SECOND EDITION

Biomedical Photonics Handbook

Volume I

Fundamentals, Devices,
and Techniques

Biomedical Photonics Handbook, Second Edition

Volume I: Fundamentals, Devices, and Techniques

Volume II: Biomedical Diagnostics

Volume III: Therapeutics and Advanced Biophotonics

SECOND EDITION

Biomedical Photonics Handbook

Volume I

Fundamentals, Devices, and Techniques

Edited by

Tuan Vo-Dinh

Duke University
Durham, North Carolina, USA

CRC Press
Taylor & Francis Group
Boca Raton London New York

CRC Press is an imprint of the
Taylor & Francis Group, an **informa** business

CRC Press
Taylor & Francis Group
6000 Broken Sound Parkway NW, Suite 300
Boca Raton, FL 33487-2742

First issued in paperback 2019

© 2015 by Taylor & Francis Group, LLC
CRC Press is an imprint of Taylor & Francis Group, an Informa business

No claim to original U.S. Government works

ISBN-13: 978-1-4200-8512-9 (hbk)
ISBN-13: 978-0-367-37848-6 (pbk)

Library of Congress Cataloging-in-Publication Data

Biomedical photonics handbook / edited by Tuan Vo-Dinh. -- Second edition.
    p. ; cm.
    Includes bibliographical references and indexes.
    Summary: "Biomedical photonics is defined as the science of harnessing light and other forms of radiant energy to address problems in medicine and biology. The field has experienced explosive growth due to the noninvasive or minimally invasive nature and cost-effectiveness of photonic modalities in medical diagnostics and therapy. The first volume of the Biomedical Photonics Handbook, Second Edition focuses on the fundamentals and advanced optical techniques and devices. It is an authoritative reference source for those involved in the research, teaching, learning, and practice of medical technologies"--Provided by publisher.
    ISBN 978-1-4398-0444-5 (set : alk. paper) -- ISBN 978-1-4200-8512-9 (v. 1 : hardcover : alk. paper) -- ISBN 978-1-4200-8514-3 (v. 2 : hardcover : alk. paper) -- ISBN 978-1-4200-8516-7 (v. 3 : hardcover : alk. paper)
    I. Vo-Dinh, Tuan, editor. II. Title: Fundamentals, devices, and techniques. III. Title: Biomedical diagnostics. IV. Title: Therapeutics and advanced biophotonics.
    [DNLM: 1. Diagnostic Imaging--instrumentation. 2. Diagnostic Imaging--methods. 3. Biosensing Techniques--instrumentation. 4. Biosensing Techniques--methods. 5. Photons--diagnostic use. WN 150]

R857.O6
610'.28--dc23   2014008504

**Visit the Taylor & Francis Web site at
http://www.taylorandfrancis.com**

**and the CRC Press Web site at
http://www.crcpress.com**

*Inspired by the love and
infinite patience of
my wife, Kim-Chi, and
my daughter, Jade*

*This book is dedicated to the
memory of my parents,
Vo Dinh Kinh and Dang Thi Dinh*

Contents

SECTION I   Photonics and Tissue Optics

SECTION II   Basic Instrumentation

SECTION III   Photonic Detection and Imaging Techniques

SECTION IV   Spectroscopic Data

Preface

In the tradition of the *Biomedical Photonics Handbook*, the second edition is intended to serve as an authoritative reference source for a broad audience involved in the research, teaching, learning, and practice of medical technologies. Biomedical photonics is defined as the science that harnesses light and other forms of radiant energy to provide the solution of problems arising in medicine and biology. This research field has recently experienced an explosive growth due to its noninvasive or minimally invasive nature and the cost-effectiveness of photonic modalities in medical diagnostics and therapy.

The field of biomedical photonics did not emerge as a well-defined, single research discipline like chemistry, physics, or biology. Its development and growth have been shaped by the convergence of three scientific and technological revolutions of the twentieth century: the *quantum theory revolution*, the *technology revolution*, and the *genomics revolution*.

The quantum theory of atomic phenomena provides a fundamental framework for molecular biology and genetics because of its unique understanding of electrons, atoms, molecules, and light itself. Out of this new scientific framework emerged the discovery of the structure of DNA, the molecular nature of cell machinery, and the genetic cause of diseases, all of which form the basis of molecular medicine. The formulation of the quantum theory not only gave birth to the field of molecular spectroscopy but also led to the development of a powerful set of photonics tools—lasers, scanning tunneling microscopes, and near-field nanoprobes—for exploring nature and understanding the cause of disease at the fundamental level.

Advances in technology also played, and continue to play, an essential role in the development of biomedical photonics. The invention of the laser was an important milestone. Laser is now the light source most widely used to excite tissues for disease diagnosis as well as to irradiate tumors for tissue removal in interventional surgery (*optical scalpels*). The microchip is another important technological development that has significantly accelerated the evolution of biomedical photonics. While the laser has provided a new technology for excitation, the miniaturization and mass production of integrated circuits, sensor devices, and their associated electronic circuitry made possible the development of the microchip, which has radically transformed the ways detection and imaging of molecules, tissues, and organs can be performed in vivo and ex vivo. Recently, nanotechnology, which involves research on materials and species at length scales between 1 and 100 nm, has been revolutionizing important areas in biomedical photonics, especially diagnostics and therapy at the molecular and cellular level. The combination of photonics and nanotechnology has already led to a new generation of devices for probing the cell machinery and elucidating intimate life processes occurring at the molecular level that were heretofore invisible to human inquiry. This will open the possibility of detecting and manipulating atoms and molecules using nanodevices, which have the potential for a wide variety of medical uses at the cellular level. The marriage of electronics, biomaterials, and photonics is expected to revolutionize many areas of medicine in the twenty-first century.

A wide variety of biomedical photonic technologies have already been developed for clinical monitoring of early disease states or physiological parameters such as blood pressure, blood chemistry, pH, temperature, and the presence of pathological organisms or biochemical species of clinical importance. Advanced optical concepts using various spectroscopic modalities (e.g., fluorescence, scattering, reflection, and optical coherence tomography) are emerging in the important area of functional imaging. Many photonic technologies originally developed for other applications (e.g., lasers and sensor systems in defense, energy, and aerospace) have now found important uses in medical applications. From the brain to the sinuses to the abdomen, precision navigation and tracking techniques are critical to position medical instruments precisely within the three-dimensional surgical space. For instance, optical stereotactic systems are being developed for brain surgery, and flexible micronavigation devices are being engineered for medical laser ablation treatments.

With the completion of the sequencing of the human genome, one of the greatest impacts of genomics and proteomics is the establishment of an entirely new approach to biomedical research. With whole-genome sequences and new automated, high-throughput systems, photonic technologies such as biochips and microarrays can address biological and medical problems systematically and on a large scale in a massively parallel manner. They provide the tools to study how tens of thousands of genes and proteins work together in interconnected networks to orchestrate the chemistry of life. Specific genes have been deciphered and linked to numerous diseases and disorders, including breast cancer, muscle disease, deafness, and blindness. Furthermore, advanced biophotonics has contributed dramatically to the field of diagnostics, therapy, and drug discovery in the postgenomic area. Genomics and proteomics present the drug discovery community with a wealth of new potential targets. Biomedical photonics can provide tools capable of identifying specific subsets of genes encoded within the human genome that can cause the development of diseases. Photonic techniques based on molecular probes are being developed to identify the molecular alterations that distinguish a diseased cell from a normal cell. Such technologies will ultimately aid in characterizing and predicting the pathologic behavior of that diseased cell, as well as the cell's responsiveness to drug treatment. Information from the human genome project will one day make personal, molecular medicine an exciting reality.

The second edition of this handbook is intended to present the most recent scientific and technological advances in biomedical photonics, as well as their practical applications, in a single source. The three-volume handbook represents the collective work of over 150 scientists, engineers, and clinicians. It includes many new topics and chapters such as fiber-optics probes design, laser and optical radiation safety, photothermal detection, multidimensional fluorescence imaging, surface plasmon resonance imaging, molecular contrast optical coherence tomography, multiscale photoacoustics, polarized light for medical diagnostics, quantitative diffuse reflectance imaging, interferometric light scattering, nonlinear interferometric vibrational imaging, nanoscintillator-based therapy, SERS molecular sentinel nanoprobes, and plasmonic coupling interference nanoprobes.

The three-volume handbook includes 71 chapters grouped in 8 sections:

1. Volume I: *Biomedical Photonics Handbook*, Second Edition: *Fundamentals, Devices, and Techniques*
2. Volume II: *Biomedical Photonics Handbook*, Second Edition: *Biomedical Diagnostics*
3. Volume III: *Biomedical Photonics Handbook*, Second Edition: *Therapeutics and Advanced Biophotonics*

In Volume I, Section I (Photonics and Tissue Optics) contains introductory chapters on the fundamental optical properties of tissue, light–tissue interactions, and theoretical models for optical imaging. Section II (Basic Instrumentation) deals with basic instrumentation and hardware systems and contains chapters on lasers and excitation sources, basic optical instrumentation, optical fibers, probe designs, laser use, and optical radiation safety. Section III (Photonic Detection and Imaging Techniques) deals with methodologies and contains chapters on various detection techniques and systems (such as lifetime imaging, microscopy, two-photon detection, photothermal detection, interferometry, Doppler imaging, light scattering, and thermal imaging). Finally, Section IV (Spectroscopic Data) provides a comprehensive compilation of useful information on spectroscopic data of biologically and medically relevant species for over 1000 compounds and systems.

In Volume II, Section I (Biomedical Analysis, Sensing, and Imaging) contains chapters describing in vitro diagnostics (e.g., glucose diagnostics, in vitro instrumentation, biosensors, surface plasmon resonance, and flow cytometry) and in vivo diagnostics (optical coherence tomography, polarized light diagnostics, functional imaging and photon migration spectroscopy, and multiscale photoacoustics). Section II (Biomedical Diagnostics and Optical Biopsy) is mainly devoted to novel optical techniques for cancer diagnostics, often referred to as *optical biopsy* (such as fluorescence, scattering, reflectance, interferometric light scattering, optoacoustics, and ultrasonically modulated optical imaging).

In Volume III, Section I (Therapeutic and Interventional Techniques) covers photodynamic therapy as well as various laser-based treatment techniques that are applied to different organs and disease endpoints (such as dermatology, pulmonology, neurosurgery, ophthalmology, otolaryngology, gastroenterology, and dentistry). There are several chapters dealing with nanotechnology for theranostics, that is, the modality combining diagnostics and therapy. Section II (Advanced Biophotonics and Nanophotonics) is devoted to the most recent advances in methods and instrumentation for biomedical and biotechnology applications. This section contains chapters on emerging photonic technologies (e.g., biochips, nanosensors, quantum dots, molecular probes, molecular beacons, molecular sentinels, plasmonic coupling nanoprobes, bioluminescent reporters, optical tweezers) that are being developed for gene expression research, gene diagnostics, protein profiling, and molecular biology investigations as well as for early diagnostics of disease biomarkers for *new medicine.*

The goal of the second edition of this handbook is to provide a comprehensive forum that integrates interdisciplinary research and development of interest to scientists, engineers, manufacturers, teachers, students, and clinical providers. Each chapter provides introductory material with an overview of the topic of interest as well as a collection of published data with an extensive list of references for further details. The handbook is designed to present the most recent advances in instrumentation and methods as well as clinical applications in important areas of biomedical photonics. Because light is rapidly becoming an important diagnostic tool and a powerful weapon in the armory of the modern physician, it is our hope that this handbook will stimulate a greater appreciation of the usefulness, efficiency, and potential of photonics in medicine.

Tuan Vo-Dinh
Duke University
Durham, North Carolina

Acknowledgments

The completion of this work has been made possible with the assistance of many friends and colleagues. I wish to express my gratitude to members of the Scientific Advisory Board of the first edition. Their thoughtful suggestions and useful advice in the planning phase of the first edition have been important in achieving the breadth and depth of this handbook. It is a great pleasure for me to acknowledge, with deep gratitude, the contribution of over 150 contributors for the 71 chapters in this three-volume handbook. I wish to thank my coworkers at Duke University and the Oak Ridge National Laboratory, and many colleagues in academia, federal laboratories, and industry, for their kind help in reading and commenting on various chapters of the manuscript. My gratitude is extended to all my present and past students, postdoctoral associates, colleagues, and collaborators, who have been traveling with me on this exciting journey of discovery with the ultimate vision of bringing research at the intersection of photonics and medicine to the service of society.

I gratefully acknowledge the support of the US Department of Energy Office of Biological and Environmental Research, the National Institutes of Health, the Defense Advanced Research Projects Agency, the Department of the Army, the Army Medical Research and Materiel Command, the Department of Justice, the Federal Bureau of Investigation, the Office of Naval Research, the Environmental Protection Agency, the Fitzpatrick Foundation, the R. Eugene and Susie E. Goodson Endowment Fund, and the Wallace Coulter Foundation.

The completion of this work has been made possible with the love, encouragement, and inspiration of my wife, Kim-Chi, and my daughter, Jade.

Editor

Tuan Vo-Dinh is R. Eugene and Susie E. Goodson Distinguished Professor of Biomedical Engineering, professor of chemistry, and director of the Fitzpatrick Institute for Photonics at Duke University. A native of Vietnam and a naturalized US citizen, he completed high school education in Saigon (now Ho Chi Minh City). He continued his studies in Europe, where he received his BS in physics in 1970 from EPFL (Ecole Polytechnique Federal de Lausanne) in Lausanne and his PhD in physical chemistry in 1975 from ETH (Swiss Federal Institute of Technology) in Zurich, Switzerland. Before joining Duke University in 2006, Dr. Vo-Dinh was director of the Center for Advanced Biomedical Photonics, group leader of Advanced Biomedical Science and Technology Group, and a corporate fellow, one of the highest honors for distinguished scientists at Oak Ridge National Laboratory (ORNL). His research has focused on the development of advanced technologies for the protection of the environment and the improvement of human health. His research activities involve biophotonics, plasmonics, nanobiotechnology, laser spectroscopy, molecular imaging, medical theranostics, cancer detection, nanosensors, chemical sensors, biosensors, and biochips.

Dr. Vo-Dinh has authored over 350 publications in peer-reviewed scientific journals. He is the author of a textbook on spectroscopy and the editor of six books. He holds over 37 US and international patents, 5 of which have been licensed to private companies for commercial development. Dr. Vo-Dinh has presented over 200 invited lectures at international meetings in universities and research institutions. He has chaired over 30 international conferences in his field of research and served on various national and international scientific committees. He also serves the scientific community through his participation in a wide range of governmental and industrial boards and advisory committees.

Dr. Vo-Dinh has received seven R&D 100 Awards for Most Technologically Significant Advance in Research and Development for his pioneering research and inventions of innovative technologies. He has received the Gold Medal Award, Society for Applied Spectroscopy (1988); the Languedoc–Roussillon Award, France (1989); the Scientist of the Year Award, ORNL (1992); the Thomas Jefferson Award, Martin Marietta Corporation (1992); two Awards for Excellence in Technology Transfer, Federal Laboratory Consortium (1995, 1986); the Inventor of the Year Award, Tennessee Inventors Association (1996); the Lockheed Martin Technology Commercialization Award (1998); the Distinguished Inventors Award, UT-Battelle (2003); and the Distinguished Scientist of the Year Award, ORNL (2003). In 1997, he was presented the Exceptional Services Award for distinguished contribution to a healthy citizenry from the US Department of Energy. In 2011, he received the Award for Spectrochemical Analysis from the American Chemical Society (ACS) Division of Analytical Chemistry.

Contributors

Gregory Bearman
ANE Image
Pasadena, California

Moshe Ben-David
Department of Biomedical
  Engineering
Tel-Aviv University
Tel-Aviv, Israel

Nada N. Boustany
Department of Biomedical
  Engineering
Rutgers, The State University of
  New Jersey
Piscataway, New Jersey

Laurent Cognet
Institut d'Optique Graduate
  School
Centre National de la Recherche
  Scientifique and University
  of Bordeaux
Talence, France

David Cuccia
Modulated Imaging
Irvine, California

Brian M. Cullum
Department of Chemistry and
  Biochemistry
University of Maryland,
  Baltimore County
Baltimore, Maryland

Chen Y. Dong
Department of Physics
National Taiwan University
Taipei, Taiwan

Christopher Dunsby
Department of Physics
and
Department of Experimental
  Medicine
Imperial College London
London, United Kingdom

Mikella E. Farrell
US Army Research Laboratory
Adelphi, Maryland

Paul French
Physics Department
Imperial College London
London, United Kingdom

Israel Gannot
Department of Biomedical
  Engineering
Tel-Aviv University
Tel-Aviv, Israel

and

ECE Department
Johns Hopkins University
Baltimore, Maryland

Jay P. Gore
Purdue University
West Lafayette, Indiana

Petr Herman
Institute of Physics
Charles University
Prague, Czech Republic

Ellen Holthoff
US Army Research Laboratory
Adelphi, Maryland

Poorya Hosseini
Department of Mechanical
  Engineering
Massachusetts Institute of
  Technology
Cambridge, Massachusetts

Ilko K. Ilev
Center for Devices and
  Radiological Health
US Food and Drug
  Administration
Silver Spring, Maryland

Ramesh Jaganathan
Advanced Biomedical Science
  and Technology Group
Oak Ridge National Laboratory
Oak Ridge, Tennessee

Paul M. Kasili
Advanced Biomedical Science
  and Technology Group
Oak Ridge National Laboratory
Oak Ridge, Tennessee

Joseph R. Lakowicz
Department of Biochemistry
    and Molecular Biology
University of Maryland School
    of Medicine
Baltimore, Maryland

Robert J. Landry
Center for Devices and
    Radiological Health
US Food and Drug
    Administration
Silver Spring, Maryland

Marcus Larsson
Department of Biomedical
    Engineering
Linköping University
Linköping, Sweden

Richard Levenson
Department of Pathology &
    Laboratory Medicine
UC Davis Medical Center
University of California, Davis
Sacramento, California

Brahim Lounis
Institut d'Optique Graduate
    School
Centre National de la Recherche
    Scientifique and University
    of Bordeaux
Talence, France

Barry R. Masters
Department of Ophthalmology
University of Bern
Bern, Switzerland

James McGinty
Physics Department
Imperial College London
London, United Kingdom

Joel Mobley
Department of Physics and
    Astronomy
University of Mississippi
Oxford, Mississippi

Gert E. Nilsson
WheelsBridge AB
Linköping, Sweden

Stephen J. Norton
Department of Biomedical
    Engineering
Duke University
Durham, North Carolina

T. Joshua Pfefer
Center for Devices and
    Radiological Health
US Food and Drug
    Administration
Silver Spring, Maryland

E. Göran Salerud
Department of Biomedical
    Engineering
Linköping University
Linköping, Sweden

Peter T.C. So
Department of Mechanical
    Engineering
Massachusetts Institute of
    Technology
Cambridge, Massachusetts

Joon Myong Song
Advanced Biomedical Science
    and Technology Group
Oak Ridge National Laboratory
Oak Ridge, Tennessee

David L. Stokes
Advanced Biomedical Science
    and Technology Group
Oak Ridge National Laboratory
Oak Ridge, Tennessee

Dimitra N. Stratis-Cullum
Electro-Optics and Photonics
    Division
US Army Research Laboratory
Adelphi, Maryland

Tomas Strömberg
Department of Biomedical
    Engineering
Linköping University
Linköping, Sweden

Nitish V. Thakor
Department of Biomedical
    Engineering
Johns Hopkins University
    School of Medicine
Baltimore, Maryland

and

Singapore Institute for
    Neurotechnology
National University of
    Singapore
Singapore

Valery V. Tuchin
Research-Educational Institute
    of Optics and Biophotonics
Saratov State University
and
Institute of Precise Mechanics
    and Control
Russian Academy of Science
Saratov, Russia

and

Optoelectronics and
    Measurement Techniques
    Laboratory
University of Oulu
Oulu, Finland

Urs Utzinger
Department of Biomedical
    Engineering
University of Arizona
Tucson, Arizona

Tuan Vo-Dinh
Department of Biomedical
  Engineering
and
Department of Chemistry
Duke University
Durham, North Carolina

Karin Wårdell
Department of Biomedical
  Engineering
Linköping University
Linköping, Sweden

Tony Wilson
University of Oxford
Oxford, United Kingdom

Lisa X. Xu
Purdue University
West Lafayette, Indiana

Dmitry A. Zimnyakov
Physics Department
Yuri Gagarin State Technical
  University of Saratov
Saratov, Russia

MATLAB Statement

MATLAB® is a registered trademark of The MathWorks, Inc. For product information, please contact:

The MathWorks, Inc.
3 Apple Hill Drive
Natick, MA 01760-2098 USA
Tel: 508-647-7000
Fax: 508-647-7001
E-mail: info@mathworks.com
Web: www.mathworks.com

1

Biomedical Photonics: A Revolution at the Interface of Science and Technology

Tuan Vo-Dinh
Duke University

1.1 Introduction

Light has played an important role in medicine throughout human history. In prehistoric times, the healing power of light was often attributed to mythological, religious, and supernatural powers. The history of light therapy dates back to the ancient Egyptians, Hindus, Romans, and Greeks, all of whom created temples to worship the therapeutic powers of light, especially sunlight, for healing the body as well as the mind and the soul. In Hindu mythology, Dhanvantari, originally a sun god, is physician of the gods and a teacher of healing arts to humans. In ancient Egyptian religion, sun worship was common with many deities associated with the sun, such as Ra, a major god identified primarily as the midday sun. In Greek mythology, Apollo, the god of healing, who taught medicine to man, is also called the sun god or the *god of light*. These and other mythological figures are testaments to humankind's recognition of the mystical healing power of light since the dawn of time.

The contribution of light to medicine has evolved throughout human history with the advent of science and technology. In the seventeenth century, the invention of the microscope by Dutch investigators was critical to the development of biological and biomedical research for the next 200 years. Cell theory emerged in the 1830s, when German scientists M. J. Schleiden and Theodor Schwann looked into their microscopes and identified the cell as the basic unit of plant and animal tissue and metabolism [1]. The microscope provided the central observation tool for a new style of research, out of which emerged the germ theory of disease, developed by Robert Koch and Louis Pasteur in the 1870s (Figure 1.1).

In the fall of 1895, the German physicist Wilhelm Roentgen, working with a standard piece of laboratory equipment, discovered a new type of radiation, the x-rays. This discovery extended the range of

FIGURE 1.1   Portrait of Louis Pasteur. Louis Pasteur used the microscope, which provided the central observation tool for a new style of research, out of which emerged the germ theory of disease. (Painting by Albert Edelfelt, Musee d'Orsay, Paris, France. Copyright Artists Right Society (ARS), New York.)

electromagnetic radiation well beyond its conventional limits. Furthermore, this discovery led to the development of a powerful new technique that uses x-rays to look into the intact body for disease diagnosis.

These examples are just some of the numerous cases where scientific discoveries and technological advances in photonics have opened new horizons to medicine and provided critical tools to investigate molecules, analyze tissue, and diagnose diseases.

1.2 Biomedical Photonics: A Definition

The field of biomedical photonics is often not well defined because it is a unique field that has emerged from research conducted at the intersection of the physical, biological, and medical sciences and engineering. Therefore, it is useful to provide here some definition of the field.

A related term that has commonly been used is *biomedical optics*. Let us examine the similarity and difference between biomedical photonics and biomedical optics. According to the *Merriam-Webster* dictionary, the term *optical* is defined as "of or relating to vision," or "of, relating to, or being objects that emit light in the visible range of frequencies." Thus, by general definition, the field of optics involves *optical* light or *visible* light, which is a particular type of electromagnetic radiation that can be seen and sensed by the human eye. On the other hand, the field of photonics, which involves photons, particles conceptually developed by Albert Einstein to describe the quanta of energy in the entire spectrum of electromagnetic radiation, is broader than the field of optics. We tend to think of optical radiation as *light*, but the rainbow of colors that make up optical or visible light is just a very small part of a much broader range of the energy range of the photon.

Photonics includes both optical and nonoptical technologies that deal with electromagnetic radiation, which is the energy propagated through space by electric and magnetic fields. The electromagnetic spectrum is the extent of that energy, ranging from gamma rays and x-rays throughout ultraviolet (UV), visible, infrared (IR), terahertz, microwave, and radio-frequency energy.

Biomedical photonics, therefore, may be defined as the science and technology that uses the entire range of electromagnetic radiation beyond visible light for medical applications. This field involves investigating, generating, and harnessing light and other forms of radiant energy whose quantum unit is the photon. The science includes the use of light absorption, emission, transmission, scattering, amplification, and detection using a wide variety of methods and technologies, such as lasers and other light sources, fiber optics, electrooptical instrumentation, microelectromechanical systems, detectors, and nanosystems for medical applications. The range of applications of biomedical photonics extends from medical diagnostics to therapy and disease prevention.

1.3 Scientific and Technological Revolutions Shaping Biomedical Photonics

The field of biomedical photonics did not emerge as a well-defined, single research discipline like chemistry, physics, or biology. Its development and growth have been shaped by the convergence of three scientific and technological revolutions of the twentieth century (Figure 1.2):

- The quantum theory revolution (1900–1950s)
- The technology revolution (1940s–1950s)
- The genomics revolution (1950s–2000)

Technological progress is usually represented as an *S curve*, rising slowly at first, then more and more rapidly until it approaches natural limits, and then tending to level off to reach theoretical limits. Some examples of scientific discoveries and technological achievements are shown in Figure 1.2. These important revolutions are still progressing today and have led to major scientific developments, such as quantum computing (based on quantum entanglement phenomena), information technology (high-speed data transmission using fiber-optic and ultrashort light pulses), and personalized medicine (individualized diagnostics and therapy based on genomic characteristics of patients). The following sections discuss the three revolutions of the twentieth century that have shaped the growth and development of biomedical photonics.

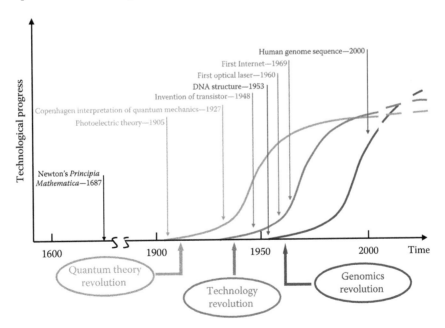

FIGURE 1.2 Three revolutions of the twentieth century that shaped biomedical photonics. Technological progress is usually represented as an *S curve*, rising slowly at first, then more and more rapidly until it approaches natural limits, and then tending to level off to reach theoretical limits.

1.3.1 Quantum Theory Revolution: A Historic Evolution of the Concept of Light

The field of photonics has significantly benefited from the development of quantum theory. With the advent of quantum theory, scientific fields such as molecular spectroscopy and photonic technologies (such as lasers, optical biopsy, optical tweezers, and near-field probes) have provided powerful tools to diagnose diseases noninvasively, interrogate the cell at the molecular level, and fight diseases at the gene level. The quantum theory of atomic phenomena provides a fundamental framework for molecular biology and genetics because of its unique understanding of electrons, atoms and molecules, and light itself. Out of this new scientific framework emerged the discovery of the DNA structure, the molecular nature of cell machinery, and the genetic cause of diseases, all of which form the basis of molecular medicine.

The quantum theory is one of the most amazing discoveries of the twentieth century and brought about a monumental paradigm shift. The concept of a paradigm was proposed by Thomas S. Kuhn in his seminal book *The Structure of Scientific Revolutions* [2]. According to Kuhn's thesis, scientists conduct their research within the framework of a collective background of shared assumptions, which make up a paradigm. During any given period, the scientific community in a given field of research has a prevailing paradigm that shapes, defines, and directs research activities in that field. People often become attached to their paradigms until a paradigm shift occurs when a major revolutionary discovery triggers a drastic change in beliefs, like the dramatic upheavals occurring in revolutions.

A dramatic paradigm shift occurred in the seventeenth century with Rene Descartes' mechanistic philosophy and Isaac Newton's scientific revolution. Descartes, a French mathematician and philosopher, established the firm belief in scientific knowledge, which forms the basis of Cartesian philosophy and the worldview derived from it. According to Bertrand Russell, such a shift in thought "had not happened since Aristotle... There is a freshness about [Descartes'] work that is not to be found in any eminent previous philosopher since Plato" [3]. For Descartes, the essence of human nature lies in thought, and all things we conceive clearly and distinctly are true. Descartes' rational philosophy integrated a complete mechanistic interpretation of physics, biology, psychology, and medicine. His celebrated statement *Cogito, ergo sum* (*I think, therefore I exist*) has profoundly influenced Western civilization for many centuries.

With Newton, a new agenda for scientific research in optics, mechanics, astronomy, and a wide variety of other fields was born. Newton's *Principia Mathematica Philosophica Naturalis*, a series of three books completed in 1687, laid the foundation for our understanding of the underlying physics of the world, which shaped the history of science and remained the main paradigm for the classical worldview for over two centuries. Newton's work concluded the intellectual quest that extended back through Galileo, Kepler, and Copernicus and ultimately back to Aristotle. Since 300 BC, Aristotle's work—which encompassed logic, physics, cosmology, psychology, natural history, anatomy, metaphysics, ethics, and aesthetics—had represented the culmination of the Hellenic Enlightenment Age and the source of science and higher learning for the following 2000 years. In the Aristotelian worldview, light was not considered among one of the four basic elements—air, earth, fire, and water—which made up the physical universe. In Newton's work, by contrast, light plays an important role in a series of three books called *Opticks*, which describes in detail a wide variety of light phenomena (such as the refraction of light, the nature of white light colors, thin-film phenomena) and optical instruments (such as the microscope and the telescope). Newton performed groundbreaking experiments using optical instruments demonstrating that light was actually a mixture of colors by using a glass prism to separate the colors. In 1865, the British physicist James Clerk Maxwell developed the theory of light propagation by unifying the theories that describe the forces of electricity and magnetism.

It is interesting that the quest for understanding light and color perception is pervasive in many fields of the humanities beyond the realm of physics. The German poet Johann Wolfgang von Goethe expressed his views on color perception in his work *Theory of Colours*, which was published in 1810. The German philosopher Arthur Schopenhauer also discussed the phenomenon of visual perception in a treatise *On Vision and Colors* in 1816 and in a later work *Theoria colorum Physiologica, eademque primaria (Fundamental Physiological Theory of Color)* in 1830. The Austrian–British philosopher, Ludwig Wittgenstein, whose interest involved the philosophy of the mind, logic, and language, published in 1950 a collection of notes *Remarks on Colour*, which presented his views delineating the differences between the scientific basis of Newton's optics and Goethe's phenomenology of color.

Then a series of unexpected discoveries concerning the nature of light itself brought into question the underlying reality of the Newtonian worldview and set the stage for yet another monumental paradigm shift, the twentieth-century revolution in quantum physics launched by Albert Einstein. A phenomenon called the *photoelectric effect* raised intriguing questions about the exact nature of light. The photoelectric effect, discovered by Heinrich Hertz, dealt with the apparent paradox that when light irradiates certain materials, an electric current is produced, but only above a certain frequency (i.e., energy). Increasing the intensity of light that has a frequency below the requisite threshold will not induce a current. In 1901, the German physicist Max Planck suggested that light came only as discrete packets of energy. But it was Einstein, working as a clerk in the Swiss patent office after graduating from the Swiss Federal Institute of Technology in Zurich, known as ETH (*Eidgenossische Technische Hochschule*), who provided a comprehensive explanation of the photoelectric effect in a paper published in 1905 and launched the field of quantum mechanics. Einstein called the particles of light *quanta* (after the Latin *quantus* for *how much?*), hence, the origin of the term *quantum theory*. Einstein showed that light consists neither of continuous waves nor of small, hard particles. Instead, it exists as bundles of wave energy called photons. Each photon has an energy that corresponds to the frequency of the waves in the bundle. The higher the frequency (the bluer the color), the greater the energy carried by that bundle.

Einstein also published another extraordinary paper in 1905 that drastically redirected modern physics. This paper dealt with special relativity, or the physics of bodies moving uniformly relative to one another. Here again, light took a central role, being the ultimate reference element having the highest speed in the universe. By postulating that nothing can move faster than light, Einstein reformulated Newtonian mechanics, which contains no such limitation. This is the heart of his celebrated formula $E = mc^2$, which equates mass (m) and energy (E), and makes the speed of light (c), a constant in the equation. Einstein's theory of relativity shattered the classical worldview, which is based on the 3D space of classical Euclidean geometry, Newtonian mechanics, and the concept of absolute space and time. In Einstein's worldview, the universe has no privileged frames of references, no master clock, and no absolute time, because the velocity of light is the ultimate limit in speed, constant in all directions. It is interesting that Einstein's Nobel Prize was awarded in 1921 not for his theories on relativity (the Nobel committee thought them too speculative at the time), but for his quantum theory on the photoelectric effect (Figure 1.3).

Acceptance of quantum theory was further reinforced by research conducted by Ernest Rutherford and Niels Bohr, using radioactive emission as a tool to investigate the structure of the atom. Essentially, quantum theory replaced the mechanical model of the atom with one where atoms and all material objects are not sharply defined entities in the physical world but rather *fuzzy clouds* with dual wave–particle nature. According to quantum theory, all objects, even the subatomic particles (electrons, protons, neutrons in the nucleus), are entities that have a dual aspect. Light and matter exhibit this duality, sometimes behaving as electromagnetic waves, sometimes as particles, called photons. The existence of any molecule or object can be defined by a *probability wave*, which predicts the likelihood of finding the object at a particular place within specific limits. The mathematical formulations developed by Werner Heisenberg, Erwin Schrodinger, and Paul Dirac from 1926 through 1933 firmly established the theoretical foundation of the quantum theory.

The rules that govern the subatomic world of blurry particles and nuclear forces are called quantum mechanics. In quantum mechanics, the state of a molecule is described by a wave function ψ, a function of the position and spins of all electrons and nuclei and the presence of any external field. The probability of finding the molecule at a particular position or spin is represented by the square of the amplitude of wave function ψ. In other words, at the subatomic level, matter does not exist with

FIGURE 1.3 Photo of Albert Einstein. Einstein launched the field of quantum mechanics, which provides the theoretical foundation for molecular spectroscopy and photonics. He called the particles of light *quanta* (after the Latin *quantus* for *how much?*), hence, the origin of the term *quantum theory*. Einstein showed that light consists neither of continuous waves nor of small, hard particles. Instead, it exists as bundles of wave energy called photons. (Courtesy of Bildarchiv ETH-Bibliothek, Zurich, Switzerland.)

certainty at definite places, but rather shows *probabilities to exist*. Even Einstein could hardly accept the fundamental nature of probability in the quantum concept of reality, saying "God does not play dice with the universe." By the 1930s, with such bizarre quantum entities as *spin* formulated by the Austrian–Swiss physicist Wolfgang Pauli (the *Pauli exclusion principle*), the worldview of quantum theory had completed its paradigm shift. Following a historic meeting in 1927, scientists acknowledged that a complete understanding of physical reality lies beyond the capabilities of rational thought, a conclusion known as the *Copenhagen interpretation of quantum mechanics* [4]. The quantum worldview implies that the structure of matter is often not mechanical or visible, and that the reality of the world cannot be explained by the physical perceptions of the human senses. Coming into full circle, a seed of this new scientific concept of the twenty-first century rejoined an important aspect of Plato's philosophy in 400 BC, which referred to the concrete objects of the visible world as imperfect copies of the forms that they *partake of*. Similarities between quantum physics and many metaphysical concepts in Eastern philosophies and Western religions have been a topic of great interest [5].

Quantum theory also had a profound effect on many fields beyond science, such as art. It is noteworthy that modern art, with its seemingly strange distortions of visual reality, also appeared in the 1930s. It was no coincidence that during the quantum revolution in science, Cubist and Surrealist art abolished realistic shapes referenced in fixed space and fixed time. Pablo Picasso's renderings of the human face with their multifaceted perspectives often reflected the multidimensional nature of reality (Figure 1.4); Salvador Dali's vision of melting clocks evoked the elasticity of a relativistic time; Henry Moore's sculptures that reshape the physical form of the human body marked a departure from traditional geometric forms (Figure 1.5); and Joan Miro's paintings often instilled a feeling of cosmic interrelationship of space–time of the new physics. The Euclidean geometry of physical reality was further shattered by the creative movement of abstract art, which surged in the early 1910s with Wassily Kandinsky and Piet Mondrian and culminated in the 1950s with Jackson Pollock and Mark Rothko of the so-called New York School of abstract expressionism, where defined shapes and forms were replaced by shades, colors, and lines of fields, which were perhaps the visual equivalents of molecular waveforms, probability clouds, and energy fields of the new physics. The interdependence of science, technology, and art was well expressed in 1948 by Pollock in the avant-garde review *Possibilities*: "It seems to me that modern

FIGURE 1.4   **(See color insert.)** Pablo Picasso, *Portrait of Dora Maar Seated*, 1937. Quantum theory also had a profound effect on many fields beyond science, such as art. It was no coincidence that during the quantum revolution in science, Cubist and Surrealist art abolished realistic shapes referenced in fixed space and fixed time. Pablo Picasso's renderings of the human face with their multifaceted perspectives often reflected the dual nature of reality. (Courtesy of Musee Picasso, Paris, France. Copyright Artists Right Society (ARS), New York.)

FIGURE 1.5   Henry Moore's sculpture: Reclining Figure (1931). It is noteworthy that modern art, with its seemingly strange distortions of visual reality, also appeared in the epoch of quantum theory. Henry Moore's sculptures that reshape the physical form of the human body marked a departure from geometric forms. (With permission from The Henry Moore Foundation.)

painting cannot express our era (the airplane, the atomic bomb, the radio) through forms inherited from the Renaissance and from any other culture of the past. Each era finds its own technique." [6].

At the same time that Niels Bohr introduced his theory of complementarity on the dual wave–particle nature of light, the Swiss psychologist Carl Jung proposed his theory of synchronicity, an *acausal connecting* phenomenon where an event in the physical world concurs somehow with a psychological state of mind. In Jung's philosophy, synchronistic experiences are associated with the relativity of space and time and a degree of unconsciousness. Synchronicity reflected an almost mystical connection between the personal psyche and the material world, which are essentially different forms of energy [7,8]. Complementarity and synchronicity are concepts belonging to two different fields, but both concepts, with their nature of duality, reflected the twentieth-century *zeitgeist* of the quantum reality in the psyche and the physical world.

Quantum theory revolutionized an important field of photonic research, molecular spectroscopy. Molecular spectroscopy, which deals with the study of the interaction of light with matter, has been a cornerstone of the renaissance in biomedical photonics since the mid-1950s, mainly because of the foundations of quantum theory. *How do we learn about the molecules that make up the cell, the tissues, and the organs when most of them are too small to be seen even with the most powerful microscopes?* There is no simple answer to this question, but a large amount of information has come from measurements using various techniques of molecular spectroscopy.

Classical concepts of the world provide no simple way to explain the interaction of light with matter. The most satisfactory and complete description of the absorption and emission of light by matter is based on time-dependent wave mechanics. It analyzes what happens to the wave functions of molecules in the presence of light. Light is a rapidly oscillating electromagnetic field. Matter (including living species) is made up of molecules, which contain distributions of charges and spins that give matter its electrical and magnetic properties. These distributions are altered when a molecule is exposed to light.

The discovery of quantum theory has not only given birth to the new field of molecular spectroscopy but also led to the development of a powerful set of photonic tools for exploring nature and understanding the cause of disease at the fundamental level. In essence, our current knowledge of how molecules bind together, how the building blocks of DNA cause a cell to grow, and how disease progresses on the molecular level has its fundamental basis in quantum theory.

With molecular spectroscopy, light can be used in many different ways to analyze complex biological systems and understand nature at the molecular level. Light at a certain wavelength λ (or frequency $\nu = c/\lambda$) is used to irradiate the sample (e.g., a bodily fluid, a tissue, or an organ) in a process called excitation. Then some of the properties of the light that emerges from the sample are measured and analyzed. Some analyses deal with the fraction of the incident radiation absorbed by a sample: the techniques involved are called absorption spectroscopies (e.g., UV, visible, IR absorption techniques). Other analyses examine the incident radiation dissipated and reflected back from the samples (elastic scattering techniques). Alternatively, it is possible to measure the light emitted and scattered by a sample, which occurs at wavelengths different from the excitation wavelength: in this case, the techniques involved are fluorescence, phosphorescence, Raman scattering, and inelastic scattering. Other specialized techniques can also be used to detect specific properties of emitted light (circular dichroism, polarization, lifetime, etc.).

The range of wavelengths used in molecular spectroscopy to study biological molecules is quite extensive. Molecular spectroscopic techniques have led to the development of a wide variety of practical techniques for minimally invasive monitoring of disease. For example, Chance et al. have developed and used near-IR absorption techniques to monitor physiological processes and brain function noninvasively [9]. Today, a wide variety of molecular spectroscopic techniques including fluorescence, Raman scattering, and bioluminescence are being developed for cancer diagnosis, disease monitoring, and drug discovery [10–25]. Table 1.1 summarizes the different types of spectroscopies and the wavelength of the electromagnetic radiation.

1.3.2 Technology Revolution

In Western civilization, science has often been associated with fundamental research and theoretical studies, whereas technology and engineering have often been viewed as originating from applied and experimental studies. Although the value of applied science was recognized very early on by the Greek mathematician Heron, who founded the first College of Technology in Alexandria in 105 BC [26], science and technology have remained largely separate since the early times of Hellenic Greece. Aristotelian tradition held that the laws that govern the universe could be understood by pure thought,

TABLE 1.1 Various Types of Spectroscopic Techniques for Biological and Biomedical Applications

Spectral Region	Wavelength (cm)	Energy (kcal mol^{-1})	Techniques	Properties
γ-ray	10^{-11}	3×10^8	Mossbauer	
X-ray	10^{-8}	3×10^5	X-ray diffraction	Structure
			X-ray scattering	Structure
UV	10^{-5}	3×10^2	UV absorption	Electronic
Visible	6×10^{-5}	5×10^3	Visible absorption	
			Luminescence	Electronic states
IR	10^{-3}	3×10^0	IR absorption	Vibrations
			IR emission	
Microwave	10^{-1}	3×10^{-2}	Microwave	Rotations of molecules
	10^0	10^{-3}	Electron paramagnetic resonance	
Radio frequency	10	3×10^{-4}	NMR	Nuclear spin

without requiring any experimental observation. In the seventeenth century, science and technology started to become interdependent. Galileo Galilei became a pivotal figure in the scientific revolution with his experimental observation of the movement of bodies by rolling balls of different weights down a slope. He also improved an important optical instrument, the telescope, which led to his revolutionary discoveries in astronomy. The development and use of the microscope also provided another revealing example of the interdependence between science and technology. It is very clear that it was indeed the discovery of a practical instrument that led to fundamental discoveries supporting the germ theory of diseases two centuries later. The seventeenth century witnessed the rise and spread of experimental science. This interdependence began to emerge when people believed that science should be useful and applied. Francis Bacon advocated his empirical, inductive method of investigation and envisioned a role for experiments designed to test nature and verify a hypothesis using an inductive process.

Descartes proposed a completely mechanical view of the world, a physical universe of objects that moves as a clockwork according to the laws of physics and the principles of geometry, which represented a radical departure from the more abstract worldview that dated back to Aristotle [1]. Being the father of the rationalistic mathematical philosophy, where rational thought and reason are the most important sources and tests of truth, Descartes was also influential in advocating what he called *practical philosophy* and the idea that knowledge should be applied *for the general good of all men*. Opposing Cartesian rationalism based on pure thought and reason, British philosopher David Hume formulated his philosophy of empiricism, which regards empirical observation by the senses as the only reliable source of knowledge. Shortly thereafter, German philosopher Immanuel Kant built a bridge between pure thought and sensory perception and reconciliated the philosophical divide between rationalists and empiricists (the so-called battle of Descartes vs. Hume). In Kantian philosophy, it is the structure of consciousness, through an activity of thought called *synthesis*, that turns appearances observed by the senses into objects and perceptions, without which there would be nothing. Kant found roles for both the empiricist and the rationalist elements by including a rational element in his theory of knowledge (the 12 rational concepts of the understanding) with the publication in 1781 of his influential work *The Critique of Pure Reason* [27]. In the nineteenth century, the interdependence between science and technology became more important, culminating in important technological discoveries of electricity and electromagnetic induction by Michael Faraday in 1831, the first electric telegraph in 1837 by Charles Wheaton, and the creation of the incandescent light bulb in 1879 by Thomas Edison in New Jersey and Joseph Swan in England. The discovery of the light bulb foreshadowed the next revolution in photonics in the twentieth century. Today, new photonic devices, such as the light-emitting diode (LED), a semiconductor light source with low energy consumption, long lifetime, and small size, will carry the photonic revolution into the twenty-first century. The following sections discuss some three important examples of technological developments, that is, the laser, the microchip, and nanotechnology, which have greatly influenced the field of biomedical photonics.

1.3.2.1 From Lasers to Medical Treatments

The invention of the laser was an important milestone in the development of biomedical photonics. The laser is becoming the most widely used light source to excite tissues for disease diagnosis as well as to irradiate tumors for tissue removal in interventional treatment. The word *laser* is an acronym for *light amplification by stimulated emission of radiation*. Einstein, who postulated photons and the phenomenon of stimulated emission, can be considered the grandfather of the laser. After Arthur Schawlow and Charles Townes published their paper on the possibility of laser action in the IR and visible spectrum [28], it was not long before many researchers began seriously considering practical devices. The first successful optical laser constructed by Maiman in 1960 consisted of a ruby crystal surrounded by a helicoidal flash tube enclosed within a polished aluminum cylindrical cavity cooled by forced air [29]. Light was amplified (laser action) within a ruby cylinder, which formed a Fabry–Perot cavity by optically polishing the ends to be parallel to within a third of a wavelength of light. Each end was coated with evaporated silver; one end was made less reflective to allow some radiation to escape as the laser beam.

Lasers are now widely used as excitation light sources in disease diagnostics and imaging and as optical scalpels in interventional surgery. Lasers, which are ideal light sources because of their monochromaticity and intensity, can be coupled to optical fibers inserted into an endoscope for in vivo diagnosis of diseases (Figure 1.6). They also have the advantages of increased precision and reduced rates of infection and bleeding. Computers and robotic systems are used to control the intensity and direction of the laser beam, reducing human error. Nowadays, lasers are commonly used to perform surgery on many organs ranging from the skin, the stomach, to the brain. Laser procedures can be used to remove wrinkles, tattoos, birthmarks, tumors, and warts. Other types of growths can also be removed. Lasers are used to treat eye conditions. In some individuals, vision problems can be corrected with laser surgery. Lasers can help treat some forms of glaucoma and eye problems related to diabetes. Lasers are

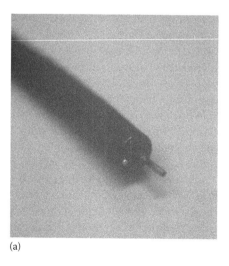 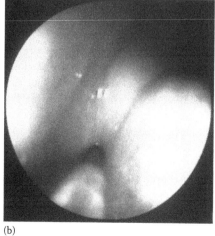

(a)   (b)

FIGURE 1.6 (See color insert.) Fiber-optic probe for in vivo laser-induced diagnostics of cancer. Laser-induced fluorescence (LIF) has been used for gastrointestinal (GI) endoscopy examinations to directly diagnose cancer of patients without requiring physical biopsy. The LIF measurement was completed in approximately 0.6 s for each tissue site. The fiber-optic probe was inserted into the biopsy channel of an endoscope (a). The fiber probe lightly touched the surface of the GI tissue being monitored (b).

being incorporated into surgical procedures for other parts of the body as well. These include the heart, the prostate gland, and the throat. Lasers are also used to open clogged arteries and remove blockages caused by tumors. Knives and scalpels may be completely eliminated someday.

1.3.2.2 From Microchips to Optical Imaging

The advent of the microchip is another important development that has significantly affected the evolution of biomedical photonics. Whereas the discovery of the laser has provided a new technology for excitation, the miniaturization and mass production of sensor devices and their associated electronic circuitry have radically transformed the ways detection and imaging of molecules, tissues, and organs can be performed in vivo and ex vivo.

Microchip technology comes from the development and widespread use of large-scale integrated circuits, consisting of hundreds of thousands of components packed onto a single tiny chip that can be mass-produced for a few cents each (Figure 1.7). This technology has enabled the fabrication of microelectronic circuitries (microchips) and photonic detectors such as photodiode arrays (PDA), charge-coupled device (CCD) cameras, and complementary metal oxide silicon (CMOS) sensor arrays in large numbers at sufficiently low costs to open a mass market and permit the widespread use of these devices in biomedical spectroscopy and molecular imaging. The miniaturization and evolution of integrated circuit technology continues to exemplify a phenomenon known as *Moore's law.* The observation made by Intel Corporation founder Gordon Moore in 1965 was that the number of components on the most complex integrated circuit chip would double each year for the next 10 years. Moore's law has come to refer to the continued chip size reduction and exponential decrease in the cost per function that can be achieved on an integrated circuit.

The miniaturization of high-density optical sensor arrays is critical to the development of innovative high-resolution imaging methods at the cellular or molecular scales that are capable of identifying and characterizing premalignant abnormalities or other early cellular changes. Photonic imaging detectors provide novel solutions for in vivo microscopic imaging sensors or microscopic implanted devices with high spatial, contrast, and temporal resolution.

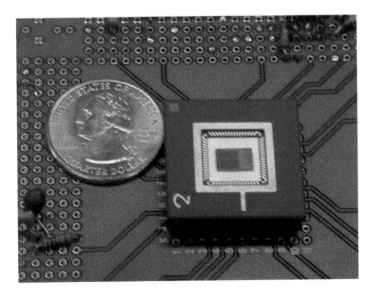

FIGURE 1.7 An optical sensor array microchip based on CMOS technology (ORNL photo). Microchip technology has enabled the fabrication of microelectronic circuitries (microchips) and photonic detectors such as PDA, CCD cameras, and CMOS sensor arrays in large numbers at sufficiently low costs to open a mass market and permit the widespread use of these devices in biomedical spectroscopy, molecular imaging, and clinical diagnostics.

During the past decade, there has been an explosion of research in biomedical photonics, resulting in scores of publications, conventions, and manufacturers offering new products in the field. Sensor miniaturization has enabled significant advances in medical imaging technologies over the last 25 years in such areas as magnetic resonance imaging (MRI), computed tomography (CT), nuclear medicine, and optical and ultrasound imaging of diseases. The development of multichannel sensor technologies has led to the development of novel photonic imaging technologies that exploit our current knowledge of the genetic and molecular origins of important diseases such as cancer and heart diseases. Those molecular biological discoveries have great implications for improving disease prevention, detection, and targeted therapy.

1.3.2.3 From Nanotechnology to Single-Cell Medicine

Nanotechnology, which involves research on and development of materials and species at length scales between 1 and 100 nm, has been revolutionizing important areas in biomedical photonics, especially diagnostics and therapy at the molecular and cellular level. The combination of molecular nanotechnology and photonics opens the possibility of detecting and manipulating atoms and molecules using nanodevices, which have the potential for a wide variety of medical uses at the cellular level.

Today, the amount of research in biomedical science and engineering at the molecular level is growing exponentially because of the availability of new investigative nanotools. These new analytical tools are capable of probing the nanometer world and will make it possible to characterize the chemical and mechanical properties of cells, discover novel phenomena and processes, and provide science with a wide range of tools, materials, devices, and systems with unique characteristics. The marriage of electronics, biomaterials, and photonics is expected to revolutionize many areas of medicine in the twenty-first century. The futuristic vision of nanorobots moving through bloodstreams armed with bioreceptor-based nanoprobes and nanolaser beams that recognize and kill specific cancer cells while sparing healthy cells might someday no longer be the *stuff of dream*.

The combination of photonics and nanotechnology has already led to a new generation of devices for probing the cell machinery and elucidating intimate life processes occurring at the molecular level that were heretofore invisible to human inquiry. Tracking biochemical processes within intracellular environments can now be performed in vivo with the use of fluorescent molecular probes (Figure 1.8) [30] and nanosensors (Figure 1.9). With powerful microscopic tools using near-field optics, scientists are now able to explore the biochemical processes and submicroscopic structures of living cells at unprecedented resolutions. It is now possible to develop nanocarriers for targeted delivery of drugs that have their shells conjugated with antibodies for targeting antigens and fluorescent chromophores for in vivo tracking.

The possibility of fabricating nanoscale components has recently led to the development of devices and techniques that can measure fundamental parameters at the molecular level. With *optical tweezer* techniques, small particles may be trapped by radiation pressure in the focal volume of a high-intensity, focused beam of light. This technique, also called *optical trapping*, may be used to move cells or subcellular organelles around at will by the use of a guided, focused light beam [31]. For example, a bead coated with an immobilized, caged bioactive probe could be inserted into a tissue or even a cell and moved around to a strategic location by an optical trapping system. The cage could then be photolyzed by multiphoton uncaging in order to release and activate the bioactive probe. Optical tweezers can also be used precisely to determine the mechanical properties of single molecules of collagen, an important tissue component and a critical factor in diagnosing cancer and the aging process [32]. The optical tweezer method uses the momentum of focused laser beams to hold and stretch single collagen molecules bound to polystyrene beads. The collagen molecules are stretched through the beads using the optical laser tweezer system, and the deformation of the bound collagen molecules is measured as the relative displacement of the microbeads, which are examined by optical microscopy. Ingenious optical trapping systems have also been used to measure the force exerted by individual motor proteins [33].

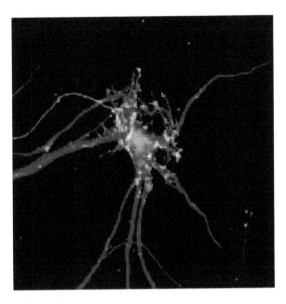

FIGURE 1.8 **(See color insert.)** Tracking biochemical processes in nerve cells using GFP. Tracking biochemical processes can now be performed with the use of fluorescent probes. For example, there is a great interest to understand the origin and movement of neurotrophic factors such as brain-derived neurotrophic factor (BDNF) between nerve cells. This figure illustrates the use of BDNF tagged with GFP to follow synaptic transport from axons to neurons to postsynaptic cells. (Adapted from Kohara K. et al., *Science*, 291, 2419, 2001.)

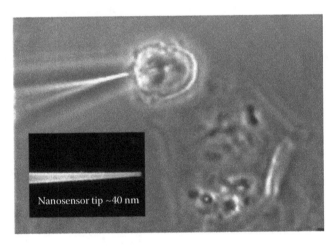

FIGURE 1.9 **(See color insert.)** A fiber-optic nanosensor with antibody probe used for detecting biochemicals in a single cell. The combination of photonics and nanotechnology has led to a new generation of devices for probing the cell machinery and elucidating intimate life processes occurring at the molecular level that were heretofore invisible to human inquiry. The insert (lower left) shows a scanning electron photograph of a nanofiber with a 40 nm diameter. The small size of the probe allowed manipulation of the nanoprobe at specific locations within a single cell. (Adapted from Vo-Dinh, T. et al., *Nat. Biotechnol.*, 18, 764, 2000; Cullum, B.M. and Vo-Dinh, T., *Trends Biotechnol.*, 18, 388, 2000.)

Molecular nanoprobes allow detection of the earliest signs of disease at the genetic, molecular, and cellular levels, leading to early and effective therapy.

An important advance in the field of biosensors has been the development of optical nanobiosensors, which have dimensions on the nanometer (nm)-size scale. Typical tip diameters of the optical fibers used in these sensors range between 20 and 100 nm. Using these nanobiosensors, it has become possible to probe individual chemical species in specific locations throughout a living cell. Figure 1.9 shows a photograph of a fiber-optic nanoprobe with antibodies targeted to benzopyrene tetrol (BPT) and designed to detect BPT in a single cell [34]. An important advantage of the optical sensing modality is the capability to measure biological parameters in a noninvasive or minimally invasive manner due to the very small size of the nanoprobe. Following measurements using the nanobiosensor, cells have been shown to survive and undergo mitosis [35].

These photonic technologies are just some examples of a new generation of nanophotonic tools that have the potential to drastically change our fundamental understanding of the life process itself. They could ultimately lead to the development of new modalities of early diagnostics and medical treatment and prevention beyond the cellular level to that of individual organelles and even DNA, the building block of life.

1.3.3 Genomics Revolution

James Watson and Francis Crick's publication of the helix structure of DNA in 1953 can be considered as the first landmark achievement that launched the genomics revolution of the twenty-first century [36]. Almost 50 years after this landmark discovery, the completion of the sequencing of the human genome marked the second major achievement in the area of molecular genetics. The draft of the published sequence encompasses 90% of the human genome's euchromatic portion, which contains the most genes. Figure 1.10 illustrates the scientific progress achieved in genomics and proteomics research since the discovery of the DNA structure.

It is useful to get an overview of this remarkable twenty-first-century achievement in molecular biology, known as the human genome project (HGP). The HGP traces its roots to an initiative in the US Department of Energy (DOE). In 1986, DOE announced its human genome initiative, believing that precise knowledge of a reference human genome sequence would be critical to its mission to pursue a deeper understanding of the potential health and environmental risks posed by energy production and use. Shortly thereafter, DOE and the National Institutes of Health (NIH) teamed up to develop a plan for a joint HGP that officially began in 1990.

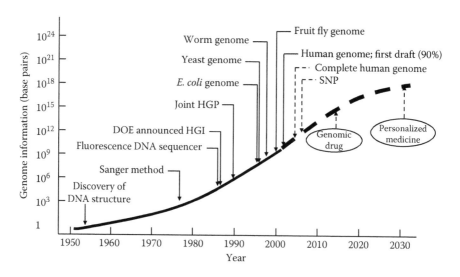

FIGURE 1.10 Advances in the genomics revolution since the discovery of the DNA helix structure. The advent of genomics has provided new information, new approaches, and new technologies for the drug discovery process. Potential applications of this knowledge are numerous and include drug discovery and customized medicine.

From the beginning of the HGP project, photonic technologies provided the critical tools for accelerating the DNA sequencing process. In 1986, Leroy Hood and co-workers described a new technique for DNA sequencing whereby four fluorescent dyes were attached to the DNA instead of using radioactive labels and reading the x-ray films [37]. This photonic detection scheme—which allowed the DNA to run down only a single lane and be illuminated by a laser and the result from four fluorescent labels to be read into a computer—has been the workhorse of the sequencing efforts.

In June 2000, scientists announced biology's most spectacular achievement: the generation of a working draft sequence of the entire human genome [38,39]. The draft encompasses 90% of the human genome's euchromatic portion, which contains the most genes. In constructing the working draft, the 16 genome sequencing centers produced over 22.1 billion bases of raw sequence data, comprising overlapping fragments totaling 3.9 billion bases and providing sevenfold coverage (sequenced seven times) of the human genome. One of the greatest impacts of knowing the genome sequence is the establishment of an entirely new approach to biomedical research. In the past, researchers studied one gene or a few genes at a time. With whole-genome sequences and new automated, high-throughput technologies, they can address biological and medical problems systematically and on a large scale in a massively parallel manner. They are now able to investigate a large number of genes in a genome, or various gene products in a particular tissue or organ or tumor. They can also study how tens of thousands of genes and proteins work together in interconnected networks to orchestrate the chemistry of life. Specific genes have been pinpointed and associated with numerous diseases and disorders, including breast cancer, muscle disease, deafness, and blindness.

Applications of this knowledge are numerous and include drug discovery and personalized medicine. Personalized medicine is a new area of health care customized to each individual, which is enabled by substantially lowering the cost of genome sequencing of an individual or a patient. Advances in photonics have contributed dramatically to the field of drug discovery in the postgenomic area. Genomics and proteomics present the drug discovery community with a wealth of new potential targets. In the pregenomics area, the basic approach to discovering new drugs mainly involved research efforts focused on individual drug targets. Nonsystematic methods were generally used to find potential drug targets. Phenotype analyses were performed by examining the differential expression of some proteins in diseased versus normal tissue to identify proteins associated with a specific disease process. When a hopeful compound was successfully synthesized, it was evaluated using time-consuming animal studies and, later, human clinical tests.

The advent of genomics has provided new information, new approaches, and new technologies for the drug discovery process. The field of optogenomics opens new approaches to investigate molecular pathways associated with neuron activity and brain function [40]. By using the ability to predict all the possible protein-coding regions in experimental systems, biomedical studies and analyses can be expanded beyond the most abundant and best characterized proteins of the cells to discover new drug targets. Most drugs today are based on about 500 molecular targets; knowledge of the genes involved in diseases, disease pathways, and drug-response sites will lead to the discovery of thousands of new targets. The increasing number of potential targets requires new instrumental approaches for rapid development of new drugs. One area of great interest is the use of high-throughput cell-based assays, which have the potential to provide a fundamental platform, the living cell, to characterize, analyze, and screen a drug target in situ.

Combined with high-throughput techniques, fluorescence technologies based on a wide variety of reporter gene assays, ion channel probes, and fluorescent probes have provided powerful tools for cell-based assays for drug discovery. An exciting new advance in fluorescent probes for biological studies has been the development of naturally fluorescent proteins for use as fluorescent probes. The jellyfish *Aequorea victoria* produces a naturally fluorescent protein known as green fluorescent protein (GFP), which was discovered by Martin Chalfie, Osamu Shimomura, and Roger Y. Tsien. Active research is under way on the applications of GFP-based assays, which use the GFP gene as a reporter of expression allowing analysis of intracellular signaling pathways using libraries of cell lines engineered to report on key cellular processes [41].

Another important advance in photonic technologies that has contributed to the dramatic growth of cell-based assays is the development of advanced imaging systems that have the combined capability of high-resolution, high-throughput, and multispectral detection of fluorescent reporters. Fluorescent reporters allow the development of live cell assays with the ability for in vivo sensing of individual biological responses across cell populations, tracking the transport of biological species within intracellular environments, and monitoring multiple responses from the same cell [42].

Cellular biosensors [43], extrinsic cellular sensors based on site-specific labeling of recombinant proteins [44,45] and cell-based assays for screening regulated genes [46] are also important photonic tools recently developed for understanding cell regulation and for drug discovery. Advanced photonic techniques using time-resolved and phase-resolved detection, polarization, and lifetime measurements further extend the usefulness of cell-based assays. Novel classes of labels using inorganic fluorophors based on quantum dots [47] or surface-enhanced Raman scattering (SERS) plasmonic nanoprobes provide unique possibilities for multiplex diagnostics and therapy [48–51]. Today, single-molecule detection techniques using various photonic modalities provide the ultimate tools to elucidate cellular processes at the molecular level.

1.4 Conclusion and Future Perspectives

As we enter the new millennium, it appears that we are witnessing a new paradigm shift in the contribution of science and technology to the development of human knowledge and the advancement of civilization. In past centuries, it was usually a discovery in basic science that led to new technology development. But today, we see that often technology creates new devices and instruments that drive scientific discoveries. As we create the tools to peer at the stars, look into the inner world of atoms, or decipher the genetic code of human cells in order to understand and ultimately eliminate the cause of diseases, we are witnessing a shift from a *science-driven* process to a *technology-driven* process in the evolution of human knowledge. Neither science nor technology holds a privileged role in the quest for knowledge. Rather, both science and technology will contribute equally to human development and societal progress as the *yin* and *yang* of life, reflecting the ever-pervasive duality of being.

Biomedical photonics is a relatively new field at the interface of science and technology that has the potential to revolutionize medicine as we know it. The rapid development of laser and imaging technology has yielded powerful tools for the study of disease on all scales, from single molecules to tissue materials and whole organs. These emerging techniques find immediate applications in biological and medical research. For example, laser microscopes permit spectroscopic and force measurements on single protein molecules, laser sources provide access to molecular dynamics and structure to perform *optical biopsy* noninvasively and almost instantaneously, while optical coherence tomography allows visualization of tissue and organs. Using genetic promoters to drive luciferase expression, bioluminescence methods can generate molecular light switches, which serve as functional indicator lights reporting cellular conditions and responses in the living animal. This technique could allow rapid assessment of and response to the effects of antitumor drugs, antibiotics, or antiviral drugs.

Developments in quantum chemistry, molecular genetics, high-speed supercomputers as well as the Internet, personal smart phones, and social media have created unparalleled capabilities for modeling complex biological systems and understanding diseases as well as for collecting, storing, connecting, and mining large amount of medical information globally in the field of so-called big data, a term underlining the exploding growth of data in the world. The Internet, born from a research program called ARPAnet initiated in 1969 by the US government's Advanced Research Project Agency (ARPA) and designed to connect mainframe computers at different universities around the country using a common language and a common protocol, is now becoming the indispensable superhighway of data transfer. In the area of supercomputing, the world's fastest supercomputer in the world, named Titan and located at the Oak Ridge National Laboratory, is already capable of 20 petaflops in 2013 (20,000 trillion calculations per second). Such a phenomenal computing speed that is expected to increase exponentially with exascale computing will provide the critical tool to treat medical data in real time. Information from the HGP will make molecular medicine an exciting reality. Current research has indicated that many diseases such as cancer occur as the result of the gradual buildup of genetic changes in single cells. Biomedical photonics can provide tools capable of identifying specific subsets of genes encoded within the human genome that can cause the development of cancer. Photonic techniques are being developed to identify the molecular alterations that distinguish a cancer cell from a normal cell. Such technologies will ultimately aid in characterizing and predicting the pathologic behavior of that cancer cell, as well as the cell's responsiveness to drug treatment.

Biomedical photonic technologies are definitely bringing a bright future to biomedical research, as they are capable of yielding the critical information bridging molecular structure and physiological function, which is the most important process in the understanding, treatment, and prevention of disease. As this handbook explores in the following sections, it is our hope that light is rapidly finding its place as an important diagnostic tool and a powerful weapon in the arsenal of the modern physician.

Acknowledgments

This work was sponsored by the US Department of Energy (DOE) Office of Biological and Environmental Research, under contract DEAC05-00OR22725 with UT-Batelle, LLC, the National Institutes of Health, the Coulter Foundation, the Department of the Army, and the Defense Advanced Research Project Agency (DARPA).

References

1. McClellan, III J. E. and Dorn, H., *Science and Technology in World History*, The John Hopkins University Press, London, U.K., 1999.
2. Kuhn, T. S., *The Structure of Scientific Revolutions*, University of Chicago Press, Chicago, IL, 1962.
3. Russell, B., *A History of Western Philosophy*, Simon & Schuster, New York, 1972.
4. Zukav, G., *The Dancing Wu Li Masters*, Bantam Books, New York, 1980.
5. Capra, F., *The Tao of Physics*, Shambhala, Berkeley, CA, 1975.
6. Ferrier, J. L., *Art of the 20th Century*, Chene-Hachette, Paris, France, 1999.
7. Jung, C. G., *L'Homme a la Decouverte de Son Ame*, Petite Bibliotheque Payot, Paris, France, 1962.
8. Shlain, L., *Arts and Physics*, Quill Publishers, New York, 1991.
9. Chance, B., Leigh, J. S., Miyake, H., Smith, D. S., Nioka, S., Greenfeld, R., Finander, M. et al., Comparison of time-resolved and time-unresolved measurements of deoxyhemoglobin in brain, *Proceedings of the National Academy of Sciences of the United States of America* 85(14), 4971–4975, 1988.
10. Anderssonengels, S., Johansson, J., Svanberg, K., and Svanberg, S., Fluorescence imaging and point measurements of tissue—Applications to the demarcation of malignant-tumors and atherosclerotic lesions from normal tissue, *Photochemistry and Photobiology* 53(6), 807–814, 1991.
11. Vo-Dinh, T., Panjehpour, M., Overholt, B. F., Farris, C., Buckley, F. P., and Sneed, R., In-vivo cancer-diagnosis of the esophagus using differential normalized fluorescence (Dnf) indexes, *Lasers in Surgery and Medicine* 16(1), 41–47, 1995.
12. Panjehpour, M., Overholt, B. F., Vo-Dinh, T., Haggitt, R. C., Edwards, D. H., and Buckley, F. P., Endoscopic fluorescence detection of high-grade dysplasia in Barrett's esophagus, *Gastroenterology* 111(1), 93–101, 1996.
13. Alfano, R. R., Tata, D. B., Cordero, J., Tomashefsky, P., Longo, F. W., and Alfano, M. A., Laser-induced fluorescence spectroscopy from native cancerous and normal tissue, *IEEE Journal of Quantum Electronics* 20(12), 1507–1511, 1984.
14. Richards-Kortum, R. and Sevick-Muraca, E., Quantitative optical spectroscopy for tissue diagnosis, *Annual Review of Physical Chemistry* 47, 555–606, 1996.

15. Lam, S. and Palcic, B., Autofluorescence bronchoscopy in the detection of squamous metaplasia and dysplasia in current and former smokers, *Journal of the National Cancer Institute* 91(6), 561–562, 1999.

16. Wagnieres, G. A., Studzinski, A. P., and VandenBergh, H. E., Endoscopic fluorescence imaging system for simultaneous visual examination and photodetection of cancers, *Review of Scientific Instruments* 68(1), 203–212, 1997.

17. Manoharan, R., Shafer, K., Perelman, L., Wu, J., Chen, K., Deinum, G., Fitzmaurice, M. et al. Raman spectroscopy and fluorescence photon migration for breast cancer diagnosis and imaging, *Photochemistry and Photobiology* 67(1), 15–22, 1998.

18. Mahadevan-Jansen, A., Mitchell, M. F., Ramanujam, N., Malpica, A., Thomsen, S., Utzinger, U., and Richards-Kortum, R., Near-infrared Raman spectroscopy for in vitro detection of cervical precancers, *Photochemistry and Photobiology* 68(1), 123–132, 1998.

19. Fujimoto, J. G., Boppart, S. A., Tearney, G. J., Bouma, B. E., Pitirs, C., and Brezinski, M. E., High-resolution in vivo intra-arterial imaging with optical coherence tomography, *Heart* 82, 128–133, 1999.

20. Boas, D. A., Oleary, M. A., Chance, B., and Yodh, A. G., Detection and characterization of optical inhomogeneities with diffuse photon density waves: A signal-to-noise analysis, *Applied Optics* 36(1), 75–92, 1997.

21. Bigio, I. J., Bown, S. G., Briggs, G., Kelley, C., Lakhani, S., Pickard, D., Ripley, P. M., Rose, I. G., and Saunders, C., Diagnosis of breast cancer using elastic-scattering spectroscopy: Preliminary clinical results, *Journal of Biomedical Optics* 5(2), 221–228, 2000.

22. Contag, C. H., Jenkins, D., Contag, F. R., and Negrin, R. S., Use of reporter genes for optical measurements of neoplastic disease in vivo, *Neoplasia* 2(1–2), 41–52, 2000.

23. Mourant, J. R., Bigio, I. J., Boyer, J., Conn, R. L., Johnson, T., and Shimada, T., Spectroscopic diagnosis of bladder cancer with elastic light scattering, *Lasers in Surgery and Medicine* 17(4), 350–357, 1995.

24. Mycek, M. A., Schomacker, K. T., and Nishioka, N. S., Colonic polyp differentiation using time-resolved autofluorescence spectroscopy, *Gastrointestinal Endoscopy* 48(4), 390–394, 1998.

25. Farkas, D. L. and Becker, D., Applications of spectral imaging: Detection and analysis of human melanoma and its precursors, *Pigment Cell Research* 14(1), 2–8, 2001.

26. Grun, B., *The Timetables of History*, Touchtone Book, New York, 1982.

27. Lavine, T. Z., *From Socrates to Sartre: The Philosophical Quest*, Bantam Books, New York, 1989.

28. Bertolotti, M., *Masers and Lasers: An Historical Approach*, Adam Hilger Ltd., Bristol, U.K., 1983.

29. Maiman, T. H., Did Maiman really invent the ruby-laser—Reply, *Laser Focus World* 27(3), 25, 1991.

30. Kohara, K., Kitamura, A., Morishima, M., and Tsumoto, T., Activity-dependent transfer of brain-derived neurotrophic factor to postsynaptic neurons, *Science*, 291, 2419–2423, 2001.

31. Askin, A., Dziedzic, J. M., and Yamane, T., Optical trapping and manipulation of single cells using infrared laser beam, *Nature* 330, 769, 1987.

32. Luo, Z. P., Bolander, M. E., and An, K. N., A method for determination of stiffness of collagen molecules, *Biochemical and Biophysical Research Communications* 232(1), 251–254, 1997.

33. Kojima, H., Muto, E., Higuchi, H., and Yanagido, T., Mechanics of single kinesin molecules measured by optical trapping nanometry, *Biophysical Journal* 73(4), 2012–2022, 1997.

34. Vo-Dinh, T., Alarie, J. P., Cullum, B. M., and Griffin, G. D., Antibody-based nanoprobe for measurement of a fluorescent analyte in a single cell, *Nature Biotechnology* 18(7), 764–767, 2000.

35. Cullum, B. M. and Vo-Dinh, T., The development of optical nanosensors for biological measurements, *Trends in Biotechnology* 18(9), 388–393, 2000.

36. Watson, J. D. and Crick, F. H. C., Molecular structure of DNA, *Nature* 171, 737, 1953.

37. Smith, L. M., Sanders, J. Z., Kaiser, R. J., Hughes, P., Dodd, C., Connell, C. R., Heiner, C., Kent, S. B. H., and Hood, L. E., Fluorescence detection in automated DNA-sequence analysis, *Nature* 321(6071), 674–679, 1986.

38. Venter, J. C., Adams, M. D., Myers, E. W., Li, P. W., Mural, R. J., Sutton, G. G., Smith, H. O. et al., The sequence of the human genome, *Science* 291(5507), 1304–1351, 2001.

39. Lander, E. S., Linton, L. M., Birren, B., Nusbaum, C., Zody, M. C., Baldwin, J., Devon, K. et al., Initial sequencing and analysis of the human genome, *Nature* 409(6822), 860–921, 2001.

40. Lief Fenno, L., Yizhar, O., and Deisseroth, K., The development and application of optogenetics, *Annual Review of Neuroscience* 34, 389, 2011.

41. Chalfie, M., Lighting up life, *Proceedings of the National Academy of Sciences* 106, 10073, 2009.

42. Thomas, N., Cell-based assays-seeing the light, *Drug Discover World* 2, 25, 2001.

43. Durick, K. and Negulescu, P., Cellular biosensors for drug discovery, *Biosensors and Bioelectronics* 16(7–8), 587–592, 2001.

44. Nakanishi, J., Nakajima, T., Sato, M., Ozawa, T., Tohda, K., and Umezawa, Y., Imaging of conformational changes of proteins with a new environment/sensitive fluorescent probe designed for site specific labeling of recombinant proteins in live cells, *Analytical Chemistry* 73(13), 2920–2928, 2001.

45. Whitney, M., Rockenstein, E., Cantin, G., Knapp, T., Zlokarnik, G., Sanders, P., Durick, K., Craig, F. F., and Negulescu, P. A., A genome-wide functional assay of signal transduction in living mammalian cells, *Nature Biotechnology* 16(13), 1329–1333, 1998.

46. Michalet, X., Pinaud, F. F., Bentolila, L. A., Tsay, G. M., Doose, S., Li, J. J., Sundaresan, G., Wu, A. M., Gambhir, S. S., and Weiss, S., Quantum dots for live cells, in vivo imaging, and diagnostics, *Science* 307, 538, 2005.

47. Vo-Dinh, T., Wang, H. N., and Scaffidi, J., Plasmonic nanoprobes for SERS biosensing and bioimaging, *Journal of Biophotonics* 3, 89, 2010.

48. Wang, H. N. and Vo-Dinh, T., Plasmonic coupling interference (PCI) nanoprobes for nucleic acid detection, *Small* 7, 3067, 2011.

49. Yuan, H., Liu, Y., Fales, A. M., Li, Y. L., Liu, J., and Vo-Dinh, T., Quantitative surface-enhanced resonant Raman scattering multiplexing of biocompatible gold nanostars for in vitro and ex vivo detection, *Analytical Chemistry* 85, 208, 2013.

50. Vo-Dinh, T., Fales, A. M., Griffin, G. D., Khoury, C. G., Liu, Y., Ngo, H., Norton, S. J., Register, J. K., Wang, H. N., Yuan, Y., Plasmonic nanoprobes: From chemical sensing to medical diagnostics and therapy, *Nanoscale* 5, 10127, 2013.

I

Photonics and Tissue Optics

2

Optical Properties of Tissue

Joel Mobley
University of Mississippi

Tuan Vo-Dinh
Duke University

Valery V. Tuchin
*Saratov State University
and
Russian Academy of Science
and
University of Oulu*

2.1 Introduction

The electromagnetic (EM) spectrum provides a diverse set of tools for probing, manipulating, and interacting with biological systems. A great variety of EM phenomena are utilized to diagnose and treat disease and to advance basic research in the life sciences. The focus of this chapter is the propagation of light in tissues. The main purpose is to provide tables of the optical properties of tissues and introduce the concepts necessary for understanding and using them, emphasizing how these properties fit into the description of light transport in tissues.

The term *light* is often used to refer to the visible region of the EM spectrum that occurs in the wavelength (λ) range from 780 to 390 nm (Figure 2.1). In this chapter, we will be dealing with a broader region of the spectrum, and we will use *light* to refer to (vacuum) wavelengths ranging from 30,000 to 100 nm.* This spectral range includes the far-, middle-, and near-infrared (IR) bands, the visible band, and the ultraviolet (UV) A, B, and C bands. A region of the spectrum that is of great importance in biomedical photonics is the *therapeutic window* (aka *diagnostic window*) from 1300 to 600 nm, in which light is not strongly absorbed by tissues. (This window is discussed in Section 2.4.3.1.)

In physics, phenomena are often described from either *classical* or *quantum* viewpoints. The term *classical* refers to theories that describe phenomena on the human scale such as Newton's laws of motion. Quantum theory describes the physics at small scales in the realm of molecules and atoms. In the classical theory,

---

* Light waves can also be described in terms of their frequency ν, which is related to wavelength by the speed of light c, $\nu = c/\lambda$. Our spectral range of interest in terms of frequency is $1 \times 10^{13} - 3 \times 10^{15}$ Hz. However, it is customary to refer to light in terms of its vacuum wavelength and frequency is rarely used in this context.

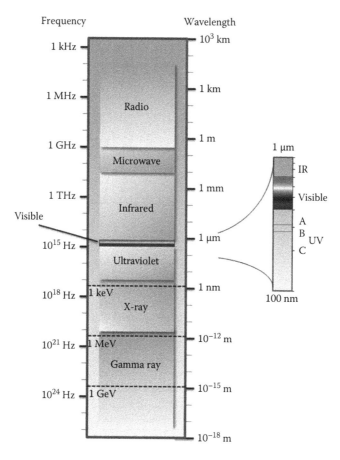

FIGURE 2.1 The electromagnetic spectrum. This figure shows the spectrum in terms of frequency on the left axis and wavelength on the right axis. The region discussed in this chapter is expanded on the right and is traditionally labeled by wavelength. This subset of the spectrum includes the UV-C (100–280 nm), UV-B (280–315 nm), and UV-A (315–400 nm) bands, the visible region (400–760 nm), and a portion of the IR-A band (760–1400 nm). An important region of this spectrum is therapeutic/diagnostic window (see Figure 2.16), which extends from around 600 to 1300 nm.

light is considered to be an oscillating EM field, and the energy carried by these EM waves can be arbitrarily small. However, one of the great insights of quantum theory is that light is granular—light waves are made up of photons, packets (i.e., *quanta*) of energy. Each photon carries an energy that is inversely proportional to the wavelength of the light, $E = hc/\lambda$, where h is Planck's constant ($6.626 \times 10^{-34}\ J\ s$) and c is the speed of light (2.998×10^8 m/s in vacuo). At the most fundamental level, light and matter interact by exchanging photons. The classical description of light and its quantum corrections are both important to the study of tissue optics. The interactions of light with objects, such as the scattering from cellular organelles, are well described by the classical viewpoint. The photon picture must be used to describe the exchange of light energy with molecules and atoms, phenomena such as absorption and fluorescence. As discussed in Chapter 1, the concept of photon exchange is essential for understanding molecular spectroscopy.*

---

* Since this handbook deals with biomedical *photonics*, it should be noted that the term photonics is often used to refer to all electromagnetic wave phenomena, whether or not the quantum nature of light (photon) is a necessary component of its description. There is a parallel with the use of the term *electronics*, which deals with electrical circuits; in many applications, the electric current flowing in a device can be thought of as a continuous flow of charge, and the fact that the current is actually composed of electrons, elementary particles that are governed by the laws of quantum mechanics, is not essential to understanding how a circuit works.

Most tissues are strong scatterers of light, and multiple scattering effects are essential features of light propagation in biomaterials. Because of the central role of multiple scattering, the direct application of the fundamental equations of electromagnetism to the tissue optics problem can be cumbersome. In place of the direct approach, the essential physics of tissue optics are described by *radiation transport theory* (RTT), which explicitly ignores wave phenomena (such as polarization and interference), in favor of tracking the transport of light energy through a medium. As discussed later, the RTT model does implicitly incorporate elements of the classical and quantum descriptions of light. However, the RTT model is not specific to light and has other important applications to phenomena such as neutron transport.

In the electromagnetic description, light waves are self-sustaining, interdependent oscillations of electric and magnetic fields (Figure 2.2). In addition to wavelength, two important parameters of EM waves are phase and polarization. The *phase* is a general property of a wave and gives rise to important effects such as interference and diffraction. The phase describes what part of an oscillating cycle a wave is in (e.g., node or antinode). *Polarization* refers to the axis parallel to the orientation of the electric field vector of the wave and is perpendicular to the direction of propagation. Two waves must have some overlap in their polarizations for interference to occur. Since RTT deals with the transport of light energy, it explicitly ignores the phase and polarization of the waves, although wave properties are implicitly included through the material parameters (such as scattering properties) used in the fundamental equations. The validity of RTT for strongly scattering optical media has been established empirically, and the theoretical link between RTT and the fundamental laws of electromagnetism has also been demonstrated (Ishimaru 1978). Since RTT provides the simplest framework for accurately predicting the dynamics of light in tissues, building a more complicated description of tissue optics directly from the fundamental laws of electromagnetism is unnecessary (Ishimaru 1978). Nevertheless, some attempt will be made to distinguish between phenomena that are inherently EM or wavelike in nature versus those that are part of the more empirical RTT model.

The remainder of this chapter provides an introduction to the fundamental photophysical concepts and the basic theoretical models for describing light propagation in tissues. We examine in some detail RTT along with the important tissue properties required by the model (μ_a, μ_s, g, and n). We will briefly discuss the Monte Carlo (MC) method for numerically performing practical transport simulations as well as the Kubelka–Munk (K–M) theory. Finally, we provide a representative listing of the relevant optical properties of tissues: the index of refraction, n; the absorption coefficient, μ_a; the scattering coefficient, μ_s; and the average cosine of scatter, g.

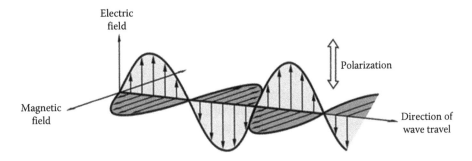

FIGURE 2.2 Electromagnetic waves. This figure is a representation of an EM wave traveling through space. The wave consists of mutually dependent, self-perpetuating oscillations of electric and magnetic fields. The axis parallel to the electric field vectors defines the axis of polarization. In the case shown, the polarization is linear and in the vertical direction. (Adapted from Hewitt, P.G., *Conceptual Physics*, Addison-Wesley, Boston, MA, 2009.)

2.2 Fundamental Optical Properties

This section discusses three photophysical processes that affect light propagation in tissues: refraction, scattering, and absorption. The fundamental parameters related to these processes are

- The *index of refraction, $n(\lambda)$*
- The *scattering cross section, σ_s*
- The *differential scattering cross section, $d\sigma_s/d\Omega$*
- The *absorption cross section, σ_a*

From the latter three, the parameters essential to the RTT model (μ_a, μ_s, and g) will emerge. For the purposes of this chapter, we will assume that the media considered are isotropic, which does not detract from the essential physics involved.

2.2.1 Refraction

2.2.1.1 Index of Refraction

The index of refraction $n(\lambda)$ (a.k.a. the refractive index) is the most fundamental of the four optical properties as it applies to a purely homogenous medium. It governs the speed of light in the medium and how light behaves at boundaries between media. The index of refraction can also be extended to apply to heterogeneous materials, as will be discussed later. In general, the refractive index $n(\lambda)$ depends on the wavelength and is defined in terms of the phase velocity of light in the medium,

$$c_m(\lambda) = \frac{c}{n(\lambda)}, \tag{2.1}$$

where $c = 2.998 \times 10^8\ m/s$ is the speed of light in vacuum ($n_{vacuum} = 1$). The index of refraction in most biologic media runs between 1.3 and 1.7, which implies a reduction in the speed of light of about 20%–40% relative to c.* The wavelength of light in the medium is given in terms of the vacuum wavelength λ as

$$\lambda_m = \frac{\lambda}{n(\lambda)}. \tag{2.2}$$

Even though both the speed and the wavelength of light depend on the refractive index, its frequency, $\nu = c/\lambda = c_m/\lambda_m$, is independent of the medium ($\nu_m = \nu_{vacuum} = \nu$). The prevailing convention is to identify a spectral component of light by its vacuum wavelength, λ. When the actual wavelength in a material is used, it will be denoted by the addition of a subscript (e.g., λ_m).

2.2.1.2 Refraction and Reflection at an Interface

When a light wave encounters a boundary between two media with different refractive indices, the path of the light is affected as shown in Figure 2.3. Light that gets transmitted into the second medium will be refracted, that is, *bent* into a new direction of propagation. Light that is reflected leaves at the same

---

* Note that the index of refraction can be defined as a complex quantity, $\tilde{n}(\lambda) = n(\lambda) - i\alpha(\lambda)$, where the real part, $n(\lambda)$, is the index defined earlier and the imaginary part, $\alpha(\lambda)$, is the attenuation. The parameter $\alpha(\lambda)$ encompasses the attenuation of a wave due to absorption and scattering. However, in this work, the absorption is considered a separate parameter and so the *complex index of refraction* $\tilde{n}(\lambda)$ will not be considered any further.

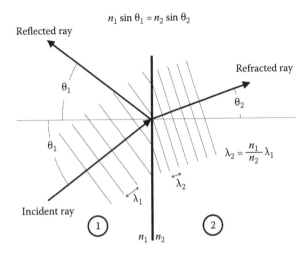

$$n_1 \sin\theta_1 = n_2 \sin\theta_2$$

Reflected ray

θ_1

Refracted ray

θ_2

θ_1

$\lambda_2 = \dfrac{n_1}{n_2}\lambda_1$

λ_2

λ_1

Incident ray

1   2

$n_1 | n_2$

FIGURE 2.3 Refraction of light at a planar boundary. The incoming and reflected waves are in medium 1 and the transmitted wave is in medium 2. The refraction angle θ_2 is given in terms of the angle of incidence θ_1 and the refractive indices, n_1 and n_2, by Snell's Law. The angle of reflection is equal to the angle of incidence. The lines represent wave crests and are separated by the wavelength in each respective medium (none are shown for the reflected wave). In this example, medium two has the larger refractive index and thus the lower speed of light of the two media.

angle relative to the boundary as the incident wave (see Figure 2.3). The amount of light reflected by and transmitted through a boundary depends on the refractive indices of the two materials, the angle of incidence, and the polarization of the incoming wave. The relation between the angle of incidence θ_1 and the angle of refraction θ_2 for the transmitted light is given by Snell's Law,

$$\sin\theta_2 = \frac{n_1}{n_2}\sin\theta_1, \tag{2.3}$$

where n_1 and n_2 are the refractive indices of the two media. There are also situations where no light will be reflected (Brewster's angle) or transmitted (total internal reflection). The refractive indices also determine the amount of energy transmitted across and reflected from a boundary. For normal incidence ($\theta_1 = 0°$) onto a planar interface, the fraction of the incident energy that is transmitted (at $\theta_2 = 0°$) into the second medium is given by

$$T = \frac{4n_1n_2}{\left(n_1+n_2\right)^2}. \tag{2.4}$$

The fraction of the incident energy that gets reflected is

$$R = 1-T = \frac{\left(n_1-n_2\right)^2}{\left(n_1+n_2\right)^2}. \tag{2.5}$$

This reflection from the interface is also known as the Fresnel reflection.* More general forms for transmission, reflection, and refraction, which are dependent on angle and polarization, can be found in Hecht (2002).

---

* This is distinct from the diffuse reflectance that will be discussed in Section 2.3. In our context, diffuse reflectance originates below the surface of the second medium.

2.2.2 Scattering

2.2.2.1 Scattering at a Localized Inclusion

Scattering occurs when light encounters a compact object with a refractive index distinct from its surroundings. In the scattering process, some fraction of the incident light is redirected over a range of angles (Figure 2.4). An important source of scattering in tissues is the refractive index mismatch between cellular organelles and the cytoplasm that envelopes them. In biomedical photonics, scattering processes are important in both diagnostic and therapeutic applications.

- *Diagnostic applications:* Scattering depends on the size, morphology, and structure of the components of tissues (e.g., lipid membranes, nuclei, collagen fibers). Abnormalities in these components and/or their distributions due to disease can affect the scattering properties of a tissue, providing a potential means for detecting pathologies, especially in imaging applications.
- *Therapeutic applications:* Scattering signals can be used to determine optimal light dosimetry (e.g., during laser-based treatment procedures) and provide useful feedback during therapy.

Scattering is most simply described by considering the incident light as a plane wave (i.e., a wave that is of uniform amplitude in planes perpendicular to the direction of propagation). In practical terms, the plane wave is assumed to be much wider than the transverse dimensions of the scattering particle. In principle, given the refractive indices of the two materials, the size and shape of the scatterer, and the wavelength, the scattered radiation can be calculated using the classical description of light. Scattering properties are quantified by the *scattering cross section*. The energy carried by a light wave is described in terms of the power P (the rate at which energy is transported, typically expressed in *watts*, W) and the intensity, I (the rate at which energy passes through a unit area and often expressed in *watts per square meter*, W/m²). For a monochromatic plane wave that has an intensity I_0 encountering the scattering object, some amount of power P_{scatt} is scattered. The ratio of the power (energy per time) scattered out of the plane wave to the incident intensity (energy per time per area) has the dimensions of an area and is known as the *scattering cross section*,

$$\sigma_s\left(\hat{\mathbf{s}}\right) = \frac{P_{scatt}}{I_0},\qquad(2.6)$$

where $\hat{\mathbf{s}}$ is the propagation direction of the plane wave relative to the scatterer (Figure 2.5). The scattering cross section is equivalent to the area that an object would have to *cut out* from the uniform plane wave in order to remove the observed amount of scattered power, P_{scatt}. The area defined by the cross section

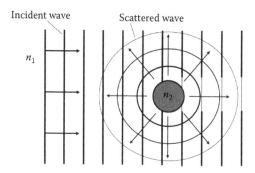

FIGURE 2.4 Light scattering. Schematic diagram of the scattering of light from a particle embedded in a host medium.

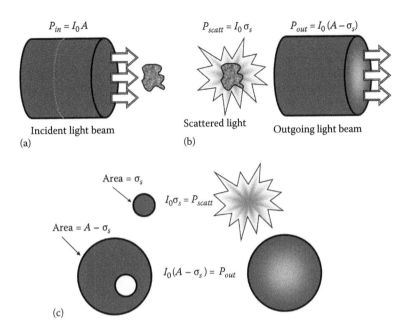

FIGURE 2.5 Scattering cross section. The scattering cross section expresses the proportionality between the intensity (power per area) of incident wave and the amount of power that is scattered from it. (a) Before encountering the scatterer, the beam has a uniform power $P_{in} = I_0 A$, where I_0 is the intensity of the beam and A is the cross-sectional area. (b) After encountering the particle, some of the energy gets scattered out of the beam, and the beam intensity is no longer uniform. (c) The amount of power scattered is equivalent to the power contained within a piece of the incoming beam with area σ_s, which is the scattering cross section.

does not necessarily represent the physical area the particle projects; it merely describes how efficiently the scatterer removes energy from the incident wave. In fact, the scattering cross section can be larger or smaller than the physical projected area of the object. In the quantum picture, the cross section is related to the likelihood that a photon will be scattered out of a beam of known intensity. In principle, scattering depends on the polarization of the incoming wave, but Equation 2.6 can be considered to represent an average over orthogonal polarization states.

The angular distribution of the scattered radiation is given by the *differential cross section* (Figure 2.6),

$$\frac{d\sigma_s}{d\Omega}(\hat{\mathbf{s}}, \hat{\mathbf{s}}'), \tag{2.7}$$

where $\hat{\mathbf{s}}'$ defines the axis of a cone of solid angle $d\Omega$ originating at the scatterer. In the quantum picture, the differential cross section determines the likelihood that a photon will be scattered in a particular direction.

For our purposes, we will assume that the scattering cross section is independent of the relative orientation of the incident light and the scatterer. This is equivalent to the assumption that the object is spherically symmetric, an approximation that is valid within the context of the RTT model. Thus, the scattering cross section can be written as a constant for a given wavelength,

$$\sigma_s(\hat{\mathbf{s}}) = \sigma_s. \tag{2.8}$$

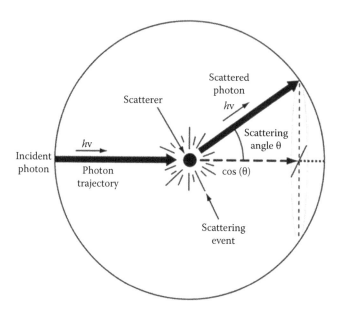

FIGURE 2.6 Angular distribution of scattering. The differential scattering cross section expresses the angular distribution of the scattered light relative to the incident light (represented here as a photon). For our purposes, we assume that the scatter does not depend on the azimuthal angle.

The spherical assumption also implies that the differential cross section will depend only on the relative orientation of the incident and scattered light,

$$\frac{d\sigma_s}{d\Omega}\left(\hat{\mathbf{s}},\hat{\mathbf{s}}'\right)=\frac{d\sigma_s}{d\Omega}\left(\hat{\mathbf{s}}\cdot\hat{\mathbf{s}}'\right), \tag{2.9}$$

where
$\hat{\mathbf{s}}\cdot\hat{\mathbf{s}}'=\cos\theta$
θ is the relative angle between the incident and scattering directions

In general, $d\sigma_s/d\Omega$ also depends on the polarization states of the incoming and outgoing waves.

A medium containing a uniform distribution of identical scatterers is characterized by the *scattering coefficient*,

$$\mu_s=\rho\sigma_s, \tag{2.10}$$

where ρ is the number density of scatterers. The scattering coefficient represents the combined scattering cross-sectional area for all the inclusions within a unit volume of the medium. The *scattering mean free path*

$$l_s=\frac{1}{\mu_s} \tag{2.11}$$

is the distance over which incident light gets reduced to 37% (e^{-1}) of its initial intensity. In the photon picture, it represents the average distance a photon travels between consecutive scattering events.

In mammalian tissues, scattering is often the dominant mechanism affecting light propagation. In many tissues, a very short pulse of collimated injected light can rapidly disperse and become diffuse in a fraction of a millimeter. Such tissues are labeled as turbid (e.g., cloudy) and act as light diffusers. This concept of diffuse light is central in tissue optics. Diffusion describes a process in which each member of a group of objects (e.g., photons, molecules) moves in random *steps* independent of the others. This random motion tends to erase any structure or order that existed in the initial state of the objects. As a simple example, consider putting a drop of dye into a glass of still water. What starts out as a single drop begins to contort into intricate swirling patterns. With time, however, the dye disperses uniformly throughout the water, resulting in a homogenous mixture. There is no residual hint of the order of the initial drop or the swirls that had once existed—those will have been erased over time by the random motion of the dye molecules. For an example from optics, consider an incandescent light bulb whose brightness is produced by a glowing wire filament. For a clear bulb, the filament, its length, and geometric shape are plain to see. If the bulb's glass is frosted instead of clear, the emitted light is uniformly bright in all directions, and there is little hint of what is inside the bulb. With the clear bulb, an orderly progression of light carries the image of the filament to your eye. The frosted bulb diffuses the light, effectively erasing the detailed information about what is inside. One of the challenges of biomedical photonics is to find ways of deriving diagnostic information from light in spite of the diffusive effects of the tissues themselves. (Such techniques are described elsewhere in this handbook.) Even with the randomizing/diffusing effects of turbid media, the distribution and evolution of the light are described by the deterministic equations of RTT (as discussed in Section 2.3).

In practice, scattering is classified into three categories defined by the ratio of the linear dimensions of the scatterer (e.g., radius) relative to the wavelength, a/λ: (1) the Rayleigh limit, where the object is small compared to the wavelength; (2) the Mie regime, where the size of the object and the wavelength are comparable; and (3) the geometric limit, where the scattering object is large compared to the wavelength. In the geometric limit, the scattering can be understood to a good approximation using ray optics (a.k.a. geometric optics) although diffraction is also a consideration. To understand the inherent scattering properties of tissues, Rayleigh and Mie are the important regimes and are discussed further here.*

2.2.2.2 Rayleigh Limit

The *Rayleigh limit* applies when the scatterers are much smaller than the wavelength. Such structures include cellular components like membranes, the smaller cellular organelles, and extracellular components such as the banded ultrastructure of collagen fibrils. In the classical picture, the implication of the small size-to-wavelength ratio is that, at any moment in time, the electric field of the light wave is uniform over the volume of the scatterer. This condition gives rise to a dipole moment in the scatterer, a slight spatial shift between the positive and negative charges within it. This dipole moment oscillates at the same frequency as the incident light and as a consequence gives off dipole radiation. For a spherical particle of radius a, the differential cross section is (Jackson 1999)

$$\frac{d\sigma_s}{d\Omega} = 8\pi^4 n_m^4 \left(\frac{n_s^2 - n_m^2}{n_s^2 + 2n_m^2}\right)^2 \frac{a^6}{\lambda^4}\left(1 + \cos^2\theta\right), \tag{2.12}$$

where
    θ is the angle between the direction of the incoming wave and the outgoing direction of interest
    n_m and n_s are the refractive indices of the medium and the scatterer, respectively (Figure 2.7)
    λ is the vacuum wavelength

This form of the differential cross section is averaged over polarization states of the incoming and scattered waves, as is appropriate for use in the RTT model. The scattering is strongest in the forward and

---

* The scattering processes mentioned in this section are *elastic*, which is when the wavelength of the scattered light is the same as the incident light. In the quantum picture, the energies of the incoming and scattered photons are the same. Inelastic scattering processes such as Raman scattering involve wavelength/energy shifts.

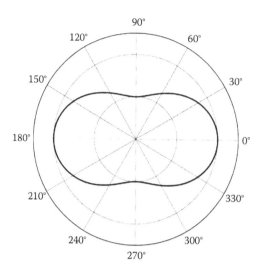

FIGURE 2.7   Rayleigh scattering. The distribution of the Rayleigh scattered light from a plane wave as a function of angle ($\theta = 0°$ is the forward direction). This is for unpolarized incident light.

rearward directions and reduced to half of the maximum laterally. It is also very sensitive to radius (a^6) and wavelength (λ^{-4}). Rayleigh scattering in the atmosphere is responsible for the blue color of the sky.

2.2.2.3 Mie Regime

Scattering of light by spherical structures comparable to the wavelength is described by the *Mie theory*.* Various subcellular structures such as mitochondria and nuclei have sizes on the order of hundreds of nanometers to a few microns, which are comparable to the wavelengths of light and are considered to fall within the Mie regime. Because the scatterer and the wavelength are of the same order, the light wave (specifically its electric field) varies over the volume of the scatterer and thus elicits a more complex response. The charges in the object oscillate at the frequency of the incoming field, and they reradiate waves with a more complex spatial variation. The result is a more varied scattering pattern than the Rayleigh type. There can be peaks and nulls in the scattering at specific wavelengths and in the angular distribution of the scattered light. Scattering behavior in the Mie regime will depend on the detailed shape of the particle among other factors. As in the Rayleigh limit, the cross section in the Mie regime can be calculated using the classical model but may be modulated by quantum effects.

2.2.3 Absorption

2.2.3.1 Absorption Processes

Absorption is a process involving the extraction of energy from light by a molecular or atomic species. In biomedical photonics, absorption processes are important in both diagnostics and therapeutic applications:

- *Diagnostic applications:* Transitions between two energy levels of a molecule are well defined at specific wavelengths (i.e., photon energies). The absorption spectrum of a molecule can serve as its fingerprint for diagnostic purposes. The absorbed light can also be re-emitted as fluorescence that can also be used as a molecular fingerprint.
- *Therapeutic applications:* Absorption of light energy can produce physical and chemical effects in tissues for treatment purposes.

---

* Note that the Mie scattering theory was specifically developed for spherical particles, although the label *Mie regime* is used to refer to scattering processes where the ratio of particle size to wavelength is on the order of one.

Absorption processes involve an important concept in quantum theory, the *energy level*, which is a quantum state of an atom or molecule. The shift of a molecular or atomic species from one energy state to another is called a *transition*. A transition from a lower to a higher energy level involves the absorption of the photon energy equal to the difference in energy, $E_{ph} = \Delta E$, between the two levels:

$$\frac{hc}{\lambda} = E_{final} - E_{initial}.$$ (2.13)

A drop from a higher energy level to a lower level is called a decay and is accompanied by a release of energy equal to the energy difference between the two levels. This release of energy may occur without emitting light (e.g., through a collision) or by the emission of a photon (e.g., luminescence).

In the quantum model, individual photons are absorbed by atoms and molecules in specific transitions, and the photon energy is used to increase their internal energy states. The regions of the spectrum where this occurs are known as *absorption bands*, and these bands are specific to particular molecular/atomic species. In general, there are three basic types of absorption processes: (1) electronic, (2) vibrational, and (3) rotational. In general, $\Delta E_{elec.} > \Delta E_{vibr.} > \Delta E_{rot.}$.

Absorption between electronic levels: In thermal equilibrium at room temperature, the states of a group of molecules are distributed across the lowest vibrational and rotational bands of the electronic ground state, S_o. When a molecule absorbs energy, it is excited from S_o to some vibrational level of one of the electronic singlet states, S_n, in the manifold S_1, ..., S_n. The intensity of the absorption (i.e., fraction of ground-state molecules promoted to the electronic excited state) depends on the intensity of the incident light (i.e., number of photons per unit area and time) and the probability of the transition with photons of a specific energy. A term often used to characterize the intensity of an absorption (or an induced emission) band is the oscillator strength, f, which may be defined from the integrated absorption spectrum by the relationship

$$f = 4.315\times10^{-9} \int \varepsilon_v dv,$$ (2.14)

where ε_v is the *molar extinction coefficient* at the frequency v.

Oscillator strengths of unity or near unity correspond to strongly allowed transitions, whereas smaller values of f indicate the smaller transition matrix elements of *forbidden* transitions. Depending on the types of species, electronic transitions have energies equivalent to those of photons from the UV through the visible to the IR range of the spectrum.

Absorption involving vibrational levels: Vibrational levels characterize the different states of vibration of the molecule. The vibrations in various degrees of freedom are quantized, giving rise to a series of energy levels for each vibrational mode of a molecule. Typically, vibrational transitions involve photons from the IR range of the spectrum.

Absorption involving rotational levels: Rotational levels represent the different states of rotational motion of a molecule, which are also quantized. Rotational transitions involve photon energies found in the far IR to submillimeter regions of the spectrum.

Molecular orbitals: The electronic energy levels of molecules are described by molecular orbitals. When a molecule undergoes an electronic transition, an electron is transferred from one molecular orbital to another. Often, the excitation is considered to be localized to a particular bond or groups of atoms. There are several types of molecular orbitals of importance in biological compounds including: π bonding orbitals, π^* antibonding orbitals, and n nonbonding (lone-pair) orbitals. The π^* antibonding orbital is less stable and has a higher energy than the π bonding orbital. An electronic transition involving the transfer of an electron from a π orbital to a π^* orbital is called a $\pi\pi^*$ transition.

Alternatively, the transfer of an electron from an n orbital to a π^* orbital is called and $n\pi^*$ transition. In transition metal complexes (e.g., metal porphyrins), the electronic energy levels may also be described by molecular orbitals formed between the metal and ligands. The term *chromophore* describes the subgroup within a molecule with an electronic transition in the visible range and is responsible for the color of a substance.

The probability of transitions between different states or energy levels is governed by quantum mechanical rules, which depend on the detailed structure of the molecules. Some transitions are said to be *allowed*, which means that they are very likely to occur, and give rise to very strong absorption bands. Others are *forbidden* transitions, which are not always literally forbidden but unlikely to occur resulting in weak absorption bands.

2.2.3.2 Absorption Cross Section and Coefficient

For a localized absorber, the absorption cross section σ_a can be defined in the same manner as for scattering,

$$\sigma_a = \frac{P_{abs}}{I_0}, \tag{2.15}$$

where P_{abs} is the amount of power absorbed out of an initially uniform plane wave of intensity I_0. As with σ_s, we make the approximation that the cross section is independent of the relative orientation of the incident light and the absorber. A medium with a uniform distribution of identical absorbing particles can be characterized by the *absorption coefficient*

$$\mu_a = \rho \sigma_a, \tag{2.16}$$

where ρ is the number density of absorbers. The reciprocal,

$$l_a = \frac{1}{\mu_a}, \tag{2.17}$$

is the *absorption mean free path, or absorption length,* and is the distance over which incident light gets reduced to 37% (e^{-1}) of its initial intensity. In the photon picture, it represents the average distance a photon travels before being absorbed. For a medium, the absorption coefficient can be defined by the following relation:

$$dI = -\mu_a I dz, \tag{2.18}$$

where dI is the differential (infinitesimal) change of intensity of a collimated light beam traversing an infinitesimal path dz through a medium with absorption coefficient μ_a. Integrating over the thickness z yields the *Beer–Lambert law,*

$$I = I_0 e^{-\mu_a z}, \tag{2.19}$$

which can also be expressed as

$$I = I_0 e^{-\varepsilon_\lambda a z}, \tag{2.20}$$

where

    ε_λ is the *molar extension coefficient* (cm^2/mol^1) at wavelength λ
    a (mol/cm^3) is the molar concentration of the absorption species
    z is the thickness (cm)

The molar extinction coefficient ε_λ (which is equal to μ_a/a) is a measure of the *absorbing power* of the species. This parameter is widely used in chemistry to express material absorption properties. Another quantity that is commonly used to describe an absorbing medium is the *transmission, T,* which is defined as the ratio of transmitted intensity I to incident intensity I_0:

$$T = \frac{I}{I_0}. \tag{2.21}$$

The *attenuation,* also called *absorbance (A)* or *optical density* (OD), of a medium is given by

$$A = \mathrm{OD} = \log_{10}\left(\frac{I_0}{I}\right) = -\log_{10}(T). \tag{2.22}$$

The variation of ε_λ with wavelength constitutes an absorption spectrum. A plot of A versus wavelength is a common way of displaying absorption spectra.

Once absorbed by a molecular species, the light energy can be dissipated by emitting a photon or nonradiatively by exchanging kinetic energy with other internal degrees of freedom of the absorbing or external species (e.g., *heating* the medium). The quantum description of the absorption and emission processes (e.g., fluorescence, phosphorescence) of molecules is further discussed elsewhere in this handbook. The most common situation is a combination of the two processes, where a small amount of the absorbed energy is dissipated nonradiatively and the majority is emitted as a photon as the molecule cascades through intermediate levels in the transition back down to the ground state. This process of emission, known as luminescence, is further broken down into fluorescence (prompt emission from a singlet state) and phosphorescence (delayed emission from a triplet state). In fluorescence, the delay between absorption and emission is typically on the order of nanoseconds, whereas in phosphorescence, emission follows much later (from milliseconds to hours). As far as photon transport in tissues is concerned, fluorescence from excited singlet states is a much more common phenomenon, although triplet states are often involved in photodynamic therapy. Under special conditions, the absorption process can also occur by the *simultaneous* extraction of two or more photons. This multiphoton absorption has important application in biomedical optics and is discussed elsewhere in this handbook.

2.3 Light Transport in Tissues

As media for light propagation, most human tissues are considered turbid (i.e., *cloudy* and/or opaque). Turbid tissues have heterogeneous structures and corresponding spatial variations in their optical properties. The spatial distribution and density of these fluctuations make these tissues strong scatterers of light. In the absence of absorption, a significant fraction of the light launched into these tissues undergoes multiple scattering, generating a diffuse light field. As optical media, these tissues have been successfully modeled as two-component systems consisting of (1) randomly distributed scattering and absorbing *particles* within (2) a homogeneous background medium. In spite of the complexities of actual tissues, this simple two-component model has proven to give a satisfactory description of optical transport in turbid tissues in most cases. In this section, we will examine the solutions from Radiation Transport Theory (RTT) in limiting cases. The Monte Carlo and Kubelka–Munk approaches to the light transport problem are also discussed, followed by a brief introduction to time-resolved propagation.

2.3.1 Preliminaries to RTT

2.3.1.1 Interference and the Diffuse Light Approximation

Light is an electromagnetic wave and thus undergoes both constructive and destructive interference. If light from two sources or apertures overlaps on a screen, they can display bright and dark regions due to this effect. To be strict, any attempt to track the detailed distribution of light energy in tissues must take the interferences into account. In RTT, the interference effects are ignored. If we consider each scatterer in a tissue to be an independent source of light, RTT tells us to treat each one as if the others do not exist—where the light overlaps, their energies add up, not their vector amplitudes. As far as the total energy is concerned, both the wave picture and RTT yield the same result. The justification of RTT lies in the random nature of the scatterers. Since the positions of the scatterers are not ordered, the interference patterns will be random as well and may even change with time. With RTT, the average energy in any region is the important quantity. It is not the microscopic details of the light field that are important, only the general flow of light energy through the tissue.

With lasers, the light entering the tissue is coherent and it can maintain most of its coherence upon exiting despite the random scattering. (The speckle pattern of laser light reflected off a surface is a manifestation of its coherence and is due to the interference of the light reaching your eye from different parts of the laser spot.) The speckle pattern that emerges from the tissue provides important information that is exploited in diagnostic techniques as discussed elsewhere. Still, even though the radiation transport approach ignores coherence, it provides an accurate and efficient way to describe the evolution of the energy distribution of light in turbid media. Figure 2.8 illustrates the diffusing properties of tissues.

The following simple example illustrates how the RT approach relates to the usual EM theory of light. In EM theory, the fundamental equations are linear in the electric and magnetic field amplitudes. This linear condition implies that the field at a point in a medium is the summation of the field vectors contributed from each individual light source. Consider the total electric field at a point due to two scatterers of light (Figure 2.9),

$$\mathbf{E}_{total}(\mathbf{r},t) = \mathbf{E}_1(\mathbf{r},t) + \mathbf{E}_2(\mathbf{r},t), \tag{2.23}$$

where $\mathbf{E}_1(\mathbf{r},t)$ and $\mathbf{E}_2(\mathbf{r},t)$ are the electric field components of the two scattered waves. (To keep things simple, we will ignore the incident field and assume the fields have harmonic time dependence.)

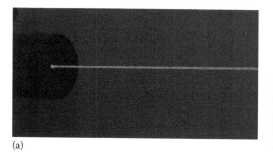

(a)

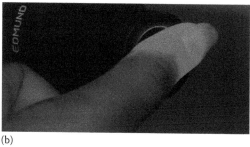

(b)

FIGURE 2.8 Light diffusion in the tissue. (a) Photograph of a beam produced by a helium–neon laser. The beam is only a few millimeters in cross section and is well collimated. (b) The diffuse glow of the light from a beam-expanded laser projected through a thumb. In passing through the thumb, the light is rendered diffuse due to multiple scattering and spreads throughout the tissue. Note that it is not strongly absorbed. When this demonstration is repeated with a green laser at 532 nm, there is no diffuse glow as this wavelength is strongly absorbed in the tissue.

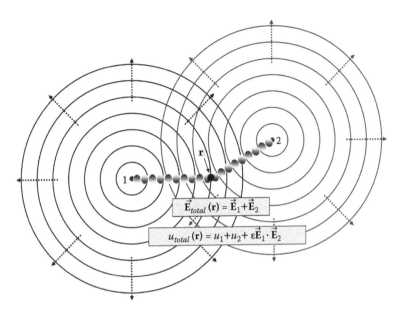

FIGURE 2.9 Superposition of EM waves. The EM field $\mathbf{E}_{total}$ at point $\mathbf{r}$ is the vector sum of the two fields from the respective sources, 1 and 2. The energy density u_{total} of the electric field point $\mathbf{r}$ due to these two waves includes an interference term to account constructive or destructive interference. In the RTT model, the interference term is ignored.

To calculate the energy per unit volume u associated with the electric field of the scattered light requires taking the square of its amplitude,

$$u(\mathbf{r}) = \frac{\varepsilon}{2}\mathbf{E}_{total}(\mathbf{r})\cdot\mathbf{E}_{total}(\mathbf{r}) = \frac{\varepsilon}{2}\left[E_1^2(\mathbf{r}) + E_2^2(\mathbf{r}) + 2\mathbf{E}_1(\mathbf{r})\cdot\mathbf{E}_2(\mathbf{r})\right]$$
$$= u_1(\mathbf{r}) + u_2(\mathbf{r}) + \varepsilon\mathbf{E}_1(\mathbf{r})\cdot\mathbf{E}_2(\mathbf{r}), \tag{2.24}$$

where ε is the *permittivity* of the medium. (Note that the permittivity ε used in this section is distinct from the molar extension coefficient ε_λ defined in Equation 2.20). The $u_1 = \varepsilon E_1^2/2$ and $u_2 = \varepsilon E_2^2/2$ terms are the contributions of each respective source as if the other source did not exist. However, the quantity expressed in Equation 2.24 also includes the term $2\mathbf{E}_1 \cdot \mathbf{E}_2$, which accounts for the interference of the two sources. If the polarization vectors of the two electric fields overlap $\left(|\mathbf{E}_1 \cdot \mathbf{E}_2| > 0\right)$, they can interfere constructively $\left(\mathbf{E}_1 \cdot \mathbf{E}_2 > 0\right)$ or destructively $\left(\mathbf{E}_1 \cdot \mathbf{E}_2 < 0\right)$ at point $\mathbf{r}$.* Therefore, the total energy density at $\mathbf{r}$ can be greater than or less than the sum of the energies of the individual sources acting alone. Still, energy must be conserved—the total energy emitted by each source and the total energy in the combined fields (with interference accounted for) calculated by integrating u over all of space are the same. In this way, the interference term has no net contribution, and the constructive and destructive interferences balance out. If the analysis is confined to some region of space, the average over that region is

$$u_{avg} = \frac{\varepsilon}{2}\overline{\mathbf{E}_{total}\cdot\mathbf{E}_{total}} = \frac{\varepsilon}{2}\left[\overline{E_1^2} + \overline{E_2^2} + 2\overline{\mathbf{E}_1 \cdot \mathbf{E}_2}\right]. \tag{2.25}$$

---

* The dot product of the two vectors is proportional to the cosine of their relative angle of orientation. If the two are perpendicular ($\theta = \pm 90°$), the product is zero.

If the region is large enough, the interference contributions average away,

$$u_{avg} = \frac{\varepsilon}{2}\left[\overline{\mathbf{E}_1^2} + \overline{\mathbf{E}_2^2}\right] = \overline{u}_1 + \overline{u}_2, \tag{2.26}$$

and

$$\overline{\mathbf{E}_1 \cdot \mathbf{E}_2} = 0. \tag{2.27}$$

For a greater number of scatterers,

$$\mathbf{E}_{total} = \sum_{j=1}^{N} \mathbf{E}_j, \tag{2.28}$$

which upon squaring has an interference term

$$\sum_{j=1}^{N}\sum_{\substack{m=1 \\ m \neq j}}^{N} \mathbf{E}_j \cdot \mathbf{E}_m. \tag{2.29}$$

Again, by considering a sufficiently large region, the interference contributions will average to zero:

$$\overline{\sum_{j=1}^{N}\sum_{\substack{m=1 \\ m \neq j}}^{N} \mathbf{E}_j \cdot \mathbf{E}_m} \approx 0. \tag{2.30}$$

The size of the region necessary for the interferences to zero out depends on how the scatterers are distributed, the nature of the scattering process itself, and the coherence of the light sources. By considering only the spatially averaged energy, the wave nature of the light can be ignored and one need track only the movement of energy through the system. This can be a simpler task, since the energies are scalar quantities (i.e., not vectors) and have no phase (i.e., they are only positive quantities). In turbid tissue, the multiple scattering due to the randomly positioned inclusions prevents the formation of any orderly (i.e., nonrandom) pattern of interference in the field. In the grand scheme of things, the contributions of the interference terms are not important to the task of tracking the flow of energy through the tissue. RTT will provide the equations that describe how this light energy moves through turbid media.

2.3.2 Radiation Transport Theory Approach

2.3.2.1 Basic Parameters

Multiple scattering in turbid media can create a very complex light field but also leads to a less demanding method of following the evolution of the light energy. The idea is to track the energy in the spreading light field using the RTT, discarding the wave-specific features of the light. The fundamental quantity that replaces the EM field in RTT model is the specific intensity, $I(\mathbf{r}, \hat{\mathbf{s}}, t)$. The following relation defines the specific intensity:

$$dP = I(\mathbf{r}, \hat{\mathbf{s}}, t) d\omega da, \tag{2.31}$$

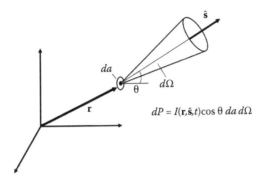

FIGURE 2.10 Specific intensity. The light power passing through a surface element *da* inside a cone of solid angle $d\Omega$ directed at an angle θ to the surface normal $\hat{n}$ is proportional to the specific intensity $I(\mathbf{r}, \hat{s}, t)$.

where dP is the light power at time t and at point $\mathbf{r}$ directed in a cone of solid angle $d\omega$ oriented in the direction defined by the unit vector $\hat{s}$, from a surface area da normal to $\hat{s}$ (Figure 2.10). Thus, $I(\mathbf{r}, \hat{s}, t)$ is the light power per unit area per unit solid angle. Since the energy in a monochromatic light field is proportional to the number of photons the field contains, the specific intensity is representative of the number of photons per second flowing from point $\mathbf{r}$ within the solid angle cone.

The medium through which the light energy propagates is characterized by three parameters:

1. The *absorption coefficient*, μ_a
2. The *scattering coefficient*, μ_s
3. The *scattering phase function* (SPF), $p(\hat{s} \cdot \hat{s}')$

where the unit vectors $\hat{s}'$ and $\hat{s}$ mark the directions of the incoming and scattered light, respectively. (The function $p(\hat{s} \cdot \hat{s}')$ does not have anything to do with the phase of propagating waves; the *phase* label is a historical legacy [Ishimaru 1978].) The coefficients μ_s and μ_a were defined earlier in Equations 2.10 and 2.16, respectively. The *attenuation coefficient* includes both the absorption and scattering coefficients and is given by

$$\mu_t = \mu_a + \mu_s. \tag{2.32}$$

The total mean free path is defined through the total attenuation coefficient as

$$l_t = \frac{1}{\mu_a + \mu_s} = \left(\frac{1}{l_a} + \frac{1}{l_s} \right)^{-1}, \tag{2.33}$$

where l_a and l_s are the mean free paths defined earlier. In this relation, the shorter mean free path is weighed more heavily in the total. For photons, l_t represents the average distance of propagation before an absorption or scattering event.

2.3.2.2 Scattering Phase Function

In RTT, the particles are assumed to be isotropic, and so the SPF is expressed in terms of $\hat{s} \cdot \hat{s}' = \cos \theta$. For light energy incident on the scatterer from the direction $\hat{s}'$, the SPF, $p(\hat{s} \cdot \hat{s}')$, describes the fraction of the energy that is scattered in the $\hat{s}$ direction. The SPF can be expressed in terms of the differential scattering cross section,

$$p(\hat{s} \cdot \hat{s}') = \frac{1}{\sigma_s} \frac{d\sigma_s}{d\Omega} (\hat{s} \cdot \hat{s}'). \tag{2.34}$$

An important constant derived from $p(\hat{s}\cdot\hat{s}')$ is the cosine-weighted average of the scattering, known as the *average cosine of scatter*:

$$g \equiv \int_{4\pi} p(\hat{s}\cdot\hat{s}')\hat{s}\cdot\hat{s}'d\Omega' = 2\pi \int_0^\pi p(\cos\theta)\cos\theta \sin\theta \, d\theta. \qquad (2.35)$$

The parameter g is a measure of scatter retained in the forward direction following a scattering event. For a Rayleigh scatterer, the SPF varies as $1+\cos^2\theta = 1+(s\cdot s')^2$, and its average cosine of scatter, g, is zero. This is because backward and forward scattering are equally likely. A scatterer with $g > 0$ is more likely to forward-scatter a photon, while a negative g indicates a preference for backward scattering. Most of the scattering processes of interest in tissue optics are within the Mie regime, and the specific SPF can be difficult to calculate without detailed knowledge of the system. An approximate SPF that is often used is the Henyey–Greenstein function (Henyey et al. 1941) (Figure 2.11):

$$p_{HG}(\cos\theta) = \frac{1}{4\pi}\frac{1-g^2}{\left(1+g^2-2g\cos\theta\right)^{3/2}}. \qquad (2.36)$$

This function is convenient to use since it is parameterized by g. For tissues, g ranges from 0.4 to 0.99. Such values indicate that scattering is strongly forward-peaked. For example, for $g = 0.6$, almost 90% of the scattered energy is within a 90° cone centered on the forward $(\theta = 0)$ direction; for $g = 0.99$, 90% of the energy is scattered to within a 5° forward cone. It is also normalized when integrated over the full 4π steradians of solid angle. Even though RTT explicitly ignores the wave nature of light, electromagnetic wave phenomena are implicitly included through the scattering-related material parameters, as these are calculated in principle using EM wave theory. The absorption depends on the molecular species in the tissue and is governed by quantum rules. Another parameter of interest is the albedo Λ,

$$\Lambda \equiv \frac{\sigma_s}{\sigma_a + \sigma_s} = \frac{\mu_s}{\mu_a + \mu_s}, \qquad (2.37)$$

which is the fraction of the total cross section that is due to scattering.

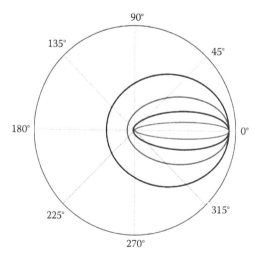

FIGURE 2.11 Henyey–Greenstein phase function. Plots of the angular dependence of the Henyey–Greenstein SPF for $g = 0.2, 0.4, 0.6,$ and 0.8, shown from outside to inside, respectively. The plots have been normalized to unity.

2.3.2.3 Radiation Transport Equation

The fundamental equation that describes light propagation in the RTT model is the *RT equation* (also known as the Boltzmann equation), which describes the fundamental dynamics of the specific intensity $I(\mathbf{r},\hat{s},t)$. Light is treated as a collection of localized, incoherent photons. (Polarization can be considered, but it is not necessary.) Consider a small packet of light energy defined by its position, $\mathbf{r}(t)$, and its direction of propagation given by the unit vector $\hat{s}$. Imagine traveling alongside the light packet as it propagates, keeping a count of the number of photons. During a short time interval dt, the packet will lose energy due to absorption and scattering out of the direction $\hat{s}$. However, the packet will also gain energy from photons scattered into the $\hat{s}$-directed packet from other directions. Finally, if we happen to be on top of a source emitting in the $\hat{s}$ direction (e.g., the end of an optical fiber), the packet may pick up more photons. These processes are related through the *radiation transport equation*:

$$\frac{1}{c_m}\frac{d}{dt}I\big(\mathbf{r}(t),\hat{s},t\big)=-(\mu_a+\mu_s)I\big(\mathbf{r}(t),\hat{s},t\big)+\frac{\mu_s}{4\pi}\int p(\hat{s}\cdot\hat{s}'')I\big(\mathbf{r}(t'),\hat{s}'',t\big)d\Omega'+Q\big(\mathbf{r}(t),\hat{s},t\big), \quad (2.38)$$

where
  c_m is the speed of light in the medium
  $Q\big(\mathbf{r}(t),\hat{s},t\big)$ is the source term

Instead of following a packet, we might also count the number of photons passing through a fixed location. Mathematically, this change in viewpoint is achieved by replacing the total time derivative with partial derivatives, $d/dt \to \partial/\partial t + c\hat{s}\cdot\vec{\nabla}$. Upon substitution, the usual form of the RT equation is obtained:

$$\frac{1}{c_m}\frac{\partial I(\mathbf{r},\hat{s},t)}{\partial t}=-\hat{s}\cdot\vec{\nabla}I(\mathbf{r},\hat{s},t)-(\mu_a+\mu_s)I(\mathbf{r},\hat{s},t)+\frac{\mu_s}{4\pi}\int_{4\pi} p(\hat{s}\cdot\hat{s}'')I(\mathbf{r},\hat{s}'',t)d\Omega'+Q(\mathbf{r},\hat{s},t). \quad (2.39)$$

In this form, the dynamics driving the change in $I(\mathbf{r},\hat{s},t)$ with time can be understood by considering the physical processes represented by each term on the right-hand side of Equation 2.39. The $-\hat{s}\cdot\vec{\nabla}I(\mathbf{r},\hat{s},t)$ term is the spatial derivative of the intensity along the $\hat{s}$ direction. The specific intensity will increase with time if this spatial derivative in the $\hat{s}$ direction is decreasing—that is, photons will *flow* from regions of high intensity to low intensity along the line parallel to $\hat{s}$. The $-(\mu_a+\mu_s)I(\mathbf{r},\hat{s},t)$ term on the right-hand side will always decrease the value of $I(\mathbf{r},\hat{s},t)$, due to scattering and absorption losses. The integral term will increase $I(\mathbf{r},\hat{s},t)$ due to photons scattered into the $\hat{s}$ cone, as will any source of light $Q(\mathbf{r},\hat{s},t)$ at $\mathbf{r}$ radiating in the $\hat{s}$ direction.

An important simplification of the RT equation is the steady-state form, which applies when the specific intensity does not change with time. The mathematical implication of the *steady-state* condition is that $\partial I(\mathbf{r},\hat{s},t)/\partial t=0$. In physical terms, the losses and gains in the medium are balanced such that the specific intensity at any point is constant in time. The steady state is applicable to many practical situations, even when the light source is pulsed or modulated. The steady-state RT equation in a source-free region is

$$\hat{s}\cdot\vec{\nabla}I(\mathbf{r},\hat{s})=-(\mu_a+\mu_s)I(\mathbf{r},\hat{s})+\frac{\mu_s}{4\pi}\int_{4\pi} p(\hat{s}\cdot\hat{s}')I(\mathbf{r},\hat{s}')d\Omega'. \quad (2.40)$$

As an example, consider a uniform light pulse that is 1 μs in duration and is incident on a 1 cm thick sample of liver ($n_{liver} = 1.38$). Over the 1 μs interval, light could traverse this sample more than 20,000 times. Within the duration of the pulse, the light field in the tissue would effectively be in a steady state.

2.3.3 Limiting Cases

Although conceptually straightforward, the direct analytical solution of the RT equation for many problems of interest is impractical. Most of the complications arise from dealing with the boundaries at tissue interfaces and geometric aspects of the tissues and light sources. In this section, the important concepts regarding the light distributions are discussed, and limiting cases of the RTT model are examined. The discussion will mainly be concerned with idealized cases in order to illustrate some basic properties of the diffuse field.

2.3.3.1 Incident and Diffuse Light

The specific intensity is zero until light is injected by an external source (e.g., a laser) or an embedded source (e.g., fiber-optic probe). In the medium, it can be expressed as

$$I(\mathbf{r},\hat{s},t) = I_c(\mathbf{r},\hat{s},t) + I_d(\mathbf{r},\hat{s},t),$$ (2.41)

where
$I_c(\mathbf{r},\hat{s},t)$ is the *primary* component which is the unscattered light
$I_d(\mathbf{r},\hat{s},t)$ is the *diffuse* component representing the scattered light

The single RT equation can be expressed as two separate equations: one involving only the primary light,

$$\frac{1}{c_m}\frac{\partial I_c(\mathbf{r},\hat{s},t)}{\partial t} + \hat{s}\cdot\vec{\nabla}I_c(\mathbf{r},\hat{s},t) = -(\mu_a+\mu_s)I_c(\mathbf{r},\hat{s},t),$$ (2.42)

and a second one involving both light components (primary and diffuse),

$$\frac{1}{c_m}\frac{\partial I_d(\mathbf{r},\hat{s},t)}{\partial t} + \hat{s}\cdot\vec{\nabla}I_d(\mathbf{r},\hat{s},t) = -(\mu_a+\mu_s)I_d(\mathbf{r},\hat{s},t) + \frac{\mu_s}{4\pi}\int_{4\pi}p(\hat{s},\hat{s}')I_d(\mathbf{r},\hat{s}',t)d\Omega'$$

$$+ \frac{\mu_s}{4\pi}\int_{4\pi}p(\hat{s},\hat{s}')I_c(\mathbf{r},\hat{s}',t)d\Omega'.$$ (2.43)

The last term in Equation 2.43 can be written as a source term for the diffuse component equation,

$$Q_c(\mathbf{r},\hat{s},t) \equiv \frac{\mu_s}{4\pi}\int_{4\pi}p(\hat{s},\hat{s}')I_c(\mathbf{r},\hat{s}',t)d\Omega'.$$ (2.44)

2.3.3.1.1 Solution for the Primary Light

If the incident field is a plane wave, the solution to Equation 2.40 for the primary (unscattered) field takes the form of the Beer–Lambert law (as shown earlier in Equation 2.19). For example, a planar source of peak intensity I_0 launched into a semi-infinite medium normal to its surface ($\hat{s}_1=\hat{z}$ and z is the depth into the tissue), the source term takes the form

$$= I_0(1-R)f(c_mt-z)e^{-(\mu_a+\mu_s)z},$$ (2.45)

where $f(c_m t - z)$ is a positive definite, normalized function describing the time envelope of the light pulse (valid for $z > 0$ and $t > 0$.) In the steady state, the plane wave solution is simply

$$I_c(z) = I_0(1-R)e^{-(\mu_a + \mu_s)z}, \tag{2.46}$$

where R is the reflectance of the interface. Note that for a plane wave, the specific intensity and intensity are related as

$$I(\mathbf{r}, \hat{s}, t) = I(z, t)\frac{\delta(\theta)}{\sin\theta}\delta(\phi), \tag{2.47}$$

where the delta functions are as defined in Ishimaru (1978). The behavior of the primary field is straightforward in most cases, and its main contribution to tissue optics is as a source for the diffuse field. In situations where absorption is relatively weak, studying light transport in tissues means tracking the distribution of the diffuse field, and thus determining I_d is the principal goal. The diffuse field is the more penetrating component of the light and is therefore of primary importance. The diffuse light gives rise to important phenomena such as diffuse reflectance. Diffuse reflectance occurs when light that penetrates into the tissue undergoes multiple scattering inside the tissue and exits back through the surface (ordinary reflectance is a surface property). This diffusely reflected light can contain information about the bulk of the medium because it has collectively sampled an extended volume of the tissue. Changes in the diffuse reflectance over time can be used to track the dynamics of certain absorbing species in the tissue, such as oxygenated hemoglobin. (Note that in other contexts, *diffuse reflectance* refers to the reflection from a rough surface or interface, not subsurface structures.)

To introduce some of the important aspects of the diffuse field and the primary light, we will examine the two limiting cases—$\mu_a \gg \mu_s$ (absorption-dominant limit) and $\mu_s \gg \mu_a$ (scattering-dominant limit)—and briefly discuss Monte Carlo calculations and Kubelka–Munk theory.

2.3.3.2 Absorption-Dominant Limit

In the case where absorption is the dominant process, the light does not penetrate the tissue deeply enough for significant scattering to occur, and the diffuse field will remain small compared to the primary one. Using Equation 2.43, one can find an approximate solution to I_d by considering only the contributions of singly scattered light from the primary field $(\mu_s/4\pi)\int_{4\pi} p(\hat{s}, \hat{s}')I_c(\mathbf{r}, \hat{s}')d\Omega'$ and ignoring the contributions of rescattered light (dropping the next to last term in Equation 2.43). Along a straight-line path of length s parallel to the direction $\hat{s}$, the equation becomes

$$\frac{d}{ds}I_d(\mathbf{r}, \hat{s}) = -(\mu_a + \mu_s)I_d(\mathbf{r}, \hat{s}) + \frac{\mu_s}{4\pi}\int_{4\pi} p(\hat{s}, \hat{s}')I_c(\mathbf{r}, \hat{s}')d\Omega', \tag{2.48}$$

where
$\mathbf{r}$ points to the path
$\mathbf{r}_0$ is the position of the path's origin

It is convenient to divide the diffuse field into components resulting from forward scattering and rearward scattering. When the primary field is planar and obeys Beer's law, a solution for diffuse field in the forward direction ($\theta = 0°$) is

$$I_d(z, \theta = 0°) = \frac{1}{4\pi}I_0\mu_t z e^{-\mu_t z}p(\cos\theta = 1), \tag{2.49}$$

where p is the SPF. In this case, the forward diffuse field peaks at a depth of $1/\mu_t$. In the backscatter direction, at a particular depth, only the primary light beyond that depth can contribute. The result is

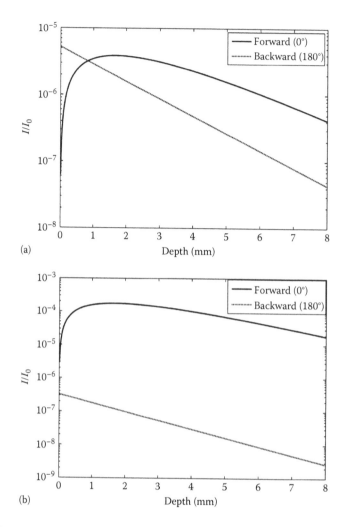

FIGURE 2.12 Absorption-dominant case. The forward- ($\theta = 0°$) and backscattered ($\theta = 180°$) diffuse light in the absorption-dominant regime ($\mu_a \gg \mu_s$) for $\mu_t \cong \mu_a = 0.6$ mm^{-1}. (a) $g = 0.0$ (isotropic scattering) and (b) $g = 0.8$. The mean free path in both cases is 1.67 mm.

the rearward field has no peak and falls off exponentially with depth. Figure 2.12 plots the forward- and backscattered fields ($\theta = 0°$ and $\theta = 180°$, respectively) for two values of g.

2.3.3.3 Scattering-Dominant Limit: The Diffusion Approximation

Scattering is the dominant transport process when the absorption is sufficiently low to permit significant penetration of light into the tissue. This case, known as the *diffusion limit*, is important because light spreads through the tissue, although it is dispersed in a random fashion. In the diffusion process, *particles* (in our case photons) moving through the medium do so in a series of steps of random length and direction (i.e., random walk). Each step begins with a scattering event, and the photon is equally likely to emerge in any direction. Note that the scattering in tissues is not isotropic but has a bias for the forward direction. This is seemingly at odds with the isotropic scattering that comes along with the diffusion limit. To reconcile these requires redefining the step size. It turns out that a single step in an isotropic scattering process can be mimicked by a sufficient number of directional scattering steps. The number of steps necessary depends on the average cosine of scatter. The coefficient describing this effective isotropic scattering is the reduced scattering coefficient μ_s'. The reduced scattering coefficient is related to the previously defined scattering parameters as follows:

$$\mu_s' = (1-g)\mu_s. \tag{2.50}$$

On average, the number of directional scattering events that are equivalent to a single isotropic scattering event is given by $1/(1-g)$. For example, in a medium with $g = 0.75$, it takes an average of four scattering events for a population of photons to disperse isotropically. In tissues, one encounters values of g from 0.4 to >0.99, which results in isotropic dispersion of photons after as few as 2 steps to over 100 steps. The photons can also be absorbed as they propagate, and the absorption properties of the medium are encompassed by the usual absorption coefficient, μ_a.

When the number of photons undergoing the random walk is large, the density of photons can be described by a continuous function whose dynamics are described by the diffusion equation. In the diffusion limit, the properties of the medium are contained in the diffusion constant,

$$D = \frac{c_m}{3\left(\mu_a + (1-g)\mu_s\right)}, \tag{2.51}$$

which has dimensions of length squared over time. The quantity

$$\mu_t' \equiv \mu_a + (1-g)\mu_s \tag{2.52}$$

is called the transfer attenuation of the medium. The diffusion coefficient can also be written as

$$D = \frac{1}{3}c_m l_t', \tag{2.53}$$

where $l_t' = 1/\mu_t'$ is the effective mean free path. The diffusion coefficient is the tissue property that governs the spatial width of the photon distribution. For the following examples, the *widths* of the photon distributions are proportional to $\sqrt{D}$.

2.3.3.3.1 Diffusion Equation

When the scattering processes are dominant, the angular dependence of the specific intensity is well approximated by the first-order expansion in the unit vector $\hat{s}$,

$$I_d(\mathbf{r},\hat{s},t) \cong \frac{1}{4\pi}\Phi_d(\mathbf{r},t) + \frac{3}{4\pi}F_d(\mathbf{r},t)\hat{s}_f \cdot \hat{s}, \tag{2.54}$$

where

$$\Phi_d(\mathbf{r},t) = \int_{4\pi} I_d(\mathbf{r},\hat{s},t)\,d\Omega \tag{2.55}$$

is the total intensity at point **r**, known as the fluence rate. The fluence rate can also be written as $\Phi_d(\mathbf{r},t) = h\nu c_m \eta_d(\mathbf{r},t)$, where $\eta_d(\mathbf{r},t)$ is the photon density. The second term on the right side of Equation 2.54 includes

$$\mathbf{F}_d(\mathbf{r},t) = F_d(\mathbf{r},t)\hat{\mathbf{s}}_f = \int_{4\pi} I_d(\mathbf{r},\hat{\mathbf{s}},t)\hat{\mathbf{s}}\,d\Omega, \tag{2.56}$$

the net intensity vector. Similarly, the net intensity vector is proportional to the photon current density, $\mathbf{J}(\mathbf{r},t) = \mathbf{F}(\mathbf{r},t)/h\nu$. Integrating all the terms in the RT equation over the entire 4π of solid angle results in a new relation expressed in terms of $\Phi_d(\mathbf{r},t)$ and $\mathbf{F}_d(\mathbf{r},t)$:

$$\frac{1}{c_m}\frac{\partial}{\partial t}\Phi_d(\mathbf{r},t) + \vec{\nabla}\cdot\mathbf{F}_d(\mathbf{r},t) = -\mu_a\Phi_d(\mathbf{r},t) + Q, \tag{2.57}$$

where Q represents a source. This equation provides an expression of the divergence of $\mathbf{F}_d$ in terms of the time rate of change of the total intensity Φ_d.* The next step is to define a second relation between $\mathbf{F}_d$ and Φ_d and to use this new relation to eliminate $\mathbf{F}_d$. The lowest-order approximation is obtained by using Fick's law,

$$c_m\mathbf{F}_d(\mathbf{r},t) = -D\vec{\nabla}\Phi_d(\mathbf{r},t). \tag{2.58a}$$

Fick's law asserts that the net flow of photons through a point in space is in the direction that the photon density decreases most steeply. Essentially, the net flow is from high photon concentrations to low ones. Substituting Equation 2.58a into Equation 2.57 yields the diffusion equation

$$\frac{\partial}{\partial t}\Phi_d(\mathbf{r},t) = D\nabla^2\Phi_d(\mathbf{r},t) - \mu_a c_m\Phi_d(\mathbf{r},t) + Q. \tag{2.59}$$

(A higher-order approximation for the diffusion problem can be derived from the RT equation in which the constants D, μ'_{tr}, and g arise naturally in the process. The resulting relation is called the telegrapher's equation. However, the telegrapher's equation has not been found to offer any significant advantages over the diffusion equation for light propagation in turbid tissues (Das et al. 1997).)

Solving the diffusion equation analytically can be impractical for many cases of interest. To gain some insight into the nature of diffuse light fields, it is instructive to look at solutions to the diffusion equation in some ideal cases, two of which are examined in the following. These can be used as starting points for solving for the fluence rate in a more realistic scenario. The numerical MC method, an effective computational technique for solving diffusion problems, is described later in this section.

---

* The essential step is the calculation of the integral over the SPF:

$$\frac{\mu_s}{4\pi}\iint_{4\pi} p(\hat{\mathbf{s}},\hat{\mathbf{s}}')I_d(\mathbf{r},\hat{\mathbf{s}}',t)\,d\Omega'd\Omega = \frac{\mu_s}{4\pi}\int_{4\pi}\left[\int_{4\pi} p(\hat{\mathbf{s}},\hat{\mathbf{s}}')\,d\Omega\right]I_d(\mathbf{r},\hat{\mathbf{s}}',t)\,d\Omega'$$

$$= \mu_s\int_{4\pi} I_d(\mathbf{r},\hat{\mathbf{s}}',t)\,d\Omega'$$

$$= \mu_s\Phi_d(\mathbf{r},t) \tag{2.58b}$$

2.3.3.3.2 *Time-Dependent Solutions*

In the following, time-dependent solutions of Equation 2.59 are given for the case in which material boundaries can be neglected. A source is used to inject light at $t = 0$ but is removed immediately after $(Q = 0, t > 0)$. Following a single injection of N photons into the medium at point $\mathbf{r}_0$ at time $t = 0$, the solution is

$$\Phi_d(\mathbf{r}, t) = hvc_m \frac{Ne^{-\mu_a c_m t}}{(4\pi Dt)^{3/2}} e^{-r^2/4Dt}, \tag{2.60}$$

where $r = |\mathbf{r} - \mathbf{r}_0|$ (Figure 2.13). The spherically symmetric distribution of the light is Gaussian and has an effective radius (e^{-1} radius) of $r_E = 2\sqrt{Dt}$. The numerator of the first factor, $Ne^{-\mu_a c_m t}$, contains the total number of photons that exist at time t, which is decreasing in time due to absorption. The denominator, $(4\pi Dt)^{3/2} = \pi^{3/2}r_E^3$, is proportional to the cube of the effective radius. The denominator is thus proportional to the volume of a sphere of radius r_E and is essentially a volumetric normalization factor. At any time, about 40% of the photons will lie within this sphere, and almost 90% will lie within the sphere of radius $3r_E$. The photon density itself falls to half of its maximum inside the effective radius:

$$r_{1/2} = 1.67\sqrt{Dt}. \tag{2.61}$$

Toward the center of the distribution, the local photon density is decreasing with time, while farther out, the density is increasing with time since the net flow is from higher to lower concentrations. The boundary between the two regions occurs at the radius:

$$r_{\substack{growth \\ boundary}} = \sqrt{6Dt\left(1 + \frac{2}{3}\mu_a c_m t\right)}. \tag{2.62}$$

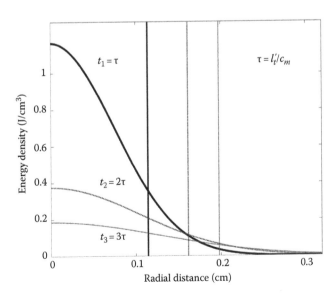

FIGURE 2.13 Scattering-dominant case—time dependent. The photon density in the time-dependent diffusion process due to the point injection of photons at $t = 0$. The material parameters are $\mu_s = 50$ cm^{-1}, $\mu_a = 1$ cm^{-1}, and $g = 0.8$. The three curves represent the solution for time $t_1 = \tau$, $t_2 = 2\tau$, and $t_3 = 3\tau$, where $\tau = l_t'/c_m$ is the time it takes for light to propagate one mean free path, which is 0.09 cm for this case.

This radius defines the boundary between the region of space from which energy is escaping and the surrounding region that sees a net increase in light energy with time. Similar solutions are available for other source geometries. For a line source of photons, injected at $t = 0$, the time-dependent solution is

$$hvc_m \frac{N_L e^{-\mu_a c_m t}}{4\pi Dt} e^{-\rho^2/4Dt},$$ (2.63)

where

    ρ is the perpendicular distance from the source line to a point in the medium
    N_L is the number of photons emitted *per unit of length* of the source

Here, the distribution has cylindrical symmetry, and a little more than 60% of the energy is within the cylinder of $\rho_{1/e} = \sqrt{4Dt}$. The growth boundary occurs at

$$\rho_{growth\ boundary} = \sqrt{4Dt(1 + \mu_a c_m t)}.$$ (2.64)

2.3.3.3.3 Steady-State Solutions

In the steady-state regime in a source-free region, the diffusion equation becomes

$$\nabla^2 \Phi_d(\mathbf{r}) - \kappa_d^2 \Phi_d(\mathbf{r}) = 0,$$ (2.65)

where

$$\frac{1}{\kappa_d} = \sqrt{\frac{D}{\mu_a c_m}} = \sqrt{\frac{1}{3\mu_a(\mu_a + (1-g)\mu_s)}}$$ (2.66)

is the diffusion length. A point source of light at $r = 0$, injecting photons at a rate to maintain the steady state, has a photon density away from the source of

$$\frac{\Phi_d(\mathbf{r})}{hvc_m} = \frac{N\kappa_d^2}{4\pi} \frac{e^{-\kappa_d r}}{r}.$$ (2.67)

In this case, about 25% of the photons are inside the sphere of radius $1/\kappa_d$. A sphere of radius $4/\kappa_d$ contains 90% of the photons (see Figure 2.14). For a line source, the photon density is

$$\frac{\Phi_d(\mathbf{r})}{hvc_m} = \frac{N_L \kappa_d^2}{2\pi} K_0(\kappa_d \rho),$$ (2.68)

where

    $K_0(\kappa_d \rho)$ is the zero-order modified Bessel function of the second kind
    N_L is the number of photons per unit length

For large values of $\kappa_d \rho$, $K_0(\kappa_d \rho) \sim e^{-\kappa_d \rho}/\sqrt{\kappa_d \rho}$.
    A cylinder of radius $1/\kappa_d$ contains 44% of the photons and a cylinder of radius $3.2/\kappa_d$ holds 90%.

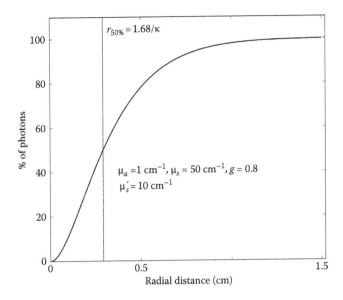

FIGURE 2.14 Scattering-dominant case—steady state. The fraction of photons within the given radius for the point source solution of the steady-state diffusion problem. The mean free path is 0.09 cm.

2.3.4 Numerical Approach: Monte Carlo Simulations

Monte Carlo (MC) methods are a broad class of computational techniques that employ random sampling. The MC simulations represent stochastic processes whose dynamics are governed by random steps and can provide a method of solution for practical problems in tissue optics. Among the first large-scale applications of MC methods were RT problems involving neutrons, and thus the MC approach is well suited to problems involving light transport in turbid media. The method is named for the administrative area of Monaco well known for its casinos. The random behavior in games of chance is similar to how an MC simulation selects variable values at random to simulate the dynamics of a model. Consider rolling a pair of dice—a number between 2 and 12 must result, but any prediction is only a guess. However, the probability of rolling a seven is six times higher than rolling a two, and with a large number of rolls, the results will approach this ratio. This is analogous to what occurs with individual photons inside tissues, whose likelihood of being absorbed or scattered is governed by probability.

The MC simulation is a very important tool in tissue optics and can be used for any albedo and SPF. In a simulation, single photons are traced through the sample step by step, and the distribution of light in the system is built from these single-photon trajectories. The algorithms for implementing the basic elements of an MC simulation are relatively simple. The parameters of each step (length, direction, weighting factor, etc.) are calculated from probability distributions that are randomly sampled. As the number of photons grows, an increasingly accurate solution to the RT problem is obtained. The actual number of photons that are necessary for a result of practical value (i.e., a good statistical sample) depends on the specific physical quantities that are to be determined and the parameters of the simulation. As few as 3,000 photons may be adequate for determining diffuse reflectance from a sample, while more than 100,000 may be required for a complex 3D simulation. (Wang et al. 1995)

In the following, the method of randomly choosing the step size is illustrated as a simple example of how parameters are generated in an MC simulation. The exponential decrease in source intensity with depth due to scattering is expressed using Beer's law,

$$\frac{I(L)}{I_0} = e^{-\mu_s L}, \tag{2.69}$$

which also represents the fraction of unscattered photons remaining after traveling a distance L. For an individual photon, this can be interpreted as the probability $p_L \left(0 \le p_L \le 1 \right)$ that a photon will be scattered somewhere between 0 and L,

$$p_L = e^{-\mu_s L}. \tag{2.70}$$

This relation is inverted to determine a step size L in terms of the probability,

$$L = -\frac{\ln p}{\mu_s}. \tag{2.71}$$

The probability is then treated as a random number drawn from a uniform distribution $f(p) = 1, \left(0 < p \le 1 \right)$, equally likely to have any value between 0 and 1. The expectation value of $\ln p$ over the uniform distribution is –1, which implies that the average value of the step length will converge to the mean free path, $\bar{L} = 1/\mu_t$, as the number of steps grows larger. Other parameters associated with a photon's path and ultimate fate can be determined in a similarly straightforward manner. MC methods have been used for various transport processes as well as for emission in tissues (Gardner et al. 1996).

2.3.5 Kubelka–Munk Model

For optically inhomogeneous media, the expression for quantitative analysis can be derived from an equation developed for optical studies on surfaces with diffuse reflectance (Vo-Dinh 1984). Schuster first conceived the continuum model and derived the basic differential equations for diffuse scattering materials to describe the behavior of radiation from a star in a foggy atmosphere (Schuster 1905). Schuster's model described two diffuse fluxes traveling in opposite directions through the gas, each contributing backscattered flux to each other. Kubelka and Munk further improved the model and derived differential equations similar to those developed by Schuster with the exception of the emission terms (Kubelka and Munk 1931). Later, Kubelka developed a set of useful equations that provided the foundations for many quantitative studies of absorption, scattering, and luminescence processes in diffuse scattering media. These equations are known as the K–M equations (Kubelka 1948):

$$-\frac{di}{dx} = -\left(S + K \right) i + Sj, \tag{2.72}$$

$$\frac{dj}{dx} = -\left(S + K \right) j + Si, \tag{2.73}$$

where
    i is the intensity of light propagating inside the sample in the forward (transmitted) direction
    j is the intensity of light propagating in the backscattered direction
    S is the scatter coefficient per unit thickness
    K is the absorption coefficient per unit thickness
    x is the distance from the nonilluminated side

The K–M equations simply describe the fact that the light beam traveling in the transmitted direction (i) decreases in intensity due to the absorption (K) and scattering (S) processes and gains intensity from the scattering process that occurs in the beam coming from the other direction (j).

The basic assumptions in the K–M model are that the sample is a planar, homogeneous, and ideal diffuser illuminated on one side with diffuse monochromatic light. The K–M model also assumes that the reflection and absorption processes occur at infinitesimal distances and are constant over the area under illumination and over the thickness of the sample. It is also implicitly assumed in the 1D approximation of the K–M model that the reflected light beam is normal to the sample surface.

The general solutions in i and j for the K–M equations are given by

$$i = A \sinh(bSx) - B \cosh(bSx) \tag{2.74}$$

$$j = (aA - bB)\sinh(bSx) - (aB - bA)\cosh(bSx), \tag{2.75}$$

where

$$a = 1 + \frac{K}{S} \tag{2.76}$$

$$b = \sqrt{a^2 - 1}. \tag{2.77}$$

The arbitrary constants A and B can be eliminated using the following conditions:

$$i = I_0 T; \quad j = 0 \quad \text{for} \quad x = 0, \tag{2.78}$$

$$i = I_0; \quad j = I_0 R \quad \text{for} \quad x = X. \tag{2.79}$$

The transmittance T and reflectance R are then given by

$$T = \frac{b}{a \sinh(bSX) + b \cosh(bSX)}, \tag{2.80}$$

$$R = \frac{\sinh(bSX)}{a \sinh(bSX) + b \cosh(bSX)}, \tag{2.81}$$

where X is the thickness of the sample.

The solution for T and R can be used to determine the coefficient SX.

For a scattering medium that does not absorb light (i.e., when $K = 0$), the transmission T and reflectance R become

$$T_0 = \frac{1}{1 + SX}, \tag{2.82}$$

$$R_0 = \frac{SX}{1 + SX}. \tag{2.83}$$

The value of SX can be determined using these equations. In general, T_0 can be derived from measurement of the absorbance $A_0 = -\log(T_0)$.

There has been a great interest in using the K–M model for investigating optical properties of diffuse media (Agati et al. 1993; Beuthan et al. 1996; Bjorn 1996; Cheong et al. 1990; Christy et al. 1995; Durkin et al. 1994; Hoffmann et al. 1998; Koukoulas and Jordan 1997; Loyalka and Riggs 1995; Mandelis and Grossman 1992; Molenaar et al. 1999; Philips-Invernizzi et al. 2001; Ragain and Johnston 1998, 2001; Rundlof and Bristow 1997; Tsuchikawa and Tsutsumi 1999; Vargas and Niklasson 1997; Waters 1994). A method utilizing the K–M theory was developed to allow the quantification of tissue reflectance spectra to study in vivo kinetic changes in the oxygen saturation of hemoglobin and myoglobin (Hoffmann et al. 1998). The K–M model was used for investigating the quantitative optical biopsy of tissues ex vivo (Beuthan et al. 1996). Later models use four fluxes, which, in addition to the diffuse forward and backward fluxes, also involve collimated forward and backward fluxes. Multiple-flux models provide data that are in good agreement with experimental results. However, the applicability of the K–M models is limited to simple slab geometries (Beuthan et al. 1996).

2.3.6 Effective Index of Refraction

Turbid tissues are heterogeneous and their optical properties depend on their composition and physical structure. In many cases, composite media can also be characterized by a single *effective* index of refraction that governs the speed of light in the tissue. This is the speed that governs the propagation of the ballistic component of light that is discussed in the next section. In this section, the effective index is described in the context of a two-component medium of scattering inclusions embedded in a host. The effective index arises from the interference between the source and scattered fields. For a plane wave source, this interference occurs when a component of the scattered field forms a planar wavefront. (The forward scattered light is the most important in this respect, although in theory there can be contributions from doubly backscattered light and the like in media with monosized spheres.) This interference manifests itself as a phase shift in the coherent field, and hence a shift in the time it takes light to propagate through the medium relative to the host alone. In this manner, the speed of light in the medium is affected by the scatterers. If we consider a slab with no embedded scatterers, the photon propagation time through the tissue is

$$\tau_0 = \frac{n_m}{c} L, \tag{2.84}$$

where
    L is the width of the tissue sample
    n_m is the index of refraction for the single medium

In a two-component medium, consisting of a host with index n_{host} and embedded scatterers with index n_{sc}, the question becomes how to define an index of refraction characteristic of the composite medium. A commonly used approximation is to compute the effective index of refraction as the volume-weighted average of its constituents,

$$n_{eff} = (1 - V_f)n_{host} + V_f n_{sc}, \tag{2.85}$$

where V_f is the volume fraction of the scatterers embedded in the medium. This can also be written as

$$n_{eff} = n_{host}\left(1 + V_f \delta\right), \tag{2.86}$$

where $n_{sc}/n_{host} = 1 + \delta$ and δ is the fractional difference between the two indices. This approximation has proven to be highly accurate in the long wavelength limit (Liu et al. 1991). As the size-to-wavelength ratio grows, the volume-weighted approximation gives way to a method that explicitly involves the scattered field. When the scatterers are irregularly shaped and/or have a wide variation of sizes, the volume-weighted average approach may remain an adequate approximation for calculating n_{eff}. For spherical scatterers, the impact of the interference on the effective refractive index for the medium can be significant, even when the scatterers are randomly dispersed in the host. This dependence of the effective light speed on the scatterers is a manifestation of the wave nature of light and cannot be derived from RTT. For more details on the role of scattering in determining the index of refraction, see Chapter 31 of Feynman et al. (1963) or Chapter 10 of Jackson (1999).

There are several theoretical formulations for defining the speed of directional light transport in a multiple scattering medium. In tissues, with their irregular structures, the singly scattered field in the forward direction is the source of the effect. Nevertheless, the contribution of multiply scattered fields on the effective index can be observed when working with more idealized structures such as tissue phantoms. Derivations of effective wavenumbers for coherent multiple scattering are discussed in Lloyd and Berry (1967). A formulation of the effective index adequate for illustrative purposes is given by Watermann and Truell (1961), which builds on the earlier work of Foldy (1945) and Lax (1951). The interference of the coherent scatter and the incident field gives rise to the following relation for the effective index of refraction of the medium:

$$n_{eff}^2 = n_{host}^2 \left(\left(1 + V_f \frac{3}{2k^2 a^3} f(\theta) \right)^2 - \left(V_f \frac{3}{2k^2 a^3} f(\pi) \right)^2 \right), \tag{2.87}$$

where
$k = 2\pi/\lambda_m = 2\pi n_{host}/\lambda$ is the wavenumber in the host
a is the scatterer radius
$f(\theta)$ is the scattering amplitude

$f(\theta)$ is related to the differential cross section as

$$|f(\theta)|^2 = \frac{d\sigma}{d\Omega}(\theta). \tag{2.88}$$

If we examine the case of scattering in the Rayleigh limit, where $|f(0)| = |f(\pi)|$, the relation becomes

$$n_{eff}^2 = n_{host}^2 \left(1 + V_f \frac{3}{k^2 a^3} f(0) \right). \tag{2.89}$$

Plugging in the explicit form of $f(\theta)$ for Rayleigh scattering and performing some substitutions yields

$$= n_{host}^2 \left(1 + V_f \delta \frac{6 + 3\delta}{3 + 2\delta + \delta^2} \right), \tag{2.90}$$

where $\delta = (n_{sc} - n_{host})/n_{host}$ as before. For small $V_f \delta$, the effective index can be approximated to lowest order in δ as

$$n_{eff} = n_{host} \left(1 + V_f \delta \right), \tag{2.86, 2.91}$$

which is equivalent to the volume-weighted result. This approximation is linear in the volume fraction, which is indicative of the single-scattering limit. The approximate relation performs quite well over realistic ranges of δ and V_f. Refractive indices for turbid tissue constituents range from 1.33 (water) to above 1.6 (melanin particles), so 0.2 could be considered a relatively large value for δ. For Rayleigh scatterers, using reasonable values for V_f and δ, the volume-weighted approach can be accurate to at least the 1% level, if not better.

2.3.7 Time-Resolved Propagation of Light Pulses

A pulse of light launched into a slab of turbid tissue can be broken into two components: (1) *source* light (i.e., that retains the shape and directionality of the original pulse) and (2) the diffuse component. The first type is also called the *ballistic* component, as it follows a straight path through the tissue, and is the first light to emerge from the far side of the slab. The speed of the ballistic photons is governed by the effective index of refraction, $c_{slab} = c/n_{eff}$, and so is not devoid of the influence of the scatterers. Following the ballistic component, the next to emerge is the early diffuse light consisting of *snake* photons. The snake photons forward-scatter through the medium in nearly a straight line, with paths resembling the shape of a crawling snake. If a detector with sufficient time resolution is used to record the emerging light, the snake photons can be isolated from the rest of the diffuse light. The bulk of the diffuse light cloud follows, consisting of the photons that have taken longer random walks through the slab.

The predictable path of the ballistic photon group makes it useful for absorption mapping and tomographic imaging (see Figure 2.15). The snake photons can also be used for mapping and tomography, as their paths are nearly straight. In scattering-dominant regimes, the snake components have the advantage over the ballistic photons because the ballistic component is more strongly attenuated as thickness increases. The remaining diffuse photons take random paths that can only be characterized statistically. The diffuse photons carry useful information about the physiological state of the tissues to the surface of the body, and their broad spatial sampling can be an advantage in some situations. There are techniques that can also make use of diffuse light for imaging, as discussed elsewhere in this handbook.

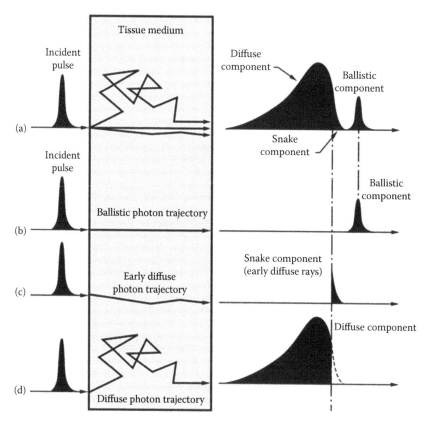

FIGURE 2.15 Time resolved propagation and photon trajectories in turbid tissue. (a) On the left is a pulse of light that is launched into the tissue. On the right is the distribution of the light that emerges (enlarged in scale). The distribution is broken out into the contributions from (b) ballistic photons, (c) early diffuse "snake" photons, and (d) the remaining diffuse photons. (Adapted from Das, B.B. et al., *Rep. Prog. Phys.*, 60, 227, 1997.)

The use of time-resolved techniques to investigate photon transport processes requires specific conditions, depending on the applications. If the aim is to be able to separate the ballistic, snake, and diffuse components in a through-transmission experiment, then the pulse width of light should be short, typically on the order of ps (10^{-12} s) or less. For a single pulse of light to fit within a 1 cm thick liver sample, it must be shorter than 50 ps (5×10^{-11} s). In order to differentiate the diffuse and ballistic components from this sample, a pulse would need to be significantly shorter than this.

Another issue for pulse propagation is dispersion, the change in the pulse shape of the ballistic component. In this case, the notion of a single velocity is somewhat ill-defined. As many as five different velocities (Brillouin 1960) have been defined for pulse transmission through dispersive media. Only two of these are relevant for our purposes, the phase velocity and the group velocity. Before now, the light speeds we have discussed are equivalent to the phase velocity, given by

$$c_m(\lambda) = \frac{c_0}{n(\lambda)}. \tag{2.1, 2.92}$$

Phase velocity is the speed of the crests of the wave for each individual wavelength component in the pulse. The phase velocity can be written as $1/(k_m/\omega)$, where $\omega = 2\pi\nu$ and $k_m = 2\pi/\lambda_m$, or more simply as $\nu\lambda_m$. For a pulse of light typical of tissue studies, the speed of the envelope (i.e., pulse shape) is given by the group velocity. The group velocity is defined as

$$c_m^{group} = \frac{1}{dk_m/d\omega} \tag{2.93}$$

$$= \frac{c_0}{n} \frac{1}{1-(\lambda/n)(dn/d\lambda)}. \tag{2.94}$$

In regions of *normal dispersion*, where $dn/d\lambda < 0$, the group velocity is a good measure of the speed of the pulse envelope. Note that the group velocity is always less than the phase velocity in this regime. In regions of anomalous dispersion, where $dn/d\lambda > 0$, this simple interpretation of the group velocity is obscured by the strong absorption. Absorption always accompanies anomalous dispersion resulting in a greatly diminished pulse amplitude and a distorted pulse shape. The definition of the *group index of refraction* follows from Equation 2.94,

$$n_g = n\left(1 - \frac{\lambda}{n}\frac{dn}{d\lambda}\right). \tag{2.95}$$

A specific type of pulsed light finding applications in biomedicine is supercontinuum light (Kano and Harnaguchi 2007; Unterhuber et al. 2004). These pulses consist of a very large collection of wavelengths and can be extremely short in duration (Ell et al. 2001). Assigning a single speed to such pulses can require a more detailed approach if there is strong dispersion.

2.4 Tissue Properties

In this section, we present the values for the optical properties of tissues and tissue constituents and provide some background and discussion. The index of refraction is the most fundamental optical property of a material and values for tissues are offered in Table 2.1. Following this, the scattering and absorption properties of tissues and their underlying physical origins are discussed. This includes an account of the absorption characteristics of various biomolecules and molecular subgroups and their

TABLE 2.1 Index of Refraction for Various Tissues, Tissue Components, and Biofluids

Tissue	Description	Wavelength (nm)	Index of Refraction	References
Stratum corneum		—	1.55	Duck (1990)
Stratum corneum	n_g	1300	1.51 ± 0.02	Tearney et al. (1995)
Epidermis	n_g	1300	1.34 ± 0.01	Tearney et al. (1995)
Dermis	n_g	1300	1.41 ± 0.03	Tearney et al. (1995)
Dermis	n_g	1300	1.400 ± 0.007	Tearney et al. (1995)
Stratum corneum	Palm of hand, $(nn_g)^{1/2}$	1300	1.47 ± 0.01	Knuttel et al. (2000)
Epidermis	Palm of hand, granular layer, $(nn_g)^{1/2}$	1300	1.43 ± 0.02	Knuttel et al. (2000)
Epidermis	Palm of hand, basal layer, $(nn_g)^{1/2}$	1300	1.34 ± 0.02	Knuttel et al. (2000)
Epidermis	Volar side of lower arm, $(nn_g)^{1/2}$	1300	1.36 ± 0.01	Knuttel et al. (2000)
Upper dermis	Palm of hand, $(nn_g)^{1/2}$	1300	1.41 ± 0.03	Knuttel et al. (2000)
Upper dermis	Volar side of lower arm, $(nn_g)^{1/2}$	1300	1.43 ± 0.02	Knuttel et al. (2000)
Stratum corneum	Dorsal surface of thumb, $(nn_g)^{1/2}$	980	1.50 ± 0.02	Alexandrov et al. (2003)
Air/skin interface	Volar side of a thumb, $(nn_g)^{1/2}$	980	1.56	Zvyagin et al. (2003)
SC/epidermis interface	Volar side of a thumb, $(nn_g)^{1/2}$	980	1.34	Zvyagin et al. (2003)
Epidermis, excised (fresh)	12 females, 27–63 years; 10 Caucasians (a), 2 African Americans (b)		(Uncert = ± 0.006)	
s and p polarization	(a)	325	1.489(s) 1.486(p)	Ding et al. (2006)
	(a and b)	442	1.449(s) 1.447(p)	Ding et al. (2006)
	(a and b)	532	1.448(s) 1.446(p)	Ding et al. (2006)
	(a)	633	1.433(s) 1.433(p)	Ding et al. (2006)
	(a)	850	1.417(s) 1.416(p)	Ding et al. (2006)
	(a)	1064	1.432(s) 1.428(p)	Ding et al. (2006)
	(a)	1310	1.425(s) 1.421(p)	Ding et al. (2006)
	(a)	1557	1.404(s) 1.400(p)	Ding et al. (2006)
Dermis, excised (fresh)	(a)	325	1.401(s) 1.403(p)	Ding et al. (2006)
	(a and b)	442	1.395(s) 1.400(p)	Ding et al. (2006)
	(a and b)	532	1.378(s) 1.381(p)	Ding et al. (2006)
	(a)	633	1.396(s) 1.393(p)	Ding et al. (2006)
	(a)	850	1.384(s) 1.389(p)	Ding et al. (2006)
	(a)	1064	1.375(s) 1.385(p)	Ding et al. (2006)
	(a)	1310	1.358(s) 1.364(p)	Ding et al. (2006)
	(a)	1557	1.363(s) 1.367(p)	Ding et al. (2006)
Porcine skin				
	$(nn_g)^{1/2}$	1300	1.415	Knuttel et al. (2004)
	Treated with detergent, $(nn_g)^{1/2}$	1300	1.365	Knuttel et al. (2004)
Dermis		325	1.393	Ding et al. (2005); Ma et al. (2005)
		442	1.376	Ding et al. (2005); Ma et al. (2005)

TABLE 2.1 (continued)   Index of Refraction for Various Tissues, Tissue Components, and Biofluids

Tissue	Description	Wavelength (nm)	Index of Refraction	References
		532	1.359	Ding et al. (2005); Ma et al. (2005)
		633	1.354	Ding et al. (2005); Ma et al. (2005)
		850	1.364	Ding et al. (2005); Ma et al. (2005)
		1064	1.36	Ding et al. (2005); Ma et al. (2005)
		1310	1.357	Ding et al. (2005); Ma et al. (2005)
		1557	1.361	Ding et al. (2005); Ma et al. (2005)
Rat skin		456–1064	1.42	Muller et al. (1995)
Mouse skin		456–1064	1.4	Muller et al. (1995)
Human brain				
Gray matter		456–1064	1.36	Muller et al. (1995)
White matter		456–1064	1.38	Muller et al. (1995)
White and gray		456–1064	1.37	Muller et al. (1995)
Human aorta				
Normal intima		456–1064	1.39	Muller et al. (1995)
Normal media		456–1064	1.38	Muller et al. (1995)
Normal adventitia		456–1064	1.36	Muller et al. (1995)
Calcified intima		456–1064	1.39	Muller et al. (1995)
Calcified media		456–1064	1.53	Muller et al. (1995)
Human bladder				
Mucous		456–1064	1.37	Muller et al. (1995)
Wall		456–1064	1.4	Muller et al. (1995)
Integral		456–1064	1.38	Muller et al. (1995)
Human colon				
Muscle		456–1064	1.36	Muller et al. (1995)
Submucous		456–1064	1.36	Muller et al. (1995)
Mucous		456–1064	1.38	Muller et al. (1995)
Integral		456–1064	1.36	Muller et al. (1995)
Human esophagus				
Mucous		456–1064	1.37	Muller et al. (1995)
Human fat				
Subcutaneous		456–1064	1.44	Muller et al. (1995)
Abdominal		456–1064	1.46	Muller et al. (1995)
Mesenteric		1300	1.467 ± 0.008	Tearney et al. (1995)
Porcine fat				
		488	1.510 ± 0.002	Cheng et al. (2002)
		632.8	1.492 ± 0.003	Cheng et al. (2002)
		1079.5	1.482 ± 0.002	Cheng et al. (2002)
		1341.4	1.478 ± 0.004	Cheng et al. (2002)
		632.8	1.493 ± 0.005	Li et al. (1996)

(continued)

TABLE 2.1 (continued)   Index of Refraction for Various Tissues, Tissue Components, and Biofluids

Tissue	Description	Wavelength (nm)	Index of Refraction	References
Bovine fat		633	1.455	Bolin et al. (1989)
Human heart				
Trabecula		456–1064	1.4	Muller et al. (1995)
Myocardial		456–1064	1.38	Muller et al. (1995)
Myocardial	Left ventricle	1300	1.382 ± 0.007	Tearney et al. (1995)
Human femoral vein		456–1064	1.39	Muller et al. (1995)
Kidney				
Human		456–1064	1.37	Muller et al. (1995)
Human		633	1.417	Bolin et al. (1989)
Canine		633	1.4	Bolin et al. (1989)
Porcine		633	1.39	Bolin et al. (1989)
Bovine		633	1.39	Bolin et al. (1989)
Liver				
Human		456–1064	1.38	Muller et al. (1995)
Human		633	1.367	Bolin et al. (1989)
Canine		633	1.38	Bolin et al. (1989)
Porcine		633	1.39	Bolin et al. (1989)
Bovine		633	1.39	Bolin et al. (1989)
Lung				
Human		456–1064	1.38	Muller et al. (1995)
Canine		633	1.38	Bolin et al. (1989)
Porcine		633	1.38	Bolin et al. (1989)
Human muscle				
		456–1064	1.37	Muller et al. (1995)
Myocardial	Left ventricle	1300	1.382 ± 0.007	Tearney et al. (1995)
Canine muscle				
		633	1.4	Bolin et al. (1989)
Bovine muscle				
		633	1.41	Bolin et al. (1989)
		592(560–640)	1.382 ± 0.004	Dirckx et al. (2005)
Ovine muscle				
	Polarized parallel to fibers	488	1.404 ± 0.003	Cheng et al. (2002)
		632.8	1.389 ± 0.002	Cheng et al. (2002)
		1079.5	1.378 ± 0.004	Cheng et al. (2002)
		1341.4	1.375 ± 0.003	Cheng et al. (2002)
	Polarized perpendicular to fibers	488	1.402 ± 0.002	Cheng et al. (2002)
		632.8	1.389 ± 0.002	Cheng et al. (2002)
		1079.5	1.375 ± 0.003	Cheng et al. (2002)
		1341.4	1.373 ± 0.003	Cheng et al. (2002)

TABLE 2.1 (continued)   Index of Refraction for Various Tissues, Tissue Components, and Biofluids

Tissue	Description	Wavelength (nm)	Index of Refraction	References
Porcine muscle				
	Polarized parallel to fibers	488	1.402 ± 0.002	Cheng et al. (2002)
		632.8	1.381 ± 0.002	Cheng et al. (2002)
		1079.5	1.372 ± 0.003	Cheng et al. (2002)
		1341.4	1.370 ± 0.003	Cheng et al. (2002)
		632.8	1.380 ± 0.007	Li et al. (1996)
	Polarized perpendicular to fibers	488	1.399 ± 0.002	Cheng et al. (2002)
		632.8	1.379 ± 0.002	Cheng et al. (2002)
		1079.5	1.370 ± 0.002	Cheng et al. (2002)
		1341.4	1.367 ± 0.003	Cheng et al. (2002)
	Polarized parallel to fibers	632.8	1.380 ± 0.007	Li et al. (1996)
	Polarized perpendicular to fibers	632.8	1.460 ± 0.008	Li et al. (1996)
Rat muscle	Wistar Han species, dehydrated tissue monitored by sample weight			
Abdominal wall	Mass = 0.1623 g	589	1.398	Oliveira et al. (2009)
	Mass = 0.1455 g	589	1.3995	Oliveira et al. (2009)
	Mass = 0.1361 g	589	1.4105	Oliveira et al. (2009)
	Mass = 0.1252 g	589	1.42	Oliveira et al. (2009)
	Mass = 0.1144 g	589	1.4295	Oliveira et al. (2009)
	Mass = 0.1053 g	589	1.441	Oliveira et al. (2009)
	Mass = 0.0955 g	589	1.4525	Oliveira et al. (2009)
	Mass = 0.0860 g	589	1.464	Oliveira et al. (2009)
	Mass = 0.0747 g	589	1.4785	Oliveira et al. (2009)
	Mass = 0.0654 g	589	1.491	Oliveira et al. (2009)
	Mass = 0.0551 g	589	1.5035	Oliveira et al. (2009)
Spleen				
	Human	456–1064	1.37	Muller et al. (1995)
	Canine	633	1.4	Bolin et al. (1989)
	Porcine	633	1.4	Bolin et al. (1989)
Human stomach				
	Muscle	456–1064	1.39	Muller et al. (1995)
	Mucous	456–1064	1.38	Muller et al. (1995)
	Integral	456–1064	1.38	Muller et al. (1995)
Rat mesentery	Gel–liquid phase change between 38°C and 42°C			
Mainly phospholipid bilayers	25°C	850	1.4245	Haruna et al. (2000)
	30°C	850	1.4239	Haruna et al. (2000)

(continued)

TABLE 2.1 (continued)   Index of Refraction for Various Tissues, Tissue Components, and Biofluids

Tissue	Description	Wavelength (nm)	Index of Refraction	References
	35°C	850	1.4223	Haruna et al. (2000)
	38°C	850	1.4216	Haruna et al. (2000)
	40°C	850	1.4186	Haruna et al. (2000)
	42°C	850	1.4027	Haruna et al. (2000)
	44°C	850	1.4016	Haruna et al. (2000)
	46°C	850	1.4	Haruna et al. (2000)
	48°C	850	1.3986	Haruna et al. (2000)
Porcine small intestine				
		488	1.391 ± 0.002	Cheng et al. (2002)
		632.8	1.373 ± 0.002	Cheng et al. (2002)
		1079.5	1.361 ± 0.003	Cheng et al. (2002)
		1341.4	1.359 ± 0.004	Cheng et al. (2002)
Human blood		633	1.4	Bolin et al. (1989)
Whole blood diluted in water				
	20%	633	1.35	Sardar et al. (1998)
	40%	633	1.35	Sardar et al. (1998)
	60%	633	1.36	Sardar et al. (1998)
	100% (extrapolated)	633	1.38	Sardar et al. (1998)
Plasma		488	1.350 ± 0.002	Cheng et al. (2002)
		632.8	1.345 ± 0.002	Cheng et al. (2002)
		1079.5	1.332 ± 0.003	Cheng et al. (2002)
		1341.4	1.327 ± 0.004	Cheng et al. (2002)
Whole blood		488	1.395 ± 0.003	Cheng et al. (2002)
		632.8	1.373 ± 0.004	Cheng et al. (2002)
		1079.5	1.363 ± 0.004	Cheng et al. (2002)
		1341.4	1.360 ± 0.005	Cheng et al. (2002)
Dry red blood cells				
	Healthy patients, $n = 7$, fixed RBC	550	1.61–1.66	Mazarevica et al. (2002)
	Diabetic patients, $n = 9$, fixed RBC	550	1.56–1.62	Mazarevica et al. (2002)
	Healthy patients, $n = 7$, intact RBC	550	1.57–1.61	Mazarevica et al. (2002)
	Diabetic patients, $n = 9$, intact RBC	550	1.61–1.64	Mazarevica et al. (2002)
Human hemoglobin				
Oxygenated	Fresnel reflectance, integrating sphere			
	Hb 287 g/L	250	1.47 ± 0.03	Friebel et al. (2005, 2006)
		300	1.441 ± 0.03	Friebel et al. (2005, 2006)
		400	1.409 ± 0.03	Friebel et al. (2005, 2006)
		500	1.413 ± 0.03	Friebel et al. (2005, 2006)

TABLE 2.1 (continued)   Index of Refraction for Various Tissues, Tissue Components, and Biofluids

Tissue	Description	Wavelength (nm)	Index of Refraction	References
		589	1.406 ± 0.03	Friebel et al. (2005, 2006)
		700	1.404 ± 0.03	Friebel et al. (2005, 2006)
		800	1.40 ± 0.03	Friebel et al. (2005, 2006)
		900	1.401 ± 0.03	Friebel et al. (2005, 2006)
		1000	1.401 ± 0.03	Friebel et al. (2005, 2006)
		1100	1.40 ± 0.03	Friebel et al. (2005, 2006)
	Hb 165 g/L	250	1.435 ± 0.03	Friebel et al. (2005, 2006)
		300	1.405 ± 0.03	Friebel et al. (2005, 2006)
		400	1.383 ± 0.03	Friebel et al. (2005, 2006)
		500	1.383 ± 0.03	Friebel et al. (2005, 2006)
		589	1.375 ± 0.03	Friebel et al. (2005, 2006)
		700	1.374 ± 0.03	Friebel et al. (2005, 2006)
		800	1.37 ± 0.03	Friebel et al. (2005, 2006)
		900	1.369 ± 0.03	Friebel et al. (2005, 2006)
		1000	1.37 ± 0.03	Friebel et al. (2005, 2006)
		1100	1.369 ± 0.03	Friebel et al. (2005, 2006)
	Hb 104 g/L	250	1.416 ± 0.03	Friebel et al. (2005, 2006)
		300	1.389 ± 0.03	Friebel et al. (2005, 2006)
		400	1.367 ± 0.03	Friebel et al. (2005, 2006)
		500	1.363 ± 0.03	Friebel et al. (2005, 2006)
		589	1.357 ± 0.03	Friebel et al. (2005, 2006)
		700	1.356 ± 0.03	Friebel et al. (2005, 2006)
		800	1.353 ± 0.03	Friebel et al. (2005, 2006)
		900	1.352 ± 0.03	Friebel et al. (2005, 2006)

(continued)

TABLE 2.1 (continued)   Index of Refraction for Various Tissues, Tissue Components, and Biofluids

Tissue	Description	Wavelength (nm)	Index of Refraction	References
		1000	1.353 ± 0.03	Friebel et al. (2005, 2006)
		1100	1.352 ± 0.03	Friebel et al. (2005, 2006)
	Hb 46 g/L	250	1.398 ± 0.03	Friebel et al. (2005, 2006)
		300	1.373 ± 0.03	Friebel et al. (2005, 2006)
		400	1.354 ± 0.03	Friebel et al. (2005, 2006)
		500	1.348 ± 0.03	Friebel et al. (2005, 2006)
		589	1.343 ± 0.03	Friebel et al. (2005, 2006)
		700	1.341 ± 0.03	Friebel et al. (2005, 2006)
		800	1.338 ± 0.03	Friebel et al. (2005, 2006)
		900	1.338 ± 0.03	Friebel et al. (2005, 2006)
		1000	1.338 ± 0.03	Friebel et al. (2005, 2006)
		1100	1.337 ± 0.03	Friebel et al. (2005, 2006)
		633	1.3750 ± 0.0003	Friebel et al. (2005, 2006)
		633	1.3600 ± 0.0003	Friebel et al. (2005, 2006)
	Hb 165 g/L	633	1.3750 ± 0.0003	Friebel et al. (2005, 2006)
	Hb 104 g/L	633	1.3600 ± 0.0003	Friebel et al. (2005, 2006)
Glycated hemoglobin	Hb 140 g/L			
	Glucose 40–400 mg/dL	820	1.382–1.415	Tuchin et al. (2004)
	Glucose 400–800 mg/dL	820	1.415–1.385	Tuchin et al. (2004)
Porcine hemoglobin				
Oxygenated	37°C, Hb 93 g/L	800	1.392 ± 0.001	Faber et al. (2004)
Deoxygenated	37°C, Hb 93 g/L	800	1.388 ± 0.002	Faber et al. (2004)
Cytoplasm	—		1.350–1.367	Duck (1990)
Cervical epithelium	Histology, cytometry and modeling, N = 20			
Cellular nuclei		Far vis/NIR	1.380–1.391	Arifler et al. (2003)
		Far vis/NIR	1.366–1.376	Arifler et al. (2003)
		Far vis/NIR	1.406–1.419	Arifler et al. (2003)
		Far vis/NIR	1.416–1.434	Arifler et al. (2003)

TABLE 2.1 (continued) Index of Refraction for Various Tissues, Tissue Components, and Biofluids

Tissue	Description	Wavelength (nm)	Index of Refraction	References
		Far vis/NIR	1.395–1.411	Arifler et al. (2003)
		Far vis/NIR	1.420–1.439	Arifler et al. (2003)
Human eye				
Aqueous humor		—	1.336	Duck (1990)
Cornea		—	1.376	Duck (1990)
Cornea, fibrils		—	1.47	Duck (1990)
Cornea, ground substance		—	1.35	Duck (1990)
Lens, Surface		—	1.386	Duck (1990)
Center		—	1.406	Duck (1990)
Vitreous humor		—	1.336	Duck (1990)
Tears		—	1.3361–1.3379	Duck (1990)
Sclera		442–1064	1.47–1.36	Muller et al. (1995)
Cornea		550	1.3771	Drexler et al. (1998); Lin et al. (2004)
		589	1.380 ± 0.005	Drexler et al. (1998); Lin et al. (2004)
	n_g	855	1.3817 ± 0.0021	Drexler et al. (1998); Lin et al. (2004)
	n_g	1270	1.389 ± 0.004	Drexler et al. (1998); Lin et al. (2004)
	n_g	1270	1.386	Drexler et al. (1998); Lin et al. (2004)
	n_g	1270	1.390 ± 0.005	Drexler et al. (1998); Lin et al. (2004)
	n_g	1270	1.3838 ± 0.0021	Drexler et al. (1998); Lin et al. (2004)
Cornea, fibrils		589	1.411 ± 0.004	Leonard et al. (1997)
Cornea, extra fibrillar		589	1.365 ± 0.009	Leonard et al. (1997)
Bovine eye				
Stroma		589	1.375	Farrell et al. (2000); Meek et al. (2003)
Hydrated fibrils		589	1.413	Farrell et al. (2000); Meek et al. (2003)
Hydrated extrafibrillar matrix		589	1.359	Farrell et al. (2000); Meek et al. (2003)
Dry collagen		589	1.547	Farrell et al. (2000); Meek et al. (2003)
Dry extrafibrillar material		589	1.485	Farrell et al. (2000); Meek et al. (2003)
Solvent (salt solution)		589	1.335	Farrell et al. (2000); Meek et al. (2003)
Hydrated stroma	$H = 3–8$, $H = 3.2$—Physiological hydration		$1.335 + 0.04/(0.22 + 0.24H)$	Farrell et al. (2000); Meek et al. (2003)

(continued)

TABLE 2.1 (continued)   Index of Refraction for Various Tissues, Tissue Components, and Biofluids

Tissue	Description	Wavelength (nm)	Index of Refraction	References
	H (hydration) = 3	589	1.378	Farrell et al. (2000); Meek et al. (2003)
	H (hydration) = 8		1.354	Farrell et al. (2000); Meek et al. (2003)
Cornea, fibrils		589	1.413 ± 0.004	Leonard et al. (1997)
Cornea, extra fibrillar		589	1.357 ± 0.009	Leonard et al. (1997)
Calf cornea	Normal, n_g, H = 1.5–5	820	(1.324 ± 0.002)	Kim et al. (2004)
	H = 5.3d − 0.67, d = corn. stroma thick (mm)		+ (0.272 ± 0.009)/ (H + 1)	
	Hydrated, H = 1.5, n_g	820	1.433 ± 0.006	Kim et al. (2004)
	Hydrated, H = 5, n_g	820	1.369 ± 0.004	Kim et al. (2004)
Rabbit eye				
Cornea, fibrils		589	1.413 ± 0.004	Leonard et al. (1997)
Cornea, extra fibrillar		589	1.357 ± 0.01	Leonard et al. (1997)
Trout eye				
Cornea, fibrils		589	1.418 ± 0.004	Leonard et al. (1997)
Cornea, extra fibrillar		589	1.364 ± 0.009	Leonard et al. (1997)
Human tooth				
Enamel		220	1.73	Duck (1990)
Enamel		400–700	1.62	Duck (1990)
Apatite		—	>1.623	Duck (1990)
Dentin matrix		Visible	1.553 ± 0.001	Grisimov (1994)
Enamel	n_g	856	1.62 ± 0.02	Wang et al. (1999)
Dentin	n_g	856	1.50 ± 0.02	Wang et al. (1999)
Enamel		842—858	1.65	Ohmi et al. (2000)
Dentin		842—858	1.54	Ohmi et al. (2000)
Human hair				
	Black	850	1.59 ± 0.08	Wang et al. (1995)
	Red	850	1.56 ± 0.01	Wang et al. (1995)
(Roughness layer width, air inclusion)	304 nm, 0.6%	400—600	1.45	Chan et al. (2006)
	299.5 nm, 0.1%	400—600	1.46	Chan et al. (2006)
	308.7 nm, 0.8%	400—600	1.47	Chan et al. (2006)
	273.7 nm, 2.2%	400—600	1.46	Chan et al. (2006)
	327.5 nm, 4.7%	400—600	1.5	Chan et al. (2006)
	359.7 nm, 5.7%	400—600	1.5	Chan et al. (2006)
Nail		842—858	1.51	Ohmi et al. (2000)
Collagen, Type 1				
Dry	n_g	850	1.53 ± 0.02	Wang et al. (1996)
Fully hydrated	n_g	850	1.43 ± 0.02	Wang et al. (1996)

roles in shaping tissue optical properties. The tissue data are provided in Table 2.2 in terms of the scattering and absorption coefficients, μ_s and μ_a, and the average cosine of scatter g.

2.4.1 Refractive Index

The index of refraction determines the speed of light in the medium relative to the speed of light in vacuum, $c_n = c/n$. Changes in the refractive index between two media give rise to scattering, refraction (Snell's law, Equation 2.3), and reflection. Since tissues are heterogeneous in composition, the refractive indices for the various tissue constituents can be used to calculate scattering properties, to determine reflection and transmission at boundaries, or to derive an *effective* index of refraction for the tissue as a whole. In turbid tissues, the index reported is the *effective* index. The refractive indices of various tissues and tissue components are given in Table 2.1.

Water makes up a significant portion of most tissues and has an index $n = 1.33$. This sets the minimum values for biofluids and soft-tissue constituents. Among the other soft-tissue components, melanin particles, found mostly in the epidermal layer of the skin (see Section 2.4.3.2), are at the high end of the refractive index scale, with a reported value above 1.6. Indices of refraction for whole tissues themselves—including parts of the brain, aorta, lung, stomach, kidney, and bladder—fall in the 1.36–1.40 range. Extracellular fluids and intracellular cytoplasm fall in the range of 1.35–1.38. The index of refraction for fatty tissues is around 1.45. The membranes that enclose cells and subcellular organelles are composed largely of lipids, and the index mismatch between the cytoplasm and these lipid structures is the origin of much of the scattering in cellular tissues. One of the hard tissues in the table, tooth enamel has an index reported in the range of 1.62–1.73. Values for various bones and their substructures are not widely reported.

2.4.2 Scattering Properties

Scattering occurs when light encounters an abrupt, localized change in the refractive index due to embedded or floating inclusions. Scattering is the phenomenon that renders light diffuse in turbid tissues. In the cell, the important scatterers are the subcellular organelles. The characteristic sizes of these organelles overlap much of the wavelength range of interest, since their dimensions can range from about 0.1 to 10 μm. As scatterers, most of these structures fall in the Mie regime, exhibiting highly anisotropic forward-directed scattering patterns. The scattering coefficients (μ_s) and average cosines of scatter (g) for various tissues are given in Table 2.2.

The mitochondria are the dominant scatterers among the organelles. These subcellular structures manufacture the large majority of the cell's ATP, which is the energy currency of the cellular realm. They are oblong and vary in size depending on cell type, with transverse dimensions of 0.5–2 μm. In addition to being enclosed in a bilipid membrane, mitochondria also have an internal bilipid membrane with folds that run throughout the interior. This structure endows this organelle with a high refractive index relative to the surrounding cytosol, which in turn gives rise to the high degree of scattering. The largest of the organelles is the cell nucleus, with a diameter in the 4–6 μm range. The nucleus takes up about 10% of the cell volume. The endoplasmic reticulum and Golgi apparatus are among the larger organelles. The submicron vesicles pinched off from these structures transport their various molecular products through the cell. These smaller vesicle-type organelles include lysosomes and peroxisomes that are 0.25–0.5 μm in diameter. Human cells themselves vary in shape and size among the different tissue types with sizes from 4 to >100 μm with a typical size near 10 μm. (Axons of nerve cells can be more than a meter in length yet small in the transverse dimensions.) An isolated cell can be a strong scatterer, but within a tissue, the scattering is largely subcellular in origin. In skin, melanosomes are important scattering structures ranging in size from 100 nm to 2 μm. These contain melanin particles, which are strung together as beads. (Melanin is discussed further in Section 2.4.3.2.) In blood, the disk-shaped erythrocytes are the

TABLE 2.2   Absorption Coefficient μ_a, the Scattering Coefficient μ_s, and the Average Cosine of Scatter g for Various Tissues, Tissue Components, and Biofluids

Tissue	Description	Wavelength (nm)	μ_a (cm^{-1})	μ_s (cm^{-1})	g	Notes	References
Skin, in vitro	Stratum corneum	193	6000				Cheong et al. (1990)
	Stratum corneum	250	1150	2600	0.9	Data from plots	van Gemert et al. (1989)
		308	600	2400	0.9	"	van Gemert et al. (1989)
		337	330	2300	0.9	"	van Gemert et al. (1989)
		351	300	2200	0.9	"	van Gemert et al. (1989)
		400	230	2000	0.9	"	van Gemert et al. (1989)
	Epidermis	250	1000	2000	0.69	"	van Gemert et al. (1989)
		308	300	1400	0.71	"	van Gemert et al. (1989)
		337	120	1200	0.72	"	van Gemert et al. (1989)
		351	100	1100	0.72	"	van Gemert et al. (1989)
		415	66	800	0.74	"	van Gemert et al. (1989)
		488	50	600	0.76	"	van Gemert et al. (1989)
		514	44	600	0.77	"	van Gemert et al. (1989)
		585	36	470	0.79	"	van Gemert et al. (1989)
		633	35	450	0.8	"	van Gemert et al. (1989)
		800	40	420	0.85	"	van Gemert et al. (1989)
	Dermis	250	35	833	0.69	"	van Gemert et al. (1989)
		308	12	583	0.71	"	van Gemert et al. (1989)
		337	8.2	500	0.72	"	van Gemert et al. (1989)
		351	7	458	0.72	"	van Gemert et al. (1989)
		415	4.7	320	0.74	"	van Gemert et al. (1989)
		488	3.5	250	0.76	"	van Gemert et al. (1989)
		514	3	250	0.77	"	van Gemert et al. (1989)
		585	3	196	0.79	"	van Gemert et al. (1989)
		633	2.7	187.5	0.8	"	van Gemert et al. (1989)
		800	2.3	175	0.85	"	van Gemert et al. (1989)
	Epidermis	577	19	480	0.787		Kienle et al. (1994, 1995)
		585	19	470	0.79		Kienle et al. (1994, 1995)
		590	19	460	0.8		Kienle et al. (1994, 1995)
		595	19	460	0.8		Kienle et al. (1994, 1995)
		600	19	460	0.8		Kienle et al. (1994, 1995)

Tissue					Comments	Reference
Dermis	517	2.2	210	0.787		Kienle et al. (1994, 1995)
	585	2.2	205	0.79		Kienle et al. (1994, 1995)
	590	2.2	200	0.8		Kienle et al. (1994, 1995)
	595	2.2	200	0.8		Kienle et al. (1994, 1995)
	600	2.2	200	0.8		Kienle et al. (1994, 1995)
Skin, blood	517	354	468	0.995		Kienle et al. (1994, 1995)
	585	191	467	0.995		Kienle et al. (1994, 1995)
	590	694	66	0.995		Kienle et al. (1994, 1995)
	595	434	65	0.995		Kienle et al. (1994, 1995)
	600	254	64	0.995		Kienle et al. (1994, 1995)
Dermis, leg	635	1.8 ± 0.2	244 ± 21	0.68		Marchesini et al. (1989)
Dermis	749	0.24 ± 0.19				Troy et al. (1996)
	789	0.75 ± 0.06				Troy et al. (1996)
	836	0.98 ± 0.15				Troy et al. (1996)
Dermis	633	<10		0.97		Laufer et al. (1998); Simpson et al. (1998)
Dermis	700	2.7 ± 1.0				Graaff et al. (1993)
Dermis	633	1.9 ± 0.6				Graaff et al. (1993)
Dermis	633	1.5				Laufer et al. (1998); Simpson et al. (1998)
Skin and underlying tissues	633	3.1	70.7	0.8	Including vein wall (leg)	Tuchin (1995)
Caucasian male skin (n = 3)	500	5.1				Chan et al. (1996a,b)
	810	0.26				Chan et al. (1996a,b)
Cascasian male skin (n = 3)	500	15.3			Mass loaded 0.1 kg/cm^2	Chan et al. (1996a,b)
	810	0.63			"	Chan et al. (1996a,b)
Cascasian male skin (n = 3)	500	13.6			Mass loaded 1.0 kg/cm^2	Chan et al. (1996a,b)
	810	0.57			"	Chan et al. (1996a,b)
Caucasian female skin (n = 3)	500	5.2				Chan et al. (1996a,b)
	810	0.97				Chan et al. (1996a,b)
Caucasian female skin (n = 3)	500	7.4			Mass loaded 0.1 kg/cm^2	Chan et al. (1996a,b)
	810	1.4			"	Chan et al. (1996a,b)

(continued)

TABLE 2.2 (continued)   Absorption Coefficient μ_a, the Scattering Coefficient μ_s, and the Average Cosine of Scatter g for Various Tissues, Tissue Components, and Biofluids

Tissue	Description	Wavelength (nm)	μ_a (cm⁻¹)	μ_s (cm⁻¹)	g	Notes	References
	Caucasian female skin ($n = 3$)	500	10			Mass loaded 1.0 kg/cm²	Chan et al. (1996a,b)
		810	1.7			"	Chan et al. (1996a,b)
	Hispanic male skin ($n = 3$)	500	3.8				Chan et al. (1996a,b)
		810	0.87				Chan et al. (1996a,b)
	Hispanic male skin ($n = 3$)	500	5.1			Mass loaded 0.1 kg/cm²	Chan et al. (1996a,b)
		810	0.93			"	Chan et al. (1996a,b)
	Hispanic male skin ($n = 3$)	500	6.2			Mass loaded 1.0 kg/cm²	Chan et al. (1996a,b)
		810	0.87			"	Chan et al. (1996a,b)
	Caucasian skin ($n = 21$)	400	3.76 ± 0.35				Bashkatov et al. (2005)
		500	1.19 ± 0.16				Bashkatov et al. (2005)
		600	0.69 ± 0.13				Bashkatov et al. (2005)
		700	0.48 ± 0.11				Bashkatov et al. (2005)
		800	0.43 ± 0.11				Bashkatov et al. (2005)
		900	0.33 ± 0.02				Bashkatov et al. (2005)
		1000	0.27 ± 0.03				Bashkatov et al. (2005)
		1100	0.16 ± 0.04				Bashkatov et al. (2005)
		1200	0.54 ± 0.04				Bashkatov et al. (2005)
		1300	0.41 ± 0.07				Bashkatov et al. (2005)
		1400	1.64 ± 0.31				Bashkatov et al. (2005)
		1500	1.69 ± 0.35				Bashkatov et al. (2005)
		1600	1.19 ± 0.22				Bashkatov et al. (2005)
		1700	1.55 ± 0.28				Bashkatov et al. (2005)
		1800	1.44 ± 0.22				Bashkatov et al. (2005)
		1900	2.14 ± 0.28				Bashkatov et al. (2005)
		2000	1.74 ± 0.29				Bashkatov et al. (2005)
	Epidermis (tissue slabs)	370	1.35 ± 0.16		0.8		Salomatina et al. (2006)
		420	1.20 ± 0.12		0.8		Salomatina et al. (2006)
		470	0.84 ± 0.06		0.8		Salomatina et al. (2006)
		488	0.76 ± 0.07		0.8		Salomatina et al. (2006)
		514	0.63 ± 0.07		0.8		Salomatina et al. (2006)
		520	0.60 ± 0.07		0.8		Salomatina et al. (2006)

570	0.39 ± 0.08	0.8	Salomatina et al. (2006)
620	0.28 ± 0.07	0.8	Salomatina et al. (2006)
633	0.26 ± 0.07	0.8	Salomatina et al. (2006)
670	0.26 ± 0.08	0.8	Salomatina et al. (2006)
720	0.24 ± 0.07	0.8	Salomatina et al. (2006)
770	0.19 ± 0.06	0.8	Salomatina et al. (2006)
820	0.15 ± 0.06	0.8	Salomatina et al. (2006)
830	0.14 ± 0.06	0.8	Salomatina et al. (2006)
870	0.10 ± 0.05	0.8	Salomatina et al. (2006)
920	0.07 ± 0.04	0.8	Salomatina et al. (2006)
970	0.06 ± 0.03	0.8	Salomatina et al. (2006)
1020	0.04 ± 0.03	0.8	Salomatina et al. (2006)
1064	0.02 ± 0.02	0.8	Salomatina et al. (2006)
1070	0.02 ± 0.02	0.8	Salomatina et al. (2006)
1120	0.02 ± 0.02	0.8	Salomatina et al. (2006)
1170	0.06 ± 0.04	0.8	Salomatina et al. (2006)
1220	0.07 ± 0.04	0.8	Salomatina et al. (2006)
1270	0.06 ± 0.04	0.8	Salomatina et al. (2006)
1320	0.11 ± 0.05	0.8	Salomatina et al. (2006)
1370	0.56 ± 0.14	0.8	Salomatina et al. (2006)
1420	2.36 ± 0.35	0.8	Salomatina et al. (2006)
1470	2.96 ± 0.42	0.8	Salomatina et al. (2006)
1520	1.89 ± 0.29	0.8	Salomatina et al. (2006)
1570	1.01 ± 0.20	0.8	Salomatina et al. (2006)
Dermis (tissue slabs)			
370	0.98 ± 0.14	0.8	Salomatina et al. (2006)
420	0.85 ± 0.11	0.8	Salomatina et al. (2006)
470	0.43 ± 0.06	0.8	Salomatina et al. (2006)
488	0.36 ± 0.05	0.8	Salomatina et al. (2006)
514	0.31 ± 0.04	0.8	Salomatina et al. (2006)
520	0.30 ± 0.04	0.8	Salomatina et al. (2006)
570	0.22 ± 0.03	0.8	Salomatina et al. (2006)
620	0.15 ± 0.02	0.8	Salomatina et al. (2006)

(continued)

TABLE 2.2 (continued)   Absorption Coefficient μ_a, the Scattering Coefficient μ_s, and the Average Cosine of Scatter g for Various Tissues, Tissue Components, and Biofluids

Tissue	Description	Wavelength (nm)	μ_a (cm^{-1})	μ_s (cm^{-1})	g	Notes	References
		633	0.15 ± 0.02		0.8		Salomatina et al. (2006)
		670	0.15 ± 0.02		0.8		Salomatina et al. (2006)
		720	0.15 ± 0.02		0.8		Salomatina et al. (2006)
		770	0.13 ± 0.02		0.8		Salomatina et al. (2006)
		820	0.11 ± 0.02		0.8		Salomatina et al. (2006)
		830	0.11 ± 0.02		0.8		Salomatina et al. (2006)
		870	0.09 ± 0.02		0.8		Salomatina et al. (2006)
		920	0.08 ± 0.02		0.8		Salomatina et al. (2006)
		970	0.08 ± 0.02		0.8		Salomatina et al. (2006)
		1020	0.07 ± 0.02		0.8		Salomatina et al. (2006)
		1064	0.05 ± 0.02		0.8		Salomatina et al. (2006)
		1070	0.05 ± 0.02		0.8		Salomatina et al. (2006)
		1120	0.06 ± 0.02		0.8		Salomatina et al. (2006)
		1170	0.12 ± 0.02		0.8		Salomatina et al. (2006)
		1220	0.13 ± 0.02		0.8		Salomatina et al. (2006)
		1270	0.10 ± 0.02		0.8		Salomatina et al. (2006)
		1320	0.15 ± 0.03		0.8		Salomatina et al. (2006)
		1370	0.48 ± 0.04		0.8		Salomatina et al. (2006)
		1420	1.76 ± 0.18		0.8		Salomatina et al. (2006)
		1470	2.19 ± 0.20		0.8		Salomatina et al. (2006)
		1520	1.41 ± 0.11		0.8		Salomatina et al. (2006)
		1570	0.85 ± 0.07		0.8		Salomatina et al. (2006)
Subcutaneous fat (tissue slabs)		370	1.18 ± 0.21		0.8		Salomatina et al. (2006)
		420	1.65 ± 0.33		0.8		Salomatina et al. (2006)
		470	0.75 ± 0.09		0.8		Salomatina et al. (2006)
		488	0.63 ± 0.08		0.8		Salomatina et al. (2006)
		514	0.47 ± 0.07		0.8		Salomatina et al. (2006)
		520	0.44 ± 0.07		0.8		Salomatina et al. (2006)
		570	0.31 ± 0.09		0.8		Salomatina et al. (2006)
		620	0.15 ± 0.03		0.8		Salomatina et al. (2006)
		633	0.14 ± 0.03		0.8		Salomatina et al. (2006)

Tissue	Wavelength	Value	g	Reference
	670	0.13 ± 0.03	0.8	Salomatina et al. (2006)
	720	0.12 ± 0.02	0.8	Salomatina et al. (2006)
	770	0.11 ± 0.02	0.8	Salomatina et al. (2006)
	820	0.10 ± 0.02	0.8	Salomatina et al. (2006)
	830	0.10 ± 0.02	0.8	Salomatina et al. (2006)
	870	0.09 ± 0.02	0.8	Salomatina et al. (2006)
	920	0.09 ± 0.02	0.8	Salomatina et al. (2006)
	970	0.09 ± 0.03	0.8	Salomatina et al. (2006)
	1020	0.08 ± 0.02	0.8	Salomatina et al. (2006)
	1064	0.07 ± 0.02	0.8	Salomatina et al. (2006)
	1070	0.07 ± 0.02	0.8	Salomatina et al. (2006)
	1120	0.08 ± 0.02	0.8	Salomatina et al. (2006)
	1170	0.14 ± 0.03	0.8	Salomatina et al. (2006)
	1220	0.15 ± 0.03	0.8	Salomatina et al. (2006)
	1270	0.10 ± 0.03	0.8	Salomatina et al. (2006)
	1320	0.12 ± 0.03	0.8	Salomatina et al. (2006)
	1370	0.27 ± 0.04	0.8	Salomatina et al. (2006)
	1420	0.93 ± 0.14	0.8	Salomatina et al. (2006)
	1470	1.08 ± 0.18	0.8	Salomatina et al. (2006)
	1520	0.70 ± 0.12	0.8	Salomatina et al. (2006)
	1570	0.43 ± 0.07	0.8	Salomatina et al. (2006)
Infiltrative basal cell carcinoma	370	0.68 ± 0.08	0.8	Salomatina et al. (2006)
	420	0.67 ± 0.11	0.8	Salomatina et al. (2006)
	470	0.33 ± 0.04	0.8	Salomatina et al. (2006)
	488	0.29 ± 0.05	0.8	Salomatina et al. (2006)
	514	0.26 ± 0.06	0.8	Salomatina et al. (2006)
	520	0.25 ± 0.06	0.8	Salomatina et al. (2006)
	570	0.20 ± 0.07	0.8	Salomatina et al. (2006)
	620	0.15 ± 0.06	0.8	Salomatina et al. (2006)
	633	0.15 ± 0.05	0.8	Salomatina et al. (2006)
	670	0.14 ± 0.05	0.8	Salomatina et al. (2006)
	720	0.13 ± 0.05	0.8	Salomatina et al. (2006)

(continued)

TABLE 2.2 (continued) Absorption Coefficient μ_a, the Scattering Coefficient μ_s, and the Average Cosine of Scatter g for Various Tissues, Tissue Components, and Biofluids

Tissue	Description	Wavelength (nm)	μ_a (cm⁻¹)	μ_s (cm⁻¹)	g	Notes	References
		770	0.11 ± 0.04		0.8		Salomatina et al. (2006)
		820	0.09 ± 0.04		0.8		Salomatina et al. (2006)
		830	0.09 ± 0.04		0.8		Salomatina et al. (2006)
		870	0.07 ± 0.03		0.8		Salomatina et al. (2006)
		920	0.06 ± 0.03		0.8		Salomatina et al. (2006)
		970	0.08 ± 0.03		0.8		Salomatina et al. (2006)
		1020	0.07 ± 0.03		0.8		Salomatina et al. (2006)
		1064	0.08 ± 0.04		0.8		Salomatina et al. (2006)
		1070	0.08 ± 0.04		0.8		Salomatina et al. (2006)
		1120	0.10 ± 0.06		0.8		Salomatina et al. (2006)
		1170	0.16 ± 0.07		0.8		Salomatina et al. (2006)
		1220	0.17 ± 0.09		0.8		Salomatina et al. (2006)
		1270	0.18 ± 0.12		0.8		Salomatina et al. (2006)
		1320	0.27 ± 0.15		0.8		Salomatina et al. (2006)
		1370	0.69 ± 0.27		0.8		Salomatina et al. (2006)
		1420	2.21 ± 0.46		0.8		Salomatina et al. (2006)
		1470	2.75 ± 0.54		0.8		Salomatina et al. (2006)
		1520	1.90 ± 0.47		0.8		Salomatina et al. (2006)
		1570	1.12 ± 0.31		0.8		Salomatina et al. (2006)
Nodular basal cell carcinoma		370	0.87 ± 0.29		0.8		Salomatina et al. (2006)
		420	0.73 ± 0.20		0.8		Salomatina et al. (2006)
		470	0.40 ± 0.12		0.8		Salomatina et al. (2006)
		488	0.34 ± 0.12		0.8		Salomatina et al. (2006)
		514	0.28 ± 0.11		0.8		Salomatina et al. (2006)
		520	0.27 ± 0.11		0.8		Salomatina et al. (2006)
		570	0.18 ± 0.09		0.8		Salomatina et al. (2006)
		620	0.13 ± 0.06		0.8		Salomatina et al. (2006)
		633	0.12 ± 0.06		0.8		Salomatina et al. (2006)
		670	0.09 ± 0.05		0.8		Salomatina et al. (2006)
		720	0.07 ± 0.04		0.8		Salomatina et al. (2006)
		770	0.04 ± 0.03		0.8		Salomatina et al. (2006)

Tissue	Wavelength	Value	g	Reference
	820	0.02 ± 0.02	0.8	Salomatina et al. (2006)
	830	0.02 ± 0.01	0.8	Salomatina et al. (2006)
	870	0.01 ± 0.01	0.8	Salomatina et al. (2006)
	920	0.01 ± 0.00	0.8	Salomatina et al. (2006)
	970	0.01 ± 0.01	0.8	Salomatina et al. (2006)
	1020	0.00 ± 0.00	0.8	Salomatina et al. (2006)
	1064	0.00 ± 0.00	0.8	Salomatina et al. (2006)
	1070	0.00 ± 0.00	0.8	Salomatina et al. (2006)
	1120	0.00 ± 0.00	0.8	Salomatina et al. (2006)
	1170	0.01 ± 0.01	0.8	Salomatina et al. (2006)
	1220	0.02 ± 0.01	0.8	Salomatina et al. (2006)
	1270	0.01 ± 0.01	0.8	Salomatina et al. (2006)
	1320	0.05 ± 0.01	0.8	Salomatina et al. (2006)
	1370	0.32 ± 0.03	0.8	Salomatina et al. (2006)
	1420	1.46 ± 0.20	0.8	Salomatina et al. (2006)
	1470	1.86 ± 0.16	0.8	Salomatina et al. (2006)
	1520	1.19 ± 0.07	0.8	Salomatina et al. (2006)
	1570	0.67 ± 0.04	0.8	Salomatina et al. (2006)
Squamous cell carcinoma	370	0.94 ± 0.20	0.8	Salomatina et al. (2006)
	420	1.21 ± 0.23	0.8	Salomatina et al. (2006)
	470	0.41 ± 0.06	0.8	Salomatina et al. (2006)
	488	0.34 ± 0.05	0.8	Salomatina et al. (2006)
	514	0.32 ± 0.04	0.8	Salomatina et al. (2006)
	520	0.32 ± 0.04	0.8	Salomatina et al. (2006)
	570	0.29 ± 0.04	0.8	Salomatina et al. (2006)
	620	0.14 ± 0.02	0.8	Salomatina et al. (2006)
	633	0.13 ± 0.02	0.8	Salomatina et al. (2006)
	670	0.11 ± 0.02	0.8	Salomatina et al. (2006)
	720	0.09 ± 0.02	0.8	Salomatina et al. (2006)
	770	0.07 ± 0.02	0.8	Salomatina et al. (2006)
	820	0.05 ± 0.02	0.8	Salomatina et al. (2006)
	830	0.05 ± 0.02	0.8	Salomatina et al. (2006)

(continued)

TABLE 2.2 (continued) Absorption Coefficient μ_a, the Scattering Coefficient μ_s, and the Average Cosine of Scatter g for Various Tissues, Tissue Components, and Biofluids

Tissue	Description	Wavelength (nm)	μ_a (cm⁻¹)	μ_s (cm⁻¹)	g	Notes	References
		870	0.04 ± 0.01		0.8		Salomatina et al. (2006)
		920	0.03 ± 0.01		0.8		Salomatina et al. (2006)
		970	0.04 ± 0.02		0.8		Salomatina et al. (2006)
		1020	0.04 ± 0.02		0.8		Salomatina et al. (2006)
		1064	0.04 ± 0.02		0.8		Salomatina et al. (2006)
		1070	0.04 ± 0.02		0.8		Salomatina et al. (2006)
		1120	0.04 ± 0.02		0.8		Salomatina et al. (2006)
		1170	0.10 ± 0.03		0.8		Salomatina et al. (2006)
		1220	0.11 ± 0.03		0.8		Salomatina et al. (2006)
		1270	0.11 ± 0.03		0.8		Salomatina et al. (2006)
		1320	0.17 ± 0.04		0.8		Salomatina et al. (2006)
		1370	0.43 ± 0.05		0.8		Salomatina et al. (2006)
		1420	1.70 ± 0.12		0.8		Salomatina et al. (2006)
		1470	2.35 ± 0.21		0.8		Salomatina et al. (2006)
		1520	1.50 ± 0.15		0.8		Salomatina et al. (2006)
		1570	0.92 ± 0.12		0.8		Salomatina et al. (2006)
Skin, ex vivo	Caucasian dermis (n = 12)	633	0.33 ± 0.09		0.9		Laufer et al. (1998); Simpson et al. (1998)
		700	0.19 ± 0.06		0.9		Laufer et al. (1998); Simpson et al. (1998)
		900	0.13 ± 0.07		0.9		Laufer et al. (1998); Simpson et al. (1998)
	Negroid dermis (n = 5)	633	2.41 ± 1.53		0.9		Laufer et al. (1998); Simpson et al. (1998)
		700	1.49 ± 0.88		0.9		Laufer et al. (1998); Simpson et al. (1998)
		900	0.45 ± 0.18		0.9		Laufer et al. (1998); Simpson et al. (1998)
	Subdermis (n = 12)	633	0.13 ± 0.05		0.9	Primarily globular fat cells	Laufer et al. (1998); Simpson et al. (1998)

Tissue	Wavelength	Value		Description	Reference
	700	0.09 ± 0.03	0.9	"	Laufer et al. (1998); Simpson et al. (1998)
	900	0.12 ± 0.04	0.9	"	Laufer et al. (1998); Simpson et al. (1998)
Back of knee (F, 51 years old)	1460	17.88 ± 1.12		Moderate inflammation in dermis	Troy et al. (2001)
	1600	5.35 ± 0.24		"	Troy et al. (2001)
	2200	7.46 ± 0.56		"	Troy et al. (2001)
Back of knee (F, 51 years old)	1460	18.70 ± 1.13		Moderate inflammation in dermis	Troy et al. (2001)
	1600	5.46 ± 0.27		"	Troy et al. (2001)
	2200	8.86 ± 0.46		"	Troy et al. (2001)
Lower back, right side (F, 66 years old)	1460	16.01 ± 0.56		Mild solar damage	Troy et al. (2001)
	1600	4.91 ± 0.10		"	Troy et al. (2001)
	2200	10.94 ± 0.23		"	Troy et al. (2001)
Lower back, right side (F, 66 years old)	1460	12.65 ± 0.96		Mild solar damage	Troy et al. (2001)
	1600	3.86 ± 0.28		"	Troy et al. (2001)
	2200	8.58 ± 0.55		"	Troy et al. (2001)
Shin (F, 67 years old)	1460	16.58 ± 3.26		Mild solar damage, chronic inflammation	Troy et al. (2001)
	1600	5.15 ± 0.60		"	Troy et al. (2001)
	2200	9.65 ± 1.17		"	Troy et al. (2001)
Shin (F, 67 years old)	1460	18.07 ± 0.42		Mild solar damage, chronic inflammation	Troy et al. (2001)
	1600	5.60 ± 0.17		"	Troy et al. (2001)
	2200	11.26 ± 0.16		"	Troy et al. (2001)
Thigh (M, 64 years old)	1000	0.69 ± 0.01		Mild chronic dermatitis	Troy et al. (2001)
	1460	16.64 ± 0.95		"	Troy et al. (2001)
	1600	4.96 ± 0.27		"	Troy et al. (2001)
	2200	13.04 ± 2.36		"	Troy et al. (2001)
Lower thigh (M, 75 years old)	1000	0.83 ± 0.03		Normal skin	Troy et al. (2001)
	1460	19.06 ± 1.22		"	Troy et al. (2001)

(continued)

TABLE 2.2 (continued)   Absorption Coefficient μ_a, the Scattering Coefficient μ_s, and the Average Cosine of Scatter g for Various Tissues, Tissue Components, and Biofluids

Tissue	Description	Wavelength (nm)	μ_a (cm⁻¹)	μ_s (cm⁻¹)	g	Notes	References
		1600	5.75 ± 0.27			"	Troy et al. (2001)
		2200	11.92 ± 0.41			"	Troy et al. (2001)
	Lower thigh (M, 75 years old)	1000	0.85 ± 0.02			Normal skin	Troy et al. (2001)
		1460	18.03 ± 2.01			"	Troy et al. (2001)
		1600	5.61 ± 0.56			"	Troy et al. (2001)
		2200	11.85 ± 0.83			"	Troy et al. (2001)
	Groin, left side (F, 42 years old)	1000	0.80 ± 0.01			Mild chronic inflammation	Troy et al. (2001)
		1460	20.49 ± 0.89			"	Troy et al. (2001)
		1600	5.85 ± 0.14			"	Troy et al. (2001)
		2200	12.46 ± 0.42			"	Troy et al. (2001)
	Groin, left side (F, 42 years old)	1000	0.77 ± 0.03			Mild chronic inflammation	Troy et al. (2001)
		1460	20.24 ± 1.04			"	Troy et al. (2001)
		1600	5.76 ± 0.28			"	Troy et al. (2001)
		2200	12.71 ± 0.58			"	Troy et al. (2001)
	Posterior thigh (M, 33 years old)	1000	0.82 ± 0.02			Mild chronic dermatitis	Troy et al. (2001)
		1460	19.01 ± 1.28			"	Troy et al. (2001)
		1600	5.81 ± 0.33			"	Troy et al. (2001)
		2200	11.13 ± 1.21			"	Troy et al. (2001)
	Axillary, right side (F, 52 years old)	1000	0.97 ± 0.08			Mild perivascular chronic inflammation	Troy et al. (2001)
		1460	21.39 ± 1.25			"	Troy et al. (2001)
		1600	6.17 ± 0.30			"	Troy et al. (2001)
		2200	12.53 ± 0.84			"	Troy et al. (2001)
	Back of thigh, upper left (M, 37 years old)	1000	0.82 ± 0.02			Mild chronic dermatitis	Troy et al. (2001)
		1460	23.31 ± 0.71			"	Troy et al. (2001)
		1600	6.68 ± 0.11			"	Troy et al. (2001)
		2200	15.19 ± 1.37			"	Troy et al. (2001)
	Scalp (M, 70 years old)	1000	1.04 ± 0.02			Mild chronic dermatitis, solar elastosis	Troy et al. (2001)
		1460	15.95 ± 0.99			"	Troy et al. (2001)

Tissue	λ (nm)	Value	Notes	Reference
	1600	5.09 ± 0.23	"	Troy et al. (2001)
	2200	12.65 ± 0.52	"	Troy et al. (2001)
Scalp (M, 61 years old)	1000	0.79 ± 0.02	Mild chronic dermatitis, solar elastosis	Troy et al. (2001)
	1460	16.47 ± 1.05	"	Troy et al. (2001)
	1600	5.11 ± 0.24	"	Troy et al. (2001)
	2200	13.30 ± 1.48	"	Troy et al. (2001)
Scalp/facial tissue (F, 68 years old)	1000	1.06 ± 0.03	Mild solar damage, chronic inflammation	Troy et al. (2001)
Scalp/facial tissue (F, 68 years old)	1460	12.81 ± 1.84	"	Troy et al. (2001)
	1600	4.26 ± 0.50	"	Troy et al. (2001)
	2200	11.32 ± 1.52	"	Troy et al. (2001)
Scalp/facial tissue (F, 53 years old)	1000	1.32 ± 0.05	Severe solar damage, mild chronic inflammation	Troy et al. (2001)
	1460	12.68 ± 5.07	"	Troy et al. (2001)
	1600	4.31 ± 1.34	"	Troy et al. (2001)
	2200	11.33 ± 3.05	"	Troy et al. (2001)
Scalp/facial tissue (F, 53 years old)	1000	1.55 ± 0.02	Mild chronic inflammation	Troy et al. (2001)
	1460	16.13 ± 1.38	"	Troy et al. (2001)
	1600	5.38 ± 0.31	"	Troy et al. (2001)
	2200	13.84 ± 1.02	"	Troy et al. (2001)
Scalp/facial tissue (F, 53 years old)	1000	1.53 ± 0.02	Mild solar damage	Troy et al. (2001)
	1460	16.82 ± 1.13	"	Troy et al. (2001)
	1600	5.57 ± 0.19	"	Troy et al. (2001)
	2200	13.46 ± 0.58	"	Troy et al. (2001)
Abdomen (F, 52 years old)	1000	0.88 ± 0.03	Mild chronic inflammation	Troy et al. (2001)
	1460	18.21 ± 2.51	"	Troy et al. (2001)
	1600	5.74 ± 0.68	"	Troy et al. (2001)
	2200	11.33 ± 0.76	"	Troy et al. (2001)
Abdomen (F, 52 years old)	1000	0.94 ± 0.02	Mild chronic inflammation	Troy et al. (2001)
	1460	18.46 ± 1.64	"	Troy et al. (2001)

(continued)

TABLE 2.2 (continued)   Absorption Coefficient μ_a, the Scattering Coefficient μ_s, and the Average Cosine of Scatter g for Various Tissues, Tissue Components, and Biofluids

Tissue	Description	Wavelength (nm)	μ_a (cm^{-1})	μ_s (cm^{-1})	g	Notes	References
		1600	5.76 ± 0.31			"	Troy et al. (2001)
		2200	13.72 ± 0.52			"	Troy et al. (2001)
	Subcutaneous ($n = 6$)	400	2.26 ± 0.24				Bashkatov et al. (2005)
		500	1.49 ± 0.06				Bashkatov et al. (2005)
		600	1.18 ± 0.02				Bashkatov et al. (2005)
		700	1.11 ± 0.05				Bashkatov et al. (2005)
		800	1.07 ± 0.11				Bashkatov et al. (2005)
		900	1.07 ± 0.07				Bashkatov et al. (2005)
		1000	1.06 ± 0.06				Bashkatov et al. (2005)
		1100	1.01 ± 0.05				Bashkatov et al. (2005)
		1200	1.06 ± 0.07				Bashkatov et al. (2005)
		1300	0.89 ± 0.07				Bashkatov et al. (2005)
		1400	1.08 ± 0.03				Bashkatov et al. (2005)
		1500	1.05 ± 0.02				Bashkatov et al. (2005)
		1600	0.89 ± 0.04				Bashkatov et al. (2005)
		1700	1.26 ± 0.07				Bashkatov et al. (2005)
		1800	1.21 ± 0.01				Bashkatov et al. (2005)
		1900	1.62 ± 0.06				Bashkatov et al. (2005)
		2000	1.43 ± 0.09				Bashkatov et al. (2005)
Skin, *in vivo*	Dermis	660	0.07–0.20				Graaff et al. (1993)
	Skin	633	0.62				Dognitz et al. (1998)
		700	0.38				Dognitz et al. (1998)
	Skin (0–1 mm)	633	0.67				Kienle et al. (1994, 1995)
	Skin (1–2 mm)	633	0.026				Kienle et al. (1994, 1995)
	Skin (>2 mm)	633	0.96				Kienle et al. (1994, 1995)
	Epidermis, forearm	633	8				Kienle et al. (1994, 1995)
	Dermis, forearm	633	0.15				Kienle et al. (1994, 1995)
	Forearm (5 subj., 14 meas.)	800	0.23 ± 0.04				Troy et al. (1996)
	Arm	633	0.17 ± 0.01				Doornbos et al. (1999)
		660	0.128 ± 0.005				Doornbos et al. (1999)
		700	0.090 ± 0.002				Doornbos et al. (1999)

Tissue		Wavelength	Value	Notes	Reference
	Foot sole	633	0.072 ± 0.002		Doornbos et al. (1999)
		660	0.053 ± 0.003		Doornbos et al. (1999)
		700	0.037 ± 0.002		Doornbos et al. (1999)
	Forehead	633	0.090 ± 0.009		Doornbos et al. (1999)
		660	0.052 ± 0.003		Doornbos et al. (1999)
		700	0.024 ± 0.002		Doornbos et al. (1999)
	Forearm (light skin, $n = 7$):	590	2.372 ± 0.282	Skin temperature, 22°C	Khalil et al. (2003)
		750	0.966 ± 0.110		Khalil et al. (2003)
		950	0.981 ± 0.073		Khalil et al. (2003)
		590	2.869 ± 0.289	Skin temperature, 38°C	Khalil et al. (2003)
		750	1.157 ± 0.106		Khalil et al. (2003)
		950	1.135 ± 0.123		Khalil et al. (2003)
	Abdominal	810	0.014	Parallel to body axis	Nickell et al. (2000)
		810	0.07	Perp. to body axis	Nickell et al. (2000)
Brain/head, in vitro	White matter (F, 32 years old)	415	2.1	24 h post mortem, data from plots	Sterenborg et al. (1989)
		488	1	"	Sterenborg et al. (1989)
		630	0.2	"	Sterenborg et al. (1989)
		800–1100	0.2–0.3	"	Sterenborg et al. (1989)
	White matter (F, 63 years old)	488	2.7	30 h post mortem, data from plots	Sterenborg et al. (1989)
		633	0.9	"	Sterenborg et al. (1989)
		800–1100	1.0–1.5	"	Sterenborg et al. (1989)
	Gray matter (M, 71 years old)	514	19.5	24 h post mortem, data from plots	Sterenborg et al. (1989)
		585	14.5	"	Sterenborg et al. (1989)
		630	4.3	"	Sterenborg et al. (1989)
		800–1100	~1.0	"	Sterenborg et al. (1989)
	Glioma (M, 65 years old)	415	16.6	4 h post mortem, data from plots	Sterenborg et al. (1989)
		488	12.5	"	Sterenborg et al. (1989)
		630	3	"	Sterenborg et al. (1989)
		800–1100	1	"	Sterenborg et al. (1989)

(continued)

TABLE 2.2 (continued)   Absorption Coefficient μ_a, the Scattering Coefficient μ_s, and the Average Cosine of Scatter g for Various Tissues, Tissue Components, and Biofluids

Tissue	Description	Wavelength (nm)	μ_a (cm⁻¹)	μ_s (cm⁻¹)	g	Notes	References
Melanoma (M, 71 years old)		585	2			24 h post mortem, data from plots	Sterenborg et al. (1989)
		630	20			″	Sterenborg et al. (1989)
		800	8			″	Sterenborg et al. (1989)
		900	4			″	Sterenborg et al. (1989)
		1100	2			″	Sterenborg et al. (1989)
White matter		633	2.2 ± 2	532 ± 41	0.82 ± 0.01	Freshly resected slabs	Tuchin (2007)
		1064	0–7.2	469 ± 34	0.870 ± 0.007	″	Tuchin (2007)
Gray matter		633	2.7 ± 2	354 ± 37	0.940 ± 0.004	″	Tuchin (2007)
		1064	0–10	134 ± 14	0.900 ± 0.007	″	Tuchin (2007)
Adult head, scalp and skull		800	0.4				Tuchin et al. (1999)
Adult head, gray matter		800	0.25				Tuchin et al. (1999)
Adult head, white matter		800	0.05				Tuchin et al. (1999)
White matter ($n = 7$)		360	2.53 ± 0.55	402.0 ± 91.9	0.702 ± 0.093		Schwarzmaier et al. (1997); Yaroslavsky et al. (1996)
		640	0.8	408.2 ± 88.5	0.84 ± 0.05		Schwarzmaier et al. (1997); Yaroslavsky et al. (1996)
		860	0.97 ± 0.40	353.1 ± 68.1	0.871 ± 0.028		Schwarzmaier et al. (1997); Yaroslavsky et al. (1996)
		1060	1.08 ± 0.51	299.5 ± 70.1	0.889 ± 0.010		Schwarzmaier et al. (1997); Yaroslavsky et al. (1996)
White matter		800	0.80 ± 0.16	140.0 ± 1.4	0.95 ± 0.02		Roggan et al. (1994)
		1064	0.40 ± 0.08	110 ± 11	0.95 ± 0.02		Roggan et al. (1994)
White matter		1064	1.6	513	0.96		Schwarzmaier et al. (1997)
White matter, coagulated ($n = 7$)		360	8.30 ± 3.65	604.2 ± 131.5	0.800 ± 0.099	Coagulation: 2 h 80°C	Schwarzmaier et al. (1997); Yaroslavsky et al. (1996)
		860	1.7 ± 1.3	417.0 ± 272.5	0.922 ± 0.025	″	Schwarzmaier et al. (1997); Yaroslavsky et al. (1996)

Tissue	Wavelength (nm)				Remarks	Reference
	1060	2.15 ± 1.34	363.3 ± 226.8	0.930 ± 0.015	"	Schwarzmaier et al. (1997); Yaroslavsky et al. (1996)
White matter, coagulated	800	0.90 ± 0.18	170 ± 10.2	0.94 ± 0.02	Homogenized tissue, coagulation (75°C)	Roggan et al. (1994)
	1064	0.5 ± 0.1	130 ± 9.1	0.93 ± 0.02	"	Roggan et al. (1994)
Gray matter (n = 7)	360	3.33 ± 2.19	141.3 ± 42.6	0.818 ± 0.093		Schwarzmaier et al. (1997); Yaroslavsky et al. (1996)
	640	0.17 ± 0.26	90.1 ± 32.5	0.89 ± 0.04		Schwarzmaier et al. (1997); Yaroslavsky et al. (1996)
	1060	0.56 ± 0.70	56.8 ± 18.0	0.90 ± 0.05		Schwarzmaier et al. (1997); Yaroslavsky et al. (1996)
Gray matter	1064	1.9	267	0.95		Schwarzmaier et al. (1997); Yaroslavsky et al. (1996)
Gray matter, coagulated (n = 7)	360	9.39 ± 1.70	426 ± 122	0.868 ± 0.031	Coagulation: 2 h 80°C	Schwarzmaier et al. (1997); Yaroslavsky et al. (1996)
	740	0.45 ± 0.27			"	Schwarzmaier et al. (1997); Yaroslavsky et al. (1996)
	1100	1.00 ± 0.45	179.8 ± 32.6	0.954 ± 0.001	"	Schwarzmaier et al. (1997); Yaroslavsky et al. (1996)
Brain tumor, astrocytoma grade III WHO	400	10	84	0.9		Willmann et al. (1999)
	633	6.3 ± 1.6	67 ± 8	0.883 ± 0.011		Willmann et al. (1999)
	700	4	50	0.88		Willmann et al. (1999)
	800	3	50	0.88		Willmann et al. (1999)

(continued)

TABLE 2.2 (continued)   Absorption Coefficient μ_a, the Scattering Coefficient μ_s, and the Average Cosine of Scatter g for Various Tissues, Tissue Components, and Biofluids

Tissue	Description	Wavelength (nm)	μ_a (cm⁻¹)	μ_s (cm⁻¹)	g	Notes	References
	White matter	450	0.14	42	0.78		Yaroslavsky et al. (2002)
		510	0.1	42.6	0.81		Yaroslavsky et al. (2002)
		630	0.08	40.9	0.84		Yaroslavsky et al. (2002)
		670	0.07	40.1	0.85		Yaroslavsky et al. (2002)
		850	0.1	34.2	0.88		Yaroslavsky et al. (2002)
		1064	0.1	29.6	0.89		Yaroslavsky et al. (2002)
	Gray matter	450	0.07	11.7	0.88		Yaroslavsky et al. (2002)
		510	0.04	10.6	0.88		Yaroslavsky et al. (2002)
		630	0.02	9	0.89		Yaroslavsky et al. (2002)
		670	0.02	8.4	0.9		Yaroslavsky et al. (2002)
		1064	0.05	5.7	0.9		Yaroslavsky et al. (2002)
	White matter	456	0.81	92.3	0.92		Gottschalk (1992)
		514	0.5	104.5	0.93		Gottschalk (1992)
		630	0.15	38.6	0.86		Gottschalk (1992)
		675	0.07	43.6	0.87		Gottschalk (1992)
		1064	0.16	51.3	0.95		Gottschalk (1992)
	Gray matter	456	0.9	68.6	0.95		Gottschalk (1992)
		514	1.17	57.8	0.97		Gottschalk (1992)
		630	0.14	47.3	0.93		Gottschalk (1992)
		675	0.06	36.4	0.91		Gottschalk (1992)
		1064	0.19	26.7	0.96		Gottschalk (1992)
	Gray matter ($n = 7$)	400	2.6 ± 0.6	128.5 ± 18.4	0.87 ± 0.02	Data from plots	Yaroslavsky et al. (2002)
		500	0.5 ± 0.2	109.9 ± 13.0	0.88 ± 0.01	"	Yaroslavsky et al. (2002)
		600	0.3 ± 0.1	94.1 ± 13.5	0.89 ± 0.02	"	Yaroslavsky et al. (2002)
		700	0.2 ± 0.1	84.1 ± 12.0	0.90 ± 0.02	"	Yaroslavsky et al. (2002)
		800	0.2 ± 0.1	77.0 ± 11.0	0.90 ± 0.02	"	Yaroslavsky et al. (2002)
		900	0.3 ± 0.2	67.3 ± 9.6	0.90 ± 0.02	"	Yaroslavsky et al. (2002)
		1000	0.6 ± 0.3	61.6 ± 5.7	0.90 ± 0.02	"	Yaroslavsky et al. (2002)
		1100	0.5 ± 0.3	55.1 ± 6.5	0.90 ± 0.02	"	Yaroslavsky et al. (2002)
	Gray matter, coagulated ($n = 7$)	400	7.5 ± 0.4	258.6 ± 18.8	0.78 ± 0.04	"	Yaroslavsky et al. (2002)
		500	1.8 ± 0.2	326.5 ± 7.7	0.85 ± 0.03	"	Yaroslavsky et al. (2002)

Tissue	Wavelength				Reference
	600	0.7 ± 0.0	319.0 ± 15.2	0.87 ± 0.03	Yaroslavsky et al. (2002)
	700	0.7 ± 0.1	319.0 ± 7.5	0.88 ± 0.03	Yaroslavsky et al. (2002)
	800	0.8 ± 0.1	252.7 ± 18.3	0.87 ± 0.02	Yaroslavsky et al. (2002)
	900	0.9 ± 0.1	214.6 ± 10.3	0.87 ± 0.02	Yaroslavsky et al. (2002)
	1000	1.4 ± 0.2	191.0 ± 18.7	0.88 ± 0.03	Yaroslavsky et al. (2002)
	1100	1.5 ± 0.2	186.6 ± 13.5	0.88 ± 0.03	Yaroslavsky et al. (2002)
White matter (n = 7)	400	3.1 ± 0.2	413.5 ± 21.4	0.75 ± 0.03	Yaroslavsky et al. (2002)
	500	0.9 ± 0.1	413.5 ± 43.9	0.80 ± 0.02	Yaroslavsky et al. (2002)
	600	0.8 ± 0.1	413.5 ± 21.4	0.83 ± 0.02	Yaroslavsky et al. (2002)
	700	0.8 ± 0.1	393.1 ± 30.9	0.85 ± 0.02	Yaroslavsky et al. (2002)
	800	0.9 ± 0.1	364.5 ± 28.6	0.87 ± 0.01	Yaroslavsky et al. (2002)
	900	1.0 ± 0.1	329.5 ± 35.0	0.88 ± 0.01	Yaroslavsky et al. (2002)
	1000	1.2 ± 0.2	305.4 ± 15.9	0.88 ± 0.01	Yaroslavsky et al. (2002)
	1100	1.0 ± 0.2	283.2 ± 22.2	0.88 ± 0.00	Yaroslavsky et al. (2002)
White matter, coagulated (n = 7)	410	8.7 ± 1.7	568.7 ± 111.9	0.83 ± 0.03	Yaroslavsky et al. (2002)
	510	2.9 ± 0.6	513.2 ± 116.9	0.87 ± 0.02	Yaroslavsky et al. (2002)
	610	1.7 ± 0.4	500.2 ± 129.9	0.90 ± 0.02	Yaroslavsky et al. (2002)
	710	1.4 ± 0.5	475.2 ± 108.3	0.91 ± 0.01	Yaroslavsky et al. (2002)
	810	1.5 ± 0.5	440.0 ± 114.3	0.92 ± 0.01	Yaroslavsky et al. (2002)
	910	1.7 ± 0.6	407.4 ± 92.8	0.93 ± 0.01	Yaroslavsky et al. (2002)
	1010	1.9 ± 0.6	367.7 ± 95.5	0.93 ± 0.00	Yaroslavsky et al. (2002)
	1100	2.4 ± 0.5	358.4 ± 81.6	0.93 ± 0.00	Yaroslavsky et al. (2002)
Astrocytoma (tumor, WHO grade II, n = 4)	400	18.8 ± 11.3	198.4 ± 55.6	0.88 ± 0.05	Yaroslavsky et al. (2002)
	490	2.5 ± 0.9	158.5 ± 53.7	0.93 ± 0.03	Yaroslavsky et al. (2002)
	600	1.2 ± 0.7	132.4 ± 49.0	0.96 ± 0.02	Yaroslavsky et al. (2002)
	700	0.5 ± 0.3	113.2 ± 41.8	0.96 ± 0.02	Yaroslavsky et al. (2002)
	800	0.7 ± 0.2	96.7 ± 41.8	0.96 ± 0.01	Yaroslavsky et al. (2002)
	900	0.3 ± 0.2	86.4 ± 34.6	0.96 ± 0.01	Yaroslavsky et al. (2002)
	1000	0.5 ± 0.3	79.0 ± 34.2	0.96 ± 0.00	Yaroslavsky et al. (2002)
	1100	0.6 ± 0.2	73.8 ± 29.6	0.96 ± 0.00	Yaroslavsky et al. (2002)

(continued)

TABLE 2.2 (continued) Absorption Coefficient μ_a, the Scattering Coefficient μ_s, and the Average Cosine of Scatter g for Various Tissues, Tissue Components, and Biofluids

Tissue	Description	Wavelength (nm)	μ_a (cm^{-1})	μ_s (cm^{-1})	g	Notes	References
	Cerebellum ($n = 7$)	400	4.7 ± 0.8	276.7 ± 19.1	0.80 ± 0.03	"	Yaroslavsky et al. (2002)
		500	1.4 ± 0.2	277.5 ± 32.6	0.85 ± 0.02	"	Yaroslavsky et al. (2002)
		600	0.8 ± 0.2	272.1 ± 12.3	0.87 ± 0.02	"	Yaroslavsky et al. (2002)
		700	0.6 ± 0.1	266.8 ± 12.1	0.89 ± 0.01	"	Yaroslavsky et al. (2002)
		800	0.6 ± 0.1	250.3 ± 17.2	0.90 ± 0.01	"	Yaroslavsky et al. (2002)
		900	0.7 ± 0.1	229.6 ± 15.8	0.90 ± 0.01	"	Yaroslavsky et al. (2002)
		1000	0.8 ± 0.1	215.4 ± 14.7	0.90 ± 0.01	"	Yaroslavsky et al. (2002)
		1100	0.7 ± 0.1	202.1 ± 13.9	0.90 ± 0.01	"	Yaroslavsky et al. (2002)
	Cerebellum, coagulated ($n = 7$)	400	19.3 ± 7.7	560.0 ± 25.5	0.61 ± 0.01	"	Yaroslavsky et al. (2002)
		500	5.1 ± 1.7	512.2 ± 47.8	0.77 ± 0.02	"	Yaroslavsky et al. (2002)
		600	2.9 ± 1.4	458.2 ± 65.6	0.78 ± 0.01	"	Yaroslavsky et al. (2002)
		700	1.7 ± 0.4	489.9 ± 70.1	0.85 ± 0.01	"	Yaroslavsky et al. (2002)
		800	1.1 ± 0.2	458.2 ± 54.0	0.87 ± 0.02	"	Yaroslavsky et al. (2002)
		900	1.1 ± 0.3	458.2 ± 65.6	0.89 ± 0.02	"	Yaroslavsky et al. (2002)
		1000	1.0 ± 0.4	419.1 ± 49.4	0.90 ± 0.03	"	Yaroslavsky et al. (2002)
		1100	1.1 ± 0.5	428.5 ± 40.0	0.91 ± 0.03	"	Yaroslavsky et al. (2002)
	Meningioma (tumor, $n = 6$)	410	4.1 ± 0.5	197.4 ± 19.8	0.88 ± 0.02	"	Yaroslavsky et al. (2002)
		490	1.3 ± 0.2	188.2 ± 18.8	0.93 ± 0.01	"	Yaroslavsky et al. (2002)
		590	0.7 ± 0.2	171.1 ± 12.7	0.95 ± 0.01	"	Yaroslavsky et al. (2002)
		690	0.3 ± 0.1	155.5 ± 15.6	0.95 ± 0.01	"	Yaroslavsky et al. (2002)
		790	0.2 ± 0.1	141.3 ± 14.2	0.96 ± 0.01	"	Yaroslavsky et al. (2002)
		910	0.2 ± 0.1	125.4 ± 9.4	0.95 ± 0.00	"	Yaroslavsky et al. (2002)
		990	0.4 ± 0.2	119.6 ± 8.9	0.95 ± 0.01	"	Yaroslavsky et al. (2002)
		1100	0.6 ± 0.2	116.8 ± 8.6	0.97 ± 0.00	"	Yaroslavsky et al. (2002)
	Pons ($n = 7$)	400	3.1 ± 0.7	163.5 ± 15.3	0.89 ± 0.02	"	Yaroslavsky et al. (2002)
		500	0.9 ± 0.3	133.7 ± 19.2	0.91 ± 0.01	"	Yaroslavsky et al. (2002)
		600	0.6 ± 0.2	109.4 ± 18.5	0.91 ± 0.01	"	Yaroslavsky et al. (2002)
		700	0.5 ± 0.2	93.5 ± 20.9	0.91 ± 0.01	"	Yaroslavsky et al. (2002)
		800	0.6 ± 0.3	83.6 ± 21.0	0.91 ± 0.01	"	Yaroslavsky et al. (2002)
		900	0.7 ± 0.3	74.8 ± 18.7	0.92 ± 0.01	"	Yaroslavsky et al. (2002)
		1000	1.0 ± 0.4	69.9 ± 17.5	0.91 ± 0.01	"	Yaroslavsky et al. (2002)

Tissue	Wavelength			Reference
Pons, coagulated (*n* = 7)	1100	0.9 ± 0.4	64.0 ± 17.8	0.92 ± 0.01	Yaroslavsky et al. (2002)
	410	17.2 ± 1.6	685.7 ± 63.7	0.85 ± 0.02	Yaroslavsky et al. (2002)
	510	8.5 ± 0.8	627.5 ± 73.6	0.89 ± 0.01	Yaroslavsky et al. (2002)
	610	7.7 ± 0.5	530.0 ± 100.0	0.89 ± 0.01	Yaroslavsky et al. (2002)
	710	6.9 ± 0.6	402.5 ± 67.7	0.89 ± 0.00	Yaroslavsky et al. (2002)
	810	6.5 ± 0.6	329.7 ± 55.4	0.89 ± 0.01	Yaroslavsky et al. (2002)
	910	5.9 ± 1.0	276.0 ± 46.4	0.88 ± 0.00	Yaroslavsky et al. (2002)
	1010	5.7 ± 1.0	241.6 ± 34.4	0.88 ± 0.00	Yaroslavsky et al. (2002)
	1100	6.5 ± 0.9	221.1 ± 31.5	0.88 ± 0.00	Yaroslavsky et al. (2002)
Thalamus (*n* = 7)	410	3.2 ± 1.0	146.7 ± 49.4	0.86 ± 0.03	Yaroslavsky et al. (2002)
	510	0.9 ± 0.3	188.0 ± 31.9	0.87 ± 0.03	Yaroslavsky et al. (2002)
	610	0.6 ± 0.2	176.3 ± 34.5	0.88 ± 0.02	Yaroslavsky et al. (2002)
	710	0.5 ± 0.3	169.0 ± 28.7	0.89 ± 0.03	Yaroslavsky et al. (2002)
	810	0.7 ± 0.3	158.5 ± 35.3	0.89 ± 0.02	Yaroslavsky et al. (2002)
	910	0.7 ± 0.3	155.4 ± 22.3	0.90 ± 0.02	Yaroslavsky et al. (2002)
	1010	0.8 ± 0.3	139.3 ± 34.9	0.90 ± 0.02	Yaroslavsky et al. (2002)
	1100	0.8 ± 0.3	146.0 ± 36.6	0.91 ± 0.02	Yaroslavsky et al. (2002)
Thalamus, coagulated (*n* = 7)	400	15.0 ± 3.3	391.1 ± 56.1	0.83 ± 0.04	Yaroslavsky et al. (2002)
	500	4.2 ± 0.9	399.9 ± 67.7	0.90 ± 0.01	Yaroslavsky et al. (2002)
	600	1.6 ± 0.6	365.7 ± 43.2	0.92 ± 0.01	Yaroslavsky et al. (2002)
	700	1.4 ± 0.3	327.0 ± 30.6	0.92 ± 0.01	Yaroslavsky et al. (2002)
	800	1.1 ± 0.3	286.0 ± 33.8	0.93 ± 0.01	Yaroslavsky et al. (2002)
	900	1.1 ± 0.3	267.4 ± 31.6	0.93 ± 0.01	Yaroslavsky et al. (2002)
	1000	1.4 ± 0.4	233.8 ± 39.7	0.93 ± 0.01	Yaroslavsky et al. (2002)
	1100	1.5 ± 0.4	223.6 ± 32.1	0.94 ± 0.01	Yaroslavsky et al. (2002)
Gray matter (*n* = 25)	400	9.778			Gebhart et al. (2006)
	418	14.873			Gebhart et al. (2006)
	428	16.722			Gebhart et al. (2006)
	450	5.161			Gebhart et al. (2006)
	488	2.272			Gebhart et al. (2006)
	500	2.206			Gebhart et al. (2006)
	550	2.955			Gebhart et al. (2006)

(continued)

TABLE 2.2 (continued)   Absorption Coefficient μ_a, the Scattering Coefficient μ_s, and the Average Cosine of Scatter g for Various Tissues, Tissue Components, and Biofluids

Tissue	Description	Wavelength (nm)	μ_a (cm⁻¹)	μ_s (cm⁻¹)	g	Notes	References
		600	1.46				Gebhart et al. (2006)
		632	0.925				Gebhart et al. (2006)
		670	0.809				Gebhart et al. (2006)
		700	0.733				Gebhart et al. (2006)
		750	0.599				Gebhart et al. (2006)
		800	0.507				Gebhart et al. (2006)
		830	0.485				Gebhart et al. (2006)
		850	0.472				Gebhart et al. (2006)
		870	0.479				Gebhart et al. (2006)
		900	0.503				Gebhart et al. (2006)
		950	0.521				Gebhart et al. (2006)
		1000	0.585				Gebhart et al. (2006)
		1064	0.502				Gebhart et al. (2006)
		1100	0.502				Gebhart et al. (2006)
		1150	0.815				Gebhart et al. (2006)
		1200	1.01				Gebhart et al. (2006)
		1250	0.865				Gebhart et al. (2006)
		1300	0.894				Gebhart et al. (2006)
	White matter ($n = 19$)	400	9.134				Gebhart et al. (2006)
		418	13.603				Gebhart et al. (2006)
		428	15.417				Gebhart et al. (2006)
		450	3.958				Gebhart et al. (2006)
		488	1.869				Gebhart et al. (2006)
		500	1.834				Gebhart et al. (2006)
		550	2.584				Gebhart et al. (2006)
		600	1.175				Gebhart et al. (2006)
		632	0.801				Gebhart et al. (2006)
		670	0.711				Gebhart et al. (2006)
		700	0.674				Gebhart et al. (2006)
		750	0.649				Gebhart et al. (2006)
		800	0.622				Gebhart et al. (2006)

830	0.626	Gebhart et al. (2006)
850	0.643	Gebhart et al. (2006)
870	0.666	Gebhart et al. (2006)
900	0.684	Gebhart et al. (2006)
950	0.785	Gebhart et al. (2006)
1000	0.883	Gebhart et al. (2006)
1064	0.752	Gebhart et al. (2006)
1100	0.762	Gebhart et al. (2006)
1150	1.135	Gebhart et al. (2006)
1200	1.42	Gebhart et al. (2006)
1250	1.268	Gebhart et al. (2006)
1300	1.274	Gebhart et al. (2006)
Glioma ($n = 39$)		
400	12.393	Gebhart et al. (2006)
418	17.496	Gebhart et al. (2006)
428	16.124	Gebhart et al. (2006)
450	4.891	Gebhart et al. (2006)
488	2.592	Gebhart et al. (2006)
500	2.352	Gebhart et al. (2006)
550	2.768	Gebhart et al. (2006)
600	1.149	Gebhart et al. (2006)
632	0.846	Gebhart et al. (2006)
670	0.741	Gebhart et al. (2006)
700	0.709	Gebhart et al. (2006)
750	0.679	Gebhart et al. (2006)
800	0.656	Gebhart et al. (2006)
830	0.662	Gebhart et al. (2006)
850	0.67	Gebhart et al. (2006)
870	0.685	Gebhart et al. (2006)
900	0.707	Gebhart et al. (2006)
950	0.768	Gebhart et al. (2006)
1000	0.938	Gebhart et al. (2006)
1064	0.822	Gebhart et al. (2006)

(continued)

TABLE 2.2 (continued) Absorption Coefficient μ_a, the Scattering Coefficient μ_s, and the Average Cosine of Scatter g for Various Tissues, Tissue Components, and Biofluids

Tissue	Description	Wavelength (nm)	μ_a (cm^{-1})	μ_s (cm^{-1})	g	Notes	References
		1100	0.831				Gebhart et al. (2006)
		1150	1.231				Gebhart et al. (2006)
		1200	1.518				Gebhart et al. (2006)
		1250	1.379				Gebhart et al. (2006)
		1300	1.412				Gebhart et al. (2006)
	Dura mater ($n = 8$)	400	3.08 ± 0.15			<24 h post mortem	Genina et al. (2005)
		450	1.51 ± 0.08			"	Genina et al. (2005)
		500	1.09 ± 0.05			"	Genina et al. (2005)
		550	1.10 ± 0.05			"	Genina et al. (2005)
		600	0.80 ± 0.04			"	Genina et al. (2005)
		650	0.70 ± 0.04			"	Genina et al. (2005)
		700	0.74 ± 0.04			"	Genina et al. (2005)
	Scalp and skull	800	0.4				Okada et al. (1997)
	Cerebrospinal fluid	800	0.01				Okada et al. (1997)
	Scalp ($n = 3$)	805	0.52 ± 0.04				Bashkatov et al. (2006)
		900	0.40 ± 0.02				Bashkatov et al. (2006)
		950	0.39 ± 0.03				Bashkatov et al. (2006)
		1000	0.33 ± 0.03				Bashkatov et al. (2006)
		1100	0.19 ± 0.04				Bashkatov et al. (2006)
		1200	0.65 ± 0.04				Bashkatov et al. (2006)
		1300	0.50 ± 0.07				Bashkatov et al. (2006)
		1400	1.98 ± 0.31				Bashkatov et al. (2006)
		1430	2.19 ± 0.29				Bashkatov et al. (2006)
		1500	2.04 ± 0.35				Bashkatov et al. (2006)
		1600	1.43 ± 0.22				Bashkatov et al. (2006)
		1700	1.87 ± 0.28				Bashkatov et al. (2006)
		1800	1.73 ± 0.22				Bashkatov et al. (2006)
		1900	2.57 ± 0.28				Bashkatov et al. (2006)
		1930	2.52 ± 0.25				Bashkatov et al. (2006)
		2000	2.09 ± 0.29				Bashkatov et al. (2006)
	Skull bone ($n = 8$)	801	0.11 ± 0.02				Bashkatov et al. (2006)

Tissue		Wavelength (nm)	Value	Reference
		900	0.15 ± 0.02	Bashkatov et al. (2006)
		980	0.23 ± 0.03	Bashkatov et al. (2006)
		1000	0.22 ± 0.03	Bashkatov et al. (2006)
		1100	0.16 ± 0.03	Bashkatov et al. (2006)
		1180	0.67 ± 0.07	Bashkatov et al. (2006)
		1200	0.67 ± 0.07	Bashkatov et al. (2006)
		1300	0.54 ± 0.05	Bashkatov et al. (2006)
		1400	2.43 ± 0.24	Bashkatov et al. (2006)
		1465	3.33 ± 0.31	Bashkatov et al. (2006)
		1500	3.13 ± 0.26	Bashkatov et al. (2006)
		1600	2.47 ± 0.40	Bashkatov et al. (2006)
		1700	2.77 ± 0.46	Bashkatov et al. (2006)
		1740	2.98 ± 0.54	Bashkatov et al. (2006)
		1800	2.97 ± 0.62	Bashkatov et al. (2006)
		1900	4.39 ± 1.33	Bashkatov et al. (2006)
		1930	4.97 ± 1.52	Bashkatov et al. (2006)
		2000	4.47 ± 1.18	Bashkatov et al. (2006)
Brain/Head, in vivo	Cortex, frontal lobe	674	<0.2	Bevilacqua et al. (1999)
		811	<0.1	Bevilacqua et al. (1999)
		849	<0.1	Bevilacqua et al. (1999)
		956	0.15 ± 0.10	Bevilacqua et al. (1999)
	Cortex, temporal lobe	674	0.2 ± 0.1	Bevilacqua et al. (1999)
		811	0.2 ± 0.1	Bevilacqua et al. (1999)
		849	<0.1	Bevilacqua et al. (1999)
		956	0.25 ± 0.10	Bevilacqua et al. (1999)
	Astrocytoma of optic nerve	674	1.4 ± 0.3	Bevilacqua et al. (1999)
		811	1.2 ± 0.3	Bevilacqua et al. (1999)
		849	0.9 ± 0.3	Bevilacqua et al. (1999)
		956	1.5 ± 0.3	Bevilacqua et al. (1999)
	Normal optic nerve	674	0.6 ± 0.3	Bevilacqua et al. (1999)
		849	0.8 ± 0.3	Bevilacqua et al. (1999)
		956	0.7 ± 0.3	Bevilacqua et al. (1999)

(continued)

TABLE 2.2 (continued)   Absorption Coefficient μ_a, the Scattering Coefficient μ_s, and the Average Cosine of Scatter g for Various Tissues, Tissue Components, and Biofluids

Tissue	Description	Wavelength (nm)	μ_a (cm⁻¹)	μ_s (cm⁻¹)	g	Notes	References
	Skull	674	0.5 ± 0.2				Bevilacqua et al. (1999)
		849	0.5 ± 0.2				Bevilacqua et al. (1999)
		956	0.5 ± 0.2				Bevilacqua et al. (1999)
	Cerebellar white matter	674	2.5 ± 0.5				Bevilacqua et al. (1999)
		849	0.95 ± 0.20				Bevilacqua et al. (1999)
		956	0.9 ± 0.2				Bevilacqua et al. (1999)
	Medulloblastoma	674	2.6 ± 0.5				Bevilacqua et al. (1999)
		849	1.0 ± 0.2				Bevilacqua et al. (1999)
		956	0.75 ± 0.20				Bevilacqua et al. (1999)
	Cerebellar white matter with scar tissue	674	<0.2				Bevilacqua et al. (1999)
		849	<0.2				Bevilacqua et al. (1999)
	Normal cortex, temporal and frontal lobe	674	>0.2		0.92	Measurements during brain surgery	Bevilacqua et al. (1997)
		849	>0.2		0.92	"	Bevilacqua et al. (1997)
		956	>0.2		0.92	"	Bevilacqua et al. (1997)
	Normal optic nerve	674	0.60 ± 0.25		0.92	"	Bevilacqua et al. (1997)
		849	0.75 ± 0.25		0.92	"	Bevilacqua et al. (1997)
		956	0.65 ± 0.25		0.92	"	Bevilacqua et al. (1997)
	Astrocytoma of optical nerve	674	1.6 ± 1.0		0.92	"	Bevilacqua et al. (1997)
		849	1.1 ± 1.0		0.92	"	Bevilacqua et al. (1997)
		950	1.8 ± 1.0		0.92	"	Bevilacqua et al. (1997)
Female breast, in vitro	Fatty normal (n = 23)	749	0.18 (0.16)	8.48 ± 3.43		Excised, kept in saline, 37°C	Troy et al. (1996)
		789	0.08 (0.10)	7.67 ± 2.57		"	Troy et al. (1996)
		836	0.11 (0.10)	7.27 ± 2.40		"	Troy et al. (1996)
	Fibrous normal (n = 35)	749	0.13 (0.19)	9.75 ± 2.27		Excised, kept in saline, 37°C	Troy et al. (1996)
		789	0.06 (0.12)	8.94 ± 2.45		"	Troy et al. (1996)
		836	0.05 (0.08)	8.10 ± 2.21		"	Troy et al. (1996)
	Infiltrating carcinoma (n = 48)	749	0.15 (0.14)	10.91 ± 5.59		Excised, kept in saline, 37°C	Troy et al. (1996)
		789	0.04 (0.08)	10.12 ± 5.05		"	Troy et al. (1996)
		836	0.10 (0.19)			"	Troy et al. (1996)

Tissue	Wavelength (nm)	Value	g	Notes	Reference
Mucinous carcinoma (n = 3)	749	0.26 (0.20)		Excised, kept in saline, 37°C	Troy et al. (1996)
	789	0.016 (0.072)		"	Troy et al. (1996)
	836	0.023 (0.108)		"	Troy et al. (1996)
Ductal carcinoma, in situ (n = 5)	749	0.076 (0.068)		Excised, kept in saline, 37°C	Troy et al. (1996)
	789	0.023 (0.034)		"	Troy et al. (1996)
	836	0.039 (0.068)		"	Troy et al. (1996)
Grandular tissue (n = 3)	540	3.58 ± 1.56		Homogenized tissue	Peters et al. (1990)
	700	0.47 ± 0.11		"	Peters et al. (1990)
	900	0.62 ± 0.05		"	Peters et al. (1990)
Fatty tissue (n = 7)	540	2.27 ± 0.57		Homogenized tissue	Peters et al. (1990)
	700	0.70 ± 0.08		"	Peters et al. (1990)
	900	0.75 ± 0.08		"	Peters et al. (1990)
Fibrocystic tissue (n = 8)	540	1.64 ± 0.66		Homogenized tissue	Peters et al. (1990)
	700	0.22 ± 0.09		"	Peters et al. (1990)
	900	0.27 ± 0.11		"	Peters et al. (1990)
Fibroadenoma (n = 6)	540	4.38 ± 3.14		Homogenized tissue	Peters et al. (1990)
	700	0.52 ± 0.47		"	Peters et al. (1990)
	900	0.72 ± 0.53		"	Peters et al. (1990)
Carcinoma (n = 9)	540	3.07 ± 0.99		Homogenized tissue	Peters et al. (1990)
	700	0.45 ± 0.12		"	Peters et al. (1990)
	900	0.50 ± 0.15		"	Peters et al. (1990)
Carcinoma	580	4.5 ± 0.8			Key et al. (1991)
	850	0.4 ± 0.5			Key et al. (1991)
	1300	0.5 ± 0.8			Key et al. (1991)
Healthy tissue adjacent to carcinoma	580	2.6 ± 1.1			Key et al. (1991)
	850	0.3 ± 0.2			Key et al. (1991)
	1300	0.8 ± 0.6			Key et al. (1991)
Fatty tissue	700		0.95 ± 0.02		Key et al. (1991)
Fibroglandular tissue	700		0.92 ± 0.03		Key et al. (1991)

(continued)

TABLE 2.2 (continued)   Absorption Coefficient μ_a, the Scattering Coefficient μ_s, and the Average Cosine of Scatter g for Various Tissues, Tissue Components, and Biofluids

Tissue	Description	Wavelength (nm)	μ_a (cm⁻¹)	μ_s (cm⁻¹)	g	Notes	References
	Carcinoma (central part)	700			0.88 ± 0.03		Key et al. (1991)
	Fatty tissue	625	0.06 ± 0.02				Zhadin et al. (1998)
	Benign tumor	625	0.33 ± 0.06				Zhadin et al. (1998)
	Fatty tissue	625	0.06 ± 0.02				Das et al. (1997)
	Benign tumor	625	0.33 ± 0.06				Das et al. (1997)
	Invasive ductal carcinoma (n = 10)	450	2.55 ± 0.30			9 in group 55–65 year, 1 in 1–35 year	Ghosh et al. (2001)
		460	2.62 ± 0.34			"	Ghosh et al. (2001)
		470	2.44 ± 0.25			"	Ghosh et al. (2001)
		480	2.32 ± 0.26			"	Ghosh et al. (2001)
		490	2.23 ± 0.25			"	Ghosh et al. (2001)
		500	2.22 ± 0.22			"	Ghosh et al. (2001)
		510	2.16 ± 0.24			"	Ghosh et al. (2001)
		520	2.12 ± 0.22			"	Ghosh et al. (2001)
		530	2.07 ± 0.22			"	Ghosh et al. (2001)
		540	1.99 ± 0.21			"	Ghosh et al. (2001)
		550	2.13 ± 0.23			"	Ghosh et al. (2001)
		560	2.09 ± 0.21			"	Ghosh et al. (2001)
		570	2.09 ± 0.25			"	Ghosh et al. (2001)
		580	2.07 ± 0.21			"	Ghosh et al. (2001)
		590	2.01 ± 0.22			"	Ghosh et al. (2001)
		600	1.90 ± 0.19			"	Ghosh et al. (2001)
		610	1.82 ± 0.18			"	Ghosh et al. (2001)
		620	1.71 ± 0.18			"	Ghosh et al. (2001)
		630	1.64 ± 0.17			"	Ghosh et al. (2001)
		640	1.55 ± 0.17			"	Ghosh et al. (2001)
		650	1.48 ± 0.15			"	Ghosh et al. (2001)
		633			0.96 ± 0.01		Ghosh et al. (2001)
		633			0.86 ± 0.02		Ghosh et al. (2001)
	Adjacent healthy tissue (n = 10)	450	1.45 ± 0.22			9 in group 55–65 year, 1 in 1–35 year	Ghosh et al. (2001)

Tissue	λ (nm)	Value		Technique	Reference
	460	1.48 ± 0.21		"	Ghosh et al. (2001)
	470	1.42 ± 0.21		"	Ghosh et al. (2001)
	480	1.35 ± 0.19		"	Ghosh et al. (2001)
	490	1.26 ± 0.21		"	Ghosh et al. (2001)
	500	1.24 ± 0.21		"	Ghosh et al. (2001)
	510	1.23 ± 0.19		"	Ghosh et al. (2001)
	520	1.19 ± 0.18		"	Ghosh et al. (2001)
	530	1.14 ± 0.17		"	Ghosh et al. (2001)
	540	1.19 ± 0.22		"	Ghosh et al. (2001)
	550	1.16 ± 0.26		"	Ghosh et al. (2001)
	560	1.14 ± 0.17		"	Ghosh et al. (2001)
	570	1.13 ± 0.16		"	Ghosh et al. (2001)
	580	1.17 ± 0.17		"	Ghosh et al. (2001)
	590	1.07 ± 0.17		"	Ghosh et al. (2001)
	600	1.00 ± 0.12		"	Ghosh et al. (2001)
	610	0.95 ± 0.12		"	Ghosh et al. (2001)
	620	0.89 ± 0.11		"	Ghosh et al. (2001)
	630	0.82 ± 0.07		"	Ghosh et al. (2001)
	640	0.79 ± 0.08		"	Ghosh et al. (2001)
	650	0.74 ± 0.08		"	Ghosh et al. (2001)
	633		0.88 ± 0.01		Ghosh et al. (2001)
	633		0.76 ± 0.01		Ghosh et al. (2001)
Female breast, in vivo	Normal (30 Japanese F, avg'd over ages)	753	0.046 ± 0.014		Suzuki et al. (1996)
	Normal (6 subj., 26–43 years old)	800	0.017–0.045		Heusmann et al. (1996)
	Normal (6 subj.)	580	0.70 ± 0.12	Transmission	Key et al. (1991)
	Normal (6 subj.)	780	0.23 ± 0.02	"	Key et al. (1991)
	Normal (6 subj.)	850	0.27 ± 0.03	"	Key et al. (1991)
	Breast cancer (5 subj., relapsed cancer)	630	0.305 ± 0.160	HPD (72 h)	Driver et al. 1991

(continued)

TABLE 2.2 (continued)   Absorption Coefficient μ_a, the Scattering Coefficient μ_s, and the Average Cosine of Scatter g for Various Tissues, Tissue Components, and Biofluids

Tissue	Description	Wavelength (nm)	μ_a (cm^{-1})	μ_s (cm^{-1})	g	Notes	References
	Normal (56 year)	674	0.04				Fishkin et al. (1997); Tromberg et al. (1997)
		811	0.035				Fishkin et al. (1997); Tromberg et al. (1997)
		849	0.035				Fishkin et al. (1997); Tromberg et al. (1997)
		956	0.085				Fishkin et al. (1997); Tromberg et al. (1997)
	Fibroadenoma w/ductal hyperplasia (56 year)	674	0.055				Fishkin et al. (1997); Tromberg et al. (1997)
		811	0.06				Fishkin et al. (1997); Tromberg et al. (1997)
		849	0.055				Fishkin et al. (1997); Tromberg et al. (1997)
		956	0.12				Fishkin et al. (1997); Tromberg et al. (1997)
	Normal (27 year)	674	0.035				Fishkin et al. (1997); Tromberg et al. (1997)
		811	0.03				Fishkin et al. (1997); Tromberg et al. (1997)
		849	0.038				Fishkin et al. (1997); Tromberg et al. (1997)
		956	0.09				Fishkin et al. (1997); Tromberg et al. (1997)
	Fluid-filled cyst (27 year)	674	0.07				Fishkin et al. (1997); Tromberg et al. (1997)
		811	0.07				Fishkin et al. (1997); Tromberg et al. (1997)
		849	0.08				Fishkin et al. (1997); Tromberg et al. (1997)
		956	0.16				Fishkin et al. (1997); Tromberg et al. (1997)
	Papillary cancer (55 year)	690	0.084 ± 0.014				Fantini et al. (1998)

Tissue	Type						References
	Normal (67 year)	825	0.085 ± 0.017				Fantini et al. (1998)
		674	0.057				Holboke et al. (2000)
		782	0.05				Holboke et al. (2000)
		803	0.047				Holboke et al. (2000)
		849	0.054				Holboke et al. (2000)
	Ductal carcinoma in situ (67 year)	674	0.17				Holboke et al. (2000)
		782	0.18				Holboke et al. (2000)
		803	0.15				Holboke et al. (2000)
		849	0.21				Holboke et al. (2000)
Aorta, in vitro	Normal	308	33			6 h post mortem	Tuchin (2007)
	Normal, coagulated	308	44			6 h post mortem	Tuchin (2007)
	Fibrous plaque	308	24			6 h post mortem	Tuchin (2007)
	Fibrous plaque, coagulated	308	34			6 h post mortem	Tuchin (2007)
	Normal	1064	0.53 ± 0.09	239 ± 45	0.9	Post mortem	Tuchin (2007)
	Coagulated	1064	0.46 ± 0.18	293 ± 73	0.9		Tuchin (2007)
	Fibro-fatty	355	17.7				Tuchin (2007)
		532	3.6				Tuchin (2007)
		1064	0.09				Tuchin (2007)
	Aorta	633	0.52	316	0.87		Tuchin (2007)
		1064	0.5	239	0.9		Tuchin (2007)
		1064	0.7				Tuchin (2007)
		1320	2.2	233	0.9		Tuchin (2007)
		1320	4.3				Tuchin (2007)
	Aorta	470	5.3 ± 0.9				Cilesiz et al. (1993); Lin et al. (1996); Tuchin (1994)
		476	5.1 ± 0.9				Cilesiz et al. (1993); Lin et al. (1996); Tuchin (1994)
		488	4.5 ± 0.9				Cilesiz et al. (1993); Lin et al. (1996); Tuchin (1994)

(continued)

TABLE 2.2 (continued) Absorption Coefficient μ_a, the Scattering Coefficient μ_s, and the Average Cosine of Scatter g for Various Tissues, Tissue Components, and Biofluids

Tissue	Description	Wavelength (nm)	μ_a (cm^{-1})	μ_s (cm^{-1})	g	Notes	References
		514.5	3.7 ± 0.9				Cilesiz et al. (1993); Lin et al. (1996); Tuchin (1994)
		580	2.8 ± 0.9				Cilesiz et al. (1993); Lin et al. (1996); Tuchin (1994)
		600	2.6 ± 0.9				Cilesiz et al. (1993); Lin et al. (1996); Tuchin (1994)
		633	2.6 ± 0.9				Cilesiz et al. (1993); Lin et al. (1996); Tuchin (1994)
		1064	2.7 ± 0.5				Cilesiz et al. (1993); Lin et al. (1996); Tuchin (1994)
	Intima	476	14.8	237	0.81	Frozen sections	Keijzer et al. (1989a,b)
		580	8.9	183	0.81	"	Keijzer et al. (1989a,b)
		600	4	178	0.81	"	Keijzer et al. (1989a,b)
		633	3.6	171	0.85	"	Keijzer et al. (1989a,b)
	Intima	1064	2.3	165	0.97		Muller et al. (1995)
	Media	476	7.3	410	0.89	Frozen sections	Keijzer et al. (1989a,b)
		580	4.8	331	0.9	"	Keijzer et al. (1989a,b)
		600	2.5	323	0.89	"	Keijzer et al. (1989a,b)
		633	2.3	310	0.9	"	Keijzer et al. (1989a,b)
	Media	1064	1	634	0.96		Muller et al. (1995)
	Adventitia	476	18.1	267	0.74	Frozen sections	Keijzer et al. (1989a,b)
		580	11.3	217	0.77	"	Keijzer et al. (1989a,b)
		600	6.1	211	0.78	"	Keijzer et al. (1989a,b)
		633	5.8	195	0.81	"	Keijzer et al. (1989a,b)
	Adventitia	1064	2	484	0.97		Muller et al. (1995)
Aorta, in vivo	Advanced fibrous atheroma	355	16.5 ± 1.7			Photoacoustic method	Oraevsky et al. (1997)
		532	3.53 ± 0.53			"	Oraevsky et al. (1997)

Tissue	Condition						References
Blood		1064	0.15 ± 0.07		0.99	"	Oraevsky et al. (1997)
	Oxygenated, Hct = 0.41	685	2.65	1413	0.995		Cheong et al. (1990)
	Oxygenated, Hct = 0.41	665	1.3	1246	0.992		Cheong et al. (1990)
	Oxygenated, Hct = 0.41	960	2.84	505			Cheong et al. (1990)
	Oxygenated, Hct = 0.4	810	4.5				Tuchin (2007)
		1064	3				Tuchin (2007)
	Deoxygenated, Hct = 0.41	665	4.87	509	0.995		Cheong et al. (1990)
		960	1.68	668	0.992		Cheong et al. (1990)
	Deoxygenated, Hct = 0.4	810	4.5				Tuchin (2007)
		1064	0.3				Tuchin (2007)
	Partially oxygenated, Hct = 0.47	450	381	2940	0.9972		Tuchin (2007)
		488	133	3190	0.9987		Tuchin (2007)
		514	116	3320	0.9988		Tuchin (2007)
		577	301	3140	0.9977		Tuchin (2007)
		630	14.3	3660	0.9976		Tuchin (2007)
		760	15.5	2820	0.9972		Tuchin (2007)
	Oxygenation >98%, Hct = 0.45–0.46	633	15.5	644.7	0.982	Data from plots	Yaroslavsky et al. (1996, 1999)
		710	4.0 ± 0.8	737 ± 75	0.986 ± 0.006	"	Yaroslavsky et al. (1996, 1999)
		765	5.3 ± 0.6	725 ± 75	0.991 ± 0.002	"	Yaroslavsky et al. (1996, 1999)
		810	6.5 ± 0.5	690 ± 80	0.989 ± 0.002	"	Yaroslavsky et al. (1996, 1999)
		865	7.2 ± 0.3	649 ± 25	0.990 ± 0.001	"	Yaroslavsky et al. (1996, 1999)
		910	8.9 ± 0.4	649 ± 25	0.992 ± 0.002	"	Yaroslavsky et al. (1996, 1999)
		965	9.3 ± 0.6	650 ± 25	0.991 ± 0.001	"	Yaroslavsky et al. (1996, 1999)
		1010	8.3 ± 0.4	645 ± 25	0.992 ± 0.001	"	Yaroslavsky et al. (1996, 1999)

(continued)

TABLE 2.2 (continued)   Absorption Coefficient μ_a, the Scattering Coefficient μ_s, and the Average Cosine of Scatter g for Various Tissues, Tissue Components, and Biofluids

Tissue	Description	Wavelength (nm)	μ_a (cm^{-1})	μ_s (cm^{-1})	g	Notes	References
		1065	5.6 ± 0.3	645 ± 25	0.992 ± 0.001	"	Yaroslavsky et al. (1996, 1999)
		1110	4.2 ± 0.3	630 ± 20	0.993 ± 0.001	"	Yaroslavsky et al. (1996, 1999)
		1165	4.1 ± 0.7	655 ± 15	0.993 ± 0.001	"	Yaroslavsky et al. (1996, 1999)
		1210	5.5 ± 0.5	654 ± 20	0.995 ± 0.001	"	Yaroslavsky et al. (1996, 1999)
	Oxygenation >98%, Hct = 0.38	633	15.2 ± 0.6	400 ± 30	0.971 ± 0.001	"	Yaroslavsky et al. (1996, 1999)
		633	16.1 ± 0.6	4130 ± 170		"	Yaroslavsky et al. (1996, 1999)
		633	16.3 ± 0.5	2390 ± 160	0.9962 ± 0.0001		Yaroslavsky et al. (1996, 1999)
	Oxygenated flowing blood, Hct = 0.1	633	2.10 ± 0.02	773 ± 5	0.994 ± 0.010	Close to physiological flow; oxygen sat. 98%	Roggan et al. (1999)
	Oxygenated flowing blood, Hct = 0.2	633	4	800	0.989	Close to physiological flow; oxygen sat. 98%, data from plots	Roggan et al. (1999)
	Oxygenated flowing blood, Hct = 0.3	633	5.4	890		"	Roggan et al. (1999)
	Oxygenated flowing blood, Hct = 0.4	633	6.6	850		"	Roggan et al. (1999)
	Oxygenated flowing blood, Hct = 0.5	633	7.7	840		"	Roggan et al. (1999)
	Oxygenated flowing blood, Hct = 0.6	633	10	840		"	Roggan et al. (1999)
	Oxygenated flowing blood, Hct = 0.7	633	12.2	950		"	Roggan et al. (1999)
	Flowing blood, 24.5°C	633	2.9–3.5				Nilsson et al. (1997)
	Flowing blood, 24.5°C–42°C	633	3.0–4.0				Nilsson et al. (1997)
	Flowing blood, 47°C	633	4.5				Nilsson et al. (1997)
	Flowing blood, 54.3°C	633	6.3				Nilsson et al. (1997)

Tissue		Wavelength			g	Notes	Reference
	Oxygenation >99%, Hct = 0.421	260	375.5 ± 9.0	631.5 ± 57.6	0.784 ± 0.030		Friebel et al. (2006)
		350	368.1	559.5	0.852		Friebel et al. (2006)
		375	338.6 ± 4.2	542.8 ± 66.5	0.872 ± 0.007		Friebel et al. (2006)
		415	782.5 ± 62.9	390.3 ± 61.2	0.668 ± 0.008		Friebel et al. (2006)
		450	263	682.6	0.923		Friebel et al. (2006)
		490	106.8	793.8	0.962		Friebel et al. (2006)
		520	120.4 ± 6.9	766.2 ± 42.4	0.967 ± 0.009		Friebel et al. (2006)
		540	232.3	655.6	0.945		Friebel et al. (2006)
		555	178.9	709.3	0.953		Friebel et al. (2006)
		575	231.6	658	0.952		Friebel et al. (2006)
		585	160.2 ± 10.3	751.7 ± 46.1	0.955 ± 0.007		Friebel et al. (2006)
		620	4.14	905.3	0.974		Friebel et al. (2006)
		630	2.51 ± 0.09	894.6 ± 28.6	0.975 ± 0.004		Friebel et al. (2006)
		670	1.22	892.3	0.976		Friebel et al. (2006)
		700	1.25	879.3	0.976		Friebel et al. (2006)
		750	1.99	840.8	0.975		Friebel et al. (2006)
		780	2.85	821.5	0.975		Friebel et al. (2006)
		800	3.27 ± 0.12	809.9 ± 66.4	0.975 ± 0.003		Friebel et al. (2006)
		830	4.9	798.7	0.975		Friebel et al. (2006)
		850	4.65	799.5	0.975		Friebel et al. (2006)
		870	5.1	784.4	0.974		Friebel et al. (2006)
		900	5.43	751.4	0.973		Friebel et al. (2006)
		950	6.15 ± 0.35	712.0 ± 69.8	0.971 ± 0.002		Friebel et al. (2006)
		980	6.79	685.9	0.97		Friebel et al. (2006)
		1000	6.51	680.8	0.97		Friebel et al. (2006)
		1050	4.91 ± 0.12	661.3 ± 12.8	0.9699 ± 0.0006		Friebel et al. (2006)
		1100	3.74	639.5	0.97		Friebel et al. (2006)
Misc. tissues, in vitro	Lung	515	25.5 ± 3.0	356 ± 39		Frozen sections	Marchesini et al. (1989)
		635	8.1 ± 2.8	324 ± 46	0.75	"	Marchesini et al. (1989)
	Lung	1064	2.8	39			Muller et al. (1995)
	Muscle	515	11.2 ± 1.8	530 ± 44	0.91	Frozen sections	Marchesini et al. (1989)

(continued)

TABLE 2.2 (continued)   Absorption Coefficient μ_a, the Scattering Coefficient μ_s, and the Average Cosine of Scatter g for Various Tissues, Tissue Components, and Biofluids

Tissue	Description	Wavelength (nm)	μ_a (cm⁻¹)	μ_s (cm⁻¹)	g	Notes	References
Muscle		1064	2	215	0.96		Muller et al. (1995)
Meniscus		360	13			Frozen, thawed	Tuchin (2007)
		400	4.6			"	Tuchin (2007)
		488	1			"	Tuchin (2007)
		514	0.73			"	Tuchin (2007)
		630	0.36			"	Tuchin (2007)
		800	0.52			"	Tuchin (2007)
		1064	0.34			"	Tuchin (2007)
Uterus		635	0.35 ± 0.10	394 ± 91	0.69	Frozen sections	Marchesini et al. (1989)
Uterus, postmenopausal		630	0.515 ± 0.054				Madsen et al. (1994)
Uterus, premenopausal		630	0.193 ± 0.013				Madsen et al. (1994)
		630	0.314 ± 0.030				Madsen et al. (1994)
		630	0.213 ± 0.024				Madsen et al. (1994)
		630	0.197 ± 0.030				Madsen et al. (1994)
Uterus, fibroid		630	0.082 ± 0.008				Madsen et al. (1994)
Bladder, integral		633	1.4	88	0.96		Cheong et al. (1990)
Bladder, integral		633	1.4	29.3	0.91	Excised, kept in saline	Muller et al. (1995)
Bladder, mucous		1064	0.7	7.5	0.85		Muller et al. (1995)
Bladder, wall		1064	0.9	54.3	0.85		Muller et al. (1995)
Bladder, integral		1064	0.4	116	0.9		Muller et al. (1995)
Heart, endocardial		1060	0.07	136	0.97	Excised, kept in saline	Cheong et al. (1990)
Heart, epicardial		1060	0.35	167	0.98		Muller et al. (1995)
Heart, myocardial		1060	0.3	177.5	0.96		Muller et al. (1995)
Heart, epicardlial		1060	0.21	127.1	0.93		Muller et al. (1995)
Heart, aneurysm		1060	0.4	137	0.98		Muller et al. (1995)
Heart, trabecula		1064	1.4	424	0.97		Muller et al. (1995)
Heart, myocardial		1064	1.4	324	0.96		Muller et al. (1995)
Heart, myocardial		1060	0.52				Hourdakis et al. (1995)
Kidney, pars conv.		1064	2.4	72	0.86		Muller et al. (1995)
Kidney, medulla ren		1064	2.1	77	0.87		Muller et al. (1995)
Femoral vein		1064	3.2	487	0.97		Muller et al. (1995)

Tissue	λ (nm)				Comments	Reference
Liver	515	18.9 ± 1.7	285 ± 20			Marchesini et al. (1989)
Liver	635	2.3 ± 1.0	313 ± 136	0.68		Marchesini et al. (1989)
Liver	1064	0.7	356	0.95		Marchesini et al. (1989)
Liver	630	3.2	414	0.95		Hourdakis et al. (1995)
Colon, muscle	1064	3.3	238	0.93		Muller et al. (1995)
Colon, submucous	1064	2.3	117	0.91		Muller et al. (1995)
Colon, mucous	1064	2.7	39	0.91		Muller et al. (1995)
Colon, integral	1064	0.4	261	0.94		Muller et al. (1995)
Esophagus	633	0.4				Tuchin (2007)
Esophagus, mucous	1064	1.1	83	0.86		Muller et al. (1995)
Fat, subcutaneous	1064	2.6	29	0.91		Muller et al. (1995)
Fat, abdominal	1064	3	37	0.91		Muller et al. (1995)
Prostate	850	0.6 ± 0.2	100 ± 20	0.94 ± 0.02	Shock frozen sections, 0.5–3 h post mortem	Muller et al. (1995)
	980	0.4 ± 0.2	90 ± 20	0.95 ± 0.02	"	Muller et al. (1995)
	1064	0.3 ± 0.2	80 ± 20	0.95 ± 0.02	"	Muller et al. (1995)
Prostate, coagulated	850	7.0 ± 0.2	230 ± 30	0.94 ± 0.02	Water bath (75°C, 10 min), 0.5–3 h post mortem	Muller et al. (1995)
	980	5.0 ± 0.2	190 ± 30	0.95 ± 0.02	"	Muller et al. (1995)
	1064	4.0 ± 0.2	180 ± 30	0.95 ± 0.02	"	Muller et al. (1995)
Prostate, normal ($n = 3$)	695	0.8	330 ± 30	0.95		Dickey et al. (2001)
Prostate, normal	1064	1.5 ± 0.2	47 ± 13	0.862	Freshly excised	Tuchin (2007)
Prostate, coagulated	1064	0.8 ± 0.2	80 ± 12	0.861	Water bath (70°C,10 min)	Tuchin (2007)
Spleen	1064	6	137	0.9		Muller et al. (1995)
Stomach, muscle	1064	3.3	29.5	0.87		Muller et al. (1995)
Stomach, mucous	1064	2.8	732	0.91		Muller et al. (1995)
Stomach, integral	1064	0.8	128	0.91		Muller et al. (1995)
Sclera	650	0.08				Svaasand et al. (1993)
Sclera ($n = 5$)	404	5.0 ± 0.5				Bashkatov (2002)
	449	3.99 ± 0.40				Bashkatov (2002)
	499	2.96 ± 0.30				Bashkatov (2002)
	549	2.26 ± 0.23				Bashkatov (2002)

(*continued*)

TABLE 2.2 (continued) Absorption Coefficient μ_a, the Scattering Coefficient μ_s, and the Average Cosine of Scatter g for Various Tissues, Tissue Components, and Biofluids

Tissue	Description	Wavelength (nm)	μ_a (cm^{-1})	μ_s (cm^{-1})	g	Notes	References
		599	1.95 ± 0.19				Bashkatov (2002)
		649	1.74 ± 0.17				Bashkatov (2002)
		699	1.67 ± 0.17				Bashkatov (2002)
		749	1.65 ± 0.17				Bashkatov (2002)
		799	1.58 ± 0.16				Bashkatov (2002)
	Tooth, dentin	633	6	1200			Zijp et al. (1991, 1997)
	Tooth, enamel	633	0.97	1.1			Zijp et al. (1991, 1997)
	Gallstones, porcinement	351	102 ± 16				Cheong et al. (1990)
		488	179 ± 28				Cheong et al. (1990)
		580	125 ± 29				Cheong et al. (1990)
		630	85 ± 11				Cheong et al. (1990)
		1060	125 ± 12				Cheong et al. (1990)
	Gallstones, cholesterol	351	85 ± 7				Cheong et al. (1990)
		488	62 ± 15				Cheong et al. (1990)
		580	36 ± 7				Cheong et al. (1990)
		630	44 ± 10				Cheong et al. (1990)
		1060	60 ± 9				Cheong et al. (1990)
Misc. tissues, in vivo	Calf (11 subj.,14 meas.)	800	0.17 ± 0.05				Matcher et al. (1997)
	Abdominal, normal tissue	674	0.0589 ± 0				Fishkin et al. (1997); Tromberg et al. (1997)
		811	0.0645 ± 0				Fishkin et al. (1997); Tromberg et al. (1997)
		849	0.069 ± 0				Fishkin et al. (1997); Tromberg et al. (1997)
		956	0.111 ± 0.015				Fishkin et al. (1997); Tromberg et al. (1997)
	Abdominal, tumor	674	0.169 ± 0.020				Fishkin et al. (1997); Tromberg et al. (1997)
		811	0.190 ± 0.015				Fishkin et al. (1997); Tromberg et al. (1997)

Tissue	λ (nm)				References
	849	0.276 ± 0.030			Fishkin et al. (1997); Tromberg et al. (1997)
	956				Fishkin et al. (1997); Tromberg et al. (1997)
Back, normal tissue	674	0.0883 ± 0.0060			Fishkin et al. (1997); Tromberg et al. (1997)
	811	0.0892 ± 0.0050			Fishkin et al. (1997); Tromberg et al. (1997)
	849	0.0915 ± 0.0030			Fishkin et al. (1997); Tromberg et al. (1997)
	956	0.127 ± 0.030			Fishkin et al. (1997); Tromberg et al. (1997)
Back, tumor	674	0.174 ± 0.020			Fishkin et al. (1997); Tromberg et al. (1997)
	811	0.177 ± 0.013			Fishkin et al. (1997); Tromberg et al. (1997)
	849	0.19 ± 0.01			Fishkin et al. (1997); Tromberg et al. (1997)
	956	0.186 ± 0.160			Fishkin et al. (1997); Tromberg et al. (1997)
Cervical stromal tissue, in vivo	849	0.34	61.1	0.9	Hayakawa et al. (2001)
Forehead, in vivo	715	0.16			Gratton et al. (1997)
Forearm, in vivo	715	0.18			Gratton et al. (1997)
	825	0.24			Gratton et al. (1997)
	825	0.16			Gratton et al. (1997)
Fat, ex vivo					
Abdominal (n = 2)	360	3.12 ± 0.78			Bashkatov et al. (2005)
	400	3.97 ± 0.99			Bashkatov et al. (2005)
	500	2.37 ± 0.59			Bashkatov et al. (2005)
	600	1.90 ± 0.47			Bashkatov et al. (2005)
	700	1.84 ± 0.46			Bashkatov et al. (2005)
	800	1.87 ± 0.47			Bashkatov et al. (2005)
	900	1.80 ± 0.45			Bashkatov et al. (2005)
	1000	1.77 ± 0.44			Bashkatov et al. (2005)
	1100	1.68 ± 0.42			Bashkatov et al. (2005)

(continued)

TABLE 2.2 (continued)   Absorption Coefficient μ_a, the Scattering Coefficient μ_s, and the Average Cosine of Scatter g for Various Tissues, Tissue Components, and Biofluids

Tissue	Description	Wavelength (nm)	μ_a (cm⁻¹)	μ_s (cm⁻¹)	g	Notes	References
		1200	1.79 ± 0.45				Bashkatov et al. (2005)
		1300	1.52 ± 0.38				Bashkatov et al. (2005)
		1400	1.75 ± 0.44				Bashkatov et al. (2005)
		1500	1.63 ± 0.41				Bashkatov et al. (2005)
		1600	1.47 ± 0.37				Bashkatov et al. (2005)
		1700	2.11 ± 0.53				Bashkatov et al. (2005)
		1800	1.92 ± 0.48				Bashkatov et al. (2005)
		1900	2.48 ± 0.62				Bashkatov et al. (2005)
		2000	2.12 ± 0.53				Bashkatov et al. (2005)
		2100	1.74 ± 0.43				Bashkatov et al. (2005)
		2200	1.65 ± 0.41				Bashkatov et al. (2005)
Mucosa	Maxillary sinus at antritis ($n = 10$)	400	4.89 ± 0.92				Bashkatov et al. (2005)
		500	1.13 ± 0.18				Bashkatov et al. (2005)
		600	0.45 ± 0.23				Bashkatov et al. (2005)
		700	0.16 ± 0.24				Bashkatov et al. (2005)
		800	0.13 ± 0.16				Bashkatov et al. (2005)
		900	0.12 ± 0.09				Bashkatov et al. (2005)
		1000	0.27 ± 0.21				Bashkatov et al. (2005)
		1100	0.16 ± 0.14				Bashkatov et al. (2005)
		1200	0.57 ± 0.31				Bashkatov et al. (2005)
		1300	0.67 ± 0.35				Bashkatov et al. (2005)
		1400	4.84 ± 1.79				Bashkatov et al. (2005)
		1500	6.06 ± 2.38				Bashkatov et al. (2005)
		1600	2.83 ± 1.01				Bashkatov et al. (2005)
		1700	2.26 ± 0.79				Bashkatov et al. (2005)
		1800	3.04 ± 1.15				Bashkatov et al. (2005)
		1900	9.23 ± 2.69				Bashkatov et al. (2005)
		2000	9.31 ± 2.28				Bashkatov et al. (2005)

Region	Tissue	Wavelength (nm)	Value		Reference
Forearm	Fat	633	0.026		Kienle et al. (1994, 1995)
	Muscle	633	0.96		Kienle et al. (1994, 1995)
Gastrointestinal tract, in vivo	Mucosa in the antrum	500	2.5 ± 0.8		Thueler et al. (2003)
		550	3.6 ± 1.3		Thueler et al. (2003)
		600	1.0 ± 0.6		Thueler et al. (2003)
		650	0.5 ± 0.5		Thueler et al. (2003)
		700	0.4 ± 0.4		Thueler et al. (2003)
		750	0.45 ± 0.45		Thueler et al. (2003)
		800	0.5 ± 0.4		Thueler et al. (2003)
		850	0.7 ± 0.5		Thueler et al. (2003)
		900	0.8 ± 0.5		Thueler et al. (2003)
	Mucosa in the fundus	500	3.3 ± 0.8		Thueler et al. (2003)
		550	4.4 ± 1.5		Thueler et al. (2003)
		600	1.8 ± 0.5		Thueler et al. (2003)
		650	0.8 ± 0.4		Thueler et al. (2003)
		700	0.7 ± 0.4		Thueler et al. (2003)
		750	0.7 ± 0.5		Thueler et al. (2003)
		800	0.7 ± 0.3		Thueler et al. (2003)
		850	0.7 ± 0.3		Thueler et al. (2003)
		900	0.8 ± 0.4		Thueler et al. (2003)
Abdomen	Muscle ($n = 1$), ex vivo	633	1.21	0.9	Laufer et al. (1998); Simpson et al. (1998)
		700	0.46	0.9	Laufer et al. (1998); Simpson et al. (1998)
		900	0.32	0.9	Laufer et al. (1998); Simpson et al. (1998)

strongest scatterers. The erythrocyte is about 2 μm thick with a diameter from 7 to 9 μm. The scattering properties of blood are dependent on the hematocrit and the degree of red cell agglomeration.

Support tissues, consisting of cells and extracellular proteins such as elastin and collagen, provide mechanical strength and durability. The scattering properties of these tissues arise from both the small-scale inhomogeneities and large-scale variations in the structures they form. The characteristic sizes of the small-scale variations are subwavelength, and the scattering is of the Rayleigh type. For example, collagen fibrils have a banded structure that produces Rayleigh scattering. This banding has a periodicity of 70 nm, which is on the order of 10 times smaller than the wavelengths in the therapeutic/diagnostic window (Saidi et al. 1995).

2.4.3 Absorption Properties

2.4.3.1 Therapeutic/Diagnostic Window

The ability of light to penetrate tissues is governed by absorption. Of primary importance is the therapeutic/diagnostic window, a region of the spectrum where the absorption of most tissues is sufficiently weak to allow significant penetration of light (see Figure 2.16). This window extends from 600 to 1300 nm, from the orange/red region of the visible spectrum into the NIR. At the short-wavelength end, the window is bounded by the absorption of hemoglobin, in both its oxygenated and deoxygenated forms. The absorption of oxygenated hemoglobin increases approximately two orders of magnitude as the wavelength shortens in the region around 600 nm. At shorter wavelengths, many biomolecules become significant absorbers (which in the UV includes DNA and the amino acids). At the IR end of the window, penetration is limited by the absorption properties of water. Within the therapeutic window, scattering is dominant over absorption, and the penetrating light can become diffuse, and may even enter into the diffusion limit. The following section discusses the absorption properties of various types of tissue components (Table 2.2).

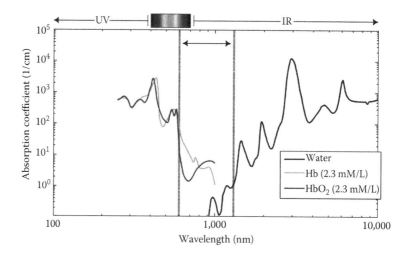

FIGURE 2.16 **(See color insert.)** The therapeutic/diagnostic window. The absorption spectra of water, deoxygenated hemoglobin (Hb), and oxygenated hemoglobin (HbO$_2$). The Hb and HbO$_2$ curves are given for a concentration of 150 g/L in water. The region of the spectrum between 600 and 1300 nm is the therapeutic/diagnostic window, where photons can achieve significant penetration in tissues. The water spectrum is from Hale and Querry (1973). The hemoglobin data are compiled in Prahl (1998). (Adapted from Jacques, S.L. and Prahl, S.A., Absorption spectra for biological tissues, available from http://omlc.ogi.edu/classroom/ece532/class3/muaspectra.html, 1998.)

2.4.3.2 Absorption Properties of Tissue Components

Absorption in tissues is a function of molecular composition. Figure 2.17 shows the chemical structures of some molecules of biological and biomedical interest. Molecules absorb photons when the photon energy matches an interval between internal energy states, subject to the quantum selection rules governing the transition. At the short-wavelength (higher photon energy) end of the spectrum, these transitions are electronic. Some of the important absorbers in the UV include DNA, the aromatic amino acids (tryptophan and tyrosine), proteins, melanins, and porphyrins (which include hemoglobin, myoglobin,

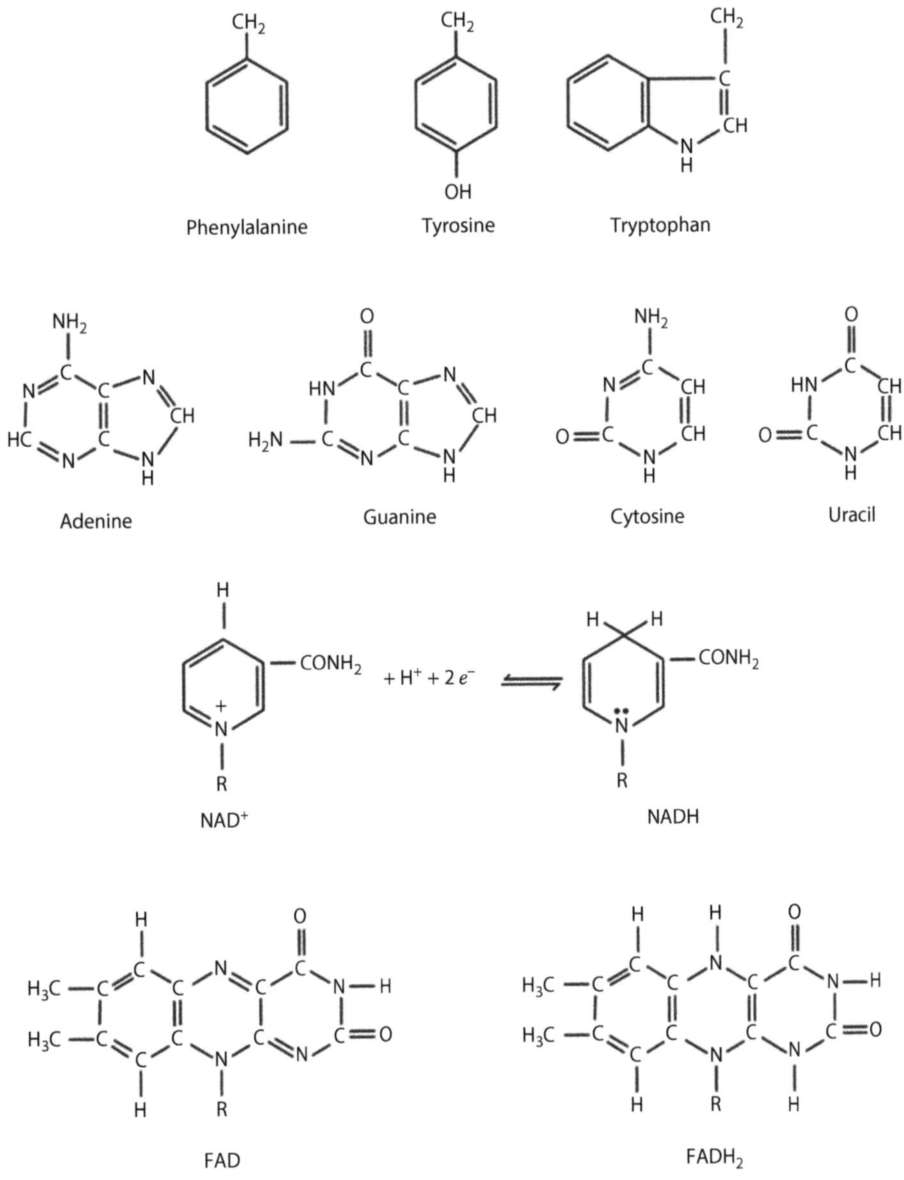

FIGURE 2.17 Chemical structures of various species of biological interest. The top row consists of the three aromatic amino acids. The second row shows the bases that are found in RNA. (In DNA, thymine takes the place of uracil.) The last two rows show molecules involved in energy metabolism in the cell.

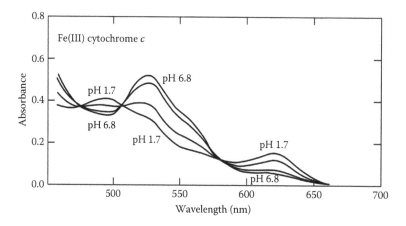

FIGURE 2.18 Absorption spectra of Fe(III) cytochrome *c* at various pH values from 6.8 to 1.7. (Adapted from Tinoco, I. Jr. et al., *Physical Chemistry: Principles and Applications in Biological Sciences*, Prentice Hall, Englewood Cliffs, NJ, 1985.)

vitamin B12, and cytochrome *c*). Figure 2.18 shows the absorption spectra of Fe(III) cytochrome *c* at various pH values from 6.8 to 1.7 (Tinoco Jr et al. 1985). Table 2.2 provides a list of absorption coefficient (μ_a) values for various tissues.

Nucleic acids: DNA has an absorption peak at 258 nm ($\varepsilon = 6.6 \times 10^3$ cm^2/mol^1), but its absorption falls significantly as the wavelength increases to 300 nm, and it has no significant activity in the visible and NIR. Figure 2.19 shows the absorption spectrum of DNA from *Escherichia coli* in the native form at 25°C (solid curve) and as an enzymatic digest of nucleotides (dashed curve) (Voet et al. 1963). RNA exhibits similar absorption properties, having a maximum absorption at 258 nm ($\varepsilon = 14.9 \times 10^3$ cm^2/mol^1). The absorption of nucleic acids arises from $n\pi^*$ and $\pi\pi^*$ transitions. The absorption spectra of the purine and pyrimidine bases (e.g., adenine, cytosine) occur between 200 and 300 nm. The absorption maxima (λ_{max}) and extinction coefficients (ε) of several purine and pyrimidine bases and their derivatives are given in Table 2.3.

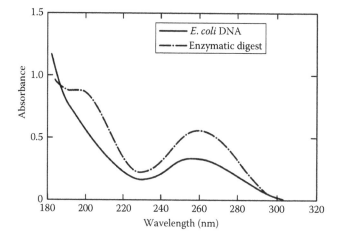

FIGURE 2.19 Absorption spectrum of DNA from *E. coli* in the native form at 25°C (solid curve) and as an enzymatic digest of nucleotides (dashed curve). (Adapted from Voet, D. et al., *Biopolymers*, 1, 93, 1963.)

TABLE 2.3 Absorption Maxima and Extinction Coefficients of Several Purine and Pyrimidine Bases and Their Derivatives

	Absorption Maximum (λ_{max}) (nm)	Extinction Coefficient (ε) (cm²/mol¹)
Adenine	260.5	13.4×10^3
Adenosine	259.5	14.9×10^3
Guanine	275	8.1×10^3
Guanosine	276	9.0×10^3
Cytidine	271	9.1×10^3
Cytosine	267	6.1×10^3
Uracil	259.5	8.2×10^3
Uridine	261.1	10.1×10^3
Thymine	264.5	7.9×10^3
Thymidine	267	9.7×10^3

Source: Campbell, I.D. and Dwek, R.A., *Biological Spectroscopy,* The Benjamin/Cummings Publishing Company, Inc., Menlo Park, CA, 1984.

Amino acids and proteins: The aromatic molecules tryptophan (Trp), tyrosine (Tyr), and phenylalanine (Phe) have local peaks in the 250–280 nm range and even larger peaks at shorter wavelengths (Figure 2.20a) (Campbell and Dwek 1984; Tinoco et al. 1985). A number of amino acid side chains absorb in the 200–230 nm range, including Trp, Tyr, and Phe. In proteins, the amino acid absorption in the region near 200 nm is overwhelmed by the much stronger absorption associated with amide backbone groups. Proteins often exhibit peaks near 280 nm due to Trp and near 200 nm due to the amide backbone groups.

Nucleotides: NADH is a dinucleotide that is the carrier molecule in the electron transport chain that is essential to the manufacture of ATP in the mitochondrion. NADH absorbs at 340 nm ($\varepsilon = 6.2 \times 10^6\,\text{mol}^{-1}$) and at 260 nm ($\varepsilon = 14 \times 10^6\,\text{mol}^{-1}$). NAD, the reduced form of NADH, absorbs only at 260 nm ($\varepsilon = 18 \times 10^6\,\text{mol}^{-1}$). Measurements of NADH absorption can be used to monitor reactions occurring in biological fluids and tissues. Figure 2.20b shows the absorption spectra of NADH and other constituents in biological tissues.

Skin: The skin is the largest organ of the body and consists of three layers: epidermis, dermis, and subcutaneous tissue. The epidermis is the outer layer; below it is the dermis, which is made up of skin appendages such as hair follicles and sweat glands, surrounded by fibrous supporting tissue and collagen. The underlying layer of subcutaneous tissue contains fat-producing cells and fibrous tissue. The outermost part of the epidermis is the corneal layer, consisting of dead skin cells, which are shed and continuously replaced by epidermal keratinocytes that originate from the basal layer. Melanocytes, which are among the basal cells of the epidermis, produce *melanin*, the protective pigment responsible for skin color. Melanocytes are stimulated by sunlight to produce melanin, which protects the skin against harmful UV radiation. The *melanins* are a group of biopolymers that provide for pigmentation and are the dominant chromophores in the skin (Prota et al. 1988). In most people, the absorption of epidermis is usually dominated by melanin. It absorbs UV radiation, dissipating the photon energy as heat, protecting the skin from UV damage. It is a polymer built by condensation of tyrosine molecules and has a broad absorption spectrum. Melanin is found in the melanosome, a membranous particle whose internal membranes are studded with melanin granules of about 10–30 nm in size. In the visible range (400–700 nm), melanin exhibits strong absorption, which decreases at longer wavelengths.

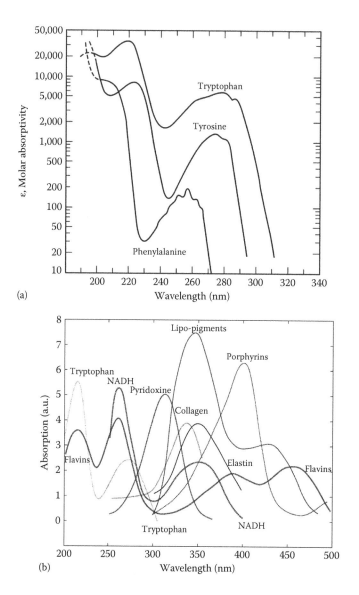

FIGURE 2.20 Absorption spectra of molecules and molecular components. (a) Absorption spectra of trypto-phan, tyrosine, and phenylalanine. (From Tinoco, I. Jr. et al., *Physical Chemistry: Principles and Applications in Biological Sciences*, Prentice Hall, Englewood Cliffs, NJ, 1985.) (b) Absorption spectra of various constitu-ents in the tissue.

Blood and hemoglobin: Knowledge about the optical properties μ_a, μ_s, and g of human blood plays an important role for many diagnostic and therapeutic applications. Hemoglobin is the molecular con-stituent of red blood cells (erythrocytes) responsible for the transport of oxygen from the lungs to the tissues. Hemoglobin consists of four polypeptide chains each with an embedded iron containing heme group. The four subunits are from the globin family of proteins. Hemoglobin exhibits several absorption bands and is one of the important molecules in defining the boundaries of the therapeutic/diagnostic window. When oxygen molecules bind to the iron, hemoglobin undergoes both a conformal change and a change in its absorption properties. When oxygenated, it is called oxyhemoglobin (HbO_2); in the reduced state, it is known as deoxyhemoglobin (Hb). The heme groups themselves consist of a porphyrin ring system that is mainly responsible for its color. This group has a high degree of electron delocalization, which results in a decrease in the energy required for an electronic transition to occur. As a result, the absorption of porphyrin occurs in the visible range. For example, the absorption spectrum of Fe(III) hemoprotein exhibits several bands at 400, 500, 530, 590, 625, and 1000 nm (Campbell and Dwek 1984). The absorption spectrum of hemoglobin changes when binding occurs. HbO_2 is a strong absorber up to 600 nm; then its absorption drops off very steeply, by almost two orders of magnitude, and remains low. The absorption of Hb, however, does not drop as dramatically; it stays relatively high although it decreases with increasing wavelengths. The isosbestic point, where the Hb and HbO_2 spectra intersect, occurs at about 805 nm (Delpy and Cope 1997). The optical properties of blood depend strongly on physiological parameters such as oxygen saturation, osmolarity, flow conditions, and hematocrit.

The integrating sphere technique and inverse MC simulations have been used to measure μ_a, μ_s, and g of circulating human blood (Roggan et al. 1999). At 633 nm, the optical properties of human blood with a hematocrit of 10% and an oxygen saturation of 98% were found to be 0.210 ± 0.002 mm^{-1} for μ_a, 77.3 ± 0.5 mm^{-1} for μ_s, and 0.994 ± 0.001 for the g factor; thus, the absorption mean free path is 13 μm and the scattering is highly directed forward. An increase of the hematocrit up to 50% leads to a linear increase of absorption and reduced scattering. Variations in osmolarity and wall shear rate led to changes of all three parameters, while variations in the oxygen saturation led only to a significant change in the absorption coefficient. A spectrum of all three parameters was measured in the wavelength range 400–2500 nm for oxygenated and deoxygenated blood; the results showed that blood absorption follows the absorption behavior of hemoglobin and water (see Figure 2.16 for Hb and HbO_2 spectra).

Chance et al. employed time-resolved spectroscopy to measure photon path lengths and thereby determine hemoglobin concentration in tissues (Chance et al. 1988). Although both deoxyhemoglobin and melanin are strong absorbers inside the diagnostic window, their relatively low concentrations in tissues explain why they do not significantly affect transmission processes. However, when light strikes a blood vessel, it encounters the strong absorption of whole blood in bulk that limits the light's penetration over a range of wavelengths.

Water: At the NIR end of the window, water is the dominant absorber due to transitions between its vibrational energy states. Within the H_2O transmission window (UV to 930 nm), the absorption of water is rather low (see Figure 2.16), yet it remains a significant contributor to the overall attenuation because of its high concentration in soft tissue.

Skull: The skull has been studied in the therapeutic window, and its absorption is quite low (0.02–0.05 mm^{-1}) compared with that of soft tissue (Firbank et al. 1993).

2.5 Conclusion

This chapter emphasizes the basic physics of light propagation in tissues with the goal of providing a fundamental framework for more in-depth study. This chapter has aimed to provide the basic context for interpreting the optical properties of tissues and to illustrate the fundamental physical processes essential for understanding photon transport in biological tissues. Many of the chapters in this handbook give detailed treatments of the wide range of phenomena that fall within the broad umbrella of biomedical photonics. More detailed treatments of various aspects of tissue optics can also be found in Chance (1989); Mishchenko et al. (2006); Splinter and Hooper (2007); Tuchin (2007); Wang and Hsin-i (2007).

2.6 Summary

- Across the spectrum, a great variety of electromagnetic wave phenomena are used in biomedicine to detect and treat disease and to advance basic research in the life sciences. In principle, the propagation of light in tissues can be described using fundamental EM theory. However, such a treatment is not practical because of the complexity of the problem. Instead, a simplified model is used which ignores wave phenomena such as polarization and interference and particle properties such as inelastic collisions. This approach, known as Radiation Transport Theory (RTT), is a general description of the flow of energy or particles in a system.
- Several properties provide the basis on which the inputs to RT theory are built:
  - The *index of refraction, $n(\lambda)$*
  - The *scattering cross section, σ_s*
  - The *differential scattering cross section, $d\sigma_s/d\Omega$*
  - The *absorption cross section, σ_a*
- The three properties of primary inputs to RT theory are
  - The *scattering coefficient, μ_s*
  - The *absorption coefficient, μ_a*
  - The *average cosine of scatter, g*
- The *index of refraction* is a fundamental property of homogeneous media. The real part of the refractive index, $n(\lambda)$, can be defined in terms of the phase velocity of light in the medium,

$$c_m(\lambda) = \frac{c}{n(\lambda)}.$$

- Refractive index mismatches on a macroscopic scale (e.g., between soft tissue and bones, skin, and skull) give rise to refraction and reflection. On the other hand, refractive index mismatches at microscopic scales (e.g., cell membrane boundaries, organelles) give rise to the scattering of light in tissue.
- The ratio of the power scattered out of a plane wave to the incident intensity is the *scattering cross section*,

$$\sigma_s(\hat{s}) = \frac{P_{scatt}}{I_0},$$

where $\hat{s}$ is the propagation direction of the plane wave relative to the scatterer.
- The *scattering coefficient* μ_s (cm⁻¹) can be expressed in terms of *particle density* ρ (cm⁻³) and scattering cross section σ_s (cm²) as

$$\mu_s = \rho\sigma_s.$$

- Scattering of light by tissue structures much smaller than the photon wavelength involves the so-called *Rayleigh limit*. Those structures include many smaller components and substructures of the cell such as organelles, vesicles, and membranes.

- When the scatterer dimensions are comparable to the wavelength, the Rayleigh approximation no longer applies, and the scattering process enters the Mie regime. Various subcellular and extracellular structures—such as mitochondria, nuclei, and collagen fibers—have sizes that are comparable to the wavelengths generally used in biomedical applications (0.5–1 μm).
- Absorption processes involve an important concept in quantum theory, the *energy level*, which is a quantum state of an atom and molecule. The energy of a particular level is determined relative to the ground level, or lowest possible energy level. The shift of a species (e.g., molecule or atom) from one energy level to another is called a *transition*. A transition of a species from a lower to a higher energy level (i.e., the transition to an excited state) requires the absorption of an amount of photon energy, hc/λ, equal to the difference in energy, ΔE, between the two levels:

$$\frac{hc}{\lambda} = \Delta E.$$

- For a localized absorber, the *absorption cross section* σ_a can be defined in the same manner as for scattering:

$$\sigma_a = \frac{P_{abs}}{I_0},$$

where P_{abs} is the amount of power absorbed out of an initially uniform plane wave of intensity (power per unit area)I_0.
- The *absorption coefficient* μ_a can be defined in terms of particle density ρ and absorption cross-section σ_a:

$$\mu_a = \rho\sigma_a.$$

- The fundamental equation that describes light propagation in the RTT model is the *radiation transport equation*, which describes the fundamental dynamics of the specific intensity. The steady-state RT equation in a source-free region is

$$\hat{\mathbf{s}} \cdot \vec{\nabla} I\left(\mathbf{r},\hat{\mathbf{s}}\right) = -\left(\mu_a + \mu_s\right) I\left(\mathbf{r},\hat{\mathbf{s}}\right) + \frac{\mu_s}{4\pi} \int_{4\pi} p\left(\hat{\mathbf{s}} \cdot \hat{\mathbf{s}}'\right) I\left(\mathbf{r},\hat{\mathbf{s}}'\right) d\Omega',$$

where $I\left(\mathbf{r},\hat{\mathbf{s}},t\right)$ is the light power per unit area per unit solid angle at point $\mathbf{r}$ directed in a cone of solid angle oriented in the direction defined by the unit vector $\hat{\mathbf{s}}$.
- The scattering phase function, $p\left(\hat{\mathbf{s}} \cdot \hat{\mathbf{s}}'\right)$, describes the fraction of light energy incident on a scatterer from the $\hat{\mathbf{s}}'$ direction that is scattered out into the $\hat{\mathbf{s}}$ direction.
- The radiation transport equation can be solved using analytical approaches based on deterministic approximations. Two common analytical approaches involve the *diffusion approximation* and the *Kubelka–Munk* model.
- In the diffusion process, photons moving through a medium do so in a series of steps of random length and direction (i.e., random walk). Each step begins with a scattering event and is equally likely to be taken in any direction. This isotropic scattering is described by the reduced scattering coefficient μ_s', which is related to the anisotropic scattering parameters:
$\mu_s' = \left(1 - g\right)\mu_s$.

- Scattering anisotropy factor *g* is defined as a cosine-weighted average of the scattering, commonly known as the *average cosine of scatter*:

$$g \equiv \int_{4\pi} p(\hat{\mathbf{s}} \cdot \hat{\mathbf{s}}')\hat{\mathbf{s}} \cdot \hat{\mathbf{s}}' d\Omega' = 2\pi \int_{0}^{\pi} p(\cos\theta)\cos\theta\sin\theta d\,\theta.$$

The average cosine of scatter is a measure of the directionality of the scattering specific to a type of tissue. For a pure Rayleigh scattering, the likelihood of scattering in a forward or rearward direction is equal and $g = 0$. When the tendency is for forward scattering, $g > 0$. A negative *g* indicates a preference for backward scattering. The *g* parameter is used with the Henyey–Greenstein SPF,

$$p_{HG}(\cos\theta) = \frac{1}{4\pi}\frac{1-g^2}{\left(1+g^2-2g\,\cos\theta\right)^{\frac{3}{2}}},$$

which provides the angular distribution of the scattered light.
- In the stochastic approach, the numerical *Monte Carlo* method is often used to calculate photon transport through the tissue.
- The ability of light to penetrate tissues depends on absorption. Within the spectral range known as the *therapeutic window* (also called the *diagnostic window*), most tissues are sufficiently weak absorbers to permit significant penetration of light. This window extends from 600 to 1300 nm, from the orange/red region of the visible spectrum into the near IR.

Acknowledgments

This work was sponsored by the US Department of Energy (DOE) Office of Biological and Environmental Research, under contract DEAC05-00OR22725 with UT-Battelle, LLC, and by the National Institutes of Health (RO1 CA88787-01). VVT was supported by grants № 224014 PHOTONICS4LIFE-FP7-ICT-2007-2, RF President's SSc-208.2008.2, and № 2.1.1/4989, 2.2.1.1/2950 of PF Program DHSP. JM acknowledges the support of the University of Mississippi.

References

Agati, G., F. Fusi, G. P. Donzelli, and R. Pratesi. 1993. Quantum yield and skin filtering effects on the formation rate of lumirubin. *Journal of Photochemistry and Photobiology B-Biology* 18(2–3):197–203.

Alexandrov, S. A., A. V. Zvyagin, K. K. Silva, and D. D. Sampson. 2003. Bifocal-optical coherence refractometry of turbid media. *Optics Letters* 28(2):117–119.

Arifler, D., M. Guillaud, A. Carraro, A. Malpica, M. Follen, and R. Richards-Kortum. 2003. Light scattering from normal and dysplastic cervical cells at different epithelial depths: Finite-difference time-domain modeling with a perfectly matched layer boundary condition. *Journal of Biomedical Optics* 8(3):484–494.

Bashkatov, A. N. 2002. Controlling of optical properties of tissues at action by osmotically active immersion liquids. PhD thesis, Saratov State University: Saratov, Russia.

Bashkatov, A. N. 2005. Optical properties of the subcutaneous adipose tissue in the spectral range 400–2500 nm. *Optics and Spectroscopy* 99(5):836–842.

Bashkatov, A. N., E. A. Genina, V. I. Kochubey, and V. V. Tuchin. 2005. Optical properties of human skin, subcutaneous and mucous tissues in the wavelength range from 400 to 2000 nm. *Journal of Physics D-Applied Physics* 38(15):2543–2555.

Bashkatov, A. N., E. A. Genina, V. I. Kochubey, and V. V. Tuchin. 2006. Optical properties of human cranial bone in the spectral range from 800 to 2000 nm. *Proceedings of the SPIE* 6163.

Beuthan, J., O. Minet, and G. Muller. 1996. Quantitative optical biopsy of liver tissue ex vivo. *IEEE Journal of Selected Topics in Quantum Electronics* 2(4):906–913.

Bevilacqua, F., D. Piguet, P. Marquet et al. 1998. In vivo local determination of tissue optical properties. *Proceedings of the SPIE* 3194.

Bevilacqua, F., D. Piguet, P. Marquet, J. D. Gross, B. J. Tromberg, and C. Depeursinge. 1999. *In vivo* local determination of tissue optical properties: Applications to human brain. *Applied Optics* 38(22):4939–4950.

Bjorn, L. O. 1996. Light propagation in biological materials and natural waters. *Scientia Marina* 60:9–17.

Bolin, F. P., L. E. Preuss, R. C. Taylor, and R. J. Ference. 1989. Refractive index of some mammalian tissues using a fiber optic cladding method. *Applied Optics* 28:2297–2303.

Brillouin, L. 1960. *Wave Propagation and Group Velocity.* New York: Academic Press.

Campbell, I. D. and R. A. Dwek. 1984. *Biological Spectroscopy.* Menlo Park, CA: Benjamin/Cummings Publishing Company, Inc.

Chan, D., B. Schulz, M. Rubhausen, S. Wessel, and R. Wepf. 2006. Structural investigations of human hairs by spectrally resolved ellipsometry. *Journal of Biomedical Optics* 11(1):6.

Chan, E., T. Menovsky, and A. J. Welch. 1996. Effect of cryogenic grinding on soft-tissue optical properties. *Applied Optics* 35(22):4526–4532.

Chan, E. K., B. Sorg, D. Protsenko, M. O'Neil, M. Motamedi, and A. J. Welch. 1996. Effects of compression on soft tissue optical properties. *IEEE Journal on Selected Topics in Quantum Electronics* 2(4):943–950.

Chance, B. ed. 1989. *Photon Migration in Tissue.* New York: Plenum Press.

Chance, B., J. S. Leigh, H. Miyake et al. 1988. Comparison of time-resolved and time-unresolved measurements of deoxyhemoglobin in brain. *Proceedings of the National Academy of Sciences of the United States of America* 85(14):4971–4975.

Cheng, S., H. Y. Shen, G. Zhang, C. H. Huang, and X. J. Huang. 2002. Measurement of the refractive index of biotissue at four laser wavelengths. *Proceedings of the SPIE* 4916.

Cheong, W.-F., S. A. Prahl, and A. J. Welch. 1990. A review of the optical properties of biological tissues. *IEEE Journal of Quantum Electronics* 26(12):2166–2185.

Christy, A. A., O. M. Kvalheim, and R. A. Velapoldi. 1995. Quantitative-analysis in diffuse-reflectance spectrometry—A modified Kubelka-Munk equation. *Vibrational Spectroscopy* 9(1):19–27.

Cilesiz, I. F. and A. J. Welch. 1993. Light dosimetry: Effects of dehydration and thermal damage on the optical properties of human aorta. *Applied Optics* 32:477–487.

Das, B. B., F. Liu, and R. R. Alfano. 1997. Time-resolved fluorescence and photon migration studies in biomedical and model random media. *Reports on Progress in Physics* 60(2):227–292.

Delpy, D. T. and M. Cope. 1997. Quantification in tissue near-infrared spectroscopy. *Philosophical Transactions of the Royal Society of London Series B-Biological Sciences* 352(1354):649–659.

Dickey, D. J., R. B. Moore, D. C. Rayner, and J. Tulip. 2001. Light dosimetry using the P3 approximation. *Physics in Medicine and Biology* 46(9):2359–2370.

Ding, H. F., J. Q. Lu, K. M. Jacobs, and X. H. Hu. 2005. Determination of refractive indices of porcine skin tissues and intralipid at eight wavelengths between 325 and 1557 nm. *Journal of the Optical Society of America A—Optics Image Science and Vision* 22(6):1151–1157.

Ding, H. F., J. Q. Lu, W. A. Wooden, P. J. Kragel, and X. H. Hu. 2006. Refractive indices of human skin tissues at eight wavelengths and estimated dispersion relations between 300 and 1600 nm. *Physics in Medicine and Biology* 51(6):1479–1489.

Dirckx, J. J. J., L. C. Kuypers, and W. F. Decraemer. 2005. Refractive index of tissue measured with conflocal microscopy. *Journal of Biomedical Optics* 10(4):8.

Dognitz, N. and G. Wagnieres. 1998. Determination of tissue optical properties by steady-state spatial frequency-domain reflectometry. *Lasers in Medical Science* 13(1):55–65.

Doornbos, R. M. P., R. Lang, M. C. Aalders, F. W. Cross, and H. J. C. M. Sterenborg. 1999. The determination of in vivo human tissue optical properties and absolute chromophore concentrations using spatially resolved steady-state diffuse reflectance spectroscopy. *Physics in Medicine and Biology* 44:967–981.

Drexler, W., C. K. Hitzenberger, A. Baumgartner, O. Findl, H. Sattmann, and A. F. Fercher. 1998. Investigation of dispersion effects in ocular media by multiple wavelength partial coherence interferometry. *Experimental Eye Research* 66(1):25–33.

Driver, I., C. P. Lowdell, and D. V. Ash. 1991. *In vivo* measurement of the optical interaction coefficients of human tumors at 630 nm. *Physics in Medicine and Biology* 36:805–813.

Duck, F. A. 1990. *Physical Properties of Tissue: A Comprehensive Reference Book*. London, U.K.: Academic Press.

Durkin, A. J., S. Jaikumar, N. Ramanujam, and R. Richardskortum. 1994. Relation between fluorescence-spectra of dilute and turbid samples. *Applied Optics* 33(3):414–423.

Ell, R., U. Morgner, F. X. Kartner et al. 2001. Generation of 5-fs pulses and octave-spanning spectra directly from a Ti:Sapphire laser. *Optics Letters* 26(6):373–375.

Faber, D. J., M. C. G. Aalders, E. G. Mik, B. A. Hooper, M. J. C. van Gemert, and T. G. van Leeuwen. 2004. Oxygen saturation-dependent absorption and scattering of blood. *Physical Review Letters* 93(2):4.

Fantini, S., S. A. Walker, M. A. Franceschini, M. Kaschke, P. M. Schlag, and K. T. Moesta. 1998. Assessment of the size, position, and optical properties of breast tumors in vivo by noninvasive optical methods. *Applied Optics* 37(10):1982–1989.

Farrell, R. A. and R. L. McCally. 2000. Corneal transparency. In *Principles and Practice of Ophthalmology*, eds. D. A. Albert and F. A. Jakobiec. Philadelphia, PA: W.B. Saunders.

Feynman, R. P., R. B. Leighton, and M. L. Sands. 1963. *The Feynman Lectures on Physics*, Vols. 1–3. Reading, MA: Addison-Wesley.

Firbank, M., M. Hiraoka, M. Essenpreis, and D. T. Delpy. 1993. Measurement of the optical properties of the skull in the wavelength range 650–950 nm. *Physics in Medicine and Biology* 38(4):503–510.

Fishkin, J. B., O. Coquoz, E. R. Anderson, M. Brenner, and B. J. Tromberg. 1997. Frequency-domain photon migration measurements of normal and malignant tissue optical properties in a human subject. *Applied Optics* 36(1):10–20.

Foldy, L. L. 1945. The multiple scattering of waves. I. General theory of isotropic scattering by randomly distributed scatterers. *Physical Review* 67(3–4):107.

Friebel, M. and M. Meinke. 2005. Determination of the complex refractive index of highly concentrated hemoglobin solutions using transmittance and reflectance measurements. *Journal of Biomedical Optics* 10(6):5.

Friebel, M. and M. Meinke. 2006. Model function to calculate the refractive index of native hemoglobin in the wavelength range of 250–1100 nm dependent on concentration. *Applied Optics* 45(12):2838–2842.

Friebel, M., A. Roggan, G. Muller, and M. Meinke. 2006. Determination of optical properties of human blood in the spectral range 250 to 1100 nm using Monte Carlo simulations with hematocrit-dependent effective scattering phase function. *Journal of Biomedical Optics* 11(3):10.

Gardner, C. M., S. L. Jacques, and A. J. Welch. 1996. Light transport in tissue: Accurate expressions for one-dimensional fluence rate and escape function based upon Monte Carlo simulation. *Lasers in Surgery and Medicine* 18(2):129–138.

Gebhart, S. C., W. C. Lin, and A. Mahadevan-Jansen. 2006. In vitro determination of normal and neoplastic human brain tissue optical properties using inverse adding-doubling. *Physics in Medicine and Biology* 51(8):2011–2027.

Genina, E. A., A. N. Bashkatov, V. I. Kochubey, and V. V. Tuchin. 2005. Optical clearing of human dura mater. *Optics and Spectroscopy* 98(3):470–476.

Ghosh, N., S. K. Mohanty, S. K. Majumder, and P. K. Gupta. 2001. Measurement of optical transport properties of normal and malignant human breast tissue. *Applied Optics* 40(1):176–184.

Gottschalk, W. 1992. Ein Messverfahren zur Bestimmung der Optischen Parameter biologischer Gevebe in vitro. Universitaet Fridriciana: Karlsruhe, Germany.

Graaff, R., A. C. M. Dassel, M. H. Koelink et al. 1993. Optical properties of human dermis in vitro and in vivo. *Applied Optics* 32:435–447.

Gratton, E., S. Fantini, M. A. Franceschini, G. Gratton, and M. Fabiani. 1997. Measurements of scattering and absorption changes in muscle and brain. *Philosophical Transactions of the Royal Society of London B* 352:727–735.

Grisimov, V. N. 1994. Refractive index of the basic substance of dentin. *Optika I Spektroskopiya* 77(2):272–273.

Hale, G. M. and M. R. Querry. 1973. Optical constants of water in the 200 nm to 200 µm wavelength region. *Applied Optics* 12:9.

Haruna, M., K. Yoden, M. Ohmi, and A. Seiyama. 2000. Detection of phase transition of a biological membrane by precise refractive-index measurement based on low-coherence interferometry. *Proceedings of the SPIE* 3915.

Hayakawa, C. K., J. Spanier, F. Bevilacqua, A. K. Dunn, J. S. You, B. J. Tromberg, and V. Venugopalan. 2001. Perturbation Monte Carlo methods to solve inverse photon migration problems in heterogeneous tissues. *Optics Letters* 26(17):1335–1337.

Hecht, E. 2002. *Optics*, 4th edn. Reading, MA: Addison-Wesley.

Henyey, L. G. and J. L. Greenstein. 1941. Diffuse radiation in the galaxy. *Astrophysical Journal* 93:70–83.

Heusmann, H., J. Kolzer, and G. Mitic. 1996. Characterization of female breast in vivo by time resolved and spectroscopic measurements in near-infrared spectroscopy. *Journal of Biomedical Optics* 1(4):425–434.

Hewitt, P. G. 2010. *Conceptual Physics*, 11th edn. Boston, MA: Addison-Wesley.

Hoffmann, J., D. W. Lubbers, and H. M. Heise. 1998. Applicability of the Kubelka-Munk theory for the evaluation of reflectance spectra demonstrated for haemoglobin-free perfused heart tissue. *Physics in Medicine and Biology* 43(12):3571–3587.

Holboke, M. J., B. J. Tromberg, X. Li, N. Shah, J. Fishkin, D. Kidney, J. Butler, B. Chance, and A. G. Yodh. 2000. Three-dimensional diffuse optical mammography with ultrasound localization in a human subject. *Journal of Biomedical Optics* 5(2):237–247.

Hourdakis, C. J. and A. A. Perris. 1995. Monte Carlo estimation of tissue optical properties for use in laser dosimetry. *Physics in Medicine and Biology* 40:351–364.

Ishimaru, A. 1978. *Wave Propagation and Scattering in Random Media*, Vols. 1 and 2. New York: Academic Press.

Jackson, J. D. 1999. *Classical Electrodynamics*, 3rd edn. Hoboken, NJ: John Wiley & Sons.

Jacques, S. L. and S. A. Prahl. 1998. Absorption spectra for biological tissues. Available from http://omlc.ogi.edu/classroom/ece532/class3/muaspectra.html (accessed February 16, 2014).

Kano, H. and H. O. Harnaguchi. 2007. Supercontinuum dynamically visualizes a dividing single cell. *Analytical Chemistry* 79(23):8967–8973.

Keijzer, M., S. L. Jacques, S. A. Prahl, and A. J. Welch. 1989. Monte-Carlo simulations for finite-diameter laser beams. *Lasers in Surgery and Medicine* 9:148–154.

Keijzer, M., R. R. Richards-Kortum, S. L. Jacques, and M. S. Feld. 1989. Fluorescence spectroscopy of turbid media: Autofluorescence of the human aorta. *Applied Optics* 28(20):4286–4292.

Key, H., E. R. Davies, P. C. Jackson, and P. N. T. Wells. 1991. Optical attenuation characteristics of breast tissues at visible and near-infrared wavelengths. *Physics in Medicine and Biology* 36(5):579–590.

Khalil, O. S., S. Yeh, M. G. Lowery, X. Wu, C. F. Hanna, S. Kantor, T. W. Jeng, J. S. Kanger, R. A. Bolt, and F. F. de Mul. 2003. Temperature modulation of the visible and near infrared absorption and scattering coefficients of human skin. *Journal of Biomedical Optics* 8(2):191–205.

Kienle, A. and R. Hibst. 1995. A new optimal wavelength for treatment of port wine stains? *Physics in Medicine and Biology* 40:1559–1576.

Kienle, A., L. Lilge, and M. S. Patterson. 1994. Investigations of multilayered tissue with *in vivo* reflectance measurements. *Proceedings of the SPIE* 2326:212–214.

Kim, Y. L., J. T. Walsh, Jr., T. K. Goldstick, and M. R. Glucksberg. 2004. Variation of corneal refractive index with hydration. *Physics in Medicine and Biology* 49(5):859–868.

Knuttel, A. and M. Boehlau-Godau. 2000. Spatially confined and temporally resolved refractive index and scattering evaluation in human skin performed with optical coherence tomography. *Journal of Biomedical Optics* 5(1):83–92.

Knuttel, A., S. Bonev, and W. Knaak. 2004. New method for evaluation of in vivo scattering and refractive index properties obtained with optical coherence tomography. *Journal of Biomedical Optics* 9(2):265–273.

Koukoulas, A. A. and B. D. Jordan. 1997. Effect of strong absorption on the Kubelka-Munk scattering coefficient. *Journal of Pulp and Paper Science* 23(5):J224–J232.

Kubelka, P. 1948. New contributions to the optics of intensely light-scattering materials. Part I. *Journal of Optical Society of America* 38:448.

Kubelka, P. and F. Munk. 1931. Ein Beitrag zur Optik der Farbanstriche. *Zeitschrift für Technische Physik* 12:59.

Laufer, J., C. R. Simpson, M. Kohl, M. Essenpreis, and M. Cope. 1998. Effect of temperature on the optical properties of *ex vivo* human dermis and subdermis. *Physics in Medicine and Biology* 43(9):2479–2489.

Lax, M. 1951. Multiple scattering of waves. *Reviews of Modern Physics* 23(4):287.

Leonard, D. W. and K. M. Meek. 1997. Refractive indices of the collagen fibrils and extrafibrillar material of the corneal stroma. *Biophysical Journal* 72(3):1382–1387.

Li, H. and S. S. Xie. 1996. Measurement method of the refractive index of biotissue by total internal reflection. *Applied Optics* 35(10):1793–1795.

Lin, R. C., M. A. Shure, A. M. Rollins, J. A. Izatt, and D. Huang. 2004. Group index of the human cornea at 1.3-mu m wavelength obtained in vitro by optical coherence domain reflectometry. *Optics Letters* 29(1):83–85.

Lin, W.-C., M. Motamedi, and A. J. Welch. 1996. Dynamics of tissue optics during laser heating of turbid media. *Applied Optics* 35(19):3413–3420.

Liu, F., K. M. Yoo, and R. R. Alfano. 1991. Speed of the coherent component of femtosecond laser-pulses propagating through random scattering media. *Optics Letters* 16(6):351–353.

Lloyd, P. and M. V. Berry. 1967. Wave propagation through an assembly of spheres. IV. Relations between different multiple scattering theories. *Proceedings of the Physical Society* 91:678–688.

Loyalka, S. K. and C. A. Riggs. 1995. Inverse problem in diffuse-reflectance spectroscopy—Accuracy of the Kubelka-Munk equations. *Applied Spectroscopy* 49(8):1107–1110.

Ma, X. Y., J. Q. Lu, H. F. Ding, and X. H. Hu. 2005. Bulk optical parameters of porcine skin dermis at eight wavelengths from 325 to 1557 nm. *Optics Letters* 30(4):412–414.

Madsen, S. J., P. Wyss, L. O. Svaasand, R. C. Haskell, Y. Tadir, and B. J. Tromberg. 1994. Determination of the optical-properties of the human uterus using frequency-domain photon migration and steady-state techniques. *Physics in Medicine and Biology* 39(8):1191–1202.

Mandelis, A. and J. P. Grossman. 1992. Perturbation theoretical approach to the generalized Kubelka-Munk problem in nonhomogeneous optical media. *Applied Spectroscopy* 46(5):737–745.

Marchesini, R., A. Bertoni, S. Andreola et al. 1989. Extinction and absorption coefficients and scattering phase functions of human tissues *in vitro*. *Applied Optics* 28:2318–2324.

Matcher, S. J., M. Cope, and D. T. Delpy. 1997. *In vivo* measurements of the wavelength dependence of tissue scattering coefficients between 760 and 900 nm measured with time-resolved spectroscopy. *Applied Optics* 36(1):386–396.

Mazarevica, G., T. Freivalds, and A. Jurka. 2002. Properties of erythrocyte light refraction in diabetic patients. *Journal of Biomedical Optics* 7(2):244–247.

Meek, K. M., S. Dennis, and S. Khan. 2003. Changes in the refractive index of the stroma and its extrafibrillar matrix when the cornea swells. *Biophysical Journal* 85(4):2205–2212.

Mishchenko, M. I., L. D. Travis, and A. A. Lacis. 2006. *Multiple Scattering of Light by Particles: Radiative Transfer and Coherent Backscattering*. New York: Cambridge University Press.

Molenaar, R., J. J. ten Bosch, and J. R. Zijp. 1999. Determination of Kubelka-Munk scattering and absorption coefficients by diffuse illumination. *Applied Optics* 38(10):2068–2077.

Muller, G. and A. Roggan, eds. 1995. *Laser-Induced Interstitial Thermotherapy*. Vol. PM25. Bellingham, WA: SPIE Press.

Nickell, S., M. Hermann, M. Essenpreis, T. J. Farrell, U. Kramer, and M. S. Patterson. 2000. Anisotropy of light propagation in human skin. *Physics in Medicine and Biology* 45(10):2873–2886.

Nilsson, A. M. K., G. W. Lucassen, W. Verkruysse, S. Andersson-Engels, and M. J. C. van Gemert. 1997. Changes in optical properties of human whole blood *in vitro* due to slow heating. *Photochemistry and Photobiology* 65(2):366–373.

Ohmi, M., Y. Ohnishi, K. Yoden, and M. Haruna. 2000. *In vitro* simultaneous measurement of refractive index and thickness of biological tissue by the low coherence interferometry. *IEEE Transactions on Biomedical Engineering* 47(9):1266–1270.

Okada, E., M. Firbank, M. Schweiger, S. R. Arridge, M. Cope, and D. T. Delpy. 1997. Theoretical and experimental investigation of near-infrared light propagation in a model of the adult head. *Applied Optics* 36(1):21–31.

Oliveira, L., A. Lage, M. P. Clemente, and V. Tuchin. 2009. Optical characterization and composition of abdominal wall muscle from rat. *Optics and Lasers in Engineering* 47(6):667–672.

Oraevsky, A. A., S. L. Jacques, and F. K. Tittel. 1997. Measurement of tissue optical properties by time-resolved detection of laser-induced transient stress. *Applied Optics* 36(1):402–415.

Peters, V. G., D. R. Wyman, M. S. Patterson, and G. L. Frank. 1990. Optical properties of normal and diseased human tissues in the visible and near infrared. *Physics in Medicine and Biology* 35:1317–1334.

Philips-Invernizzi, B., D. Dupont, and C. Caze. 2001. Bibliographical review for reflectance of diffusing media. *Optical Engineering* 40(6):1082–1092.

Prahl, S. A. 1998. Tabulated molar extinction coefficient for hemoglobin in water. Available from http://omlc.ogi.edu/spectra/hemoglobin/summary.html (accessed February 16, 2014).

Prota, G., M. D'Ischia, and A. Napolitano. 1988. The chemistry of melanins and related metabolites. In *The Pigmentary System*, eds. J. Nordlund et al., New York: Oxford University Press.

Ragain, J. C. and W. M. Johnston. 1998. Agreement of measured reflectance with Kubelka-Munk theory for human enamel and dentin. *Journal of Dental Research* 77:1433.

Ragain, J. C. and W. M. Johnston. 2001. Accuracy of Kubelka-Munk reflectance theory applied to human dentin and enamel. *Journal of Dental Research* 80(2):449–452.

Roggan, A., M. Friebel, K. Dorschel, A. Hahn, and G. Mueller. 1999. Optical properties of circulating human blood in the wavelength range 400–2500 nm. *Journal of Biomedical Optics* 4(1):36–46.

Roggan, A., O. Minet, C. Schroder, and G. Muller. 1994. The determination of optical tissue properties with double integrating sphere technique and Monte Carlo simulations. *Proceedings of the SPIE* 2100:42–56.

Rundlof, M. and J. A. Bristow. 1997. A note concerning the interaction between light scattering and light absorption in the application of the Kubelka-Munk equations. *Journal of Pulp and Paper Science* 23(5):J220–J223.

Saidi, I. S., S. L. Jacques, and F. K. Tittel. 1995. Mie and Rayleigh modeling of visible-light scattering in neonatal skin. *Applied Optics* 34(31):7410–7418.

Salomatina, E., B. Jiang, J. Novak, and A. N. Yaroslavsky. 2006. Optical properties of normal and cancerous human skin in the visible and near-infrared spectral range. *Journal of Biomedical Optics* 11(6):9.

Sardar, D. K. and L. B. Levy. 1998. Optical properties of whole blood. *Lasers in Medical Science* 13(2):106–111.

Schuster, A. 1905. Radiation through a foggy atmosphere. *Astrophysical Journal* 21:1.

Schwarzmaier, H.-J., A. N. Yaroslavsky, I. V. Yaroslavsky et al. 1997. Optical properties of native and coagulated human brain structures. *Proceedings of the SPIE* 2970:492–499.

Simpson, C. R., M. Kohl, M. Essenpreis, and M. Cope. 1998. Near-infrared optical properties of *ex vivo* human skin and subcutaneous tissues measured using the Monte Carlo inversion technique. *Physics in Medicine and Biology* 43:2465–2478.

Splinter, R. and B. A. Hooper. 2007. *An Introduction to Biomedical Optics*. New York: Taylor & Francis

Sterenborg, H. J. C. M., M. J. C. van Gemert, W. Kamphorst et al. 1989. The spectral dependence of the optical properties of human brain. *Lasers in Medical Science* 4:221–227.

Suzuki, K., Y. Yamashita, K. Ohta et al. 1996. Quantitative measurement of optical parameters in normal breast using time-resolved spectroscopy: *In vivo* results of 30 Japanese women. *Journal of Biomedical Optics* 1(3):330–334.

Svaasand, L. O., B. J. Tromberg, R. C. Haskell et al. 1993. Tissue characterization and imaging using photon density waves. *Optical Engineering* 32(2):258–266.

Tearney, G. J., M. E. Brezinski, J. F. Southern, B. E. Bouma, M. R. Hee, and J. G. Fujimoto. 1995. Determination of the refractive index of highly scattering human tissue by optical coherence tomography. *Optics Letters* 20(21):2258–2260.

Thueler, P., I. Charvet, F. Bevilacqua, M. St Ghislain, G. Ory, P. Marquet, P. Meda, B. Vermeulen, and C. Depeursinge. 2003. In vivo endoscopic tissue diagnostics based on spectroscopic absorption, scattering, and phase function properties. *Journal of Biomedical Optics* 8(3):495–503.

Tinoco Jr, I., K. Sauer, and J. C. Wang. 1985. *Physical Chemistry: Principles and Applications in Biological Sciences*. Englewood Cliffs, NJ: Prentice Hall.

Tromberg, B. J., O. Coquoz, J. Fishkin, T. Pham, E. R. Anderson, J. Butler, M. Cahn, J. D. Gross, V. Venugopalan, and D. Pham. 1997. Non-invasive measurements of breast tissue optical properties using frequency-domain photon migration. *Philosophical Transactions of the Royal Society of London B* 352:661–668.

Troy, T. L., D. L. Page, and E. M. Sevick-Muraca. 1996. Optical properties of normal and diseased breast tissues: Prognosis for optical mammography. *Journal of Biomedical Optics* 1(3):342–355.

Troy, T. L. and S. N. Thennadil. 2001. Optical properties of human skin in the near infrared wavelength range of 1000 to 2200 nm. *Journal of Biomedical Optics* 6(2):167–176.

Tsuchikawa, S. and S. Tsutsumi. 1999. Analytical characterization of reflected and transmitted light from cellular structural material for the parallel beam of NIR incident light. *Applied Spectroscopy* 53(9):1033–1039.

Tuchin, V. V. ed. 1994. *Selected Papers on Tissue Optics: Applications in Medical Diagnostics and Therapy*. Vol. MS102. Bellingham, WA: SPIE Press.

Tuchin, V. V. 1995. Fundamentals of low-intensity laser radiation interaction with biotissues: Dosimetry and diagnostical aspects. *Bulletin of the Russian Academy of Sciences, Physics Series* 59(6):120–143.

Tuchin, V. V. 2007. *Tissue Optics: Light Scattering Methods and Instruments for Medical Diagnosis*, 2nd edn., Vol. PM 166. Bellingham, WA: SPIE Press.

Tuchin, V. V., H. Podbielska, and C. K. Hitzenberger. 1999. Special section on coherence domain optical methods in biomedical science and clinics. *Journal of Biomedical Optics* 4(1):94–190.

Tuchin, V. V., R. K. Wang, E. I. Galanzha, J. B. Elder, and D. M. Zhestkov. 2004. Monitoring of glycated hemoglobin by OCT measurement of refractive index. *Proceedings of the SPIE* 5316.

Unterhuber, A., B. Povazay, K. Bizheva et al. 2004. Advances in broad bandwidth light sources for ultra-high resolution optical coherence tomography. *Physics in Medicine and Biology* 49(7):1235–1246.

van Gemert, M. J. C., J. S. Nelson, S. L. Jacques, H. J. C. M. Sterenborg, and W. M. Star. 1989. Skin optics. *IEEE Transactions on Biomedical Engineering* 36(12):1146–1154.

Vargas, W. E. and G. A. Niklasson. 1997. Applicability conditions of the Kubelka-Munk theory. *Applied Optics* 36(22):5580–5586.

Vo-Dinh, T. 1984. *Room Temperature Phosphorimetry for Chemical Analysis*. New York: Wiley.

Voet, D., W. B. Gratzer, R. A. Cox, and P. Doty. 1963. Absorption spectra of nucleotides, polynucleotides, and nucleic acids in the far ultraviolet. *Biopolymers* 1:93.

Wang, L. H., S. L. Jacques, and L. Q. Zheng. 1995. Mcml—Monte-Carlo modeling of light transport in multilayered tissues. *Computer Methods and Programs in Biomedicine* 47(2):131–146.

Wang, L. V. and W. Hsin-i. 2007. *Biomedical Optics: Principles and Imaging.* Hoboken, NJ: John Wiley & Sons.

Wang, X. J., T. E. Milner, M. C. Chang, and J. S. Nelson. 1996. Group refractive index measurement of dry and hydrated type I collagen films using optical low-coherence reflectometry. *Journal of Biomedical Optics* 1(2):212–216.

Wang, X. J., T. E. Milner, J. F. de Boer, Y. Zhang, D. H. Pashley, and J. S. Nelson. 1999. Characterization of dentin and enamel by use of optical coherence tomography. *Applied Optics* 38(10):2092–2096.

Wang, X. J., T. E. Milner, R. P. Dhond, W. V. Sorin, S. A. Newton, and J. S. Nelson. 1995. Characterization of human scalp hairs by optical low-coherence reflectometry. *Optics Letters* 20(6):524–526.

Waterman, P. C. and R. Truell. 1961. Multiple scattering of waves. *Journal of Mathematical Physics* 2(4):512–538.

Waters, D. N. 1994. Raman-spectroscopy of powders—Effects of light-absorption and scattering. *Spectrochimica Acta Part A—Molecular and Biomolecular Spectroscopy* 50(11):1833–1840.

Willmann, S., A. Terenji, I. V. Yaroslavsky, T. Kahn, P. Hering, and H.-J. Schwarzmaier. 1999. Determination of the optical properties of a human brain tumor using a new microspectrophotometric technique. *Proceedings of the SPIE* 3598:233–239.

Yaroslavsky, A. N., P. C. Schulze, I. V. Yaroslavsky, R. Schober, F. Ulrich, and H.-J. Schwarzmaier. 2002. Optical properties of selected native and coagulated human brain tissues *in vitro* in the visible and near infrared spectral range. *Physics in Medicine and Biology* 47:2059–2073.

Yaroslavsky, A. N., I. V. Yaroslavsky, T. Goldbach, and H.-J. Schwarzmaier. 1996a. Inverse hybrid technique for determining the optical properties of turbid media from integrating-sphere measurements. *Applied Optics* 35(34):6797–6809.

Yaroslavsky, A. N., I. V. Yaroslavsky, T. Goldbach, and H. J. Schwarzmaier. 1996b. The optical properties of blood in the near infrared spectra range. *Proceedings of the SPIE* 2678.

Yaroslavsky, A. N., I. V. Yaroslavsky, T. Goldbach, and H.-J. Schwarzmaier. 1999. Influence of the scattering phase function approximation on the optical properties of blood determined from integrating sphere measurements. *Journal of Biomedical Optics* 4(1):47–53.

Zhadin, N. N. and R. R. Alfano. 1998. Correction of the internal absorption effect in fluorescence emission and excitation spectra from absorbing and highly scattering media: Theory and experiment. *Journal of Biomedical Optics* 3(2):171–186.

Zijp, J. R. and J. J. Ten Bosch. 1991. Angular dependence of He-Ne laser light scattering by bovine human dentine. *Archives of Oral Biology* 36(4):283–289.

Zijp, J. R. and J. J. Ten Bosch. 1997. Anisotropy of volume-backscattered light. *Applied Optics* 36:1671–1680.

Zvyagin, A. V., K. K. Silva, S. A. Alexandrov, T. R. Hillman, J. J. Armstrong, T. Tsuzuki, and D. D. Sampson. 2003. Refractive index tomography of turbid media by bifocal optical coherence refractometry. *Optics Express* 11(25):3503–3517.

3

Light–Tissue Interactions

Valery V. Tuchin
*Saratov State University
and
Russian Academy of Science
and
University of Oulu*

3.1 Introduction

Light interactions with biological tissues and fluids can be categorized into two large classes of biological media: (1) strongly scattering (opaque) with skin, brain, vessel walls, eye sclera, blood, and lymph, and (2) weakly scattering (transparent) with cornea, crystalline lens, vitreous humor, and aqueous humor of the front chamber of the eye.[1-7] Light interactions with tissues of the first class can be described as a model of multiple scattering of scalar or vector waves in a randomly nonuniform medium with absorption. Interactions with tissues of the second class can be described as a model of single (or low-step) scattering of an ordered isotropic or anisotropic medium with closely packed scatterers made from material with absorptive properties. The surrounding medium could also be considered to have refractive and absorptive properties.

The transparency of tissues reaches its maximum in the near infrared (NIR), which is associated with the fact that living tissues do not contain strong intrinsic chromophores that can absorb radiation within this spectral range.[1-7] Light penetrates into a tissue over several centimeters, which is important for the transillumination of thick human organs (brain, breast, etc.). However, in spite of scattering decay with the increase in wavelength, tissues are still characterized by rather strong scattering of NIR radiation. This prevents one from obtaining clear images of localized inhomogeneities arising in tissues due to various pathologies, such as tumor formation, local increase in blood volume caused by a hemor-rhage, or the growth of microvessels. Therefore, attention in optical tomography and spectroscopy is focused on the development of methods for the selection of image-carrying photons or the detection of photons providing information regarding optical parameters of the scattering medium.

Methods of noninvasive optical diagnosis and spectroscopy of tissues involve two radiation regimes: continuous wave (CW) and time-resolved regimes.[1-7] Time-resolved interactions are real-ized when tissue is exposed to short laser pulses and the subsequent scattered broadened pulses are recorded (time-domain method) or by irradiation with modulated light where the depth of modula-tion of the scattered light intensity and the corresponding phase shift in the modulation frequencies are recorded (frequency-domain or phase method). The time-resolved regimes are based on the exci-tation of the photon-density wave spectrum in a strongly scattering medium, which can be described in the framework of the nonstationary radiation transfer theory (RTT). The CW regime is described by the stationary RTT.

Many optical medical technologies employ laser radiation and fiber optics; therefore, coherence of light is very important for the analysis of light interaction with tissues and cell ensembles.[2-5,7-12] This problem may be considered in terms of the loss of coherence due to the scattering of light in a randomly nonuniform medium with multiple scattering and/or the change in the statistics of speckles in the scat-tered field. The coherence of light is of fundamental importance for the selection of photons that have experienced no or a small number of scattering events, as well as for the generation of speckle-modulated fields from scattering phase objects with single and multiple scattering. Such approaches are important for coherent tomography, diffractometry, holography, photon-correlation spectroscopy, laser Doppler anemometry, and speckle-interferometry of tissues and biological flows. The use of optical sources with a short coherence length opens up new opportunities in coherent interferometry and tomography of tis-sues, organs, and blood flows, providing a specific gating of ballistic or quasi-ballistic photons.

The vector nature of light waves is important for light–tissue interaction, because in a scattering medium it is manifested as the polarization ability of an initially unpolarized incident light or the change in the character of a polarization state of an initially polarized light propagating in a medium. Similar to coherence properties of a light beam reflected from or transmitted through a biological object, the polarization properties of light can be used to select photons coming from different depths in an object.[3,5,7]

Quasi-elastic light scattering (QELS) as applied to monitoring of dynamic systems (chaotic or directed movements of tissue components or cells) is based mainly on the correlation or spectral analysis of the temporal fluctuations of the scattered light intensity.[3,5,7-10] QELS spectroscopy, also known as light-beating spectroscopy or correlation spectroscopy, is widely used for various biomedical applications, particularly for blood or lymph flow measurement and cataract diagnostics. For the study of optically thick tissue where multiple scattering prevails and photon migration (diffusion) within tissue is important for deter-mining the characteristics of fluctuations in intensity, diffusion-wave spectroscopy (DWS) is available.

Raman scattering is the basis for Raman vibrational spectroscopy. It is a great tool for studying the structure and dynamic function of biological molecules and has been used extensively for the monitor-ing and diagnosis of diseases like cataract, atherosclerotic lesions in coronary arteries, precancerous and cancerous lesions in human soft tissues, and bone and teeth pathologies.[13-15]

Light-induced fluorescence is also a powerful noninvasive method for the recognition and monitor-ing of tissue pathology.[7,16-18] Autofluorescence, fluorescence of introduced markers, time-resolved, laser-scan, and multiphoton fluorescence have been used to study human tissues and cells in situ.

Light-induced thermal effects in tissues are important for diagnostics, phototherapy, and laser surgery.[5,7,12,19-24] Optothermal (OT) spectroscopy, based on the detection of time-dependent heat generation induced in a tissue by pulsed or intensity-modulated optical radiation, is widely used in biomedicine. Among a variety of OT methods, optoacoustic (OA) and photoacoustic (PA) techniques are of great importance. They allow the estimation of optical, thermal, and acoustic properties of the tissue that depend on the peculiarities of tissue structure.

For thermal phototherapy and surgery, much higher light intensities are needed than for diagnostic purposes. Important factors that need to be considered are controllable temperature rise, and thermal and/or thermomechanical damage (coagulation, vaporization, vacuolization, pyrolysis, ablation) of a tissue.[12,20–24]

3.2 Light Interactions with a Strongly Scattering Tissue

3.2.1 Continuous-Wave Light

Biological tissues are optically inhomogeneous and absorbing media whose average refractive index is higher than that of air. This is the cause of the partial reflection of the radiation at the tissue/air interface (Fresnel reflection), while the remaining part penetrates the tissue. Multiple scattering and absorption are responsible for light beam broadening and eventual decay as it travels through a tissue, whereas bulk scattering is a major cause for the dispersion of a large fraction of radiation in the backward direction. Cellular organelles such as mitochondria, thin fibrillar structures of connective tissues, melanin granules, and red blood cells are the main scatterers in living tissues.[1–7]

Absorbed light is converted to heat or radiated in the form of fluorescence; it is also consumed in photobiochemical reactions. The absorption spectrum depends on the type of predominant absorption centers and water content of tissues (see Figure 2.16). Absolute values of absorption coefficients for typical tissues lie in the range $10^{-2}–10^4$ cm^{-1}.[1–7] In the ultraviolet (UV) and infrared (IR) ($\lambda \geq 2$ μm) spectral regions, light is readily absorbed, which accounts for the small contribution of scattering and inability of radiation to penetrate deep into tissues (only across one or more cell layers). In the wavelength range 600–1600 nm, scattering prevails over absorption and the intensity of the reflected radiation increases to 35%–70% of the total incident light (due to backscattering).

Light interaction with a multilayer and multicomponent tissue is a very complicated process (see Figure 3.1). For example, for skin, the horny layer (stratum corneum) reflects about 5%–7% of the incident light. A collimated light beam is transformed into a diffuse one by microscopic inhomogeneities at the air/horny layer interface. A major portion of reflected light results from backscattering in different skin layers (stratum corneum, epidermis, dermis, blood, and fat). The absorption of diffuse light by skin pigments is a measure of bilirubin and melanin content, and hemoglobin saturation with oxygen. These characteristics are widely used in the diagnosis of various diseases. Certain phototherapeutic and diagnostic modalities take advantage of the ready transdermal penetration of visible and NIR light into the

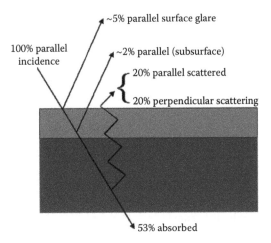

FIGURE 3.1 Simplified two-layer scattering and absorption model of skin at linear polarized incident light. (From Jacques, S.L., *Proc. SPIE*, 4707, 474, 2002.)

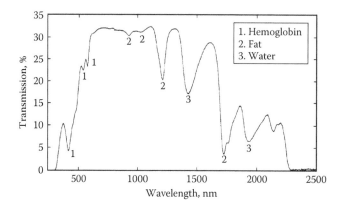

FIGURE 3.2 Transmission spectrum of a 3 mm-thick slab of female breast tissue. A spectrometer with an integrating sphere was used. The contributions of absorption bands of the tissue components (hemoglobin [1], fat [2], and water [3]) are marked. (From Marks, F.A., *Proc. SPIE*, 1641, 227, 1992.)

body in the wavelength region corresponding to the so-called therapeutic or diagnostic window, which has a few broad and narrow bands within 600–1600 nm.

Another example of heterogeneous multicomponent tissue is a female breast. The absorption bands of hemoglobin, fat, and water are clearly seen in an in vitro measured spectrum of a 3 mm slab of breast tissue presented in Figure 3.2. There is a wide window between 600 and 1150 nm and two narrow but very important windows near the central wavelengths with 1300 and 1550 nm, where the lowest percentage of light is attenuated.

A collimated (laser) beam is attenuated in a tissue layer of thickness d in accordance with the exponential law (Beer–Lambert's law):

$$I(d) = (1 - R_F)\, I_0 \exp(-\mu_t d),\tag{3.1}$$

where

$I(d)$ is the intensity of transmitted light measured using a distant photodetector with a small aperture (on line or collimated transmittance), W/cm^2

R_F is the coefficient of Fresnel reflection at the normal beam incidence, $R_F = [(n-1)/(n+1)]^2$

n is the relative mean refractive index of tissue and surrounding media

I_0 is the incident light intensity, W/cm^2

$\mu_t = \mu_a + \mu_s$ is the extinction coefficient (interaction or total attenuation coefficient), cm^{-1}

μ_a is the absorption coefficient, cm^{-1}

μ_s is the scattering coefficient, cm^{-1}

The mean free path (MFP) length between two consequent interactions with absorption or scattering is denoted by

$$l_{ph} = \mu_t^{-1}.\tag{3.2}$$

To analyze light propagation under multiple scattering conditions, it is assumed that absorbing and scattering centers are uniformly distributed across the tissue. Visible and NIR radiation is normally subject to anisotropic scattering characterized by the apparent direction of a scattered photon.

A sufficiently strict mathematical description of CW light propagation in a scattering medium is possible in the framework of stationary RTT. This theory is valid for an ensemble of scatterers located far from one another and has been successfully used to work out some practical aspects of tissue optics. The main stationary equation of RTT for monochromatic light takes the following form[1-7]:

$$\frac{\partial I(\bar{r},\bar{s})}{\partial s} = -\mu_t I(\bar{r},\bar{s}) + \frac{\mu_s}{4\pi} \int_{4\pi} I(\bar{r},\bar{s}')p(\bar{s},\bar{s}')d\Omega' + S(\bar{r},\bar{s}), \tag{3.3}$$

where

$I(\bar{r},\bar{s})$ is the radiance (or specific intensity)—average power flux density at a point $\bar{r}$ in the given direction $\bar{s}$, (W/cm² sr)

$p(\bar{s},\bar{s}')$ is the scattering phase function, sr^{-1}

$d\Omega'$ is the unit solid angle around the direction $\bar{s}'$, sr

$S(\bar{r},\bar{s})$ characterizes a radiation source inside the medium or light coming from the outside and falling on the tissue surface, for example, a laser beam; as a radiation source, internal fluorescence, excited by an external light beam, could also be considered

To characterize the relation of the scattering and absorption properties of a tissue, a parameter such as albedo is usually introduced: $\Lambda = \mu_s/\mu_t$. The albedo ranges from zero for a completely absorbing medium to unity for a completely scattering medium.

The phase function $p(\bar{s},\bar{s}')$ describes the scattering properties of the medium and is in fact the probability density function for scattering in the direction $\bar{s}'$ of a photon traveling in the direction $\bar{s}$; in other words, it characterizes an elementary scattering act. If scattering is symmetric relative to the direction of the incident wave, then the phase function depends only on the scattering angle θ (angle between directions $\bar{s}$ and $\bar{s}'$), that is, $p(\theta)$. In practice, the phase function is usually well approximated with the aid of the postulated Henyey–Greenstein function[1-7]:

$$p(\theta) = \frac{1}{4\pi} \cdot \frac{1-g^2}{\left(1+g^2 - 2g\cos\theta\right)^{3/2}}, \tag{3.4}$$

where g is the scattering anisotropy factor (mean cosine of the scattering angle θ). The value of g varies from 0 to 1: $g = 0$ corresponds to isotropic (Rayleigh) scattering and $g = 1$ to total forward scattering (Mie scattering by large particles).[27-30]

The integrodifferential equation (Equation 3.1) is frequently simplified by representing the solution in the form of spherical harmonics. Such simplification leads to a system of $(N + 1)^2$ connected differential partial derivative equations known as the P_N approximation. This system is reducible to a single differential equation on the order of $(N + 1)$. For example, four connected differential equations reducible to a single diffusion-type equation are necessary for $N = 1$.[1-7] The photon diffusion coefficient, cm²/c,

$$D = \frac{c}{3\left(\mu_s' + \mu_a\right)}, \tag{3.5}$$

and the reduced (transport) scattering coefficient, cm^{-1},

$$\mu_s' = \left(1-g\right)\mu_s, \tag{3.6}$$

are the major parameters of the diffusion equation; here c is the velocity of light in the medium.

The transport MFP of a photon (cm) is defined as

$$l_{tr} = \left(\mu'_s + \mu_a\right)^{-1}. \tag{3.7}$$

It is worthwhile to note that the transport MFP in a medium with anisotropic single scattering significantly exceeds the MFP in a medium with isotropic single scattering $l_{tr} \gg l_{ph}$. The transport MFP l_{tr} is the distance over which the photon loses its initial direction.

Based on the diffusion theory, attenuation of a wide laser beam of intensity I_0 at depths $z > l_d = 1/\mu_{eff}$ in a thick tissue may be described as[5]

$$I(z) \approx I_0 b_s \exp(-\mu_{eff} z), \tag{3.8}$$

where b_s accounts for additional irradiation of the upper layers of a tissue due to backscattering (photon recycling effect) and

$$\mu_{eff} = \left[3\mu_a\left(\mu'_s + \mu_a\right)\right]^{1/2}, \tag{3.9}$$

is the effective attenuation coefficient or inverse diffusion length, $\mu_{eff} = 1/l_d$, cm^{-1}. Consequently, the depth of light penetration into a tissue is defined as

$$l_e = l_d[\ln b_s + 1]. \tag{3.10}$$

Typically, for tissues, $b_s = 1$–5 for a beam diameter of 1–20 mm.[5,31] Thus, when wide laser beams are used for irradiation of highly scattering tissues with low absorption, CW light energy is accumulated in tissue due to high multiplicity of chaotic long-path photon migrations. A highly scattering medium works as a random cavity providing the capacity of light energy. The light power density within the superficial tissue layers may substantially (up to fivefold) exceed the incident power density and cause the overdosage during photodynamic therapy or overheating at interstitial laser thermotherapy. On the other hand, the photon recycling effect can be used for more effective irradiation of undersurface lesions at relatively small incident power densities.

3.2.2 Short Light Pulses

When probing the plane-parallel layer of a scattering medium with an ultrashort light pulse, the transmitted pulse consists of a ballistic (coherent) component, a group of photons having zigzag trajectories, and a highly intensive diffuse component (see Figure 3.3a).[1–7,16,22,32,33] Both unscattered photons and photons undergoing forward-directed single-step scattering contribute to the intensity of the ballistic component (comprised of photons traveling straight along the light beam). This component (not shown in Figure 3.3a) is subject to exponential attenuation with increasing sample thickness. This accounts for the limited utility of ballistic photons for practical diagnostic purposes in medicine.

The group of snake photons with zigzag trajectories includes photons, which experienced only a few collisions each. They propagate along trajectories that only slightly deviate from the direction of the incident beam and form the first-arriving part of the diffuse component. These photons carry information about the optical properties of the random medium.

The diffuse component is very broad and intense as it contains the bulk of incident photons after they have participated in many scattering acts and therefore migrate in different directions and have different path lengths. Moreover, the diffuse component carries information about the optical properties of

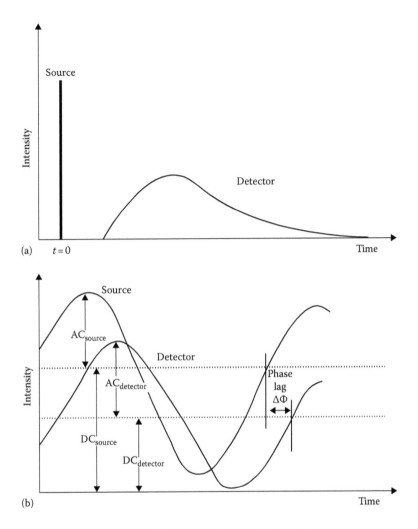

FIGURE 3.3 Schematic representation of the time evolution of the light intensity measured in response to (a) a very narrow light pulse and (b) a sinusoidally intensity-modulated light transversing an arbitrary distance in a scattering and absorbing medium. (From Fishkin, J.B. and Gratton, E., *J. Opt. Soc. Am. A*, 10, 127, 1993.)

the scattering medium, and its deformation may reflect the presence of local inhomogeneities in the medium. The resolution obtained by this method at a high light-gathering power is much lower than by the method measuring straight-passing photons. Two principal probing schemes are conceivable: one recording transmitted photons and the other taking advantage of their backscattering.

The time-dependent reflectance is defined as[32,33]

$$R(r_{sd},t) = \frac{z_0}{(4\pi D)^{3/2}} t^{-5/2} \exp\left(\frac{r_{sd}^2 + z_0^2}{2Dt}\right) \exp(-\mu_a ct), \tag{3.11}$$

where
t is the time
$z_0 = (\mu_s')^{-1}$
D is the photon diffusion coefficient, cm²/c (see Equation 3.5)

An important advantage of the pulse method is its applicability to in vivo studies due to the possibility of the separate evaluation of μ_a and μ'_s using a single measurement in the backscattering or transillumination regimes.

3.2.3 Diffuse Photon-Density Waves

When probing the plane-parallel layer of a scattering medium with an intensity-modulated light, the modulation depth of scattered light intensity $m_U \equiv AC_{detector}/DC_{detector}$ (see Figure 3.3b) and the corresponding phase shift relative to the incident light modulation phase $\Delta\Phi$ (phase lag) can be measured.[1-7,34-37] In applications to tissue spectroscopy and tomography compared with pulse measurements, this method is more simple and reliable in terms of data interpretation and noise immunity. These happen because amplitude modulation is measured at low peak powers, slow rise time, and hence smaller bandwidths than the pulse measurements need. The current measuring schemes are based on heterodyning of optical and transformed signals.[35]

The development of the theory underlying this method resulted in the discovery of a new type of waves: photon-density waves or progressively decaying waves of intensity.[1-7,34-37] Microscopically, individual photons make random migrations in a scattering medium, but collectively they form a photon-density wave at a modulation frequency ω that moves away from a radiation source (see Figure 3.3b). Photon-density waves possess typical wave properties: for example, they undergo refraction, diffraction, interference, dispersion, and attenuation.

In strongly scattering media with weak absorption far from the walls and a source or a receiver of radiation, the light distribution may be regarded as a decaying diffusion process described by the time-dependent diffusion equation for photon density. For a point light source with harmonic intensity modulation at frequency $\omega = 2\pi\nu$ placed at the point $\bar{r} = 0$, an alternating component (AC) of intensity is a going-away spherical wave with its center at the point $\bar{r} = 0$, which oscillates at a modulation frequency with modulation depth[36]

$$m_U(\bar{r},\omega) = m_I \exp\left(\bar{r}\sqrt{\frac{c\mu_a}{D}}\right)\exp\left(-\bar{r}\sqrt{\frac{\omega}{2D}}\right),\tag{3.12}$$

and undergoes a phase shift relative to the phase value at point $\bar{r} = 0$ equal to

$$\Delta\Phi(\bar{r},\omega) = \bar{r}\left(\frac{\omega}{2D}\right)^{0.5},\tag{3.13}$$

where
m_I is the intensity modulation depth of the incident light
$D = c/3(\mu'_s + \mu_a)$

The length of a photon-density wave, Λ_Φ, and its phase velocity, V_Φ, are defined by

$$\Lambda_\Phi^2 = \frac{8\pi^2 D}{\omega} \quad \text{and} \quad V_\Phi^2 = 2D\omega.\tag{3.14}$$

Measurement of $m_U(\bar{r},\omega)$ and $\Delta\Phi(\bar{r},\omega)$ allows one to separately determine the transport scattering coefficient μ'_s and the absorption coefficient μ_a and evaluate the spatial distribution of these parameters.

Keeping medical applications in mind, we can easily estimate that, for $\omega/2\pi = 500$ MHz, $\mu'_s = 15$ cm^{-1}, $\mu_a = 0.035$ cm^{-1}, and $c = (3 \times 10^{10}/1.33)$ cm/s, the wavelength is $\Lambda_\Phi \cong 5.0$ cm and the phase velocity is $V_\Phi \cong 1.77 \times 10^9$ cm/s.

3.3 Polarized Light Interaction

3.3.1 Tissue Structure and Anisotropy

The randomness of tissue structure results in fast depolarization of light propagating in tissues. However, in certain tissues (transparent tissues, such as eye tissues, cellular monolayers, mucous membrane, and superficial skin layers), the degree of polarization of transmitted or reflected light remains measurable even when the tissue is of considerable thickness. From the registered depolarization degree of initially polarized light, the transformed state of polarization, or the appearance of a polarized component in the scattered light, the information about the structure of tissues and cell ensembles can be extracted.[3,5,7,25,30,38–58] As regards practical implications, polarization techniques are believed to give rise to simplified schemes of optical medical tomography compared with time-resolved methods and also provide additional information about the structure of tissues.

Many biological tissues are optically anisotropic.[5,6,47] Tissue birefringence results primarily from the linear anisotropy of fibrous structures, which forms the extracellular media. The refractive index of a medium is higher along the length of a fiber than along its cross section. A specific tissue structure is a system composed of parallel cylinders that create a uniaxial birefringent medium with the optic axis parallel to the cylinder axes. This is called birefringence of form (Figure 3.4). A large variety of tissues such as eye cornea, tendon, cartilage, eye sclera, dura mater, muscle, artery wall, nerve, retina, bone, teeth, and myelin exhibit form birefringence. All of these tissues contain uniaxial and/or biaxial birefringent structures. For example, myocardium contains fibers oriented along two different axes. Myocardium consists mostly of cardiac muscle fibers arranged in sheets that wind around the ventricles and atria. Myocardium is typically birefringent as the refractive index along the axis of the muscle fiber is different from that in the transverse direction.

Form birefringence arises when the relative optical phase between the orthogonal polarization components is nonzero for forward-scattered light. After multiple forward-scattering events, a relative phase difference accumulates and a delay (δ_{oe}) similar to that observed in birefringent materials is introduced between orthogonal polarization components. For organized linear structures, an increase in phase delay is characterized by a difference (Δn_{oe}) in the effective refractive index for light polarized along, and perpendicular to, the long axis of the linear structures. The effect of tissue birefringence on the propagation of linearly polarized light is dependent on the angle between the incident polarization orientation and the tissue axis. Phase retardation δ_{oe} between orthogonal polarization components is proportional to the distance d traveled through the birefringent medium[59]

$$\delta_{oe} = \frac{2\pi d \Delta n_{oe}}{\lambda_0}. \tag{3.15}$$

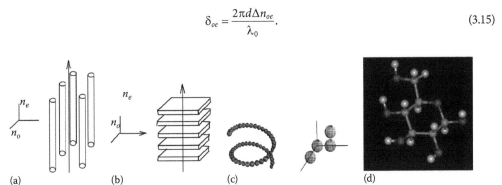

(a)   (b)   (c)   (d)

FIGURE 3.4 Examples of anisotropic models of tissues and tissue components[5]: (a) system of long dielectric cylinders; (b) system of dielectric plates; (c) chiral aggregates of particles; (d) glucose (chiral molecule) as a tissue component. (a–c: From Tuchin V.V., *Tissue Optics: Light Scattering Methods and Instruments for Medical Diagnosis*, SPIE Tutorial Texts in Optical Engineering, Vol. TT38, SPIE Press, Bellingham, WA, 2000; 2nd edn., Vol. PM166, SPIE Press, Bellingham, WA, 2007. d: Courtesy of Alex Vitkin.)

A medium of parallel cylinders is a positive uniaxial birefringent medium [$\Delta n_{oe} = (n_e - n_o) > 0$] with its optic axis parallel to the cylinder axes (Figure 3.4a). Therefore, a case defined by an incident electrical field directed parallel to the cylinder axes will be called *extraordinary*, and a case with the incident electrical field perpendicular to the cylinder axes will be called *ordinary*. The difference $(n_e - n_o)$ between the extraordinary index and the ordinary index is a measure of the birefringence of a medium comprised of cylinders. For the Rayleigh limit (when the wavelength is much larger than cylinder diameter), the form birefringence becomes[59]

$$\Delta n_{oe} = (n_e - n_o) = \frac{f_1 f_2 (n_1 - n_2)^2}{f_1 n_1 + f_2 n_2},$$ (3.16)

where
  f_1 is the volume fraction of the cylinders
  f_2 is the volume fraction of the ground substance
  n_1, n_2 are the corresponding indices

For a given index difference, maximal birefringence is expected for approximately equal volume fractions of thin cylinders and ground material. For systems with large-diameter cylinders (when the wavelength is much smaller than cylinder diameter), the birefringence goes to zero.

For tissues that could be modeled as a system of thin dielectric plates (Figure 3.4b), the form birefringence is described by the expression[5]

$$n_e^2 - n_o^2 = -\frac{f_1 f_2 (n_1 - n_2)}{f_1 n_1^2 + f_2 n_2^2},$$ (3.17)

where
  f_1 is the volume fraction occupied by the plates
  f_2 is the volume fraction of the ground substance
  n_1, n_2 are the corresponding indices

This implies that the system behaves like a negative uniaxial crystal with its optical axis aligned normally with the plate surface.

Linear dichroism (diattenuation), that is, different wave attenuation for two orthogonal polarizations, in systems formed by long cylinders or plates is defined by the difference between the imaginary parts of the effective indices of refraction. Depending on the relationship between the sizes and the optical constants of the cylinders or plates, this difference can take both positive and negative values.[60]

Reported birefringence values for tendon, muscle, coronary artery, myocardium, sclera, cartilage, and skin are on the order of 10^{-3}. A new technique—polarization-sensitive optical coherence tomography (OCT)—allows for the measurement of linear birefringence in turbid tissue with high precision. The following data have been reported using this technique: for rodent muscle, 1.4×10^{-3}; for normal porcine tendon, $(4.2 \pm 0.3) \times 10^{-3}$ and for thermally treated tendon (90°C, 20 s), $(2.24 \pm 0.07) \times 10^{-3}$; for porcine skin, (1.5×10^{-3})–(3.5×10^{-3}); for bovine cartilage, 3.0×10^{-3}; and for bovine tendon, $(3.7 \pm 0.4) \times 10^{-3}$.[54] Such birefringence provides 90% phase retardation at a depth on the order of several hundred micrometers.

The magnitude of birefringence and diattenuation are related to the density and other properties of the collagen fibers, whereas the orientation of the fast axis indicates the orientation of the

 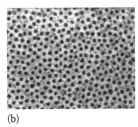 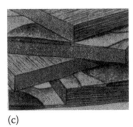 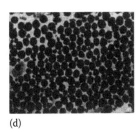

(a)   (b)   (c)   (d)

FIGURE 3.5 Electron micrographs of (a and b) the human cornea (×32,000) and (d) the sclera (×18,000)[62]: collagen fibrils have a uniform diameter and are arranged in the same direction within the lamellae; *K* is the keratocyte, (b) presents a magnified image of the middle lamella of image (a); (d) sclera collagen fibrils display various diameters, although they are quasi-ordered; (c) presents the model of lamellar-fibrillar structure of the corneal stroma. (From Hart, R.W. and Farrell, R.A., *J. Opt. Soc. Am.*, 59, 766, 1969.)

collagen fibers. The amplitude and orientation of birefringence in the skin and cartilage are not as uniformly distributed as in tendon. In other words, the densities of collagen fibers in skin and cartilage are not as uniform as in tendon, and the orientation of collagen fibers is not distributed in as orderly a fashion.

In addition to linear birefringence and dichroism (diattenuation), many tissue components show optical activity. In complex tissue structures, chiral aggregates of particles, in particular spherical particles, may be responsible for tissue optical activity (Figure 3.4c). The molecule's chirality, which stems from its asymmetric molecular structure, also results in a number of characteristic effects generically called optical activity. A well-known manifestation of optical activity is the ability to rotate the plane of linearly polarized light around the axis of propagation. The amount of rotation depends on the chiral molecular concentration, the path length through the medium, and the light wavelength. Tissues containing chiral components display optical activity (Figure 3.4d). Interest in chiral turbid media is driven by the attractive possibility of noninvasive in situ optical monitoring of glucose in diabetic patients.[61] Within turbid tissues, however, where the scattering effects dominate, the loss of polarization information is significant and the chiral effects due to the small amount of dissolved glucose are difficult to detect (see Chapter 17 of Tuchin[61]).

More sophisticated anisotropic tissue models can also be constructed. For example, the eye cornea can be represented as a system of plane anisotropic layers (plates, i.e., lamellas), each of which is composed of densely packed long cylinders (fibrils) (Figure 3.5c) with their optical axes oriented along a spiral. This fibrillar-lamellar structure of the cornea is responsible for the linear and circular dichroism and its dependence on the angle between the lamellas.

3.3.2 Polarized Light Description

Polarization refers to the pattern described by the electric field vector as a function of time at a fixed point in space. When the electric field vector oscillates in a single, fixed plane all along the beam, the light is said to be linearly polarized. This linearly polarized wave can be resolved into components parallel $E_{\parallel}$ and perpendicular $E_{\perp}$ to the scattering plane. If the plane of the electric field rotates, the light is said to be elliptically polarized, because the electric field vector traces out an ellipse at a fixed point in space as a function of time. If the ellipse happens to be a circle, the light is said to be circularly polarized. The connection between phase and polarization can be understood as follows: circularly polarized light consists of equal amounts of linear mutually orthogonal polarized components where they oscillate exactly 90° out of phase. In general, light of arbitrary elliptical polarization consists of unequal amplitudes of linearly polarized components and the electric fields for the two polarizations oscillate at the same frequency but have some constant phase difference.

Light of arbitrary polarization can be represented by four numbers known as the Stokes parameters, I, Q, U, and V: I refers to the intensity of the light; the parameters Q, U, and V represent the extent of horizontal linear polarization, 45° linear polarization, and circular polarization, respectively.[30] In terms of electric field components, the Stokes parameters are given by

$$
\begin{aligned}
I &= E_\parallel E_\parallel^* + E_\perp E_\perp^*, \\
Q &= E_\parallel E_\parallel^* - E_\perp E_\perp^*, \\
U &= E_\parallel E_\perp^* + E_\perp E_\parallel^*, \\
V &= E_\parallel E_\perp^* - E_\perp E_\parallel^*,
\end{aligned}
\tag{3.18}
$$

and the irradiance or intensity of light by

$$
I^2 \geq Q^2 + U^2 + V^2.
\tag{3.19}
$$

For an elementary monochromatic plane or spherical electromagnetic wave, Equation 3.19 is the equality.[45] For a partially polarized quasi-monochromatic light, which can be presented as a mixture of elementary waves, the Stokes parameters are sums of the respective Stokes parameters of these elementary waves, because of the fundamental property of additivity. In this case, Equation 3.19 is the inequality.[30,45]

The polarization state of the scattered light in the far zone is described by the Stokes vector connected with the Stokes vector of the incident light

$$
\mathbf{S}_s = \mathbf{M} \cdot \mathbf{S}_i,
\tag{3.20}
$$

where

$\mathbf{M}$ is the normalized 4×4 scattering matrix (intensity or Mueller's matrix)

$$
\mathbf{M} =
\begin{bmatrix}
M_{11} & M_{12} & M_{13} & M_{14} \\
M_{21} & M_{22} & M_{23} & M_{24} \\
M_{31} & M_{32} & M_{33} & M_{34} \\
M_{41} & M_{42} & M_{43} & M_{44}
\end{bmatrix},
\tag{3.21}
$$

$\mathbf{S}_i$ is the Stokes vector of incident light

In polarimetry, the Stokes vector $\mathbf{S}$ of a light beam is constructed based on six flux measurements obtained with different polarization analyzers in front of the detector:

$$
\mathbf{S} =
\begin{pmatrix}
I \\
Q \\
U \\
V
\end{pmatrix}
=
\begin{pmatrix}
I_H + I_V \\
I_H - I_V \\
I_{+45°} - I_{-45°} \\
I_R - I_L
\end{pmatrix},
\tag{3.22}
$$

where I_H, I_V, $I_{+45°}$, $I_{-45°}$, I_R, and I_L are the light intensities measured with a horizontal linear polarizer, a vertical linear polarizer, a +45° linear polarizer, a −45° linear polarizer, a right circular analyzer, and a left circular analyzer in front of the detector, respectively. Because of the relationship $I_H + I_V = I_{+45°} + I_{-45°} = I_R + I_L = I$, where I is the intensity of the light beam measured without any analyzer in front of the detector, a Stokes vector can be determined by four independent measurements.

The degree of linear polarization of scattered light is defined as

$$P_L = \frac{\left(I_\| - I_\perp\right)}{\left(I_\| + I_\perp\right)} = \frac{\left[Q_s^2 + U_s^2\right]^{1/2}}{I_s}, \tag{3.23}$$

and that of circular polarization as

$$P_C = \frac{V_s}{I_s}. \tag{3.24}$$

Elements of the light scattering matrix (LSM) depend on the scattering angle θ, the wavelength, and the geometrical and optical parameters of the scatterers. There are only 7 independent elements (of 16) in the scattering matrix of a single particle with fixed orientation and 9 relations that connect the others together. For scattering by a collection of randomly oriented scatterers, there are 10 independent parameters.

M_{11} is what is measured when the incident light is unpolarized, the scattering angle dependence of which is the phase function of the scattered light. It provides only a fraction of the information theoretically available from scattering experiments. M_{11} is much less sensitive to chirality and long-range structure than some of the other matrix elements.[30] M_{12} refers to a degree of linear polarization of the scattered light, M_{22} displays the ratio of depolarized light to the total scattered light (a good measure of scatterer nonsphericity), and M_{34} displays the transformation of 45° obliquely polarized incident light to circularly polarized scattered light (uniquely characteristic for different biological systems); the difference between M_{33} and M_{44} is a good measure of scatterer nonsphericity.

If a particle is small with respect to the wavelength of the incident light, its scattering can be described as if it were a single dipole, and the so-called Rayleigh theory is applicable under the condition that $m(2\pi a/\lambda) = 1$, where m is the relative refractive index of the scatterers, $(2\pi a/\lambda)$ is the size parameter, a is the radius of the particle, and λ is the wavelength of the incident light in a medium.[30] For NIR light and typical biological scatterers with a refractive index referring to the ground matter $m = 1.05$–1.1, the maximum particle radius is about 12–14 nm for the Rayleigh theory to remain valid. For this theory, the scattered irradiance is inversely proportional to λ^4 and increases as a^6, and the angular distribution of the scattered light is isotropic.

The Rayleigh–Gans or Rayleigh–Debye theory addresses the problem of calculating the scattering by a special class of arbitrarily shaped particles; it requires $|m - 1| \ll 1$ and $(2\pi a'/\lambda)\,|m - 1| \ll 1$, where a' is the largest dimension of the particle.[38] These conditions mean that the electric field inside the particle must be close to that of the incident field and the particle can be viewed as a collection of independent dipoles that are all exposed to the same incident field. A biological cell might be modeled as a sphere of cytoplasm with a higher refractive index ($\bar{n} = 1.37$) relative to that of the surrounding water medium ($\bar{n} = 1.35$), then $m = 1.015$, and for NIR light this theory will be valid for particles whose dimension is up to $a' = 850$–950 nm. This approximation has been applied extensively to calculations of light scattering from suspensions of bacteria.[38] It can be applicable for describing light scattering from cell components (mitochondria, lysosomes, peroxisomes, etc.) in tissues due to their small dimensions and refraction.

For describing forward direction scattering caused by large particles (on the order of 10 μm), the Fraunhofer diffraction approximation is useful.[38] According to this theory, the scattered light has the same polarization as that of the incident light and the scatter pattern is independent of the refractive index of the object. For small scattering angles, the Fraunhofer diffraction approximation can represent accurately the change in irradiance as a function of particle size. That is why this approach is applicable in the laser flow cytometry. The structure of the biological cell such as the cell membrane, its nuclear texture, and the granules in the cytoplasm can be represented by variations in optical density.

The Mie or Lorenz–Mie scattering theory is an exact solution of Maxwell's electromagnetic field equations for a homogeneous sphere.[30,38,44–46,60] In the general case, light scattered by a particle becomes elliptically polarized. For spherically symmetric particles of an optically inactive material, the Mueller scattering matrix is given by[30]

$$\mathbf{M}(\theta) = \begin{bmatrix} M_{11}(\theta) & M_{12}(\theta) & 0 & 0 \\ M_{12}(\theta) & M_{22}(\theta) & 0 & 0 \\ 0 & 0 & M_{33}(\theta) & M_{34}(\theta) \\ 0 & 0 & -M_{34}(\theta) & M_{44}(\theta) \end{bmatrix}. \tag{3.25}$$

The Mie theory has been extended to arbitrarily coated spheres and to arbitrary cylinders.[38,44–46,60] In the Mie theory, the electromagnetic fields of the incident, internal, and scattered waves are each expanded in a series. A linear transformation can be effected between the fields in each of the regions. This approach can also be used for nonspherical objects such as spheroids.[44–46] The linear transformation is called the transition matrix (T-matrix). The T-matrix for spherical particles is diagonal.

3.3.3 Single Scattering and Quasi-Ordered Tissues

Healthy tissues of the anterior human eye chamber (e.g., the cornea and lens) are highly transparent to visible light because of their ordered structure and the absence of strongly absorbing chromophors.[5,47,63] Scattering is an important feature of light propagation in eye tissues. The size of the scatterers and the distance between them are smaller than or comparable to the wavelength of visible light, and the relative refractive index of the scatterers is equally small (soft particles). Typical eye tissue models are long, round dielectric cylinders (corneal and scleral collagen fibers) (Figure 3.5) or spherical particles (lens protein structures) with a refractive index n_s; they are randomly/quasi-orderly (sclera, opaque lens) or regularly (transparent cornea and lens) distributed in the isotropic base matter with a refractive index $n_0 < n_s$. Light scattering analysis in eye tissue is often possible using a single scattering model owing to the small scattering cross section.

The corneal stroma is composed of several hundred successively stacked layers of lamellae (Figure 3.5c), which vary in width (0.5–250 μm) and thickness (0.2–0.5 μm), depending on the tissue region (three sequential lamellae are shown in Figure 3.5a). A few flat cells (keratocytes) are dispersed between the lamellae, and these occupy only 0.03–0.05 of the stromal volume. Each lamella is composed of a parallel array of collagen fibrils.

The fibrils in the human cornea have a uniform diameter of about 30.8 ± 0.8 nm with a periodicity close to two diameters, 55.3 ± 4.0 nm, and rather high regularity in the organization of fibril axes around one another (Figures 3.5a, b and 3.6a). The intermolecular spacing is 1.63 ± 0.10 nm.[64] Thus, the stroma has at least three levels of structural organization: the lamellae that lie parallel to the cornea's surface; the fibrillar structure within each lamella that consists of small, parallel collagen fibrils with uniform diameters that have some degree of order in their spatial positions; and the collagen molecular ultrastructure. In the scleral stroma, the collagen fibrils exhibit a wide range of diameters, from 25 to 230 nm, and the mean distance between fibril centers is about 285 nm (Figures 3.5d and 3.6b). Collagen intermolecular spacing is similar to that in the cornea; in bovine sclera, particularly, it is equal to 1.61 ± 0.02 nm. These fibrils are arranged in individual bundles in a parallel fashion, but more randomly than in the cornea; moreover, within each bundle, the groups of fibers are separated from each other by large empty lacunae randomly distributed in space.

Although both tissues are composed of similar molecular components, they have different micro-structures and thus very different physiological functions. The cornea is transparent, allowing for more than 90% of the incident light to be transmitted. The collagen fibrils in the cornea have a much more uniform size and spacing than those in the sclera, resulting in a greater degree of spatial order in the organization of the fibrils in the cornea compared with the sclera. The sclera of the eye is opaque to light; it scatters almost all wavelengths of visible light and thus appears white.

Light propagation in a densely packed disperse system can be analyzed using the radial distribution function $g(r)$, which statistically describes the spatial arrangement of particles in the system (Figure 3.6). The function $g(r)$ is the ratio of the local number density of the fibril centers at a distance r from a refer-ence fibril at $r = 0$ to the bulk number density of fibril centers.[63] It expresses the relative probability of finding two fibril centers separated by a distance r; thus, $g(r)$ must vanish for values of $r \leq 2a$ (where a is the radius of a fibril; fibrils cannot approach each other closer than touching). The radial distribution function of scattering centers $g(r)$ for a certain tissue may be calculated on the basis of tissue electron micrographs (see Figure 3.6).

The radial distribution function $g(r)$ was first found for the rabbit cornea by Hart and Farrell.[63] Figure 3.6a depicts a typical result for one of the cornea regions. The function $g(r) = 0$ for $r \leq 25$ nm, which is consistent with a fibril radius of 14 ± 2 nm, can be calculated from the electron micrograph. The first peak in the distribution gives the most probable separation distance, which is approximately 50 nm. The value of $g(r)$ is essentially unity for $r \geq 170$ nm, indicating that the fibril positions are correlated over no more than a few of their nearest neighbors. Therefore, a short-range order exists in the system.

Similar calculations for several regions of the human eye sclera are illustrated in Figure 3.6b. Electron micrographs from Hart and Farrell,[63] averaged for 100 fibril centers, were processed. The obtained

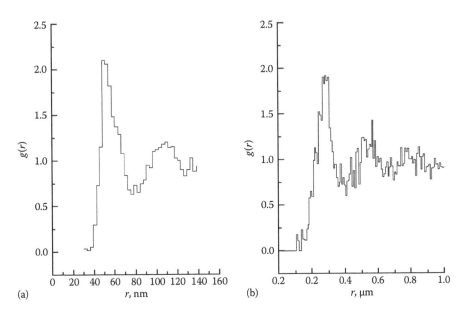

FIGURE 3.6 Experimental radial distribution function $g(r)$ measured for (a) rabbit cornea[63] and (b) human sclera[5]. $g(r)$ is proportional to the probability of particle displacement r at a certain distance from an arbitrarily fixed particle.

results present evidence of the presence of a short-range order in the sclera, although the degree of order is less pronounced than in the cornea. The function $g(r) = 0$ for $r \leq 100$ nm is consistent with the mean fibril diameter of ≈ 100 nm derived from the electron micrograph. The first peak in the distribution gives the most probable separation distance, which is approximately 285 nm. The value of $g(r)$ is essentially unity for $r \geq 750$ nm, indicating a short-range order in the system.

For an isotropic system of N identical interacting long cylinders, the scattered intensity is defined as

$$\langle I \rangle = |E_0|^2 N S_2(\theta), \tag{3.26}$$

where

    E_0 is the scattering amplitude of an isolated particle

$$S_2(\theta) = \left\{ 1 + 8\pi a^2 \rho \int_0^R \left[g(r) - 1 \right] J_0 \left(\frac{2\pi a}{\lambda} r \sin \frac{\theta}{2} \right) dr \right\} \text{ is the structure factor} \tag{3.27}$$

    a is the radius of the cylinder face
    ρ is the mean density of the cylinder faces
    J_0 is the zero-order Bessel function
    R is the distance for that $g(r) \to 1$
    θ is the scattering angle

For an isotropic system of identical spherical particles[65]

$$\langle I \rangle = |E_0|^2 N S_3(\theta), \tag{3.28}$$

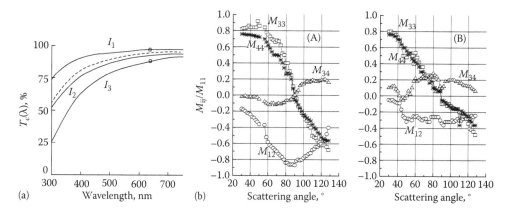

FIGURE 3.7 Single-scattering vector theory with spatial correlation of scatterers; low-scattering tissues: (a) human cornea and (b) eye lens. (a) Calculated and experimental (circles) collimated transmittance of cornea for two orthogonal polarizations of the incident light: one parallel to peripheral collagen fibers and another in perpendicular direction; (b) Human eye lens aging and cataract diagnostics: experimental angular dependence for LSM elements of (A) normal lens 5 h after the death of a 56-year-old subject and (B) cataractous lens 5 h after the death of an 88-year-old subject. (From Tuchin, V.V., *Tissue Optics: Light Scattering Methods and Instruments for Medical Diagnosis*, SPIE Tutorial Texts in Optical Engineering, Vol. TT38, SPIE Press, Bellingham, WA, 2000; 2nd edn., Vol. PM166, SPIE Press, Bellingham, WA, 2007; Calculated and measured by Maksimova, I.L. and Shubochkin, L.P., *Opt. Spectrosc.*, 70(6), 745, 1991.)

$$S_3(\theta) = \left\{ 1 + 4\pi\rho \int_0^R r^2 \left[g(r) - 1 \right] \frac{\sin qr}{qr} dr \right\}, \tag{3.29}$$

where
$q = (4\pi/\lambda)\sin(\theta/2)$
ρ is the mean density of particles
R is the distance for which $g(r) \to 1$

Quantity $S_3(\theta)$ is the 3D structure factor. This factor describes the alteration of the angle dependence of the scattered intensity that appears with a higher particle concentration.

This model can be used to describe polarized light interaction with anisotropic media showing a single scattering. The interference between elementary scattered fields, which is accounted for by structure functions $S_2(\theta)$ and $S_3(\theta)$, transforms the scattering angular dependence of Mueller matrix elements.[5,47] Figure 3.7 illustrates the application of a single-scattering vector theory with the spatial correlation of scatterers for describing polarized light transmittance by the human cornea and eye lens.

3.3.4 Vector Radiative Transfer Theory

Polarization effects of light propagation through multiply scattering media, including tissues, are fully described by the vector radiative transfer equation (VRTE). For macroscopically isotropic and symmetric plane-parallel scattering media, the VRTE can be substantially simplified as follows[5,46,47]:

$$\frac{d\mathbf{S}(\bar{r},\vartheta,\varphi)}{d\tau(\bar{r})} = -\mathbf{S}(\bar{r},\vartheta,\varphi) + \frac{\Lambda(\bar{r})}{4\pi} \int_{-1}^{+1} d(\cos\vartheta') \int_0^{2\pi} d\varphi' \bar{\mathbf{Z}}(\bar{r},\vartheta,\vartheta',\varphi-\varphi')\mathbf{S}(\bar{r},\vartheta',\varphi'), \tag{3.30}$$

where
S is the Stokes vector
$\bar{r}$ is the position vector
ϑ, φ are the angles characterizing incident direction, that is, the polar (zenith) and the azimuth angles, respectively
$d\tau(\bar{r}) = \rho(\bar{r})\langle\sigma_{ext}(\bar{r})\rangle ds$ is the optical path length element
ρ is the local particle number density
$\langle\sigma_{ext}\rangle$ is the local ensemble-averaged extinction coefficient
ds is the path length element measured along the unit vector of the direction of light propagation
Λ is the single scattering albedo
ϑ', φ' are the angles characterizing the scattering direction, that is, the polar (zenith) and the azimuth angles, respectively
$\bar{\mathbf{Z}}$ is the normalized phase matrix $\bar{\mathbf{Z}}(\bar{r},\vartheta,\vartheta',\varphi-\varphi') = \mathbf{R}(\Phi)\mathbf{M}(\theta)\mathbf{R}(\Psi)$
where
Φ and Ψ are expressed via angles ϑ, φ, ϑ', φ' characterizing the incident and scattering directions, respectively
$\mathbf{M}(\theta)$ is the single scattering Mueller matrix
θ is the scattering angle
$\mathbf{R}(\phi)$ is the Stokes rotation matrix for angle ϕ:

$$\mathbf{R}(\phi) = \begin{bmatrix} 1 & 0 & 0 & 0 \\ 0 & \cos 2\phi & -\sin 2\phi & 0 \\ 0 & \sin 2\phi & \cos 2\phi & 0 \\ 0 & 0 & 0 & 1 \end{bmatrix}. \tag{3.31}$$

Every Stokes vector and Mueller matrix are associated with a specific reference plane and coordinates. The first term on the right-hand side of VRTE (Equation 3.30) describes the change in the specific intensity vector over the distance *ds* caused by extinction and dichroism, and the second term describes the contribution of light illuminating a small volume element centered at $\bar{r}$ from all incident directions and scattered into the chosen direction. For real systems, the form of VRTE tends to be rather complex and often intractable. Therefore, a wide range of analytical and numerical techniques have been developed to solve the VRTE. Because the important property of the normalized phase matrix (Equation 3.31) is dependent on the difference of the azimuthal angles of the scattering and incident directions rather than on their specific values,[46] an efficient analytical treatment of the azimuthal dependence of the multiply scattered light, using a Fourier decomposition of the VRTE, is possible. The following techniques and their combinations can be used to solve VRTE: transfer matrix method, the singular eigenfunction method, the perturbation method, the small-angle approximation, the adding-doubling method, the matrix operator method, the invariant embedding method, and the Monte Carlo method.[3,5,7,25,40–56]

When the medium is illuminated by unpolarized light and/or only the intensity of multiply scattered light needs to be computed, the VRTE can be replaced by its approximate scalar counterpart. In that case, in Equation 3.30, the Stokes vector is replaced by its first element (i.e., radiance) (see Equation 3.22) and the normalized phase matrix by its (1,1) element (i.e., the phase function, $p(\bar{s}, \bar{s}')$) (see Equation 3.3).

The results of Monte Carlo simulations for polarized light propagation within the multiple scattering media with parameters close to those of tissues are shown in Figures 3.8 and 3.9.[5,7,47,51] These calculations clearly demonstrate that polarization properties of tissues could be dramatically transformed for the multiple scattering conditions.

For a system of small spatially uncorrelated particles, the degree of linear ($i = L$) and circular ($i = C$) polarization in the far region of the initially polarized (linearly or circularly) light transmitted through a layer of thickness *d* is defined by the relation[40]

$$P_i \cong \frac{2d}{l_s} \sin h\left(\frac{l_s}{\xi_i}\right) \cdot \exp\left(\frac{-d}{\xi_i}\right), \tag{3.32}$$

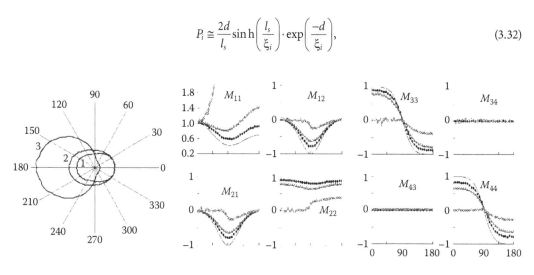

FIGURE 3.8   MC simulation: LSM elements (small particles): $\lambda = 633$ nm, $a = 50$ nm, $m = 1.2$, $f = 0.01$; single scattering (—); multiple scattering: diameter of the system—1 mm (-·-), 2 mm (-Δ-), and 20 mm (-o-); M_{12} refers to a degree of linear polarization of the SL; M_{22} displays the ratio of depolarized light to the total SL (particle nonsphericity); M_{33} and M_{44} difference is a good measure of particle nonsphericity. (From Maksimova, I.L. et al., *Opt. Spectrosc.*, 92(6), 915, 2002.)

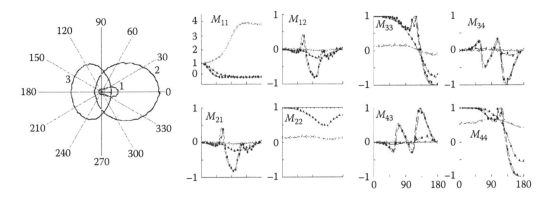

FIGURE 3.9 MC simulation: LSM elements (big particles): $\lambda = 633$ nm, $a = 300$ nm, $m = 1.2$, $f = 0.01$; single scattering (—); multiple scattering: diameter of the system—0.002 mm (–·–), 0.2 mm (–Δ–), and 2 mm (–o–); M_{12} refers to a degree of linear polarization of the SL; M_{22} displays the ratio of depolarized light to the total SL (particle nonsphericity); M_{34} displays the transformation of 45° obliquely polarized incident light to circularly polarized SL; M_{33} and M_{44} difference is a good measure of particle nonsphericity. (From Maksimova, I.L. et al., *Opt. Spectrosc.*, 92(6), 915, 2002.)

where

$l_s = 1/\mu_s$ is the scattering length

$\xi_i = (\zeta_i \cdot l_s/3)^{0.5}$ is the characteristic depolarization length for a layer of scatterers

$d > \xi_i$

$\zeta_L = l_s/[\ln (10/7)]$

$\zeta_C = l_s/(\ln 2)$

As can be seen from Equation 3.32, the characteristic depolarization length for linearly polarized light in tissues that can be represented as ensembles of Rayleigh particles is approximately 1.4 times greater than the corresponding depolarization length for circularly polarized light. One can employ Equation 3.32 to assess the depolarization of light propagating through an ensemble of large-scale spherical particles whose sizes are comparable with the wavelength of incident light (Mie scattering). For this purpose, one should replace l_s by the transport length $l_{tr} \cong 1/\mu_s'$ and take into account the dependence on the size of scatterers in ζ_L and ζ_C. With an increase in the size of scatterers, the ratio ξ_L/ζ_C changes. It decreases from ~1.4 to ~0.5 as $2\pi a/\lambda$ increases from 0 to ~4, where a is the size of scatterers and λ is the wavelength of the light in the medium, which remains virtually constant at the level of 0.5 when $2\pi a/\lambda$ grows from ~4 to 15.

The Mueller matrix for the backscattering geometry was obtained by solving a radiative transfer equation with appropriate boundary conditions.[41] Analysis of this matrix structure showed that its form coincides with the single scattering matrix for optically active spherical scatterers. Thus, different tissues or the same tissues in various pathological or functional states should display different responses to the probing with linearly and circularly polarized light. This effect can be employed both in optical medical tomography and for determining optical and spectroscopic parameters of tissues. As it follows from Equation 3.32, the depolarization length in tissues should be close to the mean transport path length l_{tr} of a photon, because this length characterizes the distance within which the direction of light propagation and, consequently, the polarization plane of linearly polarized light become totally random after many sequential scattering events.

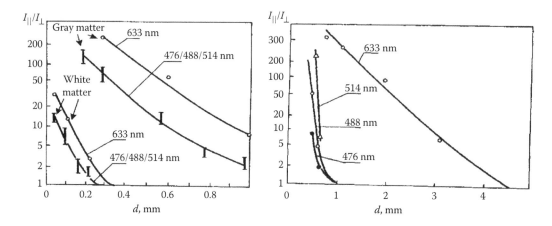

FIGURE 3.10 Tissue polarization properties. Dependence of the depolarization degree ($I_{\parallel}/I_{\perp}$) of laser radiation (He–Ne laser, λ = 633 nm and Ar laser, λ = 476/488/514 nm) on the penetration depth (d) for brain tissue (gray and white matter) and whole blood (low hematocrit).[5,39] Measurements were performed within a small solid angle (10^{-4} sr) along the axis of a laser beam 1 mm in diameter. A strong influence of fluorescence was seen for blood irradiated by Ar laser.

As the length l_{tr} is determined by the parameter g characterizing the anisotropy of scattering, the depolarization length should also substantially depend on this parameter. Correspondingly, the experimental data of Svaasand and Gomer[39] demonstrate that the depolarization length l_p of linearly polarized light, which is defined as the length within which the ratio $I_{\parallel}/I_{\perp}$ decreases down to 2, displays such a dependence. The ratio mentioned earlier varied from 300 to 1, depending on the thickness of the sample and the type of tissue (see Figure 3.10). These measurements were performed within a narrow solid angle (~10^{-4} sr) in the direction of the incident laser beam. The values of l_p considerably differed for the white matter of the brain and a tissue of the cerebral cortex: 0.23 and 1.3 mm for λ = 633 nm, respectively. Whole blood is characterized by a considerable depolarization length (about 4 mm) at λ = 633 nm, which is indicative of the dependence on the parameter g, whose value for blood exceeds the values of this parameter for tissues of many other types and can be estimated as 0.982–0.999.[5,7,10]

In contrast to depolarization, the attenuation of collimated light is determined by the total attenuation coefficient μ_t (see Equation 3.1). For many tissues, μ_t is much greater than $\mu_s' + \mu_a$. Therefore, in certain situations, it is impossible to detect pure ballistic photons (photons that do not experience scattering), but forward scattered photons retain their initial polarization and can be used for imaging purpose. It was experimentally demonstrated that laser radiation retains linear polarization on the level of $P_L \leq 0.1$ within $2.5 l_{tr}$.[25,40] Specifically, for skin irradiated in the red and NIR ranges, $\mu_a \cong 0.4$ cm^{-1}, $\mu_s' \cong 20$ cm^{-1}, and $l_{tr} \cong 0.48$ mm. Consequently, light propagating in skin can retain linear polarization within the length of about 1.2 mm. Such an optical path in a tissue corresponds to a delay time on the order of 5.3 ps, which provides an opportunity to produce polarization images of macro-inhomogeneities in a tissue with a spatial resolution equivalent to the spatial resolution that can be achieved with the selection of photons by means of more sophisticated time-resolved techniques. In addition, polarization imaging makes it possible to eliminate specular reflection from the surface of a tissue (see Figure 3.1), which enables the application of this technique for the imaging of microvessels in facile skin.[3,57,58] Polarization images show textural changes on the skin's subsurface and allows to erase melanin from such images.[25,53]

Polarization imaging is a new direction in tissue optics.[3,5,7,25,40–43,47–58] The basis for this technique is to register two-dimensional polarization patterns for the backscattering of a polarized incident narrow laser beam. The major informative images can be received using the backscattering Mueller matrix approach. To determine each of the 16 experimental matrix elements, a total of 16 images should be taken at various combinations of input and output polarization states. Spatially resolved reflectance and OCT imaging techniques combine well with the polarization method.[6,43,52,54]

3.4 Optothermal Interactions

3.4.1 Temperature Rise and Tissue Damage

When photons traveling in tissue are absorbed, heat is generated. The generated heat, described by the heat source term S at a point r, is proportional to the fluence rate of light $\phi(\mathbf{r})$ (mW/cm²) and absorption coefficient $\mu_a(r)$ at this point[3,6,20–22,66]:

$$S(r) = \mu_a(r)\ \phi(r). \tag{3.33}$$

The traditional *bioheat equation* originated from the energy balance describes the change in tissue temperature over time at a point $\mathbf{r}$ in the tissue:

$$\rho c \frac{\partial T(r,t)}{\partial t} = \nabla\left[k_m \nabla T(r,t)\right] + S(r) + \rho c w\left(T_a - T_v\right), \tag{3.34}$$

where
    ρ is the tissue density (g/cm³)
    c is the tissue-specific heat (mJ/g °C)
    $T(r,t)$ is the tissue temperature (°C) at time t
    k_m is the thermal conductivity (mW/cm °C)
    $S(r)$ is the heat source term (mW/cm³)
    w is the tissue perfusion rate (g/cm³ s)
    T_a is the inlet arterial temperature (°C)
    T_v is the outlet venous temperature (°C), all at point r in the tissue

In this equation, convection, radiation, vaporization, and metabolic heat effects are not accounted for because they are negligible in many practical cases. The source term is assumed to be stationary over the time interval of heating. The first term to the right of the equal sign describes any heat conduction (typically away from point r), and the source term accounts for heat generation due to photon absorption. In most cases of light (laser)–tissue interaction, the heat transfer caused by blood perfusion (last term) is negligible.

To solve this equation, initial and boundary conditions must be accounted for. The initial condition is the tissue temperature at $t = 0$ and the boundary conditions depend on the tissue structure and the geometry of light heating. Methods of solving the *bioheat equation* can be found in studies by Tuchin[3] and Müller and coworkers.[20–23]

Damage of a tissue results when it is exposed to high temperature for a long time period.[3,20–23,66] The damage function is expressed in terms of an Arrhenius integral:

$$\Omega(\tau) = \ln\left(\frac{C(0)}{C(\tau)}\right) = A \int_0^\tau e^{-(E_a/RT(t))} dt, \tag{3.35}$$

where

τ is the total heating time (s)
$C(0)$ is the original concentration of undamaged tissue
$C(\tau)$ is the remaining concentration of undamaged tissue after time τ
A is an empirically determined constant (s^{-1})
E_a is an empirically determined activation energy barrier (J/mol)
R is the universal gas constant (8.32 J/mol K)
T is the absolute temperature (K)

In noninvasive optical diagnostic and some photochemical applications of light, one has to keep tissue below the damaging temperature, which is also called the critical temperature T_{crit}. This temperature is defined as the temperature at which the damage accumulation rate, $d\Omega/dt$, is equal to 1.0[66]:

$$T_{crit} = \frac{E_a}{R\ln(A)}. \tag{3.36}$$

The constants A and E_a can be calculated on the basis of experimental data when tissue is exposed to a constant temperature.[21] For example, for pig skin, $A = 3.1 \times 10^{98}$ and $E_a = 6.28 \times 10^5$ J/mol, which gives $T_{crit} = 59.7°C$.

With CW light sources, the conduction of heat away from the light absorption region into the surrounding tissue increases due to the increase in temperature difference between the irradiation and the surrounding tissue. Depending on the light energy, large volumes of tissue may be damaged or may lose heat at the target tissue component. For pulsed light, little heat is usually lost during the pulse duration as light absorption is a fast process while heat conduction is relatively slow, and, therefore, more precise tissue damage is possible.

The following forms of irreversible tissue damage are expected as tissue temperature rises past T_{crit}: *coagulation* (denaturization of cellular and tissue proteins) is the basis for *tissue welding*; *vaporization* (tissue dehydration and vapor bubbles formation [*vacuolization*], $T \geq 100°C$) is the basis for mechanical destruction of tissue; and *pyrolysis* (at temperatures $T \approx 350°C–450°C$) is the basis for chemical changes in tissue due to heat. The combination of *vaporization*, *vacuolization*, and *pyrolysis* produces *thermal ablation*—the basis of laser surgical removal of tissue.

The disadvantage of thermal ablation with CW light sources is undesirable damage to surrounding tissue because of coagulation. Pulsed light can deliver sufficient energy with each pulse to ablate tissue, but the tissue should be removed within a short time period before any heat is transferred to the surrounding tissue. To achieve precise tissue cutting, lasers with a very short penetration depth and sharp focus, such as UV excimer ArF laser (193 nm), are used.

As condensed matter, tissue can undergo any noncoherent or coherent effects with laser irradiation.[67] Linear noncoherent effects exist within a wide area of pulse duration and intensities; for long pulses of 1 s, the intensity should not exceed 10 W/cm², whereas for shorter pulses of 10^{-9} s, the intensity can be up to 10^9 W/cm². Multiphoton processes may exist in relatively short pulse duration $(10^{-9}–10^{-12}$ s) at intensities of $10^9–10^{12}$ W/cm² and rather low light energies, not higher 0.1 J/cm². The linear and nonlinear coherent effects may be induced only by a very short pulse with a time duration comparable with the relaxation time of biological molecules, $\tau \leq 10^{-13}$.

3.4.2 Optothermal and Optoacoustic Effects

The time-dependent heat generated in a tissue due to interaction with pulsed or intensity-modu-lated optical radiation is known as the optothermal (OT) effect.[5,6,19,68] Such interaction also induces a number of thermoelastic effects in a tissue and, in particular, causes the generation of acoustic waves (AWs). Detection of AWs is the basis for the optoacoustic (OA) or photoacoustic (PA) method. The informative features of this method allow one to estimate the thermal, optical, and acoustic properties of a tissue, which depend on the peculiarities of tissue structure. Two main modes can be used for the excitation of a tissue's thermal response: (1) a pulse of light excites the sample and the signal is detected in the time domain with a fast detector attached to a wide-band amplifier (signal averaging and gating techniques are used to increase the signal-to-noise ratio); (2) an intensity-modulated light source, a low-frequency transducer, and phase-sensitive detection for noise sup-pression are provided.

In every case, the thermal waves generated by the heat release result in several effects that have given rise to various techniques: OA or PA; optothermal radiometry (OTR) or photothermal radiometry (PTR); photorefractive techniques; etc.[5,19,68] (see Figure 3.11).

The term OA refers primarily to the time-resolved technique utilizing pulsed lasers and measur-ing profiles of pressure in tissue, and the term PA refers primarily to spectroscopic experiments with CW-modulated light and a PA cell.

When a laser beam falls onto the sample surface and the wavelength is tuned to an absorption line of the tissue component of interest, the optical energy is absorbed by the target component and most of the energy transforms into heat. The time-dependent heating leads to all the thermal and thermoelastic effects mentioned earlier. In OA or PA techniques, a microphone or piezoelectric transducer, which is in acoustic contact with the sample, is used as a detector to measure the ampli-tude or phase of the resultant AW. In the OTR technique, distant IR detectors and array cameras are employed for estimating the sample surface temperature and its imaging. The intensity of the signals obtained with any of the OT or OA techniques depends on the amount of energy absorbed and transformed into heat and on the thermoelastic properties of the sample and its surroundings. When nonradiative relaxation is the main process in a light beam decay, and extinction is not very

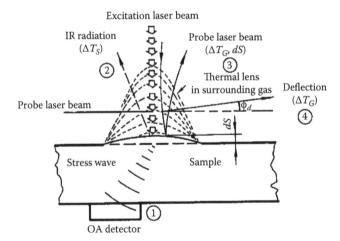

FIGURE 3.11 Schematic representation of some optothermal techniques applied to tissue study[5,68]: ΔT_S is the temperature change of a sample; ΔT_G is the temperature change of a surrounding gas; dS is the thermoelastic defor-mation; ϕ_d is the deflection angle of a probe laser beam; 1—OA technique; 2—OTR technique; 3—thermal lens technique; 4—deflection technique.

high, $\mu_a d \ll 1$ (d is the length of a cylinder within the sample occupied by a pulse laser beam), the absorbed pulse energy induces the local temperature rise defined by

$$\Delta T \cong \frac{E\mu_a d}{c_P V \rho}, \qquad (3.37)$$

where

c_P is the specific heat capacity for a constant pressure
$V = \pi w^2 d$ is the illuminated volume
w is the laser beam radius
ρ is the medium density

Suppose an adiabatic expanding of an illuminated volume over heat with a constant pressure; one can then calculate the change of the volume ΔV. This expansion induces a wave propagating in the radial direction with the sound speed v_a. The corresponding change of pressure Δp is proportional to the amplitude of mechanical oscillations

$$\Delta p \sim \left(\frac{f_a}{w}\right)\left(\frac{\beta v_a}{c_P}\right) E\mu_a, \qquad (3.38)$$

where

β is the coefficient of volumetric expansion
f_a is the frequency of the AW

Equations 3.37 and 3.38 present principles of various OT and OA techniques. The information about the absorption coefficient μ_a at the definite wavelength can be received from direct measurements of temperature change ΔT (optical calorimetry), volume change ΔV (optogeometric technique), or pressure change Δp (OA and PA techniques).

For a highly scattering tissue, measurement of the stress-wave profile and amplitude should be combined with measurement of the total diffuse reflectance in order to extract separately both absorption and scattering coefficients of the sample. The absorption coefficient in a turbid medium can be estimated from the acoustic transient profile only if the subsurface irradiance is known. For turbid media irradiated with a wide laser beam (>0.1 mm), the effect of backscattering causes a higher subsurface fluence rate compared with the incident laser fluence (see Equation 3.8).[5] Therefore, the z-axial light distribution in tissue and the corresponding stress distribution have a complex profile with a maximum at the subsurface layer. However, when the heating process is much faster than the medium expansion, the stress amplitude adjacent to the irradiated surface $\delta p(0)$ and the stress exponential tail in the depth of the tissue sample $\delta p(z)$ can be expressed as[69] (see also Equation 3.8)

$$\delta p(0) = \Gamma\mu_a E(0), \quad \text{at surface}(z=0), \qquad (3.39)$$

$$\delta p(z) = \Gamma\mu_a b_s E_0 \exp(-\mu_{eff} z), \quad \text{for } z > \frac{1}{\mu_{eff}}, \qquad (3.40)$$

where

$\Gamma = \beta v_a^2/c_T$
c_T is the specific heat of the tissue
b_s is the factor that accounts for the effect of backscattered irradiance, which increases the effective energy absorbed in the subsurface layer
μ_{eff} is defined in Equation 3.9
$E(0)$ is the subsurface irradiance
E_0 is the incident laser pulse energy at the sample surface (J/cm²)

For optically thick samples,[70]

$$E(0) \approx (1 + 7.1 R_d)E_0, \tag{3.41}$$

where R_d is the total diffuse reflection. The Grüneisen parameter Γ is a dimensionless, temperature-dependent factor proportional to the fraction of thermal energy converted into mechanical stress. For water it can be expressed with an empirical formula:

$$\Gamma = 0.0043 + 0.0053T, \tag{3.42}$$

where temperature T is measured in degrees Celsius; for $T = 37°C$, $\Gamma \approx 0.2$.

Equations 3.39 and 3.40 are strictly valid only when the heating process is much faster than the expansion of the medium. The stress is temporarily confined during laser heat deposition when the duration of the laser pulse is much shorter than the time of stress propagation across the depth of light penetration in the tissue sample. Such conditions of temporal pressure confinement in a volume of irradiated tissue enable the most efficient pressure generation (see Chapter 21 in Volume II of this handbook).[69,70]

The pulse laser heating of a tissue causes perturbations in its temperature and the corresponding modulation of its own thermal (IR) radiation. This is the basis for pulse OTR.[5,6,19,71] The maximum intensity of the thermal radiation of living objects falls within the wavelength range close to 10 μm. A detailed analysis of OTR signal formation requires knowledge on the internal temperature distribution within the tissue sample, tissue thermal diffusivity, and its absorption coefficients at the excitation μ_a and emission μ_a' (10 μm) wavelengths. Going backward, the knowledge of some of the mentioned parameters allows the reconstruction of, for example, the depth distribution of μ_a on the basis of the measured OTR signal.[71]

The surface radiometric signal $S_r(t)$ at any time t is the sum of the contributions from all depths in the tissue at time t. The radiation from deeper depths is attenuated by the IR absorption of the sample before reaching the detector. As the initial surface temperature is known, the temperature distribution in the sample depth can be extracted from $S_r(t)$ measurement.

OTR, OA, and PA transient techniques provide a convenient means for in vitro or even in vivo and in situ monitoring of human skin properties (optical absorption, thermal properties, water content) and surface concentrations of topically applied substances (drugs and sunscreen diffusion). The main difficulty of the PA method in the case of in vivo measurements is the requirement of a closed sample cell, which can guide the acoustic signal efficiently from sample to acoustical detector. The use of pulsed OA and OTR techniques is more appropriate for in vivo and in situ measurements.

For example, the photothermal flow cytometry (PTFC) technique has the ability to visualize absorbing cellular structures of moving unlabeled cells in real-time in vivo studies of circulating red and white blood cells in capillaries and lymph microvessels of rat mesentery.[72,73] The imaging of single cells in vivo is potentially important for the early diagnosis of diseases (e.g., cancer and diabetes) or for the study of the influence of various factors (e.g., drugs, nanoparticles, smoking, ionizing radiation) on individual cells.

To realize PTFC, a nonscanning fast photothermal microscopy (PTM) system was used because cells cross the area of detection in 0.1–0.01 s even in the relatively slow flow in capillaries.[53] Such a system was built on the basis of a pulsed pumping tunable optical parametric oscillator (420–570 nm, pulse width 8 ns, pulse energy 0.1–400 μJ). Laser-induced temperature-dependent variations of the refractive index in the cell were detected using a phase-contrast imaging technique with illumination by a low-energy collinear probe pulse from the Raman shifter with a wavelength of 639 nm, pulse width of 13 ns, and pulse energy of 2 nJ. The diameters of the pump- and probe-beam spots, with stable, smooth intensity profiles, ranged from 20 to 50 μm and 15 to 50 μm, respectively, and thus covered all the single cells and the entire microvessel as well. A spatial resolution of ~0.7 μm was provided. The acquisition procedure included illumination of the cell with three pulses: an initial

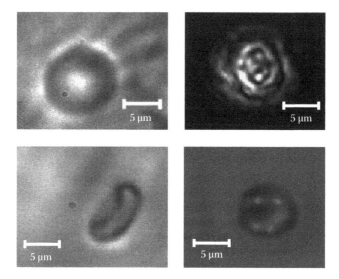

FIGURE 3.12 Photothermal flow cytometry (PTFC) technique.[73] Optical transmission (left column) and PT (right column) images in vivo of a single, moving lymphocyte (top row) and RBC (bottom row) in lymph flow in rat mesentery (vessel diameter, 105 μm; velocity, ~120 μm/s). Pump pulse parameters: wavelength, 525 nm; pulse width, 8 ns; and pulse energy, 30 and 0.5 μJ (right column, top and bottom, respectively); time delay between pump and probe pulses, 10 ns.

probe pulse followed by a 0.08 s delay to the pump pulse, and then a second probe pulse with a tunable time delay (0–5000 ns) to the pump pulse. The PT image, calculated as the difference between the two probe-pulse images, depends only on absorption contrast transformed by the pump laser pulse into refractive contrast.[73]

Using transillumination digital microscopy (TDM), red blood cells (RBCs) and lymphocytes traveling through blood and/or lymph vessels, lymphatic valves, and other mesenteric structures were imaged (Figure 3.12, left). However, because of its low absorption sensitivity, TDM was not suitable for visualizing absorbing cellular structures. In contrast, the PTFC mode (navigated by TDM) produced images of moving lymphocytes and RBCs (Figure 3.12, right), showing structures specific to PT images and associated with the spatial distribution of absorbing cellular chromophores (e.g., hemoglobin in RBCs, or cytochromes in lymphocytes). Currently, PTFC's rate of ~10 cells/s is limited by the repetition rate of the pump laser (10 Hz).

Potential applications of in vivo PTFC include (1) the identification of cells with differences in natural absorptive properties (e.g., the counting of white cells in blood flow or of rare RBCs among lymphocytes in microlymphatic vessels); (2) the monitoring of the circulation and distribution of absorbing nanoparticles used for PT probing or photosensitizing; (3) the study of laser–cell interactions; and (4) the study of the influence of different environmental factors on cells.

3.4.3 Acoustooptical Interactions

Acoustic wave (AW) and light interaction within a tissue as a heterogeneous medium is the basis for acoustooptical tomography or ultrasound (US)-modulated optical tomography, where the acoustic (US) modulation of coherent laser light traveling within a tissue is provided (see Chapter 20 in Volume II of this handbook).[74] An AW is focused into the tissue and laser light irradiates the same volume within the tissue. Any light that is encoded by the US, including both singly and multiply scattered photons, contributes to the imaging signal. Axial resolution along the acoustic axis can be achieved with US frequency sweeping and subsequent application of the Fourier transformation,[74] whereas lateral resolution can be obtained by focusing the AW into the tissue.

Three possible mechanisms have been identified for the acoustic modulation of light in scattering tissues.[74] The first mechanism is based on US-induced variations of the optical properties of a tissue caused by spatially and temporally dependent compression or rarifying of the tissue at the propagation of the AW. These variations in tissue density cause the corresponding oscillations of the optical properties of the tissue, including absorption and scattering coefficients, as well as refractive index. Accordingly, the detected intensity of light varies with the AW. However, US modulation of incoherent light has been too weak to be observed experimentally. The second mechanism is based on variations of the optical properties in response to US-induced displacement of scatterers. The displacement of scatterers, assumed to follow AW amplitudes, modulates the physical path lengths of light traveling through the acoustic field. Multiply scattered light accumulates modulated physical path lengths along its path. Consequently, the intensity of the speckles formed by the multiply scattered light fluctuates with the AW. The modulated component of the speckle pattern carries spatial information determined by the US and can be utilized for tomographic imaging. The third mechanism is caused by photon–phonon interactions, where light is considered as an ensemble of photons and AW an ensemble of phonons. The photon–phonon interactions cause a Doppler shift in the classical sense to the frequency of the photons by the acoustic frequency and its harmonics. An optical detector functions as a heterodyning device between the Doppler-shifted light and unshifted light and produces an intensity signal at the acoustic frequency and its harmonics.

Both the second and the third mechanisms require the use of coherent light and both may be associated with the speckle effect. The modulation of the speckles in the second mechanism is caused by the acoustic modulation of scatterer displacement, while the modulation of the speckles in the third mechanism is caused by the acoustic modulation of the refractive index of the tissue. The acoustic modulation of the refractive index also appears in both the first and third mechanisms. However, in the first mechanism, the variation of the refractive index causes light that may or may not be coherent to fluctuate in intensity, whereas in the second mechanism, the variation of the refractive index causes fluctuation in the phase of coherent light, which is converted to fluctuation in intensity by a square-law detector. Thus, as a result of acoustic modulation of the refractive index, the optical phase between two consecutive scattering events is modulated, multiply scattered light accumulates modulated phases along its path, and the modulated phase causes the intensity of the speckles formed by the multiply scattered light to vary with the AW.

The intensity modulation depth M is defined as the ratio between the intensity at the fundamental frequency I_1 and the unmodulated intensity I_0:

$$M = \frac{I_1}{I_0}. \tag{3.43}$$

The spectral intensity I_1 at the fundamental acoustic frequency ω_a is calculated from[74]

$$I_n = \left(\frac{1}{T_a}\right) \int_0^{T_a} \cos\left(n\omega_a\tau\right) G_1(\tau)d\tau \tag{3.44}$$

at $n = 1$; here, T_a is the acoustic period; the autocorrelation function of the scalar electric field, $E(t)$, of the scattered light $G_1(\tau)$ calculated in the approximation of weak scattering (the optical MFP [see Equation 3.7] is much longer than the optical wavelength) and weak modulation (the acoustic amplitude is much less than the optical wavelength) has a view[74]

$$G_1(\tau) = 1 - \left(\frac{1}{6}\right)\left(\frac{L}{l_{tr}}\right)\varepsilon\left[1 - \cos(\omega_a \tau)\right],$$

(3.45)

where

$\varepsilon = 6(\delta_n + \delta_d)(n_0 k_0 A)^2$

$\delta_n = (\alpha_{n_1} + \alpha_{n_2})\eta^2$

$\delta_d = 1/6$

$\alpha_{n_1} = k_a l_{tr} \tan^{-1}(k_a l_{tr})/2$

$\alpha_{n_2} = \alpha_{n_1}/[(k_a l_{tr})/\tan^{-1}(k_a l_t) - 1]$

L is the tissue slab thickness

n_0 is the background refractive index

k_0 is the optical wave vector in vacuum

A is the acoustic amplitude

k_a is the acoustic wave vector

l_t is the photon transport MFP

Parameter η is related to the adiabatic piezo-optical coefficient of the tissue $\partial n/\partial p$, the density ρ, and the acoustic velocity v_a: $\eta = (\partial n/\partial p)\rho(v_a)^2$. The parameters δ_n and δ_d (=1/6) are related to the average contributions per photon free path and per scattering event, respectively, to the ultrasonic modulation of light intensity. The contribution from the index of refraction δ_n increases with $k_a l_{tr}$ because a longer photon free path, relative to the acoustic wavelength, accumulates a greater phase modulation. By contrast, the contribution from displacement δ_d stays constant at 1/6, independent of k_a and l_{tr}. The contribution from the index of refraction above a critical point at $k_a l_{tr} = 0.559$, where contributions from refractive index and displacement are equal, increases with $k_a l_{tr}$ and significantly outmatches the contribution from displacement.

Accounting for Equation 3.45, the modulation depth of intensity fluctuations can be presented as

$$M = \left(\frac{1}{6}\right)\left(\frac{L}{l_t}\right)^2 \varepsilon \propto A^2,$$

(3.46)

This equation shows a quadratic relationship between the intensity modulation depth M and the acoustic amplitude A. Only the nonlinear terms of phase accumulation contribute to the acoustic modulation of coherent light at multiple scattering. The linear term vanishes as a result of optical random walk in scattering media. In the ballistic (nonscattering) regime, M is proportional to A due to nonaveraged contributions from the linear term of phase accumulation. In the quasi-ballistic (minimal scattering) regime, M, may show a mixed behavior with A.

3.4.4 Sonoluminescence

A sonoluminescence (SL) signal generated internally in the media with a 1 MHz CW US can be used to produce two-dimensional images of objects imbedded in turbid media.[75,76] This technique is based on the light emission phenomenon connected with the driving of small bubbles by US collapse. The bubbles start out with a radius of several microns and expand to ~50 μm, owing to a decrease in acoustic pressure in the negative half of a sinusoidal period; after the acoustic wave reaches the positive half of the period, the resulting pressure difference leads to a rapid collapse of the bubbles, accompanied by a broadband emission of light—SL. Such emission is of a short duration (in tens of picoseconds), repeatable with each cycle of sound, and has the spectrum containing molecular emission bands (with peaks near 300–500 nm) associated with the liquid, mostly water, in which the SL occurs.

SL tomography (SLT) as a new approach for optical imaging of dense turbid media (biological tissues) is described.[75,76] The major advantages of SLT include (1) high signal-to-noise ratio due to the internally generated probe optical signal; (2) high contrast of imaging; (3) good spatial resolution, which is limited by the US focal size; and (4) low cost of equipment. It was shown experimentally that there is a threshold of SL generation at applied US pressure, when the peak US pressure at the US focus was ~2 bar. The rapid increase in the intensity of SL with the acoustic pressure above the threshold indicates that the SL signal would be a sensitive measure of the local acoustic pressure.

SLT is based on several contrast mechanisms[76]: (1) for the objects with US contrast relative to the background, the SL signal originating from the object will differ from that originating from the background medium, as the SL generation is affected by the local US intensity; (2) for the objects with contrast in optical properties, the SL signal from the object is attenuated differently because the SL light must propagate through the object; (3) for objects with the ability to generate SL, the SL from the object is different, even if the local US pressure is the same.

It should be noted that the peak pressure at the US focus is typically less than ~2 bar (1.3 W/cm² in spatial-peak-temporal-peak power), which is one order in magnitude less than the safety limit set by the US Food and Drug Administration (23 bar) and two orders less than the tissue damage threshold (400 and 900 W/cm² at 1 MHz for brain and muscle, respectively).[76]

3.5 Refractive Index and Controlling of Light Interaction with Tissue

The refractive index in tissues is of great importance for light–tissue interaction. Most tissues have refractive indices for visible light in the 1.335–1.62 range (e.g., 1.55 for the stratum corneum, 1.62 for the enamel, and 1.386 at the lens surface)[5,77,78] (see Chapter 2 in this volume). It is worthwhile to note that in vitro and in vivo measurements may differ significantly. For example, the refractive index in rat mesenteric tissue in vitro was found to be 1.52 compared with only 1.38 in vivo. This difference can be accounted for by the decreased refractivity of ground matter n_0 due to impaired hydration.

The mean refractive index $\bar{n}$ of a tissue is defined by the refractive indices of its scattering centers material n_s and ground matter n_0[5,78]:

$$\bar{n} = c_s n_s + (1 - c_s) n_0,$$

$$(3.47)$$

where c_s is the volume fraction of the scatterers.

The $n_s/n_0 \equiv m$ ratio determines the scattering coefficient. For example, in a simple monodisperse model of scattering dielectric spheres[79]:

$$\mu_s' = 3.28 \pi a^2 \rho_s \left(\frac{2\pi a}{\lambda} \right)^{0.37} (m-1)^{2.09},$$

$$(3.48)$$

where
  a is the sphere radius
  ρ_s is the volume density of the spheres

Equation 3.48 is valid for noninteracting Mie scatterers, $g > 0.9$; $5 < 2\pi a/\lambda < 50$; $1 < m < 1.1$.

It follows from Equation 3.48 that even a 5% change in the refractive index of the ground matter ($n_0 = 1.35 \rightarrow 1.42$), when that of the scattering centers is $n_s = 1.47$, will cause a sevenfold decrease of μ_s'. Therefore, matching of the refractive index of the scatterers and ground material enables considerable reduction of tissue scattering. This phenomenon is very useful for improving facilities of optical tomography and for obtaining precise spectroscopic information from the depth of a tissue.

Optical parameters of a tissue, in particular the refractive index, are known to depend on water content. The refractive index of water over a broad wavelength range of 0.2–200 μm has been reported.[77] The following relation was shown to be valid for the visible and NIR wavelength range (λ in nm)[80]:

$$n_{H_2O} = \frac{1.3199 + 6878}{\lambda^2} - \frac{1.132 \times 10^9}{\lambda^4} + \frac{1.11 \times 10^{14}}{\lambda^6}. \tag{3.49}$$

For different parts of a biological cell, values of the refractive index in the NIR can be estimated as follows: extracellular fluid—$\bar{n}$ = 1.35–1.36; cytoplasm—1.360–1.375; cell membrane—1.46; nucleus—1.38–1.41; mitochondria and organelles—1.38–1.41; and melanin—1.6–1.7.[5,81] Scattering arises from a mismatch in refractive index of the components that make up the cell. In tissues, when cells are surrounded by other cells or tissue structures of similar index, certain organelles become important scatterers. For instance, the nucleus is a significant scatterer because it is often the largest organelle in the cell and its size increases relative to the rest of the cell throughout neoplastic progression. Mitochondria (500–1500 nm in diameter), lysosomes (500 nm), and peroxisomes (500 nm) are very important scatterers, whose size relative to the wavelength of light suggests that they must contribute significantly to backscattering. Melanin granular, traditionally thought of as an absorber, must be considered an important scatterer because of its size and high refractive index.[81] Structures consisting of membrane layers such as the endoplasmic reticulum or Golgi apparatus may prove significant because they contain index fluctuations of high spatial frequency and amplitude. Besides cell components, fibrous structures of tissue such as collagen and elastin must be considered important scatterers.

Refractivity measurements in a number of strongly scattering tissues at 633 nm performed with a fiber-optic refractometer have shown that fatty tissue has the largest refractive index (1.455) followed by kidney (1.418), muscular tissue (1.410), and blood and spleen (1.400).[82] The lowest refractive indices were found in lungs and liver (1.380 and 1.368, respectively) (see Chapter 2 in this volume). Refractive indices tend to decrease with increasing light wavelength from 390 to 700 nm (e.g., for bovine muscle in the range of 1.42–1.39).

It is possible to achieve a marked impairment of scattering by matching the refractive indices of scattering centers and ground matter by means of intratissue administration of the appropriate chemical agents. Experimental optical clearing in human sclera in the visible wavelength range induced by administration of x-ray contrast (verografin, trazograph), glucose, propylene glycol, polyethylene glycol, and other solutions has been described.[5,7,78,83–87] Osmotic phenomena appear to be involved when optical properties of tissues are modulated by sugar, alcohol, glycerol, and electrolyte solutions.

Experimental studies on optical clearing of normal and pathological skin and its components (epidermis and dermis) and the management of reflectance and transmittance spectra using glycerol, glycerol–water solutions, glucose, sunscreen creams, cosmetic lotions, gels, and pharmaceutical products were carried out.[5,7,78]

A marked clearing effect through the rat[85] and human[86] skin and the rabbit sclera[7,78] was observed in an in vivo tissue within a few minutes of topical application or intratissue injection of glycerol, glucose, verografin, or trazograph. In vivo reflectance spectra of the human skin with intraskin injection of 40% glucose are shown in Figure 3.13. Skin is well protected from penetration of any agent by the stratum corneum, thus preventing permeation of optical clearing agents with topical application. In addition, optical reflectance of the skin is defined mostly by the dermis. For both these reasons, the intradermis injection is useful if temporal reduction of light scattering by the skin as a whole is desirable.

Multiple scattering is a detrimental factor that limits OCT imaging performances including imaging resolution, depth, and localization (OCT is described in Chapter 10 in this volume). To improve the imaging capabilities, the multiple scattering of tissue must be reduced. The immersion technique for the application of biocompatible agents is a prospective technique for OCT because the depth and contrast

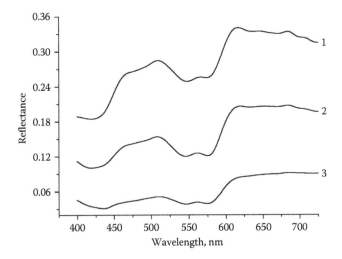

FIGURE 3.13 Reflectance spectra measured (1) before and (2) at 23 min and (3) 60 min after intraskin injection of 0.1 mL of 40% glucose into the internal side of the forearm of the male volunteer. (From Tuchin, V.V., *Tissue Optics: Light Scattering Methods and Instruments for Medical Diagnosis*, SPIE Tutorial Texts in Optical Engineering, Vol. TT38, SPIE Press, Bellingham, WA, 2000; 2nd edn., Vol. PM166, SPIE Press, Bellingham, WA, 2007; Tuchin, V.V., *Optical Clearing of Tissues and Blood*, Vol. PM 154, SPIE Press, Bellingham, WA, 2006; Tuchin, V.V., et al., *J. Tech. Phys. Lett.*, 27(6), 489, 2001.)

of OCT images can be very easily improved at immersion.[5,9,78,87–93] OCT imaging combined with OCA immersion is a useful technology for skin disease diagnosis and monitoring. To illustrate the dynamics of skin optical clearing after the application of glycerol, a set of OCT images (820 nm) of the rat skin sample at regular time intervals over a period of 40 min was recorded (Figure 3.14).[88] Both the index-matching effect, leading to the enhanced depth capability, and the localized dehydration effect, leading to the improvement of imaging contrast, are clearly evident.

The result of the OCT study is the measurement of optical backscattering or reflectance, $R(z)$, from the tissue versus the axial ranging distance, or depth, z. The reflectance depends on the optical properties of the tissue, that is, the absorption μ_a and scattering μ_s coefficients, or the total attenuation coefficient $\mu_t = \mu_a + \mu_s$. The relationship between $R(z)$ and μ_t is, however, highly complicated because of the high and anisotropic scattering of tissue. But for optical depths less than 4, reflected power can be approximately proportional to $-\mu_t z$ in an exponential scale according to the single-scattering model, that is,[9]

$$R(z) = I_0 \alpha(z) \exp(-\mu_t z), \tag{3.50}$$

where

I_0 is the optical power launched into the tissue sample
$\alpha(z)$ is the reflectivity of the sample at the depth of z

Optical depth is a measure of depth in terms of the number of MFP lengths, that is, $\mu_s z$. $\alpha(z)$ is linked to the local refractive index and the backscattering property of the sample. If $\alpha(z)$ is kept constant, μ_t can be obtained theoretically from the reflectance measurements at two different depths, z_1 and z_2:

$$\mu_t = \frac{1}{(\Delta z)} \ln \left[\frac{R(z_1)}{R(z_2)} \right], \tag{3.51}$$

where $\Delta z = |z_1 - z_2|$. As noise is inevitable in the measurement, a final result should thus be obtained by using the least-squares fitting technique to improve the accuracy of the determined value of μ_t.

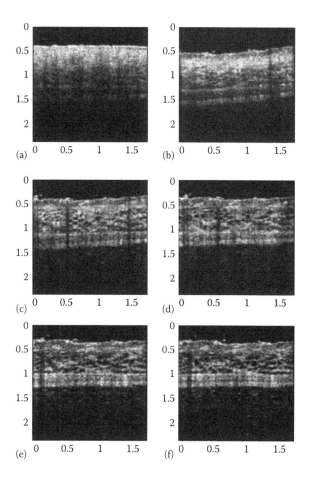

FIGURE 3.14 OCT images (λ = 820 nm) captured at (a) 0, (b) 3, (c) 10, (d) 15, (e) 20, and (f) 40 min after a topical application of 80% glycerol solution onto the rat skin. Images taken right after the rat was sacrificed, all the units presented are in millimeters, and the vertical axis presents the imaging depth. (From Wang, R.K. et al., *J. Opt. Soc. Am. B*, 18, 948, 2001.)

Optical clearing (enhancement of transmittance) ΔT by an agent application can be estimated using the following expression:

$$\Delta T = \left[\frac{(R_a - R_s)}{R_s} \right] \times 100\%,$$

(3.52)

where

R_a is the reflectance from the backward surface of the sample impregnated by an agent

R_s is that with a control sample

The possibility of in vivo diagnostics and monitoring of malignant melanoma, psoriatic erythrodermia, and observation of subepidermal blisters by controlling the scattering properties of skin through its saturation with clearing agents was demonstrated.[5,78] It should be noted that high sensitivity of OCT signal to immersion of living tissue by glucose allows monitoring its concentration in the skin at a physiological level.[61,78,90,91]

Although glycerol and glucose are effective OCAs when injected into the dermis, they do not normally penetrate so well into intact skin. In recent OCT experiments with human skin in vivo with topical application during 90–120 min of the combined lipophilic polypropylene glycol–based polymers and hydrophilic polyethylene glycol (PEG)-based polymers, both with refractive indices of 1.47, which closely matches that of skin-scattering components in SC, epidermis, and dermis, it was shown that a polymer mixture can penetrate intact skin and improve OCT images to see dermal vasculature and hair follicles more clearly.[89] This composition may have some advantages in skin optical clearing due to the hydrophilic component, which may more effectively diffuse within living epidermis and dermis; less osmotic strength may also have some advantages, but the optical clearing depth could not be improved radically in comparison with topical application of other clearing agents, such as glycerol, glucose, x-ray contrast, and propylene glycol, because of principal limitations of chemical agent diffusion through intact cell layers. Thus, to provide fast and effective optical clearing of skin, the appropriate well-known or newly developed methods of enhanced agent delivery should be applied.

The concept that index matching could improve the optical penetration depth of whole blood is proved experimentally in in vitro studies using OCT.[78,92,93] For example, for whole blood twice diluted by saline, adding 6.5% of glycerol increases the optical penetration up to 117%; optical clearing up to 150% was achieved mostly due to refractive index matching for the high–molecular weight dextran.[93]

3.6 Fluorescence

3.6.1 Fundamentals and Methods

Fluorescence arises upon light absorption and is related to an electronic transition from the excited state to the ground state of a molecule. In the case of thin samples, for example, cell monolayers or biopsies with a few micrometers in diameter, fluorescence intensity I_F is proportional to the concentration c and the fluorescence quantum yield η of the absorbing molecules.[17,94] In a scattering medium, the path lengths of scattered and unscattered photons within the sample are different, and should be accounted for.[17] However, in rather homogeneous thin samples, the linearity between I_F, c, and η is still fulfilled.

When biological objects are excited by UV light ($\lambda \leq 300$ nm), fluorescence of proteins and nucleic acids can be observed. Fluorescence quantum yields of all nucleic acid constituents, however, are around 10^{-4}–10^{-5} corresponding to lifetimes of the excited states in the picosecond time range. Autofluorescence of proteins is related to the amino acids tryptophan, tyrosin, and phenylalanine with absorption maxima at 280, 275, and 257 nm, respectively, and emission maxima between 280 (phenylalanine) and 350 nm (tryptophan). Usually, the protein spectrum is dominated by tryptophan. Fluorescence from collagen or elastin is excited between 300 and 400 nm and shows broad emission bands between 400 and 600 nm with maxima around 400, 430, and 460 nm. In particular, fluorescence of collagen and elastin can be used to distinguish various types of tissues, such as epithelial and connective tissues.[16,17,94-97]

The reduced form of coenzyme nicotinamide adenine dinucleotide (NADH) is excited selectively in a wavelength range between 330 and 370 nm. NADH is most concentrated within mitochondria, where it is oxidized within the respiratory chain located within the inner mitochondrial membrane and its fluorescence is an appropriate parameter for detection of ischemic or neoplastic tissues.[94] Fluorescence of free and protein-bound NADH has been shown to be sensitive on oxygen concentration. Flavin mononucleotide (FMN) and flavin dinucleotide (FAD) with excitation maxima around 380 and 450 nm have also been reported to contribute to intrinsic cellular fluorescence.[94]

Porphyrin molecules, for example, protoporphyrin, coproporphyrin, uroporphyrin, or hematoporphyrin, occur within the pathway of biosynthesis of hemoglobin, myoglobin, and cytochromes; thus, intrinsic fluorescence of these molecules provides information related to disease development. For example, abnormalities in heme synthesis, occurring in the cases of porphyrias and some hemolytic diseases, may enhance considerably the porphyrin level within tissues, which could be detected via tissue autofluorescence associated with porphyrin bands. Several bacteria, such as *Propionibacterium acnes* or bacteria within dental caries lesions, accumulate considerable amounts of protoporphyrin.[94] Therefore, acne or caries detection based on measurements of intrinsic fluorescence appears to be a promising method.

At present, various exogenous fluorescing dyes can be applied for probing of cell anatomy and cell physiology.[94] In humans, dyes such as fluorescein and indocyanine green are used for fluorescence angiography or blood volume determination.

Fluorescence spectra often give detailed information on fluorescent molecules, their conformation, binding sites, and interaction within cells and tissues. Fluorescence intensity can be measured either as a function of the emission wavelength or of the excitation wavelength. The fluorescence emission spectrum $I_F(\lambda)$ is specific for any fluorophore and commonly used in fluorescence diagnostics.

For many biomedical applications, an optical multichannel analyzer (OMA) (a diode array or a charge-coupled device [CCD] camera) as a detector of emission radiation is preferable, because spectra can be recorded very rapidly and repeatedly with sequences in the millisecond range. Fluorescence spectrometers for in vivo diagnostics are commonly based on fiber-optic systems. The excitation light of a lamp or a laser is guided to the tissue (e.g., some specific organ) via a fiber using appropriate optical filters. Fluorescence spectra are usually measured either via the same fiber or via a second fiber or fiber bundle in close proximity to the excitation fiber.

Various comprehensive and powerful fluorescence spectroscopies such as microspectrofluorimetry, polarization anisotropy, time-resolved with pulse excitation and frequency domain, time-gated, total internal reflection fluorescence spectroscopy and microscopy, fluorescence resonant energy transfer method, confocal laser scanning microscopy, and their combinations are available now[17,94] (see Chapter 15 in Volume II of this handbook). These methods provide the following:

- 3D topography of specimens measured in the reflection mode for morphological studies of biological samples
- High-resolution microscopy measured in the transmission mode
- 3D-fluorescence detection of cellular structures and fluorescence bleaching kinetics
- Time-resolved fluorescence kinetics
- Studies of motions of cellular structures
- Time-gated imaging in order to select specific fluorescent molecules or molecular interactions
- Fluorescence lifetime imaging
- Spectrally resolved imaging

Principles of optical clinical chemistry based on measuring the changes of fluorescence intensity, wavelength, polarization anisotropy, and lifetime are described by Lakowicz.[17] Various fluorescence techniques of selective oxygen sensing and blood glucose and blood gases detection are available.[17,61]

3.6.2 Multiphoton Fluorescence

A new direction in laser spectroscopy of biological objects is associated with multiphoton (two-, three-photon) fluorescence scanning microscopy, which makes it possible to image functional states of an object or to determine, in combination with autocorrelation analysis of the fluorescence signal, the intercellular motility in small volumes[18] (see Chapter 12 in this volume). The two-photon technique employs both ballistic and scattered photons at the wavelength of the second harmonic of incident radiation coming to a wide-aperture photodetector exactly from the focal area of the excitation beam. A unique advantage of two-photon microscopy is the possibility of investigating three-dimensional distributions of chromophores excited with UV radiation in thick samples. Such an investigation becomes possible because chromophores can be excited (e.g., at the wavelength of 350 nm) with laser radiation whose wavelength falls within the range (700 nm) where a tissue has high transparency. Such radiation can reach deep-lying layers and produces less damage in tissues. Fluorescent emission in this case lies in the visible range (>400 nm) and comparatively easily emerges from a tissue and reaches a photodetector, which registers only the legitimate signal from the focal volume without any extraneous background. Investigations of tissues and cells by means of two-photon microscopy are characterized by the following typical parameters of laser systems: the wavelength ranges from 700 to 960 nm, the pulse duration is on the order of 150 fs, the pulse repetition rate is 76–80 MHz, and the mean power is less than 10 mW. Such parameters can be achieved with mode-locked dye lasers pumped by an Nd:YAG laser or with titanium sapphire lasers pumped by an argon laser. Diode-pumped solid-state lasers also hold much promise for the purposes of two-photon microscopy.

3.7 Vibrational Energy States Excitation

Mid-infrared (MIR) and Raman spectroscopies use light-excited vibrational energy states in molecules to get information about the molecular composition, molecular structures, and molecular interactions in a sample.[13–15,98,99] In MIR spectroscopy, IR light from a broadband source (usually 2.5–25 μm or 4000–400 cm^{-1}) is directly absorbed to excite the molecules to higher vibrational states. In a Raman scattering event, light is inelastically scattered by a molecule when a small amount of energy is transferred from the photon to the molecule (or vice versa). This leads to an excitation of the molecule usually from its lowest vibrational energy level in the electronic ground state to a higher vibrational state. The energy difference between the incident and scattered photon is expressed in a wave number shift (cm^{-1}). Some vibrations can be excited by both Raman and MIR processes, while others can only be excited by either a Raman scattering or MIR absorption. Both techniques enable the recording of high-quality spectra in relative short acquisition times (30–60 s).

The MIR and Raman spectroscopy techniques are successfully applied in various areas of clinical studies, such as cancerous tissue examination, the mineralization process of bone and teeth tissue monitoring, glucose sensing in blood, noninvasive diagnosis of skin lesions on benign or malignant cells, monitoring of treatments and topically applied substances (e.g., drugs, cosmetics, moisturizers) on skin, and water exchange in human eye lens.[13–15,61,98–104]

Raman spectroscopy is widely used in biological studies, ranging from studies of purified biological compounds to investigations at the level of single cells. At present, combinations of spectroscopic techniques such as MIR and Raman with microscopic imaging techniques are explored to map molecular distributions at specific vibrational frequencies on samples to locally characterize tissues or cells.[99] Chemical imaging will become more and more important in the clinical diagnosis.

3.8 Speckles Formation

Speckle structures are produced as a result of interference of a large number of elementary waves with random phases that arise when coherent light is reflected from a rough surface or when coherent light passes through a scattering medium[3,5,7,105–109] (see Chapter 19 in this volume). Generally, there are two types of speckles: subjective speckles, which are produced in the image space of an optical system (including an eye), and objective speckles, which are formed in a free space and are usually observed on a screen placed at a certain distance from an object. As the majority of tissues are optically nonuniform, irradiation of such objects with coherent light always gives rise to the appearance of speckle structures, which either distort the results of measurements and, consequently, should be eliminated in some way, or provide new information regarding the structure and the motion of a tissue and its components.

Figure 3.15 schematically illustrates the principles of the formation and propagation of speckles. The average size of a speckle in the far-field zone is estimated as

$$d_{av} \sim \frac{\lambda}{\varphi}, \qquad (3.53)$$

where
λ is the wavelength
φ is the angle of observation

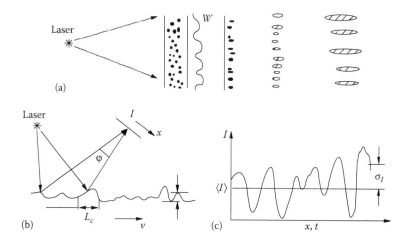

FIGURE 3.15 (a) Formation and propagation of speckles, (b) observation of speckles, and (c) intensity modulation. (From Tuchin, V.V., *Tissue Optics: Light Scattering Methods and Instruments for Medical Diagnosis*, SPIE Tutorial Texts in Optical Engineering, Vol. TT38, SPIE Press, Bellingham, WA, 2000; 2nd edn., Vol. PM166, SPIE Press, Bellingham, WA, 2007.)

Displacement of the observation point over a screen (x) or the scanning of a laser beam over an object with a certain velocity v (or an equivalent motion of the object itself with respect to the laser beam) under conditions when the observation point remains stationary gives rise to spatial or temporal fluctuations in the intensity of the scattered field. These fluctuations are characterized by the mean value of the intensity $\langle I \rangle$ and the standard deviation σ_I (see Figure 3.15b). The object itself is characterized by the standard deviation σ_h of the altitudes (depths) of inhomogeneities and the correlation length L_c of these inhomogeneities (random relief).

As many tissues and cells are phase objects, the propagation of coherent beams in tissues can be described within the framework of the model of a random phase screen. Ideal conditions for the formation of speckles, when completely developed speckles arise, can be formulated as coherent light irradiates a diffusive surface (or a transparency) characterized by Gaussian variations of optical length $\Delta L = \Delta(nh)$ and standard deviation of relief variations, $\sigma_L > \lambda$; both the coherence length of light and the sizes of the scattering area considerably exceed the differences in optical paths due to the surface relief, and many scattering centers contribute to the resulting speckle pattern.

Statistical properties of speckles can be divided into statistics of the first and second orders. Statistics of the first order describes the properties of speckle fields at each point. Such a description usually employs the intensity probability density distribution function $p(I)$ and the contrast

$$V_I = \frac{\sigma_I}{\langle I \rangle}, \quad \sigma_I^2 = \langle I \rangle^2 - \langle I \rangle^2, \tag{3.54}$$

where $\langle I \rangle$ and σ_I^2 are the mean intensity and the variance of the intensity fluctuations, respectively. In certain cases, statistical moments of higher orders are employed.

For ideal conditions, when the complex amplitude of scattered light has a Gaussian statistics, the contrast is $V_I = 1$ (developed speckles), and the intensity probability distribution is represented by a negative exponential function:

$$p(I) = \frac{1}{\langle I \rangle} \exp \left\{ -\frac{I}{\langle I \rangle} \right\}. \tag{3.55}$$

Thus, the most probable intensity value in the corresponding speckle pattern is equal to zero; that is, destructive interference occurs with the highest probability.

Partially developed speckle fields are characterized by a contrast $V_I < 1$. The contrast may be lower due to a uniform coherent or incoherent background added to the speckle field.

For phase objects with $\sigma_\phi^2 \gg 1$ and a small number of scatterers, $N = w/L_\phi$, contributing to the field at a certain point on the observation plane, the contrast of the speckle pattern is greater than unity. The statistics of the speckle field in this case is non-Gaussian and nonuniform (i.e., the statistic parameters depend on the observation angle).

The specific features of the diffraction of focused laser beams from moving phase screens underlie speckle methods of structure diagnostics and monitoring of motion parameters of tissue, blood, and lymph flows.

Statistics of the second order shows how fast the intensity changes from point to point in the speckle pattern, that is, characterizes the size and the distribution of speckle sizes in the pattern. Statistics of the second order is usually described in terms of the autocorrelation function of intensity fluctuations,

$$g_2(\Delta\xi) = \langle I(\xi + \Delta\xi)I(\xi)\rangle, \tag{3.56}$$

and its Fourier transform, representing the power spectrum of a random process; $\xi \equiv x$ or t is the spatial or temporal variable; $\Delta\xi$ is the change in variable. Angular brackets $\langle\rangle$ in Equation 3.56 stand for the averaging over an ensemble or the time.

In the elementary case when reflected light in developed speckle structures retains linear polarization, intensity distribution at the output of a dual-beam interferometer can be written as[5]

$$I(r,t) = I_r(r) + I_s(r) + 2\left[I_r(r)I_s(r)\right]^{1/2}|\gamma_{11}(\Delta t)|\cos\{\Delta\Phi_I(r) + \Delta\Psi_I(r) + \Delta\Phi_I(t)\}, \tag{3.57}$$

where

$I_r(r)$ and $I_s(r)$ are intensity distributions of the reference and signal fields, respectively
r is the transverse spatial coordinate
$\gamma_{11}(\Delta t)$ is the degree of temporal coherence of light
$\Delta\Psi_I(r)$ is the deterministic phase difference of the interfering waves
$\Delta\Phi_I(r) = \Phi_{Ir}(r) - \Phi_{Is}(r)$ is the random phase difference
$\Delta\Phi_I(t)$ is the time-dependent phase difference related to the motion of an object

In the absence of speckle modulation, the deterministic phase difference $\Delta\Psi_I(r)$ governs the formation of regular interference fringes. On the average, the output signal of a speckle interferometer reaches its maximum when the interfering fields are phase-matched ($\Delta\Psi_I(r)$ = const. within the aperture of the detector), focused laser beams are used (speckles with maximum sizes are produced), and a detector with a maximum area is employed.

3.9 Dynamic Light Scattering

3.9.1 Quasi-Elastic Light Scattering

QELS spectroscopy, photon-correlation spectroscopy, spectroscopy of intensity fluctuations, and Doppler spectroscopy are synonymous terms related to the dynamic scattering of light, which underlies a noninvasive method for studying the dynamics of particles on a comparatively large time scale.[5]

The implementation of the single-scattering mode and the use of coherent light sources are of fundamental importance in this case. The spatial scale of testing of a colloid structure (an ensemble of biological particles) is determined by the inverse of the wave vector $|\bar{s}|^{-1}$:

$$|\bar{s}| = \frac{4\pi n}{\lambda_0} \sin\left(\frac{\theta}{2}\right), \tag{3.58}$$

where

     n is the refractive index
     θ is the angle of scattering

With an allowance for self-beating due to the photomixing of the electric components of the scattered field, the intensity autocorrelation function (AF) can be measured: $g_2(\tau) = \langle I(t)I(t + \tau)\rangle$. For Gaussian statistics, this AF is related to the first-order AF by the Siegert formula:

$$g_2(\tau) = A\left[1 + \beta_{sb}|g_1(\tau)|^2\right], \tag{3.59}$$

where

     τ is the delay time
     $A = \langle i^2\rangle$ is the square of the mean value of the photocurrent, or the base line of the AF
     β_{sb} is the parameter of self-beating efficiency, $\beta_{sb} \approx 1$
     $g_1(\tau) = \exp(-\Gamma_T\tau)$ is the normalized AF of the optical field for a monodisperse system of Brownian particles
     $\Gamma_T = |\bar{s}|^2 D_T$ is the relaxation parameter
     $D_T = k_B T/6\pi\eta r_h$ is the coefficient of translation diffusion
     k_B is the Boltzmann constant
     T is the absolute temperature
     η is the absolute viscosity of the medium
     r_h is the hydrodynamic radius of a particle

Many biological systems are characterized by a bimodal distribution of diffusion coefficients, when fast diffusion (D_{Tf}) can be separated from slow diffusion (D_{Ts}) related to the aggregation of particles. The goal of QELS spectroscopy is to reconstruct the distribution of scattering particles according to size, which is necessary for the diagnosis or monitoring of a disease.

     The description of the principles and characteristics of the homodyne and heterodyne photon-correlation spectrometers, the so-called laser Doppler anemometers (LDAs), differential LDA schemes, laser Doppler microscopes, laser scanning and speckle CMOS-based full-field imagers, and review of medical applications, mainly the investigation of eye tissues (cataract diagnosis), investigation of hemodynamics of individual vessels (vessels of eye fundus or any other vessels) with the use of fiber-optic catheters, and mapping of blood microcirculation in tissues, is given by Tuchin[5,7] and Priezzhev and coworkers.[10,105–109]

3.9.2 Diffusion-Wave Spectroscopy

DWS is a new class of technique in the field of dynamic light scattering related to the investigation of the dynamics of particles within very short time intervals.[5,7,9,110–113] A fundamental difference of this method from QELS is that it is applicable in the case of dense media with multiple scattering, which is critical for tissues. In contrast to the case of single scattering, the AF of the field $g_1(\tau)$ is sensitive to the motion of a particle on the length scale on the order of $\lambda[L/l_{tr}]^{-1/2}$, which is generally much smaller than λ, because $L > l_{tr}$ (L is the total mean photon path length and l_{tr} is the transport length of a photon, $l_{tr} \approx 1/\mu_s'$). Thus, DWS AFs decay much faster than AFs employed in QELS. In recent years, DWS has come into use as a more accurate term—diffusion-correlation spectroscopy (DCS).[113]

Experimental implementation of DWS is very simple.[5,7,9,110–113] A measuring system provides irradiation of an object under study by a CW laser beam and the intensity fluctuations of the scattered radiation within a single speckle are measured with the use of a single-mode receiving fiber, photomultiplier, photon-counting system, and a fast digital correlator working in a nanosecond range. The possibilities of the DWS technique for medical applications have been demonstrated for blood flow monitoring in the human forearm. The AF slope is the indicative parameter for the determination of the blood flow velocity. The normalized AF of field fluctuations can be represented in terms of two components related to the Brownian and directed motion of scatterers (erythrocytes):

$$g_1(\tau) \approx \exp\left\{-2\left[\frac{\tau}{\tau_B^{-1}} + \left(\frac{\tau}{\tau_s^{-1}}\right)^2\right]\frac{L}{l_{tr}}\right\}, \tag{3.60}$$

where
$\tau_B^{-1} \equiv \Gamma_T = |\overline{s}|^2 D_T$
$\tau_S^{-1} \cong 0.18 G_V |\overline{s}| l_{tr}$ characterizes the directed flow
G_V is the gradient of the flow rate

This enables the expression of the slope of the AF in terms of the diffusion coefficient (characterizes blood microcirculation) and the gradient of the directed velocity of blood. The correlation function can be modified to account for other dynamic processes in addition to random and shear flow, such as turbulence.

3.10 Conclusion

This brief review on tissue optics provides information on physical optics as well as approaches in atomic and thermal physics that may be applicable to a number of different phenomena that involve light interaction with tissues. A more comprehensive and complete description of tissue optics and light/laser-tissue interactions can be found elsewhere.[2,5–7,9,20–22,47,58,61,78,114–134]

Acknowledgments

The author is thankful for the support of grants 224014 PHOTONICS4LIFE of FP7-ICT-2007-2; Russian Presidential 703.2014.2; the Government of the Russian Federation to support scientific research projects implemented under the supervision of leading scientists 14.Z50.31.0004; FiDiPro, TEKES Program (40111/11), Finland; SCOPES EC, Uzb/Switz/RF, Swiss NSF, IZ74ZO_137423/1; and 13-02-91176-NSFC_a of RFBR.

References

1. Chance B., Optical method, *Annu. Rev. Biophys. Biophys. Chem.* **20**, 1–28, 1991.
2. Müller G., Chance B., Alfano R. et al., eds., *Medical Optical Tomography: Functional Imaging and Monitoring*, Vol. IS11, SPIE Press, Bellingham, WA, 1993.
3. Tuchin V.V., ed., *Selected Papers on Tissue Optics Applications in Medical Diagnostics and Therapy*, Vol. MS102, SPIE Press, Bellingham, WA, 1994.
4. Minet O., Müller G., and Beuthan J., eds., *Selected Papers on Optical Tomography, Fundamentals and Applications in Medicine*, Vol. MS 147, SPIE Press, Bellingham, WA, 1998.
5. Tuchin V.V., *Tissue Optics: Light Scattering Methods and Instruments for Medical Diagnosis*, SPIE Tutorial Texts in Optical Engineering, Vol. TT38, SPIE Press, Bellingham, WA, 2000; 2nd edn., Vol. PM166, SPIE Press, Bellingham, WA, 2007.
6. Wang L.V. and Wu H.-I., *Biomedical Optics: Principles and Imaging*, Wiley-Intersience, Hoboken, NJ, 2007.
7. Tuchin V.V., ed., *Handbook of Optical Biomedical Diagnostics*, Vol. PM107, SPIE Press, Bellingham, WA, 2002.
8. Fercher A.F., Drexler W., Hitzenberger C.K., and Lasser T., Optical coherence tomography— Principles and applications, *Rep. Prog. Phys.* **66**, 239–303, 2003.
9. Tuchin V.V., ed., *Coherent-Domain Optical Methods: Biomedical Diagnostics, Environmental Monitoring and Material Science*, 2nd edn., Vols. 1 and 2, Springer-Verlag, Berlin, Germany, 2013.
10. Priezzhev A.V. and Asakura T., eds., Special section on optical diagnostics of biological fluids, *J. Biomed. Opt.* **4**, 35–93, 1999.
11. Tuchin V.V., Podbielska H., and Hitzenberger C.K., eds., Special section on coherence domain optical methods in biomedical science and clinics, *J. Biomed. Opt.* **4**, 94–190, 1999.
12. Katzir A., *Lasers and Optical Fibers in Medicine*, Academic Press, San Diego, CA, 1993.
13. Ozaki Y., Medical application of Raman spectroscopy, *Appl. Spectrosc. Rev.* **24**, 259–312, 1988.
14. Mahadevan-Jansen A. and Richards-Kortum R., Raman spectroscopy for detection of cancers and precancers, *J. Biomed. Opt.* **1**, 31–70, 1996.
15. Morris M.D., ed., Special section on biomedical applications of vibrational spectroscopic imaging, *J. Biomed. Opt.* **4**, 6–34, 1999.
16. Das B.B., Liu F., and Alfano R.R., Time-resolved fluorescence and photon migration studies in biomedical and random media, *Rep. Prog. Phys.* **60**, 227–292, 1997.
17. Lakowicz J.R., *Principles of Fluorescence Spectroscopy*, 2nd edn., Kluwer Academic Publishers, New York, 1999.
18. Denk W., Two–photon excitation in functional biological imaging, *J. Biomed. Opt.* **1**, 296–304, 1996.
19. Braslavsky S.E. and Heihoff K., Photothermal methods, in *Handbook of Organic Photochemistry*, Scaiano J.C., ed., CRC Press, Boca Raton, FL, 1989, pp. 327–355.
20. Müller G. and Roggan A., eds., *Laser-Induced Interstitial Thermotherapy*, SPIE Press, Bellingham, WA, 1995.
21. Welch A.J. and van Gemert M.J.C., eds., *Optical-Thermal Response of Laser-Irradiated Tissue*, 2nd edn., Springer, New York, 2011.
22. Niemz M.H., *Laser-Tissue Interactions: Fundamentals and Applications*, 3rd edn., Springer-Verlag, Berlin, Germany, 2007.
23. Jacques S.L., ed., *Laser-Tissue Interaction: Photochemical, Photothermal, Photomechanical I–XIII*, *SPIE Proceedings*, Annual collection of SPIE papers, SPIE Press, Bellingham, WA, 1990–2002.
24. Vij D.R. and Mahesh K., eds., *Lasers in Medicine*, Kluwer, Boston, MA, 2002.
25. Jacques S.L., Strength and weaknesses of various optical imaging techniques, *Proc. SPIE* **4707**, 474–488, 2002.

26. Marks F.A., Optical determination of the hemoglobin oxygenation state of breast biopsies and human breast cancer xenografts in nude mice, *Proc. SPIE* **1641**, 227–237, 1992.

27. Ishimaru A., *Wave Propagation and Scattering in Random Media*, Academic Press, New York, 1978.

28. Van de Hulst H.C., *Multiple Light Scattering*, Vol. 1, Academic Press, New York, 1980.

29. Van de Hulst H.C., *Light Scattering by Small Particles*, Dover, New York, 1981.

30. Bohren C.F. and Huffman D.R., *Absorption and Scattering of Light by Small Particles*, Wiley, New York, 1983.

31. Star W.M., Wilson B.C., and Patterson M.C., Light delivery and dosimetry in photodynamic therapy of solid tumors, in *Photodynamic Therapy: Basic Principles and Clinical Applications*, Henderson B.W. and Dougherty T.J., eds., Marcel Dekker, New York, 1992, pp. 335–368.

32. Patterson M.S., Chance B., and Wilson B.C., Time resolved reflectance and transmittance for the non-invasive measurement of tissue optical properties, *Appl. Opt.* **28**, 2331–2336, 1989.

33. Jacques S.L., Time-resolved reflectance spectroscopy in turbid tissues, *IEEE Trans. Biomed. Eng.* **36**, 1155–1161, 1989.

34. Fishkin J.B. and Gratton E., Propagation of photon-density waves in strongly scattering media containing an absorbing semi-infinite plane bounded by a straight edge, *J. Opt. Soc. Am. A* **10**, 127–140, 1993.

35. Chance B., Cope M., Gratton E., Ramanujam N., and Tromberg B., Phase measurement of light absorption and scatter in human tissue, *Rev. Sci. Instrum.* **698**, 3457–3481, 1998.

36. Rinneberg H., Scattering of laser light in turbid media, optical tomography for medical diagnostics, in *The Inverse Problem*, Lübbig H., ed., Akademie Verlag, Berlin, Germany, 1995, pp. 107–141.

37. Schmitt J.M., Knüttel A., and Knutson J.R., Interference of diffusive light waves, *J. Opt. Soc. Am. A* **9**, 1832–1843, 1992.

38. Salzmann G.C., Singham S.B., Johnston R.G., and Bohren C.F., Light scattering and cytometry, in *Flow Cytometry and Sorting*, 2nd edn., Melamed M.R., Lindmo T., and Mendelsohn M.L., eds., Wiley-Liss Inc., New York, 1990, p. 81–107.

39. Svaasand L.O. and Gomer Ch.J., Optics of tissue, in *Dosimetry of Laser Radiation in Medicine and Biology*, Mueller G.V., Sliney D.H., eds., Vol. IS5, SPIE Press, Bellingham, WA, 1989, p. 114–132.

40. Bicout D., Brosseau C., Martinez A.S., and Schmitt J.M., Depolarization of multiply scattering waves by spherical diffusers: Influence of the size parameter, *Phys. Rev. E* **49**, 1767–1770, 1994.

41. Racovic M.J., Kattavar G.W., Mehrubeoglu M., Cameron B.D., Wang L.V., Rasteger S., and Cote G.L., Light backscattering polarization patterns from turbid media: Theory and experiment, *Appl. Opt.* **38**, 3399–3408, 1999.

42. Zege E.P., Ivanov A.P., and Katsev I.L., *Image Transfer through a Scattering Medium*, Springer-Verlag, New York, 1991.

43. Wang L.V., Coté G.L., and Jacques S.L., eds., Special section on tissue polarimetry, *J. Biomed. Opt.* **7**(3), 278–397, 2002.

44. Mishchenko M.I., Hovenier J.W., and Travis L.D., eds., *Light Scattering by Nonspherical Particles*, Academic Press, San Diego, CA, 2000.

45. Mishchenko M.I., Travis L.D., and Lacis A.A., *Scattering, Absorption, and Emission of Light by Small Particles*, Cambridge University Press, Cambridge, U.K., 2002.

46. Mishchenko M.I., Travis L.D., and Lacis A.A., *Multiple Scattering of Light by Particles: Radiative Transfer and Coherent Backscattering*, Cambridge University Press, New York, 2006.

47. Tuchin V.V., Wang L.V., and Zimnyakov D.A., *Optical Polarization in Biomedical Applications*, Springer-Verlag, New York, 2006.

48. Bartel S. and Hielscher A.H., Monte Carlo simulations of the diffuse backscattering Mueller matrix for highly scattering media, *Appl. Opt.* **39**, 2000, 1580–1588.

49. Tynes H.H., Kattawar G.W., Zege E.P., Katsev I.L., Prikhach A.S., and Chaikovskaya L.I., Monte Carlo and multicomponent approximation methods for vector radiative transfer by use of effective Mueller matrix calculations, *Appl. Opt.* **40**(3), 400–412, 2001.

50. Stockford I.M., Morgan S.P., Chang P.C.Y., and Walker J.G., Analysis of the spatial distribution of polarized light backscattering, *J. Biomed. Opt.* **7**(3), 313–320, 2002.

51. Maksimova I.L., Romanov S.V., and Izotova V.F., The effect of multiple scattering in disperse media on polarization characteristics of scattered sight, *Opt. Spectrosc.* **92**(6), 915–923, 2002.

52. Wang X. and Wang L.V., Propagation of polarized light in birefringent turbid media: A Monte Carlo study, *J. Biomed. Opt.* **7**(3), 279–290, 2002.

53. Jacques S.L., Ramella-Roman J.C., and Lee K., Imaging skin pathology with polarized light, *J. Biomed. Opt.* **7**(3), 329–340, 2002.

54. de Boer J.F. and Milner T.E., Review of polarization sensitive optical coherence tomography and Stokes vector determination, *J. Biomed. Opt.* **7**(3), 359–371, 2002.

55. Gangnus S.V., Matcher S.J., and Meglinski I.V., Monte Carlo modeling of polarized light propagation in biological tissues, *Laser Phys.* **14**, 886–891, 2004.

56. Guo X., Wood M.F.G., and Vitkin I.A., Angular measurement of light scattered by turbid chiral media using linear Stokes polarimetry, *J. Biomed. Opt.* **11**, 041105, 2006.

57. Anderson R.R., Polarized light examination and photography of the skin, *Arch. Dermatol.* **127**, 1000–1005, 1991.

58. Kollias N., Polarized light photography of human skin, in *Bioengineering of the Skin: Skin Surface Imaging and Analysis*, Wilhelm K.-P., Elsner P., Berardesca E., and Maibach H.I., eds., CRC Press, Boca Raton, FL, 1997, pp. 95–106.

59. Hemenger R.P., Refractive index changes in the ocular lens result from increased light scatter, *J. Biomed. Opt.* **1**, 268–272, 1996.

60. Born M. and Wolf E., *Principles of Optics*, 7th edn., Cambridge University Press, Cambridge, U.K., 1999.

61. Tuchin V.V., ed., *Handbook of Optical Sensing of Glucose in Biological Fluids and Tissues*, CRC Press, Boca Raton, FL, 2009.

62. Kamai Y. and Ushiki T., The three-dimensional organization of collagen fibrils in the human cornea and sclera, *Invest. Ophthalmol. Vis. Sci.* **32**, 2244–2258, 1991.

63. Hart R.W. and Farrell R.A., Light scattering in the cornea, *J. Opt. Soc. Am.* **59**, 766–774, 1969.

64. Huang Y. and Meek K.M., Swelling studies on the cornea and sclera: The effect of pH and ionic strength, *Biophys. J.* **77**, 1655–1665, 1999.

65. Maksimova I.L. and Shubochkin L.P., Light-scattering matrices for a close-packed binary system of hard spheres, *Opt. Spectrosc.* **70**(6), 745–748, 1991.

66. Wright C.H.G, Barrett S.F., and Welch A.J., Laser–tissue interaction, in *Lasers in Medicine*, Vij D.R. and Mahesh K., eds., Kluwer Academic Publishers, Boston, MA, 2002, pp. 21–58.

67. Letokhov V.S., Laser biology and medicine, *Nature* **316**(6026), 325–328, 1985.

68. Tuchin V.V., *Lasers and Fiber Optics in Biomedical Science*, 2nd edn., Fizmatlit, Moscow, Russia, 2010.

69. Oraevsky A.A., Jacques S.J., and Tittel F.K., Measurement of tissue optical properties by time-resolved detection of laser-induced transient stress, *Appl. Opt.* **36**, 402–415, 1997.

70. Karabutov A.A. and Oraevsky A.A., Time-resolved detection of optoacoustic profiles for measurement of optical energy distribution in tissues, Chapter 10, in *Handbook of Optical Biomedical Diagnostics*, Vol. PM107, Tuchin V.V., ed., SPIE Press, Bellingham, WA, 2002, pp. 585–674.

71. Sathyam U.S. and Prahl S.A., Limitations in measurement of subsurface temperatures using pulsed photothermal radiometry, *J. Biomed. Opt.* **2**, 251–261, 1997.

72. Zharov V.P., Galanzha E.I., and Tuchin V.V., Integrated photothermal flow cytometry in vivo, *J. Biomed. Opt.* **10**, 647–655, 2005.

73. Zharov V.P., Galanzha E.I., and Tuchin V.V., Photothermal image flow cytometry in vivo, *Opt. Lett.* **30**(6), 107–110, 2005.

74. Wang L.V., Ultrasound-mediated biophotonics imaging: A review of acousto-optical tomography and photo-acoustic tomography, *Dis. Markers* **19**, 123–138, 2003–2004.

75. Wang L.V. and Shen Q., Sonoluminescent tomography of strongly scattering media, *Opt. Lett.* **23**(7), 561–563, 1998.
76. Shen Q. and Wang L.V., Two-dimensional imaging of dense tissue-simulating turbid media by use of sonoluminescence, *Appl. Opt.* **38**(1), 246–252, 1999.
77. Duck F.A., *Physical Properties of Tissue: A Comprehensive Reference Book*, Academic Press, London, U.K., 1990.
78. Tuchin V.V., *Optical Clearing of Tissues and Blood*, Vol. PM 154, SPIE Press, Bellingham, WA, 2006.
79. Graaff R., Aarnoudse J.G., Zijp J.R. et al., Reduced light scattering properties for mixtures of spherical particles: A simple approximation derived from Mie calculations, *Appl. Opt.* **31**, 1370–1376, 1992.
80. Kohl M., Essenpreis M., and Cope M., The influence of glucose concentration upon the transport of light in tissue-simulating phantoms, *Phys. Med. Biol.* **40**, 1267–1287, 1995.
81. Drezek R., Dunn A., and Richards-Kortum R., Light scattering from cells: Finite-difference time-domain simulations and goniometric measurements, *Appl. Opt.* **38**, 3651–3661, 1999.
82. Bolin F.P., Preuss L.E., Taylor R.C., and Ference R.J., Refractive index of some mammalian tissues using a fiber optic cladding method, *Appl. Opt.* **28**, 2297–2303, 1989.
83. Tuchin V.V., Maksimova I.L., Zimnyakov D.A. et al., Light propagation in tissues with controlled optical properties, *J. Biomed. Opt.* **2**, 401–417, 1997.
84. Liu H., Beauvoit B., Kimura M., and Chance B., Dependence of tissue optical properties on solute—Induced changes in refractive index and osmolarity, *J. Biomed. Opt.* **1**, 200–211, 1996.
85. Vargas G., Chan E.K., Barton J.K., Rylander III H.G., and Welch A.J., Use of an agent to reduce scattering in skin, *Lasers Surg. Med.* **24**, 133–141, 1999.
86. Tuchin V.V., Bashkatov A.N., Genina E.A., Sinichkin Yu.P., and Lakodina N.A., In vivo study of the human skin clearing dynamics, *J. Tech. Phys. Lett.* **27**(6), 489–490, 2001.
87. Wang R.K. and Tuchin V.V., Optical tissue clearing to enhance imaging performance for OCT, Chapter 28, in *Optical Coherence Tomography: Technology and Applications*, Drexler W. and Fujimoto J.G., eds., Springer, Berlin, Germany, 2008, pp. 851–882.
88. Wang R.K., Tuchin V.V., Xu X., and Elder J.B., Concurrent enhancement of imaging depth and contrast for Optical Coherence Tomography by hyperosmotic agents, *J. Opt. Soc. Am. B* **18**, 948–953, 2001.
89. Khan M.H., Choi B., Chess S., Kelly K.M., McCullough J., and Nelson J.S., Optical clearing of in vivo human skin: Implications for light-based diagnostic imaging and therapeutics, *Lasers Surg. Med.* **34**, 83–85, 2004.
90. Khalil O., Non-invasive glucose measurement technologies: An update from 1999 to the dawn of the New Millennium, *Diabetes Technol. Ther.* **6**(5), 660–697, 2004.
91. Larin K.V., Eledrisi M.S., Motamedi M., and Esenaliev R.O., Noninvasive blood glucose monitoring with optical coherence tomography: A pilot study in human subjects, *Diabetes Care* **25**(12), 2263–2267, 2002.
92. Brezinski M., Saunders K., Jesser C., Li X., and Fujimoto J., Index matching to improve OCT imaging through blood, *Circulation* **103**, 1999–2003, 2001.
93. Tuchin V.V., Xu X., and Wang R.K., Dynamic optical coherence tomography in optical clearing, sedimentation and aggregation study of immersed blood, *Appl. Opt.* **41**(1), 258–271, 2002.
94. Schneckenburger H., Steiner R., Strauss W., Stock K., and Sailer R., Fluorescence technologies in biomedical diagnostics, Chapter 15, in *Optical Biomedical Diagnostics*, Vol. PM107, Tuchin V.V., ed., SPIE Press, Bellingham, WA, 2002, pp. 825–874.
95. Sinichkin Yu.P., Kollias N., Zonios G., Utz S.R., and Tuchin V.V., Reflectance and fluorescence spectroscopy of human skin in vivo, Chapter 13, in *Handbook of Optical Biomedical Diagnostics*, Vol. PM107, Tuchin V.V., ed., SPIE Press, Bellingham, WA, 2002, pp. 725–785.
96. Sterenborg H.J.C.M., Motamedi M., Wagner R.F., Duvic J.R.M., Thomsen S., and Jacques S.L., In vivo fluorescence spectroscopy and imaging of human skin tumors, *Lasers Med. Sci.* **9**, 191–201, 1994.
97. Zeng H., MacAulay C., McLean D.I., and Palcic B., Spectroscopic and microscopic characteristics of human skin autofluorescence emission, *Photochem. Photobiol.* **61**, 639–645, 1995.

98. Carden A. and Morris M.D., Application of vibration spectroscopy to the study of mineralized tissues (review), *J. Biomed. Opt.* **5**, 259–268, 2000.

99. Lucassen G.W., Caspers P.J., and Puppels G.J., Infrared and Raman spectroscopy of human skin in vivo, Chapter 14, in *Handbook of Optical Biomedical Diagnostics*, Vol. PM107, Tuchin V.V., ed., SPIE Press, Bellingham, WA, 2002, pp. 787–823.

100. Hanlon E.D., Manoharan R., Koo T.-W., Shafer K.E., Motz J.T., Fitzmaurice M., Kramer J.R., Itzkan I., Dasari R.R., and Feld M.S., Prospects for in vivo Raman spectroscopy, *Phys. Med. Biol.* **45**, R1–R59, 2000.

101. Petry R., Schmitt M., and Popp J., Raman spectroscopy—A prospective tool in the life sciences, *ChemPhysChem* **4**, 14–30, 2003.

102. Skrebova Eikje N., Ozaki Y., Aizawa K., and Arase S., Fiber optic near-infrared Raman spectroscopy for clinical noninvasive determination of water content in diseased skin and assessment of cutaneous edema, *J. Biomed. Opt.* **10**, 014013-1-13, 2005.

103. Yaroslavskaya A.N., Yaroslavsky I.V., Otto C., Puppels G.J., Duindam H., Vrensen G.F.J.M., Greve J., and Tuchin V.V., Water exchange in human eye lens monitored by confocal Raman microspectroscopy, *Biophysics* **43**(1), 109–114, 1998.

104. Dainty J.C., ed., *Laser Speckle and Related Phenomena*, 2nd edn., Springer-Verlag, New York, 1984.

105. Aizu Y. and Asakura T., Bio-speckle phenomena and their application to the evaluation of blood flow, *Opt. Laser Technol.* **23**, 205–219, 1991.

106. Briers J.D., Laser Doppler and time-varying speckle: A reconciliation, *J. Opt. Soc. Am. A* **13**, 345–350, 1996.

107. Serov A., Steinacher B., and Lasser T., Full-field laser Doppler perfusion imaging and monitoring with an intelligent CMOS camera, *Opt. Express* **13**, 3681–3689, 2005.

108. Forrester K.R., Tulip J., Leonard C., Stewart C., and Bray R.C., A laser speckle imaging technique for measuring tissue perfusion, *IEEE Trans. Biomed. Eng.* **51**, 2074–2084, 2004.

109. Liu Q., Wang Z., and Luo Q., Temporal clustering analysis of cerebral blood flow activation maps measured by laser speckle contrast imaging, *J. Biomed. Opt.* **10**(2), 024019-1-7, 2005.

110. Boas D.A., Campbell L.E., and Yodh A.G., Scattering and imaging with diffusing temporal field correlations, *Phys. Rev. Lett.* **75**, 1855–1858, 1995.

111. Zimnyakov D.A., Tuchin V.V., and Yodh A.G., Characteristic scales of optical field depolarization and decorrelation for multiple scattering media and tissues, *J. Biomed. Opt.* **4**, 157–163, 1999.

112. Yu G., Lech G., Zhou C., Chance B., Mohler III E.R., and Yodh A.G., Time-dependent blood flow and oxygenation in human skeletal muscles measured with noninvasive near-infrared diffuse optical spectroscopies, *J. Biomed. Opt.* **10**(2), 024027-1-7, 2005.

113. Durduran T., Zhou C., Buckley E.M. et al., Optical measurement of cerebral hemodynamics and oxygen metabolism in neonates with congenital heart defects, *J. Biomed. Opt.* 15(3), 037004-1-10, 2010.

114. Diaspro A., ed., *Confocal and Two-Photon Microscopy: Foundations, Applications, and Advances*, Wiley-Liss Inc., New York, 2002.

115. Bouma B.E. and Tearney G.J., eds., *Handbook of Optical Coherence Tomography*, Marcel Dekker, New York, 2002.

116. Berlien H.-P. and Müller G.J., eds., *Applied Laser Medicine*, Springer-Verlag, Berlin, Germany, 2003.

117. Wilson B., Tuchin V., and Tanev S., eds., *Advances in Biophotonics*, NATO Science Series I. Life and Behavioural Sciences, Vol. 369, IOS Press, Amsterdam, the Netherlands, 2005.

118. Kishen A. and Asundi A., eds., *Photonics in Dentistry*. Series of Biomaterials and Bioengineering, Imperial College Press, London, U.K., 2006.

119. Masters B.R. and So P.T.C., eds., *Handbook of Biomedical Nonlinear Optical Microscopy*, Oxford University Press, New York, 2008.

120. Drexler W. and Fujimoto J.G., eds., *Optical Coherence Tomography: Technology and Applications*, Springer, Berlin, Germany, 2008.

121. Splinter R. and Hooper B.A., *An Introduction to Biomedical Optics*, CRC Press, Boca Raton, FL, 2007.
122. Wax A. and Backman V., eds., *Biomedical Applications of Light Scattering*, McGraw-Hill, New York, 2009.
123. Wang L., ed., *Photoacoustic Imaging and Spectroscopy*, CRC Press, Boca Raton, FL, 2009.
124. Peiponen K.-E., Myllylä R., and Priezzhev A.V., *Optical Measurement Techniques, Innovations for Industry and the Life Science*, Springer-Verlag, Berlin, Germany, 2009.
125. Baron E.D., ed., *Light-Based Therapies for Skin of Color*, Springer, London, U.K., 2009.
126. Ahluwalia G., ed., *Light Based Systems for Cosmetic Application*, William Andrew, Norwich, England, 2009.
127. Tuchin V.V., ed., *Handbook of Photonics for Biomedical Science*, CRC Press, Boca Raton, FL, 2010.
128. Pavone F.S., ed., *Laser Imaging and Manipulation in Cell Biology*, Wiley-VCH Verlag GmbH & Co. KGaA, Weinheim, Germany, 2010.
129. Boas D.A., Pitris C., and Ramanujam N., eds., *Handbook of Biomedical Optics*, CRC Press, Boca Raton, FL, 2011.
130. Popp J., Tuchin V.V., Chiou A., and Heinemann S.H., eds., *Handbook of Biophotonics*, Vol. 1: *Basics and Techniques*, Wiley-VCH Verlag GmbH & Co. KGaA, Weinheim, Germany, 2011.
131. Popp J., Tuchin V.V., Chiou A., and Heinemann S.H., eds., *Handbook of Biophotonics*, Vol. 2: *Photonics for Health Care*, Wiley-VCH Verlag GmbH & Co. KGaA, Weinheim, Germany, 2012.
132. Popp J., Tuchin V.V., Chiou A., and Heinemann S.H., eds., *Handbook of Biophotonics*, Vol. 3: *Photonics in Pharmaceutics, Bioanalysis and Environmental Research*, Wiley-VCH Verlag GmbH & Co. KGaA, Weinheim, Germany, 2012.
133. Tuchin V.V. *Dictionary of Biomedical Optics and Biophotonics*, SPIE Press, Bellingham, WA, 2012.
134. Wang R.K. and Tuchin V.V., eds., *Advanced Biophotonics: Tissue Optical Sectioning*, CRC Press, Taylor & Francis Group, London, U.K., 2013.

4

Theoretical Models and Algorithms in Optical Diffusion Tomography

Stephen J. Norton
Duke University

Tuan Vo-Dinh
Duke University

4.1 Introduction

The potential of optical tomography as a new diagnostic tool has stimulated considerable interest in the last 20 years [1–10]. Although limited to about the first 5 cm of the body, the technique offers several advantages often not available in established imaging modalities such as ultrasound, x-ray computed tomography, and magnetic resonance imaging [5,6]. These benefits include nonionizing radiation, relatively inexpensive instrumentation, and the potential for functional (i.e., spectroscopic) imaging of optical tissue properties. In functional imaging, light at specific wavelengths can be used to excite specific biological molecules of interest, such as NAD, NADH, tryptophan, and hemoglobin, in order to provide real-time, in vivo information on the functional status of tissues and organs (e.g., pH, tissue oxygenation, glucose, dysplasia, tumor). Luminescence techniques based on bioluminescence allow the monitoring of gene expression in vivo in animals for drug discovery investigations.

The use of light for diagnostics of deep tissue still presents great challenges. Electromagnetic radiation in the UV and visible wavelength range is strongly absorbed by biological species in tissue. Due to absorption by tissue chromophores, the intensity of light decreases rapidly as it penetrates deep inside tissue. And at all wavelengths, scattering of light occurs strongly in tissue. However, in the near-infrared *optical window*, where hemoglobin absorbs weakly (600–900 nm), multiple scattering in tissue dominates absorption, and penetration can be substantial, sometimes reaching 5 cm or more.

As light propagates in an optically dense medium such as tissue, it rapidly loses its coherence, and as a result, light intensity (or photon flux density or radiance) is usually the observable quantity. To a reasonable approximation, the radiance can be shown to obey a diffusion equation, which, when the source is modulated, reduces to a Helmholtz equation with a complex wave number. With a harmonically modulated source, the radiance oscillates in both space and time; this oscillation is termed a diffuse photon density wave (DPDW). Such waves, although highly damped, are diffracted and scattered by optical inhomogeneities within the tissue. A number of authors have reported both theoretical and experimental investigations of these waves [11–35].

This chapter describes several approaches to optical tomography based upon theoretical models of light propagation and scattering in tissue. We examine in some detail three optical tomography algorithms. The first algorithm employs Fourier methods to reconstruct an image from time-harmonic DPDW data recorded at one modulation frequency. This method can be regarded as a generalization of diffraction tomography (DT) developed originally for acoustical imaging. The second and third algorithms are iterative and attempt to minimize an error norm in the frequency and time domains, respectively. An overview is also given of other algorithmic approaches that have appeared in the literature in the last two decades. Experimental techniques, instrumentation, and clinical applications of optical tomography are described in Chapter 10 of this volume and Chapters 9, 18, and 24 of Volume II in this handbook. Recent reviews of optical diffusion tomography have been published by Boas et al. [8] and Arridge and coworkers [9,10].

Here, we focus on imaging algorithms for reconstructing spatial maps of optical properties of tissue by exploiting models of the interaction of light with tissue. Two basic measurement methodologies are used for imaging: (1) frequency-domain methods that employ harmonically modulated photon density waves and phase-resolved detection, which measures the phase shift of the photon density waves [17,22,24,28,36–52], and (2) *time-resolved* methods that use pulsed excitation and gated detection to examine the response of tissue to a short pulse of incident light [53–67]. Figure 4.1 shows a schematic diagram of the phase-resolved and time-resolved techniques. *Continuous-wave* imaging methods, which employ steady-state illumination, may also be classified as frequency domain, since these correspond to the zero-frequency limit [68,69].

The simplest time-resolved imaging method exploits the fact that the first-arrival photons propagate along paths that have not deviated significantly from the line joining the source and receiver. Such photons are referred to as *ballistic photons*. Photons that are scattered, predominantly in the forward direction, are sometimes referred to as *snakelike photons*, as illustrated in Figure 4.2. In the case of ballistic photons, scattering is neglected, and conventional tomographic algorithms based on straight-line

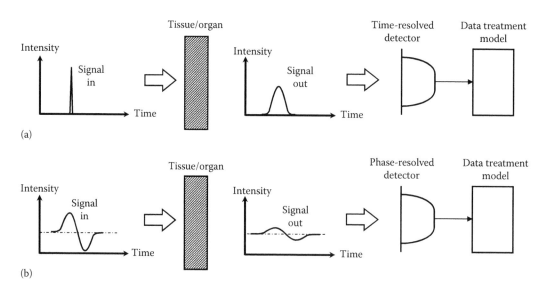

FIGURE 4.1 Schematic diagram of the (a) time-resolved and (b) phase-resolved techniques.

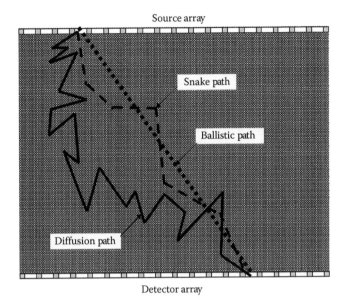

Source array

Snake path

Ballistic path

Diffusion path

Detector array

FIGURE 4.2 Schematic diagram of various photon trajectories through tissue medium.

propagation are in principle applicable. Because of the need for ultrafast laser pulses for excitation (femtosecond and picosecond time frames [70]), the very precise timing used to gate the first-arrival photons is technically challenging. In this regard, harmonic modulation of an intensity source is less demanding. Moreover, beyond a few centimeters, the relative number of ballistic photons becomes exceedingly small, so that ballistic measurements rapidly become signal-to-noise limited. Frequency-domain methods offer better signal-to-noise performance because more photons are available for measurement and synchronous detection using phase-resolved measurements can be employed to enhance the signal. In time-resolved methods, the *boxcar* or a pulse-averaging scheme is often used to improve the signal-to-noise ratio. Modern time-resolved methods typically examine the entire transient response, rather than merely the ballistic photons, and attempt to fit this response to a model based, for example, on the time-dependent diffusion equation. In this manner, scattered photons are also accounted for, and time-domain methods that employ rapid averaging of the transient signals can achieve signal-to-noise ratios competitive with frequency-domain methods. In this chapter, we describe in detail two frequency-domain imaging algorithms, followed by a time-domain imaging algorithm applicable to time-resolved measurements.

We focus on two particular formulations of the inverse problem: analytical and iterative. Here, the term *inverse problem* refers to the task of deriving the spatial distribution of optical tissue properties from the detected scattered photons. The analytical approaches are noniterative, although we distinguish them from linear inversion schemes that are purely numerical, that is, methods that solve a linear system of equations directly. These analytical approaches have largely been inspired by DT, which, in its conventional formulation, is a wave-equation-based inversion scheme [71–74]. In particular, DT provides an explicit solution to a linearized integral equation using Fourier inversion methods. Traditional DT begins with the Helmholtz equation with a real wave number, but our starting point is the diffusion equation; the latter reduces to a Helmholtz equation with a complex wave number when the source is harmonically modulated. In this case, the required generalization of DT resembles a Fourier–Laplace inversion. Markel and Schotland [51] have formulated the problem in this way, which requires an analytic continuation of the data in the Fourier domain. We describe a related approach that provides an explicit inversion formula [43], as does Markel and Schotland's method, but does not invoke an inverse Laplace transform and more explicitly defines the process of analytic continuation.

Another category of algorithms attempts to minimize iteratively a global error norm that quantifies the difference between measured and predicted data. Typically, the error norm is the mean-square error, although other norms are possible. The predicted observations are computed using a forward solution to the transport problem, but the choice of forward algorithm (e.g., finite element, finite difference, or an integral equation solution) is not relevant for our purposes, although speed and accuracy are obviously desirable traits. Iterative approaches are computationally expensive but have some advantages over analytical solutions. For example, analytical solutions, or Fourier methods, are necessarily based on a linearization procedure, which requires a weak-scattering assumption. Nonlinear iterative algorithms need make no such assumption, since these methods iterate a forward algorithm, which, in principle, need not be subject to approximations. The forward algorithm can be designed to deal with voids and complicated boundaries, as well as arbitrary source and detector-array geometries. The forward algorithm may, for example, solve the full integrodifferential transport equation [27,29,34,75–77] (the radiative transfer equation), rather than the diffusion equation, which is generally the starting point for DT-based methods. Conditions for which the diffusion approximation holds or breaks down have been examined [8–10,27,29,34,77]. Much of the early work assumed the diffusion approximation for analytic tractability and the relative simplicity of the forward algorithms. More recently, however, the trend towards formulations based on the equation of radiative transfer has increased [78–84].

In investigating iterative algorithms, we emphasize the importance of the adjoint method in computing the functional gradient (or Fréchet derivative) of the error norm. The numerical evaluation of this gradient is generally the most demanding part of the computation, but the adjoint method renders this computation vastly more efficient than a *brute-force* gradient evaluation based, for example, on finite differences. In the second algorithm described, we show how the adjoint method can be applied directly to the integrodifferential transport equation, rather than to the diffusion equation, which is the more common approach.

4.2 Photon Transport in Tissue

In this section, we review some of the basic equations obeyed by photon density waves and describe some of their properties. Most imaging algorithms begin with an equation derived from the Boltzmann transport equation (or radiative transfer equation), whose time-dependent form is given by [75,76]

$$\frac{1}{c}\frac{\partial \Psi(\boldsymbol{r},t,\boldsymbol{s})}{\partial t} + \boldsymbol{s}\cdot\nabla\Psi(\boldsymbol{r},t,\boldsymbol{s}) + \sigma_t(\boldsymbol{r})\Psi(\boldsymbol{r},t,\boldsymbol{s}) = \sigma_s(\boldsymbol{r})\int\Psi(\boldsymbol{r},t,\boldsymbol{s}')f(\boldsymbol{s}\cdot\boldsymbol{s}')d^2\boldsymbol{s}' + q(\boldsymbol{r},t,\boldsymbol{s}), \tag{4.1}$$

where

$\Psi(\boldsymbol{r},t,\boldsymbol{s})$ is the time-dependent radiance (or photon flux density) in the direction $\boldsymbol{s}$ at the point $\boldsymbol{r}$

c is the velocity of light in the medium

$\sigma_t(\boldsymbol{r}) = \sigma_s(\boldsymbol{r}) + \sigma_a(\boldsymbol{r})$, where $\sigma_s(\boldsymbol{r})$ and $\sigma_a(\boldsymbol{r})$ are the spatially dependent scattering and absorption cross sections

$f(\boldsymbol{s}\cdot\boldsymbol{s}')$ is the phase function that denotes the probability of an incident photon from direction $\boldsymbol{s}$ being scattered into the direction $\boldsymbol{s}'$

$q(\boldsymbol{r},t,\boldsymbol{s})$ is the source flux

The diffusion approximation results on performing a spherical harmonic expansion of the radiance in Equation 4.1 and retaining the lowest-order nontrivial terms in the expansion. Defining the photon density as

$$\Psi(\boldsymbol{r},t) = \int\Psi(\boldsymbol{r},t,\boldsymbol{s})d^2\boldsymbol{s}, \tag{4.2}$$

the diffusion approximation can be shown to give [75]

$$\frac{1}{c}\frac{\partial \Psi(r,t)}{\partial t} - \nabla \cdot \left[D(r)\, \nabla \Psi(r,t) \right] + \sigma_a(r)\, \Psi(r,t) = Q(r,t), \tag{4.3}$$

where

$D(r) = 1/3\left[\sigma_a(r) + \sigma_s'(r)\right]$ is the diffusion coefficient
$\sigma_s'(r) = (1-g)\sigma_s(r)$ is the reduced scattering coefficient
g is the mean cosine of the scattering angle
$Q(r,t)$ is the source density function

If we now assume that the sources in Equations 4.1 and 4.3 are both harmonically modulated at frequency ω so that $q(r,t,s) = q(r,s)\exp(-i\omega t)$ and $Q(r,t) = Q(r)\exp(-i\omega t)$, then, writing $\Psi(r,t,s) = \Psi(r,s)$ $\exp(-i\omega t)$, Equation 4.1 becomes

$$s \cdot \nabla \Psi(r,s) + \left[\sigma_t(r) - \frac{i\omega}{c} \right] \Psi(r,s) = \sigma_s(r) \int \Psi(r,s')f(s \cdot s')d^2 s' + q(r,s), \tag{4.4}$$

and, writing $\Psi(r,t) = \Psi(r)\exp(-i\omega t)$, Equation 4.3 reduces to

$$\nabla \cdot \left[D(r)\nabla \Psi(r) \right] + \left[\frac{i\omega}{c} - \sigma_a(r) \right] \Psi(r) = -Q(r). \tag{4.5}$$

Note that the radiance $\Psi(r,s)$ in Equation 4.4 and the photon density $\Psi(r)$ in Equation 4.5 are now complex quantities, since in-phase and quadrature components of the signals can be measured, corresponding to the real and imaginary parts of $\Psi(r,s)$ and $\Psi(r)$.

Now write $D(r) = D_0 + D_1(r)$ and $\sigma_a(r) = \sigma_0 + \sigma_1(r)$, where $D_1(r)$ and $\sigma_1(r)$ will denote perturbations assumed to be small compared to the background values D_0 and σ_0 if the equations are ultimately to be linearized. Then substituting into Equation 4.5 gives

$$\nabla^2 \Psi(r) + k_0^2 \Psi(r) = -\frac{1}{D_0}\nabla \cdot \left[D_1(r)\nabla \Psi(r) \right] + \frac{\sigma_1(r)}{D_0} - S(r), \tag{4.6}$$

where $S(r) \equiv Q(r)/D_0$ and the wave number is given by

$$k_0 = \left[\frac{i\omega}{cD_0} - \frac{\sigma_0}{D_0} \right]^{1/2}. \tag{4.7}$$

By definition, the wavelength and the phase velocity of the photon density wave $\Psi(r)$ are given, respectively, by [15] $\lambda = 2\pi/k_r$ and $v_p = \omega/k_r$, where k_r is the real part of Equation 4.7. The *skin depth*, or depth to which the photon density falls to e^{-1} of its incident value, is given by $\delta = 1/k_i$, where k_i is the imaginary part of Equation 4.7. If, for example, we choose a modulation frequency such that $\omega \gg \sigma_a c$, then Equation 4.7 simplifies to $k_0 = \sqrt{i\omega/cD_0} = (1+i)\sqrt{\omega/2cD_0}$, in which case, $\lambda = 2\pi\sqrt{2cD_0/\omega}$, $v_p = \sqrt{2cD_0\omega}$, and $\delta = \sqrt{2cD_0/\omega}$.

4.3 Optical Diffusion Tomography

4.3.1 Classes of Inversion Algorithms

The reconstruction of an image of optical properties requires the solution to an inverse problem. One can classify inversion algorithms as one of three types:

1. Algorithms based on linear numerical inversion methods
2. Algorithms based on Fourier-based inversion methods
3. Algorithms based on nonlinear iterative methods

The starting point for the first two methods is typically the diffusion equation, which is then transformed to an integral equation via Green's theorem (known as the Lippmann–Schwinger equation), linearized to first order in the inhomogeneity (the Born approximation), and then solved. In the first class of algorithms, a process of discretization converts the integral equation into a system of linear equations, which is then numerically solved using some type of regularized inversion scheme. Examples of this approach are singular-value decomposition and various iterative procedures such as an algebraic reconstruction technique or a conjugate-gradient algorithm using, for example, a Tikhonov regularization procedure [7,36,49,85].

In the second class of algorithms, one solves the linearized integral equation using Fourier methods. DT falls into this class; in this case, the integral equation is manipulated into a form from which the Fourier transform of the unknown optical inhomogeneity can be recovered. An inverse Fourier transform then yields the image. An aesthetically pleasing feature of Fourier-based methods is that they sometimes yield explicit inversion formulas, although the derivation of these formulas generally requires that the source and detection surfaces have simple geometries, such as lines or circles for 2D problems and planes and cylinders or spheres for 3D problems [73,86]. These inversion formulas are computationally efficient because the inversion has been performed analytically, whereas a purely numerical solution requires the equivalent of the inversion of a large matrix. Although numerical inversion algorithms allow greater freedom in formulating regularization schemes, regularization is still possible in Fourier-based methods, usually in the form of a low-pass spatial-frequency filter. This filter attenuates the higher spatial-frequency components to reduce noise sensitivity.

Imaging algorithms in the third category are based on nonlinear optimization techniques, in which a global error norm, such as the mean-square error, is iteratively minimized. This method is the most general, since, in principle, no approximations need be made for the benefit of analytical tractability. Here, the unknown inhomogeneity is sought that best predicts the data, subject perhaps to additional a priori constraints. At each iteration, the forward solution is computed on the basis of the current estimate of the unknown inhomogeneity and compared to the measurements. Normally, gradient descent methods are employed to minimize the mean-square error, although other approaches, such as simulated annealing or evolution algorithms [87], are possible.

Gaudette et al. [49] recently conducted a comprehensive review of the first class of algorithms, linear numerical schemes. In this chapter, we shall focus on the second and third classes of algorithms, Fourier methods and nonlinear iterative methods. We will discuss examples of each of these approaches from our research to illustrate both techniques.

4.3.2 Analytical and Quasi-Analytical Methods

In the category of Fourier-based methods, we include DT methods, as well as back-projection and angular spectrum techniques.

Inspired by x-ray tomography, several algorithms based on the concept of filtering and back projection have been proposed. Walker et al. [88] proposed a particularly simple back-projection algorithm in which the data, after filtering, are back-projected along the straight lines joining each source and detector. Colak et al. [38] conducted a comprehensive investigation of back-projection methods using different filtering schemes. Such methods are rapid and robust but do not take into account the diffraction or scattering of the photon density waves.

Li et al. [24] and Durduran et al. [89] examined other algorithms that rigorously take into account DPDW diffraction, by use of an angular spectrum decomposition of the scattered waves. This approach is well suited for data acquisition over planar geometries, since a spatial Fourier transform is performed over a plane (or line in 2D imaging). The angular spectrum analysis decomposes the DPDW field into spatial-frequency components, each of which propagates in different directions and attenuates at different rates. This method is particularly convenient for examining how spatial resolution changes with depth. The authors used the angular spectrum formulation to develop a quasi-analytical reconstruction algorithm based on filtering and back propagation of the DPDW field.

A closely related approach developed by a number of authors begins by linearizing the Green integral and then employing a plane-wave expansion of the Green function [46,90]. This is a first step towards a conventional DT formulation, except that in this case, the problem is complicated by the complex wave number. Along these lines, Matson [39,45,52] developed a projection slice theorem that relates the Fourier transform of the data to that of the object. However, the theorem in the form presented is not suitable for reconstructing the object directly via Fourier inversion because the complex wave number gives rise to complex spatial-frequency variables. Norton and Vo-Dinh [43] showed how an explicit solution can be derived by a process of analytical continuation that makes the spatial-frequency variables real, after which a conventional inverse Fourier transform can be performed to reconstruct the object. This approach is described in more detail in the succeeding text. A method somewhat related to that of Norton and Vo-Dinh was developed by Markel and Schotland [51], who obtained a solution to the Born integral equation based on a Fourier–Laplace inversion scheme. Their solution also implies analytically continued data. Other authors have also developed formal solutions for continuous-wave imaging (i.e., in the zero-frequency limit) based on a Laplace transform inversion [68,69], and Chen et al. [44] developed an approximate explicit solution by using a stationary-phase approximation to the Green integral.

4.3.3 Nonlinear Iterative Methods

Nonlinear iterative approaches attempt to find the inhomogeneity that best predicts the data [47,64,91,92]. In these algorithms, regularization schemes are relatively easy to implement—for example, by attaching a penalty function to the error functional, which can be designed to emphasize objects that are smooth [93] or have minimum norm or maximum entropy. Also, a priori constraints can be enforced at each iteration. Formulating the method in a maximum-likelihood or a Bayesian framework allows noise statistics to be taken explicitly into account [94–97]. Typically, the image is decomposed into pixels, each of which is described by one or more parameters, such as the diffusion or absorption coefficient at that point. A pixel basis is not necessary, however; other parameterizations are possible, such as the shape of the anomaly [50,98–101].

An essential ingredient in most of these schemes is the computation of the functional gradient (or Fréchet derivative) of the mean-square error [102]. A brute-force approach to evaluating the gradient could be performed by computing finite differences. That is, if there were N unknowns characterizing the inhomogeneity (e.g., N pixels in the image), then a minimum of N forward problems would be required to compute the N derivatives using finite differences. With the adjoint method, only two forward problems need be solved: one forward problem and its adjoint [102–109]. Norton [104], Arridge and Schweiger [105], and Dorn [106,107] appear to be among the first to have applied the adjoint method to the full radiative transfer equation. We examine this method later with a simple example of a continuous-wave inverse problem for reconstructing an image of the scattering coefficient in a 2D object. We also show how this approach can be applied to time-domain or transient data.

4.4 Algorithms for Imaging

4.4.1 Explicit Solution Based on Diffraction Tomography

We present in this section an analytical solution to the inverse diffusion problem using the methods of DT [43]. As noted, DT is a wave-equation-based inversion procedure whose starting point is a Helmholtz equation with a real wave number. As we shall see, however, the complex wave number in the Helmholtz equation (4.7) that arises from the diffusion equation can be dealt with by analytical continuation of the data in the Fourier domain; then, the conventional DT methodology can be applied.

For simplicity, we assume that the frequency ω is large compared to $\mu_a c$; although this assumption is not necessary, it simplifies the algebra. In this case, Equation 4.7 becomes simply $k_0 = \sqrt{i\omega/cD_0}$. We also assume that scattering dominates absorption, so that the parameter to be imaged is the diffusion coefficient $D(r)$ or, equivalently, the perturbation $D_1(r)$. Then Equation 4.6 reduces to

$$\nabla^2 \Psi(r) + k_0^2\, \Psi(r) = -\frac{1}{D_0}\nabla\cdot\left[D_1(r)\,\nabla\Psi(r)\right] - S(r). \tag{4.8}$$

Equation 4.8 can be converted to an integral equation using standard techniques, giving

$$\Psi(r) = \Psi_i(r) + \frac{1}{D_0}\iint \nabla\cdot\left[D_1(r')\,\nabla\Psi(r')\right]g\left(r|r'\right)\,d^3r', \tag{4.9}$$

where $g(r|r')$ is a Green function and

$$\Psi_i(r) = \iint S(r')g\left(r|r'\right)d^3r \tag{4.10}$$

is the incident field. The Green function is defined by

$$\nabla^2 g\left(r\,|\,r'\right) + k_0^2\, g\left(r|r'\right) = -\delta(r - r').$$

If the source can be approximated as a point at r_s, then letting $S(r) = \delta(r - r_s)$, Equation 4.10 reduces to $\Psi_i(r) = g(r|r_s)$. In the Born approximation, the scattered field, represented by the integral on the right-hand side of Equation 4.9, is assumed to be much smaller than the incident field, $\Psi_i(r)$. Then, to the first-order approximation in the inhomogeneity $D_1(r)$, we replace $\Psi(r)$ by $\Psi_i(r)$ in the integrand of Equation 4.9. Defining the scattered field by $\Psi_s = \Psi - \Psi_i$ and setting $r = r_d$ in Equation 4.9, where r_d is the detection point, we obtain

$$\Psi_s(r_s, r_d) = \frac{1}{D_0}\int \nabla\cdot[D_1(r)\,\nabla g(r_s|r)]\,g(r_d|r)d^3r. \tag{4.11}$$

Here, we have substituted $\Psi_i(r) = g(r|r_s) = g(r_s|r)$ in the integrand of Equation 4.9. Abbreviating $g_s = g(r_s|r)$ and $g_d = g(r_d|r)$, this integral can be rewritten using an integration by parts with the aid of the identity $\nabla\cdot[D_1\nabla g_s]\,g_d = \nabla\cdot[D_1 g_d \nabla g_s] - D_1\nabla g_d \cdot \nabla g_s$. When this is substituted into Equation 4.11, the volume integral of $\nabla\cdot[D_1 g_d \nabla g_s]$ can be converted to a surface integral over the domain of integration with the aid of the divergence theorem. This integral will vanish, since we can define $D_1(r) = 0$ on this surface. Then Equation 4.11 becomes

$$\Psi_s(r_s, r_d) = -\frac{1}{D_0}\int D_1(r)\,\nabla g(r_s|r)\cdot\nabla g(r_d|r)d^3r \tag{4.12}$$

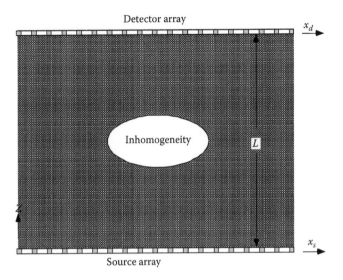

FIGURE 4.3 Simplified 2D geometry of source–detector system. The source point is confined to a line with coordinates $r_s = (x_s, 0)$ and the detector point is confined to a line with coordinates $r_d = (x_d, L)$. See text for discussion.

For simplicity, at this point, we consider a 2D problem, although the theory can be readily generalized to 3D. We then write $r = (x, z)$, with all quantities assumed to be independent of the third dimension, y. Although we will for brevity refer to the sources as *points* lying along a line array, for a true 2D problem, each *point* source is really a line source in the direction y. A similar statement can be made for the *point* detectors.

Consider now the 2D geometry shown in Figure 4.3, in which the source point is confined to a line with coordinates $r_s = (x_s, 0)$ and the detector point is confined to a line with coordinates $r_d = (x_d, L)$ parallel to, and at a distance L from, the source line. We now employ the following plane-wave expansions of the 2D Green functions:

$$g(r|r_s) = \frac{i}{4\pi} \int_{-\infty}^{\infty} \frac{dk_x}{\gamma} \exp\left[ik_x(x - x_s)\right] \exp(i\gamma z), \tag{4.13a}$$

$$g(r|r_d) = \frac{i}{4\pi} \int_{-\infty}^{\infty} \frac{dk_x}{\gamma} \exp\left[ik_x(x - x_d)\right] \exp[i\gamma(L - z)], \tag{4.13b}$$

where

$$\gamma = \sqrt{k_0^2 - k_x^2}. \tag{4.13c}$$

We now compute the 2D Fourier transform of Equation 4.12 with respect to x_s and x_d, that is,

$$\tilde{\Psi}_s(k_s, k_d) \equiv \int_{-\infty}^{\infty} \int \Psi_s(r_s, r_d) \exp\left[i(k_d x_d + k_s x_s)\right] dx_s dx_d. \tag{4.14}$$

Equations 4.13a and b are substituted into Equation 4.12 and the result into Equation 4.14. After interchanging the orders of integration and some manipulation, we obtain

$$\Psi_s(k_s,k_d) = C(k_s,k_d) \int_0^{L_x} dx \int_0^{L_z} dz\, D_1(x,z)\, \exp\big[i(k_s+k_d)x + i(\gamma_s-\gamma_d)z\big],$$ (4.15)

where

$$\gamma_s \equiv \sqrt{k_0^2 - k_s^2}$$

$$\gamma_d \equiv \sqrt{k_0^2 - k_d^2}$$

$$C(k_s,k_d) \equiv -\frac{1}{4D_0}\frac{k_sk_d-\gamma_s\gamma_d}{\gamma_s\gamma_d}\exp[i\gamma_d L].$$ (4.16)

Note that in Equation 4.15 we have limited the nonzero domain of $D_1(x,z)$ to the rectangular region $0 < x < L_x$ and $0 < z < L_z$. Note also that Equation 4.15 is in the form of a 2D Fourier transform of the inhomogeneity $D_1(x,z)$, which can be written as

$$\tilde{\Psi}_s(k_s,k_d) = C(k_s,k_d) \int_0^{L_x} dx \int_0^{L_z} dz\, D_1(x,z)\, \exp\big[i(k_x x + k_z z)\big],$$ (4.17)

with transform variables defined by

$$k_x \equiv k_s + k_d,$$ (4.18a)

$$k_z \equiv \gamma_s - \gamma_d = \sqrt{k_0^2 - k_s^2} - \sqrt{k_0^2 - k_d^2}.$$ (4.18b)

For Equation 4.17 to assume the form of an ordinary 2D Fourier transform, the transform variables k_x and k_z must both be real; however, as written, k_z in Equation 4.18b is complex, since $k_0^2 = i\omega/cD_0$. To invert Equation 4.17, we perform an analytical continuation of k_s and k_d into their respective complex planes in such a way as to make both k_x and k_z real; then, an ordinary inverse Fourier transform of Equation 4.17 will yield the reconstruction of $D_1(x,z)$, denoted $\hat{D}_1(x,z)$:

$$\hat{D}_1(x,z) = \frac{1}{(2\pi)^2}\int\int_{-\infty}^{\infty}\frac{\tilde{\Psi}_s(k_s,k_d)}{C(k_s,k_d)}\exp\big[-i(k_x x + k_z z)\big]dk_x dk_z.$$ (4.19)

The transform variables k_x and k_z in Equation 4.18 can be made real using the following change of variables:

$$k_s \equiv k_0 \sin(p-q+iv),$$ (4.20a)

$$k_d \equiv k_0 \sin(p+q+iv),$$ (4.20b)

where p, q, and v are new variables. Recalling that $k_0 \equiv \sqrt{i\omega/cD_0} = (1+i)\sqrt{\omega/2cD_0}$, substituting Equation 4.20 into Equation 4.18, and employing trigonometric identities gives

$$k_x = \sqrt{\frac{2\omega}{cD_0}}(1+i)\sin(p+iv)\cos q,$$ (4.21a)

$$k_z = \sqrt{\frac{2\omega}{cD_0}}\,(1+i)\sin(p+iv)\sin q. \tag{4.21b}$$

We now choose the variables p and v such that the imaginary part of $(1 + i)\sin(p + iv)$ vanishes. This gives the relation

$$\tan p = -\tanh v. \tag{4.22}$$

With this condition, Equations 4.21a and b become purely real. On employing Equation 4.22 to eliminate v in Equations 4.21a and b, we obtain

$$k_x = f(p)\cos q, \tag{4.23a}$$

$$k_z = f(p)\sin q, \tag{4.23b}$$

where $f(p) \equiv \sqrt{2\omega/cD_0}\,\sin 2p/\sqrt{\cos 2p}$.

Note that by forcing k_x and k_z to be real in Equation 4.19, we have made k_s and k_d complex. The implications of this will be explained shortly. In evaluating Equation 4.19, it is convenient to change the variables of integration from (k_x, k_x) to (p,q) since, in the integrand of Equation 4.19,

$$\tilde{\Psi}_s(k_s,k_d) = \tilde{\Psi}_s[k_s(p,q),k_d(p,q)] \equiv \tilde{\Psi}_s(p,q),$$

and similarly

$$C(k_s,k_d) = C[k_s(p,q),k_d(p,q)] \equiv C(p,q),$$

where the relations $k_s(p,q)$ and $k_d(p,q)$ are defined through Equations 4.21 and 4.22. From Equation 4.23, the entire Fourier plane $-\infty < k_x < \infty$ and $-\infty < k_z < \infty$ maps into the domain $0 \le p < \pi/4$ and $0 \le q < 2\pi$. Then, on changing variables to (p,q), Equation 4.19 becomes

$$\hat{D}_1(x,z) = \frac{1}{(2\pi)^2} \int_0^{\pi/4} dp\, J(p)A(p) \int_0^{2\pi} dq\, \frac{\tilde{\Psi}_s(k_s,k_d)}{C(k_s,k_d)} \exp[-if(p)(x\cos q + z\sin q)], \tag{4.24}$$

where $J(p)$ is the Jacobian of the transformation from (k_x,k_x) to (p,q), which can be shown to be $J(p) = (2\omega/cD_0)(1 + \cos^2 2p)\sin 2p/\cos^2 2p$, and $A(p)$ has been inserted to serve as a spectral apodization function, or low-pass filter, that forces the integrand to zero for large p. The simplest choice for $A(p)$ is $A(p) = 1$ for $p < p_{max}$ and $A(p) = 0$ for $p > p_{max}$. The function $A(p)$ is necessary to compensate for any exponential growth in the analytically continued data $\tilde{\Psi}_s(k_s, k_d)$ caused by noise. The low-pass filter $A(p)$ may be regarded as a form of regularization.

We conclude by discussing how measurements can be analytically continued. This is, in principle, performed by substituting the complex variables (k_s, k_d), defined by Equations 4.20 and 4.22, into the Fourier relation (4.14) and performing the integration. This will in general give rise to growing exponentials in the factor $\exp[i(k_d x_d + k_d x_d)]$, which, in the absence of noise, will be compensated for by the dying exponentials in the data $\tilde{\Psi}_s(k_s, k_d)$. If we limit the domain of integration in Equation 4.14 (which will always be the case for finite-length source and detector arrays), the integral (4.14) will remain bounded. In this process, however, noise in the data will be amplified relative to the signal, which is a reflection of the ill-conditioning of the inverse problem. As noted, the purpose of the function $A(p)$ in Equation 4.24 is to reduce this noise at the expense of some loss of resolution.

To summarize, given the measurements $\Psi_s(r_s, r_d)$, the reconstruction procedure requires the following steps. First, the modified Fourier transform of $\Psi_s(r_s, r_d)$, defined by Equation 4.14, is evaluated using the complex transform variables k_s and k_d as defined by Equations 4.20 and 4.22. This result is then substituted into Equation 4.24 to obtain the image. Figure 4.4 shows a simulated reconstruction of two-point inhomogeneities with no noise, 2% noise, and 5% noise added to the simulated data. Further details on the approach described in this section, with further discussion of the simulation, can be found in Ref. [43].

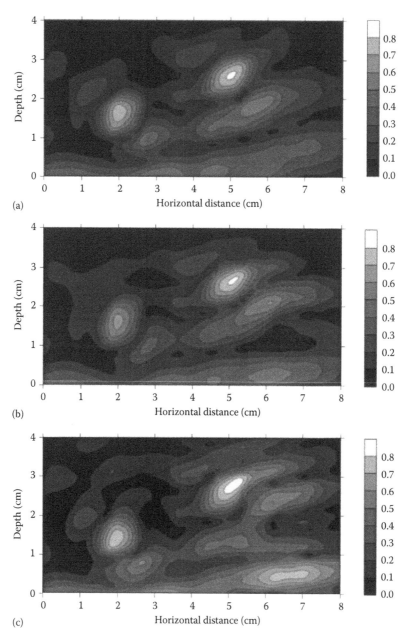

FIGURE 4.4 Simulated reconstruction of two-point inhomogeneities using theoretical algorithms: (a) With no noise, (b) 2% noise, and (c) 5% noise added to the simulated data. (From Norton, S.J. and Vo-Dinh, T., *J. Opt. Soc. Am. A,* 15, 2670, 1998.)

4.4.2 Nonlinear Iterative Algorithm: Frequency-Domain Data

The nonlinear iterative approach involves attempting to find a set of parameters that best predicts the data. For simplicity, we assume that scattering dominates absorption (i.e., $\sigma_s \gg \sigma_a$), so that the scattering cross section $\sigma_s(r)$ is the parameter of interest. Then the transport equation (4.4) becomes

$$s \cdot \nabla \Psi(r,s) + \left[\sigma_s(r) - \frac{i\omega}{c} \right] \Psi(r,s) = \sigma_s(r) \int \Psi(r,s') f(s \cdot s') \mathrm{d}^2 s' + q(r,s). \tag{4.25}$$

The adjoint equation to Equation 4.25 is

$$-s \cdot \nabla \Psi^+(r,s) + \left[\sigma_s(r) - \frac{i\omega}{c} \right] \Psi^+(r,s) = \sigma_s(r) \int \Psi^+(r,s') f(s \cdot s') \mathrm{d}^2 s' + q^+(r,s), \tag{4.26}$$

where the adjoint flux $\Psi^+(r,s)$ is defined as the solution to Equation 4.26 given the adjoint source $q^+(r,s)$, which will be defined. Equations 4.25 and 4.26 can be abbreviated as

$$L\{\Psi\} = q \tag{4.27}$$

and

$$L^+\{\Psi^+\} = q^+, \tag{4.28}$$

where L and L^+ are the linear forward and adjoint operators defined by Equations 4.25 and 4.26. One can then show that [110]

$$\iint \Psi L^+\{\Psi^+\} \mathrm{d}^3 r \mathrm{d}^2 s = \iint \Psi^+ L\{\Psi\} \mathrm{d}^3 r \mathrm{d}^2 s. \tag{4.29}$$

We define the error norm expressing the difference between the predicted flux, $\Psi(r,s)$, and the observed flux, $\Psi^{(obs)}(r,s)$, as

$$E = \frac{1}{2} \iint D(r,s) |\Psi(r,s) - \Psi^{(obs)}(r,s)|^2 \, \mathrm{d}^3 r \mathrm{d}^2 s, \tag{4.30}$$

where $D(r,s)$ is a function that restricts the domain (r,s) to *detector space*; we refer to $D(r,s)$ as the detector response function. For example, suppose we have N point detectors at $r_j, j = 1, \ldots, N$. Further, suppose that each detector has an angular response $R_j(s)$ at point r_j. Then

$$D(r,s) = \sum_{j=1}^{N} R_j(s) \delta^{(3)}(r - r_j), \tag{4.31}$$

where $\delta^{(3)}(r)$ is the 3D Dirac delta function. Substitution of Equation 4.31 into 4.30 gives the error norm

$$E = \frac{1}{2} \sum_{j=1}^{N} \int \mathrm{d}^2 s \, R_j(s) |\Psi(r_j,s) - \Psi^{(obs)}(r_j,s)|^2. \tag{4.32}$$

Our objective is to find the scattering cross section $\sigma_s(r)$ that minimizes E.

Now a small variation $\delta\sigma_s(r)$ in the cross section $\sigma_s(r)$ will give rise to a variation $\delta\Psi_s(r,s)$ in the flux and a corresponding variation δE in the error E defined by Equation 4.30:

$$\delta E = \text{Re} \iint D\left(\Psi - \Psi^{(obs)}\right)^* \delta\Psi d^3r d^2s, \tag{4.33}$$

where
    Re means real part
    * denotes the complex conjugate

The corresponding variation in the transport Equation 4.27 will be

$$\delta L\{\Psi\} + L\{\delta\Psi\} = \delta q, \tag{4.34}$$

since L is a linear operator. Normally, the source q is fixed, so that $\delta q = 0$ and Equation 4.34 gives

$$L\{\delta\Psi\} = -\delta L\{\Psi\}. \tag{4.35}$$

We wish to eliminate the unknown flux variation, $\delta\Psi$, from Equation 4.33 with the aid of Equation 4.35 to express δE explicitly in terms of the operator variation δL, which can, in turn, be related to the variation in the scattering cross section, $\delta\sigma_s$, through Equation 4.25, as we now show.

To accomplish this, define the adjoint source $q^+ \equiv D(\Psi - \Psi^{(obs)})^*$ so that Equation 4.28 becomes

$$L^+\{\Psi^+\} \equiv D\left(\Psi - \Psi^{(obs)}\right)^*. \tag{4.36}$$

Substituting Equation 4.36 into 4.33 and using Equation 4.29 gives

$$\delta E = \text{Re} \iint \delta\Psi L^+\{\Psi^+\} d^3r d^2s = \text{Re} \iint \Psi^+ L\{\delta\Psi\} d^3r d^2s. \tag{4.37}$$

Substituting Equation 4.35 into the second integral finally results in

$$\delta E = -\text{Re} \iint \Psi^+ \delta L\{\Psi\} d^3r d^2s. \tag{4.38}$$

Now, variation of the linear transport operator L defined in Equation 4.25 gives

$$\delta L\{\Psi\} = \delta\sigma_s \Psi - \delta\sigma_s \int \Psi f d^2s'. \tag{4.39}$$

Substituting Equation 4.39 into 4.38 and interchanging orders of integration gives

$$\delta E = \int \left[\nabla_s E(r)\right]\delta\sigma_s(r)d^3r, \tag{4.40}$$

where

$$\nabla_s E(r) \equiv \text{Re} \iint \Psi^+(r,s)\left[\Psi(r,s) - \int \Psi(r,s')f(s\cdot s')d^2s'\right]d^2s. \tag{4.41}$$

Here, $\nabla_s E(\mathbf{r})$ is the functional gradient (or Fréchet derivative) of the mean-square error, E, computed with respect to the scattering cross section $\sigma_s(\mathbf{r})$. Equation 4.40 may be regarded as the continuous-space analogue of a directional derivative, where the integral is analogous to a dot product between the functional gradient, $\nabla_s E(\mathbf{r})$, and the infinite-dimensional *vector* $\delta\sigma_s(\mathbf{r})$. If $\sigma_s(\mathbf{r})$ depends on a finite number of discrete parameters, Equation 4.40 reduces to the usual finite-dimensional gradient, as shown in the following. Equation 4.41 may also be interpreted as the mean-square error *sensitivity* in the sense that it tells us how sensitive a change in the error functional, E, is to a change in $\sigma_s(\mathbf{r})$ at the point $\mathbf{r}$.

Once the functional gradient has been computed, any convenient gradient descent algorithm may be employed to drive the error norm (4.32) to a minimum. Examples of descent algorithms are the method of steepest descent, the conjugate-gradient method, or various quasi-Newton methods [111–114]. The point is that any of these descent algorithms requires the gradient, which traditionally has been the most expensive part of the computation. For example, the conjugate-gradient algorithm updates the previous estimate of $\sigma_s(\mathbf{r})$, denoted $\sigma_s^{(n-1)}(\mathbf{r})$, as follows [111,112]:

$$\sigma_s^{(n)}(\mathbf{r}) = \sigma_s^{(n-1)}(\mathbf{r}) + \alpha_n f_n(\mathbf{r}),$$

where

$$f_n(\mathbf{r}) = -\nabla_s^{(n-1)} E(\mathbf{r}) + \beta_n f_{n-1}(\mathbf{r})$$

In the latter equation, the gradient $\nabla_s^{(n-1)} E(\mathbf{r})$ is computed on the basis of the estimate $\sigma_s^{(n-1)}(\mathbf{r})$, and α_n and β_n are step-size parameters that are updated at each iteration, with the initial condition $\beta_0 \equiv 0$.

4.4.2.1 Finite Number of Parameters

The finite-dimensional case can also be treated using the aforementioned results. Suppose, for example, that we represent the spatially varying cross section $\sigma_s(\mathbf{r})$ using a finite number of pixels at the points $\mathbf{r}_p$, $p = 1, \ldots, P$. More generally, we can expand $\sigma_s(\mathbf{r})$ in a set of P basis functions, as follows:

$$\sigma_s(\mathbf{r}, \mathbf{a}) = \sum_{p=1}^{P} a_p \varphi_p(\mathbf{r}), \tag{4.42}$$

where $\mathbf{a} = \{a_1, \ldots, a_P\}$ is a set of coefficients to be determined. The variation $\delta\sigma_s(\mathbf{r})$ can then be written as

$$\delta\sigma_s(\mathbf{r}, \mathbf{a}) = \sum_{p=1}^{P} \varphi_p(\mathbf{r}) \delta a_p. \tag{4.43}$$

Substituting Equation 4.43 into 4.40 gives

$$\delta E = \sum_{p=1}^{P} \left[\int [\nabla_s E(\mathbf{r})] \varphi_p(\mathbf{r}) \mathrm{d}^3 \mathbf{r} \right] \delta a_p. \tag{4.44}$$

However,

$$\delta E = \sum_{p=1}^{P} \frac{\partial E}{\partial a_p} \delta a_p. \tag{4.45}$$

Comparing Equations 4.44 and 4.45, we obtain the components of the finite-dimensional gradient of the norm E:

$$\frac{\partial E}{\partial a_p} = \int [\nabla_s E(r)]\varphi_p(r)\mathrm{d}^3 r,$$

(4.46)

where $\nabla_s E(r)$ is defined by Equation 4.41.

4.4.2.2 Adding a Regularization Term

In general, the inverse transport problem is ill-posed, and consequently ill-conditioned, because of the diffuse nature of the photon migration in tissue; thus, some regularization scheme is needed to stabilize the inversion in the presence of noisy data. One approach involves adding a penalty term to the error norm (4.30) that is designed, for example, to penalize large variations in the unknown $\sigma_s(r)$ or to emphasize smoother solutions. Regularization of this kind not only helps mitigate ill-conditioning but also serves to force uniqueness in an underdetermined problem. Equation 4.30 is then modified to read

$$E = \frac{1}{2}\iiint D\,|\Psi - \Psi^{(obs)}|^2\,\mathrm{d}^3 r\,\mathrm{d}^2 s + \frac{1}{2}\lambda\int C(\sigma_s)\mathrm{d}^3 r,$$

(4.47)

where
   $C(\sigma_s)$ is a penalty function
   λ is a parameter that controls the relative weighting of the two terms in Equation 4.47

Possible choices of $C(\sigma_s)$ are the following:

$$C(\sigma_s) = \sigma_s(r)^2$$

(4.48a)

$$C(\sigma_s) = \nabla\sigma_s(r)\cdot\nabla\sigma_s(r)$$

(4.48b)

$$C(\sigma_s) = -\sigma_s(r)\ln\sigma_s(r)$$

(4.48c)

Working through the same derivation as mentioned earlier and letting $\nabla_s E(r)$ denote the Fréchet derivative defined by Equation 4.41, the augmented norm (4.47) gives the following new Fréchet derivatives corresponding, respectively, to the earlier three penalty functions:

$$\nabla^{(1)}E(r) = \nabla_s E(r) + \lambda\sigma_s(r)$$

(4.49a)

$$\nabla_s^{(2)}E(r) = \nabla_s E(r) - \lambda\nabla^2\sigma_s(r)$$

(4.49b)

$$\nabla_s^{(3)}E(r) = \nabla_s E(r) + \lambda[1 - \ln\sigma_s(r)]$$

(4.49c)

4.4.2.3 Example: Two-Dimensional Imaging of a Scattering Cross Section

To illustrate how the aforementioned theory can be implemented, consider a simple 2D continuous-wave (steady-state) imaging problem. We divide a square imaging domain into $N \times N$ square pixels, each of which with an unknown scattering cross section denoted by σ_{mn}, with $m,n = 1, ..., N$. We wish to determine the N^2 unknowns σ_{mn} based on scattering observations on the boundary of the square domain. We consider a very simple discrete transport model for purposes of illustration; in this model, a photon incident on a pixel can be scattered in just four directions—up, down, left, and right—where each direction is described, respectively, by the unit vectors s_1, s_2, s_3, s_4. Denote the photon flux moving in the direction s_j in pixel (m,n) by $\psi_{mn}(s_j)$. This flux then obeys the following discrete transport equation:

$$s_i \cdot \nabla_d \psi_{mn}(s_i) + \sigma_{mn} \psi_{mn}(s_i) = \sigma_{mn} \sum_{j=1}^{4} f(s_i \cdot s_j) \psi_{mn}(s_j) + q_{mn}(s_i) \tag{4.50}$$

for $i = 1,\ldots,4$. This is the discretized version of Equation 4.25 in the zero-frequency limit ($\omega = 0$). Here, ∇_d is a difference operator defined in the following, $f(s_i \cdot s_j)$ is a known scattering law, and $q_{mn}(s_i)$ is a source term, also assumed known. If we assume that the pixels are of unit dimensions ($\Delta x = \Delta y = 1$), then we have for the four directions:

$$s_1 \cdot \nabla_d \psi_{mn}(s_1) = \psi_{mn}(s_1) - \psi_{m-1,n}(s_1)$$

$$s_2 \cdot \nabla_d \psi_{mn}(s_2) = \psi_{mn}(s_2) - \psi_{m+1,n}(s_2)$$

$$s_3 \cdot \nabla_d \psi_{mn}(s_3) = \psi_{mn}(s_3) - \psi_{m,n-1}(s_3)$$

$$s_4 \cdot \nabla_d \psi_{mn}(s_4) = \psi_{mn}(s_4) - \psi_{m+1,n}(s_4)$$

This is a reasonable approximation to the spatial derivatives, assuming that the flux is slowly varying on the scale of a pixel. More realistic transport models obviously exist but can be treated similarly. The forward problem can be defined as the task of computing the fluxes $\psi_{mn}(s_j)$, given the set $\{\sigma_{mn}\}$; a simple forward algorithm can be devised by iterating Equation 4.50. Our objective is to minimize the following mean-square error with respect to $\{\sigma_{mn}\}$:

$$E = \frac{1}{2} \sum_{m,n=1}^{N} \sum_{i=1}^{4} D_{mn}(s_i) \left[\psi_{mn}(s_i) - \psi_{mn}^{(obs)}(s_i) \right]^2, \tag{4.51}$$

in which the detector response function is defined by $D_{mn}(s_i) \equiv 1$ for the observed flux on the boundary of the domain, and $D_{mn}(s_i) \equiv 0$ otherwise.

We wish to show in this section how, using the adjoint method, we can compute the partial derivatives of the mean-square error $\partial E / \partial \sigma_{mn}$ in terms of known quantities. Defining for brevity $F_{ij} \equiv f(s_i \cdot s_j) - \delta_{ij}$, where δ_{ij} is the Kronecker delta, we can write Equation 4.50 in the more compact form

$$s_i \cdot \nabla_d \psi_{mn}(s_i) - \sigma_{mn} \sum_{j=1}^{4} F_{ij} \psi_{mn}(s_j) = q_{mn}(s_i). \tag{4.52}$$

The adjoint of Equation 4.52 is

$$-s_i \cdot \nabla_d^+ \psi_{mn}^+(s_i) - \sigma_{mn} \sum_{j=1}^{4} F_{ij} \psi_{mn}^+(s_j) = q_{mn}^+(s_i), \tag{4.53}$$

where the adjoint source, $q_{mn}^+(s_i)$, is to be defined, and

$$s_1 \cdot \nabla_d^+ \psi_{mn}^+(s_1) = \psi_{mn}^+(s_1) - \psi_{m+1,n}^+(s_1),$$

$$s_2 \cdot \nabla_d^+ \psi_{mn}^+(s_2) = \psi_{mn}^+(s_2) - \psi_{m-1,n}^+(s_2),$$

$$s_3 \cdot \nabla_d^+ \psi_{mn}^+(s_3) = \psi_{mn}^+(s_3) - \psi_{m,n+1}^+(s_3),$$

$$s_4 \cdot \nabla_d^+ \psi_{mn}^+(s_4) = \psi_{mn}^+(s_4) - \psi_{m,n-1}^+(s_4).$$

If we abbreviate Equations 4.52 and 4.53 by

$$L_{mn}\{\psi_{mn}(s_i)\} = q_{mn}(s_i),$$

$$L_{mn}^+\{\psi_{mn}^+(s_i)\} = q_{mn}^+(s_i),$$

then the adjoint relation analogous to Equation 4.29 is

$$\sum_{m,n=1}^{N} \sum_{i=1}^{4} \psi_{mn}(s_i) L_{mn}^+ \left\{ \psi_{mn}^+(s_i) \right\} = \sum_{m,n=1}^{N} \sum_{i=1}^{4} \psi_{mn}^+(s_i) L_{mn} \{ \psi_{mn}(s_i) \}.$$

We now define the adjoint source as

$$q_{mn}^+(s_i) \equiv D_{mn}(s_i) \left[\psi_{mn}(s_i) - \psi_{mn}^{(obs)}(s_i) \right].$$

If we follow the same procedure as described in the previous section, we obtain the following simple result for the derivatives of the mean-square error:

$$\frac{\partial E}{\partial \sigma_{mn}} = \sum_{i=1}^{4} \sum_{j=1}^{4} F_{ij} \psi_{pq}^+(s_i) \psi_{pq}(s_j), \tag{4.54}$$

where $\psi_{pq}(s_j)$ and $\psi_{pq}^+(s_i)$ are the solutions to Equations 4.52 and 4.53 computed on the basis of the current estimate of $\{\sigma_{mn}\}$. A simulation was performed by Norton [104] using these formulas to reconstruct a 64-pixel-square region with the scattering law $f(s_i \cdot s_j) = 0.1$ for $i \neq j$ and $f(s_i \cdot s_i) = 0.7$.

These formulas can be easily generalized to account for multiple sources and multiple modulation frequencies. If the modulation frequencies can be made sufficiently high so that the DPDW *skin depth* δ is comparable to the size of the region of interest, then multifrequency data can be shown to improve the conditioning of the inverse problem [115].

4.5 Nonlinear Iterative Algorithm: Time-Resolved Data

As noted, when recording transient data, it is not advantageous to restrict ourselves to the relatively small number of first-arrival photons. Arridge and coworkers [65,67,116], for example, have advocated measurement of the first few moments of the temporal response (e.g., the first moment being the mean arrival time); these measurements can then be employed in an iterative descent algorithm to minimize an appropriate error norm, similar to the scheme described earlier in this chapter. In this section, we describe a time-domain iterative algorithm designed to minimize an error norm that weights the entire transient waveform. For simplicity, we assume that the diffusion approximation holds, in which case the photon flux density, $\psi(r,t)$, obeys the time-dependent diffusion equation given by Equation 4.3:

$$\frac{1}{c}\frac{\partial\psi(r,t)}{\partial t} - \nabla\cdot\left[D(r)\,\nabla\psi(r,t)\right] + \sigma_a(r)\,\psi(r,t) = Q(r,t), \tag{4.55}$$

subject to the initial condition $\psi(r,0) = 0$. Although we employ the diffusion model (4.55) to illustrate our procedure, a similar time-domain treatment can be carried out with the time-dependent transport equation as a starting point [106].

Suppose the region of interest is illuminated by a light pulse, which diffuses through the tissue and is detected at N point detectors at r_j, $j = 1, \ldots, N$; the pulse is observed over the interval of time $[0,T]$ and the observations are denoted by $\psi^{(obs)}(r_j,t)$. The optical parameters, $D(r)$ and $\sigma_a(r)$, are then sought that minimize the time-integrated mean-square error:

$$E = \frac{1}{2}\sum_{j=1}^{N}\int_0^T w_j(t)\left[\psi(r_j,t) - \psi^{(obs)}(r_j,t)\right]^2 dt. \tag{4.56}$$

Here, the integration is over the observation interval T, and $w_j(t)$ is an arbitrary temporal weighting function. To compute the Fréchet derivative of Equation 4.56, we employ the adjoint diffusion equation, given by

$$-\frac{1}{c}\frac{\partial\psi^+(r,t)}{\partial t} - \nabla\cdot\left[D(r)\,\nabla\psi^+(r,t)\right] + \sigma_a(r)\psi^+(r,t) = Q^+(r,t), \tag{4.57}$$

where the adjoint source is defined by

$$Q^+(r,t) = \sum_{j=1}^{N} w_j(t)\left[\psi(r_j,t) - \psi^{(obs)}(r_j,t)\right]\delta(r - r_j). \tag{4.58}$$

We require that the adjoint field obey the terminal condition $\psi^+(r,T) = 0$. The reason for this will become apparent shortly. We now vary Equation 4.56, giving

$$\delta E = \sum_{j=1}^{N}\int_0^T w_j(t)\left[\psi(r_j,t) - \psi^{(obs)}(r_j,t)\right]\delta\psi(r_j,t)\,dt. \tag{4.59}$$

In view of the definition (4.58) of the adjoint source, we can write Equation 4.59 as

$$\delta E = \int d^3r \int_0^T dt\, Q^+(r,t)\delta\psi(r_j,t), \tag{4.60}$$

where the volume integral includes the spatial domain of interest. Next, substituting the left-hand side of Equation 4.57 into 4.60 and integrating by parts once with respect to time and twice with respect to space [102], we obtain

$$\delta E = \int d^3 r \int_0^T dt \left\{ \frac{1}{c} \frac{\partial \delta \psi(\mathbf{r},t)}{\partial t} - \nabla \cdot \left[D(\mathbf{r}) \nabla \delta \psi(\mathbf{r},t) \right] + \sigma_a(\mathbf{r}) \delta \psi(\mathbf{r},t) \right\} \psi^+(\mathbf{r}_j,t). \tag{4.61}$$

When integrating by parts with respect to time, we find that the integrated part vanishes on account of the initial condition $\delta \psi(\mathbf{r},0) = 0$ and the terminal condition $\psi^+(\mathbf{r},T) = 0$. Similarly, we assume that appropriate homogeneous boundary conditions hold on the surface of the domain of integration, which ensures that the boundary terms vanish when integrating by parts with respect to space. Next, we vary Equation 4.55 to obtain

$$\frac{1}{c} \frac{\partial \delta \psi(\mathbf{r},t)}{\partial t} - \nabla \cdot \left[\delta D(\mathbf{r}) \nabla \psi(\mathbf{r},t) \right] - \nabla \cdot \left[D(\mathbf{r}) \nabla \delta \psi(\mathbf{r},t) \right] + \delta \sigma_a(\mathbf{r}) \psi(\mathbf{r},t) + \sigma_a(\mathbf{r}) \delta \psi(\mathbf{r},t) = 0$$

or

$$\frac{1}{c} \frac{\partial \delta \psi(\mathbf{r},t)}{\partial t} - \nabla \cdot \left[D(\mathbf{r}) \nabla \delta \psi(\mathbf{r},t) \right] + \sigma_a(\mathbf{r}) \delta \psi(\mathbf{r},t) = \nabla \cdot \left[\delta D(\mathbf{r}) \nabla \psi(\mathbf{r},t) \right] - \delta \sigma_a(\mathbf{r}) \psi(\mathbf{r},t).$$

Substituting this into Equation 4.61 gives

$$\delta E = \int d^3 r \int_0^T dt \{ \nabla \cdot [\delta D(\mathbf{r}) \nabla \psi(\mathbf{r},t)] - \delta \sigma_a(\mathbf{r}) \psi(\mathbf{r},t) \} \psi^+(\mathbf{r}_j,t)$$

The first term can be integrated once more by parts with respect to $\mathbf{r}$ to yield

$$\delta E = -\int d^3 r \int_0^T dt \nabla \psi(\mathbf{r},t) \cdot \nabla \psi^+(\mathbf{r},t) \delta D(\mathbf{r}) - \int d^3 r \int_0^T dt \, \psi(\mathbf{r},t) \} \psi^+(\mathbf{r},t) \delta \sigma_a(\mathbf{r})$$

$$= \int d^3 r \nabla_D E(\mathbf{r}) \delta D(\mathbf{r}) + \int d^3 r \nabla_a E(\mathbf{r}) \delta \sigma_a(\mathbf{r}),$$

where the Fréchet derivatives for $D(\mathbf{r})$ and $\sigma_a(\mathbf{r})$ are, respectively, as follows:

$$\nabla_D E(\mathbf{r}) = -\int_0^T dt \, \nabla \psi(\mathbf{r},t) \cdot \nabla \psi^+(\mathbf{r},t) \tag{4.62}$$

$$\nabla_a E(\mathbf{r}) = -\int_0^T dt \, \psi(\mathbf{r},t) \, \psi^+(\mathbf{r},t) \tag{4.63}$$

When these functional derivatives are used to update the parameters $D(\mathbf{r})$ and $\sigma_a(\mathbf{r})$ in a descent algorithm of one's choice, the time-integrated error functional (4.56) can be driven to a minimum. In a manner similar to that of the previous section, a penalty term can also be added to Equation 4.56 to regularize the solution. As noted, the time-dependent residual in Equation 4.56 can be appropriately weighted through the selection of the weighting function $w_j(t)$ to control the signal-to-noise ratio. Arridge [7] argues that employing the logarithm of $\psi^{(obs)}(r,t)$ may be advantageous. This approach can be carried out in the latter case as well, after the field Equations 4.55 and 4.57 are modified by substituting $\psi(r,t) = \exp[\gamma(r,t)]$, which defines new field equations in $\gamma(r,t)$. After this logarithmic transformation, the Fréchet derivatives of $D(r)$ and $\sigma_a(r)$ can be derived in the same way as mentioned earlier. The interesting question of what constitutes sufficient data to reconstruct $D(r)$ and $\sigma_a(r)$ simultaneously is beyond the scope of our discussion but has been discussed by Arridge and Lionheart [117].

4.6 Conclusion

This chapter has examined in some detail three optical tomography algorithms. The first of these employs Fourier methods to reconstruct an image from time-harmonic DPDW data recorded at one modulation frequency. This approach may be regarded as a modification of conventional DT. The other algorithms employ a gradient descent algorithm to minimize a global error norm measuring the difference between recorded and predicted data. This approach is very general and can accommodate either time- or frequency-domain measurements. Whether the algorithm is iterative or Fourier-based, the inverse problem will inevitably be ill-conditioned because of the diffuse nature of the multiply scattered light or, equivalently, because of the smoothing effect of the forward transport operator. This implies that any inversion procedure that attempts to *undo* this smoothing will amplify noise in the data and will, as a result, require some form of regularization to mitigate the noise amplification. A variety of such schemes exist, but the price paid for regularization is some loss of spatial resolution. In this regard, it is important to note that resolution is ultimately noise limited, since the resolution typically achieved is well beyond the *diffraction limit*—that is, less than the wavelength of a photon density wave. Moreover, the size of the region probed may sometimes be less than a single DPDW wavelength, implying operation well within the near field.

In addition to reviewing two fundamental methods of image reconstruction (DT and the iterative minimization of an error norm), we have tried to provide a representative sampling of the literature in this area. We have not referenced many papers that have addressed related but more specialized topics, such as forward algorithms, more accurate photon transport models, fluorescence imaging and spectroscopic methods, effects of boundaries and unusual measurement geometries, and clinical studies. Also, in the interests of space, we have, with a few exceptions, emphasized references in archival journals rather than conference proceedings, although a great deal of work has been published in the latter (particularly the *SPIE Proceedings* [2]). The number of research papers in optical tomography has grown rapidly in the last 10 years, and one has every reason to expect that continued progress, both technical and algorithmic, will ultimately yield systems of significant clinical value.

Acknowledgments

This work was sponsored by the National Institutes of Health (Grant R01 CA88787–01) and by the US Department of Energy, Office of Biological and Environmental Research, under contract DEAC05–00OR22725 with UT-Batelle, LLC.

References

1. Singer, J.R., Grunbaum, F.A., Kohn, P., and Zubelli, J.P., Image reconstruction of the interior of bodies that diffuse radiation, *Science*, 248, 990, 1990.
2. Chance, B. et al., eds., *SPIE Proceedings*: Photon propagation in tissues, Vol. 2626, 1995; Photon propagation in tissues II, Vol. 2925, 1996; Photon propagation in tissues III, Vol. 3194, 1997; Photon propagation in tissues IV, Vol. 3566, 1998; Optical tomography and spectroscopy of tissue, Vol. 2389, 1995; Optical tomography and spectroscopy of tissue II, Vol. 2979, 1997; Optical tomography and spectroscopy III, Vol. 3597, 1999; Optical tomography and spectroscopy IV, Vol. 4250, 2001.

3. Minet, O., Muller, G., and Beuthan, J., eds., *Selected Papers on Optical Tomography, Fundamentals and Applications*, SPIE Press, Bellingham, WA, 1998.

4. Tuchin, V., *Tissue Optics*, SPIE Press, Bellingham, WA, 2000.

5. Hebden, J.C., Arridge, S.R., and Delpy, D.T., Optical imaging in medicine: I. Experimental techniques, *Phys. Med. Biol.*, 42, 825, 1997.

6. Arridge, S.R. and Hebden, J.C., Optical imaging in medicine: II. Modeling and reconstruction, *Phys. Med. Biol.*, 42, 841, 1997.

7. Arridge, S.R., Optical tomography in medical imaging, *Inverse Probl.*, 15, R41, 1999.

8. Boas, D.A., Brooks, D.H., Miller, E.L., DiMarzio, C.A., Kilmer, M., Gaudette, R.J., and Zhang, Q., Imaging the body with diffuse optical tomography, *IEEE Signal Proc. Mag.*, 18, 57, 2001.

9. Gibson, A., Hebden, J.C., and Arridge, S.R., Recent advances in diffuse optical tomography, *Phys. Med. Biol.*, 50, R1, 2005.

10. Arridge, S.R. and Schotland, J.C., Optical tomography: Forward and inverse problems, *Inverse Probl.*, 25, 1, 2009.

11. Ishimaru, A., Diffusion of light in turbid material, *Appl. Opt.*, 28, 2210, 1989.

12. Jacques, S.L., Time resolved propagation of ultrashort laser pulses within turbid tissues, *Appl. Opt.*, 28, 2223, 1989.

13. Profio, A.E., Light transport in tissue, *Appl. Opt.*, 28, 2216, 1989.

14. Arridge, S., Schweiger, M., Hiraoka, M., and Delpy, D.T., A finite element approach for modeling photon transport in tissue, *Med. Phys.*, 20, 299, 1993.

15. Tromberg, B.J., Svaasand, L.O., Tsay, T., and Haskell, R.C., Properties of photon density waves in multiple-scattering media, *Appl. Opt.*, 32, 607, 1993.

16. Knuttel, A., Schmitt, J.M., and Knutson, J.R., Spatial localization of absorbing bodies by interfering diffuse photon-density waves, *Appl. Opt.*, 32, 381, 1993.

17. Boas, D.A., O'Leary, M.A., Chance, B., and Yodh, A.G., Scattering and wavelength transduction of diffuse density waves, *Phys. Rev. E*, 47, R2999, 1993.

18. Boas, D.A., O'Leary, M.A., Chance, B., and Yodh, A.G., Scattering of diffuse photon density waves by spherical inhomogeneities within turbid media: Analytical solution and applications, *Proc. Natl. Acad. Sci. USA*, 91, 4887, 1994.

19. Feng, S., Zeng, F., and Chance, B., Photon migration in the presence of a single defect: A perturbation analysis, *Appl. Opt.*, 34, 3826, 1995.

20. Yodh, A.G. and Chance, B., Spectroscopy and imaging with diffusing light, *Phys. Today*, 48, 34, 1995.

21. Boas, D.A. and Yodh, A.G., Spatially varying dynamical properties of turbid media probed with diffusing temporal light correlation, *J. Opt. Soc. Am. A*, 14, 192, 1997.

22. Boas, D.A., O'Leary, M.A., Chance, B., and Yodh, A.G., Detection and characterization of optical inhomogeneities with diffuse photon density waves: A signal-to-noise analysis, *Appl. Opt.*, 36, 75, 1997.

23. Ostermeyer, M.R. and Jacques, S.L., Perturbation theory for diffuse light transport in complex biological tissues, *J. Opt. Soc. Am. A*, 14, 255, 1997.

24. Li, X.D., Durduran, T., Yodh, A.G., Chance, B., and Pattanayak, D.N., Diffraction tomography for biochemical imaging with diffuse-photon density waves, *Opt. Lett.*, 22, 573, 1997.

25. Sevick-Muraca, E.M., Lopez, G., Troy, T.L., Reynolds, J.S., and Hutchinson, C.L., Fluorescence and absorption contrast mechanisms for biomedical optical imaging using frequency-domain techniques, *Photochem. Photobiol.*, 66, 55, 1997.

26. Furutsu, K., Theory of a fixed scatterer embedded in a turbid medium, *J. Opt. Soc. Am. A*, 15, 1371, 1998.

27. Hielscher, A.H., Alcouffe, R.E., and Barbour, R.L., Comparison of finite-difference transport and diffusion calculations for photon migration in homogeneous and heterogeneous tissue, *Phys. Med. Biol.*, 43, 1285, 1998.

28. Ripoll, J., Nieto-Vesperinas, M., and Carminati, R., Spatial resolution of diffuse photon density waves, *J. Opt. Soc. Am. A*, 16, 1466, 1999.

29. van Rossum, M.W. and Nieuwenhuizen, T.W., Multiple scattering of classical waves: Microscopy, mesoscopy and diffusion, *Rev. Mod. Phys.*, 71, 313, 1999.

30. Jacques, S.L., Roman, J.R., and Lee, K., Imaging superficial tissues with polarized light, *Lasers Surg. Med.*, 26, 119, 2000.

31. Jacques, S.L., Ramanujam, N., Vishnoi, G., Choe, R., and Chance, B., Modeling photon transport in transabdominal fetal oximetry, *J. Biomed. Opt.*, 5, 277, 2000.

32. Lee, J. and Sevick-Muraca, E.M., Three-dimensional fluorescence enhanced optical tomography using referenced frequency-domain photon migration measurements at emission and excitation wavelengths, *J. Opt. Soc. Am. A*, 19, 759, 2002.

33. Sun, Z.G. and Sevick-Muraca, E.M., Investigation of particle interactions in dense colloidal suspensions using frequency domain photon migration: Bidisperse systems, *Langmuir*, 18, 1091, 2002.

34. Kim, A.D. and Keller, J.B., Light propagation in biological tissue, *J. Opt. Soc. Am. A*, 20, 92, 2003.

35. Elaloufi, R., Arridge, S.R., Pierrat, R., and Carminati, R., Light propagation in multilayered scattering media beyond the diffusive regime, *Appl. Opt.*, 46, 3628, 2007.

36. O'Leary, M.A., Boas, D.A., Chance, B., and Yodh, A.G., Experimental images of heterogeneous turbid media by frequency-domain diffusing-photon tomography, *Opt. Lett.*, 20, 426, 1995.

37. Reynolds, J.S., Przadka, A., Yeung, S.P., and Webb, K.J., Optical diffusion imaging: A comparative numerical and experimental study, *Appl. Opt.*, 35, 3671, 1996.

38. Colak, S.B., Papaioannou, D.G., 't Hooft, G.W., van der Mark, M.B., Schomberg, H., Paasschens, J.C.J., Melissen, J.B.M., and van Asten, N.A.A.J., Tomographic image reconstruction from optical projections in light-diffusing media, *Appl. Opt.*, 36, 180, 1997.

39. Matson, C.L., A diffraction tomographic model of the forward problem using diffuse photon density waves, *Opt. Express*, 1, 6, 1997.

40. Yao, Y., Wang, Y., Pei, Y., Wenwu, Z., and Barbour, R.L., Frequency-domain optical imaging of absorption and scattering distributions by a Born iterative method, *J. Opt. Soc. Am. A*, 14, 325, 1997.

41. Zhu, W., Wang, Y., Yao, Y., Chang, J., Graber, H.L., and Barbour, R.L., Iterative total least-squares image reconstruction algorithm for optical tomography by the conjugate gradient method, *J. Opt. Soc. Am. A*, 14, 799, 1997.

42. Lasocki, D.L., Matson, C.L., and Collins, P.J., Analysis of forward scattering of diffuse photon-density waves in turbid media: A diffraction tomography approach to an analytic solution, *Opt. Lett.*, 23, 558, 1998.

43. Norton, S.J. and Vo-Dinh, T., Diffraction tomographic imaging with photon density waves: An explicit solution, *J. Opt. Soc. Am. A*, 15, 2670, 1998.

44. Chen, B., Stamnes, J.J., and Stamnes, K., Reconstruction algorithm for diffraction tomography of diffuse photon density waves in a random medium, *Pure Appl. Opt.*, 7, 1161, 1998.

45. Matson, C.L. and Liu, H., Analysis of the forward problem with diffuse photon density waves in turbid media by use of a diffraction tomography model, *J. Opt. Soc. Am. A*, 16, 455, 1999.

46. Matson, C.L. and Liu, H., Backpropagation in turbid media, *J. Opt. Soc. Am. A*, 16, 1254, 1999.

47. Ye, J.C., Webb, K.J., Millane, R.P., and Downar, T.J., Modified distorted born iterative method with an approximate Fréchet derivative for optical diffusion tomography, *J. Opt. Soc. Am. A*, 16, 1814, 1999.

48. Braunstein, M. and Levine, R.Y., Three-dimensional tomographic reconstruction of an absorptive perturbation with diffuse photon density waves, *J. Opt. Soc. Am. A*, 17, 11, 2000.

49. Gaudette, R.J., Brooks, D.H., DiMarzio, C.A., Kilmer, M.E., Miller, E.L., Gaudette, T., and Boas, D.A., A comparison study of linear reconstruction techniques for diffuse optical tomographic imaging of absorption coefficient, *Phys. Med. Biol.*, 45, 1051, 2000.

50. Kilmer, M.E., Miller, E.L., Boas, D., and Brooks, D., A shape-based reconstruction technique for DPDW data, *Opt. Express*, 7, 481, 2000.

51. Markel, V.A. and Schotland, J.C., Inverse problem in optical diffusion tomography. I. Fourier-Laplace inversion formulas, *J. Opt. Soc. Am. A*, 18, 1336, 2001.

52. Matson, C.L., Diffraction tomography for turbid media, in *Advances in Imaging and Electron Physics*, Vol. 124, Hawkes, P., ed., Academic Press, New York, 2002.

53. Chance, B., Leigh, J.S., Miyake, H., Smith, D.S., Nioka, S., Greenfeld, R., Finander, M. et al., Comparison of time-resolved and -unresolved measurements of deoxyhemoglobin in grain, *Proc. Natl. Acad. Sci. USA*, 85, 4971, 1988.

54. Patterson, M.S., Chance, B., and Wilson, B.C., Time resolved reflectance and transmittance for the non-invasive measurement of tissue optical properties, *Appl. Opt.*, 28, 2331, 1989.

55. Benaron, D.A. and Stevenson, D.K., Optical time-of-flight and absorbance imaging of biologic media, *Science*, 259, 1463, 1993.

56. Das, B.B., Yoo, K.M., and Alfano, R.R., Ultrafast time-gated imaging in thick tissues—A step towards optical mammography, *Opt. Lett.*, 18, 1092, 1993.

57. Hee, M.R., Izatt, J.A., Swanson, E.A., and Fujimoto, J.G., Femtosecond transillumination tomography in thick issues, *Opt. Lett.*, 18, 1107, 1993.

58. Das, B.B., Barbour, R.L., Graber, H.L., Chang, J., Zevallos, M., Liu, F., and Alfano, R.R., Analysis of time-resolved data for tomographic image reconstruction of opaque phantoms and finite absorbers in diffuse media, *Proc. SPIE*, 2389(Part 1), 16, 1995.

59. Chang, J., Zhu, W., Wang, Y., Graber, H.L., and Barbour, R.L., Regularized progressive expansion algorithm for recovery of scattering media from time-resolved data, *J. Opt. Soc. Am.*, 14, 306, 1997.

60. Grosenick, D., Wabnitz, H., and Rinneberg, H., Time-resolved imaging of solid phantoms for optical mammography, *Appl. Opt.*, 36, 221, 1997.

61. Winn, J.N., Perelman, L.T., Chen, K., Wu, J., Dasari, R.R., and Feld, M.S., Distribution of the paths of early-arriving photons traversing a turbid medium, *Appl. Opt.*, 34, 8085, 1998.

62. Cai, W., Gayen, S.K., Xu, M., Zevallos, M., Alrubaiee, M., Lax, M., and Alfano, R.R., Optical tomographic image reconstruction from ultrafast time-sliced transmission measurements, *Appl. Opt.*, 39, 4237, 1999.

63. Morin, M., Mailloux, A., Painchaud, Y., and Beaudry, P., Time-resolved transmission through homogeneous scattering media: Time-response effects, *Appl. Opt.*, 38, 3681, 1999.

64. Hielscher, A.H., Klose, A.D., and Hanson, K.M., Gradient-based iterative image reconstruction scheme for time-resolved optical tomography, *IEEE Trans. Med. Imaging*, 18, 262, 1999.

65. Schmidt, F.E.W., Hebden, J.C., Hillman, M.C., Fry, M.E., Schweiger, M., Dehghani, H., Delphy, D.T., and Arridge, S.R., Multiple-slice imaging of a tissue-equivalent phantom by use of time-resolved optical tomography, *Appl. Opt.*, 39, 3380, 2000.

66. Gao, F., Poulet, P., and Yamada, Y., Simultaneous mapping of absorption and scattering coefficients from a three-dimensional model of time-resolved optical tomography, *Appl. Opt.*, 39, 5898, 2000.

67. Hebden, J.C., Veenstra, H., Dehghani, H., Hillman, E.M.C., Schweiger, M., Arridge, S.R., and Delpy, D.T., Three-dimensional time-resolved optical tomography of a conical breast phantom, *Appl. Opt.*, 40, 3278, 2001.

68. Schotland, J.C., Continuous-wave diffusion imaging, *J. Opt. Soc. Am. A*, 14, 275, 1997.

69. Cheng, X. and Boas, D.A., Diffuse optical reflection tomography with continuous-wave illumination, *Opt. Express*, 3, 118, 1998.

70. Alfano, R.R., Govindjee, W.Y.R., Becher, B., and Ebrey, T.G., Picosecond kinetics of fluorescence from chromophore of purple membrane-protein of Halobacterium-Halobium, *Biophys. J.*, 16, 541, 1976.

71. Devaney, A.J., Inversion formula for inverse scattering within the Born approximation, *Opt. Lett.*, 7, 111, 1982.

72. Devaney, A.J., A filtered backpropagation algorithm for diffraction tomography, *Ultrason. Imaging*, 4, 336, 1982.

73. Devaney, A.J. and Beylkin, G., Diffraction tomography using arbitrary transmitter and receiver surfaces, *Ultrason. Imaging*, 6, 181, 1984.

74. Devaney, A.J., Reconstructive tomography with diffracting wave fields, *Inverse Probl.*, 2, 161, 1986.

75. Ishimaru, A., *Wave Propagation and Scattering in Random Media*, IEEE Press, Piscataway, NJ, 1997.

76. Case, K.M. and Zweifel, P.F., *Linear Transport Theory*, Addison-Wesley, Reading, MA, 1967.

77. Gonzalez-Rodriguez, P. and Kim, A.D., Comparison of light scattering models for diffuse optical tomography, *Opt. Express,* 17, 8756, 2009.

78. Abdoulaev, G.S. and Hielscher, A.H., Three-dimensional optical tomography with the equation of radiative transfer, *J. Electron. Imaging*, 12, 594, 2003.

79. Aydin, E.D., de Oliveira, C.R.E., and Goddard, A.J.H., A comparison between transport and diffusion calculations using a finite element-spherical harmonics radiation transport method, *Med. Phys.*, 2, 2013, 2002.

80. Richling, S.E., Meinkohn, E., Kryzhevoi, M., and Kanschat, G., Radiative transfer with finite elements I: Basic method and tests, *Astron. Astrophys.*, 380, 776, 2001.

81. Tarvainen, T., Vauhkonen, M., Kolehmainen, V., and Kaipio, J.P., A hybrid radiative transfer-diffusion model for optical tomography, *Appl. Opt.*, 44, 876, 2005.

82. Klose, A.D. and Hielscher, A.H., Iterative reconstruction scheme for optical tomography based on the equation of radiative transfer, *Med. Phys.*, 26, 1698, 1999.

83. Klose, A.D. and Hielscher, A.H., Optical tomography using the time-independent equation of radiative transfer: Part 2. Inverse model, *J. Quant. Spectrosc. Radiat. Transf.*, 72, 715, 2002.

84. Schotland, J.C. and Markel, V.A., Fourier–Laplace structure of the inverse scattering problem for the radiative transport equation, *Inverse Probl. Imaging*, 1, 181, 2007.

85. Pogue, B., McBride, T., Prewitt, J., Osterberg, U., and Paulsen, K., Spatially variant regularization improves diffuse optical tomography, *Appl. Opt.*, 38, 2950, 1999.

86. Norton, S.J. and Linzer, M., Ultrasonic reflectivity imaging in three dimensions: Exact inverse scattering solutions for plane, cylindrical and spherical apertures, *IEEE Trans. Biomed. Eng.*, BME-28, 202, 1981.

87. Hielscher, A., Klose, A., and Beuthan, J., Evolution strategies for optical tomographic characterization of homogeneous media, *Opt. Express*, 7, 507, 2000.

88. Walker, S.A., Fantini, S., and Gratton, E., Image reconstruction from frequency-domain optical measurements in highly scattering media, *Appl. Opt.*, 36, 170, 1997.

89. Durduran, T., Culver, J.P., Holboke, M.J., Li, X.D., Zubkov, L., Chance, B., Pattanayak, D.N., and Yodh, A.G., Algorithms for 3D localization and imaging using near-field diffraction tomography with diffuse light, *Opt. Express*, 4, 247, 1999.

90. Pattanayak, D.N. and Yodh, A.G., Diffuse optical 3D-slice imaging of bounded turbid media using a new integro-differential equation, *Opt. Express*, 4, 231, 1999.

91. Bluestone, A.Y., Abdoulaev, G., Schmitz, C.H., Barbour, R.L., and Hielscher, A.H., Three-dimensional optical tomography of hemodynamics in the human head, *Opt. Express*, 9, 272, 2001.

92. Roy, R. and Sevick-Muraca, E.M., A numerical study of gradient-based nonlinear optimization methods for contrast enhanced optical tomography, *Opt. Express*, 9, 49, 2001.

93. Hiltunen, P., Calvetti, D., and Somersalo, E., An adaptive smoothness regularization algorithm for optical tomography, *Opt. Express*, 16, 19957, 2008.

94. Ye, J.C., Webb, K.J., and Bouman, C.A., Optical diffusion tomography by iterative coordinate-descent optimization in a Bayesian framework, *J. Opt. Soc. Am. A*, 16, 2400, 1999.

95. Eppstein, M.J., Dougherty, D.E., Troy, T.L., and Sevick-Muraca, E.M., Biomedical optical tomography using dynamic parameterization and Bayesian conditioning on photon migration measurements, *Appl. Opt.*, 38, 2138, 1999.

96. Milstein, A.B., Oh, S., Reynolds, J.S., Webb, K.J., Bouman, C.A., and Millane, R.P., Three dimensional Bayesian optical diffusion tomography with experimental data, *Opt. Lett.*, 27, 95, 2002.

97. Ye, J.C., Bouman, C.A., Webb, K.J., and Millane, R.P., Nonlinear multigrid algorithms for Bayesian optical diffusion tomography, *IEEE Trans. Imaging Proc.*, 10, 909, 2001.

98. Kolehmainen, V., Vauhkonen, M., Kaipio, J.P., and Arridge, S.R., Recovery of piecewise constant coefficients in optical diffusion tomography, *Opt. Express*, 7, 468, 2000.

99. Dorn, O., Shape reconstruction in scattering media with voids using a transport model and level sets, *Can. Appl. Math. Q*, 10, 239, 2004.

100. Schweiger, M., Arridge, S.R., Dorn, O., Zacharopoulos, A., and Kolehmainen, V., Reconstructing absorption and diffusion shape profiles in optical tomography using a level set technique, *Opt. Lett.*, 31, 471, 2006.

101. Zacharopoulos, A., Arridge, S.R., Dorn, O., Kolehmainen, V., and Sikora, J., 3D shape reconstruction in optical tomography using spherical harmonics and BEM, *J. Electromagn. Waves Appl.*, 20, 1827, 2006.

102. Norton, S.J., Iterative inverse-scattering algorithms: Methods for computing the Fréchet derivative, *J. Acoust. Soc. Am.*, 106, 2653, 1999.

103. Norton, S.J. and Bowler, J.R., Theory of eddy current inversion, *J. Appl. Phys.*, 73, 501, 1993.

104. Norton, S.J., A general nonlinear inverse transport algorithm using forward and adjoint flux computations, *IEEE Trans. Nucl. Sci. Eng.*, NS-44, 153, 1997.

105. Arridge, S.R. and Schweiger, M., A gradient-based optimization scheme for optical tomography, *Opt. Express*, 2, 213, 1998.

106. Dorn, O., Scattering and absorption transport sensitivity functions for optical tomography, *Opt. Express*, 7, 492, 2000.

107. Dorn, O., A transport-backtransport method for optical tomography, *Inverse Probl.*, 14, 1107, 1998.

108. Argarwal, K., Chen, L., Chen, N., and Chen, X., Multistage inversion algorithm for biological tissue imaging, *J. Biomed. Opt.*, 15, 016007, 2010.

109. Klose, A.D. and Hielscher, A.H., Optical tomography with the equation of radiative transfer, *Int. J. Numer. Methods Heat Fluid Flow*, 18, 443, 2008.

110. Bell, G.I. and Glasstone, S., *Nuclear Reactor Theory*, Van Nostrand Reinhold, New York, 1970.

111. Gill, P.E., Murry, W., and Wright, M.H., *Practical Optimization*, Academic Press, New York, 1981.

112. Luenberger, D.G., *Linear and Nonlinear Programming*, 2nd edn., Addison-Wesley, Menlo Park, CA, 1984.

113. Schweiger, M., Arridge, S.R., and Nissila, I., Gauss-Newton method for image reconstruction in diffuse optical tomography, *Phys. Med. Biol.*, 50, 2365, 2005.

114. Tarvainen, T., Vauhkonen, M., and Arridge, S.R., Image reconstruction in optical tomography using the finite element solution of the frequency domain radiative transfer equation, *J. Quant. Spectrosc. Radiat. Transf.*, 109, 2767, 2008.

115. Norton, S.J., Electromagnetic induction imaging, Oak Ridge National Laboratory Report No. K/NSP-315, Oak Ridge National Laboratory, Oak Ridge, TN, August 1995.

116. Schweiger, M. and Arridge, S.R., Comparison of two- and three-dimensional reconstruction methods in optical tomography, *Appl. Opt.*, 37, 7419, 1998.

117. Arridge, S.R. and Lionheart, W.R.B., Nonuniqueness in diffusion-based optical tomography, *Opt. Lett.*, 23, 882, 1998.

II

Basic Instrumentation

5

Basic Instrumentation in Photonics

Tuan Vo-Dinh
Duke University

This chapter presents an overview of the basic instrumentation used in photonic measurements and applications. The chapter is designed to provide an introduction to the basic setups, apparatus, and system components for readers from other research fields who wish to get further acquainted with biomedical photonics. Only basic devices and components are described in this chapter. More advanced instrumentation and specialized systems for specific applications are described in detail in Chapters 6, 9, and 10 in this handbook.

5.1 Basic Spectrometer

The detection and analysis of an optical spectrum requires a spectrophotometer. This instrument is commercially available from many manufacturers, or it may be built using standard off-the-shelf components. For routine analytical work, it is often convenient to use a commercially available instrument. For scientific investigations and special applications, it is sometimes necessary to build an apparatus designed for specific performance requirements.

5.1.1 Basic Apparatus

This section describes how various basic components can be combined to develop instrumental setups for different types of spectroscopic measurements. The common feature of spectroscopic measurements is that they all measure some spectroscopic properties that are related to the composition and structure of biochemical species in the sample of interest. There are several types of spectroscopic measurements: absorption, scattering (elastic and inelastic), and emission. A typical spectroscopic experiment that allows us to analyze complex biological systems is conceptually simple. Light at a certain wavelength λ (or frequency $v = c/\lambda$) is used to irradiate a sample of interest. This process is

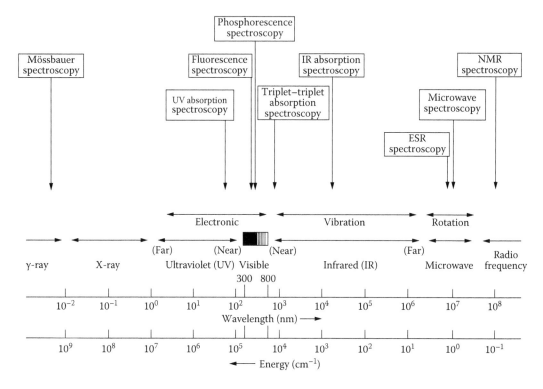

FIGURE 5.1 Various types of spectroscopies and related spectral ranges in the electromagnetic spectrum.

called *excitation*. Properties of the light that then emerges from the sample are measured and analyzed. Some properties deal with the fraction of the incident radiation absorbed by the sample: the techniques involved are collectively called absorption spectroscopy (e.g., ultraviolet [UV], visible, and infrared (IR) absorption techniques). Other properties are related to the incident radiation reflected back from the samples (elastic scattering [ES] techniques). Alternatively, one can measure the light emitted or scattered by the sample, involving processes that occur at wavelengths different from the excitation wavelength; the techniques involved are fluorescence, phosphorescence, and inelastic scattering (Raman scattering). Other specialized techniques can be used to detect specific properties of the emitted light, such as its degree of polarization and decay times.

The range of wavelengths used in various types of molecular spectroscopy to study biological molecules is quite extensive (Figure 5.1). Table 5.1 summarizes the different types of molecular spectroscopy and the associated wavelength ranges of the electromagnetic radiation.

A basic spectrophotometer generally consists of the following basic components:

1. An excitation light source
2. A dispersive device (optical filters, monochromators, or polychromators)
3. A sample to be analyzed (usually in a compartment having a sample holder)
4. A photometric detector (equipped with a readout device)

Successful application of photonic methods requires considerable attention to experimental details and a good understanding of the instrumentation performance. The recorded spectra (absorption, reflection, scattering, emission, or excitation) represent the photon emission rate or power recorded at each wavelength over a wavelength interval determined by the slit widths and dispersion of the monochromator. There are a large variety of manufacturers of spectrometers, each offering several models with different

TABLE 5.1 Various Types of Spectroscopic Techniques for Biological and Biomedical Applications

Spectral Region[a]	Wavelength[a] (cm)	Energy[a] (kcal mol^{-1})	Techniques	Properties
γ-ray	10^{-11}	3×10^8	Mossbauer	Nucleus properties
X-ray	10^{-8}	3×10^5	X-ray diffraction/scattering	Molecular structure
UV	10^{-5}	3×10^2	UV absorption	Electronic states
Visible	6×10^{-5}	5×10^3	Visible absorption	Electronic states
			Luminescence	Electronic states
IR	10^{-3}	3×10^0	IR absorption	Molecular vibrations
			IR emission	Electronic states
Microwave	10^{-1}	3×10^{-2}	Microwave	Rotations of molecules
	10^0	10^{-3}	Electron paramagnetic resonance	Nuclear spin
Radio frequency	10	3×10^{-4}	NMR	Nuclear spin

[a] Approximate values.

performance characteristics and each offering different options. Basic instrumentation components are also commercially available. For special applications, an investigator may assemble off-the-shelf components for his or her particular applications. The basic components can be adapted to design the instrument for each type of spectroscopic measurements.

5.1.2 Instrumentation for Absorption Measurements

Figure 5.2a shows a schematic arrangement of a typical instrument setup for absorption measurements. The collimated output of a light source is focused on the entrance slit of an excitation

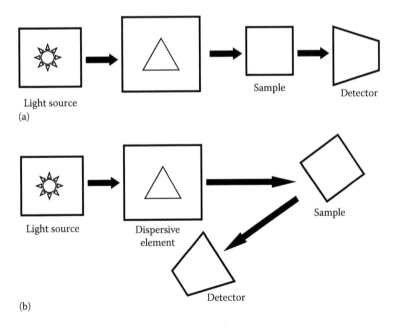

FIGURE 5.2 Various instrumental setups: (a) absorption, (b) ES.

(continued)

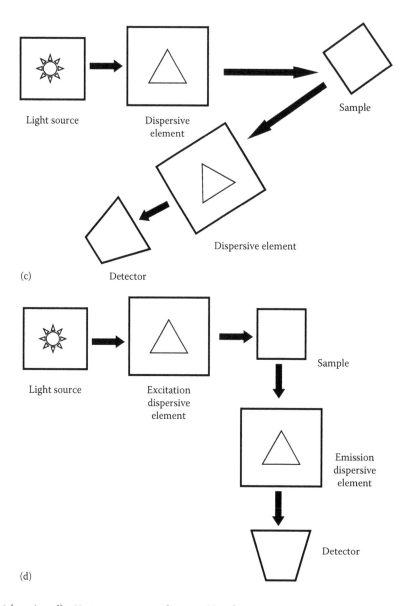

FIGURE 5.2 (continued)   Various instrumental setups: (c) inelastic scattering, and (d) emission.

monochromator for wavelength scanning. The output of the excitation monochromator is directed to the sample inside the sample compartment. The light transmitted by the sample is collected through appropriate optics and focused onto a detector. This simple instrumental setup is often used in a single-beam absorption spectrometer. Double-beam instruments include a reference beam, which is used to automatically correct intensity fluctuations in the light source in order to reduce electronic drift and lamp warm-up periods.

5.1.3 Instrumentation for Scattering Measurements

The ES technique involves detection of the backscattering of a broadband light source irradiating the sample of interest. Figure 5.2b shows a typical instrument setup that can be used for ES measurements.

A spectrometer records the backscattered light at various wavelengths and produces a spectrum that is dependent on sample structure, as well as chromophore constituents. In general, the sample is illuminated with the excitation light, which is selected with a dispersive element and then directed to a specific point location (e.g., via an optical fiber) of the sample. The scattered light is measured at the same wavelength as the excitation wavelength.

With inelastic scattering measurements, one measures the scattered light from the sample in a spectral region different from the excitation wavelength. In this case, the basic setup is similar to the ES setup but has an additional dispersive element to analyze the scattered emission from the samples (Figure 5.2c).

5.1.4 Instrumentation for Emission Measurements

Figure 5.2d shows a schematic arrangement of a typical spectrometer for emission (e.g., fluorescence, phosphorescence) measurements. The excitation light source is usually a laser or high-intensity xenon arc lamp. The collimated output of the light source is focused on the entrance slit of an excitation monochromator. The output of the excitation monochromator is directed to the sample. When a laser is used as the excitation source, the excitation monochromator is not required. The emission from the sample is collected through appropriate optics and focused onto the entrance slit of an emission monochromator. The excitation beam and the emission beam are usually focused at right angles for minimum interference from scattered light.

There are three basic classes of spectrophotometers: filter instruments, monochromator instruments, and multichannel devices. The first type of device uses optical filters, whereas the latter two systems use prisms or gratings as dispersive elements. The more expensive grating spectrometers are more versatile than filter instruments and can be used for either applied or basic studies.

Selection of an instrument appropriate to specific needs requires careful examination of many factors, including cost. The three major features to consider are the intensity of the excitation light source, the resolution and throughput of the monochromators, and the sensitivity of the detector. Although it is desirable to have an intense excitation light source, the cost–benefit ratio should be taken into consideration.

For most fluorescence measurements at room temperature, the spatial resolution need not be excessively high because the bandwidths of fluorescence spectra are usually larger than 5 nm. A monochromator with low spectral dispersion, however, would allow use of larger slit widths and consequently provide for higher radiance throughput. On the other hand, IR absorption and Raman instruments must exhibit very high spectral resolution (1 nm or less) in order to spectrally resolve the much narrower vibrational bands.

The spectrometer used for phosphorescence measurements is essentially the same as that used for fluorescence measurements. The only additional equipment required is a phosphoroscope attachment for isolating the long-lived phosphorescence from shorter-lived emissions (fluorescence, excitation scattered light). Frequently, a fluorescence spectrometer (fluorimeter) can be modified easily or equipped with a phosphoroscope to perform measurements.

The sensitivity of the spectrometer is strongly dependent on the choice of the detector, such as the photomultiplier (PM) tube or the charge-coupled device (CCD). The relative sensitivity of the PM tube or CCD can be found in data sheets from the manufacturer.

In addition to the aforementioned factors, the choice of a spectrometer also depends on other features that are more difficult to characterize, such as the reliability of electronic components, the repeatability and accuracy of scanning mechanisms, and the stray light rejection ability of the monochromators.

A cost–benefit evaluation must be made in the choice between a single-beam and a double-beam spectrometer. In the more expensive double-beam instruments, spectral fluctuations in lamp output are automatically corrected, reducing electronic drift and lamp warm-up periods. However, the performance of a single-beam instrument is adequate for most applications.

5.2 General Considerations on Instrumentation Components

5.2.1 Excitation Light Sources

UV light is generally used for excitation in many spectroscopic measurements. These UV sources may be classified into two categories, namely, line or continuum type, and can be used in a continuous wave (CW) mode or in a pulsed mode. The line sources provide sharp spectral lines, whereas continuum sources exhibit a broadband emission.

5.2.1.1 High-Pressure Arc Lamps

High-pressure arc lamps are the most commonly used radiation sources. These lamps produce an intense quasicontinuum radiation ranging from the UV (<200 nm) to the near IR (NIR) (>1000 nm) with only a few broadbands at approximately 450–500 nm. The lamps consist of two tungsten electrodes in a quartz envelope containing gases under high pressure, for example, xenon (Xe), mercury, or a Xe–mercury mixture. Lamps of this type are commercially available in a wide range of input power from a few watts to several kilowatts.

The high-pressure mercury arc lamp is similar to the high-pressure xenon lamp in appearance and performance. The spectral output of mercury lamps is of a line type, whereas that of the xenon lamps is of a continuum type. If excitation can be carried out at only one wavelength or a few fixed wavelengths of the mercury emission lines, the mercury lamp is probably the most effective radiation source. The Xe lamp, however, is more commonly used, because it provides a smoother spectral profile that is more suitable for conducting excitation spectra measurements.

The Xe arc lamp is the most versatile light source for steady-state spectrometers and has found widespread use. This lamp provides a relatively continuous light output from 250 to 700 nm. Xe arc lamps emit a continuum of light as a result of the recombination of electrons with the ionized Xe atoms. Complete separation of the electrons from the atoms yields the continuous emission. Xe lamps are available with an ellipsoidal reflector as part of the lamp itself. Parabolic reflectors in some commercially available Xe lamps collect a large solid angle of light and provide a collimated output.

The operation of high-pressure arc lamps requires special care and handling, such as reduction of the excitation stray light with a good monochromator, use of a highly regulated direct current (dc) power supply, and removal of the heat generated by the lamp output in the IR range. A warm-up period is also necessary to minimize arc wandering, because there is always some tendency for the arc to change its location inside the lamp envelope during the first half hour of operation. This arc wandering effect may cause sudden variations in the observed intensity, especially when the image of the arc is focused into a small slit aperture.

Extreme care should be exercised when inspecting high-pressure arc lamps. These lamps may explode when dropped or bumped because they are filled with gases at high pressures (~5 atm at ambient temperature and 20–30 atm at operating temperature conditions). It is recommended that special leather gloves, safety glasses, and protective headgear be used whenever the lamp housing is opened. One should not look directly at an operating Xe lamp. The extreme brightness will damage the retina, and the UV light can damage the cornea. With some older lamps, proper ventilation or use of deozonators is required to remove the ozone produced by the UV radiation of the lamp. Recently, many available Xe lamps are considered ozone-free since their operation does not generate ozone in the surrounding environment.

5.2.1.2 Low-Pressure Vapor Lamps

Low- (or medium-) pressure mercury vapor lamps are often used as line sources. They are simple to use, require little power, and offer intense UV radiation concentrated in a few lines (e.g., 253.7 nm, 365.0/265.5/366.3 nm multiplet). The mercury vapor lamps are widely used in simple filter-type spectrometers because of their low cost, intense emission characteristics, and good stability. The lamps do not need a complex power supply system and provide excellent reference light sources for calibration of spectrometers.

5.2.1.3 Incandescent Lamps

The tungsten filament incandescent lamp is the simplest continuum source. This type of incandescent lamp exhibits a smooth, continuous spectral profile determined by the blackbody radiation characteristics given by Planck's equation:

$$S_\lambda = \frac{5.8967\lambda^{-5}E_\lambda}{\exp(14{,}388/\lambda T)^{-1}}$$

where

S_λ is the spectral radiance (W cm^{-2} sr^{-1} nm^{-1})
λ is the wavelength (nm)
T is the temperature (K)
E_λ is the spectral emissivity of the filament material (dimensionless)

Since incandescent lamps usually have low UV output, they are seldom used as excitation sources, especially for luminescence measurements where samples absorb the UV. Their smooth spectral profile, however, makes them very suitable for intensity calibration procedures.

Standard incandescent lamps with calibration data provided by the National Institute of Standards and Technology (NIST) are readily available commercially. Intensity calibration data are available for the spectral range from 250 nm to 2.5 μm.

5.2.1.4 Solid-State Light Sources

Light-emitting diodes (LEDs) are solid-state light sources, which provide output over a wide range of wavelengths. These devices require little power and generate little heat. One can use a few LEDs to cover a spectral range from 400 to 700 nm. LEDs are practical light sources for many low-power photonic applications. LEDs can be amplitude modulated up to hundreds of megahertz. Another type of solid-state light source is the solid-state laser, which is described in the next section.

5.2.1.5 Lasers

Although primarily conventional light sources have been used for absorption analyses, lasers are increasingly used in luminescence and Raman measurements. The advantages offered by lasers as excitation sources include

1. Monochromaticity
2. High degree of collimation
3. High intensity
4. Phase coherence
5. Short pulse duration (with pulsed lasers)
6. Polarized radiation

Selection of laser excitation sources is determined by the wavelengths that can be matched to the absorption band of the compounds to be analyzed in order to take advantage of maximum absorption. If time-resolved measurements are performed, the pulse width of the laser is an important factor to consider. The intensity of a laser is very high at (or even near) the laser emission lines and, therefore, often interferes with the lower intensity of the emission or scattering signal being measured. A number of devices, such as a spike filter or a single monochromator, may be used to reject the Rayleigh scattered light. Notch filters, which consist of crystalline arrays of polystyrene spheres, exhibit very high-efficiency rejection of laser lines.

Many manuscripts have described the principle and applications of lasers in detail.[1-6] Some general properties of common types of lasers are in the following section.

5.2.1.5.1 General Properties of Laser

The lasing process is a stimulated emission following population inversion between different electronic levels of an active medium. Laser operation occurs inside a resonant optical cavity filled with an active medium that can be a gas, a solid crystal, a semiconductor, or a dye solution.

Some lasers are operated in the CW mode; others are pulsed. There are several methods for producing very short pulses. The simplest is to induce a population inversion for only short periods of time using pulsed electrical discharges. Other pulsing methods include Q-switching, cavity dumping, and mode locking. The Q-switching method consists of first decreasing the reflectivity or the quality, Q, of the laser cavity and then suddenly switching the Q value back to a relatively high value. Active Q-switching can be done by shutter devices, such as an electrooptic Kerr cell,[6] a Pockels cell (a device that provides a modulation linearly proportional to the applied electric field), or simply a rotating mirror. Passive Q-switching uses a saturated absorber such as gas or a dye. Cavity dumping is a pulse-forming technique for laser materials having excited-state lifetimes that are too short for Q-switching operation. Another method that produces subnanosecond pulses is the mode-locking technique. The mode-locking technique involves locking the phase of the electromagnetic longitudinal modes inside the cavity. The repetition rate of mode-locked pulses, determined by the speed of light and twice the length of the optical cavity, is too high for phosphorescence measurements. For a 1 m cavity, the time interval between two pulses is 6.6 ns. Even with a cavity design based on a folded-path mirror with a total length up to 10^3 m, the laser still produces pulses at 5.5 μs intervals.

5.2.1.5.2 Gas Lasers

The active medium of a gas laser is either an atomic (e.g., helium–neon), molecular (e.g., carbon dioxide, hydrogen cyanide, rare gas halide, water vapor), or ionized gas (e.g., argon, krypton, xenon, helium–cadmium, helium–selenium). Output powers of CW gas lasers range from milliwatts to kilowatts. Peak powers of pulsed gas lasers can reach several megawatts.

A compact, low-power gas laser often used for alignment and for calibration is the helium–neon laser. The output of this laser ranges from less than a milliwatt to several tens of milliwatts at 632.8 nm. Certain lasers also emit at 1152 nm.

Two commonly used gas lasers are the argon ion laser and the krypton ion laser. These rare gas lasers emit a series of lines from the near-UV to the IR region and have CW output ranging from tens of milliwatts to tens of watts. An ion laser that uses a metal vapor as the active medium is the helium–cadmium laser. It provides a CW output in the milliwatt range in the blue or UV regions. A CW gas laser having high power (up to several hundred watts) at 10 μm is the carbon dioxide laser.

A very practical and widely used laser is the nitrogen laser. It operates only in the pulsed mode. The nitrogen laser produces subnanosecond to tens-of-nanosecond pulses in the near-UV region (337 nm). The energy per pulse ranges from a few microjoules to a few joules.

Pulsed output in the UV region is also obtained with excimer lasers, in which the active medium is an excimer, which is a molecule that is stable only in the excited state and dissociates immediately after it emits light (e.g., krypton fluoride, xenon fluoride, xenon chloride, or argon fluoride).

In chemical lasers, the active medium, such as hydrogen fluoride or deuterium fluoride, is produced by a chemical reaction induced by electrical discharges.

Table 5.2 summarizes various characteristics of several CW and pulsed gas lasers.

5.2.1.5.3 Solid-State Lasers

Solid-state lasers can be of the pulsed type, such as the chromium-doped ruby laser, with principal output at 694 nm, or the CW type, such as the yttrium aluminum garnet (YAG) laser, with principal output at 1064 nm. Solid-state lasers are pumped optically. Optical pumping of pulsed lasers is usually achieved with a Xe flash tube. Pulsed outputs range from tenths of a joule to tens of joules. The repetition rate of these pulsed lasers is relatively low (0.1 Hz to tens of Hz). The CW YAG laser is generally used in the

TABLE 5.2 Types and Spectral Characteristics of CW and Pulsed Commercial Gas Lasers

Active Medium	Output Wavelength (nm)	CW Gas Laser	Pulsed Gas Laser
Argon	330–514	×	×
Argon fluoride	193		×
Argon–krypton	450–1,090	×	
Barium	1,500		×
Carbon dioxide	9,000–11,000	×	×
Carbon monoxide	5,000–7,000	×	×
Copper	510.5/578.2		×
Deuterium cyanide	$190 \times 10^3/195 \times 10^3$	×	
Deuterium fluoride	3,600–4,000	×	×
Fluoride	624–780		×
Gold	627.8		×
Helium–cadmium	325/442/441	×	
Helium–neon	632.8/1,152/3,391	×	
Helium–selenium	460–653	×	
Helium–xenon	3,805	×	×
Hydrogen cyanide	$311 \times 10^3/337 \times 10^3$	×	×
Hydrogen fluoride	2,600–3,000	×	×
Krypton	330–790	×	×
Lead			×
Methanol	118.8×10^3	×	×
Methyl fluoride	$190 \times 10^3 – 596 \times 10^3$	×	×
Nitrogen	337		×
Water	$(18/78/118) \times 10^3$	×	×
Xenon bromide	282		×
Xenon chloride	318		×
Xenon fluoride	351/353/354		×

pulsed mode following mode-locking, cavity dumping, or Q-switching operations. The CW output of YAG lasers is between a tenth of a watt and several watts.

Other solid-state lasers include the glass lasers, which operate only in a pulsed mode like the ruby lasers, and the F-center lasers, which provide a tunable CW output in the 2000–3000 nm spectral range.

5.2.1.5.4 Semiconductor Lasers

One important family of semiconductor lasers is the diode lasers. In a semiconductor diode laser, the front and back surfaces of the diode form the resonant cavity that generates light emission when electrons and holes recombine at the *p–n* junction. When driven by a low current, the diode emits incoherent light. Above an input current threshold, the diode begins to emit coherent laser light. The small size of semiconductor lasers allows efficient packaging. They can be operated on battery power and therefore are useful for remote operation. High-power diode lasers are generally not tunable; however, many wavelengths of interest are available for medical applications. One type of diode laser has a tunable output (the lead salt type). The other type has a fixed-wavelength output (gallium arsenide [GaAs] or gallium aluminum [GaAl] type). The lead salt lasers emit at 2000–3000 nm and are used for high-resolution spectroscopy in the IR region. The GaAs and GaAl lasers are usually employed in communications utilizing fiber-optic cables or in information processing. Diode lasers can be amplitude modulated to several gigahertz. Vertical-cavity surface-emitting lasers are semiconductor lasers that have the emission beam perpendicular from the top surface.

5.2.1.5.5 *Tunable Dye Lasers*

A dye laser consists of an organic dye solution optically pumped by a light source (flash lamp, nitrogen laser, or ion laser). Dye lasers offer many properties that make them close to being ideal light sources. Dye lasers can be tuned across a large spectral range from the UV to the NIR with a series of different dyes. The type of output desired and the absorption properties of the dye determine the choice of the pump source. Pulsed sources usually provide peak powers at levels sufficiently high for laser action in most dyes when the energy per pulse is low.

5.2.1.5.6 *Tunable Lasers with Optical Parametric Oscillators*

Optical parametric oscillators (OPOs) have extended the laser radiation from visible to the IR and have opened a new area for spectroscopists. With OPO systems, no changing of dyes solutions is needed. The operation of an OPO, in a process similar to harmonic generation, is based on the nonlinear response of a medium to a driving field (the pump laser beam) to convert photons of one wavelength to photons of other, longer wavelengths. Specifically, in the so-called parametric process, a nonlinear medium (usually a crystal) converts the high-energy photon (the pump wave) into two lower-energy photons (the signal and idler waves).

The exact wavelengths of the signal and idler are determined by the angle of the pump wave vector relative to the crystal axis. Energy can be transferred efficiently to the parametric waves if all three waves (pump, signal, idler) are traveling at the same velocity. Under most circumstances, the variation of index of refraction with crystal angle and wavelength allows this *phase matching* condition to be met only for a single set of wavelengths for a given crystal angle and pump wavelength. Thus, as the crystal rotates, different wavelengths of light are produced.[7]

When the crystal is contained in a resonant cavity, feedback causes gain in the parametric waves in a process similar to buildup in a laser cavity. Thus, light output at the resonated wavelength (and the simultaneously produced other parametric wavelength) occurs. The cavity can either be singly resonant at either the signal or idler wavelength or be doubly resonant at both wavelengths.

5.2.2 Optical Fibers and Dispersive Devices

5.2.2.1 Optical Filters

The simplest dispersive element is the optical filter, which can be used when variation of the excitation or emission wavelength is not needed. In this case, the filter is simply a single-wavelength selector device. Filters may be used to select the excitation light (excitation filters) or remove the scattered excitation light from the emission (emission filters). It is often desirable to use optical filters in addition to monochromators. In general, excitation filters with a narrow spectral bandpass are used to provide selectivity and should be able to withstand high-intensity light and, in some cases, high temperatures. Conversely, emission filters usually have a broad bandpass and are generally employed to provide maximum light throughput.

Filters may be classified into three main categories: neutral density, cutoff, and bandpass. Neutral density filters, which have a nearly constant transmission over a wide spectral range, are generally used to attenuate the light equally at all wavelengths. For example, one can use neutral density filters to strongly decrease luminescence signals and intense excitation light or to adjust or match the intensity of two signals.

Cutoff filters, which have a sharp transmission cutoff, are generally used to remove undesired radiation, such as stray light or second-order diffraction light. Long-pass (or high-pass) filters are often used to reject scattered light from the excitation and to transmit the emission of interest (e.g., fluorescence). Similarly, short-pass (or low-pass) filters are used to transmit light at the shorter wavelengths and remove light occurring at wavelengths longer than the spectral range of interest. Notch filters are used to remove light within a specific bandwidth of interest. They are often used to remove the interfering laser excitation light from the emission light (e.g., Raman or fluorescence).

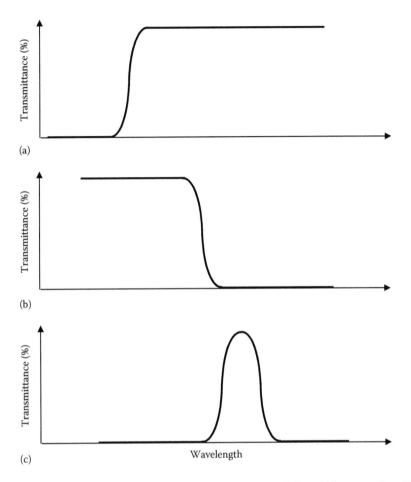

FIGURE 5.3 Schematic diagram of transmission curves of various types of filters: (a) long-pass filter, (b) short-pass filters, and (c) band-pass filter.

Band-pass filters generally serve to isolate a particular spectral region of interest for excitation or detection. Great care should be taken in the selection of band-pass filters to avoid the possibility of emission from the filter itself. When illuminated with UV light, some filters exhibit luminescence, which can interfere with the sample emission. For this reason, it is usually desirable to position the filter farther away from the sample, rather than close to the sample. Figure 5.3 shows schematic diagrams of transmission curves for various types of filters: long-pass filters (a), short-pass filters (b), and band-pass filter (c). Figure 5.4 depicts typical transmission curves of holographic and dielectric notch filters.

Filters can also be classified according to their operating principle: absorption, interference, or birefringence. Absorption filters include neutral density, cutoff, and wide band-pass filters. They may be fashioned from a variety of materials, including colored glass, gelatin, chemical solution, gas, sapphire, and alkali–halide materials. Colored glass filters are the most practical and least expensive. The spectral characteristics of the coloring inorganic species determine the transmission properties of the filters. Gelatin containing a dye is also used in absorption filters. Gelatin filters are thermally less stable than glass filters and should be used only as emission filters. Optical cells filled with a specific chemical solution or gas may be used as filters in special cases, but they are not convenient for general use.

Interference filters are generally of the narrow bandpass (10 nm) or broadband (50–80 nm) type. Carefully controlled thicknesses of vacuum-deposited layers of dielectric and metallic layers are

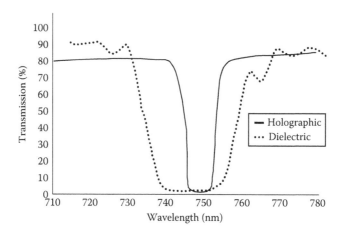

FIGURE 5.4 Typical transmission curves of holographic and dielectric notch filters.

arranged so that light outside the spectral bandpass is eliminated by destructive interference, whereas light within the bandpass is transmitted by constructive interference. Some interference filters have linearly variable transmittance across the substrate; since these filters are angle sensitive, great care should be taken when mounting them in an optical system. Interference filters, which transmit typically 30% of the intensity in the UV and 60% in the visible range, are generally used as excitation filters because of their narrow bandpass (from 50 to 1 nm). Heat from the excitation light beam is usually not a problem since the radiation that is not transmitted is rejected by reflection. Interference filters covering a broad spectral range from the UV to the IR are commercially available from many suppliers.

Dichroic filters are substrates coated with thin films to produce the desired transmission and reflection properties across the spectral range of interest. They are often used as color filters (both additive and subtractive) or as devices designed to transmit light at certain wavelengths and reflect light at other wavelength ranges. Dichroic filters are angle sensitive, so care should be taken to mount them properly in optical systems.

Birefringence filters, which are constructed with polarizers and retardation plates, may be used in polarization measurements. These filters are generally made of stretched polymer films, the absorption properties of which determine the characteristics of the filters. They usually have a bandwidth similar to that of interference filters.

5.2.2.2 Monochromators

Continuous wavelength selection is performed with monochromators, which are used to disperse polychromatic or white light into the various colors. The performance specifications of a monochromator are characterized by the spectral dispersion, the efficiency, and the stray light levels. The spectral dispersion is usually expressed in nanometers per millimeter, where the slit width is expressed in millimeters. Low stray light and high efficiency are desired qualities in selecting a monochromator. There are two types of monochromators: prism and grating monochromators.

5.2.2.2.1 Prism Monochromators

In prism monochromators, light dispersion is due to the change of the refractive index of the prism material with the wavelength of the incident light. The angular dispersion D is given by

$$D = \frac{d\theta}{d\lambda} = \frac{dn}{d\lambda}\frac{d\theta}{dn} \tag{5.1}$$

where
    θ is the angular deviation
    n is the refractive index of the prism materials
    λ is the wavelength of the light source

Prism monochromators usually produce less stray light than grating devices and are free from overlap from multiple orders, but they are less convenient to use than grating monochromators due to their nonlinear scanning dispersion.

5.2.2.2.2 Grating Monochromators

Most spectrometers are now equipped with monochromators having diffraction gratings. Gratings comprise a large number of lines, or grooves, ruled on a highly polished surface. The density and shape of its grooves determine the characteristics of a grating. Energy throughput and resolution increase with increasing number of grooves per millimeter. The width of a groove should be approximately equal to the wavelength of the light to be dispersed. The shape of the groove should be such that the maximum amount of light at a given wavelength is concentrated at only one specific angle for each order. The design and construction of the grating also determine the other properties of the monochromator, for example, reflectivity (or radiance throughout) and stray light rejection.

The general diffraction grating formula is given by

$$\sin\theta + \sin\theta' = k\frac{\lambda}{d} = kn\lambda \tag{5.2}$$

where
    θ' is the angle of incidence
    n is the number of grooves per unit length
    d is the groove spacing
    k is the dispersion order (see Figure 5.5a)

As shown in this formula, the grating disperses light because θ' is dependent on λ. A spectrum is obtained for each value of k. As a consequence, gratings also produce second- and higher-order dispersions that may overlap the lower-order emission of interest. One simple procedure to overcome this problem is to place a cutoff filter at the exit slit of the grating monochromator so that only one order is transmitted. The first-order dispersion ($k = 1$), which produces the highest throughput, is generally used for emission measurements. Higher orders, however, result in higher resolution. At the zero order ($k = 0$), there is no dispersion because all wavelengths are reflected at the same angle.

Most gratings used in modern spectrometers are of the reflection type ($\theta = \theta'$). In this case, the observation is in the direction of illumination (Littrow configuration). The grating formula then becomes

$$\sin\theta = k\frac{\lambda}{d} \tag{5.3}$$

The aforementioned relation shows that a spectral scan, which is linear with respect to the wavelength λ, can easily be produced with a simple sine-bar mechanism.

The scanning mechanisms of almost all commercial grating monochromators are readily made linear with respect to wavelength. If one prefers to record a spectrum on a linear wave-number (cm⁻¹) scale, a mechanical device that generates a reciprocal function may be attached to the monochromator. This latter device is available commercially, but it is used mostly in Raman spectrometers.

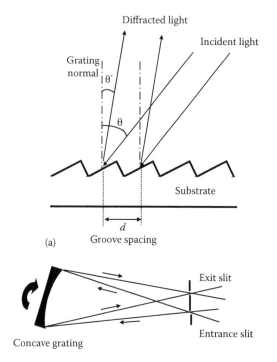

(a)   Groove spacing

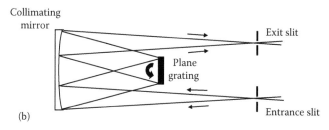

(b)

FIGURE 5.5 (a) The grating rule (θ = angle of incidence; θ' = angle of diffraction; d = groove spacing). (b) Optical configuration of planar grating and concave grating.

Three interrelated factors—spectral dispersion, resolving power, and throughput—should be considered in optimizing the conditions for a given experimental situation. The spectral dispersion, D_S, is defined by

$$D_S = \frac{d\lambda}{dl} = \frac{1}{l_f}\frac{d\Phi^{-1}}{d\lambda} = \frac{1}{l_f} = d' \tag{5.4}$$

where
  d' is the distance measured across the slit
  l_f is the focal length of the lens or mirror
  Φ is the angle of deviation (or diffraction)
  $d\Phi/d\lambda$ is the rate of change of Φ with wavelength

This equation indicates that a constant spectral dispersion and resolution are readily provided when the slit width d is kept constant. No elaborate slit program is required (as in prism monochromators) to keep the spectral resolution constant over a spectral scan.

The resolving power R, which denotes the ability of a grating monochromator to separate adjacent emission lines, is defined by

$$R = \frac{\lambda}{\Delta\lambda} = \frac{w}{\lambda}\left(\sin\theta + \sin\theta'\right) \qquad (5.5)$$

where w is the width of the grating.

By combining Equations 5.2 and 5.5, one obtains (for $k = 1$)

$$R = wn \qquad (5.6)$$

This relation shows that the resolving power is directly proportional to the grating dimension (width) and to the groove density. For example, a grating with 1200 grooves/mm and 100 mm width would have a resolving power R given by

$$R = wn = 100 \times 1{,}200 = 120{,}000 \qquad (5.7)$$

The spectral bandpass B_S is given by

$$B_S = \frac{d'}{l_f D_S} \qquad (5.8)$$

This relation indicates that a smaller slit width would yield a smaller bandpass and provide better resolution. Larger slit widths, however, yield increased signal levels and therefore higher signal-to-noise (S/N) values.

The transmission efficiency of a grating is dependent on the wavelength and on the design of the grating. Two other factors that should be taken into consideration in selecting a grating are the blaze and the blaze angle. The blaze angle is the angle at which the grooves are ruled on the grating surface, and the blaze is the wavelength at which the maximum efficiency of the grating is concentrated. In general, one selects an excitation monochromator with high efficiency in the UV range and an emission monochromator with high efficiency in the visible spectral range. For example, a grating blazed at 300 nm may be chosen for excitation, whereas a grating blazed at 500 nm may be used for emission because the fluorescence of most organic compounds occurs between 400 and 600 nm. Use of gratings with an improper blaze angle may be a source of low sensitivity for many compounds.

In many monochromators, gratings with different blazes can be snapped into and out of the grating mount. Grating monochromators are now more widely used than prism monochromators because of their higher resolution. These monochromators are convenient to use because they produce a spectral dispersion that is linear with respect to wavelength.

The light throughput for various standard gratings varies between 50% and 90% and increases with groove density. For conventionally ruled gratings, this density has a practical limit above which it may produce substantial stray light and *ghosts*. Ghosts are spurious spectral lines caused by periodic imperfections in the gratings. Stray light is defined as any light that passes through the monochromator outside the monochromator bandpass. For Raman measurements (and in some degree fluorescence measurements), which involve very low signal levels in the presence of an intense excitation source, the stray light level of the monochromator is perhaps the most critical parameter.

The advent of lasers and holography in the early 1960s has made possible the production of holographic gratings with high throughput and low stray light. Monochromators may have planar or concave gratings. Planar gratings are usually produced mechanically and may contain imperfections in some of the grooves. Concave gratings are usually produced by holographic and photoresist methods. Since they are produced optically, holographic gratings have fewer imperfections than mechanically produced (i.e., ruled) gratings. As a result of their optical configuration, a concave grating functions as both the diffracting and focusing element, resulting in one instead of three reflecting surfaces (Figure 5.5b). Fewer reflecting surfaces generally result in increased efficiency.

5.2.2.3 Tunable Filters

Tunable filters are special types of dispersive devices. These devices, such as tunable filters (acousto-optic tunable filters [AOTFs]) and liquid crystal tunable filters (LCTFs), allow the investigator to rapidly record an image at various wavelengths. An AOTF is a solid-state optical band-pass filter that can be tuned to various wavelengths within microseconds by varying the frequency of the acoustic wave propagating through the medium. The solid-state nature of the AOTF provides a high-throughput (90% diffraction efficiency) dispersive element with no moving parts, thus increasing the ruggedness of the instrumentation. AOTF devices consist of a piezoelectric transducer bonded to a birefringent crystal. The transducer is excited by a radio frequency (rf) (50–200 MHz) and generates acoustic waves in a birefringent crystal. Those waves establish a periodic modulation of the index of refraction via the elasto-optic effect. Under proper conditions, the AOTF will diffract part of the incident light within a narrow frequency range. The Bragg grating can diffract only light that enters the crystal such that its angle to the normal of the face of the crystal is within a certain range. In a noncollinear AOTF, the diffracted beam is separated from the undiffracted beam by a diffraction angle. The undiffracted beam exits the crystal at an angle equal to the incident light beam, while the diffracted beam exits the AOTF at a small angle with respect to the original beam. A detector can be placed at a distance so that the diffracted light can be monitored, while the undiffracted light does not irradiate the detector. In addition, when the incident beam is linearly polarized and aligned with the crystal axis, the polarization of the diffracted beam is rotated 90° with respect to the undiffracted beam. This can provide a second means to separate the diffracted and undiffracted beams.

LCTFs use electrically controlled liquid crystal elements that transmit a certain wavelength band while being relatively opaque to others. The LCTF is based on a Lyot filter, a device constructed of a number of static optical stages, each consisting of a birefringence retarder (quartz for LCTFs) sandwiched between two parallel polarizers. A stack of stages function together to pass a single narrow wavelength band. Tunability is provided by the partial alignment of the liquid crystals along an applied electric field between the polarizers; the stronger the field, the more the alignment and the greater the increase in retardance. Tuning times for randomly accessing wavelengths depend on the liquid crystal material used and the number of stages in the filter. At the moment, commercial devices use nematic components that result in tuning times of approximately 50–75 ms.

The operating principles of AOTFs and LCTFs and their use in imaging are described in detail in Chapter 9 in this handbook.

5.2.3 Optical Fibers

Optical fibers and waveguides are described in detail in Chapter 6 in this handbook. This section briefly mentions only the basic and salient features of optical fibers. A widely used component that provides an optical link between a spectroscopic instrument and a remotely located sample is the optical fiber. The rapid growth of fiber-optic sensing has paralleled the commercial availability of low-attenuation optical fibers. In many applications, the optical fibers comprise a core made with an optically transparent material (e.g., glass, quartz, or polymer) with a certain refractive index, n_1, surrounded by a cladding made with another material (e.g., quartz or plastic) having another refractive index, n_2. Light transmission is based on total internal reflection as depicted in Figure 5.6. Light rays that impinge on the core/cladding interface at an angle equal to or greater than the critical angle θ_c, as determined by Snell's law, can be

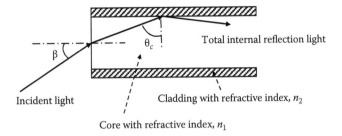

FIGURE 5.6 Total internal reflection in an optical fiber.

transmitted along the fiber by total internal reflection. Accordingly, an important fiber characteristic, the half acceptance angle β, is given by the equation

$$\sin\beta = \frac{\left(n_1^2 - n_2^2\right)^{1/2}}{n_0}$$

where n_0 is the refractive index of the medium in which the end of the fiber resides.

Low attenuation over desired spectral regions and a large β value are desirable fiber characteristics. Fibers with large values of β permit the coupling of a large amount of excitation radiation from the source at the incident end of the fiber and the efficient collection of fluorescence signal at the sensing end of the fiber.

Optical fibers can be used to transmit the excitation light to a sample and transmit the signal (reflected or scattered light) from the sample to the detector. There are several possible optical fiber configurations to perform these measurements: single-fiber system, bifurcated-fiber system, and dual-fiber system (Figure 5.7). In Figure 5.7a, a single fiber is used to transmit the excitation beam from a light source to the sample and transmit the emission from the sample to the detector. A dichroic filter is used to transmit the light at the excitation wavelength and reflect light at the emission wavelength. In Figure 5.7b, a bifurcated fiber is used, one end transmitting the excitation light to the sample and the other end transmitting the sample emission light to the detector. In Figure 5.7c, two separate parallel fibers are used, one fiber transmitting excitation light and the other fiber transmitting the emission light. In Figure 5.7d, two perpendicular fibers are used. The angle between the fibers can be varied and optimized in order to optimize the overlap of the excitation and detection volumes and to minimize scattered light.

5.2.4 Polarizers

Light can be described as consisting of two transverse electric and magnetic fields that are perpendicular to the direction of light propagation and to each other. The direction of the electric vector is used to describe the polarization of light. Most incoherent light sources consist of a large number of atomic and molecular emitters. The rays from such sources, which have electric fields with no preferred orientation, are called unpolarized. On the other hand, if a light beam (e.g., from a laser) consists of rays where the electric field vectors are oriented preferentially in the same direction, the beam is said to be linearly polarized.

Polarizers transmit light that has its electric vector aligned with the polarization axis and block light that is rotated 90°. A commonly used device is the Glan–Thompson polarizer. This device consists of a calcite prism, a birefringent material, where the refractive index is different along each optical axis of the crystal. In general, the polarization characteristics of a crystal are determined by

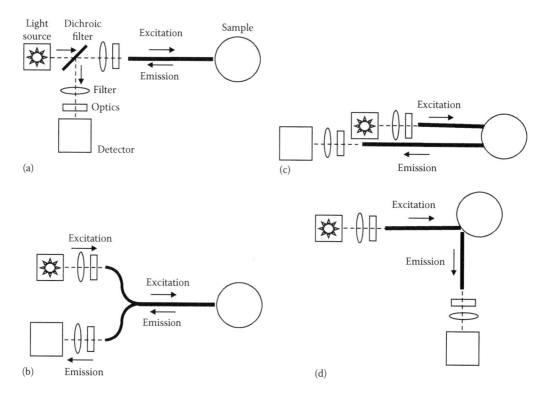

FIGURE 5.7 Various configurations using optical fibers: (a) single-fiber system, (b) bifurcated-fiber system, (c) parallel dual-fiber system, and (d) dual-fiber system.

the crystal structure and electron bonding. Calcite is widely used as a polarizing material because of its excellent transmission well into the UV range and large difference in the two index-of-refraction values. Glan–Thompson polarizers exhibit high extinction coefficients (near 10^6) and high acceptance angle (10°–15°).

Other types of polarizers include film polarizers, which are composed of materials that absorb light polarized in one direction more strongly than light of the orthogonal polarization. These are thin films of a stretched polymer (e.g., polyvinyl alcohol) that transmit the light polarized in one direction and absorb the light polarized in another direction. The film sheets are stretched to orient and align the molecules. Polarizer films are not suitable for use with intense light beams (e.g., lasers), because they absorb light and therefore can be easily damaged. Other types of polarizers, such as the Wollaston polarizers and Rochon prism polarizers, split the unpolarized light into two beams, which then must be spatially selected.

5.2.5 Detectors

Selection of a suitable detector is one of the most critical steps in the development of a spectrometer. Detectors for electromagnetic radiation can be classified into photoemissive, semiconductor, and thermal types. Photoemissive detectors are generally used in optical measurements. These devices include PM tubes, photodiodes (PDs), and imaging tubes. The PM tubes are the most commonly used because they are the most sensitive detectors for the visible and near-UV regions. There are two types of detectors, single-channel detectors and multichannel. Multichannel detectors include 1D and 2D detector arrays. Traditionally, spectroscopy has involved using a scanning monochromator and a single-element detector (e.g., PD, PM). With monochromators, only one spectral resolution element, or channel, can be monitored at a time. Detectors that permit the recording of the entire spectrum simultaneously, thus providing the multiplex advantage, are known as multichannel detectors. With multichannel systems, a complete spectrum can be recorded in the same time it takes to record one wavelength point with a scanning system.

5.2.5.1 Single-Channel Detectors

5.2.5.1.1 *Photomultipliers*

PMs are widely used for their high spectral sensitivity, wide operating range, low cost, and relatively simple electronics. A PM coupled with a monochromator is the simplest and most commonly used single-channel spectrometer. The spectral resolution and the sensitivity of the spectrometer are strongly dependent on the characteristics of the monochromator and the PM tube. Modern holographic gratings generally offer excellent stray light rejection.

The PM is a vacuum tube containing a highly sensitive surface, the photocathode, which is generally made of a metal oxide. The PM tube operates as a current source. The intensity of the current output is proportional to the intensity of the light striking the photocathode, which is a thin film of metal on the inside of the PM window. A schematic diagram of a PM tube is given in Figure 5.8. The photons cause electrons to be ejected from the photocathode surface as the result of the photoelectric effect. The photocathode is held at a high negative potential, typically −1000 to −2000 V. The photoelectrons then cascade down the dynode chain in increasing numbers as a result of secondary emission, which occurs at each dynode surface. Each dynode is an electrode coated with a material that emits several secondary electrons for each incoming electron. The electrons are accelerated down the dynode chain by a potential on the order of 100 V between two consecutive dynodes. The secondary electrons ejected from the first dynode are then accelerated onto the next dynode, where they induce further emission of secondary electrons. The secondary emission occurring at every successive dynode stage can provide an overall internal amplification factor of up to 10^8. The total number of dynodes is usually between 7 and 12, and the total high voltage between the first and last dynode is commonly between 0.7 and 1.5 kV. The total current collected at the anode is proportional to the incoming photon flux and is linear over many orders of magnitude.

The photocathode materials and the transmission characteristics of the window materials determine the spectral response of the photocathode. Only a photon with an energy greater than the work function of the cathode material results in the release of a photoelectron from the cathode. The photocathode material determines the long-wavelength cutoff, and the window material determines the short-wavelength cutoff. Figure 5.9 depicts examples of some typical cathode sensitivity responses.

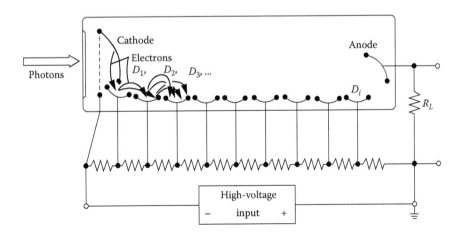

FIGURE 5.8   Schematic diagram of a photomultiplier tube. D_i, dynode i; R_L, load resistor.

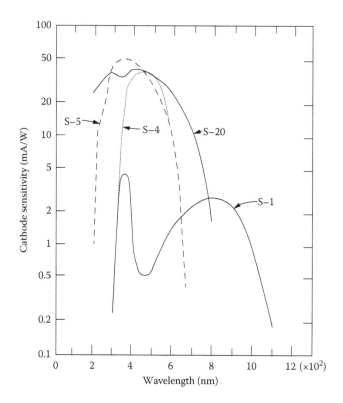

FIGURE 5.9 Examples of spectral responses curves of PMs.

The PM is equivalent to a current power supply, since all secondary electrons ejected from the dynodes are replaced from the power supply via the dynode resistor chain. The number of dynodes determines the amplification gain of the PM tube, and the gain per dynode stage is determined by the interdynode potential. The values of the dynode resistors are selected so that the current through the dynode resistor chain is about 100 times the maximum anode current. This design ensures that the anode potential remains constant during measurement. The manufacturers provide designs of typical dynode circuitry.

One limiting factor of the PM is the dark current that may be due to leakage current (imperfect insulation), ionization of residual gases inside the tube by electrons, and thermionic emission from the cathode or the dynodes. Proper preparation of PM connections significantly reduces the dark current in PM tubes. Figure 5.10 shows the five most common dynode configurations. The PM response time depends basically on the number of dynode stages and the dynode configurations. The line-focused configuration generally exhibits the fastest time response (<3 ns). In general, the capacitance between the anode and other electrodes determines the rise time of the PM tube.

Cooling the PM tube reduces the thermionic component of the dark current. With PM tubes having GaAs cathodes, cooling produces a falloff in sensitivity in the red region but increases the sensitivity over the rest of the visible spectrum. PMs require various accessories such as housing assemblies, power supplies, current stabilization devices, and magnetic shielding. Cooling equipment is usually optional. Many suppliers of PM tubes also provide the accessories and frequently offer the complete detection system in one package.

Microchannel plates (MCPs) are special PM tubes that contain numerous holes, microchannels (instead of a dynode chain), which are lined with the secondary emissive dynode material. In an MCP,

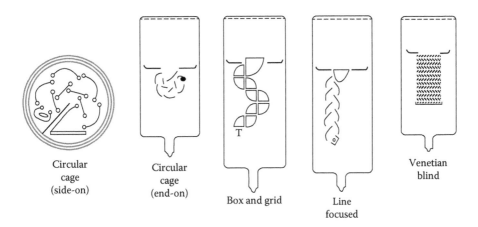

Circular cage (side-on)   Circular cage (end-on)   Box and grid   Line focused   Venetian blind

FIGURE 5.10 Typical dynode configurations of PMs.

the electrons are amplified as they drop down the voltage gradient across the MCP. MCP devices have very fast time responses and are used in the most demanding high-speed, time-resolved measurements.

5.2.5.1.2 Photodiode and Avalanche Photodiode

A simple detector is a diode, which is a semiconductor device that produces, as a result of the absorption of photons, a photovoltage or free carriers that support the conduction of photocurrent. PDs are used for the detection of optical communication signals and for the conversion of optical power to electrical power. However, the lack of high amplification limits their usefulness for high-sensitivity measurements.

An avalanche photodiode (APD) is a PD that uses avalanche breakdown to achieve internal multiplication of photocurrent. The APD is usually a silicon-based semiconductor containing a p–n junction consisting of a positively doped p region and a negatively doped n region sandwiching an area of neutral charge termed the depletion region. These diodes provide gain by the generation of electron–hole (e–h) pairs from an energetic electron that creates an *avalanche* of electrons in the substrate. Photons entering the diode first pass through the silicon dioxide layer and then through the n and p layers before entering the depletion region, where they excite free electrons and holes, which then migrate to the cathode and anode, respectively.

When a semiconductor diode has a reverse bias (voltage) applied and the crystal junction between the p and n layers is illuminated, then a current will flow in proportion to the number of photons incident upon the junction. Avalanche diodes are very similar in design to the silicon p–i–n diode; however, the depletion layer in an APD is relatively thin, resulting in a very steep localized electrical field across the narrow junction. In operation, very high reverse-bias voltages (up to 2500 V) are applied across the device. As the bias voltage is increased, electrons generated in the p layer continue to increase in energy as they undergo multiple collisions with the crystalline silicon lattice. This *avalanche* of electrons eventually results in electron multiplication that is analogous to the process occurring in one of the dynodes of a PM tube.

APDs are capable of modest gain (500–1000) but exhibit substantial dark current, which increases markedly as the bias voltage is increased. They are compact and immune to magnetic fields, require low currents, are difficult to overload, and have a high quantum efficiency (QE) that can reach 90%. Usually, APDs are run with a reverse bias that is less than the breakdown value. In this mode, they have a gain of 200–300, which is too low for photon counting. However, APDs operated in the Geiger mode (with a reverse bias slightly greater than the breakdown value) can give similar gains to those of a PM, around 10^6, and can be used for single-photon counting of low-level signals. The development of APDs as high-performance photodetectors is currently an active field of research.

5.2.5.1.3 Hybrid Detectors

Hybrid PM (HPM) tubes and hybrid PDs (HPDs) are devices that have the potential for use in single-photon counting applications. These devices have a photocathode similar to that of conventional PM tubes. Following the absorption of a photon, an electron may be emitted from the photocathode. However, unlike in PM tubes, the photoelectrons are amplified in a solid-state structure by an avalanche process, resulting in an electrical pulse. In an HPD, the electron is accelerated in a strong electric field toward a silicon sensor, which represents the anode. The electron is stopped in the depleted silicon, where a large number of e–h pairs are created, depending on its kinetic energy (10–20 keV).

5.2.5.2 Multichannel Detectors

A well-known example of a multichannel detector is the photographic plate. Other multichannel detectors include the diode array, the CCD, or the charge-injection device (CID). In multichannel spectrometers, the detector is placed at the focal plane of a polychromator, which is a monochromator with the exit slit removed. As a result, the entire emission dispersed at all wavelengths within the polychromator is detected simultaneously. The simultaneous detection of all of the dispersed emission using n spectral resolution elements reduces the measurement time by a factor of n in the case of an S/N-limited measurement or improves the S/N ratio by a factor of n in the case of a time-limited measurement.

5.2.5.2.1 Vidicons

One earlier device used in the 1970s is the vidicon, which is essentially a television-type device comprising an array of microscopic photosensitive diode junctions that are grown upon a single-silicon-crystal wafer. These diode junctions form individual microelements that are used to detect simultaneously the radiation dispersed by the polychromator. Prior to detection, each diode is charged to a preset reverse-bias potential by a fast-scanning electron beam. The electromagnetic radiation that strikes the diode causes depletion of the charge in each diode. A current proportional to the incident radiation recharges the diode to the preset potential when the electron beam scans again, thus providing information on the intensity of the incident light that has struck the diode. Consequently, an almost instantaneous recording of the spectral intensity at each diode is generated in the computer memory of the vidicon detector. Vidicons are capable of integrating radiation intensity over multiple scanning cycles because of their charge storage capabilities. Sensitivity can be further enhanced by incorporating an image-intensification section in front of the vidicon to yield a silicon-intensified target vidicon.

5.2.5.2.2 Photodiode Array

Another multichannel detector is the photodiode array (PDA), which also utilizes PDs as detection elements. Recording of the signal is performed with direct on-chip circuitry rather than with a scanning electron beam. The electronic switches needed to read from individual PDs are built right onto the chip. One diode is read at a time, and the analog signal readout through each electronic switch correlates with the amount of light intensity impinging on the PD. Most multichannel PDAs are silicon based, operating from 180 to 1100 nm. PDA systems with InGaAs detectors can be used in the NIR spectral range (800–1700 nm). Signal amplification is achieved by an MCP image intensifier. Gated detection down to 5 ns time resolution can be performed with the intensified diode array.

5.2.5.2.3 Charge-Coupled Device

More recently, multichannel devices such as CCDs are being used increasingly as a result of their high quantum yield, 2D imaging capability, and very low dark current. CCDs represent a great advance in detector instrumentation from the UV to NIR spectral range. They are very widely used because of their two-dimensionality and unique combination of sensitivity, speed, low noise, ruggedness, and durability in a relatively compact package. Because of their widespread use in spectroscopy, this section provides a detailed description of the operating principle of CCDs and the instrumental factors involved in the selection of CCDs for specific applications.

CCDs are 1D or 2D arrays of silicon PDs with metal–oxide semiconductor architectures. The detector arrays consist of individual detector elements, called pixels, which are defined by capacitors, called gates. Electrons generated by the light impinging onto the CCD charge these capacitors. Silicon exhibits an energy gap of 1.14 eV. Incoming photons with energy greater than this can excite valence electrons into the conduction band, thus creating e–h pairs. The average lifetime for these carriers is 100 μs. After this time, the e–h pair will recombine. Photons with energy from 1.14 to 5 eV generate single e–h pairs. Photons with energy of >5 eV produce multiple pairs. A 10 eV photon will produce 3 e–h pairs, on average, for every incident photon. Soft x-ray photons can generate thousands of signal electrons, making it possible for a CCD to detect single photons. For use as an IR imager, a CCD must be made of another material like germanium (band gap 0.55 eV).

The current sources (e–h pairs) produced are localized in small areas, an array of capacitors, called pixels. Common 2D CCD chips have 512 × 512 or 1024 × 1024 pixels. The charge accumulates in proportion to the light intensity impinging onto the pixel. A CCD sensor provides only one serial output, the readout register through which each capacitor can be discharged (each pixel can be read). A differential voltage is applied across each gate to perform charge transfer. The photogenerated charge is moved to the readout register by a series of parallel shifts, sequentially transferring charge from one pixel to the next within a column, until the charge finally collects in the readout register.

The charge from each row of pixels can be binned before readout to improve the S/N value. Furthermore, the dark count of CCDs is very low, especially when the detector is cooled. A CCD array can accumulate charge generated by photoelectrons almost noiselessly. However, CCD noise is produced in the act of commutating the charge out to a charge detector. Readout noise also tends to increase with increasing readout speeds; typically, the best CCD camera systems currently available give around five electrons of readout noise per pixel.

CCDs have several structures, including front-illuminated, back-illuminated, and open-electrode structures (Figure 5.11). In front-illuminated CCDs, incident photons have to penetrate a polysilicon electrode before reaching the depletion region. In back-illuminated or back-thinned CCDs, the substrate is polished and thinned to remove most of the bulk silicon substrate. Since illumination occurs from the back, the polysilicon on the front does not affect the QE of the detector. CCDs are currently the detectors with the highest QEs. Typically, a CCD array has a QE of around 40%, but back-thinning can

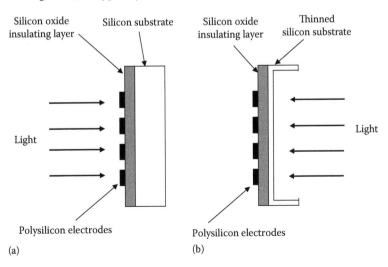

FIGURE 5.11 Structure of (a) front-illuminated and (b) back-illuminated CCDs.

increase the QE to around 80% at 600 nm. Back-thinned CCDs are usually coated with an antireflection material for enhanced response in either the UV or the NIR region. Often, reflections from boundaries of the back-thinned devices form constructive and destructive interference patterns, often referred to as the *etaloning effect*. The etaloning effect often causes an undesired oscillation superimposed on the spectrum at wavelengths longer than 650 nm.

The transmittance of the electrode depends on its thickness. Since the polysilicon electrode material does not transmit below 400 nm, some versions of front-illuminated UV CCDs have the detector coated with a phosphor, which converts UV radiation to green light. These UV CCDs can provide a 10%–15% QE response in the 120–450 nm spectral range. The coating is selected so that it does not degrade the visible and NIR response of the detector. In open-electrode CCDs, the central area of the electrode is etched to expose the underlying photosensitive silicon. These types of CCDs exhibit QEs of 30% or greater in the UV.

Cooling CCDs is often important because the operating temperature of the detector determines the dark-current noise. In general, the dark-current level is reduced by 50% for every 9°C drop in temperature. Two common cooling methods are liquid-nitrogen cooling and thermoelectric cooling using multistage Peltier devices. Liquid-nitrogen cooling provides the lowest operating temperatures, achieving typical dark-current levels of 1–3 electrons per pixel per hour at −133°C. Four-stage thermoelectrically cooled CCDs have typical operating temperatures of −73°C with typical noise levels of 1–2 electrons per pixel per minute. Thermoelectric cooling systems are often a good choice since they can offer good performance, approaching that of liquid-nitrogen cooled systems, and provide the convenience of uninterrupted operation. Whereas research spectrometers generally require cooled CCDs to detect low-level signals, conventional consumer-application CCDs are uncooled because of the high light levels and the very low bit depth of the analyzing electronics.

CCDs are often used as steady-state detectors. Since they have relatively slow time response, an intensifier stage with fast gating capability is used for time-resolved measurements. Intensified charge-coupled devices can have nanosecond gate times, which allow time-resolved measurements in combination with pulsed or gated excitation source. Electron-multiplying CCDs are systems that exhibit improved sensitivity due to a unique electron-multiplying structure built into the detector (20). Selection of CCDs is based on the spectral range of interest, resolution, and expected optical signal level. No single system can satisfy all of the possible spectral and S/N requirements in spectroscopy. The user should carefully analyze the needs and evaluate the tradeoffs involved in selecting a detector type, structure, and cooling method. Several considerations in the selection of a CCD for various applications are summarized in Table 5.3.[8]

5.2.5.2.4 Other Solid-State Detectors

Other types of solid-state detectors that have become popular include CIDs and active-pixel sensors (APSs). CIDs have pixels composed of two metal–oxide semiconductor gates that overlap and share the same row and column electrodes. The APS consists of a PD, a reset transistor, and a row-select transistor.

TABLE 5.3   Various Types of Charge-Coupled Devices and Photonics Applications

Light Level	Applications	CCD Chip Structure	Cooling Type
High	Absorption Transmission Reflection ES	Front-illuminated	Thermoelectric (two-stage)
Medium	Analytical luminescence Analytical Raman	Front-illuminated Open-electrode	Thermoelectric (four-stage)
Low	Research luminescence Research Raman	Back-illuminated	Liquid nitrogen

These devices can be highly integrated and can be manufactured using complementary metal–oxide semiconductor (CMOS) technologies.

5.2.5.2.5 CMOS Array

PDs and phototransistors are single-channel solid-state detectors. With the silicon fabrication process, the integrated electrooptic system on integrated circuit (IC) microchips, photodetector elements with amplifier circuitry, has led to the development of a new generation of multichannel (1D and 2D) detector arrays, based on the CMOS technology.

CMOS is the semiconductor technology used in the transistors that are integrated into most computer microchips. Semiconductors are made of silicon and germanium, materials that conduct electricity to some extent—more than an insulator but less than a typical conductor. Areas of these materials are doped by adding impurities, which supply extra electrons with negatively charged (N-type transistors) or positively charged carriers (P-type transistors). In CMOS technology, both kinds of transistors are used in a complementary way to form a current gate that forms an effective means of electrical control. CMOS transistors use almost no power when they are not needed.

The use of IC systems based on the CMOS technology has led to the development of extremely low-cost diagnostic sensor chips for medical applications. In CMOS, both N-type and P-type transistors are used to realize logic functions. Today, CMOS technology is the dominant semiconductor technology for microprocessors, memories, and application-specific ICs. The main advantage of CMOS is the low power dissipation. Power is dissipated only if the circuit actually switches. This design allows the integration of increased numbers of CMOS gates on an IC, resulting in much better performance.

The use of the standard CMOS process allows the production of PDs and phototransistors, as well as other numerous types of analog and digital circuitry, in a single IC chip. This feature is the main advantage of the CMOS technology compared with other detector technologies such as CCDs and CIDs. The PDs themselves are produced using the *n*-well structure that is generally used to make resistors or the body material for transistors. The capability of large-scale production using low-cost IC technology is an important advantage. The assembly process of various components is made simple by integration of several elements on a single chip. For medical applications, this advantage will allow the development of extremely low-cost, disposable biochips that can be used for in-home medical diagnosis of diseases without the need to send samples to a laboratory for analysis.[9-14] The use of CMOS sensor arrays for medical biochips is further discussed in Chapter 3 in Volume II of this handbook.

5.2.5.2.6 Streak Cameras

Streak cameras are detectors that can provide very fast time gating with temporal resolution from several picoseconds to hundreds of femtoseconds.[15] Streak cameras operate by dispersing the photoelectrons across an imaging screen at high speed using deflection plates within the detector. Streak cameras can simultaneously provide measurements of time decay at different wavelengths. In general, streak cameras have a faster response time than that of MCP PM tubes but exhibit relatively low S/N values and low dynamic ranges (3 orders of magnitude).

5.2.6 Detection Methods

5.2.6.1 Direct Current Technique

The analog or DC (direct current) method is the oldest and simplest detection technique. Since the anode is basically a current generator, a current meter can measure the anode current. It is common to measure the output by amplifying the voltage across a load resistor, which is typically between 10^5 and 10^8 Ω. The noise in the DC amplification signal is commonly reduced by a resistor–capacitor (RC) low-frequency band-pass filter. Too slow a response in the RC filter, however, necessitates very slow scanning speeds and long measurement times. A typical DC detection system involves the use of a picoammeter. The time constant for such an instrument at the highest intensity scale is a few seconds. A zero-suppression device is generally provided to compensate for dark current and zero drift.

5.2.6.2 Alternating Current Technique

The alternating current (AC) method consists of modulating the excitation signal using electronic or mechanical means. An AC amplifier is then tuned to the modulating frequency for detection. This technique rejects the noise outside the amplifier bandpass, eliminating the DC dark current. A further improvement is based on the technique of synchronous (or lock-in) detection. Stability is improved because any drift or instability of the modulating system is locked in by the amplifier. However, synchronous detection cannot eliminate some types of noise (*l/f noise*, multiplicative noise) that are amplified along with the signal.

In certain situations, it is desirable to obtain spectral information in digital form to allow data processing for correction or smoothing of spectra. One simple method to obtain data from the DC anode current uses a voltage-to-frequency converter.[16] This method is equivalent to using an analog-to-digital converter and counting the output pulses from the converter. A variety of voltage-to-frequency converters are presently commercially available.

5.2.6.3 Digital Photon Counting Technique

A truly digital technique that counts discrete photon pulses is the single-photon counting technique. Unlike the more conventional analog detection method, the single-photon counting signal output is digital in nature, producing discrete pulses of charge. The digital technique has proved to have several advantages over the analog method, especially for low-level signal detection. This method uses high-gain phototubes and high-speed electronics for detecting individual photon pulses. Unlike the previously described methods, in which the signal is analog, the PM signal output here is digital in nature. In the digital technique, discrete pulses of charge are produced at the anode, the number being proportional to the number of photons incident on the photocathode. This technique has proved to have several advantages over the conventional DC method, especially for low-level light detection.[17-20] First, the digital data can be processed directly in a manner suitable for further computer data treatment. The processing of information by digital circuitry is less susceptible to long-term drifts, which usually limit analog systems. This feature permits measurement of extremely low radiation flux. The method also makes possible the optimization of the S/N ratio by discriminating against PM dark current.

With an ideal detection device, one would expect the pulses resulting from single photoelectrons to have exactly the same pulse height. Actually, the output pulses have a certain height distribution due to thermionic emission of higher dynodes and spurious pulses generated by the PM. Electrons that do not originate at the photocathode, but from further down the dynode chain, are the main sources of PM noise or dark current. These undesired electron pulses have a height distribution that generally differs from that of the photoelectrons. For example, thermionic electrons have a pulse height distribution lower than that of photoelectrons, whereas other sources of noise—such as cosmic ray muons, after pulsing, and radioactive contamination of tube materials—generally produce pulses of higher amplitude. Most of these unwanted pulses can be eliminated by discriminatory units that select pulses within the range of height levels expected for true photosignals of interest.[21]

Whereas digital photon counting techniques are very useful for low-level signal measurements, they may not be suitable for high-intensity conditions because of pulse pileup phenomena. If two pulses arrive at the cathode at the same time or are closely spaced in time, they could be counted as a single pulse, resulting in an inaccurate count rate. The anode pulse width corresponding to a single photon for a typical PM is approximately 5 ns. This limits the maximum frequency response of the PM to 200 MHz for a periodic signal. For random events, the count rate needs to be about 100-fold less to avoid pulse pileup. Therefore, the single-photon counting technique should be employed only for photon count rates of less than 2 MHz.

5.2.6.4 Time-Resolved and Phase-Resolved Detection Methods

Various detection techniques can be designed to measure optical signals from a wide variety of spectroscopic processes, including absorption, fluorescence, phosphorescence, ES, and Raman scattering. One important parameter of these signals is the lifetime of the radiation. The lifetimes of various processes are

- Absorption: instantaneous with excitation
- Fluorescence: 10^{-10} to 10^{-8} s
- Phosphorescence: 10^{-6} to 10^{-3} s
- Scattering: almost instantaneous with excitation

Two methods of measuring signal that allow differentiation of lifetimes involve time-resolved and phase-resolved detection schemes.[22,23] The time-resolved and phase-resolved methods also improve the S/N values by differentiating the actual signal of interest from the background noise (DC signal). The use of time-resolved and phase-resolved techniques in imaging is described in Chapter 10 in this handbook.

5.2.6.4.1 Time-Resolved Detection

In the time-resolved method, a pulse excitation source is used. The width of the excitation is generally much shorter than the emission process of interest, that is, much shorter than the lifetime (decay time) of the samples. If one desires to measure the lifetime, then the time-dependent intensity is measured following the excitation pulse, and the decay time τ is calculated from the slope of a plot of log $I(t)$ versus t, or from the time at which the intensity decreases to $1/e$ of the initial intensity value I ($t = 0$). To measure the emission intensity free from the excitation pulse intensity, one can gate the detection after a delay time when the excitation pulse has decreased to zero (Figure 5.12a). Different compounds

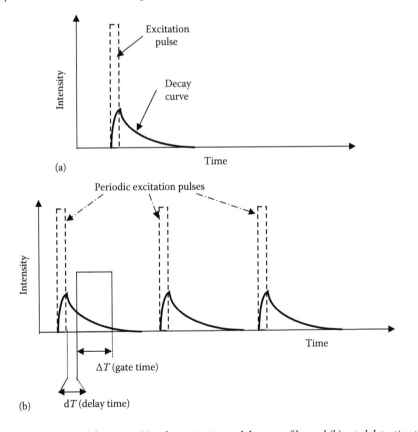

FIGURE 5.12 Time-resolved detection. (a) pulse excitation and decay profiles and (b) gated detection technique.

having different decay times can be differentiated by using different delay times (dT) and gate times (ΔT) (Figure 5.12b). An important source of noise in many measurement situations is the DC noise from the background. Improvement in the S/N ratio can be achieved by using multiple excitation pulses and applying the *boxcar* method by integrating the emission signal after each pulse (Figure 5.12b).

5.2.6.4.2 Phase-Resolved Detection

Another method that can differentiate lifetimes involves the phase-resolved techniques (often referred to as the frequency domain techniques). In this case, the sample is excited with intensity-modulated light. The intensity of the incident light changes with a very high frequency ($\omega = 2\pi f$, f being the frequency in hertz) compared with the reciprocal of the decay time τ. Following excitation with a modulated signal, the emission is also intensity modulated at the same modulation frequency. However, since the emission from the sample follows a decay profile, there is a certain delay in the emission relative to the excitation (Figure 5.13). This delay is measured as a phase shift (ϕ), which can be used to calculate the decay time. Note that at each modulation frequency ϖ, the delay is described as the phase shift ϕ_ϖ, which increases from 0° to 90° with increasing modulation frequency ϖ.

The finite time response of the sample also results in demodulation of the emission by a factor m_ϖ. This factor decreases from 1.0 to 0 with increasing modulation frequency. At low frequency, the emission closely follows the excitation. Hence, the phase angle is near zero and the modulation is near 1. As the modulation frequency is increased, the finite lifetime of the emission prevents the emission from closely following the excitation. This results in a phase delay of the emission and a decrease in the peak-to-peak amplitude of the modulated excitation (Figure 5.13).

The shape of the frequency response is determined by the number of decay times displayed by the sample. If the decay is a single exponential, the frequency is simple. One can use the phase angle or modulation at any frequency to calculate the lifetime. For single-exponential decay, the phase and modulation are related to the decay time (τ) by

$$\tan\phi_\varpi = \varpi\tau$$

and

$$m_\omega = (1 + \omega^2\tau^2)^{-1/2}$$

Therefore, one can differentiate various emissions having different decay times by selecting the phase shift optimized to the decay time of interest. This method is referred to as phase-resolved detection.

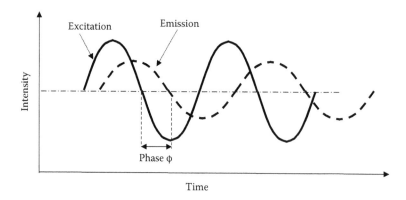

FIGURE 5.13 Phase-resolved detection.

5.2.6.5 Multispectral Imaging

Photonics diagnostic technologies can be broadly classified into two categories: (1) spectroscopic diagnostics and (2) optical imaging, often referred to as optical tomography. Spectroscopic diagnostic techniques are generally used to obtain an entire spectrum of a single sample site within a wavelength region of interest. These techniques are often referred to as point-measurement methods. On the other hand, optical imaging methods are aimed at recording a 2D image of an area of the sample of interest at one specific wavelength. A third category, which combines the two modalities, is currently receiving great interest. This category is often referred to as multispectral imaging or hyperspectral imaging.

Spectral imaging represents a hybrid modality for optical diagnostics that obtains spectroscopic information and renders it in image form. In principle, almost any spectroscopic method can also be combined with imaging. Some of these techniques use computer-based image processing in combination with microscopy. These techniques have had a major impact on research and are being implemented in cytological diagnostics. There is also considerable potential for direct clinical applications.

Instrumentation for multispectral imaging is described in detail in Chapter 9 in this handbook. Briefly, the concept of multispectral imaging is schematically illustrated in Figure 5.14. With conventional imaging, the optical emission from every pixel of an image can be recorded, but only at a specific wavelength or spectral bandpass. With conventional spectroscopy, the signal at every wavelength within a spectral range can be recorded, but for only a single analyte spot. The multispectral concept

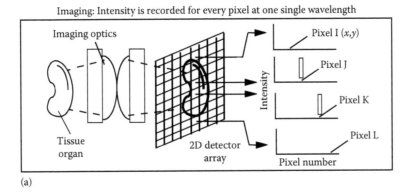

(a)

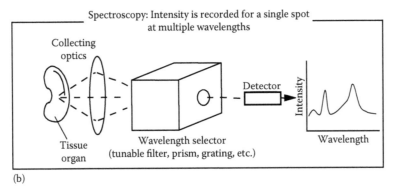

(b)

FIGURE 5.14 Principle of multispectral imaging. (a) Imaging scheme: With conventional imaging, the optical emission from every pixel of an image can be recorded, but only at a specific wavelength or spectral bandpass. (b) Spectroscopy scheme: With conventional spectroscopy, the signal at every wavelength within a spectral range can be recorded, but for only a single analyte spot.

(*continued*)

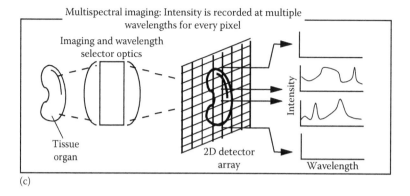

(c)

FIGURE 5.14 (continued)   Principle of multispectral imaging. (c) Multispectral imaging: The multispectral concept combines these two recording modalities and allows recording of the entire emission for every pixel on the entire image in the field of view.

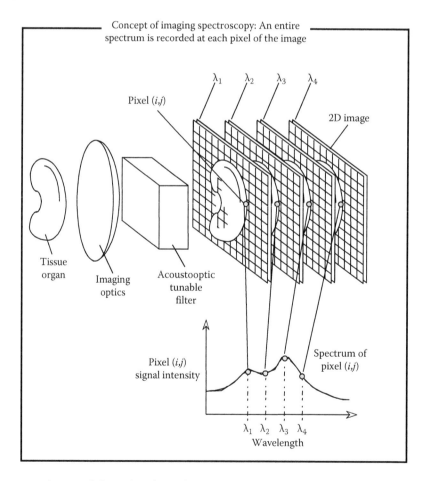

FIGURE 5.15   Multispectral data cube. The multispectral imaging approach provides a *data cube* of spectral information of the entire image at each wavelength of interest.

combines these two recording modalities and allows recording of the entire emission for every pixel on the entire image in the field of view with the use of a rapid-scanning solid-state device, such as the AOTF or an LCTF. The multispectral imaging approach provides a *data cube* of spectral information of the entire image at each wavelength of interest (Figure 5.15).

5.3 Conclusion

Photonics instrumentation includes a wide variety of basic devices available to the investigator for many applications. A simple instrumental setup can be easily assembled using off-the-shelf components. More sophisticated devices can be designed for special research requirements. Selection of components for a complete system is based on the spectral range of interest, resolution, and expected optical signal level. No single system can satisfy all of the possible spectral and S/N requirements in spectroscopy. The user should carefully analyze his/her needs and evaluate the tradeoffs involved in selecting a detector type, structure, and cooling method.

Acknowledgments

This work was sponsored by the US Department of Energy (DOE) Office of Biological and Environmental Research, under contract DEAC05-00OR22725 with UT-Batelle, LLC, by the National Institutes of Health, the Department of the Army, and by the Department of Justice, Federal Bureau of Investigation, the Coulter Foundation, and the Defense Advanced Research Project Agency.

References

1. Schawlow, A. L., Laser spectroscopy of atoms and molecules, *Science* 202(4364), 141–147, 1978.
2. Omenetto, N. (ed.), *Analytical Laser Spectroscopy*, Wiley, New York, 1979.
3. Kaminov, I. R. and Siegman, A. I., *Laser Devices and Applications*, IEEE Press, New York, 1973.
4. Kliger, D. S., *Ultrasensitive Laser Spectroscopy*, Academic Press, New York, 1983.
5. West, M. A., *Lasers in Chemistry*, Elsevier, New York, 1977.
6. Svelto, O., *Principles of Lasers*, 4th edn., Plenum Press, New York, 1998.
7. Orr, B. J., Optical parametric oscillators, in *Tunable Laser Applications*, Duarte, F. J. (ed.). Marcel Dekker, New York, 1995.
8. Gilchrist, J. R., Choosing a scientific CCD detector for spectroscopy, *Photonics Spectra* 36(3), 83–93, 2002.
9. Vo-Dinh, T., Development of a DNA biochip: Principle and applications, *Sensors and Actuators B—Chemical* B51, 52–57, 1999.
10. Vo-Dinh, T., Alarie, J. P., Isola, N., Landis, D., Wintenberg, A. L., and Ericson, M. N., DNA biochip using a phototransistor integrated circuit, *Analytical Chemistry* 71(2), 358–363, 1999.
11. Stokes, D. L., Griffin, G. D., and Vo-Dinh, T., Detection of *E. coli* using a microfluidics-based antibody biochip detection system, *Fresenius Journal of Analytical Chemistry* 369(3–4), 295–301, 2001.
12. Vo-Dinh, T., Cullum, B. M., and Stokes, D. L., Nanosensors and biochips: Frontiers in biomolecular diagnostics, *Sensors and Actuators B—Chemical* 74(1–3), 2–11, 2001.
13. Vo-Dinh, T. and Askari, M., Microarrays and biochips: Applications and potential in genomics and proteomics, *Current Genomics* 2, 399–415, 2001.
14. Vo-Dinh, T. and Cullum, B., Biosensors and biochips: Advances in biological and medical diagnostics, *Fresenius Journal of Analytical Chemistry* 366(6–7), 540–551, 2000.
15. Wiessner, A. and Staerk, H., Optical design considerations and performance of a spectro-streak apparatus for time-resolved fluorescence spectroscopy, *Review of Scientific Instruments* 64(12), 3430–3439, 1993.
16. Ingle, J. D. and Crouch, S. R., *Spectrochemical Analysis*, Prentice Hall, Englewood Cliffs, NJ, 1988.

17. Woodruff, T. A. and Malmstad, H. V., High-speed charge-to-count data domain converter for analytical measurement systems, *Analytical Chemistry* 46(9), 1162–1170, 1974.
18. Vo-Dinh, T. and Wild, U. P., High-resolution luminescence spectrometer. 1. Simultaneous recording of total luminescence and phosphorescence, *Applied Optics* 12(6), 1286–1292, 1973.
19. Vo-Dinh, T. and Wild, U. P., High resolution luminescence spectrometer. 2. Data treatment and corrected spectra, *Applied Optics* 13(12), 2899–2906, 1974.
20. Bestvater, F., Seghiri, Z., Kang, M. S., Groner, N., Li, J. Y., Im, K. B., and Wachsmuth, M., EMCCD-based spectrally resolved fluorescence correlation spectroscopy, *Optics Express* 18, 23821–23828, 2010.
21. Gustafson, T. L., Lytle, F. E., and Tobias, R. S., Sampled photon-counting with multilevel discrimination, *Review of Scientific Instruments* 49(11), 1549–1550, 1978.
22. Lakowicz, J. R., *Principles of Fluorescence Spectroscopy*, 2nd edn., Kuwer Academic, New York, 1999.
23. Suhling, K., French, P. M. W., and Phillips, D., Time-resolved fluorescence microscopy, *Photochemical and Photobiological Science* 4, 13–22, 2005.

6

Optical Fibers and Waveguides for Medical Applications

Israel Gannot
Tel-Aviv University
and
Johns Hopkins University

Moshe Ben-David
Tel-Aviv University

6.1 Introduction

Lasers for medical application are found in a wide range of electromagnetic spectrum. It starts from x-ray, goes up to UV, continues in the visible and near-infrared (NIR), and ends (at this time) in the mid-IR. This wide range of wavelengths needs to be transmitted from source (the laser) to the target tissue by a flexible device, which will enable easy manipulation of the laser beam in a medical setting.

At the beginning of laser use in medicine, the energy was transmitted by an articulated arm, which is a set of tubes connected with joints and reflecting mirrors with three plans of motion (Figure 6.1).

Although with time this set was miniaturized and became more precise and the drift of the laser beam was reduced, it is still cumbersome and it is limited to external use only. Surgical procedures are usually open procedures; however, with the introduction of the endoscope, the trend is to perform non-invasive or minimally invasive procedures. This reduces hospitalization time, causes less postoperative pain and discomfort, and is less subject to complications. In this case, the insertion of the surgical tools is done through existing openings of the body or through minor cuts made on the skin. A flexible endoscope (see Figure 6.2) could be made possible because of the optical fibers: a fiber that delivers energy and a coherent optical bundle that can deliver an image. The endoscope's working channel is used for the insertion of either a micro-surgical tool or an energy-transmitting fiber.

Optical fibers, which can transmit specific laser wavelengths, will enable minimally invasive procedures to become even more common.

Optical fibers can also transmit information from tissue back to a detector. Tissue radiates upon heating. The radiation wavelength is a function of the tissue temperature (Planck curves). This information can be transmitted by an IR-transmitting fiber, and the temperature of the tissue can be measured. One use, for example, is a feedback mechanism for tissue welding/soldering [1]. With the progress of biomedical optics and theories of photon migration in tissue, one can also measure specific

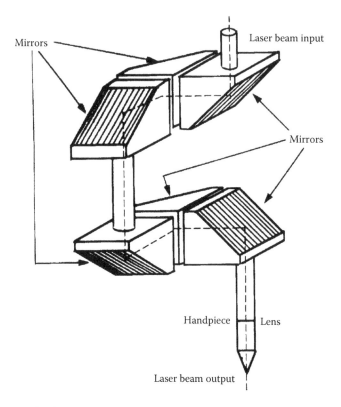

FIGURE 6.1 Laser articulated arm.

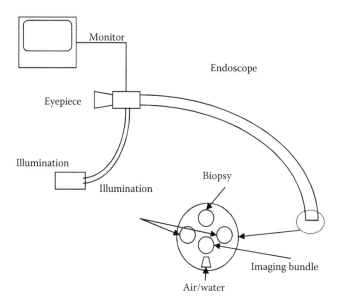

FIGURE 6.2 Schematic drawing of an endoscope.

tissue properties, through transmission and reflection. Spectroscopy of tissue can be done through evanescent waves at fibers with no cladding [2]. Fluorescence signals from tissue (either endogenous fluorophores or from exogenous materials present in tissue) can be transmitted to detectors [3] or fluorescence cameras through coherent optical bundle. These signals enable diagnosing changes in tissue, tissue transformations, and presence of tumors. These measurements can be done in a minimally invasive way through the endoscopes. The diagnosis can be immediate and less invasive and enable monitoring response to treatment.

In the next few paragraphs, we will cover the following issues:

- Theory of laser radiation propagation in optical fiber/waveguide
- Attenuation mechanisms
- Fiber and waveguide structure
- Transmission properties
- Distal tips
- Coupling devices
- Materials for making fibers and waveguides

6.2 Theory

There are two types of optical fibers that are used for medical applications. The first type is the conventional solid core fiber such as silica fibers. The second are hollow waveguides. Each type of optical fiber guides radiation according to different physical phenomena. Solid core fibers guide optical radiation by using total internal reflection, while most of the hollow waveguides guide optical radiation by pure reflection.

6.2.1 Theory of Solid Core Optical Fibers

6.2.1.1 Fiber Basics

The general structure of a solid core optical fiber (Figure 6.3) consists of a core, clad, and jacket. The optical radiation is guided through total internal reflection. Hence, the core index of refraction is slightly higher than the one of the clad. Solid core fibers can be characterized by several criteria; the most common one is their index profile. According to this criterion, there are two types of solid core fibers: step index fiber in which the refractive index profile of the core is a step function and graded index fiber in which the refractive index of the core depends on the core radius. Other criterion can be single mode or multimode. Usually, multimode fibers are used for medical applications.

6.2.1.2 Ray Theory

Ray theory is the simplest way to understand the guiding mechanism of a solid core fiber. It is applicable as long as the wavelength is much smaller than the fiber core diameter $\lambda \gg a$.

FIGURE 6.3 General structure of a solid core fiber.

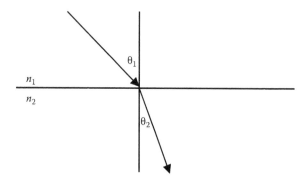

FIGURE 6.4   Snell's law.

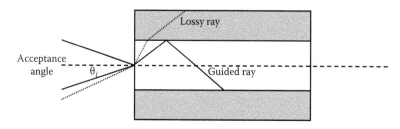

FIGURE 6.5   Guided ray and lossy ray.

When a ray impinges on a surface between two materials, it changes its angle of propagation (Figure 6.4) according to Snell's law:

$$n_1 \sin\theta_1 = n_2 \sin\theta_2 \tag{6.1}$$

where
  n is the material index of refraction
  θ is the angle of propagation

As can be seen from Snell's law, there is an angle of incidence, θ_{1c}, in which the angle of refraction equals 90°. Rays that impinge on the surface at angles above θ_{1c} will totally reflect from the surface. These rays will be guided through the fiber while other rays will be lost and decay in the clad region (Figure 6.5). Once the maximum angle of incidence is known, it is possible to determine the acceptance angle of the fiber or its numerical aperture (NA). The NA describes the ability of the fiber to collect light. The larger the NA, the greater the ability of the fiber to collect light. The NA is given by

$$\text{NA} = \sin\theta_i = \sqrt{n_1^2 - n_2^2} \tag{6.2}$$

6.2.1.3 Mode Propagation in Solid Core Optical Fibers [4]

An alternative way to ray analysis is to solve Maxwell's equation with the appropriate boundary condition. The wave equations are given by

$$\nabla^2 E - \frac{1}{c^2}\frac{\partial^2 E}{\partial t^2} = 0; \quad \nabla^2 H - \frac{1}{c^2}\frac{\partial^2 H}{\partial t^2} = 0; \quad \text{where } c^2 = \frac{1}{\mu\varepsilon} \tag{6.3}$$

This is a representation of a single-frequency wave and phase velocity. The general wave motion can be represented by $\sin(\omega t - kz)$ or $\exp[i(\omega t - kz)]$, where ω is the *temporal frequency* and k is the *spatial*

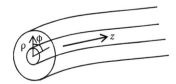

FIGURE 6.6 Cylindrical coordinates.

frequency (or *propagation vector, wave vector, wave number*). The temporal frequency denotes the number of repetitions of a wave per unit time, while the spatial frequency denotes the number of repetitions of a wave per unit distance. ω/k (or β) is known as the *phase velocity*; it is the velocity of any constant-phase point of a wave.

The wave equation in cylindrical coordinates (Figure 6.6) is given by

$$\frac{\partial^2}{\partial \rho^2} E_z + \frac{1}{\rho}\frac{\partial}{\partial \rho} E_z + \frac{1}{\rho^2}\frac{\partial^2}{\partial \phi^2} E_z + \frac{\partial^2}{\partial z^2} E_z - n^2 k_0^2 E_z = 0 \tag{6.4}$$

For a step index fiber,

$$n = \begin{cases} n_1 & \rho \le a \\ n_2 & \rho > a \end{cases}$$

It is possible to solve the wave equation by separating the variables according to the coordinates:

$$E_z\left(\rho,\phi,z\right) = F(\rho)\Phi(\phi)Z(z) \tag{6.5}$$

which yields

$$\frac{Z''}{Z} = \beta^2, \quad \Rightarrow Z = e^{i\beta z} \tag{6.6}$$

$$\frac{1}{\rho^2}\frac{\Phi''}{\Phi} = \frac{m^2}{\rho^2}, \quad \Phi = e^{im\phi} \tag{6.7}$$

and

$$\rho^2 \frac{\partial^2 F}{\partial \rho^2} + \rho \frac{\partial F}{\partial \rho} + \left(k^2\rho^2 - m^2\right)F = 0 \tag{6.8}$$

The solution to the latter equation is the famous Bessel functions. Since we would like to confine the energy only to the core, all the energy that propagates in the radial direction through the clad must decay with the distance ρ. Hence, in the core, the solution is

$$F(\rho) = AJ_m\left(k\rho\right) + A'Y_m\left(k\rho\right) \quad \rho < a \tag{6.9}$$

where

$$k^2 = n_1^2 k_0^2 - \beta^2 \tag{6.10}$$

and in the clad

$$F(\rho) = CK_m(\gamma\rho) + C'I_m(\gamma\rho) \quad \rho > a \tag{6.11}$$

In the clad, the energy decreases exponentially with the distance with a factor γ given by

$$\gamma^2 = \beta^2 - n_2^2 k_0^2 \tag{6.12}$$

It is possible to simplify the equations by setting

$$
\begin{aligned}
\rho \to \infty \quad & F(\rho) \to 0 \quad & C' = 0 \\
\rho \to 0 \quad & F(\rho) \to \text{finite} \quad & A' = 0
\end{aligned}
\tag{6.13}
$$

and thus obtaining

$$E_z = \begin{cases} AJ_m(k\rho)e^{im\phi}e^{i\beta z} & \rho \le a \\ CK_m(\gamma)e^{im\phi}e^{i\beta z} & \rho > a \end{cases} \tag{6.14}$$

and

$$H_z = \begin{cases} BJ_m(k\rho)e^{im\phi}e^{i\beta z} & \rho \le a \\ DK_m(\rho)e^{im\phi}e^{i\beta z} & \rho > a \end{cases} \tag{6.15}$$

6.2.1.4 Attenuation Mechanisms in Solid Core Fibers

There are four attenuation mechanisms in solid core fibers. The mechanisms and their cause are listed in Table 6.1.

6.2.1.4.1 Reflection

In solid core fibers, there is a difference in the index of refraction between the core and the entrance and exit surfaces. This difference causes a reflection of some of the radiation backward. The amount of energy reflected depends on the difference between the index of refraction and the angle of incidence. If the ray impinges perpendicularly to the boundary, the reflection coefficient is given by

$$R = \left(\frac{n_1 - n_2}{n_1 + n_2} \right) \tag{6.16}$$

TABLE 6.1   Loss Mechanisms in Fibers

Attenuation Mechanism	Cause
Reflection (Fresnel losses)	Difference in index of refraction at the entrance and exit surfaces of the fiber
Scattering	Intrinsic—Mie, Rayleigh, Brillouin, Raman
	Extrinsic—impurities and defects
Absorption	Intrinsic—material
	Extrinsic—impurities
Radiation	Mode coupling

6.2.1.4.2 Scattering

Scattering in solid core fibers can be divided into two categories: intrinsic scattering and extrinsic scattering. There are four intrinsic scattering mechanisms: Mie, Rayleigh, Brillouin, and Raman. The first two are caused by inhomogeneity in the index of refraction, granulation of the fiber material, or geometric faults in the fiber. These mechanisms cause elastic scattering, that is, the wavelength of the radiation does not change in the process. The attenuation is proportional to λ^{-4} and is the lower limit to the losses in the fiber.

Brillouin scattering is caused by the interaction of the photon with acoustic phonon. Raman scattering is caused by the interaction of the photon with acoustic photon. As a result of the interaction, the wavelength of the radiation changes.

Extrinsic scattering is caused by many reasons, for example, impurities in the fiber materials and faults during manufacturing. It is possible to decrease the effects of this type of scattering by closely controlling the manufacturing process of the fiber.

6.2.1.4.3 Absorption

When radiation propagates through a certain material, some of the energy is transferred to heat. That phenomenon is known as absorption. All absorption mechanisms involve moving particles (molecules or electrons) from one energy level to another. The absorption process must satisfy the following condition:

$$\lambda = \frac{hc}{\Delta E} \tag{6.17}$$

where ΔE is the difference between energy levels.

Intrinsic absorption depends on the material itself and is the one that determines the range of wavelengths that the material transmits. Electronic absorption occurs at the UV range, while multiphonon absorption occurs at longer wavelengths when the wavelength corresponds to an integer multiplication of the resonance frequency.

Extrinsic absorption is due to microscopic faults in the fibers, irregularities in the core materials, and impurities.

6.2.1.4.4 Radiation

Radiation losses are due to inappropriate coupling between the fiber and the laser beam, fiber bending, and intermodal interaction. Inappropriate coupling causes some of the rays to violate the total reflection condition and thus not to propagate through the fiber. Bending causes the rays to change their angle of incidence, and thus some of them might leave the fiber. Bending also causes coupling between low modes of propagation and higher ones. Since low modes are less attenuated, the attenuation of the fiber increases.

6.2.2 Theory of Solid Hollow Waveguides [5,6]

6.2.2.1 Hollow Waveguides Basics

The general structure of a hollow waveguide (Figure 6.7) consists of a hollow tube internally coated with a metal layer and a thin dielectric layer (the dielectric layer is optional). The optical radiation is guided through reflection from the inner layers. Hollow waveguides are characterized by the guiding mechanism of the radiation. There are two types of hollow waveguides: The first one is leaky waveguides, which have the same structure as in Figure 6.6. The second is attenuated total reflection (ATR).

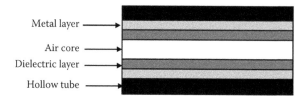

FIGURE 6.7   General structure of a hollow waveguide.

These waveguides are made of a hollow tube. The tube index of refraction is lower than the one at a region of wavelength; hence, their guiding mechanism is similar to that of solid core fibers.

6.2.2.2 Attenuation Mechanisms in Hollow Waveguides

The main attenuation mechanisms in hollow waveguides are reflection from a thin layer and scattering due to surface roughness. The first mechanism determines the wavelength that will be transmitted through the waveguide and how much it will be attenuated. In this case, the reflection coefficients are given by Fresnel equations.

The second mechanism is scattering, which is caused by the roughness of the hollow tube and the deposited layers. Figure 6.8 shows the surface of Ag and AgI layers that are deposited on a glass tube. The measurement was done using an atomic force microscope (AFM). Measurements of the surface roughness and the height distribution enable us to compute the scattering coefficient [7] and the influence of

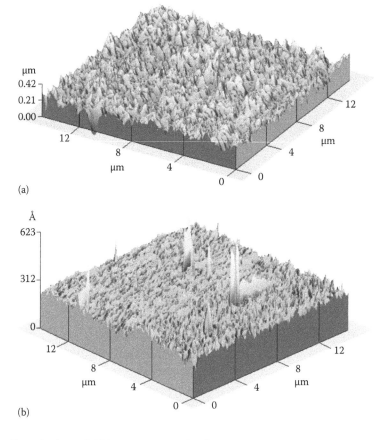

FIGURE 6.8   **(See color insert.)** AFM measurements for glass waveguide's (a) Ag and (b) AgI layers.

different parameters on the waveguide's attenuation. The main influence of scattering on the ray propagation is changing the ray's propagation angle and energy distribution. Some of the energy is reflected backward and hence is lost, and the angle of incidence of the other part is changed (usually decreased).

6.2.2.3 Ray Theory

One of the advantages of hollow waveguides is their ability to be *tuned* to almost any wavelength. The tuning is done by coating the metal layer with a thin dielectric layer, whose thickness corresponds to the desired wavelength. This can be seen from the following equation for the reflection coefficient from a thin layer:

$$r_{tot} = \frac{r_2 + r_1 e^{-2i\delta}}{1 + r_1 r_2 e^{-2i\delta}} \tag{6.18}$$

where
r_1 and r_2 are Fresnel coefficients from a layer
δ is a phase shift that determines the wavelength and is given by

$$\delta = \frac{2\pi}{\lambda} n_1 d \cos(\phi_1) \tag{6.19}$$

In order to analyze the influence of the waveguide's geometrical parameters, let us calculate the distance a ray travels between two hits on the waveguide's wall, z_i, which can be calculated using Figure 6.9 and is given by

$$z_i = 2r \tan(\phi_i) \tag{6.20}$$

where r is the waveguide's radius. The number of times it impinges on the waveguide's wall of length l, p_i, is given by

$$p_i = int\left(\frac{l}{z_i}\right) = int\left(\frac{l}{2r\tan(\phi_i)}\right) \tag{6.21}$$

The total reflection coefficient of the ray is the multiplication of all the reflections it passed on the way. It is given by

$$R_{total}(\phi_i) = R(\phi_i)^{p_i} \tag{6.22}$$

The transmission, T, and the attenuation, A, are given by

$$T_i = \frac{I_{i,out}}{I_{i,in}} \tag{6.23}$$

$$A_i = -10\log(T)$$

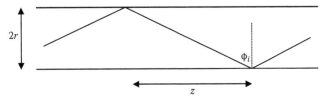

FIGURE 6.9 Tracing of one ray in the fiber/waveguide core.

More explicitly, the attenuation may be written as

$$A_i = -10 \left[\text{int}\left(\frac{1}{4r\tan(\phi_i)} \right) \log\left(R(\phi_i) \right) \right] \tag{6.24}$$

From the last equation, one can learn about the dependence of the waveguide's attenuation on its physical dimensions. It can be seen that when the waveguide's length increases or its inner radius decreases, the attenuation increases.

As can be seen from the earlier analysis, ray models explain the working principles of fibers and waveguides in a simple manner. A more complicated way is to solve Maxwell equations with the appropriate boundary conditions. This may give additional knowledge about the different modes that may propagate through the waveguide.

6.2.2.4 Multilayer Waveguides

One of the methods to overcome the attenuation mechanisms in solid core fibers and hollow waveguides is to develop a new type of fiber, a multilayer one. The general structure of a multilayer fiber is shown in Figure 6.10.

A multilayer waveguide is made of an alternating structure of two dielectric materials. Such a structure is also known as a photonic crystal. There are two ways to analyze such a structure. The first is to use the symmetry properties of the system and use quantum mechanics techniques to solve Maxwell equations [8]. This approach leads to an equation similar to Bloch equation in solid-state physics. According to the Bloch theory, there exist forbidden band gaps in which photons (electrons in solid-state physics) with certain energies cannot propagate through. In this case, the photons are *perfectly* reflected. Hence, such a device can have zero attenuation.

The second method is to solve Maxwell equation with the appropriate boundary conditions [9]. Consider the linearly polarized wave shown in Figure 6.11, impinging on a thin dielectric film between two semi-infinite transparent media. Each wave, E_{rI}, E'_{rII}, E_{rII}, and so forth, represents the resultant of all possible waves traveling in that direction, at that point in the medium. The summation process is therefore built in. The boundary conditions require that the tangential components of the electric field, **E**, and the magnetic field, **H**, be continuous across the boundaries.

By applying the boundary conditions and using Maxwell equations, it is possible to relate the fields on each side of the boundary. The relations are given by

$$\begin{bmatrix} E_I \\ H_I \end{bmatrix} = \begin{bmatrix} \cos(k_0 h) & (i\sin(k_0 h))/Y_1 \\ Y_1 i\sin(k_0 h) & \cos(k_0 h) \end{bmatrix} \begin{bmatrix} E_{II} \\ H_{II} \end{bmatrix} \tag{6.25}$$

or

$$\begin{bmatrix} E_I \\ H_I \end{bmatrix} = M_1 \begin{bmatrix} E_{II} \\ H_{II} \end{bmatrix} \tag{6.26}$$

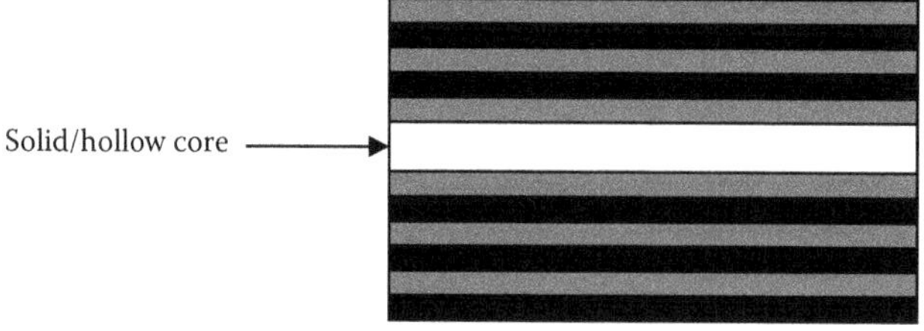

FIGURE 6.10 Cross section of a multi-layer waveguide/fiber structure.

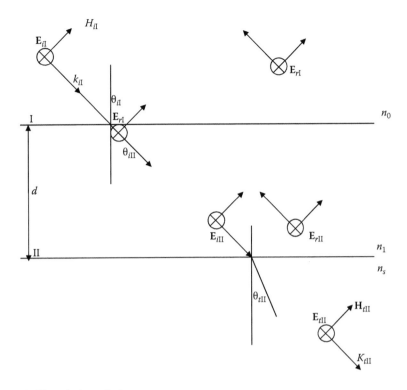

FIGURE 6.11 Fields at the boundaries.

where

$$Y_1 \equiv \sqrt{\frac{\varepsilon_0}{\mu_0}} n_1 \cos\theta_{iI} \qquad (6.27)$$

or

$$Y_1 \equiv \sqrt{\frac{\varepsilon_0}{\mu_0}} \left. n_1 \middle/ \cos\theta_{iI} \right. \qquad (6.28)$$

for TE polarization and TM polarization, respectively. The same equations hold for the second and third boundaries. It follows, therefore, that if two overlaying films are deposited on the substrate, there will be three boundaries or interfaces, and now

$$\begin{bmatrix} E_{II} \\ H_{II} \end{bmatrix} = M_{II} \begin{bmatrix} E_{III} \\ H_{III} \end{bmatrix} \qquad (6.29)$$

Multiplying both sides of this expression by M_I, we obtain

$$\begin{bmatrix} E_I \\ H_I \end{bmatrix} = M_I M_{II} \begin{bmatrix} E_{III} \\ H_{III} \end{bmatrix} \qquad (6.30)$$

In general, if p is the number of layers, each with a particular value of n and h, then the first and the last boundaries are related by

$$\begin{bmatrix} E_1 \\ H_1 \end{bmatrix} = M_1 M_{II} \ldots M_p \begin{bmatrix} E_{p+1} \\ H_{p+1} \end{bmatrix} \tag{6.31}$$

The characteristic matrix of the entire system is the result of the product (in the proper sequence) of the individual 2×2 matrices, that is,

$$M = M_1 M_{II} \ldots M_p = \begin{bmatrix} m_{11} & m_{12} \\ m_{21} & m_{22} \end{bmatrix} \tag{6.32}$$

From these equations, we can derive the reflection and transmission coefficients. Defining Y_0 and Y_s for the medium and substrate, the reflection and transmission coefficients are given by

$$r = \frac{Y_0 m_{11} + Y_0 Y_s m_{12} - m_{21} - Y_s m_{22}}{Y_0 m_{11} + Y_0 Y_s m_{12} + m_{21} + Y_s m_{22}} \tag{6.33}$$

and

$$t = \frac{2Y_0}{Y_0 m_{11} + Y_0 Y_s m_{12} + m_{21} + Y_s m_{22}} \tag{6.34}$$

To find r or t for any configuration of films, we need only compute the characteristic matrices for each film, multiply them, and then substitute the resulting matrix elements into these equations.

As an example, let us design a multilayer mirror for a laser with a wavelength of 6 μm. We will use two dielectric materials that have index of refraction, which are far apart. Such materials could be germanium ($n = 4$) and zinc selenide ($n = 2.4$). Figure 6.12 shows the reflectivity of a multilayer film made of different number of pairs of Ge and ZnSe as a function of wavelength for 0° angle of incidence. As can be seen from the figure, the more pairs there are, the more sharp is the region where the reflectivity is perfect. Attempts to create photonic crystal fibers are made by several groups in the United States, Denmark, England, Australia, and Israel [10–12].

6.2.2.5 X-Ray Waveguides

X-ray radiation is used in medicine for over a century for imaging. There are possibly new applications for x-ray, especially now after x-ray laser sources are available [13]. However, it is very hard to manipulate since they are very energetic. The conventional way to focus x-ray radiation is either coherently by using Fresnel zone plates and Bragg Fresnel lenses or incoherently by using bent crystal optics and coated fibers. The main drawback of these methods is the large spot size, which is not suitable for most applications (x-ray spectroscopy and microscopy) and especially noninvasive medical treatments for which there is a need for waveguides.

In order to overcome the drawbacks of focusing lenses and crystal, Feng et al. [14] suggested a new device composed of a low-density layer, such as carbon, which is surrounded by two high-density layers, such as nickel (Figure 6.13). They showed experimentally and theoretically that such a device may produce an x-ray beam with defined beam shape, divergence, and coherence, which corresponds to the geometrical properties of the device.

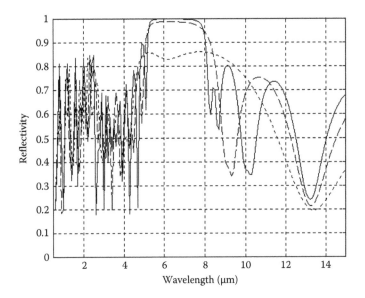

FIGURE 6.12 Transmission as a function of wavelength 6 μm optimized waveguide.

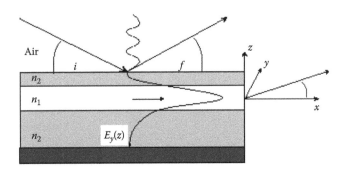

FIGURE 6.13 Single-layer x-ray waveguide.

The coupling of the x-ray beam to the device is done by grazing angle incidence. Since only a discrete set of angles can propagate through the device, the output is coherent and its spot size corresponds to the low-density layer thickness. Using such a device, Feng et al. have been producing an x-ray beam with a spot size of less than 100 Å. However, a single-layer structure also has some drawbacks; beam shape control is very difficult and the coupling efficiency is very limited. The beam shape control is difficult since it corresponds to the geometrical structure that can produce only a single spot. The coupling efficiency is poor since a single layer has a very small angular acceptance.

In order to overcome the single-layer structure drawbacks, Pfeiffer et al. [15,16] suggested a multilayer structure. The new structure (Figure 6.14) consists of one or more guiding layers. This new device enables to generate more complicated beam shapes (Figure 6.15). Moreover, the new concept enables new ways to perform x-ray spectroscopy and holography.

6.2.2.6 Coupling Devices

Coupling the laser beam to the fiber can be done in several ways. The first and most straightforward is using standard lenses. Another, more sophisticated way is using different types of tapers [17,18].

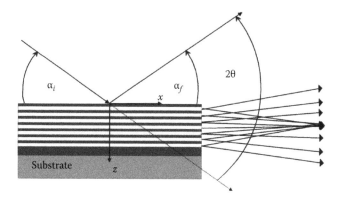

FIGURE 6.14   Multilayer x-ray waveguide. (Courtesy of Prof. Salditt group, Saarbruecken, Germany.)

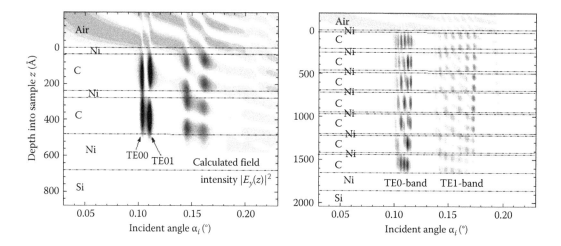

FIGURE 6.15   Field distribution of different multilayer x-ray waveguides. (Courtesy of Prof. Salditt's group, Saarbruecken, Germany.)

Coupling the laser beam into the fiber using an ordinary lens is simple and relatively straightforward. All one has to do is to make sure that the laser spot size after the lens is less than the fiber's core diameter and that the NA of the lens is suitable to that of the fiber and to align the optical axis of the lens and the fiber.

The laser spot size at the entrance to the waveguide is given by

$$\omega_0 = \frac{1.9\lambda f}{D} \tag{6.35}$$

where
  f is the focal length of the coupling lens
  D is its diameter
  λ is the laser wavelength

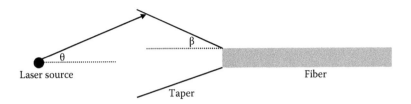

FIGURE 6.16 Tapering.

The maximum angle (i.e., the angle that determines the NA) at the entrance of the fiber is given by

$$\tan(\phi) = \frac{D}{f} \tag{6.36}$$

If the laser spot size is larger than the fiber's core diameter, then some of the energy would enter the fiber clad and will be lost. Furthermore, if the laser beam is energetic enough, it can cause damage to the fiber. If the NA of the focal lens is greater than that of the fiber, then some of the rays would not be guided through the fiber, that is, they would be lossy rays.

An alternative way to couple a laser beam into a fiber is to use a taper. A taper is a device that is attached to the fiber (see Figure 6.16). It increases the angle propagation through the waveguide and thus decreases the attenuation. Furthermore, since it changes the angle of propagation through the fiber, it also functions as a higher-mode filter or coupler, that is, it couples higher modes of propagation to lower modes of propagation.

Let us look at a laser beam with a divergence angle, θ, and a fiber that has an NA that corresponds to a certain angle, ϕ, as described in Figure 6.16. We would like to find the angle β of the taper that would enable us to couple the laser beam into the fiber successfully, that is, a ray that comes out from the laser with an angle equals to θ will propagate through the fiber.

A simple geometric calculation shows that the angle β is equal to

$$\beta = \frac{\phi - \theta}{2} \tag{6.37}$$

6.2.2.7 Distal Tips

Although coupling the laser beam into the fiber is very important, for many medical applications, controlling the beam shape at the fiber's exit is important as well. Laser beam manipulation at the end of the fiber is done using different types of tips (see Figure 6.17). The tips can be made by polishing the fiber materials (in the case of solid core fiber) or from different materials that are adhered to the fiber.

Each tip plays a role in different types of medical application. A few examples are described here. Tip c is for photodynamic therapy [19] application where radiation should diffuse to the surrounding. Tip d is for side firing of the laser beam to ablate tissue in the esophagus or colon. Tip e is created for thermotherapy (Figure 6.17) [20].

We will now examine the change of the beam shape when an input taper and a ball-shaped distal tip are added to a straight waveguide (Figure 6.18).

We have performed ray tracing on a straight waveguide and drawn the beam shape of the emerging radiation at 20 cm after waveguide output. The result is shown in Figure 6.19.

We now added an input taper and a ball-shaped distal tip and recalculated the beam shape in the new conditions. The results are shown in Figure 6.20.

The focusing of the beam is easily observed. Another advantage of the ball is sealing the waveguide and enabling work in liquid medium as may be expected within body cavities.

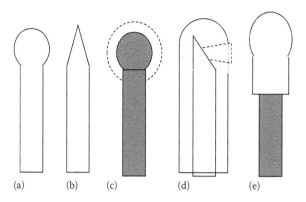

FIGURE 6.17 Types of fiber tips: (a) ball shaped, (b) tapered, (c) diffusing probe, (d) side firing, (e) metal probe.

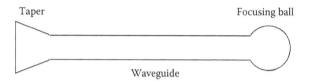

FIGURE 6.18 Waveguide with input taper and ball-shaped distal tip.

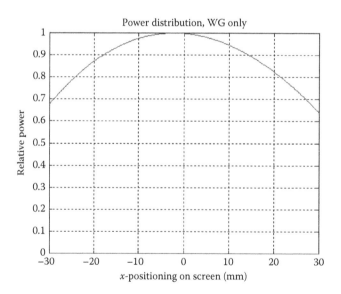

FIGURE 6.19 Energy distribution at the output of a straight waveguide.

6.2.3 Materials for Fabrication of Optical Fibers and Waveguides

6.2.3.1 Silica Fibers [21–24]

Nowadays, silica fibers constitute the backbone of optical communication and as such are manufactured by many companies. The first silica fibers were manufactured by Corning in the mid-1970s and had an attenuation of about 20 dB/km. Throughout the years, the manufacturing process had been improved and the attenuation reaches almost its theoretical limit (the typical attenuation of a silica fiber is about 0.2 dB/km). The spectral range of the fibers begins at the UV region (200 nm)

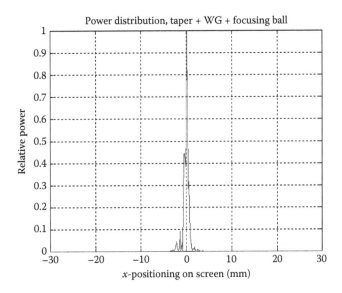

FIGURE 6.20 Beam shape at a distance of 20 cm after the addition of the input taper and distal tip.

and ends at the NIR (2000 nm). These fibers are useful for telecommunication, medical applications, power delivery, and remote sensing.

Silica fibers are solid core fibers. Their basic structure is core–clad where the core is made of silica (SiO_2) and the clad is made of silica doped with other elements such as germanium (Ge), boron (B), and fluoride (F). Doping the clad is needed in order to change its index of refraction. The fiber is coated with a plastic material (polyimide) that protects it and improves its bending capabilities. Depending on the core diameter, the fiber can transmit either single-mode or multimode radiation. The fiber can be drawn at any length, and its transmission characteristics are easily analyzed using conventional ray-tracing models.

It is possible to change the transmission characteristics by fabrication of multicomponent silica glasses in which silica comprises less than 75%/w or by high silica content fibers in which silica comprises more than 90%. It was thought that these types of fibers would have better transmission capabilities. However, impurities in these fibers increase the scattering and high OH concentration increases the absorption. Hence, the attenuation increases. Furthermore, multicomponent glasses have intrinsically lower strength and radiation hardness, while high-content silica fibers are very sensitive to moisture. These drawbacks limit their applications.

The OH concentration within the silica determines the transparency of the fiber. While low OH concentration is needed for transmission in the IR region, a higher concentration contributes to a large transmission in the UV.

6.2.3.2 Hollow Waveguides [25–39]

Hollow waveguides were initially developed for IR radiation delivery. However, they may be adapted to almost any wavelength. The basic structure of a hollow waveguide is shown in Figure 6.7. A hollow waveguide is constituted of a hollow tube (glass, fused silica, plastic, etc.), which is internally coated with a metal layer (silver, nickel, gold, etc.) and a dielectric layer (AgI, Se, Ge, polymers).

Hollow waveguides have low attenuation (less than 1 dB at the IR region), deliver high laser power (1 kW of CO_2 laser was demonstrated), and are made from nontoxic materials. Furthermore, they might be tuned to any desired wavelength by adapting the thickness of the dielectric layer and choosing the appropriate tube and inner layers. The aforementioned characteristics are the reason why hollow waveguides are suitable for many medical applications.

TABLE 6.2 Various Types of Hollow Waveguides

Tube	Metal Layer	Dielectric Layer	Group
Teflon, polyimide, fused silica, glass	Ag	AgI	Croitoru, Gannot et al.
Nickel, plastic, glass, fused silica	Ag, Ni, Al	Si, ZnS, Ge, polymers	Miyagi, Matsuura et al.
Plastic, glass, fused silica	Ag	AgI	Harrington et al.
Silver	Ag	AgI	Morrow et al.
Stainless steel	Ag, Al, Au	ZnSe, PbF$_2$	Laakman et al.

Over the years, most of the research regarding hollow waveguides was concentrated on developing hollow waveguides for the mid-IR region. The reason for choosing this wavelength range is the lack of materials and fibers, which have the desirable characteristics needed for medical applications. Till now, different hollow waveguides were developed for the IR region especially for wavelengths that are interesting for medical applications.

Several research groups around the world investigated hollow waveguides. Table 6.2 shows different hollow waveguides for different lasers. Their attenuation of a straight waveguide varies between 0.2 and 1 dB/m depending on the tube roughness and the laser wavelength.

Another type of hollow waveguide is ATR waveguides. These waveguides exploit the anomalous dispersion phenomena. Anomalous dispersion means that at a certain wavelength, the material has an index of refraction less than one. In that case, the guiding mechanism of the ATR waveguide is similar to that of a solid core fiber.

6.2.3.3 Fibers for the IR Region

Silica fibers are suitable for a wide range of wavelength from the UV to the NIR. Furthermore, their manufacturing process is well controlled, hence their low attenuation. However, this type of fiber is suitable up to 2000 nm; beyond that, there is a need for new materials and fibers. For the past three decades, intensive research has been taken place to develop optical fibers for the mid- and far-IR regions. One of the solutions, hollow waveguides, was described in the last section.

6.2.3.3.1 Glass Fibers for the IR Region

There are several glass-based fibers for the IR region. Silica-based fibers were discussed earlier. Other types are fluoride-based glass and chalcogenide fibers.

6.2.3.3.1.1 Fluoride-Based Glass [40–42]   Fluoride glass fibers are based on ZrF$_4$. These types of fibers have many advantages over silica-based glasses. Their theoretical attenuation is about 0.01 dB/km, which is an order of magnitude lower than that of silica. This is due to a shift to longer wavelength in phonon absorption. Furthermore, they possess a negative thermo-optic coefficient and are good hosts for rare earth elements. Such characteristics make these fibers a good candidate for optical fiber communications, laser amplifiers, and chemical sensing.

Over the years, different compounds of fluoride glass were investigated. The first compounds were composed of BeF$_2$. These compounds are very stable and resistant to damage. Moreover, they are transparent in a wide region of wavelengths including the UV. However, Be is a very toxic element, and the compound's hygroscopic nature makes them difficult to handle and eliminates any potential medical application.

Other fluoride compounds are heavy metal fluoride glasses that include fluorozirconate (ZrF$_4$). Such compounds are ZB—ZrF$_4$–BaF$_2$, ZBT, and ZBL, which are ZB that include ThF$_4$ and LaF$_3$, respectively. Others are ZBLA (ZrF$_4$–BaF$_2$–LaF$_3$–AlF$_3$) and ZABLAN (ZrF$_4$–BaF$_2$–LaF$_3$–AlF$_3$–NaF). The main disadvantage of ZB is its instability. Adding other components to the compound increases its stability.

TABLE 6.3 Attenuation of Chalcogenide Fibers

Chalcogenide Glass	Attenuation (dB/m)	Wavelength (μm)
Selenide	10	10.6
Sulfide	0.3	2.4
Telluride	1.5	10.6
	4	7

All fluoride glasses are transparent from the UV to the mid-IR and some of them even to the far IR. However, it is difficult to achieve the ultralow loss, which is predicted especially due to bubble formation during fiber drawing, microcrystallization, and material contamination. Typical fluoride fiber loss is in the order of 10 dB/km.

6.2.3.3.1.2 Chalcogenide Fibers [43–45] Chalcogenide glass is formed when elements from group 4 B and 5 B, such as As, Ge, Pb, Ga, and Si, are melted into silica glass. There are three groups of chalcogenide glass: sulfide, selenide, and telluride. The transparency of the glass depends on the glass composition. Sulfide glass is transparent in the visible region while incorporating selenide or telluride into the glass shifts their transparency into the IR region.

The mechanical properties also depend on the glass composition. Although the mechanical strength and thermal stability are lower than those of silica glass, they are sufficient for fiber drawing and typical fiber applications. The attenuation of chalcogenide glass also depends on the glass compound. Table 6.3 summarizes the attenuation of different kinds of chalcogenide fibers.

The main drawback of these fibers is their poisonous nature. This characteristic makes them hard to handle and limits their use in medical applications, unless protective coating is added.

6.2.3.3.2 Crystalline Fibers for the IR

There are about 100 crystalline materials that are transparent in the IR region. Most of them have similar optical characteristics but have very different mechanical properties. While transparency in a desired wavelength region is imperative, we cannot overlook the mechanical strength and durability of the fibers.

6.2.3.3.2.1 Single-Crystal Fibers Single-crystal materials have many advantages. It is possible to get a very pure material, thus overcoming scattering from impurities in the fiber material. Most of the materials have high melting point, which makes them suitable for laser power delivery. They are also chemically inert; thus, they can operate at harsh environments. An example for such material is sapphire.

However, these materials have also some drawbacks. Unlike silica fibers that can be drawn to considerable length, single-crystal fibers have to be grown from a melt; thus, long fibers are hard to manufacture. Moreover, some of the crystals are poisonous and brittle.

- *Sapphire fibers* [46–48]
  Sapphire is transparent up to 3.5 μm. These fibers are grown at very low rate (about 1 mm/min). They are made as core-only fibers and have core diameter between 0.18 and 0.55 mm. Their attenuation is 2 dB/m at 2.93 μm and maximum power delivery is 600 mJ.
  Sapphire fibers have very good mechanical properties. They have high melting point and mechanical strength. They are chemically inert, insoluble in water, and biocompatible. These properties make them good candidates for medical applications.
- *AgBr* [49]
  These fibers are manufactured either by a pulling method in which the material is heated and then pulled through a nozzle or by pressing the material through a nozzle. Fibers of up to 2 m were manufactured using these methods. The attenuation of such fibers is 1 dB/m at 10.6 μm. AgBr fiber may transmit up to 4 W of CO_2 laser.

6.2.3.3.2.2 Polycrystalline Fibers Polycrystalline fibers made of TlBr–TlI (KRS-5), KCl, AgCl, NaCl, and other compounds of heavy metals and halides are attractive because of their low theoretical attenuation. However, the attenuation of the first fibers that were manufactured was much higher, and most of the research that took place at the time concentrated on finding the origin of that loss.

Although transparency at the desired wavelength region is imperative, polycrystalline fibers have to have some more attributes. The material has to be deformed plastically in order to enable manufacturing long enough fibers. Furthermore, the crystal must be optically isotropic in order to avoid intrinsic scattering. Moreover, the material must be made out of a solid composition. Finally, the recrystallization of the material after fiber manufacturing must not impair the fiber optical characteristics. Potential materials for fiber manufacturing are thallium halides and silver halides. These fibers are made using the extrusion method.

- *Thallium halides* [50–52]
  Thallium halides have low theoretical attenuation (about 6.5 dB/km). However, the measured attenuation is much larger, between 120 and 350 dB/km. The main reason for the large attenuation is scattering, which is caused by material impurities, surface imperfection, grain boundaries, and dislocation lines. Furthermore, these fibers suffer from aging effects; they are soluble in water and sensitive to UV light.
- *Silver halide* [53–55]
  Silver halide fiber is made of $AgCl_xBr_{1-x}$. They have very low attenuation at 10.6 μm–0.15 dB/m. The main reason for the attenuation is bulk scattering, intrinsic and extrinsic absorption.

6.2.3.3.3 Liquid Core Fibers for the IR [56,57]

Liquid core fibers are hollow tubes filled with liquid that is transparent in the IR region. The only advantage of these fibers over solid core fibers is the possibility to get a more pure liquid than solid material. The liquid that is used is C_2Cl_4, which has a very high attenuation, about 100 dB/m at 3.39 μm, and CCl_4 that has an attenuation of about 4 dB/m at 2.94 μm.

6.3 Conclusions

Fibers and waveguides play a major role in the application of lasers in medicine. The variety of materials, which are used for laser transmission, is big. Each fiber or waveguide can be used in some applications but not in all. There is still a need to further improve fibers for better laser transmission at a wider range of the spectrum, smoother beam shape at the output of the fibers, higher power delivery and transmission of ultrashort energetic pulses. Most important is to have a reliable fiber that can be used safely for surgical applications.

Dedication

This chapter is dedicated to my advisor, Prof. Nathan Croitoru who passed away on February 24, 2014.

References

1. I. Cilesiz and A. Katzir, Thermal-feedback-controlled coagulation of egg white by the CO_2 laser, *Applied Optics*, 40(19), 3268–3277, July 1, 2001.
2. Y. Gotshal, R. Simhi, B.-A. Sela, and A. Katzir, Blood diagnostics using fiberoptic evanescent wave spectroscopy and neural networks analysis, *Sensors and Actuators B—Chemical*, B42(3), 157–161, August 1997.
3. I. Gannot, G. Gannot, A. Garashi, A. Gandjbakhche, A. Buchner, and Y. Keisari, Laser activated fluorescence measurements and morphological features–An in vivo study of clearance time of FITC tagged cell markers, *Journal of Biomedical Optics*, 7(1), 14–19, January 2002.

4. K. Kawano and T. Kitoh, *Introduction to Optical Waveguide Analysis*, Wiley, New York, 2001.

5. M. Ben-David, A. Inberg, I. Gannot, and N. Croitoru, The effect of scattering on the transmission of IR radiation through hollow waveguides, *Journal of Optoelectronics and Advanced Materials*, 1(3), 23–30, 1999.

6. A. Inberg, M. Ben-David, M. Oksman, A. Katzir, and N. Croitoru, Theoretical and experimental studies of infrared radiation propagation in hollow waveguides, *Optical Engineering*, 39(5), 1384–1390, 2000.

7. P. Beckman and A. Spizzichino, *The Scattering of Electromagnetic Waves from Rough Surfaces*, Pergamon Press, New York, 1963.

8. J. D. Joannopoulos, R. D. Meade, and J. N. Winn, *Photonic Crystals—Molding the Flow of Light*, Princeton University Press, Princeton, NJ, pp. 3–53, 1995.

9. I. Gannot, M. Ben-David, A. Inberg, N. Croitoru, and A. Katzir, Mid-IR optimized multi layer hollow waveguides, *BIOS*, 4253, 11–18, 2001.

10. S. D. Hart, G. R. Maskaly, B. Temelkuran, P. H. Prideaux, J. D. Joannopoulos, and Y. Fink, External reflection from omnidirectional dielectric mirror fibers, *Science*, 296, 510–513, 2002.

11. A. Bjarklev, J. Broeng, S. E. B. Libori, and E. Knudsen, Optical fibers and sensors for medical applications, *Proceedings of SPIE*, 4616, 73–80, 2002.

12. J. Knight, J. Broeng, T. A. Birks, and P. Russell, Photonic band gap guidance in optical fibers, *Science*, 282, 1476–1478, 1998.

13. F. Carroll and C. Brau, Vanderbilt MFEL program, X-ray project, http://www.vanderbilt.edu/fel/Xray/default.html (accessed March 6, 2014).

14. Y. P. Feng, S. K. Sinha, H. W. Deckman, J. B. Hastings, and D. P. Siddons, X-ray flux enhancement in thin-film waveguides using resonant beam couplers, *Physical Review Letters*, 71(4), 537–540, 1993.

15. F. Pfeiffer, X-ray waveguides, Diploma thesis, University of Munich, Munich, Germany, July 1999.

16. F. Pfeiffer, T. Salditt, P. Høghøj, I. Anderson, and N. Schell, X-ray waveguides with multiple guiding layers, *Physical Review B*, 62(24), 16939–16943, 2000.

17. M. Yaegashi, Y. Matsuura, and M. Miyagi, Hollow-tapered launching coupler for Er:YAG lasers, *Review of Laser Engineering*, 28(8), 516–519, 2000.

18. I. Ilev and R. Waynant, Uncoated hollow taper as a simple optical funnel for laser delivery, *Review of Scientific Instruments*, 70, 3840, 1999.

19. J. P. Marynissen, H. Jansen, and W. M. Star, Treatment system for whole bladder wall photodynamic therapy with in vivo monitoring and control of light dose rate and dose, *Journal of Urology*, 142(5), 1351–1355, November 1989.

20. L. Tong, D. Zhu, Q. Luo, and D. Hong, A laser pumped Nd(3+)-doped YAG fiber-optic thermal tip for laser thermotherapy, *Lasers in Surgery and Medicine*, 30(1), 67–69, 2002.

21. J. S. Sanghera and I. D. Aggarwal, *Infrared Fibers Optics*, CRC Press, Boca Raton, FL, 1998.

22. L. S. Greek, H. G. Schulze, M. W. Blades, C. A. Haynes, K.-F. Klein, and R. F. B. Turner, Fiber-optic probes with improved excitation and collection efficiency for deep-UV Raman and resonance Raman spectroscopy, *Applied Optics*, 37(1), 170–180, 1998.

23. P. Karlitschek, G. Hillrichs, and K.-F. Klein, Influence of hydrogen on the colour center formation in optical fibers induced by pulsed UV-laser radiation. 2. All-silica fibers with low-OH undoped core, *Optics Communications*, 155(4–6), 386–397, October 15, 1998.

24. K. F. Klein, R. Arndt, G. Hillrichs, M. Ruetting, M. Veidemanis, R. Dreiskemper, J. P. Clarkin, and G. W. Nelson, Optical fibers and sensors for medical applications, *Proceedings of the SPIE*, Bellingham, WA, Vol. 4253, pp. 42–49, I. Gannot, ed., 2001.

25. M. Miyagi and Y. Matsuura, Delivery of F_2-excimer laser light by aluminum hollow fibers, *Optic Express*, 6(13), 257–261, 2000.

26. Y. Matsuura, T. Yamamoto, and M. Miyagi, Hollow fiber delivery of F2-excimer laser light, *Optical Fibers and Sensors for Medical Applications*, *Proceedings of the SPIE*, Bellingham, WA, Vol. 4253, pp. 37–41, I. Gannot, ed., 2001.

27. I. Gannot, S. Schruener, J. Dror, A. Inberg, T. Ertl, J. Tschepe, G. J. Muller, and N. Croitoru, Flexible wave-guides For Er:YAG laser radiation delivery, *IEEE Transaction on Biomedical Engineering*, 42, 967, 1995.
28. I. Gannot, A. Inberg, N. Croitoru, and R. W. Waynant, Flexible waveguides for free electron laser radiation transmission, *Applied Optics*, 36(25), 6289–6293, September 1997.
29. M. Miyagi, K. Harada, Y. Aizawa, and S. Kawakami, Transmission properties of dielectric-coated metallic waveguides for infrared transmission, *Infrared Optical Materials and Fibers III*, *Proceedings of SPIE*, 484, 117–123, 1984.
30. Y. Matsuura, M. Miyagi, and A. Hongo, Dielectric-coated metallic hollow waveguide for 3 μm Er:YAG, 5 μm CO, and 10.6 μm CO_2 laser light transmission, *Applied Optics*, 29, 2213, 1990.
31. J. A. Harrington, *Infrared Fibers and Their Applications*, SPIE press, Bellingham, WA, 2004.
32. P. Bhardwaj, O. J. Gregory, C. Morrow, G. Gu, and K. Burbank, Performance of a dielectric coated monolithic hollow metallic waveguide, *Material Letters*, 16, 150–156, 1993.
33. K. D. Laakman, Hollow waveguides, U.S. Patent No. 4,652,083, 1985.
34. K. D. Laakman and M. B. Levy, U.S. Patent No. 5,500,944, 1991.
35. E. Garmire, Hollow metal waveguides with rectangular cross section for high power transmission, *Infrared Optical Materials and Fibers III, Proceedings of SPIE*, 484, 112–116, 1984.
36. T. Haidaka, T. Morikawa, and J. Shimada, Hollow core oxide glass cladding optical fibers for middle infrared region, *Journal of Applied Physics*, 52, 4467–4471, 1981.
37. R. Falciai, G. Gireni, and A. M. Scheggi, Oxide glass hollow fibers for CO_2 laser radiation transmis-sion, *Novel Optical Fiber Techniques for Medical Applications, Proceedings of SPIE*, 0494, 84–87, 1984.
38. C. C. Gregory and J. A. Harrington, Attenuation, modal, polarization properties of n < 1 hollow dielectric waveguides, *Applied Optics*, 32, 5302–5309, 1993.
39. R. Nubling and J. A. Harrington, Hollow waveguide delivery systems for high power industrial CO_2 lasers, *Applied Optics*, 34, 372–380, 1996.
40. K. Fujiura, Y. Nishida, T. Kanamori, Y. Terunuma, K. Hoshino, K. Nakagawa, Y. Ohishi, and S. Sudo, Reliability of rare-earth-doped fluoride fibers for optical fiber amplifier application, *IEEE Photonics Technology Letters*, 10(7), 946–948, July 1998.
41. P. Deng, R. Li, P. Zhang, H. Wang, and F. Gan, Defects and scattering loss in fluoride fibers: A review, *Journal of Non-Crystalline Solids*, 140(1–3), 307–313, January 2, 1992.
42. L. E. Busse and I. D. Aggarwal, Design parameters for multimode fluoride fibers: Effects of coating and bending on loss, *Optical Fiber Materials and Processing Symposium*. Material Research Society, Pittsburgh, PA, 1990, pp. 177–182.
43. B. J. Eggleton, B. Luther-Davies, and K. Richardson, Chalcogenide photonics, *Nature Photonics* 5, 141–148, 2011, doi:10.1038/nphoton.2011.309.
44. L. E. Busse, J. A. Moon, J. S. Sanghera, and I. D. Aggarwal, Mid-IR high power transmission through chalcogenide fibers: Current results and future challenges, *Proceedings of SPIE—The International Society for Optical Engineering*, Bellingham, WA, Vol. 2966, pp. 553–563, 1997.
45. S. Hocde, O. Loreal, O. Sire, B. Turlin, C. Boussard-pledel, D. Le Coq, B. Bureau et al., Biological tis-sue infrared analysis by chalcogenide glass optical fiber spectroscopy, *Biomonitoring and Endoscopy Technologies Proceeding of SPIE*, Bellingham, WA, Vol. 4158, pp. 49–56, 2000.
46. G. N. Merberg and J. A. Harrington, Optical and mechanical properties of single-crystal sapphire optical fibers, *Applied Optics*, 32, 3201, 1993.
47. P. D. Dragic, T. Hawkins, P. Foy, S. Morris, and J. Ballato, All-glass optical fibers derived from sap-phire. *Proceedings of the SPIE 8601, Fiber Lasers X: Technology, Systems, and Applications*, 86010I, February 26, 2013, doi: 10.1117/12.999361.
48. R. W. Waynant, S. Oshry, and M. Fink, Infrared measurements of sapphire fibers for medical appli-cations, *Applied Optics*, 32, 390, 1993.
49. T. J. Bridges, J. S. Hasiak, and A. R. Strand, Single crystal AgBr infrared optical fibers, *Optics Letters*, 5, 85–86, 1980.

50. M. Ikedo, M. Watari, F. Tateshi, and H. Ishiwatwri, Preparation and characterization of the TLBr-TlI fiber for a high power CO_2 laser beam, *Journal of Applied Physics*, 60, 3035–3036, 1986.

51. V. G. Artjushenko, L. N. Butvina, V. V. Vojteskhosky, E. M. Dianov, and J. G. Kolesnikov, Mechanism of optical losses in polycrystalline KRS-% fibers, *Journal of Lightwave Technology*, 4, 461–464, 1986.

52. M. Saito, M. Takizawa, and M. Miyagi, Optical and mechanical properties of infrared fibers, *Journal of Lightwave Technology*, 6, 233–239, 1988.

53. S. S. Alimpiev, V. G. Artjushenko, L. N. Butvina, S. K. Vartapetov, E. M. Dianov, Yu. G. Kolesnikov, V. I. Konov, A. O. Nabatov, S. M. Nikiforov, and M. M. Mirakjan, Polycrystalline IR fibers for laser scalpels, *International Journal of Optoelectronics*, 3(4), 333–344, 1988.

54. A. Saar, N. Barkay, F. Moser, I. Schnitzer, A. Levite, and A. Katzir, Optical and mechanical properties of silver halide fibers, *Proceedings of the SPIE*, 843, 98–104, 1987.

55. A. Saar and A. Katzir, Intrinsic losses in mixed silver halide fibers, *Proceedings of the SPIE*, 1048, 24–32, 1989.

56. H. Takahashi, I. Sugimoto, T. Takabayashi, and S. Yoshida, Optical transmission loss of liquid-core silica fibers in the infrared region, *Optics Communications*, 53, 164, 1985.

57. S. Klein, J. Meister, S. Diemer, R. Jung, W. Fuss, and P. Hering, High-power laser waveguide with a circulating liquid core for IR applications, *Specialty Fiber Optics for Biomedical and Industrial Applications, Proceedings of the SPIE*, Bellingham, WA, Vol. 2977, pp. 155–163, A. Katzir and J. A. Harrington, eds., 1997.

7

Fiber-Optic Probe Design

Urs Utzinger
University of Arizona

7.1 Introduction

The fiber-optic probe is a key element in biomedical spectroscopy and sensing applications. Optical fiber technology often is the main conduit to provide probing energy and to collect signals of interest. While optical technologies have been routinely used in the clinical environment for centuries [1–5], the integration of optical fibers has substantially expanded upon existing examination techniques and made possible new imaging techniques as well as diagnostic approaches. Fiber-optic cables provide a flexible conduit for light transport and enable the interfacing of complex light sources and advanced detection systems with the sample. With the help of fiber optics, we advance instruments into cavities and tubular structures through noninvasive and minimally invasive procedures. Through fiber optics, instruments are connected with tissue surfaces, measurements are conducted inside tissues, and optical signatures can be recorded from most living systems. Fiber-optic probes are manufactured as point sensors or imagers using cables with a diameter of half a millimeter or less, thus fitting through most needles and catheters.

Here, we review fiber-optic probe design principles and describe fiber-optic probes for fluorescence, reflectance, and Raman spectroscopy. We discuss applications for single, multiple, and imaging channels. Probe designs using Bragg gratings and interferometric techniques to measure pH, temperature, and other indirect measurands are summarized by Ronnekleiv et al. [6].

7.2 Fiber-Optic Probe and Optical Analyzer

Usually an optical system incorporates a light source, an optical analyzer with detector, and a light transport conduit, which, in many cases, is made of fiber-optic cables (Figure 7.1). Because of recent advancements in semiconductor-based integrated systems, some instrument developers incorporate sensors at the sample interface eliminating a long conduit for light analysis; however, most systems use external light sources to deliver light to the sample. When illumination and signal collection

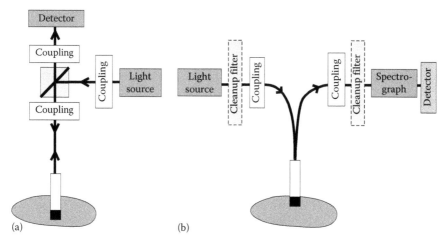

FIGURE 7.1 A fiber-optic-based spectroscopy system is based on an excitation source, a detection system, a coupling optics, and a fiber-optic probe. A flexible probe delivers illumination and collects the emitted light. (a) A fiber-optic spectroscopy system with a probe that incorporates a common optical path for illumination and detection needs a beam splitter and is illustrated in. (b) A design where the illumination and detection path are separated up to the probe tip is sown in. Because of autosignals of the optical system and limitations of the light source, often cleanup filters need to be used to achieve the desired spectral sensitivities.

occurs through the same conduit optically, a joint optical path allows for the smallest-diameter optical system (Figure 7.1a), while a separate path minimizes background signals produced in the illumination channel (Figure 7.1b). The excitation or illumination light source is typically a laser or a filtered white light source, such as a xenon or mercury lamp and more recently light-emitting diodes. Dielectric bandpass filters or monochromators can be used as filters to produce the desired spectral illumination and detection properties. When fluorescence excitation occurs with a bandpass-filtered white light source, additional filters are required to remove infrared (IR) and ultraviolet (UV) light from the beam path. Usually this is accomplished with a cold mirror in 90° configuration to protect the optical parts from excessive radiation. LEDs and laser light sources can also emit light outside the desired spectral region, and bandpass filters might be needed for them also. Pulsed light sources can be used to suppress the influence of ambient light because a gated detector collects only a fraction of the ambient light, while the average signal remains the same compared to continuously emitting sources. The coupling optics adapts the f-number of the light source to the numerical aperture (NA) of the optical fiber and guarantees optimal irradiance into the fiber. The output of a laser can be focused onto a smaller spot compared to the imaging of a light source arc onto a fiber bundle. In general, a large-diameter fiber bundle will need to be used to maximize the lamp's power usage. Usually arc lamps with lesser output power have smaller arcs with higher power densities, therefore coupling more light into a single optical fiber. The fiber-optic probe transports the remitted light from the tissue to a spectroscopic or imaging system. For most biomedical applications, exposure times are in the milliseconds to second range as modern optical analyzers (e.g., HoloSpec, Kaiser Optical Systems, Ann Arbor) and back-illuminated thinned charge-coupled devices (CCDs) with high quantum efficiencies allow efficient signal collection. Sometimes, it is necessary to place additional filter stages in front of the optical analyzer to reduce the influence of stray light originating from the illumination light source. For fluorescence applications, this filter stage holds long-pass filters, and for Raman spectroscopy, notch filters are used. To achieve the smallest probe diameters, single-fiber solutions are used in combination with a dichroic beam splitter and well-aligned coupling optics (Figure 7.1a). Single-fiber solutions are limited because of the difficulty of reducing back-scattered excitation and illumination light at the fiber coupling site and the suppression of background signals (autofluorescence and Raman scattering) induced in the fiber-optic cable by the illumination light. Nevertheless, single-fiber-based probes require a minimal amount of components and have the smallest illumination spots as well as highest light collection efficiencies.

A typical fiber probe is illustrated in Figure 7.2. A custom metal enclosure provides the interface to the sample and handling, whereas a metal coil-reinforced sleeve provides a protective enclosure to standardized interconnects. For multimode fibers, usually normed SMA-type connectors are used and for monomode fibers, FC connectors. A metal cylinder provides bifurcation of the various optical channels.

FIGURE 7.2 An example fiber-optic probe with a variety of fiber-optic tip configurations is illustrated. (Courtesy of Fibertech Optica, Kitchener, Ontario, Canada.)

Various fiber-optic configurations at the fiber-optic tip are illustrated in the same figure. Their purpose is to adjust the interrogated sample volume.

An optical fiber consists of a core, a doped cladding, and a protective jacket (Figure 7.3). Light is transmitted with total internal reflection and, when very small diameters are used, with mode confinement. Larger-diameter step index fibers (multimode fiber) do not retain polarization and the illumination pattern from the input site to the output. However, recent experiments have shown that output patterns can be generated using spatial light modulators through multimode fibers [7]. The half angle (α) of the light cone that a fiber can accept is characterized by the NA, which is defined by the difference in the refractive indices (n) of the core and the cladding materials and changes with the surrounding medium (Figure 7.3a):

$$\text{NA} = n_{\text{media}} \sin(\alpha) = \sqrt{n_{\text{core}}^2 - n_{\text{cladding}}^2} \tag{7.1}$$

For transmission in the visible wavelength range, the optical fiber core is made out of glass or plastic (e.g., acrylic or polystyrene). The doped cladding is usually made of a similar material but with lower refractive index. The optical fiber is coated with a jacket to protect it. The optical fiber is not designed to propagate

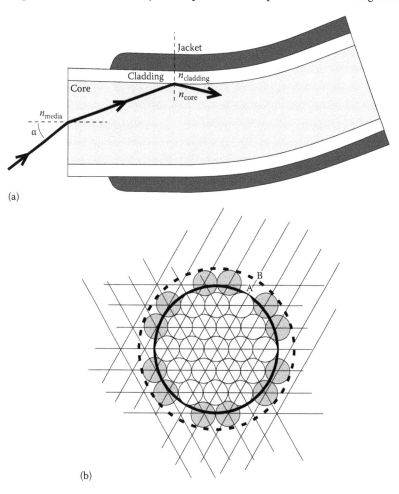

(a)

(b)

FIGURE 7.3 (a) An optical fiber consists of a core and a cladding with a lower refractive index and a rugged supportive jacket. Light is transported by total internal reflection. The light acceptance angle of the fiber is defined by the refractive indices of the media, core, and cladding. (b) In order to create flexible fiber bundles, individual fibers are hexagonal packed into a circular cross section.

light in the cladding, and if light couples into the cladding (either at the fiber end face or along the core), it creates background signals. These background signals can scatter back into the core and affect the signal of interest. When minimal background signals are critical or if the wavelength range needs to be extended to the UV and IR, high-grade fused silica is used for the core material (Figure 7.4). All-silica fibers have a doped silica cladding, while plastic clad fibers have a plastic cladding reducing costs. Optimized fiber preform

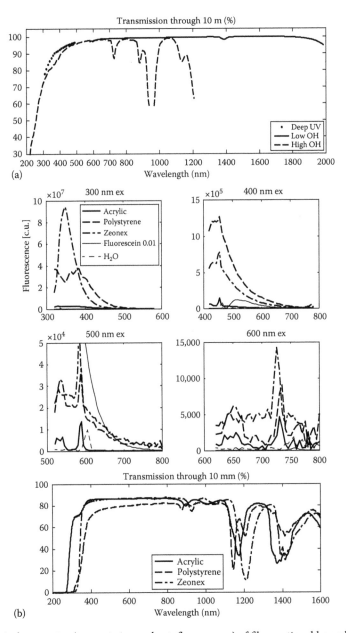

FIGURE 7.4 Optical properties (transmission and autofluorescence) of fiber-optic cables and optical materials: (a) The internal transmission of silica core fibers through a 10 m length fiber is illustrated. OH content modulates transmission in the UV and NIR. (b) Autofluorescence and transmission properties of common plastics used for lens fabrication demonstrate UV transmission and little fluorescence in the visible.

(continued)

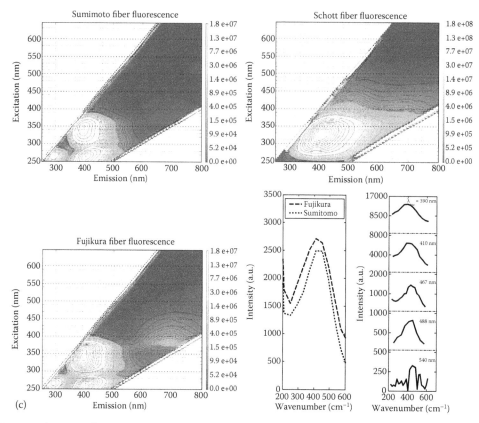

FIGURE 7.4 (continued)   Optical properties (transmission and autofluorescence) of fiber-optic cables and optical materials: (c) Autofluorescence and Raman scattering of common imaging fiber bundles show significant background in UV and consistent Raman scattering at all illumination wavelengths.

manufacturing [8–10] allows transmittance from 200 nm (solarization-resistant-UV grade fiber, Polymicro Technologies Inc., Phoenix, AZ, or CeramOptec, East Longmeadow, MA) up to 2500 nm (low-hydroxyl fiber) (Figure 7.4a), and sapphire fibers extend the transmission in the IR above 3000 nm. This allows for the application of fiber-optic probes in ultraviolet resonance Raman and IR Raman spectroscopies [11].

For spectroscopic applications, multimode fibers with a core diameter of 50–600 μm are most commonly used. Short pieces (1–2 m) of fibers with larger diameters can be produced in custom runs. The bending radius of quartz fibers resulting in no long-term defects is approximately 100 times the fiber diameter, and the momentary bending radius is approximately 50% of the long-term bending radius.

Due to bending of the fibers and defects in the fibers causing scattering, light may exit the core and hit the jacket. Most plastic jackets, such as Nylon and Tefzel®, produce significant autofluorescence when irradiated with UV light. Polyimide and metal-coated fibers, such as gold and aluminum, exhibit minimal fluorescence. During intense UV irradiation, defects may form in quartz fibers that result in increased autofluorescence in the visible [12]. The formation of these defects is partly reversible through diffusion processes in the quartz glass. Silica produces intrinsic Raman signals that will interfere with the signal of interest during Raman spectroscopy, while background signals can be subtracted to form the signal of interest by measuring a blank sample (negative standard). However, the noise associated with the background cannot be subtracted, and for large background signals, the dynamic range of the detector will need to be increased.

In order to manufacture flexible fiber-optic cables (patch cords, fiber bundles) with a large optically active area, fibers with a diameter of 100–200 μm are packed into bundles. The amount of fibers (n_{fiber})

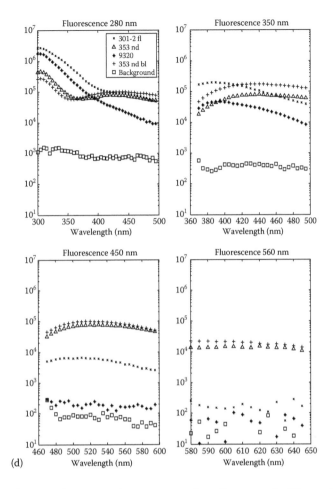

FIGURE 7.4 (continued) Optical properties (transmission and autofluorescence) of fiber-optic cables and optical materials: (d) Autofluorescence of common fiber-optic adhesives is illustrated. EPO-TEK 301 is transparent and used to bond optical elements, while EPO-TEK 353 and AngstromBond 9320 are used to embed fibers in connectors. 353ND-BL has black carbon powder mixed into the glue.

that can be packed into a round cross section is illustrated in Figure 7.3b. If fibers are packed hexagonally according to ring a, the number of fibers is determined with

$$n_{\text{fiber}} = 1 + \sum_{k=0}^{m} 6k \tag{7.2}$$

where m is the amount of rings around a central fiber. The total outer diameter (OD) is calculated by

$$\text{OD}_{\text{total}} = \text{OD}_{\text{fiber}}(1 + 2m) \tag{7.3}$$

For hexagonal packing with additional fibers, according to ring b, Equation 7.2 changes to

$$n_{\text{fiber}} = 1 + \sum_{k=0}^{m} 6k + 6m \tag{7.4}$$

and the total OD is calculated by

$$OD_{total} = OD_{fiber} 2\sqrt{1 + m + m^2} \qquad (7.5)$$

However, these calculations are based on an optimal arrangement of the fibers, which usually is not achieved under realistic manufacturing conditions.

For a given bundle cross section, the dead space (inactive area) in between the fibers increases when the fiber diameter is reduced and reaches an upper value of 25%. The inactive area additionally includes the area of the cladding and the jacket, which consumes normally more than 30% of the individual fiber cross section. If the jacket is stripped, this inactive area is reduced and the cladding accounts for approximately 17% of the fiber cross section. This leaves a total active area of approximately 60%–65% in a tightly packed fiber bundle with stripped jacket material. Melting and compressing fibers in a fiber bundle can reduce the dead space but is limited to fiber bundles with a few fibers only.

7.3 Optical Fiber Tip

To ensure optimal coupling, the end of optical fiber is either cleaved or polished. Manual polishing and inspection equipment is available through most optical component distributors. If the exit surface is polished with an oblique angle in respect to the fiber axis, the output will be deflected (Figure 7.5i through vi). If the critical angle for total internal reflection is reached, the light will leave the fiber through its cylindrical side wall (Figure 7.5iv through vi) [13–15]. The critical angle (Figure 7.5iv) for the silica–air interface is 43.3° and for the silica–water interface 66°. A fiber with a combination of a beveled and a flat polished tip [16] (formerly known as Gaser Technology [17]) can concentrate the illumination close to the fiber tip (Figure 7.5vi) as one part of the beam is exiting in the direction of the fiber axis, while the remaining part is guided sideways. To allow a larger range of steering angles (below the angle for total internal reflection), the beveled surface of the fibers can alternatively be coated (e.g., aluminum).

Several basic principles exist for sculpting the fiber tip and are illustrated in Figure 7.5a through j. Attaching a cone either eases the coupling of high-power lasers into the fiber (Figure 7.5a) or, when the cone is used in reverse order, increases the power density at the cone tip (Figure 7.5b), which can be useful for laser surgery [18]. A convex (Figure 7.5c) or concave (Figure 7.5d) surface concentrates or diverges the output of the optical fiber. An increased divergence can be useful to increase the illuminated area in front of the fiber. A convex surface focuses the output and increases the power density. Adding a sphere to the optical fiber (Figure 7.5e) enables collection of light from multiple directions due to internal reflections. If the sphere contains scattering particles, a flux meter can be constructed as the same amount of light is detected regardless of incident light direction. One can remove the fiber cladding through etching or diamond turning. If the surface of the remaining core is rough or groves are added, light will exit the core in a homogenous fashion. An example diffuser tip is illustrated in Figure 7.5f. Adding a wedge at the fiber end serves the same principle as oblique polishing (Figure 7.5g). In order to take advantage of a larger step of the refractive index at the air–glass interface as compared to air–water interface, transparent enclosure with glass plates and glass capillary tubing can be constructed as enclosures of the fiber-optic tip (Figure 7.5h and i).

Refocusing the illumination or collection beam paths decreases the sampling volume and increases collection efficiency because focusing enables a larger solid angle for signal collection. Melting the end of the fiber reshapes the exit surface. An almost spherical surface will be created by the surface tension of liquid quartz [19]. Its form is determined by the volume of melted quartz, which is related to the amount of absorbed thermal energy. A number of techniques are used to achieve this deformation: microfurnace, Bunsen microburner, electrical arc, and a CO_2 laser beam [20]. The smallest possible surface curvature is a hemisphere (Figure 7.5A and B), whereas the focusing power increases when the refractive index in the medium is decreasing. Spherical ball lenses have been used as collimators in

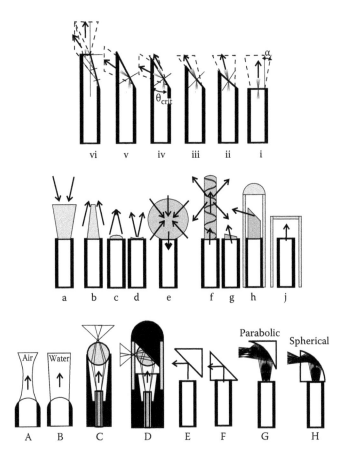

FIGURE 7.5 Fiber-tip configurations: (i–vi) Oblique polishing of the fiber tip deflects the output beam. (iv) When total internal reflection is achieved for a part of the light, the output will be split into two parts, one exiting at the front face and the other in the radial direction. (v) With steep polishing angles, complete internal reflection is achieved. (vi) A combination of front and side illumination can also be achieved and this increases power density. (a–h, j) Many options for fiber-tip sculpting exist with the goal of collecting or dispersing light at the fiber tip. (a) A taper supports the coupling of high-power laser beams into the fiber and (b) concentrates power to the taper tip when used in inverse direction. (c, d) Convex and concave front surfaces refocus or disperse the light. (e, f) A sphere or cylinder with scattering surface or embedded particles allows for homogenous illumination or light flux measurements. (g) A similar effect as shown in (ii) and (iii) can be achieved with a wedge. Capillary glass tube enclosures or glass windows allow for a glass–air interface at the fiber tip although the fiber-optic tip is immersed in water. (A–D) A spherical surface refocuses the light. (A) The fiber tip can be melted to form a sphere that allows for light concentration. (C, D) Sapphire ball lenses with high refractive indices concentrate the output of an optical fiber. (D) A half sphere allows for combined focusing and side deflection. (E, F) A coated prism serves as deflection mirror and with a nonplanar surface (G, H) can reimage and focus light.

fiber-optic connectors [21]. Sapphire spheres with a high refractive power are industrially produced in a wide range of diameters (Sandoz SA, Cugy, Switzerland; Rubis Precis, Charquemont, France). Light emitted from a fiber can be focused with this lens type into the sample (Figure 7.5C and D). The sphere and fiber can be held in a cylinder with a conic inner shape. An example fiber-optic tip with a 200 μm core diameter fiber and a sphere with a radius of 0.4 mm is illustrated in Figure 7.5C [14]. For side-looking applications, a half sphere can be implemented. Using a 45° setup and sapphire–air interface, the beam will be deflected with total internal reflection at the backside. A capillary glass tube can be used as enclosure and provides air as refractive medium.

As illustrated earlier, side looking is achieved with a beveled fiber; however, the cylindrical shape of the fiber will refocus the beam circumferentially but not axially. To avoid such deformation, a reflective surface can be introduced (Figure 7.5E and F). A prism can provide such surface. If it is attached to an outer sheet (Figure 7.5E), that sheet can be rotated and the beam is scanning the sample circumferentially without needing to rotate the optical fiber [22]. An additional concept of parabolic and spherical light concentrators [23] is illustrated in Figure 7.5G and H. The same concept can be used for smaller-diameter fibers in an imaging configuration [24]. For the manufacturing of surfaces such as the one in Figure 7.5G or H, an ultraprecision diamond cutting machine is needed (Nanotech, NH, United States).

7.4 Fiber-Optic Probe

When an optical fiber is used to interrogate a sample, three basic configurations need to be considered: (1) the illumination and detection occurs through the same fiber (Figure 7.6i), (2) the illumination and detection is accomplished with two fibers (Figure 76ii), and (3) the illumination and detection fibers include beveled or sculpted elements (Figure 7.6iii and iv).

If the surface of a sample is interrogated, the illumination and collection spot should overlap. Including a spacer at the distal end of the fibers or increasing the NA of the fibers increases this overlap. An example probe using a spacer is illustrated in Figure 7.6viii and ix and the fibers are arranged accordingly in Figure 7.6ii. In the far field, the collection efficiency for such a fiber-optic probe decreases quadratically with increased spacer thickness [25]. The spot overlap can be further improved when the spacer is replaced with a coated glass rod or a thick piece of optical fiber (Figure 7.6v). The rod (d) acts as a mixing element that homogenizes the output of the probe. Furthermore, the OD of a probe is reduced compared to using a spacer. The length of this rod should exceed OD_{fiber}/NA to allow a uniform illumination.

A better overlap is achieved with beveled fibers such as those illustrated in Figure 7.6iii, vi, vii [26]. If the probes are submersed, the beveled surface brings the sampling volume closer to the probe tip, which in turn allows collection of more light as the distance of the sample to entrance diameter of the collection fibers is reduced (Figure 7.6iii, iv, vii). The concept of dual fiber arrangement for separate illumination and collection (Figure 7.6ii) can easily extend to a multifiber design, as shown in Figure 7.6ix.

A fiber-optic probe was built to measure combined fluorescence and reflectance (Figure 7.6x) [27,28]. It included a mixing element for the fluorescence channel (Figure 7.6v) and four different source–detector separations. In the spectrograph ferrule, the detection fibers were linearly aligned, while the illumination fibers were arranged to best match the light source.

An example of a probe illuminating a ring is shown in Figure 7.6xi [29]. Multiple fibers were glued into a metallic ring. A sapphire plug with a cone-shaped deflection surface was centered onto the ring. The rotational symmetric design spreads the light from the fibers onto an elliptical spot with the long axis in circumferential direction. The inner ring of fibers could be used as collection fibers and the outer ring as illumination fibers.

7.4.1 Fiber-Optic Probe Efficiency Simulations

Several authors have investigated the effects of the fiber-optic tip configuration on collection efficiency and sampling volume. Others have evaluated how the fiber-optic probe geometry affects the spectral properties of the collected light as the influence of absorption and scattering increases with increasing length of the signals' return path.

Pfefer et al. [30] have shown the influence of the fiber diameter and shield thickness to probing depth of fluorescence remittance at 337 and 400 nm excitation. The simulation consisted of a single fiber delivering excitation and collecting fluorescence (Figure 7.6i). The Monte Carlo simulations assumed typical optical properties for esophageal tissue. The remittance increased with increasing fiber diameter and was 0.11% for 0.1 mm fiber diameter and 0.4% for 1 mm fiber diameter (fibers were placed in direct contact with the tissue). Eighty percent of the fluorescence signal originated between the tissue surface

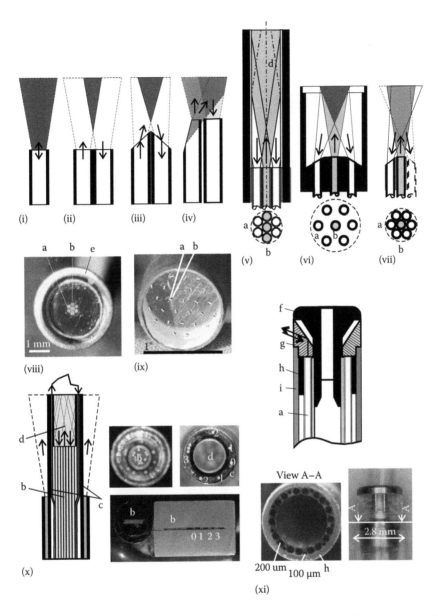

FIGURE 7.6 Fiber-optic probes: (i–iv) Multiple optical fibers are combined to create a fiber-optic probe. (i) A single fiber serves as both illuminator and collector. (ii) A fiber pair and often a six around one configuration create a fiber-optic probe where illumination and collection spots overlap in the far field of the optical system. (iii, iv) With beveled and sculpted tips, the interrogation volume is brought closer to the fiber-optic tip allowing to collect more light. (v–vii) An enclosure or assembly of multiple fibers allows for increased power delivery. (v) A cylindrical tube homogenizes the illumination and collection light, (vi, vii) whereas beveled fibers create an illumination and collection spot overlap closer to the fibers. (v–viii) The illumination power and collected light is increased by bundling several fibers together (b, illumination; a, collection). (ix) A multipixel system can be constructed using a mask holding fiber pairs. (x) An example of a combined fluorescence and reflectance measurement probe uses several illumination and collection fibers (b), an outer metal enclosure (c), a mixing element (d), and reflectance illumination and collection at several source–detector separations (0–3). (xi) A side deflecting system using a sapphire tip directs light sideways [(a) illumination and collection, (f) plug, (g) sapphire deflector, (h) metal ring, (i) outer sheet].

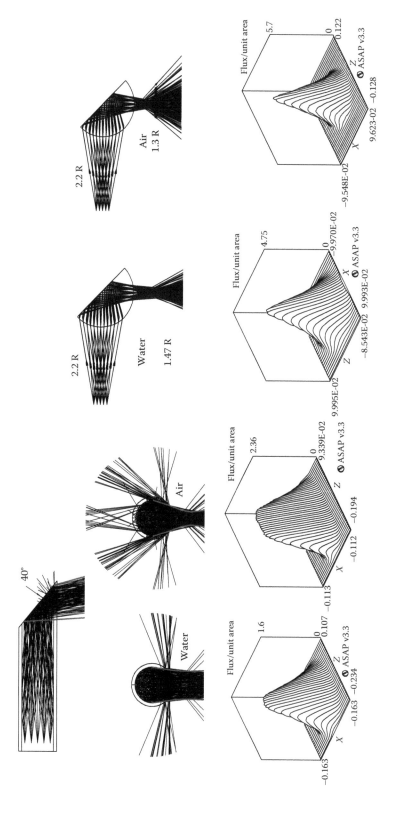

FIGURE 7.7 Sculpted fiber-optic tip simulations: Power densities were computed for beveled fibers and ball lenses attached to optical fibers. Power density is increased in water as well as in air.

and a depth of 0.175 mm when a 0.1 mm diameter fiber was used, and for a 1 mm diameter, the signal originated from a depth of 0.375 mm. Placing a shield in front of the fibers increased the probing depth further. With a 5 mm thick shield, the fluorescence was collected down to 0.425 mm (fiber diameter 0.2–1 mm), while the collected fluorescence dropped by a factor of 20 for the 0.2 mm fiber and a factor of 5 for the 0.6 mm diameter fiber. Similar results were found at 400 nm excitation wavelength.

When a tissue surface is interrogated with a fiber-optic probe, light will propagate inside the tissue and escape beyond the initial illumination spot [31,32]. In experiments with a separate excitation and fluorescence collection fiber, it was shown that when the brain is excited at 308 nm, the fluorescence ratio at 360 and 440 nm emission decreases with increased fiber separation. Furthermore, the average probing depth in esophageal tissue inside the excitation fiber area is sensitive to superficial layers; however, outside the excitation area, it originates from deeper layers [30].

Beveled fiber applications for fiber probe designs have been theoretically and experimentally analyzed by Cooney [33,34]. The sensitivity and sampling volume of beveled fibers was compared to other designs, such as a single-fiber (Figure 7.6i) flat-tipped probes (Figure 7.5j) and probes with lenses. The sensitivity and sampling volume was measured in a clear medium, and results were reported for the collection of Raman scattered light. These results are also applicable for fluorescence collection. A probe consisting of two beveled fibers (Figure 7.6iii) is at least by a factor of 1.5 more efficient than a dual fiber with a flat-tipped probe (Figure 7.6ii, NA = 0.22), and the sampling volume is smaller and located closer to the probe tip. A single-fiber probe (Figure 7.6i) is 1.8–4 times more efficient, depending on fiber diameter, than a dual fiber with a flat-tipped probe (Figure 7.6ii). The efficiency of a *Gaser-type* collection fiber (300 µm core, Figure 7.6iv) combined with a flat-tipped illumination fiber (400 µm core) has been measured by Shim et al. [16] with a sapphire Raman standard. Two configurations were considered: in the first, the light deflection angle of the *Gaser* fiber was between 13° and 32° (low deflection), and in the second, the deflection angle was between 35° and 55° (high deflection). Compared to a flat-tipped probe (Figure 7.6ii), Shim et al. found that in air, a low deflection probe measures 4 times and the high deflection probe 16 times increased signal. In water, the factors were 6.5 and 28, respectively. Similar values were found in 0.25% and 0.50% Intralipid™ solution. For a flat-tipped probe with a 400 µm illumination fiber and a 300 µm pickup fiber, maximal signal was picked up at a distance between 1.7 and 1.4 mm from the tip in water or in the Intralipid solution. The low deflection probe received the largest signal at a distance of approximately 0.5 mm and the high deflection probe at 0.25 mm from the fiber tip in water or the Intralipid solution.

Optical illumination analysis software such as ASAP (Breault Research Organization, Tucson, AZ) can be used to predict power densities at fiber-optic tips. Figure 7.7 illustrates the output of a beveled tip in water and air as well as the imaging of a fiber surface into water and air through a sapphire ball lens.

7.5 Materials for Fiber-Optic Probes

All fiber-optic probes reviewed here have a biomedical application in mind and therefore the probe tip either is in contact with tissue and body fluids or is used in a clinical environment. Therefore, it is required that the fiber-optic probe is analyzed for potential hazards before conducting clinical studies. The US Food and Drug Administration (FDA) Center for Devices and Radiological Health releases guidance documents that can be useful for such an evaluation [35]. Potential adverse events that are applicable to the use of fiber-optic probes are optical radiation hazards, thermal hazards, electrical shock hazards, clinical hazards (transmission of diseases), and material toxicity hazards. According to a general evaluation of medical devices [36], fiber-optic probes are in most cases categorized as transient surface or transient external communicating devices, and their materials should be tested for cytotoxicity, sensitization, and irritation and depend on the application, for acute systemic toxicity. If the fiber-optic probe is made of materials that have been well characterized in published literature and have a history of safe use, there is adequate justification to not conduct some or all of the suggested tests. A list of materials often used for medical devices is presented in Table 7.1.

Optical radiation hazard can be assessed by comparing the fiber-optic probe output to the threshold limit values (TLVs) or maximum permissible exposure (MPE) that has been established

TABLE 7.1 Materials Suitable for Fiber-Optic Probe Enclosures

	Fiber-Optic Application	Reference Medical Application	Biocompatibility or Approval	Useful Properties
Metals, alloys				
Stainless steel 302	Tubing, housing	Needle tubing	Literature	
Stainless steel 316	Tubing, housing	Needle tubing	Literature implants	Acid resistance
Titanium and titanium alloys	Tubing, housing		Literature implants	
Amalgam		Dental filling material	Literature dental implants	
Ceramics				
Aluminum oxide, sapphire	Scattering particles, Optical window		Literature implants	Transmission 200 nm–3 μm High thermal conductivity Almost insoluble
Barium sulfate	Scattering particle		Gastrointestinal	X-ray contrast agent
Polymers				
Acrylic[a]	Optical components			Transparent
Polyethylene, PE	Tubing, housing	Pharmaceutical bottles, catheter	Literature hip implants	
PVC	Tubing	Blood bags, cannulae	USP VI	
Teflon, PTFE	Tubing	Catheter, vascular grafts	USP VI	Temperature resistant, 230°C
PMMA	Optical components	Bone cement, blood pump	Literature optical implants	Transparent in the visible
Polycarbonate			USP VI possible bisphenol A release	Sterilizable 12°C, EtO Transparent in the visible Soft, plastic deformation
Polyester, PET		Suture, mesh, vascular grafts	Literature implants	
Polyimide	Tubing		USP VI	Temperature resistant, 220°C
Polystyrene[a]	Optical components			Transparent
Silicone rubber	Catheters, coatings, tubing		Literature implants USP VI	Temperature resistant, 200°C Flexible
Cycloolefin copolymer, topas, zeonex[a]	Optical components	Microplates for cell culture	USP VI	93% transmission visible UV transparent below 300 nm
Glue				
Cyanoacrylate (superglue)		Surgical adhesive	USP VI	Fast bonding
EPO-TEK 301[a]	Bonding of optical elements		USP VI	Transparent in the UV and visible Low autofluorescence
EPO-TEK 353,[a] 375	Sealing		USP VI	Autoclavable
AngstromBond 9320[a]	Connector		NASA	Low stress
2-TON DevCon	Sealing		Suggestive for low skin carcinogenicity	High tensile strength Water resistant Transparent

[a] Fluorescence properties shown in Figure 7.4d.

by the American Conference of Governmental Hygienists and the American National Standard Institute (ANSI) [37,38]. In contrast to those two standardization organizations, the International Commission on Non-Ionizing Radiation Protection (Oberschleissheim, Germany) makes their guidelines available online at no cost and publishes them in the *Health Physics* journal. Usually the TLVs are expressed for laser beam [39,40] and incoherent exposure of the eye and skin [41,42].

For broadband UV exposure (<400 nm), the biologically effective radiation (device emission weighted by the biologic action spectrum) should not exceed the TLV for melano-compromised skin and eye (3 mJ/cm^2). For wavelengths above 400 nm, exposure limits are governed by thermal injury to the skin and eye. Potential temperature increases should be evaluated using endpoints or by potential peak tissue temperature.

Since materials (fused silica, acrylic, polystyrene, or silicon) used for fiber manufacturing are good electrical isolators, a fiber-optic probe with minimal electrical shock hazard can be manufactured.

To avoid clinical hazards, such as the transmission of diseases, the fiber-optic probe needs to be disinfected or sterilized prior to its use with common clinical practice. If the fiber-optic probe cannot be detached from the spectroscopic equipment, the parts that may be in contact with tissue need to be soaked, for example, in a solution based on 2.4% glutaraldehyde (CIDEX, Advanced Sterilization Products, Irvine, CA) or 0.55% orthophthalaldehyde solution (CIDEX OPA). This procedure is not considered a sterilization. Low-temperature sterilization with ethylene oxide or hydrogen peroxide gas can be applied to a detachable fiber-optic probe. It is critical to consider the compatibility of the probe materials with the disinfectant. CIDEX OPA is a disinfectant that is compatible with most materials used to manufacture fiber-optic probes, including adhesives cyanoacrylate (super glue) or EPO-TEK 353 and EPO-TEK 301 (Epoxy Technology, Billerica, MA). If a fiber-optic probe is used during a surgical procedure, it can be covered with a sterile plastic drape developed to cover intraoperative ultrasound probes.

A variety of materials that can be used to enclose the fiber-optic probe have been well characterized, and their biocompatibility has been previously published in the literature. Safe choices are materials created for implants or materials that were tested for US Pharmacopeia (USP), class VI. USP, class VI, is a base requirement for medical device manufacturers. A summary of materials that could be used to create fiber-optic probes is listed in Table 7.1. One of the first alloys created for human use was stainless steel Type 302 and the corrosion-resistance-improved stainless steel Type 316 (hypodermic steel) [43]. A large variety of standard tubing diameters and wall thicknesses are available to enclose fiber-optic cables and optical elements. Aluminum oxide (Al_2O_3) is an inert bioceramic and, when grown to a crystal (sapphire), is chemically inert and almost insoluble. Sapphire has a high thermal conductivity and is optically transparent between 200 nm and 3 µm. EPO-TEK produces a variety of glues with excellent transmission and with low autofluorescence (Figure 7.4d) that can be used to bond optical elements within the fiber-optic probe. AngstromBond (Fiber Optic Center, MA, United States) manufactures adhesive to connectorize fiber-optic cables. Some exhibit also low autofluorescence characteristics (Figure 7.4d). Adhesives used to embed optical fibers in connectors are not transparent. Reflected illumination light and light coupled into the dead space between fiber bundles produce autofluorescence in the connector. Several glues from EPO-TEK were tested for USP class VI and are autoclavable (e.g., 353 and 375). Many thermoplastic polymers have been used in the body, for example, polyethylene (PE), polytetrafluoroethylene (PTFE, Teflon), polymethylmethacrylate (PMMA), and polyester (PET). These materials can be used to create flexible and heat shrink tubing or plastic enclosures for fiber-optic probes. High-temperature-resistant polymers include polyimide (220°C), PTFE (230°C), and silicone rubber (200°C). Optically transparent plastics are acrylic, polystyrene, and cyclic olefin copolymers, and their transmission and autofluorescence is illustrated in Figure 7.4b.

7.6 Fiber-Optic Cable and Fiber-Optic Probe Manufacturers

Table 7.2 contains an example list of optical fiber and component manufacturers that are relevant for custom fiber-optic probe designs. Several small companies have specialized in manufacturing custom fiber-optic probes.

TABLE 7.2   Manufacturers of Fiber-Optic Probes and Components

Company	Manufacturing of Fiber-Optic Components	Fiber-Optic Assemblies	Fiber-Optic Specialties	Location
Optical fiber and component manufacturers				
SCHOTT	Glass tubing, rods, rofiles Filters Leached image bundles	FO assemblies Cables		Massachusetts, USA; Germany
Polymicro Technologies/Molex	Optical fibers Capillary tubing Microcomponents	FO assemblies	Sculpted fiber tips Precision ferrules	Arizona, USA
CeramOptec	Optical fibers Capillary tubing Tapered fibers Collimators/focusing	Bundles FO assemblies		Massachusetts, USA; Germany
Fujikura/AFL	Optical fibers Image fiber			Japan/USA
Sumitomo Electric	Optical fibers (single mode) Image fiber			USA
Sumita Optical Glass	Fiber bundles Image fiber Light pipes			Japan
Micromaterials	Optical fibers (sapphire)		Raman probes	Florida, USA
Lumatec	Liquid light guides			Germany
Leoni Fiber Optics	Optical fibers PCF	Bundles		Germany
Corning Optical Fiber	Optical fibers			New York, USA
Fiber-optics Technology Incorporated	Optical fibers	FO assemblies Patchcords		Connecticut, USA
Fibercore	Optical fibers			England
iRphotonics/Thorlabs	Optical fibers (IR)			New Jersey, USA
NanOptics	Optical fibers (plastic)			Florida, USA
Nufern	Optical fibers			Connecticut, USA
Ofs Furukawa	Optical fibers	Patchcords		Connecticut, USA
OZ Optics	Fiber-optic couplers Attenuators Polarization maintaining components Isolators Collimators	Patchcords		Ottawa, Canada
CT Fiberoptics	Bend, tapers, glass forming	Bundles		Connecticut, USA
Gould Fiberoptics	Polarization maintaining components Attenuators Fiber-optic splitters Fiber-optic combiners Planar waveguides (splitters)	Patchcords		Maryland, USA
Industrial Fiber Optics	Optical fibers Light pipes Image fiber	Patchcords	Distributor	Arizona, USA
FluorTek	Tubing			Pennsylvania, USA
ZEUS	Tubing			South Carolina, USA

TABLE 7.2 (continued) Manufacturers for Fiber-Optic Probes and Components

Company	Manufacturing of Fiber-Optic Components	Fiber-Optic Assemblies	Fiber-Optic Specialties	Location
Assemblies				
FiberTech Optica		FO assemblies Bundles Arrays Custom probes	Raman probe	Ontario, Canada
Multimode Fiberoptics		FO assemblies Bundles Patchcords Custom proves		New Jersey, USA
Spectraconn Inc		Probes Bundles Patchcords		New Jersey, USA
Fiberoptic Engineering Corporation		FO assemblies		New Jersey, USA
Volpi		Bundles		USA/Switzerland

7.7 Fiber-Optic Light Diffuser

For photodynamic therapy, it is desirable to deliver light homogenously along the surface of a fiber-optic probe. Diffusely scattering elements can be mounted at the end of fibers or fiber bundles for light distribution. Similarly, such systems can be used to collect light independent of the light propagation direction (Figure 7.5e and f).

The scattering particles included in the diffuser elements are titanium (TiO_2), aluminum oxide (Al_2O_3), or $BaSO_4$, which are embedded in a transparent matrix such as optical glue (EPO-TEK 307, Epoxy Technology, Inc. Billerica, MA) (Table 7.1). For cylindrical illumination, tubing based on fluoropolymers (PTFE, FEP, Zeus, Orangeburg, SC) can be used as enclosure. Some oxides such as Al_2O_3 or $BaSO_4$ emit in the UV–VIS and can contribute to background signals.

An example diffuser is presented in Figure 7.8. Light exits the fiber-optic bundle and enters a turbid rod. A mirror at the end of the rod reflects unscattered light back into the rod, increasing the uniform light distribution along the optical axis [44]. The length of the probe determines the concentration of scatterers. Figure 7.8 illustrates a diffuser with a length of 30 mm and a diameter of less than 2 mm.

7.8 Fiber-Optic Probes for Reflectance Spectroscopy

The elastically scattered light that escapes the surface of a sampling volume is called reflectance [45]. The transport mean free path length of a photon in a turbid media is defined by the optical properties of the sample: the scattering and absorption coefficient as well as the scattering anisotropy. These properties depend on the chemical and structural composition of the sample. The absorption coefficient is a superposition of individual chromophores. Since the major chromophores in tissue are oxygenated and deoxygenated hemoglobin [46,47] and water, as well as other absorbers such as melanin and fat, it is possible to derive diagnostic parameters from the absorption coefficient. The structural composition of the sample affects the shape, structure, size distribution, and concentration of scattering particles. All these characteristics vary spatially and are wavelength dependent.

There are several methods to measure absorption and scattering properties in the steady state, time, and frequency domain by analyzing reflected or transmitted light. Using a steady-state approach, a spatially resolved reflectance profile can be measured when the sample is illuminated at a single point. An analytical solution to this profile can be derived from diffusion theory [48]. If the profile is measured with a fiber-optic probe, one can extract the optical properties through an inverse model approach [49].

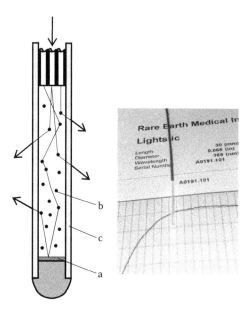

FIGURE 7.8 A cylindrical optical light diffuser with a length of 30 mm and a diameter of less than 2 mm. (a) Reflective surface, (b) scattering particle, and (c) transparent sheath.

Ideally, the profile is measured at six to nine source–detector-separated locations. For tissue, the separation distances are in the range of 2–20 mm. The fiber-optic probes used for these studies consist of a single excitation source and several spatially distributed collection fibers (Figure 7.9i and ii).

A probe with a linear alignment was constructed by Wang et al. [50,51] and places the detection fibers over a range of 1–10 mean free path lengths (Figure 7.9i). Alternatively, a circular fiber arrangement (Figure 7.9ii) by Nichols et al. [28] allows a simple calibration of the system by placing a source fiber in the center of all fibers. Measuring the spectrally resolved reflectance of all fibers simultaneously requires a dynamic range of four or more orders of magnitude. Neutral density filters in the fiber-optic path reduce the required dynamic range of the detector. Alternatively, each detection fiber is analyzed separately. Wang et al. modified their linear probe design with an oblique incident source fiber and measured the shift of the reflectance profile, which depends on the scattering properties. Fiber-optic probes that assess tissue optical properties with reflectance profiles and diffusion approximations usually require measurement locations over the range of a centimeter and more, and the assumption is made that the optical properties do not vary over this range. To overcome this limitation, Monte Carlo–trained neural networks [48,52] have been investigated. Using such approaches, fiber-optic probes with short source–detector separations (0.6–7.8 mm) have been successfully used to extract optical properties [53,54]. At a source–detector separation of 1.75 mm, the collected signal primarily depends on absorption for most common NA of optical fibers [55].

Polarization techniques have been successfully used to gate detection depth in reflectometry. This is of interest because many precancerous changes occur in the epithelial layer in the first 100–300 μm of tissue, while neovascularization is originating in deeper tissue layers. Reflectance measured in a parallel-polarized fashion (illumination and detection are linearly polarized in the same direction) contains mainly light scattered from upper tissue layers, since light from the deeper tissue layers is multiple scattered, which randomized the polarization status. Light loses its original polarization status after approximately 20 scattering mean free path lengths and becomes unpolarized in the diffusing regime [56]. Cross polarized reflectance originates from deeper tissue layers since the polarized illumination light needs to undergo several scattering events until significant components are created in perpendicular polarization direction. Subtracting the perpendicular polarized reflectance from the parallel polarized removes 90%

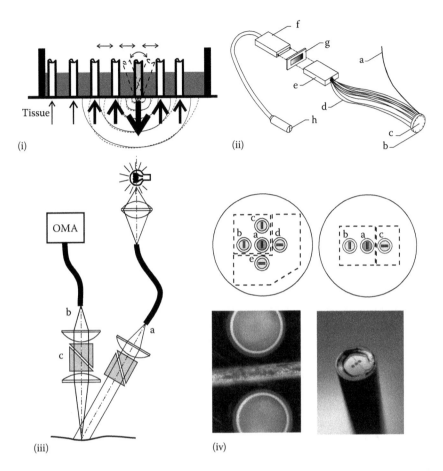

FIGURE 7.9 Reflectance measurement probes: (i, ii) the source–detector-separated reflectance profile can be measured with a linear and circular arranged fiber-optic probe. With increasing separation, the collected light intensity decreases. If the illumination angle is tilted, the apparent center of the reflectance profile shifts. (ii) If the profile is measured with a ring configuration, a center calibration fiber (c) adjusts for the system throughput. (a) Illumination fiber, (b) probe tip, (c) calibration fiber, (d) optical fibers, (e, f) linear transfer array, neutral density filter (g), and a spectrograph input ferrule (h). (iii) Polarized reflectance system with light transported from source through illumination fiber (a) and polarizer optics. Reflectance is collected (b) with an imaging system, and polarization state is analyzed in the collimated beam path (c). (iv) A frontal view of a fiber-optic-polarized reflectance probes using film polarizers (a: illumination, b–d: detection).

of light originating from deeper tissue layers [57]. Dividing this subtracted reflectance by the sum of the parallel and perpendicular polarized reflectance cancels common attenuation and the spectral characteristics of the light source and detector.

Fiber-optic probes that measure polarized reflectance consist of a polarization-filtered white light source and a polarization-filtered detection system with an optical multichannel analyzer (Figure 7.9iii). Since multimode fibers do not conserve the polarization status of light, the polarization filters need to be placed at the fiber-optic probe tip. Johnson and Mourant [58] presented a system that measures reflected light at four positions through a linear polarizer. Linear polarizing laminated film (Edmund Scientific, Barrington, NJ) was placed in front of the fibers. Suppression of different polarization modes was in the order of 98%–99%. Ideally, in a polarization-filtered probe, two or more pieces of polarization film are placed on the probe tip as shown in Figure 7.9iv. Sokolov and coworkers [59] presented in vitro results with such a fiber-optic probe.

7.9 Fiber-Optic Probes for Raman Spectroscopy

Near-IR Raman scattering can be used as a tool to perform in situ histochemical analysis [60–63]. In biological applications, approximately 10^{-10} of the incident light is Raman scattered and the Raman signal is normally six orders of magnitude weaker than typical fluorescence signals, which poses significant demands on the fiber-optic probe regarding signal collection and background suppression. Increasing the illumination source power is limited by heating hazards. The background signals originate from the laser source, the fibers, and all optical components [64]. These signals must be reduced with optical filters. Raman spectra of tissues can be recorded in vivo within a few seconds [65].

A design developed by Myrick and Angel (Figure 7.10i) [66,67] is based on Gradient-Index (GRIN) lenses allowing filters to be placed in the collimated section of the illumination and collection path. The illumination path is bandpass filtered to eliminate signals produced in the fibers and the light source at wavelengths other than the illumination wavelength. A long-pass filter or notch filter reduces specular reflections and elastically scattered light that enters into the collection path. A similar approach is shown in Figure 7.10ii [68] where dielectric filters are placed in between fibers that are held together with spring-loaded connectors. The optical fibers are attached to fiber-optic tip as illustrated in Figure 7.6v where a sapphire window isolates the probe from the surrounding environment [26,69]. Raman fiber-optic probes for submersed sensing using beveled fibers (Figure 7.10iii) were investigated by Greek et al. [15]. Mahadevan-Jansen et al. have successfully measured Raman spectra on the cervix with a fiber-optic probe (Figure 7.10iv) [70,71]. A diode laser is coupled into a single 200 µm fiber. A small-diameter dielectric filter (3–4 mm) rejects out of band light (OD5), and a mirror deflects the focused beam onto the specimen site. A quartz window provides contact between the probe and the tissue. Scattered light is imaged back on a fiber bundle. In the collimated beam path, elastically scattered and specular reflected light is rejected by a notch filter (OD6). The optics in the detection arm is 8 mm in diameter, and the whole probe diameter is less than 2 cm. A similar two-legged probe is commercially available by InPhotonics (Norwood, MA). The commercial system is further optimized by inserting a dichroic mirror that combines the excitation and collection light before focusing it onto the sample. Berger et al. [62] and Tanaka et al. [72] developed a probe design for improved signal collection using a compound parabolic concentrator (CPC) at the distal end of the probe. Small CPC dimensions with an input aperture of 0.6, an exit aperture of 2 mm, and a length of 4 mm were achieved (Figure 7.10c). The CPC improved the collection efficiency sixfold. Kaiser Optical Systems (MI, United States) produces a commercially available fiber-optic Raman probe (Figure 7.10v) [73]. Excitation light is filtered through a transmission grating and can be monitored outside the probe head. The sample spot is imaged with two microscope objectives onto a fiber-optic cable. Two Raman notch filters remove excitation light in the collimated beam path.

The scattering cross section for Raman interactions increases significantly if illumination is in resonance with the electronic transitions of the chromophore involved in the vibration. For many biologically relevant chromophores, these transitions are in the UV. Resonance enhancements are on the order of 10^4–10^5 for UV absorption bands [60]. High-throughput UV-optimized optical fibers [8] are needed for the collection and illumination path.

7.10 Fiber-Optic Imaging Probes

While most fiber-optic probes reviewed here are measuring at a single or at a few discrete locations, the main concepts can be expanded to measure multiple sites and in conjunction with an imaging system (Figure 7.11). After initial clinical trials using single-pixel fiber optical probes [74–76], multipixel probes were manufactured and clinically tested [77,78]. Those were followed with a single-pixel spectroscopy systems integrated into an endoscopic imaging system [79] where the spectral sampling point was in the middle of the field of view. Similarly, to combine complementary nature of different imaging modalities, significant work was conducted to combine multiple optical measurement techniques in a

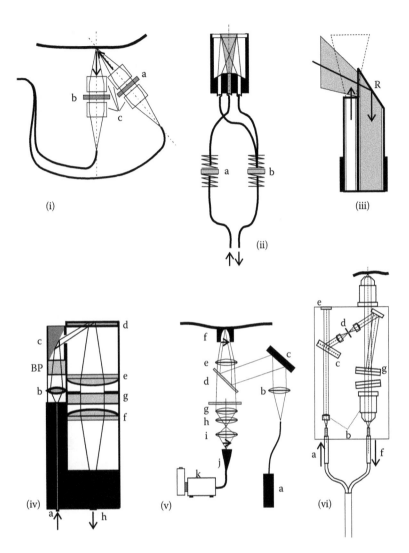

FIGURE 7.10 Fiber-optic probes for Raman spectroscopy. Because of significant instrument background, Raman probes incorporate optical filters. (i) A GRIN lens-based fiber-optic probe with a bandpass filter (a) for illumination and long-pass (b) for detection channel. (ii) A fiber-optic probe based on the design in Figure 7.6vi and in-line filters (a, b) between spring-loaded SMA connectors. (iii) Submersed fiber-optic probe with side deflection. (iv) Two separate channels are used for illumination and collection. Illumination channel (a), biconvex lens and a dielectric bandpass filter (BP) (b), transport illumination on a mirror (c), and the sample area (d). The inelastic scattered light is collected behind a quartz window and imaged with two plane convex lenses (e, f) through an aperture stop and holographic notch filter onto a flexible fiber bundle (h). (v) A Raman collection system using an optical concentrator (f), diode laser (a), collimating lens (b), reflective mirror (c), dichroic mirror (d), focusing lens (e), parabolic concentrator (f), field lenses (g, h) and transfer lens (i) to couple the Raman signal into a collection fiber bundle (j), and a high-efficiency spectrograph with a LN-cooled CCD (k). (vi) Commercial Raman probe with breadboard optics. Excitation light (a) is collimated (b) and bandpass filtered through a transmission grating (c) and aperture stop (d). The excitation intensity can be monitored (e). Raman scattered light is collected with a microscope objective and imaged on a collection fiber bundle (f). Two holographic notch filters (g) are placed in the collimated beam path to remove reflected excitation light.

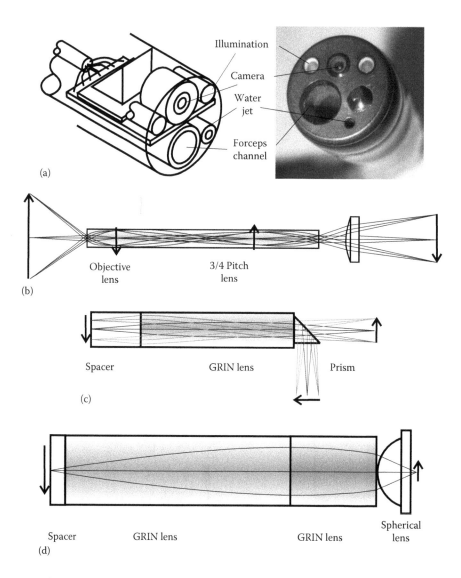

FIGURE 7.11 **(See color insert.)** Fiber-optic imaging system. (a) A modern gastroscope incorporates illumination optics, operating channel, water jet, and camera at the probe tip. (b) A similar system using GRIN lenses and a relay can be constructed for wide-field observation. (c) A single GRIN lens transfers the object under side viewing on the image plane. (d) Two-photon imaging stick lens based on GRIN lens and micro-optical component allows for high NA and observation through regular microscope objective lens when the stick is mounted in front of it.

single fiber-optic system. For example, single-point fluorescence capabilities were combined with optical coherence tomography [80,81] in a 2 mm diameter side-looking system.

A modern gastroscope is illustrated in Figure 7.11a. With the availability of integrated CMOS sensors and compact electronics manufacturing, the fiber-optic imaging bundle from early-generation endoscopes was replaced with a camera located at the endoscope tip. Illumination still occurs through fiber-optic cables and a concave lens to accommodate the large field of view of these systems. Operating channels for tissue biopsy as well as a water jet are integrated to clean the tissue surface but also the endoscope optics. The operating channel can be used to introduce fiber-optic probes for point interrogation or microendoscopes for microscopy.

To construct rigid endoscopic tips, one can incorporate gradient index optics (GRINTECH, Germany) or micro-optics (Figure 7.11b through d) [82]. Figure 7.11b illustrates a GRIN objective lens in combination with a GRIN relay lens that transfers the image onto a camera. Because of toxicity and water solubility, GRIN lenses usually need to be enclosed. In order to minimize autosignals, fiber-optic probes with GRIN lenses need to be carefully designed. A lens for a 2.3 mm diameter probe with surface chromoendoscopy capability and optical coherence tomography has been designed [83] using a design illustrated in Figure 7.11c. The chromoendoscopy channel used a 30,000 element imaging fiber bundle; a single mode fiber was added for OCT imaging [24].

An ultraslim objective lens was created for two-photon microscopy and is illustrated in Figure 7.11d. The combination with an additional optical element allowed for high NA. It was implanted transcranially in small animal studies for brain imaging [84].

While developing confocal microendoscopes, fiber-optic imaging bundles from Fujikura, Sumitomo, and Schott were analyzed for background signals and are illustrated in Figure 7.4c. Besides autofluorescence in the UV, strong Raman back-scattering signal was found throughout the spectral region [85].

References

1. Scarani P, Salvioli GP, Eusebi V. Macello Malpighi (1628–1694): A founding father of modern anatomic pathology. *The American Journal of Surgical Pathology*, 18(7):741–746; 1994.
2. von Helmholtz HLF. *Handbook of Physiological Optics*. Hamburg, Germany: Voss; 1867.
3. Shah J. Endoscopy through the ages. *BJU International* 89:645–652; 2002. doi: 10.1046/j.1464-410X.2002.02726.x.
4. Bernheim B. Organoscopy. *Annals of Surgery* 53:764; 1911.
5. Hinselman H. Verbesserung der Inspektionsmoeglichkeiten von Vulva, Vagina und Portio. *Muenchner Medizinische Wochenschrift* 73:1733; 1925.
6. Ronnekleiv E. OptoMed, assignee. Fiber optic probes. US Patent 7003184 B2; 2006.
7. Cizmar T, Dholakia K. Shaping the light transmission through a multimode optical fibre: Complex transformation analysis and applications in biophotonics. *Optics Express* 19(20):18871–18884; 2011.
8. Vydra J, Schoetz GF. Improved all-silica fibers for deep-UV applications. In: *BiOS'99 International Biomedical Optics Symposium*. International Society for Optics and Photonics, pp. 165–175; 1999.
9. Pashinin VP, Konstantinov N, Artjushenko VG, Konov VI, Silenok AS, Muller G, Schaldach B, Ulrich R. Mechanism of UV laser-induced absorption in fused silica fibers. *Fiber and Integrated Optics* 10(4):365–372; 1991.
10. Fabian H, Grzesik U, Worner KH, Henschel H, Kohn O, Schmidt HU. Radiation resistance of optical fibers: Correlation between UV attenuation and radiation-induced loss. *Proceedings of SPIE—the International Society for Optical Engineering* 1791:297–305; 1993.
11. Grant L, Schoetz G, Vydra J, Fabricant DG. Optical fiber for UV-IR broadband spectroscopy. In: *Astronomical Telescopes & Instrumentation*. International Society for Optics and Photonics, pp. 884–891; 1998.
12. Muller G, Kar H, Dorschel K, Ringelhan H. Transmission of short pulsed high power UV laser radiation through depending on pulse length, intensity and long term behaviour. *SPIE Conference on Optical Fibers in Medicine*, pp. 231–236; 1988.
13. Rol P, Niederer P. High-power laser transmission through optical fibers: Applications to ophthalmology. In: Wolbarsht M, ed. *Laser Applications in Medicine and Biology*, Vol. 5. New York: Plenum Press. pp. 141–198; 1991.
14. Rol P, Utzinger U, Beck D, Niederer P. Fiber beam shaping and ophthalmic applications. In: *International Symposium on Biomedical Optics Europe'94*. International Society for Optics and Photonics, pp. 56–62; 1995.

15. Greek L, Schulze H, Blades M, Haynes C, Klein K, Turner R. Fiber-optic probes with improved excitation and collection efficiency for deep-UV Raman and Raman resonance spectroscopy. *Applied Optics* 37(1):170–180; 1998.
16. Shim MG, Wilson BC, Marple E, Wach M. Study of fiber-optic probes for in vivo medical Raman spectroscopy. *Applied Spectroscopy* 53(6):619–627; 1999.
17. Wach ML. Visionex, assignee. Fiber optic interface for optical probes with enhanced photonic efficiency, light manipulation, and stray light rejection. US Patent 5901261; 1999.
18. Manoukian N, Kele K, Kermode JR. Fiber optic probe. US Patent 5304172; 1994.
19. Rol P. Optics for transscleral laser applications [Dissertation]. Zurich, Switzerland: Swiss Federal Institute of Technology; 1992.
20. Ward H. Molding of laser energy by shaped optic fiber tips. *Laser in Surgery and Medicine* 7:405–413; 1987.
21. Nica A. Lens coupling in fiber-optic devices: Efficiency limits. *Applied Optics* 20(18):3136–3145; 1981.
22. Bonnema GT, Cardinal KO, Williams SK, Barton JK. A concentric three element radial scanning optical coherence tomography endoscope. *Journal of Biophotonics* 2(6–7):353–356; 2009.
23. Roy A. Concentrating collectors, solar energy conversion. An introductory course. In: Dixon A, Leslie J, eds. *Solar Energy Conversion: An Introductory Course: Selected Lectures from the 5th Course on Solar Energy Conversion*. Toronto, Ontario, Canada: Pergamon, pp. 185–252; 1979.
24. Wall RA, Bonnema GT, Barton JK. Novel focused OCT-LIF endoscope. *Biomedical Optics Express* 2(3):421–430; 2011.
25. Trujillo EV, Sandison DR, Utzinger U, Ramanujam N, Mitchell MF, Richards-Kortum R. Method to determine tissue fluorescence efficiency in vivo and predict signal-to-noise ratio for spectrometers. *Applied Spectroscopy* 52(7):943–951; 1998.
26. O'Rourke PE, Toole WR. Fiber optic probe. US Patent 5774610; 1998.
27. Zuluaga AF, Utzinger U, Durkin A, Fuchs H, Gillenwater A, Jacob R, Kemp B, Fan J, Richards-Kortum R. Fluorescence excitation emission matrices of human tissue: A system for in vivo measurement and method of data analysis. *Applied Spectroscopy* 53(3):302–311; 1999.
28. Nichols MG, Hull EL, Foster TH. Design and testing of a white-light, steady-state diffuse reflectance spectrometer for determination of optical properties of highly scattering systems. *Applied Optics* 36(1):93–104; 1997.
29. Utzinger U. Selective coronary excimer laser angioplasty [Dissertation]. Zurich, Switzerland: Swiss Federal Institute of Technology; 1995.
30. Pfefer TJ, Schomacker KT, Nishioka NS. Effect of fiber optic probe design on fluorescent light propagation in tissue. *Proceedings of SPIE—The International Society for Optical Engineering* 2(14):410–416; 2001.
31. Keijzer M, Richardskortum RR, Jacques SL, Feld MS. Fluorescence spectroscopy of turbid media—Autofluorescence of the human aorta. *Applied Optics* 28(20):4286–4292; 1989.
32. Avrillier S, Tinet E, Ettori D, Tualle JM, Gelebart B. Influence of the emission-reception geometry in laser-induced fluorescence spectra from turbid media. *Applied Optics* 37(13):2781–2787; 1998.
33. Cooney TF, Skinner HT, Angel SM. Comparative study of some fiber-optic remote Raman probe designs. 2. Tests of single-fiber, lensed, and flat- and bevel-tip multi-fiber probes. *Applied Spectroscopy* 50(7):849–860; 1996.
34. Cooney TF, Skinner HT, Angel SM. Comparative study of some fiber-optic remote Raman probe designs. 1. Model for liquids and transparent solids. *Applied Spectroscopy* 50(7):836–848; 1996.
35. Electro-optical sensors for the in vivo detection of cervical cancer and its precursors: Submission guidance for an IDE/PMA, draft guidance—Not for implementation, US Department of Health and Human Services, Food and Drug Administration, Center for Devices and Radiological Health, Obstetrics and Gynecology Devices Branch, Division of Reproductive, Abdominal, Ear, Nose and Throat, and Radiological Devices, and Office of Device Evaluation. Available at http://www.fda.gov/OHRMS/DOCKETS/98fr/992211gd.pdf.

36. Biological evaluation of medical devices, FDA ODE Memo #G95-1, ISO 10993, Office of Device Evaluation, US Food and Drug Administration; 1995.

37. American Conference of Governmental Industrial Hygienists. TLVs and BEIs: Threshold limit values for chemical substances and physical agents and biological exposure indices. Cincinnati, OH: ACGIH Worldwide; 1996.

38. American National Standards Institute. *American National Standard for Safe Use of Lasers*. Vol. ANSI Z136. New York: ANSI. pp. 1–1993; 1993.

39. Guidelines on limits of exposure to laser radiation of wavelengths between 180 nm and 1,000 mu m. *Health Physics* 71(5):804–819; 1996.

40. Matthes R, Cain CP, Courant D, Freund DA, Grossman BA, Kennedy PA, Lund DJ. et al. Revision of guidelines on limits of exposure to laser radiation of wavelengths between 400 nm and 1.4 mu m. *Health Physics* 79(4):431–440; 2000.

41. McKinlay AF, Bernhardt JH, Ahlbom A, Cesarini JP, de Gruijl FR, Hietanen M, Owen R. et al. Guidelines on limits of exposure to ultraviolet radiation of wavelengths between 180 nm and 400 nm (incoherent optical radiation). *Health Physics* 87(2):171–186; 2004.

42. Guidelines on limits of exposure to broad-band incoherent optical radiation (0.38 to 3 mu m). *Health Physics* 73(3):539–554; 1997.

43. Bronzino JD, Peterson DR. *The Biomedical Engineering Handbook*. Boca Raton, FL: Taylor & Francis; 2014.

44. Sinfosky E. *High Power Diffusion Tip Fibers for Photocoagulation*. In: *Lasers and Electro-Optics Society Annual Meeting, 1996. LEOS 96. IEEE*, vol. 1, p. 263; 1996.

45. Welch AJ, van Gemert MJC, Star WM, Wilson BC. Definition and overview of tissue optics. In: Welch AJ, van Gemert MJC, eds. *Optical-Thermal Response of Laser-Irradiated Tissue*. New York: Plenum Press. pp. 15–46; 1995.

46. Zijlstra WG, Buursma, A, Meeuwsen-van der Roest WP. Absorption spectra of human fetal and adult oxyhemoglobin, de-oxyhemoglobin, carboxyhemoglobin, and methemoglobin. *Clinical Chemistry* 37(9):1633–1638; 1991.

47. Zijlstra WG, Maas AH, Moran RF. Definition, significance and measurement of quantities pertaining to oxygen carrying properties of human blood. *Scandinavian Journal of Clinical and Laboratory Investigation* 224:27–45; 1996.

48. Farrell TJ, Patterson MS, Wilson B. A diffusion theory model of spatially resolved, steady-state diffuse reflectance for the noninvasive determination of tissue optical properties in vivo. *Medical Physics* 19(4):879–888; 1992.

49. Kienle A, Lilge L, Patterson MS, Hibst R, Steiner R, Wilson BC. Spatially resolved absolute diffuse reflectance measurements for noninvasive determination of the optical scattering and absorption coefficients of biological tissue. *Applied Optics* 35(13):2304–2314; 1996.

50. Wang L, Jacques SL. Use of a laser beam with an oblique angle of incidence to measure the reduced scattering coefficient of a turbid medium. *Applied Optics* 34(13):2362–2366; 1995.

51. Lin S-P, Wang L, Jacques SL, Tittel FK. Measurement of tissue optical properties by the use of oblique-incidence optical fiber reflectometry. *Applied Optics* 36(1):136–143; 1997.

52. Wallace VP, Bamber JC, Crawford DC, Ott RJ, Mortimer PS. Classification of reflectance spectra from pigmented skin lesions—A comparison of multivariate discriminant analysis and artificial neural networks. *Physics in Medicine and Biology* 45(10):2859–2871; 2000.

53. Dam JS, Pedersen CB, Dalgaard T, Fabricius PE, Aruna P, Andersson-Engels S. Fiber-optic probe for noninvasive real-time determination of tissue optical properties at multiple wavelengths. *Applied Optics* 40(7):1155–1164; 2001.

54. Sun J, Fu K, Wang A, Lin AWH, Utzinger U, Drezek R. Influence of fiber optic probe geometry on the applicability of inverse models of tissue reflectance spectroscopy: Computational models and experimental measurements. *Applied Optics* 45(31):8152–8162; 2006.

55. Mourant JR, Bigio IJ, Jack DA, Johnson TM, Miller HD. Measuring absorption coefficients in small volumes of highly scattering media: Source-detector separations for which path lengths do not depend on scattering properties. *Applied Optics* 36(22):5655–5661; 1997.

56. Jarry G, Steimer E, Damaschini V, Epifanie M, Jurczak M, Kaiser R. Coherence and polarization of light propagating through scattering and biological tissues. *Applied Optics* 37(31):7357–7367; 1998.

57. Jacques SL, Ostermeyer MR, Wang L, Stephens DV. Polarized light transmission through skin using video reflectometry: Toward optical tomography of superficial tissue layers. *Photonics West'96*. International Society for Optics and Photonics, pp. 199–210; 1996.

58. Johnson T, Mourant J. Polarized wavelength-dependent measurements of turbid media. *Optics Express* 4(6):200–216; 1999.

59. Myakov A, Nieman L, Wicky L, Utzinger U, Richards-Kortum R, Sokolov K. Fiber optic probe for polarized reflectance spectroscopy in vivo: Design and performance. *Journal of Biomedical Optics* 7(3):388–397; 2002.

60. Carey P. *Biochemical Applications of Raman and Resonance Raman Spectroscopy*. New York: Elsevier; 1982.

61. Baraga JJ, Feld MS, Rava RP. In situ optical histochemistry of human artery using near infra-red Fourier transform Raman spectroscopy. *Proceedings of the National Academy of Science USA* 89:3473–3477; 1992.

62. Berger AJ, Itzkan I, Feld MS. Feasibility of measuring blood glucose concentration by near-infrared Raman spectroscopy. *Spectrochimica Acta Part A—Molecular Spectroscopy* 53A(2):287–292; 1997.

63. Mahadevan-Jansen A, Mitchell MF, Ramanujam N, Malpica A, Thomsen S, Utzinger U, Richards-Kortum R. Near-infrared Raman spectroscopy for in vitro detection of cervical precancers. *Photochemistry and Photobiology* 68(1):123–132; 1998.

64. de Lima CJ, Sathaiah S, Silveira L, Zangaro RA, Pacheco MT. Development of catheters with low fiber background signals for Raman spectroscopic diagnosis applications. *Artificial Organs* 24(3):231–234; 2000.

65. Shim MG, Song LM, Marcon NE, Wilson BC. In vivo near-infrared Raman spectroscopy: Demonstration of feasibility during clinical gastrointestinal endoscopy. *Photochemistry and Photobiology* 72(1):146–150; 2000.

66. Myrick ML, Angel SM, Desiderio R. Comparison of some fiber optic configurations for measurement of luminescence and Raman scattering. *Applied Optics* 29(9):1333–1344; 1990.

67. Myrick M, Angel S. Elimination of background in fiber-optic Raman measurements. *Applied Spectroscopy* 44(4):758–763; 1990.

68. Nave S, O'Rourke P, Tool W. Sampling probes enhance remote chemical analysis. *Laser Focus World* 31:83–88; 1995.

69. O'Rourke PE, Toole WR. Rugged fiber optic probe for Raman measurement. US patent 5710626; 1998.

70. Mahadevan-Jansen A, Mitchell MF, Ramanujam N, Utzinger U, Richards-Kortum R. Development of a fiber optic probe to measure NIR Raman spectra of cervical tissue in vivo. *Photochemistry and Photobiology* 68(3):427–431; 1998.

71. Utzinger U, Heintzelman DL, Mahadevan-Jansen A, Malpica A, Follen M, Richards-Kortum R. Near-infrared Raman spectroscopy for in vivo detection of cervical precancers. *Applied Spectroscopy* 55(8):955–959; 2001.

72. Tanaka K, Pacheco MTT, Brennan JF, III, Itzkan I, Berger AJ, Dasari RR, Feld MS. Compound para-bolic concentrator probe for efficient light collection in spectroscopy of biological tissue. *Applied Optics* 35(4):758–763; 1996.

73. Owen H, Battey D, Pelltier M, Slater J. *New Spectroscopic Instrument Based on Volume Holographic Optical Elements*. San Jose, CA: SPIE. pp. 260–267; 1995.

74. Ramanujam N, Mitchell MF, Mahadevan A. et al. Fluorescence spectroscopy: A diagnostic tool for Cervical Intraepithelial Neoplasia (CIN). *Gynecologic Oncology* 52:31–38; 1994.

75. Mourant JR, Bigio IJ, Boyer J, Lacey J, Johnson T, Conn RL, Bohorfousch AG, Ross AB. Diagnosis of tissue pathology with elastic scattering spectroscopy. In: *Lasers and Electro-Optics Society Annual Meeting, 1994. LEOS'94 Conference Proceedings. IEEE*, vol. 1, pp. 17–18; 1994.

76. Cothren RM, Richards-Kortum R, Sivak MV, Fitzmaurice M, Rava RP, Boyce GA, Doxtader M. et al. Gastrointestinal tissue diagnosis by laser-induced fluorescence spectroscopy at endoscopy. *Gastrointestinal Endoscopy* 36(2):105–111; 1990.

77. Pitris C, Mitchell-Follen M, Richards-Kortum R, Thomsen S, Wright T, Malpica A, Ramanujam N, Mahadevan A. Multipixel fluorescence instrumentation and clinical study for the diagnosis of the precancer and cancer of the human cervix. *American Society for Lasers in Medicine and Surgery Annual Meeting*, San Diego, CA; 1995.

78. Agrawal A. Multi-pixel fluorescence spectroscopy for the diagnosis of cervical precancer [master's thesis]. Austin, TX: The University of Texas at Austin; 1998.

79. Zeng HS, Petek M, Zorman MT, McWilliams A, Palcic B, Lam S. Integrated endoscopy system for simultaneous imaging and spectroscopy for early lung cancer detection. *Optics Letters* 29(6):587–589; 2004.

80. Tumlinson AR, Hariri LP, Utzinger U, Barton JK. Miniature endoscope for simultaneous optical coherence tomography and laser-induced fluorescence measurement. *Applied Optics* 43(1):113–121; 2004.

81. Tumlinson A, Hariri L, Barton J. Miniature endoscope for a combined OCT-LIF system. *Coherence Domain Optical Methods and Optical Coherence Tomography in Biomedicine Vii* 4956:129–138; 2003.

82. Kyrish M, Utzinger U, Descour MR, Baggett BK, Tkaczyk TS. Ultra-slim plastic endomicroscope objective for non-linear microscopy. *Optics Express* 19(8):7603–7615; 2011.

83. Wall RA, Barton JK. Fluorescence-based surface magnifying chromoendoscopy and optical coherence tomography endoscope. *Journal of Biomedical Optics* 17(8):086003; 2012.

84. Barretto RPJ, Messerschmidt B, Schnitzer MJ. In vivo fluorescence imaging with high-resolution microlenses. *Nature Methods* 6(7):U511–U561; 2009.

85. Udovich JA, Kirkpatrick ND, Kano A, Tanbakuchi A, Utzinger U, Gmitro AF. Spectral background and transmission characteristics of fiber optic imaging bundles. *Applied Optics* 47(25):4560–4568; 2008.

8

Laser and Optical Radiation Safety in Biophotonics

Robert J. Landry
US Food and Drug Administration

T. Joshua Pfefer
US Food and Drug Administration

Ilko K. Ilev
US Food and Drug Administration

8.1 Introduction

Biophotonics is an emerging field in modern biomedical technology that has opened up new horizons for many unique clinical applications in various areas ranging from minimally invasive diagnostics, bioimaging, and biosensors to biomaterial analysis and therapeutics. Recently, biophotonic techniques have been developed as alternatives to conventional medical methods for disease diagnosis, staging, interoperative monitoring, and therapeutics, as well as for drug discovery, proteomics, and environmental detection of biological agents. These techniques offer noncontact, effective, fast, and painless ways to sense and monitor various biomedical quantities. Minimally invasive biophotonic devices are rapidly finding their way into the mainstream and improving patient care and quality of life. However, in order to translate these advanced biomedical technologies from the basic *bench* laboratory studies to the patient *bedside*, a thorough evaluation of optical radiation safety issues is required. In this chapter, we will review terminology for laser and optical radiometry, optical radiation damage mechanisms, as well as specific safety standards and methodologies for evaluating photobiological effects.

Optical radiation from lasers and noncoherent sources such as light-emitting diodes (LEDs) and broadband lamps can be hazardous to the skin and eye.[1] Damage can be incurred via photothermal, photochemical, and photomechanical processes. Photothermal damage (i.e., burns) can result from a single laser pulse or continuous wave (CW) exposure. The amount of damage is dependent upon numerous factors including the area of the tissue exposed, exposure duration, and light source intensity. In a nonthermal mechanism, photon energies are high enough to break chemical bonds. This type of damage is referred to as photochemical damage. It is cumulative and related to the total exposure rather than the exposure intensity. Finally, pulsed laser radiation can be sufficiently intense to produce tissue vaporization, cavitation, and/or fragmentation of tissue. This type of damage is referred to as photomechanical damage. These mechanisms are described in greater detail in Section 8.3.

A number of laser and optical radiation safety standards have been developed to protect against eye and skin injury. The primary focus for most of these standards is accidental exposure, although there are specific standards for ophthalmic instruments where intentional ocular exposures occur. These include a general standard for ophthalmic instruments as well as specific standards for direct and indirect ophthalmoscopes, slit lamps, fundus cameras, operation microscopes, and endoilluminators.[2–8] While many of these instruments use intense light to provide for easier or more accurate diagnoses, they are also potentially hazardous.

As with other medical devices, optical systems may pose a risk of injury or death due to factors such as electrical and toxic chemicals with dye lasers, compressed and toxic gas hazards, cryogenic fluids, radio-frequency radiation, fumes, plasma emissions, explosion, and ionizing radiation emissions.[9] While these potential hazards are applicable to a variety of devices, optical devices represent unique hazards because of the interaction of light with biological tissue. For this reason, we restrict our attention in this chapter to optical radiation hazards. Specifically, we present an overview of terminology and light damage mechanisms, radiometry, laser and noncoherent optical radiation safety standards, and methodologies for evaluating optical radiation hazards.

8.2 Terminology

Relevant terminology for hazard evaluation of lasers and noncoherent light sources includes a variety of spectral and radiometric terms and quantities. Since biological effects of light are wavelength dependent, light or optical radiation must be quantitatively described with radiometric quantities. Therefore, an understanding and implementation of established terminology is critical for performing and communicating rigorous, scientifically based assessments of laser and optical radiation safety.

8.2.1 Wavelength Ranges

The portion of the electromagnetic spectrum referred to as light or optical radiation includes ultraviolet (UV), visible (VIS), and infrared (IR) bands. It covers the wavelength range of approximately 100 nm to 1 mm. Table 8.1 lists the Commission de l'Eclairage (CIE) or the International Commission on Illumination definitions for the photobiological wavelength ranges.[10] It is important to note that in some cases different organizations use different wavelength ranges. For example, the CIE upper wavelength in

TABLE 8.1 Optical Radiation Wavelength Ranges

Radiation Term	Wavelength Range
UV C	100–280 nm
UV B	280–315 nm
UV A	315–400 nm
VIS	360–400 to 760–830 nm
IR A (near)	780–1400 nm
IR B (mid)	1400–3000 nm
IR C (far)	3000 nm to 1 mm

the VIS wavelength range is 760–830 nm, while the Food and Drug Administration (FDA) Federal Laser Product Performance Standard (21 CFR 1040.10 and 21 CFR 1040.11) defines this limit as 710 nm.[11] Except under special conditions, the human eye does not easily perceive light beyond 710 nm.

It is important to note that the spectral boundaries between the UV and VIS and the VIS and IR bands are not specific wavelengths. This occurs because visual sensitivity falls off gradually rather than abruptly at a single wavelength. The range of VIS wavelengths therefore depends to some extent on the intensity of the light source. However, specific wavelength end points have been selected in the various standards for purposes of standardization.

8.2.2 Radiometric and Photometric Quantities

8.2.2.1 Radiometric Quantities

The radiometric quantities used to describe optical radiation are listed in Table 8.2. Radiant energy (Q) is energy in the form of electromagnetic waves with units of joules (J). Radiant power (Φ) is the energy flux (time rate of flow of energy emitted from a source with units of joules per second [J/s] or watts [W]).

Irradiance (E) is the radiant power that is incident on a surface per unit area of that surface and has units of W/cm² (see Figure 8.1). Radiant exposure (H) is the energy incident on a surface per unit area of surface and has units of J/cm². Thus, for a CW light source of constant power, radiant exposure is determined by multiplying the irradiance by the exposure time, t. In addition, a commonly used term in the field of optical therapeutics including low-level laser therapy (LLLT), photodynamic therapy (PDT), laser tissue ablation and surgery, and light–tissue interactions is laser radiation dose or, simply, dose. Since the optical penetration in tissue is generally superficial, the dose is defined and utilized as the quantity radiant exposure, and it has the same units of J/cm².[1,12,13]

Radiance (L) is the radiant power emitted from the surface of a source per unit area of the source and per unit solid angle in steradians (sr), as shown in Figure 8.2. It is a property of the radiating source and cannot be increased by optical elements such as lenses or mirrors. Since lasers are a point source, they are considered to have an infinite radiance. However, the reflections of a laser beam from a diffuse surface result in dispersing the radiation such that the radiance is no longer infinite.

TABLE 8.2 Selected Radiometric Quantities

Radiation Quantity	Unit
Radiant energy (Q)	Joules (J)
Radiant power (Φ)	Watts (W)
Irradiance (E)	W/cm²
Radiant exposure (H)	J/cm²
Radiance (L)	W/(cm² sr)
Time-integrated radiance (tL)	J/(cm² sr)

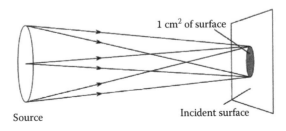

FIGURE 8.1 Sketch illustrating irradiance.

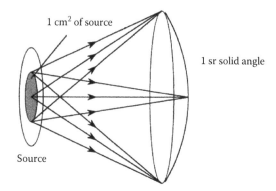

FIGURE 8.2   Sketch illustrating radiance.

Radiance is given by the following expression:

$$L = \frac{\Phi}{A \cdot \Omega},$$

(8.1)

where
   A is the area of the radiating source
   Ω is the solid angle of emission

From geometry, the solid angle Ω is given by the following expression:

$$\Omega = \frac{a}{d^2},$$

(8.2)

where a is the area on the surface of a sphere subtended by the solid angle at a distance d from the radiating source at the center of the sphere. Substituting for Ω from Equation 8.2 into 8.1 yields the following expression for radiance:

$$L = \frac{\Phi \cdot d^2}{A \cdot a}$$

(8.3)

Thus, the radiance can be determined by measuring the radiant power that is transmitted through two apertures spaced a distance d apart. We will describe its use in Section 8.6.3 in determining the radiance for a homogeneous diffuse source to better understand how to use this equation to determine the radiance for a source.

   Time-integrated radiance (tL) is the energy equivalent of radiance. While lasers emit monochromatic optical radiation, the optical radiation from noncoherent light sources is described by a spectral power distribution. For noncoherent light sources, the spectral quantities—spectral irradiance, spectral radiant exposure, etc.—are used to describe the optical radiation. For example, spectral irradiance is the irradiance per unit wavelength with units of W/(cm²·nm). A more complete description of radiometric quantities can be found in the literature.[14]

   Figure 8.3 shows spectral irradiance curves for (a) a tungsten halogen lamp and (b and c) a fluorescent lamp. Figure 8.3b is a linear plot, while Figure 8.3c is a semilog plot. From these figures, it can be seen that a semilog plot is often necessary to display all of the radiation emitted. This is particularly

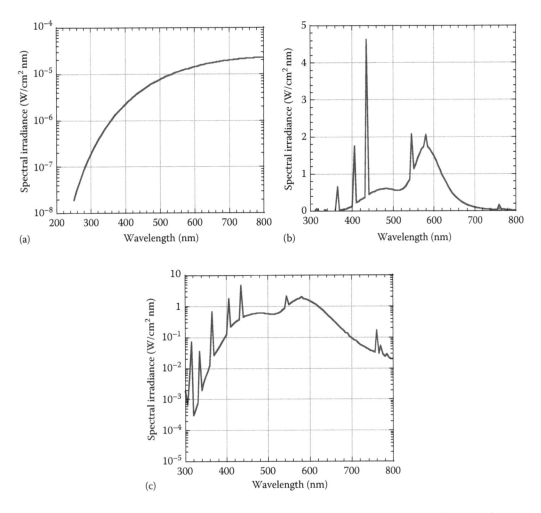

FIGURE 8.3 Plots of spectral irradiance: (a) semilog plot for a tungsten halogen lamp at a distance of 50 cm, (b) linear plot of spectral irradiance for a fluorescent lamp at a distance of 100 cm, and (c) semilog plot of spectral irradiance for the same fluorescent lamp shown in (b).

important in the UV spectral range where photobiological effects from trace amounts of UV B may dominate the bioeffect as further explained in Section 8.3.2.

8.2.2.2 Photometric Quantities

The wavelength dependence of optical radiation to produce a biological effect is described by an *action spectrum*. It is usually presented as a tabulation of the ratio of the radiation dose at specific wavelengths to the dose at the most effective wavelength for producing the biological effect. For example, the standard action spectrum for daylight vision is described by the *photopic* visual response function. This function describes the efficiency of different wavelengths of light to produce vision and is shown in Figure 8.4.[14] The semilog plot (b) shows that vision extends down to 380 nm and up to 770 nm. However, the sensitivity at these wavelengths is much reduced when compared to the sensitivity between 450 and 650 nm. The plots also show that green light is more effective in stimulating daylight vision than either red or blue light.

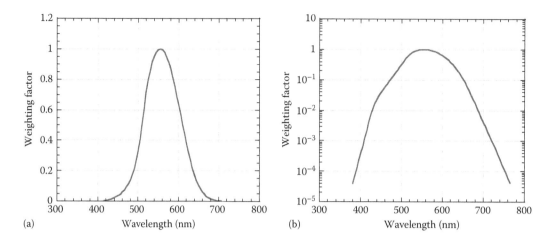

(a)   (b)

FIGURE 8.4   Plots of photopic visual response: (a) linear plot and (b) semilog plot.

When the spectral radiometric quantities are weighted with the photopic visual response function, we get the photometric quantities. For example, visually weighted radiant power becomes luminous flux, with units of lumens (Equation 8.4). In this equation, a correction factor of 683 lumens/W is used to convert the radiometric units to photometric units.[14] Illuminance is the visually weighted equivalent of irradiance with units of lumens/cm², lumens/m² = lux, or foot-candles. Luminance is the visually weighted equivalent of radiance:

$$\Phi_V = 683 \int_{380}^{770} \Phi_{e\lambda} \cdot V(\lambda) d\lambda = 683 \sum_{380}^{770} \Phi_{e\lambda} \cdot V(\lambda) \cdot \Delta\lambda \tag{8.4}$$

Equations similar to Equation 8.4 are used to convert the radiometric quantities for illuminance and luminance. A more complete description of photometric quantities can be found in the book by McCluney.[14]

When evaluating potential optical radiation hazards, as for vision, it is necessary to take the spectral weighting function into account. The resulting weighted radiometric quantities are then prefaced with the word *weighted* (e.g., weighted irradiance, weighted radiance). Sometimes it is sufficient to measure the photometric quantity to determine the potential hazards that may be associated with the exposure to VIS light as will be described in Section 8.3. All of the radiometric and photometric terms and quantities are important for evaluating laser and optical radiation hazards.

8.3 Optical Radiation Hazards and Tissue Damage Mechanisms

In the field of biophotonics, it is important to understand the diverse biological effects of light when evaluating the potential hazards associated with lasers and other light sources. As noted earlier, the biological effects of light are wavelength dependent, and intense light levels can produce damage to both skin and eyes.

8.3.1 Biological Effects of Light

Lasers and noncoherent sources of optical radiation present unique hazards, typically resulting from the absorption of light by biological tissue. Potential optical radiation hazards can be categorized by the physical origin of the damage mechanism as photochemical, photothermal, or photomechanical effects. These mechanisms are reviewed in the following sections.

8.3.2 Photochemical Damage

Photochemical damage occurs when the direct absorption of light results in a chemical or biomolecular reaction. This typically is due to the high energy level of shorter UV radiation and short wavelength VIS light in the blue and violet wavelength range. These hazards are unique when compared to other optical damage processes in that they depend on total dose. In this regard, reciprocity holds over a large time domain for this type of damage, and the effect is relatively independent of dose rate. Repeated photochemical insult can result in significant damage including carcinogenesis.

8.3.2.1 Photochemical Damage to Skin

The phototoxicity of UV radiation in skin is well known. Further, the ability of solar UV radiation to generate DNA damage resulting in skin cancer has also been recognized for some time. There are two primary mechanisms for UV damage: one governs UV A1 (340–400 nm), and another governs UV A2 (315–340 nm), UV B (280–315 nm), and UV C (100–280 nm); see Table 8.1.[15] The latter mechanism involves the absorption of radiation by DNA and RNA, causing adjacent pyrimidine bases to bond and form DNA lesions such as cyclobutane pyrimidine dimers (CPDs) and 6-4 pyrimidine photoproducts. CPDs are considered to be the primary damage mechanism because of their greater frequency and slower repair rate. Lesions that are not repaired can cause *signature* mutations in DNA sequences, such as cytosine to thymine transitions.[16] In addition to DNA damage, these wavelengths can produce erythema, melanogenesis, and thickening of skin layers.[17] It should be noted that erythema has been correlated with DNA damage.[18] Several mechanisms act to prevent the progression from DNA damage to carcinogenesis. The first is DNA repair through mechanisms such as nucleotide excision repair. If DNA is not repaired, then the p53 protein, which is activated upon UV damage, acts to initiate two other mechanisms. The first of these is an arrest of the cell growth cycle, followed by DNA repair. If, however, DNA damage is too severe for repair, p53 acts to initiate apoptosis. Failing repair and apoptosis, transmission of mutations to daughter cells may occur, resulting in transformation and carcinogenesis. The dominant damage mechanism of UV A1 radiation is not as well established as that of shorter UV wavelengths. However, it is believed that UV A1 produces reactive oxygen species through interaction with endogenous photosensitizers such as flavins and urocanic acid.[19] The ensuing production of free radicals leads to the mutagenic lesion 8-hydroxydeoxyguanine, which in turn produces DNA mutations as well as damage to membranes and other cellular constituents.[20]

8.3.2.2 Photochemical Damage to Eye

There are three primary types of photochemical hazards unique to the eye. These injury mechanisms in the cornea, lens, and retina can all be attributed to UV radiation or to VIS light. Injury to the cornea caused by UV B/C radiation is known as photokeratitis, which causes epithelial cell damage that triggers corneal inflammation, clouding, and erythema, among other symptoms. Exposure thresholds for UV B-induced corneal damage are in the 10–100 J/cm^2 range.[21] A similar effect produced in the conjunctiva, or eyelid membranes, is referred to as photoconjunctivitis.[17] This injury is essentially equivalent to a sunburn in highly sensitive tissues of the eye. While painful, these injuries tend to resolve within 24 h and do not have long-term effects.

To describe the wavelength dependence of optical radiation to produce a biological effect, an *action spectrum* is usually utilized, which, as described earlier, is presented as a tabulation of the ratio of the dose at specific wavelengths to the dose at the most effective wavelength for producing the biological effect. Prior studies have used skin erythema as a damage end point to develop an action spectrum for evaluating the potential of a UV-emitting device to cause photochemical damage in the skin.[22,23]

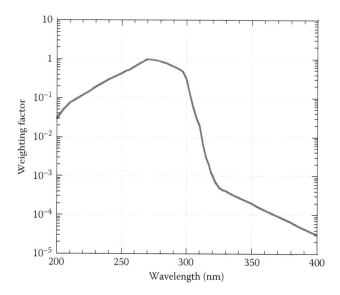

FIGURE 8.5 Plot of action spectrum for photokeratitis.

Pitts performed the pioneering work that initially established the action spectrum for the cornea.[24] Based on data for erythema and photokeratitis, the American Conference of Governmental Industrial Hygienists (ACGIH) has developed guidelines for UV safety that *should not be regarded as fine lines between safe and dangerous levels*.[25] The action spectrum for photokeratitis is shown in Figure 8.5. This action spectrum is of particular use when analyzing broadband sources of optical radiation since it can be used as a weighting function to determine the effective irradiance (E_{eff}) given by the following expression:

$$E_{eff} = \sum_{180}^{400_2} E_\lambda \cdot S(\lambda) \cdot \Delta\lambda \tag{8.5}$$

where
  E_λ is the spectral irradiance, W/(cm²·nm)
  $S(\lambda)$ is the biological weighting factor
  $\Delta\lambda$ is the summation bandwidth

ACGIH guidelines suggest that the product of E_{eff} and the exposure time should not exceed 3.0 mJ/cm².

The lens is at greater risk for damage from UV A radiation due to the greater penetration depth in this wavelength range. A UV A injury to the lens triggers a process involving alterations in lens proteins, formation of fluorescent chromophores, pigmentation, and interference with the synthesis of lens proteins (catalyze-insoluble proteins) leading to cataract formation.[26]

The third type of photochemical hazard is photoretinitis or *blue-light* photochemical injury to the retina. Photoretinitis is produced by excessive irradiation in the 400–550 nm wavelength region, with the most harmful wavelengths being those near 440 nm. Blue light at 435 nm is an order of magnitude more effective at producing photochemical injury to the retina than is radiation at 500 nm. Optical radiation at wavelengths shorter than 435 nm is even more effective in producing retinal damage in an aphakic eye (one in which the lens of the eye is missing). The injury involves a disruption in the

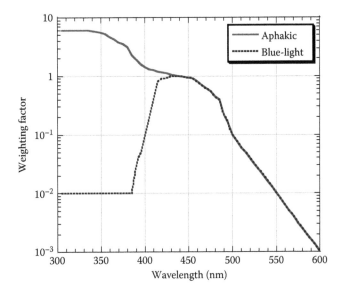

FIGURE 8.6 Plot of aphakic weighting hazard function.

recovery process of photoreceptors, leading to oxidative damage and the formation of a scotoma (an area of diminished vision in the visual field).[27] Limits for preventing photochemical damage are described in several standards including the International Electrotechnical Commission (IEC) Standard IEC 60825-1.[28] The blue-light and aphakic weighting functions are shown in Figure 8.6.

8.3.2.3 Photosensitivity

Abnormal sensitivity to light, or photosensitization, is known to result from specific diseases and the administration of exogenous compounds, including various medications. For many optical devices, these conditions would be a contraindication for use. Diseases that are known to cause photosensitivity include erythropoietic protoporphyria and lupus erythematosus.[29] While the former disease causes a buildup of the photosensitizer protoporphyrin in the skin, the mechanism of the latter is more complex, involving an abnormal and excessive physiological response to UV radiation damage. Photodamage due to absorption by exogenous chromophores is well known from its use in PDT procedures for a variety of applications, such as tumor ablation.[30] The photosensitizers used in PDT procedures, such as porfimer sodium, generate singlet oxygen free radicals, which cause cellular and vascular damage, leading to tissue necrosis. Thus, in these patients, exposure to sunlight or the use of common optical diagnostic devices, such as pulse oximeters, can be dangerous.[31] Photosensitizers can render a patient photosensitive for weeks after a procedure[32] and remain detectable in tissue for months.[33] Numerous medications can also cause photosensitivity including antihistamines, coal tar derivatives, psoralens, and tetracyclines.[34] Some patients taking these medications have extensive reactions when exposed to light, including sunburn-like rashes, hives, swelling, and blistering.[35]

8.3.3 Photothermal Damage

A common safety hazard encountered in biophotonics involves the absorption of radiant energy in tissue and/or conduction from heated surfaces, resulting in temperature rise and thermal injury or burn. Three fundamental heat transfer mechanisms affect the temperature distributions in tissue. They are convection, conduction, and radiation. However, determining the temperature distributions rapidly becomes highly complex and nonlinear when physiological and biochemical factors are considered. These factors include perfusion, phase change, transient environmental conditions, tissue morphology, and optical and thermal properties, as well as dynamic changes in these properties.

The Pennes bioheat equation is typically used to describe heat transfer in biological tissue. It is identical to the standard Fourier heat transfer equation used to calculate heat transfer in a solid body, except that it includes the effect of a generalized heat source/sink due to perfusion:

$$\rho c \frac{\partial T}{\partial \tau} = \nabla \cdot (k \nabla T) + \omega_b c_b \rho_b (T_b - T) + S \tag{8.6}$$

where
ρ is the density
c is the specific heat capacity
T is the temperature
τ is the time
k is the thermal conductivity
ω_b is the volumetric perfusion rate (1/s), the subscript b refers to blood
S is the heat source term (W/cm³)

Optical radiation is refracted, reflected, scattered, and absorbed by tissue. These complex light–tissue interactions are highly dependent on tissue optical properties such as refractive index, scattering, and absorption coefficients, and they vary with wavelength and tissue constituent. Absorbed radiation is converted to thermal energy, which results in temperature rise. The rate of heat generation produced by light propagating at any point in tissue is $S = \mu_a \phi(x,y)$, where ϕ is the local irradiance or fluence rate (W/cm²). Transfer of heat within tissue is typically dominated by conduction and convection, the latter primarily becoming significant near blood vessels and at the tissue surface.

Evaporation at the tissue surface can also act as a significant heat sink. However, the loss of water in tissue can lead to desiccation, which can in turn alter fundamental tissue properties (e.g., thermal conductivity), thus making the problem a dynamic, nonlinear one. Other dynamic changes can be produced as a function of temperature such as optical property variations.[36] Elevated temperatures may also trigger a variety of physiological effects,[37] including heat shock protein generation[38] and thermoregulatory perfusion changes.[39]

Larger increases in temperature can lead to molecular changes, most notably the denaturation of proteins such as collagen, albumin, and hemoglobin. Denaturation can bring about changes in fundamental tissue properties, such as specific heat capacity[40] and optical scattering coefficient.[41,42] Severe thermal damage can alter blood perfusion, with denaturation of vessel wall proteins or blood constituents causing temporary or permanent vessel occlusion.[43] The gold standard for identifying thermally induced tissue damage is histopathological examination. However, highly sensitive techniques for detecting thermal injury involve fluorescence and reflectance spectroscopy[44] and birefringence loss.[45] Polarization-sensitive optical coherence tomography (OCT) has also been shown to be useful for detecting thermal damage in tissue.[46]

In 1947, Henriques and Moritz[47] introduced a method for quantifying and predicting the onset and extent of thermal injury based on the Arrhenius rate process equation:

$$\frac{d\Omega(t)}{dt} = A \exp\left[-\frac{E_a}{RT(t)}\right] \tag{8.7}$$

$$\Omega(t) = \ln\left(\frac{C(t)}{C(0)}\right) = A \int_0^t \exp\left[-\frac{E_a}{RT(t)}\right] dt \qquad (8.8)$$

where

C is the concentration of living cells or molecules in the native state

R is the universal gas constant (8.31 J/mol/K)

the two primary material properties are the frequency factor or molecular collision rate, A (1/s), and the activation energy, E_a (J/mol)[35]

This equation summarizes damage processes in a single, dimensionless parameter (Ω) based on the concept that thermal damage is an exponential function of temperature and a linear function of time (exposure duration). By convention, $\Omega = 1.0$ is the threshold for observable thermal coagulation or a first-degree burn, whereas $\Omega = 10$ and $\Omega = 100$ correspond to second- and third-degree burns, respectively. An Ω value of 1.0 corresponds to a cellular damage concentration of 63%.

Each set of thermal damage parameters (A, E_a) describes the relationship between exposure duration and threshold temperature (the temperature required for coagulation) for a specific type of tissue. Figure 8.7 illustrates the relationship between threshold temperatures and exposure time, as calculated directly from thermal damage parameters for various biological tissues.[36,48–50] The data indicate that thermal damage can be produced by high-temperature exposures of less than 1 ms, such as during pulsed laser irradiation,[51] or by low-temperature, long-term exposures.[52] The 43°C safety limit specified in IEC 60601-1:2005 (Medical Electrical Equipment—Part 1: General Requirements for Basic Safety and Essential Performance)[53] is included in Figure 8.7 for comparison with the calculated threshold values at long exposure times. A more thorough discussion of rate process analyses of laser-induced thermal damage is provided by Pearce and Thomsen.[37]

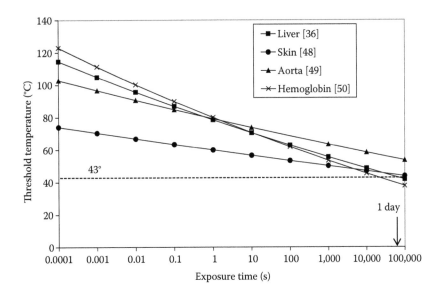

FIGURE 8.7 Relationship between exposure time and threshold temperature for thermal coagulation.

8.3.4 Photomechanical Damage

It is well established that laser irradiation can generate photomechanical effects, resulting in tissue ablation.[54,55] These effects have been successfully exploited to enable minimally invasive surgical techniques for numerous clinical applications, including laser lithotripsy and laser-assisted in situ keratomileusis (LASIK). In recent years, the use of pulsed lasers has become increasingly popular for imaging techniques, such as time-resolved fluorescence, multiphoton spectroscopy, and OCT. Therefore, the same mechanisms that have produced desirable effects during therapeutic procedures now have the potential to generate injuries during diagnostic procedures. While pulsed-laser-induced ablation incorporates a variety of highly complex phenomena that are well delineated elsewhere,[54,55] this section provides an overview of fundamental processes and considerations.

Ablation mechanisms are determined by irradiation parameters such as pulse duration (τ_p), radiant exposure, and wavelength, as well as tissue material properties, such as thermal diffusivity (α), absorption coefficient (μ_a, which is a function of wavelength), and speed of sound (σ). These parameters determine whether the irradiation event satisfies one or both of the key ablation conditions: thermal and stress confinement. A laser pulse is thermally confined when the laser pulse duration is shorter than the thermal diffusion time (τ), defined as

$$\tau = \frac{1}{4\alpha\mu_a^2} \tag{8.9}$$

Thermal confinement represents the condition in which heat diffusion from the irradiated region is minimal during the laser pulse. Stress confinement occurs when the following condition is satisfied:

$$\tau_p < \frac{1}{\alpha\mu_a} \tag{8.10}$$

This represents an irradiation event in which the pulse duration is shorter than the time required for stress waves generated by thermoelastic expansion to exit the irradiated volume. Ablation events can be placed in one of four categories: CW ablation, thermally confined ablation, stress-confined ablation, and plasma-mediated ablation. Each category is defined in terms of the parameters that cause the ablation process and the characteristic effects produced during the ablation event. CW ablation involves laser exposures for which neither stress nor thermal confinement is present. Typically, this involves lasers that provide exposure durations of hundreds of milliseconds or greater (e.g., CO_2 lasers). However, CW ablation can occur for pulses as short as several microseconds when tissue absorption coefficients are high (e.g., Er:YAG lasers). CW ablation is typically characterized by subsurface superheating and steam formation that drives the explosive ejection of tissue, followed by tissue desiccation, carbonization, and ultimately an irregular ablation crater surrounded by a wide margin of thermal damage.

A number of common pulsed lasers (e.g., the free-running Ho:YAG laser with τ_p on the order of hundreds of microseconds) produce ablation events that are thermally but not stress confined. These lasers cause rapid, localized heating that leads to limited thermal damage in adjacent tissue, although the extent of damage can expand significantly with multiple pulses. Rapid heating leads to explosive vaporization, in which expanding vapor and gas eject tissue fragments. Pressure waves are generated both at the onset and during the vaporization process. When such an event occurs in an enclosed environment (e.g., underwater), a vapor bubble can form that causes mechanical damage upon expansion followed by damage due to pressure waves generated upon bubble collapse.

Laser pulses that are both thermally and stress confined (e.g., Q-switched Nd:YAG) cause rapid thermoelastic expansion and tensile stresses. Negative stresses cause a decrease in the threshold required for cavitation, which is a process of laser-induced inner cavity formation (e.g., bubbles), and in turn, it drives material ejection and generation of shock waves. When cavitation occurs in a liquid environment away from any interface, bubbles are generated that tend to be spherical in shape. On collapse of such a highly symmetric bubble, strong pressure shock waves can be produced. This process tends to produce greater mechanical damage than nonstress-confined laser pulses but less thermal damage. For pulse durations of less than 1 ns down to 10 ps, self-focusing can concentrate radiation from a collimated beam and thus lower the ablation threshold.

Plasma-mediated ablation is produced by very short, high-intensity laser pulses (typically nano-, pico-, or femtosecond lasers) that exceed threshold irradiance for laser-induced breakdown (10^9–10^{12} W/cm^2). This type of ablation is unique in that it can occur in nonabsorbing media, given sufficient irradiance levels. Plasma generation in absorbing media involves rapid temperature rise followed by thermionic emission of free electrons, whereas in transparent media, multiphoton processes initiate ionization avalanche. In either case, the resulting plasma can reach temperatures well over 1000 K and generate hundreds of megapascals of pressure. Once formed, plasmas absorb incoming light, thus shielding deeper regions from further irradiation. While the ablation process for laser-induced plasmas is similar to that for stress-confined ablation events involving linear absorption, the mechanical effects produced are more intense (e.g., larger pressure transients, more forceful tissue ejection). Individual ablation events produced by plasma-mediated ablation tend to produce minimal collateral thermal injury. However, when a pulsed laser is used in a high-repetition-rate mode with subablation threshold energy, small increases in thermal energy can accumulate rapidly, resulting in temperature superpositioning and thermal effects that resemble CW laser processes.

8.4 Laser and Noncoherent Optical Radiation Safety Standards

8.4.1 Laser Standards

A list of laser mandatory and voluntary standards is shown in Table 8.3. All laser and laser systems marketed in the United States must be in conformance with the Federal Laser Product Performance Standard (21 CFR 1040.10 and 21 CFR 1040.11).[11] These are mandatory standards that are applicable to manufacturers. CFR 1040.10 is applicable to all laser products. CFR 1040.11 contains additional or modified requirements for specific-purpose laser products. CFR 1040.11(a) is applicable to medical laser products. CFR 1040.11(b) is applicable to surveying, leveling, and alignment laser products. CFR 1040.11(c) is applicable to demonstration (light show or display) laser products.

The basic standard, CFR 1040.10, classifies lasers and laser systems into classes according to the degree of the potential hazard associated with the levels of laser radiation that are accessible during operation. These emission levels are defined as accessible emission limits (AELs). The classification system is summarized in Table 8.4. Class I lasers do not emit levels of radiation that are known to be hazardous and are, therefore, exempt from user control measures. Class II lasers are divided into two classes. The AELs for Class IIa are the same as Class I for all wavelengths and emission durations except for emission durations ≥1000 s in the VIS wavelength range from 400 to 710 nm where the AEL is increased from 0.39 to 3.9 µW. Class II lasers have the same AELs as Class I lasers for all wavelengths and emission durations except for emission durations ≥0.25 s in the VIS wavelength range 400–710 nm where the AEL is 1 mW. Class II lasers emit accessible laser light in the VIS region, and these lasers are capable of producing eye damage through chronic exposure. In general, the human eye will blink or make another aversion response within 0.25 s when exposed to Class II laser light. This aversion reflex normally provides adequate protection. However, it may be possible to overcome a blink reflex and to stare into a Class II laser long enough to cause damage to the eye. Class II lasers have power levels less than 1 mW and are commonly found in alignment applications.

Class III lasers are further divided into two classes. Class IIIa lasers emit accessible laser light in the VIS region and are normally not hazardous when viewed momentarily with the naked eye, but they may

TABLE 8.3 Laser Standards

Standard	Notes
US 21 CFR 1040.10	US mandatory standard applicable to manufacturers. Classifies laser products and specifies testing and performance requirements.
US 21 CFR 1040.11(a)	US mandatory standard applicable to manufacturers. Specifies additional or modified requirements to CFR 1040.10 for medical laser products.
US 21 CFR 1040.11(b)	US mandatory standard applicable to manufacturers. Specifies additional or modified requirements to CFR 1040.10 for surveying, leveling, and alignment laser products.
US 21 CFR 1040.11(c)	US mandatory standard applicable to manufacturers. Specifies additional or modified requirements to CFR 1040.10 for demonstration (light show or display) laser products.
ANSI Z136.1 (2007)	US voluntary standard provides guidance for the safe use of lasers and laser systems. It is the basis for the ANSI laser standards.
ANSI Z136.3 (2011)	US voluntary standard for the safe use of lasers for diagnostic, cosmetic, preventative, and therapeutic applications.
ANSI Z136.4 (2010)	US voluntary standard for laser safety measurements.
ANSI Z136.5 (2009)	US voluntary standard for the safe use of lasers in educational institutions.
ANSI Z136.6 (2005)	US voluntary standard for the safe use of lasers outdoor (laser light shows).
IEC 60825-1 (2011)	*Safety of laser products, IEC, 2007.* The standard is a three-part standard. Two parts have requirements and specifications similar to CFR 1040.10. Standard is mandatory in countries that adopt it. The third part is informative and includes tables and MPEs. Countries can adopt all or only parts of the standard.
IEC 60825-2 (2010)	Fiber-optic communications systems.
IEC 60825-4 (2011)	Laser guards (for high-power industrial laser machines).
IEC 60825-12 (2004)	Free-space optical communications systems.
IEC 60601-2-22 (2012)	Medical laser products.

Note: These standards are reviewed and updated periodically. It is therefore important to ensure that the most recent version is being used.

TABLE 8.4 FDA Laser Standard Classification

Laser Class	Notes
Class I	Levels of laser radiation emitted are not known to be hazardous.
Class IIa	VIS light only—damage produced through chronic exposure, radiant power levels requiring more than thousands of continuous direct viewing to produce damage. Controls same as Class I.
Class II	VIS light only—damage produced through chronic exposure with radiant power levels less than 1 mW.
Class IIIa	VIS light only—not hazardous when viewed momentarily with the naked eye, may pose a hazard when viewed with optical instruments (e.g., binoculars), power levels 1–5 mW.
Class IIIb	Can cause injury by direct viewing of beam and specular reflections, power levels 5–500 mW or less than 10 J/cm² for 0.25 s pulse.
Class IV	Lasers with power levels >500 mW or greater than or equal to 10 J/cm² for 0.25 s pulse—viewing of beam and specular and diffuse reflections can cause eye and skin injuries.

pose severe eye hazards when viewed through optical instruments (e.g., microscopes and binoculars). Class IIIa lasers have power levels of 1–5 mW in the VIS wavelength range for emission durations greater than 3.8×10^{-4} s and are subject to the Class I limits for other wavelengths and emission durations. Class IIIb lasers can cause injury by direct viewing of the beam and specular reflections. The power output of Class IIIb lasers is 5–500 mW CW or, with a radiant exposure, less than 10 J/cm² for a 0.25 s pulse.

Class IV lasers include all lasers with power levels greater than 500 mW CW or, with a radiant exposure, greater than 10 J/cm^2 for a 0.25 s pulse. Viewing of the beam and of specular reflections or exposure to diffuse reflections can cause injuries to both eye and skin. The beams of Class IV lasers may start fires when they impact flammable materials. This standard also specifies measurement conditions for classification and engineering controls as well as labeling requirements.

In addition to the Federal Laser Product Performance Standard, there are a series of voluntary standards for lasers and laser systems. The American National Standards Institute (ANSI) publishes a series of laser safety standards.[56] ANSI Z136.1 (2007) is the parent document. Like the mandatory Federal laser standard, this standard contains AELs for classifying lasers that may be required if, for example, as the result of modifications, the Federal standard classification is no longer valid. It also contains maximum permissible exposure (MPE) levels. The MPE is at levels to which a person may be exposed without hazardous effects to the eye or skin. ANSI Z136.1 serves as the basis of the Z136 laser safety standards, and it provides guidance for the safe use of lasers and laser systems by defining control measures for each of the four laser classes. The other ANSI laser safety standards include standards for health care facilities, educational institutions, outdoor use, and a recommended practice for laser safety measurements.[57–60]

ANSI Z136.3, Safe Use of Lasers in Health Care Facilities, provides guidance for lasers used in diagnostic, cosmetic, preventative, and therapeutic applications. It is applicable for use in all health care facilities where bodily structure or function is changed or where symptoms are relieved.[57] ANSI Z136.5, Safe Use of Lasers in Educational Institutions, is intended for staff and students using lasers for academic instruction in universities, colleges, secondary institutions, or primary educational facilities.[58] In this standard, teaching environments include teaching laboratories, classrooms, lecture halls, science fairs, and science museums.

ANSI Z136.4, Recommended Practice for Laser Safety Measurements, is intended to assist users who have the responsibility of conducting laser hazard evaluations.[60] It includes definitions, examples, and other practical information for manufacturers, laser safety officers, technicians, and other trained laser users. ANSI Z136.6, Safe Use of Lasers Outdoors, covers product performance of lasers used outdoors including laser light shows that have been granted a variance from the provisions of the Federal Laser Product Performance Standard (21 CFR 1040).[59]

The international laser standard IEC 60825-1, published by the IEC, is similar in structure to the FDA laser standard CFR 1040.10 and applies to all laser products.[28] This standard is divided into three parts. The first two parts are mandatory in countries that adopt the standard as a national standard. The requirements and specifications in these two parts are similar to those in the FDA standard. Part three is not mandatory. It is informative and includes tables and MPEs to protect against potentially hazardous exposures. Countries can adopt the entire standard or only parts of the standard. In this regard, the United States has recognized parts of the IEC standard under Laser Notice 50 and may formally adopt parts of the IEC standard that are deemed equivalent with amendments to the FDA standard. However, where there are differences, the FDA standard will take precedence over the IEC standard.

Additionally, IEC also publishes a series of specific-purpose laser standards similar to those in CFR 1040.11. They include IEC 60825-2, Fiber-optic communications systems; IEC 60825-4, Laser guards (for high-power industrial laser machines); IEC 60825-12, Free-space optical communications systems; and IEC 60601-2-22, Medical laser products. Finally, IEC also publishes a technical report, IEC 60825-13, Safety of laser products, Part 13 Measurement for classification of laser products.[61] This technical report provides guidance on methods to perform radiometric measurements or analyses to establish the emission level of laser energy in accordance with IEC 60825-1. It includes information for calculating AELs and MPEs. The IEC standards can be obtained from the US National Committee at www.ansi.org or from the IEC at www.iec.ch or the Laser Institute of America at www.laserinstitute.org.

A number of states also regulate laser products. States that have major programs include Arizona, Illinois, Massachusetts, New York, and Texas. In addition, several states have regulations that specify persons who can perform cosmetic laser surgery. Some municipalities also have regulations addressing the use of laser pointers.

8.4.2 Standards for Broadband Sources

There are also voluntary standards for noncoherent sources of optical radiation (see Table 8.5). The American National Standards Institute/Illuminating Engineering Society of North America (ANSI/IESNA) publishes a series of standards for the photobiological safety of lamps and lamp systems. The purpose of these standards is to inform the public and OEM manufacturers about the potential optical radiation hazards that may be associated with lamps and lamp systems. Another purpose of the standards is to provide guidance for evaluating the potential hazards that may be associated with these devices.

ANSI/IESNA RP-27.1-05, Recommended Practice for Photobiological Safety for Lamps and Lamp Systems—General Requirements, is applicable to all electronically powered sources of optical radiation except for lasers and LEDs used in optical fiber communications.[62] This standard provides for the evaluation and control of optical radiation hazards for the applicable devices. ANSI/IESNA RP-27.2-00, Recommended Practice for Photobiological Safety for Lamps and Lamp Systems—Measurement Techniques, provides measurement guidance for the evaluation of potential hazards from lamps and lamp systems.[63]

ANSI/IESNA RP-27.3-07, Recommended Practice for Photobiological Safety for Lamps—Risk Group Classification and Labeling, is similar to the ANSI laser standard in that it classifies lamps into safety groups. As in the laser safety standards, the ANSI/IESNA RP-27.3 standard classifies lamps into four groups depending upon the degree of hazard associated with the lamps.[64] They include an Exempt Group, Risk Group 1 (low risk), Risk Group 2 (moderate risk), and Risk Group 3 (high risk). Lamps in the Exempt Group do not pose any photobiological hazard for the end points in the standard. Lamps in Risk Group 1 do not pose a hazard under normal behavioral limitation such as exposure to blue light for a period of 100 s. Risk Group 2 lamps do not pose a hazard due to the aversion response associated with bright light sources. Risk Group 3 lamps may pose a hazard for even momentary or brief exposures. There is also an equivalent international standard (IEC 62471) for lamps and lamp systems.[65] This standard is comprehensive and includes the requirements specified in the RP-27 series of standards cited earlier.

TABLE 8.5   Noncoherent Broadband Source Standards

Standard	Notes
ANSI/IESNA RP-27.1-05	US voluntary standard: Recommended Practice for Photobiological Safety for Lamps and Lamp Systems—General Requirements, applicable to all electronically powered sources of optical radiation except for LEDs used in optical fiber communications.
ANSI/IESNA RP-27.2-00	US voluntary standard: Recommended Practice for Photobiological Safety for Lamps and Lamp Systems—Measurement Techniques provides measurement guidance for the evaluation of potential hazards from lamps and lamp systems.
ANSI/IESNA RP-27.3-07	US voluntary standard: Recommended Practice for Photobiological Safety for Lamps—Risk Group Classification and Labeling, similar to the ANSI laser standard in that it classifies lamps into safety groups.
IEC 62471 CIE S 009:2006	International standard: contains requirements equivalent to those in the RP-27 series of standards.
ACGIH TLVs and BEIs:2013	TLVs for chemical substances and physical agents specify TLVs for both lasers and noncoherent light sources.

Note: These standards are reviewed and updated periodically. It is therefore important to ensure that the most recent version is being used.

Finally, as noted earlier, ACGIH has published threshold limit values (TLVs) for chemical substances and physical agents.[25] This publication includes TLVs for both lasers and noncoherent light sources. The ANSI/IESNA standards can be obtained from www.iesna.org, the IEC 62471 standard from www.techstreet.com, and the ACGIH standard from www.acgih.org.

8.5 Limits and Guidelines

It is important to note that many of the guidelines/limits in these standards are fundamentally identical since they are based on the same biological threshold damage data. The different standards are intended to serve different communities. MPE levels and TLVs are limits that were originally developed to provide protection from accidental exposure. However, they may be used as limits/guidance for intentional exposures such as from illumination sources or diagnostic devices.

It is important to note that the MPEs and TLVs incorporate safety factors to minimize the likelihood of hazardous exposures. The degree of the safety factor for a given bioeffect depends on the bioeffect. For example, a safety factor of about 10 is incorporated in the limit for photochemical damage to the retina, since damage to the retina is usually irreversible. The threshold for damage at which 50% of exposures would be expected to result in a minimal photochemical retinal injury is about 22 J/cm² at a wavelength of 435 nm. However, the MPE at 435 nm to prevent photochemical injury is 2.2 J/cm². On the other hand, the threshold for photokeratitis from exposure to UV radiation is about 4 mJ/cm², while the MPE to prevent a minimal injury from photokeratitis is set at 3 mJ/cm². In this case, the safety factor is much smaller than that for photochemical retinal injury because the injury for photokeratitis is temporary.

8.6 Evaluation Methods

8.6.1 General Requirements

To evaluate potential hazards to the eye or skin for unweighted biological end points, it is only necessary to determine the irradiance, radiant exposure, source radiance, or source-integrated radiance for the retinal hazards. However, to evaluate potential hazards for the weighted biological end points, it is necessary to determine weighted irradiance, weighted radiant exposure, weighted source radiance, or weighted source-integrated radiance for the retinal hazards. This is accomplished using the appropriate weighting factors tabulated in the standards. There are excellent examples in the appendices of ANSI Z136.1 for classification of laser products and hazard evaluation.[56]

The process of evaluating the biologically weighted end points for extended laser and noncoherent light sources that emit optical radiation over a wide wavelength range is also straightforward. It is accomplished by calculating a weighted radiometric quantity as noted earlier. For example, the ACGIH blue-light weighted radiance is given by the following expression:

$$L_B = \sum_{305}^{700} L_\lambda \cdot B(\lambda) \cdot \Delta\lambda \tag{8.11}$$

where
L_B is the blue-light weighted retinal radiance
L_λ is the spectral radiance
$B(\lambda)$ is the biological weighting factor at wavelength λ for photochemical injury to the retina
$\Delta\lambda$ is the wavelength summation interval, and the summation is taken over the specified wavelength range from 305 to 700 nm

The other wavelength-dependent dose-related quantities use similar expressions, all of which involve the summation of the product of the spectral quantity times the weighting function over the pertinent wavelength range (e.g., see Equation 8.5). Thus, for example, the thermal weighted radiance is determined by using the $R(\lambda)$ weighting function instead of the $B(\lambda)$ weighting function in Equation 8.11.

8.6.2 Special Requirements: Measurement Techniques for Light Hazard Evaluation

Specific averaging aperture diameters, measurement distances, and instrument acceptance angles are specified in the safety standards to assess optical hazards. Some of the measurement conditions specified in these standards may appear puzzling for a person experienced in general photometry and radiometry. However, they are needed to account for certain factors such as thermal diffusion and pupil diameter and normal eye movements that influence the degree of the light hazards as further explained in Sections 8.6.2.1 through 8.6.2.3.

8.6.2.1 Averaging Apertures

For evaluating thermal hazards to the skin, the ANSI Z136.1 laser standard specifies that a 3.5 mm averaging aperture be used. To evaluate the retinal blue-light hazard exposure limit for small sources, the ANSI/IESNA RP-27.1-05 standard for the photobiological safety of lamps and lamp systems specifies two different apertures depending upon the homogeneity of the source. A limiting aperture no greater than 2.5 cm is to be used for a source that produces a homogeneous beam, while a 7 mm diameter aperture shall be used for a beam that contains hot spots with a diameter less than 2.5 cm. The difference in the averaging apertures in the two standards reflects differences in the size of the beams to be expected at the user exposure position. In general, noncoherent light sources, such as those used for general illumination purposes, will produce larger beams at the user exposure position than those from laser beams.

It is important to note that if the beams are smaller than the averaging apertures specified, the irradiance is averaged over the entire diameter of the specified aperture. Thus, for example, the radiant power in a laser beam that is smaller than 3.5 mm is divided by the area of the 3.5 mm diameter aperture to determine the irradiance to be compared to the exposure limit.

Typically, a 7 mm averaging aperture at the cornea is specified for evaluating the potential hazards to the retina. This is intended to serve as a worst case analysis by taking into account the diameter of a dilated pupil. A normally illuminated environment results in a small pupil diameter, which limits the total radiant power that can enter the eye and thereby reduces the risk of retinal injury especially if the beam is larger than the pupil. Additionally, in some cases, such as in the ANSI/IESNA RP-27.1 standard, a 3 mm diameter averaging aperture is used for evaluating the retinal blue-light exposure hazard for nonsmall sources, while a 7 mm diameter aperture is specified to evaluate the retinal thermal hazard exposure limit for short exposure durations from 1 µs to 10 s. It is, therefore, important to use the aperture specified in the standard being used for the appropriate hazard to be evaluated.

8.6.2.2 Measurement Distances

The distance of the light source from the eye also needs to be considered when evaluating light hazards to the retina. A light source that is located too close so the eye cannot focus it on the retina presents a lower potential hazard than a light source that is within the accommodation range of the eye where it can be brought into sharp focus. The accommodation distance for a normal young adult eye is from about 20 cm to infinity. However, a highly myopic eye or a child's eye may have an accommodation distance of 10 cm or less. Thus, a conservative measurement distance of 10 cm for evaluating hazards was chosen for the ANSI and IEC laser safety standards. On the other hand, ANSI/IESNA RP-27.1 specifies that the measurement distance for the quantity radiance shall be at a minimum accessible viewing distance but no less than 20 cm. In this case, a minimum measurement distance of 20 cm was chosen since lamps are typically not used at distances less than 20 cm. Additionally, many ophthalmic instruments are used at distances closer than the accommodation distance, resulting in unusual ocular exposure geometries. Such instruments must be evaluated at the user distance as described in the paper by Landry et al.[66] It is therefore important to use the measurement distance specified in the standard being used for the specific hazard to be evaluated.

8.6.2.3 Field of View and Eye Movements

Eye movements are also important to consider because the potential hazard is greater for a stabilized eye than for a moving eye. For example, a CW point light source such as a collimated laser beam produces a nearly diffraction-limited spot on the retina of a stabilized eye with a diameter on the order of 30 µm. However, in a normal awake person, there are normal saccadic eye movements. These movements are small, jerky, involuntary eye movements, which have an average excursion of 100 µm. They can result in smearing the diffraction-limited spot into an elliptical shape with a major half-maximum axis of 150 µm and a minor half-maximum axis of 107 µm.[67] The smearing produced by normal eye movements reduces the potential optical radiation hazard when compared to that for a stabilized eye. Additionally, it is important to note that a collimated pulsed light source with a pulse width less than 250 ms will also produce a nearly diffraction-limited spot with a diameter on the order of 30 µm on the retina since the duration of the laser pulse is shorter than any significant eye movements. Therefore, specified measurement conditions for evaluating potential hazards to the eye in both laser and noncoherent optical radiation safety standards typically take eye movement into account.

When evaluating potential hazards to the retina, it is typically necessary to determine source radiance. This is achieved by defining a field of view for the measurement instrument. It is easy to show that a 30 µm spot on the retina corresponds to a field of view of 1.75 mrad. Similarly, a 180 µm spot on the retina corresponds to a field of view of 11 mrad. A field of view of 1.75 mrad is used to determine the potential hazard for a stabilized eye, while a 180 µm field of view is used to determine the potential hazard for an eye that is not stabilized. As noted earlier, it is also necessary to account for eye movements or the lack thereof to obtain an accurate determination of the potential hazard to the eye.

When a limiting field of view is used, a *field-averaged radiance* is determined instead of the actual radiance of the source. In some cases, the field-averaged radiance will be the same as the radiance of the source. In other cases, it will differ from the true radiance of the source. Methods for implementing field-averaged radiance will be described in the examples in Section 8.6.3. It is worth noting that ISO 15004-2 allows for the determination of either retinal irradiance or source radiance for the determination of the potential retinal hazards. The standard clearly specifies the retinal averaging area for the determination of retinal irradiance and a field of view for the determination of source radiance.

It is important to recognize that for a homogeneous beam, the irradiance at any point in the beam will be the same regardless of the area over which the irradiance is averaged. In such a case, the irradiance determined by measuring the radiant power in the whole beam and dividing by the area of the beam will be the same as the radiant power in any portion of the beam measured divided by the area of the portion of the beam measured. Similarly, for a homogeneous light source, even if an instrument has a field of view that is contained within the light source but is larger than the field of view specified by the standard, the same value of radiance will be measured in either case.

8.6.3 Evaluation Methods

It is important for all optical radiation safety evaluations to include the worst case maximum expected exposure duration for CW sources and the maximum expected number of pulses for pulsed light sources. A standard method for evaluating the potential optical radiation hazards to the eye for a diffuse laser beam and noncoherent light source will be reviewed because of the complexities associated with such evaluations.

8.6.3.1 Method for Evaluating Hazards to the Retina from a Diffuse Source

The light source in this example is a 440 nm wavelength 15 mW power multimode laser with a beam diameter of 3 mm that is incident upon a diffusing surface that is intended to be viewed by the eye. If we assume a Lambertian diffuse surface, the radiation will be distributed over π-steradians in accordance with the cosine of the viewing angle. The eye is located at a distance of 10 cm in a perpendicular direction from the light-diffusing surface, and the person exposed may be viewing the surface for a duration of 100 s. The hazard evaluation will be performed in accordance with the specifications in ANSI Z136.[56] It is important to note that the methods described for a diffuse source can be used for any extended or broadband source, including sources with hot spots such as those produced in some LEDs. We have chosen to use a diffuse source in this example simply to illustrate the method.

As noted earlier, the specific measurement requirements include averaging apertures and field of view for measurement instruments. In this example, the measurements of radiance are to be made with an averaging aperture of 7 mm and with an instrument that has a field of view of 11 mrad as specified in Table 8a of the standard for an exposure time of 100 s.[56] It should be noted that the 7 mm diameter aperture assumes a worst case fully dilated pupil and the 11 mrad field of view assumes an awake and task-oriented eye.

As noted in Section 8.2.2.1, the radiance of the source may be determined using Equation 8.3: $L = (\Phi \cdot d^2)/(A \cdot a)$, where Φ is the radiant power, d is the measurement distance, A is the area of the 7 mm diameter measurement aperture, and a is the area of the aperture to define the field of view of the source.

In this case, the 7 mm diameter measurement aperture is located at a distance 10 cm from the light source and over the measurement detector. The radiance is then determined by dividing the product of the radiant power transmitted through the two apertures and the square of the distance between the two apertures by the areas of the two apertures. The dual-aperture method is equivalent to measuring the irradiance $E(\Phi/A)$ at a distance d from the aperture with area, a, that defines the measurement field of view.

Establishing the measurement field of view is accomplished by placing an aperture of area a over the source as illustrated in Figure 8.8. To establish the appropriate field of view of 0.011 rad, we need to determine the diameter of the aperture d_a.

The field of view in this example is given by

$$\text{FOV} = \frac{d_a}{z} \tag{8.12}$$

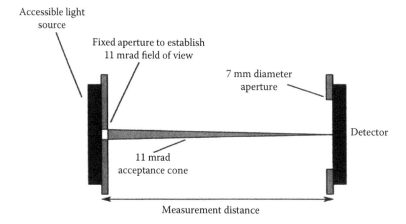

Accessible light source

Fixed aperture to establish 11 mrad field of view

7 mm diameter aperture

Detector

11 mrad acceptance cone

Measurement distance

FIGURE 8.8 Sketch of a light source that is homogeneous showing the 7 mm aperture and the 11 mrad field-of-view distance z from the aperture with area, a, that defines the measurement field of view.

where z is the measurement distance between the two apertures:

$$d_a = 0.011 \times z = 0.011 \times 100 = 1.1 \text{ mm} \tag{8.13}$$

Additionally, the irradiance (E) is measured at a distance $z = 10$ cm from the light source. Since the 1.1 mm diameter aperture blocks out a part of the beam, the measured radiant power is less than the total power by the ratio of the area of the 1.1 mm aperture diameter to the area of the 3 mm diameter beam. Thus, the measured radiant power (Φ) is reduced by a factor of 0.134 $(1.1/3)^2$. Note that the area of a 7 mm diameter aperture is 0.384 cm^2. Also, since the radiation is from a diffusing surface, it is spread over an angle of π sr. Thus, the radiation intercepted by the 7 mm diameter aperture at a distance of 10 cm is equal to the ratio of the area of the 7 mm diameter aperture to the area of the hemisphere at a distance of 10 cm. Thus, the radiant power as measured is less than the total power by the ratio of $0.384/2\pi r^2$ (6.11×10^{-4}). Thus, the total irradiance is given by

$$E = \frac{\Phi}{A} = \frac{(6.11 \times 10^{-4})(0.134)}{0.384} \Phi = 2.13 \times 10^{-4} \Phi = (2.13 \times 10^{-4})(15 \times 10^{-3}) = 3.2 \times 10^{-6} \text{ W/cm}^2 \tag{8.14}$$

An alternative method for determining the irradiance is to use an irradiance meter that limits the field of view to 11 mrad so that at 10 cm, the instrument provides a direct reading of irradiance. It is important to note that if the aperture of the irradiance meter is different than 7 mm, the reading must be corrected to account for this difference if the irradiance meter was calibrated with a different diameter aperture. This is easily accomplished by multiplying the reading obtained by the ratio of the area of the aperture of the input optics of the irradiance meter to that of the required 7 mm diameter aperture located in front of the input optics of the irradiance meter.

With the known value of irradiance, we can now determine the radiance. The formula for radiance using Equation 8.3 becomes

$$L = 3.2 \times 10^{-6} \frac{(10)^2}{\pi (0.055)^2} = 3.37 \times 10^{-2} \text{ W/(cm}^2 \text{ sr)} \tag{8.15}$$

For a 100 s exposure time, the integrated weighted radiance is 3.37 J/(cm^2 sr). This value is then compared to the MPE specified in Table 5b.[56]

The first fact to note about the MPE for this wavelength and exposure duration is that it is dependent on the source dimensions as defined in the standard by the visual angle α. So, we must first determine the visual angle of the source to the eye. It should also be noted that a minimum visual angle of $\alpha_{min} = 1.5$ mrad is used as a dividing line between an extended source and a point source. In this case, the 3 mm diameter diffuse source at a distance of 10 cm from the eye forms a visual angle of 30 mrad. Therefore, the source is considered to be an extended one. Further, since α is greater than 11 mrad, we use the MPE of 100 C_B J/(cm^2 sr) for exposure durations in the range 0.7 to 1×10^4 s. The correction factor (C_B) that takes into account the weighting factor for photochemical damage in this example is 1. Therefore, the MPE is 100 J/(cm^2 sr). For a 100 s exposure, the integrated radiance of 3.37 J/(cm^2 sr) is a factor of 30 times below the MPE.

If the source were recessed inside a cabinet with the light emitted through a glass or plastic plate as shown in Figure 8.9, the field-averaged radiance may be determined by using a lens to form an image of the light source in space outside the cabinet. The lens simply brings light collected by the lens to a focused image of the light source in the image plane of the lens. In the previous example, we could use a 20 diopter lens at a distance of 10 cm from the diffuse source. Using the lens equation, this would produce a real one-to-one image of the source at a distance of 10 cm in image space. A worst case analysis of field-averaged radiance in this example would locate the 7 mm diameter aperture directly in front of the lens and the 1.1 mm diameter aperture over the image of the source. The location of the 1.1 mm aperture

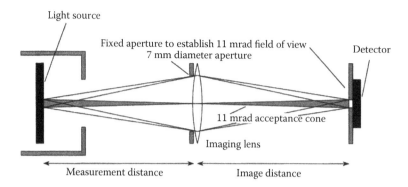

FIGURE 8.9 Sketch of homogeneous light source except for a hot spot recessed in a cabinet showing the 7 mm aperture and the 11 mrad field of view.

in the image of the light source at a distance of 10 cm from the 7 mm aperture–lens combination establishes the field of view of 11 mrad. The field-averaged radiance as presented earlier is given by dividing the product of the radiant power transmitted by the two apertures (1.1 and 7 mm diameter apertures) and the square of the distance (10 cm) between the two apertures by the areas of the two apertures used. In this case, we must also correct for the Fresnel reflection losses from the lens that is about 8% (4% per surface). This can also be easily determined by measuring the transmittance of the lens.

8.6.3.2 Evaluation Based on Photometric Measurements

In some cases, it is possible to use photometric measurements to evaluate the potential hazards from a light source. To demonstrate this, we consider the case of a 3 mm diameter diffuse emitting white LED fixation light in an ophthalmic instrument that has a radiant intensity of 0.021 mW/sr. The white light LED emits radiation only at wavelengths between 400 and 700 nm. The ocular exposure distance is 20 cm.

In this case, the radiance is given by dividing the radiant intensity by the area of the 3 mm diameter diffuse radiating surface, which is 0.071 cm^2. Thus, the radiance is 0.29 mW/(cm^2 sr). For measurements in this case, the ISO 15004-2 Ophthalmic Instrument Standard requires that an instrument with a field of view of 11 mrad be used.[62] Thus, at an ocular exposure distance of 20 cm, the field of view of the luminance meter would be limited to a maximum diameter equal to 0.011 × 200 = 2.2 mm on the radiating surface of the source. However, if the light emitted is homogeneous over the emitting surface, it is not critical that an 11 mrad field of view be used. In this case, it is only necessary that the field of view be limited to that defined by the diameter of the emitting surface. The ISO standard also specifies that a 7 mm diameter aperture shall be used to collect the luminous flux at the ocular exposure distance. In this regard, it is important to note that if the 7 mm diameter aperture is of a different diameter than that of the input optics of the luminance meter, the reading obtained must be corrected. This is necessary because the reading of the luminance meter is usually calibrated with a different input optics aperture. This correction is easily accomplished by multiplying the reading obtained by the meter by the ratio of the area of the aperture of the input optics of the luminance meter to that of the 7 mm diameter aperture.

To demonstrate the principle of the photometric measurement method, we consider the case of a green fixation light instead of a white fixation light and having the same radiance and exposure distance as in the example. A measurement of luminance as described for this device yields a value of 0.20 cd/cm^2 or 0.20 lumens/(cm^2 sr), since 1 cd/cm^2 = 1 lumen/(cm^2 sr). For a worst case analysis, we assume that all of the light is emitted at the peak wavelength of the visual response function at 553 nm. Using the relationship that 1 W at 553 nm results in 683 lumens (683 lumens/W), a radiance value of 0.29 mW/(cm^2 sr) is determined by dividing 0.20 lumens/(cm^2 sr) by 683 lumens/W. Thus, if a luminance measurement is made, it is possible to show that the radiance is a factor of about 7 times lower than the 2 mW/(cm^2 sr) limit specified in ISO 15004-2 for a Group 1 instrument.

Finally, one could also determine that the green fixation light emissions are below the limit by making a measurement of illuminance and determining the luminance. In this example, luminance may be determined by measuring the luminous flux that is transmitted through two apertures spaced a known distance, z, apart using the relationship as in the example in Section 8.6.3.1:

$$L_v = \frac{\Phi_v \cdot d^2}{A \cdot a} \tag{8.16}$$

where
L_v is the luminance
Φ_v is the luminous flux
d is the measurement distance
A is the area of the second aperture
a is the area of the first aperture

If the illuminance meter is located at a distance of 20.0 cm from the fixation light, the 11 mrad field of view requires that an aperture having a 2.2 mm diameter be placed over the emitting surface of the fixation light. The illuminance at a distance $d = 20$ cm from the light source is given by

$$E_v = \frac{\Phi_v}{A} \tag{8.17}$$

so that the formula for luminance becomes

$$L_v = E_v \frac{d^2}{a} = E_v \frac{(20)^2}{\pi(0.11)^2} \tag{8.18}$$

All that needs to be done then is to measure the illuminance that is transmitted through a 7 mm aperture at distance $z = 20$ cm using an aperture of 2.2 mm over the diffuse source.

For this example, the illuminance (E_v) measured, as in this example, is 0.19×10^{-4} lumens/cm² or 0.18 lux. Using Equation 8.18, this illuminance yields a luminance of 0.20 lumens/(cm² sr). For this green light LED, the radiance of the LED is given, as in this calculation, by the ratio of the luminance to 683 lumens/W at 553 nm yielding 0.29 mW/(cm² sr).

If the luminance of a white light source is measured as described earlier and the relative spectral power distribution is known, the spectral radiance and the total radiance of the source can be determined using the following relationships:

$$L_v = 683 \int_{380}^{770} L_\lambda \cdot V(\lambda)d\lambda = 683 \sum_{380}^{770} L_{\lambda peak} \cdot f(\lambda) \cdot V(\lambda) \cdot \Delta\lambda \tag{8.19}$$

where
L_v is the luminance of the source
L_λ is the spectral irradiance at wavelength λ of the source at distance d
$V(\lambda)$ is the visual response function
$d\lambda$ is the infinitesimally small wavelength interval
$\Delta\lambda$ is the summation interval
$L_{\lambda peak}$ is the spectral irradiance at the peak of the spectral irradiance curve and is a constant
$f(\lambda)$ is the relative spectral power distribution with f_λ being 1 at λ_{peak}

Now,

$$L_\lambda = L_{\lambda.peak} \cdot f_\lambda \qquad (8.20)$$

From these equations, and when L_v can be measured and f_λ is known, $L_{\lambda peak}$ can be determined. Once L_λ is known, the radiance (L) and the blue weighted radiance (L_B) can be determined using the following equations:

$$L = \sum_{380}^{700} L_\lambda \cdot \Delta\lambda \qquad (8.21)$$

$$L_B = \sum_{380}^{700} L_\lambda \cdot B(\lambda) \cdot \Delta\lambda \qquad (8.22)$$

L_B can then be compared to the limit.

8.6.3.3 Laser Safety Evaluation Using Advanced Noninvasive Optical Imaging Approaches

In the biophotonic field, conventional optical diagnostic/therapeutic systems typically include some basic components such as various medical lasers or noncoherent light sources and flexible fiber-optic light delivery systems for tissue manipulation, bioimaging, and/or biosensing. Such modalities have a broad area of practical biophotonic applications ranging from precise tissue ablation and optical stimulation of neurons to single cell and intracellular sensing and imaging. In order to evaluate the effectiveness and laser safety of these systems, a critical parameter that needs to be determined is the radiant exposure, H, on the target tissue (or the equivalent quantity *dose* used in the field of optical therapeutics[1,12,13]). It can be calculated using either the laser radiant power and the exposure time (for CW lasers) or the laser pulse energy (for pulsed lasers) and the laser beam spot size, which is given by the following equation[68,69]:

$$H \; [\text{J/cm}^2] = \frac{Q[\text{J}]}{A \; [\text{cm}^2]} = \frac{\Phi_a \cdot \tau_e}{\pi[d \cdot \tan\alpha + r]^2} = \frac{Q_p}{\pi[d \cdot \tan\alpha + r]^2} \qquad (8.23)$$

where Q is the laser radiant energy delivered to the irradiated tissue area A. For CW and high-repetition-rate lasers, Q is determined by the radiant power Φ_a and the exposure time τ_e, while for a single pulse and low-repetition-rate pulse lasers by the delivered laser pulse energy Q_p. The laser radiation delivery fiber acceptance angle $\alpha = \sin^{-1}(NA/n)$ is determined by the numerical aperture NA of the delivery fiber and the refraction index n of the media. d is the fiber-to-tissue distance and r is the core radius of delivery fiber (see Figure 8.10 for details).

Using Equation 8.23 for specific values of the radiant exposure, which can be controlled and precalibrated, and a laser radiation delivery fiber with known characteristics, the radiation dose is calculated using the fiber-to-tissue distance (d) that determines the laser radiation spot size. Since a small axial displacement may cause a significant dose change due to the diverging laser beam profile, precise measurement of the fiber-to-tissue distance is required. An advanced noninvasive approach to precisely measure and monitor this distance is based on optical coherence tomography (OCT) (Figure 8.10a).[68,69] When a single-mode fiber OCT probe is attached to the delivery fiber (Figure 8.10b and c), the fiber-to-tissue distance d is measured in real time with a high accuracy of a few microns. This simple noninvasive technique significantly improves the effectiveness of light–tissue interactions and the laser safety evaluation.

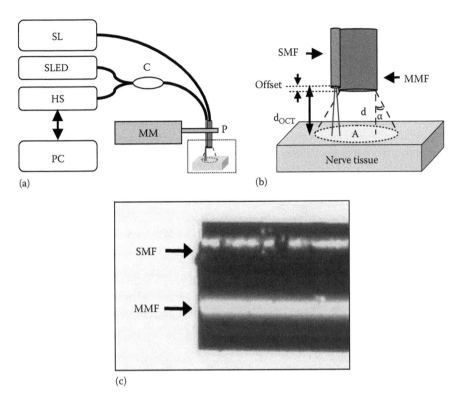

(a)

(b)

(c)

FIGURE 8.10 (a) OCT setup for precise fiber-to-tissue measurement, (b) probe design, and (c) integrated fiber probe prototype.

8.6.4 Multiple Light Sources

The evaluation of optical radiation emissions from an instrument with multiple light sources requires special consideration. For this case, it is necessary to evaluate the exposure to all consecutive or simultaneous use of light sources in 1 day that result in overlapping beams. In regard to retinal exposure, it is important to note that adjacent light sources that are not within a cone angle of 100 mrad from the eye's nodal point are considered to be independent and beams from such light sources cannot overlap on the retina. Thus, only light sources that are within a cone angle of 100 mrad from the position of the eye can produce overlapping beams on the retina. IEC 60825-1 specifies how to deal with multiple light sources that produce overlapping beams.[28] For both ocular and skin hazards, the contributions from each overlapping beam are additive for the wavelength ranges from 180 to 315, 315 to 400, 400 to 1400, and 1400 to 10^6 nm.

8.7 Radiometric Measurements

Measurements of the radiant power from point sources require locating the beam directly on the sensitive surface of the detector. Measurement of the optical radiation emissions from noncoherent broadband sources is more difficult. Ideally, such measurements would be characterized by accurate spectroradiometric measurements using a calibrated spectroradiometer with a double-grating monochromator and cosine-corrected input optics. The use of a double-grating spectroradiometer typically provides sufficient out-of-band rejection. It is beyond the scope of this paper to provide a detailed description of radiometric measurements. Such descriptions can be found elsewhere.[63]

It is important to note that in some cases, it may be possible to use relatively simple and inexpensive optical radiometric measurement instruments to determine if the optical radiation emissions from an instrument are below the limits specified in the standard. This is especially true if the optical radiation emission levels are relatively low. If the spectral distribution is known, or if it is limited by strong filtration, the initial use of a sensitive, well-characterized, broadband radiometer may even replace the need for spectroradiometry provided that a substantial tolerance exists between the measured value and the applicable limits. It should be noted that all the spectral weighting factors have values less than 1.0, except for the $A(\lambda)$ function for wavelengths less than 445 nm, which increases from 1 at 440 nm to 6 for wavelengths between 305 and 335 nm. Thus, the instrument reading would always exceed the spectrally weighted value for emissions greater than 440 nm. In addition, if the direct reading were six times lower than the limit for the retinal photochemical hazard, one could conservatively employ the broadband meter since the maximum $A(\lambda)$ weighting factor is 6.

Commercially available broadband direct reading *safety* meters can be used to directly measure one of the spectrally weighted or nonweighted quantities to assess potential optical radiation hazards to the eye or skin. Also, spot luminance meters can be used to directly measure the quantity luminance. In this regard, if the luminance of a white light source is less than 1 cd/cm², spectral data may not be needed. For example, ophthalmometers and perimeters do not emit white light exceeding 1 cd/cm². As noted earlier, the spectral irradiance of a light source can be determined from a measurement of illuminance using an illuminance meter if the relative spectral power distribution of the source is known.

Finally, it is important to estimate the uncertainty in all optical radiation measurements and in any weighted or unweighted optical radiation quantity determined. Methods for estimating the uncertainty in radiometric measurements can be found in ANSI/IESNA RP 27.2.[63]

8.8 Summary

The potential optical radiation hazards from any light source may be evaluated using the methods described and in the examples provided in standards. Alternative methods that yield equivalent results can be used. In this paper, we have described methods we believe are easier to understand and use. We have also reviewed bioeffects of optical radiation and optical radiation terminology needed to perform hazard evaluations.

Disclaimer

The mention of commercial products, their sources, or their use in connection with material reported herein is not to be construed as either an actual or implied endorsement of such products by the Department of Health and Human Services.

References

1. Sliney, D. H. and M. L. Wolbarsht, *Safety with Lasers and Other Optical Sources: A Comprehensive Handbook*. New York: Plenum Press, 1980.
2. International Standards Organization (ISO), ISO 15004-2 Ophthalmic instruments—Fundamental requirements and test methods—Part 2: Light hazard protection, 2007.
3. International Standards Organization (ISO), ISO 10943-2011 Ophthalmic instruments—Indirect ophthalmoscopes, 2011.
4. International Standards Organization (ISO), ISO 10939 Ophthalmic instruments—Slit-lamp microscopes, 2007 (Ed. 2).
5. International Standards Organization (ISO), ISO 10942 Ophthalmic instruments—Direct ophthalmoscopes, 2006 (Ed. 2).
6. International Standards Organization (ISO), ISO 10936-2 Optics and photonics—Operation microscopes used in ocular surgery, 2010 (Ed. 2).

7. International Standards Organization (ISO), ISO 15752 Ophthalmic instruments—Endoilluminators fundamental requirements and test methods for optical radiation safety, 2010.

8. International Standards Organization (ISO), ISO 10940 Ophthalmic instruments—Fundus cameras, 2010.

9. Food and Drug Administration, http://www.fda.gov/MedicalDevices/DeviceRegulationand Guidance/default.htm.

10. Commission de l'Eclairage (CIE), The International Commission on Illumination, as published in the International Lighting Vocabulary, published by CIE.

11. Code of Federal Regulations, Food and Drugs, Title 21, Parts 1040.10 & 1040.11. Washington, DC: U.S. Government Printing Office, 2007.

12. Sliney, D. H., Radiometric quantities and units used in photobiology and photochemistry: Recommendations of the Commission Internationale de l'Eclairage (International Commission on Illumination), *Photochem. Photobiol.*, 83: 425–432, 2007.

13. Huang, Y. Y. et al., Biphasic dose response in low level light therapy, *Dose Response*, 7: 358–383, 2009.

14. McCluney, W. R., *Introduction to Radiometry and Photometry*. Boston, MA: Artech House, 1994.

15. Drobetsky, E. A., J. Turcotte, and A. Chateauneuf, A role for ultraviolet A in solar mutagenesis, *Proc. Natl. Acad. Sci.*, 92: 2350–2354, 1995.

16. Matsumura, Y. and H. N. Ananthaswamy, Toxic effects of ultraviolet radiation on the skin, *Toxicol. Appl. Pharmacol.*, 195(3): 298–308, 2004.

17. Boyce, P. R., Light and health, in *Human Factors in Lighting*. New York: Macmillan, 1981.

18. Young, A. R. et al., The similarity of action spectra for thymine dimers in human epidermis and erythema suggests that DNA is the chromophore for erythema, *J. Invest. Dermatol.*, 111: 982–988, 1998.

19. Baier, J. et al., Singlet oxygen generation by UVA light exposure of endogenous photosensitizers, *Biophys. J.*, 91: 1452–1459, 2006.

20. Kvam, E. and R. M. Tyrrell, Induction of oxidative DNA base damage in human skin cells by UV and near visible radiation, *Carcinogenesis*, 18(12): 2379–2384, 1997.

21. Zuclich, J. A., Ultraviolet-induced photochemical damage in ocular tissues, *Health Phys.* 56(5): 671–682, 1989.

22. Diffey, B. L., Solar ultraviolet radiation effects on biological systems, *Phys. Med. Biol.*, 36(3): 299–328, 1991. Acknowledgments 323.

23. Anders, A. et al., Action spectrum for erythema in humans investigated with dye lasers, *Photchem. Photobiol.*, 61(2): 200–205, 1995.

24. Pitts, D. G., The human ultraviolet action spectrum, *Am. J. Optom. Physiol. Opt.*, 51: 946–960, 1974.

25. TLVs and BEIs., *American Conference of Governmental Industrial Hygienists*. Cincinnati, OH: ACGIH Worldwide, 2013.

26. Oliva, M. S. and H. A. C. Taylor, Ultraviolet radiation and the eye, *Int. Ophthalmol. Clin.*, 45(1): 1–17, 2005.

27. Ham, W. T. J., H. A. Mueller, and D. H. Sliney, Retinal sensitivity to damage from short wavelength light, *Nature*, 260(5547): 153–155, 1976.

28. International Standard, IEC 60825-1 Safety of laser products, International Electrotechnical Commission, 2011.

29. Poblete-Gutiérrez, P. et al., The porphyrias: Clinical presentation, diagnosis and treatment, *Eur. J. Dermatol.*, 16(3): 230–240, 2006.

30. Dougherty, T. J., An update on photodynamic therapy applications, *J. Clin. Laser Med. Surg.*, 20(1): 3–7, 2002.

31. Radu, A. et al., Pulse oximeter as a cause of skin burn during photodynamic therapy, *Endoscopy*, 31(9): 831–833, 1999.

32. Sibata, C. H. et al., Photodynamic therapy in oncology, *Expert Opin. Pharmacother.*, 2: 917–927, 2001.
33. Pfefer, T. J., K. T. Schomacker, and N. S. Nishioka, Long-term effects of photodynamic therapy on fluorescence spectroscopy in the human esophagus, *Photochem. Photobiol.*, 73(6): 664–668, 2001.
34. Levine, J., Medications that increase sensitivity to light: A 1990 listing, U.S. Department of Health and Human Services. Washington, DC: Government Printing Office (FDA91-8280), 1990.
35. Reid, C. D., Chemical photosensitivity: Another reason to be careful in the sun, FDAConsumer, May 1996, http://permanent.access.gpo.gov/lps1609/www.fda.gov/fdac/features/496_sun.html.
36. Kim, B., S. L. Jacques, S. Rastegar, S. Thomsen, and M. Motamedi, Nonlinear finite-element analysis of the role of dynamic changes in blood perfusion and optical properties in laser coagulation of tissue, *IEEE J. Sel. Top. Quant. Elect.*, 2(4): 922–933, 1996.
37. Pearce, J. and S. Thomsen, Rate process analysis of thermal damage, in A. J. Welch and M. J. C. van Gemert (eds.), *Optical-Thermal Response of Laser-Irradiated Tissue*. New York: Plenum Press, 1995.
38. Park, H. G. et al., Cellular responses to mild heat stress, *Cell. Mol. Life Sci.*, 62(1): 10–23, 2005.
39. Charkoudian, N., Skin blood flow in adult human thermoregulation: How it works, when it does not, and why, *Mayo Clin. Proc.*, 78(5): 603–612, 2003.
40. Si, M. S. et al., Dynamic heat capacity changes of laser-irradiated type I collagen films, *Lasers Surg. Med.*, 19(1): 17–22, 1996.
41. Pfefer, T. J. et al., Pulsed laser-induced thermal damage in whole blood, *J. Biomech. Eng.*, 122(2): 196–202, 2000.
42. Ritz, J. P. et al., Optical properties of native and coagulated porcine liver tissue between 400 and 2400 nm, *Lasers Surg. Med.*, 29(3): 205–212, 2001.
43. Barton, J. K. et al., Simultaneous irradiation and imaging of blood vessels during pulsed laser delivery, *Lasers Surg. Med.*, 24(3): 236–243, 1999.
44. Buttemere, C. R. et al., In vivo assessment of thermal damage in the liver using optical spectroscopy, *J. Biomed. Opt.*, 9(5): 1018–1027, 2004. 324 Regulation and Regulatory Science for Optical Imaging.
45. Thomsen, S., J. A. Pearce, and W. F. Cheong, Changes in birefringence as markers of thermal damage in tissues, *IEEE Trans. Biomed. Eng.*, 36(12): 1174–1170, 1989.
46. Park, B. H. et al., In vivo burn depth determination by high-speed fiber-based polarization sensitive optical coherence tomography, *J. Biomed. Opt.*, 6(4): 474–479, 2001.
47. Henriques, F. C. and A. R. Moritz, Studies in thermal injury: I. The conduction of heat to and through skin and the temperature attained therein. A theoretical and experimental investigation, *Am. J. Pathol.*, 23: 531–549, 1947.
48. Henriques, F. C., Studies in thermal injury: V. The predictability and significance of thermally induced rate processes leading to irreversible epidermal injury, *Arch. Pathol.*, 43: 489–502, 1947.
49. Agah, R. et al., Rate process model for arterial tissue thermal damage: Implications on vessel photocoagulation, *Lasers Surg. Med.*, 15: 176–184, 1994.
50. Moussa, N. A., E. N. Tell, and E. G. Cravalho, Time progression of hemolysis of erythrocyte populations exposed to supraphysiological temperatures, *ASME J. Biomech. Eng.*, 101: 213–217, 1979.
51. Pfefer, T. J. et al., Dynamics of pulsed holmium: YAG laser photocoagulation of albumen, *Phys. Med. Biol.*, 45(5): 1099–1114, 2000.
52. Wille, J. et al., Pulse oximeter-induced digital injury: Frequency rate and possible causative factors, *Crit. Care Med.*, 28(10): 3555–3557, 2000.
53. International Electrotechnical Commission (IEC), IEC 60601-1 Medical electrical equipment—Part 1: General requirements for basic safety and essential performance, 2005.
54. van Leeuwen, T. G. et al., Pulsed laser ablation of soft tissue, in A. J. Welch and M. J. C. van Gemert, (eds.), *Optical-Thermal Response of Laser-Irradiated Tissue*. New York: Plenum Press, 1995.
55. Vogel, A. and V. Venugopalan, Mechanisms of pulsed laser ablation of biological tissues, *Chem. Rev.*, 103(2): 577–644, 2003.
56. American National Standards Institute (ANSI), ANSI Z136.1 Safe use of lasers, 2007.

57. American National Standards Institute (ANSI), ANSI Z136.3 Safe use of lasers in health care facilities, 2011.

58. American National Standards Institute (ANSI), ANSI Z136.5 Safe use of lasers in educational institutions, 2009.

59. American National Standards Institute (ANSI), ANSI Z136.6 Safe use of lasers outdoors, 2005.

60. American National Standards Institute (ANSI), ANSI Z136.4 Recommended practice for laser safety measurements, 2010.

61. International Standard, IEC TR 60825-13 Safety of laser products, International Electrotechnical Commission, 2006.

62. American National Standards Institute/Illuminating Engineering Society of North America, ANSI/IESNA RP-27.1-05. Photobiological safety for lamp and lamp systems—General requirements, 2005.

63. ANSI/IESNA RP-27.2-00 Recommended practice for photobiological safety for lamps and lamp systems—Measurement techniques, 2000.

64. ANSI/IESNA RP-27.3-07 Recommended practice for photobiological safety for lamps—Risk group classification and labeling, 2007.

65. International Standard, IEC 62471 Photobiological safety of lamps and lamp systems, CIE S 009:2006.

66. Landry, R. J., R. G. Bostrom, S. A. Miller, D. Shi, and D. H. Sliney, Retinal phototoxicity: A review of standard methodology for evaluating retinal optical radiation hazards, *Health Phys.*, 100(4): 417–434, 2011.

67. Ness, J. W., H. Zwick, B. E. Stuck, D. J. Lund, B. J. Lund, J. W. Molchaney, and D. H. Sliney, Retinal image motion during deliberate fixation implications to laser safety for long duration viewing, *Health Phys.*, 78(2): 131–142, 2000.

68. Zhang, K., E. Katz, H.-D. Kim, J. Kang, and I. Ilev, Common-path optical coherence tomography guided fiber probe for spatially precise optical nerve stimulation, *Electron. Lett.*, 46: 118–120, 2010.

69. Zhang, K., E. Katz, H.-D. Kim, J. Kang, and I. Ilev, A fiber-optic nerve stimulation probe integrated with a precise common-path optical coherence tomography distance sensor, CLEO-2010 Technical Digest. Washington, DC: Optical Society of America, CTuP2, 2010.

III

Photonic Detection and Imaging Techniques

9

Biological Imaging Spectroscopy

Gregory Bearman
ANE Image

David Cuccia
Modulated Imaging

Richard Levenson
*University of
California, Davis*

9.1 Introduction

Improved detectors, new electrooptical devices, and vastly improved computational power for data analysis have fueled interest in combining biology and spectroscopy. This chapter will cover three aspects of biomedical imaging spectroscopy:

1. Data acquisition: What instruments are available for acquiring an image cube, and what are the performance trade-offs involved in choosing one over the other?
2. Data analysis: What are some of the approaches for examining very large and multivariate datasets?
3. Applications: Which current research areas in biology and medicine can exploit the power of imaging spectroscopy?

It is reasonable to wonder what distinguishes spectral imaging from standard red, green, and blue (RGB), full-color imaging. After all, our computer monitors tell us they can display 16.7 million colors—surely that should be enough. To answer this question, we need to understand the difference between *color* and spectral content. Light is composed of photons with different energies. While we can think of higher energy (shorter wavelength) photons as being *blue* and less energetic (longer wavelength) photons as being *red*, these color attributes are an artifact of the human visual system. In fact, there turns out to be no simple relationship between wavelength content of light and the color we actually perceive. This is (in part) because our eyes (and conventional color film and color digital cameras) allocate visible (VIS) light, no matter how spectrally complex, into only about three different color bins: red, green, and blue. Light with completely different spectral content can have precisely the same RGB coordinates, a phenomenon known as metamerism. For example, red light and green light can combine to form yellow. If we see a yellow object, we cannot tell if the color is spectrally pure (as it would be if it were created by a prism or rainbow) or if it arose from a mixture of red and green.

Researchers have used human color vision to interpret images since the first microscope. Although we perceive three spectral bands and cover a relatively narrow range, the human eye is quite sensitive to subtle color differences within that range. When exogenous dyes were used to differentially color cellular structures or molecules, the interpretation still relied on color vision and, more recently, on electronic color cameras. The addition of fluorescent dyes to the microscopist's tool kit began to push the limit of color vision, electronic or otherwise. The standard detection toolkit of fluorescence microscopy is an array of dichroic mirrors, filter cubes, and other filters designed to separate multiple colored probes, either in absorption or in fluorescence emission. Increasing the number of probes, as biologists want to do, creates so much spectral overlap that filtering cannot separate the probes; that is, color images of fluorescent probes that differ only slightly spectrally appear the same. In that case, we need to use some sort of spectroscopy.

Spectroscopy usually uses single-point detectors that cannot easily sample large areas or small areas at high resolution. On the other hand, *imaging* spectroscopy can spectrally image large areas, combining the function of a camera (recording spatial information) with that of a spectrometer. These devices can measure the spectral content of light at every point in the image: a 1000×1000 pixel sensor provides one million individual spectra. Once a spectral stack is acquired, mathematical approaches ranging from simple to very sophisticated can be used for analysis. Analysis of fluorescence microscopy uses spectral signatures to match each pixel with one of the known probes used in the experiment. Imaging spectroscopy tells us *what* is *where*.

Once properly calibrated, these images can be used to obtain a corrected spectrum for each image pixel, which can then be used to identify components in the target. For the geologist, imaging spectroscopy yields compositional maps of geologic sites, showing *which* minerals are *where*[1] or to determine the composition of the rain forest canopy.[2] It can detect agricultural pests,[3] drought stress, or fertilizer application levels. Spectral imaging has uses in industrial process control, in detection of ordinarily invisible bruising in fruit, in assessing the viability of transplanted organs,[4,5] in uncovering forgeries, and so on. Finally, modifications in existing designs and novel approaches have made spectral imaging easy to accomplish with a microscope; this combination has promising applications in surgical pathology and molecular biology. Fluorescent dyes have become available, which will increase the usefulness of spectral imaging in a variety of areas, including high-throughput screening, genomics, and clinical diagnostics.

9.1.1 Spectral Image Cubes

Simply put, an imaging spectrometer acquires the spectrum of each pixel in a 2D spatial scene. The easiest way to think of such a scheme is band-sequential (BSQ) imaging, in which multiple images of the same scene at different wavelengths are acquired. A key point is that the spectra be sampled densely enough to reassemble a spectrum (commensurate with the needs for analysis). A remote-sensing instrument may take hundreds of more images over the VIS to near-infrared (NIR) range. There are many technological means of obtaining these data, and this chapter will present a catalog of current technologies. The images are typically stacked in a computer, from the lowest wavelength to the highest, to create an image cube of the dataset. The spectrum of a selected pixel is obtained by skewering it in its third dimension, wavelength. While there are many ways of acquiring and storing the data, this representation is BSQ, in which the images are stacked like a deck of cards and resemble a cube with sides x, y, and λ (wavelength). Even if the data are acquired in some other fashion, they can be reconfigured into this mode. Two other data modes are band-interleaved pixel (BIP) and band-interleaved line (BIL). In BIP, the spectra of successive pixels are stored sequentially. This is advantageous for computation, as the spectrum of each pixel can be read directly, as opposed to BSQ data where one has to read in the entire cube to calculate a spectrum of any given pixel.

9.2 Instruments

Before describing specific instruments, it is worthwhile to compare spectral imaging with what can be accomplished using standard imaging systems based on conventional RGB sensors. Because most such systems rely on single-chip cameras, color images can be acquired in a single exposure, at video rate or faster. In contrast, until recently, most spectral systems require a series of exposures, so improvements in the quality or utility of the spectral data collected should be large enough to justify the potential penalties in cost and throughput and data acquisition time.

For example, while earlier systems for automated or assisted immunohistochemistry quantitation, a relatively simple problem in color analysis, used grayscale cameras and two or more color filters somewhere in the light path, other approaches exploit RGB cameras and analytical strategies of varying levels of sophistication and complexity. With automatic thresholding operations, Ruifrok[6] was able to differentiate between a diaminobenzidine (DAB) (brown) stain alone, DAB plus hematoxylin (blue), and hematoxylin alone. More recently, this group has shown that conversion of RGB images into optical density (OD) units allowed for more accurate discrimination. However, RGB sensors have intrinsically broad and overlapping regions of spectral sensitivity for their three color channels, and this adversely affects unmixing accuracy, especially when separation of similar chromogens is being attempted. Thus, for example, a dense brown stain can generate a signal in the postanalysis red channel.[7] While it may be possible to unmix red, brown, and blue using only three input images, the optimal wavelengths and bandwidths will differ from the broad channels provided by standard RGB imaging systems.

There are additional technical and practical problems with conventional color imaging. First, many color cameras use a charge-coupled device (CCD or CMOS) that produces a color-encoded analog signal that is digitized by a computer video board into R, G, and B pixel intensities. The fidelity and consistency of such a system can be variable. Section-to-section variability, along with interactions with camera controls such as automatic gain control, can induce fluctuations in the image quality. Because the color of a stained object is a product of the stain's transmittance and the camera's spectral response, it is possible that the camera could sense dyes differing in spectral properties similarly and thus be indistinguishable. Finally, the spatial resolution of single-chip color cameras is typically lower than that of monochrome cameras with the same pixel count because of the color mask and interpolation routines that merge information from three or more pixels when determining RGB and intensity values.

True imaging spectrometers, in some fashion, acquire a 3D dataset, spatial (2D), and wavelength as the third dimension. The approaches for instruments traditionally involved scanning one of the dimensions, either (1) acquiring a complete spectrum for each pixel (or line of pixels) at a single shot and then spatially scanning through the scene or, alternatively, (2) taking in the complete scene in a single exposure and then stepping through wavelengths to complete the data cube. While typically the light emerging from the imaged object is filtered for spectral content, it is also possible to control the spectral content of the illumination. Recently, other instruments have been developed that acquire both spectral and spatial information in a single exposure, although with some trade-offs (reviewed in Hagan et al.[8]).

Figure 9.1 shows an image cube and how different cuts through the data illustrate the different approaches. Some of the terminology comes from the origins of imaging spectroscopy, which involved

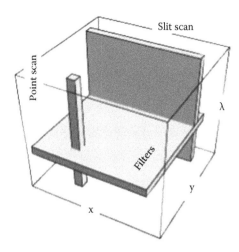

FIGURE 9.1   Schematic of various spectral imaging strategies. Not depicted are representations of newer "snapshot" approaches.

performing remote sensing from a moving platform. For example, the whiskbroom imaging spectrometer is the one in which a single point is scanned perpendicular (cross track) to the direction of motion. The spectrum of each pixel is acquired with a spectrometer and the data are taken one spectrum at a time pixel by pixel along a line. The name comes from the fact that the path of the scanned pixels resembles that of a whiskbroom in action. Similarly, a push-broom spectrometer images a slit onto a focal plane array; the spatial dimension occupies one axis of the array and the spectrum for each pixel is spread out perpendicularly to it. A complete image is acquired one line at a time as the slit is scanned in the direction of motion. In biological imaging, point-scanning and slit-scanning confocal microscopes use similar image collection geometries respectively. In addition to these techniques, which collect spectral data directly, there are other modalities that require mathematical processing of the intermediate data.

We will break up the discussion of instrument types into four general types:

1. *Spectral scanning*: These use electro-optical devices such as liquid crystal or acousto-optic tunable filters (AOTFs), as well as filter wheels that project complete images onto CCDs or other focal plane arrays. Controlling the spectral content of the illumination source rather than filtering the remitted light can also achieve spectral scanning.
2. *Spatial scanning*: These use either push-broom or whiskbroom configurations with prisms, gratings, or beam splitters to create spectral discrimination.
3. *Interferometric*: These typically (but not always) acquire a 2D image and scan optical path differences (OPDs) in some manner to obtain a complete interferogram at each pixel; the data need to be mathematically converted into spectra in wavelength space.
4. *Other approaches*: These include instruments such as the computed tomographic imaging spectrometer (CTIS) and the image slicing spectrometer.[8]

9.2.1 Spectral Scanning Instruments

Such instruments are easy to understand and have very simple optics. They consist of imaging optics, a tunable filter of some sort for spectral selection (or a tunable light source), and a camera. Since the components can be in-line or folded, such systems can be made rather compact, suitable for mounting on microscopes or other instruments. The tunable filter can be a mechanical filter wheel, a linear variable filter, or an electro-optical filter that can be tuned electronically.

9.2.1.1 Fixed Filters

The simplest implementation of an imaging spectrometer incorporates a filter wheel equipped with a set of fixed band-pass filters in a rotating mount. A variant often necessary for fluorescence imaging would substitute a set of filter cubes (combinations of dichroic mirrors, excitation and emission barrier filters) for a simple filter wheel. For applications where there are a relatively small number of wavelengths needed, preset and invariant, this can be a useful technique. For example, Speicher et al.[9] have demonstrated fluorescence-based spectral imaging with a filter system generating a combinatorial library of 27 colors, enough to paint all the human chromosomes. Furthermore, compared to other approaches, a filter wheel can be relatively inexpensive and is also quite light efficient (although the latter consideration is not straightforward and can depend on the degree of spectral cross talk between channels tolerated). These instruments have limitations. (1) They lack spectral flexibility, since only a relatively small number of wavelength choices are available in any one configuration. While one could make the filter holder larger to accommodate more filters, this increases the size and expense commensurately. (2) The performance of the filters can change unpredictably over time due to aging. (3) Switching speeds can be low. (4) Moving parts create noise and vibration. (5) There can be image registration problems due to misalignment of filters.

PIXELTEQ makes an integrated band-sequential system, SpectroCam™, with up to eight filters in a rapidly spinning filter wheel. The spectral image acquired is relatively small in terms of pixels at the time of writing (January 2013), ~1.5 MP, but, because of the sensor used, can cover a spectral range from the VIS to 1.05 μm in the NIR. While the filters can be replaced, a filter system is not as flexible as more general spectrometers. For many applications, eight filters are sufficient; for example, this is more than enough to capture ~98% of color information in a scene.[10,11] BSQ imaging using filter suffers from the fact that the optical path for each filter may be slightly different and the filters are not all coplanar. These factors create a variety of chromatic image shifts and aberrations that need to be corrected for each image; an approach to these corrections has been suggested.[12]

Several implementations of a spectral version of a Bayer filter camera have also been developed.[13-16] A Bayer color camera has a four-cell unit, containing a red filter, two green filters, and a blue filter directly over the light-sensing pixels. If, instead of R, G, and B, one had an n × m mask with custom narrowband filters over individual pixels, one now acquires up to n × m separate wavelengths per snapshot acquisition. There are fabrication and alignment issues with this approach, and the first filter-sensor construct will be fairly expensive due to start-up costs. However, cost per unit can drop rapidly as volumes ramp up. Yi et al. have reported the data acquired from such devices.[17] However, these filter-array approaches present the same sort of demosaicing issues that Bayer sensors do, except proportionately even more challenging. Remember that a Bayer filter does not actually sample all the pixels in the object plane at all three colors; it interpolates and creates an RGB image plane for each of the pixels in the unit cell; that is, it assigns a blue value even to pixels under the green filters that never saw a blue signal. The super-Bayer unit cell is bigger and same-band pixels on the object are even farther apart, increasing the chance for spatial–spectral misestimation. One solution is to simply bin the image by n, the size of the cell, which reduces the image size but avoids implementing interpolation algorithms but also leaves open the possibility of in-pixel mixing. There is no ideal solution to this, but sophisticated cross-talk corrections and interpolation schemes can help recover spatial resolution and spectral fidelity.[13]

9.2.1.2 Linear Variable Filters

A linear variable filter can also act as a spectral filtering element for an imaging spectrometer. For such filters, the transmission varies linearly along the filter; at any wavelength, λ_o (or position along the filter), the local transmission is a band-pass filter with a width that is a fixed fraction of λ_o. That width is 1%–1.5%, depending on the filter, so a typical bandwidth is ~4–10 nm over the VIS spectrum. One vendor, OCLI, has marketed a spectrometer without a grating, using a linear filter directly on top of a linear CCD detector array. There are similar versions known as circular variable filters (CVFs) in which the transmission changes with rotational angle of the filter.

An optical system with a beam waist can use a linear filter or CVF to create an imaging spectrometer by inserting the filter at the location of the minimum spot size. Since the filter's transmission is spatially dependent, a large spot size would give a large and spatially varying bandwidth, so the filter is located at a beam waist to reduce the resultant spectral smearing. In this mode, the filter acts like a filter wheel with a large number of filters. Images are acquired at each wavelength and filter translated or rotated to the next wavelength. Surface Optics Corp. (San Diego, CA) has developed an innovative variant based on time domain integration that reads out an imaging array row by row synchronized to the motion of a spinning CVF (to avoid the problem of spectral smearing). In conjunction with algorithms implemented in hardware, their instrument is capable of acquiring *and processing* 30 image cubes per second.

9.2.1.3 Tunable Filters

As the name implies, these devices can tune their spectral passband electronically and without moving parts. Advantages include quiet and vibration-free operation, switching speed, spectral selectivity, spectral purity, and flexibility. There are several important criteria that such filters need to meet. (1) Since the entire image is being filtered, the filter wavelength needs to be constant over the entire image or meet some lower limit for edge effects. (2) Introduction of the filter into the optical path cannot introduce (significant) image distortion. (3) The tuning time has to be commensurate with the dynamics of the experiment. (4) Out-of-band rejection must be sufficiently good that dim in-band signals are not contaminated by out-of-band intrusions.[18]

9.2.1.3.1 Liquid Crystal Tunable Filters

Liquid crystal tunable filters (LCTFs) use electrically controlled liquid crystal elements that transmit a certain wavelength band while being relatively opaque to others. The rejection of the unselected wavelengths, without further manipulation, is about $10^4{:}1$.[19] The band pass can be as narrow as 1 nm or even less, and the spectral range with a single device can range from 400 to 720 nm in the VIS.

Mode of action: The LCTF is based on a Lyot filter, a device constructed of a number of static optical stages, each consisting of a birefringence retarder (quartz for LCTFs) sandwiched between two parallel polarizers. A stack of stages function together to pass a single narrow wavelength band. As the incident linearly polarized light traverses the retarder, it is divided into two rays, the ordinary and extraordinary, that has different optical paths, given by

$$\Gamma(\lambda) = 2\pi \times \frac{\Delta d}{\lambda}$$

where
  Δ is the birefringence
  d is the thickness

After transmission through the retarder, only those wavelengths of light in phase are transmitted by the polarizer and passed onto the next filter stage. The transmission of a stage is

$$T(\lambda) = \cos^2\left[\frac{\Gamma(\lambda)}{2}\right]$$

as illustrated in Figure 9.2. The overlap of these continuously varying transmission curves determines which wavelengths are passed by the filter stack as a whole. To introduce tunability, a liquid crystal layer

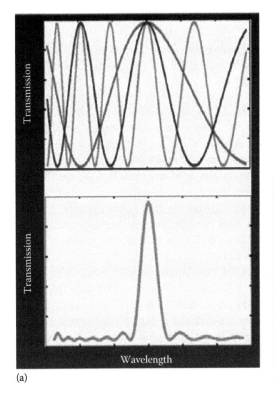

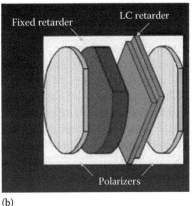

(a) (b)

FIGURE 9.2 **(See color insert.)** Liquid tunable filters: principle of operation (a) and schematic of a single stage (b), in which an LC retarder is inserted at every stage of the Lyot filter.

is added to each stage, as in Figure 9.2a, which creates minor changes in retardance affecting the position along the spectrum where the curves constructively interact. Tunability is provided by the partial alignment of the liquid crystals along an applied electric field between the polarizers; the stronger the field, the more the alignment and the greater the increase in retardance. Tuning times for randomly accessing wavelengths depend on the liquid crystal material used and the number of stages in the filter. At the moment, commercial devices use nematic components that result in tuning times of approximately 50–75 ms.

In practical terms, the VIS to NIR range can be covered from 420 to 1800 nm. This typically would require up to three separate filter units as each device only tunes over about 1 octave (twofold) spectral range.

Since the band pass is related to the number of stages in the device, any band pass can be designed and fabricated, from 16 cm^{-1} to 50 nm. The 16 cm^{-1} device has been used for Raman imaging spectroscopy.[20] The devices are rather spectrally flat over a relatively large aperture (38 mm). Like the AOTF, the LCTF is polarization sensitive, which reduces the transmission by half, unless optical means are provided to harvest both polarization states.

LCTFs work best in a collimated or telecentric optical space, as the maximum f-number that provides an off-axis shift of less than 2 nm at the filter edge is ~2.5. Since the device operation depends on interference effects, photons that are significantly off axis have a different optical path than on-axis photons, creating edge effects. However, since many optics are inherently slower than f/2.5, they can be used before optical elements. One of the authors (Bearman) has taken a number of image cubes of remote scenes with an LCTF mounted in front of a Nikon 135 mm lens operated at f/4, as have others. Similar arrangements are also available commercially from Opto-Knowledge Systems, Inc. (www.oksi.com).

The LCTF approach has been used to create image cubes for biological imaging (see Figure 9.18), confocal microscopy,[21] and agriculture and imaging archeological documents such as the Dead Sea Scrolls.[22]

9.2.1.3.2 Acousto-Optic Tunable Filter

An AOTF uses the interaction between a crystal lattice and an acoustic wave to diffract an incoming beam into a fixed wavelength, as shown in Figure 9.3. As an applied acoustic wave propagates through the crystal, it creates a grating by alternately compressing and relaxing the lattice. Those density changes create a local index of refraction changes that acts like a transmission diffraction grating, except that it diffracts one wavelength at a time, so it behaves as a tunable filter. In practice, the undiffracted zero-order beam is stopped with a beam stop and the monochromatic diffracted beam is available. Changing the frequency (and wavelength) of the acoustic wave changes the grating spacing, adjusting the wavelength of the diffracted beam. In addition, if multiple radio frequencies are launched into the crystal, then combinations of frequencies can be diffracted simultaneously; in this, it is more flexible than LCTFs, which generate only a single band pass at a time.

For VIS wavelengths in a tellurium oxide crystal, the applied acoustic wave is RF and can be switched very quickly (typically in less than 50 μs) compared to other technologies. Unlike an LCTF in which the bandwidth is fixed by the design and construction, an AOTF can vary the bandwidth by using closely spaced RFs simultaneously. There are several standard problems with AOTFs, some of which have been successfully addressed: blurred images and poor out-of-band rejection ($<10^{-3}$). The acoustic wave spreads as it propagates through the crystal, so diffracted rays leave at a variety of angles, resulting in blurred images. Use of a compensating prism[23] has significantly improved resolution. Narrowing the acceptance angle and attending to details of crystal fabrication can also overcome image blur and shift, albeit at a cost in light throughput.[24] A commercial version of such an *imaging-grade* AOTF-based system for microscopy is currently available from Gooch & Housego (England, United Kingdom).

Both the AOTF and LCTF imaging spectrometers share an important attribute: they make it easy to get very good signal-to-noise spectra. This is due to the band-sequential nature of their operation. When spanning a wavelength range, say 400–720 nm, the sample may have a considerable variation in reflectivity or transmission over that range. In addition, at the blue end of the spectrum, CCD sensitivity declines, as does the illumination intensity of many laboratory light sources. As a result, there is typically less signal in the blue relative to the red or green part of the spectrum. However, that can be compensated for by longer integration times at the wavelengths with reduced signal, something not possible with many other devices. In fact, the ideal way to operate a BSQ imaging spectrometer is to set a pixel data target value and integrate at each wavelength as long as necessary to obtain that value, maintaining the

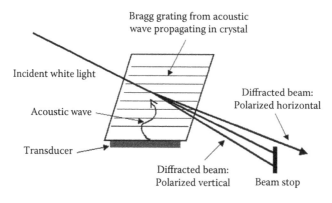

FIGURE 9.3 Acousto-optical tunable filter.

signal-to-noise ratio (SNR) at each wavelength. In that case, the *model* raw data spectrum of the target pixel would be a straight line, with the real data contained in the varying integration times for each wavelength. This is a major advantage, especially when there are no restrictions on the data acquisition time.

Spectral leakage can contaminate the acquired spectra. One advantage of the LCTF is a high rejection ratio for out-of-band transmission (10^{-3}–10^{-5}), critical for recovering spectra that can be compared with those from other laboratories or with standard spectral libraries. LCTFs can be fabricated with larger apertures than AOTFs, although that is not an issue for integration into microscopes, which do not need the large aperture. On the other hand, their major drawback is longer tuning time relative to an AOTF: ~30–50 ms vs. microseconds for the AOTF. There is a switching mode for LCTFs that is somewhat faster, around 20 ms, but that is for a limited palette of perhaps three to four wavelengths. For situations in which the integration time is photon-limited and the exposure time is ≥50 ms, the tuning time of either device becomes less of a bottleneck for data acquisition. In fact, it is usually the camera data transfer rate that dominates acquisition time when light is ample.

9.2.2 Spatial Scanning Systems

9.2.2.1 Push Broom

Lightform, Inc.[25] has developed a push-broom imaging spectrometer that is designed to mount to a C-mount camera port. It collects a slit image from the object onto a 2D camera in which the spatial information is displayed along one axis and wavelength information along the other. Wavelength dispersion is provided by a prism. This approach is well suited for scanning gels or searching an object for specific spectral features since the entire spectrum of each pixel in the slit image is available in real time. Since gels are too large to image easily, they can be mechanically scanned with this system. With this approach, the user does not have to collect an entire image cube but can assemble an image that records hits only for the spectra of interest. If there is a known spectral feature, that feature can be identified in each spectral line scan in real time and used to assemble a classified image without having to acquire the entire spectral cube for the whole image field.

Push-broom scanners have been well developed by Specim (specim.com), a Finnish company that has a wide range of instruments that cover the VIS, NIR, and SWIR. There are versions that go as far into the NIR as 2.5 μm and make them suitable for product quality measurement and geology. For microscope application, a translation stage can scan the target relative to the sensor. Stand-alone camera systems suitable for point-of-care use still require a scanner of some sort to move the slit image across the object. Specim does provide integrated solutions, but as with all physical scanners the data acquired is not truly simultaneous, although in many applications that is not important.

9.2.3 Interferometers

Rather than scanning in wavelength space, one can also scan in OPD space and capture an interferogram for each pixel, which is then inverted via the fast Fourier transform algorithm to obtain an image cube in wavelength space. Several devices have been developed and one is available commercially. Although seemingly different from instruments that acquire sequential wavelength images, many of the interferometric devices are similar in spirit and suffer from similar problems. Like the BSQ imagers, interferometric imaging spectrometers also require acquisition of many images, and sometimes an order of magnitude more images. For so many images, the data acquisition time may become limited by camera image transfer time.

Applied Spectral Imaging of Israel was perhaps the first company to make a commercially available imaging spectrometer. The device is a common-path Sagnac interferometer in which the interferogram is spread out over a 2D sensor.[26] An optical element changes the OPD in stepwise fashion, while a CCD (or other technology) focal plane array captures the resulting interference pattern at each step. Since the interferogram moves with each OPD image, object motion is a challenge for this instrument. If the object moves and the images are corrected by reregistering the spatial content to compensate, any errors will show up as incorrect interferograms and propagate into the spectra after inversion. The ASI instrument has been used for cytogenetics[27] and cell pathology,[28] to name a few applications.

Itoh et al.[29] have developed another interferometric device that uses a tilted and wedged lens array and mirrors to acquire all the necessary images at different OPDs *simultaneously* on a 2D imager. Itoh has demonstrated imaging of rapidly moving objects with this approach, a laser ablation plume and rotating (1800 RPM) targets. Since all the multiple images have to be acquired on a single detector, there is a trade-off between image size and spectral resolution.

One problem with interferometers is that of the center burst (OPD = 0), which is quite bright relative to the rest of the fringes. Since the detector is an imager, the integration time or illumination intensity has to be reduced sufficiently to avoid saturation (or blooming) for pixels at the center burst, thereby reducing the fringe contrast further out in the interferogram. The reduced fringe contrast decreases the signal and results in increased noise in the image cube in wavelength space.

There has been considerable discussion in the literature about the relative photon efficiency of interferometers compared to scanning instruments. Although on the surface the interferometer appears to have a substantial advantage over other approaches due to the fact that it collects all the spectral information simultaneously,[30] several papers[31,32] have argued that for real instruments with read noise and other noise sources, this advantage disappears in most imaging regimes. Furthermore, in the spectrally sparse scenes typical of fluorescence imaging, in which signals occupy only a fraction of the total spectral range, the ability of tunable filters to capture images only at informative wavelengths improves their performance relative to interferometer-based approaches that have to collect all wavelengths, informative or not.

9.2.4 Snapshot Methods

There are now many different optical designs for systems that can collect multispectral imaging data with a single snapshot (reviewed in Hagan et al.[8]). Such a capability can be highly advantageous for many imaging and remote-sensing tasks, including defense-related. One such approach is illustrated in Figure 9.4. With this technique, diffractive optics disperse the spectral and spatial information of each pixel onto an imaging sensor, and an image cube in wavelength space is computationally reconstructed. Since it turns out that the mathematics of the reconstruction is the same as tomographic imaging, such devices are known as CTISs. Originally proposed by several researchers[33,34] in the early 1990s, they have been further developed by Descour et al.[35] and recently commercialized by HORIBA Scientific.[36]

A diffractive optical element operating at multiple orders creates an array of images on the sensor. Development of techniques for fabricating the grating with e-beam lithography has been the main driver in development of this instrument.[37] It is important to note that each image is not simply composed of single wavelengths; that information is multiplexed over the entire array. Figure 9.4 shows how the diffractive disperser distributes the spectrum of a single pixel. Note that there is a zero-order image that can be used for focusing, a difficult task for many spectral imagers. A calibration matrix is necessary to perform the reconstruction; it is obtained by measuring the location on the image plane of pixels in the object plane at different wavelengths with a movable fiber optic coupled to a monochromator. A CTIS can operate over a large wavelength range, easily from 400 to 800 nm, and with the proper detector can operate in the IR or UV. The data from a single image can be reconstructed in a variety of ways to adjust image size and wavelength bands. For example, an image can be reconstructed with 128 × 128 pixels with 20 bands or 64 × 64 pixels and 32 bands, using the *same* dataset. The only difference between the two reconstructions is the calibration matrix.

Since it takes a single image that contains all the spectral/spatial data, it can be run at video rates,[36] assuming sufficient light and a high-speed camera (since a large pixel array is typically required by this method). This potential speed makes it suitable for studies such as endoscopy and rapid processes that other instruments cannot handle. However, as the images do require computational reconstruction, display

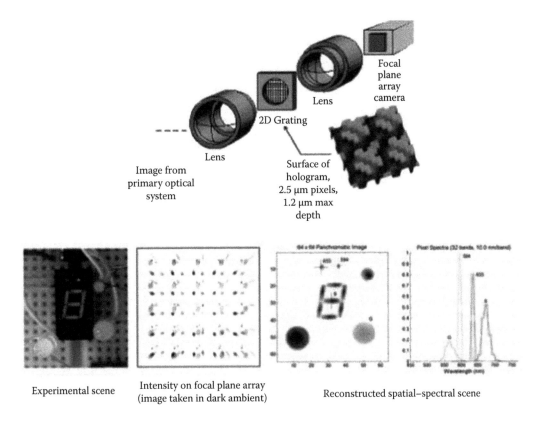

Image from primary optical system

Lens

2D Grating

Surface of hologram, 2.5 µm pixels, 1.2 µm max depth

Lens

Focal plane array camera

Experimental scene

Intensity on focal plane array (image taken in dark ambient)

Reconstructed spatial–spectral scene

FIGURE 9.4 A diffractive grating, written by e-beam lithography, multiplexes the spatial and spectral information of a single pixel onto many sensor pixels. The observed scene is bottom-left, consisting of three color LEDs, a HeNe laser spot and an 8-segment indicator. Next (2nd plot, to right) is the resulting image from the focal plane. There is a zero order image (3rd plot) that can be used for focusing as well as providing an initial starting point for the reconstruction of the image cube. Recovered spectra are shown in the final plot, bottom right.

can lag significantly behind acquisition. Alternatively, it is useful for collecting ratiometric data, since all wavelengths are acquired simultaneously. A major issue with spectral imagers has always been bandwidth—they tend to produce enormous amounts of data that present downlink or transmission problems. For example, a satellite hyperspectral imager can produce hundreds of gigabytes of data a day. In the same vein, a remotely sited or operated imaging spectrometer can easily present significant bandwidth demands for data transmission in a power-limited environment (power = bandwidth for telecommunications).

Using CTIS devices, Descour et al. have demonstrated ratiometric pH imaging with standard probes,[38] while de la Iglesia et al.[39] performed toxicology studies. In both cases, the device allowed capture of the entire spectrum of fluorescent probes at once.

The image mapping spectrometer (IMS) is another snapshot hyperspectral imaging technology capable of acquiring an entire data cube (x, y, λ) without the need for scanning.[40] This technology has been commercialized by Rebellion Photonics. Its principle of operation is shown in Figure 9.5 where an object (a fluorescently labeled cell) is imaged onto a mapping mirror by a front imaging lens. The mapping mirror is composed of hundreds to thousands of thin micromirror facets that sample individual rows of the relayed image and reflect these rows in different directions depending on the tilt of the facets. This is represented as three different directions although in practice it is many more. The reflected image associated with each direction is shown with dark space between the reflected image rows next to the mapping mirror as well as the relayed image. A collecting lens then captures each reflected image and forms a set of pupils in the collecting lens' pupil plane where each pupil corresponds to a single reflected image.

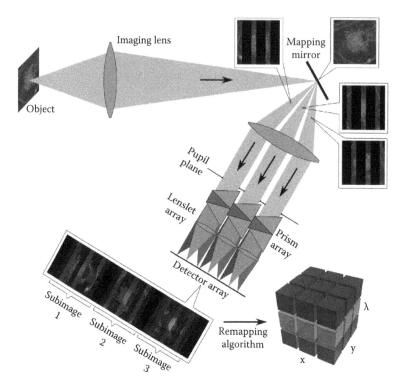

FIGURE 9.5   Layout of imaging mapping spectrometer.

A dispersion element(s) then spectrally disperses each reflected image pupil in the orthogonal direction to the image rows. A lenslet array matching the number of pupils then creates an array of subimages, which now have the dark space filled with the spectrum from each line in the redirected images. After detection by a CCD, a simple remapping algorithm can be applied to create the entire data cube at the CCD's full data rate. An example of an fluorescently labeled cell imaged using Rebellion Photonics' ARROW system based on the IMS technology is shown in Figure 9.6.

Ricoh has published an novel approach to snapshot imaging spectrometers. These devices use lenslet arrays and filter masks at an appropriate aperture to create a multispectral imager, as shown in Figure 9.7. Since the device has a set of filters in the optical path, there is a limit on the number of bands, although it is true snapshot imaging. The increasing size and decreasing cost of sensors means bigger images and more bands for this approach. There are no moving parts and the system is reasonably compact, lending itself to a handheld version.

9.2.5 Hadamard Transform Imaging Spectroscopy

Hadamard transforms have been used for spectroscopy for some time[41] and have been adapted to fluorescence microscopy. The fabrication of large-format liquid crystal spatial modulators has made this application possible as they can create the Hadamard masks rapidly and with no moving parts. In a series of papers, Jovin and colleagues have developed this implementation of imaging spectroscopy and microscopy.[42]

Like the CTIS or an interferometer, the Hadamard transform spectrometer requires computation to reconstruct the image cube in wavelength space. However, it also requires a large number of images, for example, Hanley reports acquiring 511 images in ~11 and 5 min of computation to transform the data. Increasing optical efficiency can clearly reduce some of the data time, but like the interferometer, the basic nature of the device requires many images.

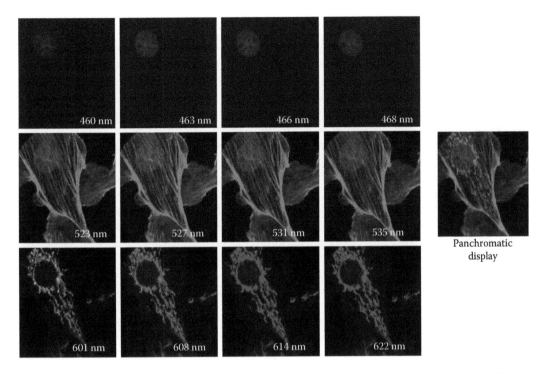

FIGURE 9.6 Different spectral planes of a (single) snapshot spectral image dataset taken of a fluorescently labeled cell using an image mapping spectrometer.

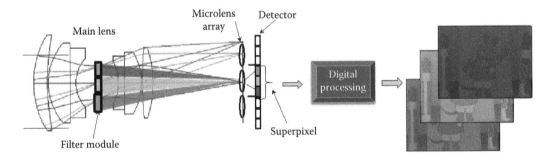

FIGURE 9.7 Overview of plenoptic camera architecture with spectral filters inserted into the main lens.

9.2.6 Fiber-Optic Image Compression

Several groups have developed fiber-optic-based systems that acquire spectral image cubes using a combination of a spectrograph, 2D detector, and a custom fiber bundle. The fiber bundle has a 2D input but the output is reordered into a line, which then serves as the input for a spectrograph. The spectrograph output is a 2D image, containing the spatial pixel in one direction, while the other dimension is the spectrum of that pixel. Myrick has constructed such a system[43] and used it to image laser ablation plumes.[44] This is in fact the same data format of a push-broom imaging spectrometer, except that the input now represents a scrambled 2D image, rather than a line imaged across an object. Similarly, Ben-Amotz has used a fiber-optic compression system for Raman imaging with a 61-bundle (pixel) image.[45] One limitation of these systems is image size; the number of pixels in the image is limited to the number of rows in the detector; Myrick reports on a 17 × 32 image that uses a detector with 544 elements in the spatial dimension.

9.2.7 Spectral Source

For microscopy bright-field applications, it is possible to accomplish spectral imaging by tuning the illumination light as well as filtering or otherwise analyzing the remitted light. Monochromators (usually relying on diffraction gratings and white light sources) are available for such purposes. However, as the name implies, monochromators provide illumination consisting only of one spectral band of light at a time. However, it is possible to create sources that are more flexible and can produce illumination of any desired pure wavelength (like a monochromator) or any selected *mixture* of pure wavelengths simultaneously, with white light output an easy option.[46] The resulting images can be collected by a high-resolution grayscale CCD camera and interpreted using appropriate algorithms and displays. It can be used to create a complete spectral image cube for a sample by taking sequential images while illuminating with a series of pure wavelengths, with greater ease and economy than by means of devices on the imaging path, such as tunable filters or interferometers. An advantage over tunable filters in some applications is that contamination of individual bands by out-of-band light is minimal. Furthermore, using software approaches such as projection pursuit vectors or principal components to define specific illuminants, the sample can be illuminated with a precisely controlled mixture of wavelengths so that the image presented to the detector is a linear superposition of the sample properties at many wavelengths. Thus, spectral discrimination that would previously have required the collection of complete spectral cubes might require acquisition of as few as one to three matched spectral images per field. Data acquisition is simplified, and, since spectral processing is being performed optically rather than computationally, both acquisition and analysis times are greatly reduced.

The ability to tune illumination has increased significantly in the last decade. Digital micromirrors can provide an ability to fine-tune the spectral output of an otherwise broad-spectrum light source.[49–51] Tidal Photonics and Gooch & Housego, for example, sell agile light sources that can provide tailored (arbitrarily shaped) spectral output. The devices put a digital multimirror (DMM) at the focal plane of an imaging spectrometer; the wavelength is spread out along x and the input slit image is along y. The DMM reimages the photons back into a light path that can be directed onto a sample. To adjust the spectral output, one turns various micromirrors off and on along the wavelength axis. For example, to reduce the intensity at some λ by half, one just turns off half the mirrors in the y direction at the x pixel corresponding to λ. This flexibility lets the user control the optical transfer function, so one can make a top hat output, for example, as in Figure 9.8.

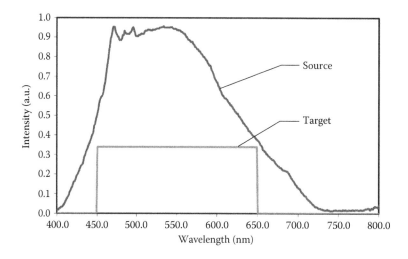

FIGURE 9.8 Spectral flattening. The SPLE is used to shape the white light spectrum of a xenon lamp (upper line) to approximate a spectrally flat top hat spectrum (lower line).

The advantage of spectrally programmable light source is that one can achieve spectral discrimination without spectrally modulating the capture light. Given that light remitted from a sample is often dim, avoiding light loss often associated with spectral filtering in the capture path is a good thing. For many situations, we have a lot of a priori knowledge about the spectral content of an image; either by prior measurement, in some basic physics, or in the case of target molecular probes, we have engineered the spectral into the scene.[47,48]

9.2.8 Multispectral Confocal Microscopy

Much biology today uses confocal microscopy as a major tool to provide high-resolution 3D imaging of cells and tissue. Considerable effort has gone into developing optics and software to provide diffraction-limited imaging in commercial instruments. Similar effort has been expended on fluorescent probes to illuminate cellular activity.

Laser scanning confocal microscopes (LSCMs) raster scan a laser spot across the object, obtaining a full image a point at a time, similar to the way a whiskbroom imaging spectrometer operates. Since a pinhole, present in the optical path to provide confocality, attenuates the signal considerably, a photo-multiplier tube (PMT) is typically used as a detector to provide sufficient gain and reasonably short pixel dwell time. To obtain an emission image cube of a fluorescently labeled sample, there are two options. Since the image is already being scanned, emitted light can be filtered prior to detection to assemble an image cube. This was performed originally by inserting an LCTF in front of the PMT in a LSCM and stepping through the emission wavelength range.[13] One could also use an AOTF as the filtering element. While serviceable, this is not a practical method, as it requires multiple scans at different wavelengths to acquire a full image cube. Aside from the significantly increased time for assembling a z-stack or a time series, the repeated scans can cause excessive photobleaching.

For an LSCM, the best data collection scheme would be to acquire the entire spectrum from each spatial pixel as it is scanned, making it in effect a whiskbroom imaging spectrometer. In this approach, the instrument collects an entire image cube in the same time as a single spatial scan. Zeiss was the first to introduce a spectral imager (the *Meta*) that obtained a high-resolution spectrum at each pixel as the LSCM scanned across the sample. Many reporter molecules could be imaged simultaneously, and overlapping spectra resolved using unmixing and classification algorithms[13,49] as well as more cleanly separate spectra in FRET experiments.[50] Now all major confocal manufacturers include multispectral capabilities, although details of the optical strategies differ.

Unlike with conventional point-scanning laser-based systems, it is feasible to add a tunable filter device to a confocal instrument that uses a Nipkow-type disk approach, since devices such as these present an entire image to a CCD. Thus, any spectroscopic device that can be interposed before a CCD camera should be compatible with this or similar confocal designs. However, any method that requires a lot of images may pose practical problems when doing real experiments due to the possibility of photobleaching and the time involved in acquiring an (x, y, z, λ, t) image stack.

9.3 Data Analysis

9.3.1 Image Analysis

A number of freestanding spectral analysis tools are available commercially or as freeware. These include ImageJ (with plugins like http://rsbweb.nih.gov/ij/plugins/spectral-unmixing-plugin.html), Research System International's ENVI™, and MultiSpec™.[51] Other programs can be assembled by the sophisticated user with resources available in such packages as MATLAB® such as the statistics, chemometrics, and image analysis toolboxes, supplemented with researcher-generated MATLAB-compatible algorithms downloadable from various Internet sites. Still others are available bundled with commercially available spectral imaging hardware (OKSI, Kairos Scientific, Spectral Dimensions, ChemIcon, etc.). These software tools will not be described further, except for some aspects to be touched upon in the following discussion.

One of the appeals to developing spectral imaging systems or applications is the richness of the datasets, comprising both spatial and spectral information, which invites the use of intriguing analytical tools. Indeed, many of the algorithms, such as automatic clustering tools, being developed for use with genomics datasets (such as the huge expression arrays) are applicable to spectral cubes, with the proviso that these methods do not encompass any of the spatial content to be found in the images. Methods attempting to link spatial and spectral data are under development.[52] Another thread in current investigations is determining how to select the minimum number of wavelengths needed to accomplish specific tasks.[53] While it may seem intuitive that more spectral data and higher spectral resolution may provide increased analytical precision, this is often not the case. Many wavelengths may be *uninformative*, and their inclusion in the dataset merely adds noise. This consideration is partially related to the so-called curse of dimensionality, which also deals with consequences of the huge internal volumes of the hyperspheres that can be used to represent high-dimensional datasets.[54] (This is more of a problem in remote sensing, in which datasets can contain images at hundreds of wavelengths.)

For the relatively simple problems posed by imaging in the VIS range and where the targets may be simply defined by fluorescent dyes or chromogens, one may be able to lower the number of wavelengths acquired to approximate the number of distinct species sought in the image. Thus, analytical techniques can be used not only to work with the datasets but also to shape how they are collected. At the limit, spectral flexibility can be used simply to provide a capability to select one or more specific wavelengths for the purposes of increasing contrast or enhancing the utility of straightforward image analysis tools. Ornberg et al.[55] describe using a tunable filter to identify optimal wavelengths for separating signal from background in samples stained with a single chromogen plus background stain.

9.3.1.1 Analysis of Spectral Images

Assuming that more than a couple of wavelengths have been collected, the task of analysis usually involves classification, unmixing, or both. Classification involves the assigning of each pixel to one or more spectrally defined classes (or to an *unclassified* class). Classification is equivalent to spectral segmentation; it is an *exclusive* operation in which a pixel or object is assigned to a *single* class using one or more of a variety of metrics. On the other hand, when pixels can be or are composed of more than one spectral class, as is often encountered when multiplexed protein or nucleic acid probes are used, then the pixels have to be *unmixed*, yielding estimates of the proportion of each class present. Overall, the steps involved typically consist of

1. Detection and/or selection of appropriate spectra for subsequent analysis
2. Spectral classification or pixel unmixing

9.3.1.1.1 Pixel Classification

There are several approaches to classifying pixels in a spectral image. The minimum squared error method compares the spectra at each pixel in the image with a set of reference spectra, choosing the most *similar* using a least-squares (Euclidean distance) criterion. This metric compares spectral means; other distance metrics such as Mahalanobis distance[56] can be used that are sensitive to higher-order statistics such as class variances. Related approaches convert spectra into n-dimensional vectors, and the angles between such vectors can be used as measures of similarity.[57] Determining which spectra to use for the classification procedure is not always straightforward. In simple cases, the reference spectra can be selected from obvious structures in the image (e.g., foci of cancer vs. normal cells) or from established spectral libraries. Alternatively, informative spectra can be extracted using statistical analysis methods, such as principal component analysis (PCA) or clustering methods.[58] Instead of using a classified pseudocolor display, spectral similarity can also be illustrated by mapping the degree of similarity using grayscale intensity. This operation can reveal otherwise unapparent morphological details.[59]

9.3.1.1.2 Pixel Unmixing

Spectral classification methods are suitable for images in which no pure spectral components are likely to exist, such as in histologically stained samples. In other types of images, such as those generated by immunofluorescence or in situ hybridization (ISH) procedures, multiple distinct spectral signals may coexist in a single pixel to form the detected signal. Spectrally mixed pixels result when objects cannot be resolved either at an object boundary (spatial mixture), when more than one object is located along the optical path (depth mixture), or when multiple probes are colocalized within a pixel. In fluorescence, due to the additive nature of the light signal, the observed spectrum is usually a linear mixture of the component spectra, weighted by the amount of each probe. A linear combination algorithm can be used to unmix the summed signal arising from the pure spectral components, to recover the weighting coefficients. Given an appropriate set of standards, the algorithm can quantitate the absolute amount of each label present.[60]

In contrast to fluorescence images, imaging multiplexed samples (such as immunohistochemistry studies) in bright field must take into account the behavior of absorbing chromophores that, rather than being additive, subtract signal from the transmitted light in a nonlinear fashion. Conversion of the bright-field image from transmittance to OD, a straightforward mathematical procedure, permits the use of the same linear unmixing algorithms that work with fluorescence.

Automated end-member detection: How does one select which spectra to use for unmixing? In many cases, this is easy. If one is doing standard immunofluorescence studies, the spectra of the fluorophores, imaged one at a time, can be stored in a spectral library and used to unmix images in which multiple fluorophores are present. But what if one does not have pure spectral species to work with, for example, if a single, multiply labeled image is available or if, to change applications, one is trying to analyze a remote scene about which there is little a priori knowledge available? There are tools that can identify the pure spectral species present in an image, without a priori knowledge, by deconstructing the spectral content into its presumed components. ENVI™ provides a tool based on convex hull analysis that considers spectral end-members (the pure species) to occupy the periphery of a data cloud all of whose mixed species must fall within, rather than on the surface. The cloud (in which each pixel's location is determined by its spectral content in n-dimensional space) is rotated randomly and projected onto a hyperplane. Pixels that repeatedly end up on the periphery after multiple projections are considered to be end-members and can be used to unmix the image. This procedure can be quite time-consuming. Another specialized utility, N-FINDR,[61] uses an analytical approach rather than multiple projections and accomplishes the same task quite efficiently.

Dimensionality reduction and automated cluster analysis: Spectral data, as noted earlier, can be expressed as points in hyperspace. Spectrally similar pixels will cluster together and algorithms, some of which are similar to those used for analyzing genetic expression arrays, can be used to identify such clusters, which might represent meaningful bases for spectral classification.[62,63] Frequently, such analysis is either impossible or inefficient when all wavelengths are included in the dataset. Because there is a great deal of covariance in typical datasets (i.e., the intensity at one wavelength predicts to a high degree the intensity at neighboring wavelengths), the number of dimensions needed to express the actual information content in a dataset is often far less than the number of dimensions in the dataset itself. PCA is one of a family of statistical tools that can identify the most informative combinations of wavelengths (by rotating the basis vectors of the original dataset) and can segregate signal from noise (with some major limitations). Typically, the dimensionality of a 25-wavelength image cube of a standard histology sample can be reduced to three or four dimensions (which are composed of linear combinations of many of the original wavelengths) while preserving virtually all of the spectral information. Clustering algorithms can then readily work on such a reduced dataset to identify meaningful spectral clusters, although some techniques, such as support vector machines,[64] are designed to use the original full feature space. A large variety of clustering methods, including iterative, analytical, neural net, fuzzy logic, and genetic algorithm-based approaches, both published and proprietary, have been developed, whose description and virtues are beyond the scope of this review. A number of these tools are available as part of the software resources identified at the beginning of this section.

9.3.1.2 More Recent Multispectral Analysis Advances

There have been a number of variations and novel approaches to analyzing spectral images. One of the more intriguing is the reformulation of spectral data (typically shown as spectral curves or conceived of as points in a multidimensional data space) into spectral phasors.[65] Converting the spectra to phasors allows them to be treated as vectors. An advantage of this technique is that binary mixtures can be visualized as lying along lines connecting the positions of the two pure spectral species, allowing analysis to be computationally very simple. Also, as with eigenvalue PCA plots, individual points in a phasor display can be mapped to pixels in the image, so spatial correlation of spectral species is easily visualized. A downloadable ImageJ plugin has been posted at http://www.staff.science.uu.nl/~ferei101/Spectral%20Phasor%20PlugIn.htm.

An adjunct to spectral hardware and software development or quality assurance projects is to have useful phantoms with well-characterized properties. An interesting absorbance microarray phantom produced by robotic printing of well-defined mixtures of absorbing and scattering dyes was described by Clarke et al.[66] and could prove to be of value.

9.3.1.3 Alternatives to *Traditional* Spectral-Based Multiplexing

Spectral imaging, with the right mix of imager and labeling approach, provides an opportunity to multiplex up to 10 fluorescently labeled probes in a single slice of tissue. This approach can be challenging, as it requires excellent control over the sample preparation and labeling methodologies, including careful balancing of fluorescent label choices and expected signal intensities in order that bright fluorophores do not overwhelm spectrally and spatially overlapping similar signals (dynamic range issues). One alternative is to spectrally image the old-fashioned way—that is, by using a conventional fluorescence microscope equipped with multiple dichroic filter cubes (10 is a current practical maximum) appropriately matched to selected fluorescent dyes, so that each signal is visualized more or less without cross talk from other dyes, in separate exposures. A tour de force, but it can take months to years to work the bugs out exposures (see work by Dragan Maric, NINDS, NIH at http://www.youtube.com/watch?v=QzcuOw6Rdv4).

Another nonspectral variant is sequential staining, imaging, bleaching, and restaining. In principle, given enough antibodies, sturdy specimens, and (a lot of) time, virtually unlimited levels of multiplexing are achievable. Some variants of this approach include the *toponomics* system developed by Schubert et al.[67,68] and one under development at General Electric.[69]

9.3.1.4 Simplified Instrumentation for Spectral Imaging

If one can characterize a scene spectrally and can identify features of interest, then it is possible to acquire spectral images in informationally efficient ways. For example, Fauch et al.[70] described a system involving a sequence of customized LED combinations that illuminate a scene with mutually orthogonal spectral functions—which have to be preestablished (see the earlier discussion on tuned or matched filtering). Higher spectral-resolution spectra can be achieved using a monochrome sensor than can be achieved with the same number of single spectral bands. This approach demonstrated good accuracy on the Munsell color chart with seven different exposures. However, one can use a standard RGB CCD or CMOS chip and as few as two double-peaked illumination sources, combined with nonnegative principal components analysis, and achieve apparently similar spectral resolution, with the trade-off being that the spatial resolution is degraded somewhat in the move from monochrome sensor to Bayer-pattern filtered RGB chip.[71]

Instead of using continuously spinning filter wheels, as in the PIXELTEQ SpectroCam device, other investigators synchronize rapidly changing spectral illumination with continuously running CCD or CMOS cameras, as described by Sun et al.[72] and Tominaga and Horiuchi.[73] Such instrumentation can be useful for in vivo imaging of in situ dynamic physiology, such as changes in blood flow or tissue oxygenation in brain regions. If combined with efficiently defined illuminants, fairly high frame rates can be acquired with straightforward instrumentation.

9.3.1.4.1 Additional Use Cases for Spectral Imaging in Clinical Practice

Some recent examples demonstrating the utility of spectral imaging for clinical specimens include the following: Gilbert and Parwani[74] showed that reactive vs. neoplastic urothelium can be distinguished using double immunostains with counterstain; Dolloff et al.[75] demonstrated the detection of apoptosis in H&E-stained specimens along with changes in spectral shape of a fluorescent reporter gene accompanying association with autophagosomes. Fiore et al.[76] found certain advantages of multispectral imaging compared to conventional RGB analysis of triple-stained prostatic adenocarcinoma, and Huang et al.[77] documented the reliability of quantitative triple staining with immunohistochemistry, again with prostatic adenocarcinoma microarray samples.

9.3.1.4.2 Novel Fluorescent Agents with Useful Spectral Properties

Quantum dot reagents continue to be developed, with advances that include smaller size, improved conjugation, improved brightness, and lower toxicity. The reader is directed to recent reviews (e.g., Dave et al.[78] and Jin and Hildebrandt[79]) that cover this rapidly evolving field. One particular example combining multiplexed QD labeling and spectral imaging to explore prostate cancer heterogeneity in clinical specimens is that of Liu et al.[80] However, progress in alternative fluorescent molecules with intriguing properties has also been achieved. One interesting approach is that devised by Guo et al.[81] that assembles oligodeoxyfluorosides into defined-sequence structures with tunable fluorescent properties. These are easily conjugated to antibodies and can emit from the violet to the red with a single broadband UV excitation and are only a few nm in size, providing less steric hindrance (better labeling in tight quarters) than provided by typically larger quantum dot reagents.

Combined spectral and spatial analysis: All the tools described are designed to work only on the spectral content of the data cube. Remarkably, the pixels could be randomly scrambled, and if their associated spectra were preserved, analysis of the resulting scrambled images by the purely spectral-based algorithms would be unaffected. Obviously, a more powerful approach would somehow combine the rich spatial information present in the images, with the spectral data. This is an evolving field with ongoing attempts to adapt remote-sensing expertise to problems in biomedical imaging.

9.4 Microscopy Applications

There are a number of areas for which spectral imaging holds out promise. This section will concentrate on applications involving microscopy and visual light, while touching on applications in other areas. In microscopy, the goal can be variously the spectral measurements of natural chromophores or environmentally sensitive indicator molecules (*imaging spectroscopy*), the detection and discrimination of multiple analytes (*multiplex imaging*), and/or the analysis of complex scenes (*spectral segmentation and morphometry*); these functions can be combined. It can thus serve as a bridge between the morphological (the traditional strength of pathology) and the molecular.

9.4.1 Imaging Spectroscopy

Conceptually, the most straightforward application of spectral imaging involves the simple acquisition of spectra from naturally occurring or adventitious chromophores within a sample. Potential uses, in biomedicine, include the characterization of different melanin moieties in normal skin, dysplastic and malignant pigmented lesions, discrimination between oxy- and deoxyhemoglobin, or the study of any pigments of interest in biological or nonbiological samples. Comparison between the acquired spectra and preexisting spectral libraries can be used to aid in the identification of specific species. An example

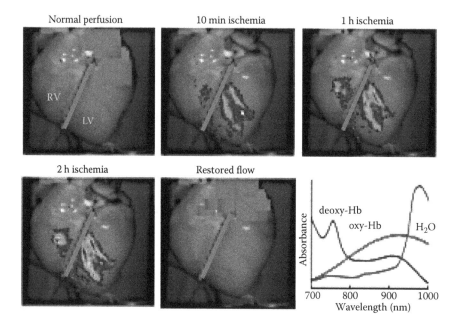

FIGURE 9.9 **(See color insert.)** Spectral detection and display of hypoxic regions of a pig heart before and after ischemia using an oxy–deoxyhemoglobin unmixing and spectral color lookup table.

of oxygenation-based studies of ischemic regions in a pig heart perfusion model is shown in Figure 9.9, which demonstrates application of macroscopic optics and a spectral range encompassing the near-IR.

Another use of spectral imaging in which the acquired spectrum has intrinsic importance is the detection of spectral shifts in (typically fluorescent) indicator dyes. Ion-sensitive dyes that shift their emission maxima in response to changing ion concentrations are well known but are not as frequently used as dyes that change their excitation profile, in part because it has been easier to switch rapidly between excitation wavelengths than to do the same on the emission side. Emission-responsive dyes, such as Indo-1, SNARF-1, Acridine Orange, and Nile Red, can be excited at a single wavelength and their emission behavior monitored either by using high-resolution spectroscopy or by detecting intensities at only two or perhaps more specific wavelength ranges. An example using propidium iodide, which is sensitive to the relative proportion of DNA and RNA in a specimen, is shown in Figure 9.10, which compares three samples of yeast under three different experimental conditions. Spectral shifts, highlighted using principal components analysis (PCA), identify yeast in each class. Such small spectral shifts are easily separated by data from an image cube, but are not separable by filters without significant cross talk.

For ratio-based ion-sensitive imaging approaches, one would ideally wish to monitor emission at a number of wavelengths simultaneously, rather than sequentially, to obtain an instantaneous pixel-by-pixel measure of ion concentration. While LCTFs can be configured to switch between wavelengths with a switching time of 1–5 ms and AOTFs in around 30 μs, these still represent serial measurements. Simultaneous measures can be achieved either by using the CTIS approach described earlier or by using beam splitters and interference filters to direct light with the desired wavelengths to one or more detectors in parallel. A commercial device that sends up to four images at different wavelengths simultaneously to a single detector is available from Roper Scientific (previously Optical Insights).

9.4.2 Multiplex Imaging, Including Immunohistochemistry and In Situ Hybridizations

Spectral imaging on an analytic level facilitates multiprobe detection techniques for proteins, RNA, and DNA. Histochemical, immunohistochemical, immunofluorescent, and fluorescent molecular probes

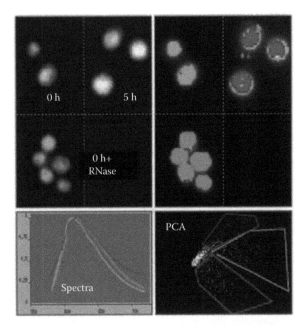

FIGURE 9.10 Spectral imaging and principal component analysis (PCA) of propidium-iodide-labeled yeast under different experimental conditions. While the raw spectra are similar, PCA enhances the statistically significant differences for visualization and quantitative analysis.

bind specifically to intra- or extracellular components and can be visualized with either fluorescence or bright-field (transmission) optics. Ideally, one would like to apply more than one specific probe at a time.

Spectral karyotyping (SKY): Pioneering work in SKY using combinatorial labeling of metaphase chromosomes[82] allowed nonambiguous identification of 27 chromosomes or chromosome pairs. Applied Spectral Imaging has commercialized this approach, which has demonstrated considerable clinical utility when applied to difficult cytogenetic problems (Figure 9.11). Similar approaches using multiple fixed filter sets (M-FISH) have been described by Ward and colleagues.[9]

Immunofluorescence: Multiprobe immunophenotyping has become widely used in evaluation of hematological malignancies, with 17 or more channels available in high-end instruments.[83]

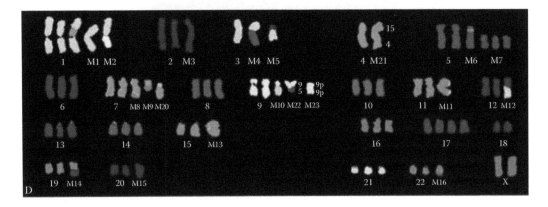

FIGURE 9.11 Spectral karyotyping of complex cancer-related chromosomal abnormalities. Multiple dyes are used to paint chromosomal regions using a spectral barcode approach resolvable via spectral imaging techniques.

The imaging approach to molecular characterization improves on flow cytometry in its ability to visualize the cells under study directly, to localize (and colocalize) cellular features, to count discrete objects on a per-cell basis, and, in tissue sections, to allow correlation with tissue microarchitecture. Using multiple labels simultaneously in the absence of spectral imaging tools is currently difficult because of the problem of spectral overlap: it is not easy to prevent signal from one dye *leaking* into the spectral channel of another, and the problem becomes intractable for conventional interference filter sets as the number of dyes is increased.[84] Another problem with immunofluorescence is the interference of autofluorescence, which can be particularly troubling when formalin-fixed tissues are being examined. Other challenging specimens include many plant samples, insects, and *Caenorhabditis elegans* nematodes. One solution to autofluorescence difficulties is to shift the excitation and emission wavelengths into the red and far-red, where autofluorescence is far less intense. If that is not possible, then spectral imaging can be used to separate the unwanted autofluorescence signal from that of specific fluorescent dyes (Figure 9.12).

Immunohistochemistry: More popular in clinical applications than immunofluorescence, IHC is widely used clinically for the detection of diagnostically or prognostically significant molecules in or on cells. In the past two decades, the technique has become central to the practice of oncologic pathology[85] since it can distinguish between look-alike lesions (e.g., mesothelioma

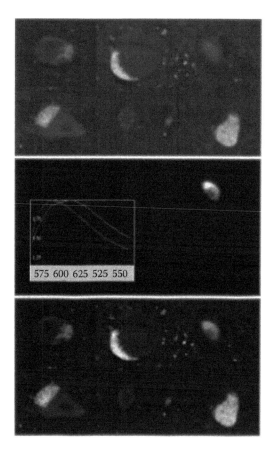

FIGURE 9.12 Detection of specific fluorescent dyes in the presence of abundant autofluorescence. Only the neuron in the top right region of the upper panel is specifically labeled with a green dye—all other signals are intrinsic brain autofluorescence. Spectral unmixing is shown in the middle panel, and the result is pseudocolored and added to the (gray) autofluorescence background in the bottom panel.

vs. carcinoma) or divine the cellular lineage of extremely undifferentiated neoplasms (lymphoma vs. other so-called *small blue cell* tumors). It can also be used to highlight the presence of otherwise easily overlooked microscopic foci of tumor, such as micrometastases lurking in lymph nodes, and can be used to measure quantitatively the levels of diagnostically or prognostically important markers such as estrogen (ER) and progesterone receptors (PR), Her2-neu, p53, ki-67, and a host of others.[86] Under some clinical circumstances and often in research situations, double- or triple-staining single slides with different chromophore-coupled antibodies may be desirable. Triple-staining procedures are not often performed because of technical difficulties; however, with the advent of programmable staining systems, complex staining protocols may become less of a hindrance. Despite the nonlinear effects of enzyme amplification, immunohistochemistry can be made quantitative, if precautions are taken.[87,88] The major problem is that it is hard to determine visually where and to what extent the different stains may physically overlap when there may be coexpression of two or more analytes in the same cellular compartment. Spectral imaging can overcome this difficulty, even in the presence of considerable spectral overlap with the chromogens. Figure 9.13 demonstrates spectral unmixing of a triple-stained breast cancer sample. This specimen was probed with an anti-PR immunostain coupled to a brown chromogen (DAB) and an anti-ER immunostain coupled to a red chromogen (Fast Red); all nuclei were counterstained with a fairly dark hematoxylin wash. The RGB image reveals how difficult it is to determine by eye which cells are expressing PR, which ER, and which both. After converting the image to OD and using previously determined spectra for linear unmixing, separate images demonstrating localization of the PR, ER, and hematoxylin stains are shown.

FISH and chromogenic ISH (CISH): ISH has proven to be an invaluable molecular tool in research and diagnosis and has enabled major strides to be taken in the fields of gene structure and expression at the level of individual cells and in complex tissues. To date, the vast majority of ISH applications have relied on fluorescence readout systems because of their sensitivity, spatial resolution, relative simplicity, and easy adaptation to multicolor and quantitative methods. As noted earlier, it can be difficult using conventional filter sets to image multiple fluors simultaneously. With spectral imaging, it is possible to visualize six or more probes simultaneously

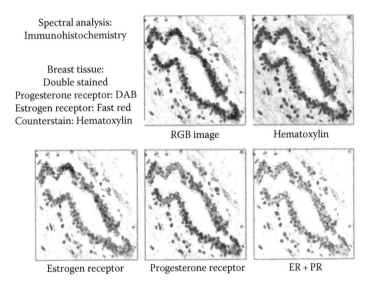

FIGURE 9.13 Multicolor immunohistochemistry can be resolved with spectral unmixing, even when absorbing color signals (in this case, brown, red, and blue) overlap spatially.

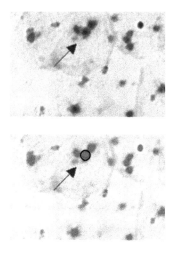

FIGURE 9.14 Spectral unmixing of chromogenic in situ hybridization (CISH) signals. In this case, the CISH spot indicated by the arrows can be unambiguously identified as being red-brown double-labeled and can be identified (or pseudocolored) after spectral unmixing, shown in the lower panel indicated by a black circle.

although similar feats can be accomplished using multiple filter cube sets and cross talk correction. As noted earlier,[30–32] issues of speed and signal-to-noise with the various approaches have aroused some degree of controversy. In any event, FISH-based techniques have proven to be somewhat problematic in the clinical arena. Drawbacks include the disadvantage that most fluorescent signals fade upon exposure to light and during storage, interference by autofluorescence (which can be severe in formaldehyde-fixed tissues), and the cost of the microscopic and imaging equipment needed (not to mention the inconvenience of having to dim the lights around the imaging station). In addition, it is difficult to combine FISH with routine histopathological stains that can reveal the morphological context of the images.

Some of these difficulties have been overcome with the development of nonfading, bright-field detection methods for ISH signals.[89–91] Signals can readily be detected in tissue sections, which can also be counterstained with hematoxylin or other general histology stains. Finally, bright-field ISH or CISH can be combined with immunohistochemistry to provide truly multiparameter molecular characterization. An example of spectrally unmixed three-color CISH is shown in Figure 9.14, which includes an example of spectrally resolving physically overlapped centromeric chromosome probes.

9.4.3 Spectral Segmentation and Morphometry

Prostate cancer cells can be spectrally detected in images of prostate biopsy tissue stained with hematoxylin and eosin (H&E). This capability could be useful, for example, in automated screening of prostate *chips* removed for benign prostatic hyperplasia. Large volumes of tissue have to be examined in a search for potentially tiny foci of clinically unsuspected cancer. Figure 9.15 demonstrates that it is possible to spectrally separate malignant and normal epithelial cells and to detect basal cells as well (these are a second cell layer found in normal prostate glands but absent in cancer). The segmentation is not perfect. Some of the imperfections (such as isolated misclassified pixels) can be suppressed using image-processing techniques. However, the limitations in the present case include the fact that the relatively unsophisticated minimum square error classification algorithm was used. More generally, it is likely that the stains, hematoxylin, and eosin—convenient, ubiquitous, and used for generations—may not be the optimal choice for purely spectral analysis of tissue, and recent advances in morphology-based tissue segmentation[92] may make this technique unnecessary.

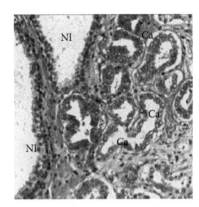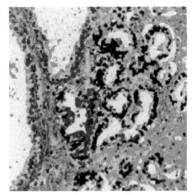

FIGURE 9.15 Hematoxylin and eosin-stained histology specimens can display useful spectral properties that allow for spectral segmentation of different tissue classes. Here prostate cancer cells (classified in black using spectral similarity tools) are spectrally distinct from adjacent normal epithelial and stromal cells. Interestingly, the malignant gland closest to the more normal regions displays intermediate (black and dark gray) spectral features.

9.5 In Vivo Imaging Spectroscopy in the Clinic

Beyond pathology, imaging spectroscopy is beginning to have an impact in medicine for in vivo tissue assessment—such areas include in vivo diagnostics, therapy or treatment monitoring, and even forensics. Fields of view and interrogation volumes of these devices range from hundreds of μm to many cm.

Generally, in vivo tissue imaging technologies are limited by length scales of tissue scattering and intrinsic chromophore absorption. In particular, tissue scattering quickly degrades spatial coherence, resulting in progressively lower-resolution methods as the interrogation depth increases; at the same time, however, multiple scattering enhances detected photon pathlengths, yielding exquisite spectral sensitivity to absorbing chromophores over larger interaction distances. In essence, targeting short interaction pathlengths, typified in microscale spectral imaging techniques, tends to yield spatially rich but spectrally sparse data with shallow interrogation depths. In comparison, macroscale techniques are often spectrally rich and can image deeper within tissues but have relatively poor spatial resolution.

The typical challenges of designing real-world medical devices are amplified in imaging spectroscopy applications due to requirement of large, coregistered datasets and associated complexity of instrumentation. Such challenges include addressing (1) possible subject motion, (2) the presence of ambient lighting, (3) the size and ergonomics of the instrumentation, and (4) the time sensitivity of medical workflows.

In the following, we touch on a range of examples, moving from micro to macro.

Multimodal in vivo confocal and multiphoton microscopy techniques have been deployed to the clinic. These devices typically employ a limited number of spectral channels to create composite representations of the distribution of tissue structures and biomolecules. Endogenous optical signals are of particular interest, as they do not require additional staining. Examples of such contrast elements include (1) fluorescence from NAD+/FAD+, collagen, elastin, and porphyrins, (2) second-harmonic generation from collagen ultrastructures, and (3) vibrational absorption from lipids, elements. Technologies successfully employed in vivo include confocal imaging,[93] coherent anti-Stokes microscopy,[94] and two-photon imaging.[95] Recently, a series of clinic-friendly multiphoton microscopes (Figure 9.16) have been commercialized by JenLab (MPTFlex, Derminspect; www.jenlab.de).

Perhaps the most prevalent example of imaging spectroscopy technology in medicine today lies hidden within the measurements of frequency-domain optical coherence tomography (FD-OCT). In this setup, a low-coherence or tunable broadband (tens to hundreds of nm) source interferometrically isolates singly scattered reflected photons, which are then spectrally resolved using either (1) spatial

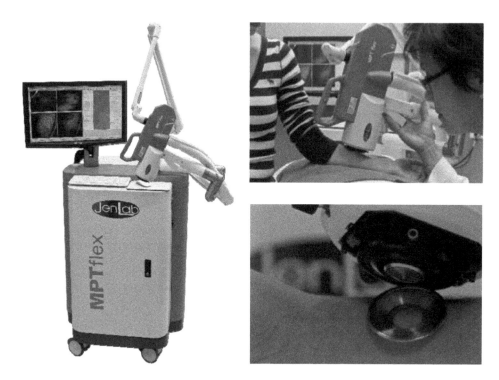

FIGURE 9.16　MPTFlex System. (Courtesy of JenLab GmbH, Saarbrücken, Germany.)

encoding with a push-broom spectrometer (Fourier domain) or, more commonly, (2) time-encoding with a high-speed spectrally swept source. The detected spectral signatures in this case represent the Fourier transform of the depth-resolved single-scattering profile. Spatial scanning of the source results in a full spectral image cube, and subsequent inverse Fourier transformation yields a final *spatial* 3D tomographic image. The most prevalent and commercially successful medical application of frequency-domain OCT has been in ophthalmology for imaging of retinal layers; however, it has been applied to a wide range of medical imaging fields, including dermatology, cardiology, and bronchoscopy. For an introduction, please see the OCT chapter later in this volume. In a similar fashion, wavelength encoding of spatial information is also harnessed in spectrally encoded confocal microscopy,[96] where wavelength division multiplexing allows simultaneous acquisition of multiple confocal pinhole spatial measurements, eliminating the need for scanning along one of the spatial axes. A recent implementation by Boudoux et al. enabled 30 fps imaging in vivo through the use of a rapidly swept source[97] and was subsequently applied to imaging of the pediatric vocal chord.[98]

Due to multiple scattering, larger-spatial scale measurements possess increased sensitivity to endogenous chromophore absorption. Examples of chromophores include hemoglobins (oxy-, deoxy-, met-, cyano-), melanins (eu-, pheo-), lipids, water, bilirubin, carotenoids, and hemosiderin (Chapter 1 in Volume II of this handbook). Devices designed to spectrally image these molecules roughly fall into two groups: (1) scanned or spatially multiplexed diffuse optical imaging or *optical biopsy* probes and (2) wide-field spectral imaging cameras. Optical spectroscopy probes are covered quite extensively in Section II of Volume II of this handbook. Spectroscopic systems that target scattering contrast have been reviewed by Qiu et al.[99] A multispectral, multidistance reflectance scanning method, laminar optical tomography, has been developed for noncontact 3D localization of chromophores[100,101] and deployed clinically for evaluation of skin lesions.[102] Krishnaswamy et al. demonstrated a broadband, dark-field scanning spectroscopic imaging system targeting fresh tissue biopsy applications.[103] A notable example of an FDA-cleared spectral probe that employs true imaging is the TVC Imaging System (InfraReDx, Inc.,

Burlington, MA). This device applies near-infrared spectroscopy to intravascular imaging to image the density and spatial distribution of lipid core plaques. Recently, it has been combined with intravascular ultrasound for a composite representation of tissue structure and composition.[104]

Wide-field imaging techniques for intraoperative guidance have historically included simple color imaging, narrowband imaging, or single-channel exogenous fluorescence. A comprehensive review is provided by DaCosta et al.,[105] which includes discussion of trade-offs that often limit practical combination of spectroscopy and imaging. A multispectral imaging system developed by Roblyer et al. strikes a practical balance between spatial and spectral diversity for detecting oral neoplasia.[106] Recently, reflectance-based spectral imaging has shown promise for surgery in the ability to resolve maps of tissue hemoglobin and oxygenation—examples include neurosurgical guidance for brain and breast tumor demarcation,[107,108] visualization of organ oxygenation during partial nephrectomy,[109] and monitoring of vascular occlusion in tissue transfer flaps.[110,111] Efforts to advance the performance of spectral imaging for intraoperative guidance often focus on identifying the spatial–spectral trade-offs to balance imaging speed and spectral specificity.[112,113] Spectral illuminants show promise in this regard and have been demonstrated for hyperspectral fluorescence endoscopy[114] and hyperspectral imaging of wounds.[115]

Spectral imaging methods have also been widely applied to skin tissue assessment. Skin lesion characterization and margin delineation have been a strong area of interest,[116,117] with commercial devices developed by Astron Clinica (SIAScope) and MELA Sciences, Inc. (MelaFind) and HyperMed (OxyView), among others. Other example applications include detection of shock,[118] assessment of burn injuries,[119] and monitoring response to phototherapies.[120,121] A review of methods and applications is provided by Kainerstorfer et al.[122]

Often, light transport model-based computational approaches are used to derive quantitative metrics such as chromophore concentration.[123,124] Alternatively, in vivo measurements can be analyzed by comparison to a grid of tabulated tissue-simulating phantom measurement results.[125] Also, calibration procedures have been developed, which address the effects of skin surface curvature on the measured intensity.[126,127]

There are limitations, however, to in vivo quantitation using *pure* reflectance spectral image cubes. Uncertainties between tissue absorption and scattering contrast limit the ability to extract quantitative chromophore concentration. For example, at a single wavelength band, higher tissue absorption causes a decrease in tissue reflectance, while higher tissue scattering causes an increase in tissue reflectance. Therefore, it is possible that high scattering can mask high absorption, resulting in a net zero observable contrast in reflectance. This effectively means that analysis of absorbing and scattering species with spectral reflectance alone must be achieved with assumptions about both (1) the presence of and (2) the specific spectral signatures of those species.[128] A new quantitative imaging method, called spatial frequency-domain imaging (SFDI), is compatible with wide-field spectral imaging and aims to alleviate these constraints.

SFDI, also referred to as modulated imaging (MI), is a noncontact optical imaging technology, which has the unique capability of spatially resolving optical absorption and scattering parameters, allowing wide-field quantitative mapping of tissue optical properties.[129–131] By separating and quantifying the multispectral absorption and scattering optical properties, SFDI removes the cross talk in reflectivity changes resulting from physically distinct contrast mechanisms. This provides a more direct assessment of tissue state via the derivation of physiologically relevant parameters, yielding insight into tissue structure and function on the mesoscale. SFDI can also be coregistered with other modalities to allow a multiscale tissue characterization.[132]

SFDI uses spatially modulated illumination for imaging of tissue constituents. Periodic illumination patterns of various spatial frequencies are projected over a large (many cm^2) area of a sample. The reflected image differs from the illumination pattern due to the optical property characteristics of the sample. Typically, sine-wave illumination patterns are used. The demodulation of these spatially modulated waves characterizes the sample modulation transfer function (MTF), which embodies the optical property information. Analytic or Monte Carlo simulation-based analysis of MTF data results in 2D maps of the quantitative absorption (μ_a) and reduced scattering (μ_s') optical properties. Mapping

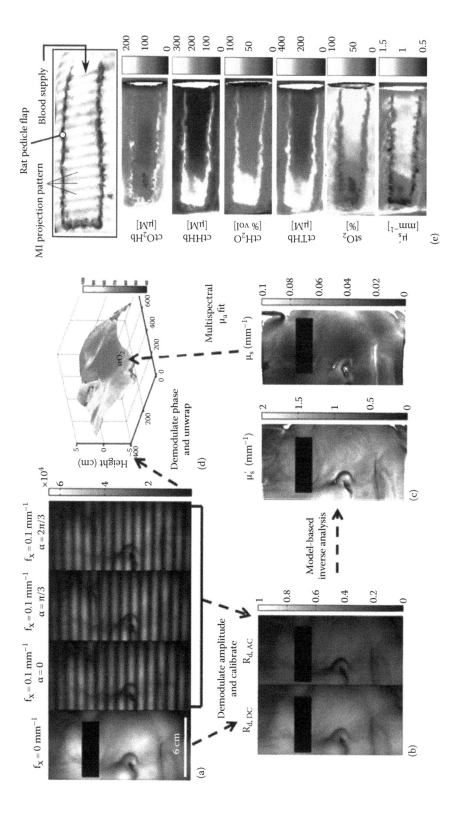

FIGURE 9.17 Flowchart of SFDI data processing and typical SFDI data products. Modulated intensity patterns (a) are projected onto the surface at each frequency (three phase images per frequency), (b) are amplitude demodulated and calibrated, and (c) fit to a multifrequency model to determine optical properties. Separately, (d) phase demodulation provides information on tissue height, which can be used for both curvature calibration and visualization. Data are processed separately for each pixel, generating spatial maps of optical properties. (e) Typical SFDI data products derived from multispectral absorption and scattering maps are shown for a rat pedicle flap, with the distal end demonstrating SFDI sensitivity to lowered perfusion (stO₂), blood pooling (ctHHb and ctTHb), edema (ctH₂O), and degradation of matrix ultrastructure/necrosis (μₛ').

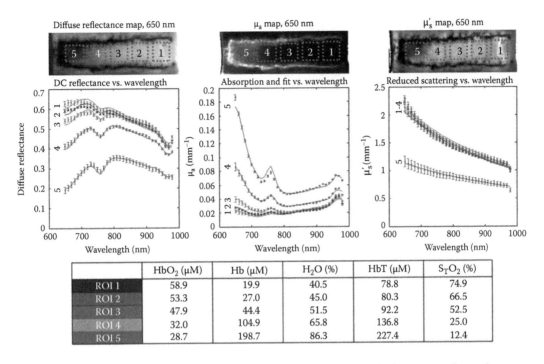

	HbO$_2$ (μM)	Hb (μM)	H$_2$O (%)	HbT (μM)	S$_T$O$_2$ (%)
ROI 1	58.9	19.9	40.5	78.8	74.9
ROI 2	53.3	27.0	45.0	80.3	66.5
ROI 3	47.9	44.4	51.5	92.2	52.5
ROI 4	32.0	104.9	65.8	136.8	25.0
ROI 5	28.7	198.7	86.3	227.4	12.4

FIGURE 9.18 Tissue flap spectral dataset of Figure 9.17e, showing the tissue 12 h after a surgery designed to cause failure in the distal (leftmost) region of the flap. Middle: region-wise average tissue diffuse reflectance (left), absorption (middle), and reduced scattering (right) spectra are plotted. Numerical labels 1–5 surrounded by dotted lines on figures (top) indicate the regions chosen for region-wise analysis. Bottom: recovered values for physiological parameters, corresponding to each numbered region. Note the gradual increase in absorption moving from proximal (right) to distal (left) regions of the flap, contrasted with the step-wise behavior of the scattering from normal to necrotic zones.

the absorption coefficient at multiple wavelengths enables quantitative spectroscopy of tissue chromophores such as oxy- and deoxyhemoglobin and water (ctO$_2$Hb, ctHHb, and ctH$_2$O) and derived physiology parameters such as tissue oxygen saturation and blood volume (stO$_2$ and ctTHb). The spatially varying phase can also be measured, yielding topological surface information. This enables visualization of the 3D tissue profile, as well as calibration data for accommodating curved surfaces in the analysis. A typical SFDI data flow is portrayed and described in Figure 9.17. Figures 9.17e shows a typical tissue flap spectral dataset acquired 12 h after a surgery designed to cause failure in the distal (leftmost) region of the flap.[133] Figure 9.18 shows a full spectral data set of the tissue flap example in Figure 9.17e.

9.6 Forensic Applications

An example of how spectral imaging can be used for forensic purposes is illustrated by the example shown in Figure 9.19. A fragment of President Lincoln's deathbed pillowcase had been preserved by the Museum of the Grand Army of the Republic (Philadelphia, PA). Dark stains on this fragment had always been assumed to represent President Lincoln's blood, but no forensic investigation had been undertaken to validate this belief. One of the authors (RL) was asked to examine the specimen with spectral imaging and to compare the spectrum with spectra acquired from aged blood stains. The resulting reflectance spectrum is shown in the right panel, overlain by spectra from blood aged 6 and 94 days. Blood stains lose spectral features as they age, so the broad and smooth spectrum from the pillowcase can be judged at least consistent with that expected from ~150-year-old blood.

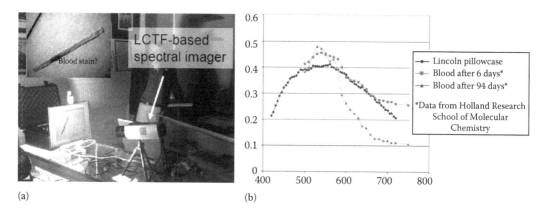

(a)   (b)

FIGURE 9.19 Spectral imaging of a fragment of the Lincoln deathbed pillowcase. (a) Shown are the fragment of pillowcase displayed in the Museum of the Grand Army of the Republic (Philadelphia, PA), the imaging setup allowing non-contact reflectance spectrum determination, and a close-up of the imaged fragment. (b) Comparison of the acquired spectrum with two other spectra of blood stains at different ages. The pillowcase spectrum is at least consistent with that of ~150-year-old blood.

Hyperspectral imaging can also be used to do optical characterization of skin bruises in the VIS/NIR spectral region.[134–136] Knowledge of the spatial, spectral, and temporal variations in bruises permits accurate modeling and age determination of bruises for forensic applications.[137,138] The VIS wavelength range for bruise detection has been explored due to the high absorption of hemoglobin and bilirubin. Also, the SWIR spectral range can be used to assess fluid accumulation (edema) and vasculature in the injured area. An example of this is shown in Figure 9.20, where a bruise is imaged on the volar side of the forearm of a male volunteer.[139] The bruise was created using a paintball gun fired at a distance of approximately

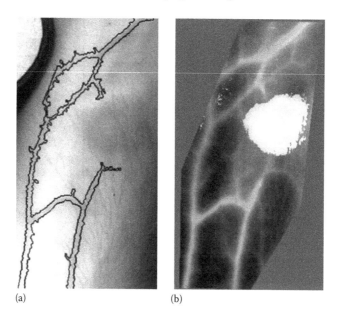

(a)   (b)

FIGURE 9.20 Paintball bruise on a male subject. Image collected within 1 h after injury. (a) Grayscale image from hyperspectral data (967 nm) with major vessel structures detected and masked out by statistical methods. Note that the bruise is shielding the vessel underneath. (b) Grayscale representation of hyperspectral composite image of bruise, demonstrating fluid accumulation (edema) in the bruised area (localized region shown as white).

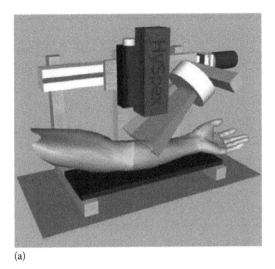

(a)

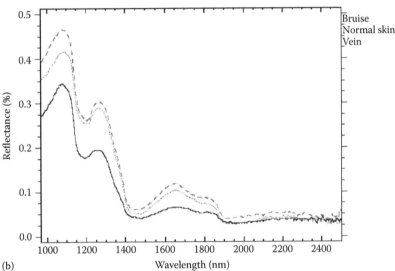

(b)

FIGURE 9.21 (a) HySpex SWIR-320m-e imaging system. (b) Reflectance spectra collected from (a). The most prominent spectral feature is water absorption.

10 m and imaged with a push-broom scanning SWIR hyperspectral camera (HySpex SWIR-320m-e, Norsk Elektro Optikk AS, Lørenskog, Norway). A hyperspectral composite image (Figure 9.20b) demonstrates fluid accumulation (edema) in the bruised area. Shown in Figure 9.21, this camera has 320 spatial pixels across the field of view and 256 spectral bands in the wavelength range 1.0–2.5 µm. Sample spectra from the image cube are shown (Figure 9.21b) for bruise, vein, and normal skin.

9.7 Conclusion

Fueled by rapid advances in instrumentation, software and algorithmic developments, novel dyes and chromogens, improvements in sample processing, and, stimulated by the genomics revolution, a need to increase throughput and multiplexing capabilities, spectral imaging is poised to make an ever-increasing contribution to biomedicine and related arts.

Disclaimer

This research was carried out partially at the Jet Propulsion Laboratory, California Institute of Technology, under a contract with the National Aeronautics and Space Administration.

References

1. Kruse, F. A., Analysis of AVIRIS data for the northern Death Valley region, California/Nevada, *Proceedings of the Second AVIRIS Workshop*, JPL publication 90-54, Pasadena, CA, 1990.
2. Johonson, L. F., F. Baret, and D. Peterson, Oregon transect: Comparison of leaf-level with canopy-level and modeled reflectance, *Summaries of the Third Annual JPL Airborne Geoscience Workshop*, JPL publication 92-14, Pasadena, CA, Vol. 1, pp. 11–13, 1992.
3. Fitzgerald, G. J., S. J. Maas, and W. R. DeTar, Early detection of spider mites in cotton using multi-spectral remote sensing, *Beltwide Cotton Conference*, Orlando, FL, 1999.
4. Abdulrauf, B. M., M. F. Stranc, M. G. Sowa, S. L. Germscheid, and H. H. Mantsch, A novel approach in the evaluation of flap failure using near IR spectroscopy and imaging, *Plastic Reconstructive Surgery* **8**, 68–72, 2000.
5. Sowa, M. G., J. R. Payette, M. D. Hewko, and H. H. Mantsch, Visible-near infrared multispectral imaging of the rat dorsal skin flap, *Journal of Biomedical Optics* **4**, 474–481, 1999.
6. Ruifrok, A. C., Quantification of immunohistochemical staining by color translation and automated thresholding, *Analytical and Quantitative Cytology and Histology* **19**, 107–113, 1997.
7. Ruifrok, A. C. and D. A. Johnston, Quantification of histochemical staining by color deconvolution, *Analytical and Quantitative Cytology and Histology* **4**, 291–299, 2001.
8. Hagan, N., R. T. Kester, L. Gao, and T. S. Tkaczyk, Snapshot advantage: A review of the light collection improvement for parallel high-dimensional measurement systems, *Optical Engineering* **51**(11), 111702, 2012.
9. Speicher, M. R., S. G. Ballard, and D. C. Ward, Karyotyping human chromosomes by combinatorial multi-fluor FISH, *Nature Genetics* **4**, 368–375, 1996.
10. Berns, R., Practical spectral imaging using a color-filter array digital camera, *Studies in Conservation* 2006. http://art-si.org/PDFs/Acquisition/TR_Practical_Spectral_Imaging.pdf (accessed February 16, 2014).
11. Berns, R. S. and F. H. Imai, The use of multi-channel visible spectrum imaging for pigment identification, in: *Preprints of the 13th Triennial Meeting of the ICOM Committee for Conservation*, ICOM, Rio de Janeiro, pp. 217–222, 2002.
12. Brauers, J. and T. Aach, Geometric calibration of lens and filter distortions for multispectral filter-wheel cameras, *IEEE Transactions on Image Processing* **20**(2), 496–505, 2011, doi: 10.1109/TIP.2010.2062193.
13. Sprigle, S., J. Caspall, L. Kong, and M. Duckworth, *Development of Handheld Erythema and Bruise Detectors*, Georgia Institute of Technology, Center for Assistive Technology and Environmental Access, Atlanta, GA, 2007. http://hdl.handle.net/1853/43222.
14. Qi, H., L. Kong, C. Wang, and L. Miao, A hand-held mosaicked multispectral imaging device for early stage pressure ulcer detection, *Journal of Medical Systems* **35**(5), 895–904, 2010, doi: 10.1007/s10916-010-9508-x.
15. Yi, D. and L. Kong, Fabrication of densely patterned micro-arrayed multichannel optical filter mosaic, *Journal of Micro/Nanolithography, MEMS, and MOEMS* **10**(3), 033020-1–033020-6, 2011, doi: 10.1117/1.3639188.
16. Yi, D. and L. Kong, Novel instrumentation of multispectral imaging technology for detecting tissue abnormity, in: A. A. S. Awwal and K. M. Iftekharuddin (eds.), Presented at the *SPIE Optical Engineering Proceedings*, San Diego, CA, Vol. 8498, 2012, doi: 10.1117/12.929398.

17. Yi, D., C. Wang, H. Qi, L. Kong, F. Wang, and A. Adibi, Real-time multispectral imager for home-based health care, *IEEE Transactions on Biomedical Engineering* **58**(3), 736–740, 2011, doi: 10.1109/TBME.2010.2077637.

18. Gat, N., Imaging spectroscopy using tunable filters: A review, *Proceedings of the SPIE* **4056**, 50–64, 2000.

19. Morris, H. R., C. C. Hoyt, and P. J. Treado, Imaging spectrometers for fluorescence and Raman micros-copy—Acousto-optic and liquid-crystal tunable filters, *Applied Spectroscopy* **48**, 857–866, 1994.

20. Morris, H. R., C. C. Hoyt, and P. Treado, Liquid crystal tunable filter Raman chemical imaging, *Applied Spectroscopy* **50**, 805–811, 1996.

21. Landsford, R., G. H. Bearman, and S. Fraser, Resolution of multiple green fluorescent protein color variants and dyes using two-photon microscopy and imaging spectroscopy, *Journal of Biomedical Optics* **6**, 311–318, 2001.

22. Bearman, G. H. and S. I. Spiro, Archeological applications of advanced imaging techniques, *Biblical Archeologist* **59**, 56–66, 1996.

23. Wachman, E. S., W. Niu, and D. L. Farkas, Imaging acousto-optic tunable filter with 0.35 micrometer spatial resolution, *Applied Optics* **35**, 5220–5226, 1996.

24. Lou Denes, CMRI, personal communication, 2001.

25. LightForm, Inc., http://www.lightforminc.com/pariss_instrumentation.html (accessed February 16, 2014).

26. Garini, Y., N. Katzir, D. Cabib, B. Buckwald, D. Soenksen, and Z. Malik, Spectral bio-imaging, in: *Fluorescence Imaging Spectroscopy and Microscopy*, Chemical Analysis Series, Vol. 137, John Wiley & Sons, Inc., New York, 1996.

27. Schröck, E., S. du Manoir, T. Veldman, B. Schoell, J. Wienberg, M. A. Ferguson-Smith, Y. Ning et al., Multicolor spectral karyotyping of human chromosomes, *Science* **26**, 494–497, 1996.

28. Malik, Z., D. Cabib, R. Buckwald, R. Talmi, Y. Garini, and S. Lipson, Fourier transform multipixel spectroscopy for quantitative cytology, *Journal of Microscopy* **182**, 133–140, 1996.

29. Itoh, K., W. Watanabe, and Y. Masuda, Parallelisms in interferometric fast spectral imaging, *Proceedings of the SPIE* **326**1, 278–288, 1998.

30. Garini, Y., A. Gil, I. Bar-Am, D. Cabib, and N. Katzir, Signal to noise analysis of multiple color fluo-rescence imaging microscopy, *Cytometry* **35**, 214–226, 1999.

31. Castleman, K. R., R. Eils, L. Morrison, J. Piper, K. Saracoglu, M. A. Schulze, and M. R. Speicher, Classification accuracy in multiple color fluorescence imaging microscopy, *Cytometry* **41**, 139–147, 2000.

32. Miller, P. and A. Harvey, Signal-to-noise analysis of various imaging systems, *Proceedings of the SPIE* **4259**, 16–21, 2001.

33. Okamato, T. and I. Yamaguchi, Simultaneous acquisition of spectral image information, *Optics Letters* **16**, 1277–1279, 1991.

34. Descour, M., Non-scanning imaging spectrometry, PhD dissertation, University of Arizona, Tucson, AZ, 1994.

35. Volin, C. E., B. K. Ford, M. R. Descour, D. W. Wilson, P. M. Maker, and G. H. Bearman, High speed spectral imager for imaging transient fluorescence phenomena, *Applied Optics* **37**, 8112–8119, 1998.

36. HORIBA Scientific, http://www.horiba.com/us/en/scientific/products/hyperspectral-imaging/details/verde-hyperspectral-camera-16537/ (accessed February 16, 2014).

37. Wilson, D., P. Maker, and R. Muller, Binary optic reflection grating for an imaging spectrometer, *Diffractive and Holographic Optics Technology III*, SPIE **2689**, 255–267, 1996.

38. Ford, B. K., C. E. Volin, S. M. Murphy, R. M. Lynch, and M. R. Descour, Computed tomography-based spectral imaging for fluorescence microscopy, *Biophysical Journal* **80**, 986–993, 2001.

39. de la Iglesia, F., J. Haskins, D. Farkas, and G. Bearman, Coherent multi-probes and quantitative spectroscopic multimode microscopy for the study of simultaneous intracellular events, *Proceedings of the International Society of Analytical Cytology, 20 Congress*, Montpellier, France, June 2000.

40. Gao, L., R. T. Kester, N. Hagen, and T. S. Tkaczyk, Snapshot Image Mapping Spectrometer (IMS) with high sampling density for hyperspectral microscopy, *Optics Express* **18**(14), 14330–14344, 2010, doi: 10.1364/OE.18.014330.

41. Treado, P. J. and M. D. Morris, Multichannel Hadamard transform Raman microscopy, *Applied Spectroscopy* **44**, 1–4, 1990.

42. Hanley, Q. S., P. J. Verveer, and T. M. Jovin, Spectral imaging in a programmable array microscope by Hadamard transform fluorescence spectroscopy, *Applied Spectroscopy* **53**, 1–10, 1999.

43. Nelson, P. M. and M. L. Myrick, Fabrication and evaluation of a dimension-reduction fiber-optic system for chemical imaging applications, *Review of Scientific Instruments* **70**, 2836–2844, 1999.

44. Nelson, P. M. and M. L. Myrick, Single-frame chemical imaging: Dimension reduction fiber-optic array improvements and application to laser-induced breakdown spectroscopy, *Applied Spectroscopy* **53**, 751–759, 1999.

45. Ma, J. and D. Ben-Amotz, Rapid micro-imaging using fiber-bundle image compression, *Applied Spectroscopy* **51**, 1845–1848, 1997.

46. Miller, P. J. and R. Levenson, Beyond image cubes: An agile lamp for practical 100% photon-efficient spectral imaging, *Proceedings of the SPIE* **4259**, 1–7, 2001.

47. Khojasteh, M. and C. MacAulay, Selective excitation light fluorescence (SELF) imaging, in: D. L. Farkas, D. V. Nicolau, and R. C. Leif (eds.), Presented at the *BiOS, SPIE*, San Francisco, CA, Vol. 7568, pp. 75680F-1–75680F-5, 2010, doi: 10.1117/12.843739.

48. Lee, M. H., H. Park, I. Ryu, and J. I. Park, Fast model-based multispectral imaging using nonnegative principal component analysis, *Optics Letters* **37**(11), 1937–1939, 2012.

49. Dickinson, M. E., C. W. Waters, and G. Bearman, Sensitive imaging of spectrally overlapping fluorochromes using the LSM 510 META. *Proceedings of the SPIE*, San Jose, CA, 2002.

50. Dinant, C., M. E. van Royen, W. Vermeulen, and A. B. Houtsmuller, Fluorescence resonance energy transfer of GFP and YFP by spectral imaging and quantitative acceptor photobleaching, *Journal of Microscopy* **231**(1), 97–104, 2008, doi: 10.1111/j.1365-2818.2008.02020.x.

51. http://www.ece.purdue.edu/~biehl/MultiSpec/.

52. Perkins, S. J., J. Theiler, S. P. Brumby, N. R. Harvey, R. B. Porter, J. J. Szymanski, and J. J. Bloch, GENIE: A hybrid genetic algorithm for feature classification in multispectral images, *Proceedings of the SPIE* **4120**, 52–62, 2000.

53. de Wolf, G. and L. J. van Vliet, Design of a four channel spectral analyzer to resolve linear combinations of two fluorescent spectra, *Proceedings of the SPIE* **3920**, 21–29, 2000.

54. Jimenez, L. O. and D. A. Landgrebe, Supervised classification in high dimensional space: Geometric, statistical, and asymptotical properties of multivariate data, *IEEE Transactions on Systems, Man, and Cybernetics, Part C: Applications and Reviews* **28**, 39–54, 1998.

55. Ornberg, R. L., B. M. Woerner, and D. A. Edwards, Analysis of stained objects in histological sections by spectral imaging and differential absorption, *Journal of Histochemistry & Cytochemistry* **47**, 1307–1314, 1999.

56. Mark, H. L. and D. Tunnell, Qualitative near-infrared reflectance analysis using Mahalanobis distances, *Analytical Chemistry* **57**, 1449–1456, 1985.

57. Kruse, F. A., A. B. Lefdoff, J. W. Boardman, K. B. Heidebrecht, A. T. Shapiro, J. P. Barloon, and A. F. Goetz, The spectral image processing system (SIPS)—Interactive visualization and analysis of imaging spectrometer data, *Remote Sensing of Environment* **44**, 145–163, 1993.

58. Harsanyi, J. C. and C. I. Chang, Hyperspectral image classification and dimensionality reduction: An orthogonal subspace projection approach, *IEEE Transactions on Geoscience and Remote Sensing* **32**, 779–785, 1994.

59. Garini, Y., N. Katzir, D. Cabib, R. A. Buckwald, D. G. Soenksen, and Z. Malik, Spectral bio-imaging, in: X. F. Wang and B. Herman (eds.), *Fluorescence Imaging Spectroscopy and Microscopy*, John Wiley & Sons, Inc., New York, 1996.

60. Farkas, D. L., C. Du, G. W. Fisher, C. Lau, W. Niu, E. S. Wachman, and R. M. Levenson, Non-invasive image acquisition and advanced processing in optical bioimaging, *Computerized Medical Imaging and Graphics* **22**, 89–102, 1998. cit_af ref_bf(Farkas, D. L. 1998 ref_num113)ref_af.

61. Winter, M. E., Fast autonomous spectral end-member determination in hyperspectral data, *Proceedings of the 13th International Conference on Applied Geologic Remote Sensing*, Vancouver, British Columbia, Canada, Vol. 2, pp. 337–344, 1999.

62. Landgrebe, D. A., Information extraction principles and methods for multispectral and hyperspectral image data, in: C. H. Chen (ed.), *Information Processing for Remote Sensing*, World Scientific, River Edge, NJ, 2000.

63. Mansfield, J. R., M. G. Sowa, J. R. Payette, B. Abdulrauf, M. F. Stranc, and H. H. Mantsch, Tissue viability by multispectral near infrared imaging: A fuzzy C-means clustering analysis, *IEEE Transactions on Medical Imaging* **17**, 1011–1018, 1998.

64. Perkins, S., N. R. Harvey, S. P. Brumby, and K. Lacker, Support vector machines for broad area feature extraction in remotely sensed images, *Proceedings of the SPIE* **4381**, 268–295, 2001.

65. Fereidouni, F., A. Bader, and H. Gerritsen, Spectral phasor analysis allows rapid and reliable unmixing of fluorescence microscopy spectral images, *Optics Express* **20**, 12729–12741, 2012.

66. Clarke, M., J. Lee, D. Samarov, D. Allen, M. Litorja, R. Nossal, and J. Hwang, Designing microarray phantoms for hyperspectral imaging validation, *Biomedical Optics Express* **3**, 1291–1299, 2012.

67. Oeltze, S., W. Freiler, R. Hillert, H. Doleisch, B. Preim, and W. Schubert, Interactive, graph-based visual analysis of high-dimensional, multi-parameter fluorescence microscopy data in toponomics, *IEEE Transactions on Visualization and Computer Graphics* **17**(12), 1882–1891, December 2011.

68. Schubert, W., A. Gieseler, A. Krusche, P. Serocka, and R. Hillert, Next-generation biomarkers based on 100-parameter functional super-resolution microscopy TIS, *New Biotechnology* **29**(5), 599–610, June 15, 2012, , ISSN 1871-6784. http://dx.doi.org/10.1016/j.nbt.2011.12.004.

69. Gerdes, M. J., C. J. Sevinsky, A. Sood, S. Adak, M. O. Bello, A. Bordwell, A. Can et al., Highly multiplexed single-cell analysis of formalin-fixed, paraffin-embedded cancer tissue, *Proceedings of the National Academy of Science of the United States of America* **110**(29), 11982–11997, July 16, 2013, doi: 10.1073/pnas.1300136110. Epub July 1, 2013.

70. Fauch, L., E. Nippolainen, V. Teplov, and A. A. Kamshilin, Re-covery of reflection spectra in a multispectral imaging system with light emitting diodes, *Optics Express* **18**, 23394–23405, 2010.

71. Lee, M., H. Park, I. Ryu, and J. Park, Fast model-based multispectral imaging using nonnegative principal component analysis, *Optics Letters* **37**, 1937–1939, 2012.

72. Sun, R., M. B. Bouchard, and E. M. Hillman, SPLASSH: Open source software for camera-based high-speed, multispectral in-vivo optical image acquisition, *Biomedical Optics Express* **1**(2), 385–397, 2010.

73. Tominaga, S. and T. Horiuchi, Spectral imaging by synchronizing capture and illumination, *Journal of the Optical Society of America* **29**, 1764–1775, 2012. http://dx.doi.org/10.1364/JOSAA.29.001764.

74. Gilbert, C. M. and A. Parwani, The use of multispectral imaging to distinguish reactive urothelium from neoplastic urothelium, *Journal of Pathology Informatics* **1**, 23, Published online October 11, 2010, doi: 10.4103/2153-3539.71064 (accessed February 16, 2014).

75. Dolloff, N. G., X. Ma, D. T. Dicker, R. C. Humphreys, L. Z. Li, and W. S. El-Deiry, Spectral imaging-based methods for quantifying autophagy and apoptosis, *Cancer Biology & Therapy* **12**(4), 349–356. Epub August 15, 2011.

76. Fiore, C., D. Bailey, N. Conlon, X. Wu, N. Martin, M. Fiorentino, S. Finn et al., Utility of multispectral imaging in automated quantitative scoring of immunohistochemistry, *Journal of Clinical Pathology* **65**(6), 496–502, Published Online First March 23, 2012, doi:10.1136/jclinpath-2012-200734.

77. Huang, W., K. Hennrick, and S. Drew, A colorful future of quantitative pathology: Validation of Vectra technology using chromogenic multiplexed immunohistochemistry and prostate tissue microarrays, *Human Pathology* **44**(1), 29–38, January 2013, doi: 10.1016/j.humpath.2012.05.009. Epub August 31, 2012.

78. Dave, S. R., C. C. White, X. Gao, and T. J. Kavanagh, Luminescent quantum dots for molecular toxicology, *Advances in Experimental Medicine and Biology* **745,** 117–137, 2012, doi: 10.1007/978-1-4614-3055-1_8.
79. Jin, Z. and N. Hildebrandt, Quantum dots for in vitro diagnostics and cellular imaging, *Trends in Biotechnology* **30**(7), 394–403, 2012.
80. Liu, J., S. K. Lau, V. A. Varma, R. A. Moffitt, M. Caldwell, T. Liu, A. N. Young et al., Molecular mapping of tumor heterogeneity on clinical tissue specimens with multiplexed quantum dots, *ACS Nano* **4**(5), 2755–2765, May 25, 2010, doi: 10.1021/nn100213v.
81. Guo, J., S. Wang, N. Dai, Y. N. Teo, E. T. Kool, Multispectral labeling of antibodies with polyfluorophores on a DNA backbone and application in cellular imaging, *Proceedings of the National Academy of Science of the United States of America* **108**(9), 3493–3498, March 1, 2011, doi: 10.1073/pnas.1017349108. Epub February 14, 2011.
82. Ried, T., M. Liyanage, S. du Manoir, K. Heselmeyer, G. Auer, M. Macville, and E. Schrock, Tumor cytogenetics revisited: Comparative genomic hybridization and spectral karyotyping, *Journal of Molecular Medicine* **75**, 801–814, 1997.
83. Chattopadhyay, P. K., S. P. Perfetto, and M. Roederer, The colorful future of cell analysis by flow cytometry, *Discovery Medicine*, **4**, 255–262, 2004. Retrieved from http://www.biomedsearch.com/nih/colorful-future-cell-analysis-by/20704956.html.
84. Brelje, T. C., M. W. Wessendorf, and R. L. Sorenson, Multicolor laser scanning confocal immunofluorescence microscopy: Practical application and limitations, in: Matsumoto, B. (ed.), *Cell Biological Applications of Confocal Microscopy*, Vol. 38, Academic Press, San Diego, CA, pp. 97–181, 1993.
85. Taylor, C. R. and R. J. Cote, Immunohistochemical markers of prognostic value in surgical pathology, *Histology and Histopathology* **12**, 1039–10552, 1997.
86. Albonico, G., P. Querzoli, S. Ferretti, E. Magri, and I. Nenci, Biophenotypes of breast carcinoma in situ defined by image analysis of biological parameters, *Pathology—Research and Practice* **192**, 117–123, 1996.
87. Zhou, R., D. L. Parker, and E. H. Hammond, Quantitative peroxidase–antiperoxidase complex-substrate mass determination in tissue sections by a dual wavelength method, *Analytical and Quantitative Cytology and Histology* **14**, 73–80, 1992.
88. Fritz, P., X. Wu, H. Tuczek, H. Multhaupt, and P. Schwarzmann, Quantitation in immunohistochemistry: A research method or a diagnostic tool in surgical pathology? *Pathologica* **87**, 300–309, 1995.
89. Speel, E. J., F. C. Ramaekers, and A. H. Hopman, Cytochemical detection systems for in situ hybridization, and the combination with immunocytochemistry: "Who is still afraid of red, green and blue?", *Histochemical Journal* **27**, 833–858, 1996.
90. Hopman, A. H., S. Claessen, and E. J. Speel, Multi-colour brightfield in situ hybridisation on tissue section, *Histochemistry and Cell Biology* **108**, 291–298, 1998.
91. Speel, E. J., A. H. Hopman, and P. Komminoth, Amplification methods to increase the sensitivity of in situ hybridization: Play card(s), *Journal of Histochemistry & Cytochemistry* **47**, 281–288, 1999.
92. Levenson, R. M., Putting the "more" back in morphology: Spectral imaging and image analysis in the service of pathology, *Archives of Pathology & Laboratory Medicine* **132**, 748–757, 2008.
93. Itzkan, I., L. Qiu, H. Fang, M. M. Zaman, E. Vitkin, I. C. Ghiran, S. Salahuddin et al., Confocal light absorption and scattering spectroscopic microscopy monitors organelles in live cells with no exogenous labels, *Proceedings of the National Academy of Sciences of the United States of America* **104**(44), 17255–17260, October 30, 2007. Epub October 23, 2007.
94. Evans, C. L., E. O. Potma, M. Puoris'haag, D. Côté, C. P. Lin, and X. S. Xie, Chemical imaging of tissue in vivo with video-rate coherent anti-Stokes Raman scattering microscopy, *Proceedings of the National Academy of Sciences of the United States of America* **102**(46), 16807–16812, 2005. Epub November 1, 2005.
95. Krasieva, T. B., C. Stringari, F. Liu, C. H. Sun, Y. Kong, M. Balu, F. L. Meyskens, E. Gratton, and B. J. Tromberg, Two-photon excited fluorescence lifetime imaging and spectroscopy of melanins in vitro and in vivo, *Journal of Biomedical Optics* **18**(3), 31107, 2013, doi: 10.1117/1.JBO.18.3.031107.

96. Tearney, G. J., R. H. Webb, and B. E. Bouma, Spectrally encoded confocal microscopy, *Optics Letters* **23**, 1152–1154, 1998.

97. Boudoux, C., S. Yun, W. Oh, W. White, N. Iftimia, M. Shishkov, B. Bouma, and G. Tearney, Rapid wavelength-swept spectrally encoded confocal microscopy, *Optics Express* **13**(20), 8214–8221, 2005.

98. Boudoux, C., S. C. Leuin, W. Y. Oh, M. J. Suter, A. E. Desjardins, B. J. Vakoc, B. E. Bouma, C. J. Hartnick, and G. J. Tearney, Optical microscopy of the pediatric vocal fold, *Archives of Otolaryngology— Head & Neck Surgery* **135**(1), 53–64, 2009, doi: 10.1001/archoto.2008.518.

99. Qiu, L., V. Turzhitsky, R. Chuttani, D. Pleskow, J. D. Goldsmith, L. Guo, E. Vitkin, I. Itzkan, E. B. Hanlon, and L. T. Perelman, Spectral imaging with scattered light: From early cancer detection to cell biology, *IEEE Journal of Selected Topics in Quantum Electronics* **18**(3), 1073–1083, 2012. Epub June 4, 2012.

100. Dunn, A. and D. Boas, Transport-based image reconstruction in turbid media with small source-detector separations, *Optics Letters* **25**(24), 1777–1779, December 15, 2000.

101. Hillman, E. M., D. A. Boas, A. M. Dale, and A. K. Dunn, Laminar optical tomography: Demonstration of millimeter-scale depth-resolved imaging in turbid media, *Optics Letters* **29**(14), 1650–1652, 2004.

102. Muldoon, T. J., S. A. Burgess, B. R. Chen, D. Ratner, and E. M. Hillman, Analysis of skin lesions using laminar optical tomography, *Biomedical Optics Express* **3**(7), 1701–1712, 2012, doi: 10.1364/ BOE.3.001701. Epub June 22, 2012.

103. Krishnaswamy, V., A. M. Laughney, K. D. Paulsen, and B. W. Pogue, Dark-field scanning in situ spectroscopy platform for broadband imaging of resected tissue, *Optics Letters* **36**(10), 1911–1913, 2011, doi: 10.1364/OL.36.001911.

104. Pu, J., G. S. Mintz, E. S. Brilakis, S. Banerjee, A. R. Abdel-Karim, B. Maini, S. Biro et al., In vivo characterization of coronary plaques: Novel findings from comparing greyscale and virtual histology intravascular ultrasound and near-infrared spectroscopy, *European Heart Journal* **33**(3), 372–383, 2012, doi: 10.1093/eurheartj/ehr387. Epub October 20, 2011.

105. DaCosta, R. S., B. C. Wilson, and N. E. Marcon, Fluorescence and spectral imaging, *Scientific World Journal* **7**, 2046–2071, 2007, doi: 10.1100/tsw.2007.308.

106. Roblyer, D., R. Richards-Kortum, K. Sokolov, A. K. El-Naggar, M. D. Williams, C. Kurachi, and A. M. Gillenwater, Multispectral optical imaging device for in vivo detection of oral neoplasia, *Journal of Biomedical Optics* **13**(2), 024019, 2008, doi: 10.1117/1.2904658.

107. Gebhart, S. C., R. C. Thompson, and A. Mahadevan-Jansen, Liquid-crystal tunable filter spectral imaging for brain tumor demarcation, *Applied Optics* **46**(10), 1896–1910, 2007.

108. Keller, M. D., S. K. Majumder, M. C. Kelley, I. M. Meszoely, F. I. Boulos, G. M. Olivares, and A. Mahadevan-Jansen, Autofluorescence and diffuse reflectance spectroscopy and spectral imaging for breast surgical margin analysis, *Lasers in Surgery and Medicine* **42**(1), 15–23, 2010, doi: 10.1002/lsm.20865.

109. Olweny, E. O., S. Faddegon, S. L. Best, N. Jackson, E. F. Wehner, Y. K. Tan, K. J. Zuzak, and J. A. Cadeddu, Renal oxygenation during robotic-assisted laparoscopic partial nephrectomy; characterization using laparoscopic digital light processing (DLP) hyperspectral imaging, *Journal of Endourology* 2012.

110. Pharaon, M. R., T. Scholz, S. Bogdanoff, D. Cuccia, A. J. Durkin, D. B. Hoyt, and G. R. Evans, Early detection of complete vascular occlusion in a pedicle flap model using quantitative spectral imaging, *Plastic and Reconstructive Surgery* **126**(6), 1924–1935, 2010, doi: 10.1097/PRS.0b013e3181f447ac.

111. Gioux, S., A. Mazhar, B. T. Lee, S. J. Lin, A. M. Tobias, D. J. Cuccia, A. Stockdale et al., First-in-human pilot study of a spatial frequency domain oxygenation imaging system, *Journal of Biomedical Optics* **16**(8), 086015, 2011, doi: 10.1117/1.3614566.

112. Corlu, A., R. Choe, T. Durduran, K. Lee, M. Schweiger, S. R. Arridge, E. M. Hillman, and A. G. Yodh, Diffuse optical tomography with spectral constraints and wavelength optimization, *Applied Optics* **44**(11), 2082–2093, 2005.

113. Mazhar, A., S. Dell, D. J. Cuccia, S. Gioux, A. J. Durkin, J. V. Frangioni, and B. J. Tromberg, Wavelength optimization for rapid chromophore mapping using spatial frequency domain imaging, *Journal of Biomedical Optics* 15(6), 061716, 2010, doi: 10.1117/1.3523373.
114. N. MacKinnon, U. Stange, P. Lane, C. MacAulay, and M. Quatrevalet, Spectrally programmable light engine for in vitro or in vivo molecular imaging and spectroscopy, *Applied Optics* 44(11), 2033–2040, 2005.
115. Xu, R. X., D. W. Allen, J. Huang, S. Gnyawali, J. Melvin, H. Elgharably, G. Gordillo et al., Developing digital tissue phantoms for hyperspectral imaging of ischemic wounds, *Biomedical Optics Express* 3(6), 1433–1445, 2012, doi: 10.1364/BOE.3.001433. Epub May 18, 2012.
116. Moncrieff, M., S. Cotton, E. Claridge, and P. Hall, Spectrophotometric intracutaneous analysis: A new technique for imaging pigmented skin lesions, *British Journal of Dermatology* 146(3), 448–457, 2002.
117. Yudovsky, D., A. Nouvong, K. Schomacker, and L. Pilon, Monitoring temporal development and healing of diabetic foot ulceration using hyperspectral imaging, *Journal of Biophotonics* 4(7–8), 565–576, 2011, doi: 10.1002/jbio.201000117. Epub April 1, 2011.
118. Gillies, R., J. E. Freeman, L. C. Cancio, D. Brand, M. Hopmeier, and J R. Mansfield, Systemic effects of shock and resuscitation monitored by visible hyperspectral imaging, *Diabetes Technology & Therapeutics* 5(5), 847–855, 2003.
119. Tehrani, H., M. Moncrieff, B. Philp, and P. Dziewulski, Spectrophotometric intracutaneous analysis: A novel imaging technique in the assessment of acute burn depth, *Annals of Plastic Surgery* 61(4), 437–440, 2008, doi: 10.1097/SAP.0b013e31815f12e6.
120. Mazhar, A., S. A. Sharif, J. D. Cuccia, J. S. Nelson, K. M. Kelly, and A. J. Durkin, Spatial frequency domain imaging of port wine stain biochemical composition in response to laser therapy: A pilot study, *Lasers in Surgery and Medicine* 44(8), 611–621, 2012, doi: 10.1002/lsm.22067. Epub August 21, 2012.
121. Saager, R. B., D. J. Cuccia, S. Saggese, K. M. Kelly, and A. J. Durkin, Quantitative fluorescence imaging of protoporphyrin IX through determination of tissue optical properties in the spatial frequency domain, *Journal of Biomedical Optics* 16(12), 126013, 2011, doi: 10.1117/1.3665440.
122. Kainerstorfer, J. M., P. D. Smith, and A. H. Gandjbakhche, Noncontact wide-field multispectral imaging for tissue characterization, *IEEE Journal of Selected Topics in Quantum Electronics* 18(4), 1343–1354, 2012, doi: 10.1109/JSTQE.2011.2175708 (accessed February 16, 2014).
123. Stamatas, G. N. and N. Kollias, In vivo documentation of cutaneous inflammation using spectral imaging, *Journal of Biomedical Optics* 12(5), 051603, September–October 2007.
124. Jacques, S. L., R. Samatham, and N. Choudhury, Rapid spectral analysis for spectral imaging, *Biomedical Optics Express* 1(1), 157–164, 2010.
125. Erickson, T. A., A. Mazhar, D. Cuccia, A. J. Durkin, and J. W. Tunnell, Lookup-table method for imaging optical properties with structured illumination beyond the diffusion theory regime, *Journal of Biomedical Optics* 15(3), 036013, 2010, doi: 10.1117/1.3431728.
126. Kainerstorfer, J. M., F. Amyot, M. Ehler, M. Hassan, S. G. Demos, V. Chernomordik, C. K. Hitzenberger, A. H. Gandjbakhche, and J. D. Riley, Direct curvature correction for noncontact imaging modalities applied to multispectral imaging, *Journal of Biomedical Optics* 15(4), 046013, 2010, doi: 10.1117/1.3470094.
127. Gioux, S., A. Mazhar, D. J. Cuccia, A. J. Durkin, B. J. Tromberg, and J. V. Frangioni, Three-dimensional surface profile intensity correction for spatially modulated imaging, *Journal of Biomedical Optics* 14(3), 034045, 2009, doi: 10.1117/1.3156840.
128. Jacques, S. L., R. Samatham, and N. Choudhury, Rapid spectral analysis for spectral imaging, *Biomedical Optics Express* 1(1), 157–164, 2010.
129. Cuccia, D. J., F. Bevilacqua, A. J. Durkin, and B. J. Tromberg, Modulated imaging: Quantitative analysis and tomography of turbid media in the spatial-frequency domain, *Optics Letters* 30(11), 1354–1356, 2005. http://dx.doi.org/10.1364/OL.30.001354.

130. Cuccia, D. J., F. Bevilacqua, A. J. Durkin, F. R. Ayers, and B. J. Tromberg, Quantitation and mapping of tissue optical properties using modulated imaging, *Journal of Biomedical Optics* **14**(2), 024012, 2009, doi: 10.1117/1.3088140.

131. Ponticorvo, A., E. Taydas, A. Mazhar, T. Scholz, H. S. Kim, J. Rimler, G. R. Evans, D. J. Cuccia, and A. J. Durkin, Quantitative assessment of partial vascular occlusions in a swine pedicle flap model using spatial frequency domain imaging, *Biomedical Optics Express* **4**(2), 298–306, 2013, http://dx.doi.org/10.1364/BOE.4.000298.

132. Balu, M., A. Mazhar, C. K. Hayakawa, R. Mittal, T. B. Krasieva, K. König, V. Venugopalan, and B. J. Tromberg, In vivo multiphoton NADH fluorescence reveals depth-dependent keratinocyte metabolism in human skin, *Biophysical Journal* **104**(1), 258–267, 2013, doi: 10.1016/j.bpj.2012.11.3809. Epub January 8, 2013.

133. Durkin, A. J., J. G. Kim, and D. J. Cuccia, Quantitative near infrared imaging of skin flaps, *ASME 2008 Third Frontiers in Biomedical Devices Conference (BIOMED2008)*, Irvine, CA, 2008. http://dx.doi.org/10.1115/BioMed2008-38100.

134. Randeberg, L.L. et al., Hyperspectral imaging of bruised skin, *Progress in Biomedical Optics and Imaging—Proceedings of SPIE*, San Jose, CA, 2006.

135. Randeberg, L. L. and J. Hernandez-Palacios, Hyperspectral imaging of bruises in the SWIR spectral region, *Photonic Therapeutics and Diagnostics VIII*, Pts 1 and 2, San Francisco, CA, Vol. 8207, 2012.

136. Randeberg, L. L., E. L. Larsen, and L. O. Svaasand, Characterization of vascular structures and skin bruises using hyperspectral imaging, image analysis and diffusion theory, *Journal of Biophotonics* **3**(1–2), 53–65, 2010.

137. Randeberg, L. et al., The optics of bruising, in: A. J. Welch and M. J. C. Gemert (eds.), *Optical-Thermal Response of Laser-Irradiated Tissue*, Springer, Dordrecht, the Netherlands, pp. 825–858, 2011.

138. Randeberg, L. L., O. A. Haugen, R. Haaverstad, and L. O. Svaasand, A novel approach to age determination of traumatic injuries by reflectance spectroscopy, *Lasers in Surgery and Medicine* **38**(4), 277–289, 2006.

139. Data provided by Lise Lyngsnes Randeberg, Julio Hernandez-Palacios, Svein-Erik Fossbråten, Lukasz Paluchowski (unpublished).

10

Lifetime-Based Imaging

Petr Herman
Charles University

Joseph R. Lakowicz
*University of Maryland
School of Medicine*

10.1 Introduction

Since the first report on the fluorescence phenomenon in 1845,[1] interest in fluorescence has rapidly increased. During the last few decades, fluorescence spectroscopy, and particularly time-resolved fluorescence, has become recognized as an important research tool in the biological sciences.[2-7] Time-resolved fluorescence is of interest due to its higher information content compared to steady-state measurements. However, the information content of the time-resolved fluorescence is usually not accessible with classical imaging techniques such as fluorescence microscopy. The situation has changed as remarkable developments in the time-resolved fluorescence methods have recently facilitated the transfer of time-resolved fluorescence from the solution spectroscopy to the field of lifetime-based sensing and imaging.[8-16]

Fluorescence emission is a radiative process that occurs on the nanosecond time scale for most fluorophores. Normally, the fluorescence is a first-order kinetic process and the intensity decay obeys the exponential law. The fluorescence lifetime τ characterizes the average amount of time that a molecule spends in the excited state following absorption of a photon. The importance of the time-resolved fluorescence for imaging can be understood from the fact that unlike intensity, the fluorescence lifetime is mostly independent of the probe concentration, photobleaching, and lightpath. All these parameters are extremely difficult to control during microscopic cellular experiments. The difficulties can be overcome by fluorescence ratiometric measurements. However, the number of fluorescent ratiometric probes is much smaller than the class of the lifetime sensors.

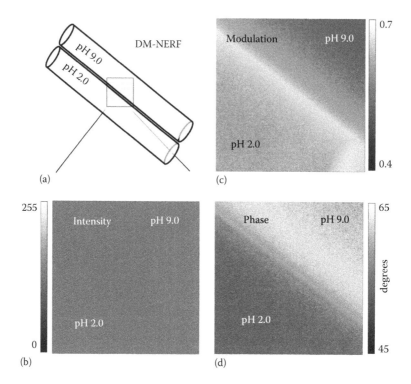

FIGURE 10.1   Demonstration of lifetime contrast in frequency domain FLIM. (a) Two capillaries filled with a buffer containing the pH-sensitive lifetime probe DM-NERF. The pH was adjusted to 9.0 and 2.0 in the upper and lower capillaries, respectively. Intensities in the both capillaries were adjusted to the same level. (b) Fluorescence intensity image does not show any contrast. (c) Modulation image. (d) Phase image. Images (c) and (d) exhibit significant contrast since modulation and phase shift depend on the fluorescence lifetime, Equation 10.9.

Lifetime-based imaging is an experimental technique in which characteristics of the fluorescence decay are measured at each spatially resolvable location within a fluorescent image. This allows generation of image contrast based on the lifetime. The concept of the lifetime contrast and its advantages for imaging applications are demonstrated in Figure 10.1. This figure shows two adjacent capillary tubes both containing the pH-sensitive fluorophore DM-NERF, but at two different pH values of 2 and 9. The steady-state intensities measured from both capillaries were identical (lower left). However, differences between the two capillaries are seen in the phase angle and modulation images (right panels), reflecting these different lifetimes. Hence, lifetime imaging can reveal spatial differences in the sample even when the steady-state intensities are identical.

For simplicity, we will not distinguish between fluorescence and phosphorescence lifetime imaging because the concepts are the same, only the time scales differ. Unless specifically noted, the same applies for fluorescence lifetime imaging (FLI) and FLI microscopy (FLIM), where the main difference between the techniques is an optical system for imaging of macroscopic and microscopic samples, respectively.

FLIM has evolved as a part of a broader discipline called fluorescence lifetime microscopy, Figure 10.2. Fluorescence lifetime microscopy includes also nonimaging applications where fluorescence decays are acquired from one or several areas of the sample.[17-27] Due to the potential of accessing intensity-independent lifetime information from subcellular volumes and due to a broad availability of lifetime probes, FLIM has rapidly gained high popularity not only in cellular biology, biophysics, and biomedical

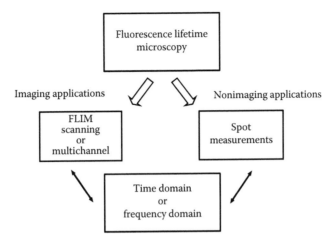

FIGURE 10.2 Fluorescence lifetime microscopy. Measurements may be done at a single location or in an imaging mode. Both TD and FD techniques can be used.

sciences, but it has been also used in forensic sciences for latent fingerprint detection[28] and in artwork analysis.[29] Various modalities and implementations of lifetime-based imaging have been previously discussed in numerous review articles.[8,9,13,15,16,30–38]

10.2 Techniques for Lifetime-Based Imaging

Spatially resolved measurements of fluorescence lifetimes can be accomplished by several means. Generally, we have two types of time-resolved fluorescence, the time-domain (TD) and the frequency-domain (FD) measurements. In principle, the two methods are equivalent. They yield the same kind of information about the examined object and they are related to each other by the Fourier transform. In practice, however, the methods use different instrumentation, the experimental data have a different appearance, and the data analysis is formally different. Depending on a specific application and detailed experimental conditions, the TD lifetime measurements can gain comparative advantage over the FD method and vice versa. However, it is mostly a personal preference or availability of instrumentation what decides whether the TD or FD approach is used.

Another aspect of FLI is the imaging modality. Images can be acquired pixel by pixel with a single-channel detector in a sequential scanning mode or line by line with an array detector in a line-scanning or push-broom configuration.[13] The scanning imaging techniques were originally developed for conventional and spectral imaging and later were successfully applied for the lifetime imaging as well. The FLIM images can also be acquired in a wide-field (WF) multichannel regime when lifetime information is simultaneously recorded from the whole field of view (from all pixels) by a 2D detector. This detector is typically an intensified CCD camera even though other spatially sensitive detectors such as multiple-anode PMT suitable for lifetime-based measurements have been reported.[39–41] In the next paragraphs, we will discuss in more detail different approaches to fluorescence lifetime microscopy and lifetime-based imaging.

10.2.1 Time-Domain

In the TD measurement, the fluorophore is excited by a short pulse of light and a time course of the fluorescence response is recorded (Figure 10.3). Because the rate of photoemission is proportional to the

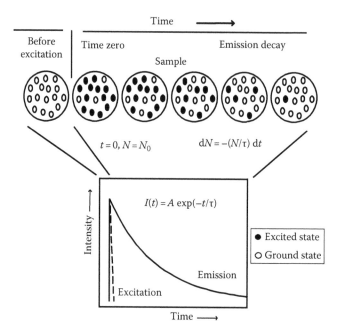

FIGURE 10.3   Concept of the emission lifetime. N, number of molecules in the excited state; τ, emission lifetime.

number of excited fluorophores, the fluorescence intensity decay $I(t)$ of a single ideal fluorophore is a monoexponential function. Often, however, the intensity decay follows a multiexponential law:

$$I(t) = \sum_i a_i \cdot e^{-t/\tau_i} \tag{10.1}$$

where
   τ_i are lifetimes
   a_i are the corresponding amplitudes

For a mixture of fluorophores, each displaying a single exponential decay, the decay time will be those of the individual fluorophores and the amplitudes will be related to the intensity and/or concentration of each species.

Several TD approaches have been used to measure fluorescence decays. They range from single-shot experiments, when the full fluorescence decay is acquired from a single excitation pulse with fast digitizers or streak cameras,[42] repetitive stroboscopic sampling, gated boxcar detection, or optical pump-probe methods,[43] to the widely used time-correlated single-photon counting (TCSPC),[44,45] which is a dominant method in nonimaging TD fluorescence spectroscopy. A number of these methods were adopted for lifetime-based imaging in a wide variety of instrumental implementations.

10.2.1.1 Time-Correlated Single-Photon Counting

A great deal of literature is available on the subject of the TCSPC. Extensive description and discussion of the method can be found, for example, in monographs by O'Connor and Philips[44] and Becker.[45] For comparison purposes with other lifetime imaging methods, it is useful to reiterate some properties and limitations of TCSPC.

The principle of the TCSPC is shown in the diagram in Figure 10.4. The method is based on repetitive measurements of the time difference between the excitation flash and a subsequent detection of the

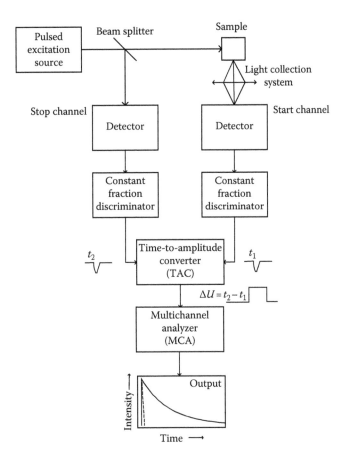

FIGURE 10.4 Schematic diagram of a TCSPC instrument.

first emission event. The time difference is typically measured by a time-to-amplitude converter (TAC), which outputs electrical pulses with amplitudes proportional to the time difference between the start and the stop pulses generated by detectors in the excitation and the emission channels. The output of the TAC is digitized and sorted by the multichannel analyzer (MCA), where a probability histogram of the count number versus time channels is built up by sampling a large number of excitation–emission events. If the emission intensity is low enough so that no more than one photon is detected per start–stop cycle, the histogram exactly reflects time evolution of the fluorescence intensity decay $I(t)$. In the opposite case, the TAC does not register the second arriving photon, and the measured intensity decay is biased to shorter times by the so-called pile-up effect. To avoid pulse pile up, the detection rate has to be limited so that detection of the two photons per excitation pulse is unlikely. An acquisition rate lower than 1–2 photons/100 excitation pulses is considered to be an upper practical limit for elimination of the pile-up effect.[44,45] Even lower acquisition rates are recommended for highly accurate measurements. At higher counting rates, the pile-up effect starts to significantly impair fluorescence decay curves and recovered lifetime parameters. Even though the pile up is predictable and it can be to some extent corrected mathematically[46–49] or suppressed by electronics discriminating the multipulse events,[45,50,51] the safest way to avoid pile-up problems is to use reasonably low counting rates.

The upper limit on the data acquisition rate, a common feature of all TAC-based TSCPC devices, places practical limitation on the 2D lifetime mapping where thousands of fluorescence decays have to be accumulated in order to obtain a lifetime image. As an example, let us assume a medium resolution 512 × 512-pixel image and a pulsed excitation with repetition rate of 80 MHz. To eliminate pile-up artifacts, the data collection rate should ideally not exceed 1% of the excitation frequency, that is, 800 kHz in the brightest pixel. However, due to the image nonuniformity, the average collection data rate might be significantly lower. Analysis of relations between number of acquired photons and accuracy of recovered lifetime parameters in TCSPC experiments has shown that very rough resolution of a single component decay needs at least 200 photons/decay.[52] Resolution of two components might require more than 10^5 counts/decay. For cuvette experiments, it is rather usual to collect 10^6–10^7 counts/decay in order to have data with a good dynamic range and signal-to-noise (S/N) ratio for multicomponent decay analysis. In FLI, we will be more likely to acquire only 10^3 counts/decay in a single pixel. This value is close to the limit where lifetime information starts to be hidden in the noise and further reduction of the counts per pixel might not be practical. Under these rather favorable conditions, the acquisition of the 512 × 512-pixel image would take more than 5 min. This time can be decreased for the cost of lower data accuracy. In our example, we used the laser repetition frequency of 80 MHz with light pulses separated by 12.5 ns. The time window between pulses in the excitation pulse train allows measurements of only lifetimes up to 2–3 ns, since fluorescence should substantially diminish before the next pulse arrives. For longer lifetimes, the excitation frequency has to be reduced and the acquisition time proportionally increases. It is evident that the TCSPC could be impractical for FLI experiments with long-lived fluorophores since the image acquisition time quickly exceeds acceptable limits. Nevertheless, with a proper fluorophore, limited spatial resolution, and limited number of counts/decay, the image acquisition time can reduce to a few seconds or several tens of seconds.[53,54] Acquisition times as short as 50 ms/image allowing *real-time* measurements of cellular dynamics were reported for an effective image size of 17 × 17 pixels.[54]

Despite the fundamental bottlenecks in the data acquisition rate, TCSPC has single-photon sensitivity, excellent time resolution, wide dynamic range, and known noise statistics. Due to its accuracy, it is suitable for determination of fractional contributions in a sample with multiple colocalized spectrally overlapping fluorescent species. Serial data acquisition with TCSPC is highly suitable for laser scanning confocal and multiphoton imaging where it can be relatively easy implemented. Commercial FLIM scanning microscopes have become available. These are reasons why the technique was employed for fluorescence lifetime microscopy or FLI by a number of researchers.[22,54–65] Applications range from a time-resolved fluorescence measurements from a single or multiple locations within a microscopic sample,[22,24] confocal scanning implementations with single[66–73] and multiphoton excitation,[54,58,61,62,74–84] to 3D imaging[57] and near-field scanning optical microscopy with fluorescence lifetime as a contrast mechanism.[85] Willemsen et al.[66] combined the TCSPC-based lifetime imaging with atomic force microscopy (AFM) in one scanning device. TCSPC has been also used for lifetime imaging in a scanning DNA sequencer.[56]

Methods have been developed to increase the data collection rate above the TCSPC limit. One of them uses parallel data acquisition with multichannel detectors. The detector could be a multianode PMT[39,41] placed in the image plane with outputs from the individual anodes routed to the multiplexed detection system. In this system, the count rate can be increased to a value close to 0.37,[86] which is a considerable improvement over single-channel TCSPC. Even though PMTs with up to 96 anodes have been reported,[87] the imaging capability and spatial resolution of FLIM instruments with the multianode PMTs is sufficient only for low spatial-resolution imaging. Multiplexed TCSPC with multichannel detector also found application in high acquisition rate scanning FLIM with spectral resolution.[62,88]

10.2.1.2 Multichannel Photon Counting

In order to overcome limitations of the TCSPC and increase the data collection rate required for 2D-lifetime scanning, a multichannel photon counting has been developed for lifetime microscopy.[23] The method utilizes simultaneous digital sampling of all photons in the emission burst while preserving their arrival-time information. Since the method does not use TAC for time measurements, the detection rate can exceed 1% limit of the TCSPC, which substantially reduces measurement time compared to the TCSPC. Modest time resolution in the nanosecond range, however, limits applicability of the technique for short-lived fluorophores.

10.2.1.3 Sampling Methods

Unlike the time-resolved cuvette experiments that are mostly performed by TCSPC, the field of the TD lifetime imaging often employs sampling methods with gated detection. The principle of the most commonly used boxcar approach[43,89] can be understood from Figure 10.5. The sample is excited by a periodic train of light pulses and the signal is recorded within a defined time interval Δt after each excitation pulse. More specifically, the detector is turned off during the fluorescence decay except for a brief period of time Δt. The fluorescence intensity in the time window of t and $t + \Delta t$ is repetitively accumulated until a desired S/N ratio is reached. The process is repeated for a set of delays t_i, until the whole fluorescence decay is sampled. It is straightforward to implement the boxcar method for the nonimaging time-resolved microscopy.[20,21] The boxcar method has been used also for scanning confocal FLIM with single-photon[90–94] or two-photon excitation[95–97] with a gated PMT as a detector. In such experiments, full images are sequentially acquired for each delay t_i, and fluorescence decays are constructed at each pixel as a function of intensity versus t_i.

10.2.1.3.1 FLIM with Gated Image Intensifier

The full advantage of gated detection in lifetime-based imaging was realized by the introduction of gated image intensifiers. Then the whole image can be sampled at once without the need of the pixel-by-pixel scanning. For given S/N ratio and image size, simultaneous sampling of all pixels brings tremendous improvement of the data acquisition time with respect to serial scanning methods. This property allowed fast lifetime mapping to the WF imaging.[98–129]

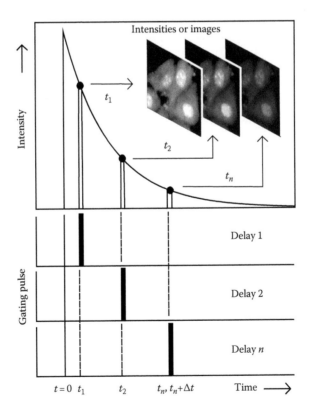

FIGURE 10.5 Gated detection. During each excitation cycle, the detector is turned on for a period of time Δt and intensity is measured. Depending on the imaging mode, the image is acquired simultaneously or pixel by pixel for each delay. The delay is changed until the whole emission decay is sampled.

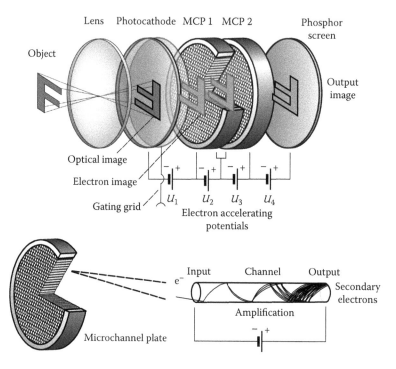

FIGURE 10.6 Schematic of a two-stage image intensifier. An image of the observed object is projected on the photocathode. Emitted primary photoelectrons are accelerated and enter pair of the MCPs where the number of electrons is amplified. Accelerated secondary electrons impinge on the phosphor screen where the bright optical image of the object is generated. The gating grid allows the gain of the intensifier to be turned *on* and *off* or continuous modulation of the gain.

An image intensifier is an optoelectronic imaging device, which amplifies brightness of an input light pattern on the photocathode while preserving the pattern. The device comprises a photocathode, one or several microchannel plates (MCPs), and a phosphorescent screen where the amplified optical image is formed. The screen is typically observed by a CCD camera for digital quantification of the image. The concept of the image intensifier is illustrated in Figure 10.6. Primary photoelectrons generated on the photocathode by an incident light are accelerated and enter the MCP consisting of a large number of metal-lined capillaries. The opposite ends of the capillaries are held at the electron accelerating potentials and their inner walls are covered by an electron emissive material. The accelerated photoelectrons therefore generate an avalanche of secondary electrons, which subsequently form a bright fluorescence spot on the phosphor screen. Since modern intensifiers exhibit practically no cross talk between adjacent channels and all components are proximity focused, the image intensifier has a very low image distortion. Depending on the number of multichannel plates, the gain of the intensifier can typically range from 10^3 for a single-stage intensifier up to 10^7 for an intensifier with three MCPs in series. Gain gating can be achieved by biasing photocathode several volts positive with respect to the MCP (by reversing the voltage U_1 in Figure 10.6) in order to prevent photoelectrons from being collected on the MCP.[130] The gate pulse then drives the photocathode more negative than MCP and the photoelectrons enter the MCP where they are amplified. Alternatively, the gating can be accomplished with improved time response by biasing an auxiliary gating grid positioned on the back of the photocathode.[99] Shutter ratios as high as 10^9–10^{12} has been reported.[130] Recent gated intensifiers can operate at high repetition frequencies with gating pulses as narrow as several hundreds of picoseconds.[131–133] Small-diameter gated intensifiers are becoming available with an exposure time as short as 50 ps. These intensifiers, however, can only operate with a very low duty cycle with repetition frequencies up to several kHz.[132,133]

While gating methods do not possess a dynamic range and the time resolution of the TCSPC with a microchannel PMT, it offers a number of important advantages for lifetime-based imaging. The boxcar acquisition method is highly configurable. The number and time width of the individual gates can be arbitrarily changed in order to match the decay characteristics of fluorophores and balance data acquisition time versus time resolution. The minimum number of time windows in order to calculate lifetime of monoexponentially decaying emission is 2.[98] Such a double-gate approach to FLI experiments[90,92,93] highly reduces data acquisition and processing time while discarding a part of information otherwise accessible by multigate methods.[125] Nevertheless, double-gate FLI is very efficient in generating a lifetime contrast even for complex decays. The lifetime contrast can be calibrated against known samples, which consequently allows extraction of biologically relevant information. Gating techniques are generally fast and allow FLIM imaging of rapid dynamic processes. A FLIM system has been reported that allows simultaneous acquisition of two gates at a rate of hundreds of frames per second.[123,134] Such system is therefore insensitive to movements of the specimen and photobleaching that may otherwise cause severe problems. Performance was demonstrated on measurements of the calcium fluxes in neonatal rat myocytes stained with the calcium probe Oregon Green 488-Bapta.[134]

Unlike the TCSPC, gating can be efficiently used with low-repetition-rate lasers. Also, multiple photons detected per laser flash are not a limitation of the technique, which can thus be used for measurements of lifetimes in the microsecond and millisecond time ranges. An important property of gating methods is the capability for background suppression. When gating is used with longer-lived fluorophores, the scattered light, reflections, autofluorescence, and short-lived fluorescence, which usually decrease image contrast in conventional fluorescence microscopy, can be easily rejected.[106,113] The normally off detector is turned on at the moment when the intensity of the short-lived components has diminished and only long-lived fluorescence is acquired. For very long-lived emission, such gating scheme has also been accomplished without expensive electronics by attaching two phase-locked choppers to a microscope.[135]

10.2.1.3.2 *Gatable CCD Cameras*

Lifetime imaging with long-lived millisecond or microsecond probes can be accomplished without expensive gatable image intensifiers. Khait et al.[136] have reported a fluorescence lifetime microscope with a millisecond temporal resolution equipped with a free-running externally synchronized CCD camera acquiring data with a speed of 33 images/s. A similar approach has been used for oxygen imaging with a millisecond phosphorescent sensor Green 2W.[137] Phosphorescence decays were sampled at preset times after the excitation flash by an externally triggered CCD camera working with an exposure time of 2.5 ms. Phosphorescence lifetime maps and a distribution of pO_2 in tissues were constructed by this method.

During the past decades, fast directly gatable CCD cameras have become available.[138–141] The technology utilizes a semiconductor-based electronic shutter integrated into the structure of the CCD chip, which allows for a short optical exposure times. The shutter rise and fall times shorter than 55 ns and with the opened/closed ratio as high as 10^4–10^5 have been reported.[138] Another approach uses control of the *charge drain* function on the interline-transfer CCD chip that allows the sensor to be toggled between an active and insensitive state within few nanoseconds and with a repetition rate up to 0.5 MHz.[141] The directly gatable cameras have been used for microsecond lifetime imaging of luminescent Pt(II)-porphyrins and ruthenium(II) complexes.[139,140] Distinguishing between 4 and 19 ns emission of sulforhodamine and quinine sulfate, respectively, was readily possible with the directly gatable camera.[141] The utilization of gain-modulated CCD and CMOS cameras suitable for FD FLIM was also reported.[142–144]

10.2.1.4 Spatially Sensitive MCP Detectors

The photoelectron emitted from the photocathode and amplified by the multichannel plate creates a localized charge on the anode centered at positions corresponding to the positions of the incident photon.

This property provides the possibility to simultaneously record the spatial and temporal information about incident photons and to use these fast detectors for a lifetime-based imaging. The temporal information is typically measured by TCSPC electronics. Several strategies for retrieving the spatial
information of the incident photons are illustrated in Figure 10.7. The spatial information can be accessed
by replacing a conventional disk anode, Figure 10.7a, by a resistive (R) or a delay line (DL) anode,
Figure 10.7b, quadrant anode (QA), Figure 10.7c, or by a set of individual anodes, Figure 10.7d. While
the DL-MCP-PMTs were used to encode a linear position across the photocathode,[55,145] the resistive[146]
or quadrant anode[70,71,147] MCP PMTs provide continuous 2D position information. The detectors with

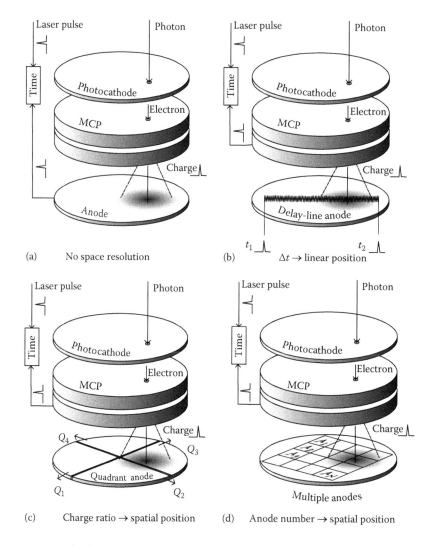

FIGURE 10.7 Principle of position-sensitive detectors for time-resolved spectroscopy. (a) Conventional MCP-
PMT without space resolution. (b) DL anode MCP-PMT with linear position resolution. The position information is derived from the difference in pulse arrival times to the terminals of the DL anode. (c) QA MCP-PMT
with the 2D spatial resolution. The position of the incident photon is calculated from the ratio of charges collected by the individual quadrants Q_1–Q_4 of the anode. (d) Multiple-anode MCP-PMT. The spatial position correlates with the electrical signal on the individual spatially separated anodes. (Revised from Kemnitz, K. et al.,
J. Flouresc., 7, 93, 1997.)

continuously coded spatial location are not multichannel devices for simultaneous signal acquisition from multiple locations, since they can register only one detection event at a time. If multiple photons simultaneously strike the photocathode at different places, the photons are registered as one detection event and an incorrect location is assigned by electronics. In order to alleviate such cases, the detection rates have to be limited to the extent as described earlier for TCSPC. From this point of view, a lifetime mapping with such detectors can be compared with a sequential random scanning of the sample. For the position-sensitive detectors, the count rate can be adjusted to the whole image, while with the nonimaging detectors, the acquisition rate should be adjusted to the pixel of maximum signal in order to avoid pile up. This could result in faster acquisition rates achievable with the position-sensitive detectors especially for images exhibiting highly different brightness across the image. With the QA detector, spatial and time resolution of 256×256 pixels and 100 ps, respectively, were achieved.[148] Nevertheless, due to advances in the image intensifier technology, which offers better spatial resolution, lower image distortion, and faster data acquisition, the importance of the position-sensitive detectors for lifetime-based imaging is decreasing.

10.2.1.5 Streak Camera

Standard FLIM methods are typically able to resolve fluorescence lifetimes longer than few hundred picoseconds. The main limitation is a time response of detectors. In order to measure faster emission rates and to improve time resolution, a streak camera technology was applied in a single-spot lifetime microscopy[27] and FLI (streakFLIM).[149,150] Modern streak cameras are optoelectronic devices with time resolution down to 200 fs and a photon counting sensitivity.[151] Their operation is based on an ultrafast photoelectron beam deflection synchronized with the excitation light pulse. The device is suitable for a single-beam or line-scanning applications since it delivers an intensity image with spatial information displayed on one and time on the other axis. The streak camera has been recently combined with a two-photon line-scanning microscope.[152,153] Performance of such streakFLIM system was demonstrated on living cells by resolution of lifetime effects caused by a single amino acid mutation in ECFP protein sequence.[153]

10.2.1.6 Multipulse Methods

Increasing the time resolution is not limited to the creation of faster electronic and optoelectronic devices. The excited state population can be controlled by intense light pulses. The availability of femtosecond lasers offers time resolution on the time scale of the pulse width. The nonlinear multipulse methods are especially suitable for confocal FLIM, since the laser light flux is concentrated in a diffraction-limited spot where a threshold for nonlinear phenomena is easier to reach.

One multipulse optical technique implemented on a scanning confocal microscope is a double-pulse FLI (DPFLIM).[154,155] The method is based on the excitation of fluorophores with ultrashort laser pulses, which populate the excited state to near the saturation value. The method does not require electronic gating of the detector and rather relies on the optical pumping of the excited state of fluorophores. Since the time resolution depends mainly on the duration of the excitation pulses, the DPFLIM can have subpicosecond time resolution. The principle of the method is schematically depicted in Figure 10.8. The sample is illuminated with intense single pulses saturating the first excited state, and the time-integrated fluorescence intensity I_1 is measured by a conventional detector. Then the measurement is repeated with saturating pulse doublets and again the total intensity I_2 is measured. The two pulses from the excitation doublet are separated by the time delay Δt. Finally, the ratio R of the two intensities yields the fluorescence lifetime τ[154]:

$$R = \frac{I_2}{I_1} = 2 - \exp\left(-\frac{\Delta t}{\tau}\right) \qquad (10.2)$$

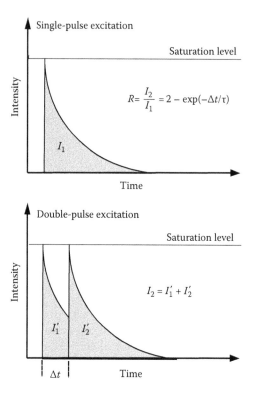

FIGURE 10.8 Principle of the optical double-pulse lifetime determination used for a confocal FLIM. (Redrawn from Muller, M. et al., *J. Microsc. Oxford*, 177, 171, 1995.)

From Figure 10.8 and Equation 10.2, it is seen that with increasing separation of the pulses, the ratio R increases until it reaches the limiting value of 2 for $\Delta t \gg \tau$. By calculation of R at each pixel, the lifetime contrast is created and fluorescence lifetime map can be constructed.

The DPFLIM was first used with a confocal microscope where saturation of the excited state can be reached without problem. With some modification, the DPFLIM is usable even with weaker nonsaturating illumination[154] and has a potential to be applied to conventional WF-FLIM.

Additional examples of multipulse nonlinear methods potentially useful for lifetime-based imaging have been outlined in the review and handbook of Lakowicz[32,89] where manipulation of the excited state population by light quenching has been proposed.

10.2.2 Frequency Domain

The history of the FD fluorometry begins in 1926 when Gaviola performed his first fluorescence lifetime measurement.[156] During the years the FD methodology rapidly evolved[89] until the last decades of the twentieth century, it was adopted for lifetime microscopy and lifetime-based imaging.[157,158] The principle of the phase and modulation measurement is schematically depicted in Figure 10.9 for a single point in the image. In FD lifetime imaging, the investigated object is exposed to the excitation light $E(t)$ harmonically modulated at an angular frequency of $\omega = 2\pi f$ with a modulation degree of $m_E = a/A$:

$$E(t) = E_0 \cdot [1 + m_E \sin(\omega t)] \tag{10.3}$$

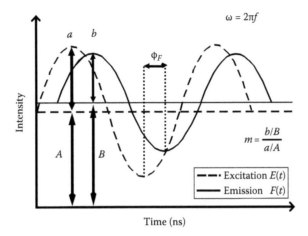

FIGURE 10.9 Concept of the FD measurements. The modulated excitation and emission light have DC components A and B, respectively, and modulation amplitudes a and b, respectively. The emission exhibits a phase lag ϕ_F with respect to the excitation light and a demodulation ratio m.

Such periodic excitation causes fluorophores to emit light $F(t)$ with the same modulation frequency ω. Due to a finite fluorescence lifetime, the emission is delayed, exhibiting a spatially dependent phase lag $\phi_F(r)$ and decreased modulation $m_F(r) = b/B$, relative to the excitation radiation:

$$F(r,t) = F_0(r) \cdot \{1 + m_F(r)\sin[\omega t - \phi_F(r)]\} \tag{10.4}$$

The spatial coordinate r is mapping position within the object from where the emission is observed. For simplicity, we set the excitation phase to zero since it can be freely manipulated, by insertion of a DL to the excitation path, and since all phases are measured relative to this phase angle. It has been shown that both the fluorescence phase $\phi_F(r)$ and the demodulation ratio $m = m_F(r)/m_E(r)$ depend on that lifetime components of the emission[159,160]:

$$\phi_F(\omega, r) = \arctan\left(\frac{N(\omega, r)}{D(\omega, r)}\right) \tag{10.5}$$

$$m(\omega, r) = \frac{m_F(\omega, r)}{m_E(\omega, r)} = \sqrt{N^2(\omega, r) + D^2(\omega, r)} \tag{10.6}$$

where

$$N(\omega, r) = \left[\sum_i \frac{a_i \omega \tau_i^2(r)}{1 + \omega^2 \tau_i^2(r)}\right] \cdot \frac{1}{\sum_i a_i \tau_i(r)} \tag{10.7}$$

$$D(\omega, r) = \left[\sum_i \frac{a_i \tau(r)}{1 + \omega^2 \tau_i^2(r)}\right] \cdot \frac{1}{\sum_i a_i \tau_i(r)} \tag{10.8}$$

For the simplest case of a monoexponential decay, the lifetime $\tau(r)$ at any location of the sample can be directly calculated from Equations 10.5 through 10.8:

$$\tau_\phi(r) = \frac{1}{\omega}\arctan(\phi_F(r)), \quad \tau_m(r) = \frac{1}{\omega}\cdot\sqrt{\frac{1}{m^2(r)}-1} \tag{10.9}$$

and $\tau = \tau_\phi = \tau_m$. Equation 10.9 can also be used to calculate the phase (τ_ϕ) and modulation (τ_m) lifetime for more complex decays. The resulting lifetimes, however, are apparent values representing complex mixing of different lifetime components.[89,161,162] For a multiexponential decay, the phase lifetime is shorter than the modulation lifetime, $\tau_\phi < \tau < \tau_m$, and such inequality indicates heterogeneous fluorescence decay.

10.2.2.1 Homodyne and Heterodyne FLIM

The theory of the phase and modulation FLIM has been discussed in a number of papers.[163–168] Briefly, the fluorescent image $F(r, t)$, Equation 10.4, of an object excited by the modulated light is projected on the photocathode of the image intensifier that has the electronic gain $G(t)$ modulated at the RF frequency $\omega + \Delta\omega$:

$$G(t) = G_0 \cdot \left\{1 + m_D \sin\left[(\omega + \Delta\omega)\cdot t - \phi_D(r)\right]\right\} \tag{10.10}$$

Inside the intensifier, frequency mixing occurs as the primary photocurrents modulated at the frequency ω interact with the electronic gain modulated at the frequency $\omega + \Delta\omega$. Due to a the long time constant of the phosphor screen caused by the millisecond phosphor lifetime, all high-frequency components become time averaged and we observe an amplified image with intensity slowly varying with the frequency $\Delta\omega$:

$$I(r,t) = I_0(r)\cdot\left\{1 + \frac{m_D \cdot m_F(r)}{2}\cos\left[\Delta\omega t - \Delta\phi(r)\right]\right\} \tag{10.11}$$

where

$$\Delta\phi(r) = \phi_F(r) - \phi_D \tag{10.12}$$

It is seen that the phase and modulation information is retained after the frequency mixing and can be extracted from Equations 10.11 and 10.12. The method is called homodyne and heterodyne for $\Delta\omega = 0$ and $\Delta\omega \neq 0$, respectively. Both methods have rapidly evolved when image intensifiers suitable for high-frequency gain modulation became available.[163,169]

The homodyne method is the dominating method used in FD WF-FLIM.[9,16,163–165,167,168,170–198] The principle of the method is schematically shown in Figure 10.10 together with the TD counterpart. The generic scheme of an FD instrument is shown in Figure 10.11. In a homodyne FLIM, the gain of the image intensifier is modulated at the same frequency as the excitation light, that is, $\Delta\omega = 0$. We observe on the phosphor screen a steady-state phase-sensitive image:

$$I(r) = I_0(r)\cdot\left\{1 + \frac{m_D \cdot m_F(r)}{2}\cos\left[\phi_F(r) - \phi_D\right]\right\} \tag{10.13}$$

The intensity $I(r)$ at any location of the phosphor screen depends on the cosine of the difference between fluorescence and detector phase angles and on the lifetime through the phase angle $\phi_F(r)$,

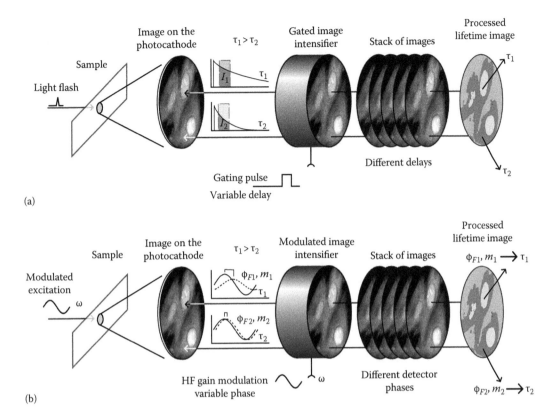

FIGURE 10.10 Schematic comparison of (a) the TD and (b) the FD methods for the WF FLI. In the TD, images are acquired for several delays of the gating pulse with respect to the excitation flash. Since the gain of the intensifier is on only during the presence of the gating pulse, fluorescence decays at different locations of the sample are simultaneously sampled and the lifetime image can be constructed on the pixel-by-pixel basis from the measured stack of the images. In the FD homodyne FLIM, the gain of the image intensifier is modulated at the same frequency as the excitation light. As a result, we observe a steady-state phase-sensitive image on the phosphor screen of the intensifier. After the stack of images is acquired at detector phases distributed between 0 and 2π, the modulation m and phase lag ϕ_F is extracted and a lifetime image is generated.

Equation 10.5, and the modulation factor $m_F(r)$, Equation 10.6. The value of ϕ_D is controlled by the apparatus and can be adjusted by a number of means. The phase-shift option can be built within the frequency synthesizer.[165] Alternatively, the phase shift can be accomplished by an external digital phase shifter[175] or passively by a DL consisting of a piece of coaxial cable inserted between the synthesizer and the image intensifier.[163] The dependence of $I(r)$ on ϕ_D for a selected location in the image is shown in Figure 10.11b. The opened circles represent the measured fluorescent intensity and the closed circles show the signal for a zero-lifetime reference (scatter), when $\phi_F = 0$. During the experiment, a stack of images is acquired for ϕ_D distributed between 0 and 2π. Since the images contain information about the $\phi_F(r)$ and $m_F(r)$ at each position, which can be extracted by the Fourier analysis or by performing a least-squares fit for each pixel in the image,[164,165] the lifetime image can be constructed. Alternatively, one can directly use the phase angle or modulation images for chemical mapping.

Heterodyne detection uses cross-correlation mixing for translation of the high-frequency fluorescence modulation to the frequency region of several Hertz or less. When used with the image intensifier or gain modulating framing camera,[199] an image with periodically oscillating intensity at the frequency $\Delta\omega$ is observed on the phosphor screen. The time-dependent image can be sampled by a CCD camera.[199] This seems to be a disadvantage since camera exposure times much shorter than $1/\Delta\omega$ is required to

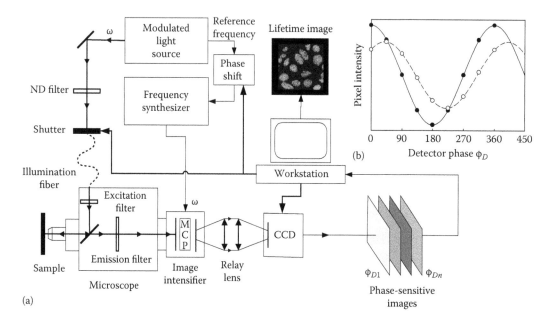

(b)

(a)

FIGURE 10.11 Homodyne FLIM. (a) Scheme of a generic instrument and (b) intensity profile through the stack of the phase-sensitive images at one arbitrarily chosen pixel as a function of a detector phase angle. Opened circles represent intensities of a fluorescent signal with phase lag $\phi_F > 0$, see Equation 10.13, closed circles show signal for a zero-lifetime reference (scatter) when $\phi_F = 0$.

properly sample the oscillations, which decreases the duty cycle for data acquisition. This is a disadvantage because measurements of dim samples may require longer integration times that may result in excessive photobleaching. To overcome this drawback, a method combining the high-frequency heterodyne mixing on the image intensifier with the low-frequency optical-homodyne image acquisition has been developed.[166,169,200] The heterodyne signal on the phosphor screen was recorded during a fraction of the heterodyne cycle by a boxcar method. This resulted in a steady-state signal on the phosphor screen integrated by the camera. The heterodyne cycle was sampled by changing delay of the acquisition window.

The heterodyne detection method does not require as strict high-frequency phase stability as the homodyne counterpart. Synchronization on the low cross-correlation frequency is much easier to achieve. However, with modern synthesizers, phase stability is not a problem. Due to the boxcar integration when only a slice of the heterodyne cycle is used for data accumulation, the acquisition times are longer with heterodyne than with homodyne detection. This disadvantage together with more complex instrumentation has caused lower use of the heterodyne WF-FLIM compared to the homodyne counterpart. Nevertheless, the heterodyne method of the lifetime determination has been successfully implemented in scanning microscopes.[201–203]

10.2.2.2 Optical Methods

Dong et al.[204] and Buehler et al.[205] have reported a unique optical cross-correlation method implemented on a scanning fluorescence lifetime microscope. Cross-correlation frequency mixing was accomplished by periodic quenching of the fluorescence by stimulated emission. The method utilizes the harmonic content[206–208] of the two pulse trains from two lasers running at repetition frequencies offset by the $\Delta\omega$ and focused in the sample. One of the laser beams excites fluorescence and the wavelength of the other is tuned to quench the emission. At the place of the beam overlap, a cross-correlation fluorescence signal at multiple harmonics of the frequency $\Delta\omega$ is created. The low-frequency fluorescence signal is detected by a standard PMT, the harmonics are electronically isolated and analyzed for demodulation and phase shift.

Optical heterodyning has two main advantages. The low-frequency fluorescence signal implies that the FLIM can be carried out without using fast photodetectors. Lifetime measurements up to 6.7 GHz have been reported with a conventional PMT.[204] Overlap of the pump and probe lasers also results in an axial sectioning since the out-of-focus signal is not cross-correlated and is rejected by the electronic filters. Because of the wavelength used in the one-photon pump-and-probing process, the spatial resolution of the technique is comparable to the confocal microscopy.

10.2.2.3 FLIM with Lock-In Amplifiers

Homodyne and heterodyne techniques are not the only detection methods for the FD FLI. Reports have been published where lifetime-based imaging was carried out with lock-in amplifiers (LIAs) in place of the detection electronics.[209,210] The LIAs are devices sensitive to both frequency and phase of the input signal and work directly with the high-frequency signal from the detector. This makes lifetime microscopes simpler; however, due to a final bandwidth of the LIAs, their usage has been limited to a frequency range up to several hundred MHz. Since LIAs for massively parallel data processing are not available, the LIAs have been utilized with scanning lifetime microscopes only. A phase sensitivity of the LIAs has been used mainly for a selective lifetime suppression in the intensity-modulated multiple wavelength scanning (IMS) technique.[209,211] Recently, dual-phase LIAs have been demonstrated to be usable for lifetime mapping.[210] The technique has been used for lifetime-based pH imaging with fluorescence probe SNAFL-2.[212]

10.2.2.4 Multifrequency FLIM

Unlike solution spectroscopy where the multifrequency instruments have been used for a long time,[213,214] FD FLI has been limited to a single modulation frequency. The main reason was a relatively long time required for a sequential acquisition of the multifrequency FLI data, which resulted in photobleaching and photodamage of the samples. Single-frequency imaging allowed generation of lifetime contrast and calculation of apparent modulation and phase lifetime images, Equation 10.9. However, this is a limitation when taking into account that fluorescence kinetics are rarely pure exponentials in a biological environment. Moreover, the heterogeneity of the decay is often the information of interest when looking for different fluorescence species within a cell. With the single-frequency FLIM, the information content of heterogeneous fluorescence decays remains inaccessible because resolution of N lifetimes needs at least N modulation frequencies.[215] With some prior information, for example, when modulation and phase lifetimes of components in a binary mixture are known, fractional contributions of both species can be determined even with a single-frequency FLIM.[216] A phasor analysis works in a similar way,[217–219] where variations in the relative contributions of mixture components to the overall signal is explored. This can be done, for example, by observing spatial variation of the mixture composition with a lifetime imaging system.

Fortunately, a solution has been found for rapid multiharmonic data acquisition. The multiharmonic method utilizes both the harmonic content of the excitation pulse train[206–208] and the harmonic content of the square-modulated gain of a detector for simultaneous data acquisition at multiple modulation frequencies. The technique was initially applied for cuvette experiments[220–223] and the single-spot heterodyne FD microscopy.[17] More recently, the multiharmonic approach has been implemented for the homodyne WF multifrequency FLIM by Squire et al. (mfFLIM)[224] and Schlachter et al. (mhFLIM)[196] FLIM. Simultaneous multiharmonic detection is achieved by mixing frequencies from the harmonic content of the emission with corresponding harmonics from the pulse-modulated gain of the image intensifier. The resulting optical image on the phosphor screen is formed as a sum of phase- and modulation-dependent images corresponding to the different harmonic frequencies. The phase and modulation is extracted at each harmonic frequency by Fourier analysis.

Compared to the serial frequency acquisition, the main advantage of the multiharmonic approach is a shorter data acquisition and a higher modulation depth. The reduction of the data acquisition time might not be as high as it would correspond to the number of simultaneously acquired frequencies. The reason is a lower duty cycle of the pulse-modulated detector.[224] Performance of the multifrequency FLIM has been demonstrated by the lifetime resolution of green fluorescent protein variants coexpressed in live cells[224] and by resolution of heterogeneous emission from human neuroblastoma cells labeled with Alexa 555 and 546 dyes.[196]

10.2.3 Three-Dimensional Wide-Field FLIM

WF microscopy suffers from the lack of sectioning capability, which is an inherent property of confocal and multiphoton imaging systems. The reason is a mixing of the signal from out-of-focus planes, which blurs the intensity image and decreases both contrast and spatial definition of observed structures. WF-FLIM suffers from the same problem. At any spatial location, the measured lifetime is an intensity-weighted average of lifetime contributions arising from the out-of-focus planes and the lifetime contrast is compromised. The improvement of the spatial and the lifetime resolution of the WF-FLIM has been the subject of a number of studies.[117,121,168,188,194,195,225–228]

10.2.3.1 Structured Illumination

One method for obtaining the optical sectioning with the WF lifetime microscope is depicted in Figure 10.12. The sectioning capability was achieved by implementation of a structured illumination,[229,230] when a grid pattern produced by a grating in the illumination lightpath is projected into the sample. Then three sets of the time-gated images are taken for three spatial positions of the grating. A simple calculation then yields a deblurred lifetime map.[117,121,227]

10.2.3.2 Spinning Disks

Sectioning capability can be introduced to the WF-FLIM by utilization of a spinning-disk confocal unit when samples are scanned simultaneously by thousands of pinholes. Parallel confocal scanning is much faster than a conventional single-beam scanning and results an image refresh rate of hundreds of frames per second. The image can therefore be conveniently viewed by a WF detector. The confocal spinning-disk technology employing a Nipkow scanner with microlenses that are used for better light efficiency is fully compatible with WF-FLIM. Integration of the spinning-disk unit was shown to considerably improve lifetime resolution and reduce artifacts, while maintaining all advantages of WF imaging.[194,195,228]

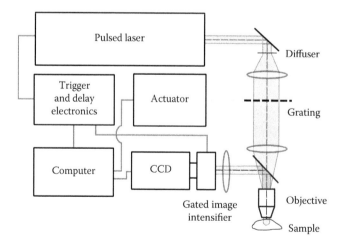

FIGURE 10.12 Scheme of a lifetime microscope with a structured illumination. A grid pattern produced by a grating in the illumination lightpath is projected on the sample. Three sets of the time-gated images are taken at three positions of the grid and a deblurred stack of intensity images is calculated. To obtain a lifetime map, the resulting stack is processed by standard methods used with the time-gated FLIM. (Revised from Cole, M.J. et al., *Opt. Lett.*, 25, 1361, 2000.)

Similar enhancement of the 3D lifetime resolution of WF-FLIM can be achieved with a multifo-cal multiphoton microscopy (MMM).[231] Multiphoton excitation and 3D sectioning is accomplished by insertion of the spinning microlens disk to the illumination path of the conventional WF microscope. This causes a near-IR excitation light from the femtosecond laser to be transformed into an array of fast-scanning foci in the object plane where multiphoton excitation occurs. The resulting two-photon image can be viewed by a conventional camera. A WF-FLIM instrument based on the MMM technique and gated image intensifier has been reported.[225]

10.2.3.3 Programmable Array Microscopes

Programmable array microscopes (PAMs) are instruments where optical sectioning is achieved by a placement of a spatial light modulator (SLM) in an image plane of the microscope. The SLM could be digital micromirror device (DMD)[232–235] or modulator based on a liquid crystal technology (LCM).[236,237] SLM can be used for generation of highly flexible programmable illumination and/or detection patterns. From this point of view, PAMs could operate the same way as confocal rotating disk microscopes; how-ever, they do not contain moving parts. Scanning pinholes can be simulated by computer-controlled spatial–temporal modulation of the LCM transmission or by rapid switching of particular micromirror elements between the on and off state. PAMs have a much greater flexibility than either the rotating disk microscopes or the single-beam scanning devices since they allow rapid random and parallel access to different regions of interest on the sample. DMD-PAMs were successfully used for an acquisition of WF optically sectioned lifetime images in combination with both homodyne FD FLIM[238,239] and TD photon counting FLIM.[235,240]

10.2.3.4 Image Deconvolution

If the point spread function (PSF) of the imaging system is known, the 3D temporal and spatial resolu-tion of the WF-FLIM can be increased by a numerical deconvolution.[168,188] Squire et al.[168] performed deconvolution with the measured PSF in the Fourier space before construction of the lifetime images. Alternatively, the FLIM data were first globally analyzed in order to separate fluorophore populations associated with different lifetimes. This allowed the separated image stacks to be individually corrected for photobleaching before the deconvolution.[188] In both cases, significant enhancement of the spatial and temporal resolution was achieved.

10.2.4 Time-Resolved Anisotropy Imaging

Fluorescence anisotropy provides information about rotational diffusion of a fluorophore and sensi-tively reflects both its motional freedom and physical properties of its microenvironment. Upon excita-tion of the fluorophore with linearly polarized light pulse, the time-resolved emission anisotropy $r(t)$ can be calculated from polarized emission decays $I_\parallel(t)$ and $I_\perp(t)$ as $r(t) = [I_\parallel(t) - I_\perp(t)]/[I_\parallel(t) + 2I_\perp(t)]$.[89] For a free rotating rigid spherical rotor and a Dirac excitation, the anisotropy decays exponentially, $r(t) = r_0 \exp(-t/\phi)$, where ϕ is a rotational correlation time that scales with the molecular volume, $\phi = \eta V/kT$, where η is the solvent viscosity. In a general case, $r(t)$ can be more complex and depends both on a detailed molecular shape[241–243] and on a spatial hindrance of its rotational motion.[244,245] Fluorescence depolarization can be also caused by a homotransfer of the electronic excitation energy (homoFRET). The emission anisotropy can therefore sensitively reflect processes such as protein conformational changes, interactions, and association into larger molecular structures. Fluorescence anisotropy can be used for monitoring of membrane properties[36,246,247] and excitation energy migration as well. Utilization of polarized fluorescence measurements for spatial mapping of these processes was therefore a logical progress in development of FLIM methods. Overview of the technique and applications in cell and tis-sue imaging can be found in recent reviews.[15,34]

The time-resolved fluorescence anisotropy imaging (rFLIM, TR-FAIM) was implemented in fre-quency[205,248,249] and TD[249–253] both in the WF[248,249,251–253] and scanning[205,249,250] imaging modes.

Applications of the time-resolved anisotropy imaging include measurements of microviscosity in live cells[205,252] or energy migration emFRET used for investigation of molecular proximity, dimerization, and aggregation of cellular fluorescent proteins.[248–250]

10.2.5 Spectral FLIM

In contrast to cuvette experiments where measurement of both spectral and lifetime information is rather common, standard FLIM instruments usually collect signal in one fixed spectral band. As a consequence, majority of valuable spectral information available from the sample emission is discarded. Both lifetime and spectral resolution might be needed for correct interpretation of fluorescence data from heterogonous samples with multiple colocalized fluorophores and for correct assignment of lifetimes to particular emitter species. FRET and autofluorescence imaging is among applications where simultaneous spectral and lifetime information is desirable.

Spectral resolution can be implemented to imaging applications by number of different ways. Some of them are reviewed and described in the biological imaging spectroscopy section of this handbook. Scanning microscopes typically use TCSPC detection for temporal resolution. A straightforward implementation of an additional spectral channel can be accomplished by a spectral splitting of the output beam by an insertion of a dichroic mirror to the data collection lightpath. Such spectral FLIM (sFLIM) was used for single molecule imaging.[59,254] Typically, 16 spectral channels can be obtained by a spectral dispersion of the output light beam across a multianode PMT array by a polychromator.[62,88,255–257] Ulrich et al. used two-photon 5D scanning microscope featuring (x, y, z, τ, λ) resolution for mapping of cellular fluorescence and autofluorescence. The spectral and time dimension was implemented in their device by an addition of a Sagnac interferometer and the TCSPC detection, respectively.[258] Praus et al.[202] used gain-modulated optical MCA (OMA) in their confocal heterodyne sFLIM device. Other implementations include 2D Hadamard transform spectral imagery[237] used in the PAM-FLIM instrument.[238] An alternative approach with a programmable DMD as a spatial illuminator was used by Bednarkiewicz et al.[235,240] for their hyperspectral sFLIM imager. The last but not the least, the push-broom technique was successfully applied for sFLIM. In this sFLIM modality, a line-illuminated strip of sample is imaged on an input slit of the imaging spectrograph. The spectrograph output is than optically coupled to the gated image intensifier viewed by a CCD camera. As a result, the x and y direction on the final image contains a spatial location along the slit and spectral information, respectively. A full spectral cube at any time slice after excitation is obtained by gating of the image intensifier and translation of the sample through the line illumination.[113,259,260] The push-broom sFLIM imagers were used, for example, for investigation of lanthanide chelates with lifetimes on a millisecond time scale[113] and for imaging of human arteries.[259]

10.3 Some Specifics of FLIM Data Analysis

Several ways have been proposed to rapidly analyze the substantial amount of data generated by lifetime imaging systems. Typically, this means to reduce a stack of images to a few images that contain spatial mapping of decay parameters, for example, lifetimes and associated amplitudes. Going even a step further, we usually want to obtain single chemical image carrying biologically relevant information. Depending on the fluorescent lifetime sensor, it can be a spatial distribution of analyte concentration, 2D map of a gas pressure, or spatial map of distances when FLIM is used for fluorescence resonance energy transfer (FRET) experiments.

10.3.1 Fast Two- and Multiple-Gate Analysis

TD lifetime mapping does not necessarily require lengthy iterative fitting of a large number of points. Fluorescence lifetime maps can be estimated from minimum of two intensity images $I_1(r)$ and $I_2(r)$

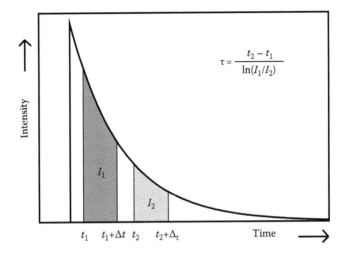

FIGURE 10.13 Principle of the two-gate method for the rapid calculation of the emission lifetime. The formula for calculation of the lifetime is valid for a monoexponentially decaying emission.

acquired in two distinct time windows.[261] Assuming windows of the equal width of Δt delayed by t_1 and t_2 relative to the excitation pulse, Figure 10.13, we can write

$$I_1(r) = \int_{t_1}^{t_1+\Delta t} a(r) \cdot \exp\left[\frac{-t}{\tau(r)}\right] dt, \quad I_2(r) = \int_{t_2}^{t_2+\Delta t} a(r) \cdot \exp\left[\frac{-t}{\tau(r)}\right] dt \tag{10.14}$$

After integration, the lifetime can be expressed as

$$\tau(r) = \frac{t_2 - t_1}{\ln[I_1(r)/I_2(r)]} \tag{10.15}$$

Equation 10.15 holds true for monoexponential decays and allows for very fast computation of the lifetime maps. To obtain good accuracy of the measured lifetime map, prior knowledge of the lifetime is essential for selecting optimal gating windows. Therefore, an effort has been invested to evaluate errors associated with the gating technique[262,263] and to develop optimal gating schemes.[264] This rapid lifetime determination technique has been extended to evaluation of double-exponential decays.[265] A rapid non-iterative approximation of multiexponential decays based on a Prony method was used for the analysis of multiple-gate FLIM data. This approach utilizing a linear least-squares approximation was shown to be more than an order of magnitude faster than nonlinear iterative methods.[125]

Despite their performance, the rapid two- or multiple-gate lifetime determination schemes typically require Dirac excitation and suffer by inability to work with a realistic impulse response function[44] in order to deconvolve emission decays. This drawback was addressed by Jo et al. who introduced a fast FLIM deconvolution analysis based on the Laguerre expansion technique. The method performs more than two orders of magnitude faster than alternative approaches, can work with as few as five gated images, and does not require infinitely short excitation pulses.[266,267]

10.3.2 Global Analysis of FLIM Data

Analysis of FLI data is a problem of the same scale as the analysis of 10^5–10^6 of time-resolved cuvette experiments. Besides the computational expenses, the number of data points on the time or frequency axis in the time or FD, respectively, has to be limited in order to collect data before irreversible processes damage the sample. Smaller number of data points and lower S/N ratio, compared to values typically obtained for cuvette measurements, make reliable multicomponent lifetime analysis from a single pixel questionable. One can increase the S/N ratio by grouping several pixels together; however, the price is a deterioration of the spatial resolution. An important possibility in the FLI data analysis is the use of prior knowledge about the sample, which leads to a dramatic improvement of the computational speed and the fit quality.[183,224,268]

One of the approaches uses the global-analysis algorithm,[159,160,269–272] which analyzes all pixels simultaneously while constraining the component lifetimes in all pixels to be equal.[183] This is an acceptable assumption for many FLI experiments where spatially distributed mixture of fluorescence species is analyzed. Such approach leads to a significant reduction of the number of fitted parameters and to improved fit reliability. It has been shown that the FD global analysis of multiple pixels can yield twice the number of lifetimes that could be recovered with ordinary single-point measurements.[183] An important consequence is a possibility of recovering two lifetimes from a single-frequency FLI experiment.[273] Examples and comparison of the single- and multifrequency global fit of FLIM data have been demonstrated on two coexpressed variants of the green fluorescence protein (GFP).[183]

A special case of the described method is called a lifetime invariant fit.[183,224] When lifetimes are a priori known, their values can be fixed during the fit and only amplitudes or intensity fractions are calculated in each pixel. In this case, the fitting process evaluates only linear parameters that are usually less correlated and can be calculated significantly faster than the nonlinear lifetime parameters. This method shares common features with the phasor analysis method in FD.[217–219] The global approach with variable lifetimes is a more general approach but the high speed of the data analysis could be of prime importance in real-time clinical applications.[167,175]

The global fit and the lifetime invariant fit are not limited to the analysis of 2D FLIM data. The concept of the invariant lifetimes has been extended to the whole sample volume and used, for example, with the WF-FLIM for the 3D deconvolution of fluorescent populations associated with phosphorylation states of the epidermal growth factor receptor ErbB1 tagged with GFP.[188]

10.4 Selected FLIM Applications

Biological and biomedical applications of the FLIM technique have been treated in many reviews.[3,10,11,33,35–38,274–276] The technique has been successfully applied in cell biology, clinical diagnostics, and analytical chemistry for determination of spatial ion and metabolite distributions, monitoring of interactions between cellular components by FRET, and for detection of abnormal tissues. In the following sections, we will briefly outline few selected biological applications.

10.4.1 Intracellular Lifetime-Based pH Imaging and Ion Mapping

Cellular functions critically depend on the maintenance of a proper intracellular environment. Changes of pH and ion concentrations can trigger signaling pathways, activate expression of genes, or influence cellular morphology. A number of lifetime probes suitable for imaging of pH distribution in cytosol and cellular compartments have been characterized[277–279] and used for imaging of living systems.[72,90,95,191,212,280–282] Lifetime-based pH imaging has been often found to be more convenient than conventional ratiometric methods due to a simpler pH calibration[90,280] and to possibly distinguish between fractions of bound and free probe.[281] This allows for elimination of the pH estimation errors caused by binding of the probe to cytosolic constituents, which is difficult to achieve with the ratiometric imaging. The potential of the lifetime-based imaging with a multiphoton excitation has been demonstrated on microbial biofilms when pH gradients up to the biofilm depth of 140 μm were resolved with the carboxyfluorescein as the pH indicator.[95]

Lifetime-based detection has been applied for imaging of cellular ions as well. Spatial calcium distributions measured with the nonratiometric lifetime probes have been reported by Lakowicz et al.,[165,283,284] concentrations of Na^+ ions in HeLa cells have been monitored with sodium green and phase/modulation fluorometry,[19] and readouts of the intracellular chloride concentrations in neural cells were performed with a clomeleon dye.[70] Optical sensors for a number of other biologically critical metal ions, for example, Zn(II), Co(II), Cd(II), Cu(II), and Ni(II) have been systematically characterized[285-287] and can be used for lifetime-based imaging.

Similarly to lifetime-based pH indicators, the intracellular lifetime response of the ion-selective fluorophores is typically different from the response in a buffer. Differences are caused mainly by interactions of probes with cellular components or due to a mismatch of intracellular viscosity, pH, and temperature.[18,107,191,288] Therefore, a careful in vivo calibration is usually necessary for the reliable estimation of the intracellular ion concentrations.

10.4.2 Lifetime-Resolved Imaging of Cellular Processes

Cellular FLIM utilizing environmental sensitivity of the emission lifetime provides more information than conventional intensity-based imaging methods. The two-photon FLIM has been used to monitor the macrophage-mediated phagocytosis with the fluorescein–BSA conjugate as an exogenous fluorescent antigen.[289] Due to intracellular proteolysis, the initial subnanosecond lifetime of the conjugate increased with time to the value close to 3.0 ns providing a monitor of the cellular proteolytic activity.

Cells contain variety of intrinsic fluorophores. One of them is the NADH, which can provide information about the oxidation–reduction status of the cells and tissues. Lifetime images of the free and protein-bound NADH autofluorescence have been first reported by Lakowicz et al.[170] Later, the FLIM of the NAD(P)H/NAD(P) pairs has revealed pathological effects induced by the exposure of cells to the strong illumination pulses.[201,290] Multiphoton FLIM of NADH has been also applied for mapping of cellular metabolism in human breast cells,[75] human dermal fibroblasts,[125] and brain tissues.[291]

Imaging of the nuclear DNA is another example of the cellular FLIM. Cell-cycle-dependent topology of the DNA has been studied with the probe YOYO-1, which exhibits base-pair-dependent lifetime.[186] Differential and simultaneous imaging of DNA and RNA in living cells has been demonstrated with SYTO13 dye and the two-photon FLIM.[292,293]

Cells have been studied also by the polarized time-resolved fluorescence. Intracellular anisotropy decays were used to map membrane fluidity and the spatial distribution of the cytoplasmic viscosity.[20,22,205,252,294] The measured microviscosities served for assessment of the spatial organization of the cytoplasm and membranes.

10.4.3 Cellular Interactions Determined by the FRET–FLIM

Due to the inverse sixth power dependence of the energy transfer efficiency on the distance between donor and acceptor molecules,[295] FRET is a powerful strategy to detect interactions, associations, and proximity of cellular components.[3,10,14,296] Pixel-by-pixel microscopic maps of the FRET efficiency can be monitored by FLI of the donor lifetime thus eliminating problems of the intensity-based measurements.

The FRET–FLIM has become widely used in cellular biology. Applications include monitoring of the phosphorylation, oligomerization, and internalization of the epidermal growth factor receptors,[179,297-299] studies related to the signal transduction and imaging of the activated protein kinase C, and its association with downstream effectors,[177,178,200] eventually topological organization of the nuclear DNA.[182,185] Time-resolved FRET microscopy was employed to study a vesicular transportation of the cholera toxin by following the energy transfer between toxin subunits,[200] to study the process of adenovirus capsid disassembly in living cells,[72] or to detect oligomerization and aggregation of alpha-synuclein molecules that are believed to play a major role in neuronal dysfunctions.[300]

A novel labeling strategy for FRET experiments involved tagging of the donor–acceptor pairs to the corresponding monoclonal antibody of the object proteins. Recently, genetically encoded fluorophores based on the GFP and its spectral variants represent an exciting possibility for in vivo protein labeling.[301] Today, the FRET experiments can take full advantage of the fusion proteins constructed as GFP chimeras.[274,302–305] FRET–FLIM was found to be useful also for the detection of an early viral infection. A plasmid construct containing donor (GFP2) and acceptor (DsRed2) sequences linked together by a cleavage site for an enterovirus 71 (EV71)-specific protease was expressed in target cells. Then on cell infection, the EV71 protease cleaved the linker that resulted in release of FRET due to separation of the donor–acceptor pair.[64]

10.4.4 Tissue Imaging and Clinical Applications

Both in vivo and ex vivo applications of the lifetime-resolved imaging at the tissue level are emerging.[35,110,128,306] Compared to the conventional fluorescence imaging that uses intensity and spectral information as an indication of a tissue abnormality, lifetime-based imaging provides additional physiochemical diagnostic information. Time-gated imaging has been used for a diagnosis of bones and teeth in order to develop diagnostic methods for bone fractures and early caries, respectively.[115,116] The diagnostic potential of the lifetime-based imaging has been demonstrated by distinguishing lifetime differences between the intact and decayed dental structures.[69,307] Additionally, multiphoton excitation has been employed for generation of the 3D lifetime images of the human-skin autofluorescence.[308] Lifetime distributions from the near-surface skin area up to the depth of ~200 µm revealed variations in metabolic states of the cells and indicated a possibility of using lifetime-based imaging for dermatological purposes. The multiphoton FLIM has proved to be a valuable tool for cancer diagnosis.[75,309] Provenzano et al. have reviewed application of this imaging technique for monitoring the tumor microenvironment and metastases.[275]

For an early clinical diagnosis of the tumor tissue in hollows organs, an endoscope equipped with a real-time lifetime-based contrast has been developed.[175] Preliminary results include endoscopic lifetime images of the excised bladder where autofluorescence of the flavin molecules has been recorded. Finally, the FD photon migration technique has been applied to tissue-simulating phantoms in order to externally image fluorescence properties inside the tissue volume for possible disease detection.[310]

10.4.5 Lifetime Imaging with Long-Lived Fluorophores

The best known application of the long-lifetime fluorophores is for oxygen imaging. Emission of many Ru(II) complexes and Pt(II) porphyrines is dynamically quenched by oxygen and can be effectively used as a lifetime oxygen sensor. Two-dimensional oxygen fluxes and partial pressure maps of surface oxygen have been monitored using such long-lived fluorophores and both the TD and FD techniques.[174,311–314] The long lifetimes enable adaptation of inexpensive LED light sources and allow for suppression of the short-lived background by proper gating of the detection. Planar chemical sensors based on the long-lived Ru(II) complexes with lifetimes responding to biologically relevant agents such as CO_2, K^+, Cl^-, NH_4^+, and glucose have been integrated on micro-plate arrays for possible clinical applications.[140,315]

10.5 Outlook

Much progress has been made in both technology and data analysis since lifetime-based imaging was first reported. During the years, the potential of the lifetime-based imaging has become well recognized and unquestionable. It seems evident that further progress of the technique is tightly linked not only to technological advances but also to advances of the probe chemistry and to introduction of new concepts in the fluorescence spectroscopy.

To answer the question where the lifetime-based imaging is heading, one has to identify current limitations of the technique and define application-driven goals. Real-world experience could be the starting point: (1) there have never been enough photons for lifetime imaging; (2) there is a need for high spatial resolution required by biological applications; (3) there is always demand for faster data acquisition and analysis to study dynamic processes; and (4) there is a strong incentive to combine multiple imaging techniques, for example, a lifetime, anisotropy, spectral, and a conventional intensity imaging in one multifunctional device. Imagers were constructed where FLIM combines with florescence correlation spectroscopy (FCS)[316] or AFM[317] in a single device. A broader spreading of the FLIM technique requires an affordable instrumentation. It seems evident that to satisfy all requirements simultaneously is not an easy task and compromises have to be made.

Our vision of the future FLIM instrument is an affordable real-time apparatus with a 3D lifetime imaging capability and the possibility of multiphoton excitation. The instrument will combine several other imaging techniques in one multifunctional device for simultaneous multidimensional imaging.

The quality and accuracy of the lifetime imaging is determined by the S/N ratio of the data. The S/N ratio can be improved mainly by acquisition of more photons per pixel. Due to the requirement for real-time operation, longer data integration time is unacceptable. Stronger excitation is also problematic due to photobleaching and sample destruction. The solution could be an introduction of brighter probes. Here, the concept of fluorescence enhancement near the surface of metallic particles could play a significant role.[5,318–322] Since the enhancement of the fluorescence intensity is caused by faster radiative rate, which is accompanied by decrease in fluorescence lifetime, the future instruments utilizing the intensity-enhanced probes should operate with a picosecond lifetime resolution.

An effort has already been made to build a real-time FLIM apparatus suitable for a clinical use and for monitoring of dynamic biological events.[122,134,167,175,180,195] Overall data acquisition rate favors WF multichannel imaging systems with an image intensifier as compared to the single-channel scanning imagers. A lack of sectioning capability of WF lifetime imagers, however, has been neutralizing this comparative advantage. At present, this drawback seems to be disappearing, and technologies utilizing confocal spinning disks,[194,195,228] MMM,[225,323] structured illumination,[117,121,227,229,230] or fast SLMs[235,238–240] have enabled efficient 3D FLI with the WF imagers. In the future, we expect to see fast 3D fluorescence lifetime imagers suitable for imaging of turbid biological tissues[310,324] or for lifetime tomography.[325]

Clinical and biological lifetime-based imaging of turbid samples will require not only a sectioning capability but also a high rejection of scattered light and suppression of autofluorescence. Two-photon exited long-lived long-wavelength fluorophores based on energy transfer[326,327] or long-lived metal–ligand complexes[328,329] will be useful with gated detection. Hardware suppression of the prompt and short-lived emission components is no longer limited to TD gated methods but has recently been introduced also to the FD fluorometry.[330,331]

It is reasonable to expect that price of the future FLI instruments will drop down, since expensive pulsed laser systems could be for routine applications replaced by pulsed or HF-modulated laser diodes or LEDs.[25,332,333] A possible way how to further reduce price of current FLIM instruments seems to be a shift to solid-state technology where expensive image intensifiers are replaced by gain-modulated lock-in CCD/CMOS chips.[334,335] A substantial reduction of hardware expense can be also achieved by utilization of standard, communications-type, radiofrequency electronics[336] or a field programmable gate arrays implemented in the digital FD FLIM.[203]

In this chapter, we have not been able to discuss all possible modalities and applications of FLIM. It is obvious that each implementation of the lifetime-based imaging has specific advantages and disadvantages and a chosen approach will always be dictated by the application. We conclude that the lifetime-based imaging has become a reliable matured technique on the way out of research laboratories. First, instruments capable of routine unsupervised operation in high-throughput screening have already been reported.[198] Nevertheless, more powerful, less expensive, and easier to operate instruments for lifetime-based imaging are still future goal.

Acknowledgments

This work was supported by GACR P205/12/0720 and Prvouk P45 (PH), and National Institutes of Health HG002655 and EB006521 (JRL).

References

1. Herschel, S. J. F. W., On a case of superficial colour presented by homogeneous liquid internally colourless, *Philosophical Transactions of the Royal Society of London* 135, 143–145, 1845.
2. D'Auria, S. and Lakowicz, J. R., Enzyme fluorescence as a sensing tool: New perspectives in biotechnology, *Current Opinion in Biotechnology* 12 (1), 99–104, 2001.
3. Wouters, F. S., Verveer, P. J., and Bastiaens, P. I., Imaging biochemistry inside cells, *Trends in Cell Biology* 11 (5), 203–211, 2001.
4. Royer, C. A. and Scarlata, S. F., Fluorescence approaches to quantifying biomolecular interactions, *Methods Enzymology* 450, 79–106, 2008.
5. Lakowicz, J. R., Ray, K., Chowdhury, M., Szmacinski, H., Fu, Y., Zhang, J., and Nowaczyk, K., Plasmon-controlled fluorescence: A new paradigm in fluorescence spectroscopy, *Analyst* 133 (10), 1308–1346, 2008.
6. Wang, Y., Shyy, J. Y., and Chien, S., Fluorescence proteins, live-cell imaging, and mechanobiology: Seeing is believing, *Annual Review of Biomedical Engineering* 10, 1–38, 2008.
7. Ciruela, F., Fluorescence-based methods in the study of protein-protein interactions in living cells, *Current Opinion in Biotechnology* 19 (4), 338–343, 2008.
8. Wang, X. F., Periasamy, A., Herman, B., and Coleman, D. M., Fluorescence Lifetime Imaging Microscopy (FLIM)—Instrumentation and applications, *Critical Reviews in Analytical Chemistry* 23 (5), 369–395, 1992.
9. Lakowicz, J. R., Fluorescence lifetime sensing creates cellular images, *Laser Focus World* 28 (5), 60–80, 1992.
10. Bastiaens, P. I. and Squire, A., Fluorescence lifetime imaging microscopy: Spatial resolution of biochemical processes in the cell, *Trends in Cell Biology* 9 (2), 48–52, 1999.
11. Ross, J. A. and Jameson, D. M., Time-resolved methods in biophysics. 8. Frequency domain fluorometry: Applications to intrinsic protein fluorescence, *Photochemical and Photobiological Science* 7 (11), 1301–1312, 2008.
12. Chang, C. W., Sud, D., and Mycek, M. A., Fluorescence lifetime imaging microscopy, *Methods in Cell Biology* 81, 495–524, 2007.
13. Niesner, R. A., Andresen, V., and Gunzer, M., Intravital two-photon microscopy: Focus on speed and time resolved imaging modalities, *Immunological Reviews* 221, 7–25, 2008.
14. Bunt, G. and Wouters, F. S., Visualization of molecular activities inside living cells with fluorescent labels, *International Review of Cytology* 237, 205–277, 2004.
15. Suhling, K., French, P. M., and Phillips, D., Time-resolved fluorescence microscopy, *Photochemical and Photobiological Science* 4 (1), 13–22, 2005.
16. van Munster, E. B. and Gadella, T. W., Fluorescence lifetime imaging microscopy (FLIM), *Advances in Biochemical Engineering Biotechnology* 95, 143–175, 2005.
17. Verkman, A. S., Armijo, M., and Fushimi, K., Construction and evaluation of a frequency-domain epifluorescence microscope for lifetime and anisotropy decay measurements in subcellular domains, *Biophysical Chemistry* 40 (1), 117–125, 1991.
18. Despa, S., Steels, P., and Ameloot, M., Fluorescence lifetime microscopy of the sodium indicator sodium-binding benzofuran isophthalate in HeLa cells, *Analytical Biochemistry* 280 (2), 227–241, 2000.
19. Despa, S., Vecer, J., Steels, P., and Ameloot, M., Fluorescence lifetime microscopy of the Na⁺ indicator Sodium Green in HeLa cells, *Analytical Biochemistry* 281 (2), 159–175, 2000.

20. Dix, J. A. and Verkman, A. S., Pyrene excimer mapping in cultured fibroblasts by ratio imaging and time-resolved microscopy, *Biochemistry* 29 (7), 1949–1953, 1990.

21. Dix, J. A. and Verkman, A. S., Mapping of fluorescence anisotropy in living cells by ratio imaging—Application to cytoplasmic viscosity, *Biophysical Journal* 57 (2), 231–240, 1990.

22. Srivastava, A. and Krishnamoorthy, G., Cell type and spatial location dependence of cytoplasmic viscosity measured by time-resolved fluorescence microscopy, *Archives of Biochemistry and Biophysics* 340 (2), 159–167, 1997.

23. Wang, X. F., Kitajima, S., Uchida, T., Coleman, D. M., and Minami, S., Time-resolved fluorescence microscopy using multichannel photon-counting, *Applied Spectroscopy* 44 (1), 25–30, 1990.

24. Keating, S. M. and Wensel, T. G., Nanosecond fluorescence microscopy—Emission kinetics of Fura-2 in single cells, *Biophysical Journal* 59 (1), 186–202, 1991.

25. Herman, P., Maliwal, B. P., Lin, H. J., and Lakowicz, J. R., Frequency-domain fluorescence microscopy with the LED as a light source, *Journal of Microscopy—Oxford* 203, 176–181, 2001.

26. Ambroz, M., MacRobert, A. J., Morgan, J., Rumbles, G., Foley, M. S., and Phillips, D., Time-resolved fluorescence spectroscopy and intracellular imaging of disulphonated aluminium phthalocyanine, *Journal of Photochemical and Photobiology B* 22 (2), 105–117, 1994.

27. Kusumi, A., Tsuji, A., Murata, M., Sako, Y., Yoshizawa, A. C., Kagiwada, S., Hayakawa, T., and Ohnishi, S., Development of a streak-camera-based time-resolved microscope fluorimeter and its application to studies of membrane fusion in single cells, *Biochemistry* 30 (26), 6517–6527, 1991.

28. Seah, L. K., Wang, P., Murukeshan, V. M., and Chao, Z. X., Application of fluorescence lifetime imaging (FLIM) in latent finger mark detection, *Forensic Science International* 160 (2–3), 109–114, 2006.

29. Comelli, D., D'Andrea, C., Valentini, G., Cubeddu, R., Colombo, C., and Toniolo, L., Fluorescence lifetime imaging and spectroscopy as tools for nondestructive analysis of works of art, *Applied Optics* 43 (10), 2175–2183, 2004.

30. Morgan, C. G. and Mitchell, A. C., Fluorescence lifetime imaging: An emerging technique in fluorescence microscopy, *Chromosome Research* 4 (4), 261–263, 1996.

31. Periasamy, A., Wang, X. F., Wodnick, P., Gordon, G. W., Kwon, S., Diliberto, P. A., and Herman, B., High-speed fluorescence microscopy: Lifetime imaging in biological sciences, *Microscopy and Microanalysis* 1 (1), 13–23, 1995.

32. Lakowicz, J. R., Emerging applications of fluorescence spectroscopy to cellular imaging: Lifetime imaging, metal–ligand probes, multi-photon excitation and light quenching, *Scanning Microscopy Supplement* 10, 213–224, 1996.

33. Tadrous, P. J., Methods for imaging the structure and function of living tissues and cells: 2. Fluorescence lifetime imaging, *Journal of Pathology* 191 (3), 229–234, 2000.

34. Levitt, J. A., Matthews, D. R., Ameer-Beg, S. M., and Suhling, K., Fluorescence lifetime and polarization-resolved imaging in cell biology, *Current Opinion in Biotechnology* 20, 28–36, 2009.

35. Elson, D., Requejo-Isidro, J., Munro, I., Reavell, F., Siegel, J., Suhling, K., Tadrous, P. et al., Time-domain fluorescence lifetime imaging applied to biological tissue, *Photochemical and Photobiological Sciences* 3 (8), 795–801, 2004.

36. de Almeida, R. F., Loura, L. M., and Prieto, M., Membrane lipid domains and rafts: Current applications of fluorescence lifetime spectroscopy and imaging, *Chemistry and Physics of Lipids* 157 (2), 61–77, 2009.

37. Wallrabe, H. and Periasamy, A., Imaging protein molecules using FRET and FLIM microscopy, *Current Opinion in Biotechnology* 16 (1), 19–27, 2005.

38. Peter, M. and Ameer-Beg, S. M., Imaging molecular interactions by multiphoton FLIM, *Biology of the Cell* 96 (3), 231–236, 2004.

39. Bonushkin, Y., Dworkin, L., Hauser, J., Kim, C., Lindgren, M., Scott, A., and Appollinari, G., Tests of a third generation multianode phototube, *Nuclear Instruments and Methods in Physics Research Section A—Accelerators Spectrometers Detectors and Associated Equipment* 381 (2–3), 349–354, 1996.

40. Salomon, M. and Williams, S. S. A., A multi-anode photomultiplier with position sensitivity, *Nuclear Instruments and Methods in Physics Research Section A—Accelerators Spectrometers Detectors and Associated Equipment* 241 (1), 210–214, 1985.

41. McLoskey, D., Birch, D. J. S., Sanderson, A., Suhling, K., Welch, E., and Hicks, P. J., Multiplexed single-photon counting. 1. A time-correlated fluorescence lifetime camera, *Review of Scientific Instruments* 67 (6), 2228–2237, 1996.

42. Vecer, J., Herman, P., and Beranek, J., Low-temperature luminescence spectra and fluorescence lifetimes of polycytidylic acid in polyalcoholic glasses, *Photochemistry and Photobiology* 57 (5), 792–795, 1993.

43. Demas, J. N., *Excited State Lifetime Measurements*, Academic Press, New York, 1983.

44. O'Connor, D. V. and Philips, D., *Time-Correlated Single Photon Counting*, Academic Press, London, U.K., 1984.

45. Becker, W., *Advanced Time-Correlated Singe Photon Counting Techniques*, Springer-Verlag, Berlin, Germany/Heidelberg, 2005.

46. Coates, P. B., The correction for photon pile-up in the measurement of radiative lifetimes, *Journal of Physics E (Series 2)* 1, 878–879, 1968.

47. Selinger, B. K. and Harris, C. M., The pile-up problem in the pulse fluorometry, in *Time-Resolved Fluorescence Spectroscopy in Biochemistry and Biology*, eds. Cundall, R. B. and Dale, R. E. Plenum Publishing Corporation, New York, 1983, pp. 115–127.

48. Harris, C. M. and Selinger, B. K., Single-photon decay spectroscopy: II. The pileup problem, *Australian Journal of Chemistry* 32, 2111–2129, 1979.

49. Williamson, J. A., Kendalltobias, M. W., Buhl, M., and Seibert, M., Statistical evaluation of dead time effects and pulse pileup in fast photon-counting—Introduction of the sequential model, *Analytical Chemistry* 60 (20), 2198–2203, 1988.

50. Meltzer, R. S. and Wood, R. M., Nanosecond time-resolved spectroscopy with a pulsed tunable dye laser and single photon time correlation, *Applied Optics* 16 (5), 1432–1434, 1977.

51. Schuyler, R. and Isenberg, I., A monophoton fluorometer with energy discrimination, *Review of Scientific Instruments* 42 (6), 813–817, 1970.

52. Kollner, M. and Wolfrum, J., How many photons are necessary for fluorescence-lifetime measurements, *Chemical Physics Letters* 200 (1–2), 199–204, 1992.

53. Schweitzer, D., Kolb, A., Hammer, M., and Thamm, E., Basic investigations for 2-dimensional time-resolved fluorescence measurements at the fundus, *International Ophthalmology* 23 (4–6), 399–404, 2001.

54. Duncan, R. R., Bergmann, A., Cousin, M. A., Apps, D. K., and Shipston, M. J., Multi-dimensional time-correlated single photon counting (TCSPC) fluorescence lifetime imaging microscopy (FLIM) to detect FRET in cells, *Journal of Microscopy* 215 (Pt 1), 1–12, 2004.

55. Kemnitz, K., Pfeifer, L., Paul, R., and Coppey-Moisan, M., Novel detectors for fluorescence imaging on the picosecond time scale, *Journal of Fluorescence* 7 (1), 93–98, 1997.

56. Lassiter, S. J., Stryjewski, W., Legendre, B. L., Jr., Erdmann, R., Wahl, M., Wurm, J., Peterson, R., Middendorf, L., and Soper, S. A., Time-resolved fluorescence imaging of slab gels for lifetime base-calling in DNA sequencing applications, *Analytical Chemistry* 72 (21), 5373–5382, 2000.

57. Schonle, A., Glatz, M., and Hell, S. W., Four-dimensional multiphoton microscopy with time-correlated single-photon counting, *Applied Optics* 39 (34), 6306–6311, 2000.

58. Barzda, V., de Grauw, C. J., Vroom, J., Kleima, F. J., van Grondelle, R., van Amerongen, H., and Gerritsen, H. C., Fluorescence lifetime heterogeneity in aggregates of LHCII revealed by time-resolved microscopy, *Biophysical Journal* 81 (1), 538–546, 2001.

59. Tinnefeld, P., Herten, D. P., and Sauer, M., Photophysical dynamics of single molecules studied by spectrally-resolved fluorescence lifetime imaging microscopy (SFLIM), *Journal of Physical Chemistry A* 105 (34), 7989–8003, 2001.

60. Tregidgo, C., Levitt, J. A., and Suhling, K., Effect of refractive index on the fluorescence lifetime of green fluorescent protein, *Journal of Biomedical Optics* 13 (3), 031218, 2008.

61. Chen, Y. and Periasamy, A., Characterization of two-photon excitation fluorescence lifetime imaging microscopy for protein localization, *Microscopy Research and Technique* 63 (1), 72–80, 2004.

62. Becker, W., Bergmann, A., Hink, M. A., Konig, K., Benndorf, K., and Biskup, C., Fluorescence lifetime imaging by time-correlated single-photon counting, *Microscopy Research and Technique* 63 (1), 58–66, 2004.

63. Bayle, V., Nussaume, L., and Bhat, R. A., Combination of novel green fluorescent protein mutant TSapphire and DsRed variant mOrange to set up a versatile in planta FRET–FLIM assay, *Plant Physiology* 148 (1), 51–60, 2008.

64. Ghukasyan, V., Hsu, Y. Y., Kung, S. H., and Kao, F. J., Application of fluorescence resonance energy transfer resolved by fluorescence lifetime imaging microscopy for the detection of enterovirus 71 infection in cells, *Journal of Biomedical Optics* 12 (2), 024016, 2007.

65. Biskup, C., Kelbauskas, L., Zimmer, T., Benndorf, K., Bergmann, A., Becker, W., Ruppersberg, J. P., Stockklausner, C., and Klocker, N., Interaction of PSD-95 with potassium channels visualized by fluorescence lifetime-based resonance energy transfer imaging, *Journal of Biomedical Optics* 9 (4), 753–759, 2004.

66. Willemsen, O. H., Noordman, O. F. J., Segering, F. B., Ruiter, A. G. T., Moers, M. H. P., and Vanhulst, N. F., Fluorescence lifetime contrast combined with probe microscopy, in *Optics at the Nanometer Scale*, eds. Nieto-Vesperinas, M. and Garcia, N. IBM, Dordrecht, the Netherlands, 1996, pp. 223–233.

67. Sasaki, K., Koshioka, M., and Masuhara, H., 3-Dimensional space-resolved and time-resolved fluorescence spectroscopy, *Applied Spectroscopy* 45 (6), 1041–1045, 1991.

68. Böhmer, M., Pampaloni, F., Wahl, M., Rahn, H. J., Erdmann, R., and Enderlein, J., Time-resolved confocal scanning device for ultrasensitive fluorescence detection, *Review of Scientific Instruments* 72 (11), 4145–4152, 2001.

69. McConnell, G., Girkin, J. M., Ameer-Beg, S. M., Barber, P. R., Vojnovic, B., Ng, T., Banerjee, A., Watson, T. F., and Cook, R. J., Time-correlated single-photon counting fluorescence lifetime confocal imaging of decayed and sound dental structures with a white-light supercontinuum source, *Journal of Microscopy* 225 (Pt 2), 126–136, 2007.

70. Jose, M., Nair, D. K., Reissner, C., Hartig, R., and Zuschratter, W., Photophysics of Clomeleon by FLIM: Discriminating excited state reactions along neuronal development, *Biophysical Journal* 92 (6), 2237–2254, 2007.

71. Jose, M., Nair, D. K., Altrock, W. D., Dresbach, T., Gundelfinger, E. D., and Zuschratter, W., Investigating interactions mediated by the presynaptic protein bassoon in living cells by Foerster's resonance energy transfer and fluorescence lifetime imaging microscopy, *Biophysical Journal* 94 (4), 1483–1496, 2008.

72. Martin-Fernandez, M., Longshaw, S. V., Kirby, I., Santis, G., Tobin, M. J., Clarke, D. T., and Jones, G. R., Adenovirus type-5 entry and disassembly followed in living cells by FRET, fluorescence anisotropy, and FLIM, *Biophysical Journal* 87 (2), 1316–1327, 2004.

73. Garcia, D. I., Lanigan, P., Webb, M., West, T. G., Requejo-Isidro, J., Auksorius, E., Dunsby, C., Neil, M., French, P., and Ferenczi, M. A., Fluorescence lifetime imaging to detect actomyosin states in mammalian muscle sarcomeres, *Biophysical Journal* 93 (6), 2091–2101, 2007.

74. Becker, W., Bergmann, A., Konig, K., and Tirlapur, U., Picosecond fluorescence lifetime microscopy by TCSPC imaging, *Multiphoton Microscopy in the Biomedical Sciences* 2 (19), 414–419 (424), 2001.

75. Bird, D. K., Yan, L., Vrotsos, K. M., Eliceiri, K. W., Vaughan, E. M., Keely, P. J., White, J. G., and Ramanujam, N., Metabolic mapping of MCF10A human breast cells via multiphoton fluorescence lifetime imaging of the coenzyme NADH, *Cancer Research* 65 (19), 8766–8773, 2005.

76. Douma, K., Megens, R. T. A., Reitsma, S., Prinzen, L., Slaaf, D. W., and Van Zandvoort, M., Two-photon lifetime imaging of fluorescent probes in intact blood vessels: A window to sub-cellular structural information and binding status, *Microscopy Research and Technique* 70 (5), 467–475, 2007.

77. Peltan, I. D., Thomas, A. V., Mikhailenko, I., Strickland, D. K., Hyman, B. T., and von Arnim, C. A., Fluorescence lifetime imaging microscopy (FLIM) detects stimulus-dependent phosphorylation of the low density lipoprotein receptor-related protein (LRP) in primary neurons, *Biochemical and Biophysical Research Communications* 349 (1), 24–30, 2006.

78. Peter, M., Ameer-Beg, S. M., Hughes, M. K., Keppler, M. D., Prag, S., Marsh, M., Vojnovic, B., and Ng, T., Multiphoton-FLIM quantification of the EGFP-mRFP1 FRET pair for localization of membrane receptor–kinase interactions, *Biophysical Journal* 88 (2), 1224–1237, 2005.

79. Murakoshi, H., Lee, S. J., and Yasuda, R., Highly sensitive and quantitative FRET–FLIM imaging in single dendritic spines using improved non-radiative YFP, *Brain Cell Biology* 36, 31–42, 2008.

80. Walczysko, P., Kuhlicke, U., Knappe, S., Cordes, C., and Neu, T. R., In situ activity of suspended and immobilized microbial communities as measured by fluorescence lifetime imaging, *Applied and Environmental Microbiology* 74 (1), 294–299, 2008.

81. Koushik, S. V. and Vogel, S. S., Energy migration alters the fluorescence lifetime of Cerulean: Implications for fluorescence lifetime imaging Forster resonance energy transfer measurements, *Journal of Biomedical Optics* 13 (3), 031204, 2008.

82. Berezovska, O., Ramdya, P., Skoch, J., Wolfe, M. S., Bacskai, B. J., and Hyman, B. T., Amyloid precursor protein associates with a nicastrin-dependent docking site on the presenilin 1-gamma-secretase complex in cells demonstrated by fluorescence lifetime imaging, *Journal of Neuroscience* 23 (11), 4560–4566, 2003.

83. Nyborg, A. C., Herl, L., Berezovska, O., Thomas, A. V., Ladd, T. B., Jansen, K., Hyman, B. T., and Golde, T. E., Signal peptide peptidase (SPP) dimer formation as assessed by fluorescence lifetime imaging microscopy (FLIM) in intact cells, *Molecular Neurodegeneration* 1, 16, 2006.

84. Parsons, M., Messent, A. J., Humphries, J. D., Deakin, N. O., and Humphries, M. J., Quantification of integrin receptor agonism by fluorescence lifetime imaging, *Journal of Cell Science* 121 (Pt 3), 265–271, 2008.

85. Kwak, E. S., Kang, T. J., and Vanden Bout, D. A., Fluorescence lifetime imaging with near-field scanning optical microscopy, *Analytical Chemistry* 73 (14), 3257–3262, 2001.

86. Suhling, K., McLoskey, D., and Birch, D. J. S., Multiplexed single-photon counting. II. The statistical theory of time-correlated measurements, *Review of Scientific Instruments* 67 (6), 2238–2246, 1996.

87. Howorth, J. R., Ferguson, I., and Wilcox, D. A., Developments in microchannel plate photomultipliers, in *Proceedings of the SPIE, Advances in Fluorescence Sensing Technology II*, ed. Lakowicz, J. R. San Jose, CA, 1995, pp. 356–362.

88. Becker, W., Bergmann, A., and Biskup, C., Multispectral fluorescence lifetime imaging by TCSPC, *Microscopic and Research Technique* 70 (5), 403–409, 2007.

89. Lakowicz, J. R., *Principles of Fluorescence Spectroscopy,* 3rd edn. Springer, New York, 2006.

90. Sanders, R., Draaijer, A., Gerritsen, H. C., Houpt, P. M., and Levine, Y. K., Quantitative pH imaging in cells using confocal fluorescence lifetime imaging microscopy, *Analytical Biochemistry* 227 (2), 302–308, 1995.

91. Sanders, R., van Zandvoort, M., Draaijer, A., Levine, Y. K., and Gerritsen, H. C., Confocal fluorescence lifetime imaging of chlorophyll molecules in polymer matrices, *Photochemistry and Photobiology* 64 (5), 817–820, 1996.

92. Buurman, E. P., Sanders, R., Draaijer, A., Gerritsen, H. C., Vanveen, J. J. F., Houpt, P. M., and Levine, Y. K., Fluorescence lifetime imaging using a confocal laser scanning microscope, *Scanning* 14 (3), 155–159, 1992.

93. Gerritsen, H. C., Sanders, R., Draaijer, A., Ince, C., and Levine, Y. K., Fluorescence lifetime imaging of oxygen in living cells, *Journal of Fluorescence* 7 (1), 11–15, 1997.

94. Roorda, R. D., Ribes, A. C., Damaskinos, S., Dixon, A. E., and Menzel, E. R., A scanning beam time-resolved imaging system for fingerprint detection, *Journal of Forensic Sciences* 45 (3), 563–567, 2000.

95. Vroom, J. M., De Grauw, K. J., Gerritsen, H. C., Bradshaw, D. J., Marsh, P. D., Watson, G. K., Birmingham, J. J., and Allison, C., Depth penetration and detection of pH gradients in biofilms by two-photon excitation microscopy, *Applied and Environmental Microbiology* 65 (8), 3502–3511, 1999.

96. de Grauw, C. J. and Gerritsen, H. C., Multiple time-gate module for fluorescence lifetime imaging, *Applied Spectroscopy* 55 (6), 670–678, 2001.

97. Sytsma, J., Vroom, J. M., De Grauw, C. J., and Gerritsen, H. C., Time-gated fluorescence lifetime imaging and microvolume spectroscopy using two-photon excitation, *Journal of Microscopy—Oxford* 191, 39–51, 1998.
98. Wang, X. F., Uchida, T., Coleman, D. M., and Minami, S., A 2-dimensional fluorescence lifetime imaging-system using a gated image intensifier, *Applied Spectroscopy* 45 (3), 360–366, 1991.
99. Oida, T., Sako, Y., and Kusumi, A., Fluorescence lifetime imaging microscopy (flimscopy). Methodology development and application to studies of endosome fusion in single cells, *Biophysical Journal* 64 (3), 676–685, 1993.
100. Cubeddu, R., Canti, G., Taroni, P., and Valentini, G., Time-gated fluorescence imaging for the diagnosis of tumors in a murine model, *Photochemical and Photobiology* 57 (3), 480–485, 1993.
101. Cubeddu, R., Taroni, P., and Valentini, G., Time-gated imaging-system for tumor-diagnosis, *Optical Engineering* 32 (2), 320–325, 1993.
102. Schneckenburger, H., Konig, K., Dienersberger, T., and Hahn, R., Time-gated microscopic imaging and spectroscopy in medical diagnosis and photobiology, *Optical Engineering* 33 (8), 2600–2606, 1994.
103. Periasamy, A., Wodnicki, P., Wang, X. F., Kwon, S., Gordon, G. W., and Herman, B., Time-resolved fluorescence lifetime imaging microscopy using a picosecond pulsed tunable dye laser system, *Review of Scientific Instruments* 67 (10), 3722–3731, 1996.
104. Hennink, E. J., deHaas, R., Verwoerd, N. P., and Tanke, H. J., Evaluation of a time-resolved fluorescence microscope using a phosphorescent Pt-porphine model system, *Cytometry* 24 (4), 312–320, 1996.
105. Cubeddu, R., Canti, G., Pifferi, A., Taroni, P., and Valentini, G., Fluorescence lifetime imaging of experimental tumors in hematoporphyrin derivative-sensitized mice, *Photochemistry and Photobiology* 66 (2), 229–236, 1997.
106. Dowling, K., Hyde, S. C. W., Dainty, J. C., French, P. M. W., and Hares, J. D., 2-D fluorescence lifetime imaging using a time-gated image intensifier, *Optics Communications* 135 (1–3), 27–31, 1997.
107. Herman, B., Wodnicky, P., Kwon, S., Periasamy, A., Gordon, G. W., Mahajan, N., and Wang, X. F., Recent developments in monitoring calcium and protein interactions in cells using fluorescence lifetime microscopy, *Journal of Fluorescence* 7 (1), 85–91, 1997.
108. Schneckenburger, H., Gschwend, M. H., Strauss, W. S. L., Sailer, R., and Steiner, R., Time-gated microscopic energy transfer measurements for probing mitochondrial metabolism, *Journal of Fluorescence* 7 (1), 9, 1997.
109. Scully, A. D., Ostler, R. B., Phillips, D., O'Neill, P., Townsend, K. M. S., Parker, A. W., and MacRobert, A. J., Application of fluorescence lifetime imaging microscopy to the investigation of intracellular PDT mechanisms, *Bioimaging* 5, 9–18, 1997.
110. Dowling, K., Dayel, M. J., Lever, M. J., French, P. M. W., Hares, J. D., and Dymoke-Bradshaw, A. K. L., Fluorescence lifetime imaging with picosecond resolution for biomedical applications, *Optics Letters* 23 (10), 810–812, 1998.
111. Dowling, K., Dayel, M. J., Hyde, S. C. W., Dainty, J. C., French, P. M. W., Vourdas, P., Lever, M. J., Dymoke-Bradshaw, A. K. L., Hares, J. D., and Kellett, P. A., Whole-field fluorescence lifetime imaging with picosecond resolution using ultrafast 10-kHz solid-state amplifier technology, *IEEE Journal of Selected Topics in Quantum Electronics* 4 (2), 370–375, 1998.
112. Konig, K., Boehme, S., Leclerc, N., and Ahuja, R., Time-gated autofluorescence microscopy of motile green microalga in an optical trap, *Cellular and Molecular Biology (Noisy-le-grand)* 44 (5), 763–770, 1998.
113. Vereb, G., Jares-Erijman, E., Selvin, P. R., and Jovin, T. M., Temporally and spectrally resolved imaging microscopy of lanthanide chelates, *Biophysical Journal* 74 (5), 2210–2222, 1998.
114. Dowling, K., Dayel, M. J., Hyde, S. C. W., French, P. M. W., Lever, M. J., Hares, J. D., and Dymoke-Bradshaw, A. K. L., High resolution time-domain fluorescence lifetime imaging for biomedical applications, *Journal of Modern Optics* 46 (2), 199–209, 1999.

115. Konig, K., Schneckenburger, H., and Hibst, R., Time-gated in vivo autofluorescence imaging of dental caries, *Cellular and Molecular Biology (Noisy-le-grand)* 45 (2), 233–239, 1999.

116. Zevallos, M. E., Gayen, S. K., Das, B. B., Alrubaiee, M., and Alfano, R. R., Picosecond electronic time-gated imaging of bones in tissues, *IEEE Journal of Selected Topics in Quantum Electronics* 5 (4), 916–922, 1999.

117. Cole, M. J., Siegel, J., Webb, S. E. D., Jones, R., Dowling, K., French, P. M. W., Lever, M. J., Sucharov, L. O. D., Neil, M. A. A., Juskaitis, R., and Wilson, T., Whole-field optically sectioned fluorescence lifetime imaging, *Optics Letters* 25 (18), 1361–1363, 2000.

118. Lee, K. C., Siegel, J., Webb, S. E., Leveque-Fort, S., Cole, M. J., Jones, R., Dowling, K., Lever, M. J., and French, P. M., Application of the stretched exponential function to fluorescence lifetime imaging, *Biophysical Journal* 81 (3), 1265–1274, 2001.

119. Requejo-Isidro, J., McGinty, J., Munro, I., Elson, D. S., Galletly, N. P., Lever, M. J., Neil, M. A. et al., High-speed wide-field time-gated endoscopic fluorescence-lifetime imaging, *Optics Letter* 29 (19), 2249–2251, 2004.

120. Munro, I., McGinty, J., Galletly, N., Requejo-Isidro, J., Lanigan, P. M., Elson, D. S., Dunsby, C., Neil, M. A., Lever, M. J., Stamp, G. W., and French, P. M., Toward the clinical application of time-domain fluorescence lifetime imaging, *Journal of Biomedical Optics* 10 (5), 051403, 2005.

121. Webb, S. E. D., Gu, Y., Leveque-Fort, S., Siegel, J., Cole, M. J., Dowling, K., Jones, R. et al., A wide-field time-domain fluorescence lifetime imaging microscope with optical sectioning, *Review of Scientific Instruments* 73 (4), 1898–1907, 2002.

122. Itoh, H., Evenzahav, A., Kinoshita, K., Inagaki, Y., Mizushima, H., Takahashi, A., Fukami, T., Hayakawa, T., and Kusumi, A., Fluorescence lifetime imaging microscopy with a high repetition gated camera and a dual-view assembly for the real time measurement, *Advances in Fluorescence Sensing Technology III* 2980, 12–19 (582), 1997.

123. Agronskaia, A. V., Tertoolen, L., and Gerritsen, H. C., Fast fluorescence lifetime imaging of calcium in living cells, *Journal of Biomedical Optics* 9 (6), 1230–1237, 2004.

124. Uchimura, T., Kawanabe, S., Maeda, Y., and Imasaka, T., Fluorescence lifetime imaging microscope consisting of a compact picosecond dye laser and a gated charge-coupled device camera for applications to living cells, *Analytical Sciences* 22 (10), 1291–1295, 2006.

125. Niesner, R., Peker, B., Schlusche, P., and Gericke, K. H., Noniterative biexponential fluorescence lifetime imaging in the investigation of cellular metabolism by means of NAD(P)H autofluorescence, *Chemphyschem* 5 (8), 1141–1149, 2004.

126. Soloviev, V. Y., Tahir, K. B., McGinty, J., Elson, D. S., Neil, M. A., French, P. M., and Arridge, S. R., Fluorescence lifetime imaging by using time-gated data acquisition, *Applied Optics* 46 (30), 7384–7391, 2007.

127. Benninger, R. K., Hofmann, O., Onfelt, B., Munro, I., Dunsby, C., Davis, D. M., Neil, M. A., French, P. M., and de Mello, A. J., Fluorescence-lifetime imaging of DNA–dye interactions within continuous-flow microfluidic systems, *Angewandte Chemie International Edition in English* 46 (13), 2228–2231, 2007.

128. Tadrous, P. J., Siegel, J., French, P. M., Shousha, S., Lalani el, N., and Stamp, G. W., Fluorescence lifetime imaging of unstained tissues: Early results in human breast cancer, *Journal of Pathology* 199 (3), 309–317, 2003.

129. Li, X., Uchimura, T., Kawanabe, S., and Imasaka, T., Use of a fluorescence lifetime imaging microscope in an apoptosis assay of Ewing's sarcoma cells with a vital fluorescent probe, *Analytical Biochemistry* 367 (2), 219–224, 2007.

130. Ushida, K., Nakayama, T., Nakazawa, T., Hamanoue, K., Nagamura, T., Mugishima, A., and Sakimukai, S., Implementation of an image intensifier coupled with a linear position-sensitive detector for measurements of absorption and emission-spectra from the nanosecond to millisecond time regime, *Review of Scientific Instruments* 60 (4), 617–623, 1989.

131. Image intensifiers, Product catalog, Hamamatsu Photonics K.K., Electron Tube Center, Iwata City, Japan, 2009, http://www.hamamatsu.com/resources/pdf/etd/II_TII0004E02.pdf.

132. Kentech Instruments Ltd., Gated image intensifiers, Wallingford, UK, 2009, http://www.kentech.co.uk.

133. LaVision GmbH, Ultra-Fast Gated Cameras, 2009, http://www.lavision.de/products/cameras/ultrafast_gated_cameras.php.

134. Agronskaia, A. V., Tertoolen, L., and Gerritsen, H. C., High frame rate fluorescence lifetime imaging, *Journal of Physics D—Applied Physics* 36 (14), 1655–1662, 2003.

135. Marriott, G., Clegg, R. M., Arndt-Jovin, D. J., and Jovin, T. M., Time resolved imaging microscopy. Phosphorescence and delayed fluorescence imaging, *Biophysical Journal* 60 (6), 1374–1387, 1991.

136. Khait, O., Smirnov, S., and Tran, C. D., Multispectral imaging microscope with millisecond time resolution, *Analytical Chemistry* 73 (4), 732–739, 2001.

137. Vinogradov, S. A., Lo, L. W., Jenkins, W. T., Evans, S. M., Koch, C., and Wilson, D. F., Noninvasive imaging of the distribution in oxygen in tissue in vivo using near-infrared phosphors, *Biophysical Journal* 70 (4), 1609–1617, 1996.

138. Reich, R. K., Mountain, R. W., McGonagle, W. H., Huang, J. C. M., Twichell, J. C., Kosicki, B. B., and Savoye, E. D., Integrated electronic shutter for back-illuminated charge-coupled-devices, *IEEE Transactions on Electron Devices* 40 (7), 1231–1237, 1993.

139. Hartmann, P. and Ziegler, W., Lifetime imaging of luminescent oxygen sensors based on all-solid-state technology, *Analytical Chemistry* 68 (24), 4512–4514, 1996.

140. Liebsch, G., Klimant, I., Frank, B., Holst, G., and Wolfbeis, O. S., Luminescence lifetime imaging of oxygen, pH, and carbon dioxide distribution using optical sensors, *Applied Spectroscopy* 54 (4), 548–559, 2000.

141. Mitchell, A. C., Wall, J. E., Murray, J. G., and Morgan, C. G., Measurement of nanosecond time-resolved fluorescence with a directly gated interline CCD camera, *Journal of Microscopy* 206 (Pt 3), 233–238, 2002.

142. Mitchell, A. C., Wall, J. E., Murray, J. G., and Morgan, C. G., Direct modulation of the effective sensitivity of a CCD detector: A new approach to time-resolved fluorescence imaging, *Journal of Microscopy* 206 (Pt 3), 225–232, 2002.

143. Lange, R. and Seitz, P., Solid-state time-of-flight range camera, *IEEE Journal of Quantum Electronics* 37 (3), 390–397, 2001.

144. Lange, R., Seitz, P., Biber, A., and Lauxtermann, S., Demodulation pixels in CCD and CMOS technologies for time-of-flight ranging, *Sensors and Camera Systems for Scientific, Industrial and Digital Photography Applications* 3965, 177–188, 2000.

145. Ainbund, M. R., Buevich, O. E., Kamalov, V. F., Menshikov, G. A., and Toleutaev, B. N., Simultaneous spectral and temporal resolution in a single photon-counting technique, *Review of Scientific Instruments* 63 (6), 3274–3279, 1992.

146. Charbonneau, S., Allard, L. B., Young, J. F., Dyck, G., and Kyle, B. J., 2-Dimensional time-resolved imaging with 100-Ps resolution using a resistive anode photomultiplier tube, *Review of Scientific Instruments* 63 (11), 5315–5319, 1992.

147. Arzhantsev, S. Y., Ainbund, M. R., Chikischev, A. Y., Koroteev, N. I., Shkurinov, A. P., Toleutaev, B. N., Turbin, E. V., Lehmann, A., Pfeifer, L., Fink, F., and Kemnitz, K., Picosecond fluorescence lifetime imaging microscopy at 1 mm space- and 10 ps time-resolution: 50 × 50 ch MCP-PMT with quadrant anode, in *Second Conference on Fluorescence Spectroscopy and Fluorescence Probes*, ed. Slavik, J. Plenum Press, Prague, Czech Republic, 1997, pp. 69–74.

148. Emiliani, V., Sanvitto, D., Tramier, M., Piolot, T., Petrasek, Z., Kemnitz, K., Durieux, C., and Coppey-Moisan, M., Low-intensity two-dimensional imaging of fluorescence lifetimes in living cells, *Applied Physics Letters* 83 (12), 2471–2473, 2003.

149. Fujiwara, M. and Cieslik, W., Fluorescence lifetime imaging microscopy: Two-dimensional distribution measurement of fluorescence lifetime, *Methods Enzymology* 414, 633–642, 2006.

150. Krishnan, R. V., Masuda, A., Centonze, V. E., and Herman, B., Quantitative imaging of protein-protein interactions by multiphoton fluorescence lifetime imaging microscopy using a streak camera, *Journal of Biomedical Optics* 8 (3), 362–367, 2003.

151. Guide to streak cameras, Hamamatsu Photonics K.K., Systems division, Hamamatsu City, Japan, 2008, http://www.hamamatsu.com/resources/pdf/sys/e_streakh.pdf.

152. Krishnan, R. V., Saitoh, H., Terada, H., Centonze, V. E., and Herman, B., Development of a multi-photon fluorescence lifetime imaging microscopy system using a streak camera, *Review of Scientific Instruments* 74 (5), 2714–2721, 2003.

153. Krishnan, R. V., Biener, E., Zhang, J. H., Heckel, R., and Herman, B., Probing subtle fluorescence dynamics in cellular proteins by streak camera based fluorescence lifetime imaging microscopy, *Applied Physics Letters* 83 (22), 4658–4660, 2003.

154. Muller, M., Ghauharali, R., Visscher, K., and Brakenhoff, G., Double-pulse fluorescence lifetime imaging in confocal microscopy, *Journal of Microscopy—Oxford* 177, 171–179, 1995.

155. Buist, A. H., Müller, M., Gijsbers, E. J., Brakenhoff, G. J., Sosnowski, T. S., Norris, T. B., and Squier, J., Double-pulse fluorescence lifetime measurements, *Journal of Microscopy* 186, 212–220, 1997.

156. Gaviola, Z., Ein fluorometer, apparat zur messung von fluoreszenzabklingungszeiten, *Zeitschrift fur Physik* 42, 853–861, 1926.

157. Murray, J. G., Cundall, R. B., Morgan, C. G., Evans, G. B., and Lewis, C., A single-photon-counting fourier-transform microfluorometer, *Journal of Physics E—Scientific Instruments* 19 (5), 349–355, 1986.

158. Wang, X. F., Uchida, T., and Minami, S., A fluorescence lifetime distribution measurement system based on phase-resolved detection using an image dissector tube, *Applied Spectroscopy* 43 (5), 840–845, 1989.

159. Gratton, E., Limkeman, M., Lakowicz, J. R., Maliwal, B. P., Cherek, H., and Laczko, G., Resolution of mixtures of fluorophores using variable-frequency phase and modulation data, *Biophysical Journal* 46 (4), 479–486, 1984.

160. Lakowicz, J. R., Laczko, G., Cherek, H., Gratton, E., and Limkeman, M., Analysis of fluorescence decay kinetics from variable-frequency phase shift and modulation data, *Biophysical Journal* 46 (4), 463–477, 1984.

161. Kilin, S. F., The duration of photo- and radioluminescence of organic compounds, *Optics and Spectroscopy* 12, 414–416, 1962.

162. Spencer, R. D. and Weber, G., Measurement of subnanosecond fluorescence lifetimes with a cross-correlation phase fluorometer, *Annals of the New York Academy of Science* 158, 361–376, 1969.

163. Lakowicz, J. R. and Berndt, K. W., Lifetime-selective fluorescence imaging using an Rf phase-sensitive camera, *Review of Scientific Instruments* 62 (7), 1727–1734, 1991.

164. Szmacinski, H., Lakowicz, J. R., and Johnson, M. L., Fluorescence lifetime imaging microscopy: Homodyne technique using high-speed gated image intensifier, *Methods in Enzymology* 240, 723–748, 1994.

165. Lakowicz, J. R., Szmacinski, H., Nowaczyk, K., Lederer, W. J., Kirby, M. S., and Johnson, M. L., Fluorescence lifetime imaging of intracellular calcium in COS cells using Quin-2, *Cell Calcium* 15 (1), 7–27, 1994.

166. Gadella, T. W. J., Jovin, T. M., and Clegg, R. M., Fluorescence Lifetime Imaging Microscopy (FLIM)—Spatial-resolution of microstructures on the nanosecond time-scale, *Biophysical Chemistry* 48 (2), 221–239, 1993.

167. Schneider, P. C. and Clegg, R. M., Rapid acquisition, analysis, and display of fluorescence lifetime-resolved images for real-time applications, *Review of Scientific Instruments* 68 (11), 4107–4119, 1997.

168. Squire, A. and Bastiaens, P. I., Three dimensional image restoration in fluorescence lifetime imaging microscopy, *Journal of Microscopy* 193 (Pt 1), 36–49, 1999.

169. Clegg, R. M., Feddersen, B., Gratton, E., and Jovin, T. M., Time resolved imaging fluorescence microscopy, in *Time-Resolved Laser Spectroscopy in Biochemistry III*, ed. Lakowicz, J. R., Los Angeles, CA, 1992, pp. 448–460.

170. Lakowicz, J. R., Szmacinski, H., Nowaczyk, K., and Johnson, M. L., Fluorescence lifetime imaging of free and protein-bound NADH, *Proceedings of the National Academy of Sciences of the United States of America* 89 (4), 1271–1275, 1992.

171. Szmacinski, H. and Lakowicz, J. R., Fluorescence lifetime-based sensing and imaging, *Sensors and Actuators B—Chemical* 29 (1–3), 16–24, 1995.

172. Mizeret, J., Wagnieres, G., Studzinski, A., Shangguan, C., and van den Bergh, H., Endoscopic tissue fluorescence life-time imaging by frequency domain light-induced fluorescence, in *Proceedings of the SPIE, Optical Biopsies*, Vol. 2627, eds. Cubeddu, R., Mordon, S. R., and Szvanberg, K., Barcelona, Spain, 1995, pp. 40–48.

173. Gadella, B. M., Vanhoek, A., and Visser, A. J. W. G., Construction and characterization of a frequency-domain fluorescence lifetime imaging microscopy system, *Journal of Fluorescence* 7, 35–43, 1997.

174. Hartmann, P., Ziegler, W., Holst, G., and Lubbers, D. W., Oxygen flux fluorescence lifetime imaging, *Sensors and Actuators B—Chemical* 38 (1–3), 110–115, 1997.

175. Wagnieres, G., Mizeret, J., Studzinski, A., and Bergh, H. v. d., Frequency-domain fluorescence lifetime imaging for endoscopic clinical cancer photodetection: Apparatus design and preliminary results, *Journal of Fluorescence* 7, 75–83, 1997.

176. Pepperkok, R., Squire, A., Geley, S., and Bastiaens, P. I., Simultaneous detection of multiple green fluorescent proteins in live cells by fluorescence lifetime imaging microscopy, *Current Biology* 9 (5), 269–272, 1999.

177. Ng, T., Squire, A., Hansra, G., Bornancin, F., Prevostel, C., Hanby, A., Harris, W. et al., Imaging protein kinase Calpha activation in cells, *Science* 283 (5410), 2085–2089, 1999.

178. Ng, T., Shima, D., Squire, A., Bastiaens, P. I., Gschmeissner, S., Humphries, M. J., and Parker, P. J., PKCalpha regulates beta1 integrin-dependent cell motility through association and control of integrin traffic, *The EMBO Journal* 18 (14), 3909–3923, 1999.

179. Wouters, F. S. and Bastiaens, P. I., Fluorescence lifetime imaging of receptor tyrosine kinase activity in cells, *Current Biology* 9 (19), 1127–1130, 1999.

180. Holub, O., Seufferheld, M. J., Gohlke, C., Govindjee, G., and Clegg, R. M., Fluorescence lifetime imaging (FLI) in real-time—A new technique in photosynthesis research, *Photosynthetica* 38 (4), 581–599, 2000.

181. Holub, O., Seufferheld, M. J., Gohlke, C., Govindjee, G., Heiss, G. J., and Clegg, R. M., Fluorescence lifetime imaging microscopy of *Chlamydomonas reinhardtii*: Non-photochemical quenching mutants and the effect of photosynthetic inhibitors on the slow chlorophyll fluorescence transient, *Journal of Microscopy* 226 (Pt 2), 90–120, 2007.

182. Murata, S., Herman, P., Lin, H. J., and Lakowicz, J. R., Fluorescence lifetime imaging of nuclear DNA: Effect of fluorescence resonance energy transfer, *Cytometry* 41 (3), 178–185, 2000.

183. Verveer, P. J., Squire, A., and Bastiaens, P. I., Global analysis of fluorescence lifetime imaging microscopy data, *Biophysical Journal* 78 (4), 2127–2137, 2000.

184. Verveer, P. J., Wouters, F. S., Reynolds, A. R., and Bastiaens, P. I., Quantitative imaging of lateral ErbB1 receptor signal propagation in the plasma membrane, *Science* 290 (5496), 1567–1570, 2000.

185. Murata, S., Herman, P., and Lakowicz, J. R., Texture analysis of fluorescence lifetime images of nuclear DNA with effect of fluorescence resonance energy transfer, *Cytometry* 43 (2), 94–100, 2001.

186. Murata, S., Herman, P., and Lakowicz, J. R., Texture analysis of fluorescence intensity and lifetime images af AT- and GC-rich regions in nuclei, *Journal of Histochemistry and Cytochemistry* 49, 1443–1452, 2001.

187. Ng, T., Parsons, M., Hughes, W. E., Monypenny, J., Zicha, D., Gautreau, A., Arpin, M., Gschmeissner, S., Verveer, P. J., Bastiaens, P. I., and Parker, P. J., Ezrin is a downstream effector of trafficking PKC-integrin complexes involved in the control of cell motility, *The EMBO Journal* 20 (11), 2723–2741, 2001.

188. Verveer, P. J., Squire, A., and Bastiaens, P. I., Improved spatial discrimination of protein reaction states in cells by global analysis and deconvolution of fluorescence lifetime imaging microscopy data, *Journal of Microscopy* 202 (Pt 3), 451–456, 2001.

189. Tertoolen, L. G., Blanchetot, C., Jiang, G., Overvoorde, J., Gadella, T. W., Jr., Hunter, T., and Hertog Jd, J., Dimerization of receptor protein-tyrosine phosphatase alpha in living cells, *BMC Cell Biology* 2 (1), 8, 2001.

190. Harpur, A. G., Wouters, F. S., and Bastiaens, P. I., Imaging FRET between spectrally similar GFP molecules in single cells, *Nature Biotechnology* 19 (2), 167–169, 2001.

191. Lin, H. J., Herman, P., and Lakowicz, J. R., Fluorescence lifetime-resolved pH imaging of living cells, *Cytometry A* 52 (2), 77–89, 2003.

192. Van Munster, E. B. and Gadella, T. W., Jr., phiFLIM: A new method to avoid aliasing in frequency-domain fluorescence lifetime imaging microscopy, *Journal of Microscopy* 213 (Pt 1), 29–38, 2004.

193. van Munster, E. B. and Gadella, T. W., Jr., Suppression of photobleaching-induced artifacts in frequency-domain FLIM by permutation of the recording order, *Cytometry A* 58 (2), 185–194, 2004.

194. van Munster, E. B., Goedhart, J., Kremers, G. J., Manders, E. M., and Gadella, T. W., Jr., Combination of a spinning disc confocal unit with frequency-domain fluorescence lifetime imaging microscopy, *Cytometry A* 71 (4), 207–214, 2007.

195. Buranachai, C., Kamiyama, D., Chiba, A., Williams, B. D., and Clegg, R. M., Rapid frequency-domain FLIM spinning disk confocal microscope: Lifetime resolution, image improvement and wavelet analysis, *Journal of Fluorescence* 18 (5), 929–942, 2008.

196. Schlachter, S., Elder, A. D., Esposito, A., Kaminski, G. S., Frank, J. H., van Geest, L. K., and Kaminski, C. F., mhFLIM: Resolution of heterogeneous fluorescence decays in widefield lifetime microscopy, *Optics Express* 17 (3), 1557–1570, 2009.

197. Matthews, S. M., Elder, A. D., Yunus, K., Kaminski, C. F., Brennan, C. M., and Fisher, A. C., Quantitative kinetic analysis in a microfluidic device using frequency-domain fluorescence lifetime imaging, *Analytical Chemistry* 79 (11), 4101–4109, 2007.

198. Esposito, A., Dohm, C. P., Bahr, M., and Wouters, F. S., Unsupervised fluorescence lifetime imaging microscopy for high content and high throughput screening, *Molecular and Cellular Proteomics* 6 (8), 1446–1454, 2007.

199. Itoh, H., Evenzahav, A., Kinoshita, K., Inagaki, Y., Mizushima, H., Takahashi, A., Hayakawa, T., and Kinosita, K., Use of a gain modulating framing camera for time-resolved imaging of cellular phenomena, *Optical Tomography and Spectroscopy of Tissue: Theory, Instrumentation, Model, and Human Studies II, Proceedings of* 2979, 733–740 (864), 1997.

200. Bastiaens, P. I. and Jovin, T. M., Microspectroscopic imaging tracks the intracellular processing of a signal transduction protein: Fluorescent-labeled protein kinase C beta I, *Proceedings of the National Academy of Science USA* 93 (16), 8407–8412, 1996.

201. Konig, K., So, P. T., Mantulin, W. W., Tromberg, B. J., and Gratton, E., Two-photon excited lifetime imaging of autofluorescence in cells during UVA and NIR photostress, *Journal of Microscopy* 183 (Pt 3), 197–204, 1996.

202. Praus, P., Gaskova, D., Kocisova, E., Chaloupka, R., Stepanek, J., Bok, J., Rejman, D., Rosenberg, I., Turpin, P. Y., and Sureau, F., Spectral decomposition of intracellular complex fluorescence using multiple-wavelength phase modulation lifetime determination: Technical approach and preliminary applications, *Biopolymers* 67 (4–5), 339–343, 2002.

203. Colyer, R. A., Lee, C., and Gratton, E., A novel fluorescence lifetime imaging system that optimizes photon efficiency, *Microscopic and Research Technique* 71 (3), 201–213, 2008.

204. Dong, C. Y., So, P. T., French, T., and Gratton, E., Fluorescence lifetime imaging by asynchronous pump-probe microscopy, *Biophysical Journal* 69 (6), 2234–2242, 1995.

205. Buehler, C., Dong, C. Y., So, P. T., French, T., and Gratton, E., Time-resolved polarization imaging by pump-probe (stimulated emission) fluorescence microscopy, *Biophysical Journal* 79 (1), 536–549, 2000.

206. Merkelo, H. S., Hartman, S. R., Mar, T., Singhal, G. S., and Govindjee, G., Mode-locked lasers: Measurements of very fast radiative decay in fluorescent systems, *Science* 164, 301–303, 1969.

207. Gratton, E. and Delgado, R. L., Use of synchrotron radiation for the measurement of fluorescence lifetimes with subpicosecond resolution, *Review of Scientific Instruments* 50, 789–790, 1979.

208. Gratton, E. and Lopez-Delgado, R., Measuring fluorescence decay times by phase-shift and modulation using the high harmonic content of pulsed light sources, *Nuovo Cimento B* 56, 110–124, 1980.

209. Carlsson, K. and Liljeborg, A., Confocal fluorescence microscopy using spectral and lifetime information to simultaneously record four fluorophores with high channel separation, *Journal of Microscopy—Oxford* 185, 37–46, 1997.

210. Carlsson, K. and Liljeborg, A., Simultaneous confocal lifetime imaging of multiple fluorophores using the intensity-modulated multiple-wavelength scanning (IMS) technique, *Journal of Microscopy* 191 (2), 119–127, 1998.

211. Aslund, N. and Carlsson, K., Confocal scanning microfluorometry of dual-labeled specimens using 2 excitation wavelengths and lock-in detection technique, *Micron* 24 (6), 603–609, 1993.

212. Carlsson, K., Liljeborg, A., Andersson, R. M., and Brismar, H., Confocal pH imaging of microscopic specimens using fluorescence lifetimes and phase fluorometry: Influence of parameter choice on system performance, *Journal of Microscopy* 199 (Pt 2), 106–114, 2000.

213. Gratton, E. and Limkeman, M., A continuously variable frequency cross-correlation phase fluorometer with picosecond resolution, *Biophysical Journal* 44 (3), 315–324, 1983.

214. Lakowicz, J. R. and Maliwal, B. P., Construction and performance of a variable-frequency phase-modulation fluorometer, *Biophysical Chemistry* 21 (1), 61–78, 1985.

215. Weber, G., Resolution of the fluorescence lifetimes in a heterogeneous system by phase and modulation measurements, *Journal of Physical Chemistry* 85 (8), 949–953, 1981.

216. Kremers, G. J., van Munster, E. B., Goedhart, J., and Gadella, T. W., Jr., Quantitative lifetime unmixing of multiexponentially decaying fluorophores using single-frequency fluorescence lifetime imaging microscopy, *Biophysical Journal* 95 (1), 378–389, 2008.

217. Redford, G. I. and Clegg, R. M., Polar plot representation for frequency-domain analysis of fluorescence lifetimes, *Journal of Fluorescence* 15 (5), 805–815, 2005.

218. Clayton, A. H., Hanley, Q. S., and Verveer, P. J., Graphical representation and multicomponent analysis of single-frequency fluorescence lifetime imaging microscopy data, *Journal of Microscopy* 213 (Pt 1), 1–5, 2004.

219. Hanley, Q. S. and Clayton, A. H., AB-plot assisted determination of fluorophore mixtures in a fluorescence lifetime microscope using spectra or quenchers, *Journal of Microscopy* 218 (Pt 1), 62–67, 2005.

220. Feddersen, B. A., Piston, D. W., and Gratton, E., Digital parallel acquisition in frequency-domain fluorimetry, *Review of Scientific Instruments* 60 (9), 2929–2936, 1989.

221. Mitchell, G. W. and Swift, K., 48000 MHF: A dual-domain Fourier transform fluorescence lifetime spectrofluorometer, in *Proceedings of the SPIE, Time-Resolved Laser Spectroscopy in Biochemistry II*, ed. Lakowicz, J. R., Los Angeles, CA, 1990, pp. 270–274.

222. Mitchell, G. W., Picosecond multiharmonic fourier fluorometer, U.S. Patent 4,937,457, 1990.

223. Watkins, A. N., Ingersoll, C. M., Baker, G. A., and Bright, F. V., A parallel multiharmonic frequency-domain fluorometer for measuring excited-state decay kinetics following one-, two-, or three-photon excitation, *Analytical Chemistry* 70 (16), 3384–3396, 1998.

224. Squire, A., Verveer, P. J., and Bastiaens, P. I., Multiple frequency fluorescence lifetime imaging microscopy, *Journal of Microscopy* 197 (Pt 2), 136–149, 2000.

225. Straub, M. and Hell, S. W., Fluorescence lifetime three-dimensional microscopy with picosecond precision using a multifocal multiphoton microscope, *Applied Physics Letters* 73 (13), 1769–1771, 1998.

226. Siegel, J., Elson, D. S., Webb, S. E. D., Parsons-Karavassilis, D., Leveque-Fort, S., Cole, M. J., Lever, M. J. et al. Whole-field five-dimensional fluorescence microscopy combining lifetime and spectral resolution with optical sectioning, *Optics Letters* 26 (17), 1338–1340, 2001.

227. Cole, M. J., Siegel, J., Webb, S. E., Jones, R., Dowling, K., Dayel, M. J., Parsons-Karavassilis, D. et al., Time-domain whole-field fluorescence lifetime imaging with optical sectioning, *Journal of Microscopy* 203 (Pt 3), 246–257, 2001.

228. Grant, D. M., Elson, D. S., Schimpf, D., Dunsby, C., Requejo-Isidro, J., Auksorius, E., Munro, I. et al., Optically sectioned fluorescence lifetime imaging using a Nipkow disk microscope and a tunable ultrafast continuum excitation source, *Optics Letters* 30 (24), 3353–3355, 2005.

229. Neil, M. A. A., Juskaitis, R., and Wilson, T., Method of obtaining optical sectioning by using structured light in a conventional microscope, *Optics Letters* 22 (24), 1905–1907, 1997.

230. Neil, M. A. A., Squire, A., Juskaitis, R., Bastiaens, P. I. H., and Wilson, T., Wide-field optically sectioning fluorescence microscopy with laser illumination, *Journal of Microscopy—Oxford* 197, 1–4, 2000.

231. Bewersdorf, J., Pick, R., and Hell, S. W., Multifocal multiphoton microscopy, *Optics Letters* 23 (9), 655–657, 1998.

232. Heintzmann, R., Hanley, Q. S., Arndt-Jovin, D., and Jovin, T. M., A dual path programmable array microscope (PAM): Simultaneous acquisition of conjugate and non-conjugate images, *Journal of Microscopy* 204 (Pt 2), 119–135, 2001.

233. Liang, M., Stehr, R. L., and Krause, A. W., Confocal pattern period in multiple-aperture confocal imaging systems with coherent illumination, *Optics Letters* 22 (11), 751–753, 1997.

234. Hanley, Q. S., Verveer, P. J., Gemkow, M. J., Arndt-Jovin, D., and Jovin, T. M., An optical sectioning programmable array microscope implemented with a digital micromirror device, *Journal of Microscopy* 196 (Pt 3), 317–331, 1999.

235. Bednarkiewicz, A., Bouhifd, M., and Whelan, M. P., Digital micromirror device as a spatial illuminator for fluorescence lifetime and hyperspectral imaging, *Applied Optics* 47 (9), 1193–1199, 2008.

236. Smith, P. J., Taylor, C. M., Shaw, A. J., and McCabe, E. M., Programmable array microscopy with a ferroelectric liquid-crystal spatial light modulator, *Applied Optics* 39 (16), 2664–2669, 2000.

237. Hanley, Q. S., Verveer, P. J., and Jovin, T. M., Spectral imaging in a programmable array microscope by hadamard transform fluorescence spectroscopy, *Applied Spectroscopy* 53 (1), 1–10, 1999.

238. Hanley, Q. S., Arndt-Jovin, D. J., and Jovin, T. M., Spectrally resolved fluorescence lifetime imaging microscopy, *Applied Spectroscopy* 56 (2), 155–166, 2002.

239. Hanley, Q. S., Lidke, K. A., Heintzmann, R., Arndt-Jovin, D. J., and Jovin, T. M., Fluorescence lifetime imaging in an optically sectioning programmable array microscope (PAM), *Cytometry A* 67 (2), 112–118, 2005.

240. Bednarkiewicz, A. and Whelan, M. P., Global analysis of microscopic fluorescence lifetime images using spectral segmentation and a digital micromirror spatial illuminator, *Journal of Biomedical Optics* 13 (4), 041316, 2008.

241. Belford, G. G., Belford, R. L., and Weber, G., Dynamics of fluorescence polarization in macromolecules, *Proceedings of the National Academy of Sciences USA* 69 (6), 1392–1393, 1972.

242. Small, E. W. and Isenberg, I., Hydrodynamic properties of a rigid molecule: Rotational and linear diffusion and fluorescence anisotropy, *Biopolymers* 16 (9), 1907–1928, 1977.

243. Thomas, J. C., Allison, S. A., Appellof, C. J., and Schurr, J. M., Torsion dynamics and depolarization of fluorescence of linear macromolecules. II. Fluorescence polarization anisotropy measurements on a clean viral phi 29 DNA, *Biophysical Chemistry* 12 (2), 177–188, 1980.

244. Holowka, D., Wensel, T., and Baird, B., A nanosecond fluorescence depolarization study on the segmental flexibility of receptor-bound immunoglobulin E, *Biochemistry* 29 (19), 4607–4612, 1990.

245. Kinosita, K. J., Kawato, S., and Ikegami, A., A theory of fluorescence polarization decay in membranes, *Biophysical Journal* 20, 289–305, 1977.

246. Herman, P., Konopasek, I., Plasek, J., and Svobodova, J., Time-resolved polarized fluorescence studies of the temperature adaptation in *Bacillus subtilis* using DPH and TMA-DPH fluorescent probes, *Biochimica et Biophysica Acta* 1190 (1), 1–8, 1994.

247. Herman, P., Malinsky, J., Plasek, J., and Vecer, J., Pseudo real-time method for monitoring of the limiting anisotropy in membranes, *Journal of Fluorescence* 14 (1), 79–85, 2004.

248. Clayton, A. H., Hanley, Q. S., Arndt-Jovin, D. J., Subramaniam, V., and Jovin, T. M., Dynamic fluorescence anisotropy imaging microscopy in the frequency domain (rFLIM), *Biophysical Journal* 83 (3), 1631–1649, 2002.

249. Lidke, D. S., Nagy, P., Barisas, B. G., Heintzmann, R., Post, J. N., Lidke, K. A., Clayton, A. H., Arndt-Jovin, D. J., and Jovin, T. M., Imaging molecular interactions in cells by dynamic and static fluorescence anisotropy (rFLIM and emFRET), *Biochemical Society Transactions* 31 (Pt 5), 1020–1027, 2003.

250. Gautier, I., Tramier, M., Durieux, C., Coppey, J., Pansu, R. B., Nicolas, J. C., Kemnitz, K., and Coppey-Moisan, M., Homo-FRET microscopy in living cells to measure monomer–dimer transition of GFP-tagged proteins, *Biophysical Journal* 80 (6), 3000–3008, 2001.

251. Siegel, J., Suhling, K., Leveque-Fort, S., Webb, S. E. D., Davis, D. M., Phillips, D., Sabharwal, Y., and French, P. M. W., Wide-field time-resolved fluorescence anisotropy imaging (TR-FAIM): Imaging the rotational mobility of a fluorophore, *Review of Scientific Instruments* 74 (1), 182–192, 2003.

252. Suhling, K., Siegel, J., Lanigan, P. M., Leveque-Fort, S., Webb, S. E., Phillips, D., Davis, D. M., and French, P. M., Time-resolved fluorescence anisotropy imaging applied to live cells, *Optics Letters* 29 (6), 584–586, 2004.

253. Spitz, J. A., Yasukuni, R., Sandeau, N., Takano, M., Vachon, J. J., Meallet-Renault, R., and Pansu, R. B., Scanning-less wide-field single-photon counting device for fluorescence intensity, lifetime and time-resolved anisotropy imaging microscopy, *Journal of Microscopy* 229 (Pt 1), 104–114, 2008.

254. Knemeyer, J. P., Herten, D. P., and Sauer, M., Detection and identification of single molecules in living cells using spectrally resolved fluorescence lifetime imaging microscopy, *Analytical Chemistry* 75 (9), 2147–2153, 2003.

255. Bird, D. K., Eliceiri, K. W., Fan, C. H., and White, J. G., Simultaneous two-photon spectral and lifetime fluorescence microscopy, *Applied Optics* 43 (27), 5173–5182, 2004.

256. Becker, W., Bergmann, A., Biscotti, G., and Ruck, A., Advanced time-correlated single photon counting technique for spectroscopy and imaging in biomedical systems, *Commercial and Biomedical Applications of Ultrafast Lasers IV* 5340, 104–112 (188), 2004.

257. Spriet, C., Trinel, D., Waharte, F., Deslee, D., Vandenbunder, B., Barbillat, J., and Heliot, L., Correlated fluorescence lifetime and spectral measurements in living cells, *Microscopic and Research Technique* 70 (2), 85–94, 2007.

258. Ulrich, V., Fischer, P., Riemann, I., and Konigt, K., Compact multiphoton/single photon laser scanning microscope for spectral imaging and fluorescence lifetime imaging, *Scanning* 26 (5), 217–225, 2004.

259. De Beule, P., Owen, D. M., Manning, H. B., Talbot, C. B., Requejo-Isidro, J., Dunsby, C., McGinty, J. et al., Rapid hyperspectral fluorescence lifetime imaging, *Microscopic and Research Technique* 70 (5), 481–484, 2007.

260. Owen, D. M., Auksorius, E., Manning, H. B., Talbot, C. B., de Beule, P. A., Dunsby, C., Neil, M. A., and French, P. M., Excitation-resolved hyperspectral fluorescence lifetime imaging using a UV-extended supercontinuum source, *Optics Letters* 32 (23), 3408–3410, 2007.

261. Woods, R. J., Scypinski, S., Love, L. J. C., and Ashworth, H. A., Transient digitizer for the determination of microsecond luminescence lifetimes, *Analytical Chemistry* 56 (8), 1395–1400, 1984.

262. Ballew, R. M. and Demas, J. N., An error analysis of the rapid lifetime determination method for the evaluation of single exponential decays, *Analytical Chemistry* 61 (1), 30–33, 1989.

263. Waters, P. D. and Burns, D. H., Optimized gated detection for lifetime measurement over a wide-range of single exponential decays, *Applied Spectroscopy* 47 (1), 111–115, 1993.

264. Chan, S. P., Fuller, Z. J., Demas, J. N., and DeGraff, B. A., Optimized gating scheme for rapid lifetime determinations of single-exponential luminescence lifetimes, *Analytical Chemistry* 73 (18), 4486–4490, 2001.

265. Sharman, K. K., Periasamy, A., Ashworth, H., Demas, J. N., and Snow, N. H., Error analysis of the rapid lifetime determination method for double-exponential decays and new windowing schemes, *Analytical Chemistry* 71 (5), 947–952, 1999.

266. Jo, J. A., Fang, Q., Papaioannou, T., and Marcu, L., Novel ultra-fast deconvolution method for fluorescence lifetime imaging microscopy based on the Laguerre expansion technique, *Conference Proceedings of the IEEE Engineering in Medicine and Biology Society* 2, 1271–1274, 2004.

267. Jo, J. A., Fang, Q. Y., and Marcu, L., Ultrafast method for the analysis of fluorescence lifetime imaging microscopy data based on the Laguerre expansion technique, *IEEE Journal of Selected Topics in Quantum Electronics* 11 (4), 835–845, 2005.

268. Pelet, S., Previte, M. J., Laiho, L. H., and So, P. T., A fast global fitting algorithm for fluorescence lifetime imaging microscopy based on image segmentation, *Biophysical Journal* 87 (4), 2807–2817, 2004.

269. Beechem, J. M., Knutson, J. R., Ross, J. B. A., Turner, B. W., and Brand, L., Global resolution of heterogeneous decay by phase modulation fluorometry—Mixtures and proteins, *Biochemistry* 22 (26), 6054–6058, 1983.

270. Knutson, J. R., Beechem, J. M., and Brand, L., Simultaneous analysis of multiple fluorescence decay curves—A global approach, *Chemical Physics Letters* 102 (6), 501–507, 1983.

271. Beechem, J. M., Ameloot, M., and Brand, L., Global and target analysis of complex decay phenomena, *Analytical Instrumentation* 14 (3–4), 379–402, 1985.

272. Beechem, J. M. and Brand, L., Global analysis of fluorescence decay—Applications to some unusual experimental and theoretical-studies, *Photochemistry and Photobiology* 44 (3), 323–329, 1986.

273. Verveer, P. J. and Bastiaens, P. I., Evaluation of global analysis algorithms for single frequency fluorescence lifetime imaging microscopy data, *Journal of Microscopy* 209 (Pt 1), 1–7, 2003.

274. Bastiaens, P. I. and Pepperkok, R., Observing proteins in their natural habitat: The living cell, *Trends in Biochemical Sciences* 25 (12), 631–637, 2000.

275. Provenzano, P. P., Eliceiri, K. W., and Keely, P. J., Multiphoton microscopy and fluorescence lifetime imaging microscopy (FLIM) to monitor metastasis and the tumor microenvironment, *Clinical and Experimental Metastasis* 26 (4), 357–370, 2009.

276. Festy, F., Ameer-Beg, S. M., Ng, T., and Suhling, K., Imaging proteins in vivo using fluorescence lifetime microscopy, *Molecular Biosystems* 3 (6), 381–391, 2007.

277. Szmacinski, H. and Lakowicz, J. R., Optical measurements of pH using fluorescence lifetimes and phase-modulation fluorometry, *Analytical Chemistry* 65 (13), 1668–1674, 1993.

278. Lin, H.-J., Szmacinski, H., and Lakowicz, J. R., Lifetime-based pH sensors: Indicators for acidic environments, *Analytical Biochemistry* 269, 162–167, 1999.

279. Lin, H.-J., Herman, P., Kang, J.-S., and Lakowicz, J. R., Fluorescence lifetime characterization of novel low pH probes, *Analytical Biochemistry* 294, 118–125, 2001.

280. Sanders, R., Gerritsen, H. C., Draaijer, A., Houpt, P. M., Van Veen, S. J. F., and Levine, Y. K., Confocal fluorescence lifetime imaging of pH in single cells, in *Proceedings of the SPIE, Time-Resolved Laser Spectroscopy in Biochemistry IV*, ed. Lakowicz, J. R., Los Angeles, CA, 1994, pp. 56–62.

281. Srivastava, A. and Krishnamoorthy, G., Time-resolved fluorescence microscopy could correct for probe binding while estimating intracellular pH, *Analytical Biochemistry* 249 (2), 140–146, 1997.

282. Andersson, R. M., Carlsson, K., Liljeborg, A., and Brismar, H., Characterization of probe binding and comparison of its influence on fluorescence lifetime of two pH-sensitive benzo[c]xanthene dyes using intensity-modulated multiple-wavelength scanning technique, *Analytical Biochemistry* 283 (1), 104–110, 2000.

283. Lakowicz, J. R., Szmacinski, H., Nowaczyk, K., and Johnson, M. L., Fluorescence lifetime imaging of calcium using Quin-2, *Cell Calcium* 13 (3), 131–147, 1992.

284. Lakowicz, J. R., Szmacinski, H., and Johnson, M. L., Calcium imaging using fluorescence lifetime and long-wavelength probes, *Journal of Fluorescence* 2 (1), 47–62, 1992.

285. Thompson, R. B., Jr, W. O. W., Maliwal, B. R., Fierke, C. A., and Frederickson, C. J., Fluorescence microscopy of stimulated Zn(II) release from organotypic cultures of mammalian hippocampus using a carbonic anhydrase-based biosensor system, *Journal of Neuroscience Methods* 96, 35–45, 2000.

286. Thompson, R. B., Maliwal, B. R., and Fierke, C. A., Selectivity and sensitivity of fluorescence lifetime-based metal ion biosensing using a carbonic anhydrase transducer, *Analytical Biochemistry* 267, 185–195, 1999.

287. Birch, D. J. S., Holmes, A. S., and Darbyshire, M., Intelligent sensor for metal ions based on fluorescence resonance energy transfer, *Measurements Science and Technology* 6, 243–247, 1995.

288. Oliver, A. E., Baker, G. A., Fugate, R. D., Tablin, F., and Crowe, J. H., Effects of temperature on calcium-sensitive fluorescence probes, *Biophysical Journal* 78, 2116–2126, 2000.

289. French, T., So, P. T., Weaver, D. J., Jr., Coelho-Sampaio, T., Gratton, E., Voss, E. W., Jr., and Carrero, J., Two-photon fluorescence lifetime imaging microscopy of macrophage-mediated antigen processing, *Journal of Microscopy* 185 (Pt 3), 339–353, 1997.

290. Paul, R. J., Oxygen concentration and the oxidation *Proceedings SPIE* reduction state of yeast: Determination of free/bound NADH and flavins by time-resolved spectroscopy, *Naturwissenschaften* 83, 32–35, 1996.

291. Chia, T. H., Williamson, A., Spencer, D. D., and Levene, M. J., Multiphoton fluorescence lifetime imaging of intrinsic fluorescence in human and rat brain tissue reveals spatially distinct NADH binding, *Optics Express* 16 (6), 4237–4249, 2008.

292. van Zandvoort, M. A. M. J., de Grauw, C. J., Gerritsen, H. C., Broers, J. L. V., Egbrink, M. G. A. O., Ramaekers, F. C. S., and Slaaf, D. W., Discrimination of DNA and RNA in cells by a vital fluorescent probe: Lifetime imaging of SYTO13 in healthy and apoptotic cells, *Cytometry* 47 (4), 226–235, 2002.

293. van Zandvoort, M. A. M. J., Engels, W., de Grauv, C. J., Gerritsen, H. C., and Slaaf, D. W., Lifetime imaging of the vital DNA/RNA probe SYTO13 in healthy and apoptotic cells, *Biomedical Nanotechnology Architectures and Applications* 4626, 464–472 (584), 2002.

294. Kuimova, M. K., Yahioglu, G., Levitt, J. A., and Suhling, K., Molecular rotor measures viscosity of live cells via fluorescence lifetime imaging, *Journal of American Chemical Society* 130 (21), 6672–6673, 2008.

295. Förster, T., Zwischenmolekulare energiewanderung und fluoreszenz, *Annals of Physics* 2, 57–75, 1948.

296. Dumas, D., Gaborit, N., Grossin, L., Riquelme, B., Gigant-Huselstein, C., De Isla, N., Gillet, P., Netter, P., and Stoltz, J. F., Spectral and lifetime fluorescence imaging microscopies: New modalities of multiphoton microscopy applied to tissue or cell engineering, *Biorheology* 41 (3–4), 459–467, 2004.

297. Gadella, T. W., Jr. and Jovin, T. M., Oligomerization of epidermal growth factor receptors on A431 cells studied by time-resolved fluorescence imaging microscopy. A stereochemical model for tyrosine kinase receptor activation, *Journal of Cell Biology* 129 (6), 1543–1558, 1995.

298. MartinFernandez, M., Tobin, M., Clarke, D., Gregory, C., and Jones, G., A high sensitivity time-resolved microfluorimeter for real-time cell biology, *Review of Scientific Instruments* 67, 3716–3721, 1996.

299. Webb, S. E., Roberts, S. K., Needham, S. R., Tynan, C. J., Rolfe, D. J., Winn, M. D., Clarke, D. T., Barraclough, R., and Martin-Fernandez, M. L., Single-molecule imaging and fluorescence lifetime imaging microscopy show different structures for high- and low-affinity epidermal growth factor receptors in A431 cells, *Biophysical Journal* 94 (3), 803–819, 2008.

300. Klucken, J., Outeiro, T. F., Nguyen, P., McLean, P. J., and Hyman, B. T., Detection of novel intracellular alpha-synuclein oligomeric species by fluorescence lifetime imaging, *FASEB Journal* 20 (12), 2050–2057, 2006.

301. Pollok, B. and Heim, R., Using GFP in FRET-based applications, *Trends in Cell Biology* 9 (2), 57–60, 1999.

302. Calleja, V., Ameer-Beg, S. M., Vojnovic, B., Woscholski, R., Downward, J., and Larijani, B., Monitoring conformational changes of proteins in cells by fluorescence lifetime imaging microscopy, *Biochemical Journal* 372 (Pt 1), 33–40, 2003.

303. Kinoshita, K., Goryo, K., Takada, M., Tomokuni, Y., Aso, T., Okuda, H., Shuin, T., Fukumura, H., and Sogawa, K., Ternary complex formation of pVHL, elongin B and elongin C visualized in living cells by a fluorescence resonance energy transfer-fluorescence lifetime imaging microscopy technique, *FEBS Journal* 274 (21), 5567–5575, 2007.

304. Lee, J. D., Huang, P. C., Lin, Y. C., Kao, L. S., Huang, C. C., Kao, F. J., Lin, C. C., and Yang, D. M., In-depth fluorescence lifetime imaging analysis revealing SNAP25A-rabphilin 3A interactions, *Microscopy and Microanalysis* 14 (6), 507–518, 2008.

305. Lee, J. D., Chang, Y. F., Kao, F. J., Kao, L. S., Lin, C. C., Lu, A. C., Shyu, B. C., Chiou, S. H., and Yang, D. M., Detection of the interaction between SNAP25 and rabphilin in neuroendocrine PC12 cells using the FLIM/FRET technique, *Microscopic Research and Technique* 71 (1), 26–34, 2008.

306. Siegel, J., Elson, D. S., Webb, S. E., Lee, K. C., Vlandas, A., Gambaruto, G. L., Leveque-Fort, S. et al., Studying biological tissue with fluorescence lifetime imaging: Microscopy, endoscopy, and complex decay profiles, *Applied Optics* 42 (16), 2995–3004, 2003.

307. Birmingham, J. J., Frequency-domain lifetime imaging methods at unilever research, *Journal of Fluorescence* 7, 45–54, 1997.

308. Masters, B. R., So, P. T., and Gratton, E., Multiphoton excitation fluorescence microscopy and spectroscopy of in vivo human skin, *Biophysical Journal* 72 (6), 2405–2412, 1997.

309. Skala, M. C., Riching, K. M., Bird, D. K., Gendron-Fitzpatrick, A., Eickhoff, J., Eliceiri, K. W., Keely, P. J., and Ramanujam, N., In vivo multiphoton fluorescence lifetime imaging of protein-bound and free nicotinamide adenine dinucleotide in normal and precancerous epithelia, *Journal of Biomedical Optics* 12 (2), 024014, 2007.

310. Sevick-Muraca, E. M., Reynolds, J. S., Troy, T. L., Lopez, G., and Paithankar, D. Y., Fluorescence lifetime spectroscopic imaging with measurements of photon migration, *Annals of the New York Academic Sciences* 838, 46–57, 1998.

311. Holst, G., Kohls, O., Klimant, I., Konig, B., Kuhl, M., and Richter, T., A modular luminescence lifetime imaging system for mapping oxygen distribution in biological samples, *Sensors and Actuators B—Chemical* 51 (1–3), 163–170, 1998.

312. Holst, G. and Grunwald, B., Luminescence lifetime imaging with transparent oxygen optodes, *Sensors and Actuators B—Chemical* 74 (1–3), 78–90, 2001.

313. Morgan, C. G., Mitchell, A. G., Murray, J. G., and Wall, E. J., New approach to lifetime-resolved luminescence imaging, *Journal of Fluorescence* 7, 65–73, 1997.

314. Gerritsen, H. C. and de Grauw, K., Fluorescence lifetime imaging of oxygen in dental biofilm, *Laser Microscopy* 1 (35), 70–78 (134), 2000.

315. Wolfbeis, O. S., Kilmant, I., Werner, T., Huber, C., Kosch, U., Krause, C., Neurauter, G., and Dürkop, A., Set of luminescence decay time based chemical sensors for clinical applications, *Sensors and Actuators B* 51, 17–24, 1998.

316. Petrasek, Z., Krishnan, M., Monch, I., and Schwille, P., Simultaneous two-photon fluorescence correlation spectroscopy and lifetime imaging of dye molecules in submicrometer fluidic structures, *Microscopic and Research Technique* 70 (5), 459–466, 2007.

317. Micic, M., Hu, D., Suh, Y. D., Newton, G., Romine, M., and Lu, H. P., Correlated atomic force microscopy and fluorescence lifetime imaging of live bacterial cells, *Colloids and Surface B: Biointerfaces* 34 (4), 205–212, 2004.

318. Gryczynski, I., Malicka, J., Gryczynski, Z., and Lakowicz, J. R., Radiative decay engineering 4. Experimental studies of surface plasmon-coupled directional emission, *Analytical Biochemistry* 324 (2), 170–182, 2004.

319. Lakowicz, J. R., Radiative decay engineering: Biophysical and biomedical applications, *Analytical Biochemistry* 298 (1), 1–24, 2001.

320. Lakowicz, J. R., Radiative decay engineering 3. Surface plasmon-coupled directional emission, *Analytical Biochemistry* 324 (2), 153–169, 2004.

321. Lakowicz, J. R., Radiative decay engineering 5: Metal-enhanced fluorescence and plasmon emission, *Analytical Biochemistry* 337 (2), 171–194, 2005.

322. Lakowicz, J. R., Shen, Y., D'Auria, S., Malicka, J., Fang, J., Gryczynski, Z., and Gryczynski, I., Radiative decay engineering. 2. Effects of Silver Island films on fluorescence intensity, lifetimes, and resonance energy transfer, *Analytical Biochemistry* 301 (2), 261–277, 2002.

323. Leveque-Fort, S., Fontaine-Aupart, M. P., Roger, G., and Georges, P., Fluorescence-lifetime imaging with a multifocal two-photon microscope, *Optics Letters* 29 (24), 2884–2886, 2004.

324. Ramanujan, V. K., Zhang, J. H., Biener, E., and Herman, B., Multiphoton fluorescence lifetime contrast in deep tissue imaging: Prospects in redox imaging and disease diagnosis, *Journal of Biomedical Optics* 10 (5), 051407, 2005.

325. Godavarty, A., Sevick-Muraca, E. M., and Eppstein, M. J., Three-dimensional fluorescence lifetime tomography, *Medical Physics* 32 (4), 992–1000, 2005.

326. Lakowicz, J. R., Piszczek, G., and Kang, J. S., On the possibility of long-wavelength long-lifetime high-quantum-yield luminophores, *Analytical Biochemistry* 288 (1), 62–75, 2001.

327. Maliwal, B. P., Gryczynski, Z., and Lakowicz, J. R., Long-wavelength long-lifetime luminophores, *Analytical Chemistry* 73 (17), 4277–4285, 2001.

328. Terpetschnig, E., Szmacinski, H., Malak, H., and Lakowicz, J. R., Metal–ligand complexes as a new class of long-lived fluorophores for protein hydrodynamics, *Biophysical Journal* 68 (1), 342–350, 1995.

329. Terpetschnig, E., Szmacinski, H., and Lakowicz, J. R., Long-lifetime metal–ligand complexes as probes in biophysics and clinical chemistry, *Methods Enzymology* 278, 295–321, 1997.

330. Lakowicz, J. R., Gryczynski, I., Gryczynski, Z., and Johnson, M. L., Background suppression in frequency-domain fluorometry, *Analytical Biochemistry* 277 (1), 74–85, 2000.

331. Herman, P., Maliwal, B. P., and Lakowicz, J. R., Real-time background suppression during frequency domain lifetime measurements, *Analytical Biochemistry* 309 (1), 19–26, 2002.

332. Elder, A. D., Frank, J. H., Swartling, J., Dai, X., and Kaminski, C. F., Calibration of a wide-field frequency-domain fluorescence lifetime microscopy system using light emitting diodes as light sources, *Journal of Microscopy* 224 (Pt 2), 166–180, 2006.

333. Herman, P. and Vecer, J., Frequency domain fluorometry with pulsed light-emitting diodes, *Annals of the New York Academy of Sciences* 1130, 56–61, 2008.

334. Esposito, A., Oggier, T., Gerritsen, H. C., Lustenberger, F., and Wouters, F. S., All-solid-state lock-in imaging for wide-field fluorescence lifetime sensing, *Optics Express* 13 (24), 9812–9821, 2005.

335. Esposito, A., Gerritsen, H. C., Oggier, T., Lustenberger, F., and Wouters, F. S., Innovating lifetime microscopy: A compact and simple tool for life sciences, screening, and diagnostics, *Journal of Biomedical Optics* 11 (3), 34016, 2006.

336. Booth, M. J. and Wilson, T., Low-cost, frequency-domain, fluorescence lifetime confocal microscopy, *Journal of Microscopy* 214 (Pt 1), 36–42, 2004.

FIGURE 1.4 Pablo Picasso, *Portrait of Dora Maar Seated*, 1937. Quantum theory also had a profound effect on many fields beyond science, such as art. It was no coincidence that during the quantum revolution in science, Cubist and Surrealist art abolished realistic shapes referenced in fixed space and fixed time. Pablo Picasso's renderings of the human face with their multifaceted perspectives often reflected the dual nature of reality. (Courtesy of Musee Picasso, Paris, France. Copyright Artists Right Society (ARS), New York.)

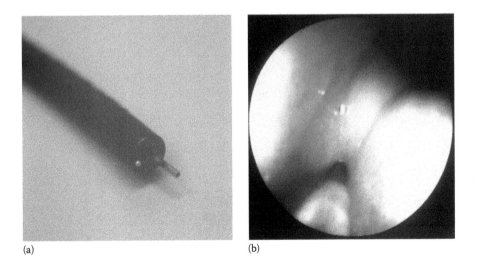

(a)   (b)

FIGURE 1.6 Fiber-optic probe for in vivo laser-induced diagnostics of cancer. Laser-induced fluorescence (LIF) has been used for gastrointestinal (GI) endoscopy examinations to directly diagnose cancer of patients without requiring physical biopsy. The LIF measurement was completed in approximately 0.6 s for each tissue site. The fiber-optic probe was inserted into the biopsy channel of an endoscope (a). The fiber probe lightly touched the surface of the GI tissue being monitored (b).

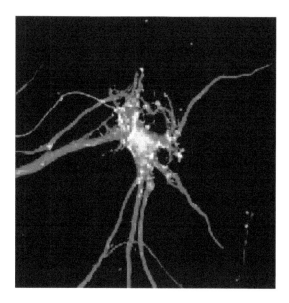

FIGURE 1.8 Tracking biochemical processes in nerve cells using GFP. Tracking biochemical processes can now be performed with the use of fluorescent probes. For example, there is a great interest to understand the origin and movement of neurotrophic factors such as brain-derived neurotrophic factor (BDNF) between nerve cells. This figure illustrates the use of BDNF tagged with GFP to follow synaptic transport from axons to neurons to postsynaptic cells. (Adapted from Kohara K. et al., *Science*, 291, 2419, 2001.)

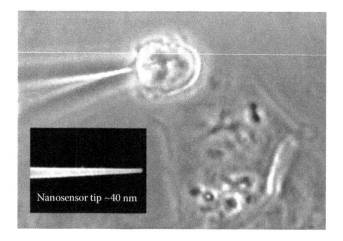

Nanosensor tip ~40 nm

FIGURE 1.9 A fiber-optic nanosensor with antibody probe used for detecting biochemicals in a single cell. The combination of photonics and nanotechnology has led to a new generation of devices for probing the cell machinery and elucidating intimate life processes occurring at the molecular level that were heretofore invisible to human inquiry. The insert (lower left) shows a scanning electron photograph of a nanofiber with a 40 nm diameter. The small size of the probe allowed manipulation of the nanoprobe at specific locations within a single cell. (Adapted from Vo-Dinh, T. et al., *Nat. Biotechnol.*, 18, 764, 2000; Cullum, B.M. and Vo-Dinh, T., *Trends Biotechnol.*, 18, 388, 2000.)

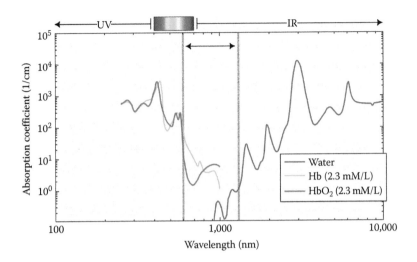

FIGURE 2.16 The therapeutic/diagnostic window. The absorption spectra of water, deoxygenated hemoglobin (Hb), and oxygenated hemoglobin (HbO₂). The Hb and HbO₂ curves are given for a concentration of 150 g/L in water. The region of the spectrum between 600 and 1300 nm is the therapeutic/diagnostic window, where photons can achieve significant penetration in tissues. (The water spectrum is from Hale and Querry (1973). The hemoglobin data are compiled in Prahl (1998). (Adapted from Jacques, S.L. and Prahl, S.A., Absorption spectra for biological tissues, available from http://omlc.ogi.edu/classroom/ece532/class3/muaspectra.html, 1998.)

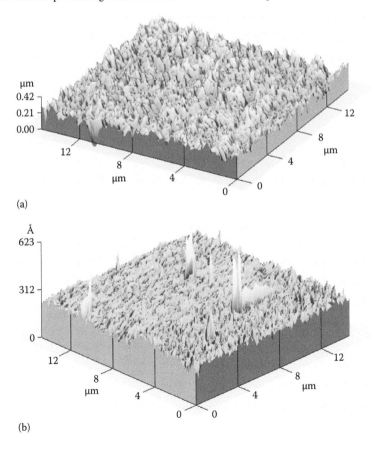

FIGURE 6.8 AFM measurements for glass waveguide's (a) Ag and (b) AgI layers.

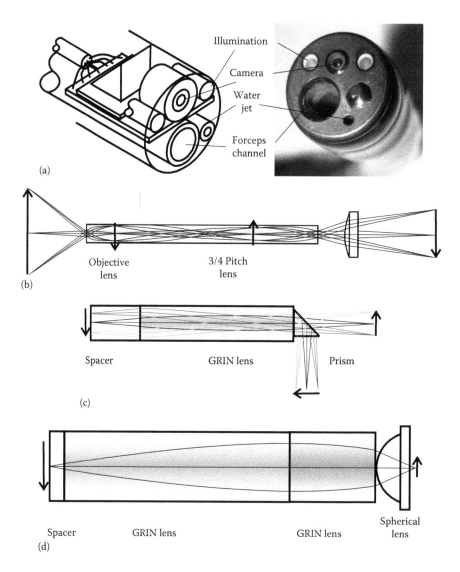

FIGURE 7.11 Fiber-optic imaging system. (a) A modern gastroscope incorporates illumination optics, operating channel, water jet, and camera at the probe tip. (b) A similar system using GRIN lenses and a relay can be constructed for wide-field observation. (c) A single GRIN lens transfers the object under side viewing on the image plane. (d) Two-photon imaging stick lens based on GRIN lens and micro-optical component allows for high NA and observation through regular microscope objective lens when the stick is mounted in front of it.

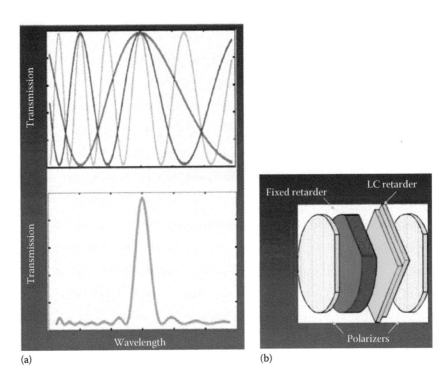

FIGURE 9.2 Liquid tunable filters: principle of operation (a) and schematic of a single stage (b), in which an LC retarder is inserted at every stage of the Lyot filter.

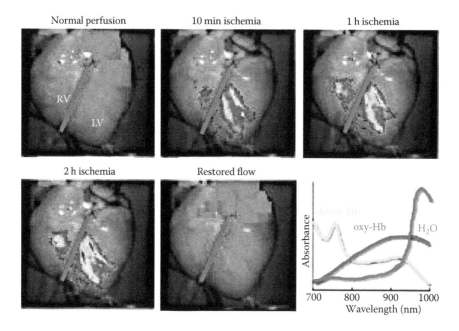

Normal perfusion   10 min ischemia   1 h ischemia

RV

LV

2 h ischemia   Restored flow

oxy-Hb   H₂O

Absorbance

700   800   900   1000
Wavelength (nm)

FIGURE 9.9 Spectral detection and display of hypoxic regions of a pig heart before and after ischemia using an oxy–deoxyhemoglobin unmixing and spectral color lookup table.

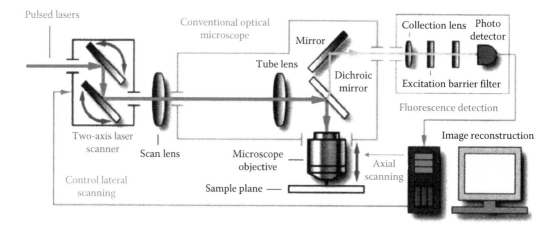

Pulsed lasers

Conventional optical microscope

Mirror

Collection lens   Photo detector

Tube lens

Dichroic mirror

Excitation barrier filter

Two-axis laser scanner

Scan lens

Fluorescence detection

Microscope objective

Axial scanning

Image reconstruction

Control lateral scanning

Sample plane

FIGURE 12.4 Schematic of a two-photon laser scanning fluorescence microscope. (From Kim, D., Ultrafast optical pulse manipulation in three dimensional-resolved microscope imaging and microfabrication. PhD, Department of Mechanical Engineering, Massachusetts Institute of Technology, 2009.)

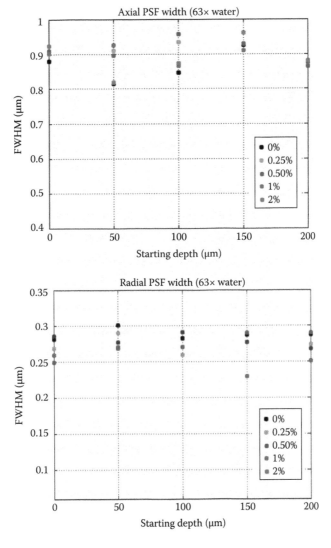

FIGURE 12.5 Averaged widths of the radial and axial PSFs measured in multiple scattering media with different scattering coefficients containing Liposyn III at concentrations of 0%, 0.25%, 0.5%, 1%, and 2%. Data were taken with a Zeiss 63×, water-immersion, C-Apochromat objective (*NA* 1.2).

(a) (b) (c)

FIGURE 12.6 3-D reconstructed two-photon images of dermal structures in a mouse ear tissue specimen. The three images show distinct structural layers: (a) epidermal keratinocytes, (b) basal cells, and (c) collagen/elastin fibers.

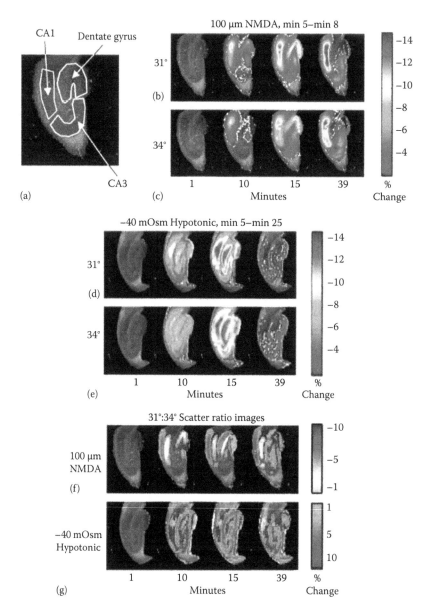

FIGURE 14.8 Single-angle and dual-angle scattering images of hippocampal slices under osmotic stress and subjected to NMDA. In each case, the color bars indicate the percent change in image intensity. In regions of the slice where the change is less than ±2%, the grayscale images of the hippocampus are shown. (a) The scatter image of a hippocampal slice is shown. The three main regions of the hippocampus are CA1, CA3, and the dentate gyrus. Also shown are single-angle scatter images at 31° (b) and 34° (c) of an NMDA-treated slice. In both (b) and (c), there is a large change in the CA1 region. There is also a significant change in the dentate gyrus that fades by minute 39 of the experiment. Single-angle scattering images at 31° (d) and 34° (e) are shown for hypotonic treatment. In both (d) and (e), there is a large change indicative of cellular swelling in the CA1 region. There is also a significant change in the dentate gyrus. Dual-angle scatter ratio images are shown for NMDA (f) and hypotonic (g) treatments. The NMDA-treated slices in (f) undergo a relatively larger change in the dual-angle scatter ratio in CA1. In CA1, the scatter ratio change is negative, possibly indicating particle shrinkage. In (g) hypotonic treatment, the magnitude of the change in the scatter ratio is less in CA1. In addition, the overall location of the scatter change is more spread out than for NMDA treatment. In both (f) and (g), the white matter areas reveal a positive change in the scatter ratio. (Reprinted from Johnson, L.J. et al., *J. Neurosci. Methods*, 98, 21, 2000. With permission.)

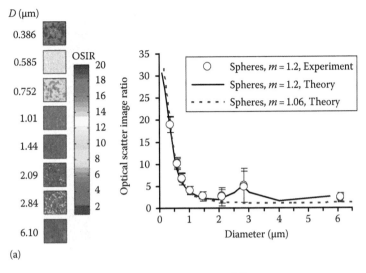

(a)

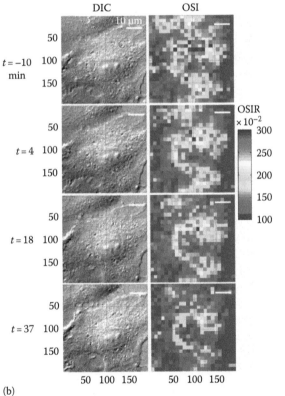

(b)

FIGURE 14.10 (a) OSI images and measurement of the OSIR in aqueous suspensions of polystyrene spheres ($m = 1.2$). The OSIR is a measure of wide to narrow angle scatter. Experimental data (open circles) and theoretical predictions (solid line, $m = 1.2$; dashed line, $m = 1.06$) are shown. The experimental data points show the mean pixel value and standard deviation in the scatter images displayed to the left of the graph. (Reprinted from Boustany, N.N. et al., *Opt. Lett.*, 26(14), 1063, 2001. With permission.) (b) Representative cell undergoing apoptosis after treatment with 2 μm STS. The cell was imaged in DIC (left panels) and OSI (right panels) at different time points. C, cytoplasm; N, nucleus. Images are displayed at times $t = -10$, 4, 18, and 37 min from STS addition at $t = 0$. The ratiometric scattering images show a decreasing scatter ratio within the cytoplasm (C).

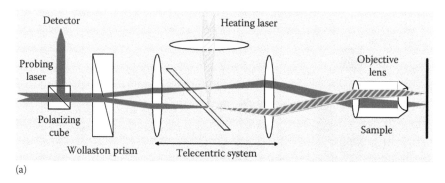

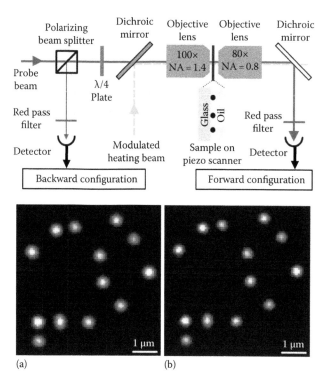

(a)

PIC image

(b)

FIGURE 16.1 (a) Experimental scheme of the photothermal interference contrast (PIC) microscope. (b) PIC image of a $10 \times 10 \ \mu m^2$ region of sample containing isolated 5 nm AuNPs.

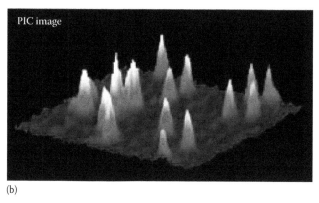

1 μm

1 μm

(a)   (b)

FIGURE 16.2 Experimental schemes of the photothermal heterodyne imaging (PHI) microscope and PHI images of the region of a sample containing isolated 10 nm AuNPs detected in the backward (a) and forward (b) configurations (see text).

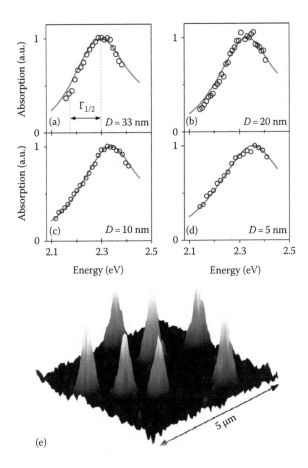

FIGURE 16.3 (a–d) Normalized absorption spectra of single AuNPs with diameters ranging from 33 to 5 nm. The extracted width at half-maximum is shown on the first spectrum. A broadening of the plasmon resonance with decreasing size is readily visible and can be compared with simulations based on Mie theory using a size-dependent modification in the dielectric constant of gold. (e) PHI image of a sample containing isolated 5.3 nm silver NPs excited at 405 nm.

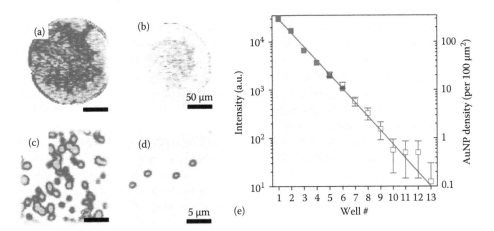

FIGURE 16.4 Direct imaging (no silver enhancement) of four wells of a DNA microarray by PHI at low (a and b) and high (c and d) resolutions. (e) Signal (left axis) vs. well number in the low-resolution (full squares) and high-resolution (open squares) regimes. The corresponding AuNPs density measured in each well is shown on the right axis.

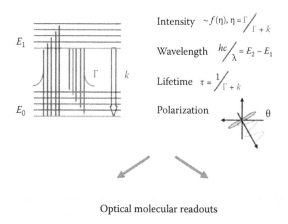

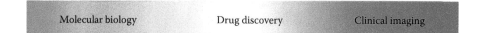

FIGURE 18.1 Overview of MDFI and metrology. (Adapted from Talbot, C. et al., Fluorescence lifetime imaging and metrology for biomedicine, Chapter 6, in: Tuchin, V., ed., *Handbook of Photonics for Biomedical Science*, CRC Press, Boca Raton, FL, 2010, pp. 159–196.)

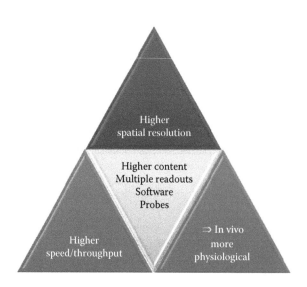

FIGURE 18.2 Frontiers of (fluorescence) bioimaging.

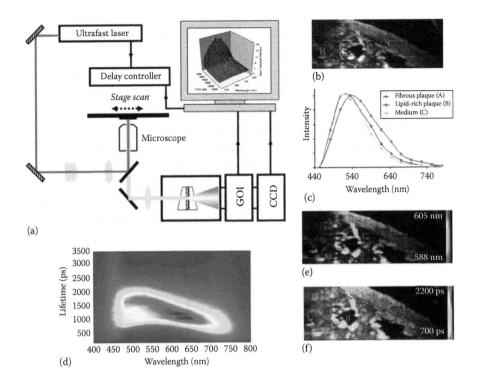

FIGURE 18.3 (a) Experimental setup for line-scanning hyperspectral FLIM, (b) integrated intensity image of sample of frozen human artery exhibiting atherosclerosis, (c) time-integrated spectra of sample regions corresponding to medium and fibrous and lipid rich plaques, (d) autofluorescence lifetime–emission matrix, (e) map of time-integrated central wavelength, and (f) spectrally integrated lifetime map of sample autofluorescence. (Adapted from De Beule, P. et al., *Microsc. Res. Tech.*, 70, 481, 2007.)

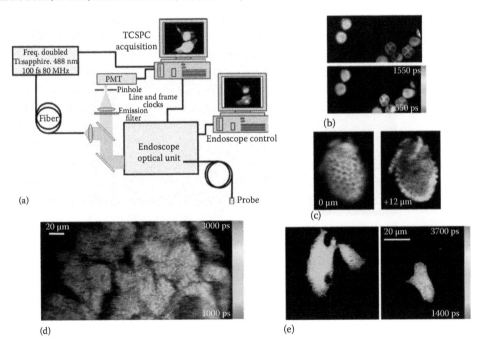

FIGURE 18.4 Confocal endomicroscope FLIM: (a) shows the experimental schematic; (b) shows intensity and corresponding optically sectioned FLIM images of stained pollen grain; (c) shows two optically sectioned FLIM images of the same pollen grain; (d) FLIM image of unlabeled rat tissue; (e) shows confocal endomicroscope FLIM images acquired in 1 s of live Cos-7 cells expressing either GFP (left) or EGP linked to RFP (right). (Adapted from Kennedy, G.T. et al., *J. Biophotonics*, 2009; Kumar, S. et al., *ChemPhysChem*, 12, 627, 2011.)

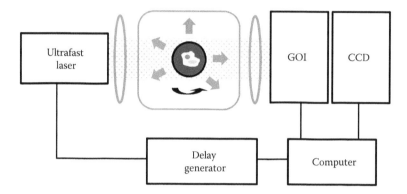

FIGURE 18.5 Schematic of tomoFLIM setup. (GOI is gated optical image intensifier.)

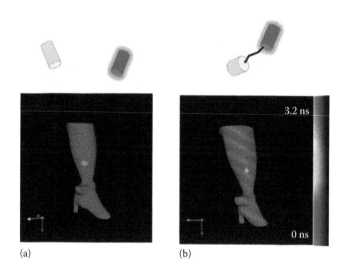

(a) (b)

FIGURE 18.6 3-D diffuse fluorescence lifetime tomography of live mouse transfected with (a) EGFP, mCherry unlinked, and (b) EGFP–mCherry directly linked. (Adapted from McGinty, J. et al., *Biomed. Opt. Express*, 2, 1907, 2011.)

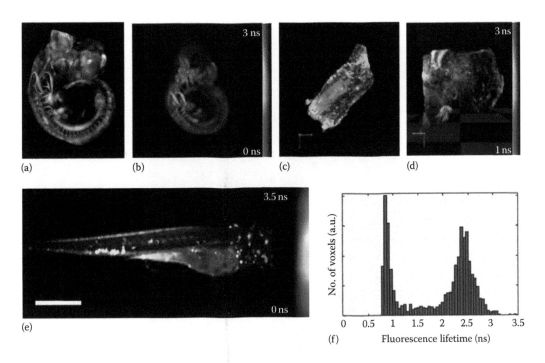

(a)   (b)   (c)   (d)

(e)   (f)

FIGURE 18.7 3-D OPT fluorescence (a) intensity and (b) FLIM images of mouse embryo cleared with BABB and with neurofilament labeled with Alexa-488 from [171]; (c) OPT autofluorescence image of portal vein branch in liver cleared with BABB but unstained showing fluorescence (acquired >500 nm with excitation at 460 nm) in green and white light absorption in red; (d) FLIM OPT of autofluorescence from unstained lung tumor cleared with BABB and excited at 480 nm; (e) FLIM OPT of live lysC:GFP transgenic zebrafish embryo 3 days postfertilization showing a single frame from 3-D fluorescence lifetime reconstruction (scale bar 500 μm) and (f) fluorescence lifetime histogram showing two clear populations corresponding to GFP and autofluorescence. (Adapted from McGinty, J. et al., *Biomed. Opt. Express*, 2, 1340, 2011.)

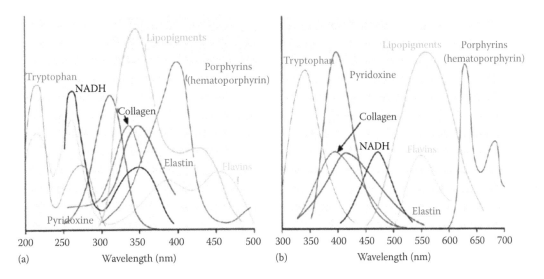

(a)   (b)

FIGURE 18.8 (a) Excitation and (b) emission spectra of main endogenous tissue fluorophores. (Adapted from Wagnieres, G.A. et al., *Photochem. Photobiol.*, 68, 603, 1998.)

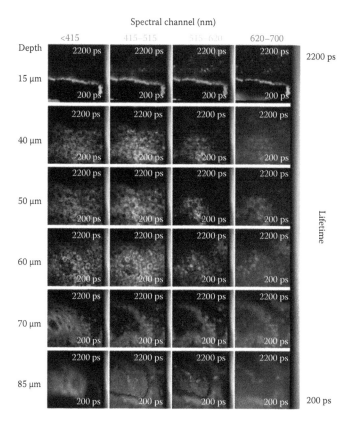

FIGURE 18.9 Multispectral optically sectioned fluorescence lifetime data stack of normal human skin acquired in vivo.

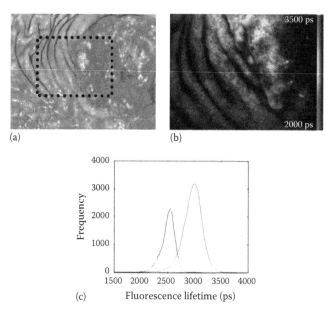

FIGURE 18.10 Wide-field time-gated FLIM with picosecond pulsed excitation at 355 nm applied to a freshly resected partial gastrectomy specimen containing a moderately differentiated intestinal-type adenocarcinoma: (a) white light image of the macroscopic specimen with area of FLIM outlined; (b) intensity-weighted false-color FLIM image with scale bar representing 1 cm; (c) autofluorescence lifetime histogram from the normal and cancerous regions of interest. (Adapted from McGinty, J. et al., *Biomed. Opt. Express*, 1, 627, 2010.)

11

Confocal Microscopy

Tony Wilson
University of Oxford

11.1 Introduction

It is probably fair to say that the development and wide commercial availability of the confocal microscope have been one of the most significant advances in light microscopy in the recent past. The main reason for the popularity of these instruments derives from their ability to permit the structure of thick specimens of biological tissue to be investigated in three dimensions by resorting to a scanning approach together with a novel (confocal) optical system.

The traditional wide-field conventional microscope is a parallel processing system that images the entire object field simultaneously. This is quite a severe requirement for the optical components, but we can relax this requirement if we no longer try to image the whole object at once. The limit of this relaxation is to require an image of only one object point at a time. In this case, all that we ask of the optics is to provide a good image of one point. The price that we pay is that we must scan in order to build up an image of the entire field. The answer to the question whether this price is worth paying will, to some extent, depend on the application in question.

A typical arrangement of a scanning confocal optical microscope is shown in Figure 11.1 in which the system is built around a host conventional microscope. The essential components are some form of mechanism for scanning the light beam (usually from a laser) relative to the specimen and appropriate photodetectors to collect the reflected or transmitted light.[1] Most of the early systems were analog in nature; however, it is now universal, thanks to the serial nature of the image formation, to use a computer to drive the microscope and to collect, process, and display the image.

In the beam scanning confocal configuration of Figure 11.1 the scanning is typically achieved by using vibrating galvanometer-type mirrors or acousto-optic beam deflectors. The use of the latter gives the possibility of TV-rate scanning, whereas vibrating mirrors are often relatively slow when imaging an extended region of the specimen, although significantly higher scanning speeds are achievable over smaller scan regions. Note that other approaches to scanning may be implemented,

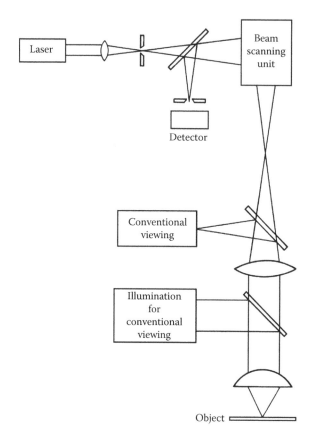

FIGURE 11.1   Schematic diagram of a confocal microscope.

such as specimen scanning and lens scanning. These methods, although not generally available commercially, do have advantage in certain specialized applications.[1,2] This chapter is necessarily limited in length, but much additional material may be found in other sources[1-7] in which a detailed list of references may be found.

11.2 Image Formation in Scanning Microscopes

We will not discuss the fine detail of the optical properties of confocal systems because this is already widely available in the literature. The essence is shown in Figure 11.1, where the confocal optical system consists simply of a point source of light that is then used to probe a single point on the specimen. The strength of the reflected or fluorescence radiation from the single object point is then measured via a point, pinhole detector. The confocal—point source and point detector—optical system therefore merely produces an *image* of a single object point and hence some form of scanning is necessary to produce an image of an extended region of the specimen. However, the use of single-point illumination and single-point detection results in novel imaging capabilities that offer significant advantages over those possessed by conventional wide-field optical microscopes. In essence, these are enhanced lateral resolution and, perhaps more importantly, a unique depth discrimination or optical sectioning property. It is this latter property that leads to the ability to obtain three-dimensional images of volume specimens.

The improvement in lateral resolution may at first seem implausible. However, it can be explained simply by a principle given by Lukosz,[8] which states, in essence, that resolution can be increased at the

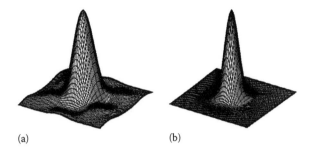

(a)   (b)

FIGURE 11.2   The point spread functions of (a) conventional and (b) confocal microscopes showing the improvement in lateral resolution that may be obtained in the confocal case.

expense of field of view. The field of view can then be increased by scanning. One way of taking advantage of Lukosz's principle is to place a very small aperture extremely close to the object. The resolution is now determined by the size of the aperture rather than the radiation. In the confocal microscope we do not use a physical aperture in the focal plane but, rather, use the back-projected image of a point detector in conjunction with the focused point source. Figure 11.2 indicates the improvement in lateral resolution that may be achieved.

The confocal principle, which was first described by Minsky,[9] was introduced in an attempt to obtain an image of a slice within a thick specimen that was free from the distracting presence of out-of-focus information from surrounding planes. The confocal optical system fulfils this requirement; its inherent optical sectioning or depth discrimination property has become the major motivation for using confocal microscopes, and is the basis of many of the novel imaging modes of these instruments.

The origin of the depth discrimination property may be understood very easily from Figure 11.3, where we show a reflection-mode confocal microscope and consider the imaging of a specimen with a rough surface. The full lines show the optical path when an object feature lies in the focal plane of the lens. At a later scan position, the object surface is supposed to be located in the plane of the vertical dashed line. In this case, simple ray tracing shows that the light reflected back to the detector pinhole arrives as a defocused blur, only the central portion of which is detected, and contributes to the image. In this way the system discriminates against features that do not lie within the focal region of the lens.

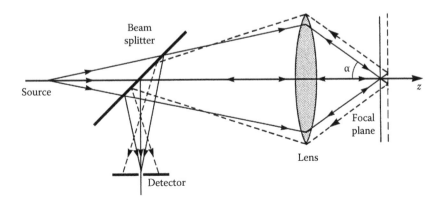

FIGURE 11.3   The origin of the optical sectioning or depth discrimination property of the confocal optical system.

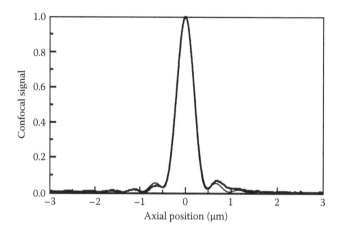

FIGURE 11.4 The variation in detected signal as a plane reflector is scanned axially through focus. The measurement was taken with a 1.3 numerical aperture objective and 633 nm radiation.

A very simple method of demonstrating the effect and giving a measure of its strength is to scan a perfect reflector axially through focus and measure the detected signal strength. Figure 11.4 shows a typical response. These responses are frequently termed the $V(z)$ by analogy with a similar technique in scanning acoustic microscopy, although the correspondence is not perfect. A simple paraxial theory models this response as:

$$I(u) = \left[\frac{\sin(u/2)}{u/2} \right]^2 \tag{11.1}$$

where u is a normalized axial coordinate related to real axial distance, z, via:

$$u = \frac{8\pi}{\lambda} nz \sin^2 \left(\frac{\alpha}{2} \right) \tag{11.2}$$

where
$\quad \lambda$ is the wavelength
$\quad n\sin \alpha$ is the numerical aperture

As a measure of the strength of the sectioning, we can choose the full width at half intensity of the $I(u)$ curves. Figure 11.5 shows this value as a function of numerical aperture for the specific case of imaging with red light from a helium neon laser. These curves were obtained using a high aperture theory, which is more reliable than Equation 11.1 at the highest values of numerical aperture. We note, of course, that these numerical values refer to nonfluorescence imaging. The qualitative explanation of optical sectioning, of course, carries over to the fluorescence case, but the actual value of the optical sectioning strength is different in the fluorescence case.

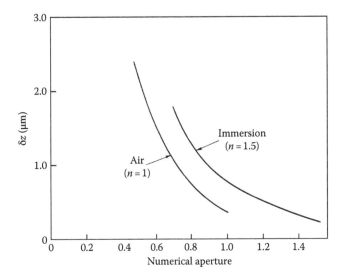

FIGURE 11.5 The optical sectioning width as a function of numerical aperture. The curves are for red light (0.6328 μm wavelength). δz is the full width at the half-intensity points of the curves of $I(u)$ against u.

11.3 Applications of Depth Discrimination

This property is one of the major reasons for the popularity of confocal microscopes, so it is worthwhile, at this point, to review briefly some of the novel imaging techniques that have become available with confocal microscopy.

Figure 11.6 illustrates the essential effect: Figure 11.6a shows a conventional image of a planar microcircuit that has deliberately been mounted with its normal at an angle to the optic axis. We see that only one portion of the circuit, running diagonally, is in focus. Figure 11.6b shows the corresponding confocal image: here the discrimination against detail outside the focal plane is clear. The areas that were out of focus in Figure 11.6a have been rejected. Furthermore, the confocal image appears to be in focus throughout the visible band, which illustrates that the sectioning property is stronger than the depth of focus.

This suggests that, if we try to image a thick translucent specimen, we can arrange, by the choice of our focal position, to image detail exclusively from one specific region. In essence, we can section the specimen optically without resorting to mechanical means. Figure 11.7 shows an idealized schematic of the process. The portion of the beehive-shaped object that we see is determined by the focus position. In this way, it is possible to take a through-focus series of images and obtain data about the three-dimensional structure of the specimen. If we represent the volume image by $I(x, y, z)$, then, ideally, by focusing at a position $z = z|_1$, we obtain the image $I(x, y, z|_1)$. This, of course, is not strictly true in practice because the optical section is not infinitely thin.

It is clear that the confocal microscope allows us to form high-resolution images with a depth of focus sufficiently small that all the detail that is imaged appears in focus. This suggests immediately that we can extend the depth of focus of the microscope by adding together (integrating) the images taken at different focal settings without sacrificing the lateral resolution. Mathematically, this extended-focus image is given by:

$$I_{EF} = I(x, y, z)dz \tag{11.3}$$

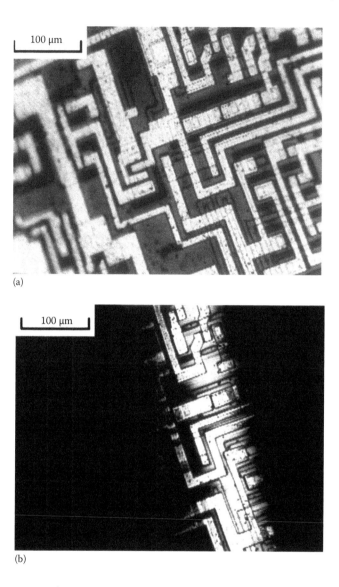

(a)

(b)

FIGURE 11.6   (a) Conventional scanning microscope image of a tilted microcircuit: the parts of the object outside the focal plane appear blurred. (b) Confocal image of the same microcircuit: only the part of the specimen within the focal region is imaged strongly.

As an alternative to the extended-focus method, we can form an auto-focus image by scanning the object axially and, instead of integrating, selecting the focus at each picture point by recording the maximum in the detected signal. Mathematically, this might be written:

$$I_{AF}(x,y) = I(x,y,z_{max})$$ (11.4)

where z_{max} corresponds to the focus setting giving the maximum signal. The images obtained are somewhat similar to the extended focus and, again, substantial increases in depth of focus may be obtained. We can go one step further in this case and turn the microscope into a noncontacting surface profilometer. Here we simply display z_{max}.

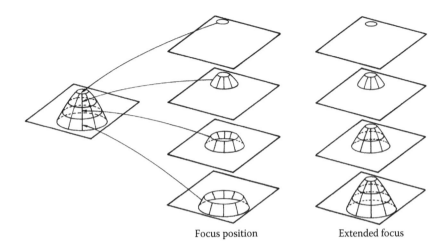

Focus position   Extended focus

FIGURE 11.7 An idealization of the optical sectioning property showing the ability to obtain a through-focus series of images, which may then be used to reconstruct the original volume object at high resolution.

It is clear by now that the confocal method gives us a convenient tool for studying three-dimensional structures in general. We essentially record the image as a series of slices and play it back in any desired fashion. Naturally, in practice it is not as simple as this, but we can, for example, display the data as an x–z image rather than an x–y image. This is somewhat similar to viewing the specimen from the side. As another example, we might choose to recombine the data as stereo pairs by introducing a slight lateral offset to each image slice as we add them up. If we do this twice, with an offset to the left in one case and the right in another, we obtain, very simply, stereo pairs. Mathematically, we form images of the form:

$$\int I(x \pm \gamma z, y, z)dz \tag{11.5}$$

where γ is a constant. In practice it may not be necessary to introduce offsets in both directions to obtain an adequate stereo view.

All that we have said so far on these techniques has been by way of simplified introduction. In particular we have not presented any fluorescence images, because these will be dealt with adequately later. The key point is that, in both bright-field and fluorescence modes, the confocal principle permits the imaging of specimens in three dimensions. Of course, the situation is more involved than we have implied. A thorough knowledge of the image formation process, together with the effects of lens aberrations and absorption, is necessary before accurate data manipulation can take place.

In conclusion to this section, it is important to emphasize that the confocal microscope does not produce three-dimensional images. The opposite is true: it essentially produces very high-quality two-dimensional images of a (thin) slice within a thick specimen. A three-dimensional rendering of the entire volume specimen may then be generated by suitably combining a number of these two-dimensional image slices from a through-focus series of images.

11.4 Fluorescence Microscopy

We now turn our attention to confocal fluorescence microscopy because this is the imaging mode usually employed in biological applications. Although we have introduced the confocal microscope in terms of bright-field imaging, the comments concerning the origin of the optical sectioning, etc. carry over directly to the fluorescence case. However, the numerical values describing the strength of the optical sectioning are, of course, different and we will return to this point later.

If we assume that the fluorescence in the object destroys the coherence of the illuminating radiation and produces an incoherent fluorescent field proportional to the intensity of the incident radiation, $I(v, u)$, then we may write the effective intensity point spread function, which describes image formation in the incoherent confocal fluorescence microscope, as

$$I(v,u)I\left(\frac{v}{\beta},\frac{u}{\beta}\right)$$

(11.6)

where the optical coordinates u and v are defined relative to the primary radiation and

$$\beta = \frac{\lambda_2}{\lambda_1}$$

is the ratio of the fluorescence to the primary wavelength. We note that

$$v = \frac{2\pi}{\lambda_1 rn\sin\alpha}$$

where r denotes the actual radial distance.

This suggests that the imaging performance depends on the value of β. In order to illustrate this, Figure 11.8 shows the variation in detected signal strength as a perfect fluorescent sheet through focus. This serves to characterize the strength of the optical sectioning in fluorescence microscopy in the same way that the mirror was used in the bright-field case. We note that the half width of these curves is essentially proportional to β and so for optimum sectioning the wavelength ratio should be as close to unity as possible.

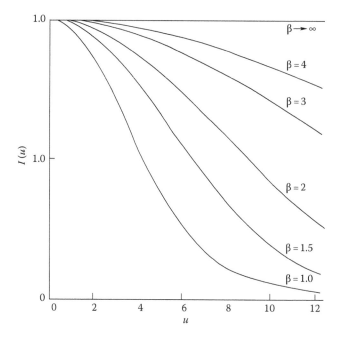

FIGURE 11.8 The detected signal as a perfect planar object is scanned axially through focus for a variety of fluorescent wavelengths. If we measure the sectioning by the halfwidth of these curves, the strength of the sectioning is essentially proportional to β.

We have just discussed what we might call one-photon fluorescence microscopy in the sense that a fluorophore is excited by a single photon of a particular wavelength. It then returns to the ground state and emits a photon at the (slightly longer) fluorescence wavelength. It is this radiation that is detected via the confocal pinhole. Recently, however, much interest has centered on two-photon excitation fluorescence microscopy.[10,11] This process relies on the simultaneous absorption of two, longer wavelength photons, following which a single fluorescence photon is emitted. The excitation wavelength is typically twice that used in the one-photon case.

The beauty of the two-photon approach lies in the quadratic dependence of the fluorescence intensity on the intensity of the illumination. This leads to fluorescence emission, which is always confined to the region of focus. In other words, the system possesses an inherent optical sectioning property. Other benefits of two-photon fluorescence over the single-photon case include the use of red or infrared lasers to excite ultraviolet dyes, confinement of photo-bleaching to the focal region (the region of excitation), and the reduced effects of scattering and greater penetration. However, it should also be remembered that, compared with single photon excitation, the fluorescence yield of many fluorescent dyes under two-photon excitation is relatively low.[12]

In order to make a theoretical comparison between the one- and two-photon modalities, we shall assume that the *emission* wavelength λ_{em} is the same irrespective of the mode of excitation. Since the wavelength required for single-photon excitation is generally shorter than the emission wavelength, we may write it as $\gamma\lambda_{em}$ where $\gamma < 1$. Because two-photon excitation requires the simultaneous absorption of two photons of half the energy, we will assume that the excitation wavelength in this case may be written as $2\gamma\lambda_{em}$, which has been shown to be a reasonable approximation for many dyes.[13]

If we now introduce optical coordinates u and v normalized in terms of λ_{em} we may write the effective point spread functions in the one-photon confocal and two-photon case as

$$I_{1p\text{-}conf} = I\left(\frac{v}{\gamma},\frac{u}{\gamma}\right)I(v,u) \tag{11.7}$$

and

$$I_{2p} = I^2\left(\frac{v}{2\gamma},\frac{u}{2\gamma}\right) \tag{11.8}$$

respectively. We note that, although a pinhole is not usually employed in two-photon microscopy, it is perfectly possible to include one if necessary. In this confocal two-photon geometry the effective point spread function becomes

$$I_{2p\text{-}conf} = I^2\left(\frac{v}{2\gamma},\frac{u}{2\gamma}\right)I(v,u) \tag{11.9}$$

If we now look at Equations 11.7 and 11.8 in the $\gamma = 1$ limit, we find

$$I_{1p\text{-}conf} = I^2(v,u) \tag{11.10}$$

and

$$I_{2p} = I^2\left(\frac{v}{2},\frac{u}{2}\right) \tag{11.11}$$

We now see that, because of the longer excitation wavelength used in two-photon microscopy, the effective point spread function is twice as large as that of the one-photon confocal in both the lateral and axial directions. The situation is somewhat improved in the confocal two-photon case, but it is worth remembering that the advantages of two-photon excitation microscopy are accompanied by a reduction in optical performance compared to the single-photon case.

From the practical point of view, the two-photon approach has certain very important advantages over the one-photon excitation in terms of image contrast when imaging through scattering media apart from the greater depth of penetration afforded by the longer wavelength excitation. In a single-photon confocal case, it is quite possible that the desired fluorescence radiation from the focal plane may be scattered after generation in such a way that it is not detected through the confocal pinhole. Further, since the fluorescence is generated throughout the entire focal volume, it is also possible that undesired fluorescence radiation, which was not generated within the focal region, may be scattered so as to be detected through the confocal pinhole.

In either case this leads to a reduction in image contrast. The situation is, however, completely different in the two-photon case. Here the fluorescence is generated only in the focal region and not throughout the focal volume. Furthermore, because all the fluorescence is detected via a large area detector—no pinhole is involved—it is not so important if further scattering events take place. This leads to high-contrast images, which are less sensitive to scattering. This is particularly important for specimens that are much more scattering at the fluorescence ($\lambda/2$) wavelength than the excitation (λ) wavelength.

11.5 Optical Architectures

In addition to the generic architecture of Figure 11.1, a number of alternative implementations of confocal microscopes have been recently developed in order to reduce the alignment tolerances of the architecture of Figure 11.1 as well as to increase image acquisition speed. Because great care is required to ensure that the illumination and detector pinholes lie in equivalent positions in the optical system, a reciprocal geometry confocal system has been developed that uses a single pinhole to launch light into the microscope as well as to detect the returning confocal image signal. A development on this is to replace the physical pinholes with a single mode optical fiber. However, these systems are essentially developments of the traditional confocal architecture. We will now describe two recent attempts to develop new confocal architectures. The first system is based on the pinhole source/detector concept, whereas the second approach describes a simple modification to a conventional microscope so as to permit optically sectioned images to be obtained.

11.5.1 The Aperture Mask System

The question of system alignment and image acquisition speed have already been answered to a certain extent by Petrán et al., who originally developed the tandem scanning (confocal) microscope[14]; this was later reconfigured into a one-sided disk embodiment (see, for example, Corle and Kino[4]).

The main component of the system is a disk containing many pinholes. Each pinhole acts as the illumination and the detection pinhole, so the system acts rather like a large number of parallel, reciprocal-geometry, confocal microscopes, each imaging a specific point on the object. However, because it is important that no light from a particular confocal system enter an adjacent system, i.e., no cross talk between adjacent confocal systems, the pinholes in the disk must be placed far apart—typically 10 pinhole diameters apart—which has 2 immediate consequences. First, only a small amount—typically 1%—of the available light is used for imaging and, second, the wide spacing of the pinholes means that the object is only sparsely probed. In order to probe, and hence image, the whole object, it is necessary to arrange the pinhole apertures in a series of Archimedean spirals and to rotate the disk. These systems have been developed and are capable of producing high-quality images without the need to use laser illumination in real time with TV rate as well as higher imaging speeds.

In order to make greater use of the available light in this approach we must, inevitably, place the pinholes closer together. This, of course, leads to cross talk between the neighboring confocal systems and thus a method must be devised to prevent this. In order to do this, we replace the Nipkow disk of the tandem scanning microscope with an aperture mask consisting of many pinholes placed as close together as possible. This aperture mask has the property that any of its pinholes can be opened and closed independently of the others in any desired time sequence. This might be achieved, for example, by using a liquid crystal spatial light modulator. Because we require no cross talk between the many parallel confocal systems, it is necessary to use a sequence of openings and closings of each pinhole completely uncorrelated with the openings and closings of all the other pinholes.

Many such ortho-normal sequences are in the literature. However, they all require the use of positive and negative numbers and, unfortunately, we cannot have negative intensity of light! The pinhole is either open, which corresponds to 1, or closed, which corresponds to 0. No position can correspond to –1. The way out of the dilemma is to not obtain the confocal signal directly. In order to use a particular ortho-normal sequence, $b_i(t)$, of plus and minus ones for the ith pinhole, we must add a constant shift to the desired sequence in order to make a sequence of positive numbers that can be encoded in terms of pinhole opening and closing.

Thus we encode each of the pinhole openings and closings as $(1 + b_i(t))/2$, which will correspond to open (1) when $b_i(t) = 1$ and to close (0) when $b_i(t) = -1$. The effect of adding the constant offset to the desired sequence is to produce a composite image that will be partly confocal due to the $b_i(t)$ terms and partly conventional due to the constant term. The method of operation is now clear. We first take an image with the pinholes encoded as we have just discussed and so obtain a composite conventional plus confocal image. We then switch all the pinholes to the open state to obtain a conventional image. It is then a simple matter to subtract the two images in real time in a computer to produce the confocal image.

Although this approach may be implemented using a liquid crystal spatial light modulator, it is cheaper and simpler merely to impress the correlation codes photolithographically on a disk and rotate it so that the transmissivity at any picture point varies according to the desired ortho-normal sequence. A blank sector may be used to provide the conventional image. If this approach is adopted, then all that is required is to replace the single-sided Nipkow disk of the tandem scanning microscope with a suitably encoded aperture disk, as shown in Figure 11.9.[15] Figure 11.10 shows a through-focus series of images of a fly's eye; Figure 11.11 shows three-dimensional representation of the fly's eye obtained with this white light real-time confocal system.

FIGURE 11.9 A typical aperture mask. When light passes through the encoded region of the disk, a composite conventional and confocal image is obtained. Light passing through the unobstructed sector provides a conventional image.

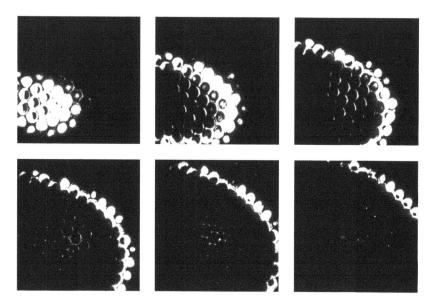

FIGURE 11.10 A through-focus series of images of a portion of a fly's eye. Each optical section was recorded in real time using standard nonlaser illumination.

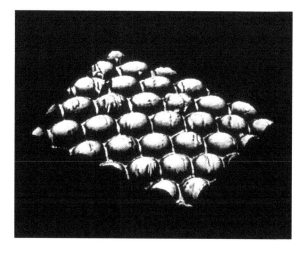

FIGURE 11.11   A computer generated three-dimensional representation of the fly's eye.

11.5.2 The Use of Structured Illumination to Achieve Optical Sectioning

The motivation for the development we have just described was to produce a light-efficient, real-time, three-dimensional imaging system. The approach was to start with the traditional confocal microscope design—point source and point detector—and to engineer a massively parallel, light-efficient three-dimensional imaging system (see also Liang et al.[16] and Hanley et al.[17]).

An alternative approach is to realize that the conventional light microscope already possesses many desirable properties: real-time image capture, standard illumination, ease of alignment, etc. However it does not produce optically sectioned images in the sense usually understood in confocal microscopy. In order to see how this deficiency may be corrected via a simple modification of the illumination system, let us look at the theory of image formation in a conventional fluorescence microscope and ask in what way the image changes as the microscope is defocused.

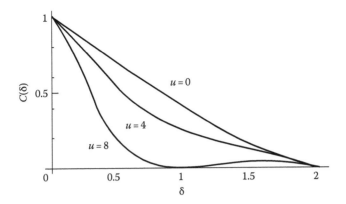

FIGURE 11.12 The transfer function for a conventional fluorescence microscope for a number of values of defocus (measured in optical units). The response for all non-zero spatial frequencies decays with defocus.

We know that in a confocal microscope the image signal from all object features attenuates with defocus and that this does not happen in a conventional microscope. However, when we look closely at the image formation process we find that it is only the zero spatial frequency (constant) component that does not change with defocus (Figure 11.12); all other spatial frequencies actually do attenuate with defocus to a greater or lesser extent. Figure 11.13 shows the image of a single spatial frequency one-dimensional bar pattern object for increasing degrees of defocus. When the specimen is imaged in focus, a good image of the bar pattern is obtained. However, with increasing defocus the image becomes progressively poorer and weaker until it eventually disappears, leaving a uniform gray level. This suggests a simple way to perform optical sectioning in a conventional microscope.

If we simply modify the illumination path of the microscope so as to project onto the object the image of a one-dimensional, single spatial frequency fringe pattern, then the image we see through the microscope will consist of a sharp image of those parts of the object where the fringe pattern is in focus, but an out-of-focus blurred image of the rest of the object. In order to obtain an optically sectioned image, it is necessary to remove the blurred out-of-focus portion as well as the fringe pattern from the in-focus optical section. There are many ways to do this—one of the simplest involves simple processing of three images taken at three different spatial positions of the fringe pattern. The out-of-focus regions remain fairly constant between these images and the relative spatial shift of the fringe pattern allows the three images to be combined in such a way as to remove the fringes. This permits us to retrieve an optically sectioned image as well as a conventional image in real time.[18]

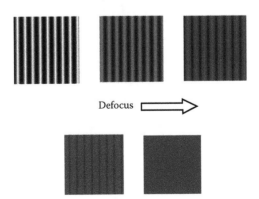

FIGURE 11.13 The image of a one-dimensional single spatial frequency bar pattern for varying degrees of defocus. For sufficiently large values of defocus, the bar pattern is not imaged at all.

The approach involves processing three conventional microscope images, so the image formation is fundamentally different from that of the confocal microscope. However, the depth discrimination or optical sectioning strength is very similar and this approach, which requires very minimal modifications to the instrument, has been used to produce high-quality three-dimensional images of volume objects directly comparable to those obtained with confocal microscopes. A schematic of the optical arrangement is shown in Figure 11.14a together with experimentally obtained axial responses in Figure 11.14b. These responses are substantially similar to those obtained in the true confocal case.

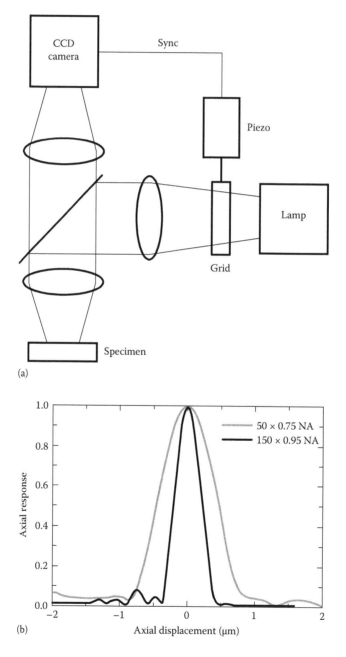

FIGURE 11.14 (a) Optical system of the structured illumination microscope together with experimentally obtained axial responses in (b), which confirm the optical sectioning ability of the instrument.

As an example of the kinds of images that can be obtained with this type of microscope, Figure 11.15 shows two images of a spiracle of a head louse. The first is an auto-focus image of greatly extended depth of field constructed from a through-focus series of images. The second is a conventional image taken at mid-focus. The dramatic increase in depth of field is clear when compared with a mid-focus conventional image. These images were taken using a standard microscope illuminator as light source. Indeed, the system is so light efficient that good quality optical sections of transistor specimens have been obtained using simply a candle as light source.

Imaging using fluorescence light is also possible using this technique. However, an alternative approach that does not require a physical grid is possible if a laser is used as the light source. In this system, the laser illumination is split into two beams that are allowed to interfere at an angle in the fluorescent specimen. This has the effect of directly *writing* a one-dimensional fringe pattern in the specimen. Spatial shift of the fringe pattern is achieved by varying the phase of one of the interfering beams.

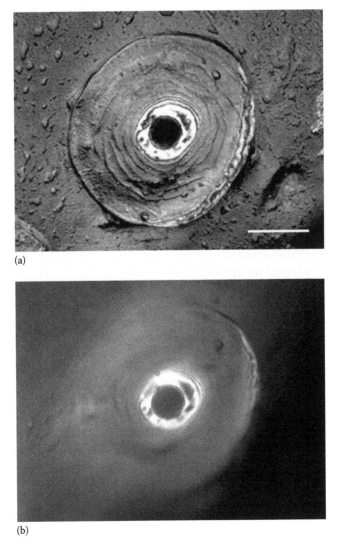

(a)

(b)

FIGURE 11.15 Two images of the region around the spiracle of a head louse. As in the previous figure, (a) is an autofocus image and (b) shows a mid-focus conventional image. Scale bar = 10 μm.

As before, three images are taken from which the optically sectioned image and the conventional image may be obtained.[19] The beauty of this approach is that no imaging optics are required at the illuminating wavelength and system alignment is trivial.

11.6 Aberration Correction

As we have seen, the most important feature of the confocal microscope is its optical sectioning property, which permits volume structures to be rendered in three dimensions via a suitable stack of through-focus images. However, in order to achieve the highest performance, it is necessary that the optical resolution be the same at all depths within the optical section. Unfortunately, there are fundamental reasons why this cannot be achieved in general. One is that high aperture microscope objectives are often designed to give optimum performance when imaging features located just below the cover glass. Another is more fundamental and is due to specimen-induced refractive index mismatches. These could be caused by a variety of reasons, such as the use of an oil immersion objective to image into a watery specimen, or may be due to refractive index inhomogeneities within the specimen. In many cases they cannot be removed by system design and a new approach must be taken. An adaptive method of correction in which the aberrations are measured using a wavefront sensor and removed with an adaptive wavefront corrector, such as a deformable mirror, is required.

A number of methods of wavefront sensing have already been developed for astronomical applications. Unfortunately these approaches are not necessarily appropriate for microscopy. Further, because the optical sectioning property of the confocal system limits the region of the specimen to be imaged to that lying close to the focal plane, it is clearly desirable that the wavefront sensor possess a similar optical sectioning property.

A very powerful method of describing a particular wavefront aberration is in terms of a superposition of aberration modes such as those described by the Zernike circle polynomials.[20] Mathematically speaking, the wavefront aberration is represented by a truncated series of Zernike polynomials (modes). These polynomials, which are two-dimensional functions, are orthogonal over the unit circle and hence are particularly useful in optics where apertures and lenses are typically circular in cross section. Their real power, however, lies in the fact that real-world aberrations can often be described by only a few low-order Zernike modes. We have shown, for example, that the aberrations introduced when focusing into a biological specimen are dominated by a few lower order modes.[21] It is therefore attractive to consider basing an adaptive system on Zernike polynomials since it will be possible to simplify the overall system design because the Zernike modes of interest could be measured directly. Equally importantly, because only a small number of modes need to be measured, the complexity of the wavefront sensor would be reduced.

The operation of the modal wavefront sensor is based upon the concept of *wavefront biasing*. This process involves the combination of certain amounts of the Zernike aberration mode being measured (the *bias mode*) with the input wavefront. Conceptually, this can be done by including appropriately shaped or etched glass plates in the optical beam path, although in practice other methods are used. First, the beam containing the input wavefront passes through a beamsplitter to create two identical beams. In the first beam, a positive amount of the bias mode is added to the wavefront, which is then focused by a lens onto a pinhole. In the second beam, an equal but opposite negative amount of the bias mode is added to the wavefront. This is again focused onto a pinhole. Behind each pinhole lies a photodetector that measures the optical power passing through the pinhole. The output signal of the sensor, which is taken to be the difference between the two photodetector signals, is found to be sensitive to the amount of the particular *bias* Zernike mode present in the input wavefront.[22]

One method of implementation of this wavefront sensor permits the entire Zernike modal content of a wavefront to be measured simultaneously. This is achieved by using a computer-generated binary phase hologram, which produces a diffraction pattern consisting of a number of spots created with the appropriate bias aberrations. A pair of spots in the diffraction pattern corresponds to the two spots produced in the positively and negatively biased paths for a given bias mode. An array of pinholes and photodetectors is positioned behind the diffraction pattern and, as before, the output signal for each mode is taken as difference between the detector signals from oppositely biased spots.

Another implementation of the modal wavefront sensor involves the sequential application of bias modes using an adaptive element. Adaptive optics systems use some form of adaptive wavefront shaping element in order to remove unwanted aberrations from the wavefront. Suitable devices include membrane or bimorph deformable mirrors and liquid crystal spatial light modulators. Effectively, the device flattens the aberrated wavefront by adding an equal but opposite aberration to it. Naturally, a device with such capabilities can also be used to apply the bias aberrations in a wavefront sensor sequentially. The sensor in this case comprises simply the biasing element, one lens, and a single pinhole/photodetector. The positive and negative bias aberrations for each mode are applied in turn and the corresponding detector signals are measured. Because the adaptive element applying the bias modes can be the same device used for correction of the wavefront, the resulting adaptive optics system can operate without a separate wavefront sensor. Considering also that only a single photodetector is required, this facilitates the design of a simple, low-cost adaptive optics system.

The practicability of this new approach has been demonstrated in a confocal microscope and has been used to measure and correct aberrations in the imaging of biological specimens.[23] Figure 11.16 shows the uncorrected and corrected x–y and x–z images of a specimen of fluorescently labeled mouse intestine.

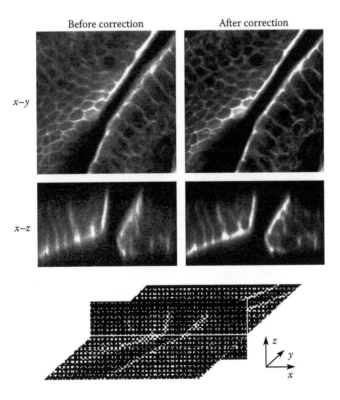

FIGURE 11.16 Confocal microscope scans before and after aberration correction. x–y (lateral) and x–z (axial) scans of a fluorescently labeled section of mouse intestine specimen are shown. The schematic beneath illustrates the relative three-dimensional orientation of the x–y and x–z scans. The image dimensions in the x and y directions are 80 μm; the z dimension is 15 μm.

11.7 Summary

We have discussed the origin of the optical sectioning property in the confocal microscope in order to introduce the range of imaging modes to which this unique form of microscope leads. A range of optical architectures has also been described. By far the most universal is that shown in Figure 11.1 where a confocal module is integrated around a conventional optical microscope. Other, more recent real-time implementations have also been described. A number of practical aspects of confocal microscopy have not been discussed. This is because they are readily available elsewhere, such as advice on the correct choice of detector pinhole size, or because they are still the focus of active research, such as the development of new contrast mechanisms for achieving enhanced three-dimensional resolution such as the stimulated emission depletion method (STED)[24] or 4-Pi.[25] These are still under development but offer great promise.

References

1. Wilson, T. and Sheppard, C.J.R., *Theory and Practice of Scanning Optical Microscopy*, Academic Press, London, U.K., 1984.
2. Wilson, T. (ed.), *Confocal Microscopy*, Academic Press, London, U.K., 1990.
3. Pawley, J.B. (ed.), *Handbook of Biological Confocal Microscopy*, Plenum Press, New York, 1995.
4. Corle, T.R. and Kino, G.K., *Confocal Scanning Optical Microscopy and Related Imaging Systems*, Academic Press, New York, 1996.
5. Gu, M., *Principles of Three-Dimensional Imaging in Confocal Microscopes*, World Scientific, Singapore, Singapore, 1996.
6. Masters, B.R., *Confocal Microscopy*, SPIE, Bellingham, WA, 1996.
7. Diaspro, A., *Confocal and Two-Photon Microscopy*, Wiley-Liss, New York, 2002.
8. Lukosz, W., Optical systems with resolving power exceeding the classical limit, *J. Opt. Soc. Am.*, 56, 1463, 1966.
9. Minsky, M., Microscopy apparatus, US Patent 3,013,467, December 19, 1961 (filed November 7, 1957).
10. Denk, W., Strikler, J.H., and Webb, W.W., Two-photon laser scanning fluorescence microscopy, *Science*, 248, 73, 1990.
11. Göppert-Mayer, M., Über Elementarakte mit zwei Quantensprungen, *Ann. Phys. (Leipzig)*, 5, 273, 1931.
12. Xu, C. and Webb, W.W., Measurement of two-photon cross-sections of molecular fluorophores with data from 690 nm to 1050 nm, *J. Opt. Soc. Am.*, B13, 481, 1996.
13. Xu, C. and Webb, W.W., Multiphoton excitation cross-sections of molecular fluorophores, *Bioimaging*, 4, 198, 1996.
14. Petrán, M., Hadravsky, M., Egger, M.D., and Galambos, R., Tandem-scanning reflected-light microscopy, *J. Opt. Soc. Am.*, 58, 661, 1968.
15. Juskaitis, R., Wilson, T., Neil, M.A.A., and Kozubek, M., Efficient real-time confocal microscopy with white light sources, *Nature*, 383, 804, 1996.
16. Liang, M.H., Stehr, R.L., and Krause, A.W., Confocal pattern period in multi-aperture confocal imaging systems with coherent illumination, *Opt. Lett.*, 22, 751, 1997.
17. Hanley, Q.S., Verveer, P.J., Gemkow, M.J., Arndt-Jovin, D., and Jovin, T.M., An optical sectioning programmable array microscope implemented with a digital micromirror device, *J. Microsc.*, 196, 317, 1999.
18. Neil, M.A.A., Juskaitis, R., and Wilson, T., Method of obtaining optical sectioning by using structured light in a conventional microscope, *Opt. Lett.*, 22, 1905, 1997.
19. Neil, M.A.A., Juskaitis, R., and Wilson, T., Real time 3D fluorescence microscopy by two beam interference illumination, *Opt. Commun.*, 153, 1, 1998.

20. Born, M. and Wolf, E., *Principles of Optics*, Pergamon Press, Oxford, U.K., 1975.
21. Booth, M.J., Neil, M.A.A., and Wilson, T., Aberration correction for confocal imaging in refractive-index-mismatched media, *J. Microsc.*, 192, 90, 1998.
22. Neil, M.A.A., Booth, M.J., and Wilson, T., New modal wave-front sensor: A theoretical analysis, *J. Opt. Soc. Am.*, 17, 1098, 2000.
23. Booth, M.J, Neil, M.A.A., Juskaitis, R., and Wilson, T., Adaptive aberration correction in a confocal microscope, *Proc. Natl. Acad. Sci. U.S.A.*, 99, 5788, 2002.
24. Hell, S. and Wichmann, J., Breaking the diffraction resolution limit by stimulated emission: Stimulated-emission-depletion microscopy, *Opt. Lett.*, 19, 870, 1994.
25. Hell, S. and Stelzer, E.H.K., Fundamental resolution improvement with a 4Pi-confocal fluorescence microscope using two-photon excitation, *Opt. Commun.*, 93, 277, 1992.

12

Two-Photon Excitation Fluorescence Microscopy

Peter T.C. So
Massachusetts Institute of Technology

Poorya Hosseini
Massachusetts Institute of Technology

Chen Y. Dong
National Taiwan University

Barry R. Masters
University of Bern

12.1 Introduction

Maria Göppert-Mayer first predicted the possibility of multiphoton excitation process in her doctoral dissertation presented in Göttingen.[1] Experimental verification of multiphoton processes was not realized until 1963 when Kaiser and Garret first observed two-photon excitation of $CaF_2:Eu^{2+}$ fluorescence.[2] Later, Kaiser and Garret further demonstrated two-photon excitation fluorescence from organic molecules. Three-photon excitation was subsequently reported by Singh and Bradley.[3] Multiphoton excitation of biochemical molecules provides complementary spectroscopic information to standard one-photon studies.[4–7]

The utilization of nonlinear optical processes to provide image contrast for microscopic studies was realized by pioneers such as Freund, Hellwarth, and Christensen.[8] Nonlinear optical imaging has inherent 3-D resolution, which was first suggested by the Oxford group including Sheppard et al.[9,10] They predicted that optical emissions that have quadratic or higher-order dependence on the excitation power will be confined to the focal plane of the microscope objective. The utilization of this principle for 3-D imaging was realized in the seminal work of Denk et al.[11] Denk and coworkers further demonstrated the importance of this approach for limiting the photodamage and for extending the penetration depth in biological specimens. This work revolutionized the application of 3-D microscopy in biomedical imaging. The applications of other nonlinear optical mechanisms in biological imaging, such as second- and third-harmonic generation, sum frequencies generation, and coherent anti-Stokes Raman scattering, soon followed.[12–14] This chapter will focus on multiphoton fluorescence microscopy, the most widely used nonlinear microscopy approach, and its applications in biology and medicine.

The subject of multiphoton microscopy has been covered in a number of other excellent reviews providing different perspectives on this subject.[15–18]

This chapter contains a discussion of the basic principles of multiphoton excitation and image formation, a review of two-photon instrumentation, and fluorescent probe choices. Since multiphoton microscopy is finding its most important applications in tissue imaging, a discussion on the strengths and the limitations of this technology for deep tissue imaging is provided. Finally, this chapter will conclude with a review on successful tissue imaging applications using two-photon microscopy.[19–21]

12.2 Basic Principles of Multiphoton Excitation and Image Formation

12.2.1 Physics of Multiphoton Excitation

The theory of two-photon excitation was predicted by Göppert-Mayer in 1931.[1] The basic physics of this phenomenon has also been described in a number of standard texts on quantum mechanics.[22] Fluorescence excitation is an interaction between the fluorophore and an excitation electromagnetic field and is described by a time-dependent Schrödinger equation where the Hamiltonian contains an electric dipole interaction term: $\vec{E}_\gamma \cdot \vec{r}$, where $\vec{E}_\gamma$ is the electric field vector of the photons and $\vec{r}$ is the position operator. This equation can be solved by perturbation theory. The first-order solution corresponds to the one-photon excitation with transition probability P:

$$P \sim \left| \left\langle f \left| \vec{E}_\gamma \cdot \vec{r} \right| i \right\rangle \right|^2 \tag{12.1}$$

where
$|i\rangle$ denotes the ground electronic state
$|f\rangle$ denotes the excited electronic state

The n-photon transitions are represented by the nth-order solutions. In the case of two-photon excitation, the transition probability between the molecular ground state $|i\rangle$ and the excited state $|f\rangle$ is represented by

$$P \sim \left| \sum_m \frac{\left\langle f \left| \vec{E}_\gamma \cdot \vec{r} \right| m \right\rangle \left\langle m \left| \vec{E}_\gamma \cdot \vec{r} \right| i \right\rangle}{\varepsilon_\gamma - \varepsilon_m} \right|^2 \tag{12.2}$$

where
ε_γ is the photonic energy associated with the electric field vector $\vec{E}_\gamma$
the summation is over all intermediate states m with energy ε_m

Therefore, two-photon excitation of molecules is a nonlinear process involving the absorption of two photons whose combined energy is sufficient to induce a molecular transition to an excited electronic state. Conventional one-photon technique uses ultraviolet or visible light to excite fluorescent molecules of interest. Excitation occurs when the absorbed photon energy matches the energy gap between the ground and excited state. The same transition can be excited by a two-photon process where two less energetic photons are simultaneously absorbed. Quantum mechanically, the first photon excites the molecule to a virtual intermediate state, and the molecule is eventually brought to the final excited state by the absorption of a second photon. A comparison between one- and

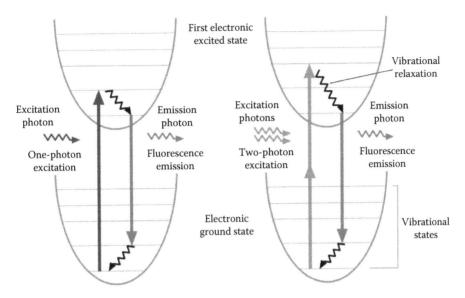

FIGURE 12.1 Jablonski diagrams for one-photon (left) and two-photon (right) excitations. Excitations occur between the ground state and the vibrational levels of the first electronic excited state. After either excitation process, the fluorophore relaxes to the lowest energy level of the first excited electronic states via vibrational processes. The subsequent fluorescence emission process for both relaxation modes is the same.

two-photon absorption is shown in Figure 12.1. In the absence of major vibrational perturbations, inductive effects, or solvent relaxation, the probability of the transition depends on the magnitude of the overlap integrals between the ground, intermediate, and excited states of a particular fluorophore. Choosing fluorophores with higher multiphoton cross sections allows more efficient excitation with lower photon dose.

A molecular state can be described by an eigenfunction with definite parity. For a molecular state with even parity, there is no change in the sign of the wavefunction when there are sign changes in the spatial coordinates. For a molecular state with odd parity, the sign of the wavefunction changes when the sign changes in the spatial coordinates. Since the dipole operator has odd parity (i.e., absorbing one photon changes the parity of the state), the one-photon transition couples initial and final states with opposite parity. On the other hand, the two-photon moment only allows transition between two states with the same parity.[22,23] In general, an n-photon process couples states with equal parity, if n is even, and opposite parity, if n is odd.

Finally, it should be noted that while the absorption cross sections for one-, two-, and multiphoton excitation of a given fluorophore are different, the molecule rapidly relaxes to the same vibrational level in the excited electronic state. The excited state residence time (fluorescence lifetime) and the fluorescence decay processes depend only on the molecular structure and its microenvironment. Therefore, the fluorescence quantum yield and emission spectra is independent of the initial excitation process.

12.2.2 Imaging Properties of Two-Photon Microscopy

One of the most important attribute of two-photon microscopy is its inherent 3-D sectioning capability. The sectioning capability of this method originates from the quadratic and higher-order dependence of the fluorescence signal upon the excitation intensity distribution.

Consider the intensity distribution at the focal point of a high numerical aperture objective. The spatial profile of the diffraction limited focus for an objective with numerical aperture, $NA = \sin(\alpha)$, is

$$I(u,v) = \left| 2 \int_0^1 J_0(v\rho) e^{-(i/2)u\rho^2} \rho\, d\rho \right|^2 \tag{12.3}$$

where
   J_0 is the zeroth-order Bessel function
   λ is wavelength of the excitation light
   $u = 4k\sin^2(\alpha/2)z$ and $v = k\sin(\alpha)r$ are the respective dimensionless axial and radial coordinates normalized to wave number $k = 2\pi/\lambda$[24,25]

The point spread function (PSF) of the fluorescence signal is a quadratic function of the excitation profile, which is just $I^2(u/2, v/2)$. In contrast, one-photon fluorescence PSF has a functional form of $I(u,v)$. For n-photon excitation, the fluorescence intensity profile is in general $I^n(u/n, v/n)$.

The difference of the fluorescence PSF resulting from one- and two-photon excitation is profound. The contribution of fluorescence signal from each axial plane can be computed for one- and two-photon excitation, as shown in Figure 12.2. Assuming negligible attenuation, the total fluorescence generated is equal at each axial plane for one-photon microscopy. In contrast, two-photon fluorescence falls off quickly from the focal plane, resulting in the localization of the signal to the focal region. For higher-photon processes, the localization property is even more pronounced. Qualitatively, the localization of multiphoton excitation can be understood by realizing that fluorescence generation is only appreciable at a region of high spatial photon density at the focal point of the microscope objective. The localization of the multiphoton excitation can be further appreciated by examining the dimensions of fluorescence PSF. Figure 12.3 illustrates the radial and axial PSF for two-photon microscopy at 960 nm. The two-photon photo-interaction volume is on the order of 0.1 fL.

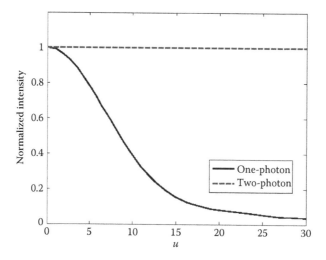

FIGURE 12.2 Total fluorescence generated at a given z-plane is calculated. This quantity is plotted as a function of its distance from the focal plane. For one-photon excitation, equal fluorescence intensity is observed in all planes and there is no depth discrimination. For two-photon excitation, the integrated intensity decreases rapidly away from the focal plane. In this figure, u is dimensionless optical axial coordinate as conventionally defined. (From Wilson, T. and Sheppard, C.J.R., *Theory and Practice of Scanning Optical Microscopy*, Academic Press, New York, 1984.)

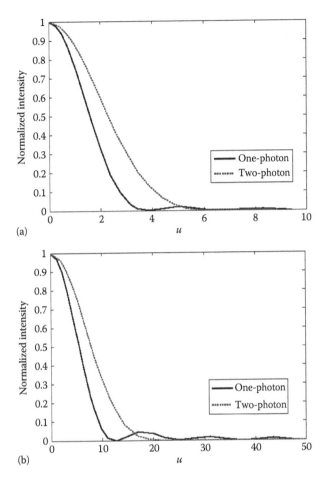

(a)

(b)

FIGURE 12.3 A comparison of the one- and two-photon PSFs in the (a) radial and (b) axial directions. In these figures, v and u are dimensionless optical coordinates along radial and axial directions as conventionally defined. (From Wilson, T. and Sheppard, C.J.R., *Theory and Practice of Scanning Optical Microscopy*, Academic Press, New York, 1984.)

The axial depth discrimination of multiphoton microscopy greatly improves image contrast as compared with wide-field fluorescence microscopy, especially for thick specimens. This localization further reduces the photo-interaction volume and greatly reduces photobleaching and photodamage in thick sample. Further, since photo-interaction occurs in a femtoliter size volume, localized photo-chemical reactions can be initiated.[26]

An important advantage of multiphoton excitation is the use of infrared light to excite fluorophores with one-photon excitation in the ultraviolet and blue-green spectral range. For one-photon excitation, the excitation spectral band overlaps the fluorescence emission band. Excitation light cannot be eliminated without losing a fraction of the fluorescence photons; microscope sensitivity is reduced. For two-photon excitation, the excitation spectrum is completely separated from the emission spectrum and very efficient filters can be used to eliminate the excitation with a minimal loss of the fluorescence. For weak fluorescence specimen in one-photon excitation, noise originated from Raman scattering is stoke-shifted into the emission band and can sometimes obscure the fluorescence signal. In multiphoton excitation, the Raman signal is red-shifted from the excitation light and is further removed from the emission band. Hyper-Raman and hyper-Rayleigh scattering can still occur in the emission band for multiphoton excitation, but these processes are typically weak.[27]

12.3 Experimental Considerations of Multiphoton Microscopy

12.3.1 Instrument Design of Multiphoton Microscopy

Multiphoton excitation efficiency increases with the spatial and temporal density of the excitation photons. Spatial localization of photons is relatively straightforward by using high numerical aperture optics. Temporal localization of photons is comparatively more difficult and has not been efficiently achieved until the advent of femtosecond pulsed lasers. The typical cross section for two-photon absorption is on the order of 1 GM, which is equivalent to 10^{-50} cm^4 s, a unit adopted in honor of Göppert-Mayer. The need for temporal localization of photons using pulsed laser can be observed by considering the absorption efficiency of a two-photon probe with cross section, δ. Two-photon absorption efficiency can be measured by n_a, the number of photons absorbed per fluorophore per pulse:

$$n_a \approx \frac{p_0^2 \delta}{\tau_p f_p^2} \left(\frac{(NA)^2}{2\hbar c\lambda} \right)^2 \tag{12.4}$$

where

    τ_p is the pulse duration
    λ is the excitation wavelength
    p_0 is the average laser intensity
    f_p is the laser's repetition rate
    NA is the numerical aperture of the focusing objective
    $\hbar$ is Planck's constant
    c is the speed of light[11]

Equation 12.4 shows that for the same average laser power and repetition frequency, the excitation probability is increased by increasing the NA of the focusing lens, which increases spatial localization of the laser excitation. The absorption efficiency further improves linearly by reducing the pulse width of the laser, indicating the importance of temporal localization.

For multiphoton excitation, femtosecond, picosecond, and continuous-wave laser sources have been used. The most commonly used laser for multiphoton microscopy is femtosecond Ti:sapphire lasers. These lasers characteristically produce 100 fs pulse train at 100 MHz repetition rate. The tuning range of Ti:sapphire systems extends from 680 to 1050 nm. Cr:LiSAF and pulse-compressed Nd:YLF lasers are some of the other femtosecond lasers used in multiphoton microscopy.[28]

Multiphoton excitation can also be generated using ps light sources, although at a lower excitation efficiency. Two-photon excitation has been achieved using mode-locked Nd:YAG (~100 ps), picosecond Ti:sapphire lasers, and pulsed dye lasers (~1 ps). Two-photon excitation using continuous-wave lasers has also been demonstrated using ArKr laser and Nd:YAG laser.[29]

Comparing different laser sources, for fixed δ, p_0, and NA, Equation 12.4 implies that the difference in excitation efficiency per unit time for pulsed and continuous-wave lasers for a two-photon process is $\sqrt{f_p \tau_p}$. This factor is typically 300 for femtosecond Ti:sapphire laser; two-photon excitation of a fluorophore that typically requires 1 mW of power using a femtosecond Ti:sapphire laser will require 300 mW using a continuous-wave laser.

The generalization of Equation 12.4 to the multiphoton case can be expressed as

$$n_a \propto \frac{1}{\tau_p^{n-1}} \tag{12.5}$$

One can conclude that multiphoton excitation becomes increasingly difficult for nonfemtosecond light sources with increasing order of excitation process.

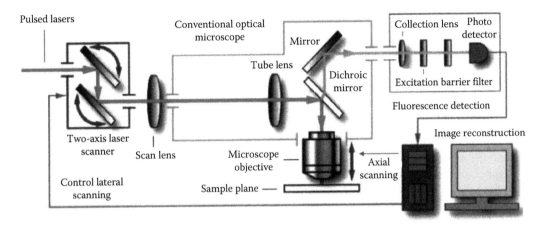

FIGURE 12.4 (See color insert.) Schematic of a two-photon laser scanning fluorescence microscope. (From Kim, D., Ultrafast optical pulse manipulation in three dimensional-resolved microscope imaging and microfabrication. PhD, Department of Mechanical Engineering, Massachusetts Institute of Technology, 2009.)

Since multiphoton excitation generates signal from a single point, the production of 3-D images requires raster scanning of this excitation volume in 3-D. In a typical multiphoton microscope, x–y raster scanning uses galvanometer-driven scanner. After beam power control and pulse-width compensation optics, x–y scanner deflects the excitation light into a fluorescence microscope. A typical multiphoton microscope uses epiluminescence geometry (Figure 12.4). The scan lens is positioned such that the x–y scanner is at its eye point while the field aperture plane is at its focal point. For a telecentric microscope, this arrangement ensures that angular scanning of the excitation light is converted to linear translation of the objective focal point. A tube lens is positioned to recollimate the excitation light directed toward the infinity-corrected objective via a dichroic mirror. The scan lens and the tube lens function together as a beam expander that overfills the back aperture of the objective lens to ensure excitation light is diffraction limited. Typically, high numerical aperture objectives are used to maximize excitation efficiency. An objective positioner translates the focal point axially for 3-D imaging.

In the emission path, the fluorescence emission is collected by the imaging objective and is transmitted through the dichroic mirror. Additional barrier filters are used to further attenuate the scattered excitation light and to select the emission band of interest. The fluorescence signal is subsequently directed toward the detector system. A number of photodetectors have been used in two-photon microscope including photomultiplier tubes (PMTs), avalanche photodiodes, and charge-coupled device (CCD) cameras. PMTs are the most common implementation because they are robust and low cost, have large active area, and have relatively good sensitivity. PMT and avalanche photodiode systems further allow the use of ultralow noise and high-sensitivity single-photon counting electronic circuitry. Electronic signal from the detector is converted to digital signal and is recorded by the data acquisition computer. The control computer further permits the archival of the image data and the 3-D rendering of these images.

12.3.2 Two-Photon Fluorescent Probes and Their Biological Applications

The design of a multiphoton imaging experiment requires not only high-sensitivity optical instrumentation but also specific and efficient fluorophores. Most fluorophores can be excited in two-photon mode at approximately twice their one-photon absorption wavelength; similarly, n-photon excitation of fluorophores typically occurs at approximately n times their one-photon excitation wavelength. However, one- and multiphoton absorption processes actually have very different quantum mechanical selection rules. For example, a fluorophore's two-photon excitation spectrum scaled to half the wavelength is in general not equivalent to its one-photon excitation spectrum. Determining the multiphoton absorption spectra of fluorophores is required to optimize multiphoton imaging. Significant progress in this area has been made.[30–32]

Multiphoton imaging can be based on both endogenous and exogenous fluorophores. On one hand, endogenous fluorophores make multiphoton imaging easier by alleviating the difficulty of in vivo labeling of biological specimen, especially in thick tissues. On the other hand, multiphoton imaging based on endogenous fluorophores is difficult as endogenous probes typically have low extinction coefficients and low quantum efficiencies.

Considering imaging based on endogenous probes, most proteins are fluorescent due to the presence of tryptophan and tyrosine. Two-photon-induced fluorescence from tryptophan and tyrosine in proteins has been studied extensively.[33–35] While these fluorophores are occasionally used for multiphoton imaging, the application of these amino acid probes is not common due to the lack of efficient laser source for two-photon excitation of these probes. Endogenous β-nicotinamide adenine dinucleotide phosphate (NAD(P)H) is the most common endogenous fluorophores used for in vivo cellular imaging.[35] The production of NAD(P)H is associated with the cellular metabolism and the intracellular redox state.[36] Flavoproteins are another endogenous marker in cellular cytoplasm and are present almost exclusively in the mitochondria. The multiphoton excitation spectrum of NAD(P)H and flavin mononucleotide (FMN) has been determined.[31] Endogenous fluorophores are also present in the extracellular matrix; the primary fluorophores are collagen and elastin. Tropocollagen and collagen fibers typically have absorption spectra similar to amino acids; their fluorescent emission is typically weak in the excitation wavelength range of Ti:sapphire lasers.[37–39] Elastin fiber fluorescence, however, can be readily excited and visualized using two-photon excitation.[40–42] While fluorescence imaging based on collagen is difficult, recent studies have shown that collagen can be imaged based on second harmonic generation due to its chiral crystalline structure.[43–49]

Tissue imaging is one of the most important application areas of multiphoton microscopy. While cellular systems can be effectively labeled using standard organic dyes,[30–32] effective uniform labeling of tissues with exogenous organic probes is extremely difficult due to limited diffusion and differential partitioning in 3-D tissue components. The invention of fluorescence protein technology presents a powerful molecular biology solution to this dilemma.[50] By transfecting/inflecting cells and tissues with fluorescent protein vectors, one can monitor the activation of a particular gene, the trafficking of specific proteins, and the changes in local cellular biochemical environment. After the introduction of green fluorescent proteins, fluorescence proteins of a wide range of colors including blue, cyan, yellow, and red have been created.[51,52] Two-photon absorption spectra for many fluorescent proteins have been determined.[53,54]

While many existing one-photon fluorophores can be used for multiphoton imaging, they are not necessarily optimized. Optimizing two-photon absorption properties in fluorophores has two important consequences. The development of efficient fluorophores can reduce the excitation laser intensity required for imaging and thus can reduce system cost and specimen photodamage. Alternatively, with high two-photon cross sections, significant excitation can be achieved with the more economical continuous-wave lasers, thus reducing the cost of two-photon systems that typically use femtosecond Ti:sapphire lasers. Searches for molecules with high two-photon absorption cross sections have led to the discovery of semiconductor nanocrystals (quantum dots) as one of the best candidates for multiphoton microscopy. Some of these quantum dots have been proven to have two-photon cross sections as high as 47,000 GM.[55] Conjugated polymers are another attractive option because of high fluorescence quantum yields, large extinction coefficients, and efficient energy-transfer properties. Novel fabrication techniques have resulted in polymer nanoparticles that are smaller in size with a two-photon cross section comparable to that of quantum dots.[56] Recent searches for molecules with high two-photon absorption cross sections have led to the synthesis of molecules with two-photon cross sections over 1000 GM.[57]

12.4 Optimization of Multiphoton Microscopy for Deep Tissue Imaging

Deep tissue imaging is an area where multiphoton microscopy has found the most important and diverse applications. An understanding of how tissue optics affects light transmission and image formation properties is essential to optimize the use of multiphoton microscopy for deep tissue imaging. It is also important to consider photodamage mechanisms in tissues. Finally, some of the most promising deep tissue imaging applications of multiphoton microscopy will be reviewed.

12.4.1 Effect of Tissue Optical Properties on Multiphoton Microscopy Efficiency and Image Formation

On one level, light propagation through tissues can be characterized by multiple scattering of light in a homogeneous medium. In the multiple scattering regime, the modeling light propagation requires a knowledge of the scattering and absorption coefficients of the tissue. The scattering coefficient measures the propensity of the tissue to deflect photons from their path. The absorption coefficient measures the propensity of the tissue to absorb photons. To quantify scattering, in addition to the scattering coefficient, it is also important to quantify the directional change of the photon after scattering. The g-factor measures the cosine of the average angular change of the photon direction after scattering. Assuming small scattering and absorption coefficients, the light attenuation through a thickness of tissue can be modeled as an exponential function:

$$I(z) = I_0 e^{-[(1-g)\mu_s + \mu_a]z} \tag{12.6}$$

where

$I(z)$ is the intensity at depth z
I_0 is the intensity at tissue surface
μ_s is the scattering coefficient
μ_a is the absorption coefficient
g is the g-factor

The scattering coefficient, absorption coefficient, and the g-factor for typical tissues are 20–200 mm^{-1}, 0.1–1 mm^{-1}, and 0.7–0.9, respectively. Since the fluorescence signal depends quadratically on the excitation power, the fluorescence signal has a mean free path of 25–250 μm.

In a homogenous medium, these phenomenological parameters fairly well characterize the microscopic light–tissue interactions. However, real tissues are not homogenous and there are mesoscopic variations in tissue optical properties that can also have dramatic effects in light propagation. For example, dermal system consists of multiple structural layers, each with very distinct scattering coefficients, absorption coefficients, and g-factors. Further, these layers also have very different indices of refraction. An understanding of light propagation in tissues will require knowing the geometries of these mesoscopic inhomogeneities.

Effective multiphoton imaging of deep tissue requires efficient transmission of excitation light to the focal volume inside thick tissues and the efficient detection of the fluorescence signal generated. The imaging depth of two-photon microscopy can be limited by the efficacy of delivering femtosecond laser pulses into tissues. Assuming that the image PSF is invariant with depth, excitation light penetration is limited by power delivery and pulse broadening. In terms of power delivery, average laser power as a function of depth is an exponential function of depth and is governed by Equation 12.6. Similar to the attenuation of the excitation light, fluorescence signal generated is also reduced due to scattering and absorption effects. In fact, since scattering and absorption efficiencies are decreasing functions of wavelength, one expects attenuation of the fluorescence signal by the surrounding tissue to be more severe than the exciting light. As compared with confocal microscopy, multiphoton microscopy is more efficient in signal detection inside a multiple scattering medium. Since the mean free path of visible photons is less than 100 μm in typical tissues, the emitted photon will encounter one or more scattering events in thick tissues. However, many of these scattered photons can be collected due to the *g*-factors of most tissues that are close to unity (primarily forward scattering). In confocal microscopy, these scattered photons are lost at the confocal aperture. In two-photon microscopy, as long as the trajectories of these scattered photons remain within the collection solid angle of the objective lens, these photons are retained using a large area detector such as a PMT tube. Because these scattered photons have more divergent light paths, detection efficiency can be improved by increasing the solid angles of all optics in the emission light path and the physical size of the detector.[58]

From Equation 12.5, one concludes that efficient multiphoton excitation requires ultrafast laser pulses. Laser pulses from Ti:sapphire lasers, which are typically on the order of 150 fs, can be broadened in the sample reducing the multiphoton excitation efficiency. This degradation is particularly important for higher-order multiphoton processes. Laser pulse dispersion in microscope objectives is significant and pulse width can easily broaden to beyond 500 fs. The effect of pulse broadening in the objective can be negated by the use of dispersion compensation optics to maintain a short pulse width. However, further pulse dispersion may occur in the tissues due to scattering and index of refraction heterogeneity in tissues. The severity of this broadening effect as femtosecond pulses travel through tissues has not been extensively studied.

In addition to the reduction in excitation and detection efficiency, light propagation through tissues may also result in contrast and resolution losses. While signal strength reduction can be compensated by increasing laser power, as long as photodamage and photobleaching are not too severe, contrast and resolution loss can result in irreversible loss of image information content. The major factors that can be responsible for resolution degradation are the scattering of the excitation and emission light, and spherical aberration. Spherical aberration is caused by mismatch in the indices of refraction between the sample and the immersion medium of the objective.

Scattering of the excitation and the emission photons from the femtoliter excitation volume may lead to a degradation of the image resolution. To quantify this factor, Dunn et al. pioneered the study of the scattering effects on resolution degradation.[59] Using Monte Carlo simulations, Dunn and coworker showed that the loss of signal intensity was due more to the scattering of the excitation photons rather than the emission photons. More importantly, this group further found that the signal strength is the limiting factor in two-photon microscope image quality. Their simulation showed that resolution did not change as the image depth was increased as measured by the full width at half maximum (FWHM) of the image PSF. Their simulation was further supported by experimental measurements, and that is in agreement with Centonze and White.[60] Dong and coworkers[61] further measured two-photon microscopy PSF directly by imaging 100 nm spheres in agarose gels with varying scattering coefficients. It is found that scattering coefficient up to physiological range has minimal or no effect on both the lateral and axial FWHMs of the PSF for both oil- and water-immersion objectives (Figure 12.5).

Considering the effect of spherical aberration, Dong et al. showed that the water-immersion objective suffers no significant PSF degradation in thick samples. On the other hand, the oil-immersion objective suffered significant degradation due the spherical aberration introduced by the index of refraction mismatch. The lateral and axial FWHMs were broadened. One may conclude that for typical tissue thickness on the order of a few hundred microns, resolution loss due to scattering is insignificant but index refraction mismatch can be a major factor. The use of water-immersion objectives is preferred in biological tissues in general, given better index matching. However, Dong et al. also studied the performance of oil versus water immersion objectives for imaging more optically heterogeneous biological tissues such as excised human skin. Human skin is an optically heterogeneous layered structure with significantly different index of refraction in each layer. It is found that there is significant spherical aberration in the imaging of dermal structures for both oil- and water-immersion

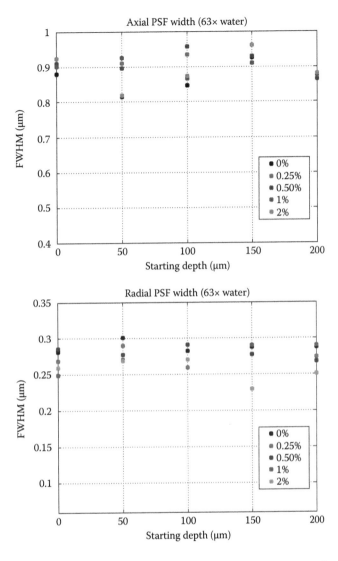

FIGURE 12.5 **(See color insert.)** Averaged widths of the radial and axial PSFs measured in multiple scattering media with different scattering coefficients containing Liposyn III at concentrations of 0%, 0.25%, 0.5%, 1%, and 2%. Data were taken with a Zeiss 63×, water-immersion, C-Apochromat objective (*NA* 1.2).

objectives. These objectives perform equivalently because neither of these objectives can match the varying index of refraction in the skin and both suffer from spherical aberration. For optically hetero-geneous samples, it is best to choose an immersion media that best matches the average index of the sample. For dermal structure, glycerine-immersion objective may potentially outperform either oil- or water-immersion lenses.

12.4.2 Photodamage Mechanisms in Tissues

Multiphoton microscopy is particularly suitable for in vivo tissue imaging due to its minimization of specimen photodamage. The region of photodamage is restricted to within the axial depth of the PSF. For a specimen with thickness comparable to the thickness of the PSF, the use of multiphoton imaging does not significantly improve specimen viability. However, the decrease in photodamage is substantial for 3-D imaging of thick tissues. Photodamage is approximately reduced by a factor equal to the ratio of the sample thickness to the axial dimension of the PSF. For 3-D imaging of a 100 μm thick specimen using multiphoton microscopy, the specimen is exposed to a light dosage about 100× less than confocal microscopy.

A number of embryology studies demonstrate well the minimal invasive nature of multiphoton microscopy. Long-term monitoring of *Caenorhabditis elegans* and hamster embryos using confocal microscopy have not been successful due to photodamage-induced developmental arrest. However, these developing embryos can be repeatedly imaged using multiphoton microscopy for hours without appreciable damage.[62-64] Further, the hamster embryos that were reimplanted after the imaging experiments were successfully developed into normal adults.

One should note that multiphoton microscopy can still cause considerable photodamage in the focal volume where photochemical interaction occurs. Multiphoton photodamage proceeds through three main mechanisms. (1) Oxidative photodamage can be caused by two- or higher-photon excitation of endogenous and exogenous fluorophores similar to ultraviolet exposure. Tissue fluorophores act as photosensitizers in the photo-oxidative process.[65,66] Photoactivation of these fluorophores results in the formation of reactive oxygen species that trigger the subsequent biochemical damage cascade in cells. Current studies found that the degree of photodamage follows a quadratic dependence on excitation power indicating that two-photon process is the primary damage mechanism.[67-71] Experiments have also been performed to measure the effect of laser pulse width on cell viability. Results indicate that the degree of photodamage is proportional to two-photon-excited fluorescence generation independent of pulse width. Therefore, using shorter pulse width for more efficient two-photon excitation also produces greater photodamage.[68,70] Flavin-containing oxidases have been identified as one of the primary endogenous targets for photodamage.[71] (2) One- and multiphoton absorption of the high-power infrared radiation can also produce thermal damage. The thermal effect resulting from two-photon absorption by water has been estimated to be on the order of 1 mK for typical excitation power.[16,72] Therefore, heating damage due to multiphoton water absorption is insignificant. However, there can be appreciable heating due to one-photon absorption in the presence of a strong infrared absorber such as melanin.[73,74] Thermal damage has been observed in the basal layer of human skin.[75] (3) Photodamage may also be caused by mechanisms resulting from the high peak power of the femtosecond laser pulses. There are indications that dielectric breakdown occasionally occurs.[67,76] In certain biological applications, this high peak power is highly advantageous in creating efficient submicrometer modifications in the biological components such as cell membrane or neurons. Perforation of the plasma membrane for in vitro transfection of cells with foreign DNA[77] or performing high-speed laser surgery in severing axons (axotomy) are two such examples.[78]

12.4.3 Tissue Level Applications of Two-Photon Microscopy

Two-photon microscopy is well suited for imaging in highly scattering specimens. A recent comparison study has convincingly demonstrated that two-photon microscopy is a superior method for this application.[60]

Multiphoton tissue imaging has been successfully applied to study the physiology of many tissue types such as the cornea structure of rabbit eyes,[75,79] the light-induced calcium signals in salamander retina,[80] the toxin effect on human intestinal mucosa,[81] and the metabolic processes of pancreatic islets.[82,83] The combination of multiphoton imaging and magnetic resonance imaging was used to investigate the heterogeneous microscopic structure of mammalian tongue tissue to obtain spatial information on two different length scales.[84]

Today, multiphoton microscopy is particularly widely used in three areas: neurobiology,[85] embryology, and dermatology. In neurobiology studies, multiphoton microscopy has been applied to study

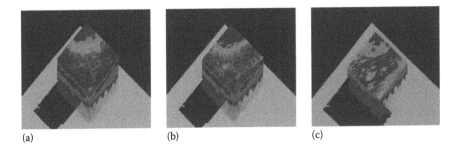

(a)   (b)   (c)

FIGURE 12.6 **(See color insert.)** 3-D reconstructed two-photon images of dermal structures in a mouse ear tissue specimen. The three images show distinct structural layers: (a) epidermal keratinocytes, (b) basal cells, and (c) collagen/elastin fibers.

the neuron structure and function in intact brain slices,[86] and to study the role of calcium signaling in dendritic spine function,[87-95] neural plasticity,[96] and hemodynamics[97] in living animals. Hyman and coworkers further applied this technology to study the formation of β-amyloid plaques associated with Alzheimer's disease.[98,99] In embryology studies, multiphoton imaging has been used to examine calcium passage during sperm–egg fusion,[100] the origin of bilateral axis in sea urchin embryos,[101] cell fusion events in *C. elegans* hypodermis,[62,63] and hamster embryo development.[64] Multiphoton microscopy has been used extensively in dermatology studies. Masters et al. employed multiphoton to image the autofluorescence of in vivo human skin down to a depth of 200 μm.[102] Cellular strata in the epidermis, including the corneum, the spinosum, and the basal layer, can be clearly resolved based on NAD(P)H fluorescence in the cytoplasm. NAD(P)H fluorescence is seen to be more pronounced in the basal layer due to its relatively higher metabolic activity. Below the epidermis, the dermal structure is visible due to elastin fluorescence and second-harmonic signal generated from the collagen matrix (Figure 12.6). Recently, two-photon microscopy has been further applied to study the transport properties of the skin,[103,104] facilitating the development of transdermal drug delivery technology.

12.5 Conclusion

Multiphoton fluorescence microscopy is becoming one of the most important recent inventions in biological imaging. This technology enables noninvasive study of biological specimen in three dimensions with submicron resolution. Two-photon excitation of fluorophores results from the simultaneous absorption of two photons. This excitation process has a number of unique advantages such as reduced specimen photodamage and enhanced penetration depth. It also produces higher-contrast images and is a novel method to trigger localized photochemical reactions. As the technology of multiphoton microscopy reaches maturity and becomes commercially available, the range of applications in biology and medicine is rapidly increasing. The suitability of multiphoton microscopy for studying the physiology of a variety of tissues has been clearly demonstrated. In the near future, other promising emerging applications of multiphoton microscopy include the use of multiphoton excitation for single molecular analysis for pharmaceutical drug discovery and the use of multiphoton processes for nanosurgery and nanoprocessing in cells and tissues.[105,106]

Acknowledgments

P.T.C. So acknowledges kind supports from the National Science Foundation grant MCB-9604382, the National Institutes of Health (NIH) grant R29GM56486-01, and the American Cancer Society grant RPG-98-058-01-CCE. C.Y. Dong acknowledges fellowship support from the NIH grant 5F32CA75736-02.

References

1. Göppert-Mayer, M., Uber Elementarakte mit zwei Quantensprungen, *Ann. Phys. (Leipzig)* 5, 273–294, 1931.
2. Kaiser, W. and Garrett, C. G. B., Two-photon excitation in $CaF_2:Eu^{2+}$, *Phys. Rev. Lett.* 7, 229–231, 1961.
3. Singh, S. and Bradley, L. T., Three-photon absorption in naphthalene crystals by laser excitation, *Phys. Rev. Lett.* 12, 162–164, 1964.
4. McClain, W. M., Excited state symmetry assignment through polarized two-photon absorption studies in fluids, *J. Chem. Phys.* 55, 2789, 1971.
5. Friedrich, D. M. and McClain, W. M., Two-photon molecular electronic spectroscopy, *Annu. Rev. Phys. Chem.* 31, 559–577, 1980.
6. Friedrich, D. M., Two-photon molecular spectroscopy, *J. Chem. Educ.* 59, 472, 1982.
7. Birge, R. R., Two-photon spectroscopy of protein-bound fluorophores, *Acc. Chem. Res.* 19, 138–146, 1986.
8. Hellwarth, R. and Christensen, P., Nonlinear optical microscopic examination of structures in polycrystalline ZnSe, *Opt. Commun.* 12, 318–322, 1974.
9. Sheppard, C. J. R., Kompfner, R., Gannaway, J., and Walsh, D., The scanning harmonic optical microscope, in *IEEE/OSA Conference on Laser Engineering and Applications*, Washington, DC, 1977, p. 100D.
10. Gannaway, J. N. and Sheppard, C. J. R., Second harmonic imaging in the scanning optical microscope, *Opt. Quant. Electron.* 10, 435–439, 1978.
11. Denk, W., Strickler, J. H., and Webb, W. W., Two-photon laser scanning fluorescence microscopy, *Science* 248(4951), 73–76, 1990.
12. Barad, Y., Eisenberg, H., Horowitz, M., and Silberberg, Y., Nonlinear scanning laser microscopy by third harmonic generation, *Appl. Phys. Lett.* 70(8), 922–924, 1997.
13. Squier, J. A., Muller, M., Brakenhoff, G. J., and Wilson, K. R., Third harmonic generation microscopy, *Opt. Exp.* 3(9), 315–324, 1998.
14. Campagnola, P. J., Wei, M.-D., Lewis, A., and Loew, L. M., High-resolution nonlinear optical imaging of live cells by second harmonic generation, *Biophys. J.* 77, 3341–3349, 1999.
15. Williams, R. M., Piston, D. W., and Webb, W. W., Two-photon molecular excitation provides intrinsic 3-dimensional resolution for laser-based microscopy and microphotochemistry, *FASEB J.* 8(11), 804–813, 1994.
16. Denk, W. J., Piston, D. W., and Webb, W. W., Two-photon molecular excitation laser-scanning microscopy, in *Handbook of Biological Confocal Microscopy*, 2 edn., Pawley, J. B. (ed.), Plenum Press, New York, 1995, pp. 445–458.
17. Hell, S. W., Nonlinear optical microscopy, *Bioimaging* 4, 121–123, 1996.
18. So, P. T., Dong, C. Y., Masters, B. R., and Berland, K. M., Two-photon excitation fluorescence microscopy, *Annu. Rev. Biomed. Eng.* 2, 399–429, 2000.
19. König, K., Multiphoton microscopy in life sciences, *J. Microsc.* 200(2), 83–104, 2000.
20. Helmchen, F. and Denk, W., New developments in multiphoton microscopy, *Curr. Opin. Neurobiol.* 12(5), 593–601, 2002.
21. Svoboda, K. and Yasuda, R., Principles of two-photon excitation microscopy and its applications to neuroscience, *Neuron* 50(6), 823–839, 2006.
22. Baym, G., *Lectures on Quantum Mechanics*, The Benjamin/Cummins Publishing Company, Menlo Park, CA, 1973.
23. Callis, P. R., The theory of two-photon induced fluorescence anisotropy, in *Nonlinear and Two-Photon-Induced Fluorescence*, Lakowicz, J. (ed.), Plenum Press, New York, 1997, pp. 1–42.
24. Sheppard, C. J. R. and Gu, M., Image formation in two-photon fluorescence microscope, *Optik* 86, 104–106, 1990.

25. Gu, M. and Sheppard, C. J. R., Comparison of three-dimensional imaging properties between two-photon and single-photon fluorescence microscopy, *J. Microsc.* 177, 128–137, 1995.

26. Denk, W., Two-photon scanning photochemical microscopy: Mapping ligand-gated ion channel distributions, *Proc. Natl. Acad. Sci. U.S.A.* 91(14), 6629–6633, 1994.

27. Xu, C., Shear, J. B., and Webb, W. W., Hyper-Rayleigh and hyper-Raman scattering background of liquid water in two-photon excited fluorescence detection, *Anal. Chem.* 69(7), 1285–1287, 1997.

28. Wokosin, D. L., Centonze, V. E., White, J., Armstrong, D., Robertson, G., and Ferguson, A. I., All-solid-state ultrafast lasers facilitate multiphoton excitation fluorescence imaging, *IEEE J. Sel. Top. Quant. Electron.* 2, 1051–1065, 1996.

29. Hell, S. W., Booth, M., and Wilms, S., Two-photon near- and far-field fluorescence microscopy with continuous-wave excitation, *Opt. Lett.* 23, 1238–1240, 1998.

30. Xu, C., Guild, J., Webb, W. W., and Denk, W., Determination of absolute two-photon excitation cross sections by in situ second-order autocorrelation, *Opt. Lett.* 20(23), 2372–2374, 1995.

31. Xu, C. and Webb, W. W., Measurement of two-photon excitation cross sections of molecular fluorophores with data from 690 to 1050 nm, *J. Opt. Soc. Am. B* 13(3), 481–491, 1996.

32. Xu, C., Zipfel, W., Shear, J. B., Williams, R. M., and Webb, W. W., Multiphoton fluorescence excitation: New spectral windows for biological nonlinear microscopy, *Proc. Natl. Acad. Sci. U.S.A.* 93(20), 10763–10768, 1996.

33. Lakowicz, J. R. and Gryczynski, I., Tryptophan fluorescence intensity and anisotropy decays of human serum albumin resulting from one-photon and two-photon excitation, *Biophys. Chem.* 45, 1–6, 1992.

34. Lakowicz, J. R., Kierdaszuk, B., Callis, P., Malak, H., and Gryczynski, I., Fluorescence anisotropy of tyrosine using one- and two-photon excitation, *Biophys. Chem.* 56, 263–271, 1995.

35. Kierdaszuk, B., Malak, H., Gryczynski, I., Callis, P., and Lakowicz, J. R., Fluorescence of reduced nicotinamides using one- and two-photon excitation, *Biophys. Chem.* 62, 1–13, 1996.

36. Masters, B. R. and Chance, B., Redox confocal imaging: Intrinsic fluorescent probes of cellular metabolism, in *Fluorescent and Luminescent Probes for Biological Activity*, Mason, W. T. (ed.), Academic Press, London, U.K., 1999, pp. 361–374.

37. LaBella, F. S. and Gerald, P., Structure of collagen from human tendon as influence by age and sex, *J. Gerontol.* 20, 54–59, 1965.

38. Dabbous, M. K., Inter- and intramolecular cross-linking in tyrosinase-treated tropocollagen, *J. Biol. Chem.* 241, 5307–5312, 1966.

39. Hoerman, K. C. and Balekjian, A. Y., Some quantum aspects of collagen, *Fed. Proc.* 25, 1016–1021, 1966.

40. LaBella, F. S., Studies on the soluble products released from purified elastic fibers by pancreatic elastase, *Arch. Biochem. Biophys.* 93, 72–79, 1961.

41. Thomas, J., Elsden, D. F., and Partridge, S. M., Degradation products from elastin, *Nature* 200, 651–652, 1963.

42. LaBella, F. S. and Lindsay, W. G., The structure of human aortic elastin as influence by age, *J. Gerontol.* 18, 111–118, 1963.

43. Stoller, P., Kim, B. M., Rubenchik, A. M., Reiser, K. M., and Da Silva, L. B., Polarization-dependent optical second-harmonic imaging of a rat-tail tendon, *J. Biomed. Opt.* 7(2), 205–214, 2002.

44. Theodossiou, T., Rapti, G. S., Hovhannisyan, V., Georgiou, E., Politopoulos, K., and Yova, D., Thermally induced irreversible conformational changes in collagen probed by optical second harmonic generation and laser-induced fluorescence, *Lasers Med. Sci.* 17(1), 34–41, 2002.

45. Campagnola, P. J., Clark, H. A., Mohler, W. A., Lewis, A., and Loew, L. M., Second-harmonic imaging microscopy of living cells, *J. Biomed. Opt.* 6(3), 277–286, 2001.

46. Theodossiou, T., Georgiou, E., Hovhannisyan, V., and Yova, D., Visual observation of infrared laser speckle patterns at half their fundamental wavelength, *Lasers Med. Sci.* 16(1), 34–39, 2001.

47. Kim, B. M., Eichler, J., Reiser, K. M., Rubenchik, A. M., and Da Silva, L. B., Collagen structure and nonlinear susceptibility: Effects of heat, glycation, and enzymatic cleavage on second harmonic signal intensity, *Lasers Surg. Med.* 27(4), 329–335, 2000.

48. Freund, I., Deutsch, M., and Sprecher, A., Connective tissue polarity: Optical second-harmonic microscopy, crossed-beam summation, and small-angle scattering in rat-tail tendon, *Biophys. J.* 50(4), 693–712, 1986.

49. Roth, S. and Freund, I., Optical second-harmonic scattering in rat-tail tendon, *Biopolymers* 20(6), 1271–1290, 1981.

50. Chalfie, M., Tu, Y., Euskirchen, G., Ward, W. W., and Prasher, D. C., Green fluorescent protein as a marker for gene expression, *Science* 263, 802–805, 1994.

51. Tsien, R. Y., The green fluorescent protein, *Annu. Rev. Biochem.* 67, 509–544, 1998.

52. Baird, G. S., Zacharias, D. A., and Tsien, R. Y., Circular permutation and receptor insertion within green fluorescent proteins, *Proc. Natl. Acad. Sci. U.S.A.* 96(20), 11241–11246, 1999.

53. Larson, D. R., Zipfel, W. R. et al., Water-soluble quantum dots for multiphoton fluorescence imaging in vivo, *Science* 300(5624), 1434–1436, 2003.

54. Rahim, N. A. A., McDaniel, W. et al., Conjugated polymer nanoparticles for two photon imaging of endothelial cells in a tissue model, *Adv. Mater.* 21(34), 3492–3496, 2009.

55. Niswender, K. D., Blackman, S. M., Rohde, L., Magnuson, M. A., and Piston, D. W., Quantitative imaging of green fluorescent protein in cultured cells: Comparison of microscopic techniques, use in fusion proteins and detection limits, *J. Microsc.* 180(Pt 2), 109–116, 1995.

56. Potter, S. M., Wang, C. M., Garrity, P. A., and Fraser, S. E., Intravital imaging of green fluorescent protein using two-photon laser-scanning microscopy, *Gene* 173, 25–31, 1996.

57. Albota, M., Beljonne, D., Bredas, J.-L., Ehrlich, J. E., Fu, J.-Y., Heikal, A. A., Hess, S. E. et al., Design of organic molecules with large two-photon absorption cross sections, *Science* 281(5383), 1653–1656, 1998.

58. Oheim, M., Beaurepaire, E., Chaigneau, E., Mertz, J., and Charpak, S., Two-photon microscopy in brain tissue: Parameters influencing the imaging depth, *J. Neurosci. Methods* 111(1), 29–37, 2001.

59. Dunn, A. K., Wallace, V. P., Coleno, M., Berns, M. W., and Tromberg, B. J., Influence of optical properties on two-photon fluorescence imaging in turbid samples, *Appl. Opt.* 39, 1194–1201, 2000.

60. Centonze, V. E. and White, J. G., Multiphoton excitation provides optical sections from deeper within scattering specimens than confocal imaging, *Biophys. J.* 75(4), 2015–2024, 1998.

61. Dong, C.-Y., Koenig, K., and So, P., Characterizing point spread functions of two-photon fluorescence microscopy in turbid medium, *J. Biomed. Opt.* 8(3), 450–459, 2003.

62. Mohler, W. A., Simske, J. S., Williams-Masson, E. M., Hardin, J. D., and White, J. G., Dynamics and ultrastructure of developmental cell fusions in the *Caenorhabditis elegans* hypodermis, *Curr. Biol.* 8(19), 1087–1090, 1998.

63. Mohler, W. A. and White, J. G., Stereo-4-D reconstruction and animation from living fluorescent specimens, *Biotechniques* 24(6), 1006–1010, 1012, 1998.

64. Squirrell, J. M., Wokosin, D. L., White, J. G., and Bavister, B. D., Long-term two-photon fluorescence imaging of mammalian embryos without compromising viability, *Nat. Biotechnol.* 17(8), 763–767, 1999.

65. Keyse, S. M. and Tyrrell, R. M., Induction of the heme oxygenase gene in human skin fibroblasts by hydrogen peroxide and UVA (365 nm) radiation: Evidence for the involvement of the hydroxyl radical, *Carcinogenesis* 11(5), 787–791, 1990.

66. Tyrrell, R. M. and Keyse, S. M., New trends in photobiology. The interaction of UVA radiation with cultured cells, *J. Photochem. Photobiol. B* 4(4), 349–361, 1990.

67. Konig, K., So, P. T. C., Mantulin, W. W., Tromberg, B. J., and Gratton, E., Two-photon excited lifetime imaging of autofluorescence in cells during UVA and NIR photostress, *J. Microsc.* 183(Pt 3), 197–204, 1996.

68. König, K., So, P. et al., Two-photon excited lifetime imaging of autofluorescence in cells during UVA and NIR photostress, *J. Microsc.* 183(Pt 3), 197, 1996.

69. Tirlapur, U. K. and König, K., Targeted transfection by femtosecond laser, *Nature* 418(6895), 290–291, 2002.
70. Yanik, M. F., Cinar, H. et al., Neurosurgery: Functional regeneration after laser axotomy, *Nature* 432(7019), 822, 2004.
71. Konig, K., Becker, T. W., Fischer, P., Riemann, I., and Halbhuber, K.-J., Pulse-length dependence of cellular response to intense near-infrared laser pulses in multiphoton microscopes, *Opt. Lett.* 24(2), 113–115, 1999.
72. Sako, Y., Sekihata, A., Yanagisawa, Y., Yamamoto, M., Shimada, Y., Ozaki, K., and Kusumi, A., Comparison of two-photon excitation laser scanning microscopy with UV-confocal laser scanning microscopy in three-dimensional calcium imaging using the fluorescence indicator Indo-1, *J. Microsc.* 185(Pt 1), 9–20, 1997.
73. Koester, H. J., Baur, D., Uhl, R., and Hell, S. W., Ca²⁺ fluorescence imaging with pico- and femtosecond two-photon excitation: Signal and photodamage, *Biophys. J.* 77(4), 2226–2236, 1999.
74. Hockberger, P. E., Skimina, T. A., Centonze, V. E., Lavin, C., Chu, S., Dadras, S., Reddy, J. K., and White, J. G., Activation of flavin-containing oxidases underlies light-induced production of H_2O_2 in mammalian cells, *Proc. Natl. Acad. Sci. U.S.A.* 96(11), 6255–6260, 1999.
75. Schonle, A. and Hell, S. W., Heating by absorption in the focus of an objective lens, *Opt. Lett.* 23(5), 325–327, 1998.
76. Jacques, S. L., McAuliffe, D. J., Blank, I. H., and Parrish, J. A., Controlled removal of human stratum corneum by pulsed laser, *J. Invest. Dermatol.* 88(1), 88–93, 1987.
77. Pustovalov, V. K., Initiation of explosive boiling and optical breakdown as a result of the action of laser pulses on melanosome in pigmented biotissues, *Kvantovaya Elektronika* 22(11), 1091–1094, 1995.
78. Buehler, C., Kim, K. H., Dong, C. Y., Masters, B. R., and So, P. T. C., Innovations in two-photon deep tissue microscopy, *IEEE Eng. Med. Biol. Mag.* 18(5), 23–30, 1999.
79. Piston, D. W., Masters, B. R., and Webb, W. W., Three-dimensionally resolved NAD(P)H cellular metabolic redox imaging of the in situ cornea with two-photon excitation laser scanning microscopy, *J. Microsc.* 178(Pt 1), 20–27, 1995.
80. Denk, W. and Detwiler, P. B., Optical recording of light-evoked calcium signals in the functionally intact retina, *Proc. Natl. Acad. Sci. U.S.A.* 96(12), 7035–7040, 1999.
81. Riegler, M., Castagliuolo, I., So, P. T., Lotz, M., Wang, C., Wlk, M., Sogukoglu, T. et al., Effects of substance P on human colonic mucosa in vitro, *Am. J. Physiol.* 276(6 Pt 1), G1473–G1483, 1999.
82. Bennett, B. D., Jetton, T. L., Ying, G., Magnuson, M. A., and Piston, D. W., Quantitative subcellular imaging of glucose metabolism within intact pancreatic islets, *J. Biol. Chem.* 271(7), 3647–3651, 1996.
83. Piston, D. W., Knobel, S. M., Postic, C., Shelton, K. D., and Magnuson, M. A., Adenovirus-mediated knockout of a conditional glucokinase gene in isolated pancreatic islets reveals an essential role for proximal metabolic coupling events in glucose-stimulated insulin secretion, *J. Biol. Chem.* 274(2), 1000–1004, 1999.
84. Napadow, V. J., Chen, Q., Mai, V., So, P. T. C., and Gilbert, R. J., Quantitative analysis of three-dimensional-resolved fiber architecture in heterogeneous skeletal muscle tissue using NMR and optical imaging methods, *Biophys. J.* 80(6), 2968–2975, 2001.
85. Fetcho, J. R. and O'Malley, D. M., Imaging neuronal networks in behaving animals, *Curr. Opin. Neurobiol.* 7(6), 832–838, 1997.
86. Denk, W., Delaney, K. R., Gelperin, A., Kleinfeld, D., Strowbridge, B. W., Tank, D. W., and Yuste, R., Anatomical and functional imaging of neurons using 2-photon laser scanning microscopy, *J. Neurosci. Methods* 54(2), 151–162, 1994.
87. Yuste, R. and Denk, W., Dendritic spines as basic functional units of neuronal integration, *Nature* 375(6533), 682–684, 1995.
88. Yuste, R., Majewska, A., Cash, S. S., and Denk, W., Mechanisms of calcium influx into hippocampal spines: Heterogeneity among spines, coincidence detection by NMDA receptors, and optical quantal analysis, *J. Neurosci.* 19(6), 1976–1987, 1999.

89. Denk, W., Sugimori, M., and Llinas, R., Two types of calcium response limited to single spines in cerebellar Purkinje cells, *Proc. Natl. Acad. Sci. U.S.A.* 92(18), 8279–8282, 1995.

90. Svoboda, K., Denk, W., Kleinfeld, D., and Tank, D. W., In vivo dendritic calcium dynamics in neocortical pyramidal neurons, *Nature* 385(6612), 161–165, 1997.

91. Svoboda, K., Helmchen, F., Denk, W., and Tank, D. W., Spread of dendritic excitation in layer 2/3 pyramidal neurons in rat barrel cortex in vivo, *Nat. Neurosci.* 2(1), 65–73, 1999.

92. Helmchen, F., Svoboda, K., Denk, W., and Tank, D. W., In vivo dendritic calcium dynamics in deep-layer cortical pyramidal neurons, *Nat. Neurosci.* 2(11), 989–996, 1999.

93. Shi, S. H., Hayashi, Y., Petralia, R. S., Zaman, S. H., Wenthold, R. J., Svoboda, K., and Malinow, R., Rapid spine delivery and redistribution of AMPA receptors after synaptic NMDA receptor activation, *Science* 284(5421), 1811–1816, 1999.

94. Mainen, Z. F., Malinow, R., and Svoboda, K., Synaptic calcium transients in single spines indicate that NMDA receptors are not saturated, *Nature* 399(6732), 151–155, 1999.

95. Maletic-Savatic, M., Malinow, R., and Svoboda, K., Rapid dendritic morphogenesis in CA1 hippocampal dendrites induced by synaptic activity, *Science* 283(5409), 1923–1927, 1999.

96. Engert, F. and Bonhoeffer, T., Dendritic spine changes associated with hippocampal long-term synaptic plasticity, *Nature* 399(6731), 66–70, 1999.

97. Kleinfeld, D., Mitra, P. P., Helmchen, F., and Denk, W., Fluctuations and stimulus-induced changes in blood flow observed in individual capillaries in layers 2 through 4 of rat neocortex, *Proc. Natl. Acad. Sci. U.S.A.* 95(26), 15741–15746, 1998.

98. Christie, R. H., Bacskai, B. J., Zipfel, W. R., Williams, R. M., Kajdasz, S. T., Webb, W. W., and Hyman, B. T., Growth arrest of individual senile plaques in a model of Alzheimer's disease observed by in vivo multiphoton microscopy, *J. Neurosci.* 21(3), 858–864, 2001.

99. Bacskai, B. J., Kajdasz, S. T., Christie, R. H., Carter, C., Games, D., Seubert, P., Schenk, D., and Hyman, B. T., Imaging of amyloid-beta deposits in brains of living mice permits direct observation of clearance of plaques with immunotherapy, *Nat. Med.* 7(3), 369–372, 2001.

100. Jones, K. T., Soeller, C., and Cannell, M. B., The passage of Ca^{2+} and fluorescent markers between the sperm and egg after fusion in the mouse, *Development* 125(23), 4627–4635, 1998.

101. Summers, R. G., Piston, D. W., Harris, K. M., and Morrill, J. B., The orientation of first cleavage in the sea urchin embryo, *Lytechinus variegatus*, does not specify the axes of bilateral symmetry, *Dev. Biol.* 175(1), 177–183, 1996.

102. Masters, B. R., So, P. T., and Gratton, E., Multiphoton excitation fluorescence microscopy and spectroscopy of in vivo human skin, *Biophys. J.* 72(6), 2405–2412, 1997.

103. Yu, B., Dong, C. Y., So, P. T. C., Blankschtein, D., and Langer, R., In vitro visualization and quantification of oleic acid induced changes in transdermal transport using two-photon fluorescence microscopy, *J. Invest. Dermatol.* 117(1), 16–25, 2001.

104. Grewal, B. S., Naik, A., Irwin, W. J., Gooris, G., de Grauw, C. J., Gerritsen, H. G., and Bouwstra, J. A., Transdermal macromolecular delivery: Real-time visualization of iontophoretic and chemically enhanced transport using two-photon excitation microscopy, *Pharm. Res.* 17(7), 788–795, 2000.

105. Tirlapur, U. K. and Konig, K., Femtosecond near-infrared laser pulse induced strand breaks in mammalian cells, *Cell Mol. Biol. (Noisy-le-grand)* 47 Online Pub, OL131–OL134, 2001.

106. Konig, K., Multiphoton microscopy in life sciences, *J. Microsc.* 200(Pt 2), 83–104, 2000.

107. Wilson, T. and Sheppard, C. J. R., *Theory and Practice of Scanning Optical Microscopy*, Academic Press, New York, 1984.

108. Kim, D., Ultrafast optical pulse manipulation in three dimensional-resolved microscope imaging and microfabrication, Ph.D., Department of Mechanical Engineering, Massachusetts Institute of Technology, Cambridge, MA, 2009.

13

Laser Doppler Perfusion Monitoring and Imaging

Tomas Strömberg
Linköping University

Karin Wårdell
Linköping University

Marcus Larsson
Linköping University

E. Göran Salerud
Linköping University

13.1 Introduction

Microcirculation—including capillaries, small arteries (arterioles), small veins (venules), and shunting vessels (arteriovenous anastomosis)—comprises the blood vessels of the most peripheral part of the vascular tree [1]. In the skin, the microvascular network is composed of different compartments, each with a different anatomy and function. The most superficial layer of the skin, the epidermis, is avascular, while the dermal papillae host the capillaries that are mainly responsible for the exchange of oxygen and metabolites with its surrounding tissue. Therefore, the blood perfusion through the capillaries is frequently referred to as the nutritive blood flow. Although this nutritive blood flow is low or irregular during resting conditions, it is of vital importance for maintaining the minute metabolic requirements of the skin. In the deeper dermal structures, the arterioles, venules, and shunting vessels reside. The main role of these vessels is to feed and drain the capillary network by adjusting their peripheral resistance and to promote the maintenance of an adequate body temperature by dissipating heat to the environment through the modulation of shunting vessels.

Microvascular blood perfusion possesses both temporal fluctuations and spatial variability. Temporal fluctuations may be either rhythmic, as in the case of vasomotion [2–4], or more stochastic [5], and are of either local or central origin, while spatial variability originates in the anatomical heterogeneity of the microvascular network [6]. Perfusion of blood in the skin is influenced not only by external factors, such as heat and topically applied vasoactive substances, but also by the intake of beverages and drugs as well as by smoking and mental stimuli. In addition, microvascular blood perfusion is regulated via the autonomous nervous system and through vasoactive agents released by the endocrine glands into the blood stream.

A number of common diseases—including various inflammatory conditions, allergic reactions, and tumors—influence the blood perfusion. In addition, impaired circulation in association with diabetes and peripheral vascular disease may lead to the formation of ulcers and ultimately to tissue necrosis. Consequently, the perfusion of blood in the skin as well as in other life-sustaining tissues needs to be investigated by the use of noninvasive methods having a minimal influence on the tissue under study. In addition, such methods should preferably record the vital status of the microcirculatory perfusion in real time and be applicable both in experimental and clinical settings. In order to evaluate both the dynamics and spatial variability of microcirculatory perfusion, the methods need to record in more than one dimension.

Many methods for the assessment of microvascular blood perfusion have been presented in the literature, including photoelectric plethysmography [7,8], thermal [9] and radioisotope [10,11] clearance, orthogonal polarization spectral imaging [12,13], laser speckle contrast analysis (LASCA) [14], optical Doppler tomography (ODT) [15], and video-photometric capillaroscopy [16,17]. However, no single methods have or can fulfill the requirements of noninvasiveness and applicability to a large number of tissues as well as offering the possibility of recording the tissue perfusion continuously or visualizing the results in terms of perfusion images. Laser Doppler flowmetry (LDF) is the technology having the best opportunity to fulfill those requirements.

The first measurements of microvascular blood flow employing the Doppler shift of monochromatic light (named after the Austrian scientist Johan Christian Doppler [18]) were reported by Riva et al. [19], who studied the blood cell velocity in the rabbit eye. Some years later, the blood velocity in exposed microvessels was investigated using a laser Doppler microscope [20,21]. In vivo evaluation of skin microcirculation by the use of coherent light scattering was first demonstrated by Stern [22]. In his setup, in which a He–Ne laser beam illuminates the skin, a portion of the laser light is scattered both by moving blood cells and by static tissue structures, while the remaining light is elastically scattered by static tissue structures alone. If the backscattered light illuminates the surface of a photodetector, a fluctuating speckle pattern will appear as a consequence of the optical mixing of waves with slightly different frequencies. These fluctuations are observed in the detected photocurrent signal as small superimposed waves with a frequency-range similar to the audio-frequency spectrum. The analysis of the photocurrent revealed a clear shift of the spectral content toward lower frequencies following the occlusion of the brachial artery by use of a pressure cuff placed around the upper arm and with the laser beam directed toward the fingertip. This fundamental discovery was soon implemented in a portable fiber-optic-based laser Doppler flowmeter [23], providing results of assessing tissue blood perfusion comparable with those obtained by the ^{133}Xe-clearance technique.

As this early prototype was used more extensively, laser-mode interference noise proved to be a major limitation of the system. A solution to this problem was to introduce a differential detector system [24], rejecting the common-mode modal interference signal while amplifying the sum of the uncorrelated blood perfusion–related signals from each detector. A signal processor using the first unnormalized moment of the photocurrent as a continuous output signal was introduced [25], and a theory that describes quasi-elastic light scattering in perfused tissue was proposed by Bonner and Nossal [26]. In the context of LDF, this quantity is generally referred to as the perfusion [27]. Although LDF does not selectively record the movement of red blood cells (RBCs), it is assumed that the RBCs in the microcirculation represent the major influence on the LDF perfusion.

In the early 1980s, laser Doppler perfusion monitoring (LDPM) was introduced as a diagnostic tool and became available to a wide range of researchers and clinicians. The ease of use of this new technology quickly created an extensive interest in recording microvascular perfusion in a wide variety of disciplines, and to date, several thousands of publications in the scientific literature cite its use [28–30].

One of the most remarkable features of microvascular blood perfusion is its substantial temporal and spatial variability [2,5]. Consequently, the comparatively small sampling volume (about 1 mm³) of fiber-optic-based LDPM instruments constitutes one of the handling aspects of this technology, especially if perfusion estimates of large tissue areas or organs are to be predicted. The large spatial variability in tissue blood perfusion creates gross differences in perfusion readings even for recordings at juxtaposed sites. This disadvantage of LDPM started the development of multichannel instruments [31] and the laser Doppler perfusion imaging (LDPI) technology at the end of the 1980s [32–34]. In LDPI, an airborne laser beam successively probes a large number of adjacent tissue volumes, and a data set of perfusion values, a perfusion image, is generated and displayed as pseudo color-coded. The commercialization phase of the LDPI technology started in the early 1990s, and several hundred publications have already verified its usefulness, particularly in dermatology [35], wound healing [36], and burns treatment [37].

LDPM and LDPI constitute a versatile pair of related noninvasive medical technologies that facilitate the study of both the temporal and spatial variabilities of tissue blood perfusion. Both technologies are easy to use, but the design of a study and interpretation of the results require the establishment of a well-prepared protocol and a thorough understanding of microcirculation as well as the influencing factors. Consequently, guidelines have been developed to assist the safe and reliable use of both LDPM [38] and LDPI [39].

13.2 Theory

Coherent light directed toward a tissue will be scattered by moving objects and by static tissue structures as the photons migrate through the tissue in a random pattern. Scattering in moving objects (predominantly RBCs) changes the frequency of the light according to the Doppler principle, while light scattered in static structures alone remains unshifted in frequency (Figure 13.1). If the backscattered light is detected by a photodetector, optical mixing of light shifted and unshifted in frequency will result in a stochastic photocurrent with a power spectral density depending on the number of RBCs as well as on their shape and the velocity distribution within the measurement volume. In the following section, we will present the classical theory [40], derived in the time domain, which relates the photocurrent power spectrum $P(\omega)$ to the properties of the blood cells in the illuminated volume, and an alternative theory, derived in the frequency domain, which relates $P(\omega)$ to the optical spectrum containing the distribution of Doppler shifts.

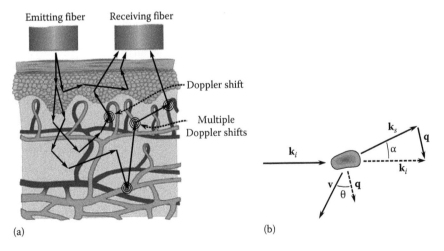

FIGURE 13.1 (a) In general, most photons have been scattered many times only in static tissue. Some photons are scattered once in an RBC and obtain a single Doppler shift; others experience multiple Doppler shifts. (b) The single scattering event.

13.2.1 Single Scattering Event

The starting point of this derivation is the interaction of a single photon and a scattering object, generally referred to as a *single scattering event* (Figure 13.1). If the incident electromagnetic wave $\mathbf{E}_i$ propagating in the direction $\mathbf{k}_i$ is denoted $\mathbf{E}_i = E_{i0}e^{i(\omega t + \mathbf{k}_i \mathbf{r})}$, where E_{i0} is the amplitude, ω is the angular frequency, t is the time, and r represents the spatial coordinates, the scattered wave $\mathbf{E}_s$ in the direction of $\mathbf{k}_s$ can be written in the form

$$\mathbf{E}_s = E_{s0}e^{i((\omega - \mathbf{k}_i\mathbf{v})t - \mathbf{k}_i\mathbf{r}_0 - \mathbf{k}_s(\mathbf{r}_1 - \mathbf{v}t))} = E_{s0}e^{i\omega t}e^{-i(\mathbf{k}_i - \mathbf{k}_s)\mathbf{v}t}e^{-i(\mathbf{k}_i\mathbf{r}_0 + \mathbf{k}_s\mathbf{r}_1)}, \tag{13.1}$$

where

E_{s0} is the amplitude of the scattered wave

$\mathbf{v}$ is the particle velocity

$e^{-i(\mathbf{k}_i\mathbf{r}_0 + \mathbf{k}_s\mathbf{r}_1)}$ is a time-invariant phase factor

Defining the scattering vector as $\mathbf{q} = \mathbf{k}_i - \mathbf{k}_s$ and neglecting the time-invariant phase factor, the scattered wave can be written as being composed of the product of the carrier frequency factor and the time-variant phase factor according to

$$\mathbf{E}_s = E_{s0}e^{i\omega t}e^{-i\mathbf{q}\mathbf{v}t}. \tag{13.2}$$

If the scattering object is stationary ($\mathbf{v} = 0$), if ($\mathbf{q} = 0$), or if $\mathbf{v}$ and $\mathbf{q}$ are perpendicular, the wave remains unshifted in frequency. Since $|\mathbf{k}_i| \approx |\mathbf{k}_s|$, the magnitude of the scattering vector equals $|\mathbf{q}| = 2|\mathbf{k}|\sin(\alpha/2) = 4\pi/\lambda_t \sin(\alpha/2)$, where α is the scattering angle and λ_t the wavelength in tissue. The frequency-determining argument of the time-variant phase factor $\mathbf{q}\mathbf{v} = qv\cos(\theta)$ constitutes the angular Doppler frequency ω_D, which amounts to $\omega_D = 4\pi/\lambda_t \sin(\alpha/2)v\cos(\theta)$, where θ is the angle between $\mathbf{v}$ and $\mathbf{q}$.

13.2.2 Detection

When a laser beam impinges on the tissue surface, the individual photons migrate through the tissue matrix in a random pattern to a depth determined by the optical properties of the tissue. This migration process implies a multitude of single scattering events with or without associated frequency shifts (Figure 13.1). A portion of all photons injected return to the tissue surface, where they become available for detection. This backscattered light is Doppler-broadened by an amount dependent on the average individual RBC velocity and scattering vectors. Therefore, the expectation value of the time-variant phase factor $\langle e^{-i\mathbf{q}\mathbf{v}\tau} \rangle$ according to Equation 13.2 needs to be determined in order to calculate the tissue perfusion.

When the backscattered light impinges on the surface of a photodetector, a fluctuating speckle pattern is formed. The frequency and magnitude of intensity fluctuations in the individual coherence areas of this speckle pattern are related to the average speed and number of moving objects within the scattering volume. In the photodetector, the instantaneous light intensity is converted to a photocurrent. If the average and the fluctuating portions of the photocurrent produced by coherence area j are denoted as $\langle i_c \rangle$ and $\Delta i_{cj}(t)$, respectively, the total photocurrent $i(t)$ proportional to the instantaneous light intensity can be expressed as [41]

$$i(t) = \sum_{j=1}^{K} \left(\langle i_c \rangle + \Delta i_{cj}(t) \right), \tag{13.3}$$

where K is the total number of coherence areas on the detector surface and all noise factors are neglected. Since $i(t)$ is a stochastic process, its characteristics may be described by the autocorrelation function

$$\left\langle i(0)i^*(\tau)\right\rangle = \sum_{j=1}^{K}\left(\left\langle i_c\right\rangle + \Delta i_{cj}(0)\right)\sum_{l=1}^{K}\left(\left\langle i_c^*\right\rangle + \Delta i_{cl}^*(\tau)\right)$$

$$= \sum_{j=1}^{K}\sum_{l=1}^{K}\left\langle i_c\right\rangle\left\langle i_c^*\right\rangle + \sum_{j=1}^{K}\sum_{l=1}^{K}\Delta i_{cj}(0)\Delta i_{cl}^*(\tau) \tag{13.4}$$

$$= K^2\left\langle i_c^2\right\rangle + K\left\langle\Delta i_c(0)\Delta i_c^*(\tau)\right\rangle,$$

where $\Delta i_c(\tau)$ represents the average fluctuating portion of the photocurrent produced by a single coherence area. Equation 13.4 is valid under the assumption that the fluctuating portions of the photocurrents produced by the individual coherence areas are independent, that is, all elements $\Delta i_{cj}(0)\Delta i_{cl}^*(\tau)$ are nonzero only for $j = l$, and each fluctuating component has an average value equal to zero. From Equation 13.4, it can be concluded that the Doppler-related portion of the photocurrent autocorrelation function scales linearly with the number of coherence areas, while the stationary part scales with the square of the number of coherence areas. Furthermore, the performance of the entire system is uniquely described by the statistics of the photocurrent produced by a single coherence area.

Since the photodetector is a square-law detector, the photocurrent produced by a single coherence area is related to the electromagnetic field $\mathbf{E}(t)$ as $i_c(t) \propto \mathbf{E}(t)\mathbf{E}^*(t)$. Hence, the autocorrelation of the photocurrent produced by a single coherence area can be expressed as

$$\left\langle i_c(0)i_c^*(\tau)\right\rangle \propto \left\langle \mathbf{E}(0)\mathbf{E}^*(0)\mathbf{E}(\tau)\mathbf{E}^*(\tau)\right\rangle, \tag{13.5}$$

where the $\mathbf{E}$-vector represents the sum of the electromagnetic field vectors impinging on the actual coherence area. This sum can be split up into two parts, of which one is the sum of all field vectors representing photons that have undergone a Doppler shift and the other is the sum of the field vectors representing photons with no frequency shift. Evaluating Equation 13.5 thus results in 16 terms, of which 10 have an expectation value that equals zero because their fluctuating parts are independent and uncorrelated. Three terms include components that are time invariant and thus represent the average photocurrent produced by the coherence area in question. Two terms represent *heterodyne* mixing of beams shifted in frequency with beams not shifted in frequency, while one term represents *homodyne* mixing of beams that have all undergone at least one frequency shift. The overall autocorrelation function of the photocurrent produced by a single coherence area can therefore be written in the form

$$\left\langle i_c(0)i_c^*(\tau)\right\rangle = \underbrace{2i_{Re}i_{Sc} + i_{Re}^2 + i_{Sc}^2}_{\text{Stationary}} + \underbrace{i_{Re}i_{Sc}\left(\left\langle e^{i\mathbf{q}\mathbf{v}\tau}\right\rangle + \left\langle e^{-i\mathbf{q}\mathbf{v}\tau}\right\rangle\right)}_{\text{heterodyne-mixing}} + \underbrace{i_{Sc}^2\left\langle\frac{1}{S^2}\sum_{k=1,k\neq l}^{S}\sum_{l=1}^{S}e^{i(\mathbf{q}_k\mathbf{v}_k - \mathbf{q}_l\mathbf{v}_l)\tau}\right\rangle}_{\text{homodyne-mixing}}, \tag{13.6}$$

where
   i_{Re} and i_{Sc} represent the average currents produced, respectively, by photons not shifted in frequency and those shifted in frequency within a single coherence area
   S denotes the total number of photons with Doppler shifts

$$\left\langle e^{i\mathbf{q}\mathbf{v}\tau}\right\rangle = \left\langle (1/S)\sum_{k=1}^{S}e^{i(\mathbf{q}_k\mathbf{v}_k\tau)}\right\rangle$$

The relationship between i_{Re} and i_{Sc} and the RBC concentration (c_{RBC}), as a tissue fraction, can be obtained from the path length distribution ($N(\rho)$) for detected photons and the scattering coefficient of RBCs ($\mu_{s,RBC}$). The Beer–Lambert law gives the probability P_{Sc} for at least one Doppler shift for a photon with path length ρ:

$$P_{Sc} = \left(1 - e^{-c_{RBC}\mu_{s,RBC}\rho}\right).$$

(13.7)

This gives $i_{Re} = \int N(\rho)e^{-\mu_s c_{RBC}\rho}d\rho$ and $i_{Sc} = \int N(\rho)(1 - e^{-\mu_s c_{RBC}\rho})d\rho$ If $\mu_s c_{RBC}\rho$ is small, $i_{Re} = \int N(\rho)d\rho$ and $i_{Sc} = \mu_{s,RBC}c_{RBC}\int N(\rho)\rho d\rho = \mu_{s,RBC}c_{RBC}\langle\rho\rangle i_{Re}$. Hence, if $\mu_{s,RBC}c_{RBC}\langle\rho\rangle$ is small, the homodyne term in Equation 13.6 may be disregarded, which has been assumed using the classical LDF theory. However, the validity of this assumption has been questioned [42].

13.2.3 LDF Theory in the Time Domain

We will show that the quantity $\int \omega P(\omega)d\omega$ is proportional to c_{RBC} and the average scalar velocity $\langle v\rangle$, while $\int P(\omega)d\omega$ is proportional to c_{RBC} alone. The assumptions are as follows: (1) c_{RBC} is low (only heterodyne detection) and (2) uniformly distributed in the tissue (only single Doppler shifts), (3) the 3D RBC velocity distribution $N_0(\mathbf{v})$ is independent of spatial coordinates, (4) individual RBCs can move independently. This derivation starts with the calculation of $\langle e^{i\mathbf{qv}\tau}\rangle_{\mathbf{v}}$ for the actual 3D velocity distribution $N_0(\mathbf{v})$. Then the expectation value of this quantity $\langle\langle e^{i\mathbf{qv}\tau}\rangle_{\mathbf{v}}\rangle_{\mathbf{q}}$ with respect to the scattering vector distribution $S_0(\mathbf{q}(\alpha))$ is calculated. According to the Wiener–Khintchine theorem [43], the Fourier transform of $\langle\langle e^{i\mathbf{qv}\tau}\rangle_{\mathbf{v}}\rangle_{\mathbf{q}}$ gives the photocurrent power spectral density. These successive calculations allow the recorded power spectral density to be linked to the RBC velocity distribution for the actual laser wavelength.

13.2.3.1 Derivation of $\langle e^{i\mathbf{qv}\tau}\rangle_{\mathbf{v}}$

The expectation value of $e^{i\mathbf{qv}t}$ with reference to $\mathbf{v}$ for a fixed value of $\mathbf{q}$ can be calculated according to

$$\left\langle e^{i\mathbf{qv}\tau}\right\rangle_{\mathbf{v}} = \int_{\mathbf{v}} N_0(\mathbf{v})e^{i\mathbf{qv}\tau}d\mathbf{v}.$$

(13.8)

With the 1D velocity distribution $N(\mathbf{v}) = 4\pi\mathbf{v}^2 N_0(\mathbf{v})$ (assumption 3) and transformation to spherical coordinates, Equation 13.8 can be written in the form

$$\left\langle e^{i\mathbf{qv}\tau}\right\rangle_{\mathbf{v}} = \int_{\mathbf{v}=0}^{\infty} N(\mathbf{v})\frac{\sin(\mathbf{qv}\tau)}{\mathbf{qv}\tau}d\mathbf{v}.$$

(13.9)

13.2.3.2 Derivation of $\langle\langle e^{i\mathbf{qv}\tau}\rangle_{\mathbf{v}}\rangle_{\mathbf{q}}$

The expectation value of $\langle e^{i\mathbf{qv}\tau}\rangle_{\mathbf{v}}$ with reference to $\mathbf{q}$ can be calculated according to

$$\left\langle\left\langle e^{i\mathbf{qv}\tau}\right\rangle_{\mathbf{v}}\right\rangle_{\mathbf{q}} = \int_{\alpha=0}^{\pi} \left\langle e^{i\mathbf{qv}\tau}\right\rangle_{\mathbf{v}} S_0(\mathbf{q}(\alpha))d\alpha.$$

(13.10)

Assuming that $S_0(\mathbf{q}(\alpha))$ possesses circular symmetry with respect to $\mathbf{k}_i$, the 1D scattering vector distribution $S(\mathbf{q}(\alpha))$ may be written in the form

$$S(\mathbf{q}(\alpha)) = \frac{2S_0(\mathbf{q}(\alpha))}{\sin(\alpha)}. \tag{13.11}$$

Inserting Equations 13.9 and 13.11 in Equation 13.10 yields

$$\left\langle \left\langle e^{i\mathbf{q}\mathbf{v}\tau} \right\rangle_{\mathbf{v}} \right\rangle_{\mathbf{q}} = \int\limits_{\alpha=0}^{\pi}\int\limits_{v=0}^{\infty} N(\mathbf{v})\frac{\sin(\mathbf{q}\mathbf{v}\tau)}{2\mathbf{q}\mathbf{v}\tau}S(\mathbf{q}(\alpha))\sin(\alpha)dvd\alpha. \tag{13.12}$$

Equation 13.12 relates the RBC 1D velocity ($N(\mathbf{v})$) and scattering vector ($S(\mathbf{q}(\alpha))$ distributions to the measurable photocurrent autocorrelation function produced by the heterodyne mixing term in Equation 13.6.

13.2.3.3 Power Spectral Density

The power spectral density $P_c(\omega)$ of the photocurrent produced by the heterodyne mixing term in a single coherence area can thus be calculated by applying the Wiener–Khintchine theorem to Equation 13.6, neglecting the stationary and the homodyne mixing terms:

$$P_c(\omega) \propto c_{RBC}i_{Re}^2 \int\limits_{\mathbf{v}=0}^{\infty} N(\mathbf{v}) \int\limits_{\alpha=0}^{\pi} \sin(\alpha)S(\mathbf{q}(\alpha)) \int\limits_{\tau=-\infty}^{\infty} \frac{\sin(\mathbf{q}\mathbf{v}\tau)}{\mathbf{q}\mathbf{v}\tau}e^{i\omega\tau}d\tau d\alpha d\mathbf{v}. \tag{13.13}$$

Using the fact that $\mathbf{q}=4\pi/\lambda_t \sin(\alpha/2)$, Equation 13.13 can, by way of pure mathematical manipulations, be reduced to

$$P_c(\omega) \propto c_{RBC}i_{Re}^2 \int\limits_{\mathbf{v}=(\lambda_t\omega/4\pi)}^{\mathbf{v}_{max}} \frac{N(\mathbf{v})}{\mathbf{v}} \int\limits_{\mathbf{q}=(\omega/\mathbf{v})}^{4\pi/\lambda_t} S(\mathbf{q})d\mathbf{q}d\mathbf{v}. \tag{13.14}$$

After normalization with the total light intensity in order to render the output signal independent of the power of the laser and taking into account that stationary components scale with K^2, while the Doppler-related portion of the signal scales with K, according to Equation 13.4, the total photocurrent power spectral density $P(\omega)$ corresponding to the heterodyne mixing term can be written in the form

$$P(\omega) \propto \frac{c_{RBC}}{K} \int\limits_{\mathbf{v}=(\lambda_t\omega/4\pi)}^{\mathbf{v}_{max}} \frac{N(\mathbf{v})}{\mathbf{v}} \int\limits_{\mathbf{q}=\omega/\mathbf{v}}^{4\pi/\lambda_t} S(\mathbf{q})d\mathbf{q}d\mathbf{v}. \tag{13.15}$$

Integration limits of Equation 13.14 indicate that in order for an RBC single scattering event to contribute to the photocurrent power spectral density at the angular frequency ω, the associated scattering vector $\mathbf{q}$ needs to have a magnitude of at least $\mathbf{q} = \omega/\mathbf{v}$. In addition, the magnitude of the RBC velocity needs to be higher than $\mathbf{v} = \lambda_t\omega/4\pi$.

As a final step, we can now calculate $\int \omega^n P(\omega)d\omega$ and demonstrate that this quantity scales linearly with the RBC concentration for $n = 0$ and the product of the RBC speed and concentration for $n = 1$. An arbitrary velocity distribution can be written as the sum of its individual parts $N(\mathbf{v}) = \sum_{\mathbf{v}_0} N_{\mathbf{v}_0} \delta(\mathbf{v} - \mathbf{v}_0)$. The contribution to $\int \omega^n P(\omega) d\omega$ from $N(\mathbf{v}) = N_{\mathbf{v}_0} \delta(\mathbf{v} - \mathbf{v}_0)$ can then be written in the form

$$
\begin{aligned}
\int \omega^n P(\omega) d\omega &\propto \frac{c_{RBC}}{K} \int_{\omega=0}^{\infty} \omega^n \int_{v=\lambda_t \omega/4\pi}^{v_{max}} \frac{N(\mathbf{v})}{\mathbf{v}} \int_{q=\omega/v}^{4\pi/\lambda_t} S(\mathbf{q}) d\mathbf{q}\, dv\, d\omega \\
&= \frac{c_{RBC} N_{\mathbf{v}_0}}{K} \int_{\omega=0}^{\infty} \omega^n \int_{v=\lambda_t \omega/4\pi}^{\infty} \frac{\delta(\mathbf{v}-\mathbf{v}_0)}{\mathbf{v}} \int_{q=\omega/v}^{4\pi/\lambda_t} S(\mathbf{q}) d\mathbf{q}\, dv\, d\omega \\
&= \frac{c_{RBC} N_{\mathbf{v}_0}}{K} \int_{x=0}^{\infty} (x \mathbf{v}_0)^n \int_{v=\lambda_t x \mathbf{v}_0/4\pi}^{\infty} \frac{\delta(\mathbf{v}-\mathbf{v}_0)\mathbf{v}_0}{\mathbf{v}} \int_{q=x\mathbf{v}_0/v}^{4\pi/\lambda_t} S(\mathbf{q}) d\mathbf{q}\, dv\, dx \\
&= \frac{c_{RBC} N_{\mathbf{v}_0} \mathbf{v}_0^n}{K} \underbrace{\int_{x=0}^{\infty} x^n \int_{q=x}^{4\pi/\lambda_t} S(\mathbf{q}) d\mathbf{q}\, dx}_{\text{constant}},
\end{aligned}
\tag{13.16}
$$

and the contribution from the complete velocity spectrum yields

$$
\int \omega^n P(\omega) d\omega \propto \frac{c_{RBC}}{K} \underbrace{\int_{v=0}^{\infty} \mathbf{v}^n N(\mathbf{v}) d\mathbf{v}}_{\langle \mathbf{v}^n \rangle} \propto c_{RBC} \langle \mathbf{v}^n \rangle.
\tag{13.17}
$$

This processor algorithm scales with the RBC concentration ($n = 0$) and with the product of RBC concentration and speed ($n = 1$) for an arbitrary velocity distribution.

13.2.4 LDF Theory in the Frequency Domain

The analogous processing of the laser Doppler signal has proven successful since the early 1980s. However, with the introduction of digital signal processing, new analysis schemes have been proposed. To fully understand the basic ideas of these approaches, it is favorable to describe the origin and processing of the laser Doppler signal in the frequency domain. According to Forrester [44] and Larsson [45], the power spectral density $P(\omega)$ of the photocurrent produced by a single coherence area is given by

$$
P(\omega) = \int_{-\infty}^{\infty} I(\beta) I(\beta + \omega) d\beta = (I * I)(\omega),
\tag{13.18}
$$

where
  * denotes cross correlation
  $I(\beta)$ is the optical spectrum containing the distribution of Doppler shifts

This equation provides a direct and intuitive link between the detected signal and the Doppler shifts occurring when moving RBCs scatter light, without any approximations regarding homodyne and/or heterodyne detection. Several authors have demonstrated the validity of these equations by comparing measurements on well-known optical phantoms with computer-based Monte Carlo simulations [46–50].

Introducing f_D, the fraction of Doppler-shifted photons, the optical spectrum can be expressed as

$$I(\beta) = I_{tot}\left((1 - f_D)\delta(\beta - \beta_c) + f_D H(\beta)\right), \tag{13.19}$$

where
   δ is the Dirac delta function
   β_c is the laser frequency
   $H(\beta)$ is the frequency distribution of the detected Doppler-shifted photons
   $I_{tot} = \int I(\beta)d\beta$ is the total amount of detected photons

For frequencies $\omega > 0$, Equation 13.18 becomes

$$P_D(\omega) = QI_{tot}^2 \left(\underbrace{2f_D(1 - f_D)H(\beta_c + \omega)}_{\text{heterodyne}} + \underbrace{f_D^2 \int_{-\infty}^{\infty} H(\beta)H(\beta + \omega)d\omega\,d\beta}_{\text{homodyne}} \right), \tag{13.20}$$

where Q is a constant that mainly depends on instrument amplification and detector sensitivity, coherence length, and the number of detected speckles. Based on Equation 13.20, the LDF concentration ($n = 0$) and perfusion ($n = 1$) can be expressed as

$$\frac{\int_0^{\infty} \omega^n P(\omega)d\omega}{I_{tot}^2} = \langle \omega^n \rangle \frac{\int_0^{\infty} P(\omega)d\omega}{I_{tot}^2}$$

$$= Q\langle \omega^n \rangle \left(2f_D - f_D^2\right). \tag{13.21}$$

Accordingly, the LDF perfusion ($n = 1$) will depend linearly on the RBC flow velocity since $\langle \omega^n \rangle \propto \langle \mathbf{v}_{RBC}^n \rangle$. As demonstrated earlier, the fraction of Doppler-shifted photons can be approximated with $f_D = c_{RBC}\mu_{s,RBC}\langle \rho \rangle$ for $f_D \ll 1$ (i.e., close to zero). For these levels of Doppler-shifted photons, Equation 13.21 becomes

$$\frac{\int_0^{\infty} \omega^n P(\omega)d\omega}{I_{tot}^2} \approx 2Q\langle \omega^n \rangle f_D \approx 2Q\langle \omega^n \rangle c_{RBC}\mu_{s,RBC}\langle \rho \rangle \propto \langle \omega^n \rangle c_{RBC}. \tag{13.22}$$

This demonstrates how the LDF concentration and perfusion measures scale with the tissue fraction of moving RBCs. For real tissue, however, f_D often becomes too large for this assumption to hold true. In this regime, the LDF concentration measure rapidly becomes nonlinear, displaying an almost $1 - \exp(-2c_{RBC}\mu_{s,RBC}\langle \rho \rangle)$ relationship to the RBC tissue fraction. The LDF perfusion measure exhibits a significantly higher degree of linearity to the tissue fraction of RBC as the degree of multiple Doppler shifts (affects $\langle \omega^n \rangle$ but not $\langle \mathbf{v}_{RBC}^n \rangle$), which increases with f_D, has a counteracting effect.

The aforementioned theory contains simplifications that can be overcome by using Monte Carlo simulations. This technique provides a tool for calculating the detected optical spectrum $I(\beta)$ based on knowledge on the tissue optical properties, the distribution of RBCs, and the RBC velocities. To make use of this technique in a real application, the inverse problem needs to be solved, that is, calculating the RBC concentration and velocity based on a detected LDF power spectrum $P(\omega)$. By using presimulated data, Fredriksson et al. have demonstrated how this can be achieved [46,51–54]. This approach does not only allow for estimating the tissue perfusion and RBC concentration in a known sampling volume but also allow for resolving the tissue perfusion in a few velocity regions. By assuming that the capillaries mainly contain low-velocity RBCs, the capillary blood flow and the arteriolar/venular can be studied separately.

13.2.5 Measurement Volume

In LDF, there is a distinct difference between measurement and sampling volume. Whereas the sampling volume only represents the volume where detected photons have propagated ($H(\mathbf{r})$), the measurement volume also takes into account the spatial distribution of blood concentration ($c_{RBC}(\mathbf{r})$) and flow velocity ($\mathbf{v}(\mathbf{r})$). The output signal from the laser Doppler flowmeter has been proposed to be [27]

$$\int_{\mathbf{r}^3} H(\mathbf{r})\mathbf{v}(\mathbf{r})c_{RBC}(\mathbf{r})d\mathbf{r}^3, \tag{13.23}$$

where $H(\mathbf{r})$ can be estimated for a specific combination of probe design and tissue optical properties by use of Monte Carlo simulation. This allows for a derivation of the median sampling depth [55]. However, Equation 13.23 fails to correctly include multiple Doppler shifts as well as speckle effects present in LDPI systems. To overcome this, Fredriksson et al. [52] have suggested an improved analysis scheme based on Monte Carlo simulated data. Their work includes a thorough analysis of measurement depths and volumes for different types of tissues, probe configurations, and LDPI setups.

13.3 Instrumentation

This section describes the instrumentation principles of LDPM and LDPI. In most LDPM devices, the light is brought to and from the tissue by optical fibers. LDPM is well suited for continuous real-time monitoring of tissue blood perfusion at a single site. In LDPI, an airborne laser beam scans the tissue and successively records the tissue perfusion from a number of sites; from these data, a 2D image of the microvascular perfusion is generated. LDPI is primarily intended for studies of the spatial variability of tissue blood perfusion.

13.3.1 Laser Doppler Perfusion Monitoring

13.3.1.1 First Experimental Setup

Following the demonstration of backscattered Doppler-broadened light from a perfused tissue [22], the first prototype of a laser Doppler flowmeter was constructed [56]. This early experimental setup utilized as light source a 15 mW He–Ne laser, emitting temporally coherent light at 632.8 nm. The large laser and a photomultiplier tube were positioned about 1 m above the tissue. In order to limit stray light, the returning light passed through a 2 mm aperture positioned at the surface of the tissue. A 0.5 mm pinhole was placed in front of the detector in such a way that only a single coherence area was selected. A flow parameter was obtained by the use of an algorithm calculating the root mean square (rms) bandwidth of the unnormalized second moment of the Doppler signal power spectral density. With measurements on skin subjected to ultraviolet-induced erythema, a linear relationship with a first-order correlation coefficient of 88% between this flow parameter and values obtained by the ^{133}Xe-clearance technique could be demonstrated.

13.3.1.2 LDPM Devices

Although the early experimental prototypes worked well in the laboratory, they proved to be too bulky for use in the clinical environment. LDPM devices, in which optical fibers guided the light to and from the tissue and a semiconductor diode replaced the photomultiplier tube, were therefore developed [23]. In LDPM, step-index or graded-index optical fibers are used as light guides. In modern equipment, these fibers have a core diameter ranging from 50 to 250 µm. The light transmitting fiber guides the light to the probe, attached to the tissue by double adhesive tape. One or more fibers pick up a portion of the backscattered and Doppler-broadened light and bring it back to the photodetector unit in the instrument. In the probe tip, the fibers are typically positioned with a core center spacing of 0–500 µm. In the tissue, the photons migrate in random pathways from the transmitting to the receiving fiber. Widening the distance between the fiber tips in the probe tends to increase the average path length of the detected photons as well as the measurement depth [55]. Both of those quantities are dependent on the optical properties of the tissue. Consequently, perfusion recorded in different organs cannot be directly compared, and no universal absolute flow unit has so far been presented for LDF applications. Depending on the intended use of the instrument, the probes may be designed for attachment to the tissue surface or as small-diameter needle probes for the recording of deep muscle perfusion [57,58]. Since skin perfusion is especially sensitive to the ambient temperature, probe holders incorporating a thermostatic element that allows the skin temperature to be clamped to a preset value have been developed [59].

Incorporating optical fibers as light guides between the tissue and the device, however, generally implies that the number of coherence areas on the photodetector surface increases and renders the system less coherent. According to Equation 13.6, this reduces the ratio of the photocurrent produced by the Doppler signal to the total light intensity by an amount proportional to the number of coherence areas. When fiber-optic-based devices were first tested, the superimposition of large spikes on the output signal at regular intervals significantly distorted the measurement. Closer analysis [60] identified optical mixing of longitudinal laser modes on the detector surface, giving rise to audio-range frequencies sweeping through the bandwidth of the device, as the origin of the problem. Since these signals are of common-mode origin, their adverse effect on the signal-to-noise ratio increases with the number of coherence areas. This problem was overcome by the introduction of a differential detector system [25] that effectively suppressed the common-mode signals while not affecting the difference of the uncorrelated Doppler signals from the two detectors. Reducing the diameter of the fiber renders the detected signal more coherent and suppresses the influence of laser-mode artifacts [61] as well as the adverse effects of fiber motion artifacts [59,62–64].

As the LDPM technology matured, a diode laser replaced the He–Ne laser as the light source [65,66], making a range of wavelengths available. The most commonly used wavelength today (780 nm) offers a slightly deeper penetration as well as reduced dependence on skin color and is also close to the isobestic points, thus eliminating the dependence on oxygen saturation [27]. In order to avoid beat notes in the Doppler signal caused by mode interference, the diode lasers need to be thermally stabilized.

In the signal processing unit, a high-pass filter with a cutoff frequency set to 20 Hz extracts the Doppler signal from the average photocurrent and pulse-synchronous light intensity fluctuations. This signal is amplified and normalized by the average photocurrent in order to make it independent of the light intensity. By use of the algorithm derived in Equation 13.18 (with $n = 1$), an output signal that scales with perfusion can be implemented by calculating the instantaneous power after feeding the normalized Doppler signal through an ω-weighing filter. The upper bandwidth limit is generally set to a value between 3 and 23 kHz [27], depending on the expected highest RBC velocity. This output signal can be fed to a recorder or a display for continuous tracking of the perfusion or to a computer for further postprocessing analysis [67–69]. In order to obtain an output signal that scales linearly with perfusion in tissues with such a high RBC concentration that the multiple generation of Doppler shifts cannot be neglected, a linearizer circuit may be utilized [70]. In this linearizer circuit, the output signal, implemented according to Equation 13.18 ($n = 1$), is multiplied by a compensation factor derived directly from the RBC concentration signal (Equation 13.18 with $n = 0$).

Double wavelength LDPM devices have been designed with the aim of penetrating different tissue depths and discriminating between superficial and global blood perfusion [71]. Green laser light

(543 nm), with a short penetration depth, has been tested especially with regard to its ability to selectively record nutritive blood flow, but with negative results [72]. Modulation of the average measurement depth can also be attained by widening the distance between the transmitting and receiving fibers [73], but no clear indication has been obtained as to whether this approach can discriminate between capillary and global tissue perfusion [74]. In order to suppress the adverse effect of the high spatial resolution of LDPM standard probes, integrating probes that record the average blood perfusion have been designed and tested with positive results [75]. The use of laser diodes integrated in the probe with photodetectors further eliminates the need for fiber optics and makes the device less sensitive to movement artifacts [66].

13.3.1.3 Recording Tissue Perfusion

In their present design, LDPM devices are generally better suited for measurements where the probe is attached to a fixed tissue position than for measurements at multiple sites with intermittent movement of the probe [59]. The reason for this constraint is that moving the probe from one skin site to another may disturb the microcirculation under study but also that rotational effects of the probe may reveal a different recording site in comparison to the previous expected, therefore creating high spatial variability effects. A typical registration demonstrating (1) the influence of a temporary occlusion of forearm skin perfusion, (2) the resulting hyperemic response following release of the occlusion, and (3) the slow increase in perfusion caused by heating the skin may serve as an illustrative example (Figure 13.2).

The recording in Figure 13.2a was performed in accordance with the following recommended protocol [38]. Prior to recording, the test subject refrains from smoking and from the intake of food and beverages for 2 h and rests in a relaxed position for at least 20 min. The preparations start with positioning the probe in a probe holder attached to the skin and by placing a pressure cuff around the upper arm. Initially, the resting perfusion is recorded for about a minute. When the pressure cuff is inflated to above systolic pressure, the perfusion value is reduced to a stable level representing *biological zero*—which does not generally coincide with the zero output level of the instrument [76]. The predominant contribution to biological zero has been demonstrated to be Brownian motion arising from interstitial tissue [77]. When the pressure in the cuff is again released, the perfusion rapidly and intermittently increases to a peak value above the resting perfusion level. This phenomenon, generally referred to as *reactive hyperemia*, represents a rapid flow of blood into the fully dilated microvascular bed. The perfusion value then returns to preocclusion levels. A successive increase in skin perfusion is then recorded as the skin temperature increases (Figure 13.2b).

In the interpretation of the results of this recording, the biological zero has generally been taken as the zero perfusion level [78,79] from which the actual perfusion is calculated. Because the LDPM output signal is expressed in arbitrary units, it is generally better to present recorded alterations in blood

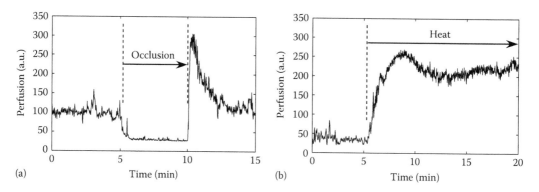

FIGURE 13.2 LDPM perfusion recordings on healthy forearm skin during occlusion (a) and local heating to 44°C (b), respectively.

perfusion as percentages of the resting value. In many situations, the latency time from the onset of stimuli to the appearance of the peak perfusion value may provide useful information [80]. Most analysis and interpretation of the LDPM output signal try to identify amplitude and temporal changes related to some physiological context, although more advanced signal processing on the LDPM output signal includes ordinary spectral analysis [81], wavelet [67,82–84], fractal dimensions [85], and empirical mode decomposition [86] techniques.

13.3.2 Laser Doppler Perfusion Imaging

13.3.2.1 From Monitoring to Imaging

When repetitive recordings in adjacent sites on a tissue with seemingly homogenous perfusion are made, identical perfusion values are generally not recorded. By comparing biopsies with results obtained using LDPM, Braverman et al. [6] demonstrated that the heterogeneous skin perfusion pattern recorded by LDPM at adjacent sites coincides with the underlying microvascular architecture. High, pulsatile perfusion values were found directly over ascending arterioles, whereas low values correlated with areas containing only capillaries and postcapillary venules. Topographic maps generated by successively moving the LDPM probe in 1 mm steps further demonstrated that high-perfusion spots correlated well with the presence of arterioles and low-perfusion areas with avascular zones [87]. To visualize this spatial variability, tissue perfusion needs to be either investigated by use of a multipoint LDPM system [31] or presented as a 2D flow image [88], rather than by a curve trace displaying temporal variations at a single site alone.

13.3.2.2 LDPI Devices

In LDPI, 2D mapping of tissue blood perfusion is performed by moving the laser beam in a raster scan over a predetermined area with no need for physical contact with the tissue (Figure 13.3). The light is guided from a low-power laser operating in the visible range to the tissue via a moving mirror system. A portion of the backscattered, Doppler-broadened light is brought back via the same mirror system to detectors placed in the proximity of the laser. This arrangement facilitates automatic focusing on and tracking of the laser spot as the beam moves over the tissue surface, thereby effectively suppressing the adverse effects of ambient stray light. Further rejection of ambient light is attained by utilizing a differential detector technology similar to that used in LDPM. In LDPI, both the laser and detector units are placed in a camera-like scanner head positioned 0.1–1 m above the tissue surface.

Two fundamentally different LDPI systems—one of which utilizes a stepwise scanning beam [32,34], while the other is based on a continuously moving beam [33]—have been presented in the literature.

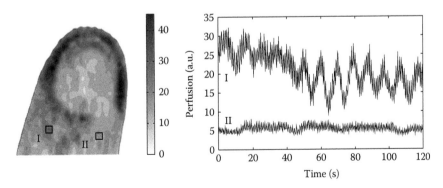

FIGURE 13.3 LDPI image of a fingertip with continuous recordings made at two skin sites. The continuous recordings comprise the average of four adjacent sites. The image is interpolated from a 40 × 40 image. (From Wårdell, K. and Nilsson, G.E., *Microvasc. Res.*, 52(2), 171, 1996. With permission.)

In stepwise scanning LDPI, the beam is stopped at each measurement site, while the backscattered, Doppler-broadened light is sampled. This arrangement avoids the interference caused by a laser beam moving in relation to the tissue, which may generate Doppler components not related to tissue perfusion. The exact duration of the beam stop is determined by the trade-off between image integrity and maximum permissible time for capturing the image. If the beam is arrested for 50 ms at each measurement site, a 64 × 64 measurement site image is generally captured in about 4 min.

In the LDPI system based on a continuously moving beam, the backscattered Doppler-broadened light is captured *on the fly*. This method allows more points to be sampled per unit time, and a full-format image may typically include 256 × 256 true measurement sites. The relative motion between the laser beam and the tissue when the backscattered light is sampled may, however, result in the generation of non-perfusion-related Doppler components and other intensity fluctuations in the backscattered light [33]. Those Doppler components are generally of low frequency, and their adverse influence on the perfusion signal can be reduced by increasing the lower cutoff frequency of the instrument bandwidth or by reducing the scanning speed. Increasing the lower cutoff frequency, however, also tends to reject the low-frequency Doppler components of the blood perfusion-related signal [89], while increased scanning speed distorts the image [90] and reduced scanning speed prolongs the image-capturing process.

In contrast to the situation in LDPM, with the use of LDPI devices, the number of coherence areas and thus the inherent amplification factor change as the distance between the detector and object is altered. This difficulty has been overcome by introducing as a light source a slightly divergent laser beam maintaining a fixed solid angle under which the laser spot on the tissue is seen from the detector, thereby keeping the number of coherence areas constant, according to Equation 13.4 [34]. It has been demonstrated empirically that the distance-dependent amplification factor of LDPI systems utilizing a collimated beam can also be reduced by using an algorithm that normalizes the perfusion signal with the light intensity raised to the power of β, where β is close to unity (rather than 2, as suggested by Equation 13.14 and 13.20).

LDPI systems utilize a signal processor similar to that in LDPM and can generate full-format images of objects from the size of a fingertip to a complete torso.

The intensity of the backscattered light can be used for discriminating the object from its background. Therefore, whenever possible, the object should be placed on a light-absorbing background to facilitate this discrimination process.

In LDPI, all parameters controlling image size, resolution, scanning speed, background threshold, etc., are set in the system software. The image can generally be displayed in an individual color scale, in which the lowest and highest perfusion values constitute the end points of the scale. This option is probably the best choice for color representation if only the relative variation in perfusion within a single image is to be investigated. If several images are to be compared, they have to be set to the same color scale before a comparison is made. Basic image-processing routines are available in the LDPI system software packages, and the images can readily be exported to more sophisticated packages for further analysis as needed.

13.3.2.3 Monitoring Mode

One important feature of LDPI is the capability of operating the system in a monitoring mode. In this mode, a small image comprising, for example, 4 × 4 measurement sites is continuously updated, and the result can be displayed either as a series of mini-images or as a time trace continuously displaying the average perfusion (Figure 13.3). This feature makes the system less dependent on the exact measurement site than single-point LDPM [91].

13.3.2.4 High-Resolution LDPI

Lateral resolution of standard LDPI is limited by the effective diameter of the laser beam (approximately 0.6–0.8 mm) and by the step length in the stepwise scanning LDPI. By keeping the distance between the scanner head and the tissue constant, focusing the laser beam onto one spot on the tissue surface, and reducing the step length, the resolution can be significantly increased and full-format images of tissue sites less than 1 cm² generated [92]. This arrangement also reduces the solid angle under which the laser spot is seen on the tissue surface, thereby increasing the average size of a coherence area on the detector surface and the inherent system amplification factor.

13.3.2.5 Recording an Image

In addition to the procedure recommended when using LDPM, the following precautions should be noted when recording a perfusion image by LDPI. The subject needs to be placed in a comfortable and relaxing position during image recording in order to avoid unintentional influence on microcirculation. If the subject moves during image recording, the image will have isolated horizontal stripes, indicating a movement artifact. Likewise, temporal fluctuations on a time scale shorter than the image recording time will result in the appearance of horizontal perfusion stripes in the image with no correspondence in the microvascular architecture. A liquid film on the skin surface may cause direct reflections of the laser light to the detector at a single spot, mimicking a (falsely) reduced tissue perfusion. The analysis of LDPI images fall into two categories: either a field of view (FOV), represented of a complete image or subimage, is selected and the mean or median and deviation parameters are calculated and compared with other recordings when the physiology is manipulated, or the recorded image is exported to an ordinary image-processing software [93] for more sophisticated analysis.

13.3.3 Performance Check and Calibration

The algorithm for LDF devices based on ω-weighing filtering of the power spectral density of the unnormalized Doppler signal was derived under assumptions that are not necessarily fulfilled in in vitro and in vivo situations. Therefore, flow simulators, with separate controls for the speed and concentration of RBCs or other scatterers, and in vivo models have been used in the evaluation of the processor algorithm.

Most flow simulator experiments demonstrate that a linear relationship between RBC speed and LDF output signal can be obtained, provided that compatibility with the instrument bandwidth [25,64] is maintained. In the evaluation of the bandwidth requirements, however, it must be borne in mind that the size of the moving scatterer substantially influences the scattering angle distribution and thus the Doppler shift generated for a given speed according to Equation 13.15. For low c_{RBC}, the probability for multiple Doppler shift generation and homodyne mixing is low. This implies that the LDF output signal also scales linearly with c_{RBC} [25]. A flow simulator based on rotating discs with embedded scatterers has been constructed [94] and used to assess the influence of optical properties and separation of the transmitting and receiving fiber on measurement depth [95].

In vivo calibration involves comparison of results obtained by LDF devices and other methods for assessment of the microvascular perfusion. Since no *gold standard* for recording tissue blood perfusion exists, the correlation of LDF data and reference method data is highly dependent on whether the two methods sample the same microvascular compartment and the degree of noninvasiveness of the reference method. Moving the LDPM probe from one position to an adjacent site while the global flow through an organ is used as the independent reference generally reveals a linear relationship with global flow at each site, albeit with a difference in scale factors caused by the heterogeneity of the tissue perfusion [96]. One of the most promising animal models for evaluation of LDPM performance is the feline intestine because of its uniform and easily controlled blood perfusion. In this model, a specific segment of the intestine is drained through a single vein, the blood flow of which can be accurately determined by a drop-counting technique [97]. A linear relationship between LDPM output signal and intestinal blood perfusion has been demonstrated within the range from 0 to 300 mL⁻¹ 100 g⁻¹ [98].

Since the processing algorithm of LDPI is identical to that of LDPM, it is expected and has been demonstrated [25] that results similar to those for LDPM are obtained when using LDPI. In addition to those performance tests, the uniformity in sensitivity over the entire image plane and the dependence on the distance between the scanner head and tissue also needs to be controlled.

The day-to-day sensitivity of LDPM instruments including the probe is checked by placing the probe tip in a suspension of latex spheres in random motion [59]. The well-defined mobility pattern of such particles should result in a stable and reproducible output signal. By use of those *motility standards*, two LDPM devices with identical specifications can be calibrated to the same sensitivity. A calibration standard that includes latex spheres undergoing Brownian motion while diffusing through a porous polyethylene material has proven to give Doppler spectra closely resembling those obtained from human skin [99]. In LDPI, the day-to-day calibration procedure can, in accordance with the procedure used in LDPM, be made by way of a two-point calibration method based on a calibration box incorporating a transparent container with *motility standard* and a white surface *zero reference*, giving the high and zero calibration points, respectively [39].

13.4 Applications

Although LDF is a relatively recent medical technology, many reports are available for various research applications in the scientific literature. The relatively slow introduction of the technology in clinical routines is probably due to the fact that tissue perfusion is poorly predictable at any given tissue site and time. Furthermore, the natural variability of tissue perfusion in terms of both temporal fluctuations and spatial heterogeneity makes it difficult to define any thumb rules for normal conditions. Instead, stimuli–response experiments have to be applied in order to challenge the tissue and reveal possible malfunctioning of the most minute vessels of the vascular tree. One of the main features of LDF, in both monitoring and imaging, is that time traces and images can easily be recorded, while the interpretation of the results implies a thorough understanding of microcirculation and its regulation. Some examples of applications for LDPM and LDPI are briefly given in the following sections.

13.4.1 LDPM Applications

Experimental investigations and clinical trials aiming at evaluating the LDPM's practical usefulness have been undertaken on many organs, both in humans during clinical studies and in animal models. This ranges over a variety of disciplines. The most commonly investigated organs include the skin [23,100–105], the kidney [91,106,107], the liver [108,109], the intestines [110,111], and the brain [112–114]. Many studies are performed with surface probes attached to the tissue of study. With the use of needle probes, however, the perfusion can furthermore be continuously monitored with minial tissue trauma, for example, in skeletal muscle and tendons [57,58,115], in the myocardium during open heart surgery [116], and in bone tissue [117–122]. Additional probe designs adapted for neurosurgical instruments, such as stereotactic systems, allow for recording of microvascular perfusion during implantation of deep brain stimulation electrodes [123,124].

Among the clinical applications, evaluation of the skin tissue perfusion in association with Raynaud's phenomenon [125–127], diabetes microangiopathy [128–132], plastic surgery and flap monitoring [34,133–137], peripheral vascular disease [80,138–142], and thermal injury [143–145] is frequently represented. Bowel ischemia and gastric blood perfusion [146–150] and pharmacological trials [151,152] are other commonly represented areas in the literature. A thorough overview of many of those applications can be found in a review paper [29] and in two books on LDF [28,30].

13.4.2 LDPI Applications

Following the introduction of LDPI in the early 1990s, this emerging technology was soon evaluated in numerous applications in which the spatial heterogeneity of tissue perfusion is significant.

In irritant [153,154] and allergy [155–157] patch testing, perfusion images visualize and quantify the erythema produced by a compound applied topically to the skin. The LDPI technology produces data that are well in accordance with visual scoring, in the assessment of both skin irritation [158] and allergy [15]. It is therefore possible to grade irritation and type irritant reactions for individual compounds. In allergy patch testing, the optimal dose, influence of the vehicle, and the application and reaction time for different allergens can be readily detected and quantified. Malignant skin tumors are highly vascularized compared to benign lesions, and LDPI has been used to characterize the perfusion pattern of lesions such as basal cell carcinoma [158], malignant melanomas, and naevi [160]. Furthermore, LDPI has been used in the assessment of basal cell carcinoma perfusion in conjunction with cryotherapy and photodynamic therapy [161]. This report suggests that LDPI can be used, not only to follow up on the healing process but also to find recurrent or residual tumor growth. LDPI has further been used to assess the depth and the healing process of burns [162,163], and a recommendation has been issued to include LDPI as a clinical tool in the evaluation of severely burned patients [37]. Characterization of perfusion in leg ulcers in combination with capillary microscopy has been performed by Gschwandtner et al. [36,164,165], while perfusion in and around pressure sores was investigated by Mayrovitz et al. [166–169]. Other applications of LDPI include assessment of diabetes neuropathy following contralateral cooling [170] and endothelial cell malfunction investigated by iontophoresis infusion of acetylcholine and nitroprusside in diabetes [171] and Alzheimer's disease [172]; angiogenesis and growth factor research [173,174] of psoriasis [175,176], scleroderma [177], and Raynaud's disease [178]; and flap surveillance [179], microdialysis [180], and phototesting [181]. Recent studies also present the possibility to use LDPI for intraoperative human brain mapping [182]. Perfusion images of the cortical surface were captured during awake surgery while the patient performed predetermined tasks.

References

1. Rhodin, J.A.G., Anatomy of the microcirculation, in *Microcirculation: Current Physiologic, Medical and Surgical Concepts*, R.M. Effros, H. Schmid-Schoenbein, and J. Dietzel, eds., 1981, New York: Academic Press, pp. 11–17.
2. Tenland, T. et al., Spatial and temporal variations in human skin blood flow. *International Journal of Microcirculation: Clinical & Experimental*, 1983, **2**(2): 81–90.
3. Colantuoni, A., S. Bertuglia, and M. Intaglietta, Variations of rhythmic diameter changes at the arterial microvascular bifurcations. *Pflugers Archiv—European Journal of Physiology*, 1985, **403**(3): 289–295.
4. Colantuoni, A., S. Bertuglia, and M. Intaglietta, Microvascular vasomotion: Origin of laser Doppler flux motion. *International Journal of Microcirculation: Clinical & Experimental* (Sponsored by the European Society for Microcirculation), 1994, **14**(3): 151–158.
5. Salerud, E.G. et al., Rhythmical variations in human skin blood flow. *International Journal of Microcirculation: Clinical & Experimental*, 1983, **2**(2): 91–102.
6. Braverman, I.M., A. Keh, and D. Goldminz, Correlation of laser Doppler wave patterns with underlying microvascular anatomy. *Journal of Investigative Dermatology*, 1990, **95**(3): 283–286.
7. Hertzman, A.B. and C.R. Spealman, Observation on the finger volume pulse recorded photoelectrically. *American Journal of Physiology*, 1937, **119**: 334.
8. Challoner, A.V.J., Photoelectric plethysmography for estimating cutaneous blood flow, in *Non-Invasive Physiological Measurements*, P. Rolfe, ed., 1979, London, U.K.: Academic Press, pp. 125–151.
9. Holti, G. and K.W. Mitchell, Estimation of the nutrient skin blood flow using a non-invasive segmented thermal probe, in *Non-Invasive Physiological Measurements*, P. Rolfe, ed., 1979, London, U.K.: Academic Press, pp. 113–123.
10. Kety, S.S., Measurement of regional circulation by the local clearance of radioactive sodium. *American Heart Journal*, 1949, **38**: 321.

11. Sejrsen, P., Measurement of cutaneous blood flow by freely diffusable radioactive isotopes: Methodological studies on the washout of krypton-85 and xenon-133 from the cutaneous tissue in man. *Danish Medical Bulletin*, 1971, **18**(Suppl. 3): 9–38.

12. Groner, W. et al., Orthogonal polarization spectral imaging: A new method for study of the micro-circulation. *Nature Medicine*, 1999, **5**(10): 1209–1212.

13. Nadeau, R.G. and W. Groner, The role of a new noninvasive imaging technology in the diagnosis of anemia. *Journal of Nutrition*, 2001, **131**(5): 1610S–1614S.

14. Briers, J.D. and S. Webster, Laser speckle contrast analysis (LASCA): A nonscanning, full-field technique for monitoring capillary blood flow. *Journal of Biomedical Optics*, 1996, **1**(2): 174–179.

15. Chen, Z. et al., Optical Doppler tomographic imaging of fluid flow velocity in highly scattering media. *Optics Letters*, 1997, **22**(1): 64–66.

16. Bollinger, A. et al., Red blood cell velocity in nailfold capillaries of man measured by a television microscopy technique. *Microvascular Research*, 1974, **7**: 61.

17. Wayland, H. and P.C. Johnson, Erythrocyte velocity measurements in microvessels by a two-slit photometric method. *American Journal of Physiology*, 1967, **22**: 333.

18. Doppler, J.C., Über das farbige Licht der Doppelsterne und einiger anderer Gestirne des Himmels (On the coloured light from double stars and some other heavenly bodies). *Abhandlungen der Königlich Böhmischen Gesellschaft der Wissenschaften* (*Treatises of the Royal Bohemian Society of Science*), 1842, **5**(2): 465.

19. Riva, C., B. Ross, and G.B. Benedek, Laser Doppler measurements of blood flow in capillary tubes and retinal arteries. *Investigative Ophthalmology*, 1972, **11**(11): 936–944.

20. Mishina, H., T. Koyama, and T. Asakura, Velocity measurements of blood flow in the capillary and vein using the laser Doppler microscope. *Applied Optics*, 1974, **14**: 2326–2327.

21. Einav, S. et al., Measurement of blood flow in vivo by Laser Doppler Anemometry through a micro-scope. *Biorheology*, 1975, **12**(3–4): 203–205.

22. Stern, M.D., In vivo evaluation of microcirculation by coherent light scattering. *Nature*, 1975, **254**(5495): 56–58.

23. Holloway, G.A., Jr. and D.W. Watkins, Laser Doppler measurement of cutaneous blood flow. *Journal of Investigative Dermatology*, 1977, **69**(3): 306–309.

24. Nilsson, G.E., T. Tenland, and P.Å. Öberg, Evaluation of a laser Doppler flowmeter for measurement of tissue blood flow. *IEEE Transactions on Biomedical Engineering*, 1980, **27**(10): 597–604.

25. Nilsson, G.E., T. Tenland, and P.Å. Öberg, A new instrument for continuous measurement of tissue blood flow by light beating spectroscopy. *IEEE Transactions on Biomedical Engineering*, 1980, **27**(1): 12–19.

26. Bonner, R. and R. Nossal, Model for laser Doppler measurements of blood flow in tissue. *Applied Optics*, 1981, **20**: 2097–2107.

27. Leahy, M.J. et al., Principles and practice of the laser-Doppler perfusion technique. *Technology & Health Care*, 1999, **7**(2–3): 143–162.

28. Shepherd, A.P. and P.Å. Öberg, eds., *Laser-Doppler Blood Flowmetry*, 1990, Boston, MA: Kluwer Academic Publishers.

29. Öberg, P.A., Laser Doppler flowmetry. *Critical Reviews in Biomedical Engineering*, 1990, **18**(2): 125–163.

30. Belcaro, G. et al., eds., *Laser Doppler*, 1994, London, U.K.: Med-Orion Publishing Co.

31. Hill, S.A. et al., Microregional blood flow in murine and human tumours assessed using laser Doppler microprobes. *British Journal of Cancer—Supplement*, 1996, **27**: S260–S263.

32. Nilsson, G.E., A. Jakobsson, and K. Wårdell, Imaging of tissue blood flow by coherent light scatter-ing. *IEEE 11th Annual EMBS Conference*, Seattle, WA, 1989, pp. 391–392.

33. Essex, T.J. and P.O. Byrne, A laser Doppler scanner for imaging blood flow in skin. *Journal of Biomedical Engineering*, 1991, **13**(3): 189–194.

34. Wårdell, K., A. Jakobsson, and G.E. Nilsson, Laser Doppler perfusion imaging by dynamic light scattering. *IEEE Transactions on Biomedical Engineering*, 1993, **40**(4): 309–316.

35. Sommer, A., Laser Doppler imaging in dermatology—Experimental and clinical applications—Time to replace the dermatologist's eye?, 2001. Maastricht, the Netherlands: University of Maastricht.

36. Gschwandtner, M.E. et al., Microcirculation in venous ulcers and the surrounding skin: Findings with capillary microscopy and a laser Doppler imager. *European Journal of Clinical Investigation*, 1999, **29**(8): 708–716.

37. Pape, S.A., C.A. Skouras, and P.O. Byrne, An audit of the use of laser Doppler imaging (LDI) in the assessment of burns of intermediate depth. *Burns*, 2001, **27**(3): 233–239.

38. Bircher, A. et al., Guidelines for measurement of cutaneous blood flow by laser Doppler flowmetry. A report from the Standardization Group of the European Society of Contact Dermatitis. *Contact Dermatitis*, 1994, **30**(2): 65–72.

39. Fullerton, A. et al., Guidelines for visualisation of cutaneous blood flow by laser Doppler perfusion imaging. *Contact Dermatitis*, 2002, **46**(3): 129–140.

40. Nilsson, G.E., A. Jakobsson, and K. Wårdell, Tissue perfusion monitoring and imaging by coherent light scattering, in *Biooptics: Optics in Biomedicine and Environmental Studies*, O.D.D. Soares and A.M. Scheggi, eds., 1992, Bellingham, WA: SPIE—The International Society for Optical Engineering, pp. 90–109.

41. Berne, B.J. and R. Pecora, *Dynamic Light Scattering*, 1976, New York: Wiley.

42. Serov, A., W. Steenbergen, and F. de Mul, Method for estimation of the fraction of Doppler-shifted photons in light scattered by mixture of moving and stationary scatterers. *Proceedings of SPIE—The International Society for Optical Engineering*, Saratov, Russia, 2000, Vol. 4001, pp. 178–189.

43. Cummins, H.Z. and H.L. Swinney, Light beating spectroscopy, in *Progress in Optics*, E. Wolf, ed., 1970, Amsterdam, the Netherlands: North-Holland Publishing Co., pp. 133–200.

44. Forrester, A.T., Photoelectric mixing as a spectroscopic tool. *Journal of the Optical Society of America*, 1961, **51**(3): 253–259.

45. Larsson, M., *Influence of Optical Properties on Laser Doppler Flowmetry*, 2004, Linköping, Sweden: Department of Biomedical Engineering, Linköpings Universitet.

46. Fredriksson, I., M. Larsson, and T. Stromberg, Absolute flow velocity components in laser Doppler flowmetry, in *Optical Diagnostics and Sensing VI*, G.L. Cote, A.V. Priezzhev, eds., 2006, Bellingham, WA: SPIE.

47. Kienle, A. et al., Determination of the scattering coefficient and the anisotropy factor from laser Doppler spectra of liquids including blood. *Applied Optics*, 1996, **35**(19): 3404–3412.

48. De mul, F.F.M. et al., Laser-Doppler velocimetry and Monte-Carlo simulations on models for blood perfusion in tissue. *Applied Optics*, 1995, **34**(28): 6595–6611.

49. Larsson, M., W. Steenbergen, and T. Strömberg, Influence of optical properties and fiber separation on laser Doppler flowmetry. *Journal of Biomedical Optics*, 2002, **7**(2): 236–243.

50. Larsson, M. and T. Stromberg, Toward a velocity-resolved microvascular blood flow measure by decomposition of the laser Doppler spectrum. *Journal of Biomedical Optics*, 2006, **11**(1): 014024.

51. Fredriksson, I., M. Larsson, and T. Stromberg, Optical microcirculatory skin model: Assessed by Monte Carlo simulations paired with in vivo laser Doppler flowmetry. *Journal of Biomedical Optics*, 2008, **13**(1): 014015.

52. Fredriksson, I., M. Larsson, and T. Stromberg, Measurement depth and volume in laser Doppler flowmetry. *Microvascular Research*, 2009, **78**(1): 4–13.

53. Fredriksson, I., M. Larsson, and T. Stromberg, Inverse Monte Carlo method in a multilayered tissue model for diffuse reflectance spectroscopy. *Journal of Biomedical Optics*, 2012, **17**(4): 047004–047012.

54. Fredriksson, I., M. Larsson, M., and T. Strömberg, Model-based quantitative laser Doppler flowmetry in skin. *Journal of Biomedical Optics*, 2010, **15**(5): 057002, 2010.

55. Jakobsson, A. and G.E. Nilsson, Prediction of sampling depth and photon pathlength in laser Doppler flowmetry. *Medical & Biological Engineering & Computing*, 1993, **31**(3): 301–307.

56. Stern, M.D. et al., Continuous measurement of tissue blood flow by laser-Doppler spectroscopy. *American Journal of Physiology*, 1977, **232**(4): H441–H448.
57. Salerud, E.G. and P.Å. Öberg, Single-fiber laser Doppler flowmetry: A method for deep tissue perfusion measurements. *Medical & Biological Engineering & Computing*, 1987, **25**(3): 329–334.
58. Kvernebo, K., L.E. Staxrud, and E.G. Salerud, Assessment of human muscle blood perfusion with single-fiber laser Doppler flowmetry. *Microvascular Research*, 1990, **39**(3): 376–385.
59. Nilsson, G.E., Perimed's LDV flowmeter, in *Laser-Doppler Flowmetry*, A.P. Shepherd and P.Å. Öberg, eds., 1990, Boston, MA: Kluwer Academic Publishers, pp. 57–72.
60. Watkins, D. and G.A. Holloway, An instrument to measure cutaneous blood flow using the Doppler shift of laser light. *IEEE Transactions on Biomedical Engineering*, 1978, **BME-25**: 28–33.
61. Holloway, G.A., Medpacific's LDV blood flowmeter, in *Laser-Doppler Blood Flowmetry*, A.P. Shepherd and P.Å. Öberg, eds., 1990, Boston, MA: Kluwer Academic Publishers, pp. 47–56.
62. Gush, R.J. and T.A. King, Investigation and improved performance of optical fibre probes in laser Doppler blood flow measurement. *Medical & Biological Engineering & Computing*, 1987, **25**(4): 391–396.
63. Newson, T.P. et al., Laser Doppler velocimetry: The problem of fibre movement artefact. *Journal of Biomedical Engineering*, 1987, **9**(2): 169–172.
64. Obeid, A.N. et al., A critical review of laser Doppler flowmetry. *Journal of Medical Engineering & Technology*, 1990, **14**(5): 178–181.
65. Kolari, P., Optoelectronic Doppler velocimetry based on semiconductor laser diode for measurements of cutaneous blood flow. *International Journal of Microcirculation: Clinical & Experimental*, 1984, **3**: 476.
66. deMul, F.F.M. et al., Mini laser-Doppler (blood) flow monitor with diode laser source and detection integrated in the probe. *Applied Optics*, 1984, **23**: 2970–2973.
67. Kvernmo, H.D. et al., Spectral analysis of the laser Doppler perfusion signal in human skin before and after exercise. *Microvascular Research*, 1998, **56**(3): 173–182.
68. Kano, T. et al., Fundamental patterns and characteristics of the laser-Doppler skin blood flow waves recorded from the finger and toe. *Journal of the Autonomic Nervous System*, 1993, **45**(3): 191–199.
69. Kano, T. et al., Effects of neural blockade and general anesthesia on the laser-Doppler skin blood flow waves recorded from the finger or toe. *Journal of the Autonomic Nervous System*, 1994, **48**(3): 257–266.
70. Nilsson, G.E., Signal processor for laser Doppler tissue flowmeters. *Medical & Biological Engineering & Computing*, 1984, **22**(4): 343–348.
71. Duteil, L., J. Bernengo, and W. Schalla, A double wavelength laser Doppler system to investigate skin microcirculation. *IEEE Transactions on Biomedical Engineering*, 1985, **BME-32**(6): 439–447.
72. Gush, R.J. and T.A. King, Discrimination of capillary and arterio-venular blood flow in skin by laser Doppler flowmetry. *Medical & Biological Engineering & Computing*, 1991, **29**(4): 387–392.
73. Johansson, K. et al., Influence of fibre diameter and probe geometry on the measuring depth of laser Doppler flowmetry in the gastrointestinal application. *International Journal of Microcirculation: Clinical & Experimental*, 1991, **10**(3): 219–229.
74. Gush, R.J., T.A. King, and M.I. Jayson, Aspects of laser light scattering from skin tissue with application to laser Doppler blood flow measurement. *Physics in Medicine & Biology*, 1984, **29**(12): 1463–1476.
75. Salerud, E.G. and G.E. Nilsson, Integrating probe for tissue laser Doppler flowmeters. *Medical & Biological Engineering & Computing*, 1986, **24**(4): 415–419.
76. Wahlberg, E. et al., Effects of local hyperemia and edema on the biological zero in laser Doppler fluxmetry (LD). *International Journal of Microcirculation: Clinical & Experimental*, 1992, **11**(2): 157–165.
77. Kernick, D.P., J.E. Tooke, and A.C. Shore, The biological zero signal in laser Doppler fluxmetry—Origins and practical implications. *Pflugers Archiv—European Journal of Physiology*, 1999, **437**(4): 624–631.

78. Colantuoni, A., S. Bertuglia, and M. Intaglietta, Biological zero of laser Doppler fluxmetry: Microcirculatory correlates in the hamster cheek pouch during flow and no flow conditions. *International Journal of Microcirculation: Clinical & Experimental*, 1993, **13**(2): 125–136.

79. Abbot, N.C. and J.S. Beck, Biological zero in laser Doppler measurements in normal, ischaemic and inflamed human skin. *International Journal of Microcirculation: Clinical & Experimental*, 1993, **12**(1): 89–98.

80. Kvernebo, K., Laser Doppler flowmetry in evaluation of lower limb atherosclerosis. *Journal of the Oslo City Hospitals*, 1988, **38**(11–12): 127–136.

81. Hsiu, H. et al., Spectral analysis on the microcirculatory laser Doppler signal at the acupuncture point. *Annual International Conference of the IEEE Engineering in Medicine and Biology Society*, Vancouver, British Columbia, Canada, 2008, Vol. 2008, pp. 1084–1086.

82. Stefanovska, A., M. Bracic, and H.D. Kvernmo, Wavelet analysis of oscillations in the peripheral blood circulation measured by laser Doppler technique. *IEEE Transactions on Biomedical Engineering*, 1999, **46**(10): 1230–1239.

83. Veber, M. et al., Wavelet analysis of blood flow dynamics: Effect on the individual oscillatory components of iontophoresis with pharmacologically neutral electrolytes. *Physics in Medicine and Biology*, 2004, **49**(8): N111–N117.

84. Assous, S. et al., S-transform applied to laser Doppler flowmetry reactive hyperemia signals. *IEEE Transactions on Biomedical Engineering*, 2006, **53**(6): 1032–1037.

85. Carolan-Rees, G. et al., Fractal dimensions of laser Doppler flowmetry time series. *Medical Engineering and Physics*, 2002, **24**(1): 71–76.

86. Humeau, A. et al., Localization of transient signal high-values in laser Doppler flowmetry signals with an empirical mode decomposition. *Medical Physics*, 2009, **36**(1): 18–21.

87. Braverman, I.M. and J.S. Schechner, Contour mapping of the cutaneous microvasculature by computerized laser Doppler velocimetry. *Journal of Investigative Dermatology*, 1991, **97**(6): 1013–1018.

88. Wardell, K., I.M. Braverman, D.G. Silverman, and G.E. Nilsson, Spatial heterogeneity in normal skin perfusion recorded with laser Doppler imaging and flowmetry. *Microvascular Research*, 1994, **48**(1), 26–38.

89. Arildsson, M., G.E. Nilsson, and K. Wårdell, Critical design parameters in laser Doppler perfusion imaging. *SPIE Proceedings, Optical Diagnostics of Living Cells and Biofluids*, San Jose, CA, 1996, Vol. 2678, pp. 401–408.

90. Mack, G.W., Assessment of cutaneous blood flow by using topographical perfusion mapping techniques. *Journal of Applied Physiology*, 1998, **85**(1): 353–359.

91. Wårdell, K. and G.E. Nilsson, Duplex laser Doppler perfusion imaging. *Microvascular Research*, 1996, **52**(2): 171–182.

92. Linden, M., H. Golster, S. Bertuglia, A. Colantuoni, F. Sjoberg, and G. Nilsson, Evaluation of enhanced high-resolution laser Doppler imaging in an in vitro tube model with the aim of assessing blood flow in separate microvessels. *Microvascular Research*, 1998, **56**(3), 261–270.

93. Ilias, M.A. et al., Assessment of pigmented skin lesions in terms of blood perfusion estimates. *Skin Research and Technology*, 2004, **10**(1): 43–49.

94. Steenbergen, W. and F.F.M. deMul. New optical tissue phantom, and its use for studying laser Doppler blood flowmetry. *SPIE Proceedings, Optical and Imaging Techniques for Biomonitoring*, San Jose, CA, 1997, doi:10.1117/12.297935.

95. Larsson, M., W. Steenbergen, and T. Strömberg. Influence of tissue phantom optical properties and emitting—Receiving fiber distance on laser Doppler flowmetry. *SPIE Proceedings, Optical Diagnostics of Biological Fluids V*, San Jose, CA, 2000, doi:10.1117/12.387125.

96. Shepherd, A.P. et al., Evaluation of an infrared laser-Doppler blood flowmeter. *American Journal of Physiology*, 1987, **252**(6 Pt 1): G832–G839.

97. Ahn, H. et al., Evaluation of laser Doppler flowmetry in the assessment of intestinal blood flow in cat. *Gastroenterology*, 1985, **88**(4): 951–957.

98. Ahn, H. et al., In vivo evaluation of signal processors for laser Doppler tissue flowmeters. *Medical & Biological Engineering & Computing*, 1987, **25**(2): 207–211.

99. Liebert, A., M. Leahy, and R. Maniewski, A calibration standard for laser-Doppler perfusion measurements. *Review of Scientific Instruments*, 1995, **66**(11): 5169–5173.

100. Holloway, G.A., Cutaneous blood flow responses to injection trauma measured by laser Doppler velocimetry. *Journal of Investigative Dermatology*, 1980, **74**(1): 1–4.

101. Nilsson, G.E., U. Otto, and J.E. Wahlberg, Assessment of skin irritancy in man by laser Doppler flowmetry. *Contact Dermatitis*, 1982, **8**(6): 401–406.

102. Engelhart, M. and J.K. Kristensen, Evaluation of cutaneous blood flow responses by [133]Xenon wash-out and a laser-Doppler flowmeter. *Journal of Investigative Dermatology*, 1983, **80**(1): 12–15.

103. Wahlberg, J.E., Skin irritancy from alkaline solutions assessed by laser Doppler flowmetry. *Contact Dermatitis*, 1984, **10**(2): 111.

104. Wahlberg, J.E. and E. Wahlberg, Patch test irritancy quantified by laser Doppler flowmetry. *Contact Dermatitis*, 1984, **11**(4): 257–258.

105. Fischer, J.C., P.M. Parker, and W.W. Shaw, Laser Doppler flowmeter measurements of skin perfusion changes associated with arterial and venous compromise in the cutaneous island flap. *Microsurgery*, 1985, **6**(4): 238–243.

106. Stern, M.D. et al., Measurement of renal cortical and medullary blood flow by laser-Doppler spectroscopy in the rat. *American Journal of Physiology*, 1979, **236**(1): F80–F87.

107. Roman, R.J. and C. Smits, Laser-Doppler determination of papillary blood flow in young and adult rats. *American Journal of Physiology*, 1986, **251**(1 Pt 2): F115–F124.

108. Arvidsson, D., H. Svensson, and U. Haglund, Laser-Doppler flowmetry for estimating liver blood flow. *American Journal of Physiology*, 1988, **254**(4 Pt 1): G471–G476.

109. Pedrosa, M.E., E.F. Montero, and A.J. Nigro, Liver microcirculation after selective denervation. *Microsurgery*, 2001, **21**(4): 163–165.

110. Feld, A.D. et al., Laser Doppler velocimetry: A new technique for the measurement of intestinal mucosal blood flow. *Gastrointestinal Endoscopy*, 1984, **30**(4): 225–230.

111. Ahn, H. et al., Assessment of blood flow in the small intestine with laser Doppler flowmetry. *Scandinavian Journal of Gastroenterology*, 1986, **21**(7): 863–870.

112. Bogaert, L. et al., The effects of LY393613, nimodipine and verapamil, in focal cerebral ischaemia. *European Journal of Pharmacology*, 2001, **411**(1–2): 71–83.

113. Fabricius, M. et al., Laminar analysis of cerebral blood flow in cortex of rats by laser-Doppler flowmetry: A pilot study. *Journal of Cerebral Blood Flow & Metabolism*, 1997, **17**(12): 1326–1336.

114. Skarphedinsson, J.O. et al., Relative cerebral ischemia in SHR due to hypotensive hemorrhage: Cerebral function, blood flow and extracellular levels of lactate and purine catabolites. *Journal of Cerebral Blood Flow & Metabolism*, 1989, **9**(3): 364–372.

115. Astrom, M. and N. Westlin, Blood flow in the human Achilles tendon assessed by laser Doppler flowmetry. *Journal of Orthopaedic Research*, 1994, **12**(2): 246–252.

116. Karlsson, M.G. et al., Myocardial perfusion monitoring during coronary artery bypass using an electrocardiogram-triggered laser Doppler technique. *Medical & Biological Engineering & Computing*, 2005, **43**(5): 582–588.

117. Hellem, S. et al., Measurement of microvascular blood flow in cancellous bone using laser Doppler flowmetry and ^{133}Xe-clearance. *International Journal of Oral Surgery*, 1983, **12**(3): 165–177.

118. Swiontkowski, M.F. et al., Laser Doppler flowmetry for bone blood flow measurement: Correlation with microsphere estimates and evaluation of the effect of intracapsular pressure on femoral head blood flow. *Journal of Orthopaedic Research*, 1986, **4**(3): 362–371.

119. Swiontkowski, M.F. et al., Laser Doppler flowmetry for clinical evaluation of femoral head osteonecrosis. Preliminary experience. *Clinical Orthopaedics & Related Research*, 1987, **218**: 181–185.

120. Swiontkowski, M.F., K. Hagan, and R.B. Shack, Adjunctive use of laser Doppler flowmetry for debridement of osteomyelitis. *Journal of Orthopaedic Trauma*, 1989, **3**(1): 1–5.

121. Swiontkowski, M.F., Surgical approaches in osteomyelitis: Use of laser Doppler flowmetry to determine nonviable bone. *Infectious Disease Clinics of North America*, 1990, **4**(3): 501–512.
122. Hobbs, C.M. and P.E. Watkins, Evaluation of the viability of bone fragments. *Journal of Bone & Joint Surgery—British Volume*, 2001, **83**(1): 130–133.
123. Wardell, K. et al., Intracerebral microvascular measurements during deep brain stimulation implantation using laser Doppler perfusion monitoring. *Stereotactic and Functional Neurosurgery*, 2007, **85**(6): 279–286.
124. Wardell, K., Zsigmond, P., Richter, J., and Hemm, S., Relationship between laser Doppler signals and anatomy during deep brain stimulation electrode implantation toward the ventral intermediate nucleus and subthalamic nucleus. *Neurosurgery*, 2013, **72**(2 Suppl. Operative): ons127–ons140; discussion ons140.
125. Engelhart, M. and J.K. Kristensen, Raynaud's phenomenon: Blood supply to fingers during indirect cooling, evaluated by laser Doppler flowmetry. *Clinical Physiology*, 1986, **6**(6): 481–488.
126. Gush, R.J., L.J. Taylor, and M.I. Jayson, Acute effects of sublingual nifedipine in patients with Raynaud's phenomenon. *Journal of Cardiovascular Pharmacology*, 1987, **9**(5): 628–631.
127. Anderson, M.E. et al., Non-invasive assessment of digital vascular reactivity in patients with primary Raynaud's phenomenon and systemic sclerosis. *Clinical & Experimental Rheumatology*, 1999, **17**(1): 49–54.
128. Tooke, J.E. et al., Skin microvascular autoregulatory responses in type I diabetes: The influence of duration and control. *International Journal of Microcirculation: Clinical & Experimental*, 1985, **4**(3): 249–256.
129. Rayman, G. et al., Impaired microvascular hyperaemic response to minor skin trauma in type I diabetes. *British Medical Journal (Clinical Research Ed.)*, 1986, **292**(6531): 1295–1298.
130. Westerman, R.A. et al., Non-invasive tests of neurovascular function: Reduced responses in diabetes mellitus. *Neuroscience Letters*, 1987, **81**(1–2): 177–182.
131. Caballero, A.E. et al., Microvascular and macrovascular reactivity is reduced in subjects at risk for type 2 diabetes. *Diabetes*, 1999, **48**(9): 1856–1862.
132. Khan, F. et al., Impaired skin microvascular function in children, adolescents, and young adults with type 1 diabetes. *Diabetes Care*, 2000, **23**(2): 215–220.
133. Larrabee, W.F., Jr. et al., Skin flap tension and wound slough: Correlation with laser Doppler velocimetry. *Otolaryngology—Head & Neck Surgery*, 1982, **90**(2): 185–187.
134. Svensson, H. et al., Detecting arterial and venous obstruction in flaps. *Annals of Plastic Surgery*, 1985, **14**(1): 20–23.
135. Svensson, H., J. Holmberg, and P. Svedman, Interpreting laser Doppler recordings from free flaps. *Scandinavian Journal of Plastic & Reconstructive Surgery & Hand Surgery*, 1993, **27**(2): 81–87.
136. Banic, A., G.H. Sigurdsson, and A.M. Wheatley, Continuous perioperative monitoring of microcirculatory blood flow in pectoralis musculocutaneous flaps. *Microsurgery*, 1995, **16**(7): 469–475.
137. Cheng, M.H. et al., Devices for ischemic preconditioning of the pedicled groin flap. *Journal of Trauma-Injury Infection & Critical Care*, 2000, **48**(3): 552–557.
138. Belcaro, G. et al., Laser Doppler flux in normal and arteriosclerotic carotid artery wall. *Vasa*, 1996, **25**(3): 221–225.
139. Kvernebo, K., C.E. Slagsvold, and T. Gjolberg, Laser Doppler flux reappearance time (FRT) in patients with lower limb atherosclerosis and healthy controls. *European Journal of Vascular Surgery*, 1988, **2**(3): 171–176.
140. Kvernebo, K., C.E. Slasgsvold, and E. Stranden, Laser Doppler flowmetry in evaluation of skin post-ischaemic reactive hyperaemia. A study in healthy volunteers and atherosclerotic patients. *Journal of Cardiovascular Surgery*, 1989, **30**(1): 70–75.
141. Ghajar, A.W. and J.B. Miles, The differential effect of the level of spinal cord stimulation on patients with advanced peripheral vascular disease in the lower limbs. *British Journal of Neurosurgery*, 1998, **12**(5): 402–408.

142. Duan, J. et al., Hyperhomocysteinemia impairs angiogenesis in response to hindlimb ischemia. *Arteriosclerosis, Thrombosis & Vascular Biology*, 2000, **20**(12): 2579–2585.
143. Micheels, J., B. Alsbjorn, and B. Sorensen, Clinical use of laser Doppler flowmetry in a burns unit. *Scandinavian Journal of Plastic & Reconstructive Surgery*, 1984, **18**(1): 65–73.
144. Alsbjorn, B., J. Micheels, and B. Sorensen, Laser Doppler flowmetry measurements of superficial dermal, deep dermal and subdermal burns. *Scandinavian Journal of Plastic & Reconstructive Surgery*, 1984, **18**(1): 75–79.
145. Schiller, W.R. et al., Laser Doppler evaluation of burned hands predicts need for surgical grafting. *Journal of Trauma-Injury Infection & Critical Care*, 1997, **43**(1): 35–39; discussion 39–40.
146. Lunde, O.C., Endoscopic laser Doppler flowmetry in evaluation of human gastric blood flow. *Journal of the Oslo City Hospitals*, 1988, **38**(11–12): 113–126.
147. Lunde, O.C. and K. Kvernebo, Gastric blood flow in patients with gastric ulcer measured by endoscopic laser Doppler flowmetry. *Scandinavian Journal of Gastroenterology*, 1988, **23**(5): 546–550.
148. Lunde, O.C., K. Kvernebo, and S. Larsen, Evaluation of endoscopic laser Doppler flowmetry for measurement of human gastric blood flow: Methodologic aspects. *Scandinavian Journal of Gastroenterology*, 1988, **23**(9): 1072–1078.
149. Lunde, O.C., K. Kvernebo, and S. Larsen, Effect of pentagastrin and cimetidine on gastric blood flow measured by laser Doppler flowmetry. *Scandinavian Journal of Gastroenterology*, 1988, **23**(2): 151–157.
150. Krohg-Sorensen, K. and O.C. Lunde, Perfusion of the human distal colon and rectum evaluated with endoscopic laser Doppler flowmetry: Methodologic aspects. *Scandinavian Journal of Gastroenterology*, 1993, **28**(2): 104–108.
151. Trimarco, B. et al., Long-term reduction of peripheral resistance with celiprolol and effects on left ventricular mass. *Journal of International Medical Research*, 1988, **16**(Suppl 1): 62A–72A.
152. Iabichella, M.L. et al., Calcium channel blockers blunt postural cutaneous vasoconstriction in hypertensive patients. *Hypertension*, 1997, **29**(3): 751–756.
153. Fullerton, A. et al., The calcipotriol dose–irritation relationship: 48 Hour occlusive testing in healthy volunteers using Finn Chambers. *British Journal of Dermatology*, 1998, **138**(2): 259–265.
154. Issachar, N. et al., Correlation between percutaneous penetration of methyl nicotinate and sensitive skin, using laser Doppler imaging. *Contact Dermatitis*, 1998, **39**(4): 182–186.
155. Bjarnason, B. and T. Fischer, Objective assessment of nickel sulfate patch test reactions with laser Doppler perfusion imaging. *Contact Dermatitis*, 1998, **39**(3): 112–118.
156. Fischer, T., A. Dahlen, and B. Bjarnason, Influence of patch-test application tape on reactions to sodium dodecyl sulfate. *Contact Dermatitis*, 1999, **40**(1): 32–37.
157. Bjarnason, B., E. Flosadottir, and T. Fischer, Assessment of budesonide patch tests. *Contact Dermatitis*, 1999, **41**(4): 211–217.
158. Fullerton, A. and J. Serup, Laser Doppler image scanning for assessment of skin irritation, in *Irritant Dermatitis: New Clinical and Experimental Aspects* (*Current Problems in Dermatology*), P. Elsner and H. Maibach, eds., 1995, Basel, Switzerland: Karger, pp. 159–168.
159. Wang, I. et al., Superficial blood flow following photodynamic therapy of malignant non-melanoma skin tumours measured by laser Doppler perfusion imaging. *British Journal of Dermatology*, 1997, **136**(2): 184–189.
160. Stücker, M. et al., Blood flow compared in benign melanocytic naevi, malignant melanomas and basal cell carcinomas. *Clinical and Experimental Dermatology*, 1999, **24**(2): 107–111.
161. Enejder, A.M. et al., Blood perfusion studies on basal cell carcinomas in conjunction with photodynamic therapy and cryotherapy employing laser-Doppler perfusion imaging. *Acta Dermato-Venereologica*, 2000, **80**(1): 19–23.
162. Kloppenberg, F.W., G.I. Beerthuizen, and H.J. ten Duis, Perfusion of burn wounds assessed by laser doppler imaging is related to burn depth and healing time. *Burns*, 2001, **27**(4): 359–363.
163. Droog, E.J., W. Steenbergen, and F. Sjöberg, Measurement of depth of burns by laser Doppler perfusion imaging. *Burns*, 2001, **27**(6): 561–568.

164. Gschwandtner, M.E. et al., Laser Doppler imaging and capillary microscopy in ischemic ulcers. *Atherosclerosis*, 1999, **142**(1): 225–232.

165. Gschwandtner, M.E. et al., Microcirculation is similar in ischemic and venous ulcers. *Microvascular Research*, 2001, **62**: 226–235.

166. Mayrovitz, H.N., J. Smith, and M. Delgado, Variability in skin microvascular vasodilatory responses assessed by laser-Doppler imaging. *Ostomy Wound Management*, 1997, **43**(9): 66–70.

167. Mayrovitz, H.N. et al., Heel blood perfusion responses to pressure loading and unloading in women. *Ostomy Wound Management*, 1997, **43**(7): 16–20.

168. Mayrovitz, H.N. and J. Smith, Heel-skin microvascular blood perfusion responses to sustained pressure loading and unloading. *Microcirculation*, 1998, **5**(2–3): 227–233.

169. Mayrovitz, H.N., J. Macdonald, and J.R. Smith, Blood perfusion hyperaemia in response to graded loading of human heels assessed by laser-Doppler imaging. *Clinical Physiology*, 1999, **19**(5): 351–359.

170. Bornmyr, S. et al., Cutaneous vasomotor responses in young type I diabetic patients. *Journal of Diabetes & Its Complications*, 1997, **11**(1): 21–26.

171. Morris, S.J., A.C. Shore, and J.E. Tooke, Responses of the skin microcirculation to acetylcholine and sodium nitroprusside in patients with NIDDM. *Diabetologia*, 1995, **38**(11): 1337–1344.

172. Algotsson, A. et al., Skin vessel reactivity is impaired in Alzheimer's disease. *Neurobiology of Aging*, 1995, **16**(4): 577–582.

173. Couffinhal, T. et al., Mouse model of angiogenesis. *American Journal of Pathology*, 1998, **152**(6): 1667–1679.

174. Rivard, A. et al., Age-dependent impairment of angiogenesis. *Circulation*, 1999, **99**(1): 111–120.

175. Krogstad, A.L., G. Swanbeck, and B.G. Wallin, Axon-reflex-mediated vasodilatation in the psoriatic plaque? *Journal of Investigative Dermatology*, 1995, **104**(5): 872–876.

176. Krogstad, A.L. et al., Capsaicin treatment induces histamine release and perfusion changes in psoriatic skin. *British Journal of Dermatology*, 1999, **141**(1): 87–93.

177. Aghassi, D., T. Monoson, and I. Braverman, Reproducible measurements to quantify cutaneous involvement in scleroderma. *Archives of Dermatology*, 1995, **131**(10): 1160–1166.

178. Picart, C. et al., Systemic sclerosis: Blood rheometry and laser Doppler imaging of digital cutaneous microcirculation during local cold exposure. *Clinical Hemorheology & Microcirculation*, 1998, **18**(1): 47–58.

179. Eichhorn, W. et al., Laser Doppler imaging of axial and random pattern flaps in the maxillo-facial area. A preliminary report. *Journal of Cranio-Maxillo-Facial Surgery*, 1994, **22**(5): 301–306.

180. Anderson, C., T. Andersson, and K. Wårdell, Changes in skin circulation after insertion of a microdialysis probe visualized by laser Doppler perfusion imaging. *Journal of Investigative Dermatology*, 1994, **102**(5): 807–811.

181. Ilias, M.A. et al., Phototesting based on a divergent beam—A study on normal subjects. *Photodermatology, Photoimmunology and Photomedicine*, 2001, **17**: 189–196.

182. Raabe, A. et al., Laser Doppler imaging for intraoperative human brain mapping. *Neuroimage*, 2009, **44**(4): 1284–1289.

14

Light Scattering Spectroscopy and Imaging of Cellular and Subcellular Events

Nada N. Boustany
*Rutgers, The State
University of New Jersey*

Nitish V. Thakor
*Johns Hopkins University
School of Medicine*

14.1 Introduction

The optical analysis of thin biological specimens, such as cells and tissue slices, plays an important role in many clinical and biological diagnostic studies, including the evaluation of biopsies of disease tissues and the examination of live tissue metabolism in real time. Several optical techniques are available to study these ex vivo biological specimens and serve as probes of human disease and biological function. Diagnostic tests may be based on the observation of fluorescently labeled molecules involved in cellular metabolism. They can also be based on the observation of intrinsic fluorescence, elastic scattering, and Raman scattering from natural biomolecules. Typically, fluorescence and Raman scattering techniques are used to identify or localize specific biochemical entities. However, an important aspect of tissue diagnosis is based on assessing cellular and subcellular morphology. The morphological analysis of tissues, cells, and subcellular organelles is the primary objective of biomedical optical techniques based on elastic light scattering.

Alterations in tissue morphology and composition will result in detectable changes in the way light is transmitted, refracted, diffracted, or reflected from a given tissue specimen. Experimental measurements, such as light scattering intensity and its angular dependence, can be used to infer changes in size/shape or refractive index of the specimen under study. Biological analysis methods based on elastic light scattering include microscopic imaging techniques, such as dark field, phase contrast, or differential interference contrast (DIC), as well as quantitative spectroscopic methods, in which the sample is not visualized. In phase contrast and DIC microscopy, variations in the refractive index of the tissue are utilized to optically manipulate the scattered wavefronts and produce a high-resolution image of the biological specimen under study. These types of microscopic images are widely used to visualize the morphology of cells in culture or track cell movement, for example. In these cases, morphological analysis of the tissue stems from direct observation by the user. On the other hand, spectroscopy-based techniques are usually utilized in applications such as flow cytometry that require automated quantification of cellular and subcellular morphology, without visualization of the specimen. Such techniques are very useful in cell and tissue screening procedures in both clinical and biological studies. Emerging techniques, which combine imaging and spectroscopy, are also being utilized to localize and quantify the scattering information within a tissue slice or a monolayer of cells.

The constituent parts of cells and biological tissue, such as organelles or connective tissue fibers, are often at the limit of the resolution of optical microscopes. Alterations in organelle or subtissue morphology can be important indicators of underlying biochemical activity in living cells. While these changes could be quantified by electron microscopy in fixed tissue, greater insight about a biological process can be gained from minimally invasive techniques, that require minimal sample preparation, and thus suited for live tissue monitoring. Quantitative light scattering techniques are an ideal tool to address this problem and complement the other microscopic modalities. Light scattering methods are noninvasive and sensitive to changes in the dimension and optical properties of particles with size on the order of the wavelength. Light scattering measurements have had a significant impact in medicine and biology. Applications in flow cytometry include cell diagnosis and differential blood cell count [1–4] and human and bacterial cell response to various agents [5–9]. In static suspensions, light scattering has been used to monitor platelet aggregation [10–12], the mitochondrial permeability transition [13–16], and the optical properties of normal and tumor cells for future tissue diagnosis [17–19]. Optical analysis of thin biological specimens may involve monolayers of cells in culture and tissue slices. These ex vivo experimental models are used extensively to study important biological processes and are crucial to advancing our understanding of biological processes at the cellular and molecular level in a controlled laboratory environment. These biological specimens are usually thin enough to be dominated by single scattering, as opposed to studies of turbid whole-tissue samples, where multiple scattering prevails.

The uses of light scattering techniques to analyze thin biological specimens are the subject of this chapter. In particular, this chapter focuses on applications of quantitative light scattering, where a specific light scattering parameter is measured, such as intensity of the scattered light or angular dependence of the scattered light. Methods for data acquisition and interpretation are discussed, as well as ongoing work in this field.

14.2 Brief Theoretical Overview

A very brief overview of the light scattering problem is given here. The general treatment of light scattering by a single particle can be found in Refs. [20,21], to which we refer the reader for further details.

14.2.1 General Formulation of Scattering by a Single Particle

The geometry of the scattering problem for an arbitrary scatterer is shown in Figure 14.1a. The scatterer is a particle of arbitrary size and a refractive index, m, relative to the surrounding medium. $m = n_p/n_m$, where n_p is the particle's refractive index and n_m is the surrounding medium's refractive index. Given a plane wave of intensity I_o and wavelength λ, incident on this particle, the far-field scattered wave will be a spherical wave originating at the particle. The intensity of the scattered wave, I_s, at any distance, r, in the far-field can be written as $I_s = I_o F(\theta, \phi)/k^2 r^2$, where the scattering direction is defined by the angles θ and ϕ and k is the wavenumber with $k = 2\pi/\lambda$. The angle θ is the angle between the incident direction and

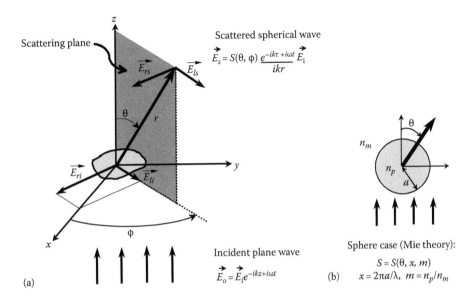

FIGURE 14.1   Relationship between incident and scattered fields. (a) The field, E_s, scattered by an arbitrary particle is related to the incident field, E_i, by way of a complex amplitude function, S, which depends on the scattering angles, θ and ϕ, the geometry of the scatterer, and the refractive index of the scatterer compared to the surrounding medium. (b) For a sphere with symmetry around ϕ, S is a function of θ, x, and m, where x is a normalized size parameter equal to the ratio of the sphere circumference to the wavelength of the light, $x = 2\pi a/\lambda$, and m is the ratio of sphere's refractive index to medium's refractive index, $m = n_p/n_m$.

the scattered direction and ϕ is the azimuthal angle of scatter. The scattering cross section of the particle is defined as $C_{sca} = 1/k^2 \int F(\theta, \phi) d\omega$, where $d\omega$ is a solid angle differential element with $d\omega = \sin\theta \, d\theta \, d\phi$. The function $F/(C_{sca}k^2)$ is the nondimensional phase function, whose integral over solid angle is equal to 1.

   Solving the scattering problem typically consists of solving for $F(\theta, \phi)$, which can be used to calculate the scattered light intensity and all directions. Solving for $F(\theta, \phi)$ involves solving for the electromagnetic field everywhere in space. Maxwell's equations are solved to calculate the electromagnetic fields inside the particle and in the medium outside the particle. The field outside the particle will be a superposition of the incident field and the field scattered by the particle. The boundary conditions at the particle/medium interface require that the tangential components of the electric and magnetic field be continuous at the interface. The problem is then reformulated in terms of electric fields, where the scattered field, E_s, is related to the incident field E_i by a complex amplitude function, $S(\theta, \phi)$. The function S may be represented as a matrix, such that

$$\begin{pmatrix} E_{ls} \\ E_{rs} \end{pmatrix} = \begin{pmatrix} S_2 & S_3 \\ S_4 & S_1 \end{pmatrix} \cdot \frac{e^{-ikr+ikz}}{ikr} \begin{pmatrix} E_{li} \\ E_{ri} \end{pmatrix}, \tag{14.1}$$

where

$$S = \begin{pmatrix} S_2(\theta,\phi) & S_3(\theta,\phi) \\ S_4(\theta,\phi) & S_1(\theta,\phi) \end{pmatrix}.$$

The elements of the matrix S in Equation 14.1 are complex numbers having both amplitude and phase and are functions of θ and ϕ. The subscripts s and i denote the scattered and incident components, respectively. The subscripts *l* and *r* denote parallel and perpendicular polarization of the *E* fields, respectively. The parallel and perpendicular directions are defined with respect to the scattering plane defined by the incident and scattering directions (Figure 14.1a). The function *F*, which was discussed earlier and which defines the *intensity* relationship between the incident and scattered light, can be deduced from the relationship between the incident and scattered fields given by the matrix *S*. For light of arbitrary polarization, it is common to rewrite Equation 14.1 as

$$
\begin{bmatrix} I_s \\ Q_s \\ U_s \\ V_s \end{bmatrix} = \frac{1}{k^2 r^2} \begin{bmatrix} S_{11} & S_{12} & S_{13} & S_{14} \\ S_{21} & S_{22} & S_{23} & S_{24} \\ S_{31} & S_{32} & S_{33} & S_{34} \\ S_{41} & S_{42} & S_{43} & S_{44} \end{bmatrix} \begin{bmatrix} I_i \\ Q_i \\ U_i \\ V_i \end{bmatrix}. \tag{14.2}
$$

I, Q, U, and *V* are the Stokes parameters, which are given in terms of the electric field components as

$$ I = E_l E_l^* + E_r E_r^*, $$

$$ Q = E_l E_l^* - E_r E_r^*, $$

$$ U = E_l E_r^* + E_r E_l^*, $$

$$ V = i(E_l E_r^* - E_r E_l^*). $$

The asterisk indicates the complex conjugate. The parameter *I* represents the intensity of the light. In this notation, unpolarized (or natural) light is represented by the Stokes vector (1, 0, 0, 0). The matrix in Equation 14.2 is known as the Mueller matrix. The explicit relationship between the elements of the Mueller scattering matrix and those of the matrix *S* in Equation 14.1 can be found in [21, p. 65]. The 16 elements of the scattering Mueller matrix depend on 7 independent variables corresponding to the magnitudes of the S_1, S_2, S_3, and S_4 of Equation 14.1 and the three possible independent phase differences between these S_j. Since the Mueller matrix has 16 elements and there are only 7 independent variables, there are 9 independent relationships between the 16 elements of the Mueller matrix.

The biological samples considered in this chapter will contain many scattering particles. However, the samples studied will be sufficiently thin to safely assume single scattering. This condition may be satisfied if $e^{-\mu_s z} \ll 1$, where μ_s is the tissue scattering coefficient and *z* is the sample thickness. μ_s is a function of wavelength, and $1/\mu_s$ represents the mean free path of the light before a scattering event occurs. Values of μ_s have been tabulated in the literature [22]. For example, μ_s is on the order of 100 cm^{-1} at 780 nm for biological tissue [23]. When single-particle scattering is considered, the following conditions ensue:

- Multiple scattering will be neglected.
- Each scatterer within the tissue will be exposed only to the radiation of the original incident beam.
- Light scattered from one particle will not be subjected to further scatter by another particle.

Thus, when only a single scattering event is considered for each particle making up the tissue, the total amount of scattering intensity by *N* particles will be equal to the sum of the individual scattering intensities by each of the *N* particles. If the particles in the sample are of varying size, the number density distribution of the particles may be taken into account. A case commonly considered is a distribution of spherical particle with different radii, *a*, in which case $N = \int_0^\infty N(a) \, da$.

14.2.2 Common Approximations to Solve for the Scattered Field of Biological Particles

The matrix S in Equation 14.1 is a general expression that describes light scattering from a single scatterer. To predict the scattered field, the matrix elements of S need to be determined. In general, these are complex numbers with magnitude and phase dependent on θ and ϕ, as well as the dimensions of the particle, the wavelength of light, and the refractive index ratio, $m = n_p/n_m$, between the particle (index n_p) and the surrounding medium (index n_m). The elements of S are rarely solved analytically in the general case of a scatterer with arbitrary shape and refractive index. Depending on the biological system at hand, an approximation is usually made to simplify the problem. Commonly used approximations are discussed in the following sections.

14.2.2.1 Rayleigh–Gans Theory for Scattering Particles with Refractive Index Ratio Close to 1

Qualitatively, a scattering particle may be viewed as being composed of different microscopic regions. An oscillating dipole moment is induced by the applied incident electric field in each of these microscopic regions. In turn, these driven dipoles scatter radiations in all directions. Thus, the scattered wave originating from the particle is the sum of all the dipole radiations. The angular intensity dependence of the scattered wave will therefore depend on the phase relationships between the radiated waves and the separations between the particle dipoles relative to the incident wavelength. If the particle is very small compared to the wavelength, it may be approximated by a single dipole, and one can use Rayleigh's theory of scattering. In this case, the elements of the matrix S in Equation 14.1 are $S_3 = S_4 = 0$, $S_1 = ik^3\alpha$, and $S_2 = ik^3\alpha \cos\theta$. α is the polarizability of the particle, and k is the wavenumber, $k = 2\pi/\lambda$. For a sphere with radius r and refractive index ratio, m, $\alpha = r^3(m^2 - 1)/(m^2 + 2)$, and for a homogeneous particle with refractive index ratio close to 1, $\alpha = (m^2 - 1)(V/4\pi)$, where V is the particle's volume (see Chapter 7 in Ref. [20]).

To satisfy the Rayleigh approximation, $|m|ka \ll 1$, where a is a length scale on the order of the size of the particle. A similar situation arises if the refractive index ratio, m, is close to 1, such that $|m-1| \ll 1$ and $2ka|m-1| \ll 1$. These two latter conditions imply that the field inside the particle is close to the incident field and that the particle may be assumed to be composed of volume elements dV that are subjected to the same incident field. Thus, instead of assuming that the particle is a single dipole, the particle is now composed of independently scattering dipoles corresponding to the different volume elements. These dipoles are driven by the same applied field, and the scattered wave is the sum of the waves scattered by these dipoles. In this case, $S_3 = S_4 = 0$, as for Rayleigh scattering and

$$S_1 = \frac{ik^3(m-1)}{2\pi}VR(\theta,\phi),$$

$$S_2 = \frac{ik^3(m-1)}{2\pi}VR(\theta,\phi)\cos\theta, \qquad (14.3)$$

$$\text{with } R(\theta,\phi) = \frac{1}{V}\int e^{i\delta}dV.$$

The phase δ refers to the phase of the scattered waves with respect to a common origin in the reference coordinate system. After reformulating δ in terms of the problem's geometry, R can then be integrated for a given particle shape. Calculations of R for spheres, ellipsoids, and cylinders are discussed in Ref. [20] (Chapter 7) and Ref. [21] (Chapter 6). The treatment of scattering resulting in Equation 14.3 is referred to as the Rayleigh–Gans or Rayleigh–Debye–Gans theory.

Since the refractive index of biological organelles, such as mitochondria, is typically close to that of the surrounding medium [24,25], various light scattering studies based on Rayleigh–Gans theory can be found in the literature. These include studies of scattering by bacteria [25,26], macromolecules [27,28], or nucleated lymphocytes [29]. Moreover, if the size and refractive index of the scattering particles satisfy the Rayleigh–Gans conditions, light scattering by a 3D scattering object may be approximated as Fraunhofer diffraction by a 2D aperture function [30]. This diffraction-based approach allows the Fourier optical treatment of diffraction and can be used to extract cellular geometric parameters from the diffraction pattern of biological cells [31–33].

14.2.2.2 Mie Theory for Spherical Particles of Arbitrary Size and Index

For a sphere with symmetry around ϕ (Figure 14.1b), S may be reexpressed as a function of θ, x, and m, where θ is the angle of scatter, x is a dimensionless size parameter equal to the ratio of the sphere circumference to the wavelength, $x = 2\pi a/\lambda$, a = particle radius, and λ = wavelength, and $m = n_p/n_m$ is the ratio of the particle's refractive index, n_p, to the surrounding medium's refractive index, n_m. The analytical solution for spheres was given by Mie in 1908 and can be found in Chapter 9 of Ref. [20] or Chapter 4 of Ref. [21]. For a sphere, S_3 and S_4 are 0 in Equation 14.1, while S_1 and S_2 can be expressed as infinite sums that can be calculated on a computer. Fortran computing routines to solve for the angular scattering function for a homogeneous or coated sphere can be found in Ref. [21]. Graaff et al. [34] present a simple numerical approximation of Mie scattering for $5 < x < 50$ and $1 < m < 1.1$. Mie theory, combined with a model of light propagation in a microscope with high numerical aperture (NA), was also used to predict high-resolution images of spheres [35]. Open source code that solves for Mie theory is also available online. One program, MieTab, has been published by August Miller in the Department of Physics at New Mexico State University and is still available for download as of this writing.

The existence of an analytical solution for the case of a spherical scatterer has prompted many researchers to approximate biological particles as spheres as a first-order approach to understand light scattering from cells and tissues. Despite the complicated morphologies of biological particles, studies have been successful in utilizing Mie theory to model the angular scattering response of bulk biological tissue. For example, the angular scattering functions of brain and muscle were successfully predicted by use of Mie theory, assuming that the tissue is composed of spheres with sizes distributed according to a skewed logarithmic function [36]. Similarly, a model based on Mie theory was able to adequately approximate reflectance spectra of colon tissue [37].

14.2.3 Solving the Scattering Problem for a Scatterer of Arbitrary Shape and Index

While biological particles are close to spherical in some cases, such as when considering the nuclei of certain cells, in general, this assumption is not necessarily warranted, especially when considering mitochondria, which appear rather filamentous in situ, or when considering neuronal dendritic structures or scattering collagen and elastin fibers in connective tissue. Thus, when considering these tissue components individually, light scattering studies should take into account their potential nonspherical and sometimes complicated geometries. Moreover, some tissue components, such as lipids, collagen, or melanin, have refractive index ratio $m > 1$ [38,39], making the Rayleigh–Gans approximation inapplicable. Analytical approaches, including the Wentzel–Kramers–Brillouin and equiphase-sphere approximations, have been proposed to predict scattering by particles of arbitrary shape [40] and potentially derive particle geometry from the scattering data. Numerical approaches to solving the problem of scattering by particles of arbitrary shape and refractive index have also been gradually emerging in applications of light scattering to biological systems. Two popular approaches for the study of nonspherical biological particles are the T-matrix (transition matrix) method originally developed by Waterman [41,42] and the finite-difference time-domain (FDTD) technique originally proposed by Yee [43]. With recent advances in computer hardware, numerical computations of angular scattering intensities can now be achieved on a personal computer. Details on the T-matrix with accompanying software can be found in Ref. [44], while an extensive reference on the FDTD method may be found in [45]. Both the T-matrix and FDTD methods are also discussed in detail with relevant references in Ref. [46].

In the T-matrix method, the incident and scattered fields are expanded into vector spherical wave functions. Due to the linearity of Maxwell's equations, the incident and scattered fields can then be related by means of a transition matrix (the T-matrix), which depends solely on the particle geometry and refractive index. Solving for the elements of the T-matrix allows prediction of the scattered field. For example, the T-matrix method was used to study light scattering by red blood cells [47] and cell nuclei [48]. On the other hand, the FDTD technique is based on the discretization of Maxwell's curl equations in time and space. Numerical calculation of the electric field as a function of time near the scattering particle is computed after applying the appropriate boundary conditions at the edges of each grid element in space. The near-field values thus computed are then transformed to yield the scattered far-field. The FDTD technique has been used to predict scattering by inhomogeneous cells composed of a spherical nucleus and of ellipsoidal organelles [17,39]. By using an incident time-limited pulse instead of an incident monochromatic plane wave, the frequency response of the scattered far-field response can be calculated. The pulsed FDTD approach was implemented to predict the angular scattering response of 2D models of inhomogeneous cells as a function of wavelength [49]. Application of the pseudospectral time-domain (PSTD) technique, which is less computationally intensive than the FDTD method, was also demonstrated to predict light scattering by inhomogeneous media with dimensions greater than 100 μm [50,51]. Such methods may ultimately pave the way for numerical modeling of light propagation and scattering by biological media of macroscopic dimensions.

14.3 Scattering Data Interpretation

Typically, changes in light scattering will result from changes in the scattering particle size, shape, refractive index, and concentration. Changes in these parameters may accompany important biochemical events in organelles, cells, and tissues and thus serve as diagnostic markers. If the particles that will change have already been identified, it may be feasible to predict the scattering behavior of these particles by utilizing the theoretical approaches described in Section 14.2. Combining the scattering measurements with the expected theoretical predictions will then serve to quantify the cellular or tissue events under study. In this case, the light scattering methods can be utilized to optically track known variations in cell or tissue substructure. The light scattering technique can serve to sort the data, for example, or to automate a well-understood diagnostic process.

Light scattering can also be used to probe tissue dynamics, in which the scattering sources have not yet been identified. Interpreting the scattering data from such measurements can, however, be particularly complicated since in this case, one has to solve the *inverse problem* consisting of extracting hitherto unknown tissue properties from the available scattering data. Typically, biological systems are inhomogeneous and contain particles of various geometries and optical properties, all of which could potentially result in the observed light scattering changes. As seen in the previous section, the dependence of scattering intensity on the angle of scatter is complicated. Thus, in order to extract an absolute optical or morphological parameter, extensive angular scattering data may have to be gathered such that sufficient measurements are available to fully characterize the angular scattering and cross section properties of the tissue under study. Such extensive data are rarely available experimentally, and theoretical prediction about the scatterers is often limited by the set of measurements at hand. In most cases, to identify and characterize the possible sources of the tissue scattering including organelles and other substructures, the biological system must be approximated by a model that can be easily described by the theoretical frameworks that are available. For example, the particles within the biological specimen are often assumed to be spherical or randomly oriented. While such simplifying assumptions are necessary to begin analyzing a given scattering problem, such assumptions restrict data interpretation. Refining the initial simplifying assumptions will often prove necessary to insure that the model is adequately taking into account the important variables in a given biological situation. In light of the difficulty of specifically identifying the sources of scattering change in a given tissue, light scattering techniques are best used in conjunction with other methods. For example, light scattering can be used to detect and localize the possible morphological changes, while additional biochemical manipulation can be used to modulate the scattering changes and help identify the molecules or tissue components leading to the observed change.

14.4 Methods and Applications of Light Scattering Measurements to the Study of Cells, Organelles, and in Tissue Slices

As we have briefly seen in Section 14.2, the amount of light scattered from a given particle typically depends on its size compared to the incident wavelength and on the ratio of its refractive index compared to the surrounding medium. In general, the larger the particle, and the larger its refractive index ratio, the larger the amount of scattering it will produce. Moreover, the shape and size of the particle will also affect the angular dependence of the scattered light. Thus, light scattered sideways as a fraction of the total light scattered will be larger for small spheres than for larger ones. Several techniques have been used to track such changes in the light scattering properties of biological samples. These techniques typically consist of measuring the angular scattering properties of a given sample. For example, changes in angular scattering as a function of sample composition or experimental condition can be monitored to deduce the size, shape, or index of the scatterers in each case. In some studies, the dependence of light scattering on incident wavelength or polarization is also taken into account.

Scattering intensities at multiple angles could be studied by collecting angular scattering intensity with the aid of a motorized goniometer. On the other hand, the pattern of diffraction by the sample could be analyzed utilizing Fourier optics to infer angular scattering. Quantitative transmission and reflection microscopy have also been used to generate images, in which the scattering intensity signals are measured locally in different parts of the specimen. Representative examples of these scattering measurement methods and their applications are discussed in this section.

14.4.1 Light Scattering Spectroscopy of Cells and Organelles in Suspensions

14.4.1.1 Methods to Study Scattering by Particle Suspensions

Particle suspensions are usually probed spectroscopically without optically resolving the individual scatterers. Scattering intensities at multiple angles can be collected with a rotating goniometer and a single photodetector [17,18,52,53] (Figure 14.2a) or by many detectors positioned at different angles around the specimen [28] (Figure 14.2b). The cell or organelle suspensions are usually contained in a cuvette, or the particles may be flowed in a single file through the optical analysis chamber of a flow cytometer [2] (Figure 14.2c). Scattering intensities are plotted versus angle around the specimen and could then be compared to theoretical predictions. By varying the polarization of the incident light and analyzing the polarization of the scattered light (if polarizers are used in Figure 14.2a), additional elements of the scattering Mueller matrix (see Section 14.2.1) can be measured [54]. Bohren and Huffman list all the possible relationships between the intensity of the scattered light at the different polarization and the intensity of the polarized incident beam in terms of the Mueller matrix elements, S_{ij} (Table 13.1 in Ref. [21]).

14.4.1.2 Applications

14.4.1.2.1 Flow Cytometry

Light scattering spectroscopy has extensively been employed in cell analysis, such as flow cytometry, to probe intracellular morphology. In a flow cytometer, scattering near the forward direction and side scattering near 90° are typically collected. While data at many angles are necessary to fully characterize the scatterers in the suspensions, limited measurements, such as the amount of near-forward scattered light and side-scattered light, can also provide very useful information. These two

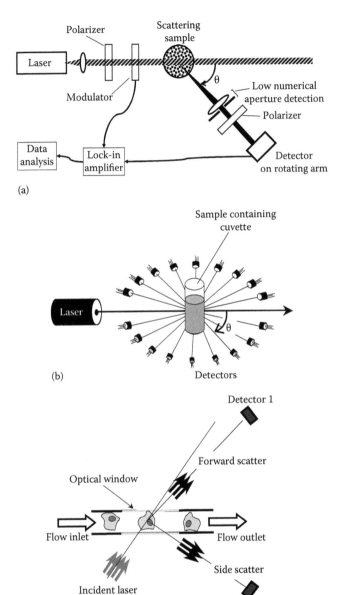

FIGURE 14.2 Schematics illustrating the principle behind experimental setups for collecting angular scattering from particle suspensions. (a) Scanning goniometer. Light scattered by the sample at a given scattering angle, θ, is collected by a low NA setup. The collection–detection setup is mounted on the arm of a rotating goniometer. The polarizers and modulator are optional. (b) Multidetector setup. Light scattered by the sample is collected as a function of scattering angle, θ, by several detectors placed around the sample. In contrast to the setup in panel A, here, the angular measurements are made simultaneously. (Adapted from Wyatt, J., *Anal. Chim. Acta*, 272, 1, 1993.) (c) Flow cytometry. The diagram shows cells flowing in single file through a flow cytometer channel, while the scattered light is collected through an optical window. Near-forward and near-90° scattering are typically collected.

measurements of forward and side scattering can be used individually or combined into a ratio to yield discriminating data that allow cell identification.

In clinical flow cytometry, cells flowing in a single file through an optical analysis chamber can be analyzed, sorted, and counted based on the ratio of forward-to-side scattering intensity. These cells may originate from a patient's blood or from a tissue biopsy. By sorting and counting the cells with specific forward-to-side scattering signatures, the status of a patient may be evaluated, and disease may be ruled in or out. For example, utilizing the elastic scattering properties of the different blood cells, human leukocytes may be sorted and counted [55]. Differentiation of leukocyte can also be improved by measuring changes in the amount of depolarization of the side-scattered light [56]. A discussion of light scattering, as it applies to flow cytometry, can be found in Ref. [2]. In this handbook, the development and applications of flow cytometry are further discussed in the accompanying chapter "Flow Cytometry with Optical Detection."

14.4.1.2.2 Angular Scattering Measurements of Isolated Mitochondria

Light scattering spectroscopy has been extensively used to study mitochondrial swelling and changes in mitochondrial matrix conformations. While change in mitochondrial morphology could be assessed by electron microscopy, dynamic studies of *viable* mitochondria typically utilize light scattering to study this organelle, whose size is close to the optical resolution of microscopes. Light scattering is a simple and convenient method, which is sensitive to changes in the size and shape of particles with dimensions on the order of the wavelength. Moreover, as an optical method, light scattering permits rapid detection commensurate with the rates at which mitochondria are expected to change. The light scattering measurements may be carried out either in a spectrophotometer with the mitochondria suspension contained in a regular cuvette or by flow cytometry. Studies on mitochondria isolated from tissue date back to the 1950s. Alterations in mitochondrial morphology measured by light scattering have been associated with mitochondrial metabolic state [57–64]. Measurements of light transmission or angular light scattering at 90° from a suspension of isolated mitochondria have long been correlated with the morphology of mitochondria in the orthodox and condensed states [14,59,60,65]. Since these early studies, light scattering has become the technique of choice to detect mitochondrial size change, and light scattering techniques have proved essential in studying the mitochondrial permeability transition [14–16,66–70].

The first scattering studies of mitochondria were interpreted by correlating the absorbance or 90° scattering intensity from the mitochondrial suspension with electron micrographs of the tested mitochondria. It was found that in most cases, mitochondrial scattering at 90° decreased as the relative number of mitochondria in the *aggregated* configuration decreased [14]. This aggregated form was typically characterized by a shrunken, electron-dense matrix space with large cristae, in contrast to the *orthodox* configuration characterized by an expanded and less electron-dense matrix space [62]. In addition, the absorbance and 90° scattering by mitochondrial suspensions were shown to decrease with increased mitochondrial swelling [71,72] (Figure 14.3). More recently, this relationship between swelling and mitochondrial absorbance and 90° scattering was utilized in the detection of mitochondrial morphology change during apoptosis. These recent studies were done utilizing flow cytometry [73] and by measuring changes in either 90° scattering or absorbance by a suspension of isolated mitochondria in a spectrophotometer cell [74–77].

Nonetheless, one should interpret single-angle scattering or absorbance measurements with great care. While under certain conditions the early scattering studies of mitochondrial have provided good correlation between light absorbance, or 90° scattering intensity, and mitochondrial morphology [14,71,72], these methods could present some shortcomings. The general relationship between transmitted light, or light scattered at one single angle, and particle volume is not always monotonic [78,79]. Moreover, changes in refractive index also contribute to the change in light scattering in addition to morphology change, thus confounding data interpretation. A study by Knight et al. [80] shows how changes in light scattered at 90° may not necessarily correlate with mitochondrial volume change and points at the difficulty in interpreting single-angle scattering data. Thus, additional validation by means

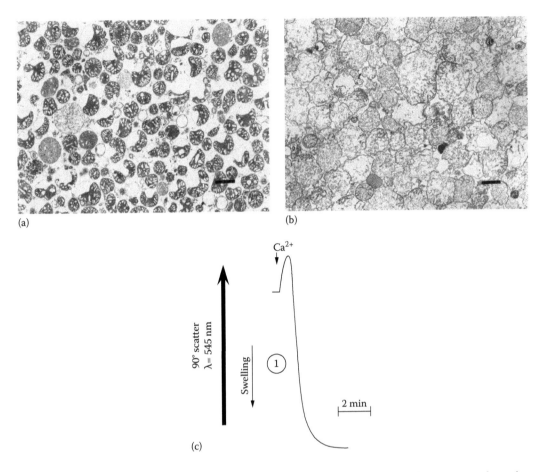

FIGURE 14.3 Electron micrographs and absorbance measurements from mitochondrial suspensions subjected to calcium overload. (a) Isolated liver mitochondria suspended in calcium-free incubation medium. (b) The medium was supplemented with 150 μm Ca^{2+}. (c) The decrease in the measured 90° light scattering at 545 nm correlates with mitochondrial swelling upon addition of 150 μm Ca^{2+} to the control medium. Bar = 1 μm. (Reprinted from Petronilli, V. et al., *J. Biol. Chem.*, 268(29), 21939, 1993. With permission.)

of electron microscopy, for example, will prove necessary to correctly infer the particles' morphologic configurations from absorbance or single-angle scattering measurements.

14.4.1.2.3 Angular Scattering Measurements of Cellular Suspensions

With the recent applications of diffuse light scattering techniques to the diagnosis of tissue in vivo (see accompanying chapter on elastic scattering and diffuse reflectance), there has been an increased interest in studying the scattering properties of cells, organelles, and subtissue structure. Scattering parameters can be used to define the morphological organization of biological tissue and to better understand the different sources of scatter that contribute to the bulk tissue signal. Scattering from cell suspensions was used by Mourant et al. to show that cells have a broad distribution of scatterer sizes [18]. Significant cell scatter was shown to originate from particles between 0.2 and 1 μm, and that small tissue particles are expected to contribute to wide angle scattering, while larger particles will contribute mainly to forward directed light scattering. Moreover, the nuclei angular scattering spectrum most closely resembled that of the whole cells (Figure 14.4). In this study, the authors assumed that the cells are comprised of spherical scatterers. They used Mie theory to analyze the angular scattering distributions. Further studies by the same group have shown that cell suspensions do not depolarize light significantly [81].

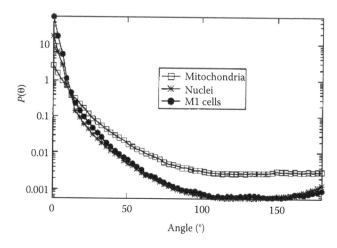

FIGURE 14.4 Normalized angular scattering measurements, $P(\theta)$, of fibroblast cells (M1 cells), isolated fibroblast nuclei and isolated fibroblast mitochondria. Values below 9° and 168° were extrapolated. (Reprinted from Mourant, J.R. et al., *Appl. Opt.*, 30(16), 3586, 1998. With permission.)

These results indicated that subcellular scatterers did not deviate much from sphericity. Several studies have focused on determining the specific size distributions of scatterers and the organelles contributing the total light scattering signal collected from cells [53,82–84]. In this context, a pertinent study emphasizes the effect of scattering cross section on the experimentally measured signal. Thus, while small scattering objects on the order of half a micron might be abundant in cells, their contribution to the measured scattering signal remains small compared to larger scattering sources such as mitochondria and lysosomes, which are less abundant but have significantly higher scattering cross sections [85]. The results of this study also demonstrate that once weighted by scattering cross section, different particle size distribution functions (e.g., log-normal or exponential) can result in largely similar total light scattering signal as a function of particle diameter.

Angular scattering measurements of cell suspensions were also utilized as experimental validation in the construction of a cell model based on the FDTD technique [17] (see Section 14.2.3). In contrast to the study by Mourant et al. [18] discussed previously, the cell model in this case does not assume a distribution of spherical scatters. Instead, the model considered a 15 μm diameter spherical cell containing a nucleus with subnuclear refractive index variations and ellipsoidal organelles. In particular, such optical cell models were used to explain the effect of adding acetic acid to cells, suggesting that acetic acid increases the frequency of fluctuations in nuclear refractive index. Acetic acid was also found to increase the amplitude of these index variations. Acetic acid addition is very relevant to cancer diagnosis. Topical application of acetic acid to tissue is a very common method used by colposcopists to enhance contrast between normal and diseased regions of the cervical epithelium. Thus, by understanding how different conditions may change the optical scattering properties of the cells under study, cell modeling, together with scattering studies of cell suspensions, represents an important set of data, which will undoubtedly be helpful when optimizing and designing current and future optical diagnostic tools based on light scattering measurements.

14.4.1.2.4 Angular Scattering Measurements of Bacteria, Macromolecules, or Vesicles in Suspension

Scattering spectroscopy of cells and organelles has direct applications to understanding the scattering properties of biological tissues. It is important to note that the methods described here for collecting angular scattering data from particle suspensions may also have other biologically relevant applications. In particular, angular scattering was used to identify bacteria [26]. The state of polarization of the light scattered by bacteria was also shown to be sensitive to very small changes in bacterial structure [54].

Angular scattering was also used to characterize the size of macromolecules in suspension [28]. In addition, in a system where angular scattering distributions were measured as a function of time, the dynamics of time-varying systems were characterized. Thus, time-dependent angular scattering measurements were used to track the polymerization of microtubules as well as dynamic changes in the size of chromaffin granules subjected to osmotic stress [27].

14.4.2 Light Scattering Spectroscopy of Cellular Monolayers and Thin Tissue Slices

14.4.2.1 Methods for Collecting Angular Scattering Measurements by Diffraction

Another way to infer angular scatter is by analyzing the sample's diffraction pattern with Fourier optics. The principle behind this method is shown in Figure 14.5. In this setup, the sample is illuminated by a plane wave of light obtained by a collimated laser beam, for example. The light scattered by the sample is collected by a lens, whose NA will determine the highest angle of scatter that can be collected in by the setup. As shown in Figure 14.5, the diffraction pattern of the sample is formed in the back focal plane, F, of the collection lens. Since the incident laser beam is collimated, the diffraction pattern is generated from light scattered by the sample. The angles of scatter are mapped onto the plane F in increasing order, moving radially away from the optical axis. The laser light that is transmitted without being scattered by the sample will be focused in the center of the plane F and can be subtracted by a beam block at this point. Usually the diffraction pattern in F is reimaged by a second lens onto a photodetector array, such as a charge-coupled device (CCD) camera [31–33,86,87]. Changes in angular scattering by the sample can be studied by analyzing the diffraction pattern of the sample. Since the cell sample can be placed on a microscope slide in this diffraction-based setup, this method is particularly useful for analyzing cells in a monolayer, as opposed to in suspension as in the previous section.

As for the angular scattering measurements of particle suspensions, the diffraction technique is of spectroscopic nature: the diffraction pattern corresponds to scattering by the entire sample region illuminated by the laser beam. The size of this illuminated region can be a few hundreds of microns in width. To correlate the angular scattering pattern with a specific region of the sample, Valentine et al. [86] used a microscope condenser in the illumination path and were able to set the laser beam diameter to 70–100 μm, thus selectively analyzing small regions of a porcine skin specimen (Figure 14.6). In that system, the optical microscope was also equipped with a beam splitter after the collection lens, such

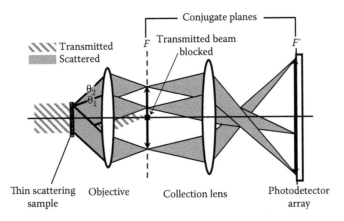

FIGURE 14.5 Experimental setup for imaging the diffraction pattern of cells or tissue slices plated on a microscope slide. F is the objective's back focal plane. F and F' are conjugate Fourier planes. The scattered light (gray beam), which forms the diffraction pattern of the sample, is reimaged onto a photodetector array, while the transmitted light (cross-hatched beam) is blocked at the center of the plane F.

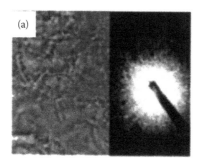 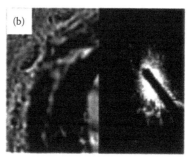

FIGURE 14.6 Images and diffraction patterns of two porcine skin specimens, 20 μm thin. (a) The tissue region sampled is homogeneous, and the diffraction pattern is isotropic. (b) A region in the vicinity of a hair results in an anisotropic scattering pattern. The black rod in the diffraction images corresponds to the transmitted light beam block. The scattering angle, θ, increases in the radial direction. Transmitted light and light scattered at 0° are focused and blocked in the center of the diffraction pattern. (Reprinted from Valentine, M.T. et al., *Opt. Lett.*, 26(12), 890, 2001. With permission.)

that an image of the sampled 70–100 μm region could be collected simultaneously with an image of its diffraction pattern on two separate cameras. Microscopes combining this methodology with elastic scattering spectroscopy [88] or Raman microspectroscopy [89] have also been described.

14.4.2.2 Applications of Diffraction to Cellular Analysis

To measure the diameters of the nucleus and cytoplasm of stained cervical cells, Turke et al. analyzed angular scattering by modeling the scatter as a Fraunhofer diffraction field from a 2D flat object [31]. The nucleated cells were modeled as two circular concentric regions having different optical densities and diameters d_N and d_C corresponding to the nucleus and cytoplasm, respectively. In some cases, the effects of offsetting the nucleus from the center of the cell were also considered. The solution of the Fraunhofer diffraction model gives the radial dependence of the light intensity in the diffraction pattern of the sample. By analyzing only one radial scan of the diffraction pattern, nuclear and cytoplasmic diameters were calculated and compared with the actual dimensions of normal, dysplastic, and cancerous cells. The results showed that the correct nuclear diameter was inferred in more than 80% of the 378 cells tested. A similar approach was taken by Burger et al. [32] to extract nuclear and cytoplasmic diameters from the Fraunhofer diffraction pattern of nucleated cell models and nucleated erythrocytes. The cellular dimensions inferred from the light scattering analysis in the diffraction pattern matched the microscopically determined dimensions of the cytoplasm and nucleus very well. An additional radial scan of the diffraction pattern taken at 90° to the first one also helped differentiate the major and minor axis diameters of these elliptical erythrocytes. More recently, modeling of light scattering as a Fraunhofer diffraction pattern was used to monitor the rounding of initially elongated cells in response to follicle-stimulating hormone [90] and the changes in cell diameter at the onset of apoptosis [91]. A study of light scattering by bacterial colonies utilizes scalar diffraction theory. In this study, the colonies, which are tested in situ on agar culture plates, are modeled as multilayered Gaussian-shaped elevations on the plate and light propagation is modeled using scalar diffraction theory in the Fresnel approximation [92]. This system was used to successfully identify and discriminate between multiple bacterial species [87].

14.4.2.3 Techniques Other Than Diffraction Pattern Analysis to Study Scattering of Cellular Monolayers and Thin Tissue Slices

Angular scattering measurements from thin tissue slices can be made by directly measuring the intensity of the forward scattered light from a tissue slice mounted on a cover slip. Direct measurements of angular light scattering were used to extract information about order and spacing between the collagen fibers in cartilage. In particular, the average scattering angle from 40 μm thin cartilage slices was shown to decrease as a function of the aggregate compressive modulus (measured separately from the bulk samples prior to slicing) [93]. This relationship between angular scattering and modulus could be explained by a theoretical analysis relating average scattering angle, compressive modulus, and a short-range order parameter. From this analytical model, this short-range order parameter, which describes the spatial correlation length between the collagen fibers, was found to be 8.2 μm.

Other scattering spectroscopic techniques of monolayers of cells and thin tissue slices have also been described. By analyzing the refractive index fluctuations of mouse liver tissue obtained by phase contrast microscopy, Schmitt and Kumar [94] showed that the number density of scattering particles in the tissue decreases as a function of increasing diameter and follows an inverse power law similar to that which describes the volume fractions of subunits of a fractal object. In this study, the scatterers were assumed to be spherical and were described using Mie theory.

Angle-dependent low coherence interferometry (aLCI) was used to measure the backscatter intensity as a function of angle from monolayers of cultured HT29 epithelial cells [95]. As for optical coherence tomography, this technique offers a depth resolution defined by the coherence length of the light source and allows scattering measurements from a specific point within the penetration depth of the sample. When probing a point close to the sample surface, or a thin monolayer of cells, the angular scattering is dominated by single scattering and can be analyzed using Mie theory. Assuming spherical scatterers, the angle-dependent low coherence method was used to extract nuclear diameter and nuclear refractive index. Once the nuclear contribution to the angular scattering is subtracted, the remaining angular scattering spectrum can be analyzed to extract information about subcellular organelles smaller than the nucleus. This study yielded a subcellular scatterer size distribution similar to that previously measured by Schmitt and Kumar [94], where the number density of tissue scatterers followed an inverse power law as a function of scatterer diameter. Improvements in the aLCI method include the use of the T-matrix method to infer the aspect ratio of subcellular scatters such as nuclei as opposed to assuming subcellular spherical particles as was done in the initial aLCI implementations [48]. While aLCI has been extensively applied for in vivo cancer diagnosis [96], recent implementations in cultured cells included observation of subcellular changes during stem cell differentiation [97] and cancer cell death [98].

14.4.3 Combining Light Scattering Spectroscopy and Imaging of Tissue Slices and Cellular Monolayers

14.4.3.1 Transmission and Reflectance Images of Brain Slices

The techniques described in the two previous subsections probe the angular dependence of light scattering by the specimen under study, but without imaging the sample itself. The scattering intensities are typically collected from an ensemble of particles, and information about the location of the scattering sources within the specimen is not always saved. Cells often respond differentially to a given treatment, and tracking the location of the scattering change within a monolayer of cells or within a tissue slice or cellular monolayer could provide a better understanding of the basic dynamics of a time-dependent biological process. To this end, *imaging* methods that are based on scattering signals can be used to record relative changes in light scattering within the full field of view. Scattering information can be collected directly from bright-field and dark-field microscopic images collected at a specific wavelength. The imaging wavelength is typically chosen in the red or near infrared, both of which penetrate tissue deeper than shorter wavelengths. The intensity distributions within transmission and reflectance microscopy images depend on the way the light is absorbed or scattered by the tissue under study. Changes in the biological composition of the tissue can affect absorbance and transmission through the specimen and can therefore be used to track subtissue dynamics. Transmission and reflectance imaging have been used to map neural activity within brain slices by differentiating the response of the various neuronal layers within the slice in response to a stimulus. Intrinsic optical signals such as transmittance or reflectance are often recorded in conjunction with fluorescent images using voltage-sensitive dyes or mitochondrial potential dyes, for example. These fluorescent images help correlate the optical scattering signal with biochemical and electrophysiological properties of the different tissue regions. A short review of the application of intrinsic optical imaging techniques to brain slices can be found in Aitken et al. [99]. Studies of intrinsic optical signals from brain slices are increasingly popular. Transmitted and reflected light signals were measured in the CA1 region of the hippocampus as a function of hypotonic stress [99,100], neural stimulation [99], or spreading depression [99]. Changes in light transmission were also compared between the CA1 and CA3 regions of the hippocampus as a function of N-methyl-D-aspartate (NMDA) and kainate-mediated excitotoxic injury [101]. Changes in transmitted light and mitochondrial depolarization measured with the fluorescent dye rhodamine 123 were used to track the spatiotemporal dynamics of hypoxia and spreading depression within hippocampal slices [102]. Using a fiber-optic excitation/collection bundle, reflected light was measured in vivo from the cat hippocampus and correlated with evoked potentials [103]. Brain slices are advantageous compared to isolated neuronal cultures in that they preserve the physiological relationship between neurons and glia, as well as the connectivity between different neuronal regions. However, to adequately preserve these relationships and maintain an experimentally viable tissue, the slices are often several microns in thickness. As a result, multiple scattering may affect the optical signals detected, and the single-particle scattering approaches presented in Section 14.2 may not be applicable. Moreover, since most studies do not record the full angular scattering response from the tissue, the optical signal interpretation becomes limited.

14.4.3.2 Dual-Angle Scattering Imaging of Brain Slices

To explore angular scattering by brain slices, Johnson et al. measured light scattered by hippocampal slices after illuminating the tissue at two different angles (Figure 14.7) [104]. These brain slices were still 310 μm thick and multiple scattering may not be negligible. However, by comparing the individual images collected at each of the two illumination angles and their ratio, Johnson et al. were able to differentiate the hippocampal response to hypotonic stress from the response to NMDA-mediated injury (Figure 14.8) [104]. When each illumination angle was considered individually, both hypotonic and NMDA treatments showed a relative decrease in the intensity of the light scattered by the CA1 region. However, when the ratio of the images collected at the two angles was considered, changes could only be measured after the NMDA treatment. Several scattering components, such as dendritic processes, axonal varicosities, or cellular organelles, can contribute to transmission, reflection, and scattering measurements of brain slices [105]. A study by Johnson et al. suggests that intrinsic optical signals measured by the dual-angle scattering technique are correlated with mitochondrial swelling during NMDA-mediated neuronal injury [106]. As such, despite the difficulty in fully interpreting the scattering response of relatively thick tissue slices, transmission and reflectance imaging of brain slices remain very valuable in that they provide a simple method to record spatiotemporal dynamics that reflect morphological change and that can be measured simultaneously in the whole preparation, unlike focal electrode recordings.

14.4.3.3 Optical Scatter Imaging of Cellular Monolayers

Boustany et al. demonstrated an optical scatter imaging (OSI) technique that produces images that directly encode a morphometric parameter within the full field of view of the microscope [107]. This OSI method combines Fourier filtering with central dark-field microscopy to detect alterations in the size of particles with wavelength-scale dimensions. A *scatter ratio* image is generated by taking the ratio of images collected at high and low NA in central dark-field microscopy. Such an image spatially encodes the ratio of wide to narrow angle scatter (or optical scatter image ratio [OSIR]) and hence provides a measure of local particle size. Figure 14.9 shows the OSI microscopy setup. The specimens are mounted on the stage of an inverted microscope, which can also be fitted with an epifluorescence and DIC imaging capabilities. The microscope condenser is adjusted to central Kohler illumination, with a condenser NA of 0.03 (condenser front aperture closed). For illumination, light from the microscope's Halogen lamp

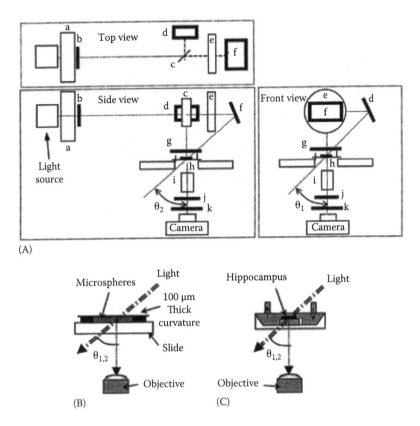

FIGURE 14.7 (A) Setup for dual-angle scattering images of hippocampal slices. (a) Shutter, (b) infrared filter, (c) beam splitter, (d) mirror, (e) shutter, (f) mirror, (g) polarizer, (h) specimen, (i) low NA objective, (j) polarizer, and (k) interference filter. (B) Stage design for experiments on microsphere suspensions. (C) Stage design for brain slice experiment. $\theta_{1,2}$ represent the two scattering angles, 31° and 34°, for which the images were acquired. (Reprinted from Johnson, L.J. et al., *J. Neurosci. Methods*, 98, 21, 2000. With permission.)

is filtered to yield an incident red beam, $\lambda = 630 \pm 5$ nm. The images are collected with a 60× oil immersion objective, NA = 1.4, and displayed on a CCD camera. In a Fourier plane conjugate to the back focal plane of the objective, a beam stop is placed in the center of an iris with variable diameter. As the inset in Figure 14.9 shows, the variable iris collects light scattered within a solid angle, bound by $2° < \theta < 10°$ for low NA and $2° < \theta < 67°$ for high NA. Two sequential dark-field images are acquired at high and low NA by manually adjusting the diameter of the variable iris. The scatter ratio image is obtained by dividing the background-subtracted high NA image by the background-subtracted low NA image.

OSI was validated on sphere suspensions and live cells [107]. Figure 14.10a shows the mean pixel value and standard deviation of OSI images collected from polystyrene sphere suspensions and plotted against sphere diameter (open circles). The optical scatter images of the suspensions are displayed to the left of the graph. The solid line represents OSIR predictions as calculated from Mie theory for $m = 1.2$ and shows excellent agreement with experiment. For comparison, the dashed line shows the theoretical prediction for $m = 1.06$. The OSIR parameter has the advantage of decreasing monotonically over a large range of sphere diameters from 0.2 to 1.5 μm. In addition, the OSIR is not significantly sensitive to refractive index changes for spheres in this size range and therefore reflects changes in morphology rather than composition.

The method was applied to cells, which naturally contain scatterers of varying size, such as the nucleus (4–15 μm) and mitochondria (0.5–2 μm). As expected, particle size variation was seen across

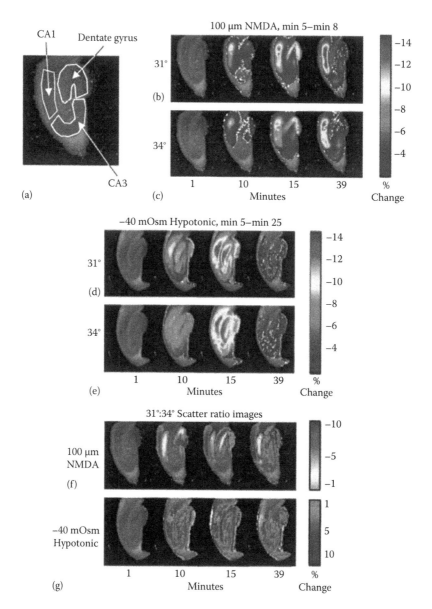

FIGURE 14.8   (See color insert.) Single-angle and dual-angle scattering images of hippocampal slices under osmotic stress and subjected to NMDA. In each case, the color bars indicate the percent change in image intensity. In regions of the slice where the change is less than ±2%, the grayscale images of the hippocampus are shown. (a) The scatter image of a hippocampal slice is shown. The three main regions of the hippocampus are CA1, CA3, and the dentate gyrus. Also shown are single-angle scatter images at 31° (b) and 34° (c) of an NMDA-treated slice. In both (b) and (c), there is a large change in the CA1 region. There is also a significant change in the dentate gyrus that fades by minute 39 of the experiment. Single-angle scattering images at 31° (d) and 34° (e) are shown for hypotonic treatment. In both (d) and (e), there is a large change indicative of cellular swelling in the CA1 region. There is also a significant change in the dentate gyrus. Dual-angle scatter ratio images are shown for NMDA (f) and hypotonic (g) treatments. The NMDA-treated slices in (f) undergo a relatively larger change in the dual-angle scatter ratio in CA1. In CA1, the scatter ratio change is negative, possibly indicating particle shrinkage. In (g) hypotonic treatment, the magnitude of the change in the scatter ratio is less in CA1. In addition, the overall location of the scatter change is more spread out than for NMDA treatment. In both (f) and (g), the white matter areas reveal a positive change in the scatter ratio. (Reprinted from Johnson, L.J. et al., *J. Neurosci. Methods*, 98, 21, 2000. With permission.)

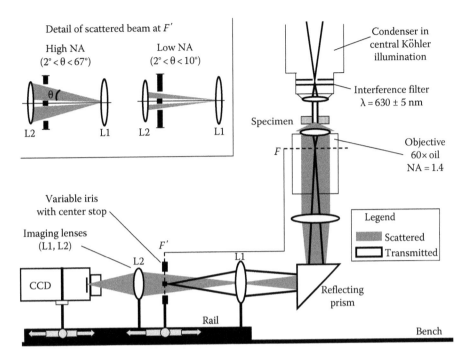

FIGURE 14.9 Apparatus for OSI of cellular monolayers. A collimated beam, λ = 630 nm (±10 nm), illuminates a sample mounted on the stage of an inverted microscope. The scattered light is collected by an oil immersion objective, NA = 1.4, and measured on a CCD camera. *F* is the objective's back focal plane. *F* and *F'* are conjugate Fourier planes. The scattered light (gray beam) is used to image the specimen on the CCD, while the transmitted light (black ray traces) is blocked at *F'*. In this setup, two images are acquired sequentially by manually varying the aperture diameter in the plane *F'*. The inset shows the scattered angles passed at the high and low NA settings of the variable diameter iris. (Reprinted from Boustany, N.N. et al., *Opt. Lett.*, 26(14), 1063, 2001. With permission.)

the cell. Figure 14.10b (top panels) shows DIC and OSI images of a normal endothelial cell. The nucleus region (N) and the cytoplasm (C) containing the mitochondria are clearly differentiated in the OSI image despite a much larger pixel size than the accompanying DIC image. In the OSI image, due to the nonlinear inverse relationship between OSIR and particle diameter, regions with small particles (C) appear brighter than regions with large particles (N). Moreover, since the value of the OSIR parameter is directly encoded at each pixel, the pixel histogram may be used to infer particle size distributions within the sample after converting the OSIR values to particle diameter using Mie theory [108]. By monitoring the cells after treatment with 2 μm staurosporine (STS) (Figure 14.10b, *t* > 0), the OSI method revealed very early, apoptosis-induced subcellular changes, which were not apparent in conventional DIC images [107]. Further studies aimed at understanding the biological sources of the scattering changes measured at the onset of apoptosis revealed that these changes were modulated by Bcl-xL [109,110] and were spatially registered with mitochondria. [111]. In a separate study, the OSI method was also used to quantify calcium-induced changes in mitochondrial shape resulting from the mitochondrial permeability transition [112]. Recent improvements in the OSI method include the use of a spatial light modulator to implement 2D Gabor filters [113,114].

Several studies have also incorporated light scattering spectroscopic data into other imaging modalities. Notably, spectroscopic optical coherence tomography and microscopy (SOCT and SOCM) were used to infer scatterer size after spectral analysis of the OCT or OCM data. These methodologies were used to identify different cell types within biological tissue sections [115], detect contrast agents [116,117], or locate strong scattering centers, such as nuclei, in the subcellular environment [118]. Several methods

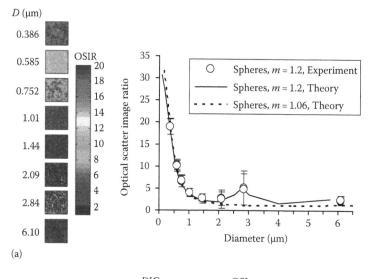

(a)

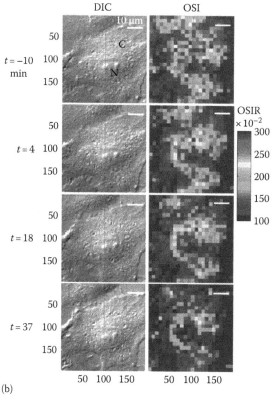

(b)

FIGURE 14.10 **(See color insert.)** (a) OSI images and measurement of the OSIR in aqueous suspensions of poly-styrene spheres ($m = 1.2$). The OSIR is a measure of wide to narrow angle scatter. Experimental data (open circles) and theoretical predictions (solid line, $m = 1.2$; dashed line, $m = 1.06$) are shown. The experimental data points show the mean pixel value and standard deviation in the scatter images displayed to the left of the graph. (Reprinted from Boustany, N.N. et al., *Opt. Lett.*, 26(14), 1063, 2001. With permission.) (b) Representative cell undergoing apoptosis after treatment with 2 μm STS. The cell was imaged in DIC (left panels) and OSI (right panels) at different time points. C, cytoplasm; N, nucleus. Images are displayed at times $t = -10$, 4, 18, and 37 min from STS addition at $t = 0$. The ratiometric scattering images show a decreasing scatter ratio within the cytoplasm (C).

can be used to analyze the scattering spectra obtained by SOCT/SOCM. These include an analysis based on scatter theory that will result in images that spatially encode scatterer size or spacing between scatterers [115]. Another method consists of dividing the spectrum into three bandwidths assigned to red, green, and blue and mapping the spectrum's centroid to a color at each location in the image [115]. Confocal light absorption and scattering spectroscopy (CLASS) combines confocal reflectance microscopy with scattering spectroscopy to produce images that map out scatterer size within the subcellular environment while achieving confocal optical sectioning [119,120] (cohCLASS). Models based on Mie theory or the T-matrix formalism discussed in Section 14.2 are used to infer particle size in CLASS. Partial wave spectroscopy was used to collect and analyze spectral fluctuations in back-reflected light from a sample illuminated with an incoherent source at low NA. The spectral fluctuations are thought to result from nanoscale axial fluctuations in refractive index at every location sampled laterally within the samples. As such, these refractive index fluctuations were mapped into a single parameter, disorder strength (L_D), which has been used successfully to image and differentiate normal and precancerous cells [121,122]. Finally, recent work in quantitative phase microscopy has been specifically aimed at extracting the scattering properties of cells and thin tissues with potentially important applications in histopathology and cell analysis. The angular scattering properties of the whole-tissue cross section may be inferred using a Fourier-based postprocessing of the field amplitude and phase collected in the image plane [123]. In addition, the scattering mean free path, l_s, can be inferred using Beer's law from measurement of the unscattered light and knowledge of the tissue section's thickness [123]. By spatially averaging the field amplitude and phase across small 2D regions as opposed to the whole-tissue slice, the scattering mean free path and anisotropy factor, g, can be mapped in two dimensions across the whole sample [124]. To this end, a scattering-phase theorem is used to derive the 2D spatial distributions of the scattering parameters l_s and g directly from the spatial variations in phase and phase gradient measured across the sample [125,126].

14.5 Summary and Conclusion

Light scattering techniques present simple and effective methods to detect subtle morphological changes, in situ, for biological particles with wavelength-scale dimensions, such as organelles or connective tissue fibers. The particles being probed need not always be individually resolved and measured by traditional morphometric methods, thus avoiding a tedious process of image recognition, particle sizing and counting. Light scattering techniques complement other microscopic methods and, in contrast to electron microscopy, require no potentially damaging cell preparation procedures.

Unlike fluorescence labeling, which results in illuminating specific biochemical markers, methods based on light scattering do not require labeling. As a result, scattering is not biochemically specific and may originate from different tissue and cellular structures, often requiring further biochemical elucidation. Nonetheless, despite having limited specificity, light scattering data have an important diagnostic value. One of the strengths of light scattering techniques is that they can reveal the presence of cellular and subcellular morphological dynamics noninvasively. Morphological information complements biochemical data and could be particularly valuable if the biochemical events and time sequence underlying a given biological behavior are not yet known to allow specific biochemical manipulation. Moreover, under certain conditions, in which only a few parameters are known to vary, scattering data can be used to efficiently automate cell differentiation and sorting during high-throughput cell analysis, as in flow cytometry.

Combining theory and experiment, various researchers have been thriving to provide novel approaches to solve the *inverse problem* and infer, from a limited light scattering data set, the morphologic and optical properties of cells and organelles. These studies are invaluable in transforming hitherto phenomenological results into fully elucidated data that could be used to design optical instruments with a high impact on biology and medicine.

References

1. Darynkiewicz, Z., Juan, G., Li, X., Goreczyca, W., Murakami, T., and Traganos, F., Cytometry in cell necrobiology: Analysis of apoptosis and accidental cell death. *Cytometry*, 1997, 27: 1–20.
2. Salzman, G.C., Sigham, S.B., Johnston, R.G., and Bohren, C.F., Light scattering and cytometry, in *Flow Cytometry and Sorting*, 2nd edn., M.R. Melamed, T. Lindmo, and M.L. Mendelsohn, eds., 1990, Wiley-Liss: New York, Chapter 5, pp. 81–107.
3. Lizard, G., Fournel, S., Genestier, L., Dhedin, N., Chaput, C., Flacher, M., Mutin, M., Panaye, G., and Revillard, J.-P., Kinetics of plasma membrane and mitochondrial alterations in cells undergoing apoptosis. *Cytometry*, 1995, 21: 275–283.
4. Ost, V., Neukammer, J., and Rinneberg, H., Flow cytometric differentiation of erythrocytes and leukocytes in dilute whole blood by light scattering. *Cytometry*, 1998, 32: 191–197.
5. Weston, K.M., Alsalami, M., and Raison, R., Cell membrane changes induced by the cytolytic peptide, melittin, are detectable by 90° laser scatter. *Cytometry*, 1994, 15: 141–147.
6. Conville, P.S., Witebsky, F.G., and MacLowry, J.D., Antomicrobial susceptibilies of mycobacteria as determined by differential light scattering and correlation with results from multiple reference laboratories. *Journal of Clinical Microbiology*, 1994, 32(6): 1554–1559.
7. Anderson, A.M., Angyal, G.N., Weaver, C.M., Felkner, I.C., Wolf, W.R., and Worthy, B.E., Potential application of laser/microbe bioassay technology for determining water-soluble vitamins in foods. *Journal of AOAC International*, 1993, 76(3): 682–690.
8. Lavergne-Mazeau, F., Maftah, A., Cenatiempo, Y., and Julien, R., Linear correlation between bacterial overexpression of recombinant peptides and cell light scatter. *Applied and Environmental Microbiology*, 1996, 62: 3042–3046.
9. Smeraldi, C., Berardi, E., and Porro, D., Monitoring of peroxisome induction and degradation by flow cytometric analysis of hansenula polymorpha cells grown in methanol and glucose media: Cell volume refractive index and FITC retention. *Microbiology*, 1994, 140: 3161–3166.
10. Hubbell, J.A., Pohl, P.I., and Wagner, W.R., The use of laser light scattering and controlled shear in platelet aggregometry. *Thrombosis and Haemostasis*, 1991, 65(5): 601–607.
11. Tohgi, H., Takashi, H., Watanabe, K., and Hiroyuki, K., Development of large platelet aggregates from small aggregates as determined by laser light scattering: Effect of aggregant concentration and antiplatelet medication. *Thrombosis and Haemostasis*, 1996, 75(5): 838–843.
12. Ozaki, Y., Satoh, K., Yatomi, Y., Yamamoto, T., and Shirasawa, Y., Detection of platelet aggregates with particle counting method using light scattering. *Analytical Biochemistry*, 1994, 218: 284–294.
13. Tedeshi, H. and Harris, D., Some observations on the photometric estimation of mitochondrial volume. *Biochimica et Biophysica Acta*, 1958, 28: 392–401.
14. Hunter, D.R. and Haworth, R.A., The Ca^{2+}-induced membrane transition in mitochondria. *Archives of Biochemistry and Biophysics*, 1979, 195(2): 453–459.
15. Bernardi, P., Vassanelli, S., Veronese, P., Colonna, R., Szabo, I., and Zoratti, M., Modulation of the mitochondrial permeability transition pore. *Journal of Biological Chemistry*, 1992, 267(5): 2934–2939.
16. Kristal, B.S. and Dubinsky, J.M., Mitochondrial permeability transition in the central nervous system: Induction by calcium cycling-dependent and -independent pathways. *Journal of Neurochemistry*, 1997, 69(2): 524–538.
17. Drezek, R., Dunn, A., and Richards-Kortum, R., Light scattering from cells: Finite-difference time-domain simulations and goniometric measurements. *Applied Optics*, 1999, 38(16): 3651–3661.
18. Mourant, J.R., Freyer, J.P., Hielscher, A.H., Eick, A.A., Chen, D., and Johnson, T.M., Mechanisms of light scattering from biological cells relevant to noninvasive optical-tissue diagnostics. *Applied Optics*, 1998, 37(16): 3586–3593.
19. Ramachandran, J., Powers, T.M., Carpenter, S., Garcia-Lopez, A., Freyer, J.P., and Mourant, J.R., Light scattering and microarchitectural differences between tumorigenic and non-tumorigenic cell models of tissue. *Optics Express*, 2007, 15(7): 4039–4053.

20. VandeHulst, H.C., *Light Scattering by Small Particles*, 1981, Dover: New York.
21. Bohren, C.F. and Huffman, D.R., *Absorption and Scattering of Light by Small Particles*, 1983, John Wiley & Sons: New York.
22. Cheong, W.F., Prahl, S.A., and Welch, A.J., A review of the optical properties of biological tissues. *IEEE Journal of Quantum Electronics*, 1990, 26(12): 2166–2185.
23. Beauvoit, B., Evans, S.M., Jenkins, T.W., Miller, E.E., and Chance, B., Correlation between the light scattering and the mitochondrial content of normal tissues and transplantable rodent tumors. *Analytical Biochemistry*, 1995, 226: 167–174.
24. Beuthan, J., Minet, O., Helfmann, J., Herrig, M., and Muller, G., The spatial variation of the refractive index in biological cells. *Physics in Medicine and Biology*, 1996, 41: 369–382.
25. Koch, A.L., Some calculations on the turbidity of mitochondria and bacteria. *Biochimica et Biophysica Acta*, 1961, 51: 429–441.
26. Wyatt, P.J., Differential light scattering: A physical method for identifying living bacterial cells. *Applied Optics*, 1968, 7(10): 1879–1896.
27. Morris, S.J., Shultens, H.A., Hellweg, M.A., Striker, G., and Jovin, T.M., Dynamics of structural changes in biological particles from rapid light scattering measurements. *Applied Optics*, 1979, 18(3): 303–311.
28. Wyatt, P.J., Light scattering and the absolute characterization of macromolecules. *Analytica Chimica Acta*, 1993, 272: 1–40.
29. Sloot, P.M.A., Hoekstra, A.G., and Figdor, C.G., Osmotic response of lymphocytes measured by means of forward light scattering: Theoretical considerations. *Cytometry*, 1988, 9: 636–641.
30. Evans, E., Comparison of the diffraction theory of image formation with the three-dimensional, first Born scattering approximation in lens systems. *Optics Communications*, 1970, 2(7): 317–320.
31. Turke, B., Seger, G., Achatz, M., and Seelen, W.V., Fourier optical approach to the extraction of morphological parameters from the diffraction pattern of biological cells. *Applied Optics*, 1978, 17(17): 2754–2761.
32. Burger, D.E., Jett, J.H., and Mullaney, P.F., Extraction of morphological features from biological models and cells by Fourier analysis of static light scatter measurements. *Cytometry*, 1982, 2(5): 327–336.
33. Banada, P.P., Guo, S.L., Bayraktar, B., Bae, E., Rajwa, B., Robinson, J.P., Hirleman, E.D., and Bhunia, A.K., Optical forward-scattering for detection of *Listeria* monocytogenes and other *Listeria* species. *Biosensors and Bioelectronics*, 2007, 22(8): 1664–1671.
34. Graaff, R., Aarnoudse, J.G., Zijp, J.R., Sloot, P.M.A., de Mul, F.F., Grieve, J., and Kolink, M.H., Reduced light scattering properties for mixtures of spherical particles: A simple approximation derived from Mie calculation. *Applied Optics*, 1992, 31(10): 1370–1376.
35. Ovryn, B. and Izen, S.H., Imaging of transparent spheres through a planar interface using a high-numerical-aperture optical microscope. *Journal of the Optical Society of America A*, 2000, 17(7): 1202–1213.
36. Schmitt, J.M. and Kumar, G., Optical scattering properties of soft tissue: A discrete particle model. *Applied Optics*, 1998, 37(13): 2788–2797.
37. Zonios, G., Perelman, L.T., Backman, V., Manoharan, R., Fitzmaurice, M., Dam, J.V., and Feld, M.S., Diffuse reflectance spectroscopy of human adenomatous colon polyps in vivo. *Applied Optics*, 1999, 38(31): 6628–6637.
38. Johnsen, S. and Widder, E.A., The physical basis of transparency in biological tissue: Ultrastructure and the minimization of light scattering. *Journal of Theoretical Biology*, 1999, 199: 181–198.
39. Dunn, A. and Richards-Kortum, R., Three-dimensional computation of light scattering from cells. *IEEE Journal of Selected Topics in Quantum Electronics*, 1996, 2(4): 898–905.
40. Li, X., Chen, Z.G., Gong, J.M., Taflove, A., and Backman, V., Analytical techniques for addressing forward and inverse problems of light scattering by irregularly shaped particles. *Optics Letters*, 2004, 29(11): 1239–1241.

41. Waterman, P.C., Matrix formulation of electro-magnetic scattering. *Proceedings of the IEEE*, 1965, 53: 805–812.
42. Waterman, P.C., Symmetry, unitarity, and geometry in electro-magnetic scattering. *Physical Review D*, 1971, 3: 825–839.
43. Yee, S.K., Numerical solution of initial boundary value problems involving Maxwell's equations in isotropic media. *IEEE Transactions on Antennas and Propagation*, 1966, 14: 302–307.
44. Barber, P.W. and Hill, S.C., *Light Scattering by Particles: Computational Methods*, 1990, World Scientific: Singapore.
45. Taflove, A. and Hagness, S.C., *Computational Electrodynamics, The Finite Difference Time Domain Method,* 3rd edn., 2005, Artech House, Inc: Norwood, MA.
46. Mishchenko, M.I., Travis, L.D., and Macke, A., T-matrix method and its applications, in *Light Scattering by Nonspherical Particles: Theory Measurements and Applications*, M.I. Mishchenko, J.W. Hovenier, and L.D. Travis, eds., 2000, Academic Press: San Diego, CA, pp. 147–173.
47. Nilsson, A.M.K., Alsholm, P.L., Karlsson, A., and Andersson-Engles, S., T-matrix computations of light scattering by red blood cells. *Applied Optics*, 1998, 37(13): 2735–2748.
48. Giacomelli, M.G., Chalut, K.J., Ostrander, J.H., and Wax, A., Application of the T-matrix method to determine the structure of spheroidal cell nuclei with angle-resolved light scattering. *Optics Letters*, 2008, 33(21): 2452–2454.
49. Drezek, R., Dunn, A., and Richards-Kortum, R., A pulsed finite-difference time-domain (FDTD) method for calculating light scattering from biological cells over broad wavelength ranges. *Optics Express*, 2000, 6(7): 148–157.
50. Tseng, S.H., Greene, J.H., Taflove, A., Maitland, D., Backman, V., and Walsh, J., Exact solution of Maxwell's equations for optical interactions with a macroscopic random medium. *Optics Letters*, 2004, 29(12): 1393–1395.
51. Tseng, S.H., Greene, J.H., Taflove, A., Maitland, D., Backman, V., and Walsh, J.T., Exact solution of Maxwell's equations for optical interactions with a macroscopic random medium: Addendum. *Optics Letters*, 2005, 30(1): 56–57.
52. Bolt, R.A. and deMul, F.F.M., Goniometric instrument for light scattering measurement of biological tissues and phantoms. *Review of Scientific Instruments*, 2002, 73(5): 2211–2213.
53. Wilson, J.D., Cottrell, W.J., and Foster, T.H., Index-of-refraction-dependent subcellular light scattering observed with organelle-specific dyes. *Journal of Biomedical Optics*, 2007, 12(1): 014010.
54. Bickel, W.S., Davidson, J.F., Huffman, D.R., and Kilkson, R., Application of polarization effects in light scattering: A new biophysical tool. *Proceedings of National Academy of Sciences USA*, 1976, 73(2): 486–490.
55. Salzman, G.C., Crowell, J.M., Martin, J.C., Trujillo, T.T., Romero, A., Mullaney, P.F., and Labauve, P.M., Cell classification by laser light scattering: Identification and separation of unstained leukocytes. *Acta Cytologica (Praha)*, 1975, 19: 374–377.
56. deGrooth, B.G., Terstappen, L.W.M.M., Puppels, G.J., and Greve, J., Light-scattering polarization measurements is a new parameter in floe cytometry. *Cytometry*, 1987, 8(6): 539–544.
57. Lehninger, A.L., Reversal of thyroxine-induced swelling of rat liver mitochondria by adenosine triphosphate. *Journal of Biological Chemistry*, 1959, 234(8): 2187–2195.
58. Packer, L., Metabolic and structural states of mitochondria. *Journal of Biological Chemistry*, 1960, 235(1): 242–249.
59. Hackenbrock, C.R., Ultrastructural bases for metabolically linked mechanical activity in mitochondria I. *Journal of Cell Biology*, 1966, 30: 269–297.
60. Packer, L., Energy-linked low amplitude mitochondrial swelling, in *Methods in Enzymology*, Vol. 10, R.W. Estabrook and M.E. Pullman, eds., 1967, Academic Press: New York, pp. 685–689.
61. Harris, R.A., Asbell, M.A., Asai, J., Jolly, W.W., and Green, D.E., The conformational basis of energy transduction in membrane systems. V. Measurement of configurational changes by light scattering. *Archives of Biochemistry and Biophysics*, 1969, 132: 545–560.

62. Hunter, D.R., Haworth, R.A., and Southward, J.H., Relationship between configuration, function, and permeability in calcium-treated mitochondria. *Journal of Biological Chemistry*, 1976, 251(16): 5069–5077.

63. Halestrap, A.P., Regulation of mitochondrial metabolism through changes in matrix volume. *Biochemical Society Transactions*, 1994, 22(2): 522–529.

64. Territo, P.R., French, S.A., Dunleavy, M.C., Evans, F.J., and Balaban, R.S., Calcium activation of heart mitochondrial oxidative phosphorylation. *Journal of Biological Chemistry*, 2001, 276(4): 2586–2599.

65. HunterJr, F.E. and Smith, E.E., Measurement of mitochondrial swelling and shrinking—High amplitude, in *Methods in Enzymology*, R.W. Estabrook and M.E. Pullman, eds., 1967, Academic Press: New York, pp. 689–696.

66. Bernardi, P., Modulation of the mitochondrial cyclosporin A-sensitive permeability transition pore by the proton electrochemical gradient. *Journal of Biological Chemistry*, 1992, 267(13): 8834–8839.

67. Petronilli, V., Constantini, P., Scorrano, L., Colonna, R., Passamonti, S., and Bernardi, P., The voltage sensor of the mitochondrial permeability transition pore is tuned by the oxidation–reduction state of vicinal thiols. *Journal of Biological Chemistry*, 1994, 269(24): 16638–16642.

68. Hoek, J.B., Farber, J.L., Thomas, A.P., and Wang, X., Calcium ion-dependent signaling and mitochondrial dysfunction: Mitochondrial calcium uptake during hormonal stimulation in intact liver cells and its implication for the mitochondrial permeability transition. *Biochimica et Biophysica Acta*, 1995, 1271: 93–102.

69. Constantini, P., Chernyak, B.V., Petronilli, V., and Bernardi, P., Modulation of the mitochondrial permeability transition pore by pyridine nucleotides and dithiol oxidation at two separate sites. *Journal of Biological Chemistry*, 1996, 271(12): 6746–6751.

70. Scorrano, L., Petronilli, V., and Bernardi, P., On the voltage dependence of the mitochondrial permeability transition pore. *Journal of Biological Chemistry*, 1997, 272(19): 12295–12299.

71. Pfeiffer, D.R., Kuo, T.H., and Chen, T.T., Some effect of Ca^{2+}, Mg^{2+}, and Mn^{2+} on the ultrastructure, light scattering properties, and malic enzyme activity of adrenal cortex mitochondria. *Archives of Biochemistry and Biophysics*, 1976, 176(2): 556–563.

72. Petronilli, V., Cola, C., Massari, S., Colonna, R., and Bernardi, P., Physiological effectors modify voltage sensing by the cyclosporin A-sensitive permeability transition pore of mitochondria. *Journal of Biological Chemistry*, 1993, 268(29): 21939–21945.

73. Vander-Heiden, M.G., Chandel, N.S., Williamson, E.K., Schumacker, P.T., and Thompson, C.B., Bcl-xL regulates the membrane potential and volume homeostasis of mitochondria. *Cell*, 1997, 91: 627–637.

74. Zamzami, N., Susin, S.A., Marchetti, P., Hirsh, T., Gomez-Monterrey, I., Castedo, M., and Kroemer, G., Mitochondrial control of apoptosis. *The Journal of Experimental Medicine*, 1996, 183: 1533–1544.

75. Jurgensmeier, J.M., Xie, Z., Deveraux, Q., Ellerby, L., Bredesen, D., and Reed, J.C., Bax directly induces release of cytochrome c from isolated mitochondria. *Proceedings of the National Academy of Sciences USA*, 1998, 95: 4997–5002.

76. Narita, M., Shimizu, S., Ito, T., Chittenden, T., Lutz, R.J., Matsuda, H., and Tsujimoto, Y., Bax interacts with the permeability transition pore to induce permeability transition and cytochrome c release in isolated mitochondria. *Proceedings of the National Academy of Sciences USA*, 1998, 95: 14681–14686.

77. Finucane, D.M., Bossy-Wetzel, E., Waterhouse, N.J., Cotter, T.G., and Green, D.R., Bax-induced caspase activation and apoptosis via cytochrome c release from mitochondria is inhibitable by Bcl-xL. *Journal of Biological Chemistry*, 1999, 274(4): 2225–2233.

78. Bryant, F.D., Latimer, P., and Seiber, B.A., Changes in total light scattering and absorption caused by changes in particle conformation—A test of theory. *Archives of Biochemistry and Biophysics*, 1969, 135: 109–117.

79. Latimer, P. and Pyle, B.E., Light scattering at various angles. Theoretical predictions of the effects of particle volume change. *Biophysical Journal*, 1972, 12: 764–773.

80. Knight, V.A., Wiggins, P.M., Harvey, J.D., and O'Brien, J.A., The relationship between the size of mitochondria and the intensity of light that they scatter in different states. *Biochimica et Biophysica Acta*, 1981, 637: 146–151.

81. Mourant, J.R., Johnson, T.M., and Freyer, J.P., Characterizing mammalian cells and cell phantoms by polarized backscattering fiber-optic measurements. *Applied Optics*, 2001, 40(28): 5114–5123.

82. Mourant, J.R., Johnson, T.M., Carpenter, S., Guerra, A., Aida, T., and Freyer, J.P., Polarized angular dependent spectroscopy of epithelial cells and epithelial cell nuclei to determine the size scale of scattering structures. *Journal of Biomedical Optics*, 2002, 7(3): 378–387.

83. Marina, O.C., Sanders, C.K., and Mourant, J.R., Correlating light scattering with internal cellular structures. *Biomedical Optics Express*, 2012, 3(2): 296–312.

84. Wilson, J.D. and Foster, T.H., Characterization of lysosomal contribution to whole-cell light scattering by organelle ablation. *Journal of Biomedical Optics*, 2007, 12(3): 3.

85. Wilson, J.D. and Foster, T.H., Mie theory interpretations of light scattering from intact cells. *Optics Letters*, 2005, 30(18): 2442–2444.

86. Valentine, M.T., Popp, A.K., Weitz, D.A., and Kaplan, P.D., Microscope-based static light-scattering instrument. *Optics Letters*, 2001, 26(12): 890–892.

87. Banada, P.P., Huff, K., Bae, E., Rajwa, B., Aroonnual, A., Bayraktar, B., Adil, A., Robinson, J.P., Hirleman, E.D., and Bhunia, A.K., Label-free detection of multiple bacterial pathogens using light-scattering sensor. *Biosensors and Bioelectronics*, 2009, 24(6): 1685–1692.

88. Cottrell, W.J., Wilson, J.D., and Foster, T.H., Microscope enabling multimodality imaging, angle-resolved scattering, and scattering spectroscopy. *Optics Letters*, 2007, 32(16): 2348–2350.

89. Smith, Z.J. and Berger, A.J., Construction of an integrated Raman- and angular-scattering microscope. *Review of Scientific Instruments*, 2009, 80(4): 044302-1–044301-8.

90. Schiffer, Z., Ashkenazy, Y., Tirosh, R., and Deutsch, M., Fourier analysis of light scattered by elongated scatterers. *Applied Optics*, 1999, 38(16): 3626–3635.

91. Shiffer, Z., Zurgil, N., Shafran, Y., and Deutsch, M., Analysis of laser scattering pattern as an early measure of apoptosis. *Biochemical and Biophysical Research Communications*, 2001, 289: 1320–1327.

92. Bae, E., Bai, N., Aroonnual, A., Robinson, J.P., Bhunia, A.K., and Hirleman, E.D., Modeling light propagation through bacterial colonies and its correlation with forward scattering patterns. *Journal of Biomedical Optics*, 2010, 15(4): 0445001-1–0445001-11.

93. Kovach, I.S. and Athanasiou, K.A., Small-angle HeNe laser light scatter and the compressive modulus of articular cartilage. *Journal of Orthopaedic Research*, 1997, 15: 437–441.

94. Schmitt, J.M. and Kumar, G., Turbulent nature of refractive index variations in biological tissue. *Optics Letters*, 1996, 21(16): 1310–1312.

95. Wax, A., Yang, C., Backman, V., Badizadegan, K., Boone, C.W., Dasari, R.R., and Feld, M.S., Cellular organization and substructure measured using angle-resolved low-coherence interferometry. *Biophysical Journal*, 2002, 82: 2256–2264.

96. Zhu, Y.Z., Terry, N.G., and Wax, A., Angle-resolved low-coherence interferometry: An optical biopsy technique for clinical detection of dysplasia in Barrett's esophagus. *Expert Review of Gastroenterology and Hepatology*, 2012, 6(1): 37–41.

97. Chalut, K.J., Kulangara, K., Wax, A., and Leong, K.W., Stem cell differentiation indicated by noninvasive photonic characterization and fractal analysis of subcellular architecture. *Integrative Biology*, 2011, 3(8): 863–867.

98. Chalut, K.J., Ostrander, J.H., Giacomelli, M.G., and Wax, A., Light scattering measurements of subcellular structure provide noninvasive early detection of chemotherapy-induced apoptosis. *Cancer Research*, 2009, 69(3): 1199–1204.

99. Aitken, P.G., Fayuk, D., Somjen, G.G., and Turner, D.A., Use of intrinsic optical signals to monitor physiological changes in brain tissue slices. *Methods: A Companion to Methods in Enzymology*, 1999, 18: 91–103.

100. Andrew, R.D., Lobinowich, M.E., and Osehobo, E.P., Evidence against volume regulation by cortical brain cells during acute osmotic stress. *Experimental Neurology*, 1997, 143: 300–312.

101. Andrew, R.D., Adams, J.R., and Polischuk, T.M., Imaging NMDA- and kainate-induced intrinsic optical signals from the hippocampal slice. *Journal of Neurophysiology*, 1996, 76(4): 2707–2717.

102. Bahar, S., Fayuk, D., Somjen, G.G., Aitken, P.G., and Turner, D.A., Mitochondrial and intrinsic optical signals imaged during hypoxia and spreading depression in rat hippocampal slices. *Journal of Neurophysiology*, 2000, 84: 311–324.

103. Rector, D.M., Poe, G.R., Kristensen, M.P., and Harper, R.M., Light scattering changes follow evoked potentials from hippocampal Schaeffer collateral stimulation. *Journal of Neurophysiology*, 1997, 78: 1707–1713.

104. Johnson, L.J., Hanley, D.F., and Thakor, N.V., Optical light scatter imaging of cellular and subcellular morphology changes in stressed rat hippocampal slices. *Journal of Neuroscience Methods*, 2000, 98: 21–31.

105. Andrew, R.D., Jarvis, C.R., and Obeidat, A.S., Potential sources of intrinsic optical signals imaged in live brain slices. *Methods: A Companion to Methods in Enzymology*, 1999, 18: 185–196.

106. Johnson, L.J., Chung, W., Hanley, D.F., and Thakor, N.V., Optical scatter imaging detects mitochondrial swelling in living tissue slices. *Neuroimage*, 2002, 17(3): 1649–1657.

107. Boustany, N.N., Kuo, S.C., and Thakor, N.V., Optical scatter imaging: Subcellular morphometry in situ with Fourier filtering. *Optics Letters*, 2001, 26(14): 1063–1065.

108. Zheng, J.Y. and Boustany, N.N., Alterations in the characteristic size distributions of subcellular scatterers at the onset of apoptosis: Effect of Bcl-x(L) and Bax/Bak. *Journal of Biomedical Optics*, 2010, 15(4): 045002.

109. Boustany, N.N., Tsai, Y.C., Pfister, B., Joiner, W.M., Oyler, G.A., and Thakor, N.V., BCL-x(L)-dependent light scattering by apoptotic cells. *Biophysical Journal*, 2004, 87(6): 4163–4171.

110. Zheng, J., Tsai, Y., Kadimcherla, P., Zhang, R., Shi, J., Oyler, G.A., and Boustany, N.N., The C-terminal transmembrane domain of Bcl-xL mediates changes in mitochondrial morphology. *Biophysical Journal*, 2008, 94(1): 286–297, PMCID:PMC2134878.

111. Pasternack, R.M., Zheng, J.-Y., and Boustany, N.N., Optical scatter changes at the onset of apoptosis are spatially associated with mitochondria. *Journal of Biomedical Optics Letters*, 2010, 15(4): 040504.

112. Boustany, N.N., Drezek, R., and Thakor, N.V., Calcium-induced alterations in mitochondrial morphology quantified in situ with optical scatter imaging. *Biophysical Journal*, 2002, 83: 1691–1700.

113. Pasternack, R.M., Qian, Z., Zheng, J., Metaxas, D.N., White, E., and Boustany, N.N., Measurement of subcellular texture by optical fourier filtering with a micromirror device. *Optics Letters*, 2008, 33(19): 2209–2211.

114. Pasternack, R.M., Qian, Z., Zheng, J., Metaxas, D.N., White, E., and Boustany, N.N., Measurement of subcellular texture by optical fourier filtering with a micromirror device—Erratum. *Optics Letters*, 2009, 34: 1939.

115. Oldenburg, A.L., Xu, C.Y., and Boppart, S.A., Spectroscopic optical coherence tomography and microscopy. *IEEE Journal of Selected Topics in Quantum Electronics*, 2007, 13(6): 1629–1640.

116. Xu, C.Y., Ye, J., Marks, D.L., and Boppart, S.A., Near-infrared dyes as contrast-enhancing agents for spectroscopic optical coherence tomography. *Optics Letters*, 2004, 29(14): 1647–1649.

117. Oldenburg, A.L., Hansen, M.N., Ralston, T.S., Wei, A., and Boppart, S.A., Imaging gold nanorods in excised human breast carcinoma by spectroscopic optical coherence tomography. *Journal of Materials Chemistry*, 2009, 19(35): 6407–6411.

118. Vinegoni, C., Ralston, T., Tan, W., Luo, W., Marks, D.L., and Boppart, S.A., Integrated structural and functional optical imaging combining spectral-domain optical coherence and multiphoton microscopy. *Applied Physics Letters*, 2006, 88(5): 053901-1–053901-3.

119. Itzkan, I., Qiu, L., Fang, H., Zaman, M.M., Vitkin, E., Ghiran, L.C., Salahuddin, S. et al. Confocal light absorption and scattering spectroscopic microscopy monitors organelles in live cells with no exogenous labels. *Proceedings of the National Academy of Sciences USA*, 2007, 104(44): 17255–17260.

120. Fang, H., Qiu, L., Vitkin, E., Zaman, M.M., Andersson, C., Salahuddin, S., Kimerer, L.M. et al. Confocal light absorption and scattering spectroscopic microscopy. *Applied Optics*, 2007, 46(10): 1760–1769.

121. Subramanian, H., Pradhan, P., Liu, Y., Capoglu, I.R., Li, X., Rogers, J.D., Heifetz, A., Kunte, D., Roy, H.K., Taflove, A., and Backman, V., Optical methodology for detecting histologically unapparent nanoscale consequences of genetic alterations in biological cells. *Proceedings of the National Academy of Sciences USA*, 2008, 105(51): 20118–20123.

122. Damania, D., Subramanian, H., Tiwari, A.K., Stypula, Y., Kunte, D., Pradhan, P., Roy, H.K., and Backman, V., Role of cytoskeleton in controlling the disorder strength of cellular nanoscale architecture. *Biophysical Journal*, 2010, 99(3): 989–996.

123. Ding, H.F., Nguyen, F., Boppart, S.A., and Popescu, G., Optical properties of tissues quantified by Fourier-transform light scattering. *Optics Letters*, 2009, 34(9): 1372–1374.

124. Ding, H.F., Wang, Z., Liang, X., Boppart, S.A., Tangella, K., and Popescu, G., Measuring the scattering parameters of tissues from quantitative phase imaging of thin slices. *Optics Letters*, 2011, 36(12): 2281–2283.

125. Wang, Z., Ding, H.F., and Popescu, G., Scattering-phase theorem. *Optics Letters*, 2011, 36(7): 1215–1217.

126. Xu, M., Scattering-phase theorem: Anomalous diffraction by forward-peaked scattering media. *Optics Express*, 2011, 19(22): 21643–21651.

15

Tissue Viability Imaging

Gert E. Nilsson
WheelsBridge AB

15.1 Introduction

The microcirculation of the skin comprises capillaries, small arteries (arterioles), small veins (venules), and arteriovenous anastomosis (AVA) shunts [1]. The blood perfusion through the capillaries supplies the tissue with oxygen, nutrients, and water while removing waste metabolites. While the capillary bed is fed by the arterioles and drained by the venules, the AVA shunts play an important role in the thermal regulation of the body. Most of these vessels reside in the upper layer of the dermis—just below the epidermis—and can be probed by the use of light in the visible and near-infrared region. An excessive amount of red blood cells (RBCs) in the skin microvascular network due to vasodilatation results in *erythema*, while vasoconstriction of the microvessels causes *blanching*. Naked eye observation has, until recently, been the main method of determining *erythema* and *blanching* in skin testing. These evaluations are generally performed by a skilled dermatologist employing a five-point severity scale to quantify the local skin reaction to a topically applied substance. Since naked eye observation is a highly subjective method, it can be difficult to attain reproducibility and to compare results reported by different researchers performing their studies in different laboratories. Consequently, there is a need for more objective, quantitative, and versatile methods in the assessment of skin *erythema* and *blanching* caused by the intake of vasoactive drugs, applications of agents on the skin surface, environmental factors, or disease. Since skin microcirculation is sensitive to applied pressure and heat, such methods should preferably be noninvasive and designed for remote use without touching the skin. As skin microcirculation further possesses substantial spatial variability [2], imaging techniques are to be preferred to single-point measurements. As skin testing frequently involves scanning a panel of individuals in rapid sequence, ease of use and productivity are the design criteria of major concern. The emerging technology of digital camera polarization spectroscopy—tissue viability imaging (TiVi)—fulfills these requirements [3].

Since advanced digital cameras were introduced in the 1990s, these devices have developed quickly and are increasingly used as part of advanced imaging systems in a variety of medical and other applications. To reduce the influence of specular reflections of the flash in the surface of the object, a cross-polarization filter technique is frequently used [4]. A photo is captured through red, green, and blue filters attached to the camera photodetector array, and the corresponding color photo—combined from the three color planes—is stored electronically in the camera memory. By transferring this photo to a computer for further information processing, images can be generated that visualize the concentration of specific chromophores in tissue such as the hemoglobin molecules in the RBCs.

This chapter first introduces a theoretical background to TiVi based on a two-layer model for diffuse light scattering in tissue, and then in Section 15.3, the implementation of a versatile device using standard digital cameras equipped with cross-polarization filters is described in further detail. The performance of this device has been evaluated by way of Monte Carlo simulations and by using a tubing system mimicking the microcirculation in the skin. In Section 15.5, examples of early in vivo applications of TiVi in human skin are addressed and toolboxes designed for special applications are described. Finally, some of the advantages and drawbacks of TiVi as compared to those of competing technologies are discussed.

15.2 Theory

Linearly polarized broadband light from the camera flash or a separate light source is partly reflected or absorbed by the upper layer of the skin and partly diffusely scattered in the dermal layers where the microvascular network is located (Figure 15.1). Most of the directly reflected light or light that suffers only a few scattering events preserves its state of linear polarization, while the light diffusely scattered in the tissue successively becomes randomly polarized before it is either absorbed or backscattered [5]. The backscattered linearly polarized light directed toward the photodetector array in the camera is effectively blocked by a filter with a polarization direction perpendicular to that of the linearly polarized light illuminating the skin. A portion of the randomly polarized light emanating from the deeper tissue layers passes through this filter and reaches the detector. The green component of the light reaching the detector is attenuated due to a high absorption in RBCs, while the red component is virtually unaltered because of its relatively low absorption in RBCs [6–8]. The surrounding tissue absorbs green and red light to approximately the same amount. Numerical data for the scattering and absorption coefficients for dermal tissue, deoxygenated RBCs, oxygenated RBCs, and epidermis are listed in Table 15.1. The values

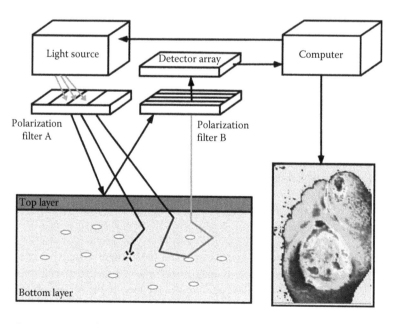

FIGURE 15.1 Operating principle of TiVi. Black arrows indicate linearly polarized light. Gray arrows indicate randomly polarized light.

TABLE 15.1 Average Approximate Absorption (μ_a) and Scattering (μ_s) Coefficients in cm^{-1} for Various Chromophores in Human Skin Tissue for the Red ($\Delta\lambda_r$) and Green ($\Delta\lambda_g$) Wavelength Regions

	Tissue	RBC_{oxy}	RBC_{deoxy}	Epidermis
$\mu_a(\Delta\lambda_r)$	2.7	3.5	25	35
$\mu_s(\Delta\lambda_r)$	187	861	861	450
$\mu_a(\Delta\lambda_g)$	3	177	201	40
$\mu_s(\Delta\lambda_g)$	223	920	920	535

are to be regarded as guideline values only, because exact quantitative determination of tissue optical properties is cumbersome and the results vary among investigators.

TiVi takes advantage of this wavelength dependence in RBC absorption. By first separating the photo color matrixes and then applying an algorithm in which a fraction of the value of each element in the green color matrix is subtracted from the corresponding value in the red color matrix, normalized with the value in the red color matrix, an output matrix representing the local RBC concentration is generated.

A two-layer model is used to theoretically analyze the spectral content of the backscattered light at different skin RBC concentrations (Figure 15.1). The thin model top layer comprises the epidermis (approximately 50–100 µm on forearm skin) that primarily acts as an absorption filter at the actual wavelengths. The model bottom layer is infinitely deep and is composed of a uniform mixture of tissue and RBCs at different concentrations. The amount of diffusely backscattered light from the bottom layer can be described by the Kubelka–Munk theory [9].

The wavelength-dependent intensity of backscattered light from the skin with a polarization perpendicular to that of the incident polarized light is described by the equation [4]

$$I_{per}(\Delta\lambda) \propto k_0\left(\Delta\lambda\right)I_0 T_{epid}(\Delta\lambda)R_d(\Delta\lambda) \tag{15.1}$$

where
  $k_0(\Delta\lambda)$ represents the fraction of the total light intensity I_0 within the wavelength interval $\Delta\lambda$, incident on the skin
  $T_{epid}(\Delta\lambda)$ represents the transmission properties of the epidermal layer (model top layer)
  $R_d(\Delta\lambda)$ represents the fraction of diffusely reflected light within the wavelength interval $\Delta\lambda$, from the dermis (model bottom layer)

The *tissue viability index* ($TiVi_{index}$) representing the local RBC concentration is defined as

$$TiVi_{index} = k_{gain}\left(\frac{I_{per}(\Delta\lambda_r) - k_1 I_{per}(\Delta\lambda_g)}{I_{per}(\Delta\lambda_r)}\right) \tag{15.2}$$

where
  $I_{per}(\Delta\lambda_r)$ and $I_{per}(\Delta\lambda_g)$ represent the wavelength-dependent intensity of backscattered light in the red and green wavelength regions, respectively, with a polarization perpendicular to that of the incident polarized light
  k_1 is a constant that can be fitted for best algorithm performance
  k_{gain} is the gain factor

For the digital cameras employed, $\Delta\lambda_r$ and $\Delta\lambda_g$ are typically in the order of 100 nm. Inserting Equation 15.1 into 15.2, cancelling the I_0 factor, and setting $k = k_1 k_0(\Delta\lambda_g)/k_0(\Delta\lambda_r)$, the algorithm expands to

$$TiVi_{index} = k_{gain} \frac{T_{epid}(\Delta\lambda_r)R_d(\Delta\lambda_r) - kT_{epid}(\Delta\lambda_g)R_d(\Delta\lambda_g)}{T_{epid}(\Delta\lambda_r)R_d(\Delta\lambda_r)} \tag{15.3}$$

By this arrangement, $TiVi_{index}$ becomes independent of the total light intensity I_0 and relates only to the diffusely backscattered light from the bottom layer, the amount of which to some degree is modulated by the absorption in the model top layer. Since the numerator approaches zero for bloodless tissue, $TiVi_{index}$ becomes close to zero for low RBC concentrations. The exact tuning is balanced by fitting the constant k. First considering a thin epidermal layer with low melanin content, the value of T_{epid} can be set to 1 and Equation 15.3 is reduced to

$$TiVi_{index} = k_{gain}\left(\frac{R_d(\Delta\lambda_r) - kR_d(\Delta\lambda_g)}{R_d(\Delta\lambda_r)}\right) \tag{15.4}$$

In Equation 15.4, $TiVi_{index}$ depends only of the fraction of backscattered light $R_d(\Delta\lambda_r)$ and $R_d(\Delta\lambda_g)$ within the red and green wavelength intervals. These fractions can be linked to the RBC concentration by the use of the Kubelka–Munk [9] theory and its modification [10] for an infinite uniform tissue layer (model bottom layer):

$$R_d(\Delta\lambda) = 1 + \frac{K(\Delta\lambda)}{S(\Delta\lambda)} - \left(\frac{K(\Delta\lambda)^2}{S(\Delta\lambda)^2} + 2\frac{K(\Delta\lambda)}{S(\Delta\lambda)}\right)^{1/2} \tag{15.5}$$

where K and S are Kubelka–Munk absorption and scattering coefficients. Using a derivation from diffusion theory [10,11], and assuming isotropic scattering, it can be shown that

$$\frac{\mu_a(\Delta\lambda)}{\mu_s(\Delta\lambda)} = \frac{3K(\Delta\lambda)}{8S(\Delta\lambda)} \tag{15.6}$$

where
  $\mu_a(\Delta\lambda)$ is the absorption coefficient
  $\mu_s(\Delta\lambda)$ is the scattering coefficient

This implies that the diffusely reflected light can be expressed as

$$R_d(\Delta\lambda) = 1 + \frac{8\mu_a(\Delta\lambda)}{3\mu_s(\Delta\lambda)} - \left(\frac{(8\mu_a(\Delta\lambda))^2}{(3\mu_s(\Delta\lambda))^2} + 2\frac{8\mu_a(\Delta\lambda)}{3\mu_s(\Delta\lambda)}\right)^{1/2} \tag{15.7}$$

The total absorption and scattering coefficients can each be considered to be composed of two parts—one that relates to the RBC ($\mu_{aRBC}(\Delta\lambda)$ and $\mu_{sRBC}(\Delta\lambda)$) and one that relates to the remaining tissue ($\mu_{aTISSUE}(\Delta\lambda)$ and $\mu_{sTISSUE}(\Delta\lambda)$). The total tissue absorption $\mu_a(\lambda)$ and scattering coefficient $\mu_s(\lambda)$ can be regarded as a linear combination of these two parts [3]:

$$\mu_a(\lambda) = RBC_f\mu_{aRBC}(\lambda) + (1 - RBC_f)\mu_{aTISSUE}(\lambda) \tag{15.8}$$

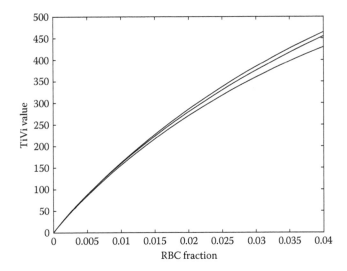

FIGURE 15.2 Simulated $TiVi_{index}$ versus RBC_f for 0% (lower curve), 70% (middle curve), and 100% (upper curve) oxygen saturation using values given in Table 15.1.

$$\mu_s(\lambda) = RBC_f \mu_{sRBC}(\lambda) + (1 - RBC_f)\mu_{sTISSUE}(\lambda) \tag{15.9}$$

where RBC_f represents the fraction of RBCs occupying the tissue volume. Equations 15.4 through 15.9 thus relate the $TiVi_{index}$ and fractions of backscattered light ($R_d(\Delta\lambda_r)$ and $R_d(\Delta\lambda_g)$) to the RBC concentration in dermal tissue. With data from Table 15.1 inserted in the equations, the theoretical relationship between tissue relative fraction of RBC and $TiVi_{index}$ is plotted for 0%, 70%, and 100% oxygen saturation, respectively, in Figure 15.2.

The influence of a thin melanin layer in the epidermis can be modeled as indicated in Equation 15.1, by way of an absorption filter with the transmission function

$$T_{epid}(\Delta\lambda) = e^{-2\mu_{aEPID}(\Delta\lambda)x} \tag{15.10}$$

where
μ_{aEPID} is the wavelength-dependent absorption coefficient of the epidermal layer
x is the layer thickness

Inserting Equation 15.10 into 15.3 yields

$$TiVi_{index} = k_{gain} \frac{R_d(\Delta\lambda_r) - ke^{-2(\mu_{aEPID}(\Delta\lambda_g)-\mu_{aEPID}(\Delta\lambda_r))x}R_d(\Delta\lambda_g)}{R_d(\Delta\lambda_r)} \tag{15.11}$$

In Caucasian forearm skin and in the skin of the back where most skin testing is performed, the epidermal layer thickness amounts to 50–100 μm, resulting in a numerical value of $e^{-2(\mu_{aEPID}(\Delta\lambda_g)-\mu_{aEPID}(\Delta\lambda_r))x}$ of about 0.90–0.95 based on values inserted from Table 15.1. If the melanin content in the epidermal layer is higher or if melanin is present in the subepidermal layers as well, a greater influence of melanin on the $TiVi_{index}$ must, however, be expected. The influence of epidermal layer thickness on the $TiVi_{index}$ is calculated from Equation 15.11 and displayed for values of x ranging from 50 to 100 μm (oxygenated blood) in Figure 15.3.

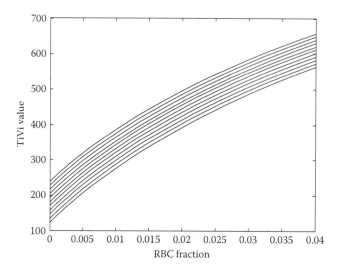

FIGURE 15.3   Simulated $TiVi_{index}$ versus RBC_f for epidermal layer thickness ranging from 50 μm (lower curve) to 100 μm (upper curve) using values given in Table 15.1.

It can be concluded from Figure 15.3 that a change in epidermal layer thickness offsets the $TiVi_{index}$, while the gain factor is affected to a minor extent only.

15.3 Instrumentation

Based on the theory for the detection of diffusely backscattered light in tissue, a system for visualization of RBC concentration in dermal tissue was developed by way of polarization light spectroscopy imaging (TiVi600, WheelsBridge AB, Linköping, Sweden [12]). The system comprises a digital camera (Canon PowerShot G9/G10 or Canon EOS 450D, Canon Inc., Japan) equipped with polarization filters and controlled remotely by a USB-connected interface with a portable PC (Dell Precision, M70, Dell Inc., Round Rock, TX). Camera settings, image acquisition, and processing, as well as generation of result diagrams are made by the use of the dedicated TiVi600 system software based on MATLAB® (Matlab, MathWorks Inc., Natick Inc., United States). The photo captured by the camera is sent to the PC where the red and green color planes are separated. Each element in these color planes has a value between 0 and 255, representative of the intensity of the actual color. Each plane includes maximally 3000 × 4000 elements corresponding to a lateral resolution of about 50 μm at an object–camera distance of 20 cm and a field of view of 15 × 20 cm². By substituting $R_d(\Delta\lambda_r)$ and $R_d(\Delta\lambda_g)$ in Equation 15.4 with the red ($\mathbf{M}_{red}$) and green ($\mathbf{M}_{green}$) color planes, respectively, the total output matrix $\mathbf{M}$ can be calculated as

$$\mathbf{M} = k_{gain}\frac{\mathbf{M}_{red} - k\mathbf{M}_{green}}{\mathbf{M}_{red}} \tag{15.12}$$

To compensate for the nonlinear feature of the algorithm, each element in the $\mathbf{M}_{out}$—matrix is further processed in accordance with

$$\mathbf{M}_{out} = \mathbf{M}e^{p*M} \tag{15.13}$$

where p is an empirical factor fit to produce the best linear performance over the physiological RBC interval (0%–4%).

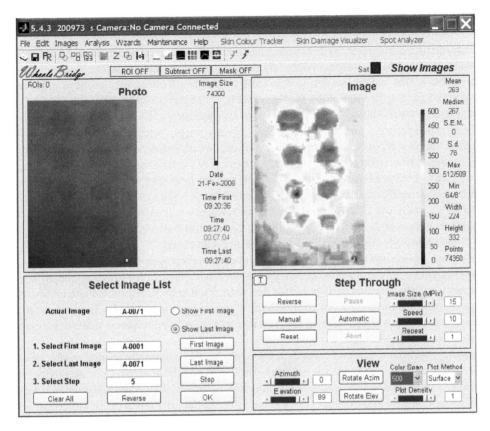

FIGURE 15.4 TiVi600 user interface displaying a photo and corresponding TiVi image following topical application of MN (10 mmol) in eight patches on the skin of the back.

$\mathbf{M}_{out}$ is a 2D array of $TiVi_{index}$ values displayed in the user interface next to the corresponding photo (Figure 15.4). A pseudocolor scale complying with that used in other imaging techniques—with blue and red color representing low and high values, respectively—is used to display the TiVi image. Each pixel corresponds to a TiVi value linearly related to local RBC concentration at the actual location.

In many applications, photos are recorded in sequence, thereby allowing studies of dynamic skin reactions following topical or systemic application of a substance that causes vasodilatation or vasoconstriction in the skin microcirculation. Such a sequence of individual TiVi images can be compiled into a video clip and played back at a rate set by the operator, thereby allowing for visualizing of a specific skin reaction in compressed time mode. By the use of integrated wizards and statistic measures, *erythema* and *blanching* reactions can be quantified and visualized. The entire process of assessing a specific skin reaction thereby becomes increasingly independent of the investigator. Data reduction is attained by reducing images to time traces that in turn can be reduced to indexes representative of an entire experiment. The maximal image acquisition rate is currently 12 images per minute and the maximal resolution is 12 million pixels. This high lateral resolution renders it possible to zoom in on a small skin area with still adequate resolution for detailed analysis of local *erythema* and *blanching*.

15.4 Evaluation

The linear relationship between the TiVi value generated and the actual concentration of RBCs was verified by way of a fluid model composed of tightly wound latex tubing simulating the blood vessels of the microcirculation. The system was perfused by fresh human blood mixed with saline to produce RBC concentrations ranging from 0% to 4% in steps of 0.2%, which was considered to cover the physiological range of skin tissue. Maximal oxygen saturation (approximately 100%) of the mixture was provided by stirring and exposure to ambient air [13], while minimal oxygen saturation (approximately 0%) was attained by bubbling N_2 gas through the blood samples. Five photos were captured for each RBC fraction and cropped to a region of interest covering the center of the fluid model. The average $TiVi_{index}$ values were then plotted as a function of the actual RBC fraction from 0% to 4%. For 0% and 100% oxygen saturation, the correlation coefficient was calculated to 0.998 ($n = 4$) and 0.997 ($n = 20$), respectively. The average $TiVi_{index}$ value calculated for the lower oxygen saturation was on average 91.5% of that calculated for the higher oxygen saturation, corresponding to a deviation of less than 3.9% within the physiological range (70%–100% oxygen saturation) well in accordance with theory.

The average penetration depth of red and green photons in skin tissue returning to the skin surface was investigated by the use of a Monte Carlo simulation model [14] adapted to accept linearly as well as cross-polarized light. By using cross-polarized detection, the average penetration depth increased from 345 to 482 μm for red wavelengths and from 316 to 387 μm for green wavelengths. This increase in depth sensitivity has been confirmed in enhanced visual scoring of skin *erythema* as well using a polarized light visualization system [15].

The average systematic drift in TiVi-system sensitivity over a 3-month time period, while the systems were moved in between laboratories, is reported to be 0.27%, and the discrepancy in sensitivity between different units is limited to 4.1%, due to offset rather than gain deviation when measurements were performed using a paper of uniform pink color as object [16]. Spatial variation in image uniformity is below 3.08% and 1.93% in the corners and the center of an individual image, respectively. Since the intensity of the flash is generally much higher than that of ambient light, virtually no influence on the image of ambient light can be demonstrated under ordinary daylight conditions. The distance dependence was assessed by recording images alternatively at a distance of 15 and 25 cm and calculating the TiVi values produced by erythema-inducing methyl nicotinate (MN) in forearm skin. No significant difference in the average TiVi value at the two distances could be demonstrated ($n = 25$). In contrast to what is the case with laser Doppler devices, TiVi is further not sensitive to movement artifacts, because no Doppler components are recorded. Since all image points are recorded simultaneously, there is no interpretation ambiguity regarding spatial and temporal variations in skin microcirculation, as may be the case with raster scan laser Doppler perfusion imaging (LDPI) devices. On the other hand, since TiVi is based on spectroscopy, it is sensitive to all objects of a specific color.

15.5 Applications

Establishment of healthy skin RBC concentration reference values is important for the design of versatile test procedures for the assessment of skin damage caused by, for instance, vibrating tools, chemical exposure, or peripheral vascular disease. One of the first applications of the emerging technology of TiVi therefore was to demonstrate how TiVi maps spatial and temporal variations in healthy Caucasian skin RBC concentration at the dorsal side of the hand at rest and during postocclusive hyperemia [17]. A lower skin RBC concentration (179–184 TiVi units [tvu]) was observed at the back of the hand and base of the thumb compared with areas adjacent to the nailfold region of the fingers (190–213 tvu) when recording was made at an ambient temperature of 21°C–23°C. Values in the same range were recorded from the volar side forearm skin (192 ± 19 tvu) and at the skin of the back (178 ± 17 tvu) in the resting state. The short-term variation (within 70 s) was 2% in all areas of the dorsal side of the hand, while day-to-day variations were in the range of 5%–7% in the back of the hand and up to 10% in areas adjacent to the nailfold region. In the postocclusive hyperemia phase, up to a 60% increase in skin RBC concentration was observed in the early part of the reactive hyperemia phase. This increase in skin RBC concentration successively decreased but remained about 18% above the preocclusive level after 30 min.

To demonstrate the potential of TiVi in the objective and operator-independent assessment of skin *blanching*, TiVi was used for the quantification of human skin *blanching* with the Minolta chromameter CR200 (Minolta, Tokyo, Japan) as an independent colorimeter reference method [18]. Desoximetasone gel 0.05% (Topicort®, TaroPharma, Taro Pharmaceuticals USA, Inc., Hawthorne, New York) was applied topically on the volar side of the forearm under occlusion for 6 h in healthy adults. The relative uncertainty in the *blanching* estimate produced by TiVi was about 5% and similar to that of the chromameter operated by a single user and taking the a* parameter as a measure of *blanching* [19]. In a separate study, the induction of *blanching* in the occlusion phase was mapped using a transparent occlusion cover. The successive induction of skin *blanching* during the occlusion phase, which is a measure of skin permeability, can thereby be mapped using TiVi. After an occlusion time of about 6 h, the RBC concentration was reduced to approximately 80% of the initial value.

To study the barrier function of the skin, MN, which is a water-soluble compound that rapidly penetrates the skin, may be used [20]. In order to minimize environmental influence, the MN solution is applied to the skin under occlusion using Finn Chambers on Scanpor® application systems (Epicutan, Tuusula, Finland). The average maximum change in recorded TiVi value (128 ± 28 tvu) was found to be about 70% of the TiVi value recorded from unprovoked skin. The average variability in *erythema* intensity between different patches on the skin of the back of the same individual, following a 30 s application time of MN (10 mmol), was calculated to be 22.1% (var. coeff.). The average time to reach 20%, 80%, and 100% of the end point *erythema* was calculated to be 114 ± 31, 264 ± 47, and 400 ± 65.2 s, respectively. The *erythema* extension generally spread outside the patch areas in which the MN was applied after about 5 min following the removal of occlusion and start of measurement, possibly caused by axon reflexes, histamine release, or lateral diffusion of the compound. The time course of *erythema* intensity induction can be modulated by lowering of fat content in topical formulations. A benzyl nicotinate formulation containing 10% fat induced *erythema* more rapidly and with higher intensity than the formulation with higher fat content (50%–80%) as demonstrated by TiVi [21]. In the same study, a corticosteroid (betamethasone 17-valerate) in vehicles with different lipid content was used to induce skin *blanching*. A significant reduction in lag time to the appearance of *blanching* as measured with TiVi was demonstrated when the lipid content was reduced from 80% to 10%.

In UVB phototesting of the skin, TiVi has been used to measure the *erythema* profile induced by a divergent UVB beam, from which the minimal *erythema* dose can be determined. TiVi has further been used to map and quantify the local vasodilatation and vasoconstriction produced by acetylcholine and sodium nitroprusside as well as by noradrenaline and phenylephrine when these vasoactive substances were forced to migrate into the skin by way of iontophoresis [22].

In a comparative study, responses of laser Doppler line scanner (LDLS) [23], laser speckle perfusion imaging [24] (LSPI), and TiVi were investigated during the reactive hyperemic phase following the release of a pressure cuff inflated to 90 and 130 mmHg and applied around the upper arm [25]. The time course for perfusion tracks recorded by LDSL and LSPI were similar with a reduction during the occlusion phase followed by a distinct reactive hyperemia peak (pressure 130 mmHg) and no reactive hyperemia (pressure 90 mmHg) that was reduced to pre-occlusion values after about 2 min. The time course of the TiVi track differed from this pattern and demonstrated a continuous increase in *erythema* (pressure 90 mmHg) and a steady RBC concentration level (pressure 130 mmHg) during the occlusion phase. This difference in track signature is explained by the fact that for the lower pressure, the venous return is blocked while the arteries supplying the skin with RBCs are still open, while at the higher cuff pressure, the RBC volume is not altered during the occlusion phase. Following topical application of MN, all methods displayed distinct and transient *erythema*. When clobetasol propionate solution (Dermovate®, 0.5 mg/mL, GlaxoSmithKline Ltd, London, United Kingdom) was topically applied on the skin to produce local vasoconstriction, both LDLS and LSPI proved to be insensitive to the action of the vasoconstrictor, while TiVi showed clear boundaries of the reaction, thus demonstrating its ability to map both *erythema* and *blanching*.

To prepare TiVi for clinical use, toolboxes with specific applications in mind have been developed.

Skin damage at the worksite can occur as a consequence of exposure to aggressive chemicals (skin irritation causing *erythema*) or following the use of vibrating tools (vasospasm causing vasoconstriction). By displaying the fraction of, for example, a hand that possesses *erythema* (RBC concentration above a user-selected threshold value) in red color, the healing process and/or progressions of contact dermatitis can be evaluated over time. The *Skin Damage Visualizer* toolbox was developed for direct assessment of skin damage at the worksite or in occupational medicine applications in general.

The color of the skin changes over time in many diseases and conditions and can thus be a useful indicator for assessment of the development and progression of skin disease or a healing process. What is generally referred to as skin color is the spectral content of backscattered light following illumination by a broadband light source. Consequently, the skin color observed is dependent not only on the optical properties of the skin but also on the spectral signature of the illuminating light. If a digital camera is used to capture a photo of the skin, the spectral sensitivity of the photodetector array further influences the color of the photos recorded. Consequently, skin color cannot be regarded as an absolute quantity. Changes in spectral content of the diffusely backscattered light from the skin under illumination by a light source with a fixed spectral signature, however, carries useful information about changing skin conditions. The *Skin Color Tracker* toolbox makes it possible to follow a preselected color (or a range of colors) over time and thereby to track, for example, the progressive development of scaling tissue in psoriasis. These color changes are minute and cannot be readily verified by naked eye observation alone.

Facial spots including acne appear most frequently during puberty with a varying degree of severity. About 90% of the adolescent population is believed to be affected by this skin disease. Numerous skin care products are commercially available for the treatment of spots, and new and more effective products are continuously released on the market. To evaluate the performance of these skin care products, the product candidates are generally tested using a panel of volunteers with spots of varying severity. The result is assessed by naked eye observation at which both the intensity and the extension of the spot *erythema* are evaluated. These evaluations are generally performed by a skilled dermatologist, and a severity score on a scale from one to five may be used to quantify the findings. Because of the unavoidable investigator-dependent element of these naked eye inspection procedures, results vary among observers and interlaboratory comparison of data may be difficult to perform. The *Spot Analyzer* toolbox renders this process fully automatic and investigator independent. Results are delivered in terms of diagrams displaying the intensity and extension of the spot *erythema*. The results from individual sessions can be integrated into a project window that displays the overall result of a test panel project.

15.6 Summary and Conclusion

This chapter deals with an emerging technology—TiVi—intended for quantitative analysis of skin *erythema* (vasodilatation) and *blanching* (vasoconstriction). TiVi was designed with ease of use in mind and to be productive in gaining investigator-independent data in terms of images visualizing the spatial distribution of RBC concentration in tissue. Due to its ability to reduce data from a sequence of images to curves and indexes, TiVi is a versatile tool in skin testing, including evaluation of new skin care products, assessment of pharmaceuticals, and in grading the performance of sensitive skin to various kinds of challenges. In contrast to the situation with LDPI technologies, TiVi measures only the concentration of RBCs and not their velocity, thereby making the images closer related to that which can be observed by the naked eye. Since LDPI is based on the Doppler principle, that is, measurement of frequency shifts generated by an object in motion, it is sensitive to all movements independent of whether they are caused by RBCs or other objects in motion. Being a spectroscopic method, TiVi is not plagued by movement artifacts and measurements of RBC concentration can be made on objects in motion as well. TiVi is, however, sensitive to the color of the object, and the amount of melanin in the epidermal layer will cause an offset in the $TiVi_{index}$ value as demonstrated in Figure 15.3. This disadvantage can be eliminated by recording only alterations in skin RBC concentration, since these generally appear on a time scale much shorter than that at which the melanin content of the skin is altered. In practice, this is achieved by subtracting—pixel by pixel—the first image from the remaining TiVi images in a sequence of images. Alternatively, the offset in $TiVi_{index}$ value can be determined from a skin site outside the region of interest and subtracted from all the remaining pixels in the TiVi image. Neither TiVi nor LDPI can display the result in absolute units, because the scattering volume is determined by the tissue optical properties and the exact location of the microvascular bed, both of which are unknown quantities. In LDPI, this represents a significant problem, since the size of the laser beam spot on the skin surface as seen from the detector unit directly determines the number of coherence areas on the detector surface and consequently also the inherent system amplification factor [26]. Both methods are therefore better suited for the comparison of alterations in microvascular RBC concentration and perfusion, respectively, as recorded before and after a stimuli–response experiment or by comparing the results from a region of interest to those from a control site. TiVi further has the additional advantage of capturing an image instantaneously, thereby avoiding misinterpretation of temporal variability as spatial heterogeneity in the microcirculation, which is the case with raster scanning LDPI. Since TiVi includes no moving parts or temperature-sensitive components, long-term stability has proven excellent and frequent calibration of the system is not necessary.

References

1. Ryan T.J. 1985. Dermal vasculature, in *Methods in Skin Research* (eds. Skerrow D. and Skerrow C.J.), John Wiley & Sons Ltd, Chichester, U.K., pp. 527–558.
2. Wårdell K., Barverman I.M., Silverman D.G., and Nilsson G.E. 1994. Spatial heterogeneity in normal skin perfusion recorded with laser Doppler imaging and flowmetry. *Microvasc Res*, 48, 26–38.
3. O'Doherty J., Henricson J., Anderson C., Leahy M.J., Nilsson G.E., and Sjöberg F. 2007. Subepidermal imaging using polarized light spectroscopy for assessment of skin microcirculation. *Skin Res Technol*, 13, 472–484.
4. Jacques S.L., Ramella-Roman J.C., and Lee K. 2002. Imaging skin pathology with polarized light. *J Biomed Opt*, 7, 329–340.
5. Zimnyakov D.A. and Sinichkin Y.P. 2000. A study of polarization decay as applied to improved imaging in scattering media. *J Opt Pure Appl Opt*, 2, 200–208.
6. Mobley J. and Vo-Dinh T. 2003. Optical properties of tissue, in *Biomedical Photonics Handbook* (ed. Vo-Dinh T.), CRC Press LLC, Boca Raton, FL.
7. Megliniski I.V. and Matcher S.J. 2003. Computer simulation of the skin reflectance spectra. *Comput Methods Programs Biomed*, 70, 179–186.
8. Wray S., Cope M., Delpy D.T., Wyatt J.S., and Reynolds E.O.R. 1988. Characterization of the near infrared absorption spectra of cytochrome aa3 and haemoglobin for the non-invasive monitoring of cerebral oxygen. *Biochim Biophys Acta*, 933(1), 184–192.
9. Ishimaru A. 1978. *Wave Propagation and Scattering in Random Media*, Vol. 1: *Single Scattering and Transport Theory*, Academic Press, London, U.K.
10. Brinkworth B.J. 1964. A diffusion model of the transport of radiation from a point source in the lower atmosphere. *Br J Appl Phys*, 15, 733–741.
11. Fabbri F., Francesschini M.A., and Fantini S. 2003. Characterization of spatial and temporal variations in the optical properties of tissue-like media with diffuse reflectance imaging. *Appl Opt*, 42(16), 3063–3072.
12. Wheelsbridge AB. www.wheelsbridge.se.
13. Nilsson G.E., Tenland T., and Öberg P.Å. 1980. Evaluation of a laser Doppler flowmeter for measurement of tissue blood flow. *IEEE Trans Biomed Eng*, BME-27(10), 597–604.
14. Ramella Roman J.C., Prahl S.A., and Jacques S.L. 2005. Three Monte Carlo programs of polarized light transport into scattering media: Part 1. *Opt Exp*, 13, 4420–4438.

15. Farage M. 2008. Enhancement of visual scoring of skin irritant reactions using cross-polarized light and parallel-polarized light. *Contact Dermatitis*, 58, 147–155.
16. Nilsson G.E., Zhai H., Chan H.P., Farahmand S., and Maibach H.I. 2009. Cutaneous bioengineering instrumentation standardization: The tissue viability imager. *Skin Res Technol*, 15, 6–13.
17. Zhai H., Chan H.P., Farahmand S., Nilsson G.E., and Maibach H.I. 2009. Tissue viability imaging: Mapping skin erythema. *Skin Res Technol*, 15, 14–19.
18. Zhai H., Chan H.P., Farahmand S., Nilsson G.E., and Maibach H.I. 2009. Comparison of tissue viability imaging and colorimetry: Skin blanching. *Skin Res Technol*, 15, 20–23.
19. Westerhof W. 2006. Colorimetry, in *Handbook of Non-Invasive Methods and the Skin*, 2nd edn. (eds. Serup J., Jemec G.B.E., and Grove G.L.), CRC Press, Boca Raton, FL, pp. 635–647.
20. Magnusson B.M., Nilsson G.E., and Anderson, C. 2006. The polarization spectroscopy camera—A promising tool for assessment of erythematous reactions to topically applied agents, *Perspectives in Percutaneous Perfusion*, La Grande Motte, France, April 18–22.
21. Wirén K., Frithiof H., Sjöqvist C., and Lodén M. March 2009. Enhancement of bioavailability by lowering of fat content in topical formulations. *Br J Dermatol*, 160(3), 552–556.
22. Henricson J., Nilsson A., Tesselar E., Nilsson G., and Sjöberg F. 2009. Tissue viability imaging: Microvascular response to vasoactive drugs induced by microdialysis. *Microvasc Res*, 78(2), 199–205.
23. Moor Instruments. http://www.moor.co.uk/.
24. Briers J.D. 2001. Laser doppler, speckle and related techniques for blood perfusion mapping and imaging. *Physiol Meas*, 22, R35–R66.
25. O'Doherty J., Clancy N.T., Enfield J.G., McNamara P., and Leahy M. 2009. Comparison of instruments for investigation of microcirculatory blood flow and red blood cell concentration. *J Biomed Opt*, 14(3), 034025.
26. Nilsson G.E., Salerud E.G., Strömberg N.O.T., and Wårdell K. 2003. Laser doppler perfusion monitoring and imaging, in *Biomedical Photonics Handbook* (ed. Vo-Dinh T.), CRC Press, Boca Raton, FL, pp. 15.1–15.24.

16

Photothermal Detection and Tracking of Individual Nonfluorescent Nanosystems

Laurent Cognet
*Centre National de la
Recherche Scientifique
and University of Bordeaux*

Brahim Lounis
*Centre National de la
Recherche Scientifique and
University of Bordeaux*

16.1 Introduction

In the fast-evolving field of nano-biosciences, simple and sensitive methods for the detection and characterization of the matter at the nanoscale are needed. Electron microscopy or near-field methods such as scanning tunneling microscopy and atomic force microscopy provide atomic resolution but need extensive sample preparations and are not fully suitable in complex environments such as biological matter. Far-field optical methods have the advantage to be noncontact and minimally invasive. The most commonly used optical techniques are based on luminescence. The ultimate sensitivity in the optical imaging of live cells is achieved by single-molecule detection (SMD). First, by removing the average inherent to ensemble measurements, SMD yields a measure of the distribution of molecular properties, which is of importance in biological systems that display static or time-dependent heterogeneity. Thus, subpopulations, even minor ones, can be studied and statistical correlations between different distributions of parameters can be performed on different subpopulations. As an example, one can cite the study of the diffusion of biomolecules in membranes (Schmidt et al. 1995; Sako et al. 2000; Schutz et al. 2000). They not only have revealed the existence of different subpopulations in terms of membrane diffusion in the plasma membrane of the live cells but also have permitted to study the characteristics of the diffusion of the different populations (Vrljic et al. 2002; Tardin et al. 2003) and have even revealed a functional role of this diffusion in the case of fast communication between neurons (Heine et al. 2008). Second, SMD gives access to the dynamic fluctuations of one parameter and can thus unravel all steps

501

of its temporal evolution without the need to synchronize all molecules (which is in practice impossible for live organisms). Keeping with the previous example in neurobiology, the existence of exchanges of receptors between different cellular compartments such as synaptic and extrasynaptic spaces of live neurons has been revealed for the first time (Tardin et al. 2003). Finally, SMD also provides an important advantage over ensemble measurements, namely, the possibility to track molecules with a position accuracy only limited by the signal-to-noise ratio at which the molecules are detected. This renders nanometer localizations possible far below the optical resolution (Schmidt et al. 1996; Yildiz et al. 2003). This superresolution feature is now widely applied to obtain superresolved images using photoactivable molecules (Betzig et al. 2006; Rust et al. 2006).

Yet, the intrinsically limited photostability of fluorescent nano-objects restrains the applicability of fluorescence microscopy. Single fluorophores emit a limited number of photons before photobleaching; at best, a few millions of photons are emitted in physiological mediums that correspond to maximum acquisition times of the order of seconds. Semiconductor nanocrystals constitute interesting alternative labels (Michalet et al. 2005). Indeed, although bulkier than individual fluorophores (especially after biofunctionalization [Groc et al. 2007]) and subject to blinking, they are much more photostable than organic dye molecules.

In this chapter, we will introduce optical methods to image and study nonluminescent nanolabels, such as gold nanoparticles (AuNPs), as an interesting alternative to fluorescent labels. AuNPs exhibit fascinating physical properties, either taken as individuals or in assemblies. The remarkable behavior of the individual AuNPs comes from their size-related electronic and optical properties (Kreibig and Vollmer 1995; Eustis and El-Sayed 2006), which has opened many applications in different fields such as in catalysis (Daniel and Astruc 2004), plasmonics (Ozbay 2006), biosensing (Rosi and Mirkin 2005), or biology (Schultz 2003; Hirsch et al. 2006). For applications in biosciences, researchers have used, in the past 30 years, different optical methods to detect individual AuNPs. Since they were all based on Rayleigh intensity scattering (Yguerabide and Yguerabide 1998; Schultz et al. 2000), such techniques were generally limited to the detection of relatively large particles (>40 nm), especially in highly scattering environments. Consequently, new methodologies were needed to allow the detection of very small AuNPs in complex environments.

16.2 Detection of Nonfluorescing Nano-Objects

There are traditionally several approaches to the far-field optical detection of individual nonluminescent nano-objects. They are based on either the scattered intensity (Yguerabide and Yguerabide 1998; Schultz et al. 2000) or the generation of new wavelengths by the particle, either in a linear photoluminescence process (Peyser et al. 2001) or in nonlinear processes (Liau et al. 2001). In conventional scattering-based methods, the intensity of the field elastically scattered by a single nanoparticle (NP) is usually spatially separated from the exciting field by particular geometrical arrangements (using dark-field illumination [Yguerabide and Yguerabide 2001], differential interference contrast and video enhancement [Debrabander et al. 1986], or total internal reflection [Sönnichsen et al. 2000]), and detected in the far-field region. However, the polarizability of an NP interacting with a laser field being proportional to its volume, intensity-scattering cross section of the particle, varies as D^6, while its absorption cross section varies as D^3 only and dominates for small sizes. For example, for AuNPs in water probed by 532 nm laser light, absorption is more important than scattering for diameters below ~100 nm.

To detect such small particles, one can detect the absorption or use the interference of the scattered field with a reference field (Lindfors et al. 2004) to get an interference signal that, similar to absorption, varies with the third power of the particle size. Because of the amplification by the reference wave, such interferometric techniques not only result in a high sensitivity but can also give access to both the amplitude and the phase of the scattered wave. In the following, we will focus on a class of these methods that rely on the photothermal effect. To date, they are the most sensitive ones, and they have the advantage to be applicable in highly scattering environments, making them suitable for a broad range of applications, especially in biosciences. In the next sections, we first introduce the principle of photothermal detection, then present the absorption spectroscopy of metal NP based on this method, and finally introduce several approaches for applications in biosciences.

16.3 Photothermal Interference Contrast Method

Excited near their surface plasmon resonance (SPR), metal NPs have a large absorption cross section and exhibit a fast electron–phonon relaxation time in the picosecond range (Link and El-Sayed 2003), which makes them very efficient light absorbers. The luminescence yield of these particles being extremely weak (Wilcoxon et al. 1998), almost all the absorbed energy is converted into heat. The increase in temperature induced by this absorption gives rise to a local variation of the refractive index. This photothermal effect can be used to detect and study individual NPs. Photothermal detection was proposed earlier by Tokeshi et al. (2001), using a thermal lens effect to detect very low concentrations of absorbing molecules in liquid solutions. Boyer et al. (2002) designed a highly sensitive polarization interference method, called photothermal interference contrast (PIC), for the detection of the small refractive index changes around an absorbing particle. The experiment was performed using a combination of two lasers (Figure 16.1a). The NPs were heated by the 514 nm line of an argon ion laser, whose intensity was modulated at high frequency. The horizontally polarized output of a HeNe laser (633 nm wavelength) was split into two perpendicularly polarized beams by a Wollaston prism forming the two arms of the interferometer (probe and reference). The probe beam was overlaid with the heating beam, and the different beams were sent to a microscope objective using a telecentric lens system. The probe and reference

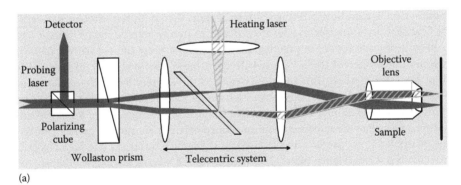

(a)

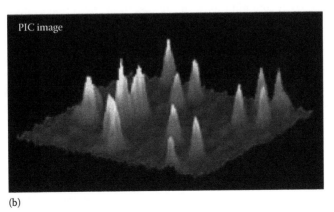

(b)

FIGURE 16.1   (See color insert.) (a) Experimental scheme of the photothermal interference contrast (PIC) microscope. (b) PIC image of a $10 \times 10\ \mu m^2$ region of sample containing isolated 5 nm AuNPs.

beams back-reflected by the sample recombined in the Wollaston prism and were reflected by a polarizing beamsplitter cube onto a fast photodiode. The PIC signals consisted of the phase difference between the two red beams measured by the variations of the red intensity at the modulation frequency of the green beam. Microscopic images were obtained by scanning the sample with respect to the three spots. With PIC, images of AuNP down to 5 nm in diameter embedded in thin polymer films were recorded with a signal-to-noise ratio larger than 10 (Figure 16.1b). It was also shown that, in addition to its intrinsic sensitivity, the photothermal method is remarkably insensitive to scattering background, even when arising from strong scatterers such as 300 nm latex beads. Using a modified PIC setup in order to 3D image NPs in thick samples, single proteins labeled with individual 10 nm AuNPs could be detected at the surface of fixed cells (Cognet et al. 2003).

The sensitivity of the PIC method, although high, could not reach the shot noise limit. Indeed, the use of high-numerical-aperture objectives induced depolarization effects that degraded the quality of the overlap between the two arms of the interferometer. As a consequence, the detection of nanometer-sized particles with PIC required relatively high laser intensities (~10 MW/cm²), which can be a serious limitation for many applications such as in biology.

16.4 Photothermal Heterodyne Imaging

A more sensitive photothermal method, called photothermal heterodyne imaging (PHI), was then developed (Berciaud et al. 2004, 2006) (alternatively called *laser-induced scattering around a nanoabsorber*, for biological applications [Blab et al. 2006; Lasne et al. 2006]). It combines a time-modulated heating beam and a nonresonant probe beam, overlapping on the sample. The probe beam produces a frequency-shifted scattered field as it interacts with the time-modulated variations of the refractive index around the absorbing NP. The scattered field is then detected through its beatnote with the probe field that plays the role of a local oscillator as in any heterodyne technique. This signal is extracted by lock-in detection. In practice, PHI can be shot noise limited since it does not suffer from the limitations inherent to PIC. Its sensitivity is more than one order of magnitude higher than that of any other method. This sensitivity enabled the unprecedented detection of individual gold clusters as small as 1.4 nm in diameter, which contain less than 100 atoms. In practice, a nonresonant probe beam (632.8 nm, HeNe, or single-frequency Ti:Sa laser) and an absorbed heating beam (532 nm, frequency-doubled neodymium-doped yttrium aluminum garnet [Nd:YAG] laser or tunable cw dye laser) are overlaid and focused on the sample by means of a high NA microscope objective (100×, NA = 1.4) (Figure 16.2a). The intensity of the heating beam is modulated at Ω ranging from typically 100 kHz to a few MHz, by an acousto-optic modulator. The PHI signal can be detected using two different configurations. In the case of the detection of the backward signal, a combination of a polarizing cube and a quarter-wave plate is used to extract the interfering probe-reflected and backward-scattered fields. In order to detect the forward signal, a second microscope objective (80×, NA = 0.8) is employed to efficiently collect the interfering probe-transmitted and forward-scattered fields. Backward or forward interfering fields are collected on fast photodiodes and fed into a lock-in amplifier in order to extract the beat signal at Ω. Integration time of 1–10 ms is typically used. Images are formed by moving the sample over the fixed laser spots by means of a 2D piezoscanner (Figure 16.2b).

The resolution of the PHI method depends on the probe and heating beam profiles and also on the dielectric susceptibility profile created around the AuNPs. Since the spatial extension of the latter is much smaller than the size of the probe beam, the transverse resolution is simply given by the product of the two beam profiles, resulting in resolutions comparable to those of confocal microscopes (Berciaud et al. 2006).

The PHI signal is proportional to the absorption cross section of the NPs. It thus scales with the volume for small-sized particles excited at given heating wavelength. Therefore, the dispersion in signals obtained from the image of a sample containing individual AuNPs with size dispersion of 10% will be equal to 30% around a mean signal (Berciaud et al. 2004, 2006).

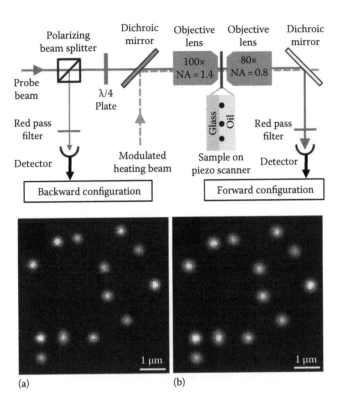

FIGURE 16.2 **(See color insert.)** Experimental schemes of the photothermal heterodyne imaging (PHI) microscope and PHI images of the region of a sample containing isolated 10 nm AuNPs detected in the backward (a) and forward (b) configurations (see text).

16.5 Single Metal Nanoparticle Absorption Spectroscopy

Noble metal NPs are strong absorbers and scatterers of visible light due to SPR. The corresponding resonant peak energies and line widths are sensitive to the NP size, shape, and local environment. For many applications where quantitative signals are needed in complex environments, it is important to understand these optical resonances with precision.

For rather large metal NPs (diameter $D \geq 20$ nm), the resonant peak energy of the SPR experiences a red shift with increasing sizes, due to retardation effects as well as increasing contributions from multipolar terms. In addition, for NPs larger than 50 nm, radiative damping of the collective electronic excitation strongly broadens the line width of the SPR. However, for NPs smaller than the electron mean free path (Kreibig and Vollmer 1995; Link and El Sayed 2000) ($D < 20$ nm), these extrinsic size effects become negligible and intrinsic size effects prevail. In a classical picture, the resulting limitation of the electron mean free path as well as interactions with the surrounding matrix (Persson 1993; Kreibig and Vollmer 1995) are responsible for additional damping at the NP surface. In the time domain, this translates into a strong reduction of the surface plasmon dephasing time down to a few fs. As a consequence, because of these additional size-dependent damping processes, the dielectric constant of small NPs can no longer be described by the bulk metal values.

Experimental studies on ensembles of metallic NPs revealed the existence of such effects (Kreibig and Vollmer 1995; Bosbach et al. 2002). However, comparisons between theory and ensemble measurements were still a matter of debate because the inhomogeneities in NPs size, shape, and local environment experimentally blur the homogeneous width of the SPR. In order to circumvent this shortcoming, PHI was used to record absorption spectra of individual AuNPs with diameters down to 5 nm with excellent

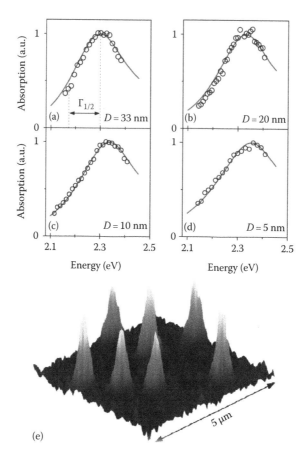

FIGURE 16.3 **(See color insert.)** (a–d) Normalized absorption spectra of single AuNPs with diameters ranging from 33 to 5 nm. The extracted width at half-maximum is shown on the first spectrum. A broadening of the plasmon resonance with decreasing size is readily visible and can be compared with simulations based on Mie theory using a size-dependent modification in the dielectric constant of gold. (e) PHI image of a sample containing isolated 5.3 nm silver NPs excited at 405 nm.

signal-to-noise ratio (Berciaud et al. 2005). Figure 16.3a through d represents the absorption spectra of four individual AuNPs with diameters of 33 and 5 nm, respectively, performed in the 515–580 nm wavelength range. The values of peak resonance energies are weakly affected by intrinsic size effects. In contrast, a significant increase in the width of the resonance clearly appears and cannot be described by Mie theory with the bulk values of the gold dielectric constant. A good agreement was found between the experimental widths and Mie simulations with size-dependent corrections (Berciaud et al. 2005).

Although the existence of intrinsic size effects in the optical response of AuNPs was unambiguously revealed, part of the damping processes are due to interband transitions. Consequently, the plasmon spectra of individual AuNPs are asymmetric, and it is delicate (even impossible for very small particles) to define a full width at half-maximum of absorption spectra. This makes it difficult to connect the widths of the plasmon resonances to the damping rate. To avoid the contribution of interband transitions in the measured spectra, one could red shift the plasmon resonance frequency, by embedding gold nanospheres in a high-index matrix or by studying the long-axis plasmon mode of gold nanorods (Klar et al. 1998) or by using core–shell NPs (Prodan et al. 2003). Another possibility is to use silver nanospheres, since their resonant energies are well separated from interband transitions. A drawback of silver particles is, however, their weak photostability due to photooxidation (Peyser et al. 2001). Interestingly, the reactivity of silver particles can be reduced by encapsulation, for example, with poly(ethylene glycol)

(PEG) (Wuelfing et al. 1998). Berciaud et al. (2006) demonstrated the detection of individual PEG-coated silver NPs with an average diameter of 5.3 nm with the PHI method (Figure 16.3e). The size distribution, however, was broad. For these experiments, a modulated diode laser emitting at 405 nm was used as the heating source with an intensity of ~50 kW/cm². As expected, the signal obtained with silver particles of 5.3 nm in diameter is about 10× higher than that of AuNPs of the same size, when identical heating intensities are used to excite the NPs at the peak of their SPR. Expected progress in the synthesis of silver NPs with narrower size distributions should permit a quantitative study of the SPR of individual silver particles much smaller than 5 nm in diameter and further applications in biosciences.

16.6 DNA Microarrays

Owing to its high sensitivity, PHI method is very promising for applications in biosciences. In a first demonstration, Blab et al. (2006) showed its potentiality to provide a new readout strategy of DNA microarrays based on AuNPs. The determination and exact quantification of gene expression are becoming increasingly important in basic pharmaceutical and clinical research. Fluorescence-based DNA assays are most widely used but suffer from the presence of autofluorescence in some biological samples and substrates, which severely interferes with the detection of the target molecules. DNA assays based on AuNP labels present a viable alternative. They commonly use AuNPs larger than 40 nm, which can be readily detected due to their strong light scattering at visible wavelengths (Yguerabide and Yguerabide 1998; Schultz et al. 2000). For increased specificity and reactivity, AuNPs smaller than 40 nm are preferred. Indeed, small AuNPs functionalized with oligonucleotides exhibit a very sharp thermal denaturation profile, and the rate of reaction on a surface is much higher than with large particles (Taton et al. 2000; Alexandre et al. 2001; Fritzsche and Taton 2003). As small AuNPs (diameter, <40 nm) barely interact with light, their direct optical detection has been impossible until recently without silver staining enhancement techniques (Taton et al. 2000; Fritzsche and Taton 2003). However, saturation at the amplification step limits the linear dynamic range as the typical size of the silver crystals is much larger than that of the AuNPs (Alexandre et al. 2001). Furthermore, spontaneous conversion of silver solution into metallic grains can occur, leading to nonspecific signals (Alexandre et al. 2001). Another alternative is the electrical detection of the AuNPs after catalytic or enzymatic deposition of the silver (Park et al. 2002; Möller et al. 2005). In this context, the possibility of detecting tiny AuNPs at the single particle level holds great promise for new and more efficient optical readout schemes of DNA assays.

PHI was thus applied on standard low-density spotted DNA microarrays in order to avoid the use of silver enhancement techniques. These arrays are well suited for routine applications as they contain a limited number of genes (usually below 1000), which enables good spotting quality and good reproducibility (Zammatteo et al. 2002). PHI provided a reliable quantification of the amount of DNA molecules in each spot of the microarrays (Figure 16.4). This determination was no longer limited by any constraints of the detection method, but only by the degree of unspecific signals on one side and by the size of the AuNPs on the other side. Indeed, as all AuNPs are detected, the lower detection of DNA depends only on unspecific DNA hybridization events and on the quality of surface treatments. Concerning the upper detection limit, the ultimate limit is given by the condition of plasmon coupling between particles. Indeed, when the average distance of the small AuNPs is comparable to their size, the optical response of AuNPs is modified (Sonnichsen et al. 2005). Furthermore, the possibility to detect much smaller AuNPs (down to 1.4 nm) (Berciaud et al. 2004) should significantly increase the dynamic range.

In addition to the high sensitivity and dynamics afforded by the present method, AuNP-based DNA arrays can be stored for long periods and measured several times. This approach thus combines the advantages of fluorescence measurements—small marker size, purely optical detection—with the high stability, specificity, and dynamic range afforded by AuNP labeling techniques. This makes photothermal approaches promising for application in biochips.

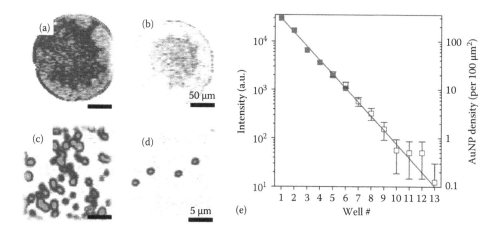

FIGURE 16.4 (See color insert.) Direct imaging (no silver enhancement) of four wells of a DNA microarray by PHI at low (a and b) and high (c and d) resolutions. (e) Signal (left axis) vs. well number in the low-resolution (full squares) and high-resolution (open squares) regimes. The corresponding AuNPs density measured in each well is shown on the right axis.

16.7 Single-Nanoparticle Tracking in Live Cells

The movements of molecules in the plasma membrane of living cells are characterized by their diversities, in both the temporal and spatial domains. Membranes exhibit a constitutive complexity, consisting of several different lipids and a great variety of proteins with highly dynamic and compartmentalized spatial distributions. These molecules explore the plasma membrane in various lateral diffusion modes and frequently interact with each other at specific locations in order to transmit information across the membrane, starting the cascades of specific signaling processes. Those molecular interactions are by nature heterogeneous, making ensemble observations of these phenomena rather challenging. The detection of individual molecules has allowed the elimination of the implicit averaging of conventional optical observations. Until now, two main approaches have been used to track individual molecules in the plasma membrane of live cells, with distinct advantages and limitations. The first one, single particle tracking (SPT), uses labels large enough to be detectable by conventional microscopes (Sheetz et al. 1989) through Rayleigh intensity scattering (~40 nm gold particles or even larger latex beads). SPT permits to follow the movement of individual molecules for very long times and possibly at very fast imaging rates (Ritchie et al. 2005). SPT, for instance, revealed barriers set for diffusion by the cytoskeleton (Kusumi and Sako 1996) and the diversity of lateral diffusion modes of receptor for neurotransmitters in live neurons (Borgdorff and Choquet 2002). However, the main drawback is the size of the beads, which might sterically hinder the interaction between the labeled molecules or alter their movements in confined environments such as synaptic clefts or endocytotic vesicles. As already mentioned in the introduction of this chapter, the second widely used technique, SMD, uses fluorescent organic dyes, autofluorescent proteins, or quantum dots. As these fluorophores are generally smaller than the target molecules, it does not have the drawback of SPT mentioned previously. The main limitation encountered in SMD studies is photobleaching, which severely limits the observation times of a single fluorophore to typically less than 1 s in live cells, and the blinking or the size of the functionalized quantum dots (Groc et al. 2007).

An experimental technique combining the advantages of SPT and SMD, namely, long observation times and small nanometer-sized labels, would thus have great potential. For biological questions, it would allow for recording the full history of proteins in cells, including intermediate states even in highly confined regions (e.g., lipid rafts or membrane protein clusters, intracellular vesicles, synapses of neurons). For live-cell imaging, PHI allows the detection of very small metal NPs (<5 nm) in highly scattering environments with intensities compatible with cell integrity. A PHI image of live neurons

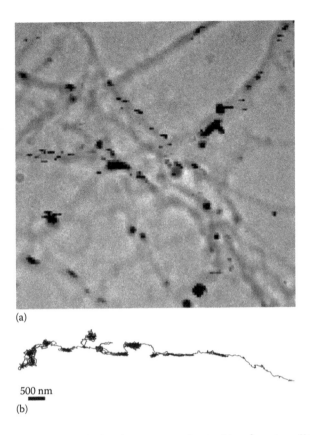

(a)

500 nm

(b)

FIGURE 16.5 (a) Superimposed bright field and PHI images of a $20 \times 20 \ \mu m^2$ portion of live cultured hippocampal neurons with GluR2 receptors labeled with 5 nm AuNPs. The PHI image (black spots and stripes) reveals the presence of point-like signals that correspond to immobile GluR2-linked AuNPs. Most NPs move during the raster scan of the sample, which produces characteristic stripe signals. (b) Trajectory of an individual receptor labeled with a 5 nm AuNP (>5 min, 9158 data points) acquired by SNaPT at video rate on a live neuron.

with 5 nm gold-labeled glutamate receptors on their outer membrane (superimposed on a white light image of the same area) is presented in Figure 16.5a. As the PHI method requires a raster scan of the sample with typically a few-millisecond integration time per point, fast imaging rates cannot be readily obtained and moving objects are not resolved during the raster scan. Consequently, mobile receptors produce stripes of signal in the image whereas immobile ones are well resolved. Lasne et al. (2006) recently designed a tracking scheme (SNaPT) based on a triangulation from three measurement points (like in the GPS scheme) to record the trajectories of single membrane proteins labeled with AuNPs in live cells at video rate. The movement of single membrane receptors could thus be recorded for several minutes (Figure 16.5b). Single-metal-NP tracking combines the advantages of small marker size with practically unlimited observation times owing to the high chemical stability of gold particles.

To estimate the local temperature rise due to laser absorption in the vicinity of the AuNPs, one can consider that the temperature inside a spherical AuNP is uniform and equal to the temperature at its surface since the thermal conductivity of metals is much higher than that of the surrounding medium. It writes $T_{surf} = \sigma_{abs} I / 4\pi \kappa a$, where I is the heating intensity, σ_{abs} is the AuNP absorption cross section, κ the thermal conductivity of the medium (water), and a the radius of the AuNPs. For 5 nm AuNPs in aqueous medium, and $I = 400 \ kW/cm^2$, one finds a rather low AuNP temperature rise of ~1.5 K. Furthermore, it decreases as the inverse of the distance from the AuNP surface. Further improvements of the SNaPT method should allow studying not only 2D movements of single molecules on cultured cells but also 3D movements in more complex system such as tissue slices. For instance, the algorithm used in SNaPT can be easily generalized for tracking movements of NPs in 3D, and we foresee that the use of near-infrared excitation of small metal NPs or nanorods would be advantageous to access deeper in the tissues.

16.8 Photothermal Absorption Correlation Spectroscopy

In most biological applications involving gold nanoprobes, the AuNPs are attached to the target molecule using ligands such as antibodies (De Mey et al. 1981), Fab fragments (Tschopp et al. 1982), DNA strands, and more recently short peptides (Levy et al. 2004). The resulting hydrodynamic radius of the functionalized probe is difficult to estimate with precision. The knowledge of this parameter is, however, crucial when proteins in confined environments of living cells are studied (Groc et al. 2007). An alternative use of PHI that allows measuring the diffusion coefficient of the NPs and/or their hydrodynamic radius in complex environments was recently proposed. This technique called photothermal absorption correlation spectroscopy (PhACS) (Octeau et al. 2009) is an analogue to fluorescence correlation spectroscopy (FCS) (Magde et al. 1972, 1974; Elson and Magde 1974), which has been essential for the investigation of fluorescent molecule photophysics, diffusion, and reaction kinetics, as well as for biological studies of molecular dynamics in live cells (Schwille et al. 1999, 2000; Wawrezinieck et al. 2005; Bacia et al. 2006; Wang et al. 2006). FCS, however, bears some limitations. The dynamic range accessible by FCS is limited by nonideal fluorophore photophysics on short time scales and photobleaching for long ones. Furthermore, one of the most significant and often overlooked factors is optical saturation, which also influences FCS measurements (Ries and Schwille 2006). With PhACS, which is based on the absorptive properties of individual nano-objects and not on fluorescence, these limitations will be overcome. PhACS thus constitutes a complementary method to FCS and should find numerous applications in analytical chemistry and biosciences.

PhACS is based on the PHI detection method and measures the time correlation function of the detected signals when the beams and sample positions are kept fixed. The signal fluctuations are recorded at high sampling rate, while NPs diffuse in the detection volume (Figure 16.6a through c). Similar to FCS experiments, the PhACS signal consists of the autocorrelation function ($G(\tau)$) of the photothermal signal (S) fluctuations (Figure 16.6d):

$$G(\tau) = \frac{\langle \delta S(t) \delta S(t+\tau) \rangle}{\langle S(t) \rangle^2} = \frac{\langle S(t) S(t+\tau) \rangle}{\langle S(t) \rangle^2} - 1 \tag{16.1}$$

where $\langle \rangle$ denotes the time average. Assuming an elliptical Gaussian-shaped observation volume and freely diffusing particles, $G(\tau)$ is given by Aragon and Pecora (1976)

$$G(\tau) = \frac{1}{N(1+\tau/\tau_D)(1+\tau/(A^2\tau_D))^{1/2}} \tag{16.2}$$

where
- N is the average number of particles in the observation volume
- A is the shape parameter of the observation volume
- τ_D is a characteristic diffusion time

PhACS provides a way to measure and quantify precisely the diffusion of absorbing NP of different sizes in environments with various viscosities. As a result, the hydrodynamic diameter of functionalized NP complexes can now be determined with nanometer precisions (Octeau et al. 2009) (Figure 16.6d).

In terms of signal dynamics, diffusion time scales and the corresponding diffusion constants, D, cover a wide range (seconds to submilliseconds to corresponding to $D \sim 0.01\text{–}100\ \mu m^2/s$). At short

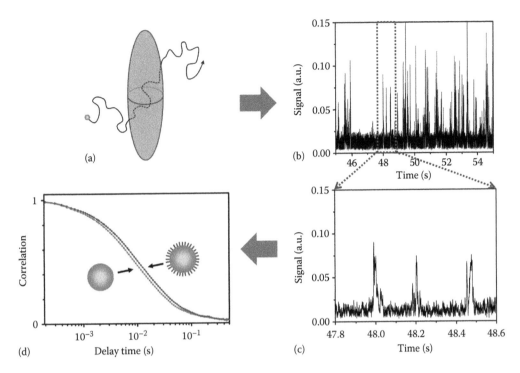

FIGURE 16.6 (a–c) Principle of PhACS: Time traces are recorded while 10 nm AuNPs diffuse (here in 80% glycerin solutions) in and out of the detection volume of PHI, which is kept immobile. PhACS curves obtained from the autocorrelation of the time traces. (d) Example of PhACS curves corresponding to bare and CALNN-coated AuNPs is shown. CALNN is a five-amino-acid long peptide, which increases the hydrodynamic radius of the AuNPs by only 0.6 nm.

time scales, it is ultimately limited by the modulation frequency of the exciting beam that can be increased to probe fast dynamics. Interestingly, the decrease in photothermal signal with the modulation frequency (Berciaud et al. 2004, 2006) can be compensated by the use of larger gold NPs (as the signal scales as $\propto d^3$) or greater excitation powers. On the longer time scales, it is not limited by photobleaching or blinking of fluorescent nano-objects like in FCS and allows exploring extremely slow diffusion dynamics not accessible by other techniques. This is because unlike FCS, PhACS correlation times arise exclusively from the diffusive properties of the NPs. As a consequence, PhACS is not sensitive to any scattering background and should thus find numerous applications in live cell studies where diffusion constants of biomolecules are extremely diverse. One can also foresee applications of PhACS in biosensing or microfluidics.

16.9 Conclusions

This chapter has introduced photothermal methods for the ultrasensitive detection of absorbing nano-objects. Applications using tiny AuNPs were proposed, such as in the field of DNA microarrays, live cell imaging, or analytical chemistry. In particular, two approaches were presented to study the dynamics of gold-labeled membrane proteins in live cells. One is based on tracking the trajectories of individual molecules (SNaPT) and the second one measures the signal fluctuations arising from diffusing NPs in the focal spot of the photothermal microscope (PhACS). Because photothermal methods bear several advantages over single-molecule fluorescence-based methods (small marker size together with practically unlimited observation times), numerous applications can be foreseen in biosciences, but also in fundamental physics.

Acknowledgments

This research was supported by CNRS (ACI Nanoscience and DRAB), Région Aquitaine, Agence Nationale pour la Recherche (ANR PNANO program), French Ministry for Education and Research (MENRT), and Human Frontiers Science Program (HFSP).

References

Alexandre, I., S. Hamels, S. Dufour, J. Collet, N. Zammatteo, F. De Longueville, J. L. Gala, and J. Remacle. 2001. Colorimetric silver detection of DNA microarrays. *Anal. Biochem.* 295(1):1–8.

Aragon, S. R. and R. Pecora. 1976. Fluorescence correlation spectroscopy as a probe of molecular dynamics. *J. Chem. Phys.* 64(4):1791–1803.

Bacia, K., S. A. Kim, and P. Schwille. 2006. Fluorescence cross-correlation spectroscopy in living cells. *Nat. Methods* 3(2):83–89.

Berciaud, S., L. Cognet, G. A. Blab, and B. Lounis. 2004. Photothermal heterodyne imaging of individual nonfluorescent nanoclusters and nanocrystals. *Phys. Rev. Lett.* 93(25):257402.

Berciaud, S., L. Cognet, P. Tamarat, and B. Lounis. 2005. Observation of intrinsic size effects in the optical response of individual gold nanoparticles. *Nano Lett.* 5(3):515–518.

Berciaud, S., D. Lasne, G. A. Blab, L. Cognet, and B. Lounis. 2006. Photothermal heterodyne imaging of individual metallic nanoparticles: Theory versus experiments. *Phys. Rev. B* 73:045424.

Betzig, E., G. H. Patterson, R. Sougrat, O. Wolf Lindwasser, S. Olenych, J. S. Bonifacino, M. W. Davidson, J. Lippincott-Schwartz, and H. F. Hess. 2006. Imaging intracellular fluorescent proteins at nanometer resolution. *Science* 313(5793):1642–1645.

Blab, G. A., L. Cognet, S. Berciaud, I. Alexandre, D. Husar, J. Remacle, and B. Lounis. 2006. Optical readout of gold nanoparticle-based DNA microarrays without silver enhancement. *Biophys. J.* 90(1):L13–L15.

Borgdorff, A. J. and D. Choquet. 2002. Regulation of AMPA receptor lateral movements. *Nature* 417(6889):649–653.

Bosbach, J., C. Hendrich, F. Stietz, T. Vartanyan, and F. Trager. 2002. Ultrafast dephasing of surface plasmon excitation in silver nanoparticles: Influence of particle size, shape, and chemical surrounding. *Phys. Rev. Lett.* 89(25):257404.

Boyer, D., P. Tamarat, A. Maali, B. Lounis, and M. Orrit. 2002. Photothermal imaging of nanometer-sized metal particles among scatterers. *Science* 297(5584):1160–1163.

Cognet, L., C. Tardin, D. Boyer, D. Choquet, P. Tamarat, and B. Lounis. 2003. Single metallic nanoparticle imaging for protein detection in cells. *Proc. Natl. Acad. Sci. USA* 100(20):11350–11355.

Daniel, M. C. and D. Astruc. 2004. Gold nanoparticles: Assembly, supramolecular chemistry, quantum-size-related properties, and applications toward biology, catalysis, and nanotechnology. *Chem. Rev.* 104(1):293–346.

Debrabander, M., R. Nuydens, G. Geuens, M. Moeremans, and J. Demey. 1986. The use of submicroscopic gold particles combined with video contrast enhancement as a simple molecular probe for the living cell. *Cell Motil. Cytoskeleton* 6(2):105–113.

De Mey, J., M. Moeremans, G. Geuens, R. Nuydens, and M. Debrabander. 1981. High-resolution light and electron-microscopic localization of tubulin with the IGS (immuno gold staining) method. *Cell Biol. Int. Rep.* 5(9):889–899.

Elson, E. L. and D. Magde. 1974. Fluorescence correlation spectroscopy. 1. Conceptual basis and theory. *Biopolymers* 13(1):1–27.

Eustis, S. and M. A. El-Sayed. 2006. Why gold nanoparticles are more precious than pretty gold: Noble metal surface plasmon resonance and its enhancement of the radiative and nonradiative properties of nanocrystals of different shapes. *Chem. Soc. Rev.* 35(3):209–217.

Fritzsche, W. and T. A. Taton. 2003. Metal nanoparticles as labels for heterogeneous, chip-based DNA detection. *Nanotechnology* 14(12):R63–R73.

Groc, L., M. Lafourcade, M. Heine, M. Renner, V. Racine, J. B. Sibarita, B. Lounis, D. Choquet, and L. Cognet. 2007. Surface trafficking of neurotransmitter receptor: Comparison between single-molecule/quantum dot strategies. *J. Neurosci.* 27(46):12433–12437.

Heine, M., L. Groc, R. Frischknecht, J.-C. Beique, B. Lounis, G. Rumbaugh, R. L. Huganir, L. Cognet, and D. Choquet. 2008. Surface mobility of postsynaptic AMPARs tunes synaptic transmission. *Science* 320(5873):201–205.

Hirsch, L., A. Gobin, A. Lowery, F. Tam, R. Drezek, N. Halas, and J. West. 2006. Metal nanoshells. *Ann. Biomed. Eng.* 34(1):15.

Klar, T., M. Perner, S. Grosse, G. von Plessen, W. Spirkl, and J. Feldmann. 1998. Surface-plasmon resonances in single metallic nanoparticles. *Phys. Rev. Lett.* 80:4249–4252.

Kreibig, U. and M. Vollmer. 1995. *Optical Properties of Metal Clusters*. Berlin, Germany: Springer-Verlag.

Kusumi, A. and Y. Sako. 1996. Cell surface organization by the membrane skeleton. *Curr. Opin. Cell Biol.* 8(4):566–574.

Lasne, D., G. A. Blab, S. Berciaud, M. Heine, L. Groc, D. Choquet, L. Cognet, and B. Lounis. 2006. Single nanoparticle photothermal tracking (SNaPT) of 5-nm gold beads in live cells. *Biophys. J.* 91(12):4598.

Levy, R., N. T. Thanh, R. C. Doty, I. Hussain, R. J. Nichols, D. J. Schiffrin, M. Brust, and D. G. Fernig. 2004. Rational and combinatorial design of peptide capping ligands for gold nanoparticles. *J. Am. Chem. Soc.* 126(32):10076–10084.

Liau, Y. H., A. N. Unterreiner, Q. Chang, and N. F. Scherer. 2001. Ultrafast dephasing of single nanoparticles studied by two-pulse second-order interferometry. *J. Phys. Chem. B* 105(11):2135.

Lindfors, K., T. Kalkbrenner, P. Stoller, and V. Sandoghdar. 2004. Detection and spectroscopy of gold nanoparticles using supercontinuum white light confocal microscopy. *Phys. Rev. Lett.* 93(3):037401.

Link, S. and M. A. El Sayed. 2000. Shape and size dependence of radiative, non-radiative and photothermal properties of gold nanocrystals. *Int. Rev. Phys. Chem.* 3(19):409–453.

Link, S., and M. A. El-Sayed. 2003. Optical properties and ultrafast dynamics of metallic nanocrystals. *Annu. Rev. Phys. Chem.* 54:331–366.

Magde, D., E. L. Elson, and W. W. Webb. 1972. Thermodynamic fluctuations in a reacting system—Measurement by fluorescence correlation spectroscopy. *Phys. Rev. Lett.* 29(11):705–708.

Magde, D., E. L. Elson, and W. W. Webb. 1974. Fluorescence correlation spectroscopy. 2. Experimental realization. *Biopolymers* 13(1):29–61.

Michalet, X., F. F. Pinaud, L. A. Bentolila, J. M. Tsay, S. Doose, J. J. Li, G. Sundaresan, A. M. Wu, S. S. Gambhir, and S. Weiss. 2005. Quantum dots for live cells, in vivo imaging, and diagnostics. *Science* 307(5709):538–544.

Möller, R., R. D. Powell, J. F. Hainfeld, and W. Fritzsche. 2005. Enzymatic control of metal deposition as key step for a low-background electrical detection for DNA chips. *Nano Lett.* 5(7):1475–1482.

Octeau, V., L. Cognet, L. Duchesne, D. Lasne, N. Schaeffer, D. G. Fernig, and B. Lounis. 2009. Photothermal absorption correlation spectroscopy. *ACS Nano* 3(2):345–350.

Ozbay, E. 2006. Plasmonics: Merging photonics and electronics at nanoscale dimensions. *Science* 311(5758):189–193.

Park, S. J., T. A. Taton, and C. A. Mirkin. 2002. Array-based electrical detection of DNA with nanoparticle probes. *Science* 295(5559):1503–1506.

Persson, B. N. J. 1993. Polarizability of small spherical metal particles: Influence of the matrix environment. *Surf. Sci.* 281:153–162.

Peyser, L. A., A. E. Vinson, A. P. Bartko, and R. M. Dickson. 2001. Photoactivated fluorescence from individual silver nanoclusters. *Science* 291(5501):103–106.

Prodan, E., C. Radloff, N. J. Halas, and P. Nordlander. 2003. A hybridization model for the plasmon response of complex nanostructures. *Science* 302(5644):419–422.

Ries, J. and P. Schwille. 2006. Studying slow membrane dynamics with continuous wave scanning fluorescence correlation spectroscopy. *Biophys. J.* 91(5):1915–1924.

Ritchie, K., X. Y. Shan, J. Kondo, K. Iwasawa, T. Fujiwara, and A. Kusumi. 2005. Detection of non-Brownian diffusion in the cell membrane in single molecule tracking. *Biophys. J.* 88(3):2266–2277.

Rosi, N. L. and C. A. Mirkin. 2005. Nanostructures in biodiagnostics. *Chem. Rev.* 105(4):1547–1562.

Rust, M. J., M. Bates, and X. W. Zhuang. 2006. Sub-diffraction-limit imaging by stochastic optical reconstruction microscopy (STORM). *Nat. Methods* 3(10):793–795.

Sako, Y., S. Minoghchi, and T. Yanagida. 2000. Single-molecule imaging of EGFR signalling on the surface of living cells. *Nat. Cell Biol.* 2(3):168–172.

Schmidt, T., G. J. Schuetz, W. Baumgartner, H. J. Gruber, and H. Schindler. 1995. Characterization of photophysics and mobility of single molecules in a fluid lipid membrane. *J. Phys. Chem.* 99:17662–17668.

Schmidt, T., G. J. Schuetz, W. Baumgartner, H. J. Gruber, and H. Schindler. 1996. Imaging of single molecule diffusion. *Proc. Natl. Acad. Sci. USA* 93(7):2926–2929.

Schultz, D. A. 2003. Plasmon resonant particles for biological detection. *Curr. Opin. Biotechnol.* 14(1):13–22.

Schultz, S., D. R. Smith, J. J. Mock, and D. A. Schultz. 2000. Single-target molecule detection with non-bleaching multicolor optical immunolabels. *Proc. Natl. Acad. Sci. USA* 97(3):996–1001.

Schutz, G. J., G. Kada, V. P. Pastushenko, and H. Schindler. 2000. Properties of lipid microdomains in a muscle cell membrane visualized by single molecule microscopy. *EMBO J.* 19(5):892–901.

Schwille, P., U. Haupts, S. Maiti, and W. W. Webb. 1999. Molecular dynamics in living cells observed by fluorescence correlation spectroscopy with one- and two-photon excitation. *Biophys. J.* 77(4):2251–2265.

Schwille, P., S. Kummer, A. A. Heikal, W. E. Moerner, and W. W. Webb. 2000. Fluorescence correlation spectroscopy reveals fast optical excitation-driven intramolecular dynamics of yellow fluorescent proteins. *Proc. Natl. Acad. Sci. USA* 97(1):151–156.

Sheetz, M. P., S. Turney, H. Qian, and E. L. Elson. 1989. Nanometre-level analysis demonstrates that lipid flow does not drive membrane glycoprotein movements. *Nature* 340(6231):284–288.

Sönnichsen, C., S. Geier, N. E. Hecker, G. Von Plessen, J. Feldmann, H. Ditlbacher, B. Lamprecht et al. 2000. Spectroscopy of single metallic nanoparticles using total internal reflection microscopy. *Appl. Phys. Lett.* 77(19):2949.

Sonnichsen, C., B. M. Reinhard, J. Liphardt, and A. P. Alivisatos. 2005. A molecular ruler based on plasmon coupling of single gold and silver nanoparticles. *Nat. Biotechnol.* 23(6):741–745.

Tardin, C., L. Cognet, C. Bats, B. Lounis, and D. Choquet. 2003. Direct imaging of lateral movements of AMPA receptors inside synapses. *EMBO J.* 22(18):4656–4665.

Taton, T. A., C. A. Mirkin, and R. L. Letsinger. 2000. Scanometric DNA array detection with nanoparticle probes. *Science* 289(5485):1757–1760.

Tokeshi, M., M. Uchida, A. Hibara, T. Sawada, and T. Kitamori. 2001. Determination of subyoctomole amounts of nonfluorescent molecules using a thermal lens microscope: Subsingle-molecule determination. *Anal. Chem.* 73(9):2112.

Tschopp, J., E. R. Podack, and H. J. Mullereberhard. 1982. Ultrastructure of the membrane attack complex of complement—Detection of the tetramolecular C9-polymerizing complex C5B-8. *Proc. Natl. Acad. Sci. USA: Biol. Sci.* 79(23):7474–7478.

Vrljic, M., S. Y. Nishimura, S. Brasselet, W. E. Moerner, and H. M. McConnell. 2002. Translational diffusion of individual class II MHC membrane proteins in cells. *Biophys. J.* 83(5):2681–2692.

Wang, Z., J. V. Shah, M. W. Berns, and D. W. Cleveland. 2006. In vivo quantitative studies of dynamic intracellular processes using fluorescence correlation spectroscopy. *Biophys. J.* 91(1):343–351.

Wawrezinieck, L., H. Rigneault, D. Marguet, and P.-F. Lenne. 2005. Fluorescence correlation spectroscopy diffusion laws to probe the submicron cell membrane organization. *Biophys. J.* 89(6):4029–4042.

Wilcoxon, J. P., J. E. Martin, F. Parsapour, B. Wiedenman, and D. F. Kelley. 1998. Photoluminescence from nanosize gold clusters. *J. Chem. Phys.* 108(21):9137–9143.

Wuelfing, W. P., S. M. Gross, D. T. Miles, and R. W. Murray. 1998. Nanometer gold clusters protected by surface-bound monolayers of thiolated poly(ethylene glycol) polymer electrolyte. *J. Am. Chem. Soc.* 120(48):12696–12697.

Yguerabide, J. and E. E. Yguerabide. 1998. Light-scattering submicroscopic particles as highly fluorescent analogs and their use as tracer labels in clinical and biological applications. *Anal. Biochem.* 262(2):137–156.

Yguerabide, J. and E. E. Yguerabide. 2001. Resonance light scattering particles as ultrasensitive labels for detection of analytes in a wide range of applications. *J. Cell. Biochem.* 37(Suppl):71–81.

Yildiz, A., J. N. Forkey, S. A. McKinney, T. Ha, Y. E. Goldman, and P. R. Selvin. 2003. Myosin V walks hand-over-hand: Single fluorophore imaging with 1.5-nm localization. *Science* 300(5628):2061–2065.

Zammatteo, N., S. Hamels, F. De Longueville, I. Alexandre, J. L. Gala, F. Brasseur, and J. Remacle. 2002. New chips for molecular biology and diagnostics. *Biotechnol. Annu. Rev.* 8:85–101.

17

Thermal Imaging for Biological and Medical Diagnostics

Jay P. Gore
Purdue University

Lisa X. Xu
Purdue University

17.1 Introduction

The change in local temperature of blood, tissue, and skin in biological systems is determined by the difference in the thermal energies received and lost by conduction transfer and advection, as well as by the thermal energy released by chemical reaction processes such as metabolism. The use of temperature (measured at a convenient location such as the ear, armpit, or mouth), as an indicator of health has been in routine use for a very long time. However, even in such use, it is widely recognized that different healthy individuals can have different temperatures. In fact, the same healthy individual can have different temperatures at different times, based on the local environmental conditions, blood perfusion rates as controlled by the nervous system, and metabolic rates as controlled by a variety of complex factors.

In addition, the single temperature measured at a chosen location does not represent the temperatures of the different parts of the body even on the skin surface. The temperature of a human being typically decreases from the core to the skin by 3°C–5°C to support the heat dissipation to the surroundings. The transport processes involving conduction and advection (blood perfusion) within the body and the thermal boundary conditions at the skin determine the local skin temperature. In addition, the sweat glands in the human skin respond to different stimuli and the surface temperature is affected by the evaporation of water at the surface.

Diagnostic techniques using single-point temperature measurements have long benefited patients. The theoretical possibility exists that temperature at multiple points, if properly interpreted, can provide information about conditions affecting blood circulation, local metabolism, sweat gland malfunction, inflammation and healing, and energy imbalances created by hyperactivity of glands such as the thyroid. Temperatures of internal surfaces such as the colon, lung, and arteries could also be indicators of the state of health. Measuring such internal surface temperatures using endoscopes is a possibility.

Thermal imaging involves measurements of surface temperature using an array of infrared sensors installed in an infrared camera. This imaging allows the simultaneous measurement of temperatures of multiple points on the skin and is also a reference for surrounding temperature. The reference measurement could be subtracted from local measurements to provide an improved estimate of the temperature. In addition, measurement of differences in temperature between two points or in different regions of the image may be better indicators of the state of health. Motivated by this, thermal imaging using infrared cameras for medical diagnostics was initiated in the early 1970s. Over the subsequent three decades the technique has received mixed reviews and produced mixed results. Although the fundamentals of the thermal imaging process are sound, the relationship to a particular anomaly is not necessarily specific; therefore, the technique has greater promise as an adjunct diagnostic tool. The technique could also become valuable for monitoring the changes in skin temperature pattern over a specified period under highly controlled conditions.

Thermal imaging was considered for early diagnosis of breast cancer based on the theoretical conjecture that a rapidly growing tumor has high metabolism and blood perfusion rates leading to an increase in the local tumor temperature. If the tumor were to be located in a region close to the skin, then the local high temperature would also appear as a high temperature spot on the skin. Pattern recognition techniques, heuristic diagnosis, and left-breast-to-right-breast temperature pattern comparisons were used to interpret analog images visually. The results of the trials were highly inconclusive and controversial.

The controversy continues today with those opposed to the technique still basing their opinions on past failures. They have not changed their opinions even though the trials were in an era before the personal computer and advances in image processing techniques that can increase measurement accuracy and yield better computerized algorithms and objective tests for specific maladies. In addition, they also discount the possible use of thermal imaging as an adjunct functional diagnosis tool, for example, to determine if a positive mammography result could be combined with a high metabolic rate indication to separate malignancies, thereby reducing the number of false biopsies. Tests conducted for use as stand-alone diagnostics are generally not indicative of the value of a method for adjunct use.

Temperature patterns on the skin are the result of biological heat and mass transfer, metabolic processes in response to specific surrounding conditions, and mental, neurological and hormonal states of the subjects. Therefore, interpretation of single thermal images must not be based on use of arbitrary pattern recognition methods because such a diagnosis can be highly nonspecific and misleading. However, interpretation using proper accounting of the biological processes that determine the skin surface temperature, as well as multiple images taken during annual examinations and of dynamic thermal response to stimuli, can yield useful results, particularly when used as a function indicator technique in conjunction with a structure indication technique such as x-ray or ultrasound imaging. Therefore, we next review the fundamentals of infrared radiation and thermal imaging, followed by abstracts of some recent work in thermal imaging for biology and diagnostics. A table of available infrared cameras is also provided, followed by two examples of the use of the bioheat transfer equation for the interpretation of thermal images.

17.2 Infrared Radiation and Thermal Imaging

Thermal imaging relies on sensing the infrared radiation emitted by all objects above absolute zero temperature. All objects emit photons as a result of transitions from a high-energy to a low-energy state. In solids, such transitions lead to a continuous distribution of energy between different wavelengths according to the Planck distribution. Objects at a relatively high temperature, such as carbon particles in a flame or a hot iron rod, emit visible light, while the emission from objects closer to normal room temperature is at much longer wavelengths. At the normal temperature of the human body, the peak of the Planck function occurs in the mid-infrared region between 9 and 10 µm wavelengths. This radiation is not visible to the human eye but, in sufficient intensity, can be felt by the human skin, one function of which is a low-sensitivity infrared array detector.

The Planck function is exponentially nonlinear in temperature. This means that the lower-temperature objects emit orders of magnitude less energy than do higher-temperature objects. Therefore, detection of infrared energy accurately is a challenging task. In addition, lower-temperature objects and their surroundings emit comparable amounts of infrared energy, which leads to difficulties in measurements. Often the signal-to-noise ratios (SNRs) are of comparable magnitude, requiring specialized background correction instrumentation, mode-locked signal processing techniques, and careful analysis of the resulting data.

The technology of infrared array detectors, associated electronics, image processing, and noise reduction has significantly improved over the last 10 years. The accuracy with which temperature and temperature changes can be measured has reached 10^{-3} K. This is better by a factor of 100 compared to the techniques used in past studies—a fact missed by many readers of the thermal imaging literature.

The infrared or thermal imaging cameras available commercially are listed in Table 17.1. The most relevant attributes of these cameras are listed in Tables 17.1 through 17.4. The most important

TABLE 17.1 Infrared Camera Manufacturers, Web Sites, Camera Models, and Array Sizes

Company	www Addresses	Camera	Array Size
FLIR	http://www.flir.com	ThermaCAM® E1	160 * 120
		ThermaCAM® SC300	160 * 120
		ThermaCAM® SC3000	320 * 240
Sierra Pacific	http://www.x20.org	PD300 IR	320 * 244
		IR 747	320 * 240
CMC Elec.	http://www.cinele.com/	TVS 8500	256 * 256
Indigo Sys.	http://www.indigosystems.com/index.html	Mirlin®-Mid	320 * 256
		Mirlin®-Uncooled	320 * 240
		Omega	160 * 120
		Alpha NIR	320 * 256
		TVS-620	320 * 240
		Phoenix-Mid	640 * 512
		Phoenix-Long	640 * 512
Infrared Sys.	http://www.infraredsys.com/	Raytheon IR500D	320 * 240
		Raytheon IRPro4	320 * 240
		Mikron 5104	320 * 240
		Mikron 7102	320 * 240
		Monroe Scientific	320 * 240
Cantronic Sys. Inc.	http://www.cantronic.com	IR805	120 * 120
		IR860	320 * 240
Ircon	http://www.ircon.com	DigiCam-IR	120 * 120
	800-323-7660	Stinger	320 * 240
Land Instruments International	http://www.landinst.com	Cyclops PPM+	120 * 120
		FTI 6	
Mikron	http://www.mikroninst.com	7200	320 * 240
		7515	320 * 240
		5104	255 * 223
		5102	255 * 223
		1100	207 * 344
		MHS500	255 * 223

TABLE 17.2   IR Camera Models, Spatial Resolution, Spectral Range, and Detector Materials

Camera	Spatial Resolution (mrad)	Spectral Range (μm)	Detector Material
ThermaCAM E1	2.6	7.5–13	FPA uncooled microbolometer
ThermaCAM SC300	2.6	7.5–13	FPA uncooled microbolometer
ThermaCAM SC3000	1.1	8–9	GaAs
PD300 IR	1.0	3.6–5	PtSi
IR 747	1.3	7.5–13	Uncooled microbolometer
TVS 8500	1.0	2.5–5.0	InSb
Mirlin-Mid		1–5.4	InSb
Mirlin-Uncooled		7.5–13.5	Uncooled microbolometer
Omega	1.7	7.5–13.5	Uncooled microbolometer
Alpha NIR	0.6	9–17	InGaAs
TVS-620	0.7	8–14	Uncooled microbolometer
Phoenix-Mid	0.25	2–5	InSb
Phoenix-Long	0.25	8–9.2	GaAs (QWIP)
Raytheon IR500D		7–14	FPA uncooled BST
Raytheon IRPro4		7–14	Uncooled BST
Mikron 5104		3–5.2	Radiometric TE MCT
Mikron 7102		8–14	Radiometric uncooled microbolometer
Monroe Scientific		2–14	Uncooled BST
IR805		8–12	Thermoelectric detector
IR860		8–14	Radiometric uncooled microbolometer
DigiCam-IR	0.25	8–12	Thermopile (TE) detector
Stinger	Not available	8–14	FPA uncooled
Cyclops PPM+	0.98	7–14	FPA uncooled
FTI 6			
7200	1.58	8–14	FPA uncooled microbolometer
7515	1.58	8–14	FPA uncooled microbolometer
5104	2.00	3–5.2	Mercury cadmium telluride
5102	1.50	8–12	Mercury cadmium telluride
1100	1.50	8–13	Mercury cadmium telluride
MHS500			Mercury cadmium telluride

parameters for biological applications are the range and accuracy. The array size and the pixel size determine the spatial resolution possible for the measurements. The wavelength sensitivity is important, even in thermal imaging, because the object imaged must have a high enough emissivity in the spectral sensitivity range of the camera. The cameras with highest sensitivity near 9 μm wavelength are the most popular for biological applications because of the Planck function's peak near this wavelength. Most biological solids also have a significant emissivity in this region, yielding a high SNR.

The cameras must be calibrated routinely using a blackbody standard source; however, the sensitivity of the detectors has long surpassed many blackbody standards used in general purpose laboratories. In addition to the blackbody calibration standard, lenses, mirrors, filters, and prisms made from infrared transmitting materials form the core tool set of a thermal imaging laboratory. Many laboratories work in the visible and the infrared portion of the light spectrum; most visible optics do not transmit infrared but most infrared optics transmit visible. Therefore, care must be taken to separate and label the optical components properly.

TABLE 17.3 Infrared Camera Models, Temperature Range, Sensitivity, Framing Rate, and Bits

Camera	Temperature Range (°C)	Sensitivity	Framing Rate (Hz)	Bits
ThermaCAM E1	−20 to 250	120 mK	60	14
ThermaCAM SC300	−40 to 500	100 mK at 30°C	50–60	14
ThermaCAM SC3000	−20 to 2000	20 mK at 30°C	50–60	14
PD300 IR	−10 to 1500	100 mK at 30°C	60	12
IR 747	−15 to 45	100 mK at 30°C		
TVS 8500	−40 to 2000	20 mK at 30°C	120	14
Mirlin-Mid	0 to 2000	25 mK	60	12
Mirlin-Uncooled	0 to 1000	100 mK	60	12
Omega	−40 to 400	40 mK	30	14
Alpha NIR			30	12
TVS-620	−20 to 900	150 mK at 30°C	30	12
Phoenix-Mid		25 mK	30	14
Phoenix-Long		30 mK	30	14
Raytheon IR500D	0 to 300	100 mK	60	12
Raytheon IRPro4	0 to 500	100 mK	30	12
Mikron 5104	−10 to 1500	100 mK at 30°C	60	12
Mikron 7102	−40 to 1500	200 mK at 30°C	60	14
Monroe Scientific	−18 to 538	100 mK at 25°C	60	10
IR805	0 to 350	200 mK	30	16
IR860	−10 to 1000	100 mK at 30°C	30	Not available
DigiCam-IR	−10 to 1100	350 mK at 30°C	60	16
Stinger	0 to 500	100 mK at 25°C	30	12
Cyclops PPM+	−15 to 300	100 mK at 30°C	60	12
FTI 6	−20 to 2000			
7200	−40 to 120	80 mK at 30°C	60	14
7515	−40 to 500	80 mK at 30°C	60	14
5104	−10 to 800	100 mK at 30°C	22	12
5102	−10 to 200	30 mK at 30°C	22	12
1100	−50 to 2000	100 mK at 30°C	4	13
MHS500	−0 to 70	100 mK	2	14

17.3 Applications of Infrared Thermal Imaging

Many applications of infrared thermal imaging have been reported in the literature.[1–5] The basic measurement involves skin temperature distribution resulting from a variety of internal and external conditions affecting the circulation and metabolic processes. Therefore, there is an existing nonspecificity in the process, which must be recognized and addressed by improving the analytical tools and recognizing that, even with the most improved tools, a fundamentally nonspecific diagnostic technique such as thermal imaging can only be used as a powerful adjunct tool. Such recognition will avoid the controversy and confusion often surrounding this issue. More scientific work is needed to distinguish the differences between enhanced blood flow because of tumor and because of stress of a polygraph test. The present problem is that proponents of thermal imaging claim that the technique can be used as a diagnostic tool in both these scenarios, while opponents of the technique claim that a health parameter as basic as temperature and its distribution on different parts of the body has no diagnostic value. Both positions are extreme because the former ignores the nonspecificity of the technique while the later ignores the fact that, along with pulse and appearance, temperature is a health diagnostic modality proven over thousands of years.

TABLE 17.4 Infrared Camera Models, Cooling Method, Cooling Temperature, and Space Required

Camera	Cooling Method	Cooling Temperature (K)	Size (mm)
ThermaCAM E1	Uncooled	—	265 * 80 * 105
ThermaCAM SC300	Uncooled	—	212 * 121 * 127
ThermaCAM SC3000	Stirling	77	220 * 135 * 130
PD300 IR	Stirling	77	222 * 127 * 140
IR 747	Uncooled	—	203 * 121 * 112
TVS 8500	Stirling	77	200 * 250 * 120
Mirlin-Mid	Stirling	77	140 * 127 * 250
Mirlin-Uncooled	Uncooled	—	102 * 115 * 203
Omega	Uncooled	—	35 * 37 * 48
Alpha NIR	Passive conduction and convection	—	53 * 64 * 95
TVS-620	Uncooled	—	115 * 217 * 142
Phoenix-Mid	Stirling	77	153 * 153 * 127
Phoenix-Long	Stirling	77	533 * 508 * 407
Raytheon IR500D	Uncooled	—	254 * 140 * 102
Raytheon IRPro4	Uncooled	—	254 * 140 * 102
Mikron 5104	Thermoelectrically cooled		203 * 90 * 220
Mikron 7102	Uncooled	—	95 * 110 * 170
Monroe Scientific	Uncooled	—	140 * 114 * 114
IR805	Uncooled	—	240 * 100 * 130
IR860	Uncooled	—	177 * 110 * 142
DigiCam-IR	Uncooled	—	240 * 100 * 130
Stinger	Uncooled	—	344 * 127 * 98
Cyclops PPM+	Uncooled	—	150 * 245 * 275
FTI 6			
7200	Uncooled	—	97 * 109 * 170
7515	Uncooled	—	97 * 109 * 170
5104			203 * 89 * 221
5102	Stirling		198 * 98 * 235
1100	Liquid nitrogen	77	135 * 159 * 228
MHS500	Thermoelectrically cooled		221 * 147 * 270

Some of the recent literature involving use of infrared thermal imaging as an experimental tool in biology and health studies is summarized briefly in the next paragraph. This summary points to the broad applications as well as the nonspecificity inherent in this measurement modality.

Barnett et al.[6] used thermal imaging to conduct a preliminary investigation of the human gingiva and suggested that, with cooling and rewarming, it can be used as an indicator of inflammatory state. Chan et al.[7] state that thermal imaging is useful for identifying the health status of the thyroid gland. Whole-body thermal images are adequate indicators of ectodermal dysplasia according to Clark and co-workers,[8] and Dickey et al.[9] report that infrared thermography can be used to measure the depth of burns in burn victims. Thermal imaging systems in the management of pain were used by Hasegawa et al.[10] Stein and co-workers[11] demonstrated the use of thermal imaging in monitoring surgical tendon repair in veterinary medicine.

One of the oldest, most controversial, but possibly most promising applications of thermal imaging involves breast cancer detection.[12] The literature in this area is briefly reviewed in Section 17.3.1. Villringer and Dirnagl[13] used infrared thermal imaging to measure brain activity. Applications of thermal imaging to internal surfaces have begun appearing, with papers on the use of infrared fiber optic catheter for thermal imaging of atherosclerotic plaque by Naghavi et al.[14]

Thermal imaging is being used in the two basic fields of biology, zoology and botany. In addition to human health studies, the applications are wide and therefore care is required in claims of specificity like those in health studies. Thermal imaging has been used in population studies of burrowing mammals by Hubbs et al.[15] Fuller and Wisniewski[16] use thermal imaging to study freezing in plants, and Jones[17] uses it to measure stomatal conductance of plants.

The most recent claims for thermal imaging involve its use in detecting anxiety by observing changes in facial temperature patterns resulting from neurological responses of the circulatory system.[18,19] Although the publicity value of such claims is high, care must be exercised in the interpretation of results based on the nonspecificity of the primary measure.

The following section describes an example of using temperature measurements in conjunction with a mathematical model. The mathematical model is then used to generate expected thermal images, which show their effectiveness in indicating high blood-perfusion and metabolic rates typically associated with a tumor. Very few studies of this nature aimed at evaluating the scientific basis for thermal imaging as a breast cancer diagnostic have been reported. Based on the results of this study, it is established that computational models of the bioheat transfer process can render thermal imaging a viable adjunct tool.

17.3.1 Calculations of Temperature Profiles in a Female Breast with and without a Tumor

17.3.1.1 Introduction

Breast cancer is one of the leading causes of death from cancer in women in the United States. One out of eight women suffers from breast cancer during her lifetime. The American Cancer Society estimated that in 2001, approximately 192,000 women in the United States would be diagnosed with breast cancer and approximately 40,600 would die as a result of this disease.[20] Early detection is considered to be the best defense against breast cancer.

Tumors grow by signaling for and receiving higher arterial blood supply, which supports the higher metabolic rates required for rapid cell division. Many diagnostic and therapeutic technologies based on these fundamentals are being considered. The differences in the energy consumption of normal and cancerous tissue lead to small but detectable local temperature changes.[21] For breast tumors near the skin surface the local temperature changes at the tumor location can be large enough to cause a local perturbation in temperature on the patient's skin.

Gautherie[22] examined 147 patients with malignant breast cancers and measured local temperatures and thermal conductivity using sterile fine-needle thermoelectric probes. He used a model based on electrical analogy to investigate correlations between thermal and geometrical parameters that appeared during heat interactions between the tumor and its surrounding tissue. The tumor metabolic heat generation rate was inferred from the model. The differences in effective thermal conductivity of healthy and cancerous tissues in vivo and in vitro as postoperative specimens were described as a function of local blood perfusion rate.

Osman and Afify[23] developed a mathematical model of the three-dimensional temperature distribution in women's breasts using a finite element method. Based on their literature survey, the blood flow to the tumor was considered to be two- to threefold that of the normal tissue. They correlated the tumor metabolic heat generation rate with the doubling time of tumor volume as measured by Gautherie et al.[24] They modeled the tumor as a point heat source with the heat generation equal to the total metabolic heat production of the neoplastic tissue. Surface temperature distributions were plotted for the malignant breasts; however, the temperature distributions inside the breasts were not discussed and compared to the measurements of Gautherie.[22]

Infrared images of skin temperature can be used as an adjunct tool for detection combined with mammography in cases in which significant thermal expression of the tumor occurs on the skin surface.

Some past studies have shown significant benefits of such a detection scheme, increasing the sensitivity rate from 85% to 95%.[25] However, many other studies were inconclusive, primarily because of incorrect measurements and qualitative interpretation. As a result, thermography is currently not used or recommended as an early detection technique. The accuracy of interpretation can be significantly enhanced if solutions of the bioheat equation are used to relate the infrared images obtained with modern high-accuracy cameras to the functional changes inside the breast. Such work is being conducted primarily outside the United States.[25-29]

17.3.1.2 Bioheat Transfer Equation

The steady-state form of the bioheat transfer equation proposed by Pennes[21] is

$$\nabla \cdot k \nabla T + \rho_b \cdot c_b \cdot \omega_b \cdot (T_a - T) + \dot{q}_m = 0 \tag{17.1}$$

where
  k denotes thermal conductivity of tissue
  ρ_b, c_b are the density and the specific heat of the blood
  ω_b (mL/s/mL) is the blood perfusion rate
  $\dot{q}_m$ is the metabolic heat generation rate (W/m³)
  T_a is the arterial blood temperature (K)
  T is the local temperature of the breast tissue

Equation 17.1 gives the quantitative relationship of the heat transfer characteristics in human tissue and includes the effects of blood perfusion (ω_b) and metabolic heat generation ($\dot{q}_m$). The temperature of the arterial blood is approximated to be the core temperature of the body.

In the bioheat equation proposed by Weinbaum and Jiji[30] the expression derived for the tensor conductivity of the tissue is a function of the local vascular geometry and flow velocity in thermally significant small vessels. The solution of this equation gives rise to an effective thermal conductivity to account for the enhanced heat transfer due to incomplete countercurrent heat exchange between paired vessels. Structural properties of the breast in terms of size and distribution of blood vessels are necessary, but not available, for the solution of this equation; therefore, Equation 17.1 was used in the present simulations.

17.3.1.3 Mathematical Model

A hemisphere of radius 9 cm was considered to match the patient in Gautherie's study.[22] The normal breast was approximated to be composed of a single layer with averaged properties. The breast with tumor included a sphere of radius 1.1 cm with its center located at 2.1 cm beneath the surface. The tumor was assigned a higher blood perfusion and metabolic heat generation rate compared to the corresponding quantities for the healthy tissue. Gautherie observed regions of increased metabolic activity around the tumor. Because metabolic heating depends on the supply of metabolites from the blood perfusion, the two quantities change in an interdependent monotonic manner. The region of abnormal metabolic rates around the tumor was modeled as 0.7 cm thick lower and upper hemispherical shells of different properties around the tumor. The two different shells are used to represent the differences in the upper and lower parts of the breast that surround the tumor.

A convective boundary condition is specified at the skin surface. Natural convection for a sphere (Nusselt number $\cong$ 2) gives a heat transfer coefficient of approximately 5 W/m² K at the surface for ambient air at 21°C. The inner boundary of the breast is the pectorals muscle, which covers the ribs and is considered as adiabatic in the present simulations.

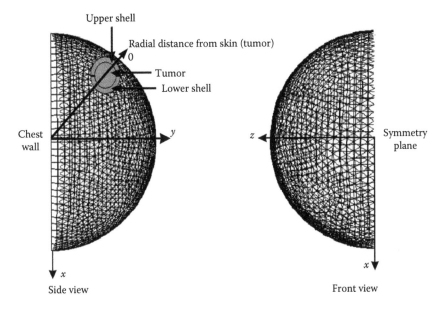

FIGURE 17.1 Numerical grid for bioheat transfer calculations in a breast model for the generation of a computational thermal image.

17.3.1.4 Solution

Figure 17.1 illustrates the computational grid created using the computer program GAMBIT 1.3. The volume is discretized into finite elements (tetrahedral and triangular). There are 96,410 cells and 238,324 faces in the grid and it satisfies a mesh independent solution. During meshing, a boundary layer was considered at the surface to account for the high temperature gradient. The grid is exported to the computer program FLUENT5 to solve for the temperature distributions. Blood perfusion and metabolic heat generation terms are added to the computations with a user-defined function written in C++. The function also increases the blood perfusion and metabolic heat generation at the tumor and the surrounding shells.

17.3.1.5 Optimum Results

The metabolic heat generation and the blood perfusion inside the normal breast were varied. Initially, the metabolic heat generation was assumed to be $\dot{q}_m = 420$ W/m^3.[31] The temperatures were calculated at three different angles, namely, 0°C, 45°C, and 90°C with respect to the x-axis. The maximum difference in temperature for these three angles was about 0.1°C.

Gautherie[22] determined the metabolic heat generation of the tumor to be 29,000 W/m^3 using the electrical analogy. Thus, for this metabolic heat generation inside the tumor, the blood perfusion rate varied. The thermal effect of higher blood perfusion is to reduce the highest temperature attained inside the tumor.

Table 17.5 gives the blood perfusion and metabolic heat generation rates used to give the best match with Gautherie's[22] temperature distributions using the least mean square error method. At the optimum, the root mean square error over the entire profile is 0.112°C for the normal breast and 0.293°C for the malignant breast. The blood perfusion rates for all domains are within the range specified by Vaupel.[32] The predicted radial temperature profile is plotted in Figure 17.2.

TABLE 17.5　Blood Perfusion and Metabolic Heat Generation Values Used for an Optimum Fit with the Gautherie Data

	$\dot{q}_m$ (W/m³)	ω_b (mL/s/mL of Tissue)
Normal	450	0.00018
Tumor	29,000	0.00900
Lower shell	11,700	0.00360
Upper shell	4,725	0.00144

Source: Data from Gautherie, M., *Ann. N. Y. Acad. Sci.,* 335, 383, 1980.

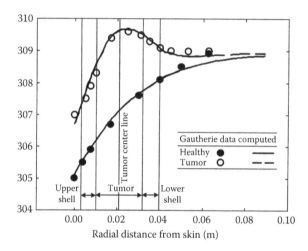

FIGURE 17.2　Verification of the numerical model parameters used in the bioheat transfer simulations.

17.3.1.6 Conclusion, Discussion, and Future Work

The calculated temperature distributions agree very well with the measurements based on the model structure and the parametric values used in the present simulation. This suggests that the tumor has approximately 1.5–2 orders of magnitude higher metabolic rate and close to 10–20 times blood perfusion rate. As long as the orders of magnitude of the metabolic rate and blood perfusion rate are correct, reasonable temperature profiles can be obtained. A more significant result is that the tumor introduces a local temperature rise on the breast surface that is accurately detectable by modern infrared cameras. Further inverse bioheat transfer calculations may provide a method of locating the tumor using the surface thermograms.

Figure 17.3 shows computational thermograms created using the results of the calculations. These data indicate that, for the tumor measured by Gautherie, the temperature differences apparent on the skin surface would be significant and detectable with thermal imaging. Furthermore, the right-hand panel shows that modern image processing techniques can enhance the thermal imaging signal significantly.

Presently, experimental verification of the combined use of thermal imaging and bioheat transfer computations is in progress in our laboratory. The initial thermal imaging results are shown in Figure 17.4. The data show that it is feasible to measure the type of temperature changes of interest in biological systems with very high accuracy and precision.

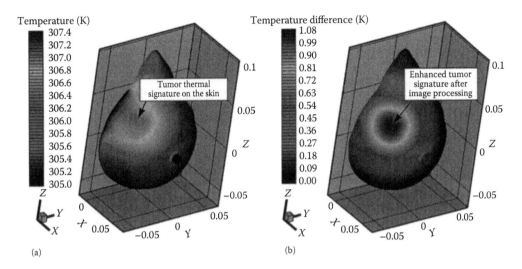

(a) (b)

FIGURE 17.3 Numerical thermal images demonstrating the fundamental ability of a thermal image to capture an abnormality and the power of image processing. (a) Skin temperature distribution of malignant breast with an embedded 2 cm diameter tumor in the upper outer quadrant and (b) enhanced tumor thermal signature on the surface of malignant breast after subtracting the normal distribution.

FIGURE 17.4 A sample infrared image demonstrating the high temperature sensitivity of modern IR cameras.

17.4 Summary and Conclusions

Infrared thermal imaging of temperature changes in a biological system is a measurement modality with a fundamental scientific basis. However, fundamental science also clearly points out the nonspecific nature of this measurement for monitoring any disease state. Therefore, this technique must be used as an adjunct to other diagnostic techniques and in conjunction with newly emerging analytical and numerical computational tools such as the one demonstrated previously.

The technique is clearly not ready for general purpose clinical applications without the establishment of a fundamental science and technology base aimed at addressing the nonspecificity issue by developing adjunct instrumentation and numerical tools. If such work is completed, thermal imaging can add a powerful measure of function in biological systems because of its fundamental basis in blood perfusion and metabolic activity.

Acknowledgments

The Indiana Twenty-First Century Research and Technology Funds (Dr. Karl Kohler, Program Director) supported this research. The contributions of PhD students Ashish Gupta and Dawn Sabados of Purdue University are gratefully acknowledged.

References

1. Incropera, F.P. and DeWitt, D.P., *Fundamentals of Heat and Mass Transfer*, John Wiley & Sons, New York, 1995.
2. Siegel, R. and Howell, J.R., *Thermal Radiation Heat Transfer*, Hemisphere, Washington, DC, 1992.
3. Modest, M.F., *Radiative Heat Transfer*, McGraw-Hill, New York, 1993.
4. DeWitt, D.P. and Nutter, G.D., *Theory and Practice of Radiation Thermometry*, John Wiley & Sons, New York, 1988.
5. Welch, A.J. and Van Gemert, M.J.C., *Optical-Thermal Response of Laser Irradiated Tissue*, Plenum Press, New York, 1995.
6. Barnett, M.L., Gilman, R.M., Charles, C.H., and Bartels, L.L., Computer-based thermal imaging of human gingiva preliminary investigation, *J. Periodontol.*, 60, 628, 1989.
7. Chan, F.H.Y., So, A.T.P., Kung, A.W.C., Lam, F.K., and Yip, H.C.L., Thyroid diagnosis by thermogram sequence analysis, *Biomed. Mater. Eng.*, 5, 169, 1995.
8. Clark, R.P., Goff, M.R., and Macdermot, K.D., Identification of functioning sweat pores and visualization of skin temperature patterns in x-linked hypohidrotic ectodermal dysplasia by whole body thermography, *Hum. Genet.*, 86, 7, 1990.
9. Dickey, F.M., Holswade, S.C., and Yee, M.L., Burn depth estimation using thermal excitation and imaging, *Proc. SPIE*, 3595, 9, 1999.
10. Hasegawa, J. and Takaya, K., Development of microcomputer-aided pyroelectric thermal imaging system and application to pain management, *Med. Biol. Eng. Comput.*, 24, 275, 1986.
11. Stein, L.E., Pijanowski, G.J., Johnson, A.L., Maccoy, D.M., and Chato, J.C.A., Comparison of steady state and transient thermography techniques using a healing tendon model, *Vet. Surg.*, 17, 90, 1988.
12. Parisky, Y.R., Skinner, K.A., Cothren, R., DeWittey, R.L., Birbeck, J.S., Conti, P.S., Rich, J.K., and Dougherty, W.R., Computerized thermal breast imaging revisited: An adjunctive tool to mammography, *Proc. IEE Eng. Med. Biol. Soc.*, 20, 919, 1998.
13. Villringer, A. and Dirnagl, U., Optical imaging of brain function and metabolism 2: Physiological basis and comparison to other functional neuroimaging methods, *Adv. Exp. Med. Biol.*, 413, 15, 1997.
14. Naghavi, M., Melling, P., Gul, K., Madjid, M., Willerson, J.T., and Casscells, W., First prototype of a 4 French 180° side-viewing infrared fiber optic catheter for thermal imaging of atherosclerotic plaque, *J. Am. Coll. Cardiol.*, 37(Suppl. 2), 3A, 2001.
15. Hubbs, A.H., Karels, T., and Boonstra, R., Indices of population size for burrowing mammals, *J. Wildl. Manage.*, 64, 296, 2000.
16. Fuller, M.P. and Wisniewski, M., The use of infrared thermal imaging in the study of ice nucleation and freezing of plants, *J. Therm. Biol.*, 23, 81, 1998.
17. Jones, H.G., Use of thermography for quantitative studies of spatial and temporal variation of stomatal conductance over leaf surfaces, *Plant Cell Environ.*, 22, 1043, 1999.
18. Kobel, J., Holowacz, W., and Podbielska, H., Thermal imaging for face recognition in optical security systems. Optical sensing for public safety, health, and security, *Proc. SPIE*, 4535, 154, 2001.
19. Pavlidis, I., Levine, J., and Baukol, P., Thermal imaging for anxiety detection, *Proceedings of the IEEE Workshop on Computer Vision beyond the Visible Spectrum: Methods and Applications*, Hilton Head Island, SC, 2000, p. 104.
20. http://www.nabco.org/.

21. Pennes, H.H., Analysis of tissue and arterial blood temperature in resting human forearms, *J. Appl. Physiol.*, 2, 93, 1948.
22. Gautherie, M., Thermopathology of breast cancer: Measurements and analysis of in vivo temperature and blood flow, *Ann. N. Y. Acad. Sci.*, 335, 383, 1980.
23. Osman, M.M. and Afify, E.M., Thermal modeling of malignant women's breast, *ASME J. Biomech. Eng.*, 110, 269, 1988.
24. Gautherie, M., Quenneville, Y., and Gros, C., Metabolic heat production, growth rate and prognosis of early breast carcinomas, *Biomedicine*, 22, 328, 1975.
25. Keyserlingk, J.R., Ahlgren, P.D., Yu, E., and Belliveay, N., Infrared imaging of the breast: Initial reappraisal using high-resolution digital technology in 100 successive cases of stage I and II breast cancer, *Breast J.*, 4, 245, 1998.
26. Ng, E.Y.K. and Sudharsan, N.M., An improved three-dimensional direct numerical modeling and thermal analysis of a female breast with tumor, *J. Eng. Med.*, 215, 25, 2001.
27. Ng, E.Y.K. and Sudharsan, N.M., Can numerical simulation adjunct to thermography be an early detection tool? *J. Thermol. Int.* [formerly *Eur. J. Thermol.*], 10, 119, 2000.
28. Sudharsan, N.M. and Ng, E.Y.K., Parametric optimization for tumor identification: Bioheat equation using ANOVA and Taguchi method, *J. Eng. Med.*, 214, 505, 2000.
29. Sudharsan, N.M., Ng, E.Y.K., and Teh, S.L., Surface temperature distribution of a breast with/without tumor, *Comput. Methods Biomech. Biomed. Eng.*, 2, 187, 1999.
30. Weinbaum, S. and Jiji, L.M., A new simplified bioheat equation for the effect of blood flow on local average tissue temperature, *ASME J. Biomech. Eng.*, 107, 131, 1985.
31. Liu, J. and Xu, L.X., Boundary information based diagnostics on the thermal states of biological bodies, *Int. J. Heat Mass Transf.*, 43, 2827, 2000.
32. Vaupel, P., *Tumor Blood Flow, Blood Perfusion and Microenvironment of Human Tumors: Implications for Clinical Radio Oncology*, Springer-Verlag, Berlin, Germany, 1998, p. 41.

18

Multidimensional Fluorescence Imaging of Biological Tissue

Christopher Dunsby
Imperial College London

James McGinty
Imperial College London

Paul French
Imperial College London

18.1 Introduction

This chapter aims to review multidimensional fluorescence imaging (MDFI) technology and its application to biological tissue, with a particular emphasis on fluorescence lifetime imaging (FLIM) of biological tissue with examples from our work at Imperial College London. Fluorescence imaging is flourishing tremendously, partly driven by advances in laser and detector technology, partly by advances in labeling technologies such as genetically expressed fluorescent proteins, and partly by advances in computational analysis techniques. Increasingly, fluorescence instrumentation is developed to provide more information than just the localization or distribution of specific fluorescent molecules. Often, fluorescence signals are analyzed to provide information on the local fluorophore environment or to contrast different fluorophores in complex mixtures—as often occur in biological tissue. This trend to higher-content fluorescence imaging increasingly exploits MDFI and measurement capabilities with instrumentation that resolves fluorescence lifetime together with other spectroscopic parameters such as excitation and emission wavelength and polarization, providing image information in two or three spatial dimensions as well as with respect to elapsed time (Figure 18.1). However, caution should be exercised when acquiring such MDFI since photobleaching or experimental considerations usually impose a limited photon *budget* and/or a maximum image acquisition time and also present significant challenges with respect to data analysis and data management. These considerations are particularly important for real-time clinical diagnostic applications, for higher-throughput assays, and for the investigation of dynamic biological systems (Figure 18.1).

Fluorescence can provide high-content optical molecular contrast [1] in single-point (cuvette-based or fiber-optic probe-based) instruments such as fluorometers as well as imaging instruments such as

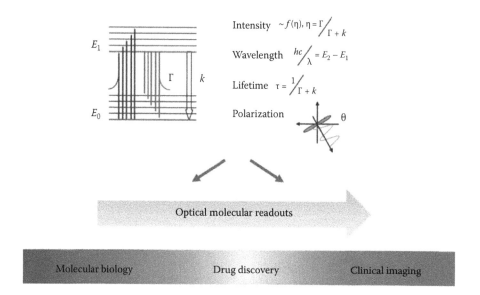

$$\text{Intensity} \quad \sim f(\eta), \eta = \frac{\Gamma}{\Gamma + k}$$

$$\text{Wavelength} \quad \frac{hc}{\lambda} = E_2 - E_1$$

$$\text{Lifetime} \quad \tau = \frac{1}{\Gamma + k}$$

$$\text{Polarization}$$

Optical molecular readouts

Molecular biology   Drug discovery   Clinical imaging

FIGURE 18.1 **(See color insert.)** Overview of MDFI and metrology. (Adapted from Talbot, C. et al., Fluorescence lifetime imaging and metrology for biomedicine, Chapter 6, in: Tuchin, V., ed., *Handbook of Photonics for Biomedical Science*, CRC Press, Boca Raton, FL, 2010, pp. 159–196.)

microscopes, multiwell plate readers, endoscopes, and tomographic instruments. Typically for cell biology, exogenous fluorescent molecules are used as *labels* to tag specific molecules of interest, while for clinical applications, it is possible to exploit the fluorescence properties of target molecules themselves to provide label-free molecular contrast, although there is an increasing trend to investigate the use of exogenous labels both to detect diseased tissue and to effect therapeutic intervention, for example, using photosensitizers for photodynamic detection (PDD) and photodynamic therapy (PDT) [2] or nanoparticles [3]. Beyond mapping distributions of fluorophores (fluorescent molecules), the use of multiple fluorophores with distinguishable spectral properties can indicate *colocalization*, which has been used to suggest interaction of labeled proteins, albeit limited to the spatial resolution of the imaging system. Higher-content analysis of fluorescence signals can provide information about the local fluorophore environment, to which the fluorescence process can be highly sensitive through its impact on the radiative and nonradiative decay rates and therefore the quantum efficiency and fluorescence lifetime. Such analysis therefore provides a means of mapping changes in molecular properties and interactions beyond the resolution (diffraction) limit.

In principle, the fluorescence intensity depends on the quantum efficiency and therefore is sensitive to the local fluorophore environment. In practice, however, it is challenging to determine the quantum efficiency from intensity measurements since this would require knowledge of the photon excitation and detection efficiencies, as well as the fluorophore concentration. These parameters are often not known in biological experiments, and quantitative measurements of intensity can be further hindered by optical scattering, internal attenuation of fluorescence (inner filter effect), and background fluorescence from other fluorophores present in a sample. More robust quantitative measurements can be made using *ratiometric* techniques. In the spectral domain, one can assume that unknown quantities such as excitation and detection efficiency, fluorophore concentration, and signal attenuation will be approximately the same for intensity measurements in two or more spectral windows and may be effectively *cancelled out* in a ratiometric measurement. This approach is used, for example, with calcium-sensing dyes that ratio measurements made at different excitation or emission wavelengths [4], and the approach has long been demonstrated to provide useful contrast between, for example, malignant and normal tissue [5,6]. Fluorescence lifetime measurements are also ratiometric, under the assumption that these unknown quantities do not change significantly during the fluorescence decay time (typically ns), such that the fluorescence signal can be compared at different delays after excitation to obtain the lifetime. Ratiometric fluorescence measurements can also be made on slower *macro* time scales, typically ranging from microseconds to seconds, to observe dynamics, for example, of calcium levels or protein transduction. Polarization provides further opportunities for ratiometric measurements, for example, to obtain information concerning molecular orientation. On the time scale of the fluorescence decay profiles, time-resolved polarization measurements can determine the molecular tumbling time—or rotational correlation time—which can be used to report molecular binding or local solvent properties.

In practice, while fluorescence excitation, emission, lifetime and polarization-based spectroscopic measurements have been widely undertaken in cuvette-based instruments such as spectrophotometers and spectrofluorometers, it has been less common to fully exploit the spectroscopic information available from fluorescence in biomedical imaging applications. This was partly due to instrumentation limitations and partly due to the challenges associated with characterizing fluorescence signals available from typically heterogeneous and often sparsely labeled biological samples. In recent years, however, the dramatic advances in laser and photonics technology, including robust tunable and ultrafast excitation sources, relatively low-cost, high-speed detection electronics, and high-performance imaging detectors, have provided biologists and medical scientists with new imaging tools and technology. These include multiphoton microscopy [7], which became widely deployed following its implementation with conveniently tunable femtosecond Ti–sapphire lasers [8] that are able to excite most of the commonly used fluorophores—including endogenous fluorophores in biological tissue that otherwise require ultraviolet excitation. For single-photon excitation, new, more convenient, excitation sources including laser diodes, light-emitting diodes (LEDs), and relatively compact diode-pumped solid-state and fiber lasers have also become available. Computer-controlled tuning facilitates automated excitation spectroscopy [9], and automated emission spectroscopy has been implemented in many commercial microscope systems. The proliferation of multiphoton microscopes also stimulated the uptake of FLIM [10,11] as a relatively straightforward and inexpensive *add-on* that provides significant new spectroscopic functionality—taking advantage of the ultrafast excitation lasers and requiring only appropriate detectors and external electronic components to implement FLIM. Progress in computers has driven the development of ever more sophisticated data analysis and signal processing techniques that make practical mathematically intensive imaging and spectroscopy techniques such as single-molecule localization, fluorescence correlation microscopy, fluorescence tomography, and multivariate analysis. These advances have been synergistic with tremendous advances in the development of fluorescence probes and labeling technologies, for which genetically expressed fluorescent proteins [12,13] provide an outstanding example. In cell biology, this is exemplified by widespread deployment of Förster resonant energy transfer (FRET) [14,15] techniques that can determine when fluorophores are located within ~10 nm of each other, providing the ultimate in colocalization and the possibility to map biomolecular processes such as signaling in live cells. In some situations, it is possible to exploit the dependence of the FRET efficiency on the distance between *FRETing* fluorophores to implement a so-called spectroscopic ruler [16,17] or to provide a quantitative readout of analytes that affect the conformation of single-molecule FRET *biosensors*, for example [18], of which the *Cameleon* FRET sensor [19] is perhaps the best known. FRET is thus a powerful tool, but detecting and quantifying the resonant energy transfer is not straightforward when implemented using intensity-based imaging, for which a number of correction calculations must be performed, for example [20]. Increasingly, FLIM and other fluorescence spectroscopic techniques are being applied to improve the reliability of FRET experiments, for example [21–23].

For medical applications, the increasing availability of suitable excitation sources, particularly the mode-locked Ti–sapphire laser, and spectroscopic techniques have prompted an increasing interest in exploiting tissue autofluorescence for label-free clinical applications. Autofluorescence signals reflect the biochemical and structural composition of the tissue and consequently are altered when tissue composition is changed by disease states such as atherosclerosis and cancer. Spectrally resolved imaging of autofluorescence has received significant attention, for example [24], and FLIM is increasingly being investigated as a means of obtaining or enhancing intrinsic autofluorescence contrast in tissues [25–27]. For biomedical

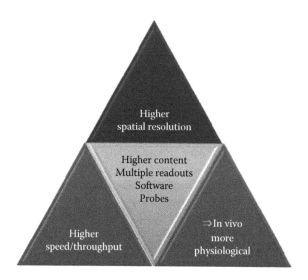

FIGURE 18.2   **(See color insert.)** Frontiers of (fluorescence) bioimaging.

research, fluorescence-based techniques applied to cell biology are being increasingly translated to studies of 3-D cell cultures, tissue sections, and live disease models, and there is also an increasing interest in applying FLIM to biophotonic devices such as microfluidic systems for lab-on-a-chip applications, for example [28,29], and to utilize FLIM and MDFI to study photonics devices, for example [30].

Looking forward, the trend to higher-content MDFI and metrology is clear, but the desire to understand the mechanisms of molecular biology and cell signaling, including fundamental disease processes, continues to drive the development of new biophotonics technology, particularly for imaging applications. As represented in Figure 18.2, key frontiers of fluorescence imaging that are being actively addressed include spatial resolution, higher speed/throughput, and increased physiological relevance. Spatial resolution has been elegantly addressed by the development of superresolution imaging techniques such as stimulated emission depletion (STED) microscopy [31,32] and stochastically switched single-molecule mapping techniques such as PALM (photoactivated localization microscopy) [33,34] and STORM (stochastic optical reconstruction microscopy) [35]. Higher-speed imaging is being addressed by the development of new instruments including high-speed photon counting electronics, electron-multiplying charge-coupled device (EM-CCD) and complementary metal oxide semiconductor (CMOS) detectors, rapid scanning confocal microscopes, and line-scanning and swept focus microscopes. In addition to permitting cell dynamics to be resolved, faster imaging technology also permits higher-throughput imaging, including of live cells. Combined with automation, this is leading to an increased interest in image-based readouts for screening applications, which is often described as high-content analysis (HCA). Following the pioneering *MitoCheck* siRNA live cell time-lapse microscopy-based screen [36], there is interest in ever more sophisticated cell-based assays including adapting MDFI approaches to use FRET to read out protein interactions in automated multiwell plate microscopy. To date, spectral ratiometric FRET readouts have been implemented in commercial plate readers, polarization FRET readouts have been proposed as a precursor to multiphoton FLIM [37] and rapid wide-field FLIM [38,39], and optically sectioned [40,41] FLIM have been implemented in multiwell plate formats. To address the frontier of physiological context, HCA is progressing to automated assays of tissue sections, and readouts of cell signaling processes are starting to be translated to in vivo measurements. Multiphoton microscopy lends itself to intravital imaging and has been particularly applied to image rodent brain function in situ for neuroscience applications. Advanced microscopy techniques are being translated to endoscopy, and new fluorescence-based molecular tomographic imaging techniques are being developed to image live disease models ranging from *Drosophila* and *Caenorhabditis elegans* to fish and mice.

In this chapter, we review different approaches to MDFI applied to biological tissue with an emphasis on FLIM and on label-free contrast based on autofluorescence, implemented in microscopes and endoscopes and with tomography.

18.2 Spectroscopic Techniques for Fluorescence Imaging and Metrology

18.2.1 Spectral Techniques

The most common and experimentally straightforward approach to fluorescence spectroscopy, including tissue spectroscopy, has been the acquisition of emission wavelength-resolved data for a fixed excitation wavelength. In its simplest form, *multispectral* data can be implemented using dichroic beamsplitters and multiple detectors, as is often applied to ratiometric imaging or colocalization studies when the spectral profiles of fluorophores are known a priori and can be used with spectral unmixing techniques, for example [42], to separate contributions from different fluorophore labels or to classify different types and states of tissue using known spectral components. It is therefore useful to study the spectral properties of tissue autofluorescence in order to establish such a priori knowledge, for which *hyperspectral* measurements that acquire the full (emission) spectral profile of a fluorescent sample are desirable. For single-channel measurements, for example, made using a spectrofluorometer, it is straightforward to use a spectrograph that can be read out using CCD, EM-CCD, or CMOS camera technology or to use a multianode photomultiplier. For hyperspectral imaging, such techniques can be implemented in laser scanning microscopes, although this leads to relatively long acquisition times owing to the sequential pixel acquisition. For faster spectrally resolved imaging, one can employ scanning filters such as acousto-optic tunable filters (AOTF) [43] or liquid crystal tunable filter (LCTF) filters [44]. Such scanning filters effectively sample the fluorescence signal, rejecting the *out-of-band* light, and so are lossy compared to spectrograph-based approaches—with the photon economy decreasing as the number of spectral bins is increased—but are faster where the number of spectral channels is small. A more photon-efficient approach is afforded by Fourier-transform spectroscopic imaging, for example, as implemented using a Sagnac interferometer [45], or by the so-called *push-broom* approach [46] implemented in a line-scanning microscope, for which the emission resulting from a line excitation is imaged to the entrance slit of a spectrograph to produce an $(x–\lambda)$ *subimage* that can be recorded on a 2-D detector. Stage scanning along the y-axis then provides the full $(x–y–\lambda)$ data cube, and the entrance slit of the spectrograph can provide optical sectioning capabilities in a manner analogous to confocal microscopy. Alternatively, one can use data compression approaches that spatially encode the light, for example, using spatial light modulator technology, to utilize Hadamard transforms [47] to increase the imaging speed. An efficient approach can also be realized using diffractive optics techniques to encode spectral and image information in a 2-D intensity distribution for single-shot data acquisition and subsequent unmixing in postprocessing [48]. In general, such approaches that require calculations to recover the desired image information can suffer from a reduction in the S/N compared to direct detection. With no a priori knowledge of tissue properties, high-content hyperspectral data sets can be analyzed using statistical multivariate techniques, for example [49], that can extract common spectral components from *training sets* of tissue fluorescence data and subsequently be applied for, for example, diagnostic purposes.

18.2.2 Fluorescence Lifetime Techniques

Spectral fluorescence data can be further complemented by fluorescence lifetime data. Typically, biological tissue exhibits complex fluorescence decay profiles corresponding to the presence of multiple fluorophore species and/or to multiple states of fluorophore species arising from variations in their interaction with the local environment. Such complex fluorescence decays are commonly fitted to

N-component multiexponential decay models, but alternative approaches have been explored, including fitting to a stretched exponential model that corresponds to a continuous lifetime distribution [50], or to a power law decay [51], or to Laguerre polynomials to generate empirical contrast [52]. In general, as more information is used to describe a fluorescence signal, then more detected photons are required to be able to make sufficiently accurate measurements of it [53,54]. Thus, fitting fluorescence decay profiles to complex models requires increased data acquisition times, which can be undesirable in terms of temporal resolution of dynamics or considerations of photobleaching or photodamage resulting from extended exposures to excitation radiation. Where quantitative analysis of complex fluorescence signals is required with short acquisition times, for example, for in vivo applications, one can forego imaging and employ single-point lifetime measurements that utilize all detected photons to represent a single complex fluorescence decay profile or one can fit fluorescence lifetime image data to a single exponential decay model for which the fluorescence lifetime obtained still provides useful contrast since it will reflect changes in the decay times or relative contributions of different components. If quantitative multicomponent FLIM is necessary, however, one can use some form of data reduction approach such as global analysis. Here, the number of photons required to be detected can be reduced using a priori knowledge or assumptions, for example, about the magnitude of one or more component lifetimes or about their relative contributions. It can sometimes be useful to assume that some parameters, such as the component lifetimes, are global, that is, they take the same value in each pixel of the image. Such global analysis, for example [55,56], is often used for the application of FLIM to FRET and may find application for autofluorescence.

In general, FLIM (and single-point fluorescence lifetime measurement) techniques are categorized as time-domain or frequency-domain techniques and, in principle, these can provide equivalent information. A further categorization of fluorescence lifetime measurement techniques can be made according to whether they are sampling techniques, using gated detection to determine the relative timing of the fluorescence compared to the excitation signal, or photon counting techniques that assign detected photons to different time bins. Almost every technique for lifetime measurement has been implemented in time-resolved fluorometers, with the early work being mainly undertaken using frequency-domain measurements, for example [57–59]. Today, time-domain techniques are more commonly applied to biological tissue, including time-correlated single-photon counting (TCSPC) [60,61], which is readily implemented using relatively low-cost electronic cards that plug directly into personal computers, and transient time-resolved detection using streak cameras [62,63] or fast A/D converters [64]. Such single-point systems often combine fluorescence spectrum and lifetime measurement and have been applied to biological tissue using pulsed excitation from nitrogen lasers, for example [65,66], or frequency-tripled Nd-doped [62] or Yb-doped lasers providing picosecond radiation at 355 nm [67]. Subsequently, lower-cost and more compact systems have been developed using picosecond gain-switched diode lasers that can provide excitation wavelengths at, for example, 375, 405, and 440 nm. More versatile systems can be realized using tunable excitation lasers such as mode-locked Ti–sapphire lasers [10,11] or, more recently, supercontinuum sources [68]. The latter multidimensional spectrofluorometer combines TCSPC fluorescence lifetime measurements with spectral (excitation and emission) and polarization-resolved measurements, which can provide a more complete analysis of the fluorescence signal and therefore of the local fluorophore environment.

It is conceptually straightforward to extend fluorescence lifetime measurement techniques to laser scanning microscopy, and FLIM was first implemented in a laser scanning FLIM microscope using TCSPC [69]. This is probably the most widely deployed approach to FLIM, although photon-binning [70] and frequency-domain techniques, for example [71,72], have also been implemented in laser scanning microscopes, and there are other line-scanning and wide-field approaches to FLIM that are reviewed in this volume and elsewhere, for example [73–75]. The optical sectioning (and improved contrast compared to wide-field imaging) conferred by laser scanning confocal/multiphoton microscopy provides improved quantitation of FLIM signals and is particularly important when imaging thick biological samples where scattering can otherwise lead to significant *out-of-plane* contributions to the detected signal. When imaging typical biological samples and assuming a single exponential decay model, laser scanning TCSPC typically requires tens of seconds to acquire sufficient photons for a single photon excited fluorescence lifetime image, and multiphoton excitation requires significantly longer acquisitions. Although recent technological advances have led to reductions in detector dead time and increased the maximum detection count rates of TCSPC it is not usually possible in practice to reach the maximum possible count rates before the onset of significant photobleaching and/or phototoxicity. The photon time-binning and frequency-domain approaches are not limited to single-photon detection and so can provide faster imaging of bright samples, but photobleaching and/or phototoxicity considerations still often limit the maximum practical imaging rates. One way to significantly increase the imaging speed of multiphoton microscopy is to use multiple excitation beams in parallel [76–78], and this approach has been applied to TCSPC FLIM using 16 parallel excitation foci and 16 detection channels [79]. In general, parallel pixel excitation and detection is applicable to all laser scanning microscopes and has been extended to optically sectioned line-scanning microscopy, for example, using a rapidly scanned multiple beam array to produce a line of fluorescence emission, which can be relayed to the input slit of a spectrograph, to facilitate *push-broom* hyperspectral imaging, or of a streak camera to implement FLIM [80]. It is also possible to rapidly scan multiple excitation beams in parallel to produce a 2-D optically sectioned fluorescence image that can be recorded using wide-field detectors. This has been implemented with single-photon excitation in spinning Nipkow disc microscopes that have been combined with wide-field time-gated detection for FLIM, for example [81–83], and with multibeam multiphoton microscopes that have also been adapted for FLIM [84–86].

Wide-field FLIM does not benefit from optical sectioning, unless combined with some other technique, but can provide imaging rates of tens to hundreds of hertz, for example [87–89], although the maximum acquisition speed is inevitably limited by the number of photons/pixel available from the (biological) sample. It is attractive for diagnostic imaging applications because of the potential for large fields of view and has been applied to FLIM of tissue autofluorescence, including via endoscopy [88,90,91]. Wide-field FLIM is most commonly implemented using modulated image intensifiers with frequency or time-domain approaches. The frequency-domain approach initially utilized sinusoidally modulated laser excitation and detection, for example [92–95], although excitation with a mode-locked pulse train is now common. In the time domain, FLIM is typically realized using time-gated microchannel plate (MCP) image intensifiers to sample the fluorescence decay profiles, for example [27,96,97]. It is possible to gate the image intensifier with sub 100 ps resolution [98], but it is usually preferable to use longer time gates to increase the detected signal [91].

18.2.3 Spectrally Resolved FLIM

It can be useful to obtain multidimensional fluorescence image information so that spectroscopic signatures can be more effectively unmixed and functional spectroscopic information correlated with sample morphology. This can be useful for FRET imaging experiments, where spectral and lifetime measurements can give complementary information [99], and for studying tissue autofluorescence, where it may be used to distinguish different metabolites or tissue matrix components. FLIM can readily be combined with multispectral imaging, for which time-resolved images are acquired in a few discrete spectral windows, for example [100], or it can be combined with hyperspectral imaging, for which the full time-resolved excitation or emission spectral profile is acquired for each image pixel. In single-beam scanning fluorescence microscopes, the fluorescence radiation can be dispersed in a spectrometer and the spectral-lifetime profiles acquired sequentially, pixel by pixel, for example, using TCSPC systems that incorporate a multianode photomultiplier [61,101]. To increase the imaging rate, it is necessary to parallelize the pixel acquisition, for example, by combining wide-field FLIM with hyperspectral imaging implemented using tunable filters or by encoding Hadamard transforms using spatial light modulator technology [102]. Hyperspectral FLIM can also be realized using the *push-broom* approach implemented in a line-scanning microscope with wide-field time-gated detection [103],

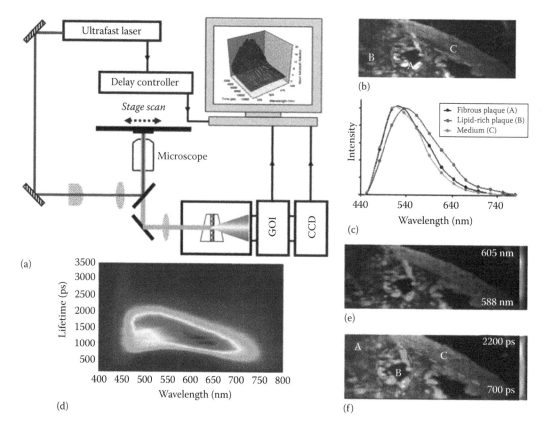

FIGURE 18.3 (See color insert.) (a) Experimental setup for line-scanning hyperspectral FLIM, (b) integrated intensity image of sample of frozen human artery exhibiting atherosclerosis, (c) time-integrated spectra of sample regions corresponding to medium and fibrous and lipid rich plaques, (d) autofluorescence lifetime–emission matrix, (e) map of time-integrated central wavelength, and (f) spectrally integrated lifetime map of sample autofluorescence. (Adapted from De Beule, P. et al., *Microsc. Res. Tech.*, 70, 481, 2007.)

as illustrated in Figure 18.3. This instrument was developed to study autofluorescence contrast in diseased tissue, and Figure 18.3 presents multidimensional fluorescence data from a hyperspectral FLIM acquisition of unstained fixed section of human cartilage.

This MDFI approach can be extended by also resolving the fluorescence signal with respect to the excitation wavelength. Using spectral selection of a fiber laser–pumped supercontinuum source to provide tunable excitation for such a line-scanning hyperspectral FLIM microscope, it was possible to acquire the fluorescence excitation–emission–lifetime (EEL) matrix for each image pixel of a sample [104]. Such a multidimensional data set enables conventional *excitation-emission matrices* (EEM) to be reconstructed for any image pixel and to obtain the fluorescence decay profile for any point in the EEM map. While such high-content multidimensional fluorescence data can provide exquisite sensitivity to variations in fluorescence emission and enhance the ability to unmix signals from different fluorophores, the size and complexity of such data sets are almost beyond the scope of ad hoc analysis by human investigators. Automated software tools are required to analyze such data in order to identify trends and fluorescence *signatures*. The combination of MDFI with image segmentation for HCA should provide powerful tools, for example, for histopathology and for screening applications.

18.2.4 Polarization Contrast

Fluorescence can also be resolved with respect to polarization, which can provide information about fluorophore orientation, rotational mobility, and resonant energy transfer [1], and polarization measurements can be combined with spectral and lifetime measurements. In general, such measurements have mainly been applied to solution phase measurements and to cell microscopy to measure the anisotropy of fluorescence emission [1] rather than to bulk tissue components where the polarization of the radiation can be scrambled through multiple scattering events. Polarization has been applied to tissue imaging, however, as a means to distinguish between less scattered and more scattered light [105], for example, from specular and subsurface backscattering, since the latter is more depolarized and therefore less attenuated by crossed polarizers in the illumination and collection paths. This has been applied to fluorescence imaging [106] where 632.8 nm excited autofluorescence images of fresh surgical specimens of bladder were ratioed with backscattered light images recorded through crossed polarizers to provide contrast indicating cancerous tissue.

18.3 MDFI Technology for In Situ Tissue Studies

This chapter primarily concerns multidimensional imaging of tissue autofluorescence, which mainly arises from cellular metabolites and tissue matrix components and offers exciting prospects for label-free clinical diagnosis, but it is important to understand that the advances in labeling techniques that have impacted cell biology are also finding many applications in tissue imaging. MDFI can be applied to histopathology to unmix multiple labels and to map the distribution of genetically expressed fluorophores in live disease models such as *C. elegans*, *Drosophila*, zebrafish, and rodents. There is also increasing interest in using fluorescence to read out protein interactions in tissue-based assays, for example, using ex vivo tissue sections, tissue cultures, or engineered tissues. For drug discovery and disease-related research, there is an increasing appreciation that in vitro cell monolayers may exhibit nonphysiological behavior that can impact the utility of cell-based assays, and this has stimulated the development of techniques to image biological processes in 3-D cell cultures, for example [107], in animal/embryo/engineered tissue samples, for example [108–110], and in live organisms. Increasingly, there is a trend to longitudinal studies in live disease models that can have the benefit of reducing the number of animals required for medical research. The combination of genetically expressed fluorophores with intravital microscopy, endoscopy, or tomography can enable the translation of assays from cell cultures to live disease models, potentially increasing the efficacy of the drug discovery pipeline and decreasing the time to failure for unsuccessful drug candidates. Ultimately, it is usually not possible to translate such label-based assays to clinical applications, although label-free MDFI readouts of metabolic pathways and tissue matrix properties do have the potential for clinical readouts. Fluorescence imaging in vivo must inevitably deal with the strong heterogeneity and optical scattering of biological tissue that make quantitative intensity measurements particularly challenging. Ratiometric readouts that do not require knowledge of probe concentrations and absolute light levels are essential and hence MDFI is an attractive approach. For tissue imaging, these can be implemented using microscopes, but these are typically limited to surface or near-surface imaging and so there is considerable interest in extending MDFI from cell-based assays to animal disease models and clinical research using endoscopy and tomographic approaches.

18.3.1 Multidimensional Fluorescence Endoscopy

Since the strong optical scattering associated with biological tissue limits the imaging depth of conventional microscopes to a few hundred μm and even multiphoton microscopes to <1 mm in most biological tissues, there is significant interest in using endoscopes to image inside animals and humans. As with microscopy, spectroscopic techniques can provide enhanced functional information compared to reflected light or intensity imaging, and autofluorescence is particularly interesting if it can afford a label-free readout of disease. Beyond diagnostic applications, MDFI via endoscopy could also permit cell signaling networks to be studied in their native context, permitting longitudinal studies that would reduce the numbers of animals required for drug discovery and providing enhanced value compared to cell-based assays in culture.

Current endoscopes may be considered in the categories of flexible video endoscopes, rigid optical endoscopes, and flexible optical endoscopes. Video endoscopes typically have a miniature CCD camera at the distal end, and the flexible section of the endoscope is essentially a cable conduit for the electronic signals, power, etc. Video endoscopes are essentially wide-field imaging instruments with an optical performance that can be considered as a wide-field microscope. Rigid optical endoscopes are typically constructed from a series of lenses enclosed in a rigid cylinder, and these relay an optical image from the distal (sample) to proximal (detector) end. They are usually employed as wide-field imaging devices with a CCD camera at the proximal end, but they can be used in scanning microscope configurations [111] and miniature rigid endoscopes have been adapted for multiphoton microscopy where a *stick lens* made from gradient index (GRIN) lenses can be employed [112]. Rigid endoscopes are widely used in orthopedic surgery, urology, and for neuroscience (imaging in rodent brains), but they are not usually suitable to study internal organs because they are not flexible enough to be passed through internal pathways, cavities, or vessels in live subjects and they are typically of limited length. For internal imaging, it is usual to employ flexible endoscopes. For intensity imaging via flexible endoscopes, video endoscopes are most commonly used, but for more sophisticated spectroscopic imaging modalities, such as hyperspectral imaging or FLIM, or for confocal or multiphoton microscopy (to provide higher resolution and optical sectioning), it is necessary to use a flexible optical endoscope. Flexible optical endoscopes can be divided into wide-field optical endoscopes, microconfocal endoscopes, and multiphoton endoscopes.

Conventional wide-field imaging can be realized endoscopically by relaying an image of the sample through a flexible optical fiber bundle from the distal end to the proximal end. The requirement for flexibility typically limits the number of optical fibers (and therefore image pixels) in a bundle to ~30,000. This is a relatively small number of pixels compared to that provided by typical CCD cameras and so such flexible optical endoscopes offer significantly fewer image resolution elements than video endoscopes or optical microscopes. The spacing between individual fiber cores (and consequent fill factor) also impacts the efficiency of light collection and the image quality, which can be degraded by cross talk between fibers within the bundle.

While wide-field endoscopy is suitable to image relatively large areas of tissue, for example, for diagnostic screening applications and optically guided biopsy, it is not able to provide depth-resolved imaging, which can be important, for example, in studying the subsurface properties of lesions, or to sufficient (subcellular) image resolution for optical biopsy to be correlated with histopathology. Accordingly, there has been considerable effort to translate the optically sectioned imaging capabilities of laser scanning confocal and multiphoton microscopes to endoscopy in order to assist in the early identification of cancerous and precancerous tissue. In general, there are two strategies to implement laser scanning confocal or multiphoton *endomicroscopy*. Confocal endomicroscopes typically utilize either a proximal scanner with an imaging fiber bundle or a distal scanner with a single optical fiber to convey the light from the sample to the (proximal) detector.

In the former case, the fiber bundle can be an array of single-mode fiber *cores* that are fabricated together to form a *coherent* bundle. A scanner at the proximal end then scans the excitation beam across the proximal end of the fiber bundle, addressing each optical fiber core sequentially, and the output at the distal end is relayed by the objective lens to scan a focused beam across the sample, for example [113,114]. The resulting fluorescence (or reflected light) is imaged back to the same fiber core, and the image of the sample is thus relayed to the proximal end of the fiber bundle. This can be imaged directly using a CCD or propagated back through the scanning system to a single detector that records the pixel information sequentially. The small size of the fiber core means that it serves as a *confocal pinhole*, which leads to optical sectioning and improved resolution and contrast compared to wide-field imaging. As with the wide-field endoscope, the limited number of fiber cores in the imaging bundle limits the image quality.

For the latter (distal scanning) approach to endomicroscopy, a miniature optical scanner is deployed at the distal end of a single-mode fiber, which serves as the *confocal pinhole* for imaging with reflected light, for example [115–117], or fluorescence, for example [118]. Considerable effort has been invested in developing miniature scanners, including microfabricated scanning mirrors or vibrating fiber tip designs, and to date, diameters of a few mm have been achieved. In general, the fiber bundle endomicroscope, which does not require a distal scanner, can be made thinner than the single-mode fiber-distal scanner approach, making it potentially more flexible and able to pass through thinner cavities, vessels, etc. Both the fiber bundle and single-fiber approaches to endomicroscopy can be adapted to multiphoton endoscopic imaging, for example [119–122], which can offer deeper penetration in biological tissue and conveniently excite autofluorescence.

To date, most endoscopy has focused on imaging tissue structure with reflected light or fluorescence. Of the spectroscopic endoscopy techniques, one gaining interest for clinical application is narrowband imaging with reflected light (NBI) [123], which has been implemented with autofluorescence imaging to obtain further contrast [124]. Spectrally resolved autofluorescence has been exploited for cancer diagnosis, for example [125], and autofluorescence has been combined with cross-polarized imaging to improve discrimination of diseased tissue, for example [126]. Multispectral confocal endomicroscopy has also been demonstrated, for example [127,128], and FLIM has been implemented in wide-field FLIM endoscopy using both rigid [26,89,90] and flexible [90] endoscopes with successful application to intravascular imaging [129] and in vivo animal studies [130].

At Imperial College London, we developed a confocal FLIM endomicroscope [131] based on a commercially available laser scanning single-photon confocal fluorescence endomicroscope [114] that is licensed for clinical use in the GI tract and for bronchoscopy. Our setup to demonstrate confocal FLIM endomicroscopy is shown in Figure 18.4. We used a *coherent* bundle of 30,000 single-mode fibers terminated with a miniature objective (Cellvizio® Mini O) to provide a lateral resolution of ~1.4 μm, a working distance of 60 μm, and a field of view of 240 μm. The excitation source was a fiber-delivered mode-locked frequency-doubled Ti–sapphire laser, and FLIM data were acquired using TCSPC to provide a fluorescence lifetime image of 350×512 pixels, which was limited by the TCSPC card memory. This microconfocal endoscope was applied to optically sectioned FLIM of stained pollen grains, label-free FLIM of autofluorescence from rat tissue, and FLIM FRET of genetically expressed fluorophores in fixed cells undergoing FRET [131]. Subsequently, this approach was applied to live cells in culture, with acquisition times ranging from 300 [132] to 2 s [133], illustrating the potential of FLIM endomicroscopy.

18.3.2 Multidimensional Fluorescence Tomography

While FLIM and FLIM FRET endomicroscopy provides a promising route to studying protein interactions in vivo, it is inevitably somewhat invasive, requires relatively large subjects, and can be impractical for time-lapse imaging studies in live subjects. As discussed elsewhere in this volume and in [134], there are a number of tomographic approaches to acquire quantitative 3-D image data in subjects ranging from embryos to rodents, for which the main challenge is the impact of the strong scattering of optical radiation in biological tissue. For relatively transparent subjects, techniques such as optical projection tomography (OPT) [135] and light sheet microscopy techniques, such as selective plane illumination microscopy (SPIM) [136] and ultramicroscopy [137], may be applied. These are essentially ballistic light imaging techniques and initially were demonstrated in conjunction with chemical clearing techniques to reduce the scattering of the optical radiation used. For live animal imaging, chemical clearing is not an option, and in scattering samples, sophisticated techniques, for example, based on inverse scattering [138], must be employed to reconstruct the 3-D image data from the diffuse light detected.

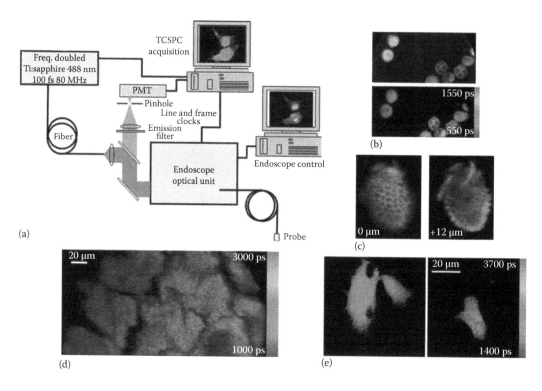

FIGURE 18.4   (See color insert.) Confocal endomicroscope FLIM: (a) shows the experimental schematic; (b) shows intensity and corresponding optically sectioned FLIM images of stained pollen grain; (c) shows two optically sectioned FLIM images of the same pollen grain; (d) FLIM image of unlabeled rat tissue; (e) shows confocal endomicroscope FLIM images acquired in 1 s of live Cos-7 cells expressing either GFP (left) or EGP linked to RFP (right). (Adapted from Kennedy, G.T. et al., *J. Biophotonics*, 2009; Kumar, S. et al., *ChemPhysChem*, 12, 627, 2011.)

There are several approaches to diffuse fluorescence tomography (DFT) that can utilize c.w. excitation and detection, for example [139,140], or time-resolved detection using frequency-domain [141,142] or time-domain [143,144] measurements, but these are generally not able to achieve image resolution close to the diffraction limit, as is possible with ballistic light imaging techniques, although impressive results are now being obtained with photoacoustic imaging [145], which has been shown to work with genetically expressed chromophores such as green fluorescent protein (GFP) [146].

To date, ballistic and DFT techniques have been largely restricted to intensity-based imaging, providing structural information and localization of specific fluorophores. Light sheet microscopy techniques have been applied to imaging live transparent small (~sub-mm) organisms such as *Drosophila melanogaster* [136], *Danio rerio* (zebrafish) [147], and several other species. The application of light sheet microscopy to imaging early embryo development with cellular resolution has been particularly successful. OPT has been applied in vivo to *D. melanogaster* [148], *C. elegans* (nematode) [149], and *D. rerio* [150]. In larger live samples that are less transparent and more highly scattering, quantitative imaging can be degraded by intensity artifacts. As discussed previously, ratiometric techniques can provide superior quantitative image data, and so it is interesting to develop spectrally resolved and FLIM tomography systems (Figure 18.5).

The spectroscopic properties of biological tissue can vary significantly through the visible spectrum, and so the fluorescence emerging from the surface of a sample can be spectrally modulated depending on the optical properties and the path length from the fluorescent source to the surface. In scattering samples, this can be exploited to determine the depth of a fluorescent target, for example [151–154].

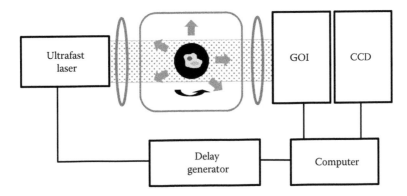

FIGURE 18.5 **(See color insert.)** Schematic of tomoFLIM setup. (GOI is gated optical image intensifier.)

Recently, a wide-field multispectral diffuse light tomography system that acquires fluorescence in a number of discrete wavelength bands [155] has been shown to yield improved spatial resolution, albeit at the cost of increased computational load. To address the latter issue, an alternative reconstruction method with improved computational cost and memory requirements for large multichannel data sets has recently been demonstrated [156]. Spectroscopic resolution can also help discriminate against unwanted background signals due to scattered excitation light and/or autofluorescence and can be used for FRET assays for biomedical research and drug discovery. However, the heterogeneity of tissue and unknown variations in spectral properties can present significant challenges for quantitative measurements of FRET in live disease models such as mice.

Since time-resolved detection is used to improve quantitative intensity imaging in highly scattering media [138], the extension to FLIM is a logical step for DFT, for example [157,158], and tomographic FLIM has been implemented in experiments based on scattering phantoms using approaches in the frequency domain, for example [159,160], and time domain, for example [161]. FLIM tomography can provide quantitative FRET readouts in highly scattering heterogeneous samples based on measurements only of the donor emission, and this could be important for drug discovery and in vivo studies of disease mechanisms. In particular, it could permit longitudinal studies in disease models, potentially decreasing the numbers of animal required for testing. Several groups are also making progress toward tomographic FLIM and FRET of live mice, for example [162–164]. At Imperial College London, we have applied the *tomoFLIM* approach depicted in Figure 18.6, using time-gated detection based on a gated image intensifier in a tomographic imaging setup in which the sample or the imaging set-up is rotated. A fiber laser–pumped supercontinuum-based light source is used to excite the fluorescence, which is imaged onto the photocathode of a gated optical intensifier (GOI), and the resulting time-gated and intensified fluorescence signals are recorded by a CCD camera. As the sample or imaging set-up is rotated, a series of time-gated images are recorded, typically at regular intervals over a full 360° rotation. We previously showed that we could reconstruct the fluorescence lifetime and other optical properties of strongly scattering samples with inclusions containing cells expressing a FRET biosensor utilizing cyan fluorescent protein (CFP) and citrine [165]. This employed a reconstruction algorithm based on the minimization of a cost functional between the measured data and a model of light diffusion in the Fourier domain [166]. Figure 18.6 illustrates the application of this time-gated DFT approach to image FRET in a live mouse [164] for which the leg muscle was transfected with enhanced green fluorescent protein (EGFP) and mCherry constructs, either separately or linked together to implement FRET. This experiment was restricted to imaging the leg because of the use of visible fluorescent proteins. Using near-infrared (NIR) fluorophores, it is possible to apply FLIM and FRET through the full body of a mouse, as has been demonstrated using capillaries of NIR dye mixtures [167]. With the prospect of

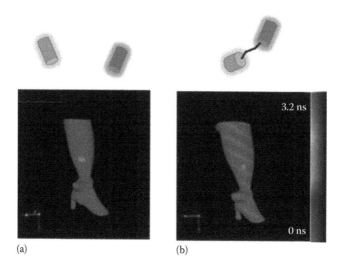

(a)   (b)

FIGURE 18.6 **(See color insert.)** 3-D diffuse fluorescence lifetime tomography of live mouse transfected with (a) EGFP, mCherry unlinked, and (b) EGFP–mCherry directly linked. (Adapted from McGinty, J. et al., *Biomed. Opt. Express*, 2, 1907, 2011.)

NIR fluorescent proteins, for example [168,169], and/or other labeling strategies, it seems likely that in vivo tomographic FLIM and FRET of live disease models, including embryos and mice, will soon be practical for biomedical research and drug discovery.

The application of spectroscopic techniques to ballistic light tomography has been more limited, perhaps because it does not offer advantages in terms of improved image quality, although multispectral imaging has been applied to separate different labels, for example [170]. To the best of our knowledge, we have developed the first ballistic light tomographic FLIM instrumentation, which is based on time-gated OPT and uses essentially the same apparatus depicted in Figure 18.5 for diffuse FLIM tomography. FLIM OPT [171] produces 3-D fluorescence lifetime images of transparent samples in the range ~0.1–1 cm. OPT is the optical equivalent of x-ray computed tomography (x-ray CT), for which a stack of image slices are reconstructed from a series of transverse images acquired at a number of projection angles, typically using the classic filtered back projection technique, which assumes parallel ray paths, and so its applicability is limited by diffraction to samples that are smaller than the depth of focus of the imaging system [135]. Due to its similarity to x-ray CT, many CT image reconstruction techniques, for example [172], can be directly applied to OPT, providing that the sample is transparent. Typically, this requires the sample to be treated (chemically cleared) to eliminate the optical scattering. For FLIM OPT, the 3-D time-gated fluorescence intensity distribution is determined for each time gate delay and then the fluorescence lifetime distribution is calculated for each voxel from the intensity decay data. Figure 18.7 shows 3-D time-integrated fluorescence intensity (a) and fluorescence lifetime (b) reconstructions of an optically cleared fixed mouse embryo imaged with FLIM OPT [171] that was fixed in formaldehyde and the neurofilament labeled with an Alexa-488-conjugated antibody. Upon excitation at 485 ± 10 nm and image reconstruction, the Alexa-488-labeled neurofilament presented a lifetime of 1360 ± 180 ps and autofluorescence, thought to be associated with the circulatory system, was also observed, presenting a fluorescence lifetime of 1030 ± 135 ps. In addition to imaging optically cleared disease models such as mouse embryos, OPT is also useful to map 3-D structures and molecular distributions in relatively large tissue samples, which would be challenging for microscopy. Figure 18.7c shows an OPT image of a histological sample in which the red-colored feature corresponds to a blood vessel (read out by its absorption), while the background tissue (colored green) structure is read out using autofluorescence. Figure 18.7d shows a further histopathology sample for which FLIM OPT is

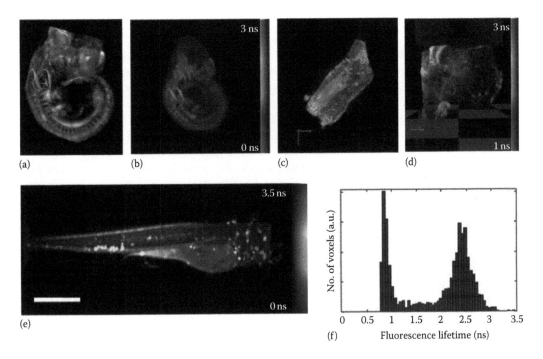

FIGURE 18.7 (See color insert.) 3-D OPT fluorescence (a) intensity and (b) FLIM images of mouse embryo cleared with BABB and with neurofilament labeled with Alexa-488 from [171]; (c) OPT autofluorescence image of portal vein branch in liver cleared with BABB but unstained showing fluorescence (acquired >500 nm with excitation at 460 nm) in green and white light absorption in red; (d) FLIM OPT of autofluorescence from unstained lung tumor cleared with BABB and excited at 480 nm; (e) FLIM OPT of live lysC:GFP transgenic zebrafish embryo 3 days postfertilization showing a single frame from 3-D fluorescence lifetime reconstruction (scale bar 500 μm) and (f) fluorescence lifetime histogram showing two clear populations corresponding to GFP and autofluorescence. (Adapted from McGinty, J. et al., *Biomed. Opt. Express*, 2, 1340, 2011.)

providing contrast based on autofluorescence lifetime [173]. Such volumetric imaging is much faster and less likely to compromise the sample structure than the conventional approach of mechanically slicing the sample and reconstructing the volumetric image from many serial sections. Thus, FLIM OPT could image 3-D fluorescence lifetime distributions in anatomically relevant samples and potentially help distinguish between extrinsic and intrinsic fluorescence signals without prior anatomical knowledge of the labeling sites, which is usually not possible with steady-state fluorescence intensity imaging.

FLIM OPT could be applied to map protein–protein interactions in 3-D using FLIM FRET to read out endpoints in optically cleared samples. Unfortunately, genetically expressed fluorescent proteins can be compromised by the optical clearing process, and so secondary labeling, for example, with antibodies to EGFP, is required when using genetically expressed fluorescent proteins for FRET. For drug discovery and biomedical research, however, it would be more useful to be able to reconstruct 3-D fluorescence lifetime distributions in live disease models to permit longitudinal studies. One way to reach this goal is to use live disease models that are transparent, and Figure 18.7e shows a FLIM OPT reconstruction of a live zebrafish embryo transfected with EGFP and immobilized in agar [150]. To extend tomoFLIM to weakly scattering disease models such as live adult zebrafish or mouse embryos is more challenging since it is necessary to address the issue of optical scattering outside the diffusion approximation, but the development of suitable reconstruction software for such samples is under way, for example [174].

18.4 MDFI of Tissue Autofluorescence

In biological tissue, autofluorescence can provide a source of label-free optical *molecular* contrast, offering the potential to distinguish healthy and diseased tissue as well as to study some of the underlying mechanisms of disease. Detecting molecular changes associated with the early manifestation of disease could permit earlier treatment and potentially improve prognosis, for example, for cancer [175]. Using spectroscopic techniques, it may be possible to detect neoplasia where the associated cellular and tissue perturbations are not observable by direct inspection and so are beyond the discrimination of conventional noninvasive diagnostic imaging techniques. In general, label-free imaging modalities are preferable for clinical imaging, particularly for diagnosis, as they avoid the issues of toxicity and pharmacokinetics associated with the administration of exogenous agents. Autofluorescence can provide label-free contrast for both single-point measurements and imaging techniques. Localized measurements are useful for *optical biopsy* in situ, while imaging could facilitate diagnostic screening and guided biopsy. In order to exploit autofluorescence for clinical applications, it is desirable to investigate the autofluorescence *signatures* of normal and diseased tissue states and to understand the molecular origins of the observed fluorescence contrast. This has motivated a number of in vitro and in vivo studies. To date, instrumentation constraints and ethical and regulatory considerations have limited the number and range of in vivo investigations, but a significant range of ex vivo studies have been undertaken.

The principal endogenous tissue fluorophores include matrix components such as collagen and elastin cross-links, metabolites such as reduced nicotinamide adenine dinucleotide (NADH), oxidized flavins (flavin adenine dinucleotide [FAD] and flavin mononucleotide [FMN]), and other endogenous fluorophores such as lipofuscin, keratin, and porphyrins. As shown in Figure 18.8 [176], these fluorophores typically have excitation maxima in the UV-A or blue (325–450 nm) spectral regions and emit Stokes-shifted fluorescence in the near-UV to the visible (390–520 nm) region of the spectrum. The autofluorescence signals detected from biological tissues will depend on the concentration and the distribution of the fluorophores present and also on the presence of chromophores (principally hemoglobin) that can absorb excitation and fluorescence light, as well as on the degree of light scattering that occurs within the tissue [24]. Autofluorescence is therefore modulated by variations in the biochemical and structural composition of the tissue and consequently can provide markers of diseases that change the tissue composition.

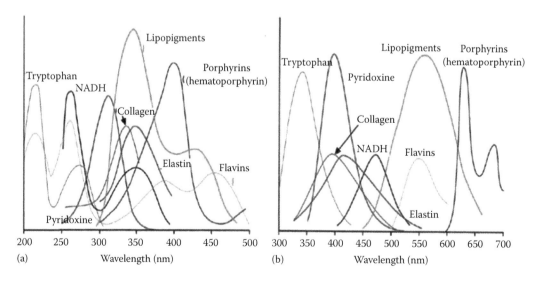

FIGURE 18.8 **(See color insert.)** (a) Excitation and (b) emission spectra of main endogenous tissue fluorophores. (Adapted from Wagnieres, G.A. et al., *Photochem. Photobiol.*, 68, 603, 1998.)

In practice, however, the dependence of autofluorescence intensity signals on fluorophore concentration, on variations in the excitation flux, detection efficiency and tissue microenvironment, and their possible attenuation by absorption and scattering within the tissue make it challenging to use autofluorescence intensity measurements as a basis to detect or study disease. Spectral information has been used to improve quantitation of autofluorescence signals, for example, through ratiometric imaging, although the heterogeneity in the distribution of tissue fluorophores and their broad, heavily overlapping, emission spectra can limit the discrimination achievable. This can be addressed using spectral unmixing approaches, including linear unmixing, for example [42], and multivariate statistical analysis, for example [49], which are being developed to identify signatures for diagnostic applications, including in combination with tissue reflectance data [177]. The complexity and heterogeneity of biological tissue, however, still present significant challenges for this approach, and fluorescence imaging tools in clinical use have been hampered by low specificity and a high rate of false-positive findings [178–180].

Increasingly, there is interest in exploiting fluorescence lifetime contrast to analyze tissue autofluorescence signals since fluorescence lifetime measurements depend on relative (rather than absolute) intensity values and FLIM is therefore largely unaffected by many factors that limit steady-state measurements. The principal tissue fluorophores exhibit characteristic lifetimes (ranging from hundreds to thousands of picoseconds) [24] that could enable spectrally overlapping fluorophores to be distinguished, and the sensitivity of fluorescence lifetime to changes in the local tissue microenvironment (e.g., pH, $[O_2]$, $[Ca^{2+}]$) [1] could provide a readout of biochemical changes indicating the onset or progression of disease. The application of time-resolved fluorescence measurements to distinguish between normal versus malignant tissue was reported as early as 1986 [62], but the complexity and performance of instrumentation associated with picosecond-resolved measurements, and particularly imaging, hindered the development of FLIM as a clinical tool. Nevertheless, *single-point measurements* of autofluorescence lifetime have revealed differences between normal and neoplastic human tissue in tumors of the esophagus [63,181,182], colon [64], brain [183,184], oral cavity [185], lung [186], breast [187,188], skin [67], and bladder [63]. Of these studies, those in Refs. [63,64,182,184,186] included in vivo measurements. Autofluorescence lifetime spectroscopy has also been applied to study atherosclerosis, for example [189,190], including in vivo [191].

There have also been a number of FLIM studies of autofluorescence of cancerous tissues. To date, most of these FLIM studies have used laser scanning microscopes and particularly multiphoton microscopes. These include ex vivo studies of brain [192], skin [193,194], cervix [195], and breast [196] and in vivo studies in animals [197] and human skin [198]. In general, multiphoton excitation provides a convenient approach to image tissue autofluorescence, and to date, the only CE-marked FLIM microscope licensed for clinical in vivo application is the DermaInspect (JenLab GmbH) multiphoton tomography system [199]. In a recent study, we modified the standard DermaInspect™ instrument to incorporate four-channel multispectral FLIM using TCSPC [200,201] and also included a spectrograph to enable hyperspectral imaging. Figure 18.9 shows a typical multispectral optically sectioned fluorescence lifetime data stack acquired in vivo of normal skin from a healthy volunteer. Such data sets from normal tissue and ex vivo samples of lesions were acquired and analyzed using automated image segmentation to identify all the pixels corresponding to intracellular fluorescence and were fitted to a double exponential decay model. Linear discriminant analysis indicated that the fitted lifetime components of the intracellular autofluorescence could be used to discriminate between normal cells and basal cell carcinoma (BCC) with an area under curve (AUC) up to 0.82 [201].

For many clinical applications such as diagnostic screening for disease, guided biopsy, or interoperative surgery, however, it is desirable to image a larger field of view than is usually possible with multiphoton microscopy. For such applications, wide-field FLIM can provide ~mm–cm fields of view at relatively high (Hz) frame rates and can be implemented in to microscopes, endoscopes, or macroscopes. To the best of our knowledge, the first application of wide-field FLIM to tissue autofluorescence utilized a wide-field frequency-domain FLIM system applied to ex vivo bladder and oral mucosa [26] and in vivo imaging of the bronchi [90]. This was followed by wide-field time-gated FLIM of ex vivo breast cancer

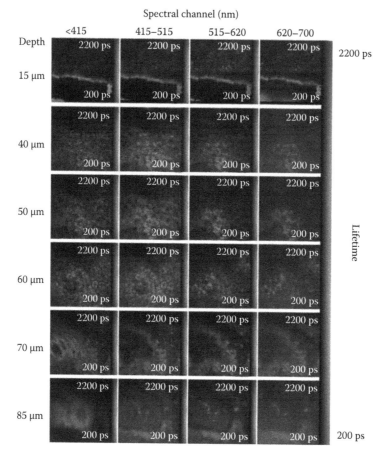

FIGURE 18.9 **(See color insert.)** Multispectral optically sectioned fluorescence lifetime data stack of normal human skin acquired in vivo.

[202] and exploratory applications of wide-field FLIM to degenerative joint disease [203], atherosclerosis [195,204,205], and skin cancer [194] and later to colon, liver, and pancreas [206]. Exemplar ex vivo auto-fluorescence lifetime images that illustrate the potential of wide-field FLIM for label-free tissue imaging are presented in Figure 18.10. These were acquired immediately after resection using a FLIM macroscope with picosecond pulse excitation at 355 nm [206]. Figure 18.10 shows a freshly resected partial gastrec-tomy specimen containing a moderately differentiated intestinal-type adenocarcinoma for which the autofluorescence lifetime images and histograms distinguish between normal mucosal tissue and the tumor. The origin for this lifetime contrast is not yet clear, but contributions arising from excitation at 355 nm can be expected from metabolites such as NADH and tissue matrix components such as collagen. To date, there are relatively few reports of in vivo clinical wide-field FLIM since [90], but a wide-field FLIM endoscope has been applied in vivo to image oral carcinoma in a hamster cheek pouch model [130] and, recently, endoscopic FLIM has been applied in vivo in a neurosurgical clinical setting to image glioblastoma multiforme (brain tumor) [207].

18.5 Conclusions

MDFI technology can provide label-free molecular contrast in biological tissue and can increase the information obtainable from exogenous probes. To realize the potential of MDFI, it is necessary to progress the various techniques described here to in vivo application for clinical and preclinical

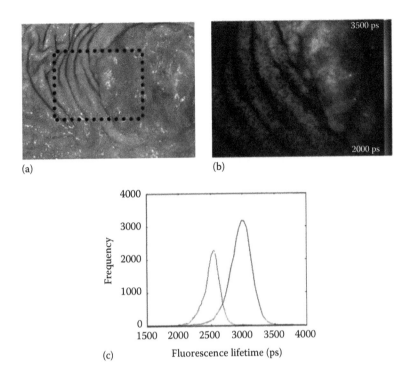

FIGURE 18.10 **(See color insert.)** Wide-field time-gated FLIM with picosecond pulsed excitation at 355 nm applied to a freshly resected partial gastrectomy specimen containing a moderately differentiated intestinal-type adenocarcinoma: (a) white light image of the macroscopic specimen with area of FLIM outlined; (b) intensity-weighted false-color FLIM image with scale bar representing 1 cm; (c) autofluorescence lifetime histogram from the normal and cancerous regions of interest. (Adapted from McGinty, J. et al., *Biomed. Opt. Express*, 1, 627, 2010.)

imaging and for drug discovery. To date, however, progress toward wide in vivo application of MDFI techniques has been slow due to the complexity and cost of the required instrumentation, its lack of commercial availability, and the relative paucity of successful case studies in terms of clinical or commercial impact. This situation is evolving with advances in compact and convenient excitation lasers, such as picosecond diode lasers and diode-pumped solid-state lasers and fiber lasers. In terms of time-resolved detectors, there are advances in modulated CMOS cameras for wide-field FLIM, for example [208], and single or multiple channel CMOS devices for on-chip measurements, for example [209], although these potentially lower-cost technologies are yet to achieve significant impact. Nevertheless, the availability of lower-cost and reliable excitation sources and detectors is in prospect and this will greatly assist the development and commercialization of clinical and preclinical instrumentation, which in turn will drive the translation of MDFI technology to industry and the clinic.

Acknowledgments

The authors gratefully acknowledge funding from the Biotechnology and Biological Sciences Research Council (BBSRC), the Department of Trade and Industry (DTI) Beacon award, the Engineering and Physical Sciences Research Council (EPSRC), the EU Framework 6 (LSHG-CT-2003-503259) and Framework 7 (EU FP7-HEALTH-2007-A, Project 201577), a Joint Infrastructure Fund Award from the Higher Education Funding Council for England (HEFCE JIF), and the Wellcome Trust.

References

1. Lakowicz, J. R., *Principles of Fluorescence Spectroscopy*, 2nd edn. Kluwer Academic Publishers, New York, 1999.
2. Ackroyd, R., Kelty, C., Brown, N., and Reed, M., The history of photodetection and photodynamic therapy, *Photochemistry and Photobiology*, 74(5), 656–669, 2001.
3. Lapotko, D., Lukianova, E., Potapnev, M., Aleinikova, O., and Oraevsky, A., Method of laser activated nanothermolysis for elimination of tumor cells, *Cancer Letters*, 239(1), 36–45, 2006.
4. Grynkiewicz, G., Poenie, M., and Tsien, R. Y., A new generation of Ca^{2+} indicators with greatly improved fluorescence properties, *Journal of Biological Chemistry*, 260(6), 3440–3450, 1985.
5. Andersson-Engels, S., Johansson, J., Stenram, U., Svanberg, K., and Svanberg, S., Malignant tumour and atherosclerotic plaque diagnosis using laser induced fluorescence, *IEEE Journal of Quantum Electronics*, 26(12), 2207–2217, 1990.
6. Koenig, F., Mcgovern, F. J., Enquist, H., Larne, R., Deutsch, T. F., and Schomacker, K. T., Autofluorescence guided biopsy for the early diagnosis of bladder cancer, *Journal of Urology*, 159, 1871–1875, 1998.
7. Denk, W., Strickler, J. H., and Webb, W. W., Two-photon laser scanning fluorescence microscopy, *Science*, 248, 73–76, 1990.
8. Denk, W. and Svoboda, K., Photon upmanship: Why multiphoton imaging is more than a gimmick, *Neuron*, 18, 351–357, 1997.
9. Dickinson, M. E., Simbuerger, E., Zimmermann, B., Waters, C. W., and Fraser, S. E., Multiphoton excitation spectra in biological samples, *Journal of Biomedical Optics*, 8(3), 329–338, 2003.
10. Sytsma, J., Vroom, J. M., De Grauw, C. J., and Gerritsen, H. C., Time-gated fluorescence lifetime imaging and microvolume spectroscopy using two-photon excitation, *Journal of Microscopy*, 191, 39–51, 1998.
11. Schönle, A., Glatz, M., and Hell, S. W., Four-dimensional multiphoton microscopy with time-correlated single photon counting, *Applied Optics*, 39, 6306–6311, 2000.
12. Tsien, R. Y., The green fluorescent protein, *Annual Review of Biochemistry*, 67, 509–544, 1998.
13. Zimmer, M., Green fluorescent protein (GFP): Applications, structure, and related photophysical behavior, *Chemical Reviews*, 102(3), 759–781, 2002.
14. Förster, T., Zwischenmolekulare Energiewanderung und Fluoreszenz, *Annals of Physics*, 2, 55–75, 1948.
15. Clegg, R. M., Holub, O., and Gohlke, C., Fluorescence lifetime-resolved imaging. Measuring lifetimes in an image. Why do it? How to do it. How to interpret it. In: Marriott, G. and Parker, I., eds., *Methods in Enzymology*. Academic Press, San Diego, CA, pp. 509–442, 2003.
16. Stryer, L., Fluorescence energy transfer as a spectroscopic ruler, *Annual Reviews in Biochemistry*, 47, 819–846, 1978.
17. dos Remedios, C. G. and Moens, P. D., Fluorescence resonance energy transfer spectroscopy is a reliable 'ruler' for measuring structural changes in proteins: Dispelling the problem of the unknown orientation factor, *Journal of Structural Biology*, 115, 175–185, 1995.
18. Miyawaki, A., Visualization of the spatial review and temporal dynamics of intracellular signaling, *Developmental Cell*, 4, 295–305, 2003.
19. Miyawaki, A., Llopis, J., Heim, R., Mccaffery, J. M., Adams, J. A., Ikura, M., and Tsien, R. Y., Fluorescent indicators for Ca^{2+} based on green fluorescent proteins and calmodulin, *Nature*, 388, 882–887, 1997.
20. Gordon, G. W., Berry, G., Liang, X. H., Levine, B., and Herman, B., Quantitative fluorescence resonance energy transfer measurements using fluorescence microscopy, *Biophysical Journal*, 74(5), 2702–2713, 1998.
21. Bastiaens, P. I. H. and Squire, A., Fluorescence lifetime imaging microscopy: Spatial resolution of biochemical processes in the cell, *Trends in Cell Biology*, 9, 48–52, 1999.

22. Jares-Erijman, E. A. and Jovin, T. M., FRET imaging, *Nature Biotechnology*, 21(11), 1387–1395, 2003.

23. Suhling, K., French, P. M. W., and Phillips, D., Time-resolved fluorescence microscopy, *Photochemical & Photobiological Sciences*, 4(1), 13–22, 2005.

24. Richards-Kortum, R. and Sevick-Muraca, E. M., Quantitative optical spectroscopy for tissue diagnosis, *Annual Review of Physical Chemistry*, 47, 555–606, 1996.

25. Das, B. B., Liu, F., and Alfano, R. R., Time-resolved fluorescence and photon migration studies in biomedical and model random media, *Reports on Progress in Physics*, 60, 227–291, 1997.

26. Mizeret, J., Wagnieres, G., Stepinac, T., and VandenBergh, H., Endoscopic tissue characterization by frequency-domain fluorescence lifetime imaging (FD-FLIM), *Lasers in Medical Science*, 12, 209–217, 1997.

27. Dowling, K., Dayel, M. J., Lever, M. J., French, P. M. W., Hares, J. D., and Dymoke-Bradshaw, A. K. L., Fluorescence lifetime imaging with picosecond resolution for biomedical applications, *Optics Letters*, 23, 810–812, 1998.

28. Benninger, R. K. P., Koc, Y., Hofmann, O., Requejo-Isidro, J., Neil, M. A. A., French, P. M. W., and Demello, A. J., Quantitative 3D mapping of fluidic temperatures within microchannel networks using fluorescence lifetime imaging, *Analytical Chemistry*, 78, 2272–2278, 2006.

29. Schaerli, Y., Wootton, R. C., Robinson, T., Stein, V., Dunsby, C., Neil, M. A. A., French, P. M. W., Demello, A. J., Abell, C., and Hollfelder, F., Continuous-flow polymerase chain reaction of single-copy DNA in microfluidic microdroplets, *Analytical Chemistry*, 81, 302–306, 2009.

30. Koeberg, M., Elson, D. S., French, P. M. W., and Bradley, D. D. C., Spatially resolved electric fields in polymer light-emitting diodes using fluorescence lifetime imaging, *Synthetic Metals*, 139, 925–928, 2003.

31. Hell, S. W. and Wichmann, J., Breaking the diffraction resolution limit by stimulated emission: Stimulated–emission-depletion fluorescence microscopy, *Optics Letters*, 19, 780–782, 1994.

32. Klar, T. A., Jakobs, S., Dyba, M., Egner, A., and Hell, S. W., Fluorescence microscopy with diffraction resolution barrier broken by stimulated emission, *Proceedings of the National Academy of Sciences of the United States of America*, 97, 8206–8210, 2000.

33. Betzig, E., Patterson, G. H., Sougrat, R., Lindwasser, O. W., Olenych, S., Bonifacino, J. S., Davidson, M. W., Lippincott-Schwartz, J., and Hess, H. F., Imaging intracellular fluorescent proteins at nanometer resolution, *Science*, 313, 1642–1645, 2006.

34. Hess, S. T., Girirajan, T. P. K., and Mason, M. D., Ultra-high resolution imaging by fluorescence photoactivation localization microscopy, *Biophysical Journal*, 91, 4258–4272, 2006.

35. Rust, M. J., Bates, M., and Zhuang, X., Sub-diffraction-limit imaging by stochastic optical reconstruction microscopy (STORM), *Nature Methods*, 3, 793–795, 2006.

36. Neumann, B., Held, M., Liebel, U., Erfle, H., Rogers, P., Pepperkok, R., and Ellenberg, J., High-throughput RNAi screening by time-lapse imaging of live human cells, *Nature Methods*, 3, 385–390, 2006.

37. Matthews, D. R., Carlin, L. M., Ofo, E., Barber, P. R., Vojnovic, B., Irving, M., Ng, T., and Ameer-Beg, S. M., Time-lapse FRET microscopy using fluorescence anisotropy, *Journal of Microscopy*, 237, 51–62, 2010.

38. Esposito, A., Dohm, C. P., Bahr, M., and Wouters, F. S., Unsupervised fluorescence lifetime imaging microscopy for high content and high throughput screening, *Molecular & Cellular Proteomics*, 6, 1446–1454, 2007.

39. Grecco, H. E., Roda-Navarro, P., Girod, A., Hou, J., Frahm, T., Truxius, D. C., Pepperkok, R., Squire, A., and Bastiaens, P. I. H., In situ analysis of tyrosine phosphorylation networks by FLIM on cell arrays, *Nature Methods*, 7, 467–472, 2010.

40. Talbot, C. B., McGinty, J., Grant, D. M., McGhee, E. J., Owen, D. M., Zhang, W., Bunney, T. D. et al., High speed unsupervised fluorescence lifetime imaging confocal multiwell plate reader for high content analysis, *Journal of Biophotonics*, 1, 514–521, 2008.

41. Alibhai, D., Kelly, D. J., Kumar, S., Warren, S., Serwa, R., Thinon, E., Alexandrov, Y. et al., Automated fluorescence lifetime imaging plate reader and its application to Forster resonant energy transfer readout of Gag protein aggregation, *Journal of Biophotonics*, 6, 398–408, 2013.

42. Chorvat Jr., D., Kirchnerova, J., Cagalinec, M., Smolka, J., Mateasik, A., and Chorvatova, A., Spectral unmixing of flavin autofluorescence components in cardiac myocytes, *Biophysical Journal*, 89, L55–L57, 2005.

43. Wachman, E. S., Niu, W. H., and Farkas, D. L., Imaging acousto-optic tunable filter with 0.35-micrometer spatial resolution, *Applied Optics*, 35, 5220–5226, 1996.

44. Farkas, D. L., Du, C. W., Fisher, G. W., Lau, C., Niu, W. H., Wachman, E. S., and Levenson, R. M., Non-invasive image acquisition and advanced processing in optical bioimaging, *Computerized Medical Imaging and Graphics*, 22, 89–102, 1998.

45. Malik, Z., Cabib, D., Buckwald, R. A., Talmi, A., Garini, Y., and Lipson, S. G., Fourier transform multipixel spectroscopy for quantitative cytology, *Journal of Microscopy-Oxford*, 182, 133–140, 1996.

46. Schultz, R. A., Nielsen, T., Zavaleta, J. R., Ruch, R., Wyatt, R., and Garner, H. R., Hyperspectral imaging: A novel approach for microscopic analysis, *Cytometry*, 43, 239–247, 2001.

47. Hanley, Q. S., Verveer, P. J., and Jovin, T. M., Spectral imaging in a programmable array microscope by Hadamard transform fluorescence spectroscopy, *Applied Spectroscopy*, 53, 1–10, 1999.

48. Volin, C. E., Ford, B. K., Descour, M. R., Garcia, J. P., Wilson, D. W., Maker, P. D., and Bearman, G. H., High-speed spectral imager for imaging transient fluorescence phenomena, *Applied Optics*, 37, 8112–8119, 1998.

49. Ramanujam, N., Mitchell, M. F., Mahadevanjansen, A., Thomsen, S. L., Staerkel, G., Malpica, A., Wright, T., Atkinson, N., and Richards-Kortum, R., Cervical precancer detection using a multivariate statistical algorithm based on laser-induced fluorescence spectra at multiple excitation wavelengths, *Photochemistry and Photobiology*, 64, 720–735, 1996.

50. Lee, K. C. B., Siegel, J., Webb, S. E. D., Lévêque-Fort, S., Cole, M. J., Jones, R., Dowling, K., Lever, M. J., and French, P. M. W., Application of the stretched exponential function to fluorescence lifetime imaging, *Biophysical Journal*, 81, 1265–1274, 2001.

51. Wlodarczyk, J. and Kierdaszuk, B., Interpretation of fluorescence decays using a power-like model, *Biophysical Journal*, 85, 589–598, 2003.

52. Jo, J. A., Fang, Q. Y., Papaioannou, T., and Marcu, L., Fast model-free deconvolution of fluorescence decay for analysis of biological systems, *Journal of Biomedical Optics*, 9, 743–752, 2004.

53. Grinwald, A., On the analysis of fluorescence decay kinetics by the method of least-squares, *Analytical Biochemistry*, 59, 583–593, 1974.

54. James, D. R. and Ware, W. R., A fallacy in the interpretation of fluorescence decay parameters, *Chemical Physics Letters*, 120, 455–459, 1985.

55. Verveer, P. J., Squire, A., and Bastiaens P. I. H., Global analysis of fluorescence lifetime imaging microscopy data, *Biophysical Journal*, 78, 2127–2137, 2000.

56. Warren, S. C., Margineanu, A., Alibhai, D., Kelly, D. J., Talbot, C., Alexandrov, Y., Munro, I., Katan, M., Dunsby, C., and French, P. M. W., Rapid global fitting of large fluorescence lifetime imaging microscopy datasets, *PLoS ONE*, 8, e70687, 2013; "FLIMfit" open source software tool available at http://www.openmicroscopy.org/site/products/partner/flimfit

57. Spencer, R. D. and Gregorio, W., Measurements of subnanosecond fluorescence lifetimes with a cross-correlation phase fluorometer, *Annals of the New York Academy of Sciences*, 158, 361–376, 1969.

58. Gratton, E. and Limkeman, M., A continuously variable frequency cross-correlation phase fluorometer with picosecond resolution, *Biophysical Journal*, 44, 315–324, 1983.

59. Lakowicz, J. R. and Maliwal, B. P., Construction and performance of a variable-frequency phase-modulation fluorometer, *Biophysical Chemistry*, 21, 61–78, 1985.

60. O'Connor, D. V. and Phillips, D., *Time-Correlated Single-Photon Counting*. Academic Press, New York, 1984.

61. Becker, W., *Advanced Time-Correlated Single Photon Counting Techniques*. Springer, Berlin, Germany, 2005.

62. Tata, D. B., Foresti, M., Cordero, J., and Tomashefpsky, P., Fluorescence polarization spectroscopy and time-resolved fluorescence kinetics of native cancerous and normal rat-kidney tissues, *Biophysical Journal*, 50, 463–469, 1986.

63. Glanzmann, T., Ballini, J.-P., Bergh, H. V. D., and Wagnieres, G., Time-resolved spectrofluorometer for clinical tissue characterization during endoscopy, *Review of Scientific Instruments*, 70, 4067–4077, 1999.

64. Mycek, M. A., Schomacker, K. T., and Nishioka, N. S., Colonic polyp differentiation using time-resolved autofluorescence spectroscopy, *Gastrointestinal Endoscopy*, 48, 390–394, 1998.

65. Fang, Q. Y., Papaioannou, T., Jo, J. A., Vaitha, R., Shastry, K., and Marcu, L., Time-domain laser-induced fluorescence spectroscopy apparatus for clinical diagnostics, *Review of Scientific Instruments*, 75, 151–162, 2004.

66. Pitts, J. D. and Mycek, M. A., Design and development of a rapid acquisition laser-based fluorometer with simultaneous spectral and temporal resolution, *Review of Scientific Instruments*, 72, 3061–3072, 2001.

67. De Beule, P. A., Dunsby, C., Galletly, N. P., Stamp, G. W., Chu, A. C., Anand, U., Anand, P., Benham, C. D., Naylor, A., and French, P. M. W., A hyperspectral fluorescence lifetime probe for skin cancer diagnosis, *Review of Scientific Instruments*, 78, 123101–123107, 2007.

68. Manning, H. B., Kennedy, G. T., Owen, D. M., Grant, D. M., Magee, A. I., Neil, M. A., Itoh, Y., Dunsby, C., and French, P. M. W., A compact, multidimensional spectrofluorometer exploiting supercontinuum generation, *Journal of Biophotonics*, 1, 494–505, 2008.

69. Bugiel, I., Konig, K., and Wabnitz, H., Investigation of cells by fluorescence laser scanning microscopy with subnanosecond time resolution, *Lasers in the Life Sciences*, 3, 47–53, 1989.

70. Buurman, E. P., Sanders, R., Draaijer, A., Gerritsen, H. C., Vanveen, J. J. F., Houpt, P. M., and Levine, Y. K., Fluorescence lifetime imaging using a confocal laser scanning microscope, *Scanning*, 14, 155–159, 1992.

71. Carlsson, K. and Liljeborg, A., Simultaneous confocal lifetime imaging of multiple fluorophores using the intensity-modulated multiple-wavelength scanning (IMS) technique, *Journal of Microscopy*, 191, 119–127, 1998.

72. So, P. T. C., French, T., Yu, W. M., Berland, K. M., Dong, C. Y., and Gratton, E., Time-resolved fluorescence microscopy using two-photon excitation, *Bioimaging*, 3, 49–63, 1995.

73. Cubeddu, R., Comelli, D., D'andrea, C., Taroni, P., and Valentini, G., Time-resolved fluorescence imaging in biology and medicine, *Journal of Physics D—Applied Physics*, 35, R61–R76, 2002.

74. Gadella, T. W. J., ed., *FRET and FLIM Imaging Techniques*. Elsevier, Amsterdam, the Netherlands, 2008.

75. Esposito, A., Gerritsen, H. C., and Wouters, F. S., Optimizing frequency-domain fluorescence lifetime sensing for high-throughput applications: Photon economy and acquisition speed, *Journal of the Optical Society of America A—Optics Image Science and Vision*, 24, 3261–3273, 2007.

76. Bewersdorf, J., Pick, R., and Hell, S. W., Multifocal multiphoton microscopy, *Optics Letters*, 23, 655–657, 1998.

77. Fittinghoff, D. N., Wiseman, P. W., and Squier, J. A., Widefield multiphoton and temporally decorrelated multiphoton microscopy, *Optics Express*, 7, 273–279, 2000.

78. Kim, K. H., Buehler, C., Bahlmann, K., Ragan, T., Lee, W. C. A., Nedivi, E., Heffer, E. L., Fantini, S., and So, P. T. C., Multifocal multiphoton microscopy based on multianode photomultiplier tubes, *Optics Express*, 15, 11658–11678, 2007.

79. Kumar, S., Dunsby, C., De Beule, P. A., Owen, D. M., Anand, U., Lanigan, P. M. P., Benninger, R. K. P. et al., Multifocal multiphoton excitation and time correlated single photon counting detection for 3-D fluorescence lifetime imaging, *Optics Express*, 15, 12548–12561, 2007.

80. Krishnan, K. V., Saitoh, H., Terada, H., Centonze, V. E., and Herman, B., Development of a multi-photon fluorescence lifetime imaging microscopy system using a streak camera, *Review of Scientific Instruments*, 74, 2714–2721, 2003.

81. Grant, D. M., Elson, D. S., Schimpf, D., Dunsby, C., Requejo-Isidro, J., Auksorius, E., Munro, I. et al., Optically sectioned fluorescence lifetime imaging using a Nipkow disk microscope and a tunable ultrafast continuum excitation source, *Optics Letters*, 30, 3353–3355, 2005.

82. Munster, E. B. V., Goedhart, J., Kremers, G. J., Manders, E. M. M., and Gadella Jr, T. W. J., Combination of a spinning disc confocal unit with frequency-domain fluorescence lifetime imaging microscopy, *Cytometry A*, 71(4), 207–214, 2007.

83. Grant, D. M., Mcginty, J., Mcghee, E. J., Bunney, T. D., Owen, D. M., Talbot, C. B., Zhang, W. et al., High speed optically sectioned fluorescence lifetime imaging permits study of live cell signaling events, *Optics Express*, 15, 15656–15673, 2007.

84. Straub, M. and Hell, S. W., Fluorescence lifetime three-dimensional microscopy with picosecond precision using a multifocal multiphoton microscope, *Applied Physics Letters*, 73, 1769–1771, 1998.

85. Leveque-Fort, S., Fontaine-Aupart, M. P., Roger, G., and Georges, P., Fluorescence-lifetime imaging with a multifocal two-photon microscope, *Optics Letters*, 29, 2884–2886, 2004.

86. Benninger, R. K. P., Hofmann, O., Mcginty, J., Requejo-Isidro, J., Munro, I., Neil, M. A., Demello, A. J., and French, P. M. W., Time-resolved fluorescence imaging of solvent interactions in microfluidic devices, *Optics Express*, 13, 6275–6285, 2005.

87. Holub, O., Seufferheld, M. J., Gohike, C., Govindjee, and Clegg, R. M., Fluorescence lifetime imaging (FLI) in real-time—A new technique in photosynthesis research, *Photosynthetica*, 38, 581–599, 2000.

88. Agronskaia, A. V., Tertoolen, L., and Gerritsen, H. C., High frame rate fluorescence lifetime imaging, *Journal of Physics D—Applied Physics*, 36, 1655–1662, 2003.

89. Requejo-Isidro, J., Mcginty, J., Munro, I., Elson, D. S., Galletly, N. P., Lever, M. J., Neil, M. A. et al., High-speed wide-field time-gated endoscopic fluorescence lifetime imaging, *Optics Letters*, 29, 2249–2251, 2004.

90. Mizeret, J., Stepinac, T., Hansroul, M., Studzinski, A., Van Den Bergh, H., and Wagnieres, G., Instrumentation for real-time fluorescence lifetime imaging in endoscopy, *Review of Scientific Instruments*, 70, 4689–4701, 1999.

91. Munro, I., Mcginty, J., Galletly, N., Requejo-Isidro, J., Lanigan, P. M. P., Elson, D. S., Dunsby, C., Neil, M. A., Lever, M. J., Stamp, G. W. H., and French, P. M. W., Toward the clinical application of time-domain fluorescence lifetime imaging, *Journal of Biomedical Optics*, 10, 051403, 2005.

92. Morgan, C. G., Mitchell, A. C., and Murray, J. G., Nanosecond time-resolved fluorescence micros-copy: Principles and practice, *Transactions of the Royal Microscopy Society*, 1, 463–466, 1990.

93. Gratton, E., Feddersen, B., and Van De Ven, M., Parallel acquisition of fluorescence decay using array detectors. In: Lakowicz, J. R., ed., *Time-Resolved Laser Spectroscopy in Biochemistry II. SPIE* vol. 1204, SPIE, Los Angeles, CA, pp. 21–25, 1990.

94. Lakowicz, J. R., Szmacinski, H., Nowaczyk, K., Berndt, K. W., and Johnson, M., Fluorescence life-time imaging, *Analytical Biochemistry*, 202, 316–330, 1992.

95. Gadella, T. W. J., Jovin, T. M., and Clegg, R. M., Fluorescence lifetime imaging microscopy (FLIM)—Spatial resolution of structures on the nanosecond timescale, *Biophysical Chemistry*, 48, 221–239, 1993.

96. Oida, T., Sako, Y., and Kusumi, A., Fluorescence lifetime imaging microscopy (flimscopy): Methodology development and application to studies of endosome fusion in single cells, *Biophysical Journal*, 64, 676–685, 1993.

97. Scully, A. D., Mac Robert, A. J., Botchway, S., O'neill, P., Parker, A. W., Ostler, R. B., and Phillips, D., Development of a laser-based fluorescence microscope with subnanosecond time resolution, *Journal of Fluorescence*, 6, 119–125, 1996.

98. Hares, J. D., Advances in sub-nanosecond shutter tube technology and applications in plasma phys-ics. In: Richardson, M. C., ed., *X Rays from Laser Plasmas, Proceedings of the SPIE* vol. 831, SPIE, Bellingham, WA, pp. 165–170, 1987.

99. Becker, W., Bergmann, A., Haustein, E., Petrasek, Z., Schwille, P., Biskup, C., Kelbauskas, L. et al., Fluorescence lifetime images and correlation spectra obtained by multidimensional time-correlated single photon counting, *Microscopy Research and Technique*, 69, 186–195, 2006.

100. Siegel, J., Elson, D. S., Webb, S. E. D., Parsons-Karavassilis, D., Lévêque-Fort, S., Cole, M. J., Lever, M. J. et al., Whole-field five-dimensional fluorescence microscopy combining lifetime and spectral resolution with optical sectioning, *Optics Letters*, 26, 1338–1340, 2001.

101. Bird, D. K., Eliceiri, K. W., Fan, C. H., and White, J. G., Simultaneous two-photon spectral and lifetime fluorescence microscopy, *Applied Optics*, 43, 5173–5182, 2004.

102. Hanley, Q. S., Arndt-Jovin, D. J., and Jovin, T. M., Spectrally resolved fluorescence lifetime imaging microscopy, *Applied Spectroscopy*, 56, 155–166, 2002.

103. De Beule, P., Owen, D. M., Manning, H. B., Talbot, C. B., Requejo-Isidro, J., Dunsby, C., Mcginty, J. et al., Rapid hyperspectral fluorescence lifetime imaging, *Microscopy Research and Technique*, 70, 481–484, 2007.

104. Owen, D. M., Auksorius, E., Manning, H. B., Talbot, C. B., De Beule, P. A., Dunsby, C., Neil, M. A., and French, P. M. W., Excitation-resolved hyperspectral fluorescence lifetime imaging using a UV-extended supercontinuum source, *Optics Letters*, 32, 3408–3410, 2007.

105. Demos, S. G. and Alfano, R. R., Optical polarization imaging, *Applied Optics*, 36, 150–155, 1997.

106. Demos, S. G., Gandour-Edwards, R., Ramsamooj, R., and Devere White, R., Spectroscopic detection of bladder cancer using near-infrared imaging techniques, *Journal of Biomedical Optics*, 9, 767–771, 2004.

107. Abbott, A., Cell culture: Biology's new dimension, *Nature*, 424, 870–872, 2003.

108. Weissleder, R. and Ntziachristos, V., Shedding light onto live molecular targets, *Nature Medicine*, 9, 123–128, 2003.

109. Oldham, M., Sakhalkar, H., Wang, Y. M., Guo, P., Oliver, T., Bentley, R., Vujaskovic, Z., and Dewhirst, M., Three-dimensional imaging of whole rodent organs using optical computed and emission tomography, *Journal of Biomedical Optics*, 12, 014009, 2007.

110. Keller, P. J., Pampaloni, F., and Stelzer, E. H. K., Life sciences require the third dimension, *Current Opinion in Cell Biology*, 18, 117–124, 2006.

111. Watson, T. F., Neil, M. A. A., Juškaitis, R., Cook, R. J., and Wilson, T., Video-rate confocal endoscopy, *Journal of Microscopy*, 207, 37–42, 2002.

112. Jung, J. C. and Schnitzer, M. J., Multiphoton endoscopy, *Optics Letters*, 28, 902–904, 2003.

113. Gmitro, A. F. and Aziz, D., Confocal microscopy through a fiber-optic imaging bundle, *Optics Letters*, 18, 565–567, 1993.

114. Laemmel, E., Genet, M., Le Goualher, G., Perchant, A., Le Gargasson, J. F., and Vicaut, E., Fibered confocal fluorescence microscopy (Cell-viZio (TM)) facilitates extended imaging in the field of microcirculation—A comparison with intravital microscopy, *Journal of Vascular Research*, 41, 400–411, 2004.

115. Dickensheets, D. L. and Kino, G. S., Micromachined scanning confocal optical microscope, *Optics Letters*, 21, 764–766, 1996.

116. Liang, C., Descour, M. R., Sung, K. B., and Richards-Kortum, R., Fiber confocal reflectance microscope (FCRM) for in-vivo imaging, *Optics Express*, 9, 821–830, 2001.

117. Tearney, G. J., Shishkov, M., and Bouma, B. E., Spectrally encoded miniature endoscopy, *Optics Letters*, 27, 412–414, 2002.

118. Seibel, E. J. and Smithwick, Q. Y. J., Unique features of optical scanning, single fiber endoscopy, *Lasers in Surgery and Medicine*, 30, 177–183, 2002.

119. Helmchen, F., Fee, M. S., Tank, D. W., and Denk, W., A miniature head-mounted two-photon microscope: High-resolution brain imaging in freely moving animals, *Neuron*, 31, 903–912, 2001.

120. Lelek, M., Suran, E., Louradour, F., Barthelemy, A., Viellerobe, B., and Lacombe, F., Coherent femtosecond pulse shaping for the optimization of a non-linear micro-endoscope, *Optics Express*, 15, 10154–10162, 2007.

121. Myaing, M. T., MacDonald, D. J., and Li, X., Fiber-optic scanning two-photon fluorescence endoscope, *Optics Letters*, 31, 1076–1078, 2006.

122. Rivera1, D. R., Brown, C. M., Ouzounov, D. G., Pavlova, I., Kobat, D., Webb, W. W., and Xu, C., Compact and flexible raster scanning multiphoton endoscope capable of imaging unstained tissue, *PNAS*, 108, 17598–17603, 2011.

123. Gono, K., Yamazaki, K., Doguchi, N., Nonami, T., Obi, T., Yamaguchi, M., Ohyama, N. et al., Endoscopic observation of tissue by narrowband illumination, *Optical Review*, 10, 211–215, 2003.

124. van den Broek, F. J. C., Fockens, P., van Eeden, S. et al., Endoscopic tri-modal imaging for surveillance in ulcerative colitis: Randomised comparison of high-resolution endoscopy and autofluorescence imaging for neoplasia detection; and evaluation of narrow-band imaging for classification of lesions, *Gut*, 57, 1083–1089, 2008.

125. Zeng, H., Weissz, A., Clinex, R., and MacAulay, C. E., *Bioimaging*, 6, 151–165, 1998.

126. Thekkek, N., Pierce, M. C., Lee, M. H., Polydorides, A. D., Flores, R. M., Anandasabapathy, S., and Richards-Kortum, R. R., Modular video endoscopy for in vivo cross-polarized and vital-dye fluorescence imaging of Barrett's-associated neoplasia, *Journal of Biomedical Optics*, 18(2), 26007, doi: 10.1117/1.JBO.18.2.026007, February 2013.

127. Jean, F., Bourg-Heckly, G., and Viellerobe, B., Fibered confocal spectroscopy and multicolor imaging system for in vivo fluorescence analysis, *Optics Express*, 15, 4008–4017, 2007.

128. Rouse, A. R. and Gmitro, A. F., Multispectral imaging with a confocal microendoscope, *Optics Letters*, 25, 1708–1710, 2000.

129. Elson, D. S., Jo, J. A., and Marcu, L., Miniaturized side-viewing imaging probe for fluorescence lifetime imaging (FLIM): Validation with fluorescence dyes, tissue structural proteins and tissue specimens, *New Journal of Physics*, 9, 127, 2007.

130. Sun, Y., Phipps, J., Elson, D. S., Stoy, H., Tinling, S., Meier, J., Poirier, B., Chuang, F. S., Farwell, D. G., and Marcu, L., Fluorescence lifetime imaging microscopy: In vivo application to diagnosis of oral carcinoma, *Optics Letters*, 34, 2081–2083, 2009.

131. Kennedy, G. T., Manning, H. B., Elson, D. S., Neil, M. A. A., Stamp, G. W., Viellerobe, B., Lacombe, F., Dunsby, C., and French, P. M. W., A fluorescence lifetime imaging scanning confocal endomicroscope, *Journal of Biophotonics*, 3, 103–107, 2010.

132. Fruhwirth, G. O., Ameer-Beg, S., Cook, R., Watson, T., Ng, T., and Festy, F., Fluorescence lifetime endoscopy using TCSPC for the measurement of FRET in live cells, *Optics Express*, 18, 11148–11158, 2010.

133. Kumar, S., Alibhai, D., Margineanu, A., Laine, R., Kennedy, G., McGinty, J., Warren, S. et al., FLIM FRET technology for drug discovery: Automated multiwell plate high content analysis, multiplexed readouts and application in situ, *ChemPhysChem*, 12, 627–633, 2011.

134. Ntziachristos, V., Ripoll, J., Wang, L. V., and Weissleder, R., Looking and listening to light: The evolution of whole-body photonic imaging, *Nature Biotechnology*, 23, 313, 2005.

135. Sharpe, J., Ahlgren, U., Perry, P., Hill, B., Ross, A., Hecksher-Sorensen, J., Baldock, R., and Davidson, D., Optical projection tomography as a tool for 3D microscopy and gene expression studies, *Science*, 296, 541–545, 2002.

136. Huisken, J., Swoger, J., Del Bene, F., Wittbrodt, J., and Stelzer, E. H. K., Optical sectioning deep inside live embryos by selective plane illumination microscopy, *Science*, 305, 1007–1009, 2004.

137. Dodt, H. U., Leischner, U., Schierloh, A., Jahrling, N., Mauch, C. P., Deininger, K., Deussing, J. M., Eder, M., Zieglgansberger, W., and Becker, K., Ultramicroscopy: Three-dimensional visualization of neuronal networks in the whole mouse brain, *Nature Methods*, 4, 331–336, 2007.

138. Gibson, A. P., Hebden, J. C., and Arridge, S. R., Recent advances in diffuse optical imaging, *Physics in Medicine and Biology*, 50, R1–R43, 2005.

139. Ntziachristos, V. and Wessleder, R., *Optics Letters*, 26, 893, 2001.

140. Deliolanis, N., Lasser, T., Hyde, D., Soubret, A., Ripoll, J., and Ntziachristos, V., Free-space fluorescence molecular tomography utilizing 360° geometry projections, *Optics Letters*, 32(4), 382–384, 2007.

141. Jiang, H., Frequency-domain fluorescent diffusion tomography: A finite-element-based algorithm and simulations, *Applied Optics*, 37, 5337–5343, 1998.

142. Milstein, A. B., Oh, S., Webb, K. J., Bouman, C. A., Zhang, Q., Boas, D. A., and Millane, R. P., Fluorescence optical diffusion tomography, *Applied Optics*, 42, 3081–3094, 2003.

143. Niedre, M. J., de Kleine, R. H., Aikawa, E., Kirsch, D. G., Weissleder, R., and Ntziachristos, V., Early photon tomography allows fluorescence detection of lung carcinomas and disease progression in mice in vivo, *Proceedings of the National Academy of Sciences of the United States of America*, 105, 19126–19131, 2008.

144. Wu, J., Perelman, L., Dasari, R. R., and Feld, M. S., Fluorescence tomographic imaging in turbid media using early-arriving photons and Laplace transforms, *Proceedings of the National Academy of Sciences of the United States of America*, 94, 8783, 1997.

145. Wang, L. V. and Hu, S., Photoacoustic tomography: In vivo imaging from organelles to organs, *Science*, 335, 1458, 2012.

146. Razansky, D., Distel, M., Vinegoni, C., Ma, R., Perrimon, N., Kçster, R. W., and Ntziachristos, V., *Nature Photonics*, 3, 412–417, 2009.

147. Huisken, J. and Stainier, D. Y. R., Even fluorescence excitation by multidirectional selective plane illumination microscopy (mSPIM), *Optics Letters*, 32, 2608–2610, 2007.

148. Vinegoni, C., Pitsouli, C., Razansky, D., Perrimon, N., and Ntziachristos, V., In vivo imaging of *Drosophila melanogaster* pupae with mesoscopic fluorescence tomography, *Nature Methods*, 5, 45–47, 2008.

149. Birk, U. J., Rieckher, M., Konstantinides, N., Darrell, A., Sarasa-Renedo, A., Meyer, H., Tavernarakis, N., and Ripoll, J., Correction for specimen movement and rotation errors for in-vivo optical projection tomography, *Biomedical Optics Express*, 1, 87–96, 2010.

150. McGinty, J., Taylor, H. B., Chen, L., Bugeon, L., Lamb, J. R., Dallman, M. J., and French, P. M. W., In vivo fluorescence lifetime optical projection tomography, *Biomedical Optics Express*, 2, 1340–1350, 2011.

151. Swartling, J., Svensson, J. et al., Fluorescence spectra provide information on the depth of fluorescent lesions in tissue, *Applied Optics*, 44, 1934–1941, 2005.

152. Zavattini, G., Vecchi, S. et al., A hyperspectral fluorescence system for 3D in vivo optical imaging, *Physics in Medicine and Biology*, 51, 2029–2043, 2006.

153. Chaudhari, A. J., Darvas, F., Bading, J. R., Moats, R. A., Conti, P. S., Smith, D. J., Cherry, S. R., and Leahy, R. M., Hyperspectral and multispectral bioluminescence optical tomography for small animal imaging, *Physics in Medicine and Biology*, 50, 5421–5441, 2005.

154. Davis, S. C., Pogue, B. W. et al., Spectral distortion in diffuse molecular luminescence tomography in turbid media, *Journal of Applied Physics*, 105, 102024, 2009.

155. Li, C., Mitchell, G. S. et al., A three-dimensional multispectral fluorescence optical tomography imaging system for small animals based on a conical mirror design, *Optics Express*, 17, 7571–7585, 2009.

156. Zacharopoulos, A. D., Svenmarker, P. et al., A matrix-free algorithm for multiple wavelength fluorescence tomography, *Optics Express*, 17, 3025–3035, 2009.

157. Oleary, M. A., Boas, D. A., Li, X. D., Chance, B., and Yodh, A. G., Fluorescence lifetime imaging in turbid media, *Optics Letters*, 21, 158, 1996.

158. Paithankar, D. Y., Chen, A. U., Pogue, B. W., Patterson, M. S., and Sevick Muraca, E. M., Imaging of fluorescent yield and lifetime from multiply scattered light reemitted from random media, *Applied Optics*, 36, 2260–2272, 1997.

159. Shives, E., Xu, Y., and Jiang, H., Fluorescence lifetime tomography of turbid media based on an oxygen-sensitive dye, *Optics Express*, 10, 1557–1562, 2002.

160. Godavarty, A., Sevick-Muraca, E. M., and Eppstein, M. J., Three-dimensional fluorescence lifetime tomography, *Medical Physics*, 32, 992–1000, 2005.

161. Kumar, A. T. N., Raymond, S. B., Dunn, A. K., Bacskai, B. J., and Boas, D. A., A time domain fluorescence tomography system for small animal imaging, *IEEE Transactions on Medical Imaging*, 27, 1152–1163, 2008.

162. Kumar, A. T. N., Chung, E., Raymond, S. B., van de Water, J., Shah, K., Fukumura, D., Jain, R. K., Bacskai, B. J., and Boas, D. A., Feasibility of in vivo imaging of fluorescent proteins using lifetime contrast, *Optics Letters*, 34, 2066–2068, 2009.

163. Nothdurft, R. E., Patwardhan, S. V., Akers, W., Ye, Y. P., Achilefu, S., and Culver, J. P., In vivo fluorescence lifetime tomography, *Journal of Biomedical Optics*, 14, 024004, 2009.

164. McGinty, J., Stuckey, D. W., Soloviev, V. Y., Laine, R., Wylezinska-Arridge, M., Wells, D. J., Arridge, S. R., French, P. M. W., Hajnal, J. V., and Sardini, A., In vivo fluorescence lifetime tomography of a FRET probe expressed in mouse, *Biomedical Optics Express*, 2, 1907–1917, 2011.

165. McGinty, J., Soloviev, V. Y. et al., Three-dimensional imaging of Förster resonance energy transfer in heterogeneous turbid media by tomographic fluorescence lifetime, *Optics Letters*, 34(18), 2772–2774, 2009.

166. Soloviev, V. Y., D'Andrea, C., Valentini, G., Cubeddu, R., and Arridge, S. R., Combined reconstruction of fluorescent and optical parameters using time-resolved data, *Applied Optics*, 48, 28, 2009.

167. Venugopal, V., Chen, J., Barroso, M., and Intes, X., Quantitative tomographic imaging of intermolecular FRET in small animals, *Biomedical Optics Express*, 3, 3161–3175, 2012.

168. Shu, X., Royant, A., Lin, M. Z., Aguilera, T. A., Lev-Ram, V., Steinbach, P. A., and Tsien, R. Y., *Science*, 324, 804–807, 2009.

169. Shcherbakova, D. M. and Verkhusha, V. V., Near-infrared fluorescent proteins for multicolor in vivo imaging, *Nature Methods*, 10, 751–754, 2013.

170. Niedworok, C. J., Schwarz, I., Ledderose, J., Giese, G., Conzelmann, K. K., and Schwarz, M. K., Charting monosynaptic connectivity maps by two-color light-sheet fluorescence microscopy, *Cell Reports*, 2, 1375–1386, 2012.

171. McGinty, J., Tahir, K. R., Laine, R., Talbot, C. B., Dunsby, C., Neil, M. A. A., Quintana, L., Swoger, J., Sharpe, J., and French, P. M. W., Fluorescence lifetime optical projection tomography, *Journal of Biophotonics*, 1, 390–394, 2008.

172. Kak, A. C. and Slaney, M., *Principles of Computerized Tomographic Imaging*. IEEE Press, New York, 1988.

173. McGinty, J., Talbot, C., Owen, D., Grant, D., Kumar, S., Galletly, N., Treanor, B. et al., Fluorescence lifetime imaging microscopy, endoscopy and tomography, In: Boas, D., Pitris, C., and Ramanujam, N., eds., *Handbook of Biomedical Optics*. CRC Press, Boca Raton, FL, pp. 589–615, 2011.

174. Soloviev, V. and Arridge, S. A., Fluorescence lifetime optical tomography in weakly scattering media in the presence of highly scattering inclusions, *Journal of the Optical Society of America A*, 28, 1513–1523, 2011.

175. American Cancer Society, *Cancer Facts and Figures 2006*. American Cancer Society, Atlanta, GA, 2006.

176. Wagnieres, G. A., Star, W. M., and Wilson, B. C., In vivo fluorescence spectroscopy and imaging for oncological applications, *Photochemistry and Photobiology*, 68, 603–632, 1998.

177. Weber, C. R., Schwarz, R. A., Atkinson, E. N., Cox, D. D., MacAulay, C., Follen, M., and Richards-Kortum, R., Model-based analysis of reflectance and fluorescence spectra for in vivo detection of cervical dysplasia and cancer, *Journal of Biomedical Optics*, 13, 064016, 2008.

178. Bard, M. P. L., Amelink, A., Skurichina, M., den Bakker, M., Burgers, S. A., van Meerbeeck, J. P., Duin, R. P. W., Aerts, J. G. J. V., Hoogsteden, H. C., and Sterenborg, H. J. C. M., Improving the specificity of fluorescence bronchoscopy for the analysis of neoplastic lesions of the bronchial tree by combination with optical spectroscopy: Preliminary communication, *Lung Cancer*, 47, 41–47, 2005.

179. Beamis, Jr., J. F., Ernst, A., Simoff, M., Yung, R., and Mathur, P., A multicenter study comparing autofluorescence bronchoscopy to white light bronchoscopy using a non-laser light stimulation system, *Chest*, 125, 148S–149S, 2004.

180. Ohkawa, A., Miwa, H., Namihisa, A., Kobayashi, O., Nakaniwa, N., Ohkusa, T., Ogihara, T., and Sato, N., Diagnostic performance of light-induced fluorescence endoscopy for gastric neoplasms, *Endoscopy*, 36, 515–521, 2004.

181. DaCosta, R. S., Wilson, B. C., and Marcon, N. E., New optical technologies for earlier endoscopic diagnosis of premalignant gastrointestinal lesions, *Journal of Gastroenterology and Hepatology*, 17(Suppl), S85–S104, 2002.

182. Pfefer, T. J., Paithankar, D. Y., Poneros, J. M., Schomacker, K. T., and Nishioka, N. S., Temporally and spectrally resolved fluorescence spectroscopy for the detection of high grade dysplasia in Barrett's esophagus, *Lasers in Surgery and Medicine*, 32, 10–16, 2003.

183. Marcu, L., Jo, J. A., Butte, P. V., Yong, W. H., Pikul, B. K., Black, K. L., and Thompson, R. C., Fluorescence lifetime spectroscopy of glioblastoma multiforme, *Photochemistry and Photobiology*, 80, 98–103, 2004.

184. Butte, P. V., Fang, Q., Jo, J. A., Yong, W. H., Pikul, B. K., Black, K. L., and Marcu, L., Intraoperative delineation of primary brain tumors using time-resolved fluorescence spectroscopy, *Journal of Biomedical Optics*, 15, 027008, 2010.

185. Chen, H. M., Chiang, C. P., You, C., Hsiao, T. C., and Wang, C. Y., Time-resolved autofluorescence spectroscopy for classifying normal and premalignant oral tissues, *Lasers in Surgery and Medicine*, 37, 37–45, 2005.

186. Uehlinger, P., Gabrecht, T., Glanzmann, T., Ballini, J.-P., Radu, A., Andrejevic, S., Monnier, P., and Wagnieres, G., In vivo time-resolved spectroscopy of the human bronchial early cancer autofluorescence, *Journal of Biomedical Optics*, 14, Article No.: 024011, 2009.

187. Pradhan, A., Das, B. B., Yoo, K. M., Cleary, J., Prudente, R., Celmer, E., and Alfano, R. R., Time-resolved UV photoexcited fluorescence kinetics from malignant and non-malignant human breast tissues, *Lasers in the Life Sciences*, 4, 225–234, 1992.

188. Jain, B., Majumder, S. K., and Gupta, P. K., Time resolved and steady state autofluorescence spectroscopy of normal and malignant human breast tissue, *Lasers in the Life Sciences*, 8, 163–173, 1998.

189. Stavridi, M., Marmarelis, V. Z., and Grundfest, W. S., Spectro-temporal studies of Xe-Cl excimer laser-induced arterial-wall fluorescence, *Medical Engineering & Physics*, 17(8), 595–601, 1995.

190. Marcu, L., Fluorescence lifetime in cardiovascular diagnostics, *Journal of Biomedical Optics*, 15, 011106, 2010.

191. Marcu, L., Jo, J. A., Fang, Q., Papaioannou, T., Reil, T., Qiao, J. H., Baker, J. D., Freischlag, J. A., and Fishbein, M. C., Detection of rupture-prone atherosclerotic plaques by time-resolved laser-induced fluorescence spectroscopy, *Arteriosclerosis (Dallas)*, 204, 156–164, 2009.

192. Kantelhardt, S. R., Leppert, J., Krajewski, J., Petkus, N., Reusche, E., Tronnier, V. M., Huttmann, G., and Giese, A., Imaging of brain and brain tumor specimens by time-resolved multiphoton excitation microscopy ex vivo, *Neuro-Oncology*, 9, 103–112, 2007.

193. Cicchi, R., Massi, D., Sestini, S., Carli, P., De Giorgi, V., Lotti, T., and Pavone, F. S., Multidimensional non-linear laser imaging of basal cell carcinoma, *Optics Express*, 15, 10135–10148, 2007.

194. Galletly, N. P., McGinty, J., Dunsby, C., Teixeira, F., Requejo-Isidro, J., Munro, I., Elson, D. S., Neil, M. A. A., Chu, A. C., French, P. M. W., and Stamp, G. W., Fluorescence lifetime imaging distinguishes basal cell carcinoma from surrounding uninvolved skin, *British Journal of Dermatology*, 159, 152–161, 2008.

195. Elson, D. S., Galletly, N., Talbot, C., Requejo-Isidro, J., McGinty, J., Dunsby, C., Lanigan, P. M. P. et al., Multidimensional fluorescence imaging applied to biological tissue. In: Geddes, C. D. and Lakowicz, J. R., eds., *Reviews in Fluorescence 2006*. Springer Science, New York, pp. 477–524, 2006.

196. Provenzano, P. P., Inman, D. R., Eliceiri, K. W., Knittel, J. G., Yan, L., Rueden, C. T., White, J. G., and Keely, P. J., Collagen density promotes mammary tumor initiation and progression, *BMC Medicine*, 6, 15, 2008.

197. Skala, M. C., Riching, K. M., Gendron-Fitzpatrick, A., Eickhoff, J., Eliceiri, K. W., White, J. G., and Ramanujam, N., In vivo multiphoton microscopy of NADH and FAD redox states, fluorescence lifetimes, and cellular morphology in precancerous epithelia, *PNAS*, 104, 19494–19499, 2007.

198. König, K., Clinical multiphoton tomography, *Journal of Biophotonics*, 1, 13–23, 2008.

199. König, K. and Riemann, I., High-resolution multiphoton tomography of human skin with subcellular spatial resolution and picosecond time resolution, *Journal of Biomedical Optics*, 8, 432–439, 2003.

200. Talbot, C. B., Patalay, R., Munro, I., Warren, S., Ratto, F., Matteini, P., Pini, R. et al., Application of ultrafast gold luminescence to measuring the instrument response function for multispectral multi-photon fluorescence lifetime imaging, *Optics Express*, 19, 13848–13861, 2011.

201. Patalay, R., Talbot, C., Alexandrov, Y., Lenz, M. O., Kumar, S., Warren, S., Munro, I. et al., Multiphoton multispectral fluorescence lifetime tomography for the evaluation of basal cell carcinomas, *PLoS ONE*, 7(9), e43460, 2012.

202. Tadrous, P. J., Siegel, J., French, P. M., Shousha, S., Lalani, E. N., and Stamp, G. W., Fluorescence lifetime imaging of unstained tissues: Early results in human breast cancer, *The Journal of Pathology*, 199, 309–317, 2003.

203. Elson, D., Requejo-Isidro, J., Munro, I., Reavell, F., Siegel, J., Suhling, K., Tadrous, P. et al., Time-domain fluorescence lifetime imaging applied to biological tissue, *Photochemical & Photobiological Science*, 3, 795–801, 2004.

204. Talbot, C., McGinty, J., McGhee, E., Owen, D., Grant, D., Kumar, S., De Beule, P. et al., Fluorescence lifetime imaging and metrology for biomedicine, Chapter 6. In: Tuchin, V., ed., *Handbook of Photonics for Biomedical Science*. CRC Press, Boca Raton, FL, pp. 159–196, 2010.

205. Sun, Y., Chaudhari, A. J., Lam, M., Xie, H., Yankelevich, D. R., Phipps, J., Liu, J., Fishbein, M. C., Cannata, J. M., Shung, K. K., and Marcu, L., Multimodal characterization of compositional, structural and functional features of human atherosclerotic plaques, *Biomedical Optics Express*, 2, 2288–2298, 2011.

206. McGinty, J., Galletly, N. P., Dunsby, C., Munro, I., Elson, D. S., Requejo-Isidro, J., Cohen, P. et al., Wide-field fluorescence lifetime imaging of cancer, *Biomedical Optics Express*, 1, 627–640, 2010.

207. Sun, Y., Hatami, N., Yee, M., Phipps, J., Elson, D. S., Gorin, F., Schrot, R. J., and Marcu, L., Fluorescence lifetime imaging microscopy for brain tumor image-guided surgery: Fluorescence lifetime imaging microscopy for brain tumor image-guided surgery, *Journal of Biomedical Optics*, 15, 056022, 2010.

208. Esposito, A., Oggier, T., Gerritsen, H. C., Lustenberger, F., and Wouters, F. S., All-solid-state lock-in imaging for wide-field fluorescence lifetime sensing, *Optics Express*, 13, 9812–9821, 2005.

209. Li, D.-U., Arlt, J., Richardson, J., Walker, R., Buts, A., Stoppa, D., Charbon, E., and Henderson, R., Real-time fluorescence lifetime imaging system with a 32×32 0.13 µm CMOS low dark-count single-photon avalanche diode array, *Optics Express*, 18, 10257–10269, 2010.

19
Speckle Correlometry

Dmitry A.
Zimnyakov
Yuri Gagarin State Technical University of Saratov

Valery V. Tuchin
Saratov State University
and
Russian Academy of Science
and
University of Oulu

19.1 Introduction

Scattering of coherent light by spatially inhomogeneous disordered media and rough surfaces leads to the formation of stochastic spatial distributions of the scattered light intensity, or *speckle* patterns. These patterns have a specific granular structure that results from the random interference of a great number of partial waves scattered by bulk or surface inhomogeneities (see Figure 19.1). The stochastic nature of scattering systems causes random phase shifts between these waves and results in statistical properties of the amplitude, phase, and intensity of speckle-modulated scattered fields.

Historically, attempts to understand how properties of observed speckles are related to the structural characteristics of a scattering system and observation conditions began with pioneer studies of laser light–matter interaction. The classic works dedicated to analysis of the statistical properties of laser speckles were made by Goodman, Dainty, Pedersen, Jakeman, Pusey, Asakura, Ioshimura, and many other researchers in the 1960s, 1970s, and early 1980s (see, for example, Refs. [1–14]). In these works, the concepts of speckle-pattern formation were based in general on the scalar approach for description of the random interference of partial waves scattered by the small-scale inhomogeneities of a scattering medium. Such an assumption leads to certain limitations in description of speckle formation by various scattering systems (especially in the case of bulk multiple scattering, where a significant change in the polarization state of propagating light occurs) but, despite these restrictions, the theory of speckle-pattern formation based on the scalar wave approach has allowed a rigid interpretation or prediction of a large number of experimentally observed phenomena. Moreover, the scalar theory can be applied for interpretation of the statistical properties of multiply scattered speckle patterns for certain observation conditions (in particular, with use of the polarization discrimination of scattered field partial components).

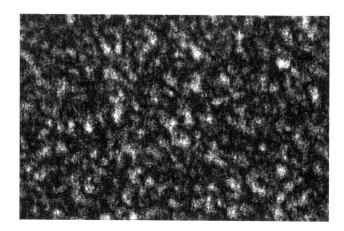

FIGURE 19.1   Speckle pattern induced in the case of laser beam scattering by a layer of human epidermis (in vitro sample).

Beginning in the mid-1980s, further development of speckle science was related to the establishment of certain analogies between classical waves propagating in disordered media and quantum-mechanical behavior of electrons forced by potential with random spatial fluctuations. The pioneering studies of interference effects were conducted by Golubentsev, Stephen, John, MacKintosh and other researchers[15-18] in the area of light propagation in random multiply scattering media. The classic examples of these effects are the manifestation of the weak localization of light in the form of coherent backscattering[19-21] or the existence of long-range correlations of scattered stochastic optical fields.[22,23] The fundamental phenomena accompanying coherent light propagation in disordered and weakly ordered systems and related to the statistical properties of multiply scattered speckles are now the object of significant research interest from many research groups. These studies resulted in modern diffusing light technologies,[24-28] widely applied in material science, biology, and medicine for probing and imaging optically dense scattering media.

Speckle patterns form when coherent light is scattered by living tissue or in vitro tissue samples and is sometimes termed biospeckles, or biospeckle phenomena. However, there is no noticeable difference between the general properties of biospeckles and the properties of speckles induced when light is scattered by a weakly ordered nonbiological system characterized by the same optical properties as the scattering tissue. The possibility of obtaining information about structure and dynamic properties of tissues by using statistical or correlation analysis of biospeckles for certain illumination and detection conditions has initiated an abundance of theoretical and experimental works dedicated to various applications of speckle technologies in biomedicine. Various applications of speckle methods in medical diagnostics (especially in the case of bioflow monitoring) are now possible, due to tremendous work by Stern, Fercher, Briers, Asakura, Aizu, Ruth, Nilsson, and many other researchers in the developing theoretical and experimental bases for biospeckle technologies.[29-42]

This chapter presents a brief review of the basic principles of speckle technologies in the context of potential applications for medical diagnostics. Typical examples of such applications are also discussed.

19.2 Statistical Properties of Speckles: Basic Principles and Results

19.2.1 First-Order Speckle Statistics

Speckle patterns result from the random interference of a great number of statistically independent waves scattered by inhomogeneities inside the scattering volume. Therefore, they may be characterized by amplitude, phase, and intensity, the values of which vary randomly from one detection point to another in the scattered optical field. The probability of determining these parameters for an arbitrarily chosen detection point within the given ranges of their possible values is characterized by the corresponding probability density functions. Knowledge of the probability density distributions of scattered-field characteristics for a given detector position allows one to evaluate the statistical moments, such as mean values and variances, for these characteristics and thus obtain the description of the first-order statistics of the observed speckle pattern. Only the intensity fluctuations can be analyzed in a conventional speckle experiment carried out without any specific interference technique. Thus, the study of the first-order statistics of speckle intensity for known illumination and observation conditions can be considered the basis for certain types of speckle technologies.

The simplest and best-known case of the first-order statistics of speckles corresponds to so-called "fully developed speckles" and can be obtained in terms of the scalar wave approach.[1-4] Let us consider a stochastic field induced by single scattering of a plane monochromatic wave by an ensemble of N statistically independent scatters. For example, if an incident collimated laser beam undergoes random spatial modulation of phase (e.g., by transmittance through a thin, inhomogeneous nonabsorbing layer in which the thickness or refractive index varies from point to point in a stochastic manner and in which the characteristic length of these variations is much larger than the wavelength used*), the number of independent scatters inside the illuminated area can be estimated as $N \approx (D/d)^2$, where D is the illuminating beam aperture and d is the correlation length of the phase fluctuations of boundary field associated with the spatial distribution of the complex amplitude of the transmitted light immediately behind the scattering object. In this case, the scattered field in an arbitrarily chosen point within the diffraction zone can be considered as the sum of partial waves scattered by statistically independent scattering sites:

$$E(\bar{r},t) = \sum_{i=1}^{N} E_i \exp\left[j\left(\omega t - \phi_i(\bar{r})\right)\right], \qquad (19.1)$$

where ω is the frequency of incident light and statistical properties of the ensemble $\phi_i(\bar{r})$ of phase shifts for each partial wave are determined by structural characteristics of the scattering object and observation conditions. In the case of a large N and

$$\sigma_\phi^2 = \left\langle \left(\phi_i - \langle\phi\rangle\right)^2 \right\rangle \gg 1,$$

the mean value of the scattered field amplitude is very close to zero and the induced speckle pattern is classified as the fully developed one. For such speckles, the probability density distributions of the real and imaginary parts of the scattered-field complex amplitude have a Gaussian form derived from the central limit theorem.[44] The probability density of the field phase is characterized by a distribution that is very close to the uniform one. Correspondingly, the intensity probability density for such fully developed speckles has the well-known negative exponential form

$$\rho(I) = \frac{1}{\langle I\rangle}\exp\left(-\frac{I}{\langle I\rangle}\right). \qquad (19.2)$$

A manifestation of such exponential statistics is the high visibility of fully developed speckle patterns. This property is manifested as the unit value of the speckle contrast, which is estimated as the ratio of the standard deviation of speckle intensity fluctuations to the mean value of speckle intensity:

$$C = \frac{\sigma_I}{\langle I\rangle} = 1.$$

---

* This scatter model is usually determined as the random phase screen (RPS) and, with certain limitations, allows the description of scattered field formation in terms of scalar diffraction theory (see Ref. [43]).

Another intrinsic property of intensity fluctuations for fully developed speckle patterns is the relation between high-order and first-order statistical moments of speckle intensity,

$$\frac{\langle I^n \rangle}{\langle I \rangle^n} = n!, \tag{19.3}$$

which can easily be obtained by calculating the statistical moments

$$\langle I^n \rangle = \int_0^\infty I^n \rho(I) di$$

using the probability density given by Equation 19.2. In the case of fully developed speckle patterns, the first-order statistics of speckle-intensity fluctuations are independent of the structural properties of the scattering object and thus cannot be used for its characterization.

This situation dramatically changes with transition from the fully developed speckle patterns to the partially developed speckles. This transition can be caused by a change in scattering or illumination and observation conditions. In particular, a decrease in the effective number of scattering sites N causes the divergence of the speckle-amplitude statistics from the Gaussian statistics; this divergence leads to formation of partially developed speckles in the diffraction zone. If deep phase modulation of an incident beam by a scattering system takes place $\left(\sigma_\phi^2 \gg 1\right)$ the contrast of partially developed speckles may significantly exceed 1; in many cases the dependence of C on the effective number of scattering sites may be approximated by the following relation:

$$C \approx \sqrt{1 + \frac{K}{N}}, \tag{19.4}$$

where the dimensionless parameter K is determined by properties of a scattering system. Analytical forms of K for various scattering systems with small numbers of N were reviewed by Jakeman.[7] In particular, for a scattering system with the Poisson-distributed number of identical scatters into a scattering volume, the parameter K is equal to 1. An ensemble of identical spherical particles that move into and out of a scattering volume can be considered as an example of such a scattering system.

Another type of scattering system that causes non-Gaussian speckles is a scattering surface consisting of equal-sized facets with statistically independent slopes. The speckle patterns produced by this scattering system are characterized by the following dependence of normalized second-order moment-of-intensity fluctuations on system parameters:

$$\frac{\langle I^2 \rangle}{\langle I \rangle^2} = 2\left(1 - N^{-1}\right) + \left[\frac{k^2 \xi^2}{4\pi P(\theta)}\right] N^{-1}, \tag{19.5}$$

where
   k is the wave number of light
   N is the number of facets inside the illuminated area
   ξ is the facet radius
   $P(\theta)$ is the probability of finding a facet facing the detector

It is easy to see that this expression may be reduced to a form similar to Equation 19.4 with K depending on k, ξ, and $P(\theta)$.

If a weak phase modulation of the illuminating beam occurs $\left(\sigma_\phi^2 \leq 1\right)$ then the dependence of the speckle contrast on N has a nonmonotonic character with the expressed maximum. The value of C

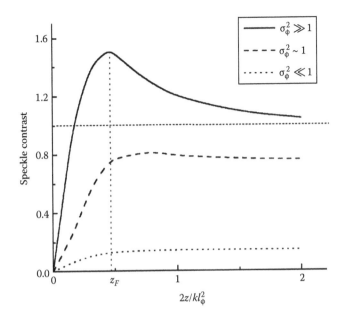

FIGURE 19.2 Typical behavior of the speckle contrast in the near-field diffraction zone in the dependence on the depth of incident wave phase modulation; l_ϕ is the correlation length of the boundary field phase fluctuations; z is the distance between scatter and observation plane.

falls to 0 with N diminishing to 0. In the case of a large, increasing N, the contrast value asymptotically approaches magnitudes that depend on the standard deviation σ_ϕ of phase fluctuations and is less than 1. In particular, theoretical analysis of similar behavior of first-order speckle statistics was given by Escamilla[45] for the case of a random phase screen with a finite correlation length of phase fluctuations and that is illuminated by the bounded monochromatic collimated beam.

Similar effects can be obtained in the case of illumination of a random phase-modulating object by a plane wave if the scattered field is observed in the near-field zone.[43] Figure 19.2 illustrates the typical behavior of the speckle contrast in the Fresnel diffraction zone for these conditions. If the illuminating wave undergoes strong phase modulation (phase fluctuations of the boundary field are characterized by $\sigma_\phi > 1$), the dependencies of C on the detection point position z behind a scattering object exhibit the nonmonotonic behavior with expressed maxima located at certain values of z_F and asymptotic decay to 1 with increasing z. Scatter parameters such as σ_ϕ and correlation length of boundary field phase fluctuations can be estimated by measuring the values of z_F and C_F.[43]

Sharp increase of the speckle contrast in the vicinity of z_F can be explained by a manifestation of the "focusing" effect. Scattering of a plane wave by an inhomogeneous medium causes the formation of a boundary field with wavefront inhomogeneities; the development of these phase inhomogeneities in the course of scattered field propagation at relatively short distances from the object leads to the appearance of a random network of caustics formed within the focusing zone behind the scattering object by structural inhomogeneities as the set of stochastically distributed small-sized lenses. For $z > z_F$, the contributing fields propagating from different inhomogeneities will overlap; this leads to interference effects. Finally, for $z > z_F$, the fully developed speckle pattern appears.

Non-Gaussian statistics of the scattered field amplitude fluctuations in the case of a small number of scatters are manifested in the peculiarities of the probability density distributions of speckle-intensity fluctuations such as, for example, the appearance of probability density bimodality for certain illumination and detection conditions.[46]

One specific case is speckle formation in the diffraction or near-field zone caused by random scatterers that exhibit the self-similarity property[47] at different length scales (fractal-like scatterers). In particular, if speckles appear when an illuminating beam is scattered by a rough surface with fractal properties, two different regimes can be considered[48]:

1. Scattering by the surfaces with fractal distribution of the relative heights measured from the baseline $D_h(\bar{r}) = \langle\{h(0) - h(\bar{r})\}^2\rangle \sim |\bar{r}|^\nu$, where the exponent ν of the surface height structure function $D_h(\bar{r})$ is related to a specific parameter of the fractal surface, such as its fractal dimension[46]; this relation can be written as $\nu = 2(2 - D)$. Theoretical estimations of the second-order statistical moment of scattered light intensity fluctuations for a scattering model made for the far diffraction zone (i.e., at distances from the scattering object that significantly exceed the ratio W^2/λ, where W is the characteristic size of illuminated area and λ is the wavelength used) as well as for the near diffraction zone (i.e., at distances that are comparable or less, W^2/λ) gives the value of $\langle I^2\rangle/\langle I\rangle^2$ close to 2, as for conventional cases of Gaussian speckles (see, for example, Refs. [1–4]).

2. Scattering by subfractal surfaces, which are characterized by fractal distributions of the local surface slope. In this case, formation of a scattered field can be described in terms of ray optics, and simple consideration leads to the power-law dependence of the normalized second-order statistical moment of scattered intensity on the illuminated area:

$$\frac{\langle I^2\rangle}{\langle I\rangle^2} \sim W^{2-\nu}, \tag{19.6}$$

where ν is the exponent of the structure function $D_s(\bar{r}) = \langle\{s(0) - s(\bar{r})\}^2\rangle^\nu$, which characterizes the properties of distribution of the surface local slope $s(\bar{r})$.

Statistical analysis of partially developed speckle patterns in the near or far diffraction zone can be applied in comparative study of the morphological differences between diseased tissues and healthy tissues. Figure 19.3 illustrates the typical shapes of histograms that characterize intensity probability distributions for partially developed speckle patterns induced by probe-light scattering in thin layers of healthy and diseased (psoriasis) human epidermis.[49]

Experiments with in vitro epidermis samples prepared as skin strips were carried out with the focused beam of an He–Ne laser; speckle patterns were caused by random interference of the forward-scattered partial components and were analyzed in the diffraction zone behind the sample. Intensity fluctuations

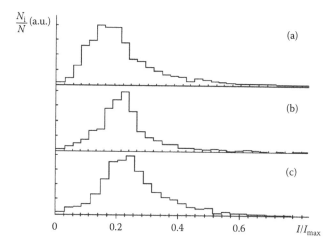

FIGURE 19.3 Histograms of intensity distributions for partially developed speckle patterns induced by focused laser beam scattering in healthy (a) and diseased (b and c, psoriasis) human epidermis layers. The thickness of the layers is about 30–40 μm. The diameter of the light spot on the sample surface is approximately 100 μm.

at the detection point were induced by the lateral scanning of the probed sample with respect to the illuminating beam. The characteristic changes in epidermal structure caused by the progress of psoriasis (e.g., the appearance of parakeratosis focuses, desquamation, and some other phenomena) are manifested as the noticeable distortion of intensity histograms and, as a result, a decrease of the contrast value for observed speckle patterns.

19.2.2 Second-Order Speckle Statistics

The statistical properties of a scattered field can also be characterized by simultaneous analysis of the correlation of the complex amplitude values for two spatially separated observation points and for different moments of time. In this way, the spatial–temporal correlation function of scattered-field fluctuations is introduced as follows:

$$G_1\left(\bar{r}_1,t_1,\bar{r}_2,t_2\right)=\left\langle E_1 E_2^*\right\rangle=\left\langle E\left(\bar{r}_1,t_1\right)E^*\left(\bar{r}_2,t_2\right)\right\rangle,\qquad(19.7)$$

where the symbol * denotes complex conjugation. For many cases, the spatial–temporal fluctuations of scattered-field amplitude can be considered the stationary random fields; this leads to the following form of $G_1\left(\bar{r}_1,t_1,\bar{r}_2,t_2\right)$

$$G_1\left(\bar{r}_1,t_1,\bar{r}_2,t_2\right)=G_1\left(\bar{r}_1,\bar{r}_2,t_1-t_2\right)=G_1\left(\bar{r}_1,\bar{r}_2,\tau\right).\qquad(19.8)$$

In a similar manner, the spatial–temporal correlation function of scattered light intensity fluctuations can be introduced:

$$G_2\left(\bar{r}_1,\bar{r}_2,\tau\right)=\left\langle I_1 I_2\right\rangle=\left\langle I\left(\bar{r}_1,t\right)I\left(\bar{r}_2,t+\Delta\tau\right)\right\rangle.\qquad(19.9)$$

Moreover, for statistically homogeneous speckle patterns, the field and intensity correlation functions depend only on $\Delta\bar{r}=\bar{r}_2-\bar{r}_1$

$$G_1\left(\bar{r}_1,\bar{r}_2,\tau\right)=G_1\left(\Delta\bar{r},\tau\right);\quad G_2\left(\bar{r}_1,\bar{r}_2,\tau\right)=G_2\left(\Delta\bar{r},\tau\right).\qquad(19.10)$$

If a scattered optical field is characterized by the Gaussian statistics of a complex amplitude that has zero mean value, then the normalized correlation functions of amplitude and intensity fluctuations $g_2(\Delta\bar{r},\tau)=\langle I_1 I_2\rangle/\langle I_1\rangle\langle I_2\rangle$ and, correspondingly, $g_1(\Delta\bar{r},\tau)=\left\langle E_1 E_2^*\right\rangle/((|E_1|^2)(|E_2|^2))^{1/2}$ are related to each other as follows[9,43]:

$$g_2\left(\Delta\bar{r},\tau\right)=1+\left|g_1\left(\Delta\bar{r},\tau\right)\right|^2.\qquad(19.11)$$

This is the well-known Siegert relation.

Three types of correlation measurements are typically considered:

- Correlations between spatially separated detectors measured with zero time delay; this type of measurement gives information about the typical size of a feature of a speckle intensity pattern that can be considered the average speckle size in the detection plane.
- Temporal intensity correlations measured at a single detection point; this type of measurement is the most typical for many practical applications and is therefore an object of further consideration.
- Simultaneous measurement of spatial–temporal correlations provided by changes in the positions of two detectors and by changes in the time delay τ.

Usually, the correlation analysis of speckle intensity fluctuations is used to study a dynamic behavior of scattering media as well as their structural characteristics; in the latter case, the dependence of speckle intensity on time may be caused by controllable motion (e.g., lateral displacement) of the probed medium with respect to the illuminating laser beam. The relationship between the dynamic properties of the scattering medium and time-dependent behavior of speckle-intensity fluctuations sufficiently depends on observation conditions. In particular, three specific zones of observation in scattered optical fields induced by probe light scattering by inhomogeneous objects with properties of the random phase screen can be mentioned:

- The near-field region in vicinity of the scattering object ($z \ll z_F$); in this case, each point of the observed speckle pattern is associated with a certain region of the phase front of the boundary field emerging from the scattering object. For the case of strong phase modulation of a probe beam by the scattering object, the average size of this region is significantly less than the phase correlation length that characterizes the scattering object as the random phase screen. Thus we should expect the temporal evolution of the observed speckle pattern to reflect the time-dependent dynamics of local fluctuations of the boundary field phase. In its turn, this dynamic is determined by local motions of scattering sites.
- The "focusing" region with $z \sim z_F$; for this detection condition, the intensity in each point of the speckle pattern is the result of superposition of partial waves coming from local areas inside the region of the boundary field phase front with a characteristic dimension of the order of ξ. Local changes in time-dependent spatial distributions of speckle intensity depend on the relative motions of spatially separated regions of the scattering object. Therefore, the relationship between the time-dependent correlation decay of intensity fluctuations and the dynamic properties of a scattering system has a more complicated form than in the case of near-field detection.
- The far field region with $z \gg z_F$; in this case, each point of the detection plane receives all contributions coming from the whole scattering region of the probed object.

The influence of detection conditions on the correlation decay of speckle intensity fluctuations is evident for speckle patterns that result from laser light scattering by moving a spatially heterogeneous "frozen" object that exhibits no mutual motions of local scattering sites and that displaces with respect to the illuminating beam as a whole. In this case, the specific property of the time-dependent spatial distributions of speckle intensity is the existence of two different types of speckle pattern evolution during the observation time; these are associated with two types of speckle motions: "boiling" motion and "translational" motion.

In the common case, the dynamic speckles translate while changing their structure and a decorrelation of intensity fluctuations occurs because of the mutual effect of the displacement and deformation of speckles. The difference between these types of speckle pattern evolution, which is intuitively understandable, can be explained in terms of an approach suggested by Yoshimura[13] and Yoshimura et al.[14] In particular, for a scattering object with properties of the random phase screen characterized by the Gaussian statistics of boundary field phase fluctuations and by the Gaussian-like form of their spatial correlation function, the normalized spatial–temporal correlation function of speckle intensity fluctuations can be expressed as

$$g_2(\Delta \bar{r}, \tau) = 1 + \exp\left(-\frac{|\Delta \bar{r}|^2}{r_s^2} + \frac{\tau_d^2}{\tau_s^2}\right) \exp\left(-\frac{(\tau - \tau_d)^2}{\tau_s^2}\right), \tag{19.12}$$

where the parameters r_s, τ_s, and τ_d are determined by the illumination and detection conditions as well as by the scattering object velocity. If intensity fluctuations are detected with two spatially separated detectors with zero time delay, then the parameter r_s can be associated with the average size of the speckles. In order to characterize speckle motion, two additional parameters, such as the correlation distance, r_c, and translation distance, r_T, can be introduced. With used notations and following Yoshimura's considerations, these parameters may be introduced as:

$$r_c = \left(\frac{1}{r_s^2} - \frac{\tau_c^2 V_s^2}{r_s^4} \right)^{-1/2} ; \quad r_T = V_s \left(\frac{1}{\tau_c^2} - \frac{V_s^2}{r_s^2} \right)^{-1/2},$$ (19.13)

where V_s is the speckle velocity. If the absolute value of ratio r_T/r_s is less than 1, then the boiling motion is dominant.

In the opposite case of $|r_T/r_T| > 1$, we can observe the more expressed translation motion of speckles. In the particular case of Gaussian beam illumination of in-plane moving scatter and free-space formation of a speckle field, this consideration leads to the relation

$$\frac{r_T}{r_s} = \frac{\{(l+z)l/a + a\}}{z},$$ (19.14)

where
  z is the distance between the scattering object and the observation plane
  l is the distance between the object and the waist plane
  a is the Rayleigh range parameter of illuminating beam evaluated as $\pi w_0^2/\lambda$ where w_0 is the waist radius

It is easy to see that if $l = 0$ (the moving object is positioned in the waist plane), then speckles observed near the scattering object ($z < a$) exhibit the dominating translation motion; in contrast, far-zone speckles are characterized by the boiling-like motion with a vanishing translation component.

The most typical speckle-correlation measurement applied to study the dynamic behavior of various scattering systems (e.g., living tissues) is single-point correlation analysis, or measurement of the time-dependent correlation decay of speckle intensity fluctuations. The basic principles of such measurements are discussed in the next section.

19.3 Temporal Correlation Analysis of Speckle Intensity Fluctuations as the Tool for Scattering Media Diagnostics

19.3.1 Single-Scattering Systems

For a single-scattering system consisting of moving scattering particles, the scattered field in the detection point can be expressed using a simple modification of Equation 19.1:

$$E(t) = \exp(j\omega t) \sum_{i=1}^{N} |E_i(t)| \exp\{-j\phi_i(t)\},$$ (19.15)

where the dependence of phase for each partial contribution on time is caused by scatter motions and therefore causes fluctuations of the scattered field amplitude in the detection point. Thus, the temporal correlation function of scattered field fluctuations in the detection point can be written as follows:

$$G_1(\tau) = \langle E(t)E^*(t+\tau)\rangle = \left\langle \exp(-j\omega\tau) \sum_{i=1}^{N}\sum_{i=1}^{N} |E_i(t)||E_i(t+\tau)| \exp\left[-j\{\phi_i(t)-\phi_i(t+\tau)\}\right]\right\rangle.$$

By using simplifying assumptions about the statistical independence of partial contributions scattered by different scattering sites and slow varying amplitudes of scattered waves, $|E_i(t) \approx E_i(t + \tau)|$, the temporal correlation function can be expressed as (see Ref. 9):

$$G_1(\tau) \approx \exp(-j\omega\tau)N\langle |E_i(t)|^2 \exp\{j\Delta\phi_i(\tau)\}\rangle. \tag{19.16}$$

Here we consider motions of scattering sites as the stationary and ergodic random process causing the increments of phase of interfering partial waves that depend only on the time delay value τ. For the zero time delay, the value of $G_1(0)$ is equal to the mean intensity of scattered light in the detection point: $G_1(0) = \langle I \rangle \approx N\langle |E_i|^2\rangle$. For simplicity, without loss of generality, we can assume the equality of contributions from each partial scattered wave and express the temporal correlation function as:

$$G_1(\tau) \approx \langle I \rangle \exp(-j\omega\tau)\langle \exp\{j\Delta\phi(\tau)\}\rangle. \tag{19.17}$$

Phase increment $\Delta\phi_i(\tau)$ for each partial wave can be expressed as $\Delta\phi_i(\tau) = \bar{q}_i\Delta\bar{r}_i(\tau)$, where $\bar{q}_i$ is the momentum transfer for each scattering event and $\Delta\bar{r}_i(\tau)$ describes the displacement of ith scattering site for the observation time τ. For the observation geometry presented in Figure 19.4, the momentum transfer module $|\bar{q}_i|$ for each scattering event can be presented as $|\bar{q}_i| \approx |\bar{q}| = (4\pi/\lambda)\sin(\theta/2)$, where θ is the scattering angle.

Finally, taking into account the property of a complex exponent with an argument as a combination of three independent random variables with the same statistical properties, the temporal correlation function of field fluctuations for single scattered dynamic speckles can be written as follows:

$$G_1(\tau) \approx \langle I \rangle \exp(-j\omega\tau)\exp\left\{-\left(4\pi\sin(\theta/2)/\lambda\right)^2 \Delta\bar{r}^2(\tau)/6\right\}, \tag{19.18}$$

where the displacement variance for scattering sites measured with the given time delay is determined by the dynamic properties of the scattering system. In particular, for ensembles of Brownian particles (this model is usual for description of the dynamical behavior of various biological systems), the value of $\Delta\bar{r}^2(\tau)$ is equal to $6D\tau$ (D is the self-diffusion coefficient of scattering particles).

In conventional schemes of scattering experiments with no use of a reference beam for detection of scattered light, the measurable parameter of a scattered optical field is the instantaneous value of

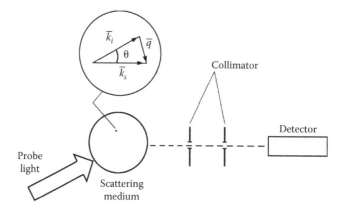

FIGURE 19.4 Typical scheme of the dynamic light scattering experiment: $\bar{k}_i$ = the wave vector of the incident wave; k_s = the wave vector of the detected scattered wave; θ = the scattering angle.

speckle intensity in the detection point. The detected intensity fluctuations are usually characterized by the intensity correlation function $G_2(\tau)$. For Gaussian scattered fields with zero mean amplitude, the normalized correlation function of intensity fluctuations $G_2(\tau) = G_2(\tau)/G_2(0)$ is related to the normalized correlation function $G_1(\tau)$ by the previously presented Siegert relation (see Equation 19.11):

$$g_2(\tau) = 1 + \beta \, | \, g_1(\tau) |^2, \tag{19.19}$$

where the parameter β depends on detection conditions. For ideal conditions (i.e., when the detector aperture is significantly smaller than the speckle size), $\beta = 1$.

Thus, for Brownian scattering systems the normalized correlation function of intensity fluctuations has the exponential form:

$$g_2(\tau) \approx 1 + \exp\left\{-2\left(\frac{4\pi \sin(\theta/2)}{\lambda}\right)^2 D\tau\right\}, \tag{19.20}$$

and the corresponding power spectrum has the Lorentzian form. The value of $\tau_0 = (4\pi^2 D/\lambda^2)^{-1} = (k^2 D)^{-1}$ is usually considered the correlation time of the field fluctuations. One of the traditional approaches to measure the properties of scattering particles (in particular, their diffusion coefficient) is the estimation of the correlation time of speckle intensity fluctuations for given scattering angle and wavelength of the probe light.

A specific case involves coherent light scattering by moving phase scatters with a fractal-like structure. In this case, the temporal intensity fluctuations detected in the fixed point of the far diffraction zone have the properties of the one-dimensional random fractal process and the corresponding fractal dimension of such a process is directly related to the fractal dimension of the scattering object (e.g., a fractal-like surface). The form of corresponding relationship depends on illumination conditions. In particular, use of a sharply focused illuminating beam causes increase of the fractal dimension of detected speckle intensity fluctuations in comparison with the fractal dimension of scattering object (the effect of "stochastization" of speckle intensity fluctuations).[50]

19.3.2 Multiple-Scattering Systems

The formation of multiply scattered dynamic speckles is a significantly more interesting case and is important for practical applications. The discrete scattering model (Equation 19.15) can be modified in order to describe dynamic speckle formation due to multiple light scattering in disordered systems; such an approach was discussed in the classic work by Maret and Wolf.[51] In this case, each partial contribution is considered as the result of sequence of n scattering events:

$$E_k(t) = \exp(j\omega t) \prod_{i=1}^{N_k} E_i \exp\left(-j\overline{q}_i \overline{r}_i(t)\right),$$

and the total scattered field in the detection point is expressed as follows:

$$E(t) = \sum_k E_k(t).$$

In further analysis, the *single-path* correlation function of field fluctuations is introduced as

$$G_2^k(\tau) = \left\langle E_k(t) E_k^*(t+\tau) \right\rangle \approx \exp(-j\omega\tau) \left\langle \prod_{i=1}^{N_k} |a_i|^2 \, \exp\left\{j\overline{q}_i \Delta\overline{r}(t)\right\} \right\rangle \tag{19.21}$$

$$\approx \exp(-j\omega\tau)\left\langle |a_i|^2\right\rangle \exp\left\{-\left\langle\overline{q}^2\right\rangle\left\langle\Delta\overline{r}^2(\tau)\right\rangle N_k/6\right\}.$$

For the discussed case, the mean value of $\bar{q}^2$ estimated for a sequence of scattering events can be expressed as $\langle \bar{q}^2 \rangle = 2k^2l/l^*$, where l is the scattering mean free path and l^* is the transport mean free path for the scattering medium. The number of scattering events for each partial contribution can be expressed as $N_k \approx s_k/l$, where s_k is the corresponding propagation path for kth partial component inside a scattering medium. Thus, the single-path correlation function has the following form:

$$G_1^k(\tau) \approx \exp(-j\omega\tau)\langle |a|^2 \rangle \exp\left\{-k^2\langle \Delta\bar{r}^2(\tau)\rangle s_k/3l^* \right\}. \tag{19.22}$$

The temporal correlation function of field fluctuations in the detection point $G_1(\tau) = \langle E(t)E^*(t+\tau)\rangle$ can be obtained by the statistical summation of the single-path correlation functions over the ensemble of partial contributions:

$$G_1(\tau) \approx \sum_k P(k)G_1^k(\tau),$$

where $P(k)$ are the statistical weights characterizing contributions of partial components to formation of a scattered field in a detection point. This expression may be modified for multiple scattering systems characterized by the continuous distribution of optical paths s by integration over the range of all possible values of s:

$$G_1(\tau) \approx \int_0^\infty \exp\left(-\frac{k^2\langle \Delta\bar{r}^2(\tau)\rangle s}{3l^*}\right)\tilde{\rho}(s)ds, \tag{19.23}$$

where $\tilde{\rho}(s)$ is the probability density of optical paths that obeys the following normalization condition: $\int_0^\infty \tilde{\rho}(s)ds = \langle I \rangle$. The normalized temporal correlation function can be introduced as $g_1(\tau) = \int_0^\infty \exp(-k^2\langle \Delta\bar{r}^2(\tau)\rangle s/3l^*)\rho(s)dx$ by using the following normalization condition: $\rho(s) = \tilde{\rho}(s)/\int_0^\infty \tilde{\rho}(s)ds$. In particular, for Brownian scattering systems, the argument of exponential kernel on the right-hand side of Equation 19.23 has the well-known form $2\tau s/\tau_0 l^*$, where τ_0 is the previously introduced correlation time of function of scattered field fluctuations for a single-scattering regime.

If a localized light source is used for illumination of a dynamic scattering system, then the correlation characteristics of scattered field fluctuations (e.g., the correlation time) strongly depend on observation conditions because of sensitivity of these characteristics to changes in path statistics characterized by variations of the pathlength density $\rho(s)$. This gives the opportunity for probing various spatially inhomogeneous dynamic media by the spatially separated light source and detector (for instance, by the probe system consisting of light-delivering and light-collecting optical fibers). Various examples of such probing technology are illustrated in Figure 19.5.

In this case, the probed scattering system consists of two layers—the superficial "static" layer consisting of motionless scatters, and underlying "dynamic" or "modulating" layer with moving scattering sites. Burned tissue with shallow necrotic layer with no blood microcirculation is a good example of a real scattering system with similar geometry. Below this layer, the living tissue is characterized by a sufficient level of blood microcirculation. For small source-detector separations, the depth of light penetration into the probed volume is small compared with the thickness of "static" layer; this scattering geometry causes the formation of dynamic speckles with large values of the correlation time. Increase of the penetration depth due to enlargement of the distance between source and detector fibers leads to faster decay of the field temporal correlation.

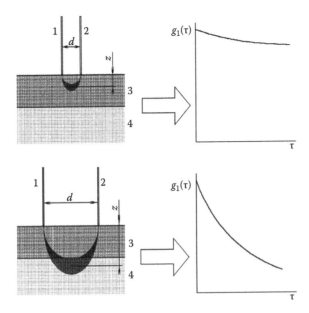

FIGURE 19.5 Location of the underlying dynamic layer (4) using measurements of the correlation decay of speckle amplitude fluctuations: 1, light-delivering fiber; 2, light-collecting fiber; and 3, superficial static layer. $g_1(\tau)$ is the normalized temporal correlation function of the scattered-field fluctuations.

Potential applications of this technique were demonstrated by Boas and Yodh for burn depth diagnostics.[52] In their experiment, an Ar-ion laser was used as an illumination source. Laser light was delivered to the probed tissue by a multimode optical fiber. Probe light scattered in the backward direction was collected by an eight-channel fiber-optic probe. The input tips of each single-mode light-collecting fiber were fixed along the detection line with 1.5 mm of separation between them. Output tips were connected with a photodetector unit through the 8×1 fiber-optic switcher. A photomultiplier tube (PMT) in the photon-counting mode was used as photodetector. Each sequence of photocounts was processed by digital correlator. This construction allowed the investigators to provide correlation analysis of speckle-intensity fluctuations for various distances between source and detector without mechanical scanning.

In vivo pig skin with thermally induced burn diseases of different levels was used to study the influence of the burned tissue thickness on parameters of normalized temporal correlation function $g_2(\tau)$ of intensity fluctuations at a given detection point. Burn diseases were induced by applying a heated metal block to the skin surface. Diseases of various levels with significantly differing thicknesses of necrotic layers can be obtained by varying the provocation time. For experimental conditions used in the study, the necrotic layer thickness was varied from ≈ 100 μm (shallow burn) to ≈ 1000 μm (deep burn, average level of disease).

The parameter most sensitive to changes of the burned tissue structure was found to be the slope of $\{g_2(\sqrt{\tau})\}$ dependencies obtained for different source-detector separations. For instance, this parameter slowly increases with an increase in source-detector separation up to the value of the order of the necrotic layer thickness. After the "critical" penetration depth comparable with the layer thickness is reached, the dramatic change in slope value occurs.

Correlation analysis of speckle intensity fluctuations associated with object scanning by the probe (for instance, as shown in Figure 19.5) can be applied for location and imaging of dynamic macro-inhomogeneities in multiple-scattering media. This possibility was demonstrated by Yodh et al.[26] in experiments with a phantom scattering object such as a resin cylinder (a static scatterer) with dynamic inclusions (a spherical cavity filled with a colloidal suspension). Titanium dioxide particles were added to resin in order to improve its scattering properties. Both scattering substances (the mixture of resin and TiO_2 particles and the colloidal suspension) were carefully prepared in order to provide equal values for the transport mean free path length. A set of data collected for different positions of a light source and a detector around the surface of the resin cylinder was used for reconstruction of the object image. The image reconstruction algorithm was based on the concept of field correlation transfer in turbid media.[53] Under these experimental conditions, this concept leads to a correlation diffusion equation similar to the light diffusion equation; the reconstructed image of the phantom object adequately characterizes the position, form, and dynamic properties of the spherical inclusion.

The possibility of recognizing various types of dynamic inhomogeneities hidden in a multiply scattering dynamic medium with use of correlation analysis of speckle intensity fluctuations was shown by Heckmeier et al.[54] They studied the dynamic light-scattering response of a multiply scattering system consisting of a Brownian medium with an embedded glass capillary in which the regular flow of a scattering substance (a suspension of polystyrene beads in water) was provided. The dependence of the decay of the temporal correlation of speckle intensity fluctuations on capillary position inside the Brownian scattering medium and flow rate was studied theoretically and experimentally. Regular flow was found to be reliably located at the depths on the order of the transport mean free path length.

Some applications of speckle correlometry techniques to in vitro tissue diagnostics are discussed in Refs. [55–59].

19.4 Angular Correlations of Multiply Scattered Light

The existence of long-range angular correlations, a fundamental property of optical fields multiply scattered by random media, was considered in terms of "angular memory" effect by Feng et al.[22] Berkovits and Feng[60] discussed the possibility of using this effect as the physical basis for tomographic imaging of optically dense disordered media. The relations between angular correlations of multiply scattered coherent light and optical properties of scattering media for the transmittance mode of light propagation were studied theoretically and experimentally by Hoover and co-workers.[61] In this study, the potential of using the angular correlation analysis for disordered scattering media characterization was investigated.

Also, Tuchin et al.[62] considered an original approach to this problem, based on the influence of angular correlation decay on an interference of optical fields induced by two illuminating coherent beams incoming in the scattering medium at different angles of incidence. In this case, the probed medium is illuminated by a spatially modulated laser beam formed by overlapping the two collimated beams. The spatial modulation of the resulting illuminating beam has the form of a regular interference pattern with the fringe spacing determined by the angle between the overlapping beams. In the absence of scattering, the angular spectra of incident beams have the Δ-like forms; the appearance of scattering causes broadening of these angular spectra and decay in the interference pattern contrast of the outgoing spatially modulated beam. Analysis of the interference pattern contrast for the outgoing beam and its dependence on the distance between the scatter and the observation plane and interference fringe period allows one to characterize the scattering properties of the probed medium.

19.5 Use of Time-Varying Speckle Contrast Analysis for Tissue Functional Diagnostics and Visualization

Analysis of speckle intensity fluctuations can be provided by acquisition of time sequences in the fixed detection point with further data processing. Another approach can be based on the statistical analysis of spatial fluctuations of time-averaged dynamic speckle patterns. Evaluation of the parameters of time-dependent intensity fluctuations (e.g., the correlation time) by use of this technique is possible in the case of ergodic dynamic speckle patterns. In this case, ensemble statistical averaging leads to the same results as temporal statistical averaging. Usually, statistical analysis of spatial fluctuations of speckle images captured with given exposure time can be carried out in the image or diffraction plane.

Two typical examples of biomedical applications of this technique can be considered; one is laser speckle contrast analysis, or the LASCA technique, developed by Fercher and Briers[33] as the tool for

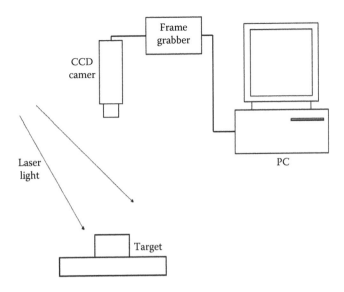

FIGURE 19.6 Basic setup for LASCA.

microcirculation analysis in upper layers of in vivo human tissues. The typical scheme of the LASCA device is very simple, as shown in Figure 19.6. The surface of probed tissue is illuminated by the broad collimated laser beam. A speckle-modulated image of the tissue surface is captured by a charge-coupled device (CCD) camera with given exposure time. In the case of expressed blood microcirculation in the upper layer of the probed tissue, the speckle pattern, which modulates the surface image, has a dynamic character due to random Doppler shifts of partial contributions scattered by moving erythrocytes. Temporal fluctuations of modulating speckles in the image plane cause the blurred captured image with a decayed value of contrast.

The contrast value for the time-averaged image depends on exposure time and falls when exposure increases. The following expression can be obtained for time-dependent contrast:

$$C(T) = \left(\frac{1}{T} \int_0^T \tilde{g}_2(\tau) d\tau \right)^{1/2}, \tag{19.24}$$

where

$\tilde{g}_2(\tau) = (G_2(\tau) - \langle I \rangle^2)/\langle I \rangle^2$

T is the exposure time

In particular, if the intensity correlation function exhibits the exponential decay: $\tilde{G}_2(\tau) = \langle I \rangle^2 \exp(-2\tau/\tau_c)$, where the correlation time $\tau_c = 1/akv$ depends on the mean velocity v of erythrocytes, wavenumber k of probe light and scattering properties of tissue described by the factor a, then the time-dependent speckle contrast and the correlation time are related through

$$C(T) = \left[\frac{\tau_c}{2T} \left\{ 1 - \exp\left(\frac{-2T}{\tau_c} \right) \right\} \right]^{1/2}. \tag{19.25}$$

If the analyzed tissue has the layered structure with significantly differing blood perfusion levels from layer to layer (e.g., human skin), then the average velocity, or mobility, of erythrocytes, which is evaluated by use of the LASCA technique, typically characterizes some average level of mobility for tissue volume with the thickness of the order of the depth of probe light penetration. This is the main disadvantage of speckle contrast analysis with broad collimated beam illumination, when a modulating speckle pattern is induced by random interference of partial components of a backscattered optical field coming from different depths of probed tissue.

In the case in which significant differences exist in the local perfusion level for various zones of the probed tissue area, the spatially inhomogeneous dynamic speckle pattern modulates the image of the tissue surface. For such an image, local estimations of exposure-dependent contrast can be used for characterization of the differences of average perfusion level from one zone of imaged surface to another. Estimated local values of the average mobility of erythrocytes can be used for reconstruction of "microcirculation maps." In the course of reconstruction of these maps, each pixel of the probed tissue surface is imaged by use of the color or gray-level scale related to the range of estimated erythrocyte mobility levels.

Briers et al.[63] demonstrated the potential of this technique for functional tissue imaging; more recent results in this field were reported by Dunn et al.[64] In this study, the LASCA technique was applied for dynamic imaging of cerebral blood flow (CBF). By illuminating the cortex with laser light and imaging the resulting speckle pattern, relative CBF images with tens of microns of spatial resolution and milliseconds of temporal resolution were obtained. This technique is easy to implement and can be used to monitor the spatial and temporal evolution of cerebral blood flow changes with high resolution in studies of cerebral pathophysiology.

The possibilities of using time-integrated speckle contrast analysis as well as speckle correlometry for studying the mechanical properties of tissue samples were demonstrated in recent works by Jacques and Kirkpatrick[65] and Duncan and Kirkpatrick.[66] This approach can be classified as the "speckle elastography of tissues."

Axial resolution in the case of time-dependent speckle contrast analysis in the image plane can be significantly improved by the use of a localized light source such as focused laser beam. Used for study of macroscopically homogeneous scattering media with thick slab geometry, this illumination scheme causes formation of statistically heterogeneous dynamic speckle patterns across the surface of the probed medium. This is because the average path of probe light propagation from the source points to the detection point strongly depends on the spatial separation between these points. The increasing distance from source and detection points causes spectral broadening of speckle intensity fluctuations from the center of the backscattered light spot on the surface of the probed medium to its edge, due to an increase in the average number of scattering events for partial contributions with large propagation paths.

The study of spatial distributions of correlation time for such statistically inhomogeneous speckle patterns allows one to obtain information not only about dynamic properties of the scattering medium, but also about conditions of probe light propagation inside the scattering volume. In particular, if the probe light travels in a layered scattering medium characterized by significant differences in dynamic properties from layer to layer, the analysis of time-averaged speckle-modulated images of the illuminated surface in the part of spatial distributions of local estimates of the time-dependent speckle contrast makes it possible to characterize the geometry of scattering system. A good example is the study of speckle contrast analysis potential for burn-depth diagnostics carried out by Sadhwani et al.[67] In this experiment, they used a burned tissue phantom composed of two layers—the static layer as the phantom of a superficial necrotic tissue layer and a modulating layer as the phantom of living tissue with an expressed microcirculation of blood (Figure 19.7).

Such scattering geometry in the case of a localized probe source leads to formation of two specific zones inside the backscattered light spot. Surrounding the source light spot, the first is characterized by the presence of a modulating static speckle pattern induced by scattering of probe light inside the static layer. The influence of partial contributions that reach the modulating layer and come back is negligibly small. The second zone, or area of dynamic speckle modulation, is located around the first. If the surface of the phantom scattering system is imaged with exposures significantly exceeding the typical value of correlation time of speckle intensity fluctuations, image for the second zone blurs with contrast value approaching 0. Thus, it is possible to estimate the thickness of the static ("necrotic") layer by measuring the inner radius of the blurred zone.

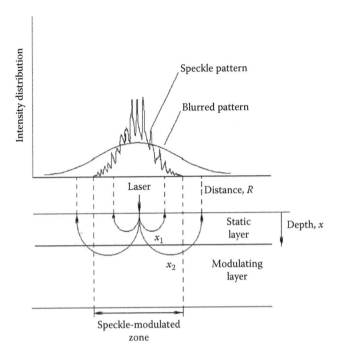

FIGURE 19.7 Formation of time-integrated speckle pattern image in the case of backward multiple scattering of the probe laser light by a two-layered medium. Photons that travel to depth x_1 in the static layer produce a speckle pattern upon returning to the surface, whereas photons that travel to depth x_2 in the modulated layer produce a time-averaged blurred pattern at the surface. The size of the speckle-modulated zone is determined by the thickness of the static layer.

The following phantom object was used for approbation of this technique. Teflon films of various thicknesses (from 0.13 to 1.3 mm) were applied as static layers. A lipid solution (Intralipid-10%) was used as the modulating medium. An He–Ne laser operating at 633 nm, used as a source of probe light, was focused to an 80 μm diameter spot on the Teflon film surface. The image of the illuminated surface was captured by CCD camera with a Nikkor 50 mm f/1.8 AF lens. The lens aperture diameter was set to minimum value (f/22) to maximize the average speckle size in the image plane. A polarizer was used between the CCD lens and the target in order to minimize the influence on image formation of specular reflection from the target surface.

Experiments with the burned tissue model have shown the strong correlation between the diameter of the nonblurred speckle pattern and the thickness of static layer. Because of the relative simplicity of the design, the inexpensive arrangement, and unsophisticated image-processing algorithm, this technique seems to be an adequate prospective tool for clinical applications.

The important question for various methods to monitor blood microcirculation on the basis of statistical, correlation, or spectral analysis of dynamic speckles is the influence of the detected light component induced by scattering of probe light by motionless scatterers existing in the scattering volume of probed tissue. For instance, this influence will be significant in the case of laser diagnostics of blood microcirculation in the dermal layer. In the case of blood flow monitoring by means of speckle contrast analysis (e.g., the LASCA technique), the presence of a static component in the image-modulating speckle pattern will cause a noticeable value of the residual speckle contrast even for large values of exposure time—values that significantly exceed the correlation time of intensity fluctuations for the dynamic component. This effect is illustrated by Figure 19.8, which presents two time-averaged dynamic speckle patterns induced by laser light scattering in human forearm skin.

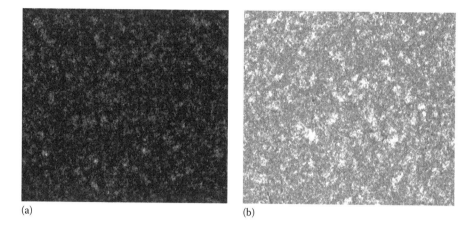

(a)   (b)

FIGURE 19.8   Time-integrated speckle patterns recorded with differing exposure times (object under study is the nail bed of a 44-year-old male). (a) 8 ms exposure time; (b) 25 ms exposure time.

The residual contrast depends on the ratio of mean intensity values of the fluctuating component and the static component of an observed speckle pattern, and on the fraction of Doppler-shifted partial contributions f in the scattered optical field. The simple analytical model that describes this ratio in the dependence on f for the case of single-point detection of fluctuating speckle intensity was considered by Serov et al.[68] They found that the modulation depth of fluctuating intensity in the detection point can be expressed as

$$\frac{\delta_I^2}{\langle I \rangle^2} = \frac{1}{N} f (2 - f),$$ (19.26)

where N is the number of speckles inside the detector aperture.

Having been verified for the case of a layered scattering model, this expression has shown excellent agreement with experimental results. Measurements of the residual contrast for in vivo tissues in combination with spatial filtration of a backscattered optical field by use of ring-like diaphragms offer the opportunity to analyze depth distributions of erythrocyte concentrations in probed tissue. The basic idea of this technique is illustrated by Figure 19.9. By using the circular and ring-like diaphragms of different radii, one can select the partial contributions of light penetrating into different depths of the tissue layer.

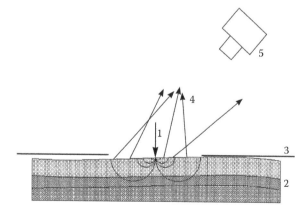

FIGURE 19.9   Use of spatial filtering of a backscattered optical field in the object plane makes it possible to improve the spatial resolution of the speckle contrast technique: 1, localized light source (focused laser beam); 2, tissue under study; 3, filtering diaphragm; 4, scattered light; and 5, CCD camera without imaging lens. (Detection of scattered light in the diffraction regime allows one to exclude the saturation of captured speckle pattern.)

19.6 Imaging of Scattering Media with Use of Partially Coherent Speckles

If a scattering medium is illuminated by partially coherent light, only partial waves with pathlength differences less than the coherence length of the illuminating light will contribute to the formation of a random interference pattern. Such discrimination of partial contributions will lead to decay of the contrast of observed speckle patterns with an increase in the spectral width of the illumination source. This effect is clearly observed in the case of multiply scattering disordered media illuminated by a broadband light source (e.g., a superluminescent diode). Qualitative analysis of the physical picture of scattered field formation due to stochastic interference of partial waves, each of them characterized by the random value of the propagation path s_i, leads us to the following condition for the visibility of an observed speckle pattern: if the mean value of the pathlength difference $\Delta s = s_i - s_j$ for two arbitrarily chosen partial contributions i and j is significantly less than the coherence length l_c of used light source:

$$\langle \Delta s \rangle = \int_0^\infty \Delta s \rho(\Delta s) d(\Delta s) \ll l_c \; [\rho(\Delta s)] \text{ is the probability density function of the pathlength difference), then}$$

the partial coherence of the illuminating beam does not strongly influence the visibility of the speckle pattern. In the opposite case of $(\Delta s) \gg l_c$ the speckle modulation of scattered field is suppressed and speckle contrast falls to zero.

The strong influence on speckle visibility of pathlength statistics, determined by the propagation conditions for partially coherent light in a multiple-scattering medium, allows one to consider the measurements of the contrast of partially coherent speckles as a technique for inhomogeneous media probing and imaging. The presence of scattering or absorbing inhomogeneities in a probed medium between the light source and detector area will distort the pathlength distribution inside the scattering volume; consequently, they will appear as changes in the contrast of speckles observed in the detection area in comparison with the homogeneous medium with the same background optical properties.

In particular, Thompson et al.[69,70] considered one version of this technique. They measured the contrast of speckles that result from propagation of laser light in the multiple-scattering slab with a strongly scattering or absorbing inclusion, because the position of the inclusion varied with respect to illuminating beam and detector (a CCD camera). A speckle-modulated image of the backward surface of the illuminated slab was formed by a lens; therefore, the observed speckles were subjective ones with the average size depending on the aperture size of the imaging lens. A polarization filter was used to improve the visibility of the observed speckles by selection of the linearly polarized component of the scattered field. The experimental results have shown that it is possible to use partially coherent speckle contrast analysis for probing and imaging macroscopically heterogeneous scattering media with the appropriately chosen spectral bandwidth for the illumination source.

The original version of an imaging technique based on observation of spatial distributions of partially coherent speckle contrast was reported by Hausler et al.[71] The schematic of the experimental setup is presented in Figure 19.10. In this case, adjustment of the reference arm of the interferometer (e.g., a Mach-Zehnder interferometer) allows one to find a region on the tissue surface where partial components of a scattered field with a certain propagation path will emerge after propagation through a scattering volume. In order to measure the speckle contrast, subtraction of two sequential images of a scattering object is used. In this case, the second image is obtained with the reference phase shifted by π.

By using this technique, the region of interest, which can be interpreted as the position of the photon horizon for the given time delay, is imaged as a border of speckle-modulated area on the image of the object surface (Figure 19.11). Varying the time delay by changing the pathlength difference between the reference and object arms of interferometer, one can study light propagation in the probed medium based on its optical properties.

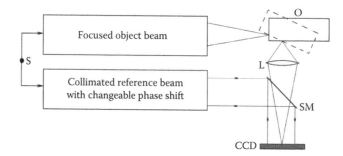

FIGURE 19.10   Schematic of low-coherence optical system for a scattering media probing with use of the *photon horizon* detection technique. S is the low-coherent light source, O is the object under study, L is the imaging lens, and SM is the semitransparent mirror.

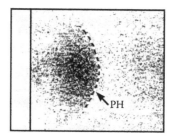

FIGURE 19.11   Visualization of the *photon horizon* (PH) with use of the speckle contrast imaging technique. The vertical line in the left part of the figure shows the front surface of the object, which is illuminated by a low-coherent beam. The speckle-modulated image is formed by light emerging from the side surface of sample.

19.7 Summary

Various speckle technologies based on analysis of decay of statistical moments of spatial, temporal, or angular fluctuations of scattered optical fields are adequately powerful tools for diagnostics and visualization of structure and dynamic properties of multiply-scattering, weakly ordered media such as human and animal tissues.

Acknowledgments

The work on this chapter was supported by grants 01-02-17493 and 00-15-96667 of the Russian Foundation for Basic Research, and by grant REC-006 of the Civil Research and Development Foundation for the Independent States of the Former Soviet Union.

References

1. Goodman, J.W., Some fundamental properties of speckles, *J. Opt. Soc. Am.*, 66, 1145, 1976.
2. Goodman, J.W., Dependence of image speckle contrast on surface roughness, *Opt. Commun.*, 14, 324, 1975.
3. Goodman, J.W., Statistical properties of laser speckle patterns, in *Laser Speckle and Related Phenomena*, 2nd edn., Dainty, J.C., Ed., Springer-Verlag, Heidelberg, Germany, 1984, p. 9.
4. Goodman, J.W., *Statistical Optics*, Wiley, New York, 1985.

5. Dainty, J.C., Ed., *Laser Speckle and Related Phenomena*, Topics in Applied Physics, Vol. 9, Springer-Verlag, Heidelberg, Germany, 1984.
6. Pedersen, H.M., Theory of speckle dependence on surface roughness, *J. Opt. Soc. Am.*, 66, 1204, 1976.
7. Jakeman, E., Speckle statistics with a small number of scatterers, *Opt. Eng.*, 23, 453, 1984.
8. Pusey, P.N., Photon correlation study of laser speckle produced by a moving rough surface, *J. Phys. D*, 9, 1399, 1976.
9. Cummins, H.Z. and Pike, E.R., Eds., *Photon Correlation and Light-Beating Spectroscopy*, NATO Advanced Study Institute Series B: Physics, Plenum Press, New York, 1974.
10. Jakeman, E. and Welford, W.T., Speckle statistics in imaging systems, *Opt. Commun.*, 21, 72, 1977.
11. Outsubo, J. and Asakura, T., Statistical properties of speckle intensity variations in the diffraction field under illumination of coherent light, *Opt. Commun.*, 14, 30, 1975.
12. Uozumi, J. and Asakura, T., First-order intensity and phase statistics of Gaussian speckles produced in the diffraction region, *Appl. Opt.*, 20, 1454, 1981.
13. Yoshimura, T., Statistical properties of dynamic speckles, *J. Opt. Soc. Am. A*, 3, 1032, 1986.
14. Yoshimura, T., Nakagawa, K., and Wakabayashi, N., Rotational and boiling motion of speckles in a two-lens imaging system, *J. Opt. Soc. Am. A*, 3, 1018, 1986.
15. Golubentsev, A.A., On the suppression of the interference effects under multiple scattering of light, *Zh. Eksp. Teor. Fiz.*, 86, 47, 1984 (in Russian).
16. Stephen, M.J., Temporal fluctuations in wave propagation in random media, *Phys. Rev. B*, 37, 1, 1988.
17. MacKintosh, F.C. and John, S., Diffusing-wave spectroscopy and multiple scattering of light in correlated random media, *Phys. Rev. B*, 40, 2382, 1989.
18. John, S., Localization of light, *Phys. Today*, 44, 32–40, May 1991.
19. Van Albada, M.P. and Lagendijk, A., Observation of weak localization of light in a random medium, *Phys. Rev. Lett.*, 55, 2692, 1985.
20. Wolf, P.-E. and Maret, G., Weak localization and coherent backscattering of photons in disordered media, *Phys. Rev. Lett.*, 55, 2696, 1985.
21. Akkermans, E., Wolf, P.E., Maynard, R., and Maret, G., Theoretical study of the coherent backscattering of light by disordered media, *J. Phys. France*, 49, 77, 1988.
22. Feng, S., Kane, C., Lee, P.A., and Stone, A.D., Correlations and fluctuations of coherent wave transmission through disordered media, *Phys. Rev. Lett.*, 61, 834, 1988.
23. Freund, I., Rosenbluh, M., and Feng, S., Memory effects in propagation of optical waves through disordered media, *Phys. Rev. Lett.*, 61, 2328, 1988.
24. Pine, D.J., Weitz, D.A., Chaikin, P.M., and Herbolzheimer, E., Diffusing-wave spectroscopy, *Phys. Rev. Lett.*, 60, 1134, 1988.
25. Boas, D.A., Campbell, L.E., and Yodh, A.G., Scattering and imaging with diffusing temporal field correlations, *Phys. Rev. Lett.*, 75, 1855, 1995.
26. Yodh, A.G., Georgiades, N., and Pine, D.J., Diffusing-wave interferometry, *Opt. Commun.*, 83, 56, 1991.
27. Kao, M.H., Yodh, A.G., and Pine, D.J., Observation of Brownian motion on the time scale of hydrodynamic interactions, *Phys. Rev. Lett.*, 70, 242, 1993.
28. Boas, D.A., Bizheva, K.K., and Siegel, A.M., Using dynamic low-coherence interferometry to image Brownian motion within highly scattering media, *Opt. Lett.*, 23, 319, 1998.
29. Stern, M.D., In vivo evaluation of microcirculation by coherent light scattering, *Nature (London)*, 254, 56, 1975.
30. Stern, M.D., Laser Doppler velocimetry in blood and multiply scattering fluids: Theory, *Appl. Opt.*, 28, 1968, 1985.
31. Briers, J.D., Wavelength dependence of intensity fluctuations in laser speckle patterns from biological specimens, *Opt. Commun.*, 13, 324, 1975.

32. Fercher, A.F., Velocity measurement by first-order statistics of time-differentiated laser speckles, *Opt. Commun.*, 33, 129, 1980.

33. Fercher, A.F. and Briers, J.D., Flow visualization by means of single-exposure speckle photography, *Opt. Commun.*, 37, 326, 1981.

34. Fujii, H., Asakura, T., Nohira, K. et al., Blood flow observed by time-varying laser speckle, *Opt. Lett.*, 10, 104, 1985.

35. Fercher, A.F., Peukert, M., and Roth, E., Visualization and measurement of retinal blood flow by means of laser speckle photography, *Opt. Eng.*, 25, 731, 1986.

36. Aizu, Y., Ogino, K., Koyama, T. et al., Evaluation of retinal blood flow using laser speckle, *J. Jpn. Soc. Laser Med.*, 8, 89, 1987.

37. Fujii, H., Nohira, K., Yamamoto, Y., Ikawa, H., and Ohura, T., Evaluation of blood flow by laser speckle image sensing, *Appl. Opt.*, 26, 5321, 1987.

38. Ruth, B., Non-contact blood flow determination using a laser speckle method, *Opt. Laser Technol.*, 20, 309, 1988.

39. Aizu, Y. and Asakura, T., Bio-speckle phenomena and their application to the evaluation of blood flow, *Opt. Laser Technol.*, 23, 205, 1991.

40. Nilsson, G., Jakobsson, A., and Wårdell, K., Tissue perfusion monitoring and imaging by coherent light scattering, *Proc. SPIE*, 1524, 90, 1991.

41. Briers, J.D. and Webster, S., Quasi real-time digital version of single-exposure speckle photography for full-field monitoring or flow fields, *Opt. Commun.*, 116, 36, 1995.

42. Dacosta, G., Optical remote sensing of heartbeats, *Opt. Commun.*, 117, 395, 1995.

43. Rhytov, S.M., Kravtsov, U.A., and Tatarsky, V.I., *Introduction to Statistical Radiophysics, Part 2, Random Fields*, Nauka Publishers, Moscow, Russia, 1978.

44. Korn, G.A. and Korn, T.M., *Mathematical Handbook for Scientists and Engineers*, McGraw-Hill, New York, 1968.

45. Escamilla, H.M., Speckle contrast from weak diffusers with a small number of correlation areas, *Opt. Acta*, 25, 777, 1978.

46. Zimnyakov, D.A. and Tuchin, V.V., About two-modality of intensity distributions of speckle fields for large-scale phase scatterers, *J. Tech. Phys. Lett.*, 21, 10, 1995.

47. Feder, J., *Fractals*, Plenum Press, New York, 1988.

48. Jakeman, E., Scattering by fractals, in *Fractals in Physics*, Pietronero, L. and Tozatti, E., Eds., North-Holland, Amsterdam, the Netherlands, 1986, p. 82.

49. Zimnyakov, D.A., Tuchin, V.V., and Utts, S.R., A study of statistical properties of partially developed speckle fields as applied to the diagnostics of structural changes in human skin, *Opt. Spectrosc.*, 76, 838, 1994.

50. Zimnyakov, D.A. and Tuchin, V.V., Fractality of speckle intensity fluctuations, *Appl. Opt.*, 35, 3328, 1996.

51. Maret, G. and Wolf, P.E., Multiple light scattering from disordered media. The effect of Brownian motions of scatterers, *Z. Phys. B*, 65, 409, 1987.

52. Boas, D.A. and Yodh, A.G., Spatially varying dynamical properties of turbid media probed with diffusing temporal light correlation, *J. Opt. Soc. Am. A*, 14, 192, 1997.

53. Ackerson, B.J., Dougherty, R.L., Reguigui, N.M., and Nobbman, U., Correlation transfer: Application of radiative transfer solution methods to photon correlation problems, *J. Thermophys. Heat Trans.*, 6, 577, 1992.

54. Heckmeier, M., Skipetrov, S.E., Maret, G., and Maynard, R., Imaging of dynamic heterogeneities in multiple-scattering media, *J. Opt. Soc. Am. A*, 14, 185, 1997.

55. Zimnyakov, D.A., Tuchin, V.V., and Mishin, A.A., Spatial speckle correlometry in applications to tissue structure monitoring, *Appl. Opt.*, 36, 5594, 1997.

56. Zimnyakov, D.A., Tuchin, V.V., and Yodh, A.G., Characteristic scales of optical field depolarization and decorrelation for multiple scattering media and tissues, *J. Biomed. Opt.*, 4, 157, 1999.

57. Zimnyakov, D.A., Maksimova, I.L., and Tuchin, V.V., Controlling of tissue optical properties. II. Coherent methods of tissue structure study, *Opt. Spectrosc.*, 88, 1026, 2000.

58. Zimnyakov, D.A. and Tuchin, V.V., Laser tomography, in *Lasers in Medicine*, Vij, D.R., Ed., Kluwer Academic, Dordrecht, the Netherlands, 2002.

59. Tuchin, V.V., Ed., *Handbook on Optical Biomedical Diagnostics*, Vol. PM107, SPIE Press, Bellingham, WA, 2002.

60. Berkovits, R. and Feng, S., Theory of speckle-pattern tomography in multiple-scattering media, *Phys. Rev. Lett.*, 65, 3120, 1990.

61. Hoover, B.G., Deslauriers, L., Grannell, S.M., Ahmed, R.E., Dilworth, D.S., Althey, B.D., and Leith, E.N., Correlations among angular wave component amplitudes in elastic multiple-scattering random media, *Phys. Rev. E*, 65, 026614, 2002.

62. Tuchin, V.V., Ryabukho, V.P., Zimnyakov, D.A., Lobachev, M.I., Lyakin, D.V., Radchenko, E.Yu., Chaussky, A.A., and Konstantinov, K.V., Tissue structure and blood microcirculation monitoring by speckle interferometry and full-field correlometry, *Proc. SPIE*, 4251, 148, 2001.

63. Briers, J.D. and Webster, S., Laser speckle contrast analysis (LASCA): A non-scanning, full-field technique for monitoring capillary blood flow, *J. Biomed. Opt.*, 1, 174, 1996.

64. Dunn, A.K., Bolay, H., Moskowitz, M.A., and Boas, D.A., Dynamic imaging of cerebral blood flow using laser speckle, *J. Cereb. Flow Metab.*, 21, 195, 2001.

65. Jacques, S.L. and Kirkpatrick, S.J., Acoustically modulated speckle imaging of biological tissues, *Opt. Lett.*, 23, 879, 1998.

66. Duncan, D.D. and Kirkpatrick, S.J., Processing algorithms for tracking speckle shifts in optical elastography of biological tissues, *J. Biomed. Opt.*, 6, 418, 2001.

67. Sadhwani, A., Schomacker, K.T., Tearney, G.T., and Nishioka, N., Determination of Teflon thickness with laser speckle. I. Potential for burn depth diagnosis, *Appl. Opt.*, 35, 5727, 1996.

68. Serov, A., Steenbergen, W., and de Mul, F., A method for estimation of the fraction of Doppler-shifted photons in light scattered by mixture of moving and stationary scatterers, *Proc. SPIE*, 4001, 178, 2000.

69. Thompson, C.A., Webb, K.J., and Weiner, A.M., Diffuse media characterization using laser speckle, *Appl. Opt.*, 36, 3726, 1997.

70. Thompson, C.A., Webb, K.J., and Weiner, A.M., Imaging in scattering media by use of laser speckle, *J. Opt. Soc. Am. A*, 14, 2269, 1997.

71. Hausler, G., Herrmann, J.M., Kummer, R., and Lindner, M.V., Observation of light propagation in volume scatterers with 10^{11}-fold slow motion, *Opt. Lett.*, 21, 1087, 1996.

IV

Spectroscopic Data

20

Spectroscopic Data in Biological and Biomedical Analysis

Dimitra N.
Stratis-Cullum
*US Army Research
Laboratory*

Mikella E. Farrell
*US Army Research
Laboratory*

Ellen Holthoff
*US Army Research
Laboratory*

David L. Stokes
*Oak Ridge National
Laboratory*

Brian M. Cullum
*University of Maryland,
Baltimore County*

Joon Myong Song
*Oak Ridge National
Laboratory*

Paul M. Kasili
*Oak Ridge National
Laboratory*

Ramesh Jaganathan
*Oak Ridge National
Laboratory*

Tuan Vo-Dinh
Duke University

20.1 Fundamental Optical Properties of Biologically Relevant Molecules

20.1.1 Raman Spectroscopy Characteristic Frequencies

Table 20.1 indicates characteristic Raman frequencies of organic compounds. This also consists of vibrational frequencies (cm^{-1}) for main chemical groups, the type of vibration energy absorbed due to vibrations of polar covalent bonds, and the functional groups that cause the vibration, and the kind of compound this characteristic vibration would be found. This table was derived from *The CRC Practical Handbook of Spectroscopy*, ed. J. W. Robinson, CRC Press, Inc., Boca Raton, FL, 1991.

20.1.2 Infrared Spectroscopy: Characteristic IR Band Positions

Table 20.2 indicates the characteristic Infrared (IR) band positions (frequencies) of organic compounds. This table consists of vibrational frequencies (cm^{-1}) of functional groups and the functional groups found within organic molecules that absorb IR energy. These frequencies and assignments have been collected from http://infrared.als.lbl.gov/IRbands.html, a website maintained by the Lawrence Berkeley National Laboratory.

20.1.3 IR Spectroscopy: Characteristic IR Absorption Frequencies

Table 20.3 consists of characteristic vibrational frequencies of main organic bonds. This table consists of the frequency range (cm^{-1}), compound type that displays the characteristic absorptions, and organic bonds involved. These frequencies and assignments have been collected from http://www.chem.ucla.edu/, a website maintained by the Chemistry Department at the University of California, Los Angeles.

20.1.4 Molecular Spectroscopy: Characteristics of Electronic Transitions between Orbitals

Table 20.4 displays the characteristics of electronic transitions between σ, η, and π orbitals. This table consists of the type of transition, Mulliken's designation, wavelength (nm), molar absorptivity (ε [L $mol^{-1}cm^{-1}$]), and examples of organic compounds with electronic transitions. This table was derived from *Spectrochemical Analysis*, J. D. Ingle, Jr., S. R. Crouch, Prentice-Hall Inc., Upper Saddle River, NJ, 1988.

20.1.5 Fluorescence of Linear Aromatics in EPA

Table 20.5 displays the fluorescence of linear aromatics in EPA (a mixture of ethanol, isopropanol, and ether) at 77 K. This table consists of aromatic compounds, overall quantum yield of fluorescence (Φ_F), excitation wavelength (λ_{ex} [nm]), emission wavelength (λ_{em} [nm]), the product of two factors, the fraction of the absorbed photons that produce triplet states and the fraction of triplet molecules that undergo fluorescence (Φ_p), and the fluorescence lifetime (τ_p[s]). This table was derived from *Spectrochemical Analysis*, J. D. Ingle, Jr., S. R. Crouch, Prentice-Hall Inc., Upper Saddle River, NJ, 1988.

TABLE 20.1   Raman Spectroscopy: Characteristic Raman Frequencies

Frequency Range (cm⁻¹)	Vibration	Compound
3400–3300	Bonded antisymmetric NH_2 stretch	Primary amines
3380–3340	Bonded OH stretch	Aliphatic alcohols
3374	CH stretch	Acetylene (gas)
3355–3325	Bonded antisymmetric NH_2 stretch	Primary amides
3350–3300	Bonded NH stretch	Secondary amines
3335–3300	≡CH stretch	Alkyl acetylenes
3300–3250	Bonded symmetric NH_2 stretch	Primary amines
3310–3290	Bonded NH stretch	Secondary amines
3190–3145	Bonded symmetric NH_2 stretch	Primary amides
3175–3154	Bonded NH stretch	Pyrazoles
3103	Antisymmetric $=CH_2$ stretch	Ethylene (gas)
3100–3020	CH_2 stretches	Cyclopropane
3100–3000	Aromatic CH stretch	Benzene derivatives
3095–3070	Antisymmetric $=CH_2$ stretch	$C=CH_2$ derivatives
3062	CH stretch	Benzene
3057	Aromatic CH stretch	Alkyl benzenes
3040–3000	CH stretch	C=CHR derivatives
3026	Symmetric $=CH_2$ stretch	Ethylene (gas)
2990–2980	Symmetric $=CH_2$ stretch	$C=CH_2$ derivatives
2986–2974	Symmetric NH_3^+ stretch	Alkyl ammonium chlorides (aq. soln.)
2969–2965	Antisymmetric CH_3 stretch	*n*-Alkanes
2929–2912	Antisymmetric CH_2 stretch	*n*-Alkanes
2884–2883	Symmetric CH_3 stretch	*n*-Alkanes
2861–2849	Symmetric CH_2 stretch	*n*-Alkanes
2850–2700	CHO group (2 bands)	Aliphatic aldehydes
2590–2560	SH stretch	Thiols
2316–2233	C≡C stretch (2 bands)	$R-C≡C-CH_3$
2301–2231	C≡C stretch (2 bands)	R-C≡C-R′
2300–2250	Pseudo-anti-symmetric N=C=O stretch	Isocyanates
2264–2251	Symmetric C≡C–C≡C stretch	Alkyl diacetylenes
2259	C≡N stretch	Cyanamide
2251–2232	C≡N stretch	Aliphatic nitriles
2220–2100	Pseudo-anti-symmetric N=C=S stretch (2 bands)	Alkyl isothiocyanates
2220–2000	C≡N stretch	Dialkyl cyanamides
2172	Symmetric C≡C–C≡C stretch	Diacetylene
2161–2134	$N^+≡C^-$ stretch	Aliphatic isonitriles
2160–2100	C≡C stretch	Alkyl acetylenes
2156–2140	C≡N stretch	Alkyl thiocyanates
2104	Antisymmetric N=N=N stretch	CH_3N_3
2094	C≡N stretch	HCN
2049	Pseudo-anti-symmetric C=C=O stretch	Ketene
1974	C≡C stretch	Acetylene (gas)
1964–1958	Antisymmetric C=C=C stretch	Allenes
1870–1840	Symmetric C=O stretch	Saturated five-membered ring cyclic anhydrides
1820	Symmetric C=O stretch	Acetic anhydride
1810–1788	C=O stretch	Acid halides
1807	C=O stretch	Phosgene

(continued)

TABLE 20.1 (continued) Raman Spectroscopy: Characteristic Raman Frequencies

Frequency Range (cm⁻¹)	Vibration	Compound
1805–1799	Symmetric C=O stretch	Noncyclic anhydrides
1800	C=C stretch	$F_2C=CF_2$ (gas)
1795	C=O stretch	Ethylene carbonate
1792	C=C stretch	$F_2C=CFCH_3$
1782	C=O stretch	Cyclobutanone
1770–1730	C=O stretch	Halogenated aldehydes
1744	C=O stretch	Cyclopentanone
1743–1729	C=O stretch	Cationic *a*-amino acids (aq. soln.)
1741–1734	C=O stretch	*O*-Alkyl acetates
1740–1720	C=O stretch	Aliphatic aldehydes
1739–1714	C=C stretch	$C=CF_2$ derivatives
1736	C=C stretch	Methylenecyclopropane
1734–1727	C=O stretch	*O*-Alkyl propionates
1725–1700	C=O stretch	Aliphatic ketones
1720–1715	C=O stretch	*O*-Alkyl formates
1712–1694	C=C stretch	RCF=CFR
1695	Nonconjugated C=O stretch	Uracil derivatives (aq. soln.)
1689–1644	C=C stretch	Monofluoroalkenes
1687–1651	C=C stretch	Alkylidene cyclopentanes
1686–1636	Amide I band	Primary amides (solids)
1680–1665	C=C stretch	Tetralkyl ethylenes
1679	C=C stretch	Methylenecyclobutane
1678–1664	C=C stretch	Trialkyl ethylenes
1676–1665	C=C stretch	*trans*-Dialkyl ethylenes
1675	Symmetric C=O stretch (cyclic dimer)	Acetic acid
1673–1666	C=N stretch	Aldimines
1672	Symmetric C=O stretch (cyclic dimer)	Formic acid (aq. soln.)
1670–1655	Conjugated C=O stretch	Uracil, cytosine, and guanine derivatives (aq. soln.)
1670–1630	Amide I band	Tertiary amides
1666–1652	C=N stretch	Ketoximes
1665–1650	C=N stretch	Semicarbazones (solid)
1663–1636	Symmetric C=N stretch	Aldazines, ketazines
1660–1654	C=C stretch	*cis*-Dialkyl ethylenes
1660–1650	Amide I band	Secondary amides
1660–1649	C=N stretch	Aldoximes
1660–1610	C=N stretch	Hydrazones (solids)
1658–1644	C=C stretch	$R_2C=CH_2$
1656	C=C stretch	Cyclohexene, cycloheptene
1654–1649	Symmetric C=C stretch (cyclic dimer)	Carboxylic acids
1652–1642	C=N stretch	Thiosemicarbazones (solid)
1650–1590	NH_2 scissors	Primary amines (weak)
1649–1625	C=C stretch	Allyl derivatives
1648–1640	N=O stretch	Alkyl nitrites
1648–1638	C=C stretch	$H_2C=CHR$
1647	C=C stretch	Cyclopropene
1638	C=O stretch	Ethylene dithiocarbonate
1637	Symmetric C=C stretch	Isoprene
1634–1622	Antisymmetric NO_2 stretch	Alkyl nitrates
1630–1550	Ring stretches (doublet)	Benzene derivatives

TABLE 20.1 (continued)   Raman Spectroscopy: Characteristic Raman Frequencies

Frequency Range (cm⁻¹)	Vibration	Compound
1623	C=C stretch	Ethylene (gas)
1620–1540	Three or more coupled C=C stretches	Polyenes
1616–1571	C=C stretch	Chloroalkenes
1614	C=C stretch	Cyclopentene
1596–1547	C=C stretch	Bromoalkenes
1581–1465	C=C stretch	Iodoalkenes
1575	Symmetric C=C stretch	1,3-Cyclohexadiene
1573	N=N stretch	Azomethane (in soln.)
1566	C=C stretch	Cyclobutene
1560–1550	Antisymmetric NO_2 stretch	Primary nitroalkanes
1555–1550	Antisymmetric NO_2 stretch	Secondary nitroalkanes
1548	N=N stretch	1-Pyrazoline
1545–1535	Antisymmetric NO_2 stretch	Tertiary nitroalkanes
1515–1490	Ring stretch	2-Furfuryl group
1500	Symmetric C=C stretch	Cyclopentadiene
1480–1470	OCH_3, OCH_2 deformations	Aliphatic ethers
1480–1460	Ring stretch	2-Furfurylidene or 2-furoyl group
1473–1446	CH_3, CH_2 deformations	*n*-Alkanes
1466–1465	CH_3 deformation	*n*-Alkanes
1450–1400	Pseudo-anti-symmetric N=C=O stretch	Isocyanates
1443–1398	Ring stretch	2-Substituted thiophenes
1442	N=N stretch	Azobenzene
1440–1340	Symmetric CO_2 stretch	Carboxylate ions (aq. soln.)
1415–1400	Symmetric CO_2 stretch	Dipolar and anionic *a*-amino acids (aq. soln.)
1415–1385	Ring stretch	Anthracenes
1395–1380	Symmetric NO_2 stretch	Primary nitroalkanes
1390–1370	Ring stretch	Naphthalenes
1385–1368	CH_3 symmetric deformation	*n*-Alkanes
1375–1360	Symmetric NO_2 stretch	Secondary nitroalkanes
1355–1345	Symmetric NO_2 stretch	Tertiary nitroalkanes
1350–1330	CH deformation	Isopropyl group
1320	Ring vibration	1,1-Dialkyl cyclopropanes
1314–1290	In-plane CH deformation	*trans*-Dialkyl ethylenes
1310–1250	Amide III band	Secondary amides
1310–1175	CH_2 twist and rock	*n*-Alkanes
1305–1295	CH_2 in-phase twist	*n*-Alkanes
1300–1280	CC bridge bond stretch	Biphenyls
1282–1275	Symmetric NO_2 stretch	Alkyl nitrates
1280–1240	Ring stretch	Epoxy derivatives
1276	Symmetric N=N=N stretch	CH_3N_3
1270–1251	In-plane CH deformation	*cis*-Dialkyl ethylenes
1266	Ring *breathing*	Ethylene oxide (oxirane)
1230–1200	Ring vibration	*para*-Disubstituted benzenes
1220–1200	Ring vibration	Mono- and 1,2-dialkyl cyclopropanes
1212	Ring *breathing*	Ethylene imine (aziridine)
1205	C_6H_5–C vibration	Alkyl benzenes
1196–1188	Symmetric SO_2 stretch	Alkyl sulfates
1188	Ring *breathing*	Cyclopropane

(*continued*)

TABLE 20.1 (continued)   Raman Spectroscopy: Characteristic Raman Frequencies

Frequency Range (cm⁻¹)	Vibration	Compound
1172–1165	Symmetric SO_2 stretch	Alkyl sulfonates
1150–950	CC stretches	*n*-Alkanes
1145–1125	Symmetric SO_2 stretch	Dialkyl sulfones
1144	Ring *breathing*	Pyrrole
1140	Ring *breathing*	Furan
1130–1100	Symmetric C=C=C stretch (2 bands)	Allenes
1130	Pseudosymmetric C=C=O stretch	Ketene
1112	Ring *breathing*	Ethylene sulfide
1111	NN stretch	Hydrazine
1070–1040	S=O stretch (1 or 2 bands)	Aliphatic sulfoxides
1065	C=S stretch	Ethylene trithiocarbonate
1060–1020	Ring vibration	*ortho*-Disubstituted benzenes
1040–990	Ring vibration	Pyrazoles
1030–1015	In-plane CH deformation	Monosubstituted benzenes
1030–1010	Trigonal ring *breathing*	3-Substituted pyridines
1030	Trigonal ring *breathing*	Pyridine
1029	Ring *breathing*	Trimethylene oxide (oxetane)
1026	Ring *breathing*	Trimethylene imine (azetidine)
1010–990	Trigonal ring *breathing*	Mono-, meta-, and 1,3,5-substituted benzenes
1001	Ring *breathing*	Cyclobutane
1000–985	Trigonal ring *breathing*	2- and 4-Substituted pyridines
992	Ring *breathing*	Benzene
992	Ring *breathing*	Pyridine
939	Ring *breathing*	1,3-Dioxolane
933	Ring vibration	Alkyl cyclobutanes
930–830	Symmetric COC stretch	Aliphatic ethers
914	Ring *breathing*	Tetrahydrofuran
906	Symmetric CON stretch	Hydroxylamine
905–837	CC skeletal stretch	*n*-Alkanes
900–890	Ring vibration	Alkyl cyclopentanes
900–850	Symmetric CNC stretch	Secondary amines
899	Ring *breathing*	Pyrrolidine
886	Ring *breathing*	Cyclopentane
877	OO stretch	Hydrogen peroxide
851–840	Symmetric CON stretch	*O*-Alkyl hydroxylamines
836	Ring *breathing*	Piperazine
835–749	C_4 skeletal stretch	Isopropyl group
834	Ring *breathing*	1,4-Dioxane
832	Ring *breathing*	Thiophene
832	Ring *breathing*	Morpholine
830–720	Ring vibration	*para*-Disubstituted benzenes
825–820	C_3O symmetrical skeletal stretch	Secondary alcohols
818	Ring *breathing*	Tetrahydropyran
815	Ring *breathing*	Piperidine
802	Ring *breathing*	Cyclohexane (chair form)
785–700	Ring vibration	Alkyl cyclohexanes
760–730	C_4O symmetrical skeletal stretch	Tertiary alcohols
750–650	C_5 symmetrical skeletal stretch	*tert*-Butyl group

TABLE 20.1 (continued) Raman Spectroscopy: Characteristic Raman Frequencies

Frequency Range (cm^{-1})	Vibration	Compound
740–585	CS stretch (1 or more bands)	Alkyl sulfides
735–690	C=S stretch	Thioamides, thioureas (solid)
733	Ring *breathing*	Cycloheptane
730–720	CCl stretch, P$_c$ conformation	Primary chloroalkanes
715–620	CS stretch (1 or more bands)	Dialkyl sulfides
709	CCl stretch	CH$_3$Cl
703	Ring *breathing*	Cyclooctane
703	Symmetric CCl$_2$ stretch	CH$_2$Cl$_2$
690–650	Pseudosymmetric N=C=S stretch	Alkyl isothiocyanates
688	Ring *breathing*	Tetrahydrothiophene
668	Symmetric CCl$_3$ stretch	CHCl$_3$
660–650	CCl stretch, P$_H$ conformation	Primary chloroalkanes
659	Symmetric CSC stretch	Pentamethylene sulfide
655–640	CBr stretch, P$_c$ conformation	Primary bromoalkanes
630–615	Ring deformation	Monosubstituted benzenes
615–605	CCl stretch, S$_{HH}$ conformation	Secondary chloroalkanes
610–590	CI stretch, P$_c$ conformation	Primary iodoalkanes
609	CBr stretch	CH$_3$Br
577	Symmetric CBr$_2$ stretch	CH$_2$BR$_2$
570–560	CCl stretch, T$_{HHH}$ conformation	Tertiary chloroalkanes
565–560	CBr stretch, P$_H$ conformation	Primary bromoalkanes
540–535	CBr stretch, S$_{HH}$ conformation	Secondary bromoalkanes
539	Symmetric CBr$_3$ stretch	CHBr
525–510	SS stretch	Dialkyl sulfides
523	CI stretch	CH$_3$I
520–510	CBr stretch, T$_{HH}$ conformation	Tertiary bromoalkanes
510–500	CI stretch, P$_H$ conformation	Primary iodoalkanes
510–480	SS stretch	Dialkyl trisulfides
495–485	CI stretch, S$_{HH}$ conformation	Secondary iodoalkanes
495–485	CI stretch, T$_{HHH}$ conformation	Tertiary iodoalkanes
484–475	Skeletal deformation	Dialkyl diacetylenes
483	Symmetric Cl$_2$ stretch	CH$_2$I$_2$
459	Symmetric CCl$_4$ stretch	CCl$_4$
437	Symmetric CI$_3$ stretch	CHI$_3$ (in soln.)
425–150	*Chain expansion*	*n*-Alkanes
355–335	Skeletal deformation	Monoalkyl acetylenes
267	Symmetric CBr$_4$ stretch	CBr$_3$ (in soln.)
200–160	Skeletal deformation	Aliphatic nitriles
178	Symmetric CI$_4$ stretch	CI$_4$ (solid)

TABLE 20.2   IR Spectroscopy: Characteristic IR Band Positions

Frequency Range (cm⁻¹)	Group
	OH stretching vibrations
3610–3645 (Sharp)	Free OH
3450–3600 (Sharp)	Intramolecular H bonds
3200–3550 (Broad)	Intermolecular H bonds
2500–3200 (Very broad)	Chelate compounds
	NH stretching vibrations
3300–3500	Free NH
3070–3350	H bonded NH
	CH stretching vibrations
3280–3340	=C–H
3000–3100	=C–H
2862–2882, 2652–2972	C–CH$_3$
2815–2832	O–CH$_3$
2810–2820	N–CH$_3$ (Aromatic)
2780–2805	N–CH$_3$ (Aliphatic)
2843–2863, 2916–2936	CH$_2$
2880–2900	CH
	SH stretching vibrations
2550–2600	Free SH
	C=N stretching vibrations
2240–2260	Nonconjugated
2215–2240	Conjugated
	C=C stretching vibrations
2100–2140	C≡CH (terminal)
2190–2260	C–C≡C–C
2040–2200	C–C≡C–C≡CH
	C=O stretching vibrations
1700–1900	Nonconjugated
1590–1750	Conjugated
~1650	Amides
	C=C stretching vibrations
1620–1680	Nonconjugated
1585–1625	Conjugated
	CH bending vibrations
1405–1465	CH$_2$
1355–1395, 1430–1470	CH$_3$
	C–O–C vibrations in esters
~1175	Formates
~1240, 1010–1040	Acetates
~1275	Benzoates
	C–O–H stretching vibrations
990–1060	Secondary cyclic alcohols
	CH out-of-plane bending vibrations in substituted ethylenic systems
905–915, 985–995	–CH=CH$_2$
650–750	–CH=CH– (*cis*)
960–970	–CH=CH– (*trans*)
885–895	C=CH$_2$

TABLE 20.3 IR Spectroscopy: Characteristic IR Absorption Frequencies

Frequency Range (cm^{-1})	Compound Type	Bond
2960–2850 (Strong) stretch	Alkanes	C–H
1470–1350 (Variable) scissoring and bending		
1380 (Medium, weak)—doublet—isopropyl, *t*-butyl	CH$_3$ umbrella deformation	
3080–3020 (Medium) stretch	Alkenes	C–H
1000–675 (Strong) bend		
3100–3000 (Medium) stretch	Aromatic rings	C–H
870–675 (Strong) bend	Phenyl ring substitution bands	
2000–1600 (Weak)—fingerprint region	Phenyl ring substitution overtones	
3333–3267 (Strong) stretch	Alkynes	C–H
700–610 (Broad) bend		
1680–1640 (Medium, weak) stretch	Alkenes	C=C
2260–2100 (Weak, strong) stretch	Alkynes	C≡C
1600, 1500 (Weak) stretch	Aromatic rings	C=C
1260–1000 (Strong) stretch	Alcohols, ethers, carboxylic acids, esters	C–O
1760–1670 (Strong) stretch	Aldehydes, ketones, carboxylic acids, esters	C=O
3640–3160 (Strong, broad) stretch	Monomeric—alcohols, phenols	O–H
3600–3200 (Broad) stretch	Hydrogen-bonded—alcohols, phenols	
3000–2500 (Broad) stretch	Carboxylic acids	
3500–3300 (Medium) stretch	Amines	N–H
1650–1580 (Medium) bend		
1340–1020 (Medium) stretch	Amines	C–N
2260–2220 (Variable) stretch	Nitriles	C≡N
1660–1500 (Strong) asymmetrical stretch	Nitro compounds	NO$_2$
1390–1260 (Strong) symmetrical stretch		

TABLE 20.4 Molecular Spectroscopy: Characteristics of Electronic Transitions between σ, η, and π Orbitals

Transition	Mulliken's Designation	Wavelength (nm)	ε (L mol^{-1} cm^{-1})	Examples
σ-σ*	$N \rightarrow V$	<200	—	Saturated hydrocarbons
π-π*	$N \rightarrow V$	200–500	≈10^4	Alkenes, alkynes, aromatics
η-σ*	$N \rightarrow Q$	160–260	10^2–10^3	H_2O, CH_3OH, CH_3Cl, CH_3NH_2
η-π*	$N \rightarrow Q$	250–600	10^1–10^2	Carbonyls, nitro, nitrate, carboxyl

TABLE 20.5 Fluorescence of Linear Aromatics in EPA[a] at 77 K

Compound	Φ_F	λ_{ex} (nm)	λ_{em} (nm)	Φ_p	τ_p (s)
Benzene	0.11	205	278	0.26	7
Naphthalene	0.29	286	321	0.1	2.6
Anthracene	0.46	365	400	<0.01	0.04
Naphthacene	0.60	390	480	—	—

[a]EPA is a mixture of ethanol, isopropanol, and ether.

TABLE 20.6   Absorption Characteristics of Common Chromophores

Group	Example	ν (10^3 cm^{-1})	λ (nm)	ε (L mol^{-1} cm^{-1})
C=C	$H_2C=CH_2$	55	182	250
		57.3	174	16,000
		58.6	170	16,500
		62	162	10,000
C≡C	$H-C≡C-CH_2-CH_3$	58	172	2,500
C=O	H_2CO	34	295	10
		54	185	Strong
C=S	$CH_3-\overset{\overset{S}{\|}}{C}-CH_3$	22	460	Weak
$-NO_2$	CH_3-NO_2	36	277	10
		47.5	210	10,000
$-N=N-$	$CH_3-N=N-CH_3$	28.8	347	15
		>38.5	<260	Strong
$-Cl$	CH_3Cl	58	172	—
$-Br$	CH_3Br	49	204	1,800
$-I$	CH_3I	38.8	258	—
		49.7	201	1,200
$-OH$	CH_3OH	55	183	200
		67	150	1,900
$-SH$	C_2H_5SH	43	232	160
$-NH_2$	CH_3NH_2	46.5	215	580
		52.5	190	3,200
$-S-$	CH_3-S-CH_3	44	228	620
		46.5	215	700
		49.3	203	2,300
C=C-C=C	$H_2C=CH-CH=CH_2$	48	209	25,000

20.1.6 Absorption Characteristics of Common Chromophores

Table 20.6 lists the absorption characteristics of common chromophores. This table consists of groups that are responsible for absorption, wavelength (λ [nm]), wavenumbers (ν [10^3 cm^{-1}]), and molar absorptivity (ε [L mol^{-1} cm^{-1}]). This table was derived from *Spectrochemical Analysis*, J. D. Ingle, Jr., S. R. Crouch, Prentice-Hall Inc., Upper Saddle River, NJ, 1988.

20.2 Spectroscopic Data for Biomedical Applications

20.2.1 UV–Visible Absorption Spectroscopy

Table 20.7 provides information on the application of UV–visible absorption technique to biomedical research.

This table covers only representative biochemical species that has been analyzed using UV–visible absorption technique.

Maximum/detection wavelength corresponds to the wavelength that was used for the detections of biochemical species in UV–visible absorption technique.

TABLE 20.7 UV–Visible Absorption

Species	Target Organ or Disease	Absorbance Maximum (nm)	Comments	Reference
Biochemical studies				
1,8-Bromo-guanosine		262		[1]
16α-Hydroxytestosterone, 16β-hydroxytestosterone		245		[2]
18β-Glycyrrhetinic acid (GA)			A novel form of adenosine-5′-phosphosulfate without an Fe–S cluster	[3]
1-Methylguanosine		210, 260		[4]
1-Methylinosine		210, 260		[4]
1-Phenylethylamine		214		[5]
2-(S-Alkylthiyl)pyrroline N-oxide		258		[6]
2,3-Dihydroxybiphenyl 1,2-dioxygenase-bound DHB		299		[7]
2,6-Dimethyldifuro-8-pyrone		306	Molar extinction coefficient = 4,700 (M−1)	[8]
2α-Hydroxytestosterone, 2β-hydroxytestosterone		245		[2]
2-Methoxy-3-isopropylpyrazine		307		[9]
2-Methylguanosine		210, 260		[4]
3,4-Methylenedioxymethamphetamine		214		[5]
3-Acetylmorphine		210		[10]
3-Methyluridine + 5-methyluridine		210, 260		[4]
3′-Phosphoadenosine-5′-phosphosulfate (PAPS)				[11]
4-Hydroxyazobenzene-2-carboxylic acid		350		[12]
5,10-Methylenetetrahydrofolate				[13]
5,6-Dihydrouridine		210, 260		[4]
6α-Hydroxytestosterone, 6β-hydroxytestosterone		245		[2]
6-Acetylmorphine		210		[10]
7α-Hydroxytestosterone		245		[2]
Acetylcodeine		210		[14]
Acetyl-CoA synthase				[15]
Adenosine		210, 260		[4]
Adenosylcobinamide (AdoCbi) amidohydrolase CbiS enzyme			Cell wall biosynthesis, spore germination	[16]
α-Human globin chain		214		[17]
Alanine racemase			Ethanol oxidation	[18]
Albonoursin		317	Molar extinction coefficient = 25,400 (M−1)	[19]
Alcohol dehydrogenase complex (ADH)			Determination of feed for farm animals	[20]

(continued)

TABLE 20.7 (continued) UV–Visible Absorption

Species	Target Organ or Disease	Absorbance Maximum (nm)	Comments	Reference
Alpha-tocopherol				[21]
Amphetamine		214		[5]
Amyloid beta-peptide	Alzheimer's disease			[22]
Amyloid beta-peptide	Alzheimer's disease			[23]
Amyloid deposit	Alzheimer's disease		Fe–S cluster transfer	[24]
Anabasine		214		[25]
Androstenedione		245		[2]
ApbC (ATPase)				[26]
ApbC/Nbp35 (Fe cluster carrier protein)				[27]
Apolipoprotein A-1 (apo A-1)			Fe uptake	[28]
Aporubredoxins			Analysis of phloem exudates	[29]
Ascorbic acid conjugates			Cellular sulfate ester cleavage	[30]
Asialotransferrin		200		[31]
AtsB (a radical SAM formylglycine-generating enzyme)			Nitrogenase activity	[32]
Azotobacter vinelandii flavodoxin hydroquinone (FIdHQ)				[33]
Benoxaprofen		305	Molar extinction coefficient = 25,000 (M−1)	[34]
Benzocaine		195		[35]
Benzoylecgonine		195		[35]
Berberine	Cortical neurons			[36]
β-Lapachone	Cancer			[37]
β-Human globin chain		214		[17]
Bile acids	Antitumor		Protein	[38]
Bis(monoacylglycerol)phosphate		231		[39]
Bis-N-nitroso-caged nitric oxides		300		[40]
β-Nicotinamide adenine dinucleotide (β-NADH)		254		[41]
Bovine β-lactoglobulin (β-LG)				[42]
Carotenoid				[43]
Carotenoid crocetin			Photoreceptor	[44]
Carprofen		240, 262, 300, 328		[34]
Channelrhodopsin-2 (ChR2)			Transcription regulation	[45]
cis-Cinnamoylcocaine, *trans*-cinnamoylcocaine		195		[35]
C-myc gene			Bacteria	[46]
Cocaine		195		[35]

TABLE 20.7 (continued) UV–Visible Absorption

Species	Target Organ or Disease	Absorbance Maximum (nm)	Comments	Reference
Codeine		214		[5]
Coenzyme B-12				[47]
Coenzyme M transferase			Antineoplastic activity	[48]
Coniferyl ferulate	Rhizome			[49]
Corticosterone		245		[2]
Cryptochrome (flavoprotein)	Circadian clock		Anti-inflammatory, antioxidant	[50]
Curcumin			S-based DNA modification	[51]
Cysteine desulfurase			Bacteria	[52]
Cysteine residues			Reactive oxygen species	[53]
Cytochrome C				[54]
Cytochrome P450 (hemoprotein)	Oxidative metabolism			[55]
Cytochrome P450 enzyme			Metal binding	[56]
Cytosine		210, 260		[4]
Dehydrated β-NADH		254		[41]
Deoxygenated hemoglobin		340, 552		[57]
Desferrioxamine (DFO)			Inhibits nitric oxide production	[58]
Dextromethorphan		220		[10]
Dianionic DHB		283, 348		[7]
Dimethylargininase	Enzyme			[59]
DinG helicase	DNA repair		Fe–S cluster in the C-terminal domain of the p58 subunit in the DNA primase, DNA synthesis	[60]
Disialotransferrin		200		[31]
DNA primase			Contains C-terminal histone acetyltransferase (HAT) domain	[61]
Elp3 subunit of the elongator complex			Antioxidant	[62]
Eosin B-protein complex		536–544		[63]
Ethyl acetate extract of *Delonix regia* Rafin flowers			Production of Fe–S proteins	[64]
Ethylmorphine		214		[5]
Fatty acid		195		[65]
Ferricytochrome C		465		[66]
Ferredoxin (BtFd)			Assembly of Fe–S clusters in plants	[67]

(continued)

TABLE 20.7 (continued)   UV–Visible Absorption

Species	Target Organ or Disease	Absorbance Maximum (nm)	Comments	Reference
Ferredoxins				[68]
Ferulic acid		340		[69]
Fibril protein	Alzheimer's disease			[70]
Fibroblast cell	Cell, antimicrobial activity		Isolated from laksa leaves	[71]
Flavonoid	DNA antioxidant			[72]
Flavonoid	DNA antioxidant			[73]
Flavonoid	UV-protective			[74]
Flavonoid				[75]
Flavonoid			UV-protective to flower parts of plants	[76]
Flavin adenine dinucleotide (FAD)	Photosynthesis		Pyrimidine synthesis	[77]
Flavoenzyme dihydroorotate dehydrogenase (DHODH)				[78]
Flavone	Plant		Free amino acid levels in tronchuda cabbage	[79]
Folding acylphosphatase		214		[80]
Free amino acids				[81]
Free-base porphyrin	Telomerase inhibitor			[82]
Gambogic acid	Antitumor			[83]
Glucose-6-phosphate dehydrogenase		214		[84]
Glucosinolate			Intracellular antioxidant	[85]
Glutathione	DNA damage			[86]
Glutathione complex				[87]
Green fluorescent protein		479		[88]
Guaifenesin		220		[10]
Guanosine		210, 260		[4]
Harmaline		254		[89]
Harmane		254		[89]
Harmine		254		[89]
Harmol		254		[89]
Harmolol		254		[89]
Hb Aubenas		214	Type of mutation: β26 Glu-Gly	[17]
Hb Chad		214	Type of mutation: α23 Glu-Lys	[17]
Hb Debrousse		214	Type of mutation: β96 Leu-Pro	[17]

TABLE 20.7 (continued) UV–Visible Absorption

Species	Target Organ or Disease	Absorbance Maximum (nm)	Comments	Reference
Hb Knossos		214	Type of mutation: β27 Ala-Ser	[17]
Hb M Iwate		214	Type of mutation: α87 His-Tyr	[17]
Hb Olympia		214	Type of mutation: β20 Val-Met	[17]
Hb Ty Gard		214	Type of mutation: β124 Pro-Gln	[17]
Heme				[90]
Heme oxygenase (HO)				[91]
Hemoglobin (Hb)	Nerve tissue			[92]
Hb				[93]
Heroin		210		[10]
Hesperetin (bioflavonoid)	Flavone, antioxidant activity		Antimicrobial peptide	[94]
Histatin				[95]
Histone H(I) chromosomal protein		210–230	Interaction with daunomycin antibiotic	[96]
HIV-1 nucleocapsid protein (NCp7)	Protein			[97]
Human aposerumtransferrin (apoTf)				[98]
Human mitochondrial branched chain amino transferase isoenzyme (hBCATm)	Mitochondria			[99]
Human serum albumin	Protein binding		Lipids, free radicals	[100]
Human serum albumin			Drug binding	[101]
Human serum albumin				[102]
Hypochlorite				[103]
Inosine		210, 260		[4]
Lacquer polysaccharide	Antioxidant activity			[104]
Lactam compound	Cancer			[105]
Leaf methanol extract	Oxidase inhibitory potential		UV filters, sunscreen	[106]
Leu-enkephalin		200		[107]
Lichen				[108]
Lidocaine		195		[35]
Lipofuscin	Retinal disease			[109]
Lipofuscin	Retinal pigment, macular degeneration			[110]
Liposomes				[111]
Low-density lipoproteins	Atherosclerosis			[112]
Lycopene	Cancer			[113]

(continued)

TABLE 20.7 (continued) UV–Visible Absorption

Species	Target Organ or Disease	Absorbance Maximum (nm)	Comments	Reference
Lycopene		471		[114]
Lysyl oxidase cofactor	Osteolathyrism			[115]
Maitotoxin		195		[116]
Melanin	Muscle tissue			[117]
Melanin				[118]
Melatonin	Melanogenesis			[119]
Melatonin	Photodegradation			[120]
Melatonin catabolite	Hormone		Fe cluster biosynthesis	[121]
Methamphetamine		214		[5]
Met-enkephalin		200		[107]
Methadone		214		[5]
Mitochondrial protein	Mitochondria			[122]
Molybdenum cofactor (Moco)			Fe–S biogenesis	[123]
Monoanionic 2,3-dihydroxybiphenyl (DHB)		305		[7]
Monsialotransferrin		200		[31]
Morphine		210		[124]
N4-Acetylcytidine		210, 260		[4]
N6-Methyladenosine		210, 260		[4]
Naproxen		230, 270, 320, 330		[34]
nfuA gene				[125]
Nicotine		214		[25]
NifU (homodimeric modular protein)	Scaffold protein			[126]
Nitrosylleghemoglobin complex	Nitrite/nitrate reduction		Bacterial sensor, regulator of organic hydroperoxide stress	[127]
Norharman		254		[89]
Nornicotine		214		[25]
OhrR			Studied peptides, Tyr-MIF-1 and MIF-1	[128]
Oxygenated hemoglobin		415, 542, 577		[57]
Pectic polysaccharides		235		[129]
Peptides	OH scavenger		ATP binding	[130]
Periplasmic heme-transport protein (HTP)				[131]
Phencyclidine		214		[132]
Phycoferritin	Nutrition analysis		Chromophore stabilization	[5]
Phytochrome			Protein interactions	[133]

TABLE 20.7 (continued) UV–Visible Absorption

Species	Target Organ or Disease	Absorbance Maximum (nm)	Comments	Reference
Phytochrome Cph1	Chromophore		Plant, *Trianthema monogyna*; pulse, *Macrotyloma uniflorum*	[134]
Plant extracts	Calcium oxalate inhibition and dissolution		Domain ability to bind physiological metals	[135]
Plant metallothionein				[136]
Plant polyphenol verbascoside	DNA protection		Estimation of postmortem interval (PMI)	[137]
Plasma				[138]
P-Nitrophenol			Antiviral, antibacterial, immune stimulating, antiallergic, anti-inflammatory, antithrombotic, antipoperoxidant	[139]
Polyphenol			Antioxidant activity	[140]
Polyphenol				[141]
Polyphenolic compounds	Antioxidant abilities			[141]
Porphyrin	Mitochondria			[142]
Proteins—MOCS1A, MOCS1B	Oxygen-sensitive Fe–S protein		Methylthiolation in tRNAs	[143]
Pseudoephedrine		220		[10]
Pseudouridine		210, 260		[4]
Radical-S-adenosylmethionine (radical-AdoMet) enzyme MiaB				[144]
Ribonuclease A (RNase A)	HIV treatment			[145]
Ribonucleotide reductase (RNR)				[146]
Serum transferrin				[147]
Silk fibroin (SF)				[148]
S-Nitrosoglutathione	Oxidizer			[149]
Staurosporine		292		[150]
Strychnine		214		[5]
Superoxide radical	Oxygen metabolism			[151]
Suprofen		300	Molar extinction coefficient = 14,300 $(M-1)$	[34]
Testosterone		245		[2]
Tetrasialotransferrin		200		[31]

(*continued*)

TABLE 20.7 (continued)   UV–Visible Absorption

Species	Target Organ or Disease	Absorbance Maximum (nm)	Comments	Reference
Thebaine		214		[5]
Thiamine (vitamin B1)			NO sensing	[152]
Tiaprofenic acid		314	Molar extinction coefficient = 14,000 (M−1)	[34]
Transcription factor NsR			Dissolution of stones and cholesterol	[153]
Tricalcium phosphate (TCP)	Human stones		Flavone constituent of brown rice and rice bran	[154]
Tricin (flavone)			tRNA modification	[155]
Trimebutine maleate		214		[156]
Trisialotransferrin		200		[31]
tRNA-modifying MiaE protein				[157]
Tumor necrosis factor alpha (TNF-alpha)	Tumor, necrosis			[158]
UDP-galactose		262		[1]
UDP-glucose		262		[1]
UDP-N-acetylgalactosamine		262		[1]
UDP-N-acetylglucosamine		262		[1]
Unfolding acylphosphatase		214		[80]
Uridine		210, 260		[4]
Vegetable oils	Antioxidant activity			[159]
Vitamin A				[160]
Xanthosine		210, 260		[4]
Tissue/biofluid studies (in vitro)				
(E)-5-(2-Bromovinyl)-2′-deoxyuridine		254	Plasma	[161]
(E)-5-(2-Bromovinyl)-2′-deoxyuridine		254	Urine	[161]
(E)-5-(2-Bromovinyl)-uracil		254	Plasma	[161]
(E)-5-(2-Bromovinyl)-uracil		254	Urine	[161]
2,8-Dihydroxyadenine		300	Urine	[162]
2-Carboxyibuprofen		214	Urine	[163]
2-Ethylidene-1,5-dimethyl-3,3-diphenylpyrrolidine		195	Urine	[164]
2-Hydroxyibuprofen		214	Urine	[163]
3-Indoxyl sulfate		220	Urine	[165]
4-Hydroxydebrisoquine		195	Urine	[166]
4-Hydroxymephenytoin		192	Urine	[167]
4-Hydroxyphenytoin		192	Urine	[167]
5-Hydroxyindole-3-acetic acid		220	Urine	[165]
5-Hydroxytryptophan		220	Urine	[165]
Acetaminophen		200	Plasma	[168]
Acetylcarnitine		260	Urine	[169]
Acetylcholinesterase (AChE)			Erythrocyte	[170]

TABLE 20.7 (continued)   UV–Visible Absorption

Species	Target Organ or Disease	Absorbance Maximum (nm)	Comments	Reference
Adenine		254	Urine	[162]
Albendazole		280	Plasma	[171]
Albendazole sulfone		280	Plasma	[171]
Albendazole sulfoxide		280	Plasma	[171]
Allopurinol		254	Urine	[162]
Amphetamine and analogs		200	Urine	[172]
Anthocyanin	Sickle cell anemia			[173]
Anthocyanins	Ocular tissue		Eye, ocular distribution	[174]
Aryl-alcohol oxidase (AAO)	Flavoenzyme		Biochemical	[175]
Atropine		200	Serum	[176]
Bambuterol		205	Plasma	[177]
Blood	Jaundice		Blood, jaundice identification	[178]
Blood	Leukemia		Blood, leukemia identification	[179]
Blood	UV radiation		Blood	[180]
Bunitrolol		210	Plasma	[181]
Carbamazepine		210	Plasma	[182]
Carbamazepine-10,11-epoxide		210	Plasma	[182]
Carnitine		260	Urine	[169]
CGP 6423, CGP 27905		200	Urine	[183]
Cicletanine		214	Plasma, urine	[184]
Cicletanine glucuronide		214	Rat urine	[185]
Cicletanine sulfate		214	Rat urine	[185]
Clenbuterol		214	Urine	[186]
Creatinine		190	Urine	[187]
Dextrorphan		200	Urine	[188]
Dihydrocodeine		210	Urine	[189]
Dihydrocodeine		210	Plasma	[190]
Dihydrocodeine-6-glucuronide		210	Urine	[189]
DNA	Oxidative damage			[191]
Ephedrine		214	Urine	[192]
Epithelial tissue			Hamster	[193]
Fluconazole		190	Plasma	[194]
Glucuronides		214	Urine	[163]
Hexanoylcarnitine		260	Urine	[169]
Hexobarbital		214	Rat plasma	[195]
Hippuric acid		190	Urine	[187]
Hippuric acid		254	Urine	[162]
Hypoxanthine		254	Urine	[162]
Ibuprofen		220	Serum	[196]
Isovalerylcarnitine		260	Urine	[169]

(continued)

TABLE 20.7 (continued)   UV–Visible Absorption

Species	Target Organ or Disease	Absorbance Maximum (nm)	Comments	Reference
Leucovorin, 5-methyl-tetrahydrofolate		289	Plasma	[197]
Levorphanol		200	Urine	[188]
Mepivacaine		215	Serum	[198]
Naproxen		220	Serum	[165]
N-Demethyldimethindene		200	Urine	[199]
N-Demethyltramadol		195	Urine	[200]
Norcodeine		210	Plasma	[190]
Nordihydrocodeine		210	Urine	[189]
Norverapamil		200	Plasma	[201]
Octanoylcarnitine		260	Urine	[169]
O-Demethyl-N-demethyltramadol		195	Urine	[200]
O-Demethyltramadol		214	Urine	[202]
O-Demethyltramadol		195	Urine	[200]
Ondansetron		254	Serum	[203]
Orotic acid		280	Urine	[204]
Oxypurinol		254	Urine	[162]
Paclitaxel		230	Plasma	[205]
Paclitaxel		230	Urine	[205]
Pentobarbital		254	Serum	[206]
Prilocaine		215	Serum	[207]
Primaquine		210	Plasma	[181]
Propionylcarnitine		260	Urine	[169]
Quinidine		220	Urine	[165]
Quinidine		210	Plasma	[190]
R-(-)-Amphetamine		195	Urine	[208]
Reduced haloperidol		200	Plasma	[209]
Salicylate		220	Urine	[165]
Salicylate		220	Serum	[165]
Salicylic acid		275	Plasma	[165]
Salicyluric acid		220	Urine	[165]
Salicyluric acid		220	Serum	[165]
Secobarbital		254	Serum	[210]
Sulfaguanidine		260	Plasma	[168]
Sulfamethoxazole		275	Plasma	[168]
Teeth	Teeth, peroxide penetration		Teeth	[211]
Terbutaline		200	Urine	[212]
Terbutaline		210	Plasma	[213]
Tolbutamide		275	Plasma	[168]
Trimethoprim		275	Plasma	[168]
Tryptophan		220	Urine	[165]
Tryptophan		220	Serum	[165]
Type I collagen			Skin	[214]
Tyrosine		220	Urine	[165]

TABLE 20.7 (continued)   UV–Visible Absorption

Species	Target Organ or Disease	Absorbance Maximum (nm)	Comments	Reference
Tyrosine		220	Serum	[165]
Uracil		280	Urine	[204]
Urate		190	Urine	[187]
Urea		190	Urine	[187]
Uric acid		220	Serum	[165]
Vanillylmandelic acid		220	Urine	[165]
Verapamil		215	Serum	[215]
Warfarin		185	Plasma	[216]
Warfarin		310	Plasma	[217]
Xanthine		254	Urine	[162]
Animal studies (in vivo)				
8-Core-modified porphyrins, 11-core-modified porphyrins	Colo-26 tumors in BALB/c mice	694		[218]
Aluminum phthalocyanine disulfonic acid		685		[219]
Ascorbate	Eye			[220]
Disulfonated aluminum phthalocyanine	Tumor-bearing mice	680–685		[221]
Epithelial tissue	Cancer		Hamster cheek	[222]
Hematoporphyrin derivative	Rat ears	500, 625		[223]
Liposomal zinc(II)-phthalocyanine	Tumor necrosis	675		[224]
Myeloperoxidase (leukocyte)	Atherosclerosis, inflammation		Mice	[225]
N-Methyl-pyrrolidone	Tumor necrosis	671		[226]
Photodynamic therapy-induced damage	Rat liver	625		[227]
Photofrin	Mouse tumor	625		[228]
Tumor cells	Dalton's lymphoma (DL)			[229]

20.2.2 Infrared Absorption Spectroscopy

Examples on applications of FTIR spectroscopy in the field of biomolecular and biomedical research are presented in Table 20.8.

Due to the extensive literature available on FTIR of biomolecules, only illustrative examples are given. Under the biomolecular section, FTIR data on side chains of amino acids (in water) are provided. In the case of proteins, classical examples on three major conformations (a, b, and random coil) are cited. Similarly, right-handed (B) and left-handed (Z) conformational states of nucleic acids are given in this table.

Few examples on phospholipids, carbohydrates, and energy metabolites are also included.

Other sections have been arranged in the alphabetical order (except biochemical species) of column B (target organ or disease).

TABLE 20.8   Infrared Absorption Spectroscopy

Species	Major IR Bands Functional Group	Shifts (cm⁻¹)	Matrix	Comments	Reference
Biochemical studies					
18 Beta-glycyrrhetinic acid (GA)					[3]
PAPS				A novel form of adenosine-5′-phosphosulfate without an Fe–S cluster	[11]
Acetylcholine receptor				Alpha helical conformation	[230]
Acetyl-CoA synthase					[15]
AdoCbi amidohydrolase CbiS enzyme					[16]
Alanine	CH₃	1470		In-plane bending (asymmetric)	[231]
Alanine	CH₃	1378		In-plane bending (symmetric)	[231]
Alanine-based peptides				Random coil and helix	[232]
Alanine racemase				Cell wall biosynthesis, spore germination	[18]
ADH				Ethanol oxidation	[20]
Alpha-tocopherol				Determination of feed for farm animals	[21]
Amyloid beta-peptide	Alzheimer's disease				[22]
Amyloid beta-peptide	Alzheimer's disease				[23]
Amyloid deposit	Alzheimer's disease				[24]
ApbC (ATPase)				Fe–S cluster transfer	[26]
ApbC/Nbp35 (Fe cluster carrier protein)					[27]
Apolipoprotein A-1 (apo A-1)					[28]
Aporubredoxins				Fe uptake	[29]
Arginine	C–N₃ H5+	1672–1673 (420–490)		Stretching (asymmetric)	[233]
Arginine	C–N₃ H5+	1633–1636 (300–340)		Stretching (symmetric)	[233]
Ascorbic acid conjugates				Analysis of phloem exudates	[30]
Asparagine	C=O	1677–1678 (310–330)		Stretching	[234]
Asparagine	NH₂	1612–1622 (140–160)		In-plane bending	[234]
Aspartic acid	C=O	1717–1788 (in COOH)		Stretching	[235]
Aspartic acid	C–O	1120–1250 (COOH) (100–200)		Stretching	[235]
Aspartic acid	COO–	1574–1579 (290–380)		Stretching (asymmetric)	[235]
Aspartic acid	COO–	1402(256)		Stretching (symmetric)	[235]
ATP, GTP, AMP, DAMP, AMP,					[236]
AtsB (a radical SAM formylglycine-generating enzyme)				Cellular sulfate ester cleavage	[32]
Azotobacter vinelandii FldHQ				Nitrogenase activity	[33]

TABLE 20.8 (continued)   Infrared Absorption Spectroscopy

Species	Major IR Bands Functional Group	Shifts (cm⁻¹)	Matrix	Comments	Reference
Berberine	Cortical neurons				[36]
Beta-amyloid				Beta sheet structure	[237]
Beta-lapachone	Cancer				[37]
Bile acids	Antitumor				[38]
Bovine beta-LG				Protein	[42]
Calf thymus DNA				Left-handed Z type	[238]
Carotenoid					[43]
Carotenoid crocetin					[44]
ChR2				Photoreceptor	[45]
Chymotrypsin				Alpha helical conformation	[239]
C-myc gene				Transcription regulation	[46]
Coenzyme B12				Bacteria	[47]
Coenzyme M transferase					[48]
Collagen IV				Triple helical structure	[240]
Coniferyl ferulate	Rhizome			Antineoplastic activity	[49]
Cryptochrome (flavoprotein)	Circadian clock				[50]
Curcumin				Anti-inflammatory, antioxidant	[51]
Cysteine (SH)	CH(COO–)	1303		In-plane bending	[241]
Cysteine (SH)	CH₂	1424		In-plane bending	[241]
Cysteine (SH)	CH₂(COO–)	1269		Wagging	[241]
Cysteine (SH)	SH	2551		Stretching	[241]
Cysteine desulfurase				S-based DNA modification	[52]
Cysteine residues				Bacteria	[53]
Cytochrome C				Reactive oxygen species	[54]
Cytochrome C oxidase				Alpha helical conformation	[242]
Cytochrome P450 (hemoprotein)	Oxidative metabolism				[55]
Cytochrome P450 enzyme					[56]
DFO				Metal binding	[58]
Dimethylargininase	Enzyme			Inhibits nitric oxide production	[59]
DinG helicase	DNA repair				[60]
DNA primase				Fe–S cluster in the C-terminal domain of the p58 subunit in the DNA primase, DNA synthesis	[61]
Elp3 subunit of the elongator complex				Contains C-terminal HAT domain	[62]
Ethyl acetate extract of *Delonix regia* Rafin flowers				Antioxidant	[64]
FAD				Characterization	[243]
Ferredoxin (BtFd)				Production of Fe–S proteins	[67]

(continued)

TABLE 20.8 (continued)   Infrared Absorption Spectroscopy

Species	Major IR Bands Functional Group	Shifts (cm⁻¹)	Matrix	Comments	Reference
Ferredoxins				Assembly of Fe–S clusters in plants	[68]
Fibril protein	Alzheimer's disease				[70]
Fibroblast cell	Cell, antimicrobial activity				[71]
Fibroin				Beta sheet structure	[244]
Flavonoid	DNA antioxidant				[72]
Flavonoid	DNA antioxidant				[73]
Flavonoid	UV-protective			UV-protective to flower parts of plants	[74]
Flavonoid				Isolated from laksa leaves	[75]
Flavonoid					[76]
FAD	Photosynthesis				[77]
DHODH				Pyrimidine synthesis	[78]
Flavone	Plant				[79]
Free amino acids				Free amino acid levels in tronchuda cabbage	[81]
Free-base porphyrin	Telomerase inhibitor				[82]
Gambogic acid	Antitumor				[83]
Glucosinolate					[85]
Glutamic acid	C–C	1074(VW), 1040(VW), 1018(VW)		Stretching	[245]
Glutamic acid	CH	1225(SH), 1187(MW), 1130(M)		In-plane bending	[245]
Glutamic acid	C–H	1260(MW)		In-plane bending	[245]
Glutamic acid	CH₂	1452(SH), 1440(S)		In-plane bending	[245]
Glutamic acid	CH₂	1359(W), 1292(W)		Wagging	[245]
Glutamic acid	CH₂	1359(W)		Twisting	[245]
Glutamic acid	CH₂	1225(SH), 1187(MW), 1130(M)		Twisting	[245]
Glutamic acid	COO–	1556–1560(450–470)		Stretching (asymmetric)	[245]
Glutamic acid	N–C	1260(MW)		Stretching	[245]
Glutamine	C=O	1668–1687(360–380, S)		Stretching	[246]
Glutamine	C–C	1052(M), 999(WM), 926(W)		Stretching	[246]
Glutamine	CH	1410(S)		In-plane bending	[246]
Glutamine	CH₂	1451(M)		In-plane bending	[246]
Glutamine	CH₂	1315(M), 1281(WM)		Wagging	[246]
Glutamine	CH₂	1256(WM), 1202(W)		Twisting	[246]
Glutamine	C–N	1410(S), 1084(W)		Stretching	[246]
Glutamine	NH₂	1586–1611 (220–240, M)		In-plane bending	[246]
Glutamine	NH₂	1410(S)		Rocking	[246]
Glutathione	DNA damage			Intracellular antioxidant	[86]

TABLE 20.8 (continued) Infrared Absorption Spectroscopy

Species	Major IR Bands Functional Group	Shifts (cm^{-1})	Matrix	Comments	Reference
Glutathione complex					[87]
Glycine	CH	1441–1446		In-plane bending	[247]
Glycine	CH$_2$	1441–1446		In-plane bending	[247]
Heme					[90]
HO					[91]
Hemoglobin (Hb)				Alpha helical conformation	[248]
Hb	Nerve tissue				[92]
Hb					[93]
Hesperetin (bioflavonoid)	Flavone, antioxidant activity				[249]
Histatin				Antimicrobial peptide	[95]
Histidine	C=C	1575(S), 1594(70, S)		Stretching	[250]
Histidine	C=N	1490(S)		Stretching	[250]
Histidine	C=N	1304(M)		Stretching	[250]
Histidine	C–C	1575(S), 1594(70, S)		Stretching	[250]
Histidine	C–C	1265(S)		Stretching	[250]
Histidine	C–C	995(M)		Stretching	[250]
Histidine	C–C	941(S)		Stretching	[250]
Histidine	CH	1490(S)		In-plane bending	[250]
Histidine	CH	1229(S)		In-plane bending	[250]
Histidine	CH	1090(S), 1106(S)		In-plane bending	[250]
Histidine	CH	995(M)		In-plane bending	[250]
Histidine	CHx	1423(M)		In-plane bending	[250]
Histidine	CHx	975(M)		In-plane bending	[250]
Histidine	C–N	1423(M)		Stretching	[250]
Histidine	C–N	1304(M)		Stretching	[250]
Histidine	C–N	1265(S)		Stretching	[250]
Histidine	C–N	1229(S)		Stretching	[250]
Histidine	C–N	1153(M), 1161(M)		Stretching	[250]
Histidine	C–N	1090(S), 1106(S)		Stretching	[250]
Histidine	C–N	975(M)		Stretching	[250]
Histidine	NH	1423(M)		In-plane bending	[250]
Histidine	NH	1229(S)			[250]
Histidine	NH	1153(M), 1161(M)		In-plane bending	[250]
Histidine	Ring	975(M)		In-plane bending	[250]
Histidine	Ring	941(S)		In-plane bending	[250]
HIV-1 nucleocapsid protein (NCp7)	Protein				[97]
Human apoTf					[98]
Human mitochondrial branched chain aminotransferase isoenzyme (hBCATm)	Mitochondria				[99]
Human serum albumin	Protein binding				[100]
Human serum albumin					[101]
Human serum albumin				Drug binding	[102]

(continued)

TABLE 20.8 (continued)   Infrared Absorption Spectroscopy

Species	Major IR Bands Functional Group	Shifts (cm⁻¹)	Matrix	Comments	Reference
Hypochlorite				Lipids, free radicals	[103]
Isoleucine	CH	1320		In-plane bending	[251]
Isoleucine	CH_2	1470		In-plane bending	[251]
Isoleucine	CH_3	1445		In-plane bending (asymmetric)	[251]
Lacquer polysaccharide	Antioxidant activity				[104]
Lactam compound	Cancer				[105]
Leaf methanol extract	Oxidase inhibitory potential				[106]
Leucine	CH_2	1470		In-plane bending	[231]
Leucine	CH_3	1445		In-plane bending (asymmetric)	[231]
Lichen				UV filters, sunscreen	[108]
Lipofuscin	Retinal disease				[109]
Lipofuscin	Retinal pigment, macular degeneration				[110]
Liposomes					[111]
Low-density lipoproteins	Atherosclerosis				[112]
Lycopene	Cancer				[113]
Lysine	NH^{3+}	1626–1629 (60–130)		In-plane bending (asymmetric)	[233]
Lysine	NH^{3+}	1526–1527 (70–100)		In-plane bending (symmetric)	[233]
Lysozyme				Helix and beta sheet	[252]
Lysyl oxidase cofactor	Osteolathyrism				[115]
Melanin	Muscle tissue				[117]
Melanin					[118]
Melatonin	Melanogenesis				[119]
Melatonin	Photodegradation				[120]
Melatonin catabolite	Hormone				[121]
Mitochondrial protein	Mitochondria			Fe cluster biosynthesis	[122]
Moco					[123]
nfuA gene				Fe–S biogenesis	[125]
NifU (homodimeric modular protein)	Scaffold protein				[126]
Nitrosylleghemoglobin complex	Nitrite/nitrate reduction				[127]
OhrR				Bacterial sensor, regulator of organic hydroperoxide stress	[128]
Oligosaccharides				Characterization	[253]
Peptides	OH scavenger			Studied peptides, Tyr-MIF-1 and MIF-1	[130]
Periplasmic HTP				ATP binding	[131]
Phenylalanine	C–C	1605 (30), 1585, 1494 (80)		Stretching	[231]

TABLE 20.8 (continued)   Infrared Absorption Spectroscopy

Species	Major IR Bands Functional Group	Shifts (cm^{-1})	Matrix	Comments	Reference
Phospholipids				Fluidity and phase behavior	[254]
Phosphatidylcholine				Chain length and hydration properties	[255]
Phosphatidylcholine				Characterization	[256]
Phospholipids bilayer	Bilayer			Structure and organization	[257]
Phycoferritin	Nutrition analysis				[132]
Phytochrome				Chromophore stabilization	[133]
Phytochrome Cph1	Chromophore			Protein interactions	[134]
Plant extracts	Calcium oxalate inhibition and dissolution			Plant, *Trianthema monogyna*; pulse, *Macrotyloma uniflorum*	[135]
Plant metallothionein				Domain ability to bind physiological metals	[136]
Plant polyphenol verbascoside	DNA protection				[137]
Plasma				Estimation of PMI	[138]
P-Nitrophenol					[139]
Poly-L-lysine				Helix-sheet transition studied by FTIR	[258]
Polynucleotide				Right-handed B type	[259]
Polynucleotide				Triple helical structure	[260]
Polyphenol				Antiviral, antibacterial, immune stimulating, antiallergic, anti-inflammatory, antithrombotic, antipoperoxidant	[140]
Polyphenol				Antioxidant activity	[141]
Polyphenolic compounds	Antioxidant abilities				[261]
Porin				Beta sheet structure	[262]
Porphyrin	Mitochondria				[142]
Proline	CH$_2$	1292(S), 1168(S)		Twisting	[263]
Proline	C–C	979(M), 945(M), 911(M)		Stretching	[263]
Proline	CH$_2$	1472(S)		In-plane bending	[263]
Proline	CH$_2$	1375(S), 1317(S)		In-plane bending	[263]
Proline	CH$_2$	1253(M), 1051(W), 1033(M)		Wagging	[263]
Proline	CH$_2$	1083(M)		Rocking	[263]
Proline	C–N	1435–1465, 979(M), 945(M), 911(M)		Stretching	[263]
Proteins—MOCS1A, MOCS1B	Oxygen-sensitive Fe–S protein				[143]
Radical-S-adenosylmethionine (radical-AdoMet) enzyme MiaB				Methylthiolation in tRNAs	[144]

(continued)

TABLE 20.8 (continued)   Infrared Absorption Spectroscopy

Species	Major IR Bands Functional Group	Shifts (cm⁻¹)	Matrix	Comments	Reference
Rec A				Alpha helical conformation	[264]
Rhodopsin				Alpha helical conformation	[265]
RNase A	HIV treatment				[145]
RNR					[146]
RNA				Triple helical structure	[266]
Serine	C–C	1364		Stretching	[241]
Serine	CH	1364			[241]
Serine	CH	1352			[241]
Serine	CH	1312			[241]
Serine	CH_2	1450			[241]
Serine	CH_2	1352		Twisting	[241]
Serine	CH_2	1312		Wagging	[241]
Serine	CH_2	1248		Twisting	[241]
Serine	CO	1030		Stretching	[241]
Serine	CO	983		Stretching	[241]
Serine	COH	1181			[241]
Serine	COO–	1352		Stretching	[241]
Serine	NH^{3+}	1030		Rocking	[241]
Serum albumin				Alpha helical conformation	[267]
Serum transferrin					[147]
SF					[148]
S-Nitrosoglutathione	Oxidizer				[149]
Soft cuticle protein				Beta sheet structure	[268]
Sugar–metal complexes				Characterization	[269]
Sugars				Determination of glucose, fructose, sucrose	[270]
Superoxide radical	Oxygen metabolism				[151]
Thiamine (vitamin B1)					[152]
Threonine	CH	1385–1420(MW), 1225–1330(MW)			[231]
Threonine	C–O	1075–1150(S), 910–950			[231]
Threonine	COH	1385–1420(MW), 1225–1330(MW)			[231]
Transcription factor NsR				NO sensing	153
Transferrin				Alpha helical conformation	[271]
TCP	Human stones			Dissolution of stones and cholesterol	[154]
Tricin (flavone)				Flavone constituent of brown rice and rice bran	[155]
tRNA-modifying MiaE protein				tRNA modification	[157]
Tryptophan	C=C pyrrole	1612(M), 1576(W)		Stretching	[272]
Tryptophan	C–C benzene	1612(M), 1576(W)		Stretching	[272]
Tryptophan	C–C benzene	1487 (S)		Stretching	[272]
Tryptophan	C–C benzene	1455(VS)		Stretching	[272]
Tryptophan	C–C benzene	1245(S)		Stretching	[272]
Tryptophan	C–C pyrrole	1412(S)		Stretching	[272]

TABLE 20.8 (continued)   Infrared Absorption Spectroscopy

Species	Major IR Bands Functional Group	Shifts (cm⁻¹)	Matrix	Comments	Reference
Tryptophan	C–C pyrrole	1352(VS)		Stretching	[272]
Tryptophan	C–C pyrrole	1245(S), 1119(M), 1010(S), 970(SH)		Stretching	[272]
Tryptophan	C–C pyrrole	1245(S), 1092(VS)		Stretching	[272]
Tryptophan	C–C pyrrole, C–N	1334(VS)		Stretching	[272]
Tryptophan	C–H benzene	1487(S)		In-plane bending	[272]
Tryptophan	C–H benzene	1455(VS)		In-plane bending	[272]
Tryptophan	C–H benzene	1412(S)		In-plane bending	[272]
Tryptophan	C–H benzene	1352(VS)		In-plane bending	[272]
Tryptophan	C–H benzene	1245(S), 970(SH), 1010(S), 1147(W), 1119(M)		In-plane bending	[272]
Tryptophan	C–H benzene	1245(S)		In-plane bending	[272]
Tryptophan	C–H pyrrole	1509(S)		In-plane bending	[272]
Tryptophan	C–H pyrrole	1064(S)		In-plane bending	[272]
Tryptophan	C–H pyrrole	1064(S)		In-plane bending	[272]
Tryptophan	C–N	1509(S)		Stretching	[272]
Tryptophan	C–N	1455(VS)		Stretching	[272]
Tryptophan	C–N	1352(VS)		Stretching	[272]
Tryptophan	C–N	1276(S)		Stretching	[272]
Tryptophan	N–C	1064(S)		Stretching	[272]
Tryptophan	N–H	1509(S)		In-plane bending	[272]
Tryptophan	N–H	1412(S)		In-plane bending	[272]
Tryptophan	N–H	1276(S)		In-plane bending	[272]
TNF-alpha	Tumor, necrosis				[158]
Tyrosine	C–C	1326, 1335 (<30), 1290–1295 (~50)		Stretching	[273]
Tyrosine	C–C ring	1614–1621 (85–150)		Stretching	[273]
Tyrosine	C–C ring	1516–1518 (340–430)		Stretching	[273]
Tyrosine	C–H	1326, 1335 (<30), 1290–1295 (~50), 1170–1179 (~50), 1100–1111 (~40)		In-plane bending	[273]
Tyrosine	C–H ring	1614–1621 (85–150)		In-plane bending	[273]
Tyrosine	C–H ring	1516–1518 (340–430)		In-plane bending	[273]
Tyrosine	CH₂	1447–1454 (~80)		In-plane bending	[273]
Tyrosine	CH₂	1376–1387 (~50)		Wagging	[273]
Tyrosine	C–O, C–C	1235–1270 (200)		Stretching	[273]
Tyrosine	COH	1169–1260 (200), 1083		In-plane bending	[273]
Unsaturated phospholipids				Conformational studies	[274]
Valine	CH	1355, 1320		In-plane bending	[231]
Valine	CH₃	1460		In-plane bending (asymmetric)	[231]
Vegetable oils	Antioxidant activity				[159]
Vitamin A					[160]

(*continued*)

TABLE 20.8 (continued)   Infrared Absorption Spectroscopy

Species	Target Organ or Disease	Major IR Bands Functional Group	Matrix	Comments	Reference
Cellular studies					
	Bacteria			Differentiation process of bacteria studied by ATR-FTIR (bacteria)	[275]
	Bacteria			Quantification from a binary mixture using FTIR (bacteria)	[276]
	Bacteria			Characterization (bacteria)	[277]
	Bacteria			Effect of antibiotics on bacteria studied by FTIR (bacteria)	[278]
	Bacteria			Identification of bacteria using FTIR (bacteria)	[279]
	BALB/c		ZnSe	Transformed by H-ras	[280]
	Bone marrow cells			Detection of hydroxyapatite (rat)	[281]
	Bronchial	820–890, O–O stretching		Detection of damage to airway cells (human)	[282]
	Cervical cancer	1025, glycogen; 1080, nucleic acids; 1080, glycogen and nucleic acids; 1240, nucleic acids; 1400, CH_3 from lipids and proteins; 1450, CH_2 from lipids and proteins		Detection of cervical cancer (human)	[283]
	Cervical cancer	Peak at 972 is a main indicator of malignancy		Detection of cervical cancer (human)	[284]
	Cervical cancer			Detection of cervical cancer (human)	[285]
	Cervical cancer			Detection of cervical cancer using ATR/FTIR study (human)	[286]
	Deinococcus radiodurans			Pharmacological application of FTIR (bacteria)	[287]
	HL60			Differentiation and apoptosis process studied by FTIR	[288]
	Keratinocytes			Lower organization of lipids compare to human skin (human)	[289]
	Leukemia		BaF2	Detection of leukemia (human)	[290]
	Leukemia			Isolated from thrombocythemic patients (human)	[291]

TABLE 20.8 (continued)   Infrared Absorption Spectroscopy

Species	Target Organ or Disease	Major IR Bands Functional Group	Matrix	Comments	Reference
	Lung cancer cells	1030, glycogen		FTIR microscopic study (human)	[292]
	Lung fibroblast			Cell cycle and death studied by synchrotron FTIR microscopy (human)	[293]
	Lymphocytes			Histocompatibility matching using FTIR microscopy (human)	[294]
	Megakaryocytes				[291]
	NIH3T3		ZnSe	Fibroblasts transformed by retrovirus	[295]
	Normal and malignant fibroblast			Characterization using FTIR microscopy (human)	[296]
	Primary cells	Normal 1082, malignant 1086	ZnSe	Cells transformed by mouse sarcoma virus (rabbit)	[297]
	Sperm, erythrocytes			FTIR microscopic study (human)	[298]
Tissue and biofluid studies (in vitro)					
Acetylcholinesterase (AChE)					[299]
Anthocyanin	Sickle cell anemia			Eye, ocular distribution	[300]
Anthocyanins	Ocular tissue			Biochemical	[301]
AAO	Flavoenzyme			Blood, leukemia identification	[302]
Blood	Jaundice			Blood	[303]
Blood	Leukemia			Blood, jaundice identification	[304]
Blood	UV radiation				[305]
DNA	Oxidative damage			Hamster	[306]
Epithelial tissue				Teeth	[307]
Teeth	Teeth, peroxide penetration			Skin	[308]
Tumor cells	DL			Erythrocyte	[309]
Type I collagen					[310]
	Alzheimer's disease			Brain tissue studied using synchrotron FTIR microscopy (human)	[214]
	Amniotic fluid			Analysis of amniotic fluid using FTIR (human)	[211]
	Blood			Analysis of blood (human)	[193]
	Brain			Enhanced resolution due to deuteration (rat)	[178]
	Brain			Characterization using IR imaging (monkey)	[180]

(continued)

TABLE 20.8 (continued) Infrared Absorption Spectroscopy

Species	Target Organ or Disease	Major IR Bands Functional Group	Matrix	Comments	Reference
	Brain			Effect of ARA-C on brain tissue studied using FTIR microscopy (rat)	[191]
	Brain tumor			Characterization of pituitary adenomas (human)	[179]
	Brain, breast			Characterization (monkey)	[175]
	Breast			Structural variation of proteins from normal to carcinomal states (human)	[170]
	Breast			Breast cancer detection using FTIR (human)	[173]
	Breast			Grading of tumors using FTIR (human)	[174]
	Breast cancer			Detection of breast cancer using ATR-FTIR (human)	[229]
	Breast implant biopsies			Characterization (human)	[311]
	Breast implants			FTIR imaging (human)	[312]
	Cartilage		CaF_2	FTIR imaging studies (bovine)	[313]
	Cartilage			FTIR imaging studies (bovine)	[314]
	Cells, tissues			Detection of trapped CO_2 (human)	[315]
	Cervical cancer			Detection of cervical cancer using FTIR microscopy (human)	[316]
	Cirrhotic liver tissue			Liver tissue analysis (human)	[317]
	Colon cancer		ZnSe	Identification of polyp and carcinoma (human)	[318]
	Colon carcinoma	995–1045, 1195–1245 phosphate symmetric and asymmetric stretching		ATR studies (human)	[319]
	Demineralized bone tissue			FTIR imaging studies (bovine)	[320]
	Dentin biopsies			Maturation of organic and mineral contents	[321]
	Follicle fluid			Characterization (human)	[322]
	Heart			Collagen deposition using FTIR microscopy (hamster UM-X 7.1)	[323]
	Heart			Characterization (human)	[324]
	Heart tissue			Detection of myocardial infarction (rat)	[325]
	Iliac crest biopsies			Osteoporosis detected by IR imaging (human)	[326]

TABLE 20.8 (continued) Infrared Absorption Spectroscopy

Species	Target Organ or Disease	Major IR Bands Functional Group	Matrix	Comments	Reference
	Iliac crest biopsies			FTIR imaging (human)	[327]
	Implanted tissue			Biodiagnostics (dog)	[328]
	Implants			Vascular healing studies using ATR-FTIR (man-made)	[329]
	Liver			Detection of hypercholesterolemia using FTIR microscopy (rabbit)	[330]
	Liver			Tissue oxygenation studied in liver transplants using NIR	[331]
	Liver and heart	2800–3000, CH_2 and CH_3 symmetric and asymmetric stretching		Detection of diabetes	[332]
	Lung	(1045, glycogen) (1467, cholesterol)		Lung cancer detection using FTIR microscopy (human)	[333]
	Lung			Detection of mercury poisoning using ATR-FTIR (mouse)	[334]
	Lymphoid tumors			Grading of lymphoid tumors using FTIR microscopy (human)	[335]
	Mineralized tissue			Maturation in horse secondary dentin (horse)	[336]
	Multiple sclerosis			Studies on multiple sclerosis using FTIR microscopy (human)	[337]
	Multiple sites			Glucose sensing using NIR transmission (human)	[338]
	Myocardial dysfunction			Detection of myocardial dysfunction (hamsters CMP)	[339]
	Niemann–Pick type C			Detection of Niemann–Pick type C (NPC) (mice BALB/c)	[340]
	Oral			Oral cell carcinoma detection using FTIR (human)	[341]
	Oral tissue	1745, triglycerides, absent in malignant samples		FTIR-FEWS studies (human)	[342]

(continued)

TABLE 20.8 (continued)   Infrared Absorption Spectroscopy

Species	Target Organ or Disease	Major IR Bands Functional Group	Matrix	Comments	Reference
	Oral/ oropharyngeal squamous cell carcinoma			Oral/oropharyngeal squamous cell carcinoma detection using FTIR microscopy and statistical analysis (human)	[343]
	Placenta			Detection of gynecological diseases (human)	[344]
	Plasma	B–H band 2493		Detection of boron after boron neutron capture therapy (human)	[345]
	Polymer implants			Biodiagnostics (man-made)	[346]
	Retinal tissue			Studies on oxidative stress using FTIR microscopy (rat)	[347]
	Saliva, urine, serum			Measuring body water (human)	[348]
	Scrapie infected			Scrapie detection in CNS using FTIR microscopy	[349]
	Serum			Quantification of serum components (human)	[350]
	Serum			Quantification of serum components (human)	[351]
	Serum			Quantification of cholesterol (human)	[352]
	Skin			Skin cancer diagnosis using FTIR-FEWS	[353]
	Skin			Basal cell carcinoma detection using FTIR (human)	[354]
	Stool			Quantifying lipid content in biological fluids (human)	[355]
	Synovial fluid			Diagnosis of arthritis using NIR (human)	[356]
	Urine			Quantification of urine components (human)	[357]
Ascorbate	Eye			Human	[358]
Epithelial tissue	Cancer			Hamster cheek	[359]
Myeloperoxidase (leukocyte)	Atherosclerosis, inflammation			Mice	[360]
Tumor cells	DL				[361]
	Brain			Noninvasive blood glucose monitoring using ATR-FTIR (human)	[222]
	Brain			Monitoring local cortical blood flow using thermal IR imaging (human)	[225]

TABLE 20.8 (continued) Infrared Absorption Spectroscopy

Species	Target Organ or Disease	Major IR Bands Functional Group	Matrix	Comments	Reference
	Brain			Thermal IR imaging (monkey)	[229]
	Lip			FEWS-FTIR study (human)	[362]
	Skin			Influence of environmental factors on skin studied using FEWS-FTIR (human)	[363]
	Skin			Cerebrovascular activity studied by NIR (human)	[364]
	Skin			Penetration depth study (human)	[365]

20.2.3 Raman Spectroscopy

Table 20.9 is meant to serve as a quick reference work that the reader can consult to find Raman-based publications regarding a variety of chemical and biochemical compounds and systems.

These works were selected based on both their application and potential significance in the field of biophotonics. As a consequence, this table includes examples of a variety of Raman-based techniques (e.g., normal Raman, resonance Raman, coherent anti-Stokes Raman spectroscopy [CARS], surface-enhanced Raman spectroscopy [SERS]).

The instrumental requirements can vary significantly for different techniques. For such details, the reader can refer to appropriate chapters of this handbook.

Because Raman band assignments can be quite extensive and matrix-dependent in Raman scattering, such details are only supplied and reported in great confidence.

This updated edition clearly demonstrates the growing potential of Raman in a variety of biomedical applications.

The authors would also like to refer the reader to the following excellent review papers:

1. Hanlon, E. B., Manoharan, R., Koo, T.-W., Shafer, K. E., Motz, J. T., Fitzmaurice, M., Kramer, J. R., Itskan, I., Dasari, R. R., and Feld, M. S., Prospects for in vivo Raman spectroscopy, *Phys. Med. Biol.* 45, R1–R59, 2000.
2. Nabiev, I., Chourpa, I., and Manfait, M., Applications of Raman and surface-enhanced Raman scattering spectroscopy in medicine, *J. Raman Spectrosc.* 25, 13–23, 1994.
3. Lawson, E. E., Barry, B. W., Williams, A. C., and Edwards, H. G. M., Biomedical applications of Raman spectroscopy, *J. Raman Spectrosc.* 28, 11–117, 1997.
4. Pappas, D., Smith, B. W., and Winefordner, J. D., Raman spectroscopy in bioanalysis, *Talanta* 51, 131–144, 2000.
5. Lin, S. Y., Li, M. J., and Cheng, W. T., FT-IR and Raman vibrational microspectroscopies used for spectral biodiagnosis of human tissues, *Spectroscopy* 21(1), 1–30, 2007.

TABLE 20.9 Raman Spectroscopy

Species	Target Organ or Disease	Excitation λ (nm)	Major Raman Shifts (cm^{-1})	Comments	Reference
Biochemical studies					
Alpha-tocopherol (aT)	Lung, oxidation of fatty acids				[366]
Aminoglutethimide	Cancer (drug)			Examined environmental and solvent effects (e.g., CCl$_4$, CHCl$_3$, and CH$_3$CN) on IR and Raman spectra of aminoglutethimide.	[367]
Amorphous calcium carbonate	Bone, biomineralization				[367]
Amsacrine (m-AMSA)	Cancer (drug)	UV		UV resonance Raman. Spectroscopically characterized.	[368]
m-AMSA	Cancer (drug)	NIR		FT-Raman. Spectroscopically characterized.	[369]
m-AMSA	Leukemia (drug)	UV	1165, 1265, 1380	UV resonance. Monitored interaction of anticancer drug, m-AMSA, with calf thymus DNA. Observed spectral changes between free and DNA-bound m-AMSA.	[370]
m-AMSA	Topoisomerase II inhibitor/cancer (drug)			SERS. Monitored intracellular interactions of m-AMSA within living K562 cancer cells.	[371]
Amyloid beta peptide	Brain, Alzheimer's disease				[372]
Amyloid fibrils	Brain, Alzheimer's disease, Creutzfeldt–Jakob disease (CJD), type 2 diabetes				[373]
Antineoplaston-A10	Cancer (drug)			Characterized with vibrational assignments. Vibrational modes similar to uracil derivatives, implying that therapy based on resemblance to pyrimidine bases.	[374]
Apatite	Cell			Micro-Raman.	[371]
Atherosclerotic lesions	Heart, atherosclerosis			Confocal Raman.	[372]
Au-imidazole	Tumors	NIR	954	NIR SERS.	[375]

Compound	Application	Spectral data	Description	Reference
Avidin			Demonstrated the importance of the side chains of biotin when interacting with avidin.	[376]
Avidin			Demonstrated the importance of the side chains of biotin when interacting with avidin.	[377]
Bacteria	Bacteria, cell component evolution			[378]
Bacteria	Bacteria, fatty acid composition			[379]
Bacteria	Bacteria, species identification		2D resonance Raman.	[380]
Bacteria	Bacterial cell wall, extracellular analysis		SERS.	[381]
Benzo(a)pyrene	Cancer		Theoretical study of the oxygenation of benzo(a)pyrene.	[382]
Biotin			Demonstrated the importance of the side chains of biotin when interacting with avidin.	[382]
Biotin			Demonstrated the importance of the side chains of biotin when interacting with avidin.	[383]
Blenoxane (bleomycins A2 and B2)	DNA/cancer (drug)		A study in drug/DNA interaction. Spectral changes observed between calf thymus DNA and bleomycin/DNA complex.	[384]
Ca phosphate	Bone, osteomyelitis			[385]
Calcium polyphosphonate (CPP) bone substitutes	Bone, mechanisms			[386]
Cationic adamantyl derivatives (Ada)	Gene splicing vector		SERS.	[387]
Cell	Blood platelet, species differentiation	1157, 1524	Laser tweezers Raman spectroscopy (LTRS).	[388]
Cell	Breast, cancer			[389]
Cell	Cell, cell growth on biopolymer surface coated with different noble metals			[390]

(continued)

TABLE 20.9 (continued) Raman Spectroscopy

Species	Target Organ or Disease	Excitation λ (nm)	Major Raman Shifts (cm⁻¹)	Comments	Reference
Cell	Cell, cellular structures			CARS.	[391]
Cell	Cell, differentiation			Provides insight into the biochemical properties of alveolar surfactant in its unperturbed cellular environment.	[392]
Cell	Cell, osteosarcoma cell			NIR SERS, measurements implying gold nanoparticles.	[393]
Cell	Cell, stem cell growth and differentiation				[394]
Cell	Epithelial tissue (human)			CARS imaging.	[395]
Cell	Keratinocytes, toxic agent exposure and response to chemical stress				[395]
Cell	Live epithelial cell, cancer			Raman tweezers, able to study cell dynamics of carcinogenesis.	[377]
Cell	Mouth, oral squamous cell carcinoma			SERS.	[377]
Cell	T cell, renal transplant rejection			Able to detect significant differences between activated and nonactivated T cells based on cell-surface receptor expression.	[396]
Cell, live	Cell, osteosarcoma cell			Raman microspectroscopy.	[397]
Cells	Cells, multidrug resistance (MDR)				[398]
Cells	Hematopoietic cell, leukemia cell, cancer			Laser trapping Raman spectroscopy, able to identify Raman markers specific to cancerous and noncancerous cell types.	[399]
Cells	T-cell activation		785, 1004, 1048, 1093, 1376, 1660		[400]
Cells, human papillomavirus (HPV)	Uterus, uterine cervical neoplasia				[401]

Cervical cell	Uterus, cervix, cancer			[402]
Cinchonine	Malaria (drug)	853, 938	Phosphate stretch band of cinchonine observed to shift upon interaction with DNA, indicative of coulombic interaction.	[403]
		1374		
cis-Dichloroplatinum derivatives	Telomerase inhibitor/cancer (drug)		Characterized cis-dichloroplatinum derivatives with N-donor ligands such as pyridine and 5-substituted quinolines.	[404]
Cisplatin	Carbon nanotubes with cisplatin, cancer			[405]
CO_2	Lung (respiratory quotient test)		Inspired, end-tidal, and mixed expired gas compositions analyzed by Raman spectroscopy.	[406]
Collagen type I	Skin, pilomatrixoma (PMX)	509, 749, 850, 960, 1246, 1665		[407]
Condensed nuclear chromatin	Cell, cell stage			[408]
Cytochrome c	Heart		SERS.	[409]
Cytochrome ĉ	Heme, NO binding		Resonance Raman.	[410]
Cytoglobin (Cgb)	Globin superfamily in mammals	572	Resonance Raman.	[411]
Dentine	Teeth, demineralization	Micro-Raman spectroscopy.		[412]
Diamine oxidase	Kidney (human)	Resonance Raman.		[413]
Dimethylcrocetin (DMCRT)	Retinoic acid nuclear receptor interaction/cancer (drug)	1210, 1165, 1541	FT–Raman, FT-SERS. Evaluated spectra for free-form DMCRT and DMCRT/ retinoic acid nuclear receptor in HL60 and K562 cancer cells.	[414]
Dinitrophenol derivative (DAMP)	Cell, lysosome		Nanoparticles used to quantify biochemicals in cellular environment.	[415]
DNA	DNA, conformation changes due to platinum antitumor drugs		Insight into structural factors involved in mechanisms underlying antitumor effects of platinum drugs.	[416]
DNA			SERS. SERRS. Studied adsorption of DNA onto substrates.	[417]

(continued)

TABLE 20.9 (continued)   Raman Spectroscopy

Species	Target Organ or Disease	Excitation λ (nm)	Major Raman Shifts (cm^{-1})	Comments	Reference
DNA (via cresyl fast violet)	HIV/AIDS	633		SERS. HIV DNA amplified (PCR) with cresyl fast violet-labeled primers and selectively hybridized to a bioassay platform.	[418]
DNA (via dye label)				SERS. Gene probes hybridized to target DNA in solution, spotted on SERS substrate.	[419]
Dopamine			950, 1006, 1258, 1378, 1508, 1603, 948, 1010, 1255, 1373, 1510	Resonance Raman.	[420]
Embryonic stem (ES) cells	Cell differentiation			CARS microscopy.	[421]
Epithelial cells	Colon, colorectal cancer				[82]
Ferrochelatase	Porphyrin, heme synthesis (mouse)			Resonance Raman.	[422]
Free-base porphyrin	DNA G-quadruplexes			UV resonance Raman, SERS, demonstrate effects of free-base porphyrins binding to DNA quadruplexes.	[423]
Gelonin	Type 1 ribosome inactivating protein/aids (drug)			Determined secondary structure of gelonin to be mainly alpha helix and beta sheet with some turn and disordered structure.	[424]
Gelonin	Type 1 ribosome inactivating protein/cancer (drug)			Determined secondary structure of gelonin to be mainly alpha helix and beta sheet with some turn and disordered structure.	[425]
Glucose				SERS, biosensor.	[426]
Glutamate	Eye, nerve cell damage, glaucoma, diabetic retinopathy		1369, 1422		[427]
Growth hormone receptor	Cell, cancer			Raman mapping.	[428]

Name	Tissue/type	Technique	Wavenumbers	Description	Ref.
Guanylate cyclase	Nitric oxide receptor/lung		474–492	Investigated interaction of YC-1 with enzyme soluble guanylate cyclase (sGC) in the presence of CO for sGC expressed by baculovirus/Sf9 cell.	[429]
Guanylate cyclase	Nitric oxide receptor/lung	UV	478, 495, 525, 567, 572, 1676	Resonance Raman. Characterized the heme domain of rat lung sGC soluble guanylate cyclase.	[430]
Guanylate cyclase	Nitric oxide receptor/lung	UV	212, 221	Resonance Raman. Identified histidine 105 in the beta 1 subunit of sGC from rat and bovine lung.	[431]
Guanylate cyclase	Nitric oxide receptor/lung	UV	497, 520, 1959	Resonance Raman. Spectra illustrated for the alpha(1) beta(1) isoform of lung sGC expressed from baculovirus.	[432]
Guanylate cyclase	Nitric oxide receptor/lung	UV	1357, 1375, 1473	Resonance Raman. Spectra reported for ferrous and ferric forms of sGC from bovine lung.	[432]
Guanylate cyclase	Nitric oxide receptor/lung	UV	203, 473, 521, 1681, 1700	Resonance Raman. Spectra reported for reduced, GO-bound, NO-bound, and oxidized NO-bound forms of sGC from bovine lung with/without guanosine-5'-triphosphate (GTP).	[433]
H^+, K^+-ATPase	Stomach			Investigated conformational changes induced by replacing K^+ ions with Na^+ ions. No significant Raman spectral changes observed.	[434]
H^+, K^+-ATPase	Stomach			Investigated secondary structure of membrane-bound regions of H^+, K^+-ATPase. Determined structure to be largely beta sheet in character.	[435]
H_2O_2	Epidermis, blood, vitiligo			FT–Raman.	[436]
Hemoglobin	Heart, blood, heart failure				[437]
Heme ci	Heme			Resonance Raman.	[143]
Heme protein	Heme protein CO activation			Investigated mechanism for heme protein activation.	[438]
Heme-PAS domain sensor protein BxRcoM-Z	Protein			Resonance Raman.	[439]

(continued)

TABLE 20.9 (continued) Raman Spectroscopy

Species	Target Organ or Disease	Excitation λ (nm)	Major Raman Shifts (cm⁻¹)	Comments	Reference
Hemoglobin	Blood, red blood cell disorders	632.8			[440]
Hemoglobin				Resonance Raman.	[441]
Hepatitis delta virus (HDV)	Ribozyme, pKa measurements				[442]
Human adult hemoglobin (HbA) group, Cycβ93	NO binding	UV		UV resonance Raman.	[443]
HAP	Inclusions (bone)			Compared biological HAP samples to ox femur.	[443]
Indoleamine 2,3-dioxygenase (IDO) enzyme	Heme enzyme			Resonance Raman.	[444]
IDO enzyme	L-Tryptophan metabolism via kynurenine pathway in nonhepatic tissue			Resonance Raman.	[445]
Insulin amyloid fibrils	Alzheimer's disease			Resonance Raman, protein-based delivery media.	[446]
Integrins	Proteins, blood clotting			SERS.	[447]
Intoplicine	Topoisomerase II inhibitor/cancer (drug)			SERS. Intoplicine found to unwind DNA and inhibit calf thymus topoisomerase II. Intoplicine spectrum eliminated when in the presence of topoisomerase II.	[448]
Lipid droplets (LDs)	Cells			CARS, variations in lipid composition and physical state between LDs contained in same cell and even within single LD.	[449]
Losartan—antihypertensive drug	Cell			Reports effects on cellular proliferation and morphology after dosing.	[450]
Lysozyme		632.8		First interpretable laser-excited spectrum of a native protein.	[451]
Matrix metalloproteinase (MMP-1)	Skin (human), IR damage, and aging			Skin aging and topical application of appropriate antioxidants represent an effective photoprotective strategy.	[452]

Media	Metabolic profiling associated with in vitro fertilization, embryo viability (human)	Allows rapid noninvasive assessment of embryonic reproductive potential before transfer. [453]
Methotrexate	Dihydrofolate reductase inhibitor/cancer (drug)	Ultrasensitive Raman difference spectroscopy illustrated light-induced reaction of methotrexate with NADPH (nicotinamide adenine dinucleotide phosphate) in dihydrofolate active site. [454]
Methotrexate	Dihydrofolate reductase inhibitor/rheumatoid arthritis (drug)	Ultrasensitive Raman difference spectroscopy illustrated light-induced reaction of methotrexate with NADPH in dihydrofolate active site. [455]
NAD(P)H cytochrome b(5) oxidoreductase	Flavoprotein	Resonance Raman. [456]
Nitric oxide synthases (NOSs)	Heme proteins	Resonance Raman. [457]
Nitrogen	Lung (respiratory quotient test)	Inspired, end-tidal, and mixed expired gas compositions analyzed by Raman spectroscopy. [458]
O_2 sensitive Fe–S protein	Human molybdenum cofactor, biosynthesis	Resonance Raman. [459]
Oxygen	Lung (respiratory quotient test)	Inspired, end-tidal, and mixed expired gas compositions analyzed by Raman spectroscopy. [460]
Phospholipids, triglycerides	Heart, atherosclerosis	[461]
Procapsid, purified recombinant core protein, purified subassemblies		Demonstrate conformational changes in procapsid. [462]
Proopiomelanocortin (POMC)-derived peptides	Skin, vitiligo, pigmentation antioxidant	FT–Raman. [463]
Protein, neuroglobin (Ngb)	Brain, hypoxic–ischemic injuries	Resonance Raman. [464]
Proteins		Elucidated kinetics and structures of proteins. [465]
Proteins		Elucidated kinetics and structures of proteins. [466]

(continued)

TABLE 20.9 (continued)   Raman Spectroscopy

Species	Target Organ or Disease	Excitation λ (nm)	Major Raman Shifts (cm^{-1})	Comments	Reference
Proteins				Elucidated kinetics and structures of proteins.	[467]
Proteins				Identified Raman vibrational bands through isotopic substitution.	[468]
Proteins				Identified Raman vibrational bands through isotopic substitution.	[469]
Proteins				Identified Raman vibrational bands through isotopic substitution.	[470]
Proteins				Identified Raman vibrational bands through isotopic substitution.	[471]
Review	Atherosclerosis				[472]
Ribozymes	Hepatitis		1528		[473]
Saturated fatty acid	Saturated fatty acid mixture, phase behavior study			FT–Raman.	[474]
Serum albumins	Albumin; exposure to proton and gamma irradiation			Confirmed radiation-induced denaturation, destruction of helical structures, and aggregation of serum albumins.	[475]
Single human lymphocytes	Lymphocyte, cancer				[476]
Spore	*Bacillus subtilis* ATCC 13933			SERS.	[477]
Sugar synthesis	Fungal beta-N0 acetylhexosaminidase				[478]
Sugars			280, 1053, 1086.5		[479]
Synthetic HAP	Teeth, decay			Helps to understand influence of fluoride application on mechanism of tooth decay.	[480]
Theraphthal	Cancer (drug-binary catalytic system)			Resonance Raman. Studied effects of molecular interactions (e.g., with Ca$^+$) and environmental factors (e.g., pH <6) on spectra of theraphthal.	[481]
Tissue	Cells, leukocytes			CARS imaging.	[482]

Tissue	Lactate production; heart failure, hypoxia, diabetic ketoacidosis			SERS.	[483]
Tissue	Lung			NIR Raman.	[484]
Tissue	Skin (mouse), cancer, radiation damage	Amide I, P=O, C–O		Assessment of radiation damage to biological samples from protons.	[485]
Triglyceride-rich lipoprotein (TGRL), very-low-density lipoproteins (VLVL)	Endothelium, lipid metabolism				[486]
Tyrosine	Human serum	245		UVRR (ultraviolet resonance Raman).	[414]
Vanadium complex	Kidney, cancer				[487]
Whole cell	Cell, cell death cycle	782, 788, 1095, 1231, 1190–1385			[488]
Whole cells	Cell culture, tumors				[489]
Whole cells	Cells, aging				[490]
	Drug			Studied kinetics and interaction mechanisms between drugs and receptors.	[491]
aT	Lung, oxidation of fatty acids				[404]
Amorphous calcium carbonate	Bone, biomineralization				[405]
m-AMSA	K562 cells/cancer (drug)			SERS. Monitored intracellular interactions of m-AMSA within living K562 cancer cells.	[369]
Amyloid beta peptide	Brain, Alzheimer's disease				[406]
Amyloid fibrils	Brain, Alzheimer's disease, CJD, type 2 diabetes				[407]
Au-imidazole	Tumors	954	NIR	NIR SERS.	[408]
Bacteriochlorophyll A	Photosynthetic bacteria			Resonance Raman. Substitution of Mg ion by Zn, Ni, Co, and Cu ions studied via Raman scattering. Molecule core size shifts caused shifts in resonance.	[492]
Ca phosphate	Bone, osteomyelitis				[409]
Calcium dipicolinate	Bacteria		UV		[493]
CPP bone substitutes	Bone, mechanisms				[410]

(continued)

TABLE 20.9 (continued)   Raman Spectroscopy

Species	Target Organ or Disease	Excitation λ (nm)	Major Raman Shifts (cm⁻¹)	Comments	Reference
Carotenoids	Liposomes			Egg phosphatidylcholine and egg phosphatidylglycerol. Observed beta carotene rapidly taken up in vitro and transported to *Gall bodies*.	[494]
Carotenoids	Lymphocytes			Lymphocytes from human blood. Observed that carotenoids concentrated in *Gall bodies*. Associated carotenoid concentrations with age, decreasing with age.	[494]
Carotenoids	Lymphocytes/lung cancer			Human. Compared carotenoid levels in lymphocytes of lung cancer patients and healthy individuals. Observed lower carotenoid levels in lung carcinoma patients.	[495]
Cationic Ada	Gene splicing vector		SERS		[411]
Chromatid	Ovary cells/mutagenesis (photo-induced)	193		Chinese hamster. Reported monitoring of mutagenesis and sister chromatin exchange induction via UV irradiation.	[496]
Chromatid	Ovary cells/mutagenesis (photo-induced)	308		Chinese hamster. Reported monitoring of mutagenesis and sister chromatin exchange induction via UV irradiation.	[496]
Chromatid	Ovary cells/mutagenesis (photo-induced)	254		Chinese hamster. Reported monitoring of mutagenesis and sister chromatin exchange induction via UV irradiation.	[496]
Cisplatin	Carbon nanotubes with cisplatin, cancer				[412]
Collagen type I	Skin, PMX		509, 749, 850, 960, 1246, 1665		[413]
Cytochrome c	Heart			SERS.	[414]

Cytochrome c'	Heme, NO binding		Resonance Raman.	[415]
Cgb	Globin superfamily in mammals	572	Resonance Raman.	[416]
Demoic acid (neurotoxin)	Phytoplankton	UV	Micro-Raman spectroscopy.	[497]
Dentine	Teeth, demineralization			[417]
Diamine oxidase	Kidney (human)		Resonance Raman.	[418]
DMCRT	HL60 and K562 cells/cancer (therapy)	1210, 1165, 1541	FT-Raman; FT-SERS. Evaluated spectra for free-form DMCRT and DMCRT/ retinoic acid nuclear receptor in HL60 and K562 cancer cells.	[378]
DNA	DNA, conformation changes due to platinum antitumor drugs		Insight into structural factors involved in mechanisms underlying antitumor effects of platinum drugs.	[419]
DNA	FD virus	UV		[498]
DNA	Virus	257		[499]
DNA	Virus	244		[499]
DNA	Virus	238		[499]
DNA	Virus	229		[499]
DNA	Virus			[499]
DNA	Virus		Resonance Raman. DNA monitored during assembly of viruses.	[500]
DNA	Virus (T7 bacteriophage)		Resonance Raman. Monitored DNA spectra for DNA during packaging into viral plasmid with little changes. Mg ions observed to influence packaging.	[501]
DNA		Visible	DNA in chromosomes. Investigation of cellular damage at different wavelengths demonstrated that 514.5 nm radiation causes damage while 660 nm radiation does not.	[502]
DNA		514.5, 660	DNA in chromosomes.	[503]
DNA		Visible	DNA in chromosomes.	[504]
DNA		Tuned	Resonance Raman. DNA preferentially detected in cells.	[505]

(continued)

TABLE 20.9 (continued)   Raman Spectroscopy

Species	Target Organ or Disease	Excitation λ (nm)	Major Raman Shifts (cm⁻¹)	Comments	Reference
DNA		Tuned		Resonance Raman. DNA preferentially detected in cells.	[505]
Dopamine		950, 1006, 1258, 1378, 1508, 1603, 948, 1010, 1255, 1373, 1510		Resonance Raman.	[420]
Ferrochelatase	Porphyrin, heme synthesis (mouse)			Resonance Raman.	[421]
Free-base porphyrin	DNA G-quadruplexes			UV Resonance Raman, SERS, demonstrate effects of free-base porphyrins binding to DNA quadruplexes.	[82]
Glucose				SERS, biosensor.	[422]
Glucose (via quinoline red)				Attached quinoline red to monitor intake of glucose into cells. Glucose intake detected by intensity changes in quinoline red Raman spectrum.	[506]
Glutamate	Eye, nerve cell damage, glaucoma, diabetic retinopathy		1369, 1422		[423]
H_2O_2	Epidermis, blood, vitiligo				[424]
Hemoglobin	Heart, blood, heart failure			FT–Raman.	[425]
Heme ci	Heme			Resonance Raman.	[426]
Heme-PAS domain sensor protein BxRcoM-Z	Protein			Resonance Raman.	[427]
Hemoglobin	Blood, red blood cell disorders	632.8			[429]
Hemoglobin	Rheumatoid arthritis	Visible		Monitoring of oxygen uptake in human cells. Concluded that patients with rheumatoid arthritis uptake oxygen faster but incompletely.	[507]

		λ	Peaks (cm⁻¹)	Notes	Ref.
Hemoglobin				Resonance Raman.	[428]
HDV	Ribozyme, pKa measurements				[430]
Hepatocytes	Liver cancer	Visible	1040, 1083, 1182, 1241, 1253	Distinguished between normal and malignant hepatocytes.	[508]
HbA group, Cycβ93	NO binding	UV		UV resonance Raman.	[431]
HAP	C4–2B cell/prostate cancer			Mineral deposition monitored for C4–2B prostate cancer cell line in promineralization media.	[509]
IDO enzyme	Heme enzyme			Resonance Raman.	[432]
IDO enzyme	L-Tryptophan metabolism via kynurenine pathway in nonhepatic tissue			Resonance Raman.	[432]
Insulin amyloid fibrils	Alzheimer's disease			Resonance Raman, protein-based delivery media.	[433]
Intoplicine	K562 cells/cancer (drug)			SERS. Intoplicine spectrum eliminated when in the presence of topoisomerase II. Intoplicine observed as complexed in nucleus and free form in cytoplasm of K562 cells.	[393]
LDs	Cells			CARS, variations in lipid composition and physical state between LDs contained in same cell, and even within single LD.	[434]
Media	Metabolic profiling associated with in vitro fertilization, embryo viability (human)			Allows rapid noninvasive assessment of embryonic reproductive potential before transfer.	[435]
NAD(P)H cytochrome b(5) oxidoreductase	Flavoprotein			Resonance Raman.	[436]
NOSs	Heme proteins			Resonance Raman.	[437]
Nucleic acids	Breast cells/cancer	257		Reported spectra for cultured normal and malignant breast cells. Differences attributed to relative amounts of nucleic acids and proteins and nuclear hypochromism.	[510]

(continued)

TABLE 20.9 (continued)   Raman Spectroscopy

Species	Target Organ or Disease	Excitation λ (nm)	Major Raman Shifts (cm^{-1})	Comments	Reference
Nucleic acids	Cervical cells/cancer	257		Reported spectra for cultured normal and malignant cervical cells. Differences attributed to relative amounts of nucleic acids and proteins and nuclear hypochromism.	[510]
Nucleic acids	T84 cancer cell	257			[511]
O$_2$ sensitive Fe–S protein	Human molybdenum cofactor, biosynthesis			Resonance Raman.	[143]
Phospholipids, triglycerides	Heart, atherosclerosis				[438]
Photosensitizers	Cancer cells			Monitored accumulation of photosensitizers in cancer cells.	[512]
Photosensitizers	Cancer cells			Monitored accumulation of photosensitizers in cancer cells.	[513]
Phthalocyanates	Cancer cells		1538	Hydrophilic phthalocyanates accumulate in lysosomes, while hydrophobic phthalocyanates accumulate in cytoplasm.	[514]
Pigments	Lung cancer	Visible		Pigments in human lymphocytes with high spatial resolution.	[515]
Pigments		Visible		Pigments in human granulocytes with high spatial resolution.	[516]
Pigments		Visible			[503]
Pigments		Visible		Pigments in human lymphocytes with high spatial resolution.	[517]
Procapsid, purified recombinant core protein, purified subassemblies				Demonstrate conformational changes in procapsid.	[439]
POMC-derived peptides	Skin, vitiligo, pigmentation antioxidant			FT–Raman.	[440]

Protein	Breast cells/cancer	257	Reported spectra for cultured normal and malignant breast cells. Differences attributed to relative amounts of nucleic acids and proteins and nuclear hypochromism.	[510]
Protein	Cervical cells/cancer	257	Reported spectra for cultured normal and malignant cervical cells. Differences attributed to relative amounts of nucleic acids and proteins and nuclear hypochromism.	[510]
Protein	FD virus	UV		[518]
Protein	Virus	257		[499]
Protein	Virus	244		[499]
Protein	Virus	238		[499]
Protein	Virus	229		[499]
Protein (active site)	Virus (HAV [hepatitis A virus] 3C)		Active site structure of HAV 3C monitored as a function of pH.	[519]
Proteins	Bacteria and viruses	231	UV resonance Raman. Obtained fluorescence-free, selective Raman detection of protein aromatic group residues in bacteria and viruses.	[520]
Proteins	Virus (P22)/replication		Protein side chain and secondary structure monitored as a virus (P22) entered a host cell, replicated components, and assembled new viruses.	[521]
Protoplasts	Bacteria	UV		[493]
Purines	Bacteria and viruses	242	UV resonance Raman. Obtained fluorescence-free, selective Raman detection of purine aromatic group residues in bacteria and viruses.	[520]
Pyrimidines	Bacteria and viruses	251	UV resonance Raman. Obtained fluorescence-free, selective Raman detection of pyrimidine aromatic group residues in bacteria and viruses.	[520]

(continued)

TABLE 20.9 (continued)   Raman Spectroscopy

Species	Target Organ or Disease	Excitation λ (nm)	Major Raman Shifts (cm⁻¹)	Comments	Reference
Review	Atherosclerosis				[441]
Saturated fatty acid	Saturated fatty acid mixture, phase behavior study			FT–Raman.	[442]
Serum albumins	Albumin; exposure to proton and gamma irradiation			Confirmed radiation-induced denaturation, destruction of helical structures, and aggregation of serum albumins.	[443]
Sugar synthesis	Fungal beta-N0 acetylhexosaminidase				[444]
Sugars		280, 1053, 1086.5			[445]
Synthetic HAP	Teeth, decay			Helps to understand influence of fluoride application on mechanism of tooth decay.	[446]
Theraphthal	A549 cells/cancer (BCS [breast conserving surgery] therapy)			Resonance Raman confocal microspectroscopy and imaging. Monitored intracellular accumulation, localization, and retention of theraphthal in living cancer cells.	[522]
Tissue	Lactate production; heart failure, hypoxia, diabetic ketoacidosis			SERS.	[447]
Tissue	Skin (mouse), cancer, radiation damage		Amide I, P=O, C–O	Assessment of radiation damage to biological samples from protons.	[448]
TGRL, VLVL	Endothelium, lipid metabolism				[449]
Tyrosine	Human serum		245	UVRR.	[450]
Vanadium complex	Kidney, cancer				[451]
	Bacteria	UV			[523]
	Bacteria	UV			[524]
	Bacteria	UV			[493]

	Bacteria	UV		Examined effects of cultural conditions on deep UV resonance Raman spectra of bacteria.	[525]
1,8-Cineole	Bacterial spores	UV			[493]
	Skin (human), pharmaceutical skin penetration				[526]
Acetone	Urine		796.5	Introduced compact, highly sensitive Raman system with integration sphere cell holder. LOD for acetone was 8 mg/dL.	[527]
Acetone	Urine		789	Human. Anti-Stokes Raman. Demonstrated quantitative analysis of acetone in urine.	[528]
Acetone	Urine	NIR	790	Human. Demonstrated quantitative analysis of acetone in urine with an LOD of 40 mg/dL.	[529]
Adenomatous and hyperplastic polyps	Colon, cancer			NIR Raman.	[530]
Adenylate	Colon	251		UV resonance Raman.	[531]
Advanced glycation end products	Eyes (human), aging and macular degeneration			Confocal Raman, able to predict chronological age of patients.	[532]
Alpha-synuclein aggregates	Fibrils, Parkinson's disease		Amide I 1667		[533]
Alumina	Bone/implants			Changes in alumina spectra investigated after implantation to establish biocompatibility.	[534]
Amiloride	Urine (drug test)			Human. SERS. Demonstrated quantitative capability in urine with colloidal silver.	[535]
Amides	Breast/cancer	1064	Normal tissue: 1078, 1300, 1445, 1651; malignant tissue: 1445, 1651; benign tumors: 1240, 1445, 1659	NIR-FT–Raman. Human.	[536]

(continued)

TABLE 20.9 (continued)   Raman Spectroscopy

Species	Target Organ or Disease	Excitation λ (nm)	Major Raman Shifts (cm⁻¹)	Comments	Reference
Amides	Skin (stratum corneum)	NIR		Human. FT-Raman. Forensic science. Demonstrated nondestructive investigation of skin samples from late Neolithic man and contemporary samples.	[537]
Amides	Various tissues	NIR		Charge coupled device (CCD). Mammalian tissues. Evaluated effects of drying, snap freezing, thawing, and formalin fixation.	[538]
Amphetamine	Urine (drug test)			Human. SERS. First phase of development of one-step drug test using selective reactive SERS substrate coating. LODs <20 ppm observed, required amide derivatization.	[539]
Amyloid plaques	Brain, Alzheimer's disease			Demonstrated that senile plaque cores are composed largely of amyloid-β and β sheets.	[540]
Apatite	Bone and enamel (animal and human)				[541]
Apatite	Bone, scaffolding				[542]
Apatite	Bone, skeletal mineralization				[543]
Aromatic amino acids	Colon/cancer	240		Human. Reported Raman spectra for normal and neoplastic colon mucosa. Neoplastic samples showed lower amino acid to nucleotide content ratio.	[544]
Aromatic amino acids	Colon/cancer	251		Human. Reported Raman spectra for normal and neoplastic colon mucosa. Neoplastic samples showed lower amino acid to nucleotide content ratio.	[544]
Atorvastatin and amlodipine	Heart (mouse), atherosclerosis				[545]
β-Amylose	Brain/Alzheimer's		1628	Human. Postmortem. Distinguished between diseased and healthy tissue.	[546]
Basal cells	Skin, basal cell carcinoma				[547]

β-Carotene	Breast/cancer	406.7		Human. Distinguished between normal and malignant breast tissue, attributing differences to relative concentrations of fatty acids and β-carotene.	[548]
β-Carotene	Breast/cancer	457.9		Human. Distinguished between normal and malignant breast tissue, attributing differences to relative concentrations of fatty acids and β-carotene.	[548]
β-Carotene	Breast/cancer	514.5		Human. Distinguished between normal and malignant breast tissue, attributing differences to relative concentrations of fatty acids and β-carotene.	[548]
Bladder tissue	Bladder, prostate, cancer			Fiber-optic Raman system.	[549]
Blood	Blood (human), hepatitis C	830	1002, 1169, 1262, 1348	NIR Raman, able to classify blood serum spectrum by identifying biochemical difference in the presence of viral infections.	[550]
Blood plasma	Cancer	488		Distinguished between cancer patients and healthy individuals by differences in carotenoid spectral regions.	[551]
Blood, tissue	Blood (animal), tissue, oxygen saturation		1355, 1378	Resonance Raman.	[552]
Blue particles???	Cancer tumors			Raman microspectroscopy with resolution better than 1 μm	[553]
Bone	Bone			Raman spectroscopic imaging.	[554]
Bone	Bone (animal), effects of gamma radiation	1064		FT-NIR–Raman.	[555]
Bone	Bone (human), craniosynostosis				[556]
Bone	Bone (human), crystallinity, osteoporosis				[557]

(*continued*)

TABLE 20.9 (continued)   Raman Spectroscopy

Species	Target Organ or Disease	Excitation λ (nm)	Major Raman Shifts (cm⁻¹)	Comments	Reference
Bone lamellar	Bone (human), strength			Raman imaging, demonstrates organization of lamellar bone from two orthogonal orientations.	[558]
Bone tibia	Bone (animal), fracture healing		958		[559]
Brain	Brain, tumor				[560]
Brain edema	Brain, edema				[561]
Breast duct epithelial	Breast, tumor	840	2600–3800		[562]
Bronchoalveolar stem cells	Lung (mouse)			Fluorescent SERS dots (FSERS-dots), labeling immunoassay technique.	[563]
Calcium HAP	Breast (chicken), malignant lesions				[564]
Calcium HAP	Breast (pig), cancer				[565]
Calcium oxalate and calcium HAP	Breast tissue, cancer			Kerr-gated Raman spectroscopy.	[566]
Calcium oxalates	Kidney stone (calcium oxalate-type)		1465, 1492	Human. Quantitative analysis performed on calcium oxalate-type kidney stones.	[567]
Calcium oxalates	Urine/stones			Human. Review.	[568]
Calcium salts	Artery (coronary)/atherosclerosis			Human. Combined intravascular ultrasound with Raman spectroscopy.	[569]
Calculi (assorted)	Urine/stones	NIR		Human. NIR-FT–Raman. Generated a Raman library and used it to correctly identify composition of various calculi.	[570]
Cancer cells	Cancer			Single-walled carbon nanotubes (SWCNTs) used as Raman tags for imaging live cancer cells.	[571]
Caries	Tooth, cavities				[572]

Substance	Tissue/Application	Wavelength	Notes	Reference
Carotenoids	Beast/cancer	488–515.5	CCD. Human. Determined 488–515 nm radiation well suited for monitoring carotenoid features due to resonance enhancement.	[573]
Carotenoids	Brain/cancer	NIR	NIR-FT–Raman. Human.	[574]
Carotenoids	Cardiovascular tissue/fatty plaques	514.5		[575]
Carotenoids	Kidney, liver, lungs, fibrotic tissue, cancer (rat and pig)			[576]
Carotenoids	Skin		Resonance Raman, noninvasive measurement of carotenoid molecules in human skin.	[577]
Cell	Cell, characterization of necrotic cell death			[578]
Cell bone	Bone, osteosarcoma		Demonstrates feasibility for in situ phenotypic differences in cells.	[579]
Ceroid	Heart (human), atherosclerotic lesions			[580]
Cholesterol	Arteries (human), arteriosclerosis			[581]
Cholesterol	Artery (coronary)/atherosclerosis		Human. Combined intravascular ultrasound with Raman spectroscopy.	[569]
Cholesterol	Eye (lens)/cataracts		Human. Studied changing cholesterol content in lens fibers via Raman scattering. Reported cholesterol and protein distribution in eye lens.	[582]
Cholesterol	Eye (lens)/cataracts		Human. Studied changing cholesterol content in lens fibers via Raman scattering. Reported cholesterol and protein distribution in eye lens.	[583]
Cholesterol	Ocular lens/cataracts	Visible		[584]
Cholesterol	Ocular lens/cataracts	Visible		[585]
Clinical drug PTH	Bone, osteoporosis	Visible		[586]
Collagen	Achilles tendon	Visible	Bovine.	[587]

(continued)

TABLE 20.9 (continued)   Raman Spectroscopy

Species	Target Organ or Disease	Excitation λ (nm)	Major Raman Shifts (cm⁻¹)	Comments	Reference
Collagen	Bladder, bladder outlet obstruction				[588]
Collagen	Bone (mouse), bone qualification				[589]
Collagen	Cervix/cancer	NIR		CCD. Human. Differentiated between precancerous and normal cervical tissues.	[590]
Collagen	Cervix/cancer	NIR		CCD. Human. Differentiated between precancerous and normal cervical tissues.	[591]
Collagen	Ocular lens/cataracts	Visible		Aggregation and pigmentation monitored.	[592]
Collagen-type 1-fibrils	Bone and teeth, biomechanics			Demonstrates that intrafibrillar mineral etches at a slower rate than extrafibrillar mineral.	[593]
Corneocytes	Skin, natural moisturizing factor-related diseases				[594]
Coronary arteries	Arteries (human), arteriosclerosis			NIRS.	[595]
Creatine	Urine	NIR	692	Human. Demonstrated quantitative analysis of creatine in urine with an LOD of 1.5 mg/dL.	[529]
Creatinine	Urine			Human. SERS needed for creatinine analysis.	[596]
Cystine	Kidney stone		498, 677	Human. Spectra of cystine-type stones almost identical to cystine powder, particularly for S–S and C–S stretches.	[597]
Cystine	Liver/cystinosis	Visible	501, 2916–2967	Human. Cystine crystals identified in liver sections.	[598]
Cystine	Spleen/cystinosis	Visible	501, 2916–2967	Human. Cystine crystals identified in spleen sections.	[598]
Cystine	Urine/stones			Human. Review.	[568]

Dentin	Teeth		FT–Raman	[599]
Dentin	Teeth, tooth hardness		FT–Raman	[600]
Dentine	Teeth	1064	Human. Microscope. Cross sections of teeth mapped by chemical functionality. Dentine and enamel distinguished.	[601]
Dentine	Teeth		Human. Studied demineralization of dentine.	[602]
Dentine	Teeth		FT–Raman	[603]
Dentine		Visible		[604]
Elastin	Achilles tendon	Visible	Bovine.	[587]
Embryo	Reproductive system, in vitro fertilization (human)		Rapid noninvasive metabolomic profiling of human embryo culture media correlated to pregnancy outcome.	[605]
ES cells	Embryo, differentiation of RNA, proteins		Ratio of RNA translated can be used to differentiate state of ES cells.	[606]
ES cells	Embryos (mouse), cellular differentiation	813		[607]
Enamel	Teeth (human), bleaching effects			[608]
Enamel	Teeth, enamel identification			[609]
Enamel	Teeth	1064	Human. Microscope. Cross sections of teeth mapped by chemical functionality. Dentine and enamel distinguished.	[601]
Epidermal phenylalanine hydroxylase (PAH)	Skin (human)		FT–Raman.	[610]
Epithelium	Bronchial, tumor		Raman microspectroscopic mapping. Confocal Raman.	[611]
Epithelium	Throat, Barrett's esophagus			[612]
Extracellular proteins	Breast/cancer	782–830	CCD. Human. Determined that NIR excitation is better than blue-green radiation for lipids.	[573]

(continued)

TABLE 20.9 (continued) Raman Spectroscopy

Species	Target Organ or Disease	Excitation λ (nm)	Major Raman Shifts (cm⁻¹)	Comments	Reference
Extracellular proteins	Breast/cancer (ductal carcinoma)	785	Normal: 1439, 1654; malignant: 1450, 1654	CCD. Human. Formalin-fixed tissue. Observed dramatic spectral differences between normal and malignant tissue samples. Diseased tissue dominated by proteins.	[613]
Extracellular proteins	Breast/cancer (fibrocystic disease)	785		CCD. Human. Formalin-fixed tissue. Observed dramatic spectral differences between normal and malignant tissue samples. Diseased tissue dominated by proteins.	[613]
Eye	Eye (human), macular pigment			Resonance Raman in vivo.	[614]
Fatty acids	Breast/cancer	406.7		Human. Distinguished between normal and malignant breast tissue, attributing differences to relative concentrations of fatty acids and β-carotene.	[548]
Fatty acids	Breast/cancer	457.9		Human. Distinguished between normal and malignant breast tissue, attributing differences to relative concentrations of fatty acids and β-carotene.	[548]
Fatty acids	Breast/cancer	514.5		Human. Distinguished between normal and malignant breast tissue, attributing differences to relative concentrations of fatty acids and β-carotene.	[548]
Femur and radius	Bone, response to mechanical loading				[615]
Femur	Bone (dog)		FT–Raman.		[616]
Femur	Bone (human), aging				[617]
Gastroesophageal junction	Esophagus, Barrett's esophagus				[618]

Analyte	Tissue/Matrix	Excitation	Raman shift	Notes	Reference
Gelatin	Achilles tendon			Bovine.	[587]
Glioblastoma	Brain (human), cancer				[619]
Glucose	Blood aqueous humor	Visible		Human. Stimulated Raman scattering. Analysis of aqueous humor precluded absorption and scattering complications associated with whole blood.	[620]
Glucose	Blood plasma			Human. Anti-Stokes Raman.	[621]
Glucose	Blood serum			Human. Anti-Stokes Raman.	[621]
Glucose	Blood serum			Human.	[622]
Glucose	Blood, type I and II diabetes				[623]
Glucose	Eye (human), cataracts	785			[624]
Glucose	Eye (human), diabetes				[625]
Glucose	Urine/diabetes		1130.5	Introduced compact, highly sensitive Raman system with integration sphere cell holder. LOD for glucose was 31 mg/dL.	[527]
Glucose	Urine/diabetes		1130	Human. Anti-Stokes Raman. Demonstrated quantitative analysis of glucose in urine.	[528]
Glucose	Urine/diabetes	NIR	1130	Human. Demonstrated quantitative analysis of glucose in urine with an LOD of 41 mg/dL.	[529]
Gonadal microliths	Testes, cancer				[626]
H_2O_2	Skin (human), vitiligo			FT-Raman, in vivo.	[627]
H_2O_2	Skin, acrofacial vitiligo			FT-Raman.	[628]
H_2O_2	Skin, vitiligo			FT-Raman, confirm that antioxidant enzymes are seriously affected in acute vitiligo.	[629]
HbO_2Sat		1360, 1375		Resonance Raman.	[630]
Heart tissue	Heart (human), calcification	830		NIR Raman.	[631]
Helical fibers	Spinal cord (guinea pigs), fibrous astroglial filaments structure analysis			CARS.	[632]

TABLE 20.9 (continued)   Raman Spectroscopy

Species	Target Organ or Disease	Excitation λ (nm)	Major Raman Shifts (cm⁻¹)	Comments	Reference
Hematin	Malaria			Polarized-resolved resonance Raman.	[633]
Hemoglobin	Blood, oxygen saturation	532	1355–1380, 1500–1650	Noninvasive and reliable in vivo SO_2 determinations in thin tissues.	[634]
Hemoglobin	Tissue	Deep violet		Resonance Raman, shock diagnosis.	[635]
Hepatocytes	Liver/cancer	Visible	1040, 1083, 1182, 1241, 1253	Distinguished between normal and malignant hepatocytes.	[508]
Human atherosclerotic lesions	Arteries (human), arteriosclerosis			NIR Raman.	[636]
Human atherosclerotic lesions	Heart (human), coronary artery disease			NIRS.	[637]
Human cortical bone	Bone (human), aging			Deep UV-Raman.	[638]
Human cortical bone	Bone, aging, fracture				[639]
Hyaluronic acid	Stratum corneum, response to moisturizing agents			Raman microspectroscopy.	[640]
HAP	Bone, osteoporosis				[641]
Hydrogen peroxide	Teeth (human), effects of tooth whitening			Bleaching strips do not change surface/subsurface hardness or chemical composition of teeth.	[642]
HAP	Bone (bovine)				[643]
Hydroxy carbonate apatite	Bone				[644]
HAP	Bone			SERS, bone-specific drug delivery device.	[645]
HAP	Bone and teeth				[646]
HAP	Bone, osteoporosis				[647]
HAPs	Bone/implants			Changes in HAP spectra investigated after implantation to establish biocompatibility.	[534]
HAPs	Bone/titanium implants	NIR	592–1048 (HAP) 960–1450 (bone)	FT-Raman. Distinguished between HAP and bone when studying HAP-treated titanium implants.	[648]
HAPs	Cardiovascular tissue/calcified tissue	514.5			[649]

HAPs	Metallic medical implants	NIR		NIR-FT–Raman. HAP powder spectra compared to spectra for HAP coatings from metallic medical implants.	[650]
Inclusions	Breast			Human. Studied the effects of silicone breast implants via Raman scattering.	[651]
Ion receptors	Bone				[652]
Lactic acid	Blood and muscle fatigue				[653]
Lamella	Bone		961, 1665	NIR Raman	[654]
L-amphetamine	Urine (drug test)			Human. SERS. Vibrational modes used to identify L-amphetamine at concentrations at the mg/mL level with colloidal silver. Spectra of drug mixtures in urine reported.	[655]
Lens	Eye (human), macular pigmentation and transparency	488		Resonance Raman.	[656]
Lens supernatants	Eye (guinea pig), cataracts				[657]
Lesions	Throat (human), Barrett's esophagus				[658]
Lesions	Throat (human), Barrett's esophagus				[659]
Lipids	Brain (human)	785			[660]
Lipids	Colon/cancer	240		Human. Reported Raman spectra for normal and neoplastic colon mucosa. Neoplastic samples showed lower amino acid to nucleotide content ratio.	[544]
Lipids	Colon/cancer	251		Human. Reported Raman spectra for normal and neoplastic colon mucosa. Neoplastic samples showed lower amino acid to nucleotide content ratio.	[544]
Lipids	Cornea	NIR		NIR-FT–Raman. Human. Investigated thermally induced molecular disorder.	[661]

(continued)

TABLE 20.9 (continued) Raman Spectroscopy

Species	Target Organ or Disease	Excitation λ (nm)	Major Raman Shifts (cm⁻¹)	Comments	Reference
Lipids	Eye (lens)			Human. Studied regional changes in the lens membrane lipid composition corresponding to lens membrane functions.	[662]
Lipids	Hair	NIR		NIR-FT–Raman. Human. Investigated thermally induced molecular disorder.	[663]
Lipids	Nail	NIR		NIR-FT–Raman. Human. Investigated thermally induced molecular disorder.	[663]
Lipids	Skin	NIR		NIR-FT–Raman. Human. Investigated thermally induced molecular disorder.	[663]
Lipids	Skin/cancer			Human. Examined 11 skin lesion types and noted changes in lipid spectral bands when compared to healthy tissue.	[664]
Lipids	Skin/cancer (basal cell carcinoma)	1064	1300, 1420–1450	Human. NIR-FT–Raman. Observed differences between normal skin and basal cell carcinoma attributed to CH_2 and $-(CH_2)(n)$-in-phase twist.	[665]
Live cancer cells				Raman microspectroscopy in vitro cellular biochemical studies.	[666]
Lung tissue	Lung (mouse), toxic chemical exposure		786	In vitro toxicological testing of pharmaceutical and in situ monitoring for growth in engineered tissues.	[667]
Lutein, zeaxanthin	Eye, macular pigment				[668]
Lutein, zeaxanthin	Retina eye, light oxidation damage			Resonance Raman.	[669]
Matrix extracellular phosphoglycoprotein (MEPE)	Teeth (human)				[670]
Mefenorex	Urine (drug test)			Human. SERS. Vibrational modes used to identify mefenorex at concentrations at the mg/mL level with colloidal silver. Spectra of drug mixtures in urine reported.	[655]

Name	Sample		Peaks (cm⁻¹)	Description	Ref.
Membraneous bone	Bone		754, 1200–1300, 1400–1470, 1600–1700, 2800–3100, 2852		[671]
Meningioma	Brain dura/central nervous system, tumor			Raman microspectroscopy, able to distinguish meningioma from dura.	[672]
Methamphetamine	Urine (drug test)			Human. SERS. First phase of development of one-step drug test using selective reactive SERS substrate coating. LODs <20 ppm observed, required amide derivatization.	[539]
Mucosal lesions	Mouth, oral lesions (human)				[673]
Mucosa tissue	Gastric system (human), carcinoma		1156, 1525, 1587, 1660		[674]
Multicomponent	Blood (whole)	NIR		CCD. Human. Performed quantitative analysis via partial least squares model.	[675]
Multicomponent	Blood serum	NIR		CCD. Human. Performed quantitative analysis via partial least squares model.	[675]
Muscle	Bladder tissue (human)			In vivo analysis of the molecular composition of normal and pathological bladders without biopsy.	[676]
Nucleic acids	Cervix/cancer	NIR		CCD. Human. Differentiated between precancerous and normal cervical tissues.	[590]
Nucleic acids	Cervix/cancer	NIR		CCD. Human. Differentiated between precancerous and normal cervical tissues.	[591]
Nucleotides	Colon/cancer	240		Human. Reported Raman spectra for normal and neoplastic colon mucosa. Neoplastic samples showed lower amino acid to nucleotide content ratio.	[544]
Nucleotides	Colon/cancer	251		Human. Reported Raman spectra for normal and neoplastic colon mucosa. Neoplastic samples showed lower amino acid to nucleotide content ratio.	[544]

(continued)

TABLE 20.9 (continued)   Raman Spectroscopy

Species	Target Organ or Disease	Excitation λ (nm)	Major Raman Shifts (cm⁻¹)	Comments	Reference
Oleic acid	Albumin, chronic wound inflammation				[677]
Organic enamel	Teeth		2941	Human. Archeological application. Samples ranging to 6000 old analyzed and dated based on the ratio of organic–inorganic band intensities.	[678]
Organs	Liver, lung, spleen; SWCNT toxicity in organs (mice)			Toxicity of SWCNTs might be due to oxidative stress.	[679]
Osteons	Bone (human)		PO_4, Amide I		[680]
Pemoline	Urine (drug test)			Human. SERS. Vibrational modes used to identify pemoline at concentrations at the mg/mL level with colloidal silver. Spectra of drug mixtures in urine reported.	[655]
Penetration enhancers	Skin (stratum corneum)			Human. Investigated mechanisms of reversible lipid disruption by drug penetration enhancers.	[681]
Penetration enhancers	Skin (stratum corneum)			Human. Investigated mechanisms of reversible lipid disruption by drug penetration enhancers.	[682]
Pentylenetetrazole	Urine (drug test)			Human. SERS. Vibrational modes used to identify pentylenetetrazole at concentrations at the mg/mL level with colloidal silver. Spectra of drug mixtures in urine reported.	[655]
Phosphates	Kidney stone (phosphate-type)	Visible		Identified phosphate-type kidney stones.	[567]
Phosphates	Teeth		960	Human. Archeological application. Samples ranging to 6000 old analyzed and dated based on ratio of organic–inorganic band intensities.	[678]
Phosphates	Urine/stones			Human. Review.	[568]

Material	Tissue/Sample	Spectroscopy	Notes	Ref.
Phospholipids	Cervix/cancer	NIR	CCD. Human. Differentiated between precancerous and normal cervical tissues.	[590]
Phospholipids	Cervix/cancer	NIR	CCD. Human. Differentiated between precancerous and normal cervical tissues.	[591]
Phospholipids	Ocular lens/cataracts	Visible		[585]
Pineal gland	Brain, nonthermal interactions with external electromagnetic field		NIR Raman.	[683]
Plaque	Artery		Human. Distinguished areas of plaque in arteries.	[684]
Plaque	Artery		Human. Distinguished areas of plaque in arteries.	[685]
Plasma	Teeth, tooth implants		Ti corrosion study in human plasma.	[686]
Poly(butylene terephthalate) (PBT)	Implants		Goat. Confocal Raman microspectroscopy. Evaluated influence of poly(ethylene oxide) (PEO) on degradation and calcification of dense PEO–PBT copolymer implants.	[687]
PEO	Implants		Goat. Confocal Raman Microspectroscopy. Evaluated influence of PEO on degradation and calcification of dense PEO–PBT copolymer implants.	[687]
Poly(methyl methacrylate) (PMMA)	Bone/implants		Changes in PMMA spectra investigated after implantation to establish biocompatibility.	[534]
Polyethylene	Bone/implants		Changes in high-density polyethylene spectra investigated after implantation to establish biocompatibility.	[534]
Polyphenolic resveratrol	Skin (pig), antioxidant		Confocal Raman.	[688]

(continued)

TABLE 20.9 (continued)   Raman Spectroscopy

Species	Target Organ or Disease	Excitation λ (nm)	Major Raman Shifts (cm⁻¹)	Comments	Reference
Polysaccharides	Skin/cancer (basal cell carcinoma)	1064	840–860	Human. NIR-FT–Raman. Observed differences between normal skin and basal cell carcinoma attributed to changes in polysaccharide structure.	[665]
Prostatic cell lines, benign and malignant (PNT1A, Lncap)	Prostate, cancer			Raman microscopy, biochemical imaging of differences between benign and malignant cell lines.	[689]
Proteins	Eye (lens)/cataracts			Human. Studied changing cholesterol content in lens fibers via Raman scattering. Reported cholesterol and protein distribution in eye lens.	[582]
Proteins	Eye (lens)/cataracts			Human. Studied changing cholesterol content in lens fibers via Raman scattering. Reported cholesterol and protein distribution in eye lens.	[583]
Proteins	Hair	NIR		NIR-FT–Raman. Human. Investigated thermally induced molecular disorder.	[663]
Proteins	Nail	NIR		NIR-FT–Raman. Human. Investigated thermally induced molecular disorder.	[663]
Proteins	Ocular lens/cataracts	Visible			[585]
Proteins	Skin	NIR		NIR-FT–Raman. Human. Investigated thermally induced molecular disorder.	[663]
Proteins	Skin/cancer			Human. Examined 11 skin lesion types and noted changes in protein spectral bands when compared to healthy tissue.	[664]
Proteins	Skin/cancer (basal cell carcinoma)	1064	1640–1680, 1220–1300, 928–940	Human. NIR-FT–Raman. Observed differences between normal skin and basal cell carcinoma attributed to amide I, amide III, nu(C–C) (valine and proline) vib. modes.	[665]
Proteins, glycogen	Bladder (human), cancer				[690]
Review	Bone				[691]

Sample	Description	Excitation (nm)	Wavenumbers	Notes	Reference
Review	Tissue, uterus, cervical precancer				[692]
Review					[693]
Silk fibroin	Biopolymers and films			In vitro approach to study rate/mechanism of fibroin degradation and other biopolymers.	[694]
Skin		785		NIR Raman, in vivo measurements.	[695]
Skin	Skin (human), hydration effects from drug penetration, Rosacea				[696]
Skin, nails	Skin (human), healthy and diseased			NIR–FT–Raman.	[697]
Soy aglycon isoflavones (SAI)	Bone (rat), eye lens, proteins			Raman spectra of isolated lens proteins (disulfide bonds), able to show increase in viability with increase in SAI.	[698]
Stratum corneum	Skin (human), atopic dermatitis and psoriasis			FT–Raman.	[699]
Stratum corneum	Skin, tumor	633	730–1170, 800–900, 1030–1130, 1200–1290, 1300–1450, 2800–2900		[700]
Sweat glands, sebaceous glands	Skin			Confocal Raman.	[701]
Synthetic HAP	Bone				[702]
Synthetic HAP	Bone, characterization of surface grafting			Development of chemical-coupled nano-biocomposites.	[703]
Teeth	Teeth (human), effects of tooth whitening		876	Hydrogen peroxide (30% soln.) demonstrates adverse effects on the mineral and organic matter of human tooth enamel.	[702]
Thioredoxin, thioredoxin reductase	Skin (human), UV damage			FT–Raman.	[704]
Tissue	Adipose tissue (pig), fatty acid content and characterization			Used for a rapid and nondestructive method to determine saturated fatty acids.	[705]

(continued)

TABLE 20.9 (continued) Raman Spectroscopy

Species	Target Organ or Disease	Excitation λ (nm)	Major Raman Shifts (cm⁻¹)	Comments	Reference
Tissue	Adrenal glands (human), neuroblastoma			Noninvasive real-time diagnostic tool for classifying pediatric tumors.	[706]
Tissue	Arteries (human), arteriosclerosis	1064		FT–Raman; able to identify and classify tissues in human carotid and distinguish layers of tissue.	[707]
Tissue	Arteries, arterial diseases			CARS imaging, label-free imaging of significant components of arterial tissues suggesting the potential of application of multimodal nonlinear microscopy to monitor onset and progression of arterial diseases.	[708]
Tissue	Bladder (human), tumor			Raman imaging.	[709]
Tissue	Brain		2400–3800		[710]
Tissue	Brain (human), cancer				[711]
Tissue	Brain (human), malignant melanoma			Raman mapping.	[712]
Tissue	Brain, bladder, cancer			Raman microspectroscopy.	[713]
Tissue	Brain, disease				[714]
Tissue	Brain, intracranial gliomas	785			[715]
Tissue	Breast (human), cancer				[716]
Tissue	Breast (human), cancer				[717]
Tissue	Breast (human), cancer		1082, 1097, 817, 1662, 1273, 1264, 1368	Used to distinguish grades and diagnosis of breast carcinoma.	[718]
Tissue	Breast (human), cancer	633			[719]
Tissue	Breast (human), cancer	633		Raman mapping for chemical/morphological understanding of breast tissue.	[720]
Tissue	Breast (human), cancer			Raman diagnostic method used for understanding biochemical/morphological changes in breast tumor tissues.	[721]

Tissue	Breast and lymph nodes (mouse), cancer	785		Demonstrate differentiation of tumor from mastitis of normal tissues.	[722]
Tissue	Breast, cancer				[723]
Tissue	Breast, carcinoma			FT–Raman.	[724]
Tissue	Bronchial (human), tumor	785	700–1800	NIR Raman.	[725]
Tissue	Cervix (human), cancer			Shows clinical potential as diagnostic tool for cervical cancer.	[726]
Tissue	Colon, cancer	785			[727]
Tissue	Coronary artery (human), atherosclerosis	830		NIR Raman.	[728]
Tissue	Coronary, coronary artery disease/ atherosclerosis		400–1800, 1400–3800	Spectral differences demonstrate chemical composition and morphology of artery wall in vivo.	[482]
Tissue	Eye (human), pterygia		1156, 1521, 1585, 1748		[729]
Tissue	Gastric system (human), cancer		1156, 1587	Confocal Raman microscopy, applied for early diagnosis of gastric cancer.	[730]
Tissue	Gastrointestinal tract (human), cancer	1064	1644	NIR Raman.	[731]
Tissue	Heart (human), tissue hypoxia cardiovascular disease (CVD)				[732]
Tissue	Kidneys, nephrocalcinosis				[727]
Tissue	Lung (human), cancer	785, 1064			[733]
Tissue	Lung, cancer	785			[714]
Tissue	Lung, cancer			Raman imaging.	[734]
Tissue	Lymph node, cancer				[735]
Tissue	Pancreas (mouse), cancer	785		Pancreatic tumors characterized by increased collagen content and decreased DNA, RNA, and lipids compared to normal tissue.	[736]
Tissue	Skin			Confocal Raman.	[737]

(continued)

TABLE 20.9 (continued) Raman Spectroscopy

Species	Target Organ or Disease	Excitation λ (nm)	Major Raman Shifts (cm⁻¹)	Comments	Reference
Tissue	Skin (human), cancer			Polarized Raman microspectroscopy; information used to elucidate the molecular mechanisms induced in basal cell carcinoma development.	[738]
Tissue	Skin (human), tumor				[677]
Tissue	Skin (mouse), spectral component differentiation				[739]
Tissue	Skin, basal cell carcinoma		Amide I, PO_2		[740]
Tissue	Skin, bullous pemphigoid (BP)		Amide I and III	Micro-Raman.	[741]
Tissue	Skin, wound repair			Confocal Raman, able to acquire detailed molecular structure information from the various proteins and their subclasses involved in wound-healing process.	[742]
Tissue	Stomach (human), cancer			FT–Raman.	[743]
Tissue	Stomach, cancer		1276, 1303	Pilot study demonstrates feasibility of discriminating normal and malignant stomach tissue.	[744]
Tissue	Thyroid, lung, oviduct, ovarian, cervical, uterine		1004, 1158, 1520, 1585, 1634		[745]
Tissue, artery	Cardiovascular system (human), atherosclerosis	NIR	961, 1071, 1439, 1451, 1663, 1655	NIR Raman, able to detect atherosclerotic plaques in carotid arterial tissue.	[746]
Tissue, cardiac valve	Cardiac valve, calcification	830	960, 1260, 1452, 1660	NIR Raman, technique used to detect calcium phosphate mineral deposition in cardiac valves.	[747]
Trabecular bone	Bone (human), osteoporosis		Carbonate, Amide I		[748]
Urates	Urine/stones			Human. Review.	[568]
Urea	Urine			Human. Urea high enough in concentration to monitor with normal Raman.	[596]

Analyte	Sample		Region	Notes	Reference
Urea	Urine	1016		Human. Anti-Stokes Raman. Demonstrated quantitative analysis of urea in urine.	[528]
Urea	Urine	1013	NIR	Human. Demonstrated quantitative analysis of urea in urine with an LOD of 4.9 mg/dL.	[529]
Uric acid	Kidney stone (uric-acid-type)		Visible	Identified uric-acid-type kidney stones.	[749]
Uric acid	Urine			Human. SERS needed for uric acid analysis.	[596]
Uric acid	Urine/stones			Human. Review.	[568]
Vitamin E	Muscle, liver, lens				[750]
Vulnerable plaque	Arteries (human), arteriosclerosis				[751]
	Artery (coronary)		NIR	CCD. Human.	[752]
	Artery		NIR	NIR-FT–Raman. Human.	[753]
	Artery		NIR	NIR-FT–Raman. Human.	[754]
	Artery		NIR	NIR-FT–Raman. Human.	[755]
	Artery		NIR	NIR-FT–Raman. Human.	[756]
	Artery (aorta)		NIR	NIR-FT–Raman. Human.	[757]
	Artery/atherosclerosis		NIR	CCD. APOE*3 Leiden transgenic mice.	[758]
	Blood		NIR	Human. FT.	[759]
	Blood serum/various cancers			Human. Raman spectra of blood serum demonstrated to be an indicator for various cancers, each with a unique Raman spectrum.	[760]
	Bone		NIR	Investigated sheep and human samples. Compared results with pure hydroxy and carbonated apatite. NIR-FT–Raman.	[761]
	Bone		NIR	Investigated sheep and human samples. Compared results with pure hydroxy and carbonated apatite. NIR-FT–Raman.	[762]
	Brain/cancer		NIR	NIR-FT–Raman. Rat.	[763]

(continued)

TABLE 20.9 (continued)　Raman Spectroscopy

Species	Target Organ or Disease	Excitation λ (nm)	Major Raman Shifts (cm⁻¹)	Comments	Reference
	Brain/cancer	NIR		NIR-FT–Raman. Human.	[764]
	Brain/Parkinson's disease	488		Monkey. Distinguished between white and gray matter.	[765]
	Breast	NIR		Human. NIR-FT–Raman. Fiber optics. Examined normal breast tissues, comparing effectiveness of Koehler laser illumination and the confocal principle.	[766]
	Breast/cancer	830		CCD. Human. Demonstrated potential of differentiating between malignant and benign lesions in breast tissues.	[767]
	Breast/cancer	NIR		CCD. Human.	[768]
	Breast/cancer	NIR		CCD. Human. Spectral imaging of lipids and proteins in thin sections of breast tissue.	[769]
	Cardiovascular tissue/calcified aorta	Visible			[770]
	Colon/cancer	NIR		CCD. Human. Determined Raman has limited utility due to low sensitivity in comparison to fluorescence and absorption. Observed excessive interference.	[768]
	Colon/cancer	UV		Human.	[771]
	Connective tissue	NIR		Human. FT.	[759]
	Cornea	Visible		Feline.	[772]
	Cornea	Visible		Monitored hydration gradients across the cornea.	[773]
	Cornea	NIR		NIR-FT–Raman. Human.	[774]
	Coronary artery/atherosclerosis	NIR		CCD. Human. Classified coronary artery lesions as non-atherosclerotic, noncalcified plaque and calcified plaque.	[775]

Coronary artery/ atherosclerosis	NIR	CCD. Human. Classified coronary artery lesions as non-atherosclerotic, noncalcified plaque and calcified plaque.	[776]
Eye	Visible	Various layers of ocular tissues investigated.	[777]
Gallstone	Visible	Human. FT–Raman. Spectral data reported for a range of gallstones.	[778]
Gallstone	NIR		[779]
Gastrointestinal cancer	NIR	CCD. Human.	[780]
Gastrointestinal cancer	NIR	Human. Raman spectroscopy described as yielding the greatest information in comparison to fluorescence, optical coherence tomography, and ultrasound.	[781]
Gynecological tissues/cancer	NIR	NIR-FT–Raman. Human.	[782]
Gynecological tissue/cancer	NIR	Human. FT. Distinguished between normal, benign, and cancerous tissues.	[783]
Hair	NIR	Human. FT–Raman spectra reported with band assignments for hair.	[784]
Larynx/cancer	830	Human. Differentiated between normal, dysplastic, and squamous cell carcinoma of the larynx. Salient spectral features observed in 850–950 and 1200–1350 cm^{-1} ranges.	[785]
Lung/cancer	1064	Human. Totally fluorescence-free Raman spectra obtained for normal and malignant lung tissues.	[786]
Mineral deposition disease	Visible		[787]
Mineral deposition disease	Visible	Human. Differentiated between mono- and dihydrate oxalates in the myocardium and the pituitary.	[788]
Mineral deposition disease/ lung	Visible		[789]

(continued)

TABLE 20.9 (continued) Raman Spectroscopy

Species	Target Organ or Disease	Excitation λ (nm)	Major Raman Shifts (cm⁻¹)	Comments	Reference
	Mineral deposits/prosthetic implants	Visible			[790]
	Mineral deposits/prosthetic implants	Visible			[791]
	Mineral deposits/prosthetic implants	Visible			[792]
	Mineral deposits/silicone breast implants	Visible			[793]
	Nail	NIR		Human. FT–Raman spectra reported with band assignments for nails.	[784]
	Ocular lens	Visible			[794]
	Ocular lens	Visible		Monitored disulfide bond formation in eye lens.	[795]
	Ocular lens	Visible			[796]
	Ocular lens	Visible			[797]
	Ocular lens	Visible			[798]
	Ocular lens	Visible		Human and rabbit.	[799]
	Ocular lens	Visible		Human.	[800]
	Ocular lens	Visible		Human.	[801]
	Ocular lens	NIR		NIR-FT–Raman.	[802]
	Oral tissues/cancer			Human. Distinguished between normal and malignant oral tissues.	[803]
	Pathological inclusions in breast tissue	Visible			[804]
	Pathological tissues (inclusions)	Visible			[805]
	Peripheral artery	NIR		CCD. Human.	[806]
	Retina/diabetes	NIR		NIR-FT–Raman.	[807]
	Skin (stratum corneum)	NIR		Human. FT–Raman. Spectra of human corneum reported with band assignments.	[808]

Skin (stratum corneum)		Snake. Modeled drug diffusion across skin.	[809]
Skin (stratum corneum)		Pig. Modeled drug diffusion across skin.	[809]
Skin (stratum corneum)	NIR	Human. FT–Raman. Compared normal, hyperkeratotic, and psoriatic stratum corneum.	[810]
Skin/cancer	NIR	NIR-FT-Raman. Human.	[811]
Skin/cancer	NIR	NIR-FT-Raman. Human.	[812]
Skin/cancer	NIR	NIR-FT-Raman. Human.	[813]
Skin/cancer (melanoma)	NIR	Human. FT.	[759]
Soft tissue/sarcoma	NIR	CCD. Human.	[768]
Stomach/cancer	NIR	Human. Malignant tissues observed to yield stringer bands related to OH, NH, C=O stretching, and H–O–H bending.	[814]
Tooth	NIR	Human. FT–Raman. Reported compositional changes in children's teeth.	[815]
Tooth dentine	NIR	Investigated sheep and human samples. Compared results with pure hydroxy and carbonated apatite. NIR-FT-Raman.	[761]
Tooth enamel	NIR	Investigated sheep and human samples. Compared results with pure hydroxy and carbonated apatite. NIR-FT-Raman.	[761]
Urinary bladder/cancer	NIR	CCD. Human. Determined Raman has limited utility due to low sensitivity in comparison to fluorescence and absorption. Observed excessive interference.	[768]
Various tissues	UV		[768]

(continued)

TABLE 20.9 (continued)   Raman Spectroscopy

Species	Target Organ or Disease	Excitation λ (nm)	Major Raman Shifts (cm^{-1})	Comments	Reference
Animal studies (in vivo)					
	Cornea			Rabbit. Confocal Raman. Fiber optics. Investigated effects of corneal dehydration via observation of changes in stretching modes of OH and CH groups.	[816]
	Eye (aqueous humor)			Rabbit. Confocal Raman. Compared rabbit and blind human eyes. Assessed biochemical properties of eye-specific Raman signatures.	[817]
	Eye (cornea)			Rabbit. Confocal Raman. Compared rabbit and blind human eyes. Assessed biochemical properties of eye-specific Raman signatures.	[817]
	Eye (lens)			Rabbit. Confocal Raman. Compared rabbit and blind human eyes. Assessed biochemical properties of eye-specific Raman signatures.	[817]
	Ocular lens			Rabbit.	[797]
	Oral tissue/cancer (dysplasia)			*Rat.* In vivo spectra obtained for normal and dysplastic rat palate tissue. Distinguished between normal tissue, low-grade dysplasia, and high-grade dysplasia/carcinoma.	[818]
Axonal myelin	Spine (guinea pig), demyelinating diseases				[819]
Tissue	Colon, cancer			SERS, conjugated nanoparticles able to target tumor biomarkers.	[820]
Bone	Bone (dog), composition	785		Raman spectroscopic diffuse tomographic imaging.	[821]
Carbonate ions	Bone, aging			Raman microscopy.	[822]
Cataract	Eye (rat), diabetes mellitus		540–1100, 2800–3730		[823]

Analyte	Sample			Technique/Notes	Reference
Cetyl trimethyl ammonium chloride (CTAC)	Eye	3300–3600		FT–Raman	[824]
Organs	Lung, liver, and spleen (mouse); nanodiamond toxicity				[825]
Sapphyrin	Pancreas (mouse), tumor			Resonance Raman	[826]
Tissue	Brain, glioblastoma				[827]
Tissue	Muscle, heart, brain, liver, spleen, kidney, cancer			NIR–FT–Raman	[828]
Clinical studies (in vivo)					
Atherosclerotic plaque	Heart (human), atherosclerosis, CVD				[829]
Beta carotene	Skin (human), IR damage			Resonance Raman.	[830]
Beta carotene, lycopene	Skin, UV damage		488, 514		[831]
Beta carotene, lycopene	Skin (human)		1160, 1525	Resonance Raman.	[832]
Beta carotene and lycopene	Skin (human), carotenoid antioxidants			Resonance Raman used to measure seasonal pigmentation response in skin.	[833]
Bone	Bone, aging			Raman microprobe.	[834]
Carotenoid	Eye (human)				[835]
Carotenoid pigment	Retina (human)			Resonance Raman.	[836]
Carotenoids	Skin/cancer			Human. Noninvasive. Determined that carotenoid concentration is lower in malignant than healthy skin tissue.	[837]
Dentin	Teeth, aging	Amide I	244	UVRRS (ultraviolet resonance Raman spectroscopy).	[838]
Dimethyl sulfoxide (DMSO)	Skin (human), barrier penetration			NIR–FT–Raman, the penetration depth of DMSO is exposure time dependent.	[839]
Epidermis	Skin (human), electrical shock				[840]
Glucose	Interstitial fluid				[841]
H_2O_2	Epidermis (human), blood, vitiligo		875, 1040	FT–Raman.	[842]

(continued)

TABLE 20.9 (continued)　Raman Spectroscopy

Species	Target Organ or Disease	Excitation λ (nm)	Major Raman Shifts (cm⁻¹)	Comments	Reference
L-Tryptophan (Trp)	Skin (human), vitiligo		930, 1050	FT–Raman, dermatology application to follow the effect of oxidative stress in the skin of patients with vitiligo.	[843]
Lung	Lung (human), cancer			NIR Raman.	[844]
Lutein, zeaxanthin	Eye (human), aging			Resonance Raman.	[845]
Lutein, zeaxanthin	Eye (human), macular degeneration			C=C region.	[846]
Lycopene	Skin (human)	514		Resonance Raman.	[847]
Lycopene, beta carotene	Skin (human), UV damage				[848]
Melanin	Skin (human)		1580, 1380	NIR Raman.	[849]
Skin	Skin (human)			Handheld Raman microspectrometer.	[850]
Skin	Skin (human), characterization				[851]
Skin	Skin (human), composition of lipids and proteins			NIR-FT–Raman.	[852]
Skin	Skin (human), pigmentation		1200–1300, 1450–2940, 1640–1680, 3250	NIR-FT–Raman.	[853]
Stratum corneum	Skin, response to moisturizers (human)			Confocal Raman.	[854]
Tissue	Breast, cancer			Tool for real-time diagnosis of tissue abnormalities.	[855]
Tissue	Skin (human), cancer				[856]
Tissue	Skin (human), chemical analysis	1064		Raman spectrum of skin reflects chemical composition of dermis.	[857]
Tissue	Skin (human), cutaneous edema			NIR Raman.	[858]

Tissue	Skin (human), penetration of topical ointment/occlusion			Sensitive and specific technique to monitor both lipid uptake and skin occlusion events following topical application.	[859]
Tissue	Skin (human), tissue characterization and clinical diagnosis			NIR Raman.	[860]
Tissue	Skin (human), tissue thickness			Confocal Raman.	[861]
trans-Retinol	Skin (human), pharmaceutical drug delivery			Effects of *trans*-retinol delivery into the skin were measured in vivo.	[862]
Urocanic acid (UCA)	Skin (human), UV damage		400–2200	Confocal Raman used for monitoring amount of *trans*-UCA in the stratum corneum to assess the efficiency of sun protective substances.	[863]
Adipose tissue		785	1077, 1265, 1303, 1444, 1658, 1755	Human. Noninvasive. Fiber-optic probe. CCD. Distinguished between fat tissue and bone. >200 mW laser power.	[538]
Artery (aorta)		810		CCD. Human.	[755]
Artery (aorta)		810		CCD. Human.	[864]
Bone		785	957, 1032, 1087, 1188, 1248, 1454, 1670	Human. Noninvasive. Fiber-optic probe. CCD. Distinguished between fat tissue and bone. >200 mW laser power.	[538]
Cervix/cancer		NIR		CCD. Human. Collected spectra via fiber-optic probe.	[865]
Eye (aqueous humor)				*Confocal Raman. Compared rabbit and blind human eyes. Assessed biochemical properties of eye-specific Raman signatures.*	[817]

(continued)

Biomedical Photonics Handbook: Fundamentals, Devices, and Techniques

TABLE 20.9 (continued)   Raman Spectroscopy

Species	Target Organ or Disease	Excitation λ (nm)	Major Raman Shifts (cm⁻¹)	Comments	Reference
	Eye (cornea)			Confocal Raman. Compared rabbit and blind human eyes. Assessed biochemical properties of eye-specific Raman signatures.	[817]
	Eye (lens)			Confocal Raman. Compared rabbit and blind human eyes. Assessed biochemical properties of eye-specific Raman signatures.	[817]
	Gastrointestinal tissues			Human. Raman spectra collected via fiber-optic probe during routine endoscopy. Subtle differences between normal and diseased tissues observed.	[866]
	Hair	785		Human. Noninvasive. Fiber-optic probe. CCD. >200 mW laser power.	[538]
	Mouth	785		Human. Noninvasive. Fiber-optic Probe. CCD. >200 mW laser power.	[538]
	Nail	NIR		Human. NIR-FT-Raman. Measured fingernails and discussed feasibility of technique for determination of metabolic disorders, drug use, poisoning, and local infections.	[867]
	Skin	785		Human. Noninvasive. Fiber-optic probe. CCD. >200 mW laser power.	[538]
	Skin/cancer	NIR		NIR-FT-Raman. Human.	[813]
		830		CCD. Human.	[526]

20.2.4 Fluorescence Spectroscopy

Table 20.10 is meant to provide an introductory level amount of information about the various fluorescence properties of many different chemical and biochemical species that have proven useful in the field of biomedical photonics.

Due to the overwhelming number of fluorescent chemical species of biomedical importance and the short time available for compiling this information, it is far from comprehensive; however, it is the hope of the authors that this table will grow with each new edition of this handbook.

In addition, the authors would like to gratefully acknowledge the efforts of several important manuscripts for providing a great deal of the information tabulated here. In particular, the authors would like to acknowledge the two very helpful review papers:

1. Richards-Kortum, R. and Sevick-Muraca, E., Quantitative optical spectroscopy for tissue diagnosis, *Annual Review of Physical Chemistry* 47, 555–606, 1996.
2. Wagnieres, G. A., Star, W. M., and Wilson, B. C., In vivo fluorescence spectroscopy and imaging for oncological applications, *Photochemistry and Photobiology* 68(5), 603–632, 1998.
3. Bachmann, L., Zezell, D. M., da Costa Ribeiro, A., Gomes, L., and Ito, A. S., Fluorescent spectroscopy of biological tissues—A review, *Applied Spectroscopy Reviews* 41, 575–590, 2006.

20.2.5 Elemental Analyses

Table 20.11 is meant to provide a quick reference of publications pertaining to the elemental analysis using a variety of techniques, for biomedical applications.

The techniques used to perform the analysis are abbreviated as follows: AAS (atomic absorption spectroscopy), AES (atomic emission spectroscopy), MP-AES (microwave plasma atomic emission spectroscopy), ICP-AES (inductively coupled plasma atomic emission spectroscopy), LIPS (laser-induced plasma spectroscopy), XRF (x-ray fluorescence), and PIXE (particle-induced x-ray emission).

This updated edition clearly demonstrates the continued importance of elemental analyses in various biomedical applications. In particular, there has been significant growth in the literature pertaining to various tissue studies.

The authors would also like to refer the reader to the following excellent review papers from which some of the information in this compilation is tabulated:

Taylor, A., Branch, S., Fisher, A., Halls, D., and White, M., Atomic spectrometry update.
Clinical and biological materials, foods, and beverages, *J. Anal. At. Spectrom.*, 16, 421–446.
Potts, P. J., Ellis, A. T., Kregsamer, P., Marshall, J., Streli, C., West, M., and Wobrauschek, P. *J. Anal. At. Spectrom.* 18, 1297–1316, 2003.
Taylor, A., Branch, S., Day, M. P., Patriarca, M., and White, M. *J. Anal. At. Spectrom.* 24, 535–579, 2009.

TABLE 20.10 Fluorescence Spectroscopy

Species	Endogenous/Exogenous Origin	Absorbance Maxima (nm)	Molar Extinction Coefficient (M−1)	Excitation λ_{ex} (nm)	Emission λ_{em} (nm)	Quantum Yield	Matrix	Lifetime t_1 (rel % t_1)	Lifetime t_2 (rel % t_2)	Lifetime t_3 (rel % t_3)	Comments	Reference
2,6-Dimethyldifluor-8-pyrone (DDP)	Endogenous fluorophore	—	—	295	395	—	Calf, rabbit, or human articular cartilage and synovial fluid	—	—	—	DDP is a hyaline cartilage specific compound present in all articular cartilage from various articulating joints/animal species; DDP level increases with articular cartilage depth and decreases with cartilage maturation.	[868]
3-Hydroxykynurenine	Endogenous fluorophore	—	—	355	>355	—	Extracts of human eye (young)	24,000	400,000	—	—	[869]
3-Hydroxykynurenine	Endogenous fluorophore	—	—	355	>355	—	Extracts of human eye (old)	0.031–0.061	—	—	—	[869]
4-Pyridoxic acid (PA)	Endogenous fluorophore	—	—	315	430	—	—	—	—	—	Autofluorescence is strongly weakened by the absorption of blood near 420 nm.	[870]
PA	Endogenous fluorophore	307	—	315	425	—	—	—	—	—	Physiologic pH.	[871]
5-Aminolaevulinic acid (ALA)	Exogenous fluorophore	—	—	635	—	—	Solution	—	—	—	Used as a photodynamic therapy agent; optimal time for fluorescence in vivo 4–6 h.	[872]
ALA	Exogenous fluorophore	—	—	514	—	—	Solution	—	—	—	Used as a photodynamic therapy agent.	[873]
5-Hydroxytriptamine	Endogenous fluorophore	—	—	366	—	—	Solution and biolgical tissues	—	—	—	Autofluorescence spectroscopy for the direct measurement of neurotransmitters.	[874]
Aluminum phthalocyanine	Exogenous fluorophore	—	—	610	>645	—	Phosphate buffered saline (PBS)	5.0 ± 0.1	—	—	—	[875]

Fluorophore	Type		Excitation	Medium	Emission				Notes	Reference
Aluminum phthalocyanine	Exogenous fluorophore	—	610	PBS + BSA (bovine serum albumin)	>645	5.5 ± 0.2 (92.0±1.0)	1.0 ± 0.2 (8±1)	—	—	[876]
Aluminum phthalocyanine	Exogenous fluorophore	—	610	0.01 mM cetyltrimethylammonium bromide (CTAB) + 0.1 M PBS	>645	6.0 ± 0.1	—	—	—	[876]
Aluminum phthalocyanine	Exogenous fluorophore	—	610	1% Triton in 0.1 M PBS	>645	5.9 ± 0.1	—	—	—	[876]
Benzoporphyrin monoacetate (BpD-MA)	Exogenous fluorophore	—	400	Solution	690	5.5	—	—	Used as a photodynamic therapy agent; optimal time for fluorescence in vivo 2–4 h; phototoxic.	[877]
Beta-carotene	Endogenous fluorophore	—	Blue	Powder/solution	520	9.6	2	0.3	—	[878]
Carious dentin	Endogenous fluorophore	—	394, 527, 629	—	700	—	—	—	—	[879]
Carious dentine	Endogenous fluorophore	—	655	—	720, 810	—	—	—	—	[879]
Carious enamel	Endogenous fluorophore	—	400	—	480, 624, 635, 690	—	—	—	—	[879]
Carious enamel	Endogenous fluorophore	—	337	—	440, 490, 590, 630	—	—	—	—	[879]
Cathepsin B	Exogenous fluorophore	—	610–650	Solution	680–720	—	—	—	Imaging for early osteoarthritis diagnosis.	[880]
Ceroid	Endogenous fluorophore	—	340–395	—	430–460	—	—	—	Physiologic pH.	[881]
Ceroid	Endogenous fluorophore	—	340–395	—	540–640	—	—	—	Physiologic pH.	[882]
Collagen	Endogenous fluorophore	—	280, 265, 330, 450	Powder	310, 385, 390, 530	—	—	—	Physiologic pH.	[883]
Collagen	Endogenous fluorophore	—	UV	Powder	Violet-blue	2.7 (25)	8.9 (35)	—	—	[884]
Collagen	Endogenous fluorophore	—	340	Powder/solution	395	9.9	5.0	0.8	—	[885]
Collagen	Endogenous fluorophore	—	270	Powder/solution	395	—	—	—	—	[885]
Collagen	Endogenous fluorophore	—	285	Powder/solution	310	—	—	—	—	[885]
Collagen	Endogenous fluorophore	330	300–370	—	350–470	—	—	—	—	[886]
Collagen	Endogenous fluorophore	—	340	—	394	—	—	—	Species observed in collagen of old humans.	[870]
Collagen	Endogenous fluorophore	—	370	—	450	—	—	—	—	[879]
Collagen	Endogenous fluorophore	—	325	—	350–410	—	—	—	Physiologic pH.	[887]

(continued)

TABLE 20.10 (continued)   Fluorescence Spectroscopy

Species	Endogenous/Exogenous Origin	Absorbance Maxima (nm)	Molar Extinction Coefficient (M-1)	Excitation I_{ex} (nm)	Emission I_{em} (nm)	Quantum Yield	Matrix	Lifetime t_1 (rel % t_1)	Lifetime t_2 (rel % t_2)	Lifetime t_3 (rel % t_3)	Comments	Reference
Collagen	Endogenous fluorophore	—	—	330, 370	375, 455	—	—	—	—	—	—	[888]
Collagen	Endogenous fluorophore	—	—	405	—	—	—	0.61	3.32	—	—	[889]
Collagen (type I)	Endogenous fluorophore	340, 500	—	340	410	—	—	—	—	—	—	[890]
Collagen (type I)	Endogenous fluorophore	340, 500	—	500	520	—	—	—	—	—	—	[890]
Collagen cross-links	Endogenous fluorophore	—	—	335, 370	390, 460	—	—	—	—	—	—	[879]
Collagen, elastin, HP	Endogenous fluorophore	325	—	325	400	—	—	—	—	—	Physiologic pH.	[891]
Collagen, elastin, LP	Endogenous fluorophore	—	—	325	400	—	—	—	—	—	Physiologic pH.	[892]
Coproporphyrin	Endogenous fluorophore	—	—	505	610	—	Methanol	—	—	—	—	[893]
Coproporphyrin IX	Exogenous fluorophore	—	—	400	650–670	—	DMSO	20 (100)	—	—	—	[884]
Dental calculus	Endogenous fluorophore	—	—	420	495, 595, 650, 695	—	—	—	—	—	Subgingival.	[879]
Dental calculus	Endogenous fluorophore	—	—	635	700, 783	—	—	—	—	—	Subgingival.	[879]
Dental calculus	Endogenous fluorophore	—	—	398, 405, 507, 538	633	—	—	—	—	—	Supragingival.	[879]
Dental calculus	Endogenous fluorophore	—	—	402, 412, 506, 577, 628	700	—	—	—	—	—	Supragingival.	[879]
Dentin level caries	Endogenous fluorophore	—	—	337	405, 435, 490, 555	—	—	—	—	—	—	[879]
Desmosine	Endogenous fluorophore	—	—	400	475	—	—	—	—	—	—	[894]
DsRed=drFP583	Exogenous fluorophore	—	72,500	558	583	68%	Powder dissolved in saline at 200 g/L; Solution	—	—	—	Photostable, intensity-dependent blinking, no pH sensitivity.	[895]
Elastin	Endogenous fluorophore	—	—	350, 410, 450	420, 500, 520	—	Powder	—	—	—	Physiologic pH.	[883]
Elastin	Endogenous fluorophore	—	—	—	—	—	Powder	2 (75)	6.7 (65)	—	—	[884]
Elastin	Endogenous fluorophore	330	—	330	405	—	—	—	—	—	—	[890]
Elastin	Endogenous fluorophore	—	—	460	520	—	Powder/solution	6.7	1.4	—	—	[885]
Elastin	Endogenous fluorophore	—	—	360	410	—	Powder/solution	7.8	2.6	0.5	—	[885]
Elastin	Endogenous fluorophore	—	—	425	490	—	Powder/solution	—	—	—	—	[885]

Name	Type						Form				Comments	Ref.
Elastin	Endogenous fluorophore	—	—	260	410	—	Powder/solution	—	—	—		[885]
Elastin	Endogenous fluorophore	348	—	300–400	330–550	—	—	—	—	—		[886]
Elastin	Endogenous fluorophore	—	—	400	475	—	Powder dissolved in saline at 200 g/L	—	—	—		[894]
Elastin/collagen cross-links	Endogenous fluorophore	—	—	420, 460	500, 540	—	—	—	—	—		[879]
Endogenous porphyrins	Endogenous fluorophore	—	—	400	610	—	Powder/solution	—	—	—		[896]
Endogenous porphyrins	Endogenous fluorophore	—	—	400	675	—	Powder/solution	—	—	—		[896]
Enhanced blue fluorescent protein (EBFP)	Exogenous fluorophore	—	31,000	380	440	—	Solution	—	—	—	Following the trafficking and function of proteins in living cells and monitoring the intracellular environment.	[897]
EBFP	Exogenous fluorophore	—	31,000	383	445	25%	Solution	—	—	—	Low photostability, dim dye, candidate protein for three-photon excitation.	[895]
Enhanced cyan fluorescent protein (ECFP)	Exogenous fluorophore	—	26,000	434	477	40%	Solution	—	—	—	Tested less suitable for single-molecule microscopy, low photostability.	[895]
ECFP	Exogenous fluorophore	—	26,000	434	477	—	Solution	—	—	—	Following the trafficking and function of proteins in living cells and monitoring the intracellular environment.	[897]
Enhanced green fluorescent protein (EGFP)	Exogenous fluorophore	—	55,000	489	508	60%	Solution	—	—	—	Single-molecule microscopy, most widely applied label due to brightness and photostability. (pH-dependent) blinking.	[895]

(continued)

TABLE 20.10 (continued) Fluorescence Spectroscopy

Species	Endogenous/Exogenous Origin	Absorbance Maxima (nm)	Molar Extinction Coefficient (M-1)	Excitation I_{ex} (nm)	Emission I_{em} (nm)	Quantum Yield	Matrix	Lifetime t_1 (rel % t_1)	Lifetime t_2 (rel % t_2)	Lifetime t_3 (rel % t_3)	Comments	Reference
Enhanced yellow fluorescent protein (EYFP)	Exogenous fluorophore	—	84,000	514	527	—	Solution	—	—	—	Following the trafficking and function of proteins in living cells and monitoring the intracellular environment.	[897]
EYFP	Exogenous fluorophore	—	84,000	514	522	61%	Solution	—	—	—	Single-molecule microscopy; (pH- and intensity-dependent) blinking.	[895]
Eosinophils—circulating	Endogenous fluorophore	—	—	370, 500	440, 550	—	—	—	—	—	Physiologic pH.	[898]
Eosinophils—granules	Endogenous fluorophore	—	—	380	520	—	—	—	—	—	Physiologic pH.	[899]
Eosinophils—granules	Endogenous fluorophore	—	—	450	520	—	—	—	—	—	Physiologic pH.	[900]
EqFP611	Exogenous fluorophore	—	78,000	559	611	45%	Solution	—	—	—	Intensity-dependent blinking, large Stokes shift, tetramer, dissociates at high dilution.	[895]
Flavin adenine dinucleotide (FAD)	Endogenous fluorophore	—	—	450	515	—	—	—	—	—	Physiologic pH.	[901]
FAD	Endogenous fluorophore	—	—	450	515	—	—	—	—	—	—	[902]
FAD	Endogenous fluorophore	—	—	370	535	—	—	—	—	—	—	[902]
FAD	Endogenous fluorophore	—	—	325	525–600	—	—	—	—	—	Physiologic pH.	[901]
FAD	Endogenous fluorophore	—	—	400	550	—	Powder dissolved in saline at 0.1 M	—	—	—	—	[894]
FAD	Endogenous fluorophore	—	—	405	—	—	—	2.94	—	—	—	[889]
Flavin mononucleotide (FMN) (riboflavin-5'-phosphate)	Endogenous fluorophore	—	—	440	520	—	Powder/solution	4.7	—	—	—	[903]
Flavins	Endogenous fluorophore	215, 260, 388, 455	—	200–500	500–600	—	—	—	—	—	—	[886]
FMN	Endogenous fluorophore	—	—	436	500	—	Water	5.2 (100)	—	—	—	[904]

Galantamine	Exogenous fluorophore	—	280	310	—	Phosphate buffer	—	—	—	—	Treatment for Alzheimer's disease.	[905]
Green fluorescent protein (GFP)	Exogenous fluorophore	7,000	395, 470	509	—	Solution	—	—	—	—	Following the trafficking and function of proteins in living cells and monitoring the intracellular environment.	[897]
Hematoporphyrin (HP)	Exogenous fluorophore	—	Visible	610	9	Solution	—	13.5	—	—	Used as a tumor marking agent; optimal time for fluorescence in vivo 2–4 h; phototoxic.	[906]
Hematoporphyrin derivative (HpD)	Exogenous fluorophore	—	400	610	2–7	Solution	—	15.5	2.5	—	Used as a photodynamic therapy agent; optimal time for fluorescence in vivo 12–72 h; phototoxic.	[907]
Hematoporphyrin monomethyl ether	Endogenous fluorophore	—	—	625–690	—	Physiological saline	—	14.6	—	—	Fluorescence lifetime imaging used as a contrast parameter between normal and malignant tissues for the photodynamic diagnosis of various cancers.	[908]
Hematoporphyrin monomethyl ether	Endogenous fluorophore	—	—	625–690	—	Human serum	—	16.6	—	—	Fluorescence lifetime imaging used as a contrast parameter between normal and malignant tissues for the photodynamic diagnosis of various cancers.	[908]
Hematoporphyrins	Endogenous fluorophore	402	300–450	610–710	—	—	—	—	—	—	—	[886]
Indocyanine green (ICG)	Exogenous fluorophore	—	780	—	—	Solution	—	—	—	—	Noninvasive fluorescence imaging of the human brain.	[909]
KAEDE	Exogenous fluorophore	98,800	508	518	80%	Solution	—	—	—	—	Emission switch from green to red after UV illumination, tetramer, quantum yield in pH resistant.	[895]

(continued)

TABLE 20.10 (continued) Fluorescence Spectroscopy

Species	Endogenous/Exogenous Origin	Absorbance Maxima (nm)	Molar Extinction Coefficient (M-1)	Excitation λ_{ex} (nm)	Emission λ_{em} (nm)	Quantum Yield	Matrix	Lifetime t_1 (rel % t_1)	Lifetime t_2 (rel % t_2)	Lifetime t_3 (rel % t_3)	Comments	Reference
KAEDE	Exogenous fluorophore	—	60,400	572	582	33%	Solution	—	—	—	Emission switch from green to red after UV illumination, tetramer, quantum yield in pH resistant.	[910]
Keratin	Endogenous fluorophore	—	—	277	382	—	30 mg/mL solution in urea	—	—	—	—	[911]
Keratin	Endogenous fluorophore	—	—	405	—	—	—	0.61	4.80	—	—	[889]
Keratin, horn	Endogenous fluorophore	—	—	370	460	—	—	—	—	—	—	[879]
Lipofuscin	Endogenous fluorophore	—	—	340–395	430–460, 540–640	—	—	—	—	—	Physiologic pH.	[901]
Lipofuscin	Endogenous fluorophore	—	—	340–395	430–460, 540–640	—	—	—	—	—	Physiologic pH.	[881]
Lipofuscin	Endogenous fluorophore	—	—	340–395	540–640	—	—	—	—	—	Physiologic pH.	[882]
Lipofuscin	Endogenous fluorophore	—	—	467	632	—	Dimethyl sulfoxide	—	—	—	Fluorescence intensity is temperature dependent.	[912]
Lipopigments	Endogenous fluorophore	345, 430	—	310–470	450–670	—	—	—	—	—	—	[886]
MAG	Exogenous fluorophore	—	41,800	492	505	80%	Solution	—	—	—	Monomeric mutant of AG.	[895]
Melanin	Endogenous fluorophore	—	—	785	NIR	—	—	—	—	—	In vivo.	[913]
Melanin	Endogenous fluorophore	—	—	785	NIR	—	—	—	—	—	In vivo.	[914]
Mono-L-aspartyl chlorin e6 (MACE)	Exogenous fluorophore	—	—	410	664	—	Solution	3.7	—	—	Used as a photodynamic therapy agent; phototoxic	[915]
mRFP1	Exogenous fluorophore	—	44,000	584	607	25%	Solution	—	—	—	Decreased photostability, monomeric dim DsRed-mutant, red-shifted relative to DsRed.	[895]

Fluorophore	Type	Absorption (nm)	ε	Excitation (nm)	Emission (nm)	Quantum yield	Lifetime (ns)			Medium	Comments	Ref
Nicotinamide adenine dinucleotide (NAD+) oxidized form	Endogenous fluorophore	260	18×10^{-3}	—	—	—	—	—	—	Powder/solution	Physiologic pH.	[916]
Nicotinamide adenine dinucleotide (NADH) reduced form	Endogenous fluorophore	—	—	350	460	—	0.6	0.2	—	—	—	[885]
NADH reduced form	Endogenous fluorophore	260, 340	14.4×10^{-3}, 6.2×10^{-3}	290	440	—	—	—	—	—	Physiologic pH.	[916]
NADH reduced form	Endogenous fluorophore	—	—	340	450	—	—	—	—	—	Physiologic pH.	[917]
NADH reduced form	Endogenous fluorophore	—	—	340	450	—	—	—	—	—	Physiologic pH.	[902]
NADH reduced form	Endogenous fluorophore	260, 350	—	250–280, 320–375	425–525	—	—	—	—	—	—	[886]
NADH reduced form	Endogenous fluorophore	—	—	350	460	—	—	—	—	—	—	[879]
NADH reduced form	Endogenous fluorophore	—	—	325	440–525	—	—	—	—	—	Physiologic pH.	[887]
NADH reduced form	Endogenous fluorophore	—	—	400	450	—	—	—	—	Powder dissolved in saline at 0.3 M	—	[894]
NADH reduced form	Endogenous fluorophore	—	—	405	—	—	0.44	—	—	—	Two-photon fluorescence lifetime imaging microscopy.	[889]
NADH reduced form	Endogenous fluorophore	—	—	780	450	—	1.3	—	—	Dulbecco's modified Eagle's medium (DMEM)	—	[918]
NADPH	Endogenous fluorophore	—	—	365	>410	—	0.45–0.60 (75–90)	1.4–2.2 (10–25)	—	Water	—	[919]
NADPH	Endogenous fluorophore	—	—	345	465	—	—	—	—	—	—	[870]
Nile blue A (NBA)	Exogenous fluorophore	—	—	620	660	—	—	—	—	Solution	Used as a photodynamic therapy agent; not phototoxic.	[920]
PAGFP	Exogenous fluorophore	—	—	505	525	—	—	—	—	Solution	Photoactivable (413 nm) mutant of EYFP.	[895]
Phenylalanine	Endogenous fluorophore	260	2.0×10^{-4}	260	280	0.04	6.8	—	—	Water	Physiologic pH.	[916]
Phenylalanine	Endogenous fluorophore	—	—	260	282	0.02	—	—	—	—	Neutral pH.	[921]
Photofrin II	Exogenous fluorophore	—	—	364	615	—	15.4 (100)	—	—	0 mM CTAB	Cationic micelles influence aggregation.	[884]
Photofrin II	Exogenous fluorophore	—	—	364	630	—	14.6 (64.3)	2.97 (17.6)	0.74 (18.1)	0 mM CTAB	Cationic micelles influence aggregation.	[922]

(continued)

TABLE 20.10 (continued)   Fluorescence Spectroscopy

Species	Endogenous/Exogenous Origin	Absorbance Maxima (nm)	Molar Extinction Coefficient (M-1)	Excitation λ_{ex} (nm)	Emission λ_{em} (nm)	Quantum Yield	Matrix	Lifetime t_1 (rel % t_1)	Lifetime t_2 (rel % t_2)	Lifetime t_3 (rel % t_3)	Comments	Reference
Photofrin II	Exogenous fluorophore	—	—	364	675	—	0 mM CTAB	14.36 (81.3)	1.95 (18.7)	—	Cationic micelles influence aggregation.	[922]
Photofrin II	Exogenous fluorophore	—	—	364	615	—	0.005 mM CTAB	15.04 (94.3)	2.44 (5.7)	—	Cationic micelles influence aggregation.	[922]
Photofrin II	Exogenous fluorophore	—	—	364	630	—	0.005 mM CTAB	14.44 (22.4)	2.37 (14.7)	0.42 (62.9)	Cationic micelles influence aggregation.	[922]
Photofrin II	Exogenous fluorophore	—	—	364	675	—	0.005 mM CTAB	14.07 (20.3)	2.21 (31.2)	0.67 (48.5)	Cationic micelles influence aggregation.	[922]
Photofrin II	Exogenous fluorophore	—	—	364	615	—	0.01 mM CTAB	14.99 (80.7)	2.27 (6.8)	0.58 (12.5)	Cationic micelles influence aggregation.	[922]
Photofrin II	Exogenous fluorophore	—	—	364	630	—	0.01 mM CTAB	14.15 (10.4)	2.2 (14.2)	0.5 (75.4)	Cationic micelles influence aggregation.	[922]
Photofrin II	Exogenous fluorophore	—	—	364	675	—	0.01 mM CTAB	13.84 (7.3)	2.43 (25.3)	0.83 (67.4)	Cationic micelles influence aggregation.	[922]
Photofrin II	Exogenous fluorophore	—	—	364	615	—	0.05 mM CTAB	13.88 (18.2)	2.47 (28.5)	0.67 (53.2)	Cationic micelles influence aggregation.	[922]
Photofrin II	Exogenous fluorophore	—	—	364	630	—	0.05 mM CTAB	12.41 (3.1)	2.66 (18.9)	0.62 (78)	Cationic micelles influence aggregation.	[922]
Photofrin II	Exogenous fluorophore	—	—	364	675	—	0.05 mM CTAB	8.17 (3.2)	2.67 (34.1)	1.06 (62.7)	Cationic micelles influence aggregation.	[922]
Photofrin II	Exogenous fluorophore	—	—	364	630	—	0.1 mM CTAB	10.5 (21.5)	3.71 (43.2)	0.96 (35.3)	Cationic micelles influence aggregation.	[922]
Photofrin II	Exogenous fluorophore	—	—	364	675	—	0.1 mM CTAB	7.55 (49.2)	3.61 (50.8)	—	Cationic micelles influence aggregation.	[922]

Compound	Type			Excitation (nm)	Emission (nm)		Medium			Comments	Ref.
Photofrin II	Exogenous fluorophore	—	—	364	630	—	1.0 mM CTAB	16.05 (89.6)	3.38 (10.4)	Cationic micelles influence aggregation.	[922]
Photofrin II	Exogenous fluorophore	—	—	364	675	—	1.0 mM CTAB	15.28 (63.1)	5.75 (36.9)	Cationic micelles influence aggregation.	[922]
Porphyrin	Endogenous fluorophore	—	—	420	640	—		—	—	—	[902]
Porphyrins	Endogenous fluorophore	—	—	400	610–675	—		—	—	—	[870]
Porphyrins	Endogenous fluorophore	—	—	405	600	—		—	—	—	[879]
Protoporphyrin IX	Exogenous fluorophore	—	—	400	650–670	—	DMSO	17 (89)	3 (11)	—	[884]
Protoporphyrin IX	Endogenous fluorophore	—	—	505	640	—	Methanol	—	—	—	[893]
Protoporphyrin photoproduct	Exogenous fluorophore	—	—	400	650–670	—	DMSO	0.7 (38)	4.5 (62)	—	[884]
Pulp level caries	Endogenous fluorophore	—	—	337	405, 435, 490, 530	—		—	—	—	[879]
Pyridoxal (PL)	Endogenous fluorophore	—	—	330	380	—		—	—	Physiologic pH.	[871]
Pyridoxal 5′-phosphate (PLP)	Endogenous fluorophore	—	—	330	400	—		—	—	Physiologic pH.	[871]
Pyridoxal 5′-phosphate (B6 derivative)	Endogenous fluorophore	—	—	330	385	—		—	—	—	[870]
Pyridoxamine (PM)	Endogenous fluorophore	326	—	335	400	—		—	—	Physiologic pH.	[871]
Pyridoxine	Endogenous fluorophore	308	—	275–350	360–470	—		—	—	—	[886]
Pyridoxine (PN)	Endogenous fluorophore	324	—	332	400	—		—	—	Physiologic pH.	[871]
Red fluorescent protein (RFP)	Exogenous fluorophore	—	22,500	558	583	—	Solution	—	—	Following the trafficking and function of proteins in living cells and monitoring the intracellular environment.	[897]
Reduced tin metalloporphyrin (Sn.NT2H₂)	Exogenous fluorophore	—	—	410	645	16	Solution	0.65	—	Used as a photodynamic therapy agent; phototoxic.	[923]
Riboflavin (B2) and its derivatives (coenzymes of FAD and FMN)	Endogenous fluorophore	—	—	370, 450	520, 530	—		—	—	—	[870]

(continued)

TABLE 20.10 (continued)   Fluorescence Spectroscopy

Species	Endogenous/Exogenous Origin	Absorbance Maxima (nm)	Molar Extinction Coefficient (M-1)	Excitation λ_{ex} (nm)	Emission λ_{em} (nm)	Quantum Yield	Matrix	Lifetime t_1 (rel % t_1)	Lifetime t_2 (rel % t_2)	Lifetime t_3 (rel % t_3)	Comments	Reference
Root carious	Endogenous fluorophore	—	—	400	480, 624, 650, 687	—	—	—	—	—	—	[879]
Sound and carious enamel	Endogenous fluorophore	—	—	405	455, 500, 582, 622	—	—	—	—	—	—	[879]
T1	Exogenous fluorophore	—	30,100	554	586	42%	Solution	—	—	—	Orange-red color, 36% brightness relative to DsRed1	[895]
Tetrasulfonated aluminum phthalocyanine (AlSPc)	Exogenous fluorophore	—	—	350	675	4	Solution	5.3	—	—	Used as a photodynamic therapy agent; optimal time for fluorescence in vivo 24–48 h; phototoxic.	[924]
Tryptophan	Endogenous fluorophore	280	5.6×10^{-3}	280	350	0.2	—	—	—	—	Physiologic pH.	[916]
Tryptophan	Endogenous fluorophore	—	—	280	350	—	—	—	—	—	Physiologic pH.	[917]
Tryptophan	Endogenous fluorophore	—	—	275	350	—	Powder/solution	2.8	1.5	—	—	[885]
Tryptophan	Endogenous fluorophore	215	—	200–240, 250–280	300–380	—	—	—	—	—	—	[886]
Tryptophan	Endogenous fluorophore	—	—	295	253	0.13	Water	3.1	—	—	Neutral pH.	[921]
Tryptophan	Endogenous fluorophore	—	—	295	345	—	—	—	—	—	—	[879]
T-sapphire	Exogenous fluorophore	—	44,000	399	511	60%	Solution	—	—	—	Largest Stokes shift reported for protein fluorophore, no pH sensitivity, fluorescence process involves an internal proton transfer.	[895]
Tyrosine	Endogenous fluorophore	275	1.4×10^{-3}	275	300	0.1	—	—	—	—	Physiologic pH.	[916]
Tyrosine	Endogenous fluorophore	—	—	270	320	—	—	—	—	—	Physiologic pH.	[879]
Tyrosine	Endogenous fluorophore	—	—	275	304	0.14	Water	3.6	—	—	Neutral pH.	[921]
Uroporphyrin III	Endogenous fluorophore	—	—	505	615	—	Methanol	—	—	—	—	[893]
Zn-protoporphyrin	Exogenous fluorophore	—	—	400	650–670	—	DMSO	2 (100)	—	—	—	[884]

Cellular studies

Species	Target Organ or Disease	Absorbance Maximum (nm)	Molar Extinction Coefficient (M-1)	Excitation λ_{ex} (nm)	Emission λ_{em} (nm)	Quantum Yield	Matrix	Lifetime t1 (rel % t1)	Lifetime t2 (rel % t2)	Lifetime t3 (rel % t3)	Comments	Reference
1-Pyreneisothiocyanate labeled anti-mouse IgG	Various cancers	—	—	Blue/visible	Visible	—	Labeled antibodies attached to surface proteins on mouse cells	20–55	—	—	Long lifetime allowed for differentiation of different types of cells via flow cytometry.	[925]
1-Pyrenesulfonyl chloride labeled anti-mouse IgG	Various cancers	—	—	Blue/visible	Visible	—	Labeled antibodies attached to surface proteins on mouse cells	20–55	—	—	Long lifetime allowed for differentiation of different types of cells via flow cytometry.	[925]
9-Acetoxy-tetra-*n*-propylporphycene (ATPPn)	Bladder cancer	—	—	Visible	Red/NIR	—	Human bladder carcinoma cell line	—	—	—	This is a recently developed photosensitizer that has shown promise based upon cellular studies with bladder carcinoma cells in vitro.	[926]
Aluminum phthalocyanine	Various cancers	—	—	610	>645	—	Cultured leukemic cells (human erythroleukemic cells)	2.2 ± 0.4 (50)	6.1 ± 0.2 (50)	—	—	[876]
Aluminum phthalocyanine	Various cancers	—	—	364	630	—	In L1210 mouse cells, 4 h following i.p. admin.	5.18 (86.33)	1.52 (13.67)	—		[875]
Aluminum phthalocyanine	Various cancers	—	—	364	630	—	In L1210 mouse cells, 12 h following i.v. admin.	5.9 (39.74)	2.87 (50.24)	0.93 (10.01)		[875]
B-lymphoma cells (CA46)	Ocular diseases	—	—	351, 488	—	—	Cultured CA46 cells	—	—	—	Ex vivo study showed that CA46 cells could be detected based on their autofluorescence.	[927]
Chlorophyll-derived photosensitizers	Lung cancer	760	—	760	NIR	—	In OAT 75 small cell lung carcinoma cells	—	—	—	Absorption maximum of the photosensitizer can shift between 20 and 100 nm depending upon local environment within the cell.	[928]

(continued)

TABLE 20.10 (continued)　Fluorescence Spectroscopy

Species	Target Organ or Disease	Absorbance Maximum (nm)	Molar Extinction Coefficient (M-1)	Excitation λ_{ex} (nm)	Emission λ_{em} (nm)	Quantum Yield	Matrix	Lifetime t1 (rel % t1)	Lifetime t2 (rel % t2)	Lifetime t3 (rel % t3)	Comments	Reference
Epithelial cells from normal, hyperplastic, and adenomatous colon	Colon cancer	—	—	488	—	—	Cultured epithelial cells	—	—	—	Differences between normal, hyperplastic, and adenomatous epithelial cells are attributed in part to differences in the intrinsic numbers of mitochondria and lysosomes.	[929]
Esophageal cells (normal and malignant)	Esophageal cancer	—	—	351	—	—	—	—	—	—	Tumoral cells had a fluorescence intensity approximately twice as that of normal cells.	[930]
Flavins	Various diseases	—	—	365	500	—	Intact yeast cells	0.2–0.35 (45)	2.0–3.0 (25)	6.0–8.0 (30)	—	[904]
Flavins	Various diseases	—	—	365	500	—	Defective yeast cells	0.30–0.50 (35)	2.0–3.0 (30)	6.0–8.0 (35)	t1 is lower but represents more fluorescence in intact cells.	[904]
Flavoproteins, oxidized	Bladder cancer	—	—	488	550–560	—	Human urothelial cells (normal and malignant) in media	—	—	—	Normal cells exhibited greater than 10 times the autofluorescence intensity of tumor cells.	[931]
Glioma, solid tumor, and invasive tumor cells	Brain cancer	—	—	710–920	—	—	Cultured mouse cell lines	—	—	—	Distinct fluorescence lifetimes of endogenous fluorophores were found in different cellular compartments in cultured cells.	[932]
GFP-labeled protein kinase B	Cells	—	7,000	395, 470	509	—	Living cells	—	—	—	Used to study intracellular signaling pathways, including the actions of insulin.	[897]
HpD	Various cancers	—	—	364	630	—	In L1210 mouse cells, 12 h following i.v. admin.	14.36 (10.83)	3.12 (20.07)	0.61 (69.1)	—	[875]
Hematoporphyrin	Various cancers	—	—	364	630	—	In L1210 mouse cells, 4 h following i.p. admin.	15.3 (45.39)	4.32 (18.16)	1.11 (36.46)	—	[875]

Hematoporphyrin aggregates	Various cancers	—	Visible	675	—	Various types of cells	—	—	—	Fluorescence emission at 675 nm was assigned to hematoporphyrin monomers.	[933]
Hematoporphyrin monomers	Various cancers	—	Visible	635	—	Various types of cells	—	—	—	Fluorescence emission at 635 nm was assigned to hematoporphyrin monomers.	[933]
Human breast cells (normal and malignant)	Breast cancer	—	UV/blue	340, 440	—	Human breast cancer cells (normal and malignant) in media	—	—	—	Ratiometric analyses of the fluorescence and autofluorescence intensity of the cells at 340 to 440 nm allow for the differentiation of normal and malignant cells.	[926]
Human osteosarcoma 143B and HeLa cells	Various cancers	—	780	450	—	—	—	1.3 (untreated) 3.5 (apoptosis)	—	Two-photon fluorescence lifetime imaging microscopy; differentiation of apoptosis from necrosis by changes in NADH fluorescence lifetime.	[918]
Human semen	Infertility	—	488	622	—	—	—	—	—	—	[934]
Human seminal plasma	Infertility	—	488	622	—	—	—	—	—	Sperm motility could be correlated to the autofluorescence emission.	[934]
Human spermatozoa	Infertility	—	488	622	—	—	—	—	—	Spermatozoa and sperm motility could be correlated to the autofluorescence emission.	[934]
Melanin	Melanoma	—	800	—	—	—	—	—	—	Selectively observable fluorescence of melanin in cells and offers the possibility of cell classification.	[935]

(continued)

TABLE 20.10 (continued)　Fluorescence Spectroscopy

Species	Target Organ or Disease	Absorbance Maximum (nm)	Molar Extinction Coefficient (M-1)	Excitation L_{ex} (nm)	Emission L_{em} (nm)	Quantum Yield	Matrix	Lifetime t_1 (rel % t_1)	Lifetime t_2 (rel % t_2)	Lifetime t_3 (rel % t_3)	Comments	Reference
Murine fibroblast cells (normal and malignant)	Various cancers	—	—	290	Blue-green	—	Murine fibroblast cell lines	—	—	—	The autofluorescence intensity of malignant cells was significantly less than that of normal cells, with tryptophan being the primary component responsible for the difference.	[936]
N-(1-pyrene) maleimide labeled anti-mouse IgG	Various cancers	—	—	Blue/visible	Visible	—	Labeled antibodies attached to surface proteins on mouse cells	20–55	—	—	Long lifetime allowed for differentiation of different types of cells via flow cytometry.	[925]
NADH reduced form and FAD	Breast cancer	—	—	780, 890	—	—	Normal and tumorous epithelium in carcinoma in situ (CIS) of the breast	—	—	—	Tumor cells produced higher intensity and had a longer fluorescence lifetime; the shift to a longer lifetime in tumor cells was independent of the free and bound state of FAD and NADH and of the excitation wavelength.	[937]
NADPH	Diabetes	—	—	370	420–480	—	Cultured 3T3-L1 adipocytes	7.23	—	—	Addition of glucose caused an increase in autofluorescence intensity and shortened the lifetime.	[938]
NADPH	Various diseases	—	—	365	450	—	Intact and defective yeast cells	0.2–0.3 (30)	1.4–2.4 (40)	6.0–8.0 (30)	Deficient cell fluorescence is four times higher.	[904]
Oral epithelial cells (normal and squamous)	Oral cancer	—	—	300	320–580	—	Several different culturing media selected to cause cell differentiation	—	—	—	No difference between the differentiated cells could be determined from the resulting spectra.	[939]

Sample	Disease/Process		Ex	Em		Conditions			Results	Ref
Oral epithelial cells (normal and squamous)	Oral cancer	—	340	360–620	—	Several different culturing media selected to cause cell differentiation	—	—	Spectral shifts occurred between the three types of cell differentiation.	[939]
Oral epithelial cells (normal and squamous)	Oral cancer	—	365	400–630	—	Several different culturing media selected to cause cell differentiation	—	—	Small changes in fluorescence intensity and spectral shape occurred between the various differentiated cells.	[939]
Oral epithelial cells (normal and squamous)	Oral cancer	—	420	440–630	—	Several different culturing media selected to cause cell differentiation	—	—	Small changes in fluorescence intensity and spectral shape occurred between the various differentiated cells.	[939]
Oral epithelial cells (normal and squamous)	Oral cancer	—	200–360	380	—	Several different culturing media selected to cause cell differentiation	—	—	Significant changes in intensity and spectral profile occurred between the various differentiated cells.	[939]
Oral epithelial cells (normal and squamous)	Oral cancer	—	240–415	450	—	Several different culturing media selected to cause cell differentiation	—	—	Significant changes in intensity and spectral profile occurred between the various differentiated cells.	[939]
Oral epithelial cells (normal and squamous)	Oral cancer	—	250–420	480	—	Several different culturing media selected to cause cell differentiation	—	—	Significant changes in intensity and spectral profile occurred between the various differentiated cells.	[939]
Oral epithelial cells (normal and squamous)	Oral cancer	—	270–480	520	—	Several different culturing media selected to cause cell differentiation	—	—	Significant changes in intensity and spectral profile occurred between the various differentiated cells.	[939]
Ovary cells	Photoconversion	—	320–400	455	—	Chinese hamster ovary cells	—	—	Optical trapping of ovary cells using 730 nm was found to induce an increase in autofluorescence intensity and a 6 nm red shift in its maxima.	[940]

(continued)

TABLE 20.10 (continued)   Fluorescence Spectroscopy

Species	Target Organ or Disease	Absorbance Maximum (nm)	Molar Extinction Coefficient (M-1)	Excitation λ_{ex} (nm)	Emission λ_{em} (nm)	Quantum Yield	Matrix	Lifetime t_1 (rel % t_1)	Lifetime t_2 (rel % t_2)	Lifetime t_3 (rel % t_3)	Comments	Reference
Phorbol-13-acetate-12-N-methyl-N-4-(N,N'-di(2hydroxyethyl)amino-7-nitrobenz-2-oxa-1,3-diazole-aminododecanoate (N-C12-Ac(13))	Cellular signal transduction	488	—	488	Visible	—	P3HR-1 Burkitt lymphoma cells	—	—	—	Used in the investigation of protein kinase C and its role in signal transduction.	[941]
Photofrin II	Various cancers	—	—	364	630	—	In L1210 mouse cells, 4 h following i.p. admin.	14.9 (40.9)	4.91 (18.32)	1.18 (41.58)	Cationic micelles influence aggregation.	[875]
Photofrin II	Various cancers	—	—	364	630	—	In L1210 mouse cells, 12 h following i.v. admin.	14.81 (7.3)	4.65 (25.49)	1.08 (67.27)	Cationic micelles influence aggregation.	[875]
Protoporphyrin	Various cancers	—	—	420	610–690	—	Incubated meerkat kidney cells dorsal skinfold of hamster	1–2	2.0–3.0	11.0–14.0		[942]
Retinal pigment epithelial (RPE) cells	Normal and diseased RPE	—	—	568	624	—	RPE cells in humans and macaques	—	—	—	In vivo imaging demonstrated that with increasing eccentricity, RPE cell density and mosaic regularity decreased, whereas RPE cell size and spacing increased.	[943]
Various types of cells	Cell proliferation	—	—	320–350	450	—	Various type of cells	—	—	—	Differentiation between slow and rapidly growing cells was possible based on ratiometric analyses.	[944]
Various types of cells	Cell proliferation	—	—	340	360–660	—	Various type of cells	—	—	—	Differentiation between slow and rapidly growing cells was possible based on ratiometric analyses.	[944]

Tissue and biofluid studies (in vitro)

Adenomatous colorectal tissue	Gastrointestinal cancers	—	290	330	—	In vitro mouse intestines	—	Mean fluorescence intensity of polyps of the small intestine and colon were both significantly higher than that of the normal mucosa.	[945]
Arterial tissue	Atherosclerosis	—	337	370–510	—	In vitro arterial tissue	—	Discrimination between stable and unstable lesions by studying the time-resolved spectra of components present in the arterial wall (collagen, lipoproteins, and cholesterol)	[946]
Beta-carotene	Atherosclerosis	—	488	Visible	—	Arterial tissue incubated with beta-carotene	—	Beta-carotene was shown to reduce the total autofluorescence intensity of atherosclerotic plaques in arterial tissues.	[947]
Bladder tissue (normal and tumorous)	Bladder cancer	—	220–500	280–700	—	In vitro bladder tissue	—	Autofluorescence can distinguish malignant from normal bladder tissue; excitation wavelengths of 280 and 330 nm are more significant for differentiation.	[948]
Blood plasma (porphyrins)	Colon cancer	—	—	615–635	—	Blood	—	The intensity of fluorescence emission was different between patients bearing colorectal cancer and blood donors.	[949]

(continued)

TABLE 20.10 (continued) Fluorescence Spectroscopy

Species	Target Organ or Disease	Absorbance Maximum (nm)	Molar Extinction Coefficient (M-1)	Excitation λ_{ex} (nm)	Emission λ_{em} (nm)	Quantum Yield	Matrix	Lifetime t_1 (rel % t_1)	Lifetime t_2 (rel % t_2)	Lifetime t_3 (rel % t_3)	Comments	Reference
Brain tissue	Brain cancer	—	—	337	360–550	—	In vitro cortex, glioblastoma, multiforme, and white matter	>3 ns at 380; <1 ns at 460	—	—	High-grade gliomas are characterized by fluorescence lifetimes that varied with emission wavelength and their emission is longer than that of normal brain tissue.	[950]
Brain tissue (tumor and epileptogenic cerebral cortex)	Brain cancer; epilepsy	—	—	337, 360, 440	—	—	In vitro pediatric brain tissue	—	—	—	Statistically significant differences found between neoplastic brain and normal gray matter, normal gray matter and differences found between neoplastic brain and normal gray matter.	[951]
Breast tissue (normal and tumorous)	Breast cancer	—	—	458	510–580	—	In vitro breast tissue	—	—	—	Discrimination between normal and neoplastic tissue reaches sensitivity and specificity of 100%.	[952]
Breast tissues (normal, benign, and malignant)	Breast cancer	—	—	325	—	—	Breast tissue	—	—	—	Intensity plots of collagen and NADPH yielded accurate classification of the different tissue types.	[953]
Bronchial tissue (healthy, metaplastic, and dysplastic/CIS)	Pre-/early cancers in the tracheobronchial tree	—	—	400–480	590	—	In vitro bronchial tissue	—	—	—	Maximize the tumor versus healthy and the tumor versus inflammatory/metaplastic contrast in the detection of pre-/early malignant lesions.	[954]
Cervical tissue (atypical)	Cervical lesions	—	—	488	500–700	—	In vitro cervical tissue	—	—	—	Significant differences of the spectral intensity profile for the high-grade squamous intraepithelial lesion (HGSIL) compared to cervical carcinoma (CC) and low-grade squamous intraepithelial lesion (LGSIL).	[955]

Sample	Disease		Excitation	Emission	Tissue			Findings	Ref.
Cervical tissue (normal and malignant)	Cervical cancer	—	325	—	In vitro cervical tissue	—	—	Principal component analysis of total fluorescence gives specificity and sensitivity over 95%.	[956]
Cervical tissues (normal, inflammatory, and dysplastic)	Cervical cancer	—	365, 440	460, 525	In vitro frozen human cervical tissues	—	—	Autofluorescence intensities are significantly lower in the epithelia of severely dysplastic tissues relative to normal and inflammatory.	[957]
Collagen	Various diseases	—	380	Visible	Cervical tissue	—	—	Increases in fluorescence emission due to NADH were found for dysplastic tissue relative to normal tissues.	[958]
Colorectal tissue	Colorectal carcinoma	—	450	550–800	In vitro *colorectal tissue*	—	—	Difference in spectral shape and autofluorescence intensity from tissue affected by adenocarcinoma as compared to healthy tissue.	[959]
Congo red	Amyloidosis	—	UV	Red	In vitro *abdominal fat pad aspirations*	—	—	Amyloid deposits showed bright red fluorescence.	[960]
Dentin (healthy and demineralized)	Tooth decay	—	488	480–520	Whole tooth	—	—	Demineralized dentin exhibited lower fluorescence intensity than healthy dentin at 529 nm.	[961]
Dentin (healthy and demineralized)	Tooth decay	—	460–480	520	Whole tooth	—	—	Demineralized dentin exhibited a more pronounced peak than healthy dentin at 520 nm.	[961]
Dentine (healthy and carious)	Tooth decay	—	488	>515	Whole tooth	—	—	Autofluorescence correlated well with the amount of demineralization.	[962]

(continued)

TABLE 20.10 (continued)   Fluorescence Spectroscopy

Species	Target Organ or Disease	Absorbance Maximum (nm)	Molar Extinction Coefficient (M-1)	Excitation λ_{ex} (nm)	Emission λ_{em} (nm)	Quantum Yield	Matrix	Lifetime t_1 (rel % t_1)	Lifetime t_2 (rel % t_2)	Lifetime t_3 (rel % t_3)	Comments	Reference
Eosinophils, circulating	Hodgkin's disease/ lymphomas	—	—	280	330	—	Eosinophils isolated from blood	—	—	—	Fluorescence emission at this wavelength can be attributed to tryptophan.	[963]
Eosinophils, circulating	Hodgkin's disease/ lymphomas	—	—	360	440	—	Eosinophils isolated from blood	—	—	—	—	[963]
Eosinophils, circulating	Hodgkin's disease/ lymphomas	—	—	380	415	—	Eosinophils isolated from blood	—	—	—	—	[963]
Eosinophils, circulating	Hodgkin's disease/ lymphomas	—	—	365	Blue-violet	—	Eosinophils isolated from blood	—	—	—	—	[963]
Eosinophils, tissue dwelling	Hodgkin's disease/ lymphomas	—	—	365	Amber-gold	—	Eosinophils isolated from various tumor tissues	—	—	—	—	[963]
Esophageal tissue (normal and tumorous)	Esophageal cancer	—	—	458	500–530, >585	—	In vitro *esophageal tissue*	—	—	—	Specific differences in the autofluorescence from Barrett's, squamous and gastric mucosa from the esophagus.	[929]
Excised and frozen colon tissue sections (normal and tumorous)	Colon cancer	—	—	351–364	Visible	—	Human colon tissue	—	—	—	This article characterizes resulting fluorescence emission from various locations in the tissue.	[964]
Excised and frozen human cervical tissues (normal and malignant)	Cervical cancer	—	—	365	460	—	Frozen cervical tissue	—	—	—	Autofluorescence emission in the epithelia showed small differences in intensity between severe dysplasia and mild dysplasia of tissues; epithelial fluorescence is less in malignant tissue than normal tissue.	[965]

Excised and frozen human cervical tissues (normal and malignant)	Cervical cancer	—	440	525	Frozen cervical tissue	—	—	Autofluorescence emission in the stroma showed small differences in intensity between inflammatory and severely dysplastic tissues; epithelial fluorescence is less in malignant tissue than normal tissue.	[966]
Excised arterial tissue	Atherosclerosis	—	458	Visible	Arterial tissue	—	—	Good differentiation of normal and atherosclerotic tissue can be seen after excitation with visible wavelengths.	[967]
Excised arterial tissue	Atherosclerosis	—	476	Visible	Arterial tissue	—	—	Good differentiation of normal and atherosclerotic tissue can be seen after excitation with visible wavelengths.	[968]
Excised arterial tissue (calcified and noncalcified)	Atherosclerosis	—	495	520–800	Arterial tissue	—	—	Characteristic spectral and temporal characteristics of 495 nm induced autofluorescence of normal intima, calcified plaques, and fibrofatty plaques were determined.	[969]
Excised arterial tissue (calcified and noncalcified)	Atherosclerosis	—	380	400–600	Arterial tissue	—	—	Differentiation of atherosclerotic lesions from normal tissue was attempted based upon autofluorescence emission (it was found that a better excitation wavelength was 325 nm).	[970]

(continued)

TABLE 20.10 (continued) Fluorescence Spectroscopy

Species	Target Organ or Disease	Absorbance Maximum (nm)	Molar Extinction Coefficient (M-1)	Excitation L_{ex} (nm)	Emission L_{em} (nm)	Quantum Yield	Matrix	Lifetime t_1 (rel % t_1)	Lifetime t_2 (rel % t_2)	Lifetime t_3 (rel % t_3)	Comments	Reference
Excised arterial tissue (calcified and noncalcified)	Atherosclerosis	—	—	450	400–600	—	Arterial tissue	—	—	—	Differentiation of atherosclerotic lesions from normal tissue was attempted based upon autofluorescence emission (it was found that a better excitation wavelength was 325 nm).	[970]
Excised arterial tissue (calcified and noncalcified)	Atherosclerosis	—	—	308	321–657	—	Arterial tissue	—	—	—	A XeCl laser was used for the simultaneous ablation and autofluorescence-based diagnosis of atherosclerosis that provided a means of differentiating normal tissue from various plaques.	[971]
Excised arterial tissue (calcified and noncalcified)	Atherosclerosis	—	—	351–364	Visible	—	Arterial tissue	—	—	—	Autofluorescence of collagen, elastin, and ceroid can be used to differentiate normal and atherosclerotic tissues.	[972]
Excised arterial tissue (calcified and noncalcified)	Atherosclerosis	—	—	476	Visible	—	Arterial tissue	—	—	—	Chemical and morphological characteristics responsible for autofluorescence signals were determined.	[973]
Excised arterial tissue (calcified and noncalcified)	Atherosclerosis	—	—	325	443	—	Arterial tissue	—	—	—	Normalized fluorescence intensities at 443 nm can be used to differentiate between normal and calcified tissues.	[967]

Sample	Disease		Excitation (nm)	Emission (nm)		Tissue type			Comments	Ref.
Excised arterial tissue (calcified and noncalcified)	Atherosclerosis	—	325	375–385 and 435–445	—	Arterial tissue filled with blood	—	—	An autoguidance system for angioplasty was investigated based upon ratiometric analyses of autofluorescence at 380 and 440 nm.	[974]
Excised arterial tissue (calcified)	Atherosclerosis	—	325	400–600 (no peak at 480)	—	Arterial tissue	—	—	Noncalcified tissue has a distinct peak at 480 nm and calcified tissue shows only a broad band emission.	[970]
Excised arterial tissue (collagen and elastin components)	Atherosclerosis	—	306–310	380	—	Arterial tissue	—	—	By ratioing tryptophan fluorescence intensity to collagen and elastin component, classification of tissue is possible.	[975]
Excised arterial tissue (noncalcified)	Atherosclerosis	—	248	370–460	—	Arterial tissue	—	—	Autofluorescence shows broad band fluorescence emission.	[976]
Excised arterial tissue (noncalcified)	Atherosclerosis	—	325	400–600 (peak at 480)	—	Arterial tissue	—	—	Noncalcified tissue has a distinct peak at 480 nm and calcified tissue shows only a broad band emission.	[970]
Excised arterial tissue (tryptophan component)	Atherosclerosis	—	306–310	340	—	Arterial tissue	—	—	By ratioing tryptophan fluorescence intensity to collagen and elastin component, classification of tissue is possible.	[977]
Excised bladder tissue (normal mucosa, flat lesions, and papillary tumors)	Bladder cancer	—	337	385 and 455	—	Bladder tissue	—	—	Ratiometric measurements of the autofluorescence intensities at 385 and 440 nm were used to distinguish malignant tumors from nonmalignant and inflamed tissue.	[978]
Excised bladder tissue (normal mucosa, flat lesions, and papillary tumors)	Bladder cancer	—	375–440 (D-light)	Visible	—	Bladder tissue	—	—	Autofluorescence imaging could distinguish between normal mucosa, flat lesions, and papillary tumors.	[979]

(continued)

TABLE 20.10 (continued) Fluorescence Spectroscopy

Species	Target Organ or Disease	Absorbance Maximum (nm)	Molar Extinction Coefficient (M-1)	Excitation λ_{ex} (nm)	Emission λ_{em} (nm)	Quantum Yield	Matrix	Lifetime t_1 (rel % t_1)	Lifetime t_2 (rel % t_2)	Lifetime t_3 (rel % t_3)	Comments	Reference
Excised brain tissue (normal and Alzheimer's)	Alzheimer's disease	—	—	647	650–850	—	Human brain tissue (temporal cortex)	—	—	—	Autofluorescence emission from temporal cortex samples was used to correctly diagnose Alzheimer's disease.	[980]
Excised brain tissue (normal and glioma)	Brain cancer	358	—	—	Visible	—	Human brain tissue	—	—	—	By combining comparative genomic hybridization and fluorescence in situ hybridization, differentiation of infiltrating gliomas has been shown to be feasible.	[981]
Excised brain tissue (normal and tumorous)	Brain cancer	—	—	337	370–500	—	In vitro brain tissue	—	—	—	Meningioma is characterized by unique fluorescence characteristics that enable discrimination of tumor and normal tissue.	[982]
Excised brain tissue (normal and tumorous)	Eye cancer	—	—	800	—	—	Choroidal tissue	—	—	—	Fluorescence of choroidal melanomas exhibited a more reddish appearance and less intensity than that of healthy tissue.	[983]
Excised brain tissue (normal and tumorous)	Brain cancer	—	—	337	460	—	Human brain tissue	—	—	—	The autofluorescence intensity of normal brain tissue was significantly greater than that of primary brain tumorous regions.	[984]
Excised breast tissue (cancer, fibrous, and adipose)	Breast cancer	—	—	532/632.8	NIR	—	Breast tissue	—	—	—	Emission intensity is considerably different in breast cancer; 632.8 nm excitation yields best contrast.	[985]

Excised breast tissue (normal and tumorous)	Breast cancer	—	488	—	—	Breast tissue	—	—	The photobleaching decay profiles of cancerous tissues were faster than those of normal tissue profiles.	[986]
Excised breast tissue (normal)	Breast cancer	—	488	530, 550 and 590	—	Normal breast tissue	—	—	Presence of three peaks due to absorption by hemoglobin at 420, 542, and 575 nm.	[987]
Excised breast tissue (tumor)	Breast cancer	—	488	530	—	Tumor breast tissue	—	—	Only a single peak is visible in the fluorescence profile.	[986]
Excised calcified arterial tissue	Atherosclerosis	—	248	397, 442, 450, 461, 528 and 558	—	Arterial tissue	—	—	Autofluorescence shows multiple, prominent atomic fluorescence lines.	[976]
Excised carotid atherosclerotic plaques	Atherosclerosis	—	337	Visible	—	Carotid arterial tissue	—	—	Autofluorescence of carotid atherosclerotic plaques was used to determine the three primary components: fibrous tissue, lipid constituents, and calcified plaque.	[988]
Excised carotid atherosclerotic plaques	Atherosclerosis	—	476	Visible	—	Carotid arterial tissue	—	—	Autofluorescence of carotid atherosclerotic plaques was used to determine the three primary components: fibrous tissue, lipid constituents, and calcified plaque.	[987]
Excised carotid atherosclerotic plaques	Atherosclerosis	—	488	Visible	—	Carotid arterial tissue	—	—	Autofluorescence of carotid atherosclerotic plaques was used to determine the three primary components: fibrous tissue, lipid constituents, and calcified plaque.	[987]

(continued)

TABLE 20.10 (continued)   Fluorescence Spectroscopy

Species	Target Organ or Disease	Absorbance Maximum (nm)	Molar Extinction Coefficient (M-1)	Excitation λ_{ex} (nm)	Emission λ_{em} (nm)	Quantum Yield	Matrix	Lifetime t_1 (rel % t_1)	Lifetime t_2 (rel % t_2)	Lifetime t_3 (rel % t_3)	Comments	Reference
Excised carotid atherosclerotic plaques	Atherosclerosis	—	—	458	Visible	—	Carotid arterial tissue	—	—	—	Autofluorescence of carotid atherosclerotic plaques was used to determine the three primary components: fibrous tissue, lipid constituents, and calcified plaque.	[987]
Excised cervical tissue (normal and tumorous)	Cervical cancer	—	—	330	385	—	Cervical tissue	—	—	—	Fluorescence intensity of normalized emission spectra is greater for normal tissue than tumor tissue.	[989]
Excised cervical tissue (normal and tumorous)	Cervical cancer	—	—	365	475	—	Cervical tissue	—	—	—	Fluorescence intensity at 475 nm increases with degree of dysplasia/malignancy.	[990]
Excised colon tissue (normal and adenomatous)	Colon cancer	—	—	330, 370, and 430	404, 480, and 680	—	Human colon tissue	—	—	—	—	[991]
Excised colon tissue (normal and adenomatous)	Colon cancer	—	—	325	350–600	—	Human colon tissue	—	—	—	370 nm was found to be the optimal excitation wavelength for discrimination of normal and adenomatous tissues.	[992]
Excised colon tissue (normal and tumorous)	Colon cancer	—	—	366	480–580	—	Human colon tissue	—	—	—	The relative amplitude of normal and timorous tissue varied significantly between 480 and 580 nm.	[993]

Sample	Condition		Excitation	Emission		Tissue				Notes	Ref.
Excised colon tissue (normal mucosa and adenomatous polyps)	Colon cancer	—	325	Visible	—	Human colon tissue	—	—	—	Autofluorescence intensity in adenomatous polyps is lower than for normal tissue and small spectral differences can be seen.	[994]
Excised heart tissue (nodal conductive and atrial endomyocardial)	Heart arrhythmia	—	308	440–500	—	Heart tissue	—	—	—	Nodal tissue demonstrated a decrease in normalized fluorescence intensity and peak width relative to atrial endomyocardial tissue.	[995]
Excised heart tissue (nodal conductive and ventricular endocardium)	Heart arrhythmia	—	308	430–550	—	Heart tissue	—	—	—	Nodal tissue demonstrated an increase in fluorescence intensity relative to atrial endomyocardial tissue.	[994]
Excised human eye lens	Vision problems	—	430–490	530–630	—	Human eye lens	—	—	—	Intrinsic eye lens transmittance can be determined based upon autofluorescence analyses.	[996]
Excised lung tissue (normal)	Lung cancer	—	488	530, 550, and 590	—	Normal lung tissue	—	—	—	Presence of three peaks due to absorption by hemoglobin at 420, 542, and 575 nm.	[986]
Excised lung tissue (tumor)	Lung cancer	—	488	530	—	Tumor lung tissue	—	—	—	Only a single peak is visible in the fluorescence profile.	[986]
Excised oral tissue (normal and tumorous)	Oral cancer	—	410	Visible	—	Oral tissue	—	—	—	Autofluorescence emission spectra excited with 410 nm light are capable of differentiating normal and abnormal tissues.	[997]

(continued)

TABLE 20.10 (continued) Fluorescence Spectroscopy

Species	Target Organ or Disease	Absorbance Maximum (nm)	Molar Extinction Coefficient (M^{-1})	Excitation I_{ex} (nm)	Emission I_{em} (nm)	Quantum Yield	Matrix	Lifetime t_1 (rel % t_1)	Lifetime t_2 (rel % t_2)	Lifetime t_3 (rel % t_3)	Comments	Reference
Excised oral tissue (normal, malignant, and dysplastic)	Oral cancer	—	—	410	635	—	Oral tissue	—	—	—	Abnormal tissues exhibited an increase in fluorescence intensity at 635 nm.	[998]
Excised oral tissue (normal, malignant, and premalignant)	Oral cancer	—	—	300	330–470	—	Oral tissue	—	—	—	Ratiometric fluorescence measurements at 330 and 470 nm allowed for the differentiation of malignant and premalignant lesions from normal tissues.	[999]
Excised skin tissue	Skin cancer	—	—	365	Visible	—	Skin tissue (147 samples)	—	—	—	Based upon the resulting autofluorescence, it was possible to differentiate nondysplastic nevi from melanomas and dysplastic nevi (tumor tissue exhibited reduced fluorescence).	[1000]
Excised skin tissue (normal and abnormal)	Nonmelanoma skin cancers	—	—	410	—	—	Skin tissue	—	—	—	Emission signal intensity is effective for detection of basal cell carcinoma, squamous cell carcinoma, and actinic keratosis.	[1001]
Excised skin tissue (tumor)	Nonmelanoma skin cancers	—	—	390	450–650	—	Skin tissue	—	—	—	Nonmelanoma skin tumors can be distinguished from healthy tissue.	[1002]

Material	Disease		Specimen	Excitation	Emission		Value	Ratio	Comments	Reference
Excised stomach tissue (normal, dysplastic, and tumor)	Stomach cancer	—	Stomach tissue	325	440 and 395	—	—	—	Normalized fluorescence intensities at both 440 and 395 nm exhibit significant differences between normal and tumorous tissues.	[1003]
Extracts of human eye (soluble and insoluble)	Cataractogenesis	—	Cortical and nuclear extracts	UV	Blue/green	—	—	—	The green to blue autofluorescence intensity ratio was found to be greater than six for cataractous tissue fractions.	[1004]
Femoral head or femur	Degenerative joint disease	—	Femoral head	410	—	—	—	—	Clear contrast of bone, degenerated cartilage, and subchondral cyst.	[1005]
Flavins	Various diseases	—	Normal rat kidney	488	500–550	—	0.357 +/- 0.018 (27)	1.220 +/- 0.035 (63)	—	[875]
Flavins	Various diseases	—	Normal rat kidney	488	600–650	—	0.204 +/- 0.011 (47)	1.0 +/- 0.006 (53)	—	[875]
Flavins	Various diseases	—	Cancerous rat kidney	488	500–550	—	0.223 +/- 0.015 (59)	1.966 +/- 0.037 (41)	—	[875]
Flavins	Various diseases	—	Cancerous rat kidney	488	600–650	—	0.236 +/- 0.014 (67)	1.963 +/- 0.037 (33)	—	[875]
Gastric mucosa (normal and tumorous)	Gastritis and gastric cancer	—	In vitro gastric mucosa	488	550, 600	—	—	—	Differentiation of normal tissue and neoplastic lesions.	[1006]
HpD	Various cancers	—	Tumor tissue in dihematoporphyrin ether (DHE)-injected rat	320	630	—	17	6	—	[1007]
HpD	Various cancers	—	Normal tissue in surrounding muscle	320	630	—	17	0.7	—	[1006]
Heart tissue	Human heart	—	In vitro heart conduction system and ventricular myocardium	320–370	420–465	—	—	—	Optical visualization to determine optical differences characteristic for heart tissues.	[1008]
Melanins derived from 3-hydroxyanthranilic acid	Colon cancer	—	Human colon tissue	324	413	—	—	—	This article characterizes the autofluorescence properties of plasma-soluble melanins.	[1009]

(continued)

TABLE 20.10 (continued)　Fluorescence Spectroscopy

Species	Target Organ or Disease	Absorbance Maximum (nm)	Molar Extinction Coefficient (M-1)	Excitation λex (nm)	Emission λem (nm)	Quantum Yield	Matrix	Lifetime t1 (rel % t1)	Lifetime t2 (rel % t2)	Lifetime t3 (rel % t3)	Comments	Reference
Melanins derived from dopa, catecholamines, catechol, and 3-hydroxykynurenine	Colon cancer	—	—	345	445	—	Human colon tissue	—	—	—	This article characterizes the autofluorescence properties of plasma-soluble melanins.	[1008]
NADH	Cervical dysplasia	—	—	380	—	—	In vitro cervical tissue	—	—	—	An increase in NADH fluorescence and a decrease in collagen fluorescence in dysplastic tissue.	[1010]
NADH	Various cancers	—	—	325	350–525	—	In vitro tissues	—	—	—	Normal tissues show higher collagen autofluorescence contribution than NADH, while cancerous tissues show increased NADH autofluorescence relative to collagen emission.	[887]
NADH	Various cancers	—	—	325	440–600	—	In vitro tissues	—	—	—	Rapidly growing cancerous tissues have higher concentrations of NADH and FAD and therefore increased autofluorescence.	[887]
NADH reduced form	Various diseases	—	—	380	Visible	—	Cervical tissue	—	—	—	Increases in fluorescence emission due to NADH were found for dysplastic tissue relative to normal tissues.	[958]
NADPH	Various diseases	—	—	320	400	—	Normal arterial wall	0.3	2	7	Calcified plaques have a higher ratio of slow (400 nm) to fast fluorescence (480 nm).	[1011]

Oral tissues (normal and tumorous)	Oral cancer	—	Various	635, 690	—	In vitro oral tissue	—	—	Prominent autofluorescence from dysplastic or malignant mucosa compared to normal mucosa.	[1012]
Oral tissues (normal and tumorous)	Oral cancer	—	375–440	>515	—	In vitro oral tissue	—	—	Autofluorescence intensity of cancerous tissue was much greater than normal mucosa.	[1011]
Ovarian tissues (normal, benign, and malignant)	Ovarian cancer	—	325	350–600	—	In vitro ovarian tissue	—	—	Discrimination of normal, benign, and malignant conditions based on autofluorescence.	[1013]
Pancreatic tissues (normal, pancreatitis, and adenocarcinoma)	Pancreatic cancer	—	355	360–700	—	Excised human pancreatic tissues and in vivo human pancreatic cancer xenografts in nude mice	—	—	Measurements were associated with NAD(P)H and collagen; the relative collagen emission from adenocarcinoma and pancreatitis tissues was larger than from normal tissues.	[1014]
Paraffin-embedded uterine cervix tissue	Alveolar rhabdomyosarcoma of the uterine cervix	—	—	—	—	In vitro uterine cervix tissue known to have tumors	—	—	Fluorescence in situ hybridization indentified specific chromosomal abnormalities.	[1015]
Prostate and periprostatic neural tissue	Prostate	—	780	380–530	—	Excised prostate and cavernous nerves from male Sprague–Dawley rats	—	—	Potential to improve the precision of nerve-sparing prostatectomy	[1016]
Rectum organ tissues (normal and malignant)	Rectal cancer	—	200–400	492, 544	—	In vitro rectal tissues	—	—	Fluorescence intensity of normal rectal tissue is higher than malignant rectal tissue; difference in fluorescence intensity at 492 and 544 nm between normal and malignant tissue.	[1017]

(continued)

TABLE 20.10 (continued)   Fluorescence Spectroscopy

Species	Target Organ or Disease	Absorbance Maximum (nm)	Molar Extinction Coefficient (M-1)	Excitation I_{ex} (nm)	Emission I_{em} (nm)	Quantum Yield	Matrix	Lifetime t_1 (rel % t_1)	Lifetime t_2 (rel % t_2)	Lifetime t_3 (rel % t_3)	Comments	Reference
Animal studies (in vivo)												
1,1'-*Bis*-(4-sulfobutyl) indo tricarbocyanine-5,5'-dicarboxylic acid diglucamide monosodium salt (SIDAG)	Breast cancer	—	—	740	750–800	—	In vivo mammary tumors in rats	—	—	—	SIDAG was enriched in a mammary tumor up to a ratio of 6:1.	[1018]
2',7'-*Bis*-(2 carboxyethyl)-5-(and-6)-carboxyfluorescein	Various cancers	—	—	465	Visible	—	Grafted tumors on mice	—	—	—	Differentiation of tumors from normal tissue was accomplished based upon intracellular pH measurements with this exogenous dye.	[1019]
Arterial tissue (atherosclerotic)	Atherosclerosis	—	—	UV/blue	410–490	—	Rabbit arteries	—	—	—	Autofluorescence analyses were capable of monitoring the disruption of atherosclerotic plaques following injection of Russell's viper venom.	[1020]
Benzoporphyrin derivatized-monoacid (BPD-MA)	Various cancers	—	—	337	380–750	—	Rat tumors (various organs)	—	—	—	BPD-MA exhibited approximately the same demarcation ability as other common photosensitizers.	[1021]
Brain tissue (normal and glioma)	Brain cancer	—	—	360	470	—	Rat brain	—	—	—	Fluorescence due to NAD(P)H; decreased autofluorescence intensity for gliomas relative to normal tissue.	[1022]
Brain tissue (normal and glioma)	Brain cancer	—	—	440	520	—	Rat brain	—	—	—	Fluorescence due to flavins; decreased autofluorescence intensity for gliomas relative to normal tissue.	[1021]

Brain tissue (normal and glioma)	Brain cancer	—	490	630	—	Rat brain	—	—	Fluorescence due to porphyrins; decreased autofluorescence intensity for gliomas relative to normal tissue.	[1021]
Buccal mucosa	Buccal cavity	—	350–410	500–520	—	In vivo buccal mucosa of dogs	—	—	Various regions showed similar emission profiles.	[1023]
Cathepsin B	Arthritis	—	610–650	680–720	—	In vivo knee joints of male nude mice	—	—	Cathepsin B fluorescent imaging showed a significant difference between the osteoarthritic and normal joints.	[880]
Cathepsin B	Atherosclerosis	—	675	710±10	—	Arterial tissue of transgenic mice	—	—	Using a tomographic imagining system, cathepsin B activity is high in inflamed atherosclerotic lesions.	[1024]
Cy5.5	Human mammary MDA-MB 468 (EGFR+) and MDA-MB-435 (EGFR−) cancer cells	—	—	—	—	Female athymic nude mice	—	—	Monitoring of epidermal growth factor receptor (EGFR) targeted therapy.	[1025]
Cy5.5	Various cancers	—	610–650	680–720	—	HP1080 fibrosarcoma overexpressing cathepsin B grown subcutaneously in the mammary fat pad of a nude mouse	—	—	The tumor demonstrated significant probe activation; background tissue exhibited minimal fluorescence.	[1026]
DsRed2	PC-3 human prostate	—	563	582	—	Male nude mice	—	—	Monitoring of tumor growth.	[1024]
DsRed2	RFP expressing B16F0 melanoma; MMT060562 mammary; Dunning and PC-3 prostate; HCT-116 colon tumors	—	563	582	—	Nude C57/B6-GFP mice with subcutaneous melanoma/orthotopic breast/orthotopic prostate/ orthotopic colon tumors	—	—	Monitoring of tumor growth and metastatic progression.	[1024]
Esophageal multispheroidal tumor (induced by trans-retinoic acid)	Esophageal cancer	—	UV/blue	340, 450, and 520	—	Rat esophagus	—	—	Autofluorescence intensities at 340, 450, and 520 nm demonstrated differences between normal and cancerous tissues.	[1027]

(continued)

TABLE 20.10 (continued) Fluorescence Spectroscopy

Species	Target Organ or Disease	Absorbance Maximum (nm)	Molar Extinction Coefficient (M-1)	Excitation λ_{ex} (nm)	Emission λ_{em} (nm)	Quantum Yield	Lifetime t_1 (rel % t_1)	Lifetime t_2 (rel % t_2)	Lifetime t_3 (rel % t_3)	Matrix	Comments	Reference
Esophageal tumor (induced by N-nitroso-N-methylbenzylamine [NMBA])	Esophageal cancer	—	—	UV/violet	380	—	—	—	—	Rat esophagus	Alteration of fluorescence emission correlated to disease progression, from normal to dysplasia, to invasive cancer.	[1028]
Foam cell lesions	Atherosclerosis	—	—	308	Visible	—	—	—	—	Hypercholesterolemic rabbit arteries	Autofluorescence from foam cell lesions exhibited red shifts and spectral broadening similar to oxidized low-density lipoproteins.	[1029]
GFP	Lymphoma cells	—	—	—	509	—	—	—	—	Lymphoma model in transgenic mice	Monitoring of the effects of genotype on therapeutic efficacy.	[1024]
GFP	MCA-MB-231human breast carcinoma	—	—	—	509	—	—	—	—	Xenograft breast cancer and metastases model in female nude mice	Detection of bone metastasis.	[1024]
Heart tissue (native and transplanted)	Heart transplant rejection	—	—	Blue	Visible	—	—	—	—	Rat heart tissue (midtransverse ventricular)	A correlation was found between the severity of tissue rejection and the autofluorescence from the heart.	[1030]
Heart tissue (normal and hypoxic)	Myocardial hypoxia	—	—	308	350–600	—	—	—	—	Mouse heart tissue	Hypoxia was found to reduce the autofluorescence between 455 and 505 nm and remove two spectral peaks at 540 and 580 nm, relative to normal tissue, leaving only a single peak at 555 nm.	[1031]
Hematoporphyrins	Various cancers	—	—	337	630	—	—	—	—	Rat tumors (various organs)	—	[1032]

Sample	Application		Excitation	Emission		Model			Findings	Reference
Joint tissue (arthritic and normal)	Arthritis	—	300/360	355–365/475–485	—	In vivo joint tissue of mice	—	—	Arthritis influences collagen/elastin and NADPH autofluorescence.	[1033]
K1735P melanoma	Skin cancer	—	UV/violet	360–700	—	Melanomas implanted intradermally in the ears of C3H/HeN mice	—	—	Resulting autofluorescence spectra showed decreases in the fluorescence intensity over the spectral range of 385–425 nm; however, no spectral differences could be determined between the normal tissue and unpigmented melanomas.	[1034]
Kidney tissue (normal and hypoxic)	Renal hypoxia	—	308	350–600	—	Mouse kidney tissue	—	—	Hypoxia was found to reduce the autofluorescence between 455 and 505 nm and remove two spectral peaks at 540 and 580 nm, relative to normal tissue, leaving only a single peak at 555 nm.	[1030]
Mammary tissue (normal and tumorous)	Breast cancer	—	632.8, 670	700, 750, and 800	—	In vivo mammary tissue of rats	—	—	Autofluorescence intensity of malignant tumors under 670 nm excitation was higher than normal tissue; intensity of benign tumors was lower than normal tissue.	[1035]
MS-2 fibrosarcoma	Various cancers	—	337	400–500	—	Implanted in a NALB-CDFI mice	—	—	Autofluorescence intensity was found to be lower in the tumor tissue than in the healthy tissue and the fast component of the biexponential fluorescence decay is significantly lower in the tumor as well.	[1036]

(continued)

TABLE 20.10 (continued)   Fluorescence Spectroscopy

Species	Target Organ or Disease	Absorbance Maximum (nm)	Molar Extinction Coefficient (M-1)	Excitation λ_{ex} (nm)	Emission λ_{em} (nm)	Quantum Yield	Matrix	Lifetime t_1 (rel % t_1)	Lifetime t_2 (rel % t_2)	Lifetime t_3 (rel % t_3)	Comments	Reference
Oral tissue (normal and malignant)	Oral cancer	—	—	410	635	—	Hamster cheek pouch (tumors induced with 7,12-dimethylbenz(a)anthracene)	—	—	—	Neoplastic lesions showed a characteristic fluorescence between 630 and 640 nm.	[1037]
Oral tissue (normal and malignant)	Oral cancer	—	—	350–370	Visible	—	Hamster cheek pouch	—	—	—	Neoplastic tissue showed an increase in fluorescence intensity relative to normal tissue with excitation between 350 and 370 nm.	[1038]
Oral tissue (normal and malignant)	Oral cancer	—	—	400–450	Visible	—	Hamster cheek pouch	—	—	—	Neoplastic tissue showed a decrease in fluorescence intensity relative to normal tissue with excitation between 400 and 450 nm, with an optimal wavelength of 410 nm.	[1037]
Oral tissue (normal, hyperplastic, papilloma, and invasive carcinoma)	Oral cancer	—	—	405	430–700	—	Hamster cheek pouch (tumors induced with 7,12-dimethylbenz(a)anthracene)	—	—	—	Ratiometric analyses of the autofluorescence emission at 530/620 and 530/630 nm provided a means of differentiating the various stages and types of tumors present.	[1039]
Pancreas tissue (normal and tumorous)	Pancreatic cancer	—	—	355	470 and 640	—	Rat pancreatic tissue (normal and tumorous)	—	—	—	—	[1040]
Pheophorbide-a (Ph-a)	Pancreatic cancer	—	—	355	680	—	Rat pancreatic tissue (normal and tumorous)	—	—	—	Normalization of Ph-a fluorescence intensity was performed via ratiometric analyses with press autofluorescent signals.	[1039]

Compound	Application		Excitation		Emission		Sample		Remarks		Ref.
Protoporphyrin IX	Skin	—	420	—	635, 707	—	Skin tissue of mice	—	Squamous cell carcinoma tissue exhibited strong red fluorescence.	—	[1041]
Protoporphyrin IX (5-aminolevulinic acid induced)	Liver cancer	—	Visible	—	635	—	Hepatic tumors in rats	—	Large quantities of PpIX was found to accumulate in both normal and tumorous liver tissues.	—	[1042]
Protoporphyrin IX (5-aminolevulinic acid induced)	Various cancers	—	405	—	635–705	—	Rat tumors (various organs)	—	Maximum buildup of PpIX occurred within 1 h.	—	[1043]
Skin tissue	Skin aging	—	295, 335	—	380	—	In vivo skin tissue of hairless mice	—	Autofluorescence of epidermal tryptophan moieties and collagen cross-links in the dermal matrix may serve as markers for skin aging, photoaging, and immediate assessment of exposure to UV A radiation.	—	[1044]
Trimethoxylated carotenoporphyrin	Various cancers	—	425	—	655 and 720	—	MS-2 fibrosarcoma in BALB/c mice	—	The greatest extent of carotenoporphyrin fluorescence occurred in the liver.	—	[1045]
Trimethylated carotenoporphyrin	Various cancers	—	425	—	655 and 720	—	MS-2 fibrosarcoma in BALB/c mice	—	The greatest extent of carotenoporphyrin fluorescence occurred in the liver.	—	[1044]
Vasculature of the heart injected with ICG	Heart	—	785	—	835	—	Heart of a human-sized pig	—	Imaging of normal vasculature, detection of the location of significant stenoses, verifying vessel potency after coronary artery bypass grafting.	—	[1023]
Vasculature of the heart injected with ICG	Heart	—	400–700	—	752	—	In vivo heart of an adult rat	—	First demonstration of vascular imaging with NIR quantum dots.	—	[1023]

(continued)

TABLE 20.10 (continued) Fluorescence Spectroscopy

Human clinical studies (in vivo)

Species	Target Organ or Disease	Absorbance Maximum (nm)	Molar Extinction Coefficient (M-1)	Excitation L_{ex} (nm)	Emission L_{em} (nm)	Quantum Yield	Matrix	Lifetime t_1 (rel % t_1)	Lifetime t_2 (rel % t_2)	Lifetime t_3 (rel % t_3)	Comments	Reference
5,10,15,20-Tetra(m-hydroxyphenyl) chlorin (mTHPC)	Various cancers	—	—	—	—	—	In vivo esophageal tissue	8.5 ± 0.8	—	—	mTHPC was found to have a fluorescence lifetime of 8.5 ns in esophageal tissue.	[975]
5-Aminolevulinic acid (5-ALA)	Brain cancer	—	—	377	460, 500	—	In vivo brain tissue	—	—	—	Species is taken up by gliomas where breakdown of the blood–brain barrier has occurred, but not in normal brain.	[1046]
Alveoli (healthy and smokers)	Lung	—	—	488	—	—	In vivo alveolar ducts and sacs	—	—	—	Alveolar macrophages were not detectable in nonsmokers, but a specific tobacco-tar-induced fluorescence was observed in smokers.	[1047]
Bladder tissue (normal and tumorous)	Bladder cancer	—	—	337	370–490	—	In vivo bladder tissue	—	—	—	Spectral and temporal differences in the autofluorescence emission were used to differentiate normal and tumorous tissues.	[1045]
Bladder tissue (normal and tumorous)	Bladder cancer	—	—	337	385–455	—	In vivo bladder tissue	—	—	—	By ratioing the autofluorescence intensity at 385 to the intensity at 455 nm, it was possible to diagnose the tissue with 98% sensitivity.	[1048]
Bladder tissue (normal, inflammatory mucosa, and neoplastic urothelial lesion)	Bladder cancer	—	—	308	Visible	—	In vivo bladder tissue	—	—	—	The shape of the autofluorescence emission of CIS tissue was significantly different from both normal and inflamed mucosa allowing a ratiometric measurement of the intensities at 360 and 440 nm to be used for diagnosis.	[1049]

Tissue	Disease				Excitation (nm)	Detection		In vivo/in vitro				Findings	Reference
Bladder tissue (normal, inflammatory mucosa and neoplastic urothelial lesion)	Bladder cancer	—	—	—	337	Visible	—	In vivo bladder tissue	—	—	—	The overall fluorescence intensity of bladder tumor tissue was found to be much less than that of normal tissue irregardless of stage.	[931]
Bladder tissue (normal, inflammatory mucosa and neoplastic urothelial lesion)	Bladder cancer	—	—	—	480	Visible	—	In vivo bladder tissue	—	—	—	The overall fluorescence intensity of bladder tumor tissue was found to be much less than that of normal tissue irregardless of stage.	[931]
Bladder tissue (normal, inflammatory mucosa and neoplastic urothelial lesion)	Bladder cancer	—	—	—	308	360 and 440	—	In vivo bladder tissue	—	—	—	Ratiometric analyses of the autofluorescence intensity at 360 to 440 nm was used to differentiate normal or inflamed tissue from neoplastic lesions.	[1050]
Bladder tissue (normal, inflammatory mucosa and neoplastic urothelial lesion)	Bladder cancer	—	—	—	337	Visible	—	In vivo bladder tissue	—	—	—	Autofluorescence intensity of tumors was significantly less than that of normal tissue.	[1050]
Bladder tissue (normal, inflammatory mucosa and neoplastic urothelial lesion)	Bladder cancer	—	—	—	488	Visible	—	In vivo bladder tissue	—	—	—	Autofluorescence intensity of tumors was significantly less than that of normal tissue.	[1050]
Bladder tissue (normal, inflammatory mucosa and neoplastic urothelial lesion)	Bladder cancer	—	—	—	337	385 and 455	—	In vivo bladder tissue	—	—	—	Ratiometric analyses based upon tissue autofluorescence were used to correctly identify bladder lesions.	[1051]
Brain tissue (normal and malignant gliomas)	Brain cancer	—	—	—	460 and 625	Visible/red/NIR	—	In vivo brain tissue	—	—	—	By combining autofluorescence and diffuse reflectance, differentiation of normal and malignant brain tissue has been shown to be feasible.	[1052]

(continued)

TABLE 20.10 (continued)   Fluorescence Spectroscopy

Species	Target Organ or Disease	Absorbance Maximum (nm)	Molar Extinction Coefficient (M-1)	Excitation I_{ex} (nm)	Emission I_{em} (nm)	Quantum Yield	Matrix	Lifetime t_1 (rel % t_1)	Lifetime t_2 (rel % t_2)	Lifetime t_3 (rel % t_3)	Comments	Reference
Brain tissue (tumor)	Brain	—	—	280	460, 500	*In vivo brain tissue*	—	—	—	—	A shift of the predominant spectral peak by 40 nm suggests radiation damage in brain tissue.	[1053]
Brain tissue exposed to 5-aminolevulinic acid	Brain cancer	—	—	405	636	—	In vivo brain tissue	—	—	—	Neoplastic cells were present in the tissue region that displayed a peak at 636 nm; no neoplastic cells were present in the region that exhibited only the excitation light peak.	[1053]
Bronchial tissue (normal and cancerous)	Bronchial carcinoma	—	—	385–465	505	—	In vivo bronchial tissue	—	—	—	Autofluorescence from tumor tissue is significantly lower than normal tissue.	[1054]
Bronchial tissue (normal and cancerous)	Endobronchial cancers	—	—	430, 430 plus 665	Green, red	—	In vivo bronchial tissue	—	—	—	The addition of backscattered red light to the tissue autofluorescence improved the contrast between healthy and diseased tissue.	[1055]
Bronchial tissue (normal and tumorous)	Lung cancer	—	—	480	Visible	—	In vivo lung tissue	—	—	—	Spectral and temporal differences in the autofluorescence emission were used to differentiate normal and tumorous tissues.	[1045]
Bronchial tissue (normal, dysplastic, and CIS)	Lung cancer	—	—	325	Green/red	—	In vivo lung tissue	—	—	—	Ratiometric measurements of lung tissue over the red and green regions of the spectrum were used to diagnose the tissue as normal, dysplastic, or CIS.	[1056]

Tissue	Application		Excitation	Emission		Sample	Notes				Ref.
Bronchial tissue (normal, metaplastic, and early cancer)	Lung cancer	—	350–495	Green–red	—	In vivo lung tissue	Absolute autofluorescence measurements allowed for the differentiation of normal, metaplastic, and early cancerous stages of lung tissue.	—	—	—	[1057]
Bronchial tissue (normal, metaplastic, and early cancer)	Lung cancer	—	400–480	600–800	—	In vivo lung tissue	Autofluorescence differences between normal, metaplastic, and dysplastic tissue were found, with the optimal excitation wavelength for differentiation being 405 nm.	—	—	—	[1058]
Bronchial tissue (normal and tumorous)	Lung cancer	—	405	430–680	—	In vivo bronchial tissue	Spectral contrast due to enhanced blood concentration just below the epithelial layers of the lesion; thickening of the epithelium in the lesions is the probable cause.	0.17 (10)	2.02 (40)	6.84 (50)	[1059]
Carotid plaques	Atherosclerotic plaque vulnerability	—	337	360–550	—	In vivo carotid plaques	Spectral intensities and time-dependent parameters at discrete emission wavelengths allow for discrimination of various compositional and pathological features associated with plaque vulnerability.	—	—	—	[1060]
Cervical tissue (normal and tumorous)	Cervical cancer	—	330	385	—	In vivo cervical tissue	The averaged normalized fluorescence intensity was found to be greater for normal tissue as compared to abnormal tissue.	—	—	—	[1061]

(continued)

TABLE 20.10 (continued)   Fluorescence Spectroscopy

Species	Target Organ or Disease	Absorbance Maximum (nm)	Molar Extinction Coefficient (M-1)	Excitation L$_{ex}$ (nm)	Emission L$_{em}$ (nm)	Quantum Yield	Matrix	Lifetime t$_1$ (rel % t$_1$)	Lifetime t$_2$ (rel % t$_2$)	Lifetime t$_3$ (rel % t$_3$)	Comments	Reference
Cervical tissue (normal, nonneoplastic, and carcinoma intraepithelial neoplasia [CIN])	Cervical cancer	—	—	337	Visible	—	In vivo cervical tissue	—	—	—	Differentiation of tissues was performed based upon changes in tissue autofluorescence related to the species collagen, oxyhemoglobin, and NAD(P)H	[1062]
Cervix	Cervical cancer	—	—	330, 440		—	In vivo	—	—	—	Multispectral digital colposcope used to measure autofluorescence of the cervix; correctly identified cervical intraepithelial neoplasia.	[1063]
Collagen	Colon cancer	—	—	337	390	—	In vivo colon tissue	—	—	—	Normal colonic tissue showed a decrease in autofluorescence intensity due to collagen relative to hyperplastic or adenomatous tissues.	[1064]
Colorectal tissue (normal, adenomatous, and cancerous)	Colorectal cancer and dysplasia	—	—	White or violet-blue	—	—	In vivo colorectal tissue	—	—	—	Cancer and adenomas with severe dysplasia showed specific differences between the fluorescence spectra as compared with normal mucosa and hyperplastic polyps.	[1065]
Endogenous skin protoporphyrins	Skin cancer	—	—	Visible	600, 620, 640 and 670	—	In vivo skin tissue	1–5	—	—	Lipophile skin bacterium, Propionibacterium acnes, that upon irradiation caused photodynamic activity were studied.	[1066]

Substance	Application		Excitation (nm)	Emission (nm)		Medium			Notes			Reference
Esophageal tissue (low risk, high-grade dysplasia, and carcinoma)	Barrett's esophagus	—	337, 400	550	—	In vivo esophageal tissue	—	—	Steady-state fluorescence was more effective than time-resolved data in diagnosing high-grade dysplasia.	—	—	[1067]
Esophageal tissue (normal and tumorous)	Esophageal cancer	—	410	450–600	—	In vivo esophageal tissue	—	—	Tumor tissue exhibited a decrease in normalized fluorescence intensity at 480 nm relative to normal tissue.	—	—	[1068]
Esophageal tissue (normal and tumorous)	Esophageal cancer	—	Violet-blue	450–700	—	In vivo esophageal tissue	—	—	Specific differences in the autofluorescence spectra of esophageal squamous cell carcinoma, adenocarcinoma of the esophagus, and adenocarcinoma of the stomach were found.	—	—	[1069]
Esophageal tissue (normal and tumorous)	Esophageal cancer	—	337	370–490	—	In vivo esophageal tissue	—	—	Spectral and temporal differences in the autofluorescence emission were used to differentiate normal and tumorous tissues.	—	—	[1045]
Flavoproteins in skin tissue	Skin cancer	—	960 (two-photon absorption)	520	—	In vivo skin tissue	—	—	Confocal microscopy was used to determine the spatial location of various autofluorescent species within skin cells.	—	—	[1070]
GB137-labeled cathepsins (cancer cells)	Various cancers	—	NIR	NIR	—	In vivo various types of cells	—	—	GB137 includes an acyloxymethyl ketone reactive moiety that targets cysteine proteases, a fluorophore, and a quencher.	—	—	[1071]

(continued)

TABLE 20.10 (continued)   Fluorescence Spectroscopy

Species	Target Organ or Disease	Absorbance Maximum (nm)	Molar Extinction Coefficient (M-1)	Excitation l_{ex} (nm)	Emission l_{em} (nm)	Quantum Yield	Matrix	Lifetime t_1 (rel % t_1)	Lifetime t_2 (rel % t_2)	Lifetime t_3 (rel % t_3)	Comments	Reference
Human cornea	Diabetes	—	—	360–370	532–630	—	In vivo human eyes	—	—	—	Autofluorescence measurements were used to distinguish healthy patients from patients with diabetes mellitus.	[1072]
Human cornea	Diabetes	—	—	400–410	532–630	—	In vivo human eyes	—	—	—	Autofluorescence measurements were used to distinguish healthy patients from patients with diabetes mellitus.	[1072]
Human cornea	Diabetes	—	—	415–425	532–630	—	In vivo human eyes	—	—	—	Autofluorescence measurements were used to distinguish healthy patients from patients with diabetes mellitus.	[1072]
Human cornea	Diabetes	—	—	425–435	532–630	—	In vivo human eyes	—	—	—	Autofluorescence measurements were used to distinguish healthy patients from patients with diabetes mellitus.	[1072]
Human cornea	Diabetes	—	—	431–441	532–630	—	In vivo human eyes	—	—	—	Autofluorescence measurements were used to distinguish healthy patients from patients with diabetes mellitus.	[1072]
Human cornea	Diabetes	—	—	435–445	532–630	—	In vivo human eyes	—	—	—	Autofluorescence measurements were used to distinguish healthy patients from patients with diabetes mellitus.	[1072]
Human cornea	Diabetes	—	—	445–455	532–630	—	In vivo human eyes	—	—	—	Autofluorescence measurements were used to distinguish healthy patients from patients with diabetes mellitus.	[1072]

Sample	Condition		Excitation	Emission		Sample			Notes	Reference
Human cornea	Diabetes	—	465–475	532–630	—	In vivo human eyes	—	—	Autofluorescence measurements were used to distinguish healthy patients from patients with diabetes mellitus.	[1072]
Human cornea	Diabetes	—	475–485	532–630	—	In vivo human eyes	—	—	Autofluorescence measurements were used to distinguish healthy patients from patients with diabetes mellitus.	[1072]
Human cornea	Diabetes	—	415–491	515–630	—	In vivo human eyes	—	—	Corneal autofluorescence was found to increase in people with diabetes mellitus.	[1073]
Human eye lens	Vision problems	—	UV/blue	495–520	—	In vivo human eyes	—	—	Lens autofluorescence could be correlated to coloration and opalescence of the lens nucleus in humans.	[1074]
Human eye lens	Vision problems	—	430–490	530–630	—	Human eye lens	—	—	Intrinsic eye lens transmittance can be determined based upon autofluorescence analyses.	[995]
ICG in saline	Posterior uveitis and central serous chorioretinopathy	—	—	—	—	Eye	—	—	ICG angiography can predict the time to growth of small choroidal melanomas by revealing characteristic patterns within their microcirculation.	[1023]
Larynx tissue (cancerous and precancerous)	Laryngeal cancer	—	374–440	Green	—	In vivo *larynx tissue*	—	—	Normal laryngeal mucosa showed green autofluorescence; moderate and high epithelial dysplasia, CIS, and cancer displayed a diminished green fluorescence.	[1075]

(continued)

TABLE 20.10 (continued) Fluorescence Spectroscopy

Species	Target Organ or Disease	Absorbance Maximum (nm)	Molar Extinction Coefficient (M-1)	Excitation λ_{ex} (nm)	Emission λ_{em} (nm)	Quantum Yield	Matrix	Lifetime t_1 (rel % t_1)	Lifetime t_2 (rel % t_2)	Lifetime t_3 (rel % t_3)	Comments	Reference
Laryngeal tissues (normal and tumorous)	Laryngeal cancer	—	—	325	Visible	—	In vivo laryngeal tissue	—	—	—	Images were obtained over several emission bands and ratiometric analyses were performed to differentiate normal tissue from carcinoma tissue.	[1076]
Laryngeal tissues (normal and tumorous)	Laryngeal cancer	—	—	375–440	Green	—	In vivo laryngeal tissue	—	—	—	Tumor autofluorescence intensity was greatly reduced relative to surrounding normal tissue.	[1077]
Laryngeal tissues (normal, dysplastic, CIS, and microinvasive lesions)	Laryngeal cancer	—	—	380–460	Light green	—	In vivo laryngeal tissue	—	—	—	Autofluorescence diagnoses were capable of distinguishing normal tissue, dysplastic tissue, CIS, and microinvasive lesions.	[1078]
Lipofuscin	Vision problems	—	—	Visible	710	—	In vivo human eyes	—	—	—	Measurements of macular pigment density could be determined based upon autofluorescence.	[1079]
NADH reduced	Colon cancer	—	—	337	460	—	In vivo colon tissue	—	—	—	—	[1065]
NADH reduced	Colon cancer	—	—	370	460	—	In vivo colon tissue	—	—	—	Ratiometric analyses of autofluorescence emission were used to identify tissues as normal, hyperplastic, or adenomatous.	[1080]
Normal, adenomatous, and cancerous colorectal tissue	Colon cancer	—	—	Blue	—	—	In vivo colorectal tissue	—	—	—	Compared with normal mucosa and hyperplastic polyps, rectal cancers, and adenomas with severe dysplasia showed specific autofluorescence differences.	[1081]

Sample	Condition		Excitation (nm)	Emission (nm)	In vivo		1.5 (peripapillary region)	5 (optical disc)		Notes	Ref.
Ocular fundus	Eye	—	457.9	—	In vivo *human ocular fundus*	—	1.5 (peripapillary region)	5 (optical disc)	—	Breathing 100% oxygen affects the fluorescence lifetimes.	[946]
Oral and pharynx tissue (normal and tumorous)	Head and neck squamous cell carcinoma	—	375–440	>515	In vivo oral and pharynx tissues	—	—	—	—	Using autofluorescence, tumors presented as darker areas with an accentuated reddish-blue color.	[1011]
Oral tissues (normal and tumorous)	Oral cancer	—	337	375–700	In vivo *oral tissue*	—	—	—	—	Comparison of principal component analysis and nonlinear algorithms component analysis and nonlinear algorithms for classifications of cancerous and normal tissue autofluorescence.	[1011]
Oral tissues (connective tissues)	Oral cancer	—	365	Visible	In vivo oral tissue	—	—	—	—	Tumor margining was performed using 365 nm light to excite the edges of the tumor.	[1082]
Oral tissues (normal)	Oral cancer	—	330	340–601	In vivo oral tissue	—	—	—	—	460/380 nm ratio increased from fibrosis to healthy mucosa to hyperkeratosis to dysplasia to cancer.	[1011]
Oral tissues (normal and lesions)	Oral cancer	—	405	—	In vivo *oral tissue*	—	—	—	—	Autofluorescence imaging coupled with objective image analysis as a noninvasive tool for the detection of oral neoplasia.	[1083]
Oral tissues (normal and premalignant lesions)	Oral cancer	—	410	633	In vivo normal verrucous hyperplasia, epithelial hyperplasia, and epithelial dysplasia	—	—	—	—	Time-resolved autofluorescence spectroscopy is used for the diagnosis of oral premalignant lesions.	[1084]

(continued)

TABLE 20.10 (continued)　Fluorescence Spectroscopy

Species	Target Organ or Disease	Absorbance Maximum (nm)	Molar Extinction Coefficient (M-1)	Excitation λ_{ex} (nm)	Emission λ_{em} (nm)	Quantum Yield	Matrix	Lifetime t_1 (rel % t_1)	Lifetime t_2 (rel % t_2)	Lifetime t_3 (rel % t_3)	Comments	Reference
Oral tissues (normal and tumorous)	Oral cancer	—	—	370	630–640	—	In vivo oral tissue	—	—	—	Differences between normal and neoplastic tissues were found based upon its autofluorescence emission.	[1085]
Oral tissues (normal and tumorous)	Oral cancer	—	—	410	630–640	—	In vivo oral tissue	—	—	—	Differences in the autofluorescence emission between normal and neoplastic tissues were found to be optimal following 410 nm excitation.	[1085]
Oral tissues (normal and tumorous)	Oral cancer	—	—	337	Visible	—	In vivo oral tissue	—	—	—	The autofluorescence intensity of contralateral sites was much greater than the abnormal sites.	[1086]
Oral tissues (normal and tumorous)	Oral cancer	—	—	410	Red/blue	—	In vivo oral tissue	—	—	—	The ratio of red fluorescence to blue fluorescence was greater in abnormal tissue than contralateral areas.	[1086]
Oral tissues (normal and tumorous)	Oral cancer	—	—	442	Red/green	—	In vivo oral tissue	—	—	—	Ratiometric images of red autofluorescence and green autofluorescence images were used to identify oral cavity neoplasia.	[1087]
Oral tissues (normal and tumorous)	Oral cancer	—	—	360	>480	—	In vivo oral tissue	—	—	—	Fluorescence photography of the tissue autofluorescence could be used to distinguish between benign and malignant oral cavity tumors.	[1088]

Sample		Type								Comments	Reference
Oral tissues (normal and tumorous)	—	Oral cancer	—	350	472	—	In vivo oral tissue	—	—	Identification of oral cavity neoplasia was demonstrated based upon the tissue autofluorescence.	[1089]
Oral tissues (normal and tumorous)	—	Oral cancer	—	380	472	—	In vivo oral tissue	—	—	Identification of oral cavity neoplasia was demonstrated based upon the tissue autofluorescence.	[1089]
Oral tissues (normal and tumorous)	—	Oral cancer	—	400	472	—	In vivo oral tissue	—	—	Identification of oral cavity neoplasia was demonstrated based upon the tissue autofluorescence.	[1089]
Oral tissues (normal and tumorous)	—	Oral cancer	—	330	380	—	In vivo oral tissue	—	—	Differentiation of normal and tumorous tissues was performed based upon a decrease in the 330 nm excitation band in tumor tissue.	[1090]
Oral tissues (normal and tumorous)	—	Oral cancer	—	340	390	—	In vivo oral tissue	—	—	Diagnosis of the stage of dysplasia was determined based upon changes in the autofluorescence emission intensity at 390 nm.	[1090]
Oral tissues (normal and tumorous)	—	Oral cancer	—	370	Visible	—	In vivo oral tissue	—	—	—	[1086]
Oral tissues (normal and tumorous)	—	Oral cancer	—	410	Visible	—	In vivo oral tissue	—	—	—	[1086]
Oral tissues (normal and tumorous)	—	Oral cancer	—	11 λ from 337–610	350–700	—	In vivo oral tissue	—	—	Correlation of early biochemical and histologic changes in oral tissue with spectral features in fluorescence, reflectance, and light scattering spectra to diagnose early stages of oral malignancies.	[1011]

(*continued*)

TABLE 20.10 (continued) Fluorescence Spectroscopy

Species	Target Organ or Disease	Absorbance Maximum (nm)	Molar Extinction Coefficient (M-1)	Excitation λ_{ex} (nm)	Emission λ_{em} (nm)	Quantum Yield	Matrix	Lifetime t_1 (rel % t_1)	Lifetime t_2 (rel % t_2)	Lifetime t_3 (rel % t_3)	Comments	Reference
Oral tissues (normal and tumorous)	Oral cancer	—	—	330	340–601	—	In vivo oral tissue	—	—	—	Partial least squares and artificial neural network classification algorithm was used to distinguish a malignant from benign lesions.	[1011]
Oral tissues (normal and tumorous)	Oral cancer	—	—	6 λ from 365–450	467–867	—	In vivo oral tissue	—	—	—	Partial least squares, artificial neural network, and emission wavelength ratio classification algorithm was used to separate healthy mucosa from oral cancer.	[1011]
Oral tissues (normal, precancerous, and cancerous)	Oral cancer	—	—	340	420–440	—	In vivo oral tissue	—	—	—	—	[1091]
Pharynx tissues (normal and tumorous)	Oral cancer	—	—	370	Visible	—	In vivo pharynx tissue	—	—	—	—	[1085]
Pharynx tissues (normal and tumorous)	Oral cancer	—	—	410	Visible	—	In vivo pharynx tissue	—	—	—	—	[1085]
Pharynx tissues (normal and tumorous)	Oropharyngeal cancer	—	—	330	380	—	In vivo pharynx tissue	—	—	—	Differentiation of normal and tumorous tissues was performed based upon a decrease in the 330 nm excitation band in tumor tissue.	[1090]
Pharynx tissues (normal and tumorous)	Oropharyngeal cancer	—	—	340	390	—	In vivo pharynx tissue	—	—	—	Diagnosis of the stage of dysplasia was determined based upon changes in the autofluorescence emission intensity at 390 nm.	[1090]

Compound	Cancer		Excitation	Emission					Tissue	Notes	Ref.
Photogen	Various cancers	—	510	Red/NIR	—	—	—	—	In vivo tissues (lungs, larynx, skin, gastric esophageal, and gynecological)	Drug accumulation studies were performed and it was found that the accumulation depended dramatically upon the tissue type and stage of cancer.	[1092]
Protoporphyrin (5-aminolevulinic acid hexylester hydrochloride-induced)	Various cancers	—	Visible	Red	—	15.9 ± 1.2	—	—	In vivo bladder tissue	PpIX was found to exhibit a mono-exponential decay of 15.9 ns in bladder tissue.	[1045]
Protoporphyrin (ALA-induced)	Various cancers	—	405	550–750	—	—	—	—	In vivo skin tissue	Diagnoses of malignant melanomas based upon multiple ratiometric measurements were performed.	[1093]
Protoporphyrin (ALA-induced)	Various cancers	—	435	550–750	—	—	—	—	In vivo skin tissue	Diagnoses of malignant melanomas based upon multiple ratiometric measurements were performed.	[1093]
Protoporphyrin (ALA-induced)	Various cancers	—	375–440	Green/red	—	—	—	—	In vivo oral tissue	Ratiometric images of both the red (PpIX) fluorescence and green (background) fluorescence were used to differentiate normal and tumorous tissues.	[1094]
Protoporphyrin (ALA-induced)	Various cancers	—	Visible	Red	—	0.230 (17)	17.1 (83)	—	In vivo tumor flank before irradiation	—	[1095]
Protoporphyrin (ALA-induced)	Various cancers	—	Visible	Red	—	0.270 (5.5)	5 (45)	—	In vivo tumor flank after irradiation	—	[1095]
Protoporphyrin (ALA-induced)	Various cancers	—	375–440	red	—	—	—	—	In vivo laryngeal tissue	PpIX fluorescence allowed for the differentiation of normal tissue from malignant neoplasms.	[1077]

(continued)

TABLE 20.10 (continued)   Fluorescence Spectroscopy

Species	Target Organ or Disease	Absorbance Maximum (nm)	Molar Extinction Coefficient (M-1)	Excitation λ_{ex} (nm)	Emission λ_{em} (nm)	Quantum Yield	Matrix	Lifetime t_1 (rel % t_1)	Lifetime t_2 (rel % t_2)	Lifetime t_3 (rel % t_3)	Comments	Reference
Protoporphyrin (ALA-induced)	Various cancers	—	—	375–440	Red	—	In vivo oral tissue	—	—	—	PpIX fluorescence allowed for the differentiation of normal tissue from malignant neoplasms and was found to have a 10:1 contrast between malignant and normal tissues.	[1096]
Protoporphyrin (IX)	Brain cancer	—	—	Blue	Red	—	In vivo *brain tissue*	—	—	—	Fluorescence intensity increased exponentially with the grades of astrocytoma.	[1040]
Reduced pyridine nucleotides in skin tissue	Skin cancer	—	—	730 (two-photon absorption)	Visible	—	In vivo skin tissue	—	—	—	Confocal microscopy was used to determine the spatial location of various autofluorescent species within skin cells.	[1070]
Skin tissue	Melanoma, evaluating skin lesions	—	—	785	NIR	—	In vivo *skin tissue of patients with nonmelanoma skin cancer*	—	—	—	NIR fluorescence is greater within the skin lesion than the surrounding normal skin.	[913]
Skin tissue	Skin aging	—	—	330, 370	375, 455	—	In vivo *skin tissue of patients with nonmelanoma skin cancer*	—	—	—	375 nm skin autofluorescence may be used as a biological marker of skin aging in vivo.	[888]
Skin tissue	Skin cancer	—	—	960 (three-photon absorption)	425	—	In vivo skin tissue	—	—	—	Confocal microscopy was used to determine the spatial location of various autofluorescent species within skin cells.	[1070]
Skin tissue	Skin cancer	—	—	325	Visible	—	In vivo skin tissue	—	—	—	Autofluorescence emission was found to correlate to photoaging of the skin.	[1097]

Sample	Application		Excitation	Emission		Description					Comments	Ref.
Skin tissue	Skin cancer	—	365	440	—	In vivo skin tissue	—	—	—		Results demonstrated that there was no difference between normal and nonmelanoma tumors.	[1098]
Skin tissue	Skin cancer	—	375	400–700 (peak at 436)	—	In vivo skin tissue	—	—	—		Results demonstrated that there was no difference between normal and nonmelanoma tumors.	[1098]
Skin tissue	Skin cancer	—	380	470	—	In vivo skin tissue	—	—	—		Autofluorescence emission resulting from a skin sample was found to strongly depend upon the absorption and scattering properties of the tissue.	[1099]
Skin tissue	Skin cancer	—	635–640	670–690	—	In vivo skin tissue of patients with nonmelanoma skin cancer	—	—	—		Carcinoma cells are characterized by increased autofluorescence in the red region of the spectrum.	[1100]
Skin tissue	Skin disorders	—	785	NIR	—	In vivo skin tissue of patients with nonmelanoma skin cancer	—	—	—		Cutaneous melanin in pigmented skin disorders emits higher NIR autofluorescence.	[914]
Skin tissue (melanocytic lesions)	Melanoma	—	—	470, 550	—	In vivo skin tissue of patients with nonmelanoma skin cancer	—	—	—		Noninvasive early detection of melanoma.	[935]
Skin tissue (normal and tumorous)	Nonmelanoma skin cancer	—	UV	—	—	In vivo skin tissue of patients with nonmelanoma skin cancer	—	—	—		Autofluorescence was more intense in tumor than in normal tissue.	[1101]
Skin tissue (palms, arms, legs, and cheeks)	Glycation	—	375	442, 460, 478, and 496	—	In vivo skin tissue of patients with nonmelanoma skin cancer	—	0.5 (78)	2.6 (18)	9.2 (3)	—	[946]

(continued)

TABLE 20.10 (continued) Fluorescence Spectroscopy

Species	Target Organ or Disease	Absorbance Maximum (nm)	Molar Extinction Coefficient (M-1)	Excitation λ_{ex} (nm)	Emission λ_{em} (nm)	Quantum Yield	Matrix	Lifetime t_1 (rel % t_1)	Lifetime t_2 (rel % t_2)	Lifetime t_3 (rel % t_3)	Comments	Reference
Skin tissue (precancerous actinic keratosis, malignant squamous cell carcinoma, and basal cell carcinoma)	Nonmelanoma skin cancer	—	—	337	350–700	—	In vivo skin tissue of patients with nonmelanoma skin cancer	—	—	—	Fluorophore contributions of normal, benign, and malignant nonmelanoma cancers are significantly different from each other.	[1102]
Skin tissue injected with ICG	Burn depth	—	—	795	835	—	In vivo skin tissue of patients with nonmelanoma skin cancer	—	—	—	NIR fluorescence to categorize the severity of burns.	[1022]
Tissues from the head and neck (normal and tumorous)	Head and neck cancers	—	—	442	480–520, >630	—	In vivo head and neck tissues	—	—	—	Diagnosis was performed by looking for brown or brownish-red areas and dark areas.	[1010]
Tongue tissue (normal and moderately differentiated squamous cell carcinoma)	Oral cancer	—	—	350	390–625	—	In vivo tongue tissue	—	—	—	Abnormal tissue displayed a reduced fluorescence emission between 400 and 450 nm relative to normal tissue.	[1103]
Tongue tissue (normal and moderately differentiated squamous cell carcinoma)	Oral cancer	—	—	410	460–675	—	In vivo tongue tissue	—	—	—	Abnormal tissue displayed a dramatic decrease in fluorescence intensity over the entire spectrum, as well as a distinct peak at approximately 630 nm.	[1103]
Tongue tissue (normal and moderately differentiated squamous cell carcinoma)	Oral cancer	—	—	460	490–720	—	In vivo tongue tissue	—	—	—	Abnormal tissue displayed a dramatic decrease in fluorescence intensity over the entire spectrum.	[1103]
Vasculature of the heart	Heart	—	—	806	—	—	In vivo human heart	—	—	—	In 5% of coronary artery bypass graft patients, NIR fluorescence analysis suggested graft failure.	[1022]

TABLE 20.11   Elemental Analyses

Species	Target Organ or Disease	Technique	Matrix	Comments	References
Biochemical studies					
K	Rat myocardium and skeletal muscle	AAS	Deionized water		[1104]
Mg	Rat myocardium and skeletal muscle	AAS	Deionized water		[1104]
Cellular studies					
B		AES	Cell suspension	The procedure is applicable to the analysis of boron in the ppm range with a high degree of precision and accuracy.	[1105]
Ca	Rat hepatocytes	AAS		Effects of hypothermia on cytosolic free calcium concentration ($[Ca2+](I)$) and total cellular calcium content.	[1106]
Zn		AAS			[1107]
Tissue and biofluid studies (in vitro)					
Ab		AAS	Blood, liver tissue		[1108]
Ag		AAS	Tissue		[1109]
Ag	Liver, kidney cortex, five brain regions: gray matter of cerebrum, white matter of cerebrum, nucleus lentiformis, cerebellum, brain stem	AAS	Tissue		[1110]
Al	Brain	XRF	Brain		[1111]
Al	Human organs/ tissue (lung, kidney, liver, hair, blood)	ICP-AES, AAS	Human organs/tissue (lung, kidney, liver, hair, blood)	Reference values human tissues/organs.	[1112]
Al		AAS	Fetal serum, amniotic fluid, and organs	Transplacental passage.	[1113]
Al		ICP-AES	Blood		[1114]
Al		ICP-AES	Human organs	Reference values study.	[1115]
Al	Alzheimer's disease	AAS	Tissue		[1116]
Al	Liver	AAS	Tissue		[1117]

(continued)

TABLE 20.11 (continued) Elemental Analyses

Species	Target Organ or Disease	Technique	Matrix	Comments	References
l	Liver	AAS	Tissue	To investigate the possible absorption and deposition of bismuth or aluminum from agents used in the treatment of peptic ulcers.	[1118]
Al	Spinal cord, brain stem, cerebellum, forebrain	AAS		The tissue distribution of Al did not follow that of essential cations as examined in this study.	[1119]
Al	Alzheimer's disease	AAS	Bone		[1120]
Al		LIPS, AAS	Teeth, bone		[1121]
Al	Brain	XRF			[1111]
Al		LIBS	Model biological tissues	Trace Al in model biological tissues.	[1122]
Al	Liver, kidney, pancreas		Animal tissues	fs-LIBS.	[1123]
Al		AES	Human serum		[1124]
As	Urine	AAS	Urine		[1125]
As	Rat kidney	AES	Rat kidney	Results indicate that arsenic accumulates in the kidney cortex synchronously over time. Arsenic also accumulated in the liver and red blood cells.	[1126]
As		AAS	Tissue		[1109]
As		ICP-AES, AAS		Various biomedical applications.	[1127]
As	Urine	AAS			[1125]
As	Breast milk	AAS		Environmental toxic exposure of very young children to A by determining As in breast milk. As found not to be excreted into breast milk at significant extent.	[1128]
B	Human organs/tissue (lung, kidney, liver, hair, blood)	ICP-AES, AAS	Human organs/tissue (lung, kidney, liver, hair, blood)	Reference values human tissues/organs.	[1112]
	Blood, urine, tissues	AE, ICP-AES		Pharmacokinetics of a compound used for BNCT was studied.	[1129]

TABLE 20.11 (continued)   Elemental Analyses

Species	Target Organ or Disease	Technique	Matrix	Comments	References
B	B	AES	Tumor, tissue, liver, skin	The procedure is applicable to the analysis of boron in the ppm range with a high degree of precision and accuracy.	[1105]
B		ICP-AES	Blood	BNCT.	[1130]
B		AES	Blood	BNCT.	[1131]
B	Melanoma	ICP-AES			[1132]
Ba	Human organs/ tissue (lung, kidney, liver, hair, blood)	ICP-AES, AAS	Human organs/tissue (lung, kidney, liver, hair, blood)	Reference values human tissues/organs.	[1112]
Ba		ICP-AES	Human organs	Reference values study.	[1115]
Ba		ICP-AES, AAS		Various biomedical applications.	[1127]
Ba	Liver	ICP-AES	Tissue		[1133]
Be	Human organs/ tissue (lung, kidney, liver, hair, blood)	ICP-AES, AAS	Human organs/tissue (lung, kidney, liver, hair, blood)	Reference values human tissues/organs.	[1112]
Be	Liver, kidney	AAS	Liver, kidney	Effects of two chelating agents on the toxicity and distribution of Be were investigated.	[1134]
Bi	Human organs/ tissue (lung, kidney, liver, hair, blood)	ICP-AES, AAS	Human organs/tissue (lung, kidney, liver, hair, blood)	Reference values human tissues/organs.	[1112]
Bi		AAS	Tissue		[1109]
Bi	Liver	AAS	Tissue	To investigate the possible absorption and deposition of bismuth or aluminum from agents used in the treatment of peptic ulcers.	[1118]
Ca	Atherosclerotic and normal tissue	LIPS		Laser angioplasty tissue characterization.	[1135]
Ca	Rat eye tissue	XRF	Rat eye tissue	Hereditary retinal degeneration.	[1136]
Ca	Heart disease	AAS	Aortic valve tissue	Surface structure of decalcified aortic valve tissue.	[1137]
Ca	Heart disease	AAS	Aortic valve tissue		[1138]
Ca		AAS	Stomach, kidneys, bone, liver		[1139]

(continued)

TABLE 20.11 (continued)   Elemental Analyses

Species	Target Organ or Disease	Technique	Matrix	Comments	References
Ca		ICP-AES	Hair		[1140]
Ca	Osteoporosis	AAS	Bone	Steroid osteoporosis treatment.	[1141]
Ca	Wilson's disease	ICP-AES	Brain tissue		[1142]
Ca	Liver	ICP-AES	Tissue		[1133]
Ca	Spinal cord, brain stem, cerebellum, forebrain	ICP-AES			[1119]
Ca	Meniscal degeneration	ICP-AES	Mensisci	Relationship between meniscal degeneration and element contents.	[1143]
Ca	Heart disease	AAS	Arterial wall		[1144]
Ca		LIPS	Cornea	Plasma emission spectra exhibited significant dependence on sample hydration. This dependence can be used for estimation of water content of irradiated model material and real cornea.	[1145]
Ca		AES	Cerebrospinal fluid		[1146]
Ca	Osteoporosis	LIPS	Hair		[1147]
Ca	Dental	LIBS	Teeth	Surface hardness studies of calcified tissues including human teeth.	[1148]
Ca	Breast cancer	XRF	Breast tissue	2D XRF mapping of invasive papillary carcinoma.	[1149]
Ca	Dental	XRF	Teeth		[1150]
Ca	Liver, kidney, pancreas		Animal tissues	fs-LIBS.	[1123]
Cd	Human organs/ tissue (lung, kidney, liver, hair, blood)	ICP-AES, AAS	Human organs/tissue (lung, kidney, liver, hair, blood)	Reference values human tissues/organs.	[1112]
Cd	Aorta	AAS	Aorta	Cadmium accumulation in aortas of smokers.	[1151]
Cd	Prostatic cancer	AAS	Tissue		[1152]
Cd		MP-AES	Blood		[1153]
Cd			Urine, tissue		[1154]
Cd		ICP-AES	Human organs	Reference values study.	[1115]
Cd		AAS	Tissue		[1109]
Cd		ICP-AES, AAS		Various biomedical applications.	[1127]
Cd	Lung	AAS	Tissue		[1155]
Cd	Liver	ICP-AES	Tissue		[1133]
Cd		AES, AAS	Blood, urine, tissue	Review.	[1156]

TABLE 20.11 (continued) Elemental Analyses

Species	Target Organ or Disease	Technique	Matrix	Comments	References
Co	Human organs/tissue (lung, kidney, liver, hair, blood)	ICP-AES, AAS	Human organs/tissue (lung, kidney, liver, hair, blood)	Reference values human tissues/organs.	[1112]
Co		ICP-AES, AAS		Various biomedical applications.	[1127]
Co	Lung	AAS	Tissue		[1155]
Co		AAS	Serum	The purpose of this study was to measure the serum cobalt levels and their correlation with clinical and radiological findings in patients with metal on metal hip articulating surfaces.	[1157]
Cr	Human organs/tissue (lung, kidney, liver, hair, blood)	ICP-AES, AAS	Human organs/tissue (lung, kidney, liver, hair, blood)	Reference values human tissues/organs.	[1112]
Cr	Rat liver, rat kidney	AAS	Tissue		[1158]
Cr	Rat tissue, blood	AAS	Rat tissue, blood		[1159]
Cr		ICP-AES	Human organs	Reference values study.	[1115]
Cr	Lung	AAS	Tissue		[1155]
Cr	Liver	ICP-AES	Tissue		[1133]
Cr		AAS	Tissue	The objective of this study was to determine whether low plasma chromium concentrations ($\leq$ to 3 nmol/L) are associated with altered glucose, insulin or lipid concentrations during pregnancy.	[1160]
Cu	Human organs/tissue (lung, kidney, liver, hair, blood)	ICP-AES, AAS	Human organs/tissue (lung, kidney, liver, hair, blood)	Reference values human tissues/organs.	[1112]
Cu	Rat liver	AAS	Rat liver	Effect of chronic exposure to excess dietary metal supplementation on liver specimens from rats.	[1161]
Cu	Rat kidney	AES	Rat kidney	Results indicate that arsenic and copper accumulate in the kidney cortex synchronously over time.	[1126]

(continued)

TABLE 20.11 (continued) Elemental Analyses

Species	Target Organ or Disease	Technique	Matrix	Comments	References
Cu	Colitis	AAS	Serum	The serum concentrations of copper remained unaltered during colitis.	[1162]
Cu	Colitis/rat colon	AAS	Tissue		[1162]
Cu	Chronic alcohol abuse	AAS	Embryo, liver	Maternal hepatic, endometrial, and embryonic levels following alcohol consumption during pregnancy in QS mice.	[1163]
Cu	Rat tissue, blood	AAS	Rat tissue, blood		[1159]
Cu	Liver disease	AAS	Tissue, plasma		[1164]
Cu	Blood plasma	ICP-AES	Plasma		[1165]
Cu	Extrahepatic biliary atresia	AAS	Tissue		[1166]
Cu		ICP-AES	Blood	Reference values for some bulk and trace elements in blood plasma and whole blood (Pb) of mothers and their newborn infants.	[1167]
Cu		ICP-AES	Liver		[1168]
Cu		ICP-AES	Blood		[1114]
Cu		ICP-AES	Human organs	Reference values study.	[1115]
Cu	Wilson's disease	AAS	Kidney, liver		[1169]
Cu		AAS	Stomach, kidneys, bone, liver		[1139]
Cu		ICP-AES	Hair		[1140]
Cu		AES	Human breast milk		[1170]
Cu	Rat small intestine	AAS	Tissue		[1171]
Cu	Wilson's disease	ICP-AES	Brain tissue		[1142]
Cu	Wilson's disease	ICP-AES	Brain tissue		[1172]
Cu	Lung	AAS	Tissue		[1155]
Cu	Liver	ICP-AES	Tissue		[1133]
Cu	Cataracts	AAS	Tissue	These data support the hypothesis that transition metal-mediated HO production may play a role in the etiology of age-related nuclear cataract.	[1173]
Cu	Liver disease	AAS	Liver, serum		[1174]
Cu	Human kidney tumor	XRF, ICP-AES			[1175]

TABLE 20.11 (continued)   Elemental Analyses

Species	Target Organ or Disease	Technique	Matrix	Comments	References
Cu	Breast cancer, stomach cancer, colon cancer	AAS	Serum, tissue	The results suggest that movements of copper out from, and of zinc into, tissues occur, and it is proposed that these changes could be a response to enhanced cytokine production rather than being a feature of any mechanism(s) for the onset of cancer.	[1176]
Cu	Heart disease	AAS	Arterial wall		[1144]
Cu	Aortoiliac occlusive disease	AAS	Tissue		[1177]
Cu		AES	Cerebrospinal fluid		[1146]
Cu	Hypoglycemia	AAS	Cerebrospinal fluid	Newborns.	[1178]
Cu	Liver and kidney	XRF	Liver and kidney tissue	Evaluation of biologically important trace metals in liver, kidney, and breast tissue.	[1179]
Cu	Liver, kidney, pancreas		Animal tissues	fs-LIBS.	[1123]
Fe	Breast milk	AAS	Breast milk		[1180]
Fe	Rat eye tissue	XRF	Rat eye tissue	Hereditary retinal degeneration.	[1136]
Fe	Beta-thalassemia				[1181]
Fe	Chronic alcohol abuse	AAS	Embryo, liver	Maternal hepatic, endometrial, and embryonic levels following alcohol consumption during pregnancy in QS mice.	[1163]
Fe	Rat tissue, blood	AAS	Rat tissue, blood		[1159]
Fe		ICP-AES	Blood	Reference values for some bulk and trace elements in blood plasma and whole blood (Pb) of mothers and their newborn infants.	[1167]
Fe		AAS	Adipose tissue		[1182]
Fe		ICP-AES	Hair		[1140]
Fe		AES	Human breast milk		[1170]
Fe	Rat small intestine	AAS	Tissue		[1171]
Fe	Wilson's disease	ICP-AES	Brain tissue		[1142]
Fe	Liver	ICP-AES	Tissue		[1133]

(continued)

TABLE 20.11 (continued) Elemental Analyses

Species	Target Organ or Disease	Technique	Matrix	Comments	References
Fe	Cataracts	AAS	Tissue	These data support the hypothesis that transition metal-mediated HO production may play a role in the etiology of age-related nuclear cataract.	[1173]
Fe	Spinal cord, brain stem, cerebellum, forebrain	ICP-AES			[1119]
Fe	Liver disease	AAS	Liver, serum		[1174]
Fe		AES	Cerebrospinal fluid		[1146]
Fe	Liver and kidney	XRF	Liver and kidney tissue	Evaluation of biologically important trace metals in liver, kidney, and breast tissue.	[1179]
Fe	Breast cancer	XRF	Breast tissue	2D XRF mapping of invasive papillary carcinoma.	[1149]
Fe	Liver, kidney, pancreas		Animal tissues	fs-LIBS.	[1123]
Fe	AES		Human serum		[1124]
Ga	Lungs, brain, testes, and ovaries	ICP-AES	Lungs, brain, testes, and ovaries	Anticancer therapy drug study, organometallic gallium (Ga) complex.	[1183]
Hg	Tissues	AAS	Tissues	Improved preparation of small biological samples for mercury analysis.	[1184]
Hg	Methylmercury exposure	AAS	Toenail, hair, blood	Toenails, an easily accessible tissue for the estimation of methylmercury exposure, have been shown to be closely correlated with the well-established samples for biomarkers, viz., blood and hair mercury.	[1185]
Hg	Connective tissue diseases		Urine		[1186]
Hg		ICP-AES, AAS	Hair		[1187]
Hg		AAS	Tissue, blood		[1188]
Hg		ICP-AES, AAS		Various biomedical applications.	[1127]
I		ICP-AES	Urine, plasma		[1189]
K	Rat eye tissue	XRF	Rat eye tissue	Hereditary retinal degeneration.	[1136]
K		AES	Blood		[1190]

TABLE 20.11 (continued)   Elemental Analyses

Species	Target Organ or Disease	Technique	Matrix	Comments	References
K	Liver	ICP-AES	Tissue		[1133]
K		AES	Cerebrospinal fluid		[1146]
K	Liver and kidney	XRF	Liver and kidney tissue	Evaluation of biologically important trace metals in liver, kidney, and breast tissue.	[1179]
K	Breast cancer	XRF	Breast tissue	2D XRF mapping of invasive papillary carcinoma.	[1149]
K	Liver, kidney, pancreas		Animal tissues	fs-LIBS.	[1123]
K		AES	Human serum		[1124]
Li	Human organs/tissue (lung, kidney, liver, hair, blood)	ICP-AES, AAS	Human organs/tissue (lung, kidney, liver, hair, blood)	Reference values human tissues/organs.	[1112]
Li		AES	Blood		[1190]
Li		ICP-AES	Human organs	Reference values study.	[1115]
Li		AES	Serum		[1191]
Li		AAS	Blood		[1192]
Li		AES	Human serum		[1124]
Mg	Human organs/tissue (lung, kidney, liver, hair, blood)	ICP-AES, AAS	Human organs/tissue (lung, kidney, liver, hair, blood)	Reference values human tissues/organs.	[1112]
Mg	Chronic alcohol abuse	AAS	Embryo, liver	Maternal hepatic, endometrial, and embryonic levels following alcohol consumption during pregnancy in QS mice.	[1163]
Mg		AES	Blood		[1190]
Mg		ICP-AES	Human organs	Reference values study.	[1115]
Mg		ICP-AES, AAS		Various biomedical applications.	[1127]
Mg		AAS	Stomach, kidneys, bone, liver		[1139]
Mg		ICP-AES	Hair		[1140]
Mg	Wilson's disease	ICP-AES	Brain tissue		[1142]
Mg	Liver	ICP-AES	Tissue		[1133]
Mg	Spinal cord, brain stem, cerebellum, forebrain	ICP-AES			[1119]
Mg	Meniscal degeneration	ICP-AES	Mensisci	Relationship between meniscal degeneration and element contents.	[1143]

(continued)

TABLE 20.11 (continued)   Elemental Analyses

Species	Target Organ or Disease	Technique	Matrix	Comments	References
Mg	Breast cancer, stomach cancer, colon cancer	AAS	Serum, tissue	The results suggest that movements of copper out from, and of zinc into, tissues occur, and it is proposed that these changes could be a response to enhanced cytokine production rather than being a feature of any mechanism(s) for the onset of cancer.	[1176]
Mg	Heart disease	AAS	Arterial wall		[1144]
Mg		AES	Cerebrospinal fluid		[1146]
Mg	Hypoglycemia	AAS	Cerebrospinal fluid	Newborns.	[1178]
Mg	Dental	LIBS	Teeth	Surface hardness studies of calcified tissues including human teeth.	[1148]
Mg	Liver, kidney, pancreas		Animal tissues	fs-LIBS.	[1123]
Mn	Human organs/ tissue (lung, kidney, liver, hair, blood)	ICP-AES, AAS	Human organs/tissue (lung, kidney, liver, hair, blood)	Reference values human tissues/organs.	[1112]
Mn	Extrahepatic biliary atresia	AAS	Tissue		[1166]
Mn		AES	Blood		[1190]
Mn		ICP-AES	Human organs	Reference values study.	[1115]
Mn		ICP-AES	Hair		[1140]
Mn	Liver	ICP-AES	Tissue		[1133]
Mn	Spinal cord, brain stem, cerebellum, forebrain	AAS			[1119]
Mn		AES	Human serum		[1124]
Mo		AAS	Tissue		[1193]
Multielement	Beta-thalassemia major	PIXE			[1194]
Na		AES	Blood		[1190]
Na	Liver	ICP-AES	Tissue		[1133]
Na		LIPS	Skin	Na line = 589 nm; ablation wavelengths = 1064, 532, 266, 213 nm.	[1195]
Na		AES	Cerebrospinal fluid		[1146]
Na	Liver, kidney, pancreas		Animal tissues	fs-LIBS.	[1123]
Na		AES	Human serum		[1124]
Ni		AAS	Blood		[1196]
Ni		ICP-AES	Human organs	Reference values study.	[1115]
Ni		ICP-AES, AAS		Various biomedical applications.	[1127]

TABLE 20.11 (continued)   Elemental Analyses

Species	Target Organ or Disease	Technique	Matrix	Comments	References
Ni		AAS	Blood, liver tissue		[1108]
Ni	Lung	AAS	Tissue		[1155]
Ni	Liver	ICP-AES	Tissue		[1133]
P	Wilson's disease	ICP-AES	Brain tissue		[1142]
P	Liver	ICP-AES	Tissue		[1133]
P	Meniscal degeneration	ICP-AES	Mensisci	Relationship between meniscal degeneration and element contents.	[1143]
P	Dental	XRF	Teeth		[1150]
P	Liver, kidney, pancreas		Animal tissues	fs-LIBS.	[1123]
Pb	Human organs/ tissue (lung, kidney, liver, hair, blood)	ICP-AES, AAS	Human organs/tissue (lung, kidney, liver, hair, blood)	Reference values human tissues/organs.	[1112]
Pb		ICP-AES	Blood		[1197]
Pb	Rat tissue, blood	AAS	Rat tissue, blood	These results show that at 5 ppm Pb exposure, brain and kidney accumulate Pb significantly, confirming that at these exposure levels, blood Pb is not a good index of tissue burden.	[1198]
Pb		MP-AES	Blood		[1199]
Pb	Osteoporosis	AAS	Bone		[1200]
Pb	Liver disease		Blood, tissue	Increased levels of lead were found in the blood of patients who consumed alcohol and those with alcoholic liver disease.	[1201]
Pb		ICP-AES	Human organs	Reference values study.	[1115]
Pb		ICP-AES, AAS		Various biomedical applications.	[1127]
Pb		PIXE	Bone		[1202]
Pb		ICP-AES	Hair		[1140]
Pb	Liver, lung, kidney, brain	AAS	Tissue	The concentrations of lead in liver, lung, kidney, brain, hair, and nails were determined in 32 deceased, long-term exposed male lead smelter workers and compared with those of 10 male controls.	[1203]
Pb		XRF	Hair, nails		[1203]
Pb	Presenile dementia	AAS	Brain tissue		[1204]

(continued)

TABLE 20.11 (continued)   Elemental Analyses

Species	Target Organ or Disease	Technique	Matrix	Comments	References
Pb	Teeth	AAS	Hard tissue		[1205]
Pb		LIPS, AAS	Teeth, bone		[1121]
Pt		ICP-AES	Blood		[1206]
		AAS	Urine		[1207]
S	Wilson's disease	ICP-AES	Brain tissue		[1142]
S	Liver	ICP-AES	Tissue		[1133]
S	Meniscal degeneration	ICP-AES	Mensisci	Relationship between meniscal degeneration and element contents.	[1143]
Sb		AAS	Tissue		[1109]
e	Rat lymphoid tissue	AAS	Rat lymphoid tissue	Effect of aging on levels of selenium in the lymphoid tissues of rats.	[1208]
Se	Rat liver	AAS	Rat liver	Effect of chronic exposure to excess dietary metal supplementation on liver specimens from rats.	[1161]
Se	Colitis	AAS	Serum	The serum concentrations of selenium also remained unaltered during colitis.	[1162]
Se	Colitis/rat colon	AAS	Tissue		[1162]
Se	Stomach cancer	AAS	Gastric tissue	The Se concentration in the biopsies of patients with gastric ulceration and cancer was significantly lower than that in patients with gastritis ($p < 0.05$) and the other conditions ($p < 0.0001$).	[1209]
Se	Liver	ICP-AES	Tissue		[1133]
Se	Human kidney tumor	XRF, ICP-AES			[1175]
Se	Cataracts	AAS	Serum, lens, and aqueous humor	Decreased Se in aqueous humor and sera of patients with senile cataract may reflect defective antioxidative defense systems that may lead to the formation of cataract.	[1210]
Se		AES	Cerebrospinal fluid		[1146]
	Liver	XRF	Tissue		[1211]
Si		AAS	Body fluids, tissue		[1212]
Si	Breast	AAS	Tissue, serum		[1213]
Sr	Human organs/ tissue (lung, kidney, liver, hair, blood)	ICP-AES, AAS	Human organs/tissue (lung, kidney, liver, hair, blood)	Reference values human tissues/organs.	[1112]
Sr		ICP-AES	Human organs	Reference values study.	[1115]
		AAS	Serum, urine, bone		[1214]

TABLE 20.11 (continued) Elemental Analyses

Species	Target Organ or Disease	Technique	Matrix	Comments	References
Sr		AAS	Blood		[1192]
Sr		LIPS, AAS	Teeth, bone		[1121]
Sr	Liver, kidney, pancreas	Animal tissues		fs-LIBS.	[1123]
Ti		AAS	Tissue		[1215]
	Brain	AAS	Tissue	Effects on distribution and lipid peroxidation in brain regions.	[1216]
V	Human organs/ tissue (lung, kidney, liver, hair, blood)	ICP-AES, AAS	Human organs/tissue (lung, kidney, liver, hair, blood)	Reference values human tissues/organs.	[1112]
V		AAS	Vaginal tissue		[1217]
V		AAS	Vaginal tissue		[1218]
V		ICP-AES, AAS		Various biomedical applications.	[1127]
V	Liver	ICP-AES	Tissue		[1133]
V		AAS	Serum		[1219]
W		ICP-AES, AAS		Various biomedical applications.	[1127]
	Human organs/ tissue (lung, kidney, liver, hair, blood)	ICP-AES, AAS	Human organs/tissue (lung, kidney, liver, hair, blood)	Reference values human tissues/organs.	[1162]
Zn	Colitis	AAS	Serum	The serum concentrations of zinc also remained unaltered during colitis.	[1162]
Zn	Colitis/rat colon	AAS	Tissue		[1162]
Zn	Chronic alcohol abuse	AAS	Embryo, liver	Maternal hepatic, endometrial, and embryonic levels following alcohol consumption during pregnancy in QS mice.	[1163]
Zn	Rat tissue, blood Beta-thalassemia major	AAS	Rat tissue, blood Urine		[1159] [1220]
Zn	Thoracic empyema	AAS	Serum		[1221]
Zn	Extrahepatic biliary atresia	AAS	Tissue		[1166]
Zn		ICP-AES	Blood	Reference values for some bulk and trace elements in blood plasma and whole blood (Pb) of mothers and their newborn infants.	[1167]
Zn		AES	Blood		[1190]
	Prostatic cancer	AAS	Tissue		[1152]
Zn		AAS	Adipose tissue		[1182]

(continued)

TABLE 20.11 (continued)   Elemental Analyses

Species	Target Organ or Disease	Technique	Matrix	Comments	References
Zn		ICP-AES	Blood		[1114]
Zn		ICP-AES	Human organs	Reference values study.	[1115]
Zn		AAS	Newborn liver		[1222]
		AAS	Stomach, kidneys, bone, liver		[1139]
Zn		ICP-AES	Hair		[1140]
Zn	Wilson's disease	ICP-AES	Brain tissue		[1142]
Zn	Liver	AAS	Tissue		[1223]
Zn	Liver	ICP-AES	Tissue		[1133]
Zn	Spinal cord, brain stem, cerebellum, forebrain	ICP-AES			[1119]
Zn	Liver disease	AAS	Liver, serum		[1174]
Zn	Human kidney tumor	XRF, ICP-AES			[1175]
Zn	Breast cancer, stomach cancer, colon cancer	AAS	Serum, tissue	The results suggest that movements of copper out from, and of zinc into, tissues occur, and it is proposed that these changes could be a response to enhanced cytokine production rather than being a feature of any mechanism(s) for the onset of cancer.	[1176]
Zn	Heart disease	AAS	Arterial wall		[1144]
Zn	Aortoiliac occlusive disease	AAS	Tissue		[1177]
Zn	Nasopharyngeal cancer	AAS	Tissue		[1224]
Zn	Breast cancer	AAS	Tissue		[1225]
Zn		AES	Cerebrospinal fluid		[1146]
Zn	Hypoglycemia	AAS	Cerebrospinal fluid	Newborns.	[1178]
Zn	Esophagus	XRF	Esophageal specimens	Zinc concentration and connection to esophageal cancer risk.	[1226]
Zn	Liver and kidney	XRF	Liver and kidney tissue	Evaluation of biologically important trace metals in liver, kidney, and breast tissue.	[1179]
Zn	Breast cancer	XRF	Breast tissue	2D XRF mapping of invasive papillary carcinoma.	[1149]
Zn	Liver, kidney, pancreas		Animal tissues	fs-LIBS.	[1123]
Zn		AES	Human serum		[1124]
Zr		ICP-AES, AAS		Various biomedical applications.	[1127]

TABLE 20.11 (continued) Elemental Analyses

Species	Target Organ or Disease	Technique	Matrix	Comments	References
Animal studies (in vivo)					
As	Rat kidney				[1227]
Cd	Tissues	XRF	Tissues		[1228]
Cu	Rat kidney				[1227]
Hg	Tissues	XRF	Tissues		[1228]
Pb	Tissues	XRF	Tissues		[1228]

References

1. Lehmann, R., Huber, M., Beck, A., Schindera, T., Rinkler, T., Houdali, B., Weigert, C., Haring, H., Voelter, W., and Schleicher, E. D., *Electrophoresis*, 21, 3010–3015, 2000.
2. Sanwald, P., Blankson, E. A., Dulery, B. D., Schoun, J., Huebert, N. D., and Dow, J., Isocratic high-performance liquid-chromatographic method for the separation of testosterone metabolites, *J. Chromatogr. B Biomed. Appl.*, 672, 207–215, 1995.
3. Zhou, N., Liang, Y. Z., and Wang, P., 18 Beta-glycyrrhetinic acid interaction with bovine serum albumin, *J. Photochem. Photobiol. A Chem.*, 185(2–3), 271–276, 2007.
4. Liebich, H. M., Lehmann, R., Xu, G., Wahl, H. G., and Haring, H., Application of capillary electrophoresis in clinical chemistry: The clinical value of urinary modified nucleosides, *J. Chromatogr. B*, 745, 189–196, 2000.
5. Bjφrnsdottir, I. and Hansen, S. H., Fast separation of 16 seizure drug substances using non-aqueous capillary electrophoresis, *J. Biochem. Biophys. Methods*, 38, 155–161, 1999.
6. Stoyanovsky, D. A., Goldman, R., Jonnalagadda, S. S., Day, B. W., Claycamp, H. G., and Kagan, V. E., Detection and characterization of the electron paramagnetic resonance-silent glutathionyl-5,5-dimethyl-1-pyrroline N-oxide adduct derived from redox cycling of phenoxyl radicals in model systems and HL-60 cells, *Arch. Biochem. Biophys.*, 330, 3–11, 1996.
7. Vaillancourt, F. H., Barbosa, C. J., Spiro, T. G., Bolin, J. T., Blades, M. W., Turner, R. F., and Eltis, L. D., Definitive evidence for monoanionic binding of 2,3-dihydroxybiphenyl to 2,3-dihydroxybiphenyl 1,2-dioxygenase from UV resonance Raman spectroscopy, UV/Vis absorption spectroscopy, and crystallography, *J. Am. Chem. Soc.*, 124, 2485–2496, 2002.
8. Gahunia, H. K., Lough, A., Vieth, R., and Pritzker, K., A cartilage derived novel compound DDP (2,6-dimethyldifuro-8-pyrone): Isolation, purification, and identification, *J. Rheumatol.*, 29, 147–153, 2002.
9. Cheng, T. B. and Reineccius, G. A., A study of factors influencing 2-methoxy-3-isopropylpyrazine production by pseudomonas-perolens using acid trap and UV spectroscopy, *Appl. Microbiol. Biotechnol.*, 36(3), 304–308, 1991.
10. Xu, X. and Stewart, J. T., *J. Liq. Chromatogr. Rel. Technol.*, 23, 1–13, 2000.
11. Kopriva, S., Fritzemeier, K., Wiedemann, G., and Reski, R., The putative moss 3′-phosphoadenosine-5′-phosphosulfate reductase is a novel form of adenosine-5′-phosphosulfate reductase without an iron-sulfur cluster, *J. Biol. Chem.*, 282(31), 22930–22938, 2007.
12. Hofstetter, H., Morpurgo, M., Hofstetter, O., Bayer, E. A., and Wilchek, M., A labeling, detection, and purification system based on 4-hydroxyazobenzene-2-carboxylic acid: An extension of avidin-biotin system, *Anal. Biochem.*, 284, 354–366, 2000.
13. Shin, H. C., Shimoda, M., and Kokue, E., Identification of 5,10-methylenetetrahydrofolate in rat Bile, *J. Chromatogr. B Biomed. Appl.*, 661, 237–244, 1994.
14. Visky, D., Kraszni, M., Hosztafi, S., and Noszal, B., *Chromatographia*, 51, 294–300, 2000.

15. Maynard, E. L., Tan, X., and Lindahl, P. A., Autocatalytic activation of acetyl-CoA synthase, *J. Biol. Inorg. Chem.*, 9(3), 316–322, 2004.

16. Woodson, J. D. and Escalante-Semerena, J. C., The cbiS gene of the archaeon *Methanopyrus kandleri* AV19 encodes a bifunctional enzyme with adenosylcobinamide amidohydrolase and alpha-ribazole-phosphate phosphatase activities, *J. Bacteriol.*, 188(12), 4227–4235, 2006.

17. Saccomani, A., Cecilia, G., Wajcman, H., and Righetti, P. G., Detection of neutral and charged mutations in α- and β-human globin chains by capillary zone electrophoresis in isoelectric, acidic buffers, *J. Chromatogr. A*, 832, 225–238, 1999.

18. Kanodia, S., Agarwal, S., Singh, P., and Bhatnagar, R., Biochemical characterization of alanine racemase—a spore protein produced by *Bacillus anthracis*, *BMB Rep.*, 42(1), 47–52, 2009.

19. Kanzaki, H., Imura, D., Nitoda, T., and Kawazu, K., Enzymatic dehydrogenation of cyclo(L-Phe-L-Leu) to a bioactive derivative, albonoursin, *J. Mol. Catal. B Enzym.*, 6, 265–270, 1999.

20. Gomez-Manzo, S., Contreras-Zentella, M., Gonzalez-Valdez, A., Sosa-Torres, M., Arreguin-Espinoza, R., and Escamilla-Marvan, E., The PQQ-alcohol dehydrogenase of *Gluconacetobacter diazotrophicus.*, *Int. J. Food Microbiol.*, 125(1), 71–78, 2008.

21. Pieszka, M., Gasior, R., and Barowicz, T., Evaluation of HPLC method for the rapid and simple determination of alpha-tocopherol acetate in feed premixes, *J. Anim. Feed Sci.*, 11(3), 527–536, 2002.

22. Meloni, G., Faller, P., and Vasak, M., Redox silencing of copper in metal-linked neurodegenerative disorders reaction of Zn(7)metallothionein-3 with Cu2+ ions, *J. Biol. Chem.*, 282(22), 16068–16078, 2007.

23. Riviere, C., Richard, T., Quentin, L., Krisa, S., Merillon, J. M., and Monti, J. P., Inhibitory activity of stilbenes on Alzheimer's beta-amyloid fibrils in vitro, *Bioorg. Med. Chem.*, 15(2), 1160–1167, 2007.

24. Damante, C. A., Osz, K., Nagy, Z., Papalardo, G., Grasso, G., Impellizzeri, G., Rizzarelli, E., and Sovago, I., The metal loading ability of beta-amyloid N-terminus: A combined potentiometric and spectroscopic study of copper(II) complexes with beta-amyloid(1–16), its short or mutated peptide fragments and its polyethylene glycol (PEG)-ylated analogue, *Inorg. Chem.*, 47(20), 9669–9683, 2008.

25. Lochmann, H., Bazzanella, A., Kropsch, S., and Bachmann, K., Determination of tobacco alkaloids in single plant cells by capillary electrophoresis, *J. Chromatogr. A*, 917, 311–317, 2001.

26. Boyd, J. M., Pierik, A. J., Netz, D. J. A., Lill, R., and Downs, D. M., Bacterial ApbC can bind and effectively transfer iron-sulfur clusters, *Biochemistry*, 47(31), 8195–8202, 2008.

27. Boyd, J. M., Drevland, R. M., Downs, D. M., and Graham, D. E., Archaeal ApbC/Nbp35 homologs function as iron-sulfur cluster carrier proteins, *J. Bacteriol.*, 191(5), 1490–1497, 2009.

28. Banerjee, S., Huber, T., and Sakmar, T. P., Rapid incorporation of functional rhodopsin into nanoscale apolipoprotein bound bilayer (NABB) particles, *J. Mol. Biol.*, 377(4), 1067–1081, 2008.

29. Bonomi, F., Iametti, S., Ferranti, P., Kurtz, D. M., Morleo, A., and Ragg, E. M., Iron priming, guides folding of denatured aporubredoxins, *J. Biol. Inorg. Chem.*, 13(6), 981–991, 2008.

30. Hancock, R. D., Chudek, J. A., Walker, P. G., Pont, S. D. A., and Viola, R., Ascorbic acid conjugates isolated from the phloem of Cucurbitaceae, *Phytochemistry*, 69(9), 1850–1858, 2008.

31. Crivellente, F., Fracasso, G., Valentini, R., Manetto, G., Riviera, A. P., and Tagliaro, F., Improved method for carbohydrate-deficient transferrin determination in human serum by capillary zone electrophoresis, *J. Chromatogr. B*, 739, 81–93, 2000.

32. Grove, T. L., Lee, K. H., St Clair, J., Krebs, C., and Booker, S. J., In vitro characterization of AtsB, a radical SAM formylglycine-generating enzyme that contains three [4Fe-4S] clusters, *Biochemistry*, 47(28), 7523–7538, 2008.

33. Lowery, T. J., Wilson, P. E., Zhang, B., Bunker, J., Harrison, R. G., Nyborg, A. C., Thiriot, D., and Watt, G. D., Flavodoxin hydroquinone reduces *Azotobacter vinelandii* Fe protein to the all-ferrous redox state with a S=0 spin state, *Proc. Natl. Acad. Sci USA*, 103(46), 17131–17136, 2006.

34. Bosca, F., Marin, L., and Miranda, M. A., Photoreactivity of the nonsteroidal anti-inflammatory 2-arylpropionic acids with photosensitizing side effects, *Photochem. Photobiol.*, 74, 637–655, 2001.

35. Lurie, I. S., Bethea, J. M., Mckibben, T. D., Hays, P. A., Pelligrini, P., Sahai, R., and Weinberger, R., Use of dynamically coated capillaries for the routine analysis of methamphetamine, amphetamine, MDA, MDMA, MDEA, and cocaine using capillary electrophoresis, *J. Forens. Sci.*, 46, 2001.

36. Chen, Y. Y., Wang, X. L., Sun, H., Xing, D. M., Hu, J., Wai, Z. H., and Du, L. J., Characterization of the transportation of berberine in *Coptidis rhizoma* extract through rat primary cultured cortical neurons, *Biomed. Chromatogr.*, 22(1), 28–33, 2008.

37. Reinicke, K. E., Bey, E. A., Bentle, M. S., Pink, J. J., Ingalls, S. T., Hoppel, C. L., Misico, R. I. et al., Development of beta-lapachone prodrugs for therapy against human cancer cells with elevated NAD(P)H: Quinone oxidoreductase 1 levels, *Clin. Cancer Res.*, 11(8), 3055–3064, 2005.

38. Criado, J. J., Manzano, J. L., and Rodriguez-Fernandez, E., New organotropic compounds—Synthesis, characterization and reactivity of Pt(II) and Au(III) complexes with bile acids: DNA interactions and 'in vitro' anticancer activity, *J. Inorg. Biochem.*, 96(2–3), 311–320, 2003.

39. Luquain, C., Laugier, C., Lagarde, M., and Pageaux, J. F., High-performance liquid chromatography determination of bis(monoacylglycerol) phosphate and other lysophospholipids, *Anal. Biochem.*, 296, 41–48, 2001.

40. Namiki, S., Kaneda, F., Ikegami, M., Arai, T., Fujimori, K., Asada, S., Hama, H., Kasuya, Y., and Goto, K., Bis-N-nitroso-caged nitric oxides: Photochemistry and biological performance test by rat aorta vasorelaxation, *Bioorg. Med. Chem.*, 7(8), 1695–1702, 1999.

41. Ma, L., Gong, X., and Yeung, E. S., Combinational screening of enzyme activity by using multiplexed capillary electrophoresis, *Anal. Chem.*, 72, 3383–3387, 2000.

42. Taheri-Kafrani, A., Bordbar, A. K., Mousavi, S. H. A., and Haertle, T., Beta-lactoglobulin structure and retinol binding changes in presence of anionic and neutral detergents, *J. Agric. Food Chem.*, 56(16), 7528–7534, 2008.

43. Maoka, T., Akimoto, N., Yim, M. J., Hosokawa, M., and Miyashita, K., New C-37 skeletal carotenoid from the clam, *Paphia amabillis*, *J. Agric. Food Chem.*, 56(24), 12069–12072, 2008.

44. Zsila, F., Bikadi, Z., and Simonyi, M., Further insight into the molecular basis of carotenoid-albumin interactions: Circular dichroism and electronic absorption study on different crocetin-albumin complexes, *Tetrahedron Asym.*, 13(3), 273–283, 2002.

45. Ritter, E., Stehfest, K., Berndt, A., Hegemann, P., and Bartl, F. J., Monitoring light-induced structural changes of channelrhodopsin-2 by UV-visible and Fourier transform infrared spectroscopy, *J. Biol. Chem.*, 283(50), 35033–35041, 2008.

46. Jain, A. A. and Rajeswari, M. R., Binding studies on peptide-oligonucleotide complex: Intercalation of tryptophan in GC-rich region of c-myc gene, *Biochim. Biophys. Acta*, 1622(2), 73–81, 2003.

47. Gray, M. J., Tavares, N. K., and Escalante-Semerena, J. C., The genome of *Rhodobacter sphaeroides* strain 2.4.1 encodes functional cobinamide salvaging systems of archaeal and bacterial origins, *Mol. Microbiol.*, 70(4), 824–836, 2008.

48. Boyd, J. A. and Ensign, S. A., Evidence for a metal-thiolate intermediate in alkyl group transfer from epoxypropane to coenzyme m and cooperative *metal-ion* binding in epoxyalkane: CoM transferase, *Biochemistry*, 44(39), 13151–13162, 2005.

49. Kong, L., Yu, Z. Y., Bao, Y. M., Su, X. Y., Zou, H. F., and Li, X., Screening and analysis of an antineoplastic compound in Rhizoma Chuanxiong by means of in vitro metabolism and HPLC-MS, *Anal. Bioanal. Chem.*, 386(2), 264–274, 2006.

50. Berndt, A., Kottke, T., Breitkreuz, H., Dvorsky, R., Hennig, S., Alexander, M., and Wolf, E., A novel photoreaction mechanism for the circadian blue light photoreceptor *Drosophila* cryptochrome, *J. Biol. Chem.*, 282(17), 13011–13021, 2007.

51. Mandeville, J. S., Froehlich, E., and Tajmir-Riahi, H. A., Study of curcumin and genistein interactions with human serum albumin, *J. Pharm. Biomed. Anal.*, 49(2), 468–474, 2009.

52. You, D. L., Wang, L. R., Yao, F., Zhou, X. F., and Deng, Z. X., A novel DNA modification by sulfur: DndA is a NifS-like cysteine desulfurase capable of assembling DndC as an iron-sulfur cluster protein in *Streptomyces lividans*, *Biochemistry*, 46(20), 6126–6133, 2007.

53. Fajardo-Cavazos, P., Rebeil, R., and Nicholson, W. L., Essential cysteine residues in *Bacillus subtilis* spore photoproduct lyase identified by alanine scanning mutagenesis, *Curr. Microbiol.*, 51(5), 331–335, 2005.

54. Thariat, J., Collin, F., Marchetti, C., Ahmed-Adrar, N. S., Vitrac, H., Jore, D., and Gardes-Albert, M., Marked difference in cytochrome c oxidation mediated by HO center dot and/or O-2(radical anion) free radicals in vitro, *Biochimie*, 90(10), 1442–1451, 2008.

55. Hawkes, D. B., Adams, G. W., Burlingame, A. L., de Montellano, P. R. O., and De Voss, J. J., Cytochrome P450(cin) (CYP176A), isolation, expression, and characterization, *J. Biol. Chem.*, 277(31), 27725–27732, 2002.

56. Woithe, K., Geib, N., Zerbe, K., Li, D. B., Heck, M., Fournier-Rousset, S., Meyer, O., Vitali, F., Matoba, N., Abou-Hadeed, K., and Robinson, J. A., Oxidative phenol coupling reactions catalyzed by OxyB: A cytochrome p450 from the vancomycin producing organism. Implications for vancomycin biosynthesis, *J. Am. Chem. Soc.*, 129(21), 6887–6895, 2007.

57. Brown, S. B., *Introduction Spectroscopy for Biochemistry*, Academic Press, London, U.K., 1980.

58. Ye, Y., Bloch, S., Xu, B., and Achilefu, S., Novel near-infrared fluorescent integrin-targeted DFO analogue, *Bioconj. Chem.*, 19(1), 225–234, 2008.

59. Stone, E. M., Costello, A. L., Tierney, D. L., and Fast, W., Substrate-assisted cysteine deprotonation in the mechanism of dimethylargininase (DDAH) from *Pseudomonas aeruginosa*, *Biochemistry*, 45(17), 5618–5630, 2006.

60. Ren, B., Duan, X. W., and Ding, H. G., Redox control of the DNA damage-inducible protein DinG helicase activity via its iron-sulfur cluster, *J. Biol. Chem.*, 284(8), 4829–4835, 2009.

61. Weiner, B. E., Huang, H., Dattilo, B. M., Nilges, M. J., Fanning, E., and Chazin, W. J., An iron-sulfur cluster in the C-terminal domain of the p58 subunit of human DNA primase, *J. Biol. Chem.*, 282(46), 33444–33451, 2007.

62. Paraskevopoulou, C., Fairhurst, S. A., Lowe, D. J., Brick, P., and Onesti, S., The elongator subunit Elp3 contains a Fe4S4 cluster and binds S-adenosylmethionine, *Mol. Microbiol.*, 59(3), 795–806, 2006.

63. Waheed, A. A. and Gupta, P. D., Application of an eosin B dye method for estimating a wide range of proteins, *J. Biochem. Biophys. Methods*, 33, 187–196, 1996.

64. Bade, J. D., Jaju, A. H., Khandagale, S. T., Aher, A. N., Bhamber, R. S., and Pal, S. C., In vitro antioxidant and free radical scavenging activity of ethyl acetate extract of *Delonix regia* Rafin flowers, *Asian J. Chem.*, 21(2), 1323–1329, 2009.

65. Li, Z., Gu, T., Kelder, B., and Kopchick, J. J., Analysis of fatty acids in mouse cells using reversed-phase high-performance liquid chromatography, *Chromatographia*, 54, 463–467, 2001.

66. Kelm, M., Dahmann, R., Wink, D., and Feelisch, M., The nitric oxide/superoxide assay. Insights into the biological chemistry of the NO/O-2. Interaction, *J. Biol. Chem.*, 272, 9922–9932, 1997.

67. Shirakawa, T., Takahashi, Y., Wada, K., Hirota, J., Takao, T., Ohmori, D., and Fukuyama, K., Identification of variant molecules of *Bacillus thermoproteolyticus* ferredoxin: Crystal structure reveals bound coenzyme A and an unexpected [3Fe-4S] cluster associated with a canonical [4Fe-4S] ligand motif, *Biochemistry*, 44(37), 12402–12410, 2005.

68. Meyer, J., Clay, M. D., Johnson, M. K., Stubna, A., Munck, E., Higgins, C., and Wittung-Stafshede, P., A hyperthermophilic plant-type [2Fe-2S] ferredoxin from *Aquifex aeolicus* is stabilized by a disulfide bond, *Biochemistry*, 41(9), 3096–3108, 2002.

69. Zupfer, J. M., Churchill, K. E., Rasmusson, D. C., and Fulcher, R. G., Variation in ferulic acid concentration among diverse barley cultivars measured by HPLC and microspectrophotometry, *J. Agric. Food Chem.*, 46, 1350–1354, 1998.

70. Panlilio, M. T. T., Espiritu, C. P., Quiming, N. S., Vergel, R. B., Reyes, M. F. G., and Villanueva, J. A., Specific binding of beta(2)-microglobulin with trypan blue, *J. Health Sci.*, 51(6), 702–707, 2005.

71. Ghosh, S. and Banthia, A. K., Biocompatibility and antibacterial activity studies of polyamidoamine (PAMAM) dendron, side chain dendritic oligourethane (SCDOU), *J. Biomed. Mater. Res. A*, 71A(1), 1–5, 2004.

72. Kanakis, C. D., Tarantilis, P. A., Polissiou, M. G., and Tajmir-Riahi, H. A., Interaction of antioxidant flavonoids with tRNA: Intercalation or external binding and comparison with flavonoid-DNA adducts, *DNA Cell Biol.*, 25(2), 116–123, 2006.

73. Kanakis, C. D., Tarantilis, P. A., Polissiou, M. G., Diamantoglou, S., and Tajmir-Riahi, H. A., Antioxidant flavonoids bind human serum albumin, *J. Mol. Struct.*, 798(1–3), 69–74, 2006.

74. Karioti, A., Kitsaki, C. K., Zygouraki, S., Ziobora, M., Djeddi, S., Skaltsa, H., and Liakopoulos, G., Occurrence of flavonoids in Ophrys (Orchidaceae) flower parts, *Flora*, 203(7), 602–609, 2008.

75. Peng, Z. F., Strack, D., Baumert, A., Subramaniam, R., Goh, N. K., Chia, T. F., Tan, S. N., and Chia, L. S., Antioxidant flavonoids from leaves of *Polygonum hydropiper* L, *Phytochemistry*, 62(2), 219–228, 2003.

76. Polster, J., Dithmar, H., Burgemeister, R., Friedemann, G., and Feucht, W., Flavonoids in plant nuclei: Detection by laser microdissection and pressure catapulting (LMPC), in vivo staining, and UV-visible spectroscopic titration, *Physiol. Plant.*, 128(1), 163–174, 2006.

77. Masuda, S., Hasegawa, K., and Ono, T., Light-induced structural changes of apoprotein and chromophore in the sensor of blue light using FAD (BLUF) domain of AppA for a signaling state, *Biochemistry*, 44(4), 1215–1224, 2005.

78. Zameitat, E., Pierik, A. J., Zocher, K., and Loffler, M., Dihydroorotate dehydrogenase from *Saccharomyces cerevisiae*: Spectroscopic investigations with the recombinant enzyme throw light on catalytic properties and metabolism of fumarate analogues, *FEMS Yeast Res.*, 7(6), 897–904, 2007.

79. Horvath, C. R., Martos, P. A., and Saxena, P. K., Identification and quantification of eight flavones in root and shoot tissues of the medicinal plant Huang-qin (*Scutellaria baicalensis* Georgi) using high-performance liquid chromatography with diode array and mass spectrometric detection, *J. Chromatogr. A*, 1062(2), 199–207, 2005.

80. Righetti, P. G., Gelfi, C., Bossi, A., Olivieri, E., Castelletti, L., Verzola, B., and Stoyanov, A. V., Capillary electrophoresis of peptides and proteins in isoelectric buffers: An update, *Electrophoresis*, 21, 4046–4053, 2000.

81. Oliveira, A. P., Pereira, D. M., Andrade, P. B., Valentao, P., Sousa, C., Pereira, J. A., Bento, A., Rodrigues, M. A., Seabra, R. M., and Silva, B. M., Free amino acids of tronchuda cabbage (*Brassica oleracea* L. var. costata DC): Influence of leaf position (internal or external) and collection time, *J. Agric. Food Chem.*, 56(13), 5216–5221, 2008.

82. Wei, C., Jia, G., Yuan, J., Feng, Z., and Li, C., A spectroscopic study on the interactions of porphyrin with G-Quadruplex DNAs, *Biochemistry*, 45(21), 6681–6691, 2006.

83. Feng, F., Liu, W. Y., Wang, Y. H., Guo, Q. L., and You, Q. D., Structure elucidation of metabolites of gambogic acid in vivo in rat bile by high-performance liquid chromatography-mass spectrometry and high-performance liquid chromatograph-nuclear magnetic resonance, *J. Chromatogr. B Anal. Technol. Biomed. Life Sci.*, 86(2), 218–226, 2007.

84. St'astna, M., Radko, S. P., and Chrambach, A., Separation efficiency in protein zone electrophoresis performed in capillaries of different diameters, *Electrophoresis*, 21, 985–992, 2000.

85. Lutfullah, G., Anjum, A. R., Ali, I., Ahmad, M., and Ahmad, I., Effects of sulfur on total glucosinolate content of in vitro regenerated plants of oilseed rape (*Brassica napus* L.), *J. Chem. Soc. Pakistan*, 29(2), 189–193, 2007.

86. Battin, E. E. and Brumaghim, J. L., Metal specificity in DNA damage prevention by sulfur antioxidants, *J. Inorg. Biochem.*, 102(12), 2036–2042, 2008.

87. Mauzeroll, J. and Bard, A. J., Scanning electrochemical microscopy of menadione-glutathione conjugate export from yeast cells, *Proc. Natl. Acad. Sci. USA*, 101(21), 7862–7867.

88. Nielsen, S. B., Lapierre, A., Andersen, J. U., Pedersen, U. V., Tomita, S., and Andersen, L. H., Absorption spectrum of the green fluorescent protein chromophore anion in vacuo, *Phys. Rev. Lett.*, 87, 228102, 2001.

89. Cheng, J. and Mitchelson, K. R., Improved separation of six harmane alkaloids by high-performance capillary electrophoresis, *J. Chromatogr. A*, 761, 297–305, 1997.

90. Ghosh, K., Thompson, A. M., Goldbeck, R. A., Shi, X. L., Whitman, S., Oh, E., Zhu, Z. W., Vulpe, C., and Holman, T. R., Spectroscopic and biochemical characterization of hemne binding to yeast Dap1p and mouse PGRMC1p, *Biochemistry*, 44(50), 16729–16736, 2005.

91. Yi, L. and Ragsdale, S. W., Evidence that the heme regulatory motifs in heme oxygenase-2 serve as a thiol/disulfide redox switch regulating heme binding, *J. Biol. Chem.*, 282(29), 21056–21067, 2007.

92. Dewilde, S., Ebner, B., Vinck, E., Gilany, K., Hankeln, T., Burmester, T., Kreiling, J. et al., The nerve hemoglobin of the bivalve mollusc *Spisula solidissima*—Molecular cloning, ligand binding studies, and phylogenetic analysis, *J. Biol. Chem.*, 281(9), 5364–5372, 2006.

93. Gladwin, M. T., Grubina, R., and Doyle, M. P., The new chemical biology of nitrite reactions with hemoglobin: R-state catalysis, oxidative denitrosylation, and nitrite reductase/anhydrase, *Acc. Chem. Res.*, 42(1), 157–167, 2009.

94. Tommasini, S., Calabro, M. L., Stancanelli, R., Donato, P., Costa, C., Catania, S., Villari, V., Ficarra, P., and Ficarra, R., The inclusion complexes of hesperetin and its 7-rhamnoglucoside with (2-hydroxypropyl)-beta-cyclodextrin, *J. Pharm. Biomed. Anal.*, 39(3–4), 572–580, 2005.

95. Kulon, K., Valensin, D., Kamysz, W., Valensin, G., Nadolski, P., Porciatti, E., Gaggelli, E., and Koztowski, H., The His-His sequence of the antimicrobial peptide demegen P-113 makes it very attractive ligand for Cu2+, *J. Inorg. Biochem.*, 102(4), 960–972, 2008.

96. Zargar, S. J. and Rabbani, A., Interaction of daunomycin antibiotic with histone H(1): Ultraviolet spectroscopy and equilibrium dialysis studies, *Int. J. Biol. Macromol.*, 30, 113–117, 2002.

97. Jenkins, L. M. M., Hara, T., Durell, S. R., Hayashi, R., Inman, J. K., Piquemal, J. P., Gresh, N., and Appella, E., Specificity of acyl transfer from 2-mercaptobenzamide thioesters to the HIV-1 nucleo-capsid protein, *J. Am. Chem. Soc.*, 129(36), 11067–11078, 2007.

98. Racine, R., Moisy, P., Paquet, F., Metivier, H., and Madic, C., In vitro study of the interaction between neptunium ions and aposerumtransferrin by absorption spectrophotometry and ultrafiltration: The case of Np(V), *Radiochim. Acta*, 91(2), 115–122, 2003.

99. Conway, M. E., Yennawar, N., Wallin, R., Poole, L. B., and Hutson, S. M., Identification of a perox-ide-sensitive redox switch at the CXXC motif in the human mitochondrial branched chain amino-transferase, *Biochemistry*, 41(29), 9070–9078, 2002.

100. Zsila, F., Fitos, I., Bikadi, Z., Simonyi, M., Jackson, H. L., and Lockwood, S. F., In vitro plasma pro-tein binding and aqueous aggregation behavior of astaxanthin dilysinate tetrahydrochloride, *Bioorg. Med. Chem. Lett.*, 14(21), 5357–5366, 2004.

101. Ouameur, A. A., Mangier, E., Diamantogiou, S., Rouillon, R., Carpentier, R., and Tajmir-Riahi, H. A., Effects of organic and inorganic polyamine cations on the structure of human serum albumin, *Biopolymers*, 73(4), 503–509, 2004.

102. Tajmir-Riahi, H. A., An overview of drug binding to human serum albumin: Protein folding and unfolding, *Sci. Iran.*, 14(2), 87–95, 2007.

103. Panasenko, O. M., Arnhold, J., and Sergienko, V. I., Impairment of membrane lipids by hypochlorite, *Biol. Membr.*, 19(5), 403–434, 2002.

104. Zou, C., Du, Y. M., Li, Y., Yang, J. H., Feng, T., Zhang, L., and Kennedy, J. F., Preparation of lacquer poly-saccharide sulfates and their antioxidant activity in vitro, *Carbohydr. Polym.*, 73(2), 322–331, 2008.

105. Lee, M. Y., Lin, H. Y., Cheng, F. W., Chiang, W. C., and Kuo, Y. H., Isolation and characterization of new lactam compounds that inhibit lung and colon cancer cells from adlay (*Coix lachryma-jobi* L. var. ma-yuen Stapf) bran, *Food Chem. Toxicol.*, 46(6), 1933–1939, 2008.

106. Akowuah, G. A., Zhari, I., Sadikun, A., and Norhayati, I., HPTLC densitometric analysis of *Orthosiphon stamineus* leaf extracts and inhibitory effect on xanthine oxidase activity, *Pharm. Biol.*, 44(1), 65–70, 2006.

107. Messana, I., Rossetti, D. V., Cassiano, L., Misiti, F., Giardina, B., and Castagnola, M., Peptide analysis by capillary (zone) electrophoresis, *J. Chromatogr. B*, 699, 149–171, 1997.

108. Rancan, F., Rosan, S., Boehm, K., Fernandez, E., Hidalgo, M. E., Quihot, W., Rubio, C., Boehm, F., Piazena, H., and Oltmanns, U., Protection against UVB irradiation by natural filters extracted from lichens, *J. Photochem. Photobiol. B Biol.*, 68(2–3), 133–139, 2002.

109. Kim, S. R., Jang, Y. P., Jockusch, S., Fishkin, N. E., Turro, N. J., and Sparrow, J. R., The all-trans-retinal dimer series of lipofuscin pigments in retinal pigment epithelial cells in a recessive Stargardt disease model, *Proc. Natl. Acad. Sci. USA*, 104(49), 19273–19278, 2007.

110. Fishkin, N. E., Sparrow, J. R., Allikmets, R., and Nakanishi, K., Isolation and characterization of a retinal pigment epithelial cell fluorophore: An all-trans-retinal dimer conjugate, *Proc. Natl. Acad. Sci. USA*, 102(20), 7091–7096, 2005.

111. Sun, C. Y., Zhang, Y. J., Fan, Y., Li, Y. J., and Li, J. H., Mannose-*Escherichia coli* interaction in the presence of metal cations studied in vitro by colorimetric polydiacetylene/glycolipid liposomes, *J. Inorg. Biochem.*, 98(6), 925–930, 2004.

112. Guarino, A. J., Lee, S. P., Tulenko, T. N., and Wrenn, S. P., Aggregation kinetics of low density lipoproteins upon exposure to sphingornyelinase, *J. Colloid Interface Sci.*, 279(1), 109–116, 2004.

113. Hackett, M. M., Lee, J. H., Francis, D., and Schwartz, S. J., Thermal stability and isomerization of lycopene in tomato oleoresins from different varieties, *J. Food Sci.*, 69(7), C536–C541, 2004.

114. Froescheis, O., Moalli, S., Liechti, H., and Bausch, J., Determination of lycopene in tissues and plasma of rats by normal-phase high-performance liquid chromatography with photometric detection, *J. Chromatogr. B*, 739, 291–299, 2000.

115. Dawson, D. A., Rinaldi, A. C., and Poch, G., Biochemical and toxicological evaluation of agent-cofactor reactivity as a mechanism of action for osteolathyrism, *Toxicology*, 177(2–3), 267–284, 2002.

116. Bouaicha, N., Ammar, M., Hennion, M. C., and Sandra, P., A new method for determination of maitotoxin by capillary zone electrophoresis with ultraviolet detection, *Toxicon*, 35, 955–962, 1997.

117. Tu, Y. G., Sun, Y. Z., Tian, Y. G., Xie, M. Y., and Chen, J., Physicochemical characterisation and antioxidant activity of melanin from the muscles of Taihe Black-bone silky fowl (Gallus gallus domesticus Brisson), *Food Chem.*, 114(4), 1345–1350, 2009.

118. Yuan, W. L., Burleigh, S. H., and Dawson, J. O., Melanin biosynthesis by Frankia strain CeI5, *Physiol. Plant.*, 131(2), 180–190, 2007.

119. Rizzi, A., Comai, S., Bertazzo, A., Costa, C. V. L., Allegri, G., and Traldi, P., An investigation on the possible role of melatonin in melanogenesis, *J. Mass Spectrom.*, 41(4), 517–526, 2006.

120. Bromme, H. J., Peschke, E., and Israel, G., Photo-degradation of melatonin: Influence of argon, hydrogen peroxide, and ethanol, *J. Pineal Res.*, 44(4), 366–372, 2008.

121. Silva, S. O., Rodrigues, M. R., Carvalho, S. R. Q., Catalani, L. H., Campa, A., and Ximenes, V. F., Oxidation of melatonin and its catabolites, N-1-acetyl-N (2)-formyl-5-methoxykynuramine and N-1-acetyl-5-methoxykynuramine, by activated leukocytes, *J. Pineal Res.*, 37(3), 171–175, 2004.

122. Wu, G., Mansy, S. S., Hemann, C., Hille, R., Surerus, K. K., and Cowan, J. A., Iron-sulfur cluster biosynthesis: Characterization of *Schizosaccharomyces pombe* Isa1, *J. Biol. Inorg. Chem.*, 7(4–5), 526–532, 2002.

123. Fischer, K., Llamas, A., Tejada-Jimenez, M., Schrader, N., Kuper, J., Ataya, F. S., Galvan, A., Mendel, R. R., Fernandez, E., and Schwarz, G., Function and structure of the molybdenum cofactor carrier protein from *Chlamydomonas reinhardtii*, *J. Biol. Chem.*, 281(40), 30186–30194, 2006.

124. Proksa, B., Separation of morphine and its oxidation products by capillary zone electrophoresis, *J. Pharm. Biomed. Anal.*, 20, 179–183, 1999.

125. Angelini, S., Gerez, C., Ollagnier-de Choudens, S., Sanakis, Y., Fontecave, M., Barras, F., and Py, B., NfuA, a new factor required for maturing Fe/S proteins in *Escherichia coli* under oxidative stress and iron starvation conditions, *J. Biol. Chem.*, 283(20), 14084–14091, 2008.

126. Smith, A. D., Jameson, G. N. L., Dos Santos, P. C., Agar, J. N., Naik, S., Krebs, C., Frazzon, J., Dean, D. R., Huynh, B. H., and Johnson, M. K., NifS-mediated assembly of [4Fe-4S] clusters in the N- and C-terminal domains of the NifU scaffold protein, *Biochemistry*, 44(39), 12955–12969, 2005.

127. Meakin, G. E., Bueno, E., Jepson, B., Bedmar, E. J., Richardson, D. J., and Delgado, M. J., The contribution of bacteroidal nitrate and nitrite reduction to the formation of nitrosylleghaemoglobin complexes in soybean root nodules, *Microbiology*, 153, 411–419, 2007.

128. Panmanee, W., Vattanaviboon, P., Poole, L. B., and Mongkolsuk, S., Novel organic hydroperoxide-sensing and responding mechanisms for OhrR, a major bacterial sensor and regulator of organic hydroperoxide stress, *J. Bacteriol.*, 188(4), 1389–1395, 2006.

129. Hotchkiss, A. T. and Hicks, K. B., Analysis of pectate lyase-generated oligogalacturonic acids by high-performance anion-exchange chromatography with pulsed amperometric detection, *Carbohydr. Res.*, 247, 1–7, 1993.

130. Hadjimitova, V., Traykov, T., and Bocheva, A., TYR-MIFS peptides as hydroxyl radical scavengers, *Comp. Rend. Acad. Bulg. Sci.*, 60(6), 687–690, 2007.

131. Tong, Y. and Guo, M. L., Cloning and characterization of a novel periplasmic heme-transport protein from the human pathogen *Pseudomonas aeruginosa*, *J. Biol. Inorg. Chem.*, 12(6), 735–750, 2007.

132. Deepa, G. L., Sashidhar, R. B., and Deshpande, V., Purification and characterization of phycoferritin from the blue-green alga, *Arthrospira* (Spirulina) *platensis*, *J. Appl. Phycol.*, 20(4), 359–366, 2008.

133. von Stetten, D., Seibeck, S., Michael, N., Scheerer, P., Mroginski, M. A., Murgida, D. H., Krauss, N. et al., Highly conserved residues Asp-197 and His-250 in Agp1 phytochrome control the proton affinity of the chromophore and Pfr formation, *J. Biol. Chem.*, 282(3), 2116–2123, 2007.

134. Hahn, J., Strauss, H. M., Landgraf, F. T., Gimenez, H. F., Lochnit, G., Schmieder, P., and Hughes, J., Probing protein-chromophore interactions in Cph1 phytochrome by mutagenesis, *FEBS J.*, 273(7), 1415–1429, 2006.

135. Das, I., Gupta, S. K., Ansari, S. A., Pandey, V. N., and Rastogi, R. P., In vitro inhibition and dissolution of calcium oxalate by edible plant *Trianthema monogyna* and pulse *Macrotyloma uniflorum* extracts, *J. Cryst. Growth*, 273(3–4), 546–554, 2005.

136. Domenech, J., Mir, G., Huguet, G., Capdevila, M., Molinas, M., and Atrian, S., Plant metallothionein domains: Functional insight into physiological metal binding and protein folding, *Biochimie*, 88(6), 583–593, 2006.

137. Zhao, C. Y., Dodin, G., Yuan, C. S., Chen, H. F., Zheng, R. L., Jia, Z. J., and Fan, B. T., In vitro, protection of DNA from Fenton reaction by plant polyphenol verbascoside, *Biochim. Biophys. Acta*, 1723(1–3), 114–123, 2005.

138. Li, W., Ke, Y., He, G. S., Xu, Y. C., and Wang, Z. Y., Study on the correlation between PMI and OD changes in rat's plasma, *Spectrosc. Spectr. Anal.*, 28(12), 2944–2946, 2008.

139. Wan, N., Gu, J. D., and Yan, Y., Degradation of p-nitrophenol by *Achromobacter xylosoxidans* Ns isolated from wetland sediment, *Int. Biodeterior. Biodegrad.*, 59(2), 90–96, 2007.

140. Bratu, M. M., Porta, S., Negreanu-Pirjol, T., Roncea, F., and Miresan, H., Capacity of Mg salts to form polyphenols out of vegetals, *Rev. Chim.*, 57(8), 823–825, 2006.

141. Perron, N. R. and Brumaghim, J. L., A review of the antioxidant mechanisms of polyphenol compounds related to iron binding, *Cell Biochem. Biophys.*, 53(2), 75–100, 2009.

142. Synytsya, A., Kral, V., Synytsya, A., Volka, K., and Sessler, J. L., In vitro interaction of macrocyclic photosensitizers with intact mitochondria: A spectroscopic study, *Biochim. Biophys. Acta*, 1620(1–3), 85–96, 2003.

143. Hanzelmann, P., Hernandez, H. L., Menzel, C., Garcia-Serres, R., Huynh, B. H., Johnson, M. K., Mendel, R. R., and Schindelin, H., Characterization of MOCS1A, an oxygen-sensitive iron-sulfur protein involved in human molybdenum cofactor biosynthesis, *J. Biol. Chem.*, 279(33), 34721–34732, 2004.

144. Hernandez, H. L., Pierrel, F., Elleingand, E., Garcia-Serres, R., Huynh, B. H., Johnson, M. K., Fontecave, M., and Atta, M., MiaB, a bifunctional radical-S-adenosylmethionine enzyme involved in the thiolation and methylation of tRNA, contains two essential [4Fe-4S] clusters, *Biochemistry*, 46(17), 5140–5147, 2007.

145. Gaudreau, S., Novetta-Dellen, A., Neault, J. F., Diamantoglou, S., and Tajmir-Riahi, H. A., 3′-Azido-3′-deoxythymidine binding to ribonuclease A: Model for drug-protein interaction, *Biopolymers*, 72(6), 435–441, 2003.

146. Seyedsayamdost, M. R., Xie, J., Chan, C. T. Y., Schultz, P. G., and Stubbe, J., Site-specific insertion of 3-aminotyrosine into subunit alpha 2 of *E coli* ribonucleotide reductase: Direct evidence for involvement of Y-730 and Y-731 in radical propagation, *J. Am. Chem. Soc.*, 129(48), 15060–15071, 2007.

147. Zhang, M. X., Gumerov, D. R., Kaltashov, I. A., and Mason, A. B., Indirect detection of protein-metal binding: Interaction of serum transferrin with In3+ and Bi3+, *J. Am. Soc. Mass Spectrom.*, 15(11), 1658–1664, 2004.

148. Li, X. G., Wu, L. Y., Huang, M. R., Shao, H. L., and Hu, X. C., Conformational transition and liquid crystalline state of regenerated silk fibroin in water, *Biopolymers*, 89(6), 497–505, 2008.

149. Tao, L. M. and English, A. M., Protein S-glutathiolation triggered by decomposed S-nitrosoglutathione, *Biochemistry*, 43(13), 4028–4038, 2004.

150. Gurley, L. R., Umbarger, K. O., Kim, J. M., Bradbury, E. M., and Lehnert, B. E., Development of a high-performance liquid-chromatographic method for the analysis of staurosporine, *J. Chromatogr. B Biomed. Appl.*, 670, 125–138, 1995.

151. Molina-Heredia, F. P., Houee-Levin, C., Berthomieu, C., Touati, D., Tremey, E., Favaudon, V., Adam, V., and Niviere, V., Detoxification of superoxide without production of H2O2: Antioxidant activity of superoxide reductase complexed with ferrocyanide, *Proc. Natl. Acad. Sci. USA*, 103(40), 14750–14755, 2006.

152. Leonardi, R., Fairhurst, S. A., Kriek, M., Lowe, D. J., and Roach, P. L., Thiamine biosynthesis in *Escherichia coli*: Isolation and initial characterisation of the ThiGH complex, *FEBS Lett.*, 539(1–3), 95–99, 2003.

153. Yuki, E. T., Elbaz, M. A., Nakano, M. M., and Moenne-Loccoz, P., Transcription factor NsrR from *Bacillus subtilis* senses nitric oxide with a 4Fe-4S cluster, *Biochemistry*, 47(49), 13084–13092, 2008.

154. Das, I. and Verma, S., Human stones: Dissolution of calcium phosphate and cholesterol by edible plant extracts and bile acids, *J. Sci. Ind. Res.*, 67(4), 291–294, 2008.

155. Cai, H., Verschoyle, R. D., Steward, W. P., and Gescher, A. J., Determination of the flavone tricin in human plasma by high-performance liquid chromatography, *Biomed. Chromatogr.*, 17(7), 435–439, 2003.

156. Li, F. and Yu, L., Determination of trimebutine maleate in rat plasma and tissues by using capillary zone electrophoresis, *Biomed. Chromatogr.*, 15, 248–251, 2001.

157. Mathevon, C., Pierrel, F., Oddou, J. L., Garcia-Serres, R., Blonclin, G., Latour, J. M., Menage, S., Gambarelli, S., Fontecave, M., and Atta, M., TRNA-modifying MiaE protein from *Salmonella typhimurium* is a nonheme diiron monooxygenase, *Proc. Natl. Acad. Sci. USA*, 104(33), 13295–13300, 2007.

158. Kobayashi, S., Ogawa, N., Fujimura, Y., Tachibana, H., and Yamada, K., A chrysanthemum flower, Shiranui Himekiku, extract enhance TNF-alpha production, *J. Jpn. Soc. Food Sci. Technol.*, 53(8), 430–436, 2006.

159. Valavanidis, A., Nisiotou, C., Papageorgiou, Y., Kremli, I., Satravelas, N., Zinieris, N., and Zygalaki, H., Comparison of the radical scavenging potential of polar and lipidic fractions of olive oil and other vegetable oils under normal conditions and after thermal treatment, *J. Agric. Food Chem.*, 52(8), 2358–2365, 2004.

160. N'Soukpoe-Kossi, C. N., Sedaghat-Herati, R., Ragi, C., Hotchandani, S., and Tajmir-Riahi, H. A., Retinol and retinoic acid bind human serum albumin: Stability and structural features, *Int. J. Biol. Macromol.*, 40(5), 484–490, 2007.

161. Olgemoller, J., Hempel, G., Boos, J., and Blaschke, G., Determination of (E)-5-(2-bromovinyl)-2′-deoxyuridine in plasma and urine by capillary electrophoresis, *J. Chromatogr. B*, 726, 261–268, 1999.

162. Wessel, T., Lanvers, C., Fruend, S., and Hempel, G., Determination of purines including 2,8-dihydroxyadenine in urine using capillary electrophoresis, *J. Chromatogr. A*, 894, 157–164, 2000.

163. Bjornsdottir, I., Kepp, D. R., Tjornelund, J., and Hansen, S. H., Separation of the enantiomers of ibuprofen and its major phase I metabolites in urine using capillary electrophoresis, *Electrophoresis*, 19, 455–460, 1998.

164. Lanz, M. and Thormann, W., Characterization of the stereoselective metabolism of methadone and its primary metabolite via cyclodextrin capillary electrophoretic determination of their urinary enantiomers, *Electrophoresis*, 17, 1945–1949, 1996.

165. Caslavska, J., Gassmann, E., and Thormann, W., Modification of a tunable UV-visible capillary electrophoresis detector for simultaneous absorbency and fluorescence detection—Profiling of bodyfluids for drugs and endogenous compounds, *J. Chromatogr. A*, 709(1), 147–156, 1995.

166. Lanz, M., Theurillat, R., and Thormann, W., Characterization of stereoselectivity and genetic polymorphism of the debrisoquine hydroxylation in man via analysis of urinary debrisoquine and 4-hydroxydebrisoquine by capillary electrophoresis, *Electrophoresis*, 18, 1875–1881, 1997.

167. Desiderio, C., Fanali, S., Kupfer, A., and Thormann, W., Analysis of mephenytoin, 4-hydroxymephenytoin and 4-hydroxyphenytoin enantiomers in human urine by cyclodextrin micellar electrokinetic capillary chromatography—Simple determination of a hydroxylation polymorphism in man, *Electrophoresis*, 15, 87–93, 1994.

168. Kunkel, A. and Watzig, H., Micellar electrokinetic capillary chromatography as a powerful tool for pharmacological investigations without sample pretreatment: A precise technique providing cost advantages and limits of detection to the low nanomolar range, *Electrophoresis*, 20, 2379–2389, 1991.

169. Vernez, L., Thormann, W., and Krahenbuhl, S., Analysis of carnitine and acylcarnitines in urine by capillary electrophoresis, *J. Chromatogr. A*, 895, 309–316, 2000.

170. Eckert, S., Eyer, P., Muckter, H., and Worek, F., Development of a dynamic model for real-time determination of membrane-bound acetylcholinesterase activity upon perfusion with inhibitors and reactivators, *Biochem. Pharmacol.*, 72(3), 358–365, 2006.

171. Prochazkova, A., Chouki, M., Theurillat, R., and Thormann, W., Therapeutic drug monitoring of albendazole: Determination of albendazole, albendazole sulfoxide, and albendazole sulfone in human plasma using nonaqueous capillary electrophoresis, *Electrophoresis*, 21, 729–736, 2000.

172. Varesio, E. and Veuthey, J.-L., Chiral separation of amphetamines by high-performance capillary electrophoresis, *J. Chromatogr. A*, 717, 219–228, 1995.

173. Mpiana, P. T., Mudogo, V., Tshibangu, D. S. T., Kitwa, E. K., Kanangila, A. B., Lumbu, J. B. S., Ngbolua, K. N., Atibu, E. K., and Kakule, M. K., Antisickling activity of anthocyanins from *Bombax pentadrum*, *Ficus capensis* and *Ziziphus mucronata*: Photodegradation effect, *J. Ethnopharmacol.*, 120(3), 413–418, 2008.

174. Matsumoto, H., Nakamura, Y., Iida, H., Ito, K., and Ohguro, H., Comparative assessment of distribution of blackcurrant anthocyanins in rabbit and rat ocular tissues, *Exp. Eye Res.*, 83(2), 348–356, 2006.

175. Ruiz-Duenas, F. J., Ferreira, P., Martinez, M. J., and Martinez, A. T., In vitro activation, purification, and characterization of *Escherichia coli* expressed aryl-alcohol oxidase, a unique H_2O_2-producing enzyme, *Protein Expr. Purif.*, 45(1), 191–199, 2006.

176. Jin, L. J., Wang, Y., Xu, R., Go, M. L., Lee, H. K., and Li, S. F. Y., Chiral resolution of atropine, homatropine and eight synthetic tropinyl and piperidinyl esters by capillary zone electrophoresis with cyclodextrin additives, *Electrophoresis*, 20, 198–203, 1999.

177. Palmarsdottir, S., Mathiasson, L., Jonsson, J. A., and Edholm, L.-E., Determination of a basic drug, bambuterol, in human plasma by capillary electrophoresis using double stacking for large volume injection and supported liquid membranes for sample pretreatment, *J. Chromatogr. B*, 688, 127–134, 1997.

178. Gunasekaran, S. and Uthra, D., FTIR and UV-visible spectral study on normal and jaundice blood samples, *Asian J. Chem.*, 20(7), 5695–5703, 2008.

179. Gunasekaran, S., Natarajan, R. K., Renganayaki, V., and Rathikha, R., FTIR and UV visible spectrophotometric approach to discriminate leukemic sera, *Asian J. Chem.*, 20(4), 2521–2530, 2008.

180. Zalesskaya, G. A., Ulashchik, V. S., Mit'kovskaya, N. P., Kuchinskii, A. V., and Laskina, O. V., Spectral signs of photochemical reactions when blood is exposed in vivo to therapeutic doses of ultraviolet radiation, *J. Appl. Spectrosc.*, 75(3), 426–432, 2008.

181. Tanaka, Y., Yanagawa, M., and Terabe, S., Separation of neutral and basic enantiomers by cyclodextrin electrokinetic chromatography using anionic cyclodextrin derivatives as chiral pseudo-stationary phases, *J. High Resolut. Chromatogr.*, 19, 421–433, 1996.

182. Kuldvee, R. and Thormann, W., Determination of carbamazepine and carbamzepine-10,11-epoxide in human serum and plasma by micellar electrokinetic capillary chromatography in the absence of electroosmosis, *Electrophoresis*, 22, 1345–1355, 2001.

183. Li, F., Cooper, S. F., and Mikkelsen, S. R., Enantioselective determination of oxprenolol and its metabolites in human urine by cyclodextrin-modified capillary zone electrophoresis, *J. Chromatogr. B*, 674, 227–285, 1995.

184. Prunonosa, J., Obach, R., Diez-Cascon, A., and Gouesclou, L., *J. Chromatogr.*, 574, 127–133, 1992.

185. Garay, R. P., Rosati, C., Fanous, K., Allard, M., Morin, E., Lamiable, D., and Vistelle, R., Evidence for (+)-cicletanine sulfate as an active natriuretic metabolite of cicletanine in the rat, *Eur. J. Pharmacol.*, 274, 175–180, 1995.

186. Altria, K. D., Goodall, D. M., and Rogan, M. M., *Electrophoresis*, 15, 824–827, 1994.

187. Alfazema, L. N., Howells, S., and Perrett, D., Optimised separation of endogenous urinary components using cyclodextrin-modified micellar electrokinetic capillary chromatography, *Electrophoresis*, 21, 2503–2508, 2000.

188. Aumatell, A. and Wells, R. J., Chiral differentiation of the optical isomers of racemethorphan and racemorphan in urine by capillary zone electrophoresis, *J. Chromatgr. Sci.*, 31, 502–508, 1993.

189. Hufschmid, E., Theurillat, R., Martin, U., and Thormann, W., *J. Chromatogr. B*, 668, 159–170, 1995.

190. Wey, A. B., Zhang, C., and Thormann, W., Head-column field-amplified sample stacking in binary system capillary electrophoresis preparation of extracts for determination of opioids in microliter amounts of body fluids, *J. Chromatogr. A*, 853, 95–106, 1999.

191. Marty, R., Ouameur, A. A., Neault, J. F., Nafisi, S., and Tajmir-Riahi, H. A., AZT-DNA interaction, *DNA Cell Biol.*, 23(3), 135–140, 2004.

192. Mazzeo, J. R., Grover, E. R., Swartz, M. E., and Petersen, J. S., Novel chiral surfactant for the separation of enantiomers by micellar electrokinetic capillary chromatography, *J. Chromatogr. A*, 680, 125–135, 1994.

193. Palmer, G. M., Marshek, C. L., Vrotsos, K. M., and Ramanujam, N., Optimal methods for fluorescence and diffuse reflectance measurements of tissue biopsy samples, *Lasers Surg. Med.*, 30(3), 191–200, 2002.

194. Heeren, F., Tanner, R., Theurillat, R., and Thormann, W., Determination of the antimycotic drug fluconazole in human plasma by micellar electrokinetic capillary chromatography with detection at 190 nm, *J. Chromatogr. A*, 745, 165–172, 1996.

195. Francotte, E., Cherkaoui, S., and Faupel, M., Separation of the enantiomers of some racemic non-steroidal aromatase inhibitors and barbiturates by capillary electrophoresis, *Chirality*, 5, 516–526, 1993.

196. Soini, H., Stefansson, M., Riekkola, M.-L., and Novotny, M. V., Maltooligosaccharides as chiral selectors for the separation of pharmaceuticals by capillary electrophoresis, *Anal. Chem.*, 66, 3477–3484, 1994.

197. Shibukawa, A., Lloyd, D. K., and Wainer, I. W., Simultaneous chiral separation of leucovorin and its major metabolite 5-methyl-tetrahydrofolate by capillary electrophoresis using cyclodextrins as chiral selectors—Estimation of the formation constant and mobility of the solute-cyclodextrin complexes, *Chromatographia*, 35, 419–429, 1993.

198. Siluveru, M. and Stewart, J. T., HPCE determination of R(+) and S(−) mepivacaine in human serum using a derivatized cyclodextrin and ultraviolet detection, *J. Pharm. Biomed. Anal.*, 15, 1751–1756, 1997.

199. Heuermann, M. and Blaschke, G., Simultaneous enantioselective determination and quantification of dimethindene and its metabolite n-demethyl-dimethindene in human urine using cyclodextrins as chiral additives in capillary electrophoresis, *J. Pharm. Biomed. Anal.*, 12, 753–760, 1994.

200. Rudaz, S., Veuthey, J. L., Desiderio, C., and Fanali, S., *J. Chromatogr. A*, 846, 227–237, 1999.

201. Dethy, J.-M., De Broux, S., Lesne, M., Longstreth, J., and Gilbert, P., Stereoselective determination of verapamil and norverapamil by capillary electrophoresis, *J. Chromatogr. B*, 654, 121–127, 1994.

202. Kurth, B. and Blaschke, G., Achiral and chiral determination of tramadol and its metabolites in urine by capillary electrophoresis, *Electrophoresis*, 20, 555–563, 1999.

203. Siluveru, M. and Stewart, J. T., Enantioselective determination of S-(+)- and R-(−)-ondansetron in human serum using derivatized cyclodextrin-modified capillary electrophoresis and solid-phase extraction, *J. Chromatogr. B.*, 691, 217–222, 1997.
204. Salerno, C., D'Eufemia, P., Celli, M., Finocchiaro, R., Crifo, C., and Giardini, O., Determination of urinary orotic acid and uracil by capillary zone electrophoresis, *J. Chromatogr. B*, 734, 175–178, 1999.
205. Hempel, G., Lehmkuhl, D., Krumpelmann, S., Blaschke, G., and Boos, J., Determination of paclitaxel in biological fluids by micellar electrokinetic chromatography, *J. Chromatogr. A*, 745, 173–179, 1996.
206. Srinivasan, K. and Bartlett, M. G., Capillary electrophoresis stereoselective determination of R-(+)- and S-(−)-pentobarbital from serum using hydroxypropyl-gamma-cyclodextrin, solid-phase extraction and ultraviolet detection, *J. Chromatogr. B.*, 703, 289–294, 1997.
207. Siluveru, M. and Stewart, J. T., Stereoselective determination of R-(−)- and S-(+)-prilocaine in human serum by capillary electrophoresis using a derivatized cyclodextrin and ultraviolet detection, *J. Chromatogr. B.*, 693, 205–210, 1997.
208. Ramseier, A., Caslavska, J., and Thormann, W., Stereoselective screening for and confirmation of urinary enantiomers of amphetamine, methamphetamine, designer drugs, methadone and selected metabolites by capillary electrophoresis, *Electrophoresis*, 20, 2726–2738, 1999.
209. Wu, S.-M., Ko, W.-K., Wu, H.-L., and Chen, S.-H., Trace analysis of haloperidol and its chiral metabolite in plasma by capillary electrophoresis, *J. Chromatogr. A*, 846, 239–243, 1999.
210. Srinivasan, K., Zhang, W., and Bartlett, M. G., Rapid simultaneous capillary electrophoretic determination of (R)- and (S)-secobarbital from serum and prediction of hydroxypropyl-gamma-cyclodextrin-secobarbital stereoselective interaction using molecular mechanics simulation, *J. Chromatogr. Sci.*, 36, 85–90, 1998.
211. Gokay, O., Mujdeci, A., and Algin, E., In vitro peroxide penetration into the pulp chamber from newer bleaching products, *Int. Endodont. J.*, 38(8), 516–520, 2005.
212. Sheppard, R. L., Tong, X., Cai, J., and Henion, J. D., Chiral separation and detection of terbutaline and ephedrine by capillary electrophoresis coupled with ion-spray mass-spectrometry, *Anal. Chem.*, 67, 2054–2058, 1995.
213. Palmarsdottir, S. and Edholm, L.-E., Enhancement of selectivity and concentration sensitivity in capillary zone electrophoresis by online coupling with column liquid-chromatography and utilizing a double stacking procedure allowing for microliter injections, *J. Chromatogr. A*, 693, 131–143, 1995.
214. Lin, Y. K. and Liu, D. C., Comparison of physical-chemical properties of type I collagen from different species, *Food Chem.*, 99(2), 244–251, 2006.
215. Soini, H., Riekkola, M.-L., and Novotny, M. V., Chiral separations of basic drugs and quantitation of bupivacaine enantiomers in serum by capillary electrophoresis with modified cyclodextrin buffers, *J. Chromatogr.*, 608, 265–274, 1992.
216. D'Hulst, A. and Verbeke, N., Separation of the enantiomers of coumarinic anticoagulant drugs by capillary electrophoresis using maltodextrins as chiral modifiers, *Chirality*, 6, 225–229, 1994.
217. Gareil, P., Gramond, J. P., and Guyon, F., Separation and determination of warfarin enantiomers in human plasma samples by capillary zone electrophoresis using a methylated beta-cyclodextrin-containing electrolyte, *J. Chromatogr.*, 615, 317–325, 1993.
218. Hilmey, D. G., Abe, M., Nelen, M. I., Stilts, C. E., Baker, G. A., Baker, S. N., Bright, F. V. et al., Water-soluble, core-modified porphyrins as novel, longer-wavelength-absorbing sensitizers for photodynamic therapy. II. Effects of core heteroatoms and meso-substituents on biological activity, *J. Med. Chem.*, 45, 449–461, 2002.
219. Cubeddu, R., Canti, G., Musolino, M., Pifferi, A., Taroni, P., and Valentini, G., In vivo absorption spectrum of disulphonated aluminium phthalocyanine in tumor bearing mice, *Fifth Biannual Meeting of the International Photodynamic Association*, Amelia Island, FL, September 21–24, 1994, Abstract 74.

220. Panjehpour, M., DeNovo, R. C., Petersen, M. G., Overholt, B. F., Bower, R., Rubinchik, V., and Kelly, B., *Lasers Surg. Med.*, 30, 26–30, 2002.

221. Cubeddu, R., Canti, G., D'Andrea, C., Pifferi, A., Taroni, P., Torricelli, A., and Valentini, G., Effects of photodynamic therapy on the absorption properties of disulphonated aluminum phthalocyanine in tumor-bearing mice, *J. Photochem. Photobiol. B*, 60, 73–78, 2001.

222. Skala, M. C., Palmer, G. M., Vrotsos, K. M., Gendron-Fitzpatrick, A., and Ramanujam, N., Comparison of a physical model and principal component analysis for the diagnosis of epithelial neoplasias in vivo using diffuse reflectance spectroscopy, *Opt. Express*, 15(12), 7863–7875, 2007.

223. Star, W. M., Versteeg, A. A. A., Van Putten, W. L. J., and Marijnissen, J. P. A., Wavelength dependence of hematoporphyrin derivative photodynamic treatment effects on rat ears, *Photochem. Photobiol.*, 52, 547–554, 1990.

224. Van Leengoed, H. L. L. M., Cuomo, V., Versteeg, A. A. C., Van der Veen, N., Jori, G., and Star, W. M., In vivo fluorescence and photodynamic activity of zinc phthalocyanine administered in liposomes, *Br. J. Cancer*, 69, 840–845, 1994.

225. Sans, M. Q., Chen, J. W., Weissleder, R., and Bogdanov, A. A., Myeloperoxidase activity imaging using Ga-67 labeled substrate, *Mol. Imaging Biol.*, 7(6), 403–410, 2005.

226. Schieweck, K., Isele, U., Capraro, H.-G., Ochsner, M., Maurer, Th., Kratz, J., Gentsch, C., Jori, G., Segalla, A., and Biolo, R., Preclinical studies with CGP 55847, Iiposomal zinc(II)-phthalocyanine, *Fifth Biannual Meeting of the International Photodynamic Association*, Amelia Island, FL, September 21–24, 1994, Abstract 8.

227. Farrell, T. J., Olivo, M. C., Patterson, M. S., Wrona, H., and Wilson, B. C., Investigation of the dependence of tissue necrosis on irradiation wavelength and time post injection using a photodynamic threshold dose model, in P. Spinelli, M. Dal Fante, and R. Marchesini (eds.), *Photodynamic Therapy and Biomedical Lasers*, Elsevier, Amsterdam, the Netherlands, pp. 830–834, 1992.

228. Potter, W. R., Bellnier, D. A., and Oseroff, A. R., Tissue concentration of sensitizer by in vivo reflectance spectroscopy, *Fifth Biannual Meeting of the International Photodynamic Association*, Amelia Island, FL, September 21–24, 1994, Abstract 46.

229. Tyagi, S., Singh, N., Singh, S. M., and Singh, U. P., Synthesis, structural, and antitumor studies of some 5-fluorocytosine and guanine complexes, *Synth. React. Inorg. Metal-Org. Chem.*, 34(3), 573–591, 2004.

230. Butler, D. H. and McNamee, M. G., FTIR analysis of nicotinic acetylcholine-receptor secondary structure in reconstituted membranes, *Biochim. Biophys. Acta*, 1150(1), 17–24, 1993.

231. Colthup, N. B., Daly, L. H., and Wiberley, S. E., *Introduction to Infrared and Raman Spectroscopy*, 2nd edn., Academic press, New York.

232. Martinez, G. and Millhauser, G., FTIR spectroscopy of alanine-based peptides—Assignment of the Amide I' modes for random coil and helix, *J. Struct. Biol.*, 114, 23–27, 1995.

233. Venyaminov, S. Y. and Kalnin, N. N., Quantitative IR spectrophotometry of peptide compounds in water (H_2O) solutions. I. Spectral parameters of amino acid residue absorption bands, *Biopolymers*, 30, 1243–1257, 1990.

234. Rahmelow, K., Hubner, W., and Ackermann, T., Infrared absorbancies of proteins side chains, *Anal. Biochem.*, 257, 1–11, 1998.

235. Pinchas, S. and Laulicht, I., *Infrared Spectra of Labelled Compounds*, Academic Press, London, U.K., 1971.

236. ElMahdaoui, L., Neault, J. F., and TajmirRiahi, H. A., Carbohydrate-nucleotide interaction. The effects of mono- and disaccharides on the solution structure of AMP, dAMP, ATP, GMP, dGMP, and GTP studied by FTIR difference spectroscopy, *J. Inorg. Biochem.*, 65(2), 123–131, 1997.

237. Szabo, Z., Klement, E., Jost, K., Zarandi, M., Soos, K., and Penke, B., An FT-IR study of the beta-amyloid conformation: Standardization of aggregation grade, *Biochem. Biophys. Res. Commun.*, 265(5), 297–300, 1999.

238. Tajmirriahi, H. A., Neault, J. F., and Naoui, M., Does DNA acid fixation produce left-handed-Z structure, *FEBS Lett.*, 370(1–2), 105–108, 1995.

239. Chang, Q. L., Liu, H. H., and Chen, J. Y., Fourier-transform infrared-spectra studies of protein in reverse micelles—Effect of Aot/Isooctane on the secondary structure of alpha-chymotrypsin, *Biochim. Biophys. Acta-Prot. Struct. Mol. Enzymol.*, 1206(2), 247–252, 1994.

240. Lee, S. M., Lin, S. Y., and Liang, R. C., Secondary conformational structure of type IV collagen in different conditions determined by Fourier transform infrared microscopic spectroscopy, *Artif. Cells Blood Immobil. Biotechnol.*, 23(2), 193–205, 1995.

241. Susi, H., Byler, D. M., and Geraaimowicz, W., Vibrational analysis of aminoacids: Cysteine, serine, beta-chloroalanine, *J. Mol. Struct.*, 102, 63–79, 1983.

242. Rich, P. R. and Breton, J., FTIR studies of the CO and cyanide adducts of fully reduced bovine cytochrome c oxidase, *Biochemistry*, 40(21), 6441–6449, 2001.

243. Birss, V. I., Hinman, A. S., McGarvey, C. E., and Segal, J., In-situ ftir thin-layer reflectance spectroscopy of flavin adenine-dinucleotide at a mercury gold electrode, *Electrochim. Acta*, 39(16), 2449–2454, 1994.

244. Um, I. C., Kweon, H. Y., Park, Y. H., and Hudson, S., Structural characteristics and properties of the regenerated silk fibroin prepared from formic acid, *Int. J. Biol. Macromol.*, 29(2), 91–97, 2001.

245. Sengupta, P. K. and Krimm, S., Vibrational analysis of peptides, polypeptides and proteins. XXI beta-calcium-poly(L-glutamate), *Bioploymers*, 23, 1565–1594, 1984.

246. Dhamelincourt, P. and Ramirez, F. J., Polarized micro-Raman and FTIR spectra of glutamine, *Appl. Spectrosc.*, 47, 446–451, 1993.

247. Laulicht, I., Pinchas, S., Samuel, D., and Wasserman, I., The infrared absorption spectrum of oxygen-18 labeled glycine, *J. Phys. Chem.*, 70, 2719–2725, 1966.

248. Chen, R. P. and Spiro, T. G., Monitoring the allosteric transition and CO rebinding in hemoglobin with time-resolved FTIR spectroscopy, *J. Phys. Chem. A*, 106(14), 3413–3419, 2002.

249. Tommasini, S., Calabro, M. L., Stancanelli, R., Donato, P., Costa, C., Catania, S., Villari, V., Ficarra, P., and Ficarra, R., The inclusion complexes of hesperetin and its 7-rhamnoglucoside with (2-hydroxypropyl)-beta-cyclodextrin, *J. Pharm. Biomed. Anal.*, 39(3–4), 572–580, 2005.

250. Hasegawa, K., Ono, T.-A., and Noguchi, T., Vibrational spectra and ab initio DFT calculations of 4-methylimidazole and its different protonation forms: Infrared and Raman markers of the protonation state of histidine side chain, *J. Phys. Chem. B*, 104, 4253–4265, 2000.

251. Overman, S. A. and Thomas, G. J., Raman markers of nonaromatic side chains in an alpha-helix assembly: ala, asp, glu, gly, ile, leu, lys, ser and val residues of phage fd subunits, *Biochemistry*, 38, 4018–4027, 1999.

252. Perez, C. and Griebenow, K., Fourier-transform infrared spectroscopic investigation of the thermal denaturation of hen egg-white lysozyme dissolved in aqueous buffer and glycerol, *Biotechnol. Lett.*, 22(23), 1899–1905, 2000.

253. Kacurakova, M. and Mathlouthi, M., FTIR and laser-Raman spectra of oligosaccharides in water: Characterization of the glycosidic bond, *Carbohydr. Res.*, 284(2), 145–157, 1996.

254. Lobau, J., Sass, M., Pohle, W., Selle, C., Koch, M. H. J., and Wolfrum, K., Chain fluidity and phase behaviour of phospholipids as revealed by FTIR and sum-frequency spectroscopy, *J. Mol. Struct.*, 481, 407–411, 1999.

255. Gauger, D. R., Selle, C., Fritzsche, H., and Pohle, W., Chain-length dependence of the hydration properties of saturated phosphatidylcholines as revealed by FTIR spectroscopy, *J. Mol. Struct.*, 565, 25–29, 2001.

256. Pohle, W., Gauger, D. R., Fritzsche, H., Rattay, B., Selle, C., Binder, H., and Bohlig, H., FTIR-spectroscopic characterization of phosphocholine-headgroup model compounds, *J. Mol. Struct.*, 563, 463–467, 2001.

257. Lewis, R. and McElhaney, R. N., The structure and organization of phospholipid bilayers as revealed by infrared spectroscopy, *Chem. Phys. Lipids*, 96(1–2), 9–21, 1998.

258. Shibata, A., Yamamoto, M., Yamashita, T., Chiou, J. S., Kamaya, H., and Ueda, I., Biphasic effects of alcohols on the phase transition of poly(L-lysine) between alpha-helix and beta sheet conformations, *Biochemistry*, 31(25), 5728–5733, 1992.

259. Dornberger, U., Spackova, N., Walter, A., Gollmick, F. A., Sponer, J., and Fritzsche, H., Solution structure of the dodecamer d-(CATGGGCC-CATG)(2) is B- DNA. Experimental and molecular dynamics study, *J. Biomol. Struct. Dyn.*, 19(1), 159–174, 2001.

260. Akhebat, A., Dagneaux, C., Liquier, J., and Taillandier, E., Triple helical polynucleotidic structures— An FTIR study of the C+.G.C triplet, *J. Biomol. Struct. Dyn.*, 10(3), 577–588, 1992.

261. Perron, N. R., Hodges, J. N., Jenkins, M., and Brumaghim, J. L., Predicting how polyphenol antioxidants prevent DNA damage by binding to iron, *Inorg. Chem.*, 47(14), 6153–6161.

262. Cabiaux, V., Oberg, K. A., Pancoska, P., Walz, T., Agre, P., and Engel, A., Secondary structures comparison of aquaporin-1 and bacteriorhodopsin: A Fourier transform infrared spectroscopy study of two-dimensional membrane crystals, *Biophys. J.*, 73(1), 406–417, 1997.

263. Herlinger, A. W. and Long, T. V., Laser-Raman and infrared spectra of aminoacids and their metal complexes. 3. Proline and bisprolinato complexes, *J. Am. Chem. Soc.*, 92, 6481–6486, 1970.

264. Butler, B. C., Hanchett, R. H., Rafailov, H., and MacDonald, G., Investigating structural changes induced by nucleotide binding to RecA using difference FTIR, *Biophys. J.*, 82(4), 2198–2210, 2002.

265. Garciaquintana, D., Garriga, P., and Manyosa, J., Quantitative characterization of the structure of rhodopsin in disk membrane by means of Fourier-transform infrared-spectroscopy, *J. Biol. Chem.*, 268(4), 2403–2409, 1993.

266. Klinck, R., Guittet, E., Liquier, J., Taillandier, E., Gouyette, C., and Huynhdinh, T., Spectroscopic evidence for an intramolecular RNA triple-helix, *FEBS Lett.*, 355(3), 297–300, 1994.

267. Bramanti, E. and Benedetti, E., Determination of the secondary structure if isomeric forms of human serum albumin by a particular frequency deconvolution to Fourier transform IR analysis, *Biopolymers*, 38(5), 639–653, 1996.

268. Iconomidou, V. A., Chryssikos, G. D., Gionis, V., Willis, J. H., and Hamodrakas, S. J., Soft-cuticle protein secondary structure as revealed by FT-Raman, ATR-FTIR and CD spectroscopy, *Insect. Biochem. Mol. Biol.*, 31(9), 877–885, 2001.

269. Bandwar, R. P., Sastry, M. D., Kadam, R. M., and Rao, C. P., Transition-metal saccharide chemistry: Synthesis and characterization of D-glucose, D-fructose, D-galactose, D- xylose, D-ribose, and maltose complexes of Co(II), *Carbohydr. Res.*, 297(4), 333–339, 1997.

270. Sivakesava, S. and Irudayaraj, J., Determination of sugars in aqueous mixtures using mid-infrared spectroscopy, *Appl. Eng. Agric.*, 16(5), 543–550, 2000.

271. Hadden, J. M., Bloemendal, M., Haris, P. I., Srai, S. K. S., and Chapman, D., Fourier-transform infrared-spectroscopy and differential scanning calorimetry of transferrins—Human serum transferrin, rabbit serum transferrin and human lactoferrin, *Biochim. Biophys. Acta Protein Struct. Mol. Enzymol.*, 1205(1), 59–67, 1994.

272. Lagant, P., Vergoten, G., and Peticolas, W. L., On the use of ultraviolet resonance Raman intensities to elaborate molecular force fields: Application to nucleic acid bases and aromatic amino acid residues models, *Biospectroscopy*, 4, 379–393, 1998.

273. Takeuchi, H. and Harada, I., Normal coordinate analysis of the indole ring, *Spectrochim. Acta*, 42, 1069–1078, 1986.

274. Chia, N. C. and Mendelsohn, R., Conformational disorder in unsaturated phospholipids by FTIR spectroscopy, *Biochim. Biophys. Acta Biomembr.*, 1283(2), 141–150, 1996.

275. Gue, M., Dupont, V., Dufour, A., and Sire, O., Bacterial swarming: A biochemical time-resolved FTIR-ATR study of *Proteus mirabilis* swarm-cell differentiation, *Biochemistry*, 40(39), 11938–11945, 2001.

276. Oberreuter, H., Mertens, F., Seiler, H., and Scherer, S., Quantification of micro-organisms in binary mixed populations by Fourier transform infrared (FT-IR) spectroscopy, *Lett. Appl. Microbiol.*, 30(1), 85–89, 2000.

277. Schultz, C. P., Liu, K., Johnston, J. B., and Mantsch, H. H., Study of chronic lymphocytic leukemia cells by FT-IR spectroscopy and cluster analysis, *Leuk. Res.*, 20(8), 649–655, 1996.

278. Zeroual, W., Manfait, M., and Choisy, C., FT-IR spectroscopy study of perturbations induced by antibiotic on bacteria (*Escherichia coli*), *Pathol. Biol.*, 43(4), 300–305, 1995.

279. van der Mei, H. C., Naumann, D., and Busscher, H. J., Grouping of oral streptococcal species using Fourier-transform infrared spectroscopy in comparison with classical microbiological identification, *Arch. Oral. Biol.*, 38(11), 1013–1019, 1993.

280. Ramesh, J., Salman, A., Hammody, Z., Cohen, B., Gopas, J., Grossman, N., and Mordechai, S., FTIR microscopic studies on normal and H-Ras oncogene transfected cultured mouse fibroblasts, *Eur. Biophys. J. Biophys. Lett.*, 30(4), 250–255, 2001.

281. Ohgushi, H., Dohi, Y., Katuda, T., Tamai, S., Tabata, S., and Suwa, Y., In vitro bone formation by rat marrow cell culture, *J. Biomed. Mater. Res.*, 32(3), 333–340, 1996.

282. Hemmingsen, A., Allen, J. T., Zhang, S. F., Mortensen, J., and Spiteri, M. A., Early detection of ozone-induced hydroperoxides in epithelial cells by a novel infrared spectroscopic method, *Free Radic. Res.*, 31(5), 437–448, 1999.

283. Neviliappan, S., Fang Kan, L., Tiang Lee Walter, T., Arulkumaran, S., and Wong, P. T., Infrared spectral features of exfoliated cervical cells, cervical adenocarcinoma tissue, and an adenocarcinoma cell line (SiSo), *Gynecol. Oncol.*, 85(1), 170–174, 2002.

284. Morris, B. J., Lee, C., Nightingale, B. N., Molodysky, E., Morris, L. J., Appio, R., Sternhell, S., Cardona, M., Mackerras, D., and Irwig, L. M., Fourier transform infrared spectroscopy of dysplastic, papillomavirus-positive cervicovaginal lavage specimens, *Gynecol. Oncol.*, 56(2), 245–249, 1995.

285. Lowry, S. R., The analysis of exfoliated cervical cells by infrared microscopy, *Cell. Mol. Biol.*, 44(1), 169–177, 1998.

286. Wong, P. T. T., Lacelle, S., Fung, M. F. K., Senterman, M., and Mikhael, N. Z., Characterization of exfoliated cells and tissues from human endocervix and ectocervix by FTIR and ATR/FTIR spectroscopy, *Biospectroscopy*, 1(5), 357–364, 1995.

287. Melin, A. M., Perromat, A., and Deleris, G., Pharmacologic application of FTIR spectroscopy: Effect of ascorbic acid-induced free radicals on *Deinococcus radiodurans*, *Biospectroscopy*, 5(4), 229–236, 1999.

288. Zhou, J., Wang, Z., Sun, S., Liu, M., and Zhang, H., A rapid method for detecting conformational changes during differentiation and apoptosis of HL60 cells by Fourier-transform infrared spectroscopy, *Biotechnol. Appl. Biochem.*, 33(2), 127–132, 2001.

289. Pouliot, R., Germain, L., Auger, F. A., Tremblay, N., and Juhasz, J., Physical characterization of the stratum corneum of an in vitro human skin equivalent produced by tissue engineering and its comparison with normal human skin by ATR-FTIR spectroscopy and thermal analysis (DSC), *Biochim. Biophys. Acta-Mol. Cell Biol. Lipids*, 1439(3), 341–352, 1999.

290. Romano, S., Monici, M., Mazzinghi, P., Bernabei, P. A., and Fusi, F., Spectroscopic study of human leukocytes, *Phys. Med.*, 13, 291–295, 1997.

291. Benedetti, E., Bramanti, E., Papineschi, F., Vergamini, P., and Benedetti, E., An approach to the study of primitive thrombocythemia (PT) megakaryocytes by means of Fourier transform infrared microspectroscopy (FT-IR-M), *Cell. Mol. Biol. (Noisy-le-grand)*, 44(1), 129–139, 1998.

292. Wang, H. P., Wang, H. C., and Huang, Y. J., Microscopic FTIR studies of lung cancer cells in pleural fluid, *Sci. Total Environ.*, 204(3), 283–287, 1997.

293. Holman, H. Y., Martin, M. C., Blakely, E. A., Bjornstad, K., and McKinney, W. R., IR spectroscopic characteristics of cell cycle and cell death probed by synchrotron radiation based Fourier transform IR spectromicroscopy, *Biopolymers*, 57(6), 329–335, 2000.

294. Wood, B. R., Tait, B., and McNaughton, D., Fourier-transform infrared spectroscopy as a tool for detecting early lymphocyte activation: A new approach to histocompatibility matching, *Hum. Immunol.*, 61(12), 1307–1314, 2000.

295. Huleihel, M., Salman, A., Erukhimovitch, V., Ramesh, J., Hammody, Z., and Mordechai, S., Novel spectral method for the study of viral carcinogenesis in vitro, *J. Biochem. Biophys. Methods*, 50(2–3), 111–121, 2002.

296. Lasch, P., Pacifico, A., and Diem, M., Spatially resolved IR microspectroscopy of single cells, *Biopolymers*, 67(4–5), 335–338, 2002.

297. Huleihel, M., Talyshinsky, M., and Erukhimovitch, V., FTIR microscopy as a method for detection of retrovirally transformed cells, *Spectroscopy*, 15(2), 57–64, 2001.

298. Wood, B. R., Quinn, M. A., Tait, B., Ashdown, M., Hislop, T., Romeo, M., and McNaughton, D., FTIR microspectroscopic study of cell types and potential confounding variables in screening for cervical malignancies, *Biospectroscopy*, 4(2), 75–91, 1998.

299. Dukor, R. K., Liebman, M. N., and Johnson, B. L., A new, non-destructive method for analysis of clinical samples with FT-IR microspectroscopy. Breast cancer tissue as an example, *Cell Mol. Biol. (Noisy-le-grand)*, 44(1), 211–217, 1998.

300. Jackson, M., Mansfield, J. R., Dolenko, B., Somorjai, R. L., Mantsch, H. H., and Watson, P. H., Classification of breast tumors by grade and steroid receptor status using pattern recognition analysis of infrared spectra, *Cancer Detect. Prev.*, 23(3), 245–253, 1999.

301. Gao, T., Feng, J., and Ci, Y., Human breast carcinomal tissues display distinctive FTIR spectra: Implication for the histological characterization of carcinomas, *Anal. Cell Pathol.*, 18(2), 87–93, 1999.

302. Ci, Y. X., Gao, T. Y., Dong, J. Q., Kan, X., and Guo, Z. Q., FTIR assessment of the secondary structure of proteins in human breast benign and malignant tissues, *Chin. Sci. Bull.*, 44(24), 2215–2221, 1999.

303. Jackson, M., Choo, L. P., Watson, P. H., Halliday, W. C., and Mantsch, H. H., Beware of connective-tissue proteins—Assignment and implications of collagen absorptions in infrared-spectra of human tissues, *Biochim. Biophys. Acta-Mol. Basis Dis.*, 1270(1), 1–6, 1995.

304. Lee, L. S., Lin, S. Y., Chi, C. W., Liu, H. C., and Cheng, C. L., Non-destructive analysis of the protein conformational structure of human pituitary adenomas using reflectance FT-IR microspectroscopy, *Cancer Lett.*, 94(1), 65–69, 1995.

305. Lester, D. S., Kidder, L. H., Levin, I. W., and Lewis, E. N., Infrared microspectroscopic imaging of the cerebellum of normal and cytarabine treated rats, *Cell. Mol. Biol. (Noisy-le-grand)*, 44(1), 29–38, 1998.

306. Lewis, E. N., Gorbach, A. M., Marcott, C., and Levin, I. W., High-fidelity Fourier transform infrared spectroscopic imaging of primate brain tissue, *Appl. Spectrosc.*, 50(2), 263–269, 1996.

307. Wetzel, D. L., Slatkin, D. N., and Levine, S. M., FT-IR microspectroscopic detection of metabolically deuterated compounds in the rat cerebellum: A novel approach for the study of brain metabolism, *Cell Mol. Biol. (Noisy-le-grand)*, 44(1), 15–27, 1998.

308. Krusejarres, J. D., Janatsch, G., Gless, U., Marbach, R., and Heise, H. M., Glucose and other constituents of blood determined by ATR-FTIR- Spectroscopy, *Clin. Chem.*, 36(2), 401–402, 1990.

309. Liu, K. Z. and Mantsch, H. H., Simultaneous quantitation from infrared spectra of glucose concentrations, lactate concentrations, and lecithin/sphingomyelin ratios in amniotic fluid, *Am. J. Obstet. Gynecol.*, 180, 696–702, 1999.

310. Choo, L. P., Wetzel, D. L., Halliday, W. C., Jackson, M., LeVine, S. M., and Mantsch, H. H., In situ characterization of beta-amyloid in Alzheimer's diseased tissue by synchrotron Fourier transform infrared microspectroscopy, *Biophys. J.*, 71(4), 1672–1679, 1996.

311. Shaw, R. A., Low-Ying, S., Leroux, M., and Mantsch, H. H., Toward reagent-free clinical analysis: Quantitation of urine urea, creatinine, and total protein from the mid-infrared spectra of dried urine films, *Clin. Chem.*, 46(9), 1493–1495, 2000.

312. Shaw, R. A., Kotowich, S., Eysel, H. H., Jackson, M., Thomson, G. T., and Mantsch, H. H., Arthritis diagnosis based upon the near-infrared spectrum of synovial fluid, *Rheumatol. Int.*, 15(4), 159–165, 1995.

313. McIntosh, L. M., Jackson, M., Mantsch, H. H., Stranc, M. F., Pilavdzic, D., and Crowson, A. N., Infrared spectra of basal cell carcinomas are distinct from non-tumor-bearing skin components, *J. Invest. Dermatol.*, 112(6), 951–956, 1999.

314. Franck, P., Sallerin, J. L., Schroeder, H., Gelot, M. A., and Nabet, P., Rapid determination of fecal fat by Fourier transform infrared analysis (FTIR) with partial least squares regression and an attenuated total reflectance accessory, *Clin. Chem.*, 42(12), 2015–2020, 1996.

315. Sukuta, S. and Bruch, R., Factor analysis of cancer Fourier transform infrared evanescent wave fiber-optical (FTIR-FEW) spectra, *Lasers Surg. Med.*, 24(5), 382–388, 1999.

316. Liu, K. Z., Shaw, R. A., Man, A., Dembinski, T. C., and Mantsch, H. H., Reagent-free, simultaneous determination of serum cholesterol in HDL and LDL by infrared spectroscopy, *Clin. Chem.*, 48(3), 499–506, 2002.

317. Gotshal, Y., Simhi, R., Sela, B. A., and Katzir, A., Blood diagnostics using fiberoptic evanescent wave spectroscopy and neural networks analysis, *Sens. Actuat. B-Chem.*, 42(3), 157–161, 1997.

318. Shaw, R. A., Kotowich, S., Leroux, M., and Mantsch, H. H., Multianalyte serum analysis using mid-infrared spectroscopy, *Ann. Clin. Biochem.*, 35, 624–632, 1998.

319. Kneipp, J., Beekes, M., Lasch, P., and Naumann, D., Molecular changes of preclinical scrapie can be detected by infrared spectroscopy, *J. Neurosci.*, 22(8), 2989–2997, 2002.

320. Jennings, G., Bluck, L., Wright, A., and Elia, M., The use of infrared spectrophotometry for measuring body water spaces, *Clin. Chem.*, 45(7), 1077–1081, 1999.

321. Homan, J. A., Radel, J. D., Wallace, D. D., Wetzel, D. L., and Levine, S. M., Chemical changes in the photoreceptor outer segments due to iron induced oxidative stress: Analysis by Fourier transform infrared (FT-IR) microspectroscopy, *Cell. Mol. Biol. (Noisy-le-grand)*, 46(3), 663–672, 2000.

322. Afanasyeva, N. I. and Bruch, R. F., Biocompatibility of polymer surfaces interacting with living tissue, *Surf. Interface Anal.*, 27(4), 204–212, 1999.

323. Nishida, Y., Yoshida, S., Li, H. J., Higuchi, Y., Takai, N., and Miyakawa, I., FTIR spectroscopic analyses of human placental membranes, *Biopolymers*, 62(1), 22–28, 2001.

324. Saini, P., Lai, J. C., and Lu, D. R., FT-IR measurement of mercaptoundecahydrododecaborate in human plasma, *J. Pharm. Biomed. Anal.*, 12(9), 1091–1095, 1994.

325. Schultz, C. P., Liu, K. Z., Kerr, P. D., and Mantsch, H. H., In situ infrared histopathology of keratinization in human oral/oropharyngeal squamous cell carcinoma, *Oncol. Res.*, 10(5), 277–286, 1998.

326. Fukuyama, Y., Yoshida, S., Yanagisawa, S., and Shimizu, M., A study on the differences between oral squamous cell carcinomas and normal oral mucosas measured by Fourier transform infrared spectroscopy, *Biospectroscopy*, 5(2), 117–126, 1999.

327. Wu, J. G., Xu, Y. Z., Sun, C. W., Soloway, R. D., Xu, D. F., Wu, Q. G., Sun, K. H., Weng, S. F., and Xu, G. X., Distinguishing malignant from normal oral tissues using FTIR fiber-optic techniques, *Biopolymers*, 62(4), 185–192, 2001.

328. Kidder, L. H., Colarusso, P., Stewart, S. A., Levin, I. W., Appel, N. M., Lester, D. S., Pentchev, P. G., and Lewis, E. N., Infrared spectroscopic imaging of the biochemical modifications induced in the cerebellum of the Niemann-Pick type C mouse, *J. Biomed. Optics*, 4(1), 7–13, 1999.

329. Bromberg, P. S., Gough, K. M., and Dixon, I. M. C., Collagen remodeling in the extracellular matrix of the cardiomyopathic Syrian hamster heart as assessed by FTIR attenuated total reflectance spectroscopy, *Can. J. Chem.-Revue Canadienne De Chim.*, 77(11), 1843–1855, 1999.

330. LeVine, S. M. and Wetzel, D. L., Chemical analysis of multiple sclerosis lesions by FT-IR microspectroscopy, *Free Radic. Biol. Med.*, 25(1), 33–41, 1998.

331. Burmeister, J. J. and Arnold, M. A., Evaluation of measurement sites for noninvasive blood glucose sensing with near-infrared transmission spectroscopy, *Clin. Chem.*, 45(9), 1621–1627, 1999.

332. Magne, D., Pilet, P., Weiss, P., and Daculsi, G., Fourier transform infrared microspectroscopic investigation of the maturation of nonstoichiometric apatites in mineralized tissues: A horse dentin study, *Bone*, 29(6), 547–552, 2001.

333. Das, R. M., Ahmed, M. K., Mantsch, H. H., and Scott, J. E., FT-IR spectroscopy of methylmercury-exposed mouse lung, *Mol. Cell. Biochem.*, 145(1), 75–79, 1995.

334. Andrus, P. G. and Strickland, R. D., Cancer grading by Fourier transform infrared spectroscopy, *Biospectroscopy*, 4(1), 37–46, 1998.

335. Yano, K., Ohoshima, S., Gotou, Y., Kumaido, K., Moriguchi, T., and Katayama, H., Direct measurement of human lung cancerous and noncancerous tissues by fourier transform infrared microscopy: Can an infrared microscope be used as a clinical tool? *Anal. Biochem.*, 287(2), 218–225, 2000.

336. Severcan, F., Toyran, N., Kaptan, N., and Turan, B., Fourier transform infrared study of the effect of diabetes on rat liver and heart tissues in the C-H region, *Talanta*, 53(1), 55–59, 2000.

337. Kitai, T., Tanaka, A., Tokuka, A., Sato, B., Mori, S., Yanabu, N., Inomoto, T. et al. Intraoperative measurement of the graft oxygenation state in living related liver transplantation by near infrared spectroscopy, *Transpl. Int.*, 8(2), 111–118, 1995.

338. Jackson, M., Ramjiawan, B., Hewko, M., and Mantsch, H. H., Infrared microscopic functional group mapping and spectral clustering analysis of hypercholesterolemic rabbit liver, *Cell. Mol. Biol. (Noisy-le-grand)*, 44(1), 89–98, 1998.

339. Lyman, D. J., Murray-Wijelath, J., Ambrad-Chalela, E., and Wijelath, E. S., Vascular graft healing. II. FTIR analysis of polyester graft samples from implanted bi-grafts, *J. Biomed. Mater. Res.*, 58(3), 221–237, 2001.

340. Lyman, D. J. and Murray-Wijelath, J., Vascular graft healing: I. FTIR analysis of an implant model for studying the healing of a vascular graft, *J. Biomed. Mater. Res.*, 48(2), 172–186, 1999.

341. Mendelsohn, R., Paschalis, E. P., and Boskey, A. L., Infrared spectroscopy, microscopy, and microscopic imaging of mineralizing tissues: Spectra-structure correlations from human iliac crest biopsies, *J. Biomed. Opt.*, 4(1), 14–21, 1999.

342. Mendelsohn, R., Paschalis, E. P., Sherman, P. J., and Boskey, A. L., IR microscopic imaging of pathological states and fracture healing of bone, *Appl. Spectrosc.*, 54(8), 1183–1191, 2000.

343. Liu, K. Z., Jackson, M., Sowa, M. G., Ju, H. S., Dixon, I. M. C., and Mantsch, H. H., Modification of the extracellular matrix following myocardial infarction monitored by FTIR spectroscopy, *Biochim. Biophys. Acta-Mol. Basis Dis.*, 1315(2), 73–77, 1996.

344. Manoharan, R., Baraga, J. J., Rava, R. P., Dasari, R. R., Fitzmaurice, M., and Feld, M. S., Biochemical analysis and mapping of atherosclerotic human artery using FT-IR microspectroscopy, *Atherosclerosis*, 103(2), 181–193, 1993.

345. Liu, K. Z., Dixon, I. M., and Mantsch, H. H., Distribution of collagen deposition in cardiomyopathic hamster hearts determined by infrared microscopy, *Cardiovasc. Pathol.*, 1999, 8(1), 41–47.

346. Thomas, N., Goodacre, R., Timmins, E. M., Gaudoin, M., and Fleming, R., Fourier transform infrared spectroscopy of follicular fluids from large and small antral follicles, *Hum. Reprod.*, 15(8), 1667–1671, 2000.

347. Magne, D., Weiss, P., Bouler, J. M., Laboux, O., and Daculsi, G., Study of the maturation of the organic (type I collagen) and mineral (nonstoichiometric apatite) constituents of a calcified tissue (dentin) as a function of location: A Fourier transform infrared microspectroscopic investigation, *J. Bone Miner. Res.*, 16(4), 750–757, 2001.

348. Paschalis, E. P., Verdelis, K., Doty, S. B., Boskey, A. L., Mendelsohn, R., and Yamauchi, M., Spectroscopic characterization of collagen cross-links in bone, *J. Bone Min. Res.*, 16(10), 1821–1828, 2001.

349. Bindig, U., Winter, H., Wasche, W., Zelianeos, K., and Muller, G., Fiber-optical and microscopic detection of malignant tissue by use of infrared spectrometry, *J. Biomed. Opt.*, 7(1), 100–108, 2002.

350. Chiriboga, L., Xie, P., Yee, H., Zarou, D., Zakim, D., and Diem, M., Infrared spectroscopy of human tissue. IV. Detection of dysplastic and neoplastic changes of human cervical tissue via infrared microscopy, *Cell Mol. Biol. (Noisy-le-grand)*, 44(1), 219–229, 1998.

351. Diem, M., Chiriboga, L., and Yee, H., Infrared spectroscopy of human cells and tissue. VIII. Strategies for analysis of infrared tissue mapping data and applications to liver tissue, *Biopolymers*, 57(2), 282–290, 2000.

352. Argov, S., Ramesh, J., Salman, A., Sinelnikov, I., Goldstein, J., Guterman, H., and Mordechai, S., Diagnostic potential of Fourier-transform infrared microspectroscopy and advanced computational methods in colon cancer patients, *J. Biomed. Opt.*, 7(2), 248–254, 2002.

353. Camacho, N. P., West, P., Torzilli, P. A., and Mendelsohn, R., FTIR microscopic imaging of collagen and proteoglycan in bovine cartilage, *Biopolymers*, 62(1), 1–8, 2001.

354. Schultz, C. P., Eysel, H. H., Mantsch, H. H., and Jackson, M., Carbon dioxide in tissues, cells, and biological fluids detected by FTIR spectroscopy, *J. Phys. Chem.*, 100(16), 6845–6848, 1996.

355. Potter, K., Kidder, L. H., Levin, I. W., Lewis, E. N., and Spencer, R. G. S., Imaging of collagen and proteoglycan in cartilage sections using Fourier transform infrared spectral imaging, *Arthritis Rheum.*, 44(4), 846–855, 2001.

356. Kidder, L. H., Kalasinsky, V. F., Luke, J. L., Levin, I. W., and Lewis, E. N., Visualization of silicone gel in human breast tissue using new infrared imaging spectroscopy, *Nat. Med.*, 3(2), 235–237, 1997.

357. Ali, S. R., Johnson, F. B., Luke, J. L., and Kalasinsky, V. F., Characterization of silicone breast implant biopsies by Fourier transform infrared mapping, *Cell. Mol. Biol. (Noisy-le-grand)*, 44(1), 75–80, 1998.

358. Brancaleon, L., Bamberg, M. P., Sakamaki, T., and Kollias, N., Attenuated total reflection-Fourier transform infrared spectroscopy as a possible method to investigate biophysical parameters of stratum corneum in vivo, *J. Invest. Dermatol.*, 116(3), 380–386, 2001.

359. George, J. S., Lewine, J. D., Goggin, A. S., Dyer, R. B., and Flynn, E. R., IR imaging of a monkey's head: Local temperature changes in response to somatosensory stimulation, *Adv. Exp. Med. Biol.*, 333, 125–136, 1993.

360. Rausch, M. and Eysel, U. T., Visualization of ICBF changes during cortical infarction thermo encephaloscopy, *Neuroreport*, 7, 2603–2606, 1996.

361. Uemura, T., Nishida, K., Ichinose, K., Shimoda, S., and Shichiri, M., Non-invasive blood glucose measurement by Fourier transform spectroscopic analysis through the mucous membrane of the lip: Application of chalcogenide optical fiber system, *Front. Med. Biol. Eng.*, 9(2), 137–153, 1999.

362. Wielgus, A. R. and Sarna, T., Ascorbate enhances photogeneration of hydrogen peroxide mediated by the iris melanin. *Photochem. Photobiol.*, 8(3), 683–691, 2008.

363. Brooks, A., Afanasyeva, N. I., Makhine, V., Bruch, R. F., Kolyakov, S. F., Artjushenko, S., and Butvina, L. N., New method for investigations of normal human skin surfaces in vivo using fiber-optic evanescent wave Fourier transform infrared spectroscopy (FEW-FTIR), *Surf. Interf. Anal.*, 27(4), 221–229, 1999.

364. Smielewski, P., Kirkpatrick, P., Minhas, P., Pickard, J. D., and Czosnyka, M., Can cerebrovascular reactivity be measured with near-infrared spectroscopy? *Stroke*, 26(12), 2285–2292, 1995.

365. Snieder, M. and Hansen, W. G., Crystal effect on penetration depth in attenuated total reflectance Fourier-transform infrared study of human skin, *Mikrochim. Acta*, 677–678, 1997.

366. Glice, M. M., Les, A., and Bajdor, K., IR, Raman and theoretical *ab initio* RHF study of aminoglutethimide—An anticancer drug, *J. Mol. Struct.*, 450, 141–153, 1998.

367. Buttler, C. A., Cooney, R. P., and Denny, W. A., Raman-spectroscopic studies of amsacrine, *Appl. Spectrosc.*, 46, 1540–1544, 1992.

368. Butler, C. A., Cooney, R. P., and Denny, W. A., Resonance Raman-study of the binding of the anticancer drug amsacrine to DNA, *Appl. Spectrosc.*, 48, 822–826, 1994.

369. Chourpa, I., Morjani, H., Riou, J. F., and Manfait, M., Intracellular interactions of antitumour drug amsacrine (m-AMSA) as revealed by surface-enhanced Raman spectroscopy, *FEBS Lett.*, 397, 61–64, 1996.

370. Michalska, D., The Raman and normal coordinate analysis of 3-(N-phenylacetylamino)-2,6-piperidinedione, antineoplastin-A10, the new antitumor drug, *Spectrochim. Acta A-Mol. Biomol. Spectrosc.*, 49, 303–314, 1993.

371. Torreggiani, A., Fagnano, C., and Fini, G., Involvement of lysine and tryptophan side-chains in the biotin-avidin interaction, *J. Raman Spectrosc.*, 28, 23–27, 1997.

372. Torreggiani, A. and Fini, G., Raman spectroscopic studies of ligand-protein interactions: The binding of biotin analogues by avidin, *J. Raman Spectrosc.*, 29, 229–236, 1998.

373. Chiang, H. P., Mou, B., Li, K. P., Chiang, P., Wang, D., Lin, S. J., and Tse, W. S., FT-Raman, FT-IR and normal-mode analysis of carcinogenic polycyclic aromatic hydrocarbons, Part II—A theoretical study of the transition states of oxygenation of benzo(a)pyrene (BaP), *J. Raman. Spectrosc.*, 32, 53–58, 2001.

374. Rajani, C., Kincaid, J. R., and Petering, D. H., A systematic approach toward the analysis of drug-DNA interactions using Raman spectroscopy: The binding of metal-free bleomycins A(2) and B-2 to calf thymus DNA, *Biopolymers*, 52, 110–128, 1999.

375. Weselucha-Birczynska, A. and Nakamoto, K., Study of the antimalarial drug cinchonine with nucleic acids by Raman spectroscopy, *J. Raman Spectrosc.*, 27, 915–919, 1996.

376. Cavigiolio, G., Bendetto, L., Boccaleri, E., Colangelo, D., Viano, I., and Osella, D., Pt(II) complexes with different N-donor ligands for specific inhibition of telomerase, *Inorg. Chim. Acta*, 305, 61–68, 2000.

377. Hoffman, G. M., Torres, A., and Forster, H. V., Validation of a volumeless breath-by-breath method for measurement of respiratory quotient, *J. Appl. Physiol.*, 75, 1903–1910, 1993.

378. Beljebbar, A., Morjani, H., Angibousat, J. F., Sockalingum, G. D., Polissiou, M., and Manfait, M., Molecular and cellular interaction of the differentiating antitumour agent dimethylcrocetin with nuclear retinoicd acid receptor as studied by near-infrared and visible SERS spectroscopy, *J. Raman Spectrosc.*, 28, 159–163, 1997.

379. Graham, D., Smith, W. E., Linacre, A. M. T., Munro, C. H., Watson, N. D., and White, P. C., Selective detection of deoxyribonucleic acids at ultralow concentrations by SERRS, *Anal. Chem.*, 69, 4703–4707, 1997.

380. Isola, N. R., Stokes, D. L., and Vo-Dinh, T., Surface-enhanced Raman gene rpobe for HIV detection, *Anal. Chem.*, 70, 1352–1356, 1998.

381. Vo-Dinh, T., Houke, K., and Stokes, D. L., Surface-enhanced Raman gene probes, *Anal. Chem.*, 66, 3379–3383, 1994.

382. Pal, B. and Bajpai, P. K., Spectroscopic characterization of gelonin—Assignments secondary structure and thermal denaturation, *Ind. J. Biochem. Biophys.*, 35, 166–171, 1998.

383. Denninger, J. W., Schelvis, J. P. M., Brandish, P. E., Zhao, Y., Babcock, G. T., and Marletta, M. A., Interaction of soluble guanylate cyclase with YC-1: Kinetic and resonance Raman studies, *Biochemistry*, 39, 4191–4198, 2000.

384. Schelvis, J. P. M., Zhao, Y., Marletta, M. A., and Babcock, G. T., Resonance Raman characterization of the heme domain of soluble guanylate cyclase, *Biochemistry*, 37, 16289–16297, 1998.

385. Zhao, Y., Schelvis, J. P. M., Babcock, G. T., Marletta, M. A., Identification of histidine 105 in the beta 1 subunit of soluble guanylate cyclase as the heme proximal ligand, *Biochemistry*, 37(13), 4502–4509, 1998.

386. Fan, B. C., Gupta, G., Danziger, R. S., Friedman, J. M., and Rousseau, D. L., Resonance Raman characterization of soluble gunylate cyclase expressed from baculovirous, *Biochemistry*, 37, 1178–1184, 1998.

387. Li, Z. Q., Li, X. Y., Sheu, F. S., Chen, D. M., and Yu, N. T., Isolation, purification and characterization of soluble guanylate cyclase from bovine lung, *Chem. Res. Chin. Univ.*, 13, 111, 1997.

388. Tomita, T., Ogura, T., Tsuyama, S., Imai, Y., and Kitagawa, T., Effects of GTP on bound nitric oxide of soluble guanylate cyclase pobed by resonance Raman spectroscopy, *Biochemistry*, 36, 10155, 1997.

389. Raussens, V., Pezolet, M., Ruysschaert, J. M., and Goormaghtigh, E., Structural difference in the H+,K+-ATPase between the E1 and E2 conformations—An attenuated total reflection infrared spectroscopy, UV circular dichroism and Raman spoectroscopy study, *Eur. J. Biochem.*, 262, 176–183, 1999.

390. Rauseens, V., De Jongh, H., Pezelot, H., Ruysschaert, J. M., and Goormaghtigh, E., Secondary structure of the intact H+,K+-ATPase and of its membrane-embedded region—An attenuated total reflection infrared spectroscopy, circular dichroism and Raman spectroscopic study, *Eur. J. Biochem.*, 252, 261–267, 1998.

391. Vogel, K. M., Spiro, T. G., Shelver, D., Thorsteinsson, M. V., and Roberts, G. P., Resonance Raman evidence for a novel charge relay mechanism of the CO-dependent heme protein transcription factor CooA, *Biochemistry*, 38, 2679–2687, 1999.

392. Walters, M. A., Leung, Y. C., Blumenthal, N. C., LeGeros, R. Z., and Konsker, K. A., A Raman and infrared spectroscopic investigation of biological hydroxyapatite, *J. Inorg. Biochem.*, 39, 193–200, 1990.

393. Morjani, H., Riou, J. F., Nabiev, I., Lavelle, F., and Manfait, M., Molecular and cellular interactions between intoplicine, DNA, and topoisomerase-II studies by surface-enhanced Raman scattering spectroscopy, *Cancer Res.*, 53, 4784–4790, 1993.

394. Lord, R. C. and Yu, N. T., Laser-excited Raman spectroscopy of biomolecules I. Native lysozyme and its constituent amino acids, *J. Mol. Biol.*, 50, 509, 1970.

395. Chen, Y. Q., Gulotta, M., Cheung, H. T. A., and Callender, R., Ligth activates reduction of metho-trexane by NADPH in the ternary complex with *Escherichia coli* dihydrofolate reductase, *Photochem. Photobiol.*, 69, 77–85, 1999.

396. Carey, P. R., Raman spectroscopy in enzymology: The first 25 years, *J. Raman Spectrosc.*, 29, 7–14, 1998.

397. Callender, R., Deng, H., and Gilmanshin, R., Raman difference studies of protein structure and folding, enzymatic catalysis and ligand binding, *J. Raman Spectrosc.*, 29, 15–21, 1998.

398. Zhao, X. and Spiro, T. G., Ultraviolet resonance Raman spectroscopy of hemoglobin with 200 and 2121 nm excitation: H-bonds of tyrosines and prolines, *J. Raman Spectrosc.*, 29, 49–55, 1998.

399. Overman, S. A. and Thomas, G. J., Raman markers of nonaromatic side chains in an alpha-helix assembly: Ala, Asp, Glu, Gly, Ile, Leu, Lys, Ser, and Val residues of phage fd subunits, *Biochemistry*, 38, 4018–4027, 1999.

400. Tsuboi, M., Overman, S. A., and Thomas, G. J., Orientation of tryptophan-26 in coat protein sub-units in the filamentous virus Ff bny polarized Raman microspectroscopy, *Biochemistry*, 35, 10403–10410, 1996.

401. Overman, S. A. and Thomas, G. J., Amide modes of the alpha-helix: Raman spectroscopy of fila-mentous virus Fd containing peptid C-13 and H-2 labels in coat protein subunits, *Biochemistry*, 37, 5654–5665, 1998.

402. Overman, S. A. and Thomas, G. J., Novel vibrational assignments for proteins from Raman spectra of viruses, *J. Raman Spectrosc.*, 29, 23–29, 1998.

403. Torreggiani, A. and Fini, G., Drug-antiserum molecular interactions: A Raman spectroscopic study, *J. Raman Spectrosc.*, 30, 295–300, 1999.

404. Beattie, J. R., Maguire, C., Gilchrist, S., Barrett, L. J., Cross, C. E., Possmayer, F., Ennis, M. et al., The use of Raman microscopy to determine and localize vitamin E in biological samples. *FASEB J.*, 21(3), 766–776, 2007.

405. Dai, L., Douglas, E. P., and Gower, L. B., Compositional analysis of a polymer-induced liquid-pre-cursor (PILP) amorphous $CaCO_3$ phase, *J. Non-Cryst. Solids*, 354(17), 1845–1854, 2008.

406. Suzuki, K., Miura, T., and Takeuchi, H., Inhibitory effect of copper(II) on zinc(II)-induced aggrega-tion of amyloid [beta]-peptide, *Biochem. Biophys. Res. Commun.*, 285(4), 991–996, 2001.

407. Gosal, W. S., Clark, A. H., Pudney, P. D. A., and Ross-Murphy, S. B., Novel amyloid fibrillar networks derived from a globular protein:B-lactoglobulin, *Langmuir*, 18(19), 7174–7181, 2002.

408. Souza, G. R., Levin, C. S., Hajitou, A., Pasqualini, R., Arap, W., and Miller, J. H., In vivo detection of gold-midazole self-assembly complexes: NIR-SERS signal reporters, *Anal. Chem.*, 78(17), 6232–6237, 2006.

409. Dion, A., Berno, B., Hall, G., and Filiaggi, M. J., The effect of processing on the structural char-acteristics of vancomycin-loaded amorphous calcium phosphate matrices, *Biomaterials*, 26(21), 4486–4494, 2005.

410. Chen, F., Wang, K., and Liu, C., Crystalline structure and its effects on the degradation of linear calcium polyphosphate bone substitute, *Appl. Surf. Sci.*, 255(2), 270–272, 2008.

411. Burckbuchler, V., Wintgens, V. R., Leborgne, C., Lecomte, S., Leygue, N., Scherman, D., Kichler, A., and Amiel, C., Development and characterization of new cyclodextrin polymer-based DNA deliv-ery systems, *Bioconj. Chem.*, 19(12), 2311–2320, 2008.

412. Tripisciano, C. and Borowiak-Palen, E., Cisplatin functionalized single-walled carbon nanotubes, *Phys. Status Solidi B-Basic Solid State Phys.*, 245(10), 1979–1982, 2008.

413. Cheng, W. T., Liu, M. T., Liu, H. N., and Lin, S. Y., Micro-Raman spectroscopy used to identify and grade human skin pilomatrixoma, *Microsc. Res. Techn.*, 68(2), 75–79, 2005.

414. Zhang, C., Smirnov, A. I., Hahn, D., and Grebel, H., Surface enhanced Raman scattering of biospecies on anodized aluminum oxide films, *Chem. Phys. Lett.*, 440(4–6), 239–243, 2007.

415. Huston, W. M., Andrew, C. R., Servid, A. E., McKay, A. L., Leech, A. P., Butler, C. S., and Moir, J. W. B., Heterologous overexpression and purification of cytochrome c', from *Rhodobacter capsulatus* and a mutant (K42E) in the dimerization region. Mutation does not alter oligomerization but impacts the heme iron spin state and nitric oxide binding properties, *Biochemistry*, 45(14), 4388–4395, 2006.

416. Sawai, H., Kawada, N., Yoshizato, K., Nakajima, H., Aono, S., and Shiro, Y., Characterization of the heme environmental structure of cytoglobin, a fourth globin in humans, *Biochemistry*, 42(17), 5133–5142, 2003.

417. Santini, A. and Miletic, V., Quantitative micro-Raman assessment of dentine demineralization, adhesive penetration, and degree of conversion of three dentine bonding systems, *Eur. J. Oral Sci.*, 116(2), 177–183, 2008.

418. Elmore, B. O., Bollinger, J. A., and Dooley, D. M., Human kidney diamine oxidase: Heterologous expression, purification, and characterization, *J. Biol. Inorg. Chem.*, 7(6), 565–579, 2002.

419. Vrána, O., Masek, V., Drazan, V., and Brabec, V., Raman spectroscopy of DNA modified by intrastrand cross-links of antitumor cisplatin, *J. Struct. Biol.*, 159(1), 1–8, 2007.

420. Barreto, W. J., Barreto, S. R. G., Santos, M. A., Schimidt, R., Paschoal, F. M. M., Mangrich, A. S., and deOliveira, L. F. C., Interruption of the MnO_2 oxidative process on dopamine and -dopa by the action of S2O32, *J. Inorg. Biochem.*, 84(1–2), 89–96, 2001.

421. Lu, Y., Sousa, A., Franco, R., Mangravita, A., Ferreira, G. C., Moura, I., and Shelnutt, J. A., Binding of protoporphyrin ix and metal derivatives to the active site of wild-type mouse ferrochelatase at low porphyrin-to-protein ratios, *Biochemistry*, 41(26), 8253–8262, 2002.

422. Shafer-Peltier, K. E., Haynes, C. L., Glucksberg, M. R., and Van Duyne, R. P., Toward a glucose biosensor based on surface-enhanced raman scattering, *J. Am. Chem. Soc.*, 125(2), 588–593, 2003.

423. Katz, A., Kruger, E. F., Minko, G., Liu, C. H., Rosen, R. B., and Alfano, R. R., Detection of glutamate in the eye by Raman spectroscopy, *J. Biomed. Opt.*, 8(2), 167–172, 2003.

424. Hasse, S., Gibbons, N. C. J., Rokos, H., Marles, L. K., and Schallreuter, K. U., Perturbed 6-tetrahydrobiopterin recycling via decreased dihydropteridine reductase in vitiligo: More evidence for H_2O_2 stress, *J. Invest. Dermatol.*, 122(2), 307–313, 2004.

425. Rodnenkov, O. V., Luneva, O. G., Ulyanova, N. A., Maksimov, G. V., Rubin, A. B., Orlov, S. N., and Chazov, E. I., Erythrocyte membrane fluidity and haemoglobin haemoporphyrin conformation: Features revealed in patients with heart failure, *Pathophysiology*, 11(4), 209–213, 2005.

426. de Vitry, C., Desbois, A., Redeker, V., Zito, F., and Wollman, F.-A., Biochemical and spectroscopic characterization of the covalent binding of heme to cytochrome b6, *Biochemistry*, 43(13), 3956–3968, 2004.

427. Marvin, K. A., Kerby, R. L., Youn, H., Roberts, G. P., and Burstyn, J. N., The transcription regulator RcoM-2 from *Burkholderia xenovorans* is a cysteine-ligated hemoprotein that undergoes a redox-mediated ligand switch, *Biochemistry*, 47(34), 9016–9028, 2008.

428. Wainwright, L. M., Wang, Y., Park, S. F., Yeh, S.-R., and Poole, R. K., Purification and spectroscopic characterization of Ctb, a group III truncated hemoglobin implicated in oxygen metabolism in the food-borne pathogen *Campylobacter jejuni*, *Biochemistry*, 45(19), 6003–6011, 2006.

429. Wood, B. R., Tait, B., and McNaughton, D., Micro-Raman characterisation of the R to T state transition of haemoglobin within a single living erythrocyte, *Biochim. Biophys. Acta (BBA)—Mol. Cell Res.*, 1539(1–2), 58–70, 2001.

430. Gong, B., Chen, J.-H., Chase, E., Chadalavada, D. M., Yajima, R., Golden, B. L., Bevilacqua, P. C., and Carey, P. R., Direct measurement of a pKa near neutrality for the catalytic cytosine in the genomic HDV ribozyme using Raman crystallography, *J. Am. Chem. Soc.*, 129(43), 13335–13342, 2007.

431. Juszczak, L. J., Manjula, B., Bonaventura, C., Acharya, S. A., and Friedman, J. M., UV resonance Raman study of beta93-modified hemoglobin A: Chemical modifier-specific effects and added influences of attached poly(ethylene glycol) chains, *Biochemistry*, 41(1), 376–385, 2002.

432. Samelson-Jones, B. J. and Yeh, S.-R., Interactions between nitric oxide and indoleamine 2,3-dioxygenase, *Biochemistry*, 45(28), 8527–8538, 2006.

433. Dzwolak, W., Insulin amyloid fibrils form an inclusion complex with molecular iodine: A misfolded protein as a nanoscale scaffold, *Biochemistry*, 46(6), 1568–1572, 2007.

434. Rinia, H. A., Burger, K. N. J., Bonn, M., and Müller, M., Quantitative label-free imaging of lipid composition and packing of individual cellular lipid droplets using multiplex CARS microscopy, *Biophys. J.*, 95(10), 4908–4914, 2008.

435. Scott, R., Seli, E., Miller, K., Sakkas, D., Scott, K., and Burns, D. H., Noninvasive metabolomic profiling of human embryo culture media using Raman spectroscopy predicts embryonic reproductive potential: A prospective blinded pilot study, *Fertil. Steril.*, 90(1), 77–83, 2008.

436. Zhu, H., Larade, K., Jackson, T. A., Xie, J. X., Ladoux, A., Acker, H., Berchner-Pfannschmidt, U. et al., NCB5OR is a novel soluble NAD(P)H reductase localized in the endoplasmic reticulum, *J. Biol. Chem.*, 279(29), 30316–30325, 2004.

437. Rousseau, D. L., Li, D., Couture, M., and Yeh, S.-R., Ligand-protein interactions in nitric oxide synthase, *J. Inorg. Biochem.*, 99(1), 306–323, 2005.

438. Brands, P. J. M., van de Poll, S. W. E., Quaedackers, J. A., Mutsaers, P. H. A., Puppels, G. J., van der Laarse, A., and de Voigt, M. J. A., Combined micro-PIXE and NIR Raman spectroscopic plaque characterisation in a human atherosclerotic aorta sample, *Nucl. Instr. Methods Phys. Res. Sect. B: Beam Interact. Mater. Atoms*, 181(1–4), 454–459, 2001.

439. Benevides, J. M., Juuti, J. T., Tuma, R., Bamford, D. H., and Thomas, G. J., Characterization of subunit-specific interactions in a double-stranded RNA virus: Raman difference spectroscopy of the 6 procapsid, *Biochemistry*, 41(40), 11946–11953, 2002.

440. Spencer, J. D., Gibbons, N. C. J., Rokos, H., Peters, E. M. J., Wood, J. M., and Schallreuter, K. U., Oxidative stress via hydrogen peroxide affects proopiomelanocortin peptides directly in the epidermis of patients with vitiligo, *J. Invest. Dermatol.*, 127(2), 411–420, 2007.

441. van de Poll, S. W. E., Romer, T. J., Puppels, G. J., and van der Laarse, A., Raman spectroscopy of atherosclerosis, *J. Cardiovasc. Risk*, 9(5), 255–261, 2002.

442. Costa, M. C., Sardo, M., Rolemberg, M. P., Ribeiro-Claro, P., Meirelles, A. J. A., Coutinho, J. A. P., and Krähenbühl, M. A., The solid-liquid phase diagrams of binary mixtures of consecutive, even saturated fatty acids: Differing by four carbon atoms, *Chem. Phys. Lipids*, 157(1), 40–50, 2009.

443. Synytsya, A., Alexa, P., de Boer, J., Loewe, M., Moosburger, M., Wurkner, M., and Volka, K., Raman spectroscopic study of serum albumins: An effect of proton- and gamma-irradiation, *J. Raman Spectrosc.*, 3(12), 1646–1655, 2007.

444. Ettrich, R., Kopecky, V., Hofbauerova, K., Baumruk, V., Novak, P., Pompach, P., Man, P. et al., Structure of the dimeric N-glycosylated form of fungal beta-N-acetylhexosaminidase revealed by computer modeling, vibrational spectroscopy, and biochemical studies, *BMC Struct. Biol.*, 7, 14, 2007.

445. Mohacek-Grosev, V., Vibrational analysis of hydroxyacetone, *Spectrochim. Acta A: Mol. Biomol. Spectrosc.*, 61(3), 477–484, 2005.

446. Gerth, H. U. V., Dammaschke, T., Schafer, E., and Zuchner, H., A three layer structure model of fluoridated, enamel containing CaF_2, Ca(OH)(2) and FAp, *Dent. Mater.*, 23(12), 1521–1528, 2007.

447. Shah, N. C., Lyandres, O., Walsh, J. T., Glucksberg, M. R., and Van Duyne, R. P., Lactate and sequential lactate-glucose sensing using surface-enhanced Raman spectroscopy, *Anal. Chem.*, 79(18), 6927–6932, 2007.

448. Synytsya, A., Alexa, P., Besserer, J., De Boer, J., Froschauer, S., Gerlach, R., Loewe, M. et al., Raman spectroscopy of tissue samples irradiated by protons, *Int. J. Radiat. Biol.*, 80(8), 581–591, 2004.

449. Chan, J. W., Motton, D., Rutledge, J. C., Keim, N. L., and Huser, T., Raman spectroscopic analysis of biochemical changes in individual triglyceride-rich lipoproteins in the pre- and postprandial state, *Anal. Chem.*, 77(18), 5870–5876, 2005.

450. Clarkson, J. and Smith, D. A., UV Raman evidence of a tyrosine in apo-human serum transferrin with a low pK(a) that is elevated upon binding of sulphate, *FEBS Lett.*, 503(1), 30–34, 2001.

451. NobIIa, P., Vieites, M., Parajón-Costa, B. S., Baran, E. J., Cerecetto, H., Draper, P., González, M. et al., Vanadium(V) complexes with salicylaldehyde semicarbazone derivatives bearing in vitro anti-tumor activity toward kidney tumor cells (TK-10): Crystal structure of [VVO2(5-bromosalicylaldehyde semicarbazone)], *J. Inorg. Biochem.*, 99(2), 443–451, 2005.

452. Wang, G. W., Yao, H. L., He, B. J., Peng, L. X., and Li, Y. Q., Raman micro-spectroscopy of single blood platelets, *Spectrosc. Spectr. Anal.*, 27(7), 1347–1350, 2007.

453. Ramser, K., Wenseleers, W., Dewilde, S., Van Doorslaer, S., and Moens, L., The combination of resonance Raman spectroscopy, optical tweezers and microfluidic systems applied to the study of various heme-containing single cells, *Spectrosc. Int. J.*, 22(4), 287–295, 2008.

454. Wang, G., Liu, X., and Ding, C., Phase composition and in-vitro bioactivity of plasma sprayed calcia stabilized zirconia coatings, *Surf. Coat. Tech.*, 202(24), 5824–5831, 2008.

455. Buschman, H. P., Deinum, G., Motz, J. T., Fitzmaurice, M., Kramer, J. R., van der Laarse, A., Bruschke, A. V., and Feld, M. S., Raman microspectroscopy of human coronary atherosclerosis: Biochemical assessment of cellular and extracellular morphologic structures in situ, *Cardiovasc. Pathol.*, 10(2), 69–82, 2001.

456. De Gelder, J., De Gussem, K., Vandenabeele, P., Vancanneyt, M., De Vos, P., and Moens, L., Methods for extracting biochemical information from bacterial Raman spectra: Focus on a group of structurally similar biomolecules—Fatty acids, *Anal. Chim. Acta*, 603(2), 167–175, 2007.

457. De Gelder, J., De Gussem, K., Vandenabeele, P., De Vos, P., and Moens, L., Methods for extracting biochemical information from bacterial Raman spectra: An explorative study on Cupriavidus metallidurans, *Anal. Chim. Acta*, 585(2), 234–240, 2007.

458. Grun, J., Manka, C. K., Nikitin, S., Zabetakis, D., Comanescu, G., Gillis, D., and Bowles, J., Identification of bacteria from two-dimensional resonant-Raman spectra, *Anal. Chem.*, 79(14), 5489–5493, 2007.

459. Jarvis, R. M., Law, N., Shadi, I. T., O'Brien, P., Lloyd, J. R., and Goodacre, R., Surface-enhanced Raman scattering from intracellular and extracellular bacterial locations, *Anal. Chem.*, 80(17), 6741–6746, 2008.

460. Brown, K. L., Palyvoda, O. Y., Thakur, J. S., Nehlsen-Cannarella, S. L., Fagoaga, O. R., Gruber, S. A., and Auner, G. W., Raman spectroscopic differentiation of activated versus non-activated T lymphocytes: An in vitro study of an acute allograft rejection model, *J. Immunol. Methods*, 340(1), 48–54, 2009.

461. Kah, J. C. Y., Kho, K. W., Lee, C. G. L., Sheppard, C. J. R., Shen, Z. X., Soo, K. C., and Olivo, M. C., Early diagnosis of oral cancer based on the surface plasmon resonance of gold nanoparticles, *Int. J. Nanomed.*, 2(4), 785–798, 2007.

462. Kano, H. and Hamaguchi, H. O., Coherent Raman imaging of human living cells using a supercontinuum light source, *Jpn. J. Appl. Phys.*, 46(10A), 6875–6877, 2007.

463. Kano, H., Molecular vibrational imaging of a human cell by multiplex coherent anti-Stokes Raman scattering microspectroscopy using a supercontinuum light source, *J. Raman Spectrosc.*, 39(11), 1649–1652, 2008.

464. Kim, B. S., Lee, C. C. I., Christensen, J. E., Huser, T. R., Chan, J. W., and Tarantal, A. F., Growth, differentiation, and biochemical signatures of rhesus monkey mesenchymal stem cells, *Stem Cells Dev.*, 17(1), 185–198, 2008.

465. Perna, G., Lastella, M., Lasalvia, M., Mezzenga, E., and Capozzi, V., Raman spectroscopy and atomic force microscopy study of cellular damage in human keratinocytes treated with HgCl$_2$, *J. Mol. Struct.*, 834–836, 182–187, 2007.

466. Swain, R. J., Kemp, S. J., Goldstraw, P., Tetley, T. D., and Stevens, M. M., Spectral monitoring of surfactant clearance during alveolar epithelial type II cell differentiation, *Biophys. J.*, 95(12), 5978–5987, 2008.

467. Tang, H. W., Yang, X. B. B., Kirkham, J., and Smith, D. A., Chemical probing of single cancer cells with gold nanoaggregates by surface-enhanced Raman scattering, *Appl. Spectrosc.*, 62(10), 1060–1069, 2008.

468. Yu, C. X., Gestl, E., Eckert, K., Allara, D., and Irudayaraj, J., Characterization of human breast epithelial cells by confocal Raman micro spectroscopy, *Cancer Detect. Prev.*, 30(6), 515–522, 2006.

469. Zhang, W. and Chu, P. K., Enhancement of antibacterial properties and biocompatibility of polyethylene by silver and copper plasma immersion ion implantation, *Surf. Coat. Technol.*, 203(5–7), 909–912, 2008.

470. Zheng, F., Qin, Y. J., and Chen, K., Sensitivity map of laser tweezers Raman spectroscopy for single-cell analysis of colorectal cancer, *J. Biomed. Opt.*, 12(3), 9, 2007.

471. Swain, R. J., Jell, G., and Stevens, M. A., Non-invasive analysis of cell cycle dynamics in single living cells with Raman micro-spectroscopy, *J. Cell. Biochem.*, 104(4), 1427–1438, 2008.

472. Chan, J. W., Taylor, D. S., Lane, S. M., Zwerdling, T., Tuscano, J., and Huser, T., Nondestructive identification of individual leukemia cells by laser trapping Raman spectroscopy, *Anal. Chem.*, 80(6), 2180–2187, 2008.

473. Krishna, C. M., Kegelaerl, G., Adt, I., Rubin, S., Kartha, V. B., Manfait, M., and Sockalingum, G. D., Combined Fourier transform infrared and Raman spectroscopic identification approach for identification of multidrug resistance phenotype in cancer cell lines, *Biopolymers*, 82(5), 462–470, 2006.

474. Mannie, M. D., McConnell, T. J., Xie, C., and Li, Y.-Q., Activation-dependent phases of T cells distinguished by use of optical tweezers and near infrared Raman spectroscopy, *J. Immunol. Methods*, 297(1–2), 53–60, 2005.

475. Jess, P. R. T., Smith, D. D. W., Mazilu, M., Dholakia, K., Riches, A. C., and Herrington, C. S., Early detection of cervical neoplasia by Raman spectroscopy, *Int. J. Cancer*, 121(12), 2723–2728, 2007.

476. Krishna, C. M., Prathima, N. B., Malini, R., Vadhiraja, B. M., Bhatt, R. A., Fernandes, D. J., Kushtagi, P., Vidyasagar, M. S., and Kartha, V. B., Raman spectroscopy studies for diagnosis of cancers in human uterine cervix, *Vib. Spectrosc.*, 41(1), 136–141, 2006.

477. Matthaus, C., Boydston-White, S., Miljkovic, M., Romeo, M., and Diem, M., Raman and infrared microspectral imaging of mitotic cells, *Appl. Spectrosc.*, 60(1), 1–8, 2006.

478. Shamsaie, A., Heim, J., Yanik, A. A., and Irudayaraj, J., Intracellular quantification by surface enhanced Raman spectroscopy, *Chem. Phys. Lett.*, 461(1–3), 131–135, 2008.

479. Konorov, S. O., Glover, C. H., Piret, J. M., Bryan, J., Schulze, H. G., Blades, M. W., and Turner, R. F. B., In situ analysis of living embryonic stem cells by coherent anti-stokes Raman microscopy, *Anal. Chem.*, 79(18), 7221–7225, 2007.

480. Chen, K., Qin, Y. J., Zheng, F., Sun, M. H., and Shi, D. R., Diagnosis of colorectal cancer using Raman spectroscopy of laser-trapped single living epithelial cells, *Opt. Lett.*, 31(13), 2015–2017, 2006.

481. Cheng, C. Y., Perevedentseva, E., Tu, J. S., Chung, P. H., Cheng, C. L., Liu, K. K., Chao, J. I., Chen, P. H., and Chang, C. C., Direct and in vitro observation of growth hormone receptor molecules in A549 human lung epithelial cells by nanodiamond labeling, *Appl. Phys. Lett.*, 90(16), 163903, 2007.

482. Chan, K. L. A., Zhang, G. J., Tomic-Canic, M., Stojadinovic, O., Lee, B., Flach, C. R., and Mendelsohn, R., A coordinated approach to cutaneous wound healing: Vibrational microscopy and molecular biology, *J. Cell. Mol. Med.*, 12(5B), 2145–2154, 2008.

483. Chowdhury, M. H., Gant, V. A., Trache, A., Baldwin, A., Meininger, G. A., and Cote, G. L., Use of surface-enhanced Raman spectroscopy for the detection of human integrins, *J. Biomed. Opt.*, 11(2), 8, 2006.

484. Etcheverry, S. B., Ferrer, E. G., Naso, L., Barrio, D. A., Lezama, L., Rojo, T., and Williams, P. A. M., Losartan and its interaction with copper(II): Biological effects, *Bioorg. Med. Chem.*, 15(19), 6418–6424, 2007.

485. Schroeder, P., Lademann, J., Darvin, M. E., Stege, H., Marks, C., Bruhnke, S., and Krutmann, J., Infrared radiation-induced matrix metalloproteinase in human skin: Implications for protection, *J. Invest. Dermatol.*, 128(10), 2491–2497, 2008.

486. Chan, J. W., Taylor, D. S., Zwerdling, T., Lane, S. M., Ihara, K., and Huser, T., Micro-Raman spectroscopy detects individual neoplastic and normal hematopoietic cells, *Biophys. J.*, 90(2), 648–656, 2006.

487. Short, M. A., Lam, S., McWilliams, A., Zhao, J. H., Lui, H., and Zeng, H. S., Development and preliminary results of an endoscopic Raman probe for potential in vivo diagnosis of lung cancers, *Opt. Lett.*, 33(7), 711–713, 2008.

488. Veilleux, I., Spencer, J. A., Biss, D. P., Cote, D., and Lin, C. P., In vivo cell tracking with video rate multimodality laser scanning microscopy, *IEEE J. Sel. Top. Quant. Electron.*, 14(1), 10–18, 2008.

489. Notingher, I., Verrier, S., Haque, S., Polak, J. M., and Hench, L. L., Spectroscopic study of human lung epithelial cells (A549) in culture: Living cells versus dead cells, *Biopolymers*, 72(4), 230–240, 2003.

490. Boydston-White, S., Romeo, M., Chernenko, T., Regina, A., Miljkovic, M., and Diem, M., Cell-cycle-dependent variations in FTIR micro-spectra of single proliferating HeLa cells: Principal component and artificial neural network analysis, *Biochim. Biophys. Acta Biomembr.*, 1758(7), 908–914, 2006.

491. Short, K. W., Carpenter, S., Freyer, J. P., and Mourant, J. R., Raman spectroscopy detects biochemical changes due to proliferation in mammalian cell cultures, *Biophys. J.*, 88(6), 4274–4288, 2005.

492. Naveke, A., Lapouge, K., Sturgis, J. N., Hartwich, G., Simonin, J., Scheer, H., and Robert, B., Resonace Raman spectroscopy of metal-substituted bacteriochlorophylls: Characterization of Raman bands sensitive to bacteriochlorin conformation, *J. Raman Spectrosc.*, 28(8), 599–604, 1997.

493. Manoharan, R., Ghiamati, E., Dalterio, R. A., Britton, K., Nelson, W. H., and Sperry, J. F., UV resonance Raman spectra of bacteria, bacterial spores, protoplasts, and calcium dipicolinate, *J. Microbiol. Methods*, 11, 1–15, 1990.

494. Ramanauskaite, R. B., SegersNolten, I. G. M. J., deGrauw, K. J., Sijtsema, N. M., vandermaas, L., Greve, J., Otto, C., and Figdor, C. G., Carotenoid levels in human lymphocytes, measured by Raman spectroscopy, *Pure Appl. Chem.*, 69, 2131–2134, 1997.

495. Schut, T. C. B., Puppels, G. J., Kraan, Y. M., Greve, J., VanderMaas, L. L. J., and Figdor, C. G., Intracellular carotenoid levels measured by Raman microspectroscopy: Comparison of lymphocytes from lung cancer patients and healthy individuals, *Int. J. Cancer*, 74, 20–25, 1997.

496. Rasmussen, R. E., Hammer-Wilson, M., and Berns, M. W., Mutation and chromatid exchange induction in Chinese hamster ovary (CHO) cells by pulsed excimer laser radiation at 193 nm and 308 nm and continuous UV radiation at 254 nm, *Photochem. Photobiol.*, 49, 413–418, 1989.

497. Yao, Y., Nelson, W. H., Hargraves, P., and Zhang, J., UV resonance Raman study of demoic acid—A marine neurotoxic amino acid, *Appl. Spectrosc.*, 51, 785–791, 1997.

498. Wen, Z. Q., Overman, S. A., and Thomas, G. J. Jr., Structure and interactions of the single-stranded DNA genome of filamentous virus FD: Investigation by ultraviolet resonance Raman spectroscopy, *Biochemistry*, 36, 7810–7820, 1997.

499. Wen, Z. Q. and Thomas, G. J. Jr., UV resonance Raman spectroscopy of DNA and protein constituents of viruses: Assignments and cross sections for excitations at 257, 244, 238, and 229 nm, *Biopolymers*, 45, 247–256, 1998.

500. Peticolas, W. L., Patapoff, T. W., Thomas, G. A., Postlewait, J., and Powell, J. W., Laser Raman spectroscopy of chromosomes in living eukaryotic cells: DNA polymorphism in vivo, *J. Raman Spectrosc.*, 27, 571–578, 1996.

501. Overman, S. A., Aubrey, K. L., Reilly, K. E., Osman, O., Hayes, S. J., Serwer, P., and Thomas, G. J., Conformation and interactions of the packaged double-stranded DNA genome of bacteriophage T7, *Biospectroscopy*, 4, S47–S56, 1998.

502. Puppels, G. J., Demul, F. F. M., Otto, C., Greve, J., Robetnicoud, M., Arndtjovin, D. J., and Jovin, T. M., Studying single living cells and chromosomes by confocal Raman microspectroscopy, *Nature*, 347, 301–303, 1990.

503. Puppels, G. J., Olminkhof, J. H., Segers-Nolten, G. M., Otto, C., de Mul, F. F., and Greve, J., Laser irradiation and Raman spectroscopy of single living cells and chromosomes: Sample degradation occurs with 514.5 nm but not with 660 nm laser light, *Exp. Cell. Res.*, 195, 361–367, 1991.

504. Puppels, G. J., Otto, C., Greve, J., Robert-Nicoud, M., and Arndt-Jovin, T. M., Raman microscpec-troscopy of low pH induced changes in DNA structure of polytene chromosomes, *Biochemistry*, 33, 3386–3395, 1994.

505. Doig, S. J. and Prendergast, F. G., Continuously tunable, quasi-continuous-wave source for ultravio-let resonanace Raman-spectroscopy, *Appl. Spectrosc.*, 49, 247–252, 1995.

506. Carey, P. R., Resonance Raman labels and Raman labels, *J. Raman Spectrosc.*, 29, 861–868, 1998.

507. Hoey, S., Brown, D. H., McConnell, A. A., Smith, W. E., Marabani, M., and Sturrock, R. D., Resonance Raman spectroscopy of hemoglobin in intact cells: A probe of oxygen uptake by erythrocytes in rheumatoid arthritis, *J. Inorg. Biochem.*, 34, 189–199, 1988.

508. Hawi, S. R., Campbell, W. B., Kajdacsy-Balla, A., Murphy, R., Adar, F., and Nithipatikom, K., Characterization of normal and malignant hepatocytes by Raman microspectroscopy, *Cancer Lett.*, 110, 35–40, 1996.

509. Lin, D. L., Tarnowski, C. P., Zhang, J., Dai, J. L., Rohn, E., Patel, A. H., Morris, M. D., and Keller, E. T., Bone metastatic LNCaP-derivative C4–2B prostate cancer cell line mineralizes in vitro, *Prostate*, 47, 212–221, 2001.

510. Yazdi, Y., Ramanujam, N., Lotan, R., Mitchell, M. F., Hittelman, W., and Richards-Kortum, R., Resonance Raman spectroscopy at 257 nm excitation of normal and malignant cultured breast and cervical cells, *Appl. Spectrosc.*, 53, 82–85, 1999.

511. Sureau, F., Chinsky, L., Amirand, C., Ballini, J. P., Duquesne, M., Laigle, A., Turpin, P. Y., and Vigny, P., An ultraviolet micro-Raman spectrometer- resonance Raman spectroscopy within signle living cells, *Appl. Spectrosc.*, 44, 1047–1051, 1990.

512. Freeman, T. L., Cope, S. E., Stringer, M. R., Cruse-Sawyer, J. E., Batchelder, D. N., and Brown, S. B., Raman spectroscopy for the determination of photosensitizer localization in cells, *J. Raman Spectrosc.*, 28, 641–643, 1997.

513. Freeman, T. L., Cope, S. E., Stringer, M. R., Cruse-Sawyer, J. E., Brown, S. B., Batchelder, D. N., and Birbeck, K., Investigation of the subcellular localization of zinc phthalocyanines by Raman map-ping, *Appl. Spectrosc.*, 52, 1257–1263, 1998.

514. Arzhantsev, S. Y., Chikishev, A. Y., Koroteev, N. I., Greve, J., Otto, C., and Sijtsema, N. M., Localization study of Co-phthalocyanines in cells by Raman micro(spectro)scopy, *J. Raman Spectrosc.*, 30, 205–208, 1999.

515. Bakker Schut, T. C., Puppels, G. J., Kraan, Y. M., Greve, J., van der Maas, L.L., and Figdor, C. G., Intracellular carotenoid levels measured by Raman microspectroscopy: Comparison of lympho-cytes from lung cancer patients and healthy individuals, *Int. J. Cancer*, 74, 20–25, 1997.

516. Puppels, G. J., Garritsen, H. S. P., Sergersnolten, G. M. J., Demul, F. F. M., and Grefe, J., Raman microscopic approach to the study of human granulocytes, *Biophys. J.*, 60, 1046–1056, 1991.

517. Puppels, G. J., Garritsen, H. S. P., Kummer, J. A., and Greve, J., Carotenoids located in human lympho-cyte subpopulations and natural killer cells by Raman microspectroscopy, *Cytometry*, 14, 251–256, 1993.

518. Grygon, C. A., Perno, J. R., Fodor, S. P. A., and Spiro, T. G., Ultraviolet resonance Raman spectros-copy as a probe of protein structure in the FD virus, *Biotechniques*, 6, 50–55, 1988.

519. Dinakarpandian, D., Shenoy, B., Pusztai-Carey, M., Malcolm, B. A., and Carey, P. C., Active site properties of the 3C proteinase from hepatitis A virus (a hybrid cisteine/serine protease) probed by Raman spectroscopy, *Biochemistry*, 38, 4943–4948, 1997.

520. Chadha, S., Manoharan, R., Moenne-Loccoz, P., Nelson, W. H., Peticolas, W. L., and Sperry, J. F., Comparison of the UV resonance Raman-spectra of bacteria, bacterial-cell walls, and ribosomes excited in the deep UV, *Appl. Spectrosc.*, 47, 38–43, 1993.

521. Tuma, R. and Thomas, G. J., Mechanisms of virus assembly probed by Raman spectroscopy: The icosahedral bacteriophage P22, *Biophys. Chem.*, 68, 17–31, 1997.

522. Feofanov, A. V., Grichine, A. I., Shitova, L. A., Karmakova, T. A., Yakubovskaya, R. I., Egret-Charlier, M., and Vigny, P., Confocal Raman microspectroscopy of thapthal in living cancer cells, *Biophys. J.*, 78, 499–512, 2000.

523. Nelson, W. H., Manoharan, R., and Sperry, J. F., UV resonance Raman studies of bacteria, *Appl. Spectrosc. Rev.*, 27, 67–124, 1992.

524. Dalterio, R. A., Nelson, W. H., Britt, D., Sperry, J., and Purcell, F. J., A resonance Raman microprobe study of chromobacteria in water, *Appl. Spectrosc.*, 40, 271–273, 1986.

525. Manoharan, R., Ghiamati, E., Chadha, S., Nelson, W. H., and Sperry, J. F., Effect of cultural conditions on deep UV resonance Raman spectra of bacteria, *Appl. Spectrosc.*, 47, 2145–2150, 1993.

526. Brennan, J. F., Wang, Y., Dassari, R. R., and Feld, S. S., Near-infrared Raman spectrometer systems for human tissue studies, *Appl. Spectrosc.*, 51, 201–208, 1997.

527. Dou, X. M., Yamaguchi, Y., Yamamoto, H., Doi, S., and Ozaki, Y., A highly sensitive compact Raman system without a spectrometer for quantitative analysis of biological samples, *Vibrat. Spectrosc.*, 14, 199–205, 1997.

528. Dou, X. M., Yamaguchi, Y., Yamamoto, H., Doi, S., and Ozaki, Y., Quantitaive analysis of metabolites in urine by anti-Stokes Raman spectroscopy, *Biospectroscopy*, 3, 113–120, 1997.

529. Dou, X., Yamagouchi, Y., Yamamato, H., Doi, S., and Ozaki, Y., Quantitative analysis of metabolites in urine using a highly precise, compact, near-infrared Raman spectrometer, *Vibrat. Spectrosc.*, 13, 83–89, 1996.

530. Williams, A. C., Edwards, H. G. M., Lawson, E. E., and Barry, B. W., Molecular interactions between the penetration enhancer 1,8-cineole and human skin, *J. Raman Spectrosc.*, 37(1–3), 361–366, 2006.

531. Molckovsky, A., Song, L.-M. W. K., Shim, M. G., Marcon, N. E., and Wilson, B. C., Diagnostic potential of near-infrared Raman spectroscopy in the colon: Differentiating adenomatous from hyperplastic polyps, *Gastrointest. Endosc.*, 57(3), 396–402, 2003.

532. Boustany, N. N., Absorption coefficient and purine photobleaching rate in colon mucosa during resonance Raman spectroscopy at 251 nm, *Appl. Opt.*, 40(34), 6396–6405, 2001.

533. Glenn, J. V., Beattie, J. R., Barrett, L., Frizzell, N., Thorpe, S. R., Boulton, M. E., McGarvey, J. J., and Stitt, A. W., Confocal Raman microscopy can quantify advanced glycation end product (AGE) modifications in Bruch's membrane leading to accurate, nondestructive prediction of ocular aging, *FASEB J.*, 21(13), 3542–3552, 2007.

534. Bertoluzza, A., Fagnano, C., Tinti, A., Morelli, M. A., Tosi, M. R., Maggi, G., and Marchetti, P. G., Raman and infrared spectroscopic study of the molecular characterization of the biocompatibility of prosthetic biomaterials, *J. Raman Spectrosc.*, 25, 109–114, 1994.

535. Calvo, N., Montres, R., and Laserna, J. J., Surface-enhanced Raman-spectrometry of ameloride on colloidal silver, *Anal. Chim. Acta*, 280, 263–268, 1993.

536. Alfono, R. R., Liu, C. H., Sha, W. L., Zhu, H. R., Akins, D. L., Cleary, J., Prudente, R., and Clemer, E., Human breast tissues studies by IR Fourier-transform Raman spectroscopy, *Lasers Life Sci.*, 4, 23–28, 1991.

537. Edwards, H. G. M., Farwell, D. W., Williams, A. C., Barry, B. W., and Rull, F., Novel spectroscopic deconvolution procedure for complex biological systems—Vibrational components in the FT-Raman spectra of ice-man and contemporary skin, *J. Chem. Soc. Faraday Trans.*, 91, 3883–3887, 1995.

538. Shim, M. G. and Wilson, B. C., The effects of ex vivo handling procedures on the near-infrared Raman spectra of normal mammalian tissues, *Photochem. Photobiol.*, 63, 662–671, 1996.

539. Sulk, R. A., Corcoran, R. C., and Carron, K. T., Surface enhanced Raman detection of amphetamine and methamphetamine by modification with 2-mercaptonicotinic acid, *Appl. Spectrosc.*, 53, 954–959, 1999.

540. Apetri, M. M., Maiti, N. C., Zagorski, M. G., Carey, P. R., and Anderson, V. E., Secondary structure of [alpha]-synuclein oligomers: Characterization by Raman and atomic force microscopy, *J. Mol. Biol.*, 355(1), 63–71, 2006.

541. Grynpas, M. D. and Omelon, S., Transient precursor strategy or very small biological apatite crystals? *Bone*, 41(2), 162–164, 2007.

542. Pasteris, J. D., Wopenka, B., Freeman, J. J., Rogers, K., Valsami-Jones, E., van der Houwen, J. A. M., and Silva, M. J., Lack of OH in nanocrystalline apatite as a function of degree of atomic order: Implications for bone and biomaterials, *Biomaterials*, 25(2), 229–238, 2004.

543. Dong, J., Atwood, C. S., Anderson, V. E., Siedlak, S. L., Smith, M. A., Perry, G., and Carey, P. R., Metal binding and oxidation of amyloid-B within isolated senile plaque cores: Raman microscopic evidence, *Biochemistry*, 42(10), 2768–2773, 2003.

544. Boustany, N., Crawford, J. M., Manoharan, R., Dasari, R. R., and Feld, M. S., Analysis of nucleotides and aromatic amino acids in normal and neoplastic colon mucosa by ultraviolet resonance Raman spectroscopy, *Lab. Invest.*, 79, 1201–1214, 1999.

545. Taddei, P., Tinti, A., Reggiani, M., and Fagnano, C., In vitro mineralization of bioresorbable poly([epsilon]-caprolactone)/apatite composites for bone tissue engineering: A vibrational and thermal investigation, *J. Mol. Struct.*, 744–747, 135–143, 2005.

546. Sajid, J., Elhaddoui, A., and Turrell, S., Fourier transform vibrational spectroscopic analysis of human cerebral tissue, *J. Raman Spectrosc.*, 28, 165–169, 1997.

547. van de Poll, S. W. E., Delsing, D. J. M., Jukema, J. W., Princen, H. M. G., Havekes, L. M., Puppels, G. J., and van der Laarse, A., Raman spectroscopic investigation of atorvastatin, amlodipine, and both on atherosclerotic plaque development in APOE*3 Leiden transgenic mice, *Atherosclerosis*, 164(1), 65–71, 2002.

548. Redd, D. C. B., Feng, Z. C., Yue, K. T., and Gansler, T. S., Raman spectroscopic characterization of human breast tissues: Implications for breast cancer diagnosis, *Appl. Spectrosc.*, 47, 787–791, 1993.

549. Nijssen, A., Schut, T. C. B., Heule, F., Caspers, P. J., Hayes, D. P., Neumann, M. H. A., and Puppels, G. J., Discriminating basal cell carcinoma from its surrounding tissue by Raman spectroscopy, *J. Invest. Dermatol.*, 119(1), 64–69, 2002.

550. Crow, P., Molckovsky, A., Stone, N., Uff, J., Wilson, B., and WongKeeSong, L. M., Assessment of fiberoptic near-infrared Raman spectroscopy for diagnosis of bladder and prostate cancer, *Urology*, 65(6), 1126–1130, 2005.

551. Larsson, K. and Hellgren, L., A study of the combined Raman and fluorescence scattering from human blood plasma, *Experientia*, 30, 481–483, 1974.

552. Saade, J., Pacheco, M. T. T., Rodrigues, M. R., and Silveira, L., Identification of hepatitis C in human blood serum by near-infrared Raman spectroscopy, *Spectrosc. Int. J.*, 22(5), 387–395, 2008.

553. Huong, P. V., New possibilities of Raman micro-spectroscopy, *Vibrat. Spectrosc.*, 11, 17–28, 1996.

554. Kubisz, L. and Polomska, M., FT NIR Raman studies on [gamma]-irradiated bone, *Spectrochim. Acta Mol. Biomol. Spectrosc.*, 66(3), 616–625, 2007.

555. Terada, N., Ohno, N., Saitoh, S., and Ohno, S., Application of, in vivo cryotechnique, to detect erythrocyte oxygen saturation in frozen mouse tissues with confocal Raman cryomicroscopy, *J. Struct. Biol.*, 163(2), 147–154, 2008.

556. Sahar, N. D., Hong, S. I., and Kohn, D. H., Micro- and nano-structural analyses of damage in bone, *Micron*, 36(7–8), 617–629, 2005.

557. Tarnowski, C. P., Ignelzi, M. A., Wang, W., Taboas, J. M., Goldstein, S. A., and Morris, M. D., Earliest mineral and matrix changes in force-induced musculoskeletal disease as revealed by Raman micro-spectroscopic imaging, *J. Bone Miner. Res.*, 19(1), 64–71, 2004.

558. Yerramshetty, J. S. and Akkus, O., The associations between mineral crystallinity and the mechanical properties of human cortical bone, *Bone*, 42(3), 476–482, 2008.

559. Kazanci, M., Wagner, H. D., Manjubala, N. I., Gupta, H. S., Paschalis, E., Roschger, P., and Fratzl, P., Raman imaging of two orthogonal planes within cortical bone, *Bone*, 41(3), 456–461, 2007.

560. Lopes, C. B., Pacheco, M. T. T., Silveira Jr, L., Duarte, J., Cangussú, M. C. T., and Pinheiro, A. L. B., The effect of the association of NIR laser therapy BMPs, and guided bone regeneration on tibial fractures treated with wire osteosynthesis: Raman spectroscopy study, *J. Photochem. Photobiol. B Biol.*, 89(2–3), 125–130, 2007.

561. Koljenovic, S., Bakker Schut, T. C., Wolthuis, R., Vincent, A. J. P. E., Hendriks-Hagevi, G., Santos, L., Kros, J. M., and Puppels, G. J., Raman spectroscopic characterization of porcine brain tissue using a single fiber-optic probe, *Anal. Chem.*, 79(2), 557–564, 2007.

562. Wolthuis, R., van Aken, M., Fountas, K., Robinson, J. S., Bruining, H. A., and Puppels, G. J., Determination of water concentration in brain tissue by Raman spectroscopy, *Anal. Chem.*, 73(16), 3915–3920, 2001.

563. Kneipp, J., Schut, T. B., Kliffen, M., Menke-Pluijmers, M., and Puppels, G., Characterization of breast duct epithelia: A Raman spectroscopic study, *Vibrat. Spectrosc.*, 32(1), 67–74, 2003.

564. Woo, M.-A., Lee, S.-M., Kim, G., Baek, J., Noh, M. S., Kim, J. E., Park, S. J. et al., Multiplex immunoassay using fluorescent-surface enhanced Raman spectroscopic dots for the detection of bronchioalveolar stem cells in murine lung, *Anal. Chem.*, 81(3), 1008, 2008.

565. Matousek, P. and Stone, N., Prospects for the diagnosis of breast cancer by noninvasive probing of calcifications using transmission Raman spectroscopy, *J. Biomed. Opt.*, 12(2), 8, 2007.

566. Stone, N. and Matousek, P., Advanced transmission Raman spectroscopy: A promising tool for breast disease diagnosis, *Cancer Res.*, 68(11), 4424–4430, 2008.

567. Kodati, V. R., Tomasi, G. E., Turmin, J. L., and Tu, A. T., Raman-spectroscopic identification of calcium-oxalate-type kidney-stone, *Appl. Spectrosc.*, 44, 1408–1411, 1990.

568. Carmona, P., Bellanato, J., and Escolar, E., Infrared and Raman spectroscopy of uranari calculi: A review, *Biospectroscopy*, 3, 331–346, 1997.

569. Romer, T. J., Brennan, J. F., Puppels, G. J., Tuinenburg, J., van Duinen, S. G., van der Laarse, A., van der Steen, A. F. W., Bom, N. A., and Bruschke, A. V. C., Intravascular ultrasound combined with Raman spectroscopy to localize and quantify cholesterol and calcium salts in atherosclerotic coronary arteries, *Arterioscl. Throm. Vasc.*, 20, 478–483, 2000.

570. Hong, T. D. N., Phat, D., Plaza, P., Daudon, M., and Dao, N. Q., Identification of urinary calculi by Raman laser fiber optics spectroscopy, *Clin. Chem.*, 38, 292–298, 1992.

571. Baker, R., Matousek, P., Ronayne, K. L., Parker, A. W., Rogers, K., and Stone, N., Depth profiling of calcifications in breast tissue using picosecond Kerr-gated Raman spectroscopy, *Analyst*, 132(1), 48–53, 2007.

572. Liu, Z., Li, X., Tabakman, S. M., Jiang, K., Fan, S., and Dai, H., Multiplexed multicolor raman imaging of live cells with isotopically modified single walled carbon nanotubes, *J. Am. Chem. Soc.*, 130(41), 13540–13541, 2008.

573. Frank, C. J., Redd, D. C. B., Gansler, T. S., and McCreery, R. L., Characterization of human breast biopsy specimens with near-IR Raman spectroscopy, *Anal. Chem.*, 66, 319–326, 1994.

574. Mizuno, A., Kitajima, H., Kawauchi, K., Muraishi, S., and Ozaki, Y., Near infrared Fourier transform Raman spectroscopic study of human brain tissues and tumours, *J. Raman Spectrosc.*, 25, 265–269, 1994.

575. Clarke, R. H., Wang, Q., and Isner, J. M., Laser Raman spectroscopy of atherosclerotic lesions in human coronary artery segments, *Appl. Opt.*, 27, 4799–4800, 1988.

576. Bergeson, S. D., Peatross, J. B., Eyring, N. J., Fralick, J. F., Stevenson, D. N., and Ferguson, S. B., Resonance Raman measurements of carotenoids using light-emitting diodes, *J. Biomed. Opt.*, 13(4), 6, 2008.

577. Ko, A. C. T., Choo-Smith, L. P., Hewko, M., Leonardi, L., Sowa, M. G., Dong, C. C. S., Williams, P., and Cleghorn, B., Ex vivo detection and characterization of early dental caries by optical coherence tomography and Raman spectroscopy, *J. Biomed. Opt.*, 10(3), 16, 2005.

578. Lorincz, A., Haddad, D., Naik, R., Naik, V., Fung, A., Cao, A., Manda, P. et al., Raman spectroscopy for neoplastic tissue differentiation: A pilot study, *J. Pediatr. Surg.*, 39(6), 953–956, 2004.

579. Kunapareddy, N., Freyer, J. P., and Mourant, J. R., Raman spectroscopic characterization of necrotic cell death, *J. Biomed. Opt.*, 13(5), 9, 2008.

580. Notingher, I., Jell, G., Lohbauer, U., Salih, V., and Hench, L. L., In situ non-invasive spectral discrimination between bone cell phenotypes used in tissue engineering, *J. Cell. Biochem.*, 92(6), 1180–1192, 2004.

581. van de Poll, S. W. E., Schut, T. C. B., van den Laarse, A., and Puppels, G. J., In situ investigation of the chemical composition of ceroid in human atherosclerosis by Raman spectroscopy, *J. Raman Spectrosc.*, 33(7), 544–551, 2002.
582. Vrensen, G. F. J. M. and Duindam, H. J., Maturation of fiber membranes in the human eye lens-ultrastructural and Raman microspectroscopic observations, *Ophthalmic Res.*, 27(S1), 78–85, 1995.
583. Yaroslavsky, I. V., Yaroslavsky, A. N., Otto, C., Puppels, G. J., Vrensen, G. F. J. M., Duindam, H. J., and Greve, J., Combined elastic and Raman light-scattering of human eye lenses, *Exp. Eye Res.*, 59, 393–399, 1994.
584. Duindam, J. J., Vrensen, G. F., Otto, C., Puppels, G. J., and Greve, J., New approach to assess the cholesterol distribution in the eye lens: Confocal Raman microspectroscopy and filipin cytochemistry, *J. Lipid Res.*, 36, 1139–1146, 1995.
585. Duindam, J. J., Vrensen, G. F., Otto, C., and Greve, J., Cholesterol, phospholipids, and protein changes in focal opacities in the human eye lens, *Invest. Opthalmaol. Vis. Sci.*, 39, 94–103, 1998.
586. van de Poll, S. W., Kastelijn, K., Bakker Schut, T. C., Strijder, C., Puppels, G. J., Pasterkamp, G., and van der Laarse, A., On-line detection of cholesterol and calcification by catheter-based Raman spectroscopy in human atherosclerotic plaque ex vivo, *J. Am. Coll. Cardiol.*, 41(6, Suppl. 1), 40–41, 2003.
587. Freshour, B. G. and Koenig, J. L., Raman scattering of collagen, gelatin, and elastine, *Bioplolymers*, 14, 379–391, 1975.
588. Pettway, G. J., Schneider, A., Koh, A. J., Widjaja, E., Morris, M. D., Meganck, J. A., Goldstein, S. A., and McCauley, L. K., Anabolic actions of PTH (1–34): Use of a novel tissue engineering model to investigate temporal effects on bone, *Bone*, 36(6), 959–970, 2005.
589. de Jong, B. W. D., Schut, T. C. B., Coppens, J., Wolffenbuttel, K. P., Kok, D. J., and Puppels, G. J., Raman spectroscopic detection of changes in molecular composition of bladder muscle tissue caused by outlet obstruction, *Vibrat. Spectrosc.*, 32(1), 57–65, 2003.
590. Mahadevan-Jensen, A. and Richards-Kortum, R., Raman spectroscopy for the detection of cancers and precancers, *J. Biomed. Opt.*, 1, 31–70, 1996.
591. Mahadevan-Jensen, A., Mitchell, M. F., Ramanujam, N., Malpica, A., Thompsen, S., Utzinger, U., and Richards-Kortum, R., NIR Raman spectroscopy for *in vitro* detection of cervical precancers, *Photochem. Photobiol.*, 68, 123–132, 1998.
592. Ozaki, Y., Kaneuchi, F., Iwamoto, T., Yoshiura, M., and Iriyama, K., Nondestructive analysis of biological materials by FT-IR-ATR 1. Direct evidence of for the existence of collagen helix in lens capsule, *Appl. Spectrosc.*, 43, 138–141, 1989.
593. Ramasamy, J. G. and Akkus, O., Local variations in the micromechanical properties of mouse femur: The involvement of collagen fiber orientation and mineralization, *J. Biomech.*, 40(4), 910–918, 2007.
594. Balooch, M., Habelitz, S., Kinney, J. H., Marshall, S. J., and Marshall, G. W., Mechanical properties of mineralized collagen fibrils as influenced by demineralization, *J. Struct. Biol.*, 162(3), 404–410, 2008.
595. Zhang, G. J., Moore, D. J., Mendelsohn, R., and Flach, C. R., Vibrational microspectroscopy and imaging of molecular composition and structure during human corneocyte maturation, *J. Invest. Dermatol.*, 126(5), 1088–1094, 2006.
596. Premasiri, W. R., Clarke, R. H., and Womble, M. E., Urine analysis by laser Raman spectroscopy, *Lasers Surg. Med.*, 28, 330–334, 2001.
597. Kodati, V. R. and Tu, A. T., Raman-spectroscopic identification of cystine-type kidney stone, *Appl. Spectrosc.*, 44, 837–839, 1990.
598. Centeno, J. A., Ishak, K., Mullick, F. G., Gahl, W. A., and O'Leary, T. J., Infrared microspectroscopy and laser Raman microprobe in the diagnosis of cystinosis, *Appl. Spectrosc.*, 48, 569–572, 1994.
599. Borges, A. F. S., Bitar, R. A., Kantovitzc, K. R., Correr, A. B., Martin, A. A., and Puppin-Rontani, R. M., New perspectives about molecular arrangement of primary and permanent dentin, *Appl. Surf. Sci.*, 254(5), 1498–1505, 2007.
600. de Puala, A. R. and Sathaiah, S., Raman spectroscopy for diagnosis of atherosclerosis: A rapid analysis using neural networks, *Med. Eng. Phys.*, 27(3), 237–244, 2005.

601. Wentrup-Byrne, E., Armstrong, C. A., Armstrong, R. S., and Collins, B. M., Fourier transform Raman microscopic mapping of the molecular components in a human tooth, *J. Raman Spectrosc.*, 28, 151–158, 1997.

602. Van der Veen, M. H. and Tenbosch, J. J., The influence of mineral loss on the auto-fluorescent behavior of in vitro demineralized dentine, *Caries Res.*, 30, 93–99, 1996.

603. Borges, A. F. S., Bittar, R. A., Pascon, F. M., Sobrinho, L. C., Martin, A. A., and Rontani, R. M. P., NaOCl effects on primary and permanent pulp chamber dentin, *J. Dent.*, 36(9), 745–753, 2008.

604. Rippon, W. P., Koenig, J. L., and Walton, A., Laser Raman spectroscopy of biopolymers and proteins, *Agri. Food Chem.*, 19, 692–697, 1971.

605. Brody, R. H., Edwards, H. G. M., and Pollard, A. M., Chemometric methods applied to the differentiation of Fourier-transform Raman spectra of ivories, *Anal. Chim. Acta*, 427(2), 223–232, 2001.

606. Notingher, I., Bisson, I., Polak, J. M., and Hench, L. L., In situ spectroscopic study of nucleic acids in differentiating embryonic stem cells, *Vibrat. Spectrosc.*, 35(1–2), 199–203, 2004.

607. Seli, E., Sakkas, D., Scott, R., Kwok, S. C., Rosendahl, S. M., and Burns, D. H., Noninvasive metabolomic profiling of embryo culture media using Raman and near-infrared spectroscopy correlates with reproductive potential of embryos in women undergoing in vitro fertilization, *Fertil. Steril.*, 88(5), 1350–1357, 2007.

608. Klocke, A., Mihailova, B., Zhang, S. Q., Gasharova, B., Stosch, R., Guttler, B., Kahl-Nieke, B., Henriot, P., Ritschel, B., and Bismayer, U., CO_2-laser-induced zonation in dental enamel: A Raman and IR microspectroscopic study, *J. Biomed. Mater. Res. B Appl. Biomater.*, 81B(2), 499–507, 2007.

609. Notingher, I., Bisson, I., Bishop, A. E., Randle, W. L., Polak, J. M. P., and Hench, L. L., In situ spectral monitoring of mRNA translation in embryonic stem cells during differentiation in vitro, *Anal. Chem.*, 76(11), 3185–3193, 2004.

610. Santini, A., Pulham, C. R., Rajab, A., and Ibbetson, R., The effect of a 10% carbamide peroxide bleaching agent on the phosphate concentration of tooth enamel assessed by Raman spectroscopy, *Dent. Traumatol.*, 24(2), 220–223, 2008.

611. Boere, I. A., Bakker Schut, T. C., van den Boogert, J., de Bruin, R. W. F., and Puppels, G. J., Use of fibre optic probes for detection of Barrett's epithelium in the rat oesophagus by Raman spectroscopy, *Vibrat. Spectrosc.*, 32(1), 47–55, 2003.

612. Schallreuter, K. U., Wazir, U., Kothari, S., Gibbons, N. C. J., Moore, J., and Wood, J. M., Human phenylalanine hydroxylase is activated by H_2O_2: A novel mechanism for increasing the l-tyrosine supply for melanogenesis in melanocytes, *Biochem. Biophys. Res. Commun.*, 322(1), 88–92, 2004.

613. Frank, C. J., McCreery, R. L., and Redd, D. C. B., Raman spectroscopy of normal and diseased human breast tissues, *Anal. Chem.*, 67, 777–783, 1995.

614. Koljenovic, S., Schut, T. C. B., van Meerbeeck, J. P., Maat, A., Burgers, S. A., Zondervan, P. E., Kros, J. M., and Puppels, G. J., Raman microspectroscopic mapping studies of human bronchial tissue, *J. Biomed. Opt.*, 9(6), 1187–1197, 2004.

615. Hogg, R. E., Anderson, R. S., Stevenson, M. R., Zlatkova, M. B., and Chakravarthy, U., In vivo macular pigment measurements: A comparison of resonance Raman spectroscopy and heterochromatic flicker photometry, *Br. J. Ophthalmol.*, 91(4), 485–490, 2007.

616. Callender, A. F., Finney, W. F., Morris, M. D., Sahar, N. D., Kohn, D. H., Kozloff, K. M., and Goldstein, S. A., Dynamic mechanical testing system for Raman microscopy of bone tissue specimens, *Vibrat. Spectrosc.*, 38(1–2), 101–105, 2005.

617. Döpner, S., Müller, F., Hildebrandt, P., and Müller, R. T., Integration of metallic endoprotheses in dog femur studied by near-infrared Fourier-transform Raman microscopy, *Biomaterials*, 23(5), 1337–1345, 2002.

618. Yerramshetty, J. S., Lind, C., and Akkus, O., The compositional and physicochemical homogeneity of male femoral cortex increases after the sixth decade, *Bone*, 39(6), 1236–1243, 2006.

619. Barr, H., Kendall, C., and Stone, N., The light solution for Barrett's oesophagus: Photodiagnosis and photodynamic therapy for columnar-lined oesophagus, *Photodiagnosis Photodyn. Ther.*, 1(1), 75–84, 2004.

620. Tarr, R. V. and Steffes, P. G., Non-invasive blood glucose measurement system and method using stimulated Raman spectroscopy, US Patent 5,243,983.

621. Dou, X. M., Yamaguchi, Y., Yamamoto, H., Uenoyama, H., and Ozaki, Y., Biological applications of anti-stokes Raman spectroscopy: Quantitative analysis of glucose in plasma and serum by a highly sensitive multichannel Raman spectrometer, *Appl. Spectrosc.*, 50, 1301–1306, 1996.

622. Koo, T.-W., Berger, A. J., Itzkan, I., Horowitz, G., and Feld, M. S., Measurement of glucose in blood serum using Raman spectroscopy, *IEEE LEOS Mag.*, 12, 18, 1998.

623. Koljenovic, S., Choo-Smith, L. P., Schut, T. C. B., Kros, J. M., van den Berge, H. J., and Puppels, G. J., Discriminating vital tumor from necrotic tissue in human glioblastoma tissue samples by Raman spectroscopy, *Lab. Invest.*, 82(10), 1265–1277, 2002.

624. Lambert, J. L., Pelletier, C. C., and Borchert, M., Glucose determination in human aqueous humor with Raman spectroscopy, *J. Biomed. Opt.*, 10(3), 8, 2005.

625. Chaiken, J., Finney, W., Knudson, P. E., Weinstock, R. S., Khan, M., Bussjager, R. J., Hagrman, D. et al., Effect of hemoglobin concentration variation on the accuracy and precision of glucose analysis using tissue modulated, noninvasive, in vivo Raman spectroscopy of human blood: A small clinical study, *J. Biomed. Opt.*, 10(3), 12, 2005.

626. Pelletier, C. C., Lambert, J. L., and Borchert, M., Determination of glucose in human aqueous humor using Raman spectroscopy and designed-solution calibration, *Appl. Spectrosc.*, 59(8), 1024–1031, 2005.

627. de Jong, B. W. D., De Gouveia Brazao, C. A., Stoop, H., Wolffenbuttel, K. P., Oosterhuis, J. W., Puppels, G. J., Weber, R. F. A., Looijenga, L. H. J., and Kok, D. J., Raman spectroscopic analysis identifies testicular microlithiasis as intratubular hydroxyapatite, *J. Urol.*, 171(1), 92–96, 2004.

628. Schallreuter, K. U. and Elwary, S., Hydrogen peroxide regulates the cholinergic signal in a concentration dependent manner, *Life Sci.*, 80(24–25), 2221–2226, 2007.

629. Schallreuter, K. U., Kruger, C., Rokos, H., Hasse, S., Zothner, C., and Panske, A., Basic research confirms coexistence of acquired Blaschkolinear Vitiligo and acrofacial Vitiligo, *Arch. Dermatol. Res.*, 299(5–6), 225–230, 2007.

630. Schallreuter, K. U., Rubsam, K., Gibbons, N. C. J., Maitland, D. J., Chavan, B., Zothner, C., Rokos, H., and Wood, J. M., Methionine sulfoxide Reductases A and B are deactivated by hydrogen peroxide (H_2O_2) in the epidermis of patients with vitiligo, *J. Invest. Dermatol.*, 128(4), 808–815, 2008.

631. Torres, I. P., Terner, J., Pittman, R. N., Somera, L. G., and Ward, K. R., Hemoglobin oxygen saturation measurements using resonance Raman intravital microscopy, *Am. J. Physiol. Heart Circ. Physiol.*, 289(1), H488–H495, 2005.

632. Otero, E. U., Sathaiah, S., Silveira, L., Pomerantzeff, P. M. A., and Pasqualucci, C. A. G., Raman spectroscopy for diagnosis of calcification in human heart valves, *Spectroscopy*, 18(1), 75–84, 2004.

633. Fu, Y., Wang, H., Shi, R., and Cheng, J.-X., Second harmonic and sum frequency generation imaging of fibrous astroglial filaments in ex vivo spinal tissues, *Biophys. J.*, 92(9), 3251–3259, 2007.

634. Frosch, T., Kustner, B., Schlucker, S., Szeghalmi, A., Schmitt, M., Kiefer, W., and Popp, J., In vitro polarization-resolved resonance Raman studies of the interaction of hematin with the antimalarial drug chloroquine, *J. Raman Spectrosc.*, 35(10), 819–821, 2004.

635. Torres, I. P., Terner, J., Pittman, R. N., Proffitt, E., and Ward, K. R., Measurement of hemoglobin oxygen saturation using Raman microspectroscopy and 532-nm excitation, *J. Appl. Physiol.*, 104(6), 1809–1817, 2008.

636. Silveira, L., Sathaiah, S., Zangaro, R. A., Pacheco, M. T. T., Chavantes, M. C., and Pasqualucci, C. A. G., Correlation between near-infrared Raman spectroscopy and the histopathological analysis of atherosclerosis in human coronary arteries, *Laser. Surg. Med.*, 30(4), 290–297, 2002.

637. Ward, K. R., Barbee, R. W., Reynolds, P. S., Torres Filho, I. P., Tiba, M. H., Torres, L., Pittman, R. N., and Terner, J., Oxygenation monitoring of tissue vasculature by resonance Raman spectroscopy, *Anal. Chem.*, 79(4), 1514–1518, 2007.

638. Nalla, R. K., Kruzic, J. J., Kinney, J. H., Balooch, M., Ager III, J. W., and Ritchie, R. O., Role of micro-structure in the aging-related deterioration of the toughness of human cortical bone, *Mater. Sci. Eng. C*, 26(8), 1251–1260, 2006.

639. Silveira, L., Sathaiah, S., Zangaro, R. A., Pacheco, M. T. T., Chavantes, M. C., and Pasqualucci, C. A., Near-infrared raman spectroscopy of human coronary arteries: Histopathological classification based on mahalanobis distance, *J. Clin. Laser Med. Surg.*, 21(4), 203–208, 2003.

640. Ritchie, R. O., Nalla, R. K., Kruzic, J. J., Ager, J. W., Balooch, G., and Kinney, J. H., Fracture and age-ing in bone: Toughness and structural characterization, *Strain*, 42(4), 225–232, 2006.

641. Chrit, L., Bastien, P., Biatry, B., Simonnet, J. T., Potter, A., Minondo, A. M., Flament, F. et al., In vitro and in vivo confocal Raman study of human skin hydration: Assessment of a new moisturizing agent, pMPC, *Biopolymers*, 85(4), 359–369, 2007.

642. Guggenbuhl, P., Filmon, R., Mabilleau, G., Baslé, M. F., and Chappard, D., Iron inhibits hydroxyapa-tite crystal growth in vitro, *Metabolism*, 57(7), 903–910, 2008.

643. Gotz, H., Duschner, H., White, D. J., and Klukowska, M. A., Effects of elevated hydrogen peroxide "strip" bleaching on surface and subsurface enamel including subsurface histomorphology, micro-chemical composition and fluorescence changes, *J. Dent.*, 35(6), 457–466, 2007.

644. Cukrowski, I., Popovic, L., Barnard, W., Paul, S. O., van Rooyen, P. H., and Liles, D. C., Modeling and spectroscopic studies of bisphosphonate-bone interactions. The Raman, NMR and crystallographic investigations of Ca-HEDP complexes, *Bone*, 41(4), 668–678, 2007.

645. Cerruti, M., Bianchi, C. L., Bonino, F., Damin, A., Perardi, A., and Morterra, C., Surface modifica-tions of bioglass immersed in tris-buffered solution. A multitechnical spectroscopic study, *J. Phys. Chem. B*, 109(30), 14496–14505, 2005.

646. Li, H., Khor, K. A., Chow, V., and Cheang, P., Nanostructural characteristics, mechanical properties, and osteoblast response of spark plasma sintered hydroxyapatite, *J. Biomed. Mater. Res. A*, 82A(2), 296–303, 2007.

647. Iafisco, M., Palazzo, B., Falini, G., Di Foggia, M., Bonora, S., Nicolis, S., Casella, L., and Roveri, N., Adsorption and conformational change of myoglobin on biomimetic hydroxyapatite nanocrystals functionalized with alendronate, *Langmuir*, 24(9), 4924–4930, 2008.

648. Otto, C., de Grauw, C. J., Duindam, J. J., Sijtsema, N. M., and Greve, J., Applications of micro-Raman imaging in biomedical research, *J. Raman Spectrosc.*, 28, 143–150, 1997.

649. Clarke, R. H., Hanlon, E. B., Isner, J. M., and Brody, H., Laser Raman spectroscopy of calcified ath-erosclerotic lesions in cardiovascular tissue, *Appl. Opt.*, 26, 3175–3177, 1987.

650. Tudor, A. M., Melia, C. D., Davies, M. C., Anderson, D., Hastings, G., Morrey, S., Domingos Santos, J., and Monteiro, F., The analysis of biomedical hydroxyapatite powders and hydroxyapatite coatings on metallic medical implants by near-IR fourier transform Raman-spectroscopy, *Spectrochim. Acta A*, 49, 675–680, 1993.

651. Frank, C. J., McCreery, R. L., Redd, D. C. B., and Gansler, T. S., Detection of silicone in lymph-node biopsy specimens by near-IR Raman-spectroscopy, *Appl. Spectrosc.*, 47, 387–390, 1993.

652. Walsh, P. J., Buchanan, F. J., Dring, M., Maggs, C., Bell, S., and Walker, G. M., Low-pressure synthe-sis and characterisation of hydroxyapatite derived from mineralise red algae, *Chem. Eng. J.*, 137(1), 173–179, 2008.

653. Parkesh, R., Mohsin, S., Lee, T. C., and Gunnlaugsson, T., Histological, spectroscopic, and sur-face analysis of microdamage in bone: Toward real-time analysis using fluorescent sensors, *Chem. Mater.*, 19(7), 1656–1663, 2007.

654. Pilotto, S., Pacheco, M. T. T., Silveira, L., Villaverde, A. B., and Zangaro, R. A., Analysis of near-infrared Raman spectroscopy as a new technique for a transcutaneous non-invasive diagnosis of blood components, *Laser. Med. Sci.*, 16(1), 2–9, 2001.

655. Ruperez, A., Montes, R., and Laserna, J. J., Identification of stimulant-drugs by surface-enhanced Raman-spectrometry on colloidal silver, *Vibrat. Spectropsc.*, 2, 145–154, 1991.

656. Hofmann, T., Heyroth, F., Meinhard, H., Fränzel, W., and Raum, K., Assessment of composition and anisotropic elastic properties of secondary osteon lamellae, *J. Biomech.*, 39(12), 2282–2294, 2006.

657. Hogg, R. E., Zlatkova, M. B., Chakravarthy, U., and Anderson, R. S., Investigation of the effect of simulated lens yellowing, transparency loss and refractive error on in vivo resonance Raman spectroscopy, *Ophthal. Physiol. Opt.*, 27(3), 225–231, 2007.

658. Gosselin, M.-È., Kapustij, C. J., Venkateswaran, U. D., Leverenz, V. R., and Giblin, F. J., Raman spectroscopic evidence for nuclear disulfide in isolated lenses of hyperbaric oxygen-treated guinea pigs, *Exp. Eye Res.*, 84(3), 493–499, 2007.

659. Wong Kee Song, L.-M., Optical spectroscopy for the detection of dysplasia in Barrett's esophagus, *Clin. Gastroenterol. Hepatol.*, 3(7, Suppl. 1), S2–S7, 2005.

660. Wong Kee Song, L.-M. and Wilson, B. C., Optical detection of high-grade dysplasia in Barrett's esophagus, *Tech. Gastrointest. Endosc.*, 7(2), 78–88, 2005.

661. Lawson, E. E., Anigbogu, A. N., Williams, A. C., Barry, B. W., and Edwards, H. G., Thermally induced moilecular disorder in human stratum corneum lipids compared with a model phospholipid system: FT Raman spectroscopy, *Spectrochim. Acta A Mol. Biomol. Spectrosc.*, 54, 543–558, 1989.

662. Borchman, D., Ozaki, Y., Lamba, O. P., Byrdwell, W. C., Czarnecki, M. A., and Yappert, M. C., Structural characterization of clear human lens lipid-membranes by near-infrared Fourier-transform Raman spectroscopy, *Curr. Eye Res.*, 14, 511–515, 1995.

663. Gniadecka, M., Faurskov Nielsen, O., Christensen, D. H., and Wulf, H. C., Structure of water, proteins, and lipids in intact human skin, hair and nail, *J. Invest. Dermatol.*, 110, 393–398, 1998.

664. Gniadecka, M., Wulf, H. C., Nielsen, O. F., Christensen, D. H., and Hercogova, J., Distinctive molecular abnormalities in benign and malignant skin lesions: Studies by Raman spectroscopy, *Photochem. Photobiol.*, 66, 418–423, 1997.

665. Gniadecka, M., Wulf, H. C., Mortensen, N. N., Nielsen, O. F., and Christensen, D. H., Diagnosis of basal cell carcinoma by Raman spectroscopy, *J. Raman Spectrosc.*, 28, 125–129, 1997.

666. Krafft, C., Neudert, L., Simat, T., and Salzer, R., Near infrared Raman spectra of human brain lipids, *Spectrochim. Acta A Mol. Biomol. Spectrosc.*, 61(7), 1529–1535, 2005.

667. Notingher, I., Jell, G., Notingher, P. L., Bisson, I., Tsigkou, O., Polak, J. M., Stevens, M. M., and Hench, L. L., Multivariate analysis of Raman spectra for in vitro non-invasive studies of living cells, *J. Mol. Struct.*, 744–747, 179–185, 2005.

668. Bhosale, P., Serban, B., Zhao, D. Y., and Bernstein, P. S., Identification and metabolic transformations of carotenoids in ocular tissues of the Japanese quail *Coturnix japonica*, *Biochemistry*, 46(31), 9050–9057, 2007.

669. Notingher, I., Selvakumaran, J., and Hench, L. L., New detection system for toxic agents based on continuous spectroscopic monitoring of living cells, *Biosens. Bioelectron.*, 20(4), 780–789, 2004.

670. Neelam, K., O'Gorman, N., Nolan, J., O'Donovan, O., Wong, H. B., Eong, K. G. A., and Beatty, S., Measurement of macular pigment: Raman spectroscopy versus heterochromatic flicker photometry, *Invest. Ophthalmol. Vis. Sci.*, 46(3), 1023–1032, 2005.

671. Liu, H., Li, W., Shi, S. T., Habelitz, S., Gao, C., and DenBesten, P., MEPE is downregulated as dental pulp stem cells differentiate, *Arch. Oral Biol.*, 50(11), 923–928, 2005.

672. Penel, G., Delfosse, C., Descamps, M., and Leroy, G., Composition of bone and apatitic biomaterials as revealed by intravital Raman microspectroscopy, *Bone*, 36(5), 893–901, 2005.

673. Koljenovic, S., Schut, T. B., Vincent, A., Kros, J. M., and Puppels, G. J., Detection of meningioma in dura mater by Raman spectroscopy, *Anal. Chem.*, 77(24), 7958–7965, 2005.

674. de Veld, D. C. G., Schut, T. C. B., Skurichina, M., Witjes, M. J. H., Van der Wal, J. E., Roodenburg, J. L. N., and Sterenborg, H., Autofluorescence and Raman microspectroscopy of tissue sections of oral lesions, *Laser. Med. Sci.*, 19(4), 203–209, 2005.

675. Berger, A. J., Koo, T.-W., Itzkan, I., Horowitz, G., and Feld, M. S., Multicomponent blood analysis by near-infrared Raman spectroscopy, *Appl. Opt.*, 38, 2916–2926, 1999.

676. Shen, A. G., Ye, Y., Zhang, J. W., Wang, X. H., Hu, J. M., Xie, W., and Shen, J., Screening of gastric carcinoma cells in the human malignant gastric mucosa by confocal Raman microspectroscopy, *Vibrat. Spectrosc.*, 37(2), 225–231, 2005.

677. de Jong, B. W. D., Bakker Schut, T. C., Wolffenbuttel, K. P., Nijman, J. M., Kok, D. J., and Puppels, G. J., Identification of bladder wall layers by Raman spectroscopy, *J. Urol.*, 168(4, Suppl. 1), 1771–1778, 2002.

678. Bertoluzza, A., Brasili, P., Castri, L., Facchini, F., Fagnano, G., and Tinti, A., Preliminary results in dating human skeletal remains by Raman spectroscopy, *J. Raman Spectrosc.*, 28, 185–188, 1997.

679. Edwards, J. V., Howley, P., Davis, R., Mashchak, A., and Goheen, S. C., Protease inhibition by oleic acid transfer from chronic wound dressings to albumin, *Int. J. Pharm.*, 340(1–2), 42–51, 2007.

680. Yang, S.-T., Wang, X., Jia, G., Gu, Y., Wang, T., Nie, H., Ge, C., Wang, H., and Liu, Y., Long-term accumulation and low toxicity of single-walled carbon nanotubes in intravenously exposed mice, *Toxicol. Lett.*, 181(3), 182–189, 2008.

681. Anigbogu, A. N. C., Williams, A. C., Barry, B. W., and Edwards, H. G. M., Fouorier-transform Raman-spectroscopy of interactions between the penetration enhancer dimethyl-sulfoxide and human stratum-corneum, *Int. J. Pharm.*, 125, 265–282, 1995.

682. Anibogu, A. N. C., Williams, A. C., Barry, B. W., and Edwards, H. G. M., in K. R. Brain, V. J. James, and K. A. Walters (eds.), *Prediction of Percutaneous Penetration: Methods, Measurements and Modelling*, Vol. 3, STS Publishing, Portland, OR, pp. 27–36, 1995.

683. Kazanci, M., Roschger, P., Paschalis, E. P., Klaushofer, K., and Fratzl, P., Bone osteonal tissues by Raman spectral mapping: Orientation-composition, *J. Struct. Biol.*, 156(3), 489–496, 2006.

684. Weinmann, P., Jouan, M., Dao, N. Q., Lacroix, B., Croiselle, C., Bonte, J. P., and Luc, G., Quantitative analysis of cholestrol and cholesteryl esters in human atherosclerotic plaques using near-infrared Raman spectroscopy, *Atherosclerosis*, 140, 81–88, 1998.

685. Romer, T. J., Brennan, J. F., Schut, T. C. B., Wolthius, R., van der Hoogen, R. C. M., Emeis, J. J., van der Laarse, A., Bruschke, A. V. G., and Puppels, G. J., Raman spectroscopy for quantifying cholesterol in intact coronary artery wall, *Atherosclerosis*, 141, 117–124, 1998.

686. Baconnier, S., Lang, S. B., Polomska, M., Hilczer, B., Berkovic, G., and Meshulam, G., Calcite microcrystals in the pineal gland of the human brain: First physical and chemical studies, *Bioelectromagnetics*, 23(7), 488–495, 2002.

687. Radder, A. M., Van Loon, J. A., Puppels, G. J., and Van Bitterswijk, C. A., Degradation and calcification of a PEO/PBT copolymer series, *J. Mater. Sci. Mater. Med.*, 6, 510–517, 1995.

688. Bozzini, B., Carlino, P., D'Urzo, L., Pepe, V., Mele, C., and Venturo, F., An electrochemical impedance investigation of the behaviour of anodically oxidised titanium in human plasma and cognate fluids, relevant to dental applications, *J. Mater. Sci. Mater. Med.*, 19(11), 3443–3453, 2008.

689. Zhang, G., Flach, C. R., and Mendelsohn, R., Tracking the dephosphorylation of resveratrol triphosphate in skin by confocal Raman microscopy, *J. Control. Release*, 123(2), 141–147, 2007.

690. Taleb, A., Diamond, J., McGarvey, J. J., Beattie, J. R., Toland, C., and Hamilton, P. W., Raman microscopy for the chemometric analysis of tumor cells, *J. Phys. Chem. B*, 110(39), 19625–19631, 2006.

691. Bazant-Hegemark, F., Edey, K., Swingler, G. R., Read, M. D., and Stone, N., Review: Optical micrometer resolution scanning for non-invasive grading of precancer in the human uterine cervix, *Technol. Cancer Res. Treat.*, 7(6), 483–496, 2008.

692. Baena, J. R. and Lendl, B., Raman spectroscopy in chemical bioanalysis, *Curr. Opin. Chem. Biol.*, 8(5), 534–539, 2004.

693. de Jong, B. W. D., Bakker Schut, T. C., Maquelin, K., van der Kwast, T., Bangma, C. H., Kok, D.-J., and Puppels, G. J., Discrimination between nontumor bladder tissue and tumor by Raman spectroscopy, *Anal. Chem.*, 78(22), 7761–7769, 2006.

694. Boskey, A. L. and Mendelsohn, R., Infrared spectroscopic characterization of mineralized tissues, *Vibrat. Spectrosc.*, 38(1–2), 107–114, 2005.

695. Taddei, P., Arai, T., Boschi, A., Monti, P., Tsukada, M., and Freddi, G., In vitro study of the proteolytic degradation of *Antheraea pernyi* silk fibroin, *Biomacromolecules*, 7(1), 259–267, 2006.

696. Li, Y. Z., Chen, R., Zeng, H. S., Huang, Z. W., Feng, S. Y., and Xie, S. S., Raman spectroscopy of Chinese human skin in vivo, *Chin. Opt. Lett.*, 5(2), 105–107, 2007.

697. Tfayli, A., Piot, O., and Manfait, M., Confocal Raman microspectroscopy on excised human skin: Uncertainties in depth profiling and mathematical correction applied to dermatological drug permeation, *J. Biophotonics*, 1(2), 140–153, 2008.

698. Edwards, H. G. M., Gniadecka, M., Petersen, S., Hart Hansen, J. P., Faurskov Nielsen, O., Christensen, D. H., and Wulf, H. C., NIR-FT Raman spectroscopy as a diagnostic probe for mummified skin and nails, *Vibrat. Spectrosc.*, 28(1), 3–15, 2002.

699. Chrit, L., Hadjur, C., Morel, S., Sockalingum, G., Lebourdon, G., Leroy, F., and Manfait, M., In vivo chemical investigation of human skin using a confocal Raman fiber optic microprobe, *J. Biomed. Opt.*, 10(4), 11, 2005.

700. Lien, T.-F., Chen, W., Hsu, Y.-L., Chen, H.-L., and Chiou, R. Y. Y., Influence of soy aglycon isoflavones on bone-related traits and lens protein characteristics of ovariectomized rats and bioactivity performance of osteoprogenitor cells, *J. Agric. Food Chem.*, 54(21), 8027–8032, 2006.

701. Wohlrab, J., Vollmann, A., Wartewig, S., Marsch, W. C., and Neubert, R., Noninvasive characterization of human stratum corneum of undiseased skin of patients with atopic dermatitis and psoriasis as studied by Fourier transform Raman spectroscopy, *Biopolymers*, 62(3), 141–146, 2001.

702. Haque, S., Rehman, I., and Darr, J. A., Synthesis and characterization of grafted nanohydroxyapatites using functionalized surface agents, *Langmuir*, 23(12), 6671–6676, 2007.

703. Caspers, P. J., Lucassen, G. W., and Puppels, G. J., Combined in vivo confocal Raman spectroscopy and confocal microscopy of human skin, *Biophys. J.*, 85(1), 572–580, 2003.

704. Jiang, T., Ma, X., Wang, Y. N., Tong, H., Shen, X. Y., Hu, Y. G., and Hu, J. M., Investigation of the effects of 30% hydrogen peroxide on human tooth enamel by Raman scattering and laser-induced fluorescence, *J. Biomed. Opt.*, 13(1), 9, 2008.

705. Nogueira, G. V., Silveira, L., Martin, A. A., Zangaro, R. A., Pacheco, M. T. T., Chavantes, M. C., and Pasqualucci, C. A., Raman spectroscopy study of atherosclerosis in human carotid artery, *J. Biomed. Opt.*, 10(3), 7, 2005.

706. Pandya, A. K., Serhatkulu, G. K., Cao, A., Kast, R. E., Dai, H., Rabah, R., Poulik, J. et al., Evaluation of pancreatic cancer with Raman spectroscopy in a mouse model, *Pancreas*, 36(2), E1–E8, 2008.

707. Lyng, F. M., Faoláin, E. Ó., Conroy, J., Meade, A. D., Knief, P., Duffy, B., Hunter, M. B., Byrne, J. M., Kelehan, P., and Byrne, H. J., Vibrational spectroscopy for cervical cancer pathology, from biochemical analysis to diagnostic tool, *Exp. Mol. Pathol.*, 82(2), 121–129, 2007.

708. Thakur, J. S., Dai, H. B., Serhatkulu, G. K., Naik, R., Naik, V. M., Cao, A., Pandya, A. et al., Raman spectral signatures of mouse mammary tissue and associated lymph nodes: Normal, tumor and mastitis, *J. Raman Spectrosc.*, 38(2), 127–134, 2007.

709. Choi, J., Choo, J., Chung, H., Gweon, D. G., Park, J., Kim, H. J., Park, S., and Oh, C. H., Direct observation of spectral differences between normal and basal cell carcinoma (BCC) tissues using confocal Raman microscopy, *Biopolymers*, 77(5), 264–272, 2005.

710. Rehman, S., Movasaghi, Z., Tucker, A. T., Joel, S. P., Darr, J. A., Ruban, A. V., and Rehman, I. U., Raman spectroscopic analysis of breast cancer tissues: Identifying differences between normal, invasive ductal carcinoma and ductal carcinoma in situ of the breast tissue, *J. Raman Spectrosc.*, 38(10), 1345–1351, 2007.

711. Krafft, C., Sobottka, S. B., Schackert, G., and Salzer, R., Near infrared Raman spectroscopic mapping of native brain tissue and intracranial tumors, *Analyst*, 130(7), 1070–1077, 2005.

712. Krafft, C., Sobottka, S. B., Schackert, G., and Salzer, R., Raman and infrared spectroscopic mapping of human primary intracranial tumors: A comparative study, *J. Raman Spectrosc.*, 37(1–3), 367–375, 2006.

713. Kawabata, T., Mizuno, T., Okazaki, S., Hiramatsu, M., Setoguchil, T., Kikuchi, H., Yamamoto, M. et al., Optical diagnosis of gastric cancer using near-infrared multichannel Raman spectroscopy with a 1064-nm excitation wavelength, *J. Gastroenterol.*, 43(4), 283–290, 2008.

714. Koljenovic, S., Schut, T. C. B., Wolthuis, R., de Jong, B., Santos, L., Caspers, P. J., Kros, J. M., and Puppels, G. J., Tissue characterization using high wave number Raman spectroscopy, *J. Biomed. Opt.*, 10(3), 11, 2005.

715. Krafft, C., Knetschke, T., Siegner, A., Funk, R. H. W., and Salzer, R., Mapping of single cells by near infrared Raman microspectroscopy, *Vibrat. Spectrosc.*, 32(1), 75–83, 2003.

716. Gniadecka, M., Nielsen, O. F., and Wulf, H. C., Water content and structure in malignant and benign skin tumours, *J. Mol. Struct.*, 661, 405–410, 2003.

717. Rabah, R., Weber, R., Serhatkulu, G. K., Cao, A., Dai, H., Pandya, A., Naik, R., Auner, G., Poulik, J., and Klein, M., Diagnosis of neuroblastoma and ganglioneuroma using Raman spectroscopy, *J. Pediatr. Surg.*, 43(1), 171–176, 2008.

718. Wang, H.-W., Le, T. T., and Cheng, J.-X., Label-free imaging of arterial cells and extracellular matrix using a multimodal CARS microscope, *Opt. Commun.*, 281(7), 1813–1822, 2008.

719. Yu, G., Xu, X. X., Niu, Y., Wang, B., Song, Z. F., and Zhang, C. P., Studies on human breast cancer tissues with Raman microspectroscopy, *Spectrosc. Spectr. Anal.*, 24(11), 1359–1362, 2004.

720. Yu, G., Xu, X. X., Lu, S. H., Zhang, C. Z., Song, Z. F., and Zhang, C. P., Confocal Raman microspectroscopic study of human breast morphological elements, *Spectrosc. Spectr. Anal.*, 26(5), 869–873, 2006.

721. Yu, G., Zhang, P., Tan, E. Z., and Zhang, C. Z., Study of human tumor tissues by Raman imaging spectra, *Spectrosc. Spectr. Anal.*, 27(2), 295–298, 2007.

722. Silveira, L., De Paula, A. R., Pasqualucci, C. A., and Pacheco, M. T. T., Independent component analysis applied to Raman spectra for classification of in vitro human coronary arteries, *Instrum. Sci. Technol.*, 36(2), 134–145, 2008.

723. Bitar, R. A., Martinho, H. D. S., Tierra-Criollo, C. J., Ramalho, L. N. Z., Netto, M. M., and Martin, A. A., Biochemical analysis of human breast tissues using Fourier-transform Raman spectroscopy, *J. Biomed. Opt.*, 11(5), 8, 2006.

724. Azrad, E., Cagnano, E., Halevy, S., Rosenwaks, S., and Bar, I., Bullous pemphigoid detection by micro-Raman spectroscopy and cluster analysis: Structure alterations of proteins, *J. Raman Spectrosc.*, 36(11), 1034–1039, 2005.

725. Hu, Y., Shen, A., Jiang, T., Ai, Y., and Hu, J., Classification of normal and malignant human gastric mucosa tissue with confocal Raman microspectroscopy and wavelet analysis, *Spectrochim. Acta A Mol. Biomol. Spectrosc.*, 69(2), 378–382, 2008.

726. Ly, E., Piot, O., Durlach, A., Bernard, P., and Manfait, M., Polarized Raman microspectroscopy can reveal structural changes of peritumoral dermis in basal cell carcinoma, *Appl. Spectrosc.*, 62(10), 1088–1094, 2008.

727. Krafft, C., Codrich, D., Pelizzo, G., and Sergo, V., Raman and FTIR imaging of lung tissue: Methodology for control samples, *Vibrat. Spectrosc.*, 46(2), 141–149, 2008.

728. Shen, A. G., Ye, Y., Wang, X. H., Chen, C. C., Zhang, H. B., and Hu, J. M., Raman scattering properties of human pterygium tissue, *J. Biomed. Opt.*, 10(2), 5, 2005.

729. Santos, L. F., Wolthuis, R., Koljenovic, S., Almeida, R. M., and Puppels, G. J., Fiber-optic probes for in vivo Raman spectroscopy in the high-wavenumber region, *Anal. Chem.*, 77(20), 6747–6752, 2005.

730. Haka, A. S., Shafer-Peltier, K. E., Fitzmaurice, M., Crowe, J., Dasari, R. R., and Feld, M. S., Diagnosing breast cancer by using Raman spectroscopy, *Proc. Natl. Acad. Sci. USA*, 102(35), 12371–12376, 2005.

731. Kalyan Kumar, K., Anand, A., Chowdary, M. V. P., Keerthi, Kurien, J., Murali Krishna, C., and Mathew, S., Discrimination of normal and malignant stomach mucosal tissues by Raman spectroscopy: A pilot study, *Vibrat. Spectrosc.*, 44(2), 382–387, 2007.

732. Liu, G., Liu, J. H., Zhang, L., Yu, F., and Sun, S. Z., Raman spectroscopic study of human tissues, *Spectrosc. Spectr. Anal.*, 25(5), 723–725, 2005.

733. Yu, G., Lu, A. J., Wang, B., Tan, E. Z., and Gao, D. W., Study on Raman linear model of human breast tissue, *Spectrosc. Spectr. Anal.*, 28(5), 1091–1094, 2008.

734. Krafft, C., Kirsch, M., Beleites, C., Schackert, G., and Salzer, R., Methodology for fiber-optic Raman mapping and FTIR imaging of metastases in mouse brains, *Anal. Bioanal. Chem.*, 389(4), 1133–1142, 2007.

735. Huang, Z., McWilliams, A., Lui, H., McLean, D., Lam, D. I. S., and Zeng, H., Near-infrared Raman spectroscopy for optical diagnosis of lung cancer, *Int. J. Cancer*, 107, 1047, 2003.

736. Palero, J. A., de Bruijn, H. S., van der Ploeg van den Heuvel, A., Sterenborg, H. J. C. M., and Gerritsen, H. C., Spectrally resolved multiphoton imaging of in vivo and excised mouse skin tissues, *Biophys. J.*, 93(3), 992–1007, 2007.

737. Brozek-Pluska, B., Placek, I., Kurczewski, K., Morawiec, Z., Tazbir, M., and Abramczyk, H., Breast cancer diagnostics by Raman spectroscopy, *J. Mol. Liq.*, 141(3), 145–148, 2008.

738. Luneva, O. G., Brazhe, N. A., Maksimova, N. V., Rodnenkov, O. V., Parshina, E. Y., Bryzgalova, N. Y., Maksimov, G. V., Rubin, A. B., Orlov, S. N., and Chazov, E. I., Ion transport, membrane fluidity and haemoglobin conformation in erythrocyte from patients with cardiovascular diseases: Role of augmented plasma cholesterol, *Pathophysiology*, 14(1), 41–46, 2007.

739. Olsen, E. F., Rukke, E. O., Flatten, A., and Isaksson, T., Quantitative determination of saturated-, monounsaturated- and polyunsaturated fatty acids in pork adipose tissue with non-destructive Raman spectroscopy, *Meat Sci.*, 76(4), 628–634, 2007.

740. Chau, A. H., Motz, J. T., Gardecki, J. A., Waxman, S., Bouma, B. E., and Tearney, G. J., Fingerprint and high-wavenumber Raman spectroscopy in a human-swine coronary xenograft in vivo, *J. Biomed. Opt.*, 13(4), 3, 2008.

741. Schallreuter, K. U. and Wood, J. M., Thioredoxin reductase—Its role in epidermal redox status, *J. Photochem. Photobiol. B Biol.*, 64, 179–184, 2001.

742. Caspers, P. J., Lucassen, G. W., Carter, E. A., Bruining, H. A., and Puppels, G. J., In vivo confocal Raman microspectroscopy of the skin: Noninvasive determination of molecular concentration profiles, *J. Invest. Dermatol.*, 116(3), 434–442, 2001.

743. Kummeling, M. T. M., de Jong, B. W. D., Laffeber, C., Kok, D.-J., Verhagen, P. C. M. S., van Leenders, G. J. L. H., van Schaik, R. H. N., van Woerden, C. S., Verhulst, A., and Verkoelen, C. F., Tubular and interstitial nephrocalcinosis, *J. Urol.*, 178(3), 1097–1103, 2007.

744. Isabelle, M., Stone, N., Barr, H., Vipond, M., Shepherd, N., and Rogers, K., Lymph node pathology using optical spectroscopy in cancer diagnostics, *Spectrosc. Int. J.*, 22(2–3), 97–104, 2008.

745. Ling, X. F., Xu, Y. Z., Weng, S. F., Li, W. H., Zhi, X., Hammaker, R. M., Fateley, W. G. et al., Investigation of normal and malignant tissue samples from the human stomach using Fourier transform Raman spectroscopy, *Appl. Spectrosc.*, 56(5), 570–573, 2002.

746. Min, Y. K., Yamamoto, T., Kohda, E., Ito, T., and Hamaguchi, H., 1064 nm Near-infrared multichannel Raman spectroscopy of fresh human lung tissues, *J. Raman Spectrosc.*, 36(1), 73–76, 2005.

747. Rocha, R., Silveira, L., Villaverde, A. B., Pasqualucci, C. A., Costa, M. S., Brugnera, A., and Pacheco, M. T. T., Use of near-infrared Raman spectroscopy for identification of atherosclerotic plaques in the carotid artery, *Photomed. Laser Surg.*, 25(6), 482–486, 2007.

748. Rocha, R., Villaverde, A. B., Pasqualucci, C. A., Silveira, L., Costa, M. S., and Pacheco, M. T. T., Identification of calcifications in cardiac valves by near infrared Raman spectroscopy, *Photomed. Laser Surg.*, 25(4), 287–290, 2007.

749. Kodati, V. R., Tu, A. T., and Turmin, J. L., Raman spectroscopic identification of uric-acid-type kidney stone, *Appl. Spectrosc.*, 44, 1134–1136, 1990.

750. McCreadie, B. R., Morris, M. D., Chen, T. C., Rao, D. S., Finney, W. F., Widjaja, E., and Goldstein, S. A., Bone tissue compositional differences in women with and without osteoporotic fracture, *Bone*, 39(6), 1190–1195, 2006.

751. Huang, C.-H., Chang, R.-J., Huang, S.-L., and Chen, W., Dietary vitamin E supplementation affects tissue lipid peroxidation of hybrid tilapia, *Oreochromis niloticus* × *O. aureus*, *Comp. Biochem. Physiol. B Biochem. Mol. Biol.*, 134(2), 265–270, 2003.

752. Brennan, J. F., Romer, T. J., Lees, R. S., Tercyak, A. M., Kramer, J. R., and Feld, M. S., Determination of human coronary artery composition by Raman spectroscopy, *Circulation*, 96, 99–105, 1997.

753. Rava, R. P., Baraga, J. J., and Feld, M. S., Near-infrared Fourier-transform Raman spectroscopy of human artery, *Spectrochim. Acta*, 47, 509–512, 1991.

754. Manoharan, R., Baraga, J. J., Feld, M. S., and Rava, R. P., Quantitative histochemical analysis of human artery using Raman spectroscopy, *J. Photochem. Photobiol. B Biol.*, 16, 211–233, 1992.

755. Baraga, J. J., Feld, M. S., and Rava, R. P., In situ optical histochemistry of human artery using near infrared Fourier transform Raman spectroscopy, *Proc. Natl. Acad. Sci. USA*, 89, 3473–3477, 1992.

756. Redd, D. C., Yue, K. T., Martini, L. G., and Kaufman, S. L., Raman spectroscopy of human atherosclerotic plaque: Implications for laser angioplasty, *J. Vasc. Interv. Radiol.*, 2, 247–252, 1991.

757. Liu, C. H., Glassman, W. L. S., Zhu, H. R., Akins, D. L., Deckelbaum, L. I., Stetz, M. L., O'Brien, K., Scott, J., and Alfono, R. R., Near-IR Fourier transform Raman spectroscopy of normal and artherosclerotic human aorta, *Lasers Life Sci.*, 4, 257–264, 1992.

758. Romer, T. J., Brennan, J. F., Puppels, G. J., Laarse, A. V. D., Princen, H. M. G., Folger, O., Buschman, H. P. J., Jukema, J. W., Havekes, L. M., and Bruschke, A. V. G., Raman spectroscopy provides chemical mappings of atherosclerotic plaques in APOE*3 Leiden transgenic mice, *J. Am. Coll. Cardiol.*, 31, 500A–501A, 1998.

759. Schrader, B., Dippel, B., Frendel, S., Keller, S., Lochte, T., Riedl, M., Sculte, R., and Tatsch, E., NIR FT Raman spectroscopy—A new tool in medical diagnostics, *J. Mol. Struct.*, 408, 247–251, 1997.

760. Gou, P., Yi, G. H., Xiong, P., Yuan, Y. L., Xie, Q., and Chen, C. X., Raman spectra of the serums from cancerous persons, *Spectrosc. Spectr. Anal.*, 20, 844–846, 2000.

761. Rehman, I., Smith, R., Hench, L. L., and Bonfield, W., Structural evaluation of human and sheep bone and comparison with synthetic hydroxyapatite by FT Raman spectroscopy, *J. Biomed. Mater. Res.*, 29, 1287–1294, 1995.

762. Smith, R. and Rehman, I., Fourier-transform Raman-spectroscopic studies of human bone, *J. Mater. Sci. Mater. Med.*, 5, 775–778, 1994.

763. Mizuno, A., Hayashi, T., Tashibu, K., Maraishi, S., Kawauchi, K., and Ozaki, Y., Near-infrared FT-Raman spectra of the rat-brain tissues, *Neurosci. Lett.*, 141, 47–52, 1992.

764. Keller, S., Schrader, B., Hoffman, A., Schrader, W., Metz, K., Rehlaender, A., Pahnke, J., Ruwe, M., and Budach, W., Application of near-infrared Fourier-transform Raman spectroscopy in medical research, *J. Raman Spectrosc.*, 25, 663–671, 1994.

765. Ong, C. W., Shen, Z. X., He, Y., Lee, T., and Tang, S. H., Raman microspectroscopy of the brain tissues in the substantia nigra and MPTP-induced Parkinson's disease, *J. Raman Spectrosc.*, 30, 91–96, 1999.

766. Dipple, B., Tatsch, E., and Schrader, B., Development of an inverted NIR-FT-Raman microscope for biomedical applications, *J. Mol. Struct.*, 408, 247–251, 1997.

767. Manoharan, R., Shafer, K., Perelman, L., Wu, J., Chen, K., Deinum, G., Fitzmauice, M. et al., Raman spectroscopy and fluorescence photon migration for breast cancer diagnosis and imaging, *Photochem. Photobiol.*, 67, 15–22, 1998.

768. Manoharan, R., Wang, Y., and Feld, M. S., Review: Histochemical analysis of biological tissues using Raman spectroscopy, *Spectrochim. Acta A*, 52, 215–249, 1996.

769. Kline, N. and Treado, P. J., Raman chemical imaging of breast tissue, *J. Raman Spectrosc.*, 28, 119–124, 1997.

770. Klug, D. D., Singleton, D. L., and Walley, V. M., Laser Raman spectrum of calcified human aorta, *Lasers Surg. Med.*, 12, 13–17, 1992.

771. Manoharan, R., Wang, Y., Dasari, R. R., Singer, S., Rava, R. P., and Feld, M. S., UV resonance Raman spectroscopy for the detection of colon cancer, *Laser. Life Sci.*, 6, 217–227, 1995.

772. Goheen, S. C., Lis, L. J., and Jauffman, J. W., Raman spectroscopy of intact feline corneal collagen, *Biochem. Biophys. Acta*, 536, 197–204, 1978.

773. Bauer, N. J. C., Wicksted, J. P., Jongsma, F. H. M., March, W. F., Hendrikse, F., and Montamedi, M., Non-invasive assessment of the hydration gradient across the cornea using confocal Raman spectroscopy, *Invest. Opthalmol. Vis. Sci.*, 39, 831–835, 1998.

774. Williams, A. C., Barry, B. W., Edwards, H. G., and Farwell, D. W., A critical comparison of some Raman spectroscopic techniques for studies of human stratum corneum, *Pharm. Res.*, 10, 1642–1647, 1993.

775. Romer, T. J., Brennan, J. F., Fitzmaurice, M., Feldstein, M. L., Deinum, G., Myles, J. L., Kramer, J. R., Less, R. S., and Feld, M. S., Histopathy of human coronary artery atherosclerosis by quantifying its chemical composition by Raman spectroscopy, *Circulation*, 97, 878–885, 1998.

776. Deinum, G., Rodriguez, D., Romer, T. J., Fitzmaurice, M., Kramer, J. R., and Feld, M. S., Histological classification of Raman spectra of human coronary atherosclerosis using principal component analysis, *Appl. Spectrosc.*, 53, 938–942, 1999.

777. Jongsma, F. H. M., Erckens, R. J., Wicksted, J. P., Bauer, N. J. C., Hendrikse, F., March, W. F., and Motamedi, M., Confocal Raman spectroscopy system for noncontact scanning of ocular tissues: An *in vitro* study, *Opt. Eng.*, 36, 3193–3199, 1997.

778. Ishida, H., Kamoto, R., Uchida, S., Ishitani, A., Ozaki, Y., Iriyama, K., Tsukie, T., Shibata, K., Ishihara, F., and Kameda, H., Raman microprobe and Fourier transform-infrared microsampling studies of the microstructure of gallstones, *Appl. Spectrosc.*, 41, 407–412, 1997.

779. Wentrup-Byrne, E., Rintoul, J., Smith, J. L., and Fredericks, P. M., Comparison of vibrational spectroscopic techniques for the characterization of human gallstones, *Appl. Spectrosc.*, 49, 1028–1036, 1995.

780. Bohorfoush, A. G., Tissue spectroscopy for gastrointestinal diseases, *Endoscopy*, 28, 372–380, 1996.

781. Barr, H., Dix, T., and Stone, N., Optical spectroscopy for the early diagnosis of gastrointestinal malignancy, *Laser Med. Sci.*, 13, 3–13, 1998.

782. Liu, C. H., Das, B. B., Sha Glassman, W. L., Tang, G. C., Yoo, K. M., Zhu, H. R., Akins, D. L. et al., Raman, fluorescence and time-resolved light scattering as optical diagnostic techniques to separate diseased and normal media, *J. Photochem. Photobiol. B Biol.*, 16, 187–209, 1992.

783. Liu, C. H., Das, B. B., Glassman, W. L. S., Tang, G. C., Yoo, K. M., Zhu, H. R., Akins, D. L. et al., Raman, fluorescence, and time-resolved light scattering as optical diagnostic techniques to separate diseased and normal biomedical media, *J. Photochem. Photobiol. B Biol.*, 16, 187–209, 1992.

784. Williams, A. C., Edwards, H. G. M., and Barry, B. W., Raman-spectra of human keratotic biopolymers—Skin, callus, hair and nail, *J. Raman Spectrosc.*, 25, 95–98, 1994.

785. Stone, N., Stavroulaki, P., Kendall, C., Birchall, M., and Barr, H., Raman spectroscopy for early detection of laryngeal malignancy: Preliminary results, *Laryngoscope*, 110, 1756–1763, 2000.

786. Kaminaka, S., Yamazaki, H., Ito, T., Kohda, E., and Hamaguchi, H. O., Near-infrared Raman spectroscopy of human lung tissues: Possibility of molecular-level cancer dignosis, *J. Raman Spectrosc.*, 32, 139–141, 2001.

787. McGill, N., Dieppe, P. A., Bowden, M., Gardiner, D. J., and Hall, M., Identification of pathological mineral deposits by Raman microspectroscopy, *Lancet*, 337, 77–78, 1991.

788. Pestaner, J. P., Mullick, F. G., Johnson, F. B., and Centeno, J. A., Calcium oxalate crystals in human pathology: Molecular analysis with the laser Raman microprobe, *Arch. Pathol. Lab. Med.*, 120, 537–540, 1993.

789. Buitveld, H., De Mul, F. F. M., and Greve, J., Identification of inclusions in lung tissues with a Raman microprobe, *Appl. Spectrosc.*, 38, 304–306, 1984.

790. Hahn, D. W., Wohlfarth, D. L., and Parks, N. L., Analysis of polyethylene wear debris using micro Raman spectroscopy: A report on the presence of beta-carotene, *J. Biomed. Mater. Res.*, 35, 31–37, 1997.

791. Wolfarth, D. L., Han, D. W., Bushar, G., and Parks, N. L., Separation and characterization of polyethylene wear debris from synovial fluid and tissue samples of revised knee replacements, *J. Biomed. Mater. Res.*, 43, 57–61, 1997.

792. Kalasinsky, V. F., Johnson, F. B., and Ferwerda, R., Fourier transform IR and Raman microspectroscopy of materials in tissue, *Cell Mol. Biol.*, 44, 141–144, 1998.

793. Luke, J. L., Kalasinsky, V. F., Turnicky, R. P., Centeno, J. A., Johnson, F. B., and Mullick, F. B., Pathological and biophysical findings associated with silicone breast implants: A study of capsular tissues from 86 cases, *Plast. Reconstr. Surg.*, 100, 1558–1565, 1997.

794. Farrell, R. and McCauley, R., On corneal transparency and its loss with swelling, *J. Opt. Soc. Am.*, 66, 342–345, 1976.

795. Yu, N. T., DeNagel, D. C., Pruett, P. L., and Kuck, J. F. R., Disulfide bond formation in the eye lens, *Proc. Natl. Acad. Sci. USA*, 82, 7965–7968, 1985.

796. Mizuno, A., Ozaki, Y., Kamada, Y., Myazaki, H., Itoh, K., and Iriyama, K., Direct measurement of Raman spectra of intact lens in a whole eyeball, *Curr. Eye Res.,* 1, 609–613, 1981.

797. Yu, N. T., Kuck, J. F. R., and Askren, C. C., Laser Raman spectroscopy of the lens *in situ*, measured in an anesthetized rabbit, *Curr. Eye Res.,* 1, 615–618, 1982.

798. Yu, N. T., DeNagel, D. C., and Kuck, J. F. R., Ocular lenses, in T. G. Spiro (ed.), *Biological Applications of Raman Spectroscopy,* Wiley, New York, pp. 47–80, 1987.

799. Bot, A. C. C., Huizinga, A., de Mul, F. F. M., Vrensen, G. F. J. M., and Greve, J.,Raman microspectroscopy of fixed rabbit and human lenses and lens slices—New potentials, *Exp. Eye Res.*, 49, 161–169, 1989.

800. Siebenga, I., Vrensen, G. F., Otto, K., Puppels, G. J., De Mul, F. F., and Greve, J., Aging and changes in protein conformation in the human lens: A Raman microspectroscopy study, *Exp. Eye Res.*, 54, 759–767, 1992.

801. Smeets, M. H., Vrensen, G. F. J. M., Otto, K., Puppels, G. J., and Greve, J., Local variations in protein structure in the human eye lens—A Raman microspectroscopy study, *Biophys. Biochim. Acta*, 1164, 236–242, 1993.

802. Nie, S. M., Bergbauer, K. L., Kuck, J. F. R., and Yu, N. T., Near-infrared Fourier-transform Raman spectroscopy in human lens research, *Exp. Eye Res.*, 51, 619–623, 1990.

803. Venkatakrishna, K., Kurien, J., Pai, K. M., Valiathan, M., Kumar, N. N., Krishna, C. M., Ullas, G., and Kartha, V. B., Optical pathology of oral tissue: A Raman spectroscopy diagnostic method, *Curr. Sci.*, 80, 665–669, 2001.

804. Schaeberle, M. D., Kalasinsky, V. F., Luke, J. L., Lewis, E. N., Levin, I. W., and Treado, P. J., Raman chemical imaging: Histopathology of inclusions in human breast tissue, *Anal. Chem.*, 66, 1829–1833, 1996.

805. Abraham, J. L. and Etz, E., Molecular microanalysis of pathological specimens *in situ* with a laser Raman microprobe, *Science*, 206, 718, 1979.

806. Salenius, J. P., Brennan, J. F., Miller, A., Wang, Y., Aretz, T., Sacks, B., Dasari, R. R., and Feld, M. S., Biochemical composition of human peripheral arteries using near infrared Raman spectroscopy, *J. Vasc. Surg.*, 27, 710–719, 1998.

807. Sebag, J., Nie, S., Reiser, K., Charles, M. A., and Yu, N. T., Raman spectroscopy of vitreous in proliferative diabetic retinopathy, *Invest. Opthalmol. Vis. Sci.*, 35, 2976–2980, 1994.

808. Williams, A. C., Edwards, H. G. M., and Barry, B. W., Fourier-transform Raman spectroscopy—A novel application for examining human stratum-corneum, *Int. J. Pharm.*, 81, R11–R14, 1992.

809. Williams, A. C., Barry, B. W., and Edwards, H. G. M., Comparison of Fourier-transform Raman spectra of mammalain and reptilian skin, *Analyst*, 119, 563–566, 1994.

810. Edwards, H. G. M., Williams, A. C., and Barry, B. W., Potential applications of FT-Raman spectroscopy for dermatological diagnostics, *J. Mol. Struct.*, 347, 379–387, 1995.

811. Barry, B. W., Edwards, H. G. M., and Williams, A. C., Fourier-transform Raman and infrared vibrational study of human skin—Assignment of spectral bands, *J. Raman Spectrosc.*, 23, 641–645, 1992.

812. Fendel, S. and Schrader, B., Investigation of skin and skin lesions by NIR-FT Raman spectroscopy, *Fresenius J. Anal. Chem.*, 360, 609–613, 1998.

813. Caspers, P. J., Lucassen, G. W., Wolthuis, R., Bruining, H. A., and Puppels, G. J., *In vitro* and *in vivo* Raman spectroscopy of human skin, *Biospectroscopy*, 4, 569–572, 1998.

814. Ling, X. F., Li, W. H., Song, Y. Y., Yang, Z. L., Xu, Y. Z., Weng, S. F., Xu, Z., Fu, X. B., Zhou, X. S., and Wu, J. G., FT-Raman spectroscopic investigation on stomach cancer, *Spectrosc. Spectr. Anal.*, 20, 692–693, 2000.

815. Hendra, P. J., Jones, C., and Warnes, G., *Fourier Transform Raman Spectroscopy: Instrumentation and Chemical Applications*, Ellis Horwood, Chichester, U.K., pp. 209–211, 1991.

816. Erckens, R. J., Bauer, N. J. C., March, W. F., Jongsma, F. H. M., Hendrikse, F., and Motamedi, M., Non-invasive in-vivo assessment of corneal dehydration in the rabbit using confocal Raman spectroscopy, *Invest. Opthalmol. Vis. Sci.*, 37, 1668–1678, 1996.

817. Bauer, N. J. C., March, W. F., Wicksted, J. P., Hendrikse, F., Jongsma, F. H. M., and Motamedi, M., In-vivo confocal Raman spectroscopy of the human eye, *Invest. Opthalmol. Vis. Sci.*, 37, 3450, 1996.

818. Schut, T. C. B., Witjes, M. J. H., Sterenborg, H. J. C. M., Speelman, O. C., Roodenburg, J. L. N., Marple, E. T., Bruining, H. A., and Puppels, G. J., *In vivo* detection of dysplastic tissue by Raman spectroscopy, *Anal. Chem.*, 72, 6010–6018, 2000.

819. Wang, H., Fu, Y., Zickmund, P., Shi, R., and Cheng, J.-X., Coherent anti-stokes Raman scattering imaging of axonal myelin in live spinal tissues, *Biophys. J.*, 89(1), 581–591, 2005.

820. Qian, X. M., Peng, X. H., Ansari, D. O., Goen, Q. Y., Chen, G. Z., Shin, D. M., Yang, L., Young, A. N., Wang, M. D., and Nie, S. M., In vivo tumor targeting and spectroscopic detection with surface-enhanced Raman nanoparticle tags, *Nat. Biotechnol.*, 26, 83, 2008.

821. Schulmerich, M. V., Cole, J. H., Dooley, K. A., Morris, M. D., Kreider, J. M., Goldstein, S. A., Srinivasan, S., and Pogue, B. W., Noninvasive Raman tomographic imaging of canine bone tissue, *J. Biomed. Opt.*, 13(2), 3, 2008.

822. Akkus, O., Adar, F., and Schaffler, M. B., Age-related changes in physicochemical properties of mineral crystals are related to impaired mechanical function of cortical bone, *Bone*, 34(3), 443–453, 2004.

823. Furic, K., Mohacek-Grosev, V., and Hadzija, M., Development of cataract caused by diabetes mellitus: Raman study, *J. Mol. Struct.*, 744–747, 169–177, 2005.

824. Baby, A. R., Lacerda, A. C. L., Velasco, M. V. R., Lopes, P. S., Kawano, Y., and Kaneko, T. M., Spectroscopic studies of stratum corneum model membrane from *Bothrops jararaca* treated with cationic surfactant, *Coll. Surf. B Biointerf.*, 50(1), 61–65, 2006.

825. Yuan, Y., Chen, Y., Liu, J.-H., Wang, H., and Liu, Y., Biodistribution and fate of nanodiamonds in vivo, *Diam. Relat. Mater.*, 18(1), 95–100, 2009.

826. Synytsya, A., Kral, V., Pouckova, P., and Volka, K., Resonance Raman and UV-visible spectroscopic studies of water-soluble sapphyrin derivative: Drug localization in tumor and normal mice tissues, *Appl. Spectrosc.*, 55(2), 142–148, 2001.

827. Amharref, N., Beljebbar, A., Dukic, S., Venteo, L., Schneider, L., Pluot, M., and Manfait, M., Discriminating healthy from tumor and necrosis tissue in rat brain tissue samples by Raman spectral imaging, *Biochim. Biophys. Acta Biomembr.*, 1768(10), 2605–2615, 2007.

828. Synytsya, A., Kral, V., Matejka, P., Pouckova, P., Volka, K., and Sessler, J. L., Biodistribution assessment of a lutetium(III) texaphyrin analogue in tumor-bearing mice using NIR Fourier-transform Raman spectroscopy, *Photochem. Photobiol.*, 79(5), 453–460, 2004.

829. Motz, J. T., Fitzmaurice, M., Miller, A., Gandhi, S. J., Haka, A. S., Galindo, L. H., Dasari, R. R., Kramer, J. R., and Feld, M. S., In vivo Raman spectral pathology of human atherosclerosis and vulnerable plaque, *J. Biomed. Opt.*, 11(2), 9, 2006.

830. Darvin, M. E., Gersonde, I., Albrecht, H., Zastrow, L., Sterry, W., and Lademann, J., In vivo Raman spectroscopic analysis of the influence of IR radiation on the carotenoid antioxidant substances beta-carotene and lycopene in the human skin. Formation of free radicals, *Laser Phys. Lett.*, 4(4), 318–321, 2007.

831. Darvin, M. E., Gersonde, I., Albrecht, H., Sterry, W., and Lademann, J., In vivo Raman spectroscopic analysis of the influence of UV radiation on carotenoid antioxidant substance degradation of the human skin, *Laser Phys.*, 16(5), 833–837, 2006.

832. Darvin, M. E., Gersonde, I., Meinke, M., Sterry, W., and Lademann, J., Non-invasive in vivo determination of the carotenoids beta-carotene and lycopene concentrations in the human skin using the Raman spectroscopic method, *J. Phys. D Appl. Phys.*, 38(15), 2696–2700, 2005.

833. Darvin, M. E., Patzelt, A., Knorr, F., Blume-Peytavi, U., Sterry, W., and Lademann, J., One-year study on the variation of carotenoid antioxidant substances in living human skin: Influence of dietary supplementation and stress factors, *J. Biomed. Opt.*, 13(4), 9, 2008.

834. Akkus, O., Polyakova-Akkus, A., Adar, F., and Schaffler, M. B., Aging of microstructural compartments in human compact bone, *J. Bone Miner. Res.*, 18(6), 1012–1019, 2003.

835. Ermakov, I., Ermakova, M., Gellermann, W., and Bernstein, P. S., Macular pigment Raman detector for clinical applications, *J. Biomed. Opt.*, 9(1), 139–148, 2004.

836. Ermakov, I. V., McClane, R. W., Gellermann, W., and Bernstein, P. S., Resonant Raman detection of macular pigment levels in the living human retina, *Opt. Lett.*, 26(4), 202–204, 2001.

837. Hata, T. R., Scholz, T. A., Ermakov, I. V., McClane, R. W., Khachik, F., Gellermann, W., and Pershing, L. K., Non-invasive Raman spectroscopic detection of carotenoids in human skin, *J. Invest. Dermatol.*, 115, 441–448, 2000.

838. Ager, J. W., Nalla, R. K., Balooch, G., Kim, G., Pugach, M., Habelitz, S., Marshall, G. W., Kinney, J. H., and Ritchie, R. O., On the increasing fragility of human teeth with age: A deep-UV resonance Raman study, *J. Bone Miner. Res.*, 21(12), 1879–1887, 2006.

839. Greve, T. M., Andersen, K. B., and Nielsen, O. F., Penetration mechanism of dimethyl sulfoxide in human and pig ear skin: An ATR-FTIR and near-FT Raman spectroscopic in vivo and in vitro study, *Spectrosc. Int. J.*, 22(5), 405–417, 2008.

840. Danielsen, L., Gniadecka, M., Thomsen, H. K., Pedersen, F., Strange, S., Nielsen, K. G., and Petersen, H. D., Skin changes following defibrillation—The effect of high voltage direct current, *Forensic Sci. Int.*, 134(2–3), 134–141, 2003.

841. Stuart, D. A., Yuen, J. M., Shah, N., Lyandres, O., Yonzon, C. R., Glucksberg, M. R., Walsh, J. T., and Van Duyne, R. P., In vivo glucose measurement by surface-enhanced Raman spectroscopy, *Anal. Chem.*, 78(20), 7211–7215, 2006.

842. Rokos, H., Moore, J., Hasse, S., Gillbro, J. M., Wood, J. M., and Schallreuter, K. U., In vivo fluorescence excitation spectroscopy and in vivo Fourier-transform Raman spectroscopy in human skin: Evidence of H_2O_2 oxidation of epidermal albumin in patients with vitiligo, *J. Raman Spectrosc.*, 35(2), 125–130, 2004.

843. Rokos, H., Wood, J. M., Hasse, S., and Schallreuter, K. U., Identification of epidermal L-tryptophan and its oxidation products by in vivo FT-Raman spectroscopy further supports oxidative stress in patients with vitiligo, *J. Raman Spectrosc.*, 39(9), 1214–1218, 2008.

844. Kaminaka, S., Ito, T., Yamazaki, H., Kohda, E., and Hamaguchi, H., Near-infrared multichannel Raman spectroscopy toward real-time in vivo cancer diagnosis, *J. Raman Spectrosc.*, 33(7), 498–502, 2002.

845. Ermakov, I. V., Sharifzadeh, M., Ermakova, M., and Gellermann, W., Resonance Raman detection of carotenoid antioxidants in living human tissue, *J. Biomed. Opt.*, 10(6), 18, 2005.

846. Ermakov, I. V., Ermakova, M. R., and Gellermann, W., Simple Raman instrument for in vivo detection of macular pigments, *Appl. Spectrosc.*, 59(7), 861–867, 2005.

847. Darvin, M. E., Gersonde, I., Albrecht, H., Meinke, M., Sterry, W., and Lademann, J., Non-invasive in vivo detection of the carotenoid antioxidant substance lycopene in the human skin using the resonance Raman spectroscopy, *Laser Phys. Lett.*, 3(9), 460–463, 2006.

848. Ermakov, I. V., Ermakova, M. R., Gellermann, W., and Lademann, J., Noninvasive selective detection of lycopene and beta-carotene in human skin using Raman spectroscopy, *J. Biomed. Opt.*, 9(2), 332–338, 2004.

849. Huang, Z. W., Lui, H., Chen, X. K., Alajlan, A., McLean, D. I., and Zeng, H. S., Raman spectroscopy of in vivo cutaneous melanin, *J. Biomed. Opt.*, 9(6), 1198–1205, 2004.

850. Lieber, C. A. and Mahadevan-Jansen, A., Development of a handheld Raman microspectrometer for clinical dermatologic applications, *Opt. Exp.*, 15(19), 11874–11882, 2007.

851. Zhao, J. H., Lui, H., McLean, D. I., and Zeng, H. S., Integrated real-time Raman system for clinical in vivo skin analysis, *Skin Res. Technol.*, 14(4), 484–492, 2008.

852. Greve, T. M., Andersen, K. B., Nielsen, O. F., ATR-FTIR, FT-NIR and near-FT-Raman spectroscopic studies of molecular composition in human skin in vivo and pig ear skin in vitro, *Spectrosc. Int. J.*, 22(6), 437–457, 2008.

853. Knudsen, L., Johansson, C. K., Philipsen, P. A., Gniadecka, M., and Wulf, H. C., Natural variations and reproducibility of in vivo near-infrared Fourier transform Raman spectroscopy of normal human skin, *J. Raman Spectrosc.*, 33(7), 574–579, 2002.

854. Crowther, J. M., Sieg, A., Blenkiron, P., Marcott, C., Matts, P. J., Kaczvinsky, R., and Rawlings, A. V., Measuring the effects of topical moisturizers on changes in stratum corneum thickness, water gradients and hydration in vivo, *Br. J. Dermatol.*, 159(3), 567–577, 2008.

855. Abramczyk, H., Placek, I., Brozek-Pluska, B., Kurczewski, K., Morawiecc, Z., and Tazbir, M., Human breast tissue cancer diagnosis by Raman spectroscopy, *Spectrosc. Int. J.*, 22(2–3), 113–121, 2008.

856. Huang, Z. W., Lui, H., McLean, D. I., Korbelik, M., and Zeng, H. S., Raman spectroscopy in combination with background near-infrared autofluorescence enhances the in vivo assessment of malignant tissues, *Photochem. Photobiol.*, 81(5), 1219–1226, 2005.

857. Naito, S., Min, Y. K., Sugata, K., Osanai, O., Kitahara, T., Hiruma, H., and Hamaguchi, H., In vivo measurement of human dermis by 1064 nm-excited fiber Raman spectroscopy, *Skin Res., Technol.*, 14(1), 18–25, 2008.

858. Eikje, N. S., Ozaki, Y., Aizawa, K., and Arase, S., Fiber optic near-infrared Raman spectroscopy for clinical noninvasive determination of water content in diseased skin and assessment of cutaneous edema, *J. Biomed. Opt.*, 10(1), 13, 2005.

859. Stamatas, G. N., de Sterke, J., Hauser, M., von Stetten, O., and van der Pol, A., Lipid uptake and skin occlusion following topical application of oils on adult and infant skin, *J. Dermatol. Sci.*, 50(2), 135–142, 2008.

860. Huang, Z. W., Zeng, H. S., Hamzavi, I., McLean, D. I., and Lui, H., Rapid near-infrared Raman spectroscopy system for real-time in vivo skin measurements, *Opt. Lett.*, 26(22), 1782–1784, 2001.

861. Egawa, M., Hirao, T., and Takahashi, M., In vivo estimation of stratum corneum thickness from water concentration profiles obtained with Raman spectroscopy, *Acta Derm. Venereol.*, 87(1), 4–8, 2007.

862. Pudney, P. D. A., Melot, M., Caspers, P. J., van der Pol, A., and Puppels, G. J., An in vivo confocal Raman study of the delivery of trans-retinol to the skin, *Appl. Spectrosc.*, 61(8), 804–811, 2007.

863. Egawa, M. and Iwaki, H., In vivo evaluation of the protective capacity of sunscreen by monitoring urocanic acid isomer in the stratum corneum using Raman spectroscopy, *Skin Res. Technol.*, 14(4), 410–417, 2008.

864. Baraga, J. J., Feld, M. S., and Rava, R. P., Rapid near-infrared Raman spectroscopy of human tissue with a spectrograph and CCD detector, *Appl. Spectrosc.*, 46(1992), 187–190, 1992.

865. Mahadevan-Jensen, A., Mitchell, M. F., Ramanujam, N., Utzinger, U., and Richards-Kortum, R., Development of a fiber optic probe to measure NIR Raman spectra of cervical tissue *in vivo*, *Photochem. Photobiol.*, 68(1998), 427–431, 1998.

866. Shim, M. G., Song, L. M. W. M., Marcon, N. E., and Wilson, B. C., In-vivo near-infrared Raman spectroscopy: Demonstration of feasibility during clinical gastrointestinal endoscopy, *Photochem. Photobiol.*, 72(2000), 146–150, 2000.

867. Schrader, B., Keller, S., Lochte, T., Fendel, S., Moore, D. S., Simon, A., Sawatski, J., NIR FT Raman-spectroscopy in medical diagnosis, *J. Mol. Struct.*, 348, 293–296, 1995.

868. Gahunia, H. K., Vieth, R., and Pritzker, K. P. H., Novel fluorescent compound (DDP) in calf, rabbit, and human articular cartilage and synovial fluid, *J. Rheumatol.*, 29, 154–160, 2002.

869. Dillon, J. and Atherton, S.J., *Photochem. Photobiol.*, 51, 465–468, 1990.

870. Giraev, K. M., Ashurbekov, N. A., and Medzhidov, R. T., Fluorescent-spectroscopic research of in vivo tissues pathological conditions, *Int. J. Mod. Phys. B*, 18, 899–910, 2004.

871. Bridges, J. W., Davies, D. S., and Williams, R. T., *Biochem. J.*, 98, 451–468, 1966.

872. Kelty, C. J., Brown, N. J., Reed, M. W. R., and Ackroyd, R., The use of 5-aminolaevulinic acid as a photosensitiser in photodynamic therapy and photodiagnosis, *Photochem. Photobiol. Sci.*, 1, 158–168, 2002.

873. Claydon, P. E. and Ackroyd, R., 5-Aminolaevulinic acid-induced photodynamic therapy and photo-detection in Barrett's esophagus, *Dis. Esophagus*, 17, 205–212, 2004.

874. Crespi, F., Croce, A. C., Fiorani, S., Masala, B., Heidbreder, C., and Bottiroli, G., Autofluorescence spectrofluorometry of central nervous system (CNS) neuromediators, *Lasers Surg. Med.*, 34, 39–47, 2004.

875. Cubeddu, R., Ramponi, R., Taroni, P., and Canti, G., *J. Photochem. Photobiol. B Biol.*, 11, 319–328, 1991.

876. Ambroz, M., MacRobert, A. J., Morgan, J., Rumbles, G., Foley, M. S. C., and Phillips, D., *J. Photochem. Photobiol. B*, 22, 105–117, 1994.

877. Kessel, D., *Photochem. Photobiol.*, 49, 579–582; Andersson-Engels, S., Johansson, J., Svanberg, K., and Svanberg, S., 1989, unpublished data.

878. Andersson-Engels, S., Baert, L., Berg, R., D'Hallewin, M. A., Johansson, J., Stenram, U., Svanberg, K., and Svanberg, S., *Proc. SPIE*, 1426, 31–43.

879. Bachmann, L., Zezell, D. M., Ribeiro, A. D., Gomes, L., and Ito, A. S., Fluorescence spectroscopy of biological tissues—A review, *Appl. Spectrosc. Rev.*, 41, 575–590, 2006.

880. Lai, W. F. T., Chang, C. H., Tang, Y., Bronson, R., and Tung, C. H., Early diagnosis of osteoarthritis using cathepsin B sensitive near-infrared fluorescent probes, *Osteoarthr. Cartil.*, 12, 239–244, 2004.

881. Eldred, G. E., Miller, G. V., Stark, W. S., and Feeney-Burns, L., *Science*, 16, 757–759, 1982.

882. Sohal, R. S., *Enzymology*, 105, 484–487, 1984.

883. Richards-Kortum, R., Rava, R. P., Baraga, J., Fitzmaurice, M., Kramer, J., and Feld, M., in R. Pratesi, (ed.), *Optronic Techniques in Diagnostic and Therapeutic Medicine*, Plenum Press, New York, 1990.

884. Konig, K., Schneckenburger, H., Hemmer, J., Tromberg, B., and Steiner, R., *Proc. SPIE*, 2135, 129–138, 1994.

885. Andersson-Engels, S., Baert, L., Berg, R., D'Hallewin, M. A., Johansson, J., Stenram, U., Svanberg, K., and Svanberg, S., *Proc. SPIE*, 1426, 31–43, 1991; Rava, R. P., Richards-Kortum, R., Fitzmaurice, M., Cothren, R., Petras, R., Sivak, M., Levin, H., and Feld, M., *Proc. SPIE*, 1426, 68–78, 1991.

886. Wagnières, G. A., Star, W. M., and Wilson, B. C., *Photochem. Photobiol.*, 68, 603–632, 1998.

887. Crowell, E., Wang, G. F., Cox, J., Platz, C. P., and Geng, L., Correlation coefficient mapping in fluorescence spectroscopy: Tissue classification for cancer detection, *Anal. Chem.*, 77, 1368–1375, 2005.

888. Na, R. H., Stender, I. M., Henriksen, M., and Wulf, H. C., Autofluorescence of human skin is age-related after correction for skin pigmentation and redness, *J. Invest. Dermatol.*, 116, 536–540, 2001.

889. Wu, Y. C. and Qu, J. A. Y., Combined depth- and time-resolved autofluorescence spectroscopy of epithelial tissue, *Opt. Lett.*, 31, 1833–1835, 2006.

890. Cheng, S.-H., Master's report, University of Texas at Austin, Austin, TX, 1992.

891. Fujimoto, D., *Biochem. Biophys. Res. Commun.*, 76, 1124–1129, 1977.

892. Eyre, D. and Paz, M., *Annu. Rev. Biochem.*, 53, 717–748, 1984.

893. Moesta, K. T., Ebert, B., Handke, T., Nolte, D., Nowak, C., Haensch, W. E., Pandey, R. K., Dougherty, T. J., Rinneberg, H., and Schlag, P. M., Protoporphyrin IX occurs naturally in colorectal cancers and their metastases, *Cancer Res.*, 61, 991–999, 2001.

894. Ueda, Y. and Kobayashi, M., Spectroscopic studies of autofluorescence substances existing in human tissue: Influences of lactic acid and porphyrins, *Appl. Opt.*, 43, 3993–3998, 2004.

895. Kohl, T. and Schwille, P., *Microscopy Techniques*, Springer-Verlag, Berlin, Germany, Vol. 95, pp. 107–142, 2005.

896. Yang, Y., Ye, T., Li, F., and Ma, P., *Lasers Surg. Med.*, 7, 528–532, 1987.

897. Tavare, J. M., Fletcher, L. M., and Welsh, G. I., Review—Using green fluorescent protein to study intracellular signalling, *J. Endocrinol.*, 170, 297–306, 2001.

898. Barnes, D., Aggarwal, S., Thomsen, S., Fitz-maurice, M., and Richards-Kortum, R., *Photochem. Photobiol.* 58, 297–303, 1993.

899. Mayeno, A. N., Hamann, K. J., and Gleich, G. J., *J. Leukoc. Biol.*, 51, 172–175, 1992.
900. Weil, G. J. and Chused, T. M., *Blood* 57, 1099–1104.
901. Tsuchida, M., Miura, T., and Aibara, K., *Chem. Phys. Lipids*, 44, 297–225, 1987.
902. Richards-Kortum, R. and Sevick-Muraca, E., *Annu. Rev. Phys. Chem.*, 47, 555–606, 1996.
903. Visser, A. J. W. G., Sanetema, J. S., and van Hoek, A., *Photochem. Photobiol.*, 39, 11–16.
904. Schneckenburger, H. and Konig, K., *Opt. Eng.*, 31, 1447–1451, 1992.
905. Malakova, J., Nobilis, M., Svoboda, Z., Lisa, M., Holcapek, M., Kvetina, J., Klimes, J., and Palicka, V., High-performance liquid chromatographic method with UV photodiode-array, fluorescence and mass spectrometric detection for simultaneous determination of galantamine and its phase I metabolites in biological samples, *J. Chromatogr. B-Anal. Technol. Biomed. Life Sci.*, 853, 265–274, 2007.
906. Andersson-Engels, S., Ankerst, J., Johansson, J., Svanberg, K., and Svanberg, S., *Lasers Med. Sci.*, 4, 115–123; Kessel, D., Byrne, C. J., and Ward, A. D., *Photochem. Photobiol.*, 53, 469–474, 1991.
907. Moan, J. and Sommer, S., *Photobiochem. Photobiophys.*, 3, 93–103, 1981; Andreoni, A. and Cubeddu, R., *Chem. Phys. Lett.*, 108, 141–144, 1984.
908. Li, B. H., Lu, Z. K., and Xie, S. S., Time-resolved fluorescence studies of hematoporphyrin monomethyl ether for photodynamic diagnosis, *Spectrosc. Spect. Anal.*, 23, 331–333, 2003.
909. Liebert, A., Wabnitz, H., Obrig, H., Erdmann, R., Moller, M., Macdonald, R., Rinneberg, H., Villringer, A., and Steinbrink, J., Non-invasive detection of fluorescence from exogenous chromophores in the adult human brain.
910. Zhang, H., Uselman, R. R., and Yee, D. Exogenous near-infrared fluorophores and their applications in cancer diagnosis: Biological and clinical perspectives, *Expert Opin. Med. Diagn.* 5, 241–251, 2011.
911. Pena, A. M., Strupler, M., Boulesteix, T., and Schanne-Klein, M. C., Spectroscopic analysis of keratin endogenous signal for skin multiphoton microscopy, *Opt. Express*, 13, 6268–6274, 2005.
912. Framme, C., Schule, G., Birngruber, R., Roider, J., Schutt, F., Kopitz, J., Holz, F. G., and Brinkmann, R., Temperature dependent fluorescence of A2-E, the main fluorescent lipofuscin component in the RPE, *Curr. Eye Res.*, 29, 287–291, 2004.
913. Huang, Z. W., Zeng, H. S., Hamzavi, I., Alajlan, A., Tan, E., McLean, D. I., and Lui, H., Cutaneous melanin exhibiting fluorescence emission under near-infrared light excitation, *J. Biomed. Opt.*, 11, 6, 2006.
914. Han, X., Lui, H., McLean, D. I., and Zeng, H., Near-infrared autofluorescence imaging of cutaneous melanins and human skin in vivo, *J. Biomed. Opt.*, 14, 024017/024011–024017/024015, 2009.
915. Aizawa, K., Okunaka, T., Ohtani, T., Kaabe, H., Yasunaka, Y., O'Hata, S., Ohtomo, N. et al., *Photochem. Photobiol.*, 46, 789–794; Kessel, D., *Photochem. Photobiol.*, 49, 447–452, 1989.
916. Campbell, I. D. and Dwek, R. A., *Biological Spectroscopy*, Benjamin Cummings, Menlo Park, CA, 1984.
917. Lakowicz, J. R., *Principles of Fluorescence Spectroscopy*, Plenum Press, New York, 1985.
918. Wang, H. W., Gukassyan, V., Chen, C. T., Wei, Y. H., Guo, H. W., Yu, J. S., and Kao, F. J., Differentiation of apoptosis from necrosis by dynamic changes of reduced nicotinamide adenine dinucleotide fluorescence lifetime in live cells, *J. Biomed. Opt.*, , 13, 9, 2008.
919. Schneckenburger, H. and Konig, K., *Opt. Eng.*, 31, 1447–1451; Lakowicz, J. R., *Principles of Fluorescence Spectroscopy*, Plenum Press, New York, 1985.
920. Lin, C.-W., Shulok, J. R., Wong, Y.-K., Schanbacher, C. F., Cinotta, L., and Foley, J. W., *Cancer Res.*, 51, 1109–1116, 1991.
921. Lakowicz, J. R., *Principles of Fluorescence Spectroscopy*, 2nd edn., Kluwer Academic, New York, 1999.
922. Cubeddu, R., Ramponi, R., and Bottiroli, G., *Chem. Phys. Lett.*, 128, 439–442, 1986.
923. Kessel, D., *Photochem. Photobiol.*, 50, 169–174, 1989.
924. Tralau, C. J., Barr, H., Sandeman, D. R., Barton, T., Lewin, M. R., and Brown, S. G., *Photochem. Photobiol.*, 46, 777–781, 1987.
925. Andeoni, A., Bottiroli, G., Colasanti, A., Giangare, M. C., Riccio, P., Roberti, G., and Vaghi, P., *J. Biochem. Biophys. Methods*, 29(2), 157–172, 1994.

926. Aicher, A., Miller, K., Reich, E., and Hautmann, R., *Opt. Eng.*, 32(2), 342–346, 1993.

927. Pantanelli, S. M., Li, Z. Q., Fariss, R., Mahesh, S. P., Liu, B. Y., and Nussenblatt, R. B., Differentiation of malignant B-lymphoma cells from normal and activated T-cell populations by their intrinsic autofluorescence, *Cancer Res.*, 69, 4911–4917, 2009.

928. Moser, J. G., Ruck, A., Schwarzmaier, H. J., and Westphalfrosch, C., *Opt. Eng.*, 31(7), 1441–1446, 1992.

929. DaCosta, R. S., Wilson, B. C., and Marcon, N. E., Optical techniques for the endoscopic detection of dysplastic colonic lesions, *Curr. Opin. Gastroenterol.*, 21, 70–79, 2005.

930. Villette, S., Pigaglio-Deshayes, S., Vever-Bizet, C., Validire, P., and Bourg-Heckly, G., Ultraviolet-induced autofluorescence characterization of normal and tumoral esophageal epithelium cells with quantitation of NAD(P)H, *Photochem. Photobiol. Sci.*, 5, 483–492, 2006.

931. Anidjar, M., Cussenot, O., Blais, J., Bourdon, O., Avrillier, S., Ettori, D., Villette, J. M., Fiet, J., Teillac, P., and LeDuc, A., *J. Urol.*, 155(5), 1771–1774, 1996.

932. Leppert, J., Krajewski, J., Kantelhardt, S. R., Schlaffer, S., Petkus, N., Reusche, E., Huttmann, G., and Giese, A., Multiphoton excitation of autofluorescence for microscopy of glioma tissue, *Neurosurgery*, 58, 759–767, 2006.

933. Seidlitz, H. K., Stettmaier, K., Wessels, J. M., and Schneckenburger, H., *Opt. Eng.*, 31(7), 1482–1486, 1992.

934. Amano, T. K. K. and Ohkawa, M., *Arch Androl.*, 36(1), 9–15, 1996.

935. Dimitrow, E., Riemann, I., Ehlers, A., Koehler, M. J., Norgauer, J., Elsner, P., Konig, K., and Kaatz, M., Spectral fluorescence lifetime detection and selective melanin imaging by multiphoton laser tomography for melanoma diagnosis, *Exp. Dermatol.*, 18, 509–515, 2009.

936. Grossman, N., Ilovitz, E., Chaims, O., Salman, A., Jagannathan, R., Mark, S., Cohen, B., Gopas, J., and Mordechai, S., *J. Biochem. Biophys. Methods*, 50(1), 53–63, 2001.

937. Conklin, M. W., Provenzano, P. P., Eliceiri, K. W., Sullivan, R., and Keely, P. J., Fluorescence lifetime imaging of endogenous fluorophores in histopathology sections reveals differences between normal and tumor epithelium in carcinoma in situ of the breast, *Cell Biochem. Biophys.*, 53, 145–157, 2009.

938. Evans, N. D., Gnudi, L., Rolinski, O. J., Birch, D. J. S., and Pickup, J. C., Glucose-dependent changes in NAD(P)H-related fluorescence lifetime of adipocytes and fibroblasts in vitro: Potential for non-invasive glucose sensing in diabetes mellitus, *J. Photochem. Photobiol. B-Biol.*, 80, 122–129, 2005.

939. Sacks, P. G., Savage, H. E., Levine, J., Kolli, V. R., Alfano, R. R., and Schantz, S. P., *Cancer Lett.*, 104(2), 171–181, 1996.

940. Konig, K., Liu, Y. G., Sonek, G. J., Berns, M. W., and Tromberg, B. J., *Photochem. Photobiol.*, 62(5), 830–835, 1995.

941. Balazs, M. S. J., Lee, W. C., Haugland, R. P., Guzikowski, A. P., Fulwyler, M. J., Damjanovich, S., Feurstein, B. G., and Pershadsingh, H. A., *J. Cell Biochem.*, 46(3), 266–276, 1991.

942. Seidlitz, H. K., Stettmaier, K., Wessels, J. M., and Schneckenburger, H., *Opt. Eng.*, 31, 1482–1486.

943. Morgan, J. I. W., Dubra, A., Wolfe, R., Merigan, W. H., and Williams, D. R., In vivo autofluorescence imaging of the human and macaque retinal pigment epithelial cell mosaic, *Invest. Ophthalmol. Vis. Sci.*, 50, 1350–1359, 2009.

944. Zhang, J. C., Savage, H. E., Sacks, P. G., Delohery, T., Alfano, R. R., Katz, A., and Schantz, S. P., *Lasers Surg. Med.*, 20(3), 319–331, 1997.

945. Banerjee, B., Henderson, J. O., Chaney, T. C., and Davidson, N. O., Detection of murine intestinal adenomas using targeted molecular autofluorescence, *Dig. Dis. Sci.*, 49, 54–59, 2004.

946. Blackwell, J., Katika, K. M., Pilon, L., Dipple, K. M., Levin, S. R., and Nouvong, A., In vivo time-resolved autofluorescence measurements to test for glycation of human skin, *J. Biomed. Opt.*, 13, 15, 2008.

947. Ye, B. Q. and Abela, G. S., *Opt. Eng.*, 32(2), 326–333, 1993.

948. Zheng, W., Lau, W., Cheng, C., Soo, K. C., and Olivo, M., Optimal excitation-emission wavelengths for autofluorescence diagnosis of bladder tumors, *Int. J. Cancer*, 104, 477–481, 2003.

949. Lualdi, M., Colombo, A., Leo, E., Morelli, D., Vannelli, A., Battaglia, L., Poiasina, E., and Marchesini, R., Natural fluorescence spectroscopy of human blood plasma in the diagnosis of colorectal cancer: Feasibility study and preliminary results, *Tumori*, 93, 567–571, 2007.

950. Marcu, L., Jo, J. A., Butte, P. V., Yong, W. H., Pikul, B. K., Black, K. L., and Thompson, R. C., Fluorescence lifetime spectroscopy of glioblastoma multiforme, *Photochem. Photobiol.*, 80, 98–103, 2004.

951. Lin, W. C., Sandberg, D. I., Bhatia, S., Johnson, M., Morrison, G., and Ragheb, J., Optical spectroscopy for in-vitro differentiation of pediatric neoplastic and epileptogenic brain lesions, *J. Biomed. Opt.*, 14, 10, 2009.

952. Hage, R., Galhanone, P. R., Zangaro, R. A., Rodrigues, K. C., Pacheco, M. T. T., Martin, A. A., Netto, M. M., Soares, F. A., and da Cunha, I. W., Using the laser-induced fluorescence spectroscopy in the differentiation between normal and neoplastic human breast tissue, *Lasers Med. Sci.*, 18, 171–176, 2003.

953. Chowdary, M. V. P., Mahato, K. K., Kumar, K. K., Mathew, S., Rao, L., Krishna, C. M., and Kurien, J., Autofluorescence of breast tissues: Evaluation of discriminating algorithms for diagnosis of normal, benign, and malignant conditions, *Photomed. Laser Surg.*, 27, 241–252, 2009.

954. Zellweger, M., Grosjean, P., Goujon, D., Monnier, P., van den Bergh, H., and Wagnieres, G., In vivo autofluorescence spectroscopy of human bronchial tissue to optimize the detection and imaging of early cancers, *J. Biomed. Opt.*, 6, 41–51, 2001.

955. Rodero, A. B., Silveira, L., Rodero, D. A., Racanicchi, R., and Pacheco, M. T. T., Fluorescence spectroscopy for diagnostic differentiation in uteri's cervix biopsies with cervical/vaginal atypical cytology, *J. Fluoresc.*, 18, 979–985, 2008.

956. Chidananda, S. M., Satyamoorthy, K., Rai, L., Manjunath, A. P., and Kartha, V. B., Optical diagnosis of cervical cancer by fluorescence spectroscopy technique, *Int. J. Cancer*, 119, 139–145, 2006.

957. Ramanujam, N., Richards-Kortum, R., Thomsen, S., Mahadevan-Jansen, A., Follen, M., and Chance, B., Low temperature fluorescence imaging of freeze-trapped human cervical tissues, *Opt. Express*, 8, 335–343, 2001.

958. Drezek, R., Sokolov, K., Utzinger, U., Boiko, I., Malpica, A., Follen, M., and Richards-Kortum, R., *J. Biomed. Opt.*, 6(4), 385–396, 2001.

959. Horak, L., Zavadil, J., Duchac, V., Javorsky, S., Kostka, F., Svec, A., and Lezal, D., Auto-fluorescence spectroscopy of colorectal carcinoma: Ex vivo study, *J. Optoelectron. Adv. Mater.*, 8, 396–399, 2006.

960. Giorgadze, T. A., Shiina, N., Baloch, Z. W., Tomaszewski, J. E., and Gupta, P. K., Improved detection of amyloid in fat pad aspiration: An evaluation of Congo red stain by fluorescent microscopy, *Diagn. Cytopathol.*, 31, 300–306, 2004.

961. van der Veen, M. H. t. B., J. J., *Eur. J. Oral Sci.*, 103(6), 375–381, 1995.

962. Banerjee, A. and Boyde, A., *Caries Res.*, 32(3), 219–226, 1998.

963. Barnes, D. A. S., Thomsen, S., Fitzmaurice, M., and Richards-Kortum, R., *Photochem. Photobiol.*, 58(2), 297–303, 1993.

964. Romer, T. J., Fitzmaurice, M., Cothren, R. M., Richardskortum, R., Petras, R., Sivak, M. V., and Kramer, J. R., *Am. J. Gastroenterol.*, 90(1), 81–87, 1995.

965. Ramanujam, N., Richards-Kortum, R., Thomsen, S., Mahadevan-Jansen, A., Follen, M., and Chance, B., *Opt. Express*, 8(6), 335–343, 2001.

966. Ramanujam, N., Richards-Kortum, R., Thomsen, S., Mahadevan-Jansen, A., Follen, M., and Chance, B., *Opt. Express*, 8(6), 335–343, 2001.

967. Lucas, A., Radosavljevk, M. J., Lu, E., and Gaffney, E. J., *Can. J. Cardiol.*, 6(6), 219–228, 1990.

968. Richards-Kortum, R., Rava, R. P., Fitzmaurice, M., Kramer, J. R., and Feld, M. S., *Appl. Cardiol., Am. Heart J.*, 122(4), 1141–1150, 1991.

969. Scheu, M., Kagel, H., Zwaan, M., Lebeau, A., and Engelhardt, R., *Lasers Surg. Med.*, 11(2), 133–140, 1991.

970. Bosshart, F. U. U., Hess, O. M., Wyser, J., Mueller, A., Schneider, J., Niederer, P., Anliker, M., and Krayenbuehl, H. P., *Cardiovasc. Res.*, 26(6), 620–625, 1992.

971. Morguet, A. J., Korber, B., Abel, B., Hippler, H., Wiegand, V., and Kreuzer, H., *Lasers Surg. Med.*, 14(3), 238–248, 1994.
972. Verbunt, R., Fitzmaurice, M. A., Kramer, J. R., Ratliff, N. B., Kittrell, C., Taroni, P., Cothren, R. M., Baraga, J., and Feld, M., *Am. Heart J.*, 123(1), 208–216, 1992.
973. Chaudhry, H. W. R.-K. R., Kolubayev, T., Kittrell, C., Partovi, F., Kramer, J. R., and Feld, M. S., *Lasers Surg. Med.*, 9(6), 572–580, 1989; Andersson-Engels, S. J., Svanberg, K., and Svanberg, S., *Photochem. Photobiol.*, 53(6), 807–814, 1991.
974. Deckelbaum, L. I., Desai, S. P., Kim, C., and Scott, J. J., *Lasers Surg. Med.*, 16(3), 226–234, 1995.
975. Baraga, J. J., Rava, R. P., Taroni, P., Kittrell, C., Fitzmaurice, M., and Feld, M. S., *Lasers Surg. Med.*, 10(3), 245–261, 1990.
976. Laufer, G., Wollenek, G., Hohla, K., Horvat, R., Henke, K. H., Buchelt, M., Wutzl, G., and Wolner, E., *Circulation*, 78(4), 1031–1039, 1988.
977. Koenig, F., McGovern, F. J., Althausen, A. F., Deutsch, T. F., and Schomacker, K. T., *J. Urol.*, 156(5), 1597–1601, 1996.
978. Frimberger, D., Zaak, D., Stepp, H., Knuchel, R., Baumgartner, R., Schneede, P., Schmeller, N., and Hofstetter, A., *Urology*, 58(3), 372–375, 2001.
979. Hanlon, E. B., Itzkan, I., Dasari, R. R., Feld, M. S., Ferrante, R. J., McKee, A. C., Lathi, D., and Kowall, N. W., *Photochem. Photobiol.*, 70(2), 236–242, 1999.
980. Burger, P. C., Minn, A. Y., Smith, J. S., Borell, T. J., Jedlicka, A. E., Huntley, B. K., Goldthwaite, P. T., Jenkins, R. B., and Feuerstein, B. G., Losses of chromosomal arms 1p and 19q in the diagnosis of oligodendroglioma. A study of paraffin-embedded sections, *Modern Pathol.*, 14, 842–853, 2001.
981. Butte, P. V., Pikul, B. K., Hever, A., Yong, W. H., Black, K. L., and Marcu, L., Diagnosis of meningioma by time-resolved fluorescence spectroscopy, *J. Biomed. Opt.*, 10, 9, 2005.
982. Schneider, M., Teuchner, K., and Leupold, D., Two-photon fluorescence of ocular melanomas. Studies on a new diagnostic method, *Ophthalmologe*, 102, 703–707, 2005.
983. Lin, W. C., Toms, S. A., Motamedi, M., Jansen, E. D., and Mahadevan-Jansen, A., *J. Biomed. Opt.*, 5(2), 214–220, 2000.
984. Demos, S. G., Bold, R., White, R. D., and Ramsamooj, R., Investigation of near-infrared autofluorescence imaging for the detection of breast cancer, *IEEE J. Sel. Top. Quant. Electron.*, 11, 791–798, 2005.
985. Gupta, S., Bhawna, Goswami, P., Agarwal, A., and Pradhan, A., Experimental and theoretical investigation of fluorescence photobleaching and recovery in human breast tissue and tissue phantoms, *Appl. Opt.*, 43, 1044–1052, 2004.
986. Alfano, R. R., Tang, G. C., Pradhan, A., Lam, W., Choy, D. S. J., and Opher, E., *IEEE J. Quant. Electron.*, 23(10), 1806–1811, 1987; Tang, G. C., Pradhan, A., Sha, W., Chen, J., Liu, C. H., Wahl, S. J., and Alfano, R. R., *Appl. Opt.*, 28(12), 2337–2342, 1989.
987. Anastassopoulou, N., Arapoglou, B., Demakakos, P., Makropoulou, M. I., Paphiti, A., and Serafetinides, A. A., *Lasers Surg. Med.*, 28(1), 67–73, 2001.
988. Richards-Kortum, R., Mitchell, M. F., Ramanujam, N., Mahadevan, A., and Thomsen, S., *J. Cell Biochem. Suppl.*, 19, 111–119, 1994.
989. Lohmann, W. M. J., Lohmann, C., and Kunzel, W., *Eur. J. Obstet. Gynecol. Reprod. Biol.*, 31(3), 249–253, 1989.
990. Richards-Kortum, R., Rava, R. P., Petras, R. E., Fitzmaurice, M., Sivak, M., and Feld, M. S., *Photochem. Photobiol.*, 53(6), 777–786, 1991.
991. Kapadia, C. R., Cutruzzola, F. W., O'Brien, K. M., Stetz, M. L., Enriquez, R., and Deckelbaum, L. I., *Gastroenterology*, 99(1), 150–157, 1990.
992. Bottiroli, G., Croce, A. C., Locatelli, D., Marchesini, R., Pignoli, E., Tomatis, S., Cuzzoni, C., Dipalma, S., Dalfante, M., and Spinelli, P., *Lasers Surg. Med.*, 16, 48–60, 1995.
993. Chwirot, B. W., Kowalska, M., Sypniewska, N., Michniewicz, Z., and Gradziel, M., *J. Photochem. Photobiol. B-Biol.*, 50(2–3), 174–183, 1999.

994. Perk, M., Flynn, G. J., Gulamhusein, S., Wen, Y., Smith, C., Bathgate, B., Tulip, J., Parfrey, N. A, and Lucas, A., *Pacing Clin. Electrophysiol.*, 16(8), 1701–1712, 1993.

995. Larsen, M. and Lundandersen, H., *Graefes Arch. Clin. Exp. Ophthalmol.*, 229(4), 363–370, 1991.

996. Ingrams, D. R., Dhingra, J. K., Roy, K., Perrault, D. F., Bottrill, I. D., Kabani, S., Rebeiz, E. E. et al., *J. Sci. Spec. Head Neck*, 19(1), 27–32, 1997.

997. Roy, K., Bottrill, I. D., Ingrams, D. R., Pankratov, M. M., Rebeiz, E. E., Woo, P., Kabani, S. et al., *SPIE*, 2395, 135–142, 1995.

998. Chen, C. T., Wang, C. Y., Kuo, Y. S., Chiang, H. H., Chow, S. N., Hsiao, I. Y., and Chang, C. P., *Proc. Natl. Sci. Counc. Repub. China, Part B, Life Sci.*, 20, 123–130, 1996.

999. Lohmann, W., Nilles, M., and Bodeker, R. H., *Naturwissenschaften*, 78, 456–457, 1991.

1000. Panjehpour, M., Julius, C. E., Phan, M. N., Vo-Dinh, T., and Overholt, S., Laser-induced fluorescence spectroscopy for in vivo diagnosis of non-melanoma skin cancers, *Lasers Surg. Med.*, 31, 367–373, 2002.

1001. Yaroslavsky, A. N., Salomatina, E., Neel, V., Anderson, R., and Flotte, T., Fluorescence polarization of tetracycline derivatives as a technique for mapping nonmelanoma skin cancers, *J. Biomed. Opt.*, 12, 014005-014001–014005-014009, 2007.

1002. Chwirot, B. W., Chwirot, S., Jedrzejczyk, W., Jackowski, M., Raczynska, A. M., Winczakiewicz, J., and Dobber, J., *Lasers Surg. Med.*, 21(2), 149–158, 1997.

1003. Yappert, M. C., Borchman, D., and Byrdwell, W. C., *Invest. Ophthalmol. Vis. Sci.*, 34(3), 630–636, 1993.

1004. Elson, D., Requejo-Isidro, J., Munro, I., Reavell, F., Siegel, J., Suhling, K., Tadrous, P. et al., *Royal Soc. Chem.*, 795–801, 2004.

1005. Silveira, L., Betiol, J. A., Silveira, F. L., Zangaro, R. A., and Pacheco, M. T. T., Laser-induced fluorescence at 488 nm excitation for detecting benign and malignant lesions in stomach mucosa, *J. Fluoresc.*, 18, 35–40, 2008.

1006. Andersson-Engels, S., Johansson, J., Stenram, U., Svanberg, K., and Svanberg, S., *IEEE J. Quant. Electron.*, 26, 2207–2217.

1007. Bagdonas, S., Zurauskas, E., Streckyte, G., and Rotomskis, R., Spectroscopic studies of the human heart conduction system ex vivo: Implication for optical visualization, *J. Photochem. Photobiol. B-Biol.*, 92, 128–134, 2008.

1008. Hegedus, Z. L. N. U., *Arch. Int. Physiol. Biochim. Biophys.*, 102(6), 311–313, 1994.

1009. Drezek, R., Sokolov, K., Utzinger, U., Boiko, I., Malpica, A., Follen, M., and Richards-Kortum, R., Understanding the contributions of NADH and collagen to cervical tissue fluorescence spectra: Modeling, measurements, and implications, *J. Biomed. Opt.*, 6, 385–396, 2001.

1010. Andersson-Engels, S., Ankerst, J., Johansson, J., Svanberg, K., and Svanberg, S., *Photochem. Photobiol.*, 57, 978–983, 1993.

1011. De Veld, D. C. G., Witjes, M. J. H., Sterenborg, H., and Roodenburg, J. L. N., The status of in vivo autofluorescence spectroscopy and imaging for oral oncology, *Oral Oncol.*, 41, 117–131, 2005.

1012. Kamath, S. D., Bhat, R. A., Ray, S., and Mahato, K. K., Autofluorescence of normal, benign, and malignant ovarian tissues: A pilot study, *Photomed. Laser Surg.*, 27, 325–335, 2009.

1013. Chandra, M., Scheiman, J., Heidt, D., Simeone, D., McKenna, B., and Mycek, M. A., Probing pancreatic disease using tissue optical spectroscopy, *J. Biomed. Opt.*, 12, 3, 2007.

1014. Rivasi, F., Botticelli, L., Bettelli, S. R., and Masellis, G., Alveolar rhabdomyosarcoma of the uterine cervix. A case report confirmed by FKHR break-apart rearrangement using a fluorescence in situ hybridization probe on paraffin-embedded tissues, *Int. J. Gynecol. Pathol.*, 27, 442–446, 2008.

1015. Yadav, R., Mukherjee, S., Hermen, M., Tan, G., Maxfield, F. R., Webb, W. W., and Tewari, A. K., Multiphoton microscopy of prostate and periprostatic neural tissue: A promising imaging technique for improving nerve-sparing prostatectomy, *J. Endourol.*, 23, 861–867, 2009.

1016. Meng, T., Xu, D. F., Xu, Y. Z., Ying, Z., Wang, D. J., and Wu, J. G., Investigation on the fluorescence spectroscopy of normal and malignant rectum tissues, *Spectrosc. Spec. Anal.*, 27, 1156–1160, 2007.

1017. Ebert, B., Sukowski, U., Grosenick, D., Wabnitz, H., Moesta, T. K., Licha, K., Becker, A., Semmler, W., Schlag, P. M., and Rinneberg, H., Near-infrared fluorescent dyes for enhanced contrast in optical mammography: Phantom experiments, *J. Biomed. Opt.*, 6, 134–140, 2001.

1018. Devoisselle, J. M., Maunoury, V., Mordon, S., and Coustaut, D., *Opt. Eng.*, 32(2), 239–243, 1993.

1019. Christov, A., Dai, E., Drangova, M., Liu, L. Y., Abela, G. S., Nash, P., McFadden, G., and Lucas, A., *Photochem. Photobiol.*, 72(2), 242–252, 2000.

1020. Andersson-Engels, S., Ankerst, J., Johansson, J., Svanberg, K., and Svanberg, S., *Photochem. Photobiol.*, 57(6), 978–983, 1993.

1021. Chung, Y. G. Schwartz, J. A., Gardner, C. M., Sawaya, R. E., and Jacques, S. L., *J. Korean Med. Sci.*, 12(2), 135–142, 1997.

1022. Fiorotti, R. C., Nicola, J. H., and Nicola, E. M. D., Native fluorescence of oral cavity structures: An experimental study in dogs, *Photomed. Laser Surg.*, 24, 22–28, 2006.

1023. Frangioni, J. V., In vivo near-infrared fluorescence imaging, *Curr. Opin. Chem. Biol.*, 7, 626–634, 2003.

1024. Koo, V., Hamilton, P. W., and Williamson, K., Non-invasive in vivo imaging in small animal research, *Cell. Oncol.*, 28, 127–139, 2006.

1025. Graves, E. E., Weissleder, R., and Ntziachristos, V., Fluorescence molecular imaging of small animal tumor models, *Curr. Mol. Med.*, 4, 419–430, 2004.

1026. Schantz, S. P. and Alfano, R. R., *J. Cell. Biochem.*, 199–204, 1993.

1027. Glasgold, R., Glasgold, M., Savage, H., Pinto, J., Alfano, R., and Schantz, S., *Cancer Lett.*, 82(1), 33–41, 1994.

1028. Oraevsky, A. A., Jacques, S. L., Pettit, G. H., Sauerbrey, R. A., Tittel, F. K., Nguy, J. H., and Henry, P. D., *Circ. Res.*, 72(1), 84–90, 1993.

1029. Morgan, D. C., Wilson, J. E., MacAulay, C. E., MacKinnon, N. B., Kenyon, J. A., Gerla, P. S., Dong, C. M. et al., *Circulation*, 100(11), 1236–1241, 1999.

1030. Shehada, R. E. N., Marmarelis, V. Z., Mansour, H. N., and Grundfest, W. S., *IEEE Trans. Biomed. Eng.*, 47(3), 301–312, 2000.

1031. Svanberg, K., Kjellen, E., Ankerst, J., Montan, S., Sjoholm, E., and Svanberg, S., *Cancer Res.*, 46(8), 3803–3808, 1986.

1032. Hansch, A., Sauner, D., Hilger, I., Frey, O., Haas, M., Malich, A., Brauer, R., and Kaiser, W. A., Noninvasive diagnosis of arthritis by autofluorescence, *Invest. Radiol.*, 38, 578–583, 2003.

1033. Sterenborg, H., Thomsen, S., Jacques, S. L., and Motamedi, M., *Melanoma Res.*, 5(4), 211–216, 1995.

1034. Fournier, L. S., Lucidi, V., Berejnoi, K., Miller, T., Demos, S. G., and Brasch, R. C., In-vivo NIR auto-fluorescence imaging of rat mammary tumors, *Opt. Express*, 14, 6713–6723, 2006.

1035. Colasanti, A., Kisslinger, A., Fabbrocini, G., Liuzzi, R., Quarto, M., Riccio, P., Roberti, G., and Villani, F., *Lasers Surg. Med.*, 26(5), 441–448, 2000.

1036. Dhingra, J. K., Zhang, X., McMillan, K., Kabani, S., Manoharan, R., Itzkan, I., Feld, M. S., and Shapshay, S. M., *Laryngoscope*, 108(4), 471–475, 1998.

1037. Coghlan, L., Utzinger, U., Drezek, R., Heintzelman, D., Zuluaga, A., Brookner, C., Richards-Kortum, R., Gimenez-Conti, I., and Follen, M., *Opt. Express*, 7(12), 436–446, 2000.

1038. Vengadesan, N., Aruna, P., and Ganesan, S., *Br. J. Cancer*, 77(3), 391–395, 1998.

1039. Tassetti, V. H. A., Sowinska, M., Evrard, S., Heisel, F., Cheng, L. Q., Miehe, J. A., Marescaux, J., and Aprahamian, M., *Photochem. Photobiol.*, 65, 997–1006, 1997.

1040. Miyoshi, N., Ogasawara, T., Nakano, K., Tachihara, R., Kaneko, S., Sano, K., Fukuda, M., and Hisazumi, H., In light of recent developments, application of fluorescence spectral analysis in tumor diagnosis, *Appl. Spectrosc. Rev.*, 39, 437–455, 2004.

1041. Svanberg, K., Liu, D. L., Wang, I., AnderssonEngels, S., Stenram, U., and Svanberg, S., *Br. J. Cancer*, 74(10), 1526–1533, 1996.

1042. Johansson, J., Berg, R., Svanberg, K., and Svanberg, S., *Lasers Surg. Med.*, 20(3), 272–279, 1997.

1043. Tian, W. D., Gillies, R., Brancaleon, L., and Kollias, N., Aging and effects of ultraviolet A exposure may be quantified by fluorescence excitation spectroscopy in vivo, *J. Invest. Dermatol.*, 116, 840–845, 2001.

1044. Nilsson, H., Johansson, J., Svanberg, K., Svanberg, S., Jori, G., Reddi, E., Segalla, A., Gust, D., Moore, A. L., and Moore, T. A., *Br. J. Cancer*, 76(3), 355–364, 1997.

1045. Glanzmann, T., Ballini, J. P., van den Bergh, H., and Wagnieres, G., *Rev. Sci. Instrum.*, 70(10), 4067–4077, 1999.

1046. Toms, S. A., Konrad, P. E., Lin, W. C., and Weil, R. J., Neuro-oncological applications of optical spectroscopy, *Technol. Cancer Res. Treat.*, 5, 231–238, 2006.

1047. Thiberville, L., Salaun, M., Lachkar, S., Dominique, S., Moreno-Swirc, S., Vever-Bizet, C., and Bourg-Heckly, G., Human in vivo fluorescence microimaging of the alveolar ducts and sacs during bronchoscopy, *Euro. Respir. J.*, 33, 974–985, 2009.

1048. Koenig, F., McGovern, F. J., Althausen, A. F., Deutsch, and T. F. Schomacker, K. T., *J. Urol.*, 156(5), 1597–1601, 1996.

1049. Anidjar, M., Ettori, D., Cussenot, O., Meria, P., Desgrandchamps, F., Cortesse, A., Teillac, P., LeDuc, A., and Avrillier, S., *J. Urol.*, 156(5), 1590–1596, 1996.

1050. Avrillier, S., Tinet, E., Ettori, D., and Anidjar, M., *Phys. Scripta*, T72, 87–92, 1997.

1051. Koenig, F., McGovern, F. J., Enquist, H., Larne, R., Deutsch, T. F., and Schomacker, K. T., *J. Urol.*, 159(6), 1871–1875, 1998.

1052. Lin, W. C., Toms, S. A., Johnson, M., Jansen, E. D., and Mahadevan-Jansen, A., *Photochem., Photobiol.*, 73(4), 396–402, 2001.

1053. Utsuki, S., Oka, H., Sato, S., Suzuki, S., Shimizu, S., Tanaka, S., and Fujii, K., Possibility of using laser spectroscopy for the intraoperative detection of nonfluorescing brain tumors and the boundaries of brain tumor infiltrates—Technical note, *J. Neurosurg.*, 104, 618–620, 2006.

1054. Huttenberger, D., Gabrecht, T., Wagnieres, G., Weber, B., Linder, A., Foth, H. J., and Freitag, L., Autofluorescence detection of tumors in the human lung—Spectroscopical measurements in situ, in an in vivo model and in vitro, *Photodiagn. Photodyn. Ther.*, 5, 139–147, 2008.

1055. Gabrecht, T., Glanzmann, T., Freitag, L., Weber, B. C., van den Bergh, H., and Wagnieres, G., Optimized autofluorescence bronchoscopy using additional backscattered red light, *J. Biomed. Opt.*, 12, 9, 2007.

1056. Lam, S., Hung, J. Y. C., Kennedy, S. M., Leriche, J. C., Vedal, S., Nelems, B., Macaulay, C. E., and Palcic, B., *Am. Rev. Respir. Dis.*, 146(6), 1458–1461, 1992.

1057. Zellweger, M., Goujon, D., Conde, R., Forrer, M., van den Bergh, H., and Wagnieres, G., *Appl. Opt.*, 40(22), 3784–3791, 2001.

1058. Zellweger, M., Grosjean, P., Goujon, D., Monnier, P., van den Bergh, H., and Wagnieres, G., *J. Biomed. Opt.*, 6(1), 41–51, 2001.

1059. Uehlinger, P., Gabrecht, T., Glanzmann, T., Ballini, J. P., Radu, A., Andrejevic, S., Monnier, P., and Wagnieres, G., In vivo time-resolved spectroscopy of the human bronchial early cancer autofluorescence, *J. Biomed. Opt.*, 14, 024011/024011–024011/024019, 2009.

1060. Marcu, L., Jo, J. A., Fang, Q. Y., Papaioannou, T., Reil, T., Qiao, J. H., Baker, J. D., Freischlag, J. A., and Fishbein, M. C., Detection of rupture-prone atherosclerotic plaques by time-resolved laser-induced fluorescence spectroscopy, *Atherosclerosis*, 204, 156–164, 2009.

1061. Richardskortum, R., Mitchell, M. F., Ramanujam, N., Mahadevan, A., and Thomsen, S., *J. Cell. Biochem.*, 111–119, 1994.

1062. Ramanujam, N., Mitchell, M.F., Mahadevan, A., Warren, S., Thomsen, S., Silva, E., and Richards-Kortum, R., *Proc. Natl Acad. Sci. USA*, 91(21), 10193–10197, 1994; Ramanujam, N., Mitchell, M. F., Mahadevan, A., Thomsen, S., Silva, E., and Richards-Kortum, R., *Gynecol. Oncol.*, 52(1), 31–38, 1994.

1063. Benavides, J. M., Chang, S., Park, S. Y., Richards-Kortum, R., Mackinnon, N., MacAulay, C., Milbourne, A., Malpica, A., and Follen, M., Multispectral digital colposcopy for in vivo detection of cervical cancer, *Opt. Express*, 11, 1223–1236, 2003.

1064. Schomacker, K. T., Flotte, J. K., Compton, C. C., Flotte, T. J., Richter, J. M., Nishioka, N. S., and Deutsch, T. F., *Lasers Surg. Med.*, 12(1), 63–78, 1992.

1065. Mayinger, B., Jordan, M., Horner, P., Gerlach, C., Muehldorfer, S., Bittorf, B. R., Matzel, K. E., Hohenberger, W., Hahn, E. G., and Guenther, K., Endoscopic light-induced autofluorescence spectroscopy for the diagnosis of colorectal cancer and adenoma, *J. Photochem. Photobiol. B-Biol.*, 70, 13–20, 2003.

1066. Konig, K., Ruck, A., and Schneckenburger, H., *Opt. Eng.*, 31(7), 1470–1474, 1992.

1067. Pfefer, T. J., Paithankar, D. Y., Poneros, J. M., Schomacker, K. T., and Nishioka, N. S., Temporally and spectrally resolved fluorescence spectroscopy for the detection of high grade dysplasia in Barrett's esophagus, *Lasers Surg. Med.*, 32, 10–16, 2003.

1068. Vo-Dinh, T., Panjehpour, M., and Overholt, B. F., *NY Acad Sci.*, New York, 116–122, 1998; Vo-Dinh, T., Panjehpour, M., Overholt, B. F., and Buckley, P., *Appl. Spectrosc.*, 51(1), 58–63, 1997.

1069. Mayinger, B., Horner, P., Jordan, M., Gerlach, C., Horbach, T., Hohenberger, W., and Hahn, E. G., *Am. J. Gastroenterol.*, 96(9), 2616–2621, 2001; Mayinger, B., Horner, P., Jordan, M., Gerlach, C., Horbach, T., Hohenberger, W., and Hahn, E. G., *Gastrointest. Endosc.*, 54(2), 195–201, 2001.

1070. Masters, B. R., So, P. T., and Gratton, E., *Biophys. J.*, 72(6), 2405–2412, 1997.

1071. Terai, T. and Nagano, T., Fluorescent probes for bioimaging applications, *Curr. Opin. Chem. Biol.*, 12, 515–521, 2008.

1072. Van Schaik, H. J., Alkemade, C., Swart, W., and Van Best, J. A., *Exp. Eye Res.*, 68(1), 1–8, 1999.

1073. van Schaik, H. J., Coppens, J., van den Berg, T., and van Best, J. A., *Exp. Eye Res.*, 69(5), 505–510, 1999.

1074. Siik, S., Chylack, L. T., Friend, J., Wolfe, J., Teikari, J., Nieminen, H., and Airaksinen, P. J., *Acta Ophthalmol. Scand.*, 77(5), 509–514, 1999.

1075. Arens, C., Dreyer, T., Glanz, H., and Malzahn, K., Springer-Verlag, 71–76, 2004.

1076. Zargi, M. S. L., Fajdiga, I., Bubnic, B., Lenarcic, J., and Oblak, P., *Eur. Arch. Otorhinolaryngol. Suppl.* 1, S113–S116, 1997.

1077. Mehlmann, N., Betz, C. S., Stepp, H., Arbogast, S., Baumgartner, R., Grevers, G., and Leunig, A., *Lasers Surg. Med.*, 25(5), 414–420, 1999.

1078. Arens, C., Malzahn, K., Dias, O., Andrea, M., and Glanz, H., *Laryngo-Rhino-Otol.*, 78(12), 685–691, 1999.

1079. De Leon, H., Ollerenshaw, J. D., Griendling, K. K., and Wilcox, J. N., *Circulation*, 104(14), 1591–1593, 2001.

1080. Cothren, R. M., Sivak, M. V., VanDam, J., Petras, R. E., Fitzmaurice, M., Crawford, J. M., Wu, J., Brennan, J. F., Rava, R. P., Manoharan, R., and Feld, M. S., *Gastrointest. Endosc.*, 44(2), 168–176, 1996.

1081. DaCosta, R. S., Andersson, H., Cirocco, M., Marcon, N. E., and Wilson, B. C., Autofluorescence characterisation of isolated whole crypts and primary cultured human epithelial cells from normal, hyperplastic, and adenomatous colonic mucosa, *J. Clin. Pathol.*, 58, 766–774, 2005.

1082. Fryen, A. G. H., lohmann, W., Dreyer, T., and Bohle, R. M., *Acta Otolaryngol. (Stockh.)*, 117(2), 316–319, 1997.

1083. Roblyer, D., Kurachi, C., Stepanek, V., Williams, M. D., El-Naggar, A. K., Lee, J. J., Gillenwater, A. M., and Richards-Kortum, R., Objective detection and delineation of oral neoplasia using autofluorescence imaging, *Cancer Prev. Res.*, 2, 423–431, 2009.

1084. Chen, H. M., Chiang, C. P., You, C., Hsiao, T. C., and Wang, C. Y., Time-resolved autofluorescence spectroscopy for classifying normal and premalignant oral tissues, *Lasers Surg. Med.*, 37, 37–45, 2005.

1085. Dhingra, J. K., Perrault, D. F., McMillan, K., Rebeiz, E. E., Kabani, S., Manoharan, R., Itzkan, I., Feld, M. S., and Shapshay, S. M., *Arch. Otolaryngol-Head Neck Surg.*, 122(11), 1181–1186, 1996.

1086. Gillenwater, A., Jacob, R., Ganeshappa, R., Kemp, B., El-Naggar, A. K., Palmer, J. L., Clayman, G., Mitchell, M. F., and Richards-Kortum, R., *Arch. Otolaryngol—Head Neck Surg.*, 124(11), 1251–1258, 1998.

1087. Kulapaditharom, B. and Boonkitticharoen, V., *Annal. Otol. Rhinol. Laryngol.*, 107(3), 241–246, 1998.

1088. Onizawa, K., Saginoya, H., Furuya, Y., and Yoshida, H., *Cancer Lett.*, 108, 61–66, 1996.

1089. Heintzelman, D. L., Utzinger, U., Fuchs, H., Zuluaga, A., Gossage, K., Gillenwater, A. M., Jacob, R., Kemp, B., and Richards-Kortum, R. R., *Photochem. Photobiol.*, 72(1), 103–113, 2000.

1090. Kolli, V. R., Shaha, A. R., Savage, H. E., Sacks, P. G., Casale, M. A., and Schantz, S. P., *Am. J. Surg.*, 170(5), 495–498, 1995.

1091. Utzinger, U., Bueeler, M., Oh, S., Heintzelman, D. L., Svistun, E. S., Abd-El-Barr, M., Gillenwater, A., and Richards-Kortum, R., Optimal visual perception and detection of oral cavity neoplasia, *IEEE Trans. Biomed. Eng.*, 50, 396–399, 2003.

1092. Chissov, V. I., Sokolov, V. V., Filonenko, E. V., Menenkov, V. D., Zharkova, N. N., Kozlov, D. N., Polivanov, I. N., Prokhorov, A. M., Pyhov, R. L., and Smirnov, V. V., *Khirurgiia (Mosk)*, 5, 37–41, 1995.

1093. Sterenborg, H., Saarnak, A. E., Frank, R., and Motamedi, M., *J. Photochem. Photobiol. B-Biol.*, 35(3), 159–165, 1996.

1094. Heyerdahl, H., Wang, I., Liu, D. L., Berg, R., AnderssonEngels, S., Peng, Q., Moan, J., Svanberg, S., and Svanberg, K., *Cancer Lett.*, 112(2), 225–231, 1997.

1095. Konig, K., Schneckenburger, H., Ruck, A., and Steiner, R., *J. Photochem. Photobiol.*, 18(2–3), 287–290, 1993.

1096. Leunig, A., Rick, K., Stepp, H., Goetz, A., Baumgartner, R., and Feyh, J., *Laryngo-Rhino-Otol*, 75(8), 459–464, 1996.

1097. Leffell, D. J., Stetz, M. L., Milstone, L. M., and Deckelbaum, L. I., *Arch. Dermatol.*, 124(10), 1514–1518, 1988.

1098. Sterenborg, H. J. C., Motamedi, M., Waner, R. F., Duvic, M., Thomsen, S., and Jacques, S. L., *Lasers Med. Sci.*, 9, 191–201, 1994.

1099. Zeng, H. S., Macaulay, C., Palcic, B., and McLean, D. I., *Phys. Med. Biol.*, 38(2), 231–240, 1993.

1100. Mateasik, A., Smolka, J., Hrin, L., and Chorvat, D., *Interperiodica*, 213–216, 2003.

1101. Brancaleon, L., Durkin, A. J., Tu, J. H., Menaker, G., Fallon, J. D., Kollias, N., In vivo fluorescence spectroscopy of nonmelanoma skin cancer, *Photochem. Photobiol.*, 73, 178–183, 2001.

1102. Rajaram, N., Kovacic, D., Migden, M. F., Reichenberg, J. S., Nguyen, T. H., and Tunnell, J. W., In vivo determination of optical properties and fluorophore characteristics of non-melanoma skin cancer, *Proceedings of the SPIE—The International Society for Optical Engineering* 2009, 7161, 716102 (716109 pp.).

1103. Zuluaga, A. F., Utzinger, U., Durkin, A., Fuchs, H., Gillenwater, A., Jacob, R., Kemp, B., Fan, J., and Richards-Kortum, R., *Appl. Spectrosc.*, 53(3), 302–311, 1999.

1104. Alkhamis, K. I., Alhadiyah, B. M., Bawazir, S. A., Ibrahim, O. M., and Alyamani, M. J., Quantification of muscle-tissue magnesium and potassium using atomic-absorption spectrometry, *Anal. Lett.*, 28(6), 1033–1053, 1995.

1105. Barth, R. F., Adams, D. M., Soloway, A. H., Mechetner, E. B., Alam, F., and Anisuzzaman, A. K. M., Determination of boron in tissues and cells using direct-current plasma atomic emission-spectroscopy, *Anal. Chem.*, 63(9), 890–893, 1991.

1106. Kim, J. S. and Southard, J. H., Alteration in cellular calcium and mitochondrial functions in the rat liver during cold preservation, *Transplantation*, 65(3), 369–375, 1998.

1107. Bax, C. M. R. and Bloxam, D. L., 2 Major pathways of zinc(Ii) acquisition by human placental syncytiotrophoblast, *J. Cell. Physiol.*, 164(3), 546–554, 1995.

1108. de Pena, Y. P., Vielma, O., Burguera, J. L., Burguera, M., Rondon, C., and Carrero, P., On-line determination of antimony(III) and antimony(V) in liver tissue and whole blood by flow injection—hydride generation—Atomic absorption spectrometry, *Talanta*, 55.

1109. da Silva, J. B. B., Bertilia, M., Giacomelli, O., de Souza, I. G., and Curtius, A. J., Iridium and rhodium as permanent chemical modifiers for the determination of Ag, As, Bi, Cd, and Sb by electrothermal atomic absorption spectrometry, *Microchem. J.*, 60, 249–257.

1110. Drasch, G., Gath, H. J., Heissler, E., Schupp, I., and Roider, G., Silver concentrations in human tissues, their dependence on dental amalgam and other factors, *J. Trace Elem. Med. Biol.*, 9(2), 82–87, 1995.

1111. Ishihara, R., Ektessabi, A. M., Hanaichi, T., Takeuchi, T., Fujita, Y., Ishihara, Y., and Ohta, T., *Int. J. PIXE*, 9(3–4), 259, 1999.

1112. Coni, E., Alimonti, A., Fornarelli, L., Beccaloni, E., Sabiioni, E., Pietra, R., Bolis, G. B., Cristallini, E., Stacchini, A., and Caroli, S., *Acta Chim. Hungarica*, 128(4–5), 563–572, 1991.

1113. Anane, R., Bonini, M., and Creppy, E. E., Transplacental passage of aluminium from pregnant mice to fetus organs after maternal transcutaneous exposure, *Hum. Exp. Toxicol.*, 16(9), 501–504, 1997.

1114. Chappuis, P., Poupon, J., and Rousselet, F., A sequential and simple determination of zinc, copper and aluminum in blood-samples by inductively coupled plasma atomic emission-spectrometry, *Clin. Chim. Acta*, 206(3), 155–165, 1992.

1115. Coni, E., Alimonti, A., Bolis, G. B., Cristallini, E., and Caroli, S., An experimental approach to the assessment of reference values for trace-elements in human organs, *Trace Elem. Electrolytes*, 11(2), 84–91, 1994.

1116. DiPaolo, N., Masti, A., Comparini, I. B., Garosi, G., DiPaolo, M., Centini, F., Brardi, S., Monaci, G., and Finato, V., Uremia, dialysis and aluminium, *Int. J. Artif. Organs*, 20(10), 547–552, 1997.

1117. Fiejka, M., Fiejka, E., and Dlugaszek, M., Effect of aluminium hydroxide administration on normal mice: Tissue distribution and ultrastructural localization of aluminium in liver, *Pharmacol. Toxicol.*, 78(3), 123–128, 1996.

1118. Gane, E., Sutton, M. M., Pybus, J., and Hamilton, I., Hepatic and cerebrospinal fluid accumulation of aluminium and bismuth in volunteers taking short course anti-ulcer therapy, *J. Gastroenterol. Hepatol.*, 11(10), 911–915, 1996.

1119. Golub, M. S., Han, B., and Keen, C. L., Developmental patterns of aluminum and five essential mineral elements in the central nervous system of the fetal and infant guinea pig, *Biol. Trace Elem. Res.*, 55(3), 241–251, 1996.

1120. Omahony, D., Denton, J., Templar, J., Ohara, M., Day, J. P., Murphy, S., Walsh, J. B., and Coakley, D., Bone aluminum content in Alzheimer's disease, *Dementia*, 6(2), 69–72, 1995.

1121. Samek, O., Beddows, D. C. S., Telle, H. H., Kaiser, J., Liska, M., Caceres, J. O., and Urena, A. G., Quantitative laser-induced breakdown spectroscopy analysis of calcified tissue samples, *Spectrochim. Acta Part B-Atom. Spectrosc.*, 56(6), 865–875, 2001.

1122. Adamson, M. D. and Rehse, S. J. *Appl. Opt.*, 46, 5844–5852, 2007.

1123. Santos, D., Samad, R. E., Trevizan, L. C., de Freitas, A. Z., Vieira, N. D., and Krug, F., *J. Appl. Spectrosc.*, 62, 1137–1143, 2008.

1124. Bianchi, F., Maffini, M., Mangia, A., Marengo, E., and Mucchino, C., *J. Pharm. Biomed. Anal.*, 43, 659–665.

1125. Becker-Ross, H., Florek, S., and Heitmann, U., *J. Anal. At. Spectrom.*, 57(2), 137.

1126. Ademuyiwa, O. and Elsenhans, B., Time course of arsenite-induced copper accumulation in rat kidney, *Biol. Trace Elem. Res.*, 74(1), 81–92, 2000.

1127. De Aza, P. N., Guitian, F., De Aza, S., and Valle, F. J., Analytical control of wollastonite for biomedical applications by use of atomic absorption spectrometry and inductively coupled plasma atomic emission spectrometry, *Analyst*, 123(4), 681–685, 1998.

1128. Concha, G., Vogler, G., Nermell, B., and Vahter, M., *Int. Arch. Occup. Environ. Health*, 71(1), 42, 1998.

1129. Yamaguchi, T., Nakajima, Y., Miyamoto, H., Mizobushi, M., Kanazu, T., Kadono, K., Nakamoto, K., and Ikeuchi, I., *J. Toxicol. Sci.*, 23(Suppl. 4), 577, 1998.

1130. Brooke, S. L., Green, S., Charles, M. W., and Beddoe, A. H., The measurement of thermal neutron flux depression for determining the concentration of boron in blood, *Phys. Med. Biol.*, 46(3), 707–715, 2001.

1131. Buchar, E., Bednarova, S., Gruner, B., Walder, P., Strouf, O., and Janku, I., Dose-dependent disposition kinetics and tissue accumulation of boron after intravenous injections of sodium mercaptoundecahydrododecaborate in rabbits, *Cancer Chemother. Pharmacol.*, 29, 450–454, 1992.

1132. Iratsuka, J., Yoshino, K., Kondoh, H., Imajo, Y., and Mishima, Y., Biodistribution of boron concentration on melanoma-bearing hamsters after administration of p-, m-, o-boronophenylalanine, *Jpn. J. Cancer Res.*, 91(4), 446–450, 2000.

1133. Galyean, M. L., Ralphs, M. H., Reif, M. N., Graham, J. D., and Braselton, W. E., Effects of previous grazing treatment and consumption of locoweed on liver mineral concentrations in beef steers, *J. Anim. Sci.*, 74(4), 827–833, 1996.

1134. Shukla, S., Sharma, P., Johri, S., and Mathur, R., *J. Appl. Toxicol.*, 18(5), 331, 1998.

1135. Lawrence, I., Deckelbaum, Scott, J. J., Stetz, M. L., O'Brien, M. O., and Baker, G., *Lasers Surg. Med.* 12, 18–24, 1992.

1136. Sergeant, C., Gouget, B., Llabador, Y., Simonoff, M., Yefimova, M., Courtois, Y., and Jeanny, J. C., *Nucl. Instrum. Meth. Phys. Res.*, 158(1–4), 344, 1999.

1137. Dahm, M., Dohmen, G., Groh, E., Krummenauer, F., Hafner, G., Mayer, E., Hake, U., and Oelert, H., Decalcification of the aortic valve does not prevent early recalcification, *J. Heart Valve Dis.*, 9(1), 21–26, 2000.

1138. Dahm, M., Prufer, D., Mayer, E., Groh, E., Choi, Y. H., and Oelert, H., Early failure of an autologous pericardium aortic heart valve (ATCV) prosthesis, *J. Heart Valve Dis.*, 7(1), 30–33, 1998.

1139. Dlugaszek, M., Fiejka, M. A., Graczyk, A., Aleksandrowicz, J. C., and Slowikowska, M., Effects of various aluminium compounds given orally to mice on Al tissue distribution anal tissue concentrations of essential elements, *Pharmacol. Toxicol.*, 86(3)

1140. Dombovari, J. and Papp, L., Comparison of sample preparation methods for elemental analysis of human hair, *Microchem. J.*, 59(2), 187–193, 1998.

1141. Emelyanov, A. V., Shevelev, S. E., Murzin, B. A., and Amosov, V. I., Efficiency of calcium and vitamin D-3 in the treatment of steroid osteoporosis in patients with hormone-dependent bronchial asthma, *Terapevticheskii Arkhiv*, 71(11), 68–69, 1999.

1142. Faa, G., Lisci, M., Caria, M. P., Ambu, R., Sciot, R., Nurchi, V. M., Silvagni, R., Diaz, A., and Crisponi, G., Brain copper, iron, magnesium, zinc, calcium, sulfur and phosphorus storage in Wilson's disease, *J. Trace Elem. Med. Biol.*

1143. Habata, T., Ohgushi, H., Takakura, Y., Tohno, Y., Moriwake, Y., Minami, T., and Fujisawa, Y., Relationship between meniscal degeneration and element contents, *Biol. Trace Elem. Res.*, 79(3), 247–256, 2001.

1144. Iskra, M., Patelski, J., and Majewski, W., Relationship of calcium, magnesium, zinc and copper concentrations in the arterial wall and serum in atherosclerosis obliterans and aneurysm, *J. Trace Elem. Med. Biol.*, 11(4), 248–252, 1997.

1145. Pallikaris, I. G., Ginis, H. S., Kounis, G. A., Anglos, D., Papazoglou, T. G., and Naoumidis, L. P., Corneal hydration monitored by laser-induced breakdown spectroscopy, *J. Refract. Surg.*, 14(6), 655–660, 1998.

1146. Walther, L. E., Streck, S., Winnefeld, K., Walther, B. W., Kolmel, H. W., and Beleites, E., Reference values for electrolytes (Na, K, Ca, Mg) and trace elements (Fe, Cu, Zn, Se) in cerebrospinal fluid, *Trace Elem. Electrolytes*, 15(4), 177–180, 1998.

1147. Ohmi, M., Nakamura, M., Morimoto, S., and Haruna, M., Nanosecond time-gated spectroscopy of laser-ablation plume of human hair to detect calcium for potential diagnoses, *Opt. Rev.*, 7(4), 353–357, 2000.

1148. Abdel-Salam, Z., Nanjing, Z., Anglos, D., and Harith, M., *Appl. Phys. B-Lasers Opt.*, 94, 141–147, 2009.

1149. Ando, M., Yamasaki, K., Ohbayashi, C., Esumi, H., Hyodo, K., Sugiyama, H., Li, G., Maksimenko, A., and Kawai, T., *Jpn. J. Appl. Phys. Part 2-Lett. Expr. Lett.*, 44, L998–L1001, 2005.

1150. Antunes, A., Salvador, V. L. R., Scapin, M. A., de Rossi, W., and Zezell, D. M., *Laser Phys. Lett.*, 2, 318–323, 2005.

1151. Abu-Hayyeh, S., Sian, M., Jones, K. G., Manuel, A., and Powell, J. T., Cadmium accumulation in aortas of smokers, *Arterioscler. Thromb. Vasc. Biol.*, 21(5), 863–867, 2001.

1152. Rys, M., Nawrocka, A. D., Miekos, E., Zydek, C., Foksinski, M., Barecki, A., and Krajewska, W. M., Zinc and cadmium analysis in human prostate neoplasms, *Biol. Trace Elem. Res.*, 59(1–3), 145–152, 1997.

1153. Bulska, E., Emteborg, H., Baxter, D. C., Frech, W., Ellingsen, D., and Thomassen, Y., Speciation of mercury in human whole-blood by capillary gas-chromatography with a microwave-induced plasma emission detector system following complexometric extraction.

1154. Chakraborty, R., Das, A. K., Cervera, M. L., and Delaguardia, M., The atomization of cadmium in graphite furnaces, *Anal. Proc.*, 32(7), 245–249, 1995.

1155. Fortoul, T. I., Osorio, L. S., Tovar, A. T., Salazar, D., Castilla, M. E., and OlaizFernandez, G., Metals in lung tissue from autopsy cases in Mexico City residents: Comparison of cases from the 1950s and the 1980s, *Environ. Health Perspect.*, 104(6), 630–632.

1156. Davis, A. C., Wu, P., Zhang, X. F., Hou, X. D., and Jones, B. T., *Appl. Spectrosc. Rev.*, 41, 35–75, 2006.

1157. Gleizes, V., Poupon, J., Lazennec, J. Y., Chamberlin, B., and Saillant, G., Advantages and limits of determinating serum cobalt levels in patients with metal on metal articulating surfaces, *Revue De Chirurgie Orthopedique Et Reparatrice De L Appareil Mote.*

1158. Anderson, R. A., Bryden, N. A., and Polansky, M. M., Lack of toxicity of chromium chloride and chromium picolinate in rats, *J. Am. Coll. Nutr.*, 16(3), 273–279, 1997.

1159. Anderson, R. A., Bryden, N. A., Polansky, M. M., and Gautschi, K., Dietary chromium effects on tissue chromium concentrations and chromium absorption in rats, *J. Trace Elem. Exp. Med.*, 9(1), 11–25, 1996.

1160. Gunton, J. E., Hams, G., Hitchman, R., and McElduff, A., Serum chromium does not predict glucose tolerance in late pregnancy, *Am J. Clin. Nutr.*, 73(1), 99–104, 2001.

1161. Aburto, E. M., Cribb, A. E., and Fuentealba, C., Effect of chronic exposure to excess dietary copper and dietary selenium supplementation on liver specimens from rats, *Am. J. Vet. Res.*, 62(9), 1423–1427, 2001.

1162. Al-Awadi, F. M., Khan, I., Dashti, H. M., and Srikumar, T. S., Colitis-induced changes in the level of trace elements in rat colon and other tissues, *Annal. Nutr. Metabol.*, 42(5), 304–310, 1998.

1163. Amini, S. A., Walsh, K., Dunstan, R., Dunkley, P. R., and Murdoch, R. N., Maternal hepatic, endometrial, and embryonic levels of Zn, Mg, Cu, and Fe following alcohol consumption during pregnancy in QS mice, *Res. Commun. Alcohol Substance.*

1164. Baker, A., Gormally, S., Saxena, R., Baldwin, D., Drumm, B., Bonham, J., Portmann, B., and Mowat, A. P., Copper-associated liver-disease in childhood, *J. Hepatol.*, 23(5), 538–543, 1995.

1165. Bamiro, F. O., Littlejohn, D., and Marshall, J., Determination of copper in blood-serum by direct-current plasma and inductively coupled plasma atomic emission-spectrometry, *J. Anal. Atomic Spectr.*, 3(1), 279–284, 1988.

1166. Bayliss, E. A., Hambidge, K. M., Sokol, R. J., Stewart, B., and Lilly, J. R., Hepatic concentrations of zinc, copper and manganese in infants with extrahepatic biliary atresia, *J. Trace Elem. Med. Biol.*, 9(1), 40–43, 1995.

1167. Bertram, C., Bertram, H. P., Schussler, M., and Pfeiffer, M., Element pattern in blood plasma and whole blood from healthy pregnant women and their newborn infants, *Trace Elem. Electrolytes*, 15(4), 190–199, 1998.

1168. Braselton, W. E., Stuart, K. J., Mullaney, T. P., and Herdt, T. H., Biopsy mineral analysis by inductively coupled plasma-atomic emission spectroscopy with ultrasonic nebulization, *J. Vet. Diagn. Invest.*, 9(4), 395–400, 1997.

1169. Deng, D. X., Ono, S., Koropatnick, J., and Cherian, M. G., Metallothionein and apoptosis in the toxic milk mutant mouse, *Lab. Invest.*, 78(2), 175–183, 1998.

1170. Dorea, J. G., Iron and copper in human milk, *Nutrition*, 16(3), 209–220, 2000.

1171. During, A., Fields, M., Lewis, C. G., and Smith, J. C., Beta-carotene 15,15'-dioxygenase activity is responsive to copper and iron concentrations in rat small intestine, *J. Am. Coll. Nutr.*, 18(4), 309–315, 1999.

1172. Faa, G., Nurchi, V., Demelia, L., Ambu, R., Parodo, G., Congiu, T., Sciot, R., Vaneyken, P., Silvagni, R., and Crisponi, G., Uneven hepatic copper distribution in Wilson's disease, *J. Hepatol.*, 22(3), 303–308, 1995.

1173. Garner, B., Davies, M. J., and Truscott, R. J. W., Formation of hydroxyl radicals in the human lens is related to the severity of nuclear cataract, *Exp. Eye Res.*, 70(1), 81–88, 2000.

1174. Hatano, R., Ebara, M., Fukuda, H., Yoshikawa, M., Sugiura, N., Kondo, F., Yukawa, M., and Saisho, H., Accumulation of copper in the liver and hepatic injury in chronic hepatitis C, *J. Gastroenterol. Hepatol.*, 15(7), 786–791, 2000.

1175. Homma, S., Sasaki, A., Nakai, I., Sagai, M., Koiso, K., and Shimojo, N., Distribution of copper, selenium, and zinc in human kidney tumors by nondestructive synchrotron-radiation x-ray- fluorescence imaging, *J. Trace Elem. Exp. Med.*

1176. Sbir, T., Tamer, L., Erkisi, M., Kekec, Y., Doran, F., Varinli, S., and Taylor, A., Copper, zinc and magnesium in serum and tissues from patients with carcinoma of breast, stomach and colon, *Trace Elem. Electrol.*, 12(3), 113–115, 1995.

1177. Jaakkola, P., Hippelainen, M., and Kantola, M., Copper and zinc concentrations of abdominal-aorta and liver in patients with infrarenal abdominal aortic-aneurysm or aortoiliacal occlusive disease, *Ann. Chir. Gynaecol.*, 83(4), 304–308, 1994.

1178. Aral, Y. Z., Gucuyener, K., Atalay, Y., Hasanoglu, A., Turkyilmaz, C., Sayal, A., and Biberoglu, G., Role of excitatory aminoacids in neonatal hypoglycemia, *Acta Paediatr. Jpn.*, 40(4), 303–306, 1998.

1179. Al-Ebraheem, A., Farquharson, M. J., and Ryan, E., *Appl. Radiat. Isotopes*, 67, 470–474, 2009.

1180. Bermejo, P., Penae, E., Dominguez, R., Brmejo, A., Fraga, J. M., and Cocho, J. A., *Talanta*, 50, 1211, 2000.

1181. Ambu, R., Crisponi, G., Sciot, R., Vaneyken, P., Parodo, G., Iannelli, S., Marongiu, et al., Uneven hepatic iron and phosphorus distribution in beta-thalassemia, *J. Hepatol.*, 23(5), 5.

1182. Burguera, J. L., Burguera, M., Carrero, P., Rivas, C., Gallignani, M., and Brunetto, M. R., Determination of iron and zinc in adipose-tissue by online microwave-assisted mineralization and flow-injection graphite- furnace atomic-absorption spectrometry.

1183. Collery, P., Domingo, J. L., and Keppler, B. K., Preclinical toxicology and tissue gallium distribution of a novel antitumour gallium compound: Tris(8- quinolinolato)gallium(III), *Anticancer Res.*, 16(2), 687–691, 1996.

1184. Adair, B. M. and Cobb, G. P., Improved preparation of small biological samples for mercury analysis using cold vapor atomic absorption spectroscopy, *Chemosphere*, 38(12), 2951–2958, 1999.

1185. Alfthan, G. V., Toenail mercury concentration as a biomarker of methylmercury exposure, *Biomarkers*, 2(4), 233–238, 1997.

1186. Arnett, F. C., Fritzler, M. J., Ahn, C., and Holian, A., Urinary mercury levels in patients with autoantibodies to U3- RNP (fibrillarin), *J. Rheumatol.*, 27(2), 405–410, 2000.

1187. Boaventura, G. R., Barbosa, A. C., and East, G. A., Multivessel system for cold-vapor mercury generation—Determination of mercury in hair and fish, *Biol. Trace Elem. Res.*, 60(1–2), 153–161, 1997.

1188. Cominos, X., Athanaselis, S., Dona, A., and Koutselinis, A., Analysis of total mercury in human tissues prepared by microwave decomposition using a hydride generator system coupled to an atomic absorption spectrometer, *Forensic Sci. Int.*, 118(1), 43–17, 2001.

1189. Agut, A., Laredo, F. G., Sanchezvalverde, M. A., Murciano, J., and Tovar, M. D., Plasma-levels and urinary-excretion of iodine after oral- administration of iohexol in dogs and cats, *Invest. Radiol.*, 30(5), 296–299, 1995.

1190. Besteman, A. D., Bryan, G. K., Lau, N., and Winefordner, J. D., Multielement analysis of whole blood using a capacitively coupled microwave plasma atomic emission spectrometer, *Microchem. J.*, 61(3), 240–246, 1999.

1191. Dol, I., Knochen, M., and Vieras, E., Determination of lithium at ultratrace levels in biological-fluids by flame atomic emission-spectrometry—Use of 1st-derivative spectrometry, *Analyst*, 117(8), 1373–1376, 1992.

1192. Matusiewicz, H., Determination of natural levels of lithium and strontium in human-blood serum by discrete injection and atomic emission-spectrometry with a nitrous-oxide acetylene flame, *Anal. Chim. Acta.*, 136(APR), 215–223, 1982.

1193. Marczenko, Z., Lobinski, R., Griepink, B., Wells, D. E., Biemann, K., Gries, W. H., Jackwerth, E., Leroy, M., Lamotte, A., Westmoreland, D. G., Zolotov, Y. A., Ballschmiter, K., Dams, R., Fuwa, K., Grasserbauer, M., Linscheid, M. W., Morita, M., Huntau, M.

1194. Afarideh, H., Amirabadi, A., HadjiSaeid, S. M., Mansourian, N., Kaviani, K., and Zibafar, E., Biomedical studies by PIXE, *Nucl. Instrum. Meth. Phys. Res. B*, 109, 270–277, 1996.

1195. Hu, X. H., Fang, Q. Y., Cariveau, M. J., Pan, X. N., and Kalmus, G. W., Mechanism study of porcine skin ablation by nanosecond laser pulses at 1064, 532, 266, and 213 nm, *IEEE J. Quant. Electron.*, 37(3), 322–328, 2001.

1196. Chakraborty, R., Das, A. K., Cervera, M. L., and delaGuardia, M., A generalized method for the determination of nickel in different samples by ETAAS after rapid microwave-assisted digestion, *Anal. Lett.*, 30(2), 283–303, 1997.

1197. Alvarado, J., Cavalli, P., Omenetto, N., Rossi, G., Ottaway, J. M., and Littlejohn, D., Direct determination of lead in whole-blood using electrothermal vaporization.

1198. Areola, O. O., Jadhav, A. L., and Williams-Johnson, M., Relationship between lead accumulation in blood and soft tissues of rats subchronically exposed to low levels of lead, *Toxic Subst. Mech.*, 18(3), 149–161, 1999.

1199. Besteman, A. D., Lau, N., Liu, D. Y., Smith, B. W., and Winefordner, J. D., Determination of lead in whole blood by capacitively coupled microwave plasma atomic emission spectrometry, *J. Anal. Atomic Spectrom.*, 11(7), 479–481, 1996.

1200. Bjora, R., Falch, J. A., Staaland, H., Nordsletten, L., and Gjengedal, E., Osteoporosis in the Norwegian moose, *Bone*, 29(1), 70–73, 2001.

1201. Castilla, L., Castro, M., Grilo, A., Guerrero, P., Lopezartiguez, M., Soria, M. L., and Martinezparra, D., Hepatic and blood lead levels in patients with chronic liver-disease, *Eur. J. Gastroenterol. Hepatol.*, 7(3), 243–249, 1995.

1202. Deibel, M. A., Savage, J. M., Robertson, J. D., Ehmann, W. D., and Markesbery, W. R., Lead determinations in human bone by particle-induced x-ray-emission (pixel) and graphite-furnace atomic-absorption spectrometry (Gfaas), *J. Radioanal. Nucl. Chem.*, 195, 83–89, 1995.

1203. Gerhardsson, L., Englyst, V., Lundstrom, N. G., Nordberg, G., Sandberg, S., and Steinvall, F., Lead in tissues of deceased lead smelter workers, *J. Trace Elem. Med. Biol.*, 9(3), 136–143, 1995.

1204. Haraguchi, T., Ishizu, H., Takehisa, Y., Kawai, K., Yokota, O., Terada, S., Tsuchiya, K. et al., Lead content of brain tissue in diffuse neurofibrillary tangles with calcification (DNTC): The possibility of lead neurotoxicity, *Neuroreport*, 12, 3887–3890, 2001.

1205. Keating, A. D., Keating, J. L., Halls, D. J., and Fell, G. S., Determination of lead in teeth by atomic-absorption spectrometry with electrothermal atomization, *Analyst*, 112(10), 1381–1385, 1987.

1206. Dinoto, V., Ni, D., Via, L. D., Scomazzon, F., and Vidali, M., Determination of platinum in human blood using inductively-coupled plasma-atomic emission-spectrometry with an ultrasonic nebulizer, *Analyst*, 120(6), 1669–1673, 1995.

1207. Canavese, C., DeCostanzi, E., Branciforte, L., Caropreso, A., Nonnato, A., Pietra, R., Fortaner, S. et al., Rubidium deficiency in dialysis patients, *J. Nephrol.*, 14(3), 169–175, 2001.

1208. Abulaban, F. S., Saadeddin, S. M., AlSawaf, H. A., and AlBekairi, A. M., Effect of ageing on levels of selenium in the lymphoid tissues of rats, *Med. Sci. Res.*, 25(5), 303–305, 1997.

1209. Burguera, J. L., Villasmil, L. M., Burguera, M., Carrero, P., Rondon, C., Delacruz, A., Brunetto, M. R., and Gallignani, M., Gastric tissue selenium levels in healthy-persons, cancer and noncancer patients with different kinds of mucosal damage, *J. Trace Elem. Med. Biol.*, 9(3), 160–164, 1995.

1210. Karakucuk, S., Mirza, G. E., Ekinciler, O. F., Saraymen, R., Karakucuk, I., and Ustdal, M., Selenium concentrations in serum, lens and aqueous-humor of patients with senile cataract, *Acta Ophthalmol. Scand.*, 73(4), 329–332, 1995.

1211. Milman, N., Laursen, J., Byg, K. E., Pedersen, H. S., Mulvad, G., and Hansen, J. C., *J. Trace Elem. Med. Biol.*, 17, 301–306, 2003.

1212. Centeno, J. A., Offiah, O. O., Rastogi, T., Dewitt, T. J., and Luke, J. L., Electrothermal atomic-absorption determination of silicon in body-fluids and tissue specimens, *Abstr. Pap. Am. Chem. Soc.*, 208, 228-ENVR, 1994.

1213. Leung, F. Y. and Edmond, P., Determination of silicon in serum and breast tissue by electrothermal atomic absorption spectrometry, *Clin. Chem.*, 42(6), 845, 1996.

1214. Dhaese, P. C., VanLandeghem, G. F., Lamberts, L. V., Bekaert, V. A., Schrooten, I., and DeBroe, M. E., Measurement of strontium in serum, urine, bone, and soft tissues by Zeeman atomic absorption spectrometry, *Clin. Chem.*, 43(1), 121–128, 1997.

1215. Jorgenson, D. S., Mayer, M. H., Ellenbogen, R. G., Centeno, J. A., Johnson, F. B., Mullick, F. G., and Manson, P. N., Detection of titanium in human tissues after craniofacial surgery, *Plast. Reconstr. Surg.*, 99(4), 976–979, 1997.

1216. Galvan-Arzate, S., Martinez, A., Medina, E., Santamaria, A., and Rios, C., Subchronic administration of sublethal doses of thallium to rats: Effects on distribution and lipid peroxidation in brain regions, *Toxicol. Lett.*, 116(1–2), 37–43, 2000.

1217. D'Cruz, O. J., Waurzyniak, B., and Uckun, F. A., Subchronic (13-week) toxicity studies of intravaginal administration of spermicidal vanadocene dithiocarbarnate in mice, *Contraception*, 64(3), 177–185, 2001.

1218. D'Cruz, O. J., Waurzyniak, B., and Uckun, F. M., Subchronic (13-week) toxicity studies of intravaginal administration of spermicidal vanadocene acetylacetonato monotriflate in mice, *Toxicology*, 170(1–2), 31–43, 2002.

1219. Heinemann, G. and Vogt, W., Quantification of vanadium in serum by electrothermal atomic absorption spectrometry, *Clin. Chem.*, 42(8), 1275–1282, 1996.

1220. Aydinok, Y., Coker, C., Kavakli, K., Polat, A., Nisli, G., Cetiner, N., Kantar, M., and Cetingul, N., Urinary zinc excretion and zinc status of patients with beta-thalassemia major, *Biol. Trace Elem. Res.*, 70(2), 165–172, 1999.

1221. Balkan, M. E. and Ozgunes, H., Serum protein and zinc levels in patients with thoracic empyema, *Biol. Trace Elem. Res.*, 54(2), 105–112, 1996.

1222. Coni, P., Ravarino, A., Farci, A. M. G., Callea, F., VanEyken, P., Sciot, R., Ambu, R. et al., Zinc content and distribution in the newborn liver, *J. Pediatr. Gastroenterol. Nutr.*, 23(2), 125–129, 1996.

1223. Gabrielson, K. L., Remillard, R. L., and Huso, D. L., Zinc toxicity with pancreatic acinar necrosis in piglets receiving total parenteral nutrition, *Vet. Pathol.*, 33(6), 692–696, 1996.

1224. Jayasurya, A., Bay, B. H., Yap, W. M., Tan, N. G., and Tan, B. K. H., Proliferative potential in nasopharyngeal carcinoma: Correlations with metallothionein expression and tissue zinc levels, *Carcinogenesis*, 21(10), 1809–1812, 2000.

1225. Jin, R. X., Bay, B. H., Tan, P. H., and Tan, B. K. H., Metallothionein expression and zinc levels in invasive ductal breast carcinoma, *Oncol. Rep.*, 6(4), 871–875, 1999.

1226. Abnet, C. C., Lai, B., Qiao, Y. L., Vogt, S., Luo, X. M., Taylor, P. R., Dong, Z. W., Mark, S. D., and Dawsey, S. M., *J. Natl. Cancer Inst.*, 97, 301–306, 2005.

1227. Ademuyiwa, O., Elsenhans, B., Nguyen, P. T., and Forth, W., Arsenic copper interaction in the kidney of the rat: Influence of arsenic metabolites, *Pharmacol. Toxicol.*, 78(3), 154–160, 1996.

1228. McNeill, F. E. and O'Meara, J. M., *Adv. X Ray Anal.*, 41, 910, 1999.

Index

Milton Keynes UK
Ingram Content Group UK Ltd.
UKHW050458071024
449327UK00015B/432